ENCYCLOPEDIA OF PHYSICS

CHIEF EDITOR
S. FLÜGGE

VOLUME XVIII/1
MAGNETISM

EDITOR
H. P. J. WIJN

WITH 150 FIGURES

SPRINGER-VERLAG
BERLIN · HEIDELBERG · NEW YORK
1968

HANDBUCH DER PHYSIK

HERAUSGEGEBEN VON

S. FLÜGGE

BAND XVIII/1

MAGNETISMUS

BANDHERAUSGEBER

H. P. J. WIJN

MIT 150 FIGUREN

SPRINGER-VERLAG
BERLIN · HEIDELBERG · NEW YORK
1968

Alle Rechte vorbehalten. Kein Teil dieses Buches darf ohne schriftliche Genehmigung des Springer-Verlages übersetzt oder in irgendeiner Form vervielfältigt werden. © by Springer-Verlag Berlin · Heidelberg 1968. Library of Congress Catalog Card Number A 56-2942. Printed in Germany.

Die Wiedergabe von Gebrauchsnamen, Handelsnamen, Warenbezeichnungen usw. in diesem Werk berechtigt auch ohne besondere Kennzeichnung nicht zu der Annahme, daß solche Namen im Sinne der Warenzeichen- und Markenschutz-Gesetzgebung als frei zu betrachten wären und daher von jedermann benutzt werden dürften

Titel-Nr. 5756

Inhaltsverzeichnis.

Paramagnetic Relaxation. By Dr. J. C. Verstelle, Kamerlingh Onnes Laboratorium der Rijksuniversiteit, Leiden (Netherlands), and Dr. D. A. Curtis, Lecturer in Applied Physics, Department of Applied Physics, University of Durham, Durham (Great Britain). (With 21 Figures) 1

   A. Introduction 1

      I. Paramagnetic relaxation 2
         a) Paramagnetism 2
         b) Phenomenological theories 4

      II. Experimental methods 9

   B. Spin-spin relaxation 13

      I. General theory 14

      II. Applications of the general theory 20
         a) $\mathscr{H}_z = \mathscr{H}_{ex} = 0$ 20
         b) $\mathscr{H}_{el} = 0$ 22
         c) $\mathscr{H}_{el} \neq 0$ 26

      III. Cross-relaxation 28

   C. Spin-lattice relaxation 30

      I. Transition probabilities and relaxation times 30

      II. Field and temperature dependence of relaxation mechanisms 39

      III. Relaxation due to spin-orbit coupling 47
         a) Rare-earth salts 47
         b) Spin-lattice relaxation of the iron group transition elements 56
         c) Numerical parameters 60
         d) Anisotropy in relaxation times 63
         e) Relaxation in S-state ions 64
         f) Crystal bonding 67

      IV. Relaxation due to alternative perturbations 70

      V. Relaxation in the phonon system 80

   General references 92

Electron Spin Resonance. By Dr. D. J. E. Ingram, Professor of Physics, University of Keele, Keele, Staffordshire (Great Britain). (With 29 Figures) 94

      I. General principles 94
      II. Parameters associated with electron resonance spectra 98
      III. Experimental technique 108
      IV. Basic absorption processes 120
      V. g-values and their theoretical interpretation 126
      VI. The Spin Hamiltonian 132
      VII. Hyperfine splitting and the Spin Hamiltonian 135
      VIII. Examples of spectrum analysis 137
      IX. General conclusion 142

Structural Information from Paramagnetic Resonance. By Dr. JOHN E. WERTZ, Professor of Physical Chemistry, Head, Chemical Physics, Department of Chemistry, University of Minnesota, Minneapolis, Minnesota (United States of America). (With 39 Figures) 145

- A. Introduction 145
    - I. Types of systems studied 145
    - II. The g-tensor 146
    - III. Hyperfine splitting 148
- B. Free radicals 151
    - I. π-radicals 151
    - II. σ-radicals 204
- C. Triplet state systems and biradicals 205
    - I. Introduction 205
    - II. Resonant behavior of triplet systems 207
    - III. Ground state triplet species 221
    - IV. Biradicals 226
- D. Defects in insulating solids 228
    - I. Electron excess centers 228
    - II. Electron-deficient centers 239
    - III. Other defects in insulators 245
- E. Defects in semiconductors 253
    - I. Compounds 255
- F. Miscellaneous topics 256
    - I. ESR of gases 256
    - II. Transition metal ions in liquid solution 257
    - III. Generation of odd-electron centers 259
    - IV. Sources of error 261
    - V. Special techniques 262
- General references 263

Austauschwechselwirkung in Isolatoren. Von Professor Dr. S. V. VONSOVSKY und Dr. B. V. KARPENKO, Institut für Metallphysik, Akademie der Wissenschaften der UdSSR, Sverdlovsk (UdSSR). (Mit 9 Figuren) 265

- Einführung 265
- A. Die Theorie des Wasserstoffmoleküls in der Heitler-London-Näherung 271
- B. Die direkte Austauschwechselwirkung 286
    - I. Die Verallgemeinerung der Heitler-London-Methode auf den Kristall 286
    - II. Das Vektormodell von DIRAC und VAN VLECK 296
    - III. Das Vektormodell bei orthogonalisierten Wellenfunktionen 321
- C. Die indirekte Austauschwechselwirkung 335
    - IV. Der Superaustausch nach KRAMERS und ANDERSON 336
        - a) Das Problem der drei Zentren und vier Elektronen. Orthogonale Bahnen . 336
        - b) Das Problem der drei Zentren und vier Elektronen. Nichtorthogonale Bahnen 359
        - c) Die Verallgemeinerung auf den Kristall 365
    - V. Der neue Lösungsweg ANDERSONs für das Problem der indirekten Wechselwirkung 368
- D. Die Austauschwechselwirkung und die Wellenfunktionen der Elektronen im Kristall 374
- Literatur 384

Magnetic Semiconductors. By Dr. SIEGFRIED METHFESSEL, IBM — Thomas J. Watson Research Center, Yorktown Heights, New York, and Dr. DANIEL C. MATTIS, Professor of Physics, Belfer Graduate School of Science, Yeshiva University, New York (United States of America). (With 52 Figures) 389

- A. Preface 389
    - I. General remarks 389
    - II. Novel features 394
    - III. Organization of this article 395
- B. Problems related to the electron band structure of magnetic materials 395
    - I. The one-electron band model of semiconductors 397
    - II. Localized electrons and magnetic moments 400
- C. Theory of electron transport properties 412
    - I. Magnetic insulators, the Mott transition, and the two types of energy gap . . 413
    - II. Band structure of rare earth semiconductors 418
    - III. Conductivity properties of magnetic semiconductors 426
- D. Theory of indirect exchange 430
- E. Electron orbitals and energies in rare earth ions 451
- F. Optical properties of crystals 463
    - I. Crystalline field effects on $4f$ electrons 463
    - II. The $5d$ orbitals in lanthanides 471
    - III. Spectroscopy of exchange interactions 480
- G. Experimental evidence for indirect exchange 487
    - I. Metallic and insulating rare earth materials 489
    - II. Magnetic interaction via semiconducting carriers 512
- H. Electrical properties of magnetic semiconductors 531
    - I. The phase diagram of correlated electrons 532
    - II. Electrical properties of $3d$-semiconductors 545
    - III. Rare earth chalcogenides 553

Sachverzeichnis (Deutsch-Englisch) 563

Subject Index (English-German) 578

Paramagnetic Relaxation.

By

J. C. VERSTELLE and D. A. CURTIS.

With 21 Figures.

A. Introduction.

1. History. The problem: "How fast does the average magnetic moment in a paramagnetic substance respond to a sudden change in the magnetic field in which the substance is placed?", is one of long standing. Already in 1920 the question was raised by LENZ[1]: "Bei zunehmender Verlangsamung der Umklappungen (of elementary magnetic moments) muß eine Verzögerung in der Einstellung des mittleren magnetischen Moments, also der Suszeptibilität, erfolgen, und es wäre interessant zu wissen ob bei den tiefsten Temperaturen der Versuche von KAMERLINGH ONNES und OOSTERHUIS nicht schon Andeutungen solcher Verzögerungen beobachtet werden können." KAMERLINGH ONNES accepted the challenge, but not before EHRENFEST[2] had given his opinion: "In general we may expect, that the corresponding retardations in the establishment of the magnetisation would show themselves most easily at very low temperatures and rapidly alternating fields. They would for instance give rise to a kind of hysteresis and a corresponding development of heat ..." Ignoring one of the fundamental laws of experimental physics, that is always to try and look for effects differing from an initial zero value, KAMERLINGH ONNES and BREIT[3] instead tried to find a decrease of the susceptibility, or dispersion, in a periodically alternating field and were forced by uncertainties in their otherwise positive results, to conclude their report with: "However it would be preposterous to conclude that the susceptibility is actually decreased by the amount found." There the matter stood for more than a decade, even after several other attempts.

With the advance of the quantum mechanical understanding of paramagnetism and a powerful impulse from the practical viewpoint, the problem of the feasibility of adiabatic demagnetization, interest returned. In 1932 a, nowadays famous, article by another Swede, WALLER[4] appeared. In this, WALLER already distinguished the two main relaxation phenomena, spin-spin relaxation and spin-lattice relaxation and for the latter found the two fundamental mechanisms the so-called direct and quasi-Raman processes. On the experimental side, in Holland, the thread left by KAMERLINGH ONNES was taken up by GORTER, who, taking heed of the suggestion by EHRENFEST, attempted to measure absorption phenomena. With his calorimetric method, that even now more than 30 years later is still in use, he was the first who unequivocally demonstrated the existence of paramagnetic relaxation[5]. After this discovery experimental results were published

---
[1] W. LENZ: Physik. Z. **21**, 613 (1920).
[2] P. EHRENFEST: Comm. Leiden, Suppl. No. 44b (1920).
[3] G. BREIT, and H. KAMERLINGH ONNES: Comm. Leiden No. 168c (1924).
[4] I. WALLER: Z. Physik **79**, 370 (1932).
[5] C. J. GORTER: Comm. Leiden, No. 241e; Physica **3**, 503 (1936); — Comm. Leiden No. 247b; Physica **3**, 1006 (1936).

in ever increasing numbers. The theoretical aspects interested men such as KRONIG[1], FIERZ[2], HEITLER, TELLER[3]. TEMPERLEY[4], and last but not least VAN VLECK[5]. The rise of paramagnetic and nuclear resonance in the late forties and early fifties drew away the attention from relaxation phenomena, but the development of the solid-state maser was the cause of a sharp increase in research, experimental as well as theoretical. Even though the flood of publications is slowly decreasing nowadays, one can hardly say that all the problems are solved. It is even safe to state that whatever sample is chosen, by measuring relaxation phenomena, one can find results that cannot be described completely by existing theories.

I. Paramagnetic relaxation.

a) Paramagnetism.

2. Curie's law. As the name paramagnetic relaxation implies, we will only consider the dynamical aspects of the magnetisation in paramagnetic substances and even narrow down the attention to the region of substances in which the equilibrium magnetisation $\boldsymbol{M}$ as a function of the magnetic field $\boldsymbol{H}$ is governed by CURIE's law:

$$\boldsymbol{M} = \chi_0 \cdot \boldsymbol{H} = \frac{C}{T} \boldsymbol{H}. \tag{2.1}$$

χ_0 being the static susceptibility and C, CURIE's constant. In general χ_0 will be a tensor. In most theories to be reviewed in this article, however, for simplicity the substances are considered to be isotropic and homogeneous so that the equilibrium magnetisation will have the same direction as the magnetic field and χ_0 is a scalar. In all these cases the z-direction of the coordinate system will be chosen along the direction of the magnetic field.

VAN VLECK[6] has shown under which conditions (2.1) can be derived. Consider a sample of a paramagnetic substance of volume V in a magnetic field H. The Hamiltonian of the system can be written as:

$$\mathcal{H}_{\text{tot}} = \mathcal{H} - V M_z H \tag{2.2}$$

M_z being the operator of the component of the magnetic moment in the direction of the field. The mean value of M_z can be calculated by making use of the statistical density matrix operator ϱ:[7]

$$\varrho = \frac{\exp[-\mathcal{H}_{\text{tot}}/kT]}{\text{Tr} \exp[-\mathcal{H}_{\text{tot}}/kT]} \tag{2.3}$$

according to:

$$\overline{M}_z = \text{Tr}\, \varrho\, M_z. \tag{2.4}$$

Making use of the operator identity for two, in general noncommuting, operators A and B:

$$\exp[-\lambda_0(A+B)] = \exp[-\lambda_0 A]\left\{1 - \int_0^{\lambda_0} d\lambda \exp[\lambda A] B \exp[-\lambda(A+B)]\right\}. \tag{2.5}$$

---

[1] R. KRONIG: Physica **6**, 33 (1939).
[2] M. FIERZ: Physica **5**, 433 (1938).
[3] W. HEITLER, and E. TELLER: Proc. Roy. Soc. (London) A **155**, 629 (1936).
[4] H. N. V. TEMPERLEY: Proc. Cambridge Phil. Soc. **35**, Pt. II, 256 (1939).
[5] J. H. VAN VLECK: J. Chem. Phys. **7**, 72 (1939); — Phys. Rev. **57**, 426 (1940).
[6] J. H. VAN VLECK: Theory of Electric and Magnetic Susceptibilities. Oxford 1932.
[7] R. C. TOLMAN: The Principles of Statistical Mechanics. Oxford: Oxford University Press 1938.

(2.4) can be written as:

$$\overline{M}_z = \frac{\operatorname{Tr} M_z \exp[-\mathcal{H}/kT]\left\{1+\int_0^{1/kT} d\lambda \exp[\lambda\mathcal{H}] V M_z H \exp[-\lambda(\mathcal{H}-VM_zH)]\right\}}{\operatorname{Tr} \exp(-\mathcal{H}/kT)\left\{1+\int_0^{1/kT} d\lambda \exp[\lambda\mathcal{H}] V M_z H \exp[-\lambda(\mathcal{H}-VM_zH)]\right\}}. \qquad (2.6)$$

Expanding the identity by iteration in a series of increasing powers of H, and retaining only terms linear in H, one finds:

$$\left.\begin{aligned}\overline{M}_z &= \operatorname{Tr} \varrho_0 M_z - VH \operatorname{Tr} \varrho_0 M_z \cdot \operatorname{Tr} \varrho_0 \int_0^{1/kT} d\lambda \exp[\lambda\mathcal{H}] M_z \exp[-\lambda\mathcal{H}] + \\ &\quad + VH \operatorname{Tr} \varrho_0 M_z \int_0^{1/kT} d\lambda \exp[\lambda\mathcal{H}] M_z \exp[-\lambda\mathcal{H}]\end{aligned}\right\} \qquad (2.7)$$

in which:

$$\varrho_0 = \exp[-\mathcal{H}/kT]/\operatorname{Tr}\exp[-\mathcal{H}/kT]. \qquad (2.8)$$

It is now supposed that we can make the high temperature approximation. VAN VLECK has shown what this implies in terms of the energy level systems of the single magnetic ions. One has to postulate the existence of a low lying set of energy levels $|m\rangle$ with spacings much smaller than kT, between which matrix elements of M_z of the form $\langle m|M_z|m'\rangle$ exist. Furthermore one has to assume that no matrix elements of M_z exist between these low energy states and states $|n\rangle$ for wich the energy difference $E_n - E_m$ is very much larger than kT and consequently will not be populated. These hypotheses can be generalized for the system under consideration. As long as the operator M_z only has matrix elements, in a representation diagonalizing $\mathcal{H}$, between states for which

$$|E_m - E_{m'}| \ll kT \qquad (2.9)$$

it is possible to expand (2.7) in a series of increasing powers of $1/kT$:

$$\overline{M}_z = \langle M_z \rangle - VH \langle M_z \rangle^2/kT + VH \langle M_z^2 \rangle/kT + \cdots \qquad (2.10)$$

in which the notation:

$$\langle A \rangle = \operatorname{Tr} \varrho_0 A \qquad (2.11)$$

is used. We have neglected the so-called "high-frequency" terms of VAN VLECK, originating from matrix elements $\langle n|M_z|m\rangle$ that lead to a temperature independent part of the susceptibility. Furthermore, it is possible to assume that the magnetic moment for zero fields will average out, implying the non-existence of a spontaneous magnetisation, so that the expression for CURIE's law is:

$$\overline{M}_z = VH \langle M_z^2 \rangle/kT \qquad (2.12)$$

and correspondingly

$$\chi_0 = V \langle M_z^2 \rangle/kT. \qquad (2.13)$$

CURIE's constant defined per unit of volume being:

$$C = \langle M_z^2 \rangle/k. \qquad (2.14)$$

3. The Hamiltonian. In the preceding section no comment has been made on the form of the Hamiltonian $\mathcal{H}_{\text{tot}}$. Its complexity, as can be guessed from consideration of what is actually happening inside a paramagnetic sample, defies the imagination. A near infinite number of magnetic ions, each with its own intricate set of energy levels, interacting with each other, vibrating in the sample under the influence of the thermal motions of its non-magnetic neighbours and the interactions with external fields and surrounding medium, is contained within the symbol, $\mathcal{H}_{tot}$. In all its generality one could try to split it up to describe at least several aspects separately:

$$\mathcal{H}_{tot} = \mathcal{H}_S + \mathcal{H}_{S,i} + \mathcal{H}_L + \mathcal{H}_{L,i} + \mathcal{H}_{L,S} + \mathcal{H}_B + \mathcal{H}_{L,B} + \mathcal{H}_{S,B}. \tag{3.1}$$

Here the subscripts S, L and B stand for the magnetic or spin system, crystal-lattice and external bath respectively, the subscript i for interaction within a system and double subscripts for interactions between systems. The remaining problem is: "How to reduce this split-up Hamiltonian even further?" The most obvious simplification will be reducing the number of the terms according to the problem under consideration. So it will be possible to isolate the system of interacting magnetic ions, described by $\mathcal{H}_S + \mathcal{H}_{S,i}$, under certain circumstances. This will be done in the description of spin-spin relaxation. In the spin-lattice relaxation theories the terms, $\mathcal{H}_L$ and $\mathcal{H}_{L,S}$ are taken into account together with $\mathcal{H}_S$ or $\mathcal{H}_S + \mathcal{H}_{S,i}$ and in some cases $\mathcal{H}_{L,i}$ and/or $\mathcal{H}_{L,B}$. The terms $\mathcal{H}_B$ and $\mathcal{H}_{S,B}$ are mostly not of any interest.

A very fruitful concept to simplify the description of the spin system is the introduction of the so-called spin Hamiltonian[1]. This expression using only spin operators, actually describes the "low-frequency" levels of VAN VLECK's hypothesis. Defining an effective spin operator $\mathbf{S}$, such that the number of these levels equals $2S+1$, one can introduce the spin Hamiltonian, whose terms are functions of the components of $\mathbf{S}$, such that the eigenvalues of the spin Hamiltonian will coincide with the ground state of the actual magnetic ions. All the excited states of the ions can be left out of consideration as the main effects of their influence on the lower states have been accounted for by parameters in the spin Hamiltonian.

Although, in principle, it is possible to give a general relaxation theory, using all the terms in the total Hamiltonian $\mathcal{H}_{tot}$, for the actual calculation of the interaction parameters, one has to fall back on methods used in the simpler theories describing one specific relaxation phenomenon. This is one reason why in this article no attempt is made to review existing general theories. Another reason for this is a rather fundamental difference in experimental methods used to measure relaxation times, the first one based on a linear approximation in the magnetic field disturbance, parallel to the existing field H, while the other is essentially non-linear in the disturbance perpendicular to the static field, as will be outlined in Sects. 4 and 5.

b) Phenomenological theories.

4. Thermodynamics. One of the first theories successfully describing some aspects of paramagnetic relaxation is one by CASIMIR and DU PRÉ[2]. In it it is assumed that both the spin system and the lattice or phonon system are separate thermodynamical systems, each in internal equilibrium, possibly with different temperatures T_S and T_L. Equilibrium in the whole system is established by spin-lattice interactions, by which energy is transferred from one system to the other. To simplify matters, the heat contact between lattice and surrounding heat bath is considered to be so strong that the lattice temperature always equals the bath temperature.

According to the first law of thermodynamics, the energy balance for the spin system now takes the form:

$$\Delta Q = dU + V M_z dH \tag{4.1}$$

---

[1] A. ABRAGAM, and M. H. L. PRYCE: Proc. Roy. Soc. (London) A **205**, 135 (1951). — B. BLEANEY, and K. W. H. STEVENS: Repts. Progr. Phys. **16**, 108 (1953).

[2] H. B. G. CASIMIR, and F. K. DU PRÉ: Physica **5**, 507 (1938).

in which U is the so-called spectroscopic energy of the spin system. Other definitions of U are possible with corresponding different expressions for the first law (4.1)[1]. Comparing (4.1) with the well known formulation of the first law:

$$\Delta Q = dU + p\, dV \qquad (4.2)$$

all the known expressions for the thermodynamical formula and quantities can be transposed into corresponding magnetic expressions by replacing p and V by M_z and H. For instance (4.1) can be written, taking $V=1$, as:

$$\Delta Q = \left(\frac{\partial U}{\partial T_S}\right)_H dT_S + \left\{\left(\frac{\partial U}{\partial H}\right)_{T_S} + M_z\right\} dH. \qquad (4.3)$$

By using the second law:

$$\Delta Q = T_S\, dS \qquad (4.4)$$

and the corresponding Maxwell-relations, (4.3) can now be transformed to:

$$\Delta Q = C_H\, dT_S + T_S \left(\frac{\partial M_z}{\partial T_S}\right)_H dH \qquad (4.5)$$

in which:

$$C_H = \left(\frac{\partial U}{\partial T_S}\right)_H \qquad (4.6)$$

is the specific heat at constant field, per unit volume of the spin system. Holding M_z constant in (4.5) one gets:

$$\Delta Q = C_M\, dT_S = C_H\, dT_S + T_S \left(\frac{\partial M_z}{\partial T_S}\right)_H \left(\frac{\partial H}{\partial T_S}\right)_{M_z} dT_S \qquad (4.7)$$

and thus, the specific heat at constant magnetisation is given by:

$$\left.\begin{array}{l} C_M = C_H + T_S \left(\dfrac{\partial M_z}{\partial T_S}\right)_H \left(\dfrac{\partial H}{\partial T_S}\right)_{M_z} \\[6pt] = C_H - T_S \left(\dfrac{\partial M_z}{\partial T_S}\right)_H^2 \Big/ \left(\dfrac{\partial M_z}{\partial H}\right)_{T_s}. \end{array}\right\} \qquad (4.8)$$

As the definition of the static susceptibility is:

$$M_z = \chi_0 H \qquad (4.9)$$

in which χ_0 is positive for paramagnetic substances and as long as paramagnetic saturation is excluded, we see that:

$$\chi_0 = \left(\frac{\partial M_z}{\partial H}\right)_{T_s} \qquad (4.10)$$

and thus by (4.8):

$$C_H \geq C_M. \qquad (4.11)$$

Using the expression:

$$\Delta Q = C_H \left(\frac{\partial T_S}{\partial M_z}\right)_H dM_z + C_M \left(\frac{\partial T_S}{\partial H}\right)_{M_z} dH \qquad (4.12)$$

and considering the spin system to be completely isolated from the lattice, we find the adiabatic susceptibility χ_{ad}:

$$\chi_{\text{ad}} = \left(\frac{\partial M_z}{\partial H}\right)_S = \frac{C_M}{C_H} \left(\frac{\partial M_z}{\partial H}\right)_{T_s} = \frac{C_M}{C_H} \chi_0. \qquad (4.13)$$

Now introduce an alternating field of vanishingly small amplitude h in the same direction as the existing static field H. Expressions (4.1), (4.3), (4.4), (4.5), (4.7) and

---

[1] L. J. F. BROER: Physica **12**, 49 (1946). — C. J. GORTER: Paramagnetic Relaxation. Amsterdam: Elsevier 1947.

(4.12) can now be written as differential equations with the time as variable, for instance (4.1) becomes:

$$\frac{d\Delta Q}{dt} = \frac{dU}{dt} + VM_z \frac{dH}{dt}. \tag{4.14}$$

Introducing the complex notation for the total field:

$$H = H_0 + h \exp[i\omega t] \tag{4.15}$$

it is now supposed that in the linear approximation in h the magnetisation can be written as:

$$M_z = M_{z_0} + m \exp[i\omega t] \tag{4.16}$$

and the spin temperature as:

$$T_S = T_L + \vartheta \exp[i\omega t]. \tag{4.17}$$

The next assumption is that the spin-lattice interaction causes a heat flow from spin system to lattice, proportional to the difference between the temperatures of both systems:

$$\Delta Q = -\alpha(T_S - T_L). \tag{4.18}$$

In the steady state, the spin system will now be described by:

$$i\omega C_H \left(\frac{\partial T_S}{\partial M_z}\right)_H m + i\omega C_M \left(\frac{\partial T_S}{\partial H}\right)_{M_z} h = -\alpha\vartheta. \tag{4.19}$$

An additional expression for ϑ is:

$$\vartheta = \left(\frac{\partial T_S}{\partial M_z}\right)_H m + \left(\frac{\partial T_S}{\partial H}\right)_{M_z} h \tag{4.20}$$

Combining (4.19) with (4.20) gives:

$$\chi(\omega) = m/h = \chi_0 (C_M + \alpha/i\omega)/(C_H + \alpha/i\omega) \tag{4.21}$$

which can be transformed into:

$$\chi(\omega) = \chi_{ad} + (\chi_0 - \chi_{ad}) \frac{1}{1 + i\omega\tau_L} \tag{4.22}$$

where:

$$\tau_L = \frac{C_H}{\alpha} \tag{4.23}$$

is called the spin-lattice relaxation time.

$\chi(\omega)$ is the complex susceptibility indicating a phase difference between m and h. Writing:

$$\chi(\omega) = \chi'(\omega) - i\chi''(\omega) \tag{4.24}$$

one finds for the real component:

$$\chi'(\omega) = \chi_{ad} + (\chi_0 - \chi_{ad}) \frac{1}{1 + \omega^2 \tau_L^2} \tag{4.25}$$

which gives the in-phase component of the susceptibility describing the dispersion effect connected with the relaxation. The imaginary part of $\chi(\omega)$:

$$\chi''(\omega) = (\chi_0 - \chi_{ad}) \frac{\omega\tau_L}{1 + \omega^2 \tau_L^2} \tag{4.26}$$

is a measure for the absorption of energy by the spin system, caused by the component 90° out of phase (Fig. 1). The energy absorption per second, per unit volume of the sample is:

$$P = -\overline{M_z \frac{dH}{dt}} = \frac{1}{2}\omega\chi''(\omega)h^2. \qquad (4.27)$$

The perfect agreement of this theory of CASIMIR and DU PRÉ with experiment in a multitude of cases justifies the simple model and the assumptions made. It is not certain, however, whether all assumptions made are necessary. In fact there is experimental evidence that the existence or non-existence of equilibrium in the

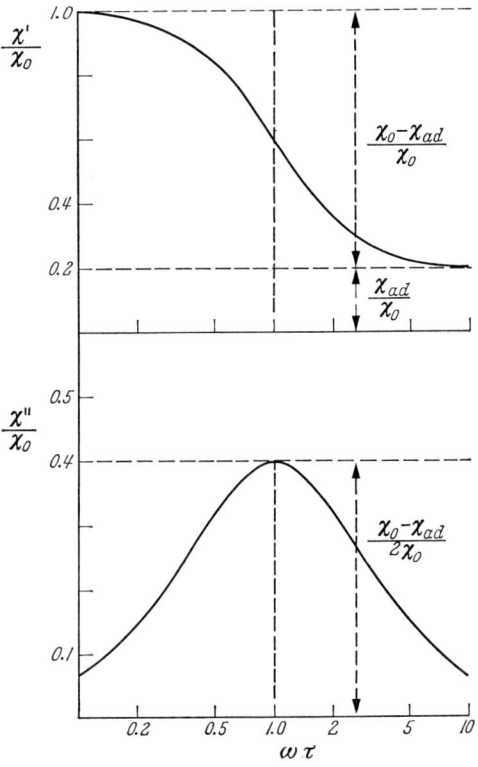

Fig. 1. Casimir and Du Pre relaxation curves for $\chi'(\omega)/\chi_0$ and $\chi''(\omega)/\chi_0$, with $\chi_{ad}/\chi_0 = 0.2$.

spin system has no influence on the spin-lattice relaxation time, although it could have a slight but hardly measurable effect on the intensity of the absorption and dispersion phenomena. This fact is of great importance, as in most microscopic theoretical calculations of the spin-lattice relaxation times the spin system is considered as an ensemble of non-interacting spins and consequently relaxation times are calculated by means of transition probabilities of isolated spins. All these theories make use, however, of the assumption that the lattice or phonon system is in equilibrium. In fact when the phonon system is not in equilibrium, this has a considerable influence on the relaxation phenomena, as for instance in the so-called phonon bottleneck effect.

According to Eq. (4.25) $\chi'(\omega)$ equals χ_{ad} in the limit of high frequencies. Actually this is a consequence of the assumption of equilibrium within the spin system, or in other words the assumption of an instantaneous establishment of this equilibrium after a change in the magnetic field. It is obvious that this last assumption can not be true. In fact it is the main purpose of experiments on spin-spin relaxation to find out how equilibrium in the spin system is established. Experimental evidence however indicates that in most cases at sufficiently low temperatures the characteristic times connected with the establishment of equilibrium in the spin system are short compared with spin-lattice relaxation times. Without making further assumptions the thermodynamical theory of CASIMIR and DU PRÉ will not be capable of describing spin-spin relaxation.

5. The Bloch equations.

Another phenomenological theory with considerable success, actually not describing paramagnetic relaxation but rather paramagnetic resonance, is one due to BLOCH[1]. As it has to take into account relaxation effects and has been the basis for a completely new method of measuring relaxation times, we will discuss it briefly.

Consider a paramagnetic sample in a magnetic field $\boldsymbol{H}$. Then the equation of motion for the magnetisation vector $\boldsymbol{M}$ will describe the precession of $\boldsymbol{M}$ around $\boldsymbol{H}$:

$$\frac{d\boldsymbol{M}}{dt} = \gamma(\boldsymbol{M} \times \boldsymbol{H}) \tag{5.1}$$

γ is called the magnetomechanical ratio. In this equation interaction between the spins themselves and between the spins and the surroundings have not been taken into account. The frequency of precession of $\boldsymbol{M}$ around $\boldsymbol{H}$ is called the Larmor frequency and is given by:

$$\omega_L = \gamma H. \tag{5.2}$$

Without interactions no change in the component of $\boldsymbol{M}$ along the $\boldsymbol{H}$-direction, taken parallel to the z-axis, can take place. BLOCH introduced two parameters to describe damping, one, such that any transverse M_x- or M_y-component will vanish with a characteristic time τ_2 and the second, such that the M_z-component will reach the equilibrium value $M_{z_0} = \chi_0 H$ with a characteristic time τ_1. The modified equations of motion, also called the Bloch equations, are:

$$\left. \begin{array}{l} \dfrac{dM_{x,y}}{dt} = \gamma(\boldsymbol{M} \times \boldsymbol{H})_{x,y} - M_{x,y}/\tau_2, \\[6pt] \dfrac{dM_z}{dt} = \gamma(M \times H)_z - (M_z - M_{z_0})/\tau_1. \end{array} \right\} \tag{5.3}$$

When a linearly polarized alternating field:

$$h_x = \tfrac{1}{2}h\{\exp[i\omega t] + \exp[-i\omega t]\}, \quad h_y = h_z = 0 \tag{5.4}$$

is applied perpendicular to $\boldsymbol{H}$, one can show that the approximate solution in the neighbourhood of resonance for the M_x-component is:

$$M_x = \tfrac{1}{2} M_{z_0} \gamma h \tau_2 (\Delta\omega\tau_2 - i) \exp[i\omega t]/(1 + (\Delta\omega\tau_2)^2 + \tfrac{1}{4}\gamma^2 h^2 \tau_1 \tau_2) \tag{5.5}$$

with the resonance frequency ω_0, shifted slightly from the Larmor frequency:

$$\omega_0^2 = \omega_L^2 + 1/\tau_2^2 \tag{5.6}$$

and

$$\Delta\omega = \omega - \omega_0. \tag{5.7}$$

The power absorbed from the alternating field near resonance now becomes:

$$P = \tfrac{1}{2}\omega\chi_x'' h^2 \approx \tfrac{1}{4} M_{z_0} H \gamma^2 h^2 \tau_2/(1 + (\Delta\omega\tau_2)^2 + \tfrac{1}{4}\gamma^2 h^2 \tau_1 \tau_2). \tag{5.8}$$

---

[1] F. BLOCH: Phys. Rev. **70**, 460 (1946).

When the term $\frac{1}{4}\gamma^2 h^2 \tau_1 \tau_2$ in the numerator is very small compared to $1+(\Delta\omega\tau_2)^2$, the power absorbed in the sample will be proportional to h^2 as expected and independent of τ_1. When, however, the amplitude of the alternating field is increased saturation occurs and in the limit of large h, P reaches the value:

$$P_{\text{sat}} \approx M_0 H/\tau_1. \tag{5.9}$$

When interpreting this result it is customary to identify τ_1 with the spin-lattice relaxation time τ_L. Some caution is warranted here. We have already seen in the Casimir-Du Pré formalism that the time dependence of the M_z-component includes spin-spin relaxation effects and so can not be describing by one single characteristic time τ_1. Only when the condition $\chi_{\text{ad}} \ll \chi_0$ is fulfilled, can the effects due to spin-spin relaxation be neglected, as these effects are proportional to χ_{ad}. This is nearly always the case in nuclear spin resonance experiments but certainly not always in the case of electron spin resonance. Another difference is the way in which the spin system is disturbed. In the non-resonance experiments, when the alternating field is in the z-direction, the frequencies used to study spin-lattice relaxation are of the order of τ_L^{-1}. The alternating field in first order cannot induce transitions between the different quantum mechanical states of the spin system but shifts the energy levels periodically. The transitions between the levels, necessary to restore equilibrium, are either induced by internal interactions and/or by the phonons in the lattice. In the resonance experiments the energy levels in first order are not shifted by the high frequency field but transitions are induced within a specific set of states, with a specific energy level spacing and connected with each other by specific matrix elements of the magnetic moment. Only when the energy absorbed in this set of levels is distributed by interactions over all the levels in the spin system, can we expect the equality of the relaxation times τ_1 and τ_L. In this case the resonant absorption has the same effect as a sudden increase of the magnetic field, both cause a rise of the temperature of the spin system. When, however, the equilibrium in the spin system is not reached and the excited set of states relax immediately to the lattice one can not expect, a priori, equality of τ_1 and τ_L.

The use of the term spin-lattice relaxation time for τ_1 almost automatically implies the use of the term spin-spin relaxation time for τ_2. The differences between the spin-spin relaxation affecting the M_z-component and BLOCH's transversal relaxation time τ_2 are evident. As τ_2 is a measure of the paramagnetic resonance linewidth, we shall not consider it further in this article. Furthermore, we shall reserve the term spin-spin relaxation for all the effects happening within the spin system itself that can be measured in the z-direction, the direction of the external constant field in which the sample is placed.

II. Experimental methods.

6. It is beyond the scope of this article to give an extensive review of the apparatus and methods used to measure paramagnetic relaxation. Some principles will be outlined and some references given. As was mentioned experimental methods to measure paramagnetic relaxation fall into two main categories, the non-linear resonance or saturation experiments and the linear non-resonance experiments.

This distinction between linear and non-linear methods is fundamental, in the sense that during saturation experiments the system under consideration i.e. the spin system, can be forced into a state which is far from the equilibrium state. When the intensity of the paramagnetic resonance line which is saturated, is taken as a measure of the temperature of the spin-system, temperatures of ten to hundred times the lattice or equilibrium temperature can easily be reached. In such cases one has to be careful in assuming the heat exchange between spin system and lattice and eventually between lattice and bath to be linearly proportional to the temperature differences. A check on the consistency of the results by measuring with widely varying saturation powers seems to be an absolute necessity.

Another matter of importance, common to all relaxation measurements and particular to spin-lattice relaxation, is the purity of the sample. Measurements of DE VRIES[1] indicate that not only magnetic impurities but also non-magnetic impurities of a few tenths of a percent can affect, by a factor of ten or more, the value of the spin-lattice relaxation time. Aging effects in some paramagnetic samples are now known to be the cause of differences, amounting to twenty percent, in the values of the magnetic specific heat and the adiabatic susceptibility found by different investigators. As these effects are very difficult to detect otherwise, it could be advisable to investigate at least two examples of the same substance, preferably of different origin.

7. The resonance saturation methods.

The resonance saturation methods can be divided in the continuous wave (c.w.) or steady state and the pulse saturation methods. In both cases essentially the same apparatus is used as in paramagnetic resonance spectroscopy, with the addition of some means to measure the power dissipated in the cavity in the steady state apparatus and the use of a high power microwave source in both cases.

The steady state saturation method is based on the solution of the Bloch-equations, (5.8) and (5.9). By increasing the power incident on the cavity and by measuring the absorbed power, either by a change in the reflected power from the cavity or by monitoring the transmitted power in a transmission cavity, one is in principle able to find the value of the relaxation time. Particularly at very low temperatures the method has the disadvantage of the amount of power that has to be dissipated and transferred to the surrounding cooling bath. This could cause an uncertainty in the temperature of the sample. On the other hand, however, do modern detection techniques and components make it possible to use such small samples, that even the power needed to induce complete saturation can be measured in milliwatts? For practical lay-out we refer to the end of this paragraph where a list of references is given.

In the pulse-saturation method a pulse of high power microwaves saturates the sample after which the absorption is monitored, using negligible power, at the same or another frequency. When the pulse ends, absorption will be at a minimum according to (5.8). As the energy in the spin system is transferred to the lattice and from there on to the bath the intensity of the monitored absorption line will increase again. The rate of increase is a measure of the time necessary to establish the equilibrium state. It is possible that different processes are involved in this return to equilibrium, which is one reason not to expect a simple exponential behaviour. Another reason could be the effect of non-linearity. Without these complications, the pulse-method would be one of the most direct ways to measure relaxation times.

It is the great practical advantage of the resonance methods that one is able to investigate relaxation effects in materials that are of interest for practical microwave applications e.g. maser materials, under almost identical circumstances as they will be used. A great drawback is the limitation to a relatively very

---

[1] A. J. DE VRIES: Thesis Leiden 1965; Physica **36**, 65 (1967).

small range of frequencies or, in other words, magnetic fields for each experimental set-up. The study of the field-dependence of the relaxation times is consequently a rather elaborate enterprise.

8. The non resonance methods. While the microwave signals in the resonance methods are used as carrier of relatively very slowly changing information, this is not the case for the other methods, where essentially a Fourier-analysis is made of the relaxation phenomena. Here the fundamental interest lies in the linear response of the material at a specific frequency to a harmonically periodic varying magnetic field of the same frequency.

The basic notion is that if a paramagnetic sample is inserted in a coil with an original inductance L_0 and an original series resistance R_0, the impedance of the coil will have the value:

$$Z = R + i\omega L = R + i\omega L_0 (1 + 4\pi q \chi(\omega))$$
$$= R_0 \left(1 + \frac{\omega L_0}{R_0} \cdot 4\pi q \chi''(\omega)\right) + i\omega L_0 \left(1 + 4\pi q \chi'(\omega)\right) \quad (8.1)$$

where $\chi(\omega) = \chi'(\omega) - i\chi''(\omega)$ and q is the filling factor.

According to this expression it would suffice to measure the impedance of the coil with and without the sample inserted to know the product $q\chi(\omega)$. However, it is not always feasible to insert or remove the sample during the experiment. Furthermore, the actual changes in the inductance of the coil are of the order of at most a percent while changes of the order of 10^{-7} can be of fundamental interest. There exist several different approaches to the problem of measuring these changes, which can be divided according to the quality factor $Q = \omega L_0/R_0$ of the coil attainable. At frequencies exceeding 10^8 Hz the quality factors of the resonance cavities that have to be used are sufficiently large (>500) to make accurate measurements of $\chi(\omega)$ possible. The relative changes of the losses in the cavity as caused by changes in $\chi''(\omega)$ are large, while the changes of the well defined resonance frequency caused by changes of $\chi'(\omega)$ can be measured accurately. The problem left, is to obtain the absolute value of $\chi(\omega)$, which is a problem common to all relaxation measurements. In most cases the solution can be found by applying a sufficiently high magnetic field perpendicular to the direction of the high frequency field. $\chi'(\omega)$ is then known to reach a value equal to the static susceptibility value χ_0. Thus, in principle the value of qL_0 in (8.1) is known and a measurement of the quality factor in this situation $(\chi''(\omega) = 0)$ is sufficient to determine the value of R_0.

At lower frequencies another approach must be followed. One method to increase the sensitivity is the null-method as provided by bridge circuits in which the sample coil is incorporated. Here three ways are open after disturbance of the initial balance of a bridge due to a change in $\chi(\omega)$: compensation for both the changes in L as well as in R, compensation for either ΔL or ΔR in which case the residual output of the bridge in general will be proportional to the non-compensated component and thirdly: no compensation, but phase sensitive detection of the quadrature components of the output that are in general proportional to the quadrature components of $\chi(\omega)$. Of all three possibilities examples can be found in literature, ranging from the old but dependable Hartshorn bridge (frequency range 10 to 10^3 Hz), as example of the first alternative, to the modern example of the third possibility covering a frequency range of about four decades (10^2 to 10^6 Hz) in which the coil that contains the sample forms a part of a simple Wheatstone bridge configuration (Fig. 2).

At sub-audio frequencies the Fourier-analysis method of measuring relaxation phenomena is not practical. Here the method of studying the response to a step in the external magnetic field is indicated, next to the pulse saturation resonance method.

In the gap left between 10^7 Hz and 10^8 Hz, stray capacitances of the leads connecting the bridge to the sample coil, which in general is placed in a cryostat, lead to experimental difficulties and impractical small values of L_0.

It is in this frequency range that the oldest methods to measure paramagnetic relaxation absorption $(\chi''(\omega))$ or dispersion $(\chi'(\omega))$ firmly hold their ground: the calorimeter method and the beat-frequency method respectively.

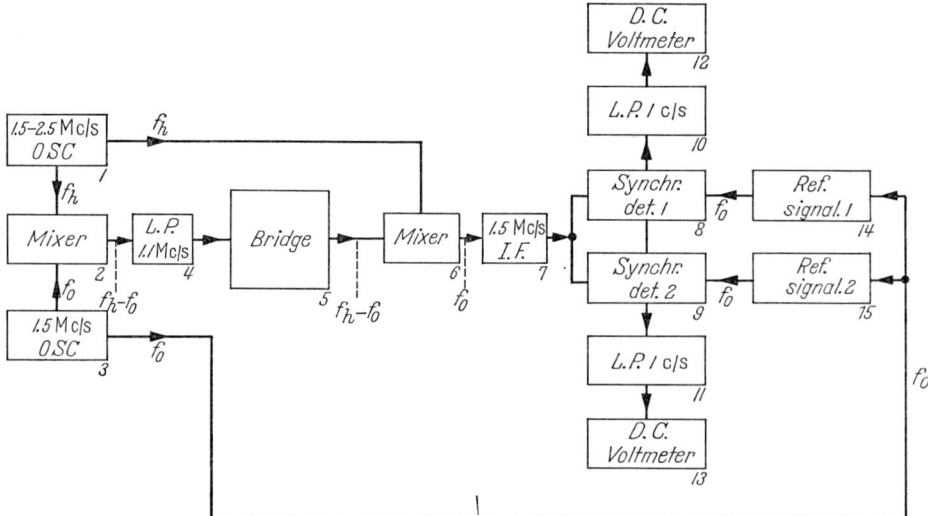

Fig. 2. Block diagram of a modern bridge method measuring system. Taken from A. J. DE VRIES, Thesis Leiden 1965.

In the first method the temperature rise of the sample which is caused by the energy absorption from the high frequency field, is measured. The energy absorption per second being given by (4.27), $\chi''(\omega)$ can be determined when the heat capacities of the calorimeter and the sample and the intensity of the high frequency field are known.

The beat frequency method is based upon the fact that a change of the inductance of a coil, which forms part of the frequency determining circuit in an oscillator, causes the frequency of the generated signal to change by an amount that is approximately proportional to $\chi'(\omega)$:

$$\omega = \frac{1}{\sqrt{CL}} \sim \frac{1}{\sqrt{CL_0}} (1 - 2\pi q \chi'(\omega)).$$

By mixing this oscillator signal with that of a constant frequency generator the low frequency beats can be measured by means of audio frequency apparatus. Modern frequency counters make it possible to measure the frequency directly with sufficient accuracy.

For all the non-resonance apparatus mentioned the sensitivity claimed is about 10^{-7} for samples with a volume of about 1 cm³. This means that a sample has to contain at least 10^{17} spins at a temperature of 1 °K to obtain useful results as compared with claims of 10^{10} to 10^{11} spins for the modern resonance saturation techniques under equivalent circumstances. This enormous difference can be traced back to two causes: the ratio of the quality factors of the resonant cavity ($\sim 10^4$) and the coils used at audio and sub-audio frequencies ($\lesssim 1$) and the ratio of the resonance and relaxation absorption coefficients ($\propto \chi''$) that are actually measured. Comparing (4.25) with (5.5...8) one finds that this last factor is of the order of $H/\Delta H$, the quotient of the external field and the linewidth of the resonance line. This ratio is of the order 10^2 for concentrated samples, to 10^3 for diluted samples.

Literature.

Steady state saturation techniques.
ESCHENFELDER, A. H., and R. T. WEIDNER: Phys. Rev. **92**, 869 (1953).
BÖLGER, B.: Proc. Koninkl. Ned. Akad. Wetenschap B **62**, 348 (1959); — Thesis Leiden 1959.

Pulse saturation techniques.
GIORDMAINE, J. A., L. E. ALSOP, F. R. NASH, and C. H. TOWNES: Phys. Rev. **109**, 302 (1958).
DAVIS jr., F. C., M. W. P. STRANDBERG, and R. L. KYHL: Phys. Rev. **111**, 1268 (1958).
BOWERS, K. D., and W. B. MIMS: Phys. Rev. **115**, 285 (1959).
DREWES, G. W. J.: Thesis Leiden 1967.

Other resonance techniques.
CASTLE jr., J. G., P. F. CHESTER, and P. E. WAGNER: Phys. Rev. **119**, 953 (1960).

Non resonance techniques.
GORTER, C. J.: Paramagnetic Relaxation. Amsterdam: Elsevier Publ. Co. 1947.
VERSTELLE, J. C., G. W. J. DREWES, and C. J. GORTER: Physica **26**, 520 (1960).
DE VRIES, A. J.: Thesis Leiden 1965, Appl. Sci. Res. **17**, 31 (1967).

B. Spin-spin relaxation.

9. Introduction. The history of spin-spin relaxation starts with the paper by WALLER on paramagnetic relaxation. He distinguished the difference from the spin-lattice relaxation and gave an estimate of the spin-spin relaxation time in zero external field. The next important development was a paper by KRONIG and BOUWKAMP[1], who predicted a Gaussian field dependence of the spin-spin relaxation time:

$$\tau_s \propto e^{\alpha H^2}. \qquad (9.1)$$

During and after the second world war BROER[2] extended WALLER's work and rejected KRONIG and BOUWKAMP's result. BROER's work was taken up by CASPERS[3] about 10 years later. CASPERS though not giving all the correct answers, has at least given clear insight into nearly all the problems connected with the spin system. Another approach was due to TJON[4] who criticised some of CASPERS' results. Very recently TERWIEL and MAZUR[5], whom we thank for their permission to use some of their, as yet unpublished work, have shown TJON's approach to be wrong in the general case as well.

Besides this sequence of fundamental work mainly of Dutch origin we should mention that of PHILIPPOT[6]. A more pragmatic viewpoint has been taken else-

---

[1] R. DE L. KRONIG, and C. J. BOUWKAMP: Physica **5**, 521 (1938).
[2] L. J. F. BROER: Physica **10**, 801 (1943).
[3] W. J. CASPERS: Physica **25**, 43, 645 (1959); **26**, 778, 809 (1960).
[4] J. A. TJON: Physica **30**, 1, 10 (1964).
[5] R. H. TERWIEL, and P. MAZUR: Physica **32**, 1813 (1966).
[6] J. PHILIPPOT: Phys. Rev. A **133**, 471 (1964).

where; several papers were published on the assumption that the spin system can be divided into several systems each with its own temperature. It is not generally recognised that Kubo and Tomita[1], in their article on the line shape in paramagnetic resonance were the first who applied this two-systems concept to derive Kronig and Bouwkamp's formula for the spin-spin relaxation time (c.f. Yokota[2]). Furthermore it is in their paper that Kubo and Tomita derive the so-called relaxation function on which the theories by Capsers, Tjon and others are based.

I. General theory.

10. The Hamiltonian. The experimental evidence for the correctness of Casimir and Du Pré's assumption concerning the equilibrium within the spin system, which relaxes as a proper thermodynamical system to the temperature of the lattice, makes it clear that the equilibrium in the spin system itself in general is reached with a characteristic time that is much shorter than the spin-lattice relaxation time. As a consequence one can assume that the spin system is isolated from its surroundings during the process of the establishment of this equilibrium. This fact, that is illustrated in formula (4.22) by the occurrence of the term χ_{ad}, justifies the simplification to two terms of the Hamiltonian (3.1) needed to describe the spin system:

$$\mathcal{H} = \mathcal{H}_S + \mathcal{H}_{S,i}. \tag{10.1}$$

Here, as was mentioned before, $\mathcal{H}_S$ is adequately described by the so-called spin Hamiltonian (Sect. 3). This Hamiltonian consists in general of several terms:

the Zeeman term, to describe the influence of the external magnetic field:

$$(\mathcal{H}_Z)_k = -\mathbf{M}_k \cdot \mathbf{H} = -g_k \beta \, \mathbf{H} \cdot \mathbf{S}_k \tag{10.2}$$

in which g_k is the effective spectroscopic Landé factor of the k-th ion and β is the Bohr magneton,

the crystalline electric field term $(\mathcal{H}_{el})_k$ which we will not specify,

several terms describing the influence of the magnetic moments of the nuclei of the magnetic ions, causing the so-called hyperfine structure splittings $(\mathcal{H}_{h.f.s})_k$. All these terms, mostly linear or quadratic in the effective spin variable $\mathbf{S}_k$ of one specific ion k are summed over all the ions in the sample to obtain:

$$\mathcal{H}_S = \mathcal{H}_Z + \mathcal{H}_{el} + \mathcal{H}_{h.f.s.}. \tag{10.3}$$

The term $\mathcal{H}_{S,i}$ accounts for the interactions between the ions, the magnetic dipole-dipole interaction $\mathcal{H}_{dip}$, and the exchange interaction $\mathcal{H}_{ex}$. $\mathcal{H}_{S,i}$ therefore consists of sums over ion pairs of biquadratic forms of the variables $\mathbf{S}_k$ and $\mathbf{S}_l$ of which typical terms are:

$$(\mathcal{H}_{dip})_{k,l} = g^2 \beta^2 \, r_{kl}^{-3} \{\mathbf{S}_k \cdot \mathbf{S}_l - 3 r_{kl}^{-2} (\mathbf{S}_k \cdot \mathbf{r}_{kl})(\mathbf{S}_k \cdot \mathbf{r}_{kl})\} \tag{10.4}$$

and

$$(\mathcal{H}_{ex})_{kl} = J_{kl} \mathbf{S}_k \cdot \mathbf{S}_l. \tag{10.5}$$

It will be convenient to split $\mathcal{H}_{S,i}$ into different terms:

$$\mathcal{H}_{S,i} = \sum_{m=-2}^{+2} \mathcal{H}_m \tag{10.6}$$

---

[1] R. Kubo, and K. Tomita: J. Phys. Soc. Japan **9**, 889 (1954).
[2] M. Yokota: J. Phys. Soc. Japan **10**, 762 (1955).

for which the following commutation properties hold:

$$[M_z, \mathcal{H}_m] = m\, g\, \beta \mathcal{H}_m: \tag{10.7}$$

$$\left.\begin{aligned}
\mathcal{H}_0 &= \sum_{k>l} a_{kl} S_{zk} S_{zl} + \sum_{k>l} b_{kl}(S_{+k}S_{-l}+S_{-k}S_{+l}), \\
\mathcal{H}_{\pm 1} &= \sum_{k>l} c_{kl}^{\pm}(S_{\pm k}S_{zl}+S_{zk}S_{\pm l}), \\
\mathcal{H}_{\pm 2} &= \sum_{k>l} d_{kl}^{\pm} S_{\pm k}S_{\pm l}
\end{aligned}\right\} \tag{10.8}$$

with

$$S_{\pm} = S_x \pm i S_y \tag{10.9}$$

and

$$\left.\begin{aligned}
a_{kl} &= J_{kl} + g^2 \beta^2 r_{kl}^{-3}(1 - 3\cos^2\vartheta_{kl}), \\
b_{kl} &= \tfrac{1}{2} J_{kl} - \tfrac{1}{4} g^2 \beta^2 r_{kl}^{-3}(1 - 3\cos^2\vartheta_{kl}), \\
c_{kl}^{\pm} &= -\tfrac{3}{2} g^2 \beta^2 r_{kl}^{-3} \sin\vartheta_{kl}\cos\vartheta_{kl}\exp(\mp i\varphi_{kl}), \\
d_{kl}^{\pm} &= -\tfrac{3}{4} g^2 \beta^2 r_{kl}^{-3} \sin^2\vartheta_{kl}\exp(\mp 2i\varphi_{kl})
\end{aligned}\right\} \tag{10.10}$$

in which ϑ_{kl} and φ_{kl} are the polar angles of the radius vector $\boldsymbol{r}_{kl}$ connecting ions k and l.

The relative magnitude of the different terms in (10.1) is of decisive importance in the general aspects of spin-spin relaxation, as they lead to very different relaxation phenomena. This can most easily be seen by considering the response of the spin system to a disturbance in the external field $\boldsymbol{H}$ in the form of a negative step of vanishingly small amplitude $\boldsymbol{h}$ at some time t_0. It is supposed that the spin system is in equilibrium at times $t < t_0$ and is isolated from the lattice, which means that energy is conserved, and thus, the expectation value of $\mathcal{H}$ is time independent during the whole process of establishment of a new equilibrium after t_0:

$$\overline{\mathcal{H}(t)} = \overline{\mathcal{H}}, \quad t > t_0. \tag{10.11}$$

Splitting the Hamiltonian into the Zeeman term $\mathcal{H}_Z$, that is first of all affected by the step in the field, and the rest of the terms $\mathcal{H}'$, this means that:

$$\overline{\mathcal{H}_Z(t)} + \overline{\mathcal{H}'(t)} = \overline{\mathcal{H}}, \quad t > t_0. \tag{10.12}$$

Now, as:

$$\mathcal{H}_Z(t) = -\boldsymbol{H}\cdot\boldsymbol{M}(t) \tag{10.13}$$

and we shall be mainly interested in the time dependence of $\overline{\boldsymbol{M}(t)}$, it follows from (10.12) that this dependence is completely governed by the specific form of $\mathcal{H}'$:

$$\boldsymbol{H}\cdot\frac{d}{dt}\overline{\boldsymbol{M}(t)} = -\frac{d}{dt}\overline{\mathcal{H}'(t)}. \tag{10.14}$$

11. The relaxation function. The arguments in the preceding section can be carried a step further to obtain a general expression for the response of the magnetic moment. We shall use the same method that we used to derive the Curie's law (Sect. 2). To simplify notation we take $t_0 = 0$ and the direction of $\boldsymbol{H}$ and $\boldsymbol{h}$ along the z-axis. For $t<0$ the equilibrium state is described by the density matrix operator $\varrho(-\infty)$:

$$\varrho(-\infty) = \frac{\exp[-\{\mathcal{H}' - (H+h)M_z\}/kT_0]}{\operatorname{Tr}\exp[-\{\mathcal{H}' - (H+h)M_z\}/kT_0]}, \quad t<0. \tag{11.1}$$

According to Curie's law, then:
$$\overline{\mathcal{H}_z(t)} = -\frac{H+h}{kT_0} \langle\langle M_z^2 \rangle\rangle, \quad t<0 \tag{11.2}$$

where we have introduced the notation:
$$\langle\langle A \rangle\rangle = \frac{\operatorname{Tr} A}{\operatorname{Tr} 1}. \tag{11.3}$$

For times $t>0$, the time dependence of $\varrho(t)$ is given by
$$\begin{aligned}\varrho(t) &= \exp[-i(\mathcal{H}'-HM_z)t/\hbar]\varrho(-\infty)\exp[i(\mathcal{H}'-HM_z)t/\hbar] \\ &= \exp[-iLt]\varrho(-\infty).\end{aligned} \tag{11.4}$$

Here we have used the quantum mechanical Liouville operator L, defined by (11.4) or by:
$$LA = \frac{1}{\hbar}[\mathcal{H}, A] \tag{11.5}$$

A being any operator. Thus:
$$\overline{\mathcal{H}_z(t)} = \operatorname{Tr} \varrho(t)\mathcal{H}_z = \operatorname{Tr} \varrho(-\infty)\mathcal{H}_z(t) \tag{11.6}$$

with
$$\mathcal{H}_z(t) = \exp[iLt]\mathcal{H}_z. \tag{11.7}$$

Using the method in deriving Curie's law (Sect. 2) one now finds, retaining only terms linear in h:
$$\overline{\mathcal{H}_z(t)} = \frac{h}{kT_0}\langle\langle \mathcal{H}_z(t) M_z \rangle\rangle - \frac{H^2}{kT_0}\langle\langle M_z^2 \rangle\rangle \tag{11.8}$$

or:
$$\overline{M_z(t)} = \frac{h}{kT_0}\langle\langle M_z M_z(t) \rangle\rangle + \frac{H}{kT_0}\langle\langle M_z^2 \rangle\rangle. \tag{11.9}$$

Defining the relaxation function $\varphi(t)$ by:
$$\varphi(t) = h^{-1}\left\{\overline{M_z(t)} - \frac{H}{kT_0}\langle\langle M_z^2 \rangle\rangle\right\} \tag{11.10}$$

it follows that:
$$\varphi(t) = \frac{1}{kT_0}\langle\langle M_z M_z(t) \rangle\rangle. \tag{11.11}$$

It is however not this relaxation function that will be used in further applications to spin-spin relaxation. By assuming that eventually for t approaching "infinity" a new equilibrium state is reached, that can be described by a new effective temperature T_S for the spin system, it can be shown that $\varphi(t)$ will reach a limiting value that is in general different from zero. Let this new equilibrium be described by the density matrix $\varrho(\infty)$:
$$\varrho(\infty) = \frac{\exp[-(\mathcal{H}'-HM_z)/kT_s]}{\operatorname{Tr}\exp[-(\mathcal{H}'-HM_z)/kT_s]}. \tag{11.12}$$

Then, as the spin system is isolated:
$$\overline{\mathcal{H}(t)} = \overline{\mathcal{H}(0)} = \overline{\mathcal{H}(\infty)} = \overline{\mathcal{H}} \tag{11.13}$$

or:
$$\operatorname{Tr} \varrho(-\infty)(\mathcal{H}'-HM_z) = \operatorname{Tr} \varrho(\infty)(\mathcal{H}'-HM_z) \tag{11.14}$$

from which follows:
$$T_S = T_0(1 + h\langle\langle \mathcal{H} M_z \rangle\rangle/\langle\langle \mathcal{H}^2 \rangle\rangle). \tag{11.15}$$

Now it is possible to define a new relaxation function $\varphi_{ad}(t)$, which has the property that it vanishes in the limit for large t:

$$\begin{aligned}\varphi_{ad}(t) &= h^{-1}[\overline{M_z(t)} - \overline{M_z(\infty)}] \\ &= \varphi(t) - (kT_0)^{-1}\langle\langle\mathcal{H} M_z\rangle\rangle^2/\langle\langle\mathcal{H}^2\rangle\rangle \\ &= (kT_0)^{-1}\{\langle\langle M_z M_z(t)\rangle\rangle - \langle\langle\mathcal{H} M_z\rangle\rangle^2/\langle\langle\mathcal{H}^2\rangle\rangle\}.\end{aligned} \quad (11.16)$$

This expression can be written in a shorter form by making use of projection operators. We define the projection operator P_A that projects any operator 0, in operator space, on to A, or, in other words, selects from the diagonal matrix elements in a representation diagonalizing A, those that are given by the operator equation:

$$P_A 0 = A \frac{\langle\langle A\, 0\rangle\rangle}{\langle\langle A^2\rangle\rangle}. \quad (11.17)$$

It is clear that according to this definition:

$$P_A^2 = P_A \quad (11.18)$$

and

$$(1 - P_A)^2 = 1 - P_A \quad (11.19)$$

so that $(1 - P_A)$ is a projection operator too that extracts the rest of the diagonal and the non-diagonal elements in the A-representation.

Now, looking in the representation diagonalizing the Hamiltonian $\mathcal{H} = \mathcal{H}' - HM_z$, one finds:

$$P_\mathcal{H} M_z(t) = P_\mathcal{H} M_z = \mathcal{H}\frac{\langle\langle \mathcal{H} M_z\rangle\rangle}{\langle\langle\mathcal{H}^2\rangle\rangle} \quad (11.20)$$

and as:

$$\langle\langle M_z M_z(t)\rangle\rangle = \langle\langle P_\mathcal{H} M_z\, P_\mathcal{H} M_z(t)\rangle\rangle + \langle\langle (1 - P_\mathcal{H}) M_z (1 - P_\mathcal{H}) M_z(t)\rangle\rangle \quad (11.21)$$

it follows that:

$$\varphi_{ad}(t) = \frac{1}{kT}\langle\langle (1 - P_\mathcal{H}) M_z (1 - P_\mathcal{H}) M_z(t)\rangle\rangle \quad (11.22)$$

where, in the approximation used, T equals either T_0 or T_S. By introducing the operators μ and $\mu(t)$:

$$\begin{aligned}\mu &= (1 - P_\mathcal{H}) M_z, \\ \mu(t) &= (1 - P_\mathcal{H}) M_z(t)\end{aligned} \quad (11.23)$$

(11.22) can be written as:

$$\varphi_{ad}(t) = (kT)^{-1}\langle\langle \mu\mu(t)\rangle\rangle. \quad (11.24)$$

12. The relation of φ_{ad} to χ_{ad}. In Sect. 4 the relation (4.13):

$$\chi_{ad}/\chi_0 = C_M/C_H$$

was derived. As the relaxation function $\varphi_{ad}(t)$ has been derived with equivalent assumptions we expect a relation between $\varphi_{ad}(t)$ and χ_{ad} to exist. We shall therefore first have to obtain expressions for the specific heats C_H and C_M. According to (4.6) the specific heat at constant field C_H is given by $C_H = (\partial U/\partial T)_H$, which becomes by using the density matrix in the high temperature limit:

$$C_H = \left\{\frac{\partial}{\partial T}\operatorname{Tr}\varrho\mathcal{H}\right\}_H = \frac{1}{kT^2}\langle\langle\mathcal{H}^2\rangle\rangle. \quad (12.1)$$

The specific heat at constant magnetisation is, c.f. (4.8):

$$C_M = C_H - T \left(\frac{\partial M_z}{\partial T}\right)_H^2 \bigg/ \left(\frac{\partial M_z}{\partial H}\right)_T$$
$$= \frac{1}{kT^2} \langle\langle \mathscr{H}^2 \rangle\rangle - \frac{1}{kT^2} \frac{\langle\langle \mathscr{H} M_z \rangle\rangle^2}{\langle\langle M_z^2 \rangle\rangle} \quad (12.2)$$

where use has been made of the fact that $\text{Tr}\mathscr{H} = 0$.

Taking (11.16) together with (4.13), (12.1) and (12.2) one finds:

$$\chi_{\text{ad}} = (kT)^{-1} \langle\langle \mu^2 \rangle\rangle = \varphi_{\text{ad}}(0), \quad (12.3)$$

which result is in accordance with the thermodynamical approach by CASIMIR and DU PRÉ.

13. The generalised master equation. Extending the results of the preceding sections a step further TERWIEL and MAZUR have derived a kind of master equation that is the basis of their theory concerning spin-spin relaxation. Use is made of the very elegant methods of ZWANZIG[1] to find the time development of the relaxation function.

Regarding the specific form of the autocorrelation function of μ in (11.24) it is realised that the relevant part of $\mu(t)$ is contained in the part that is diagonal in the representation diagonalizing μ and can be extracted by using the projection operator P_μ:

$$kT \varphi_{\text{ad}}(t) = \langle\langle \mu P_\mu \mu(t) \rangle\rangle \equiv \langle\langle \mu \mu_R(t) \rangle\rangle. \quad (13.1)$$

Now as:

$$\mu(t) = \exp[iLt] \mu(0) \quad (13.2)$$

one finds by differentiation, writing $\mu_I(t)$ for the irrelevant part of $\mu(t)$:

$$\frac{d\mu_R(t)}{dt} = i P_\mu L \mu_R(t) + i P_\mu L \mu_I(t),$$
$$\frac{d\mu_I(t)}{dt} = i(1-P_\mu) L \mu_R(t) + i(1-P_\mu) L \mu_I(t). \quad (13.3)$$

The formal solution of $\mu_I(t)$ is given by:

$$\mu_I(t) = \exp[i(1-P_\mu)Lt]\mu_I(0) + i\int_0^t d\tau \exp[i(1-P_\mu)L\tau](1-P_\mu) L \mu_R(t-\tau). \quad (13.4)$$

Substituting this in (13.3) gives:

$$\frac{d\mu_R(t)}{dt} = iP_\mu L \mu_R(t) - \int_0^t d\tau\, G(\tau) \mu_R(t-\tau) + i K(t) \mu_I(0) \quad (13.5)$$

with:

$$G(\tau) = P_\mu L \exp[i(1-P_\mu)L\tau] L P_\mu \quad (13.6)$$

and:

$$K(t) = P_\mu L \exp[i(1-P_\mu)Lt]. \quad (13.7)$$

Now as:

$$\mu_I(0) = (1-P_\mu)\mu = 0 \quad (13.8)$$

and:

$$P_\mu L P_\mu A = 0 \quad (13.9)$$

---

[1] R. ZWANZIG: Lectures in Theoretical Physics, vol. III, p. 106. New York: Interscience Publ. 1961.

for any operator A and due to the fact that:

$$P_\mu \mu_R(t-\tau) = \mu \frac{\langle\langle \mu\mu(t-\tau)\rangle\rangle}{\langle\langle \mu^2\rangle\rangle} \tag{13.10}$$

(13.5) reduces to:

$$\frac{d\varphi_{ad}(t)}{dt} = -\int_0^t d\tau \frac{\langle\langle \mu L \exp[i(1-P_\mu)L\tau]L\mu\rangle\rangle}{\langle\langle \mu^2\rangle\rangle} \varphi_{ad}(t-\tau). \tag{13.11}$$

Mention has to be made here that MAZUR and TERWIEL made use of a different operator μ'. To take account of the limiting value of $\varphi(t)$, they define:

$$\varphi_{ad}(t) = \varphi(t) - \varphi(\infty) = (kT)^{-1}\langle\langle \mu'\mu'(t)\rangle\rangle \tag{13.12}$$

with:

$$\mu' \equiv M_z - i[kT\,\varphi(\infty)]^{\frac{1}{2}} = M_z - i\frac{\langle\langle \mathscr{H} M_z\rangle\rangle}{\langle\langle \mathscr{H}^2\rangle\rangle^{\frac{1}{2}}} \tag{13.13}$$

which has to be compared with the definition in Sect. 11:

$$\mu = (1 - P_\mathscr{H})M_z = M_z - \mathscr{H}\frac{\langle\langle \mathscr{H} M_z\rangle\rangle}{\langle\langle \mathscr{H}^2\rangle\rangle} \tag{13.14}$$

The definitions are seen not to be equivalent but lead to the same results when used in the autocorrelation function $\langle\langle \mu\mu(t)\rangle\rangle$.

Although the final expression found looks promising, the problem of spin-spin relaxation is by no means solved as will be shown in the rest of the chapter. To end this part, the relation between the relaxation function in general and the complex susceptibility will be reviewed.

14. The relaxation function and complex susceptibility. The linearity assumption made in the derivation of both the relaxation function and the complex susceptibility makes it possible by the principle of superposition to find a relation connecting the two viewpoints of relaxation. Formally one can arrive at this result by regarding the delta-function as the derivative of the unit step function and using its Fourier representation. The required relation is:

$$\chi(\omega) = \chi'(\omega) - i\chi''(\omega) = -\int_0^\infty \frac{d\varphi(t)}{dt} \exp[-i\omega t]\,dt \tag{14.1}$$

or taking the real and imaginary components separately:

$$\chi'(\omega) = \chi_0 - \omega \int_0^\infty \varphi(t)\sin\omega t\,dt \tag{14.2}$$

and

$$\chi''(\omega) = \omega \int_0^\infty \varphi(t)\cos\omega t\,dt. \tag{14.3}$$

15. The Kramers-Kronig relations. As a consequence of the causality condition imposed on the relaxation-function, i.e. no response exists before a disturbance has taken place, there exist relations between $\chi'(\omega)$ and $\chi''(\omega)$, the so-called Kramers-Kronig relations:

$$\chi'(\omega) - \chi'(\infty) = 2/\pi \int_0^\infty \frac{\omega_1\chi''(\omega_1) - \omega\chi''(\omega)}{\omega_1^2 - \omega^2}\,d\omega_1 \tag{15.1}$$

and:

$$\chi''(\omega) = -2\omega/\pi \int_0^\infty \frac{\chi'(\omega_1) - \chi'(\omega)}{\omega_1^2 - \omega^2}\,d\omega_1. \tag{15.2}$$

One of the most useful relations that can be derived from the first relation is one connecting the total dispersion to the total intensity of the absorption:

$$\chi'(0) - \chi'(\infty) = 2/\pi \int_0^\infty \chi''(\omega)/\omega \, d\omega. \tag{15.3}$$

According to this relation the total intensity of the absorption over the frequency range in which spin-spin relaxation phenomena are taking place is proportional to χ_{ad}. For this reason the results of experiments are often given as a ratio of χ_{ad}.

A second conclusion from the Kramers-Kronig relations is the following constraint on the area under the dispersion curve:

$$\int_0^\infty \{\chi'(\omega) - \chi'(\infty)\} \, d\omega = 0. \tag{15.4}$$

To obtain this expression we made use of the theoretical result that the even moments of $\chi''(\omega)/\omega$ are finite and so:

$$\lim_{\omega \to \infty} \omega \, \chi''(\omega) = 0 \tag{15.5}$$

(15.4) indicates that in all physical cases $\chi'(\omega)$ will eventually reach negative values.

For the moments mentioned the following relations with the relaxation function can be derived by differentiation of $\varphi(t)$:

$$\langle \omega^n \rangle = \frac{\int_{-\infty}^\infty \{\chi(\omega)/\omega\} \omega^n \, d\omega}{\int_{-\infty}^\infty \{\chi(\omega)/\omega\} \, d\omega} = (-i)^n \frac{\left. \frac{d^n \varphi(t)}{d t^n} \right|_{t=0}}{\varphi(0)} \tag{15.6}$$

provided $\varphi(t)$ is differentiable.

The existence of a finite second moment $\langle \omega^2 \rangle$ excludes the existence of a single relaxation process of the form:

$$\varphi(t) \propto \exp[-t/\tau] \tag{15.7}$$

as can be shown by inspecting the Fourier-transform of $\varphi(t)$, although in some special cases results are found that are experimentally indistinguishable from this pure exponential decay.

II. Applications of the general theory.

The spin-spin relaxation phenomena can be divided theoretically as well as experimentally into at least three main groups: those

a) without an externally applied magnetic field and without exchange interactions,

b) without crystalline electric field splittings,

c) with all the terms in the Hamiltonian present.

a) $\mathcal{H}_z = \mathcal{H}_{ex} = 0$.

A further division has to be made according to the presence or absence of electric field splittings.

16. $\mathcal{H}_Z = \mathcal{H}_{ex} = \mathcal{H}_{el} = 0$. This case has been considered by WALLER. Later contributions to the theory have been given by BROER[1] and Miss WRIGHT[2]. The method used is that of the computation of the moments of the frequency distribution function defined by:

$$f(\omega) = \chi''(\omega)/\omega. \tag{16.1}$$

This method is in principle based on the possibility to define a function by all its moments, but has the great disadvantage that the computation of the moments higher than the second in practical cases is of prohibitive complexity. Furthermore, it has been shown by GRANT[3] that the knowledge of only a few moments does not even approximately define the wanted function, but that it is possible to find functions of completely different appearance having the same second, fourth and even sixth moments. It is therefore more convenient to postulate a specific form of $f(\omega)$, based on more or less theoretical expectations, and use the moments as parameters to fit this function to the specific case under consideration. Usually, the Gaussian form for either $f(\omega)$ or its Fourier transform $\varphi(t)$ is chosen (BROER[1]), partly because of its mathematical convenience, partly because it seems to fit the experimental results quite well (LOCHER, GORTER)[4]. Thus, as here $\chi_{ad} = \chi_0$, one choses:

$$\varphi_{ad}(t) = \chi_0 \exp[-\alpha t^2] \tag{16.2}$$

with:

$$\alpha = \tfrac{1}{2} \langle \omega^2 \rangle \tag{16.3}$$

or:

$$f(\omega)/\chi_0 = (\pi/2)^{\frac{1}{2}} \langle \omega^2 \rangle^{-\frac{1}{2}} \exp[-\omega^2/2\langle \omega^2 \rangle]. \tag{16.4}$$

$\langle \omega^2 \rangle$ is completely determined by the dipole-dipole interaction term:

$$\begin{aligned}\langle \omega^2 \rangle &= -\langle\langle M_z \ddot{M}_z(0) \rangle\rangle/\langle\langle M_z^2 \rangle\rangle = \hbar^{-2} \langle\langle [\mathcal{H},[\mathcal{H}, M_z]] M_z \rangle\rangle/\langle\langle M_z^2 \rangle\rangle \\ &= \hbar^{-2} \langle\langle [\mathcal{H}_{dip}, M_z][M_z, \mathcal{H}_{dip}] \rangle\rangle/\langle\langle M_z^2 \rangle\rangle. \end{aligned} \tag{16.5}$$

Miss WRIGHT has performed calculations of the fourth moment for ions in a simple cubic lattice and found the value in reasonable agreement with the assumption of the Gaussian function.

A remark should be made here on the use of the relaxation parameter $\varrho' = 2\pi \tau'$ in earlier publications on spin-spin relaxation experiments. All these experiments were performed on the low-frequency side of the relaxation absorption maximum and $\chi''(\omega)$ was found to be proportional to ω. In analogy with the Debye relaxation or pure exponential decay of $\varphi(t)$, the ratio was defined as the spin-spin relaxation time τ':

$$\chi''(\omega)/\chi_0 = \omega \tau'/(1 + \omega^2 \tau'^2) \approx \omega \tau'. \tag{16.6}$$

Although τ' thus defined, cannot be identified with a proper relaxation time, it has a definite meaning, as can be seen from:

$$\tau' = \frac{f(0)}{\chi_0} = (\pi/2)^{\frac{1}{2}} \langle \omega^2 \rangle^{-\frac{1}{2}} \tag{16.7}$$

and thus is a measure for the second moment of the frequency distribution function $f(\omega)$.

---
[1] L. J. F. BROER: Physica **10**, 801 (1943).
[2] A. WRIGHT: Phys. Rev. **76**, 826 (1949).
[3] W. J. C. GRANT: Physica **30**, 1433 (1964).
[4] P. R. LOCHER, and C. J. GORTER: Physica **27**, 997 (1961)

17. $\mathscr{H}_z = \mathscr{H}_{ex} = 0$, $\mathscr{H}_{el} \neq 0$. This case, which has some similarity with nuclear quadrupole resonance, has also been treated by BROER. Besides the zero-frequency band, analogous to Sect. 16, $f(\omega)$ has a second maximum centered at the frequency $\Delta E/\hbar$, where ΔE is the energy splitting caused by the crystalline field. If there exist more seperate levels, $f(\omega)$ will have more peaks, corresponding to the energy differences. No satisfactory treatment has been given yet of the width of either the zero-frequency band or the other peaks. The intensity of the different peaks can be computed from the matrix elements of the magnetic moment connecting the different levels in the single ion spin Hamiltonian representation.

b) $\mathscr{H}_{el} = 0$.

18. The interaction terms $\mathscr{H}_{dip}$ and $\mathscr{H}_{ex}$ show their different behaviour clearly when an external field is introduced. $\mathscr{H}_{ex}$ commutes with the Zeeman term $\mathscr{H}_z$. Only part of $\mathscr{H}_{dip}$ does so. The non-commuting or non-secular part of $\mathscr{H}_{dip}$ can be considered as providing a connecting mechanism between $\mathscr{H}_z$ and $\mathscr{H}_{ex}$. This view has been taken by BLOEMBERGEN and WANG[1] among others in a model in which the spin system is divided into a Zeeman-energy reservoir and an exchange-energy reservoir, each a thermodynamic system with its own temperature. The dipole-dipole energy is neglected in this model except to provide a heat link between the two systems. The exchange reservoir is considered to be in good thermal contact with the lattice. Thus the model is completely equivalent to the one proposed by CASIMIR and DU PRÉ (Sect. 4) and leads to a simple exponential decay. This simple model can be extended to the situation of a completely isolated spin system. Thus the energy fluctuations that are introduced in the Zeeman energy reservoir by the alternating field are not transported to the lattice, as in the model of BLOEMBERGEN and WANG, but will cause temperature fluctuations of the exchange reservoir. It appears that this simple model is able to provide a surprisingly good phenomenological description of the relaxation processes, even when the dipole-dipole energy is not neglected. We shall use it therefore as an illustration in the following sections to provide a clearer insight.

Extending CASIMIR and DU PRÉ's derivation to this case immediately leads to the question, how to account for the dipole-dipole energy. The secular part of $\mathscr{H}_{dip}$ that commutes with M_z can be taken together with the exchange energy. For the non-secular part we shall, following the suggestion of JEENER e.a.[2], make the choice to include it in the Zeeman system. The argument for this choice is that the non-secular part plays a major role in establishing the equilibrium situation within the Zeeman system leading, as was shown by CASPERS among others, to two resonance-like absorption peaks centred at the Larmor-frequency and at twice the Larmor frequency. With this assumption Eq. (4.12) now reads:

$$dQ = (C_H)_Z \left(\frac{\partial T_Z}{\partial M_z}\right)_H dM_z + (C_M)_Z \left(\frac{\partial T_Z}{\partial H}\right)_{M_z} dH \qquad (18.1)$$

in which the subscript Z indicates the Zeeman system. Without the assumption above $(C_M)_Z$ would vanish. Now:

$$(C_M)_Z = \frac{b_{n.s.}}{T^2} \qquad (18.2)$$

in which:

$$b_{n.s.} = 2\{\langle\langle\mathscr{H}_1 \mathscr{H}_{-1}\rangle\rangle + \langle\langle\mathscr{H}_2 \mathscr{H}_{-2}\rangle\rangle\} \qquad (18.3)$$

---

[1] N. BLOEMBERGEN, and S. WANG: Phys. Rev. **92**, 72 (1954).
[2] J. JEENER, H. EISENDRATH, and R. VAN STEENWINKEL: Phys. Rev. A **133**, 478 (1964).

and:
$$(C_H)_Z = \frac{b_{n.s.}}{T^2} + \frac{CH^2}{T^2}, \tag{18.4}$$

C being Curie's constant and
$$CH^2 = \langle\langle \mathscr{H}_Z^2 \rangle\rangle. \tag{18.5}$$

The heat exchange per second between the Zeeman system and the exchange system is again assumed to be proportional to the temperature difference:
$$\frac{dQ}{dt} = -\alpha(T_Z - T_{ex}) \tag{18.6}$$

This heat input into the exchange system will lead to a temperature rise:
$$\alpha(T_Z - T_{ex}) = C_{ex} \frac{dT_{ex}}{dt}, \tag{18.7}$$

in which:
$$C_{ex} = \frac{\langle\langle \mathscr{H}_0^2 \rangle\rangle}{T^2} = \frac{b_0}{T^2}. \tag{18.8}$$

Analogous to the rest of the derivation in Sect. 4 one now finds:
$$\frac{\chi(\omega)}{\chi_0} = \frac{(C_M)_Z + \alpha C_{ex}(\alpha + i\omega C_{ex})^{-1}}{(C_H)_Z + \alpha C_{ex}(\alpha + i\omega C_{ex})^{-1}} \tag{18.9}$$

which can be transformed to:
$$\frac{\chi(\omega)}{\chi_{ad}} = \frac{b + CH^2}{b_{n.s.} + CH^2}\left(\frac{b_{n.s.}}{b} + \frac{CH^2}{b+CH^2}\frac{b_0}{b}\frac{1}{1+i\omega\tau_S}\right) \tag{18.10}$$

in which:
$$b = b_{n.s.} + b_0, \tag{18.11}$$

$$\tau_S = \frac{b_0}{b+CH^2}\frac{(C_H)_Z}{\alpha} \tag{18.12}$$

and:
$$\chi_{ad} = \frac{b}{b+CH^2}\chi_0. \tag{18.13}$$

In the high-field limit (18.10) reduces to:
$$\frac{\chi(\omega)}{\chi_{ad}} = \frac{b_{n.s.}}{b} + \frac{b_0}{b}\frac{1}{1+i\omega\tau_S} \tag{18.14}$$

which was derived first by Caspers. He did not use the two-reservoir model.

In view of the arbitrariness of the assumption made, that is including $b_{n.s.}$ in $(C_H)_Z$, it is not certain whether (18.10) is unconditionally valid. The main features as depicted by (18.14) and in the limiting case of Bloembergen and Wang's model, where $b_{n.s.} \approx 0$ are in agreement with other theories. The two-reservoir model with the two temperatures therefore seems a legitimate assumption, as was argued by Philippot[1], who gave a thermodynamical interpretation of T_Z and T_{ex}.

Considering now the limiting cases in detail, making use of the relaxation function, we first distinguish the case:

19. $H_{ex} \gg \mathscr{H}_{dip}, \mathscr{H}_{el} = 0$. As is known so far this is the only case for which a satisfactory solution exists in terms of the generalised master equation (Sect. 13):
$$\frac{d\varphi_{ad}(t)}{dt} = -\int_0^t d\tau\, F(\tau)\, \varphi_{ad}(t-\tau). \tag{19.1}$$

---
[1] J. Philippot: Phys. Rev. A **133**, 471 (1964).

The reason for this is that $\varphi_{ad}(t)$ is a relatively slowly changing function of time, as is illustrated by expression (18.10), which reduces in this case to:

$$\frac{\chi(\omega)}{\chi_{ad}} = \frac{1}{1+i\omega\tau_S}, \tag{19.2}$$

the constant term $b_{n.s.}/b$ being approximately zero. This term, when not zero, corresponds to a rapid drop of $\varphi_{ad}(t)$ before relaxation takes place. $F(t)$ on the other hand is a rapidly decreasing function of time, so one is allowed to make the following approximation:

$$\frac{d\varphi_{ad}(t)}{dt} = -\int_0^\infty d\tau\, F(\tau)\, \varphi_{ad}(t) \tag{19.3}$$

with the solution:

$$\varphi_{ad}(t) = \chi_{ad} \exp[-t/\tau_S] \tag{19.4}$$

and:

$$\frac{1}{\tau_S} = \int_0^\infty d\tau\, F(\tau). \tag{19.5}$$

For the computation of τ_S in terms of the interactions use is made of the Fourier transform of the kernel $F(t)$:

$$\hat{F}(\omega) = \frac{1}{\pi}\int_{-\infty}^\infty d\tau\, F(\tau) \exp[-i\omega\tau], \tag{19.6}$$

so that:

$$\frac{1}{\tau_S} = \pi\hat{F}(0). \tag{19.7}$$

$\hat{F}(\omega)$ is a function peaked at multiples of the Larmor frequency, from which any but the first two, in the weak coupling limit used, can be neglected. The moments of $\hat{F}(\omega)$ are given by the derivatives of $F(t)$. Here again as in Sect. 16 the Gaussian line shape is chosen to approximate the different portions of $\hat{F}(\omega)$, resulting in:

$$\frac{1}{\tau_S} = \frac{1}{\tau_{S,1}} + \frac{1}{\tau_{S,2}} \tag{19.8}$$

with:

$$\frac{1}{\tau_{S,1}} = 2\pi\left(\frac{g\beta}{\hbar}\right)^2 \frac{\langle\langle\mathscr{H},\mathscr{H}_{-1}\rangle\rangle}{\langle\langle\mu^2\rangle\rangle} \frac{1}{(2\pi\langle\omega^2\rangle_1)^{\frac{1}{2}}} \exp[-(g\beta H)^2/2\hbar^2\langle\omega^2\rangle_1], \tag{19.9}$$

$$\frac{1}{\tau_{S,2}} = 2\pi\left(\frac{g\beta}{\hbar}\right)^2 \frac{4\langle\langle\mathscr{H}_1\mathscr{H}_{-2}\rangle\rangle}{\langle\langle\mu^2\rangle\rangle} \frac{1}{(2\pi\langle\omega^2\rangle_2)^{\frac{1}{2}}} \exp[-(2g\beta H)^2/2\hbar^2\langle\omega^2\rangle_2] \tag{19.10}$$

and

$$\langle\omega^2\rangle_{1,2} = \frac{1}{\hbar^2} \frac{\langle\langle[\mathscr{H}_0,\mathscr{H}_{1,2}][\mathscr{H}_{-1,-2},\mathscr{H}_0]\rangle\rangle}{\langle\langle\mathscr{H}_{1,2}\mathscr{H}_{-1,-2}\rangle\rangle}. \tag{19.11}$$

As nowhere in the derivation of these formula is any condition made on the magnitude of the external field, (19.9) and (19.10) are valid for all field values.

20. Exchange interaction not dominating; $\mathscr{H}_{el} = 0$. The establishment of equilibrium within the Zeeman system now is accompanied by a rapid decrease of the relaxation function in the form of damped oscillations (Fig. 3). In first order approximation the frequencies of these oscillations are exact multiplies of the Larmor frequency. The intensities of these Larmor lines in the Fourier transform $f(\omega)$ of $\varphi_{ad}^{(t)}$ are given by:

$$I_{L1} = 2\frac{\langle\langle\mathscr{H}_1\mathscr{H}_{-1}\rangle\rangle}{b} \chi_{ad} \tag{20.1}$$

and
$$I_{L2} = 2 \frac{\langle\langle \mathcal{H}_2 \mathcal{H}_{-2}\rangle\rangle}{b} \chi_{ad} \tag{20.2}$$

in the high-field limit, leaving for the intensity of the relaxation band at zero-frequency:
$$I_0 = \frac{\langle\langle \mathcal{H}_0^2 \rangle\rangle}{b} \chi_{ad}. \tag{20.3}$$

This is the same result as was found in the high-field limit for $\chi(\omega)$ (18.14). A difficulty now arises in the application of the generalised master equation as the

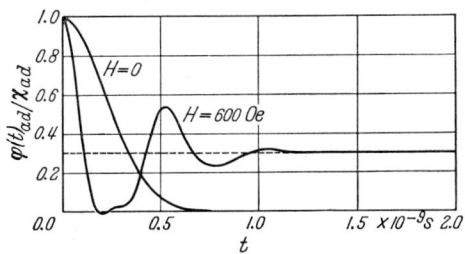

Fig. 3. The relaxation function $\varphi_{ad}(t)/\chi_{ad}$ for $CuCs_2(SO_4)_2 \cdot 6 H_2O$ for $H=0$ and $H=600$ Oe. $\tau_s \sim 8 \cdot 10^{-8}$ sec for $H=600$ Oe. Taken from P. R. LOCHER, and C. J. GORTER, Physica **27**, 997 (1961).

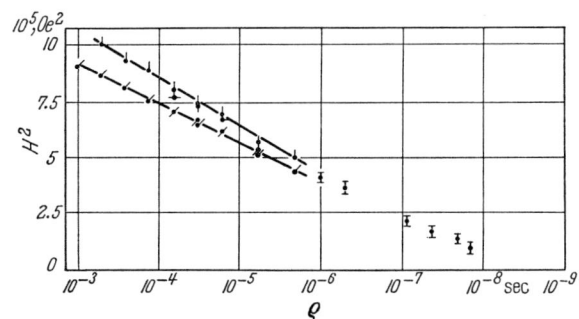

Fig. 4. The relaxation parameter $\varrho = 2\pi \tau_s$ as a function of H^2 for $CuCs_2(SO_4)_2 \cdot 6 H_2O$. The lines have been drawn through the values as measured in single crystals, upper line $H//K_2$-axis, lower line $H//K_1$-axis. The other values have been found in powdered samples. Taken from A. J. DE VRIES, Thesis Leiden 1965.

approximation to take $\varphi_{ad}(t-\tau)$ outside the integral is not allowed. Another complication is the specific form of the operator μ:
$$\mu = (1 - P_{\mathcal{H}}) M_z = \frac{b}{b + CH^2} M_z + \frac{CH^2}{b + CH^2} \left(\frac{\mathcal{H}_{ex} + \mathcal{H}_{dip}}{H} \right) \tag{20.4}$$

from which can be seen that the non-secular part of $\mathcal{H}_{dip}$ cannot be neglected, in the high-field limit. This leads to considerable complications in the computation of $\widehat{F}(\omega)$. This fact was not recognised by CASPERS and TJON in their theories on spin-spin relaxation, where they essentially took the weak-coupling limit in the previous section. Investigations by MAZUR and TERWIEL, as yet unpublished, make it clear, however, that the system will tend to equilibrium with a characteristic time τ_S, in the high field limit, analogous to (19.8)

$$\frac{1}{\tau_S} = \frac{1}{\tau_{S,1}} + \frac{1}{\tau_{S,2}} + \cdots \tag{20.5}$$

in which the leading terms again will be of the general form:

$$\frac{1}{\tau} = C_1 H^2 \exp[-C_2 H^2]. \tag{20.6}$$

The Gaussian field dependence here also stems from a line-shape assumption. Experimental results qualitatively justify this assumption exceedingly well. Measurements by De Vries[1] (to be published in Physica) (Fig. 4) on copper and cobalt tutton salts qualitatively agree with the general expression (20.6) over several orders of magnitude.

21. $\mathscr{H}_Z, \mathscr{H}_{el} = 0, \mathscr{H}_{ex} \neq 0$. When there is no external field, the form of $f(\omega)$ has again to be determined from its moments. In this case, however, it is not sufficient to determine only the second moment as the Gaussian line shape does not give a good description. In fact the second moment of $f(\omega)$ is independent of exchange (16.5). The fourth moment is increased by the presence of an exchange term in the Hamiltonian indicating a shift of the intensity from the central part of the line to the wings. In the case of extremely large exchange interaction the line is known to have a Debye or Lorentz form, corresponding to an exponential decrease of the relaxation function. A mixed Gaussian-Lorentz line form containing two parameters has been proposed to fit the experimental results (Locher and Gorter[2])

$$f(\omega) = A \frac{\tau}{1 + \omega^2 \tau^2} \exp[-\alpha^2 \tau^2 \omega^2] \tag{21.1}$$

with the normalising constant:

$$A^{-1} = \exp[\alpha^2] \operatorname{Erfc} \alpha \tag{21.2}$$

the second and fourth moments are given by:

$$\langle \omega^2 \rangle = (B-1)\tau^{-2}, \tag{21.3}$$

$$\langle \omega^4 \rangle = \{1 + B(\tfrac{1}{2}\alpha^2 - 1)\}\tau^{-4}, \tag{21.4}$$

in which

$$B^{-1} = \alpha \pi^{\tfrac{1}{2}} \exp[\alpha^2] \cdot \operatorname{Erfc} \alpha, \tag{21.5}$$

Reasonable agreement with experiment is found. Tjon has given theoretical arguments for its use.

c) $\mathscr{H}_{el} \neq 0$.

22. The complication brought about by the presence of electric field splittings is demonstrated in the work of Caspers, who extended his theory on spin-spin relaxation to this case. Another attempt to give an explanation of the experimental results is due to Bloembergen and co-workers[3] in an article on cross-relaxation phenomena. The relaxation phenomena for which these theories were extended, were first reported by De Vrijer and Gorter[4]. The effects are strongest in salts for which the dipole-dipole energy is small compared to the electric field splitting and are consequently found most easily in samples in which the concentration of magnetic ions is decreased by dilution with non-magnetic ions. Relaxation takes place in external fields from zero up to about those fields for which the Zeeman energy equals the electric field splitting. The relaxation time

---

[1] De Vries: Thesis Leiden 1965; Physica **36**, 91 (1967).
[2] P. R. Locher, and C. J. Gorter: Physica **27**, 997 (1961).
[3] N. Bloembergen, P. Shapiro, P. S. Pershan, and J. O. Artman: Phys. Rev. **114**, 445 (1959).
[4] F. W. de Vrijer, and C. J. Gorter: Physica **18**, 549 (1959).

decreases with increasing magnetic field. The intensity of the relaxation effect is sharply peaked at some intermediate field, the maximum value depending strongly on the ratio $\mathcal{H}_{dip}/\mathcal{H}_{el}$, as does the relaxation time itself. The smaller this ratio, the larger the intensity and the longer the relaxation time.

All or most of these properties can be explained by the above mentioned theories. In essence in both theories the probability is calculated for multiple-spin transitions in which Zeeman energy is approximately conserved, but in which the magnetic moment of the states between which the transition takes place is different. The excess in energy is thought to be taken up by the dipole-dipole interaction energy. An example of such a process is given in Fig. 5, where an

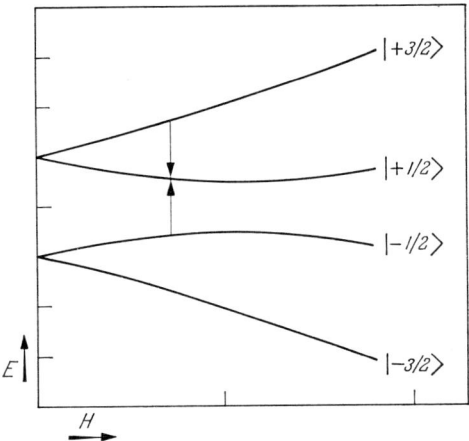

Fig. 5. Energy level scheme for an ion with $S=\frac{3}{2}$ in a trigonal electric crystal field. The states are indicated by the high field quantum numbers.

energy level diagram is drawn for an ion with effective spin $\frac{3}{2}$. As can be seen in the process of the two-spin transition, indicated by arrows, energy is conserved but the total magnetic moment, being equal to $\sum_i -\frac{\partial E_i}{\partial H}$, is changed. By means of such transitions a new equilibrium population of the levels can be established after a change of the external magnetic field, the change in magnetic moment accounting for the corresponding dispersion. The drawback of this simple model, however, is the anisotropy that has to be expected when the direction of the external field is changed with respect to the direction of the electric field gradient. The energy level diagram is then changed and the process pictured becomes increasingly improbable at the given value of the external field. This anisotropy has not been found experimentally. This could mean that the model has to be extended to include processes of more than two spins, which will be increasingly difficult to compute.

In strong magnetic fields, when the Zeeman energy is much larger than $\mathcal{H}_{el}$, the same qualitative behaviour is found as in salts without electric crystalline field splittings. This can be expected, as in this case $\mathcal{H}_{el}$ can be considered as a perturbation with a corresponding relatively small shift of the energy levels in the spin system.

The influence of exchange interactions has not yet been investigated, except in the case of extremely strong exchange, where the influence of $\mathcal{H}_{el}$ is drowned and the same effects are found as in samples without electric field splittings.

III. Cross-relaxation.

23. In the article by BLOEMBERGEN, SHAPIRO, PERSHAN and ARTMAN mentioned in the previous section a general formalism is given by which all spin-spin relaxation phenomena could be explained in terms of so-called cross-relaxation processes. The term cross-relaxation has been coined to describe effects in paramagnetic resonance saturation experiments. By cross-relaxation processes the power absorbed in a sample by saturating a specific line in the resonance spectrum is thought to be distributed over the whole energy spectrum of the spin system by spin-spin interactions. In paramagnetic resonance experiments cross-relaxation can be shown to exist by monitoring a resonance line lying close to the line which is saturated. When this last line is saturated the intensity of the first line is seen to decrease. Obviously power is transported within the energy level scheme of the spin system with the effect that the population difference of the levels connected by the transitions that cause the monitored resonance line, decreases. BLOEMBERGEN e.a. propose to evaluate the probability for these cross-relaxation transitions by finding the appropriate interaction term in the Hamiltonian that is able to cause a transition by which for example one spin in the excited state of the saturated line flips down, while another spin is excited, taking up the energy. The probability of this process is then, using first order time-dependent perturbation theory, proportional to the square of this interaction operator:

$$w \propto |\mathcal{H}_{int}|^2. \tag{23.1}$$

To account for the influence of the rest of the spin system a line shape factor $f(\Delta H)$ is introduced, so that the cross-relaxation time will be given by an expression of the form:

$$\frac{1}{\tau_{cr}} = A |\mathcal{H}_{int}|^2 f(\Delta H). \tag{23.2}$$

A being a constant independent of the frequency or the magnetic field and ΔH being the separation of the two resonance lines. Several methods to approximate the function $f(\Delta H)$ have been given.

In order to investigate the assumptions that have to be made to derive an expression like (23.2) by the methods as were used in the relaxation theory of Sects. 11 and 13 we will consider the simple case of a paramagnetic substance containing two different kinds of ions with spin $\frac{1}{2}$. The g-factors of the two spin species are considered to be slightly different:

$$\left| \frac{g^{(1)} - g^{(2)}}{g^{(1)}} \right| \ll 1. \tag{23.3}$$

Furthermore we will investigate the high-field limit, which means that the line widths of the resonance lines are much smaller than the external field H.

The Hamiltonian of the system consists of three terms: the spin Hamiltonians of both the spin species separately and an interaction term containing products of spin operators of each species.

$$\mathcal{H} = \mathcal{H}^{(1)} + \mathcal{H}^{(2)} + \mathcal{H}^{(1,2)}. \tag{23.4}$$

As is usual in the existing theories on cross-relaxation this Hamiltonian now will be strongly simplified by discarding all the non-secular terms, i.e. all the terms that do not commute with the total magnetic moment operator:

$$M_z = M_z^{(1)} + M_z^{(2)}. \tag{23.5}$$

Some justification can be found in the fact that these non-secular terms cause relaxations within the spin-systems (1) and (2) separately and contribute to cross-relaxation in higher order only. As a consequence the relaxation function:

$$\varphi(t) = \frac{1}{kT} \langle\langle M_z M_z(t) \rangle\rangle \tag{23.6}$$

with

$$M_z(t) = e^{iLt} M_z = e^{i\mathscr{H}t/\hbar} M_z e^{-i\mathscr{H}t/\hbar}$$

will be independent of time. The different components of $\varphi(t)$ however will still be time dependent:

$$\begin{aligned}\varphi^{(t)} &= \varphi^{(1)} + \varphi^{(2)} + \varphi^{(1,2)} \\ &= \frac{1}{kT} \left\{ \langle\langle M_z^{(1)} M_z^{(1)}(t) \rangle\rangle + \langle\langle M_z^{(2)} M_z^{(2)}(t) \rangle\rangle + \langle\langle M_z^{(1)} M_z^{(2)}(t) \rangle\rangle \right\}. \end{aligned} \tag{23.7}$$

This is caused by the fact that the secular interaction term $\mathscr{H}_0^{(1,2)}$ will contain a term of the form:

$$\mathscr{H}_{0,1}^{(1,2)} = \sum_{k,l} A_{kl} (S_{+k}^{(1)} S_{-l}^{(2)} + S_{-k}^{(1)} S_{+l}^{(2)}) \tag{23.8}$$

which does not commute with either $M_z^{(1)}$ or $M_z^{(2)}$.

Another assumption that has to be made is that the mixed term $\varphi^{(1,2)}$ in (23.6) can be neglected. In that case:

$$\frac{d\varphi^{(1)}(t)}{dt} = -\frac{d\varphi^{(2)}(t)}{dt}. \tag{23.9}$$

The problem now has been simplified to one analogous to the case of a spin system with large exchange interactions in Sect. 19. One has to be careful, however, to take into account the limiting values of $\varphi^{(1)}(t)$ and $\varphi^{(2)}(t)$ for $t \to \infty$. These limits can be derived from thermodynamical considerations, when it is assumed that both spin systems have come to thermal equilibrium separately. Following TERWIEL and MAZUR (c.f. Sect. 13) one can introduce the proper operators to derive the relaxation Eq. (19.3) by defining:

$$\begin{aligned}\mu^{(1)} &= M_z^{(1)} - i(kT\varphi^{(1)}(\infty))^{\frac{1}{2}}, \\ \mu^{(2)} &= M_z^{(2)} - i(kT\varphi^{(2)}(\infty))^{\frac{1}{2}}. \end{aligned} \tag{23.10}$$

The cross-relaxation time will then be given by:

$$\frac{1}{\tau_{cr}} = \int_0^\infty d\tau \frac{\langle\langle \mu^{(1)} L e^{i(1-P_{\mu^{(1)}})L\tau} L \mu^{(1)} \rangle\rangle}{\langle\langle \mu^{(1)2} \rangle\rangle} \tag{23.11}$$

Here it has been tacitly assumed that the interaction term $\mathscr{H}_{0,1}^{(1,2)}$ is much smaller then the sum of the secular interaction terms in $\mathscr{H}$, otherwise no weak coupling limit can be used. This assumption is equivalent to that by which the existence of a relaxation process is taken for granted.

(23.11) can be approximated, following the procedure given in Sect. 19 to:

$$\frac{1}{\tau_{cr}} = 2\pi \left(\frac{g^{(1)}\beta}{\hbar}\right)^2 \frac{\langle\langle \mathscr{H}_{0,1}^{(1,2)2} \rangle\rangle}{\langle\langle \mu^{(1)2} \rangle\rangle} \frac{1}{(2\pi\langle\Delta\omega^2\rangle)^{\frac{1}{2}}} \exp\left[-\frac{(g^{(1)}-g^{(2)})^2 \beta^2 H^2}{2\hbar^2 \langle\omega^2\rangle}\right] \tag{23.12}$$

with:

$$\langle\Delta\omega^2\rangle = \frac{1}{\hbar^2} \frac{\langle\langle [\mathscr{H}_{0,1}^{(1,2)}, \mathscr{H}_0][\mathscr{H}_0, \mathscr{H}_{0,1}^{(1,2)}] \rangle\rangle}{\langle\langle \mathscr{H}_{0,1}^{(1,2)2} \rangle\rangle}.$$

This is not exactly the same result as can be derived from the theory given by BLOEMBERGEN e.a. as $\langle\langle\mu^2\rangle\rangle$ is in general a function of the field. It differs by a factor $\langle\langle M_z^2\rangle\rangle/\langle\langle\mu^2\rangle\rangle$ from their formula. We could have obtained their result by not using the μ-operators, but the M_z-operators instead, which means that no account would have been taken of the fact that $\varphi^{(1)}(t)$ and $\varphi^{(2)}(t)$ do not vanish in the limit of large t.

The method to derive expression (23.12) given here can be extended to all other processes mentioned in BLOEMBERGEN's article.

In view of all the assumptions made, however, caution is warranted. The experimental results, as described in the previous chapter, for instance, indicate the importance of higher order processes. In such cases it is not justified to neglect the non-secular terms in the Hamiltonian.

C. Spin-lattice relaxation.

24. Introduction. In the remainder of the article we will consider the phenomenon of spin-lattice relaxation. Throughout most of the discussion we will assume that the spin-spin processes rapidly bring about quasi-equilibrium in the spin system and that the spin-lattice processes require longer times. We will discuss the estimation of this time from specific models. It is usual to consider the models as representing perfect paramagnetic material with no physical defects or chemical imperfections. Allowance will be made for the possible presence of these imperfections in Sects. 57 and 60 to 67.

Before we introduce a specific model, let us consider some points which are common to all models: (1) the relation between the relaxation times as measured experimentally and the transition probabilities within the spin system; (2) the properties of the wave functions of the states of an ion, in particular parity and time-reversal symmetry; (3) the properties of the phonons in the solid, in particular their matrix elements with the strain operator and their density of states; and (4) the general properties of the spin-lattice operators which can cause electron transitions.

I. Transition probabilities and relaxation times.

25. Rate equations. In Sects. 6 and 7 we listed some of the experimental techniques used to measure τ_L. As the spin system changes under the influence of the phonons, electron transitions occur between the various spin levels and we wish to know the relation between the electron populations, the transition probabilities and τ_L for each experimental method.

α) *Paramagnetic relaxation.* The relation can be obtained for the paramagnetic relaxation technique as follows. Eq. (4.23) gives the relation between τ_L, the specific heat C_H and α the heat transfer coefficient. If the total energy of the spin system is E, the energy of spin state m is E_m and the probability of occupation of state m is p_m, then:

$$dE/dt = \frac{d}{dt}\left(\sum_m p_m E_m\right) = \sum_m (dp_m/dt) E_m, \quad (25.1)$$

where:

$$E = \sum_m E_m \exp(-E_m/kT_s) \Big/ \sum_m \exp(-E_m/kT_s). \quad (25.2)$$

T_s represents the temperature of the spin system. The probabilities p_n and p_m are related by the Maxwell-Boltzmann relaxation $p_n/p_m = \exp[(E_m - E_n)/kT_s]$

Sect. 25. Rate equations.

which in the high temperature limit simplifies to $1-(E_n-E_m)/kT_s$. Now:

$$dp_m/dt = \sum_{n\neq m}(p_n W_{nm}-p_m W_{mn}), \tag{25.3}$$

where W_{mn} is the transition probability for a transition from state m to state n. When the spin system is in equilibrium with the lattice, dynamic equilibrium, represented by the principle of detailed balance, requires that the change in populations of levels m and n is zero, so $N_n W_{nm} = N_m W_{mn}$. N_m and N_n are the equilibrium populations of levels m and n and are related by the Maxwell-Boltzmann relation. Therefore:

$$W_{nm}/W_{mn} = \exp((E_n-E_m)/kT_L) \simeq 1 - \frac{E_m-E_n}{kT_L}, \tag{25.4}$$

where T_L is the lattice temperature and the simplification is in the high temperature limit. Using the ratio from Eq. (25.4) dE/dt becomes:

$$\frac{dE}{dt} = \left(\frac{1}{T_s}-\frac{1}{T_L}\right)\left(\sum_{m,n}(E_m-E_n)^2 W_{mn}\right) \bigg/ 2k\sum_m \delta_{mm}. \tag{25.5}$$

From (25.5) we can obtain α as it is given by $\alpha = (dE/dt)/(T_L-T_s)$. C_H is obtained by differentiating the total energy E, given by (25.2), with respect to T_s. From Eq. (4.23):

$$\tau_L = \sum_m E_m^2 \bigg/ \sum_{m,n}(E_m-E_n)^2 W_{mn}. \tag{25.6}$$

This equation is valid when $|T_s-T_L| \ll T_L$. During the course of the derivation it was assumed that the transitions due to spontaneous emission could be neglected and that the alternating magnetic field, necessary in order to perform paramagnetic relaxation experiments, does not induce a significant number of transitions. HEBEL and SLICHTER[1] discussed this derivation.

β) *The saturation technique.* The measurement of τ_L by the saturation technique is performed by inducing microwave transitions at one frequency between two energy levels. If sufficient power is used the populations of the energy level system are disturbed, the perturbation is observed as a decrease in the intensity of the paramagnetic resonance line. The new populations will be determined by the dynamic equilibrium between the transitions induced by the lattice and the transitions induced by the alternating magnetic field. The intensity of the paramagnetic absorption is determined by the population differences between the levels which are monitored by the signal. The total population of the levels must remain constant at N so:

$$\sum_m N_m = N, \tag{25.7}$$

where the N_m represent the populations of the levels m. As a simplification we consider that the spins in the system are isolated from each other and interact weakly with the lattice. Let the transition probability due to the lattice be W_{mn} and that due to the alternating field U_{mn}. Then the total transition probability between m and n is V_{mn}. The rate of change of population of level m is given by:

$$\frac{dN_m}{dt} = \sum_{n\neq m}(N_n V_{nm}-N_m V_{mn}), \tag{25.8}$$

where the summation extends over all the levels. In dynamic equilibrium $dN_m/dt=0$. We have N homogeneous linear equations but there are only $N-1$

---

[1] L. C. HEBEL, and C. P. SLICHTER: Phys. Rev. **113**, 1504 (1959).

independent variables. Assuming the saturating signal is applied between the ionic levels 1 and 2 and writing $N_1-N_2=\Delta_{12}$, we have:

$$\left.\begin{aligned}&N_1[W_{21}+U_{12}-(U_{12}+\sum_k W_{1k})]-\Delta_{12}(W_{21}+U_{12})+N_3 W_{31}+\cdots N_n W_{n1}=0,\\&N_1[W_{12}+U_{12}-(U_{12}+\sum_k W_{2k})]+\Delta_{12}(U_{12}+\sum_k W_{2k})+N_3 W_{32}+\cdots N_n W_{n2}=0,\\&\cdots\cdots\cdots\cdots\cdots\cdots\cdots\cdots\cdots\cdots\cdots\cdots\cdots\cdots\cdots\cdots\cdots\cdots\cdots\cdots\cdots\cdots\cdots\\&N_1(W_{1n}+W_{2n})-\Delta_{12}W_{2n}+N_{3n}W_{3n}+\cdots\cdots\cdots\cdots\cdots\cdots\cdots\cdots N_n W_{nn}=0.\end{aligned}\right\} \quad (25.9)$$

It is possible to eliminate the second equation in (25.9) by using (25.7). Then by using CRAMER'S rule the solution for Δ_{12} is:

$$\Delta_{12}=\frac{C_{22}}{\sum\limits_{k\neq 2}W_{2k}C_{2k}+U_{12}C_{21}+C_{22}}\simeq\frac{C_{22}}{\sum\limits_{k\neq 2}W_{2k}C_{2k}+U_{12}C_{21}}, \quad (25.10)$$

where C_{2k} is the cofactor of the second column element in the k-th row. The simplification in Eq. (25.10) is correct in the high temperature limit when the population differences Δ_{mn} are small. If there are only two ionic levels it can be shown that the steady state solution of Eq. (25.9) is:

$$\Delta_{12}=\frac{N_{10}-N_{20}}{1+2U_{12}/(W_{12}+W_{21})}=\frac{N_{10}-N_{20}}{1+2U_{12}\tau_L}. \quad (25.11)$$

The subscript 0 refers to the populations at equilibrium when $U_{12}=0$. Comparing Eqs. (25.10) and (25.11) we can write for τ_L the relaxation time between two ionic levels 1 and 2 in the presence of other levels (effective spin greater than $\tfrac{1}{2}$):

$$\tau_L=\frac{1}{2}\left(W_{21}+\frac{1}{C_{21}}\sum_{k=3}W_{2k}C_{2k}\right)^{-1}. \quad (25.12)$$

This is most easily demonstrated by comparing the expressions for the saturation parameter in both cases; this parameter is given by:

$$S_{12}=\Delta_{12}(U)/\Delta_{12}(U=0). \quad (25.13)$$

LLOYD and PAKE[1] discussed this derivation.

γ) *The pulse saturation technique.* The relationship:

$$\tau_L=1/(W_{12}+W_{21}) \quad (25.14)$$

used in Eq. (25.11) can be most easily derived by considering the recovery of the spin system from the microwave power saturation when the power is switched off, i.e. $U_{12}=0$. Then:

$$\begin{aligned}(N_1-N_2)-(N_{10}-N_{20})&=[(N_1-N_2)_{t=0}-(N_{10}-N_{20})]\times\\&\times\exp[-t(W_{12}+W_{21})]=A\exp(-t/\tau_L).\end{aligned} \quad (25.15)$$

τ_L is the spin-lattice relaxation time by definition. This is the solution for a two level system. Eq. (25.8), with $U_{12}=0$, represents the conditions in the pulse saturation measuring technique where the spin system is disturbed by a pulse of microwave power and the recovery of the populations to their equilibrium values is monitored by normal microwave techniques. The general solution of this

---

[1] J. P. LLOYD, and G. E. PAKE: Phys. Rev. **94**, 579 (1954).

equation for the recovery of population of level m is:

$$N_m = N_{m0} + \sum_{k}^{L-1} A_{km} \exp(-t/\tau_{Lk}), \qquad (25.16)$$

where L is the number of ionic levels. Therefore, in the general case there is more than one relaxation time.

δ) *Cross-relaxation.* Up to this point we have only discussed systems where the electron transitions exchange energy with the lattice or the electromagnetic field. However, the populations of the energy levels may alter without a change in energy of the spin system when cross-relaxation occurs. This makes the solution of the rate equations even more complicated because Eq. (25.8) becomes:

$$\mathrm{d} N_m/\mathrm{d} t = \sum_{n \neq m}(N_n V_{nm} - N_m V_{mn}) + \sum_{nop} W_{mo,np} N^{-1}(N_n N_p - N_m N_o) + \\ + \sum_{noprs} W_{mo,np,rs} N^{-2}(N_n N_p N_s - N_m N_o N_r), \qquad (25.17)$$

where $W_{mo,np}$ is the probability for cross-relaxation between the levels m, n and o, p and $W_{mo,np,rs}$ is the probability for a cross-relaxation process involving three pairs of levels; higher order processes may be included if necessary. Equations of this complexity can only be solved in approximation and we will gain no advantage attempting to find a general relation between a single parameter τ_L and the transition probabilities. SIEGMAN[1] gave a solution of Eq. (25.17) in the high temperature limit by considering the entropy of the spin system and seeking the population distribution which maximised the entropy. ARMSTRONG and SZABO[2] give a technique using matrix algebra to solve (25.17) in the high temperature limit where it is possible to linearise the deviations from the equilibrium populations.

26. Transition probabilities. In the last section we discussed the relation between the relaxation times as measured experimentally and the transition probabilities between the electron energy levels. Therefore, the problem now is to calculate the transition probabilities in the systems of interest.

By analogy with the optical case, where electron transitions are brought about by the interaction between the atoms and the photons in the space containing the atoms, we can write the transition probability $W_{\alpha\beta}$, the probability per unit time of making a transition from state α to state β of the system of spins and lattice, as:

$$W_{\alpha\beta} = \left(\frac{2\pi}{\hbar}\right) |\langle\beta| V'|\alpha\rangle|^2 \varrho(E). \qquad (26.1)$$

This is a result of first order time dependent perturbation theory [6]. V' is the first term in the series expansion of the time dependent perturbation. We shall have to go to second order theory to explain the relaxation phenomena observed so we have to add two extra terms to Eq. (26.1). These are:

$$\frac{2\pi}{\hbar} \left\{ |\langle\beta| V''|\alpha\rangle|^2 + \left| \sum_{\gamma} \frac{\langle\beta| V'|\gamma\rangle\langle\gamma| V'|\alpha\rangle}{E_\alpha - E_\gamma} \right|^2 \right\} \varrho(E_1) \varrho(E_2). \qquad (26.2)$$

V'' is the second order term in the series expansion of the perturbation used here in first order perturbation theory while V' is applied twice and so requires second order perturbation theory. The $\varrho(E)$ are the densities of the final states involved

---
[1] A. E. SIEGMAN: Phys. Rev. **119**, 562 (1960).
[2] R. A. ARMSTRONG, and A. SZABO: Can. J. Phys. **38**, 1304 (1960).

Handbuch der Physik, Bd. XVIII/1.

in the relaxation and $E_\alpha - E_\gamma$ is the energy difference between the ground state of the system and the intermediate state $|\gamma\rangle$ which is involved in the second order processes.

We have the following problems; 1) What is the perturbation between the lattice and the spins which causes them to exchange energy? 2) What are the energy states of the combined system of spins and lattice and what are their state functions? 3) Having established the state functions, the perturbation and its operator, we must evaluate the matrix elements in the expressions (26.1) and (26.2).

A qualitative answer to question 1) is that the thermal vibrations of the lattice cause the ions in the lattice to be in continual motion with respect to one another and this relative motion produces time dependent perturbations in the interionic interactions. These interactions include the crystalline field potential, the dipolar interaction, the exchange interaction, the hyperfine structure interaction and the nuclear quadrupole interaction. The last two are really interactions between the nucleus of an ion and its electron cloud but the motions of the neighbouring ions perturb the electron cloud as well as moving the nucleus. If $V(r_1^0, r_2^0, \ldots, r_n^0)$ is the equilibrium value of the interionic interaction, where the r_n^0 are the vectors specifying the interionic spacings, and $V(r_1, r_2, \ldots, r_n)$ is the value of the interaction at the r_n where $r_n^0 - r_n = \Delta r_n(t)$, which is a function of time due to the thermal vibrations, then we can write $V_{\text{perturbation}}$ as:

$$V_{\text{perturbation}} = V(r_1^0, r_2^0, \ldots, r_n^0) - V(r_1, r_2, \ldots, r_n)$$
$$= \sum_{j=1}^{n} \frac{\partial V}{\partial r_j}\bigg|_0 \Delta r_j(t) + \frac{1}{2} \sum_{\substack{j=1, l=1 \\ j>l}}^{n} \frac{\partial^2 V}{\partial r_j \, \partial r_l}\bigg|_0 \Delta r_j(t) \Delta r_l(t) + \cdots \quad (26.3)$$

Therefore:

$$V' = \sum_{j=1}^{n} \frac{\partial V}{\partial r_j}\bigg|_0 \Delta r_j(t). \quad (26.4)$$

r_j and r_l represent coordinates of r_j and r_l and when the subscript is repeated in a term summation over the coordinates is also required.

From Eq. (26.4) we see that the perturbation between the lattice and the spins can be written as a product of two terms, one representing the rate of change of the interionic interaction with ion separation and the other representing the time variation of the ionic positions. The former is, thus, a property of the ions, while the latter is a property of the thermal vibrations, which when quantised are described by phonons, and this appearance of the two properties separately allows us to make a convenient simplification in the state functions of the combined system of spins and lattice. Assuming that the interaction between the phonons and the ions is very weak the wave function $|\alpha\rangle$ describing the total system can be written as $|a\rangle |p\rangle$, the product of the wave functions for the ions and the phonons considered separately. Then the matrix element in (26.1) becomes:

$$\langle \beta | \sum_{j=1}^{n} \frac{\partial V}{\partial r_j}\bigg|_0 \Delta r_j(t) | \alpha \rangle = \sum_{j=1}^{n} \langle b | \frac{\partial V}{\partial r_j}\bigg|_0 | a \rangle \langle p'' | \Delta r_j(t) | p' \rangle. \quad (26.5)$$

This "factorisation" allows us in the simple case of a very weak perturbation V' to consider the ions and the phonons separately.

27. The Hamiltonian. The states $|a\rangle$ and $|b\rangle$ are electron states of the ion which can be determined from the following Hamiltonian:

$$\mathcal{H} = \mathcal{H}_a + \mathcal{H}_b + \mathcal{H}_c + \cdots + \mathcal{H}_i \quad (27.1)$$

where:

$$\mathcal{H}_a = \sum_k \left(\frac{p_k^2}{2m} - \frac{Ze^2}{r_k} \right) + \sum_{j<k} \frac{e^2}{r_{jk}} \tag{27.1a}$$

is the Hamiltonian of the ion with only Coulomb interactions considered;

$$\mathcal{H}_b = \sum_{j,k} (a_{jk} \boldsymbol{l}_j \cdot \boldsymbol{s}_k + b_{jk} \boldsymbol{l}_j \cdot \boldsymbol{l}_k + c_{jk} \boldsymbol{s}_j \cdot \boldsymbol{s}_k) \tag{27.1b}$$

represents the magnetic interactions between the electron spins and the orbital moments;

$$\mathcal{H}_c = \sum_k \frac{eh}{4\pi mc} (\boldsymbol{l} + 2\boldsymbol{s})_k \cdot \boldsymbol{H} + \frac{e^2 H^2}{8mc^2} \sum_k (x^2 + y^2)_k \tag{27.1c}$$

is the interaction between the electrons and an external magnetic field;

$$\mathcal{H}_d = 2\gamma \beta \beta_N \sum_k \left[\left\{ \frac{(\boldsymbol{l}_k - \boldsymbol{s}_k) \cdot \boldsymbol{I}}{r_k^3} + \frac{3(\boldsymbol{r}_k \cdot \boldsymbol{s}_k)(\boldsymbol{r}_k \cdot \boldsymbol{I})}{r_k^5} \right\} + \frac{8\pi}{3} \delta(\boldsymbol{r}_k) \boldsymbol{s}_k \cdot \boldsymbol{I} \right] \tag{27.1d}$$

is the interaction between the magnetic moment of the nucleus and the orbital and the spin moment of the electrons;

$$\mathcal{H}_e = \frac{e^2 Q}{2I(I-1)} \sum_k \left[\frac{I(I+1)}{r_k^3} - \frac{3(\boldsymbol{r}_k \cdot \boldsymbol{I})^2}{r_k^5} \right] \tag{27.1e}$$

is the nuclear quadrupole interaction, (Q is the nuclear quadrupole moment[1]);

$$\mathcal{H}_f = -\gamma \beta_N \boldsymbol{H} \cdot \boldsymbol{I} \tag{27.1f}$$

is the direct interaction of the nuclear moment with the external magnetic field (β_N is the nuclear Bohr magneton);

$$\mathcal{H}_g = 2 \sum_j J_{ij} \boldsymbol{s}_i \cdot \boldsymbol{s}_j \tag{27.1g}$$

is the exchange interaction between ion i and its neighbours;

$$\mathcal{H}_h = \sum_j \left[\frac{\boldsymbol{\mu}_i \cdot \boldsymbol{\mu}_j}{r_{ij}^3} - \frac{3(\boldsymbol{\mu}_i \cdot \boldsymbol{r}_{ij})(\boldsymbol{\mu}_j \cdot \boldsymbol{r}_{ij})}{r_{ij}^5} \right] \tag{27.1h}$$

is the dipole interaction between ion i and its neighbours;

$$\mathcal{H}_i = \sum_{n,m} A_n^m \langle r^n \rangle Y_n^m(\vartheta, \varphi) \tag{27.1i}$$

is the crystalline field interaction representing the Coulomb interaction of the neighbouring ions at the site of the magnetic ion. (The symbols in these equations have their usual meanings.) The first six terms refer to the case of a free ion and are in order of decreasing energy; the last three terms refer to interactions with other ions, (27.1g) and (27.1h) are interactions with other magnetic ions while (27.1i) represents the interaction with all the surrounding ions. The position of the latter interactions in the total Hamiltonian in order of decreasing energy depends on the particular material considered. Omitted from the Hamiltonian are terms which account for deviations in the bonding between ions from the pure ionic bond and terms which describe shielding effects.

28. Ionic energy levels. The calculation of the state functions of a paramagnetic ion in a crystal lattice can be performed by well-known techniques [7] to [12]. The influence of the crystalline field may lift the degeneracy in the energy level

---
[1] J. E. MACK: Revs. Mod. Phys. **22**, 64 (1950).

scheme of the ion in free space. We are only interested in the lowest lying levels because at the temperatures at which experimental work is normally performed (0.1 to 300 °K) most of the electrons are in these levels.

Consider an ion of the $3d$ transition group without a nuclear magnetic moment and with no interionic interactions, then the state functions are calculated from the Hamiltonian consisting of terms (27.1a), (27.1b) and (27.1i) when no external magnetic field is present. Often ions of this type are found in six-fold coordination, the six neighbours forming a distorted octahedron. In this case the large cubic field lifts some, or all, of the orbital degeneracy of the ground multiplet. Distortions from the cubic symmetry will lift some or all of the remaining degeneracy. This will always happen according to the Jahn-Teller theorem[1], which states that, except for linear complexes, a complex which has a degenerate ground state will distort so that the degeneracy is reduced. The result of such a distortion is that the energy of the complex is reduced. This is not true in the case of spin degeneracy due to time-reversal symmetry. Its effect is described by Kramers theorem[2], which states that every level of a complex with an odd number of electrons is at least twofold degenerate under the influence of external perturbations which are electrostatic in type. In the $3d$ group the influence of the spin-orbit coupling also tends to lift orbital degeneracies. Thus, for many ions, the crystalline field produces a set of levels, each level non-degenerate in the case of systems with an even number of electrons and twofold degenerate for an odd number of electrons. In the case of $3d$ ions, the crystal field acting without the first term in the perturbation (27.1b) leaves states with a definite value of the square of the orbital angular momentum but the time average of the spatial components of the orbital angular momentum is zero in the first approximation. This is the so-called quenching of the orbital angular momentum. However, introducing this term, the spin-orbit coupling, so that higher states are mixed into the ground state functions, partially lifts the quenching of the orbital momentum. This is seen in electron paramagnetic resonance (E.P.R.) work where the g-values of the ground state deviate from the free-spin value. The spin-orbit coupling also introduces some orbital angular momentum into the ground states of S-state ions. The quenching effects do not occur in the rare earth ions where the crystalline field perturbation is smaller than the spin-orbit coupling, because the magnetic electrons are in the $4f$ shell which is partially shielded from the crystalline field by other electron shells.

29. Time-reversal symmetry. The action of time-reversal on a wave function of an odd electron system produces an orthogonal wave function which is linearly independent of the first but has the same energy, therefore, the two functions represent a doubly degenerate state, the so-called Kramers degeneracy. The action of time-reversal on an operator means replacing every linear momentum $\boldsymbol{p}$ by $-\boldsymbol{p}$ and every spin angular momentum $\boldsymbol{s}$ by $-\boldsymbol{s}$. The Hamiltonian of a free ion contains terms in p^2 and $\boldsymbol{p}\cdot\boldsymbol{s}$ [3] so the action of time-reversal has no effect in this case. A purely electrostatic perturbation added to the Hamiltonian does not alter this situation but a magnetic perturbation introduces terms such as $\boldsymbol{p}\cdot\boldsymbol{A}$, $\boldsymbol{A}$ is the vector potential, which do change sign under the action of time-reversal. This lifts the degeneracy of the Kramers doublets. So transitions within the doublet can only occur under the influence of magnetic perturbations.

The spin-lattice perturbation to be considered in detail in Sect. 38 requires the modulation of the crystalline field potential to produce spin transitions by

---

[1] H. A. Jahn, and E. Teller: Proc. Roy. Soc. (London) A **161**, 220 (1937).
[2] H. A. Kramers: Proc. Koninkl. Ned. Akad. Wetenschap. **33**, 959 (1930).

means of variations of the spin-orbit coupling. Because of Kramers theorem this mechanism cannot work for time-reversal degenerate doublets so the degeneracy has to be removed first by introducing an external magnetic field. Therefore, in most of the calculations to be given later, the state functions for Kramers doublets will be given in the case of an external magnetic field being present.

Introducing the hyperfine structure interaction, (27.1d) or the dipolar interaction between neighbouring ions, (27.1h), also lifts the Kramers degeneracy. As the nuclear quadrupole interaction, (27.1e), is only present when $I \geq 1$, the Kramers degeneracy must already have been lifted by the hyperfine structure interaction. The spin-orbit interaction, the first term in (27.1b), even though it is basically an interaction between the magnetic moments of the orbit and the spin, does not reverse its sign under the action of the time-reversal operator so it can not lift the degeneracy.

A more definite demonstration that an electrostatic operator cannot cause transitions within Kramers-doublets states is as follows. Let the operator be V and assume that it is a function of electron charge and space coordinates only so that it is a real operator; let the state functions be $|\pm \tfrac{1}{2} m\rangle$ where m must be an odd number. Then, the matrix element is $\langle \tfrac{1}{2} m | V | -\tfrac{1}{2} m \rangle$ which under the action of time-reversal becomes $(-1)^m \langle -\tfrac{1}{2} m | V^* | \tfrac{1}{2} m \rangle$ which must also be equal to the Hermitean conjugate $-\langle \tfrac{1}{2} m | V | -\tfrac{1}{2} m \rangle$; we have made use of the fact that V is real to simplify this result. This is true only if they are both zero. In the applications which we shall consider there is no loss in generality in assuming that V is real.

A second interesting result of time-reversal symmetry is the so-called Van Vleck cancellation. This was first discussed by VAN VLECK [13]. We shall discuss it in more detail in Sect. 40.

30. Parity. The states usually considered in relaxation theory are from the same multiplet so they must all have the same parity. This is true in the first approximation but possible admixing from higher multiplets of different parity will give a state function with mixed parity in the next approximation. This can usually be neglected when the multiplet separations are large.

Electric dipole transitions can not occur between levels with the same parity because the sign of the perturbation operator E changes under the action of the parity operator so the matrix elements $\int \psi_1^* \mathrm{E}\, \psi_2\, dv$ when integrated over all space are zero. Therefore the spin transitions which will cause the spin-lattice transitions must be brought about by magnetic dipole transitions which can connect states with the same parity. Electric quadrupole transitions may also connect states with the same parity but these are usually neglected compared with the magnetic dipole transitions.

31. Phonons. We will discuss here only some important points from the theory of phonons because many authors have written about the theory. Among these are ZIMAN [14] and KLEMENS [15]. For every value of the wave vector $\boldsymbol{k}$ there are $3n$ modes of vibration, where n is the number of ions per unit cell. The total number of modes is $3N$, where N is the number of ions in the lattice. Each mode is represented by a normal coordinate q_i. The modes are divided into two sets of branches, the optical branches where there are finite frequencies of vibration even when the wave vector tends to zero and the acoustic branches where the frequencies tend to zero as the wave vector tends to zero. A second difference between the two sets of branches is that in the optical branch the atoms in each unit cell vibrate in opposite directions to each other while in the acoustic branch the atoms in each cell vibrate together.

Quantisation is introduced into the description of the lattice properties by introducing phonons. The total Hamiltonian for the lattice is:

$$\mathscr{H}_L = \sum_p \hbar\omega_p [N(\omega_p) + \tfrac{1}{2}] \tag{31.1}$$

where the summation is over all the phonon modes. The phonons obey Bose-Einstein statistics so the number of phonons with energy $\hbar\omega$ is given by:

$$N(\omega) = \left[\exp\left(\frac{\hbar\omega}{kT}\right) - 1\right]^{-1}. \tag{31.2}$$

This number can also be written in terms of the annihilation operator a and the creation operator a* because $N(\omega) = \mathrm{a} \cdot \mathrm{a}^*$. This notation allows us to write the matrix elements for the phonons in a simple form. The states of the lattice are described by the numbers of phonons in each mode, so a typical state vector is $|N_1, N_2, \ldots, N_i, \ldots, N_{3N}\rangle$. Then the matrix elements which make the lattice Hamiltonian diagonal and which satisfy the commutation relation $\mathrm{a} \cdot \mathrm{a}^* - \mathrm{a}^* \cdot \mathrm{a} = 1$ are:

and
$$\left. \begin{array}{l} \langle N_1, \ldots, N_i+1, \ldots, N_{3N} | \mathrm{a}^* | N_1, \ldots, N_i, \ldots, N_{3N} \rangle = (N_i+1)^{\frac{1}{2}} \\ \langle N_1, \ldots, N_i-1, \ldots, N_{3N} | \mathrm{a} | N_1, \ldots, N_i, \ldots, N_{3N} \rangle = (N_i)^{\frac{1}{2}}. \end{array} \right\} \tag{31.3}$$

The displacements $\Delta \boldsymbol{r}(t)$ must be written in terms of a and a*. The atom j at $\boldsymbol{Q}$ is displaced by $\boldsymbol{q}$ by the passage of phonons:

$$\boldsymbol{q}_{\boldsymbol{Q}\alpha} = (\hbar/2M)^{\frac{1}{2}} \sum_p (\omega_p)^{-\frac{1}{2}} \varphi_{p\alpha} (\mathrm{a}_p + \mathrm{a}_p^*) \exp(-i\boldsymbol{k} \cdot \boldsymbol{Q}), \tag{31.4}$$

where M is the lattice mass, φ is the polarisation vector, $\boldsymbol{k}$ is the wave vector, p is the phonon mode, and $\alpha = x, y, z$. Assuming that the wavelength of the lattice motion is much larger than the dimensions of a unit cell an approximate expression for the relative displacement of two neighbouring ions, $\Delta \boldsymbol{r}(t)$, is:

$$\Delta \boldsymbol{r}(t) \simeq (\hbar/2M)^{\frac{1}{2}} \sum_p (\omega_p)^{\frac{1}{2}} v^{-1} \varphi_{p\alpha} (\mathrm{a}_p + \mathrm{a}_p^*) \boldsymbol{K}_p \cdot \boldsymbol{R}_p \sin \Omega_p, \tag{31.5}$$

where $\boldsymbol{K}_p$ and $\boldsymbol{R}_p$ are the unit wave and position vectors and Ω_p is an arbitrary phase constant.

Eqs. (26.1) and (26.2) require a knowledge of the density of states of the phonons. In paramagnetic relaxation theory, as in other theories which require a knowledge of the density of states, an assumption is usually made about the frequency distribution. The most usual assumption is a Debije distribution:

$$\varrho(\omega) = \frac{1}{2\pi^2} \left(\frac{2}{v_t^3} + \frac{1}{v_l^3}\right) \omega^2 V = \hbar \varrho(E), \tag{31.6}$$

where v_l and v_t are the longitudinal and transverse velocities of sound in the material and V is the volume of the sample. This distribution was first obtained by solving the problem of the possible vibrations in an elastic sphere of isotropic material. The wavelength has to be small to neglect the boundary conditions but large compared to the interatomic spacings. The atomic structure is allowed for by remembering that only a finite number of modes of vibration can occur and this means that the distribution in Eq. (31.6) has to be cut off at a maximum frequency ω_m:

$$\omega_m = \sqrt[3]{\left\{\frac{18\pi^2 N}{V}\left(\frac{2}{v_t^3} + \frac{1}{v_l^3}\right)^{-1}\right\}}. \tag{31.7}$$

The theory is limited to the case of one atom per unit cell and only the acoustic branch of the spectrum is considered. Also the effects of dispersion are neglected. In much theoretical work on paramagnetic relaxation the distinction between the longitudinal and transverse velocities is neglected.

The analytical expressions for the lattice Hamiltonian, the number of phonons in any mode, the density of states of the phonons and the matrix elements between modes are based on simplifications. The forces between the ions are assumed to be harmonic, the structure in the unit cell is obliterated so that every point in the cell has the same motion, the possible phonon modes are given by the Debije distribution as calculated for an isotropic elastic medium with no dispersion and neglecting the optical branches, and each mode is represented by a quantised simple harmonic oscillator with a normal coordinate q_i.

32. Matrix elements. Most of the rest of this article is devoted to discussion of the spin-lattice interaction and the evaluation of (26.1) and (26.2). The main interest will be in the matrix elements of the interionic interaction, the first factor in Eq. (26.5), because most of the work will assume that the phonons are described by the Debije distribution. Only in **CV** will this assumption be dropped.

Two approaches may be made to the problem of evaluating the matrix elements of the ionic interaction. They may be evaluated from the properly determined complete wave functions and the full operator for the interaction. The second technique, in direct analogy with the use of the spin-Hamiltonian, is to evaluate them from spin state functions with an operator for the interaction so constructed that the influence of the orbital part of the state functions is included in certain constants. Thus, if $\mathscr{H}$ is the full Hamiltonian for the ion, the state functions are given by $\mathscr{H}|\psi_n\rangle = E_n|\psi_n\rangle$. Considering only the ground states of this system it is possible to derive a spin-Hamiltonian $\mathscr{H}_s$ which when it operates on the spin states $|S\rangle$ gives the same energy levels E_n. That is $\mathscr{H}_s|S\rangle = E_n|S\rangle$. Now, if V is the proper operator for the full spin-lattice interaction, it is possible to derive an effective perturbation operator V_{eff} such that the matrix elements for the transition probability between $|\psi_b\rangle$ and $|\psi_a\rangle$ and between $|S\rangle$ and $|S\pm 1\rangle$ are equal. That is $\langle \psi_b|V|\psi_a\rangle = \langle S|V_{\text{eff}}|S\pm 1\rangle$. Therefore the two procedures are closely related.

II. Field and temperature dependence of relaxation mechanisms.

How does the experimental physicist usually present his results on paramagnetic relaxation? From his measurements he obtains a parameter which he identifies as the relaxation time τ_L and he presents this as a function of temperature, magnetic field, crystal orientation and concentration of magnetic ions. Combining the considerations of the last sections with Eqs. (26.1), (26.2) and (26.5), predictions can be made about the temperature and field variation of τ_L in a simple manner. These simple predictions give no information about concentration effects and very little information about orientation effects.

33. Direct processes. When a spin flips from one state $|a\rangle$ to a neighbouring state $|b\rangle$ the energy of the spin system alters by ∂_i (the energy level separation). To conserve the total energy of the system spins plus phonons, the population of the phonon mode $\omega_i = \partial_i/\hbar$ changes by one. This is the direct process of relaxation.

α) *Non-Kramers salts.* By a non-Kramers salt we mean a salt which contains paramagnetic ions with an even number of electrons. We make the distinction between Kramers and non-Kramers salts because of the difference in the properties of their wave functions already discussed in Sect. 29. Consider the simple case of an ion with only two states $|a\rangle$ and $|b\rangle$ which are separated in energy by ∂_i (Fig. 6). The transition probability for the direct process is given by Eq. (26.1). Using the "factorisation" in Eq. (26.5) and remembering the $\omega_i^{\frac{1}{2}}$ which occurs in Eq. (31.5) we have:

$$W_{ba} = \text{constant} \times \omega_i |\langle N_1, \ldots, N_i+1, \ldots |a^*| N_1, \ldots, N_i, \ldots\rangle|^2 \varrho(\omega)\, \delta(\partial_i - \hbar\omega), \quad (33.1)$$

where we have only retained terms of interest outside the constant, and where $\delta(\partial_i - \hbar\omega)$ refers to the Dirac delta function which serves here as the mathematical expression of the requirements of the conservation of energy. Using Eqs. (31.2), (31.3) and (31.6) in (33.1) and putting all the parameters except those depending on frequency and temperature in a constant:

$$W_{ba} = \text{constant} \times [\{\exp(\hbar\omega_i/kT) - 1\}^{-1} + 1] \times \omega_i^3. \quad (33.2)$$

If $\hbar\omega_i \ll kT$ and there is no crystalline field splitting, this simplifies to:

$$W_{ba} = \text{constant} \times T \times H^2, \quad (33.3)$$

where we have also used the relation $\hbar\omega_i = g\beta H$ in order to see the magnetic field dependence. A similar expression is obtained for the probability of a spin flip in the reverse direction and by using the simplified form of Eq. (25.12) the temperature and field dependences of τ_L for the direct process are:

$$\tau_L^{-1} = \text{constant} \times T \times H^2. \quad (33.4)$$

The physical meaning of (33.4) is easily understood, lowering the temperature reduces the number of phonons available to participate in the relaxation process,

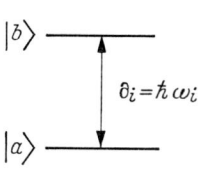

Fig. 6. Energy level scheme for a simple non-Kramers salt.

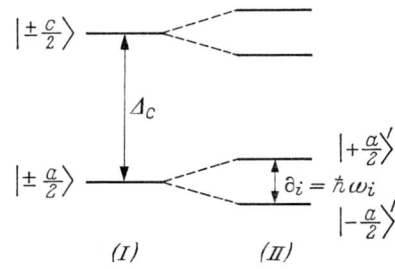

Fig. 7 I and II. Energy level scheme for a simple Kramers salt showing: (I) the unperturbed system, (II) the perturbed system.

but increasing the magnetic field increases the energy of the spin system and requires higher energy phonons to participate and they are more plentiful.

When crystalline field splittings are present we can write

$$\hbar\omega_i \simeq (\hbar^2\omega_0^2 + g^2\beta^2 H^2)^{\frac{1}{2}}$$

where $\hbar\omega_0$ represents the crystal field splitting. In this case the Zeeman effect does not make the major contribution to the level separation and the field dependency is now not necessarily H^2; it may be much weaker, or non-existent.

β) *Kramers salts.* If the two levels form a Kramers doublet, see Fig. 7, then a magnetic field is required to lift the Kramers degeneracy. The state functions become, in first order perturbation theory:

$$\begin{aligned}\left|-\tfrac{1}{2}a\right\rangle' &= \left|-\tfrac{1}{2}a\right\rangle + \Lambda\beta H \sum_c \left\{ \frac{\langle \pm\tfrac{1}{2}c|L+2S|-\tfrac{1}{2}a\rangle}{-\Delta c}\left|\pm\tfrac{1}{2}c\right\rangle \right\} \\ \left|+\tfrac{1}{2}a\right\rangle' &= \left|+\tfrac{1}{2}a\right\rangle + \Lambda\beta H \sum_c \left\{ \frac{\langle \pm\tfrac{1}{2}c|L+2S|+\tfrac{1}{2}a\rangle}{-\Delta c}\left|\pm\tfrac{1}{2}c\right\rangle \right\},\end{aligned} \quad (33.5)$$

where the $|\pm\tfrac{1}{2}c\rangle$ are the excited states, and Λ is the Landé g-factor. Because of the time-degeneracy $\left\langle +\tfrac{a}{2}\middle|V\middle|-\tfrac{a}{2}\right\rangle = 0$. Thus, the matrix elements of the ion are proportional to H^2 and so the relaxation time is given by:

$$\tau_L^{-1} = \text{constant} \times T \times H^4. \quad (33.6)$$

Of course Kramers doublets cannot be split by a crystalline field so the field dependency cannot be disturbed by this.

γ) *Experiment.* Tables 2 and 3 contain results of experiments and comparison with detailed theory. The predicted temperature dependence has been observed many times in many salts but the magnetic field dependence is not often seen. Indeed, quite often in relatively low fields the relaxation time varies with field in complete contradiction to the theory, increasing as the field increases instead of decreasing[1]. This is probably due to interionic interactions in the specimens which have not been allowed for in the simple theory. The theory to be given in Sect. 56 will try to cover these points. However, DAVIDS and WAGNER[2] have observed a TH^4 dependency in $K_3Co(CN)_6$ containing Fe^{3+}.

34. Raman processes. From (26.2) we have the Raman process whereby a spin flips from $|b\rangle$ to $|a\rangle$ and the energy is taken up in the phonon system by means of a two phonon process where the phonon ω_k is destroyed and the phonon ω_j is created such that $\partial_i = \hbar(\omega_j - \omega_k)$. In a similar manner to Eq. (33.1) we have:

$$\begin{aligned}W_{ba} = \text{constant} \times \int\int_0^{\omega_m} \omega_k \omega_j |\langle N_1, \ldots, N_k-1, \ldots, N_j+1, \ldots | a+a^* | \times \\ \times |N_1, \ldots, N_k, \ldots, N_j, \ldots\rangle|^2 \varrho(\omega_k) \varrho(\omega_j) \delta\{\partial_i - \hbar(\omega_j - \omega_k)\} d\omega_k\, d\omega_j.\end{aligned} \quad (34.1)$$

The variables in the double integral are related through the requirement of energy conservation. In the second order perturbation there are the two alternatives already mentioned in Sect. 26, the first term in the series in Eq. (26.3) may act twice, or the second term in the series may act once. In both cases the term $\Delta \mathbf{r}(t)$ occurs twice, so two phonons are involved in the process.

α) *Non-Kramers salts.* In analogous manner to the derivation of Eq. (33.2), Eq. (34.1) becomes:

$$W_{ba} = \text{constant} \times \int_0^{\omega_m} \omega_k^3 \left(\omega_k + \frac{\partial_i}{\hbar}\right)^3 \left\{ \frac{1}{\left[\exp\left(\frac{\hbar\omega_k}{kT}\right)-1\right]\left[\exp\left(\frac{\hbar\omega_k+\partial_i}{kT}\right)-1\right]} + \frac{1}{\exp\left(\frac{\hbar\omega_k}{kT}\right)-1} \right\} d\omega_k. \quad (34.2)$$

---

[1] L. C. VAN DER MAREL, J. VAN DEN BROEK, and C. J. GORTER: Physica **23**, 361 (1957).
[2] D. A. DAVIDS, and P. E. WAGNER: Phys. Rev. Letters **12**, 141 (1964).

This integral may be simplified by remembering that in many cases $\partial_i \ll \hbar\omega_k, \hbar\omega_j$ and kT. Then (34.2) becomes:

$$W_{ba} = \text{constant} \times \int_0^{\omega_m} \omega_k^6 \frac{\exp\left(\frac{\hbar\omega_k}{kT}\right)}{\left[\exp\left(\frac{\hbar\omega_k}{kT}\right)-1\right]^2} d\omega_k \tag{34.3}$$

which is one of a family of integrals which have been tabulated by ZIMAN[1] and ROGERS and POWELL[2] amongst others. The general form of the integral is:

$$I_n = \int_0^{\omega_m} \omega^n \frac{\exp(\hbar\omega/kT)}{[\exp(\hbar\omega/kT)-1]^2} d\omega \tag{34.4}$$

and in the limit of $T \gg \Theta_D$, Θ_D is the Debije temperature of the material related to the cut-off angular frequency ω_m, Eq. (31.7), by $k\Theta_D = \hbar\omega_m$, (34.4) simplifies to:

$$I_n \simeq \int_0^\infty \omega^n \exp\left(-\frac{\hbar\omega}{kT}\right) d\omega \simeq n! \left(\frac{kT}{\hbar}\right)^{n+1}. \tag{34.5}$$

In the temperature range $\Theta_D/10 \lesssim T \lesssim \Theta_D/3$ (34.4) is adequately represented by:

$$I_n = n!\left(\frac{kT}{\hbar}\right)^{n+1}\left[1 - \exp\left(-\frac{\Theta_D}{T}\right) \sum_0^n \frac{1}{n!}\left(\frac{\Theta_D}{T}\right)^n\right] \tag{34.6}$$

which goes to the value of Eq. (34.5) in the low temperature limit. When the temperature is not too far from the Debije temperature then most of the value of the integral comes from those frequencies close to ω_m so (34.4) can be written as a TAYLOR's series about ω_m:

$$I_n = \left(\frac{k\Theta_D}{\hbar}\right)^{n+1} \left\{ \frac{\exp\left(\frac{\Theta_D}{T}\right)}{\left[\exp\left(\frac{\Theta_D}{T}\right)-1\right]^2} \cdot \frac{1}{n+1} + \right.$$
$$\left. + \frac{\Theta_D}{T} \frac{\exp\left(\frac{\Theta_D}{T}\right)\left[\exp\left(\frac{\Theta_D}{T}\right)+1\right]}{\left[\exp\left(\frac{\Theta_D}{T}\right)-1\right]^3} \frac{1}{(n+1)(n+2)} \right\}. \tag{34.7}$$

This series is not correct when $T \ll \Theta_D$, but it is a good approximation over most of the rest of the temperature range. In the high temperature limit, $T \gg \Theta_D$, Eqs. (34.4) and (34.7) tend towards the value:

$$I_n = (k/\hbar)^{n+1} \Theta_D T^2 f(n), \tag{34.8}$$

where $f(n) \simeq n^{-1}$. The exact value of $f(n)$ depends on the approximation made, that is, is the TAYLOR's series stopped at the first or the second term? BLACKMAN[3] gives a series for the specific case of $n=4$ in the integral.

From the solutions to the integral at low temperatures Eq. (34.3) gives:

$$W_{ba} = \text{constant} \times T^7, \tag{34.9}$$

---

[1] J. M. ZIMAN: Proc. Roy. Soc. (London) A **226**, 436 (1954).
[2] W. M. ROGERS, and R. L. POWELL: Tables of Transport Integrals. National Bureau of Standards Circular No. 595. Washington D.C. 1958.
[3] M. BLACKMAN: Handbuch der Physik, Bd. 7/1, S. 325. Berlin-Göttingen-Heidelberg: Springer 1955.

while at very high temperatures:

$$W_{ba} = \text{constant} \times T^2. \tag{34.10}$$

Using the simple rate equations for transitions between two levels the expressions for τ_L in the two limits are:

$$\tau_L^{-1} = \text{constant} \times T^7 \quad \text{and} \quad \tau_L^{-1} = \text{constant} \times T^2. \tag{34.11}$$

By assuming that there are no direct transitions from $|b\rangle$ or $|a\rangle$ to $|c\rangle$ we are really limiting the phonons to energies less than the energy difference Δ_c, that is $k\Theta_D < \Delta_c$. We have also tacitly assumed that the energy difference $E_\alpha - E_\gamma$ in (26.2) is given by Δ_c to a good approximation, that is the phonon energies are neglected with respect to the energy separations between the ground and excited states of the spin system. This is the so-called adiabatic approximation [13].

β) *Kramers salts.* In a Kramers salt we have to allow for the time-reversal degeneracy which, as will be shown in Sect. 40, means that this adiabatic approximation is no longer applicable. In this case the phonon energies have to be included in (26.2). This is the non-adiabatic approximation. Then the transition probability becomes:

$$W_{ba} = \text{constant} \times \int_0^{\omega_m} \frac{\omega_k^8 \exp\left(\frac{\hbar \omega_k}{kT}\right) d\omega_k}{\left[\exp\left(\frac{\hbar \omega_k}{kT}\right) - 1\right]^2} \tag{34.12}$$

so the expressions for τ_L in the low and high temperature limits are:

$$\tau_L^{-1} = \text{constant} \times T^9 \quad \text{and} \quad \tau_L^{-1} = \text{constant} \times T^2. \tag{34.13}$$

Eqs. (34.3) and (34.12) do not show any magnetic field dependence but more detailed calculations, to be discussed in Sect. 56, show that there is some field dependence.

35. Orbach processes. Consider now the three energy levels in Fig. 8, where $\partial_i \ll kT \ll \Delta_c$. The transition probability for transitions from $|c\rangle$ to $|a\rangle$ is given by (33.2) which we rewrite as:

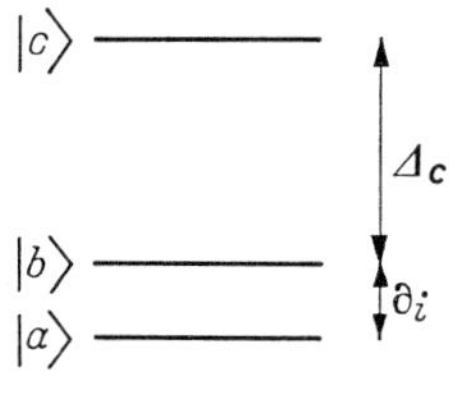

$$W_{ca} = \text{constant} \times \frac{\Delta_c^3}{\hbar^3} \times ((\exp(\Delta_c/kT) - 1)^{-1} + 1), \tag{35.1}$$

where $\hbar \omega_i = \Delta_c$. Because Δ_c is large and cannot be influenced by external magnetic fields to any significant extent (35.1) can be further simplified by including the term $\Delta_c/\hbar^3$ in the constant. In analogous manner to the derivation of (33.2), and the related expression (35.1), we derive:

Fig. 8. Energy level scheme to demonstrate the Orbach process.

$$W_{bc} = \text{constant} \times \left(\exp\left(\frac{\Delta_c}{kT}\right) - 1\right)^{-1}. \tag{35.2}$$

Remembering the relative magnitudes of the energies quoted above, (35.1) and (35.2) reduce to:

$$W_{ca} = \text{constant} \quad \text{and} \quad W_{bc} = \text{constant} \times \exp(-\Delta_c/kT). \tag{35.3}$$

The constants are of similar magnitudes so if the relaxation from $|b\rangle$ to $|a\rangle$ is via $|c\rangle$ only, the relaxation time is given by:

$$\tau_L^{-1} = \text{constant} \times \exp(-\Delta_c/kT). \tag{35.4}$$

Table 1. *Some theoretical predictions for τ_L as a function of T and H.*

Entry	Energy Levels	Expression for τ_L	Simplified Expression for τ_L	Conditions	Equation Nos. in Text	Reference	Comments
1	$\overline{}^{s}_{\partial}$ $\overline{}_{s}$	const. $\times \partial^3 \coth \dfrac{\partial}{2kT}$	const. $\times \partial^2 T$	$\partial \ll kT, kT_s$ $T \ll \Theta_D$	(33.4), (39.4), (39.5)	a	Direct processes. In entry 2 $H_\parallel$ and $H_\perp$ may require different $\varDelta$'s according to the energy levels involved. $T \gtrsim \partial/k$, spontaneous emission dominates. Reference b.
2	K.d. $\overline{}$ $\varDelta$ $\overline{}_{\partial}^{\partial}$ K.d.	const. $\times \dfrac{\partial^3 H^2}{\varDelta^2} \coth \dfrac{\partial}{2kT}$	const. $\times \dfrac{H^4 T}{\varDelta^2}$	$\partial = g\beta H \ll kT, kT_s$ $T \ll \Theta_D$ $\varDelta \gg \partial$	(33.6), (39.6), (39.7)	a, b	
3a			const. $\times \dfrac{T^2}{\varDelta^2}$		(34.11), (40.2)	a	Raman processes. Entries 3a to 3d, the term linear in strain used in second order perturbation theory. Entries 3e and 3f, the second order term in strain used in first order perturbation theory. Entries 3c and 3d are Brons-Van Vleck type formulae. 3d is difficult to use because μ and μ' are difficult to evaluate.
3b		const. $\times \dfrac{I_6}{\varDelta^2}$			(34.11), (40.2), (40.3)		
3c	$\overline{}^{s}$ $\varDelta$ $\overline{}^{s}_{\partial}$ $\overline{}_{s}$		const. $\times \dfrac{T^7}{\varDelta^2} \times \dfrac{(H^2 + H^2_{\rm hfs} + H^2_{\rm dip})}{(H^2 + H^2_{\rm hfs} + \frac{1}{2} H^2_{\rm dip})}$	$\partial \ll kT, \varDelta, E_{\rm ph}$ $T \gtrsim \Theta_D$, (3a), (3e) $T \ll \Theta_D$, (3b), (3c), (3d), (3f) $g_\perp = 0$, (3c) $H_{\rm int} \simeq H$, (3c) and (3d) $g_\perp \ne 0$, (3d)	(56.2), $H^2_{\rm hfs} = 0$	c	
3d			const. $\times \dfrac{T^7}{\varDelta^2} \times \dfrac{(H^2 + \mu H^2_{\rm hfs} + \mu' H^2_{\rm dip})}{(H^2 + H^2_{\rm hfs} + \frac{1}{2} H^2_{\rm dip})}$		(56.3)		
3e		const. $\times I_6$	const. $\times T^2$		(34.11)	d	
3f			const. $\times T^7$		(34.11), (40.10)		

Orbach processes.

4a	K.d. ⟨∂⟩ s / s	$\text{const.} \times \dfrac{I_8}{\Delta^4}$		(34.13), (40.2)	a	Raman processes. Entries 4a to 4d, the linear term in strain used in second order perturbation theory. Entry 4e, the second order strain term used in first order theory. Entries 4d and 4e, a magnetic field removes the Kramers degeneracy. When $H_{\text{dip}} \cong H$ entry 4b has Brons-Van Vleck dependencies similar to entries 3c and 3d.
4b	K.d. ⟨Δ⟩ s / s	$\text{const.} \times \dfrac{T^9}{\Delta^4}$	$\partial \ll kT$ $T \gtrsim \Theta_D$, (4a) $T \ll \Theta_D$, (4b) to (4e) $E_{\text{ph}} \ll \Delta$, (4a), (4b), (4d) and (4e) $E_{\text{ph}} \gg \Delta$, (4c)	(34.13), (40.2), (40.7)	a	
4c	Δ ⟨∂⟩ s / s	$\text{const.} \times I_4$		(49.3)	e	
4d	K.d. ⟨∂⟩ s / s	$\text{const.} \times \dfrac{H^2 I_6}{\Delta^4}$		(40.8)	d	
4e	⟨∂⟩ Δ	$\text{const.} \times \dfrac{H^2 I_6}{\Delta^2}$		(40.11)	d	
5	— s —	$\text{const.} \times \dfrac{\Delta^3}{\exp\left(\dfrac{\Delta}{kT}\right) - 1}$	$E_{\text{opt}} \gg kT$	(62.4)	f	Raman processes using optical mode phonons.
6	⟨∂⟩ Δ ⟨∂⟩	$\text{const.} \times \Delta^3 \exp\left(-\dfrac{\Delta}{kT}\right)$	$\partial \ll kT \ll \Delta$	(35.4), (40.4)	a, g	Orbach processes. The energy levels may be from Kramers or non-Kramers salts.
7	⟨∂⟩ Δ	$\text{const.} \times \dfrac{\Delta^3 \exp\left(\dfrac{\Delta}{kT}\right)}{\exp\left(\dfrac{\Delta}{kT}\right) - 1}$	$\partial \ll kT \ll \Delta$	(49.1)	g	

In this table, except for entry 5, the phonons are assumed to be described by a Debije spectrum.

E_{ph} is the energy of the phonons in the mode most densely occupied.
E_{opt} is the energy of the phonons in the optical mode involved in the relaxation.
I_n is the integral given by equation (34.4).
H_{int} is the magnitude of the internal fields in the sample.
s in the second column is a non-degenerate state.
K.d. in the second column is a Kramers doublet.
The other symbols have their normal meanings.

References:

[a] R. Orbach: Proc. Roy. Soc. (London) A **264**, 458 (1961).
[b] R. H. Ruby, H. Benoit, and C. D. Jeffries: Phys. Rev. **127**, 51 (1962).
[c] R. Orbach: Proc. Roy. Soc. (London) A **264**, 485 (1961).
[d] P. L. Scott, and C. D. Jeffries: Phys. Rev. **127**, 32 (1962).
[e] R. Orbach, and M. Blume: Phys. Rev. Letters **8**, 478 (1962).
[f] B. I. Kochelaev: Soviet Phys. JETP **10**, 171 (1960).
[g] A. A. Manenkov, and A. M. Prokhorov: Soviet Phys. JETP **15**, 951 (1962).

Therefore the Orbach relaxation mechanism in the simple form derived here shows an exponential dependence on temperature. The same expression is obtained for both Kramers and non-Kramers salts because the only transitions which we have allowed in our derivation are those between levels which are not degenerate with respect to time-reversal.

Table 1 contains a list of possible relaxation mechanisms and their dependence on temperature and magnetic field. Most of the entries in the table can be checked by similar considerations to those in this section. The considerations of this section do not give any indication of the form and magnitude of the spin-lattice interaction and this will be discussed in several of the remaining sections of this article.

36. Extremes of temperature and magnetic field. At temperatures above the Debije temperature the spin-lattice relaxation rate is inversely proportional to T^2, Eqs. (34.11) and (34.13). This has been observed experimentally many times. As the material starts to approach its melting point this dependency ends because diffusion of the ions starts. The influence of diffusion on the relaxation of nuclei has been considered by many workers (see for example EISENSTADT and REDFIELD[1]) but it is unlikely to be observed in electron spin systems because of the shortness of the other time constants at these temperatures.

At very low temperatures, experiments have been performed down to 0.08 °K[2,3] the direct relaxation process gives way to spontaneous emission of phonons by the spins. The number of phonons at angular frequency ω is given by (31.2) which goes to zero as $kT \ll \hbar\omega$, $\hbar\omega$ is the energy level splitting. Therefore, the chances of stimulated emission and absorption become negligible compared to spontaneous emission. The coefficient of spontaneous emission is not the coefficient calculated for a single spin in free space. In spin-lattice relaxation experiments the emission rate is enhanced for two reasons. Firstly, the spins are not in empty space but in a coil or resonant cavity where the energy density of the field per unit frequency range and unit volume is increased. Secondly, there are many spins within a space whose dimensions are smaller than the velocity of light divided by the frequency width of the transition. This causes coherence effects which increase the spontaneous emission rate, but not far enough to make the rate significant at any temperature above approximately 1 °K[4]. The spontaneous emission is temperature independent. A similar mechanism is the low temperature limit of the "inverted" Orbach mechanism, when the relaxation in the two step process is limited by spontaneous emission.

Another important phenomenon at low temperature is the increase in the lifetime of the phonons which produces the so-called phonon bottleneck. This is considered in **CV**.

Little work has been performed on relaxation at high magnetic fields, i.e. above 25,000 oersteds, but it is expected that no new effects will appear. Indeed, there is likelihood of better agreement between theory and experiment because under the influence of high fields the ions become more independent of each other. The higher the field, the more slowly do the spins reach equilibrium with each other and so their interaction with the lattice, during short intervals of time, may be regarded as that of truly isolated spins. At high fields the direct process may

---

[1] M. EISENSTADT: Phys. Rev. **132**, 630 (1963); **133**, A 191 (1964). — M. EISENSTADT, and A. G. REDFIELD: Phys. Rev. **132**, 635 (1963).

[2] A. M. PROKHOROV, and V. B. FEDOROV: Proceedings of the Internat. Conference on Magnetism, Nottingham, p. 449. The Institute of Physics and the Physical Society 1964.

[3] A. R. MIEDEMA, and K. W. MESS: Physica **30**, 1849 (1964).

[4] B. BÖLGER: Thesis University of Leiden 1959.

dominate at higher temperatures than normally observed. Large H increases the number of phonons which can participate in the direct relaxation process while low T means that only the lower energy phonons are excited and so the numerous high energy phonons which take part in the Raman process do not exist. The ratio H/T is also significant in the phenomenon of paramagnetic saturation and when it becomes large there may be very little susceptibility left in which to observe any relaxation phenomena.

We delay until Sect. 56 the discussion of low field work because this involves the treatment of paramagnetic ions relaxing while under the influence of one another. Our work at the moment is concerned with isolated ions.

III. Relaxation due to spin-orbit coupling.

a) Rare-earth salts.

37. The crystalline field. It is possible in rare-earth salts to make considerable progress towards evaluating a value of τ_L using a phenomenological spin-lattice theory. This approach, even with its assumptions, gives adequate order of magnitude values for τ_L.

In rare-earth salts, J, the total angular momentum quantum number, is a good quantum number and is used to describe the eigenfunctions of the electron energy levels. This occurs because the spin-orbit coupling is stronger than the electric crystalline field interaction in these salts. To establish the energy levels and the wave functions required in the calculation we use crystal field theory. The electric field at the position of the paramagnetic electron is found by solving LAPLACE's equation $\nabla^2 V = 0$. The solution for the electrostatic potential V can be expanded in a series of Legendre polynomials:

$$V = \sum_{m \leq n} V_n^m. \qquad (37.1)$$

Fortunately the number of terms which we need in this series is limited by the symmetry of the potential, and the symmetry and parity of the wave functions of the paramagnetic ion. Following ORBACH's line of approach [16] we shall consider the rare-earth ethyl sulphates as a specific example where the rare-earth ion has nine water molecules as nearest neighbours. The site symmetry is C_{3h}. Because of the parity of the wave-functions all the terms with odd n may be neglected as their matrix elements with the potential are zero. In this case the potential changes sign under the parity operation while the product of the wave functions in the integral does not, thus, integrating over all space gives zero. Further, the values of n are limited to $2l$ or less where l is the angular momentum quantum number which for rare-earths is 3, so only terms with $n = 0, 2, 4, 6$, can occur. The matrix elements for terms with higher n are zero. This occurs because the wave functions can also be expanded in a series of spherical harmonics, and when this is done non-zero values of the integral only occur when the above condition is satisfied. Finally the values of m are limited to 0 and ± 6 by the symmetry of the site because m occurs in the term $\exp(im\varphi)$ which must have a definite value for φ and $\varphi + k2\pi$, k is any integer. Therefore, the final expression for the potential is:

$$V = A_2^0 \langle r^2 \rangle Y_2^0(\vartheta, \varphi) + A_4^0 \langle r^4 \rangle Y_4^0(\vartheta, \varphi) + A_6^0 \langle r^6 \rangle Y_6^0(\vartheta, \varphi) + \\ + (A_6^{-6} Y_6^{-6}(\vartheta, \varphi) + A_6^6 Y_6^6(\vartheta, \varphi)) \langle r^6 \rangle, \qquad (37.2)$$

where A_n^m is a crystalline field parameter, $\langle r^n \rangle$ is the mean of the n-th power of the f shell electron radius, Y_n^m is the spherical harmonic of degree n and azimuthal quantum number m. The term with $m=n=0$ is ignored because it is a constant which shifts all the levels by the same amount. To ease the calculation the concept of an operator equivalent as developed by ELLIOTT and STEVENS[1] is used. The Wigner-Eckhart theorem in group theory[2] states that the matrix elements of operators, transforming in the same manner under rotations, are proportional to one another within an irreducible manifold of states. This allows us to replace the operators involving x, y, z, in the expression for the electrostatic potential by equivalent operators involving J_x, J_y, J_z. ELLIOTT and STEVENS have tabulated the proportionality constants necessary for this substitution. Now the energy levels and wave functions of the paramagnetic ion in the static crystalline field can be calculated. We assume that the phonon properties are described by Eqs. (31.1) to (31.7).

38. The spin-lattice interaction. Having established the energy levels in the combined electron-phonon system and the required wave functions we must discuss the perturbation which causes the electron transitions. The thermal motion of the atoms in the solid causes variations in the strength of the crystalline field at the site of the paramagnetic ion, these variations modulate the orbital component of the angular momentum of the electrons and this modulation is transferred to the spin of the electrons by the spin-orbit coupling. Various aspects of this mechanism were discussed by HEITLER and TELLER[3], FIERZ[4], KRONIG[5], VAN VLECK [13] and many other workers.

The magnitude of the orbit-lattice interaction can be estimated by adapting the static theory of the crystalline field to the dynamic case. The interaction can be written as a series for the strain in the paramagnetic complex XY_9, Y represents here a molecule of water:

$$V_{OL}(j) = \sum_{n,m} V_n^m(j) + \sum_{n,m} \frac{\partial V_n^m(j)}{\partial Q}\bigg|_0 \varepsilon_j Q + \frac{1}{2!} \sum_{n,m} \frac{\partial^2 V_n^m(j)}{\partial Q \, \partial Q'}\bigg|_0 \varepsilon_j Q \varepsilon_j Q' + \cdots . \quad (38.1)$$

$V_{OL}(j)$ is the orbit-lattice potential at $\boldsymbol{Q}$, the site of the j-th ion, $V_n^m(j)$ is the crystalline field potential at the site of j-th ion and Q and Q' are coordinates of $\boldsymbol{Q}$. $\boldsymbol{q} = \varepsilon_j \boldsymbol{Q}$ defines the strain at the site of the ion but here all directional properties of the strain tensor are neglected and the scalar $\varepsilon_j \simeq \partial q/\partial Q$ is used instead. The same limitations apply to the possible values of n in this case as in the static case but consideration of the twenty-four normal vibrational modes* of the paramagnetic cluster XY_9 shows that all of the V_n^m transform as one or another of the irreducible representations of the group of the vibrational modes and no more of the V_n^m can be eliminated.

39. Direct processes.

α) *Non-Kramers salts.* Eq. (38.1) is equivalent to (26.3). The first term is the static crystalline field while the second term is V', the spin-lattice perturbation to be used in Eq. (26.1). Consider again Fig. 6 which represents the states of a simple non-Kramers salt. The transition probability for the direct process is given

---

* Each member Y of the XY_9 cluster is regarded as a single particle. The three translational modes and the three rotational modes are ignored because they leave the relative positions of the members of the cluster unchanged.

[1] R. J. ELLIOTT, and K. W. H. STEVENS: Proc. Roy. Soc. (London) A **215**, 437 (1952).
[2] E. P. WIGNER: Group Theory and its Application to the Quantum Mechanics of Atomic Spectra. Academic Press 1959.
[3] W. HEITLER, and E. TELLER: Proc. Roy. Soc. (London) A **155**, 629 (1936).
[4] M. FIERZ: Physica **5**, 433 (1938).
[5] R. DE L. KRONIG: Physica **6**, 33 (1939).

Direct processes.

by Eq. (33.1). The constant in this equation contains the matrix elements of the ion states and the numerical constants of the phonon matrix elements and the Debije distribution. It is usual to assume that $\varepsilon_j Q \dfrac{\partial V_n^m(j)}{\partial Q} = \varepsilon_j V_n^m(j)$, see Sect. 44, then Eq. (33.1) is:

$$W_{ba} = \frac{2\pi}{\hbar} \int \left| \sum_{n,m,j} \langle a|V_n^m(j)|b\rangle \langle \ldots N_i+1, \ldots |\varepsilon_j|\ldots N_i, \ldots\rangle \right|^2 \times \\ \times \frac{1}{2\pi^2} \cdot \frac{\omega^2 V d\omega}{\hbar v^3} \delta(\partial_i - \hbar\omega). \qquad (39.1)$$

The summation is over the lattice sites and the integral is over the phonon distribution. The Dirac delta function restricts the integral to having a value at the phonon energy ∂_i only. The phonon matrix element is in terms of the strain operator. This is related to the matrix elements of the creation and annihilation operators, Eq. (31.3), by using (31.4) which gives the displacement q of the ion j from its site at Q. Differentiating (31.4) with respect to Q gives ε_j in terms of the operators a and a*. In the summation over the lattice sites only those with the spin in state $|b\rangle$ contribute if all dipolar forces are neglected. Then the expression for W_{ba} becomes:

$$W_{ba} = (3/2\pi \varrho v^5 \hbar)(\partial_i/\hbar)^3 \left|\langle a|\sum_{n,m}V_n^m|b\rangle\right|^2 N_b(N_i+1), \qquad (39.2)$$

where N_b is the number of atoms in the state $|b\rangle$, v is the velocity of sound in the crystal and ϱ is the density of the crystal. While simplifying this expression the relation $\omega = vk$ was used, which is true for long wavelength acoustic phonons. Its use is not really an extra assumption because the phonon structure is approximated by the Debije distribution.

The reverse process also occurs with an electron being excited from $|a\rangle$ to $|b\rangle$. The probability W_{ab} is given by:

$$W_{ab} = (3/2\pi \varrho v^5 \hbar)(\partial_i/\hbar)^3 \left|\langle b|\sum_{n,m}V_n^m|a\rangle\right|^2 N_a N_i. \qquad (39.3)$$

These two transition probabilities are used in the rate equations for the electron levels $|a\rangle$ and $|b\rangle$, and if they are the lowest levels of the rare earth ion their populations can be assumed to be independent of the other levels which are effectively unpopulated at low temperatures. The solution to the rate equation gives the following expression for the spin-lattice relaxation time τ_L:

$$\tau_L^{-1} = \frac{3\coth(\partial_i/2kT)(\partial_i/\hbar)^3}{(2\pi\hbar\varrho v^5)}\left|\langle a|\sum_{n,m}V_n^m|b\rangle\right|^2. \qquad (39.4)$$

Eq. (39.4) reduces to:

$$\tau_L^{-1} = \frac{3\partial_i^2 kT}{(\pi\hbar^4\varrho v^5)}\left|\langle a|\sum_{n,m}V_n^m|b\rangle\right|^2 \qquad (39.5)$$

when $\partial_i \ll kT$, which is the complete form of Eq. (33.4).

The direct process exchanges energy between the spins and the low frequency phonons in the tail of the frequency distribution. In this case some of the simplifications in the phonon structure are justified because only accoustic, long wavelength phonons participate. This process extends to very low temperatures because it only uses the low energy phonons which can be excited at these temperatures. If there are dipolar interactions between the spins, the possibility of multiple spin flips arises but imposing a magnetic field much larger than the internal fields effectively isolates the spins and so (39.4) can be expected to hold.

β) *Kramers salts.* Using the state functions calculated by first order perturbation theory, Eq. (33.5), the expression for τ_L for a Kramers salt with energy levels as depicted in Fig. 7 is:

$$\tau_L^{-1} = \frac{3\partial_i^2 (4A^2\beta^2)}{2\pi\varrho v^5 \hbar} \frac{|\mathbf{H}\cdot\langle -\tfrac{1}{2}c|\mathbf{J}|-\tfrac{1}{2}a\rangle|^2}{A_c^2} \times \\ \times \left|\langle -\tfrac{1}{2}a|\sum_{n,m} V_n^m|+\tfrac{1}{2}c\rangle\right|^2 \coth(\partial_i/2kT),\quad (39.6)$$

which when $g\beta H \ll kT$ reduces to:

$$\tau_L^{-1} = \frac{3(4A^2\beta^2)(g^2\beta^2)H^4 kT}{(\pi\varrho v^5 \hbar^4 A_c^2)} |\boldsymbol{\eta}\cdot\langle -\tfrac{1}{2}c|\mathbf{J}|-\tfrac{1}{2}a\rangle|^2 \left|\langle -\tfrac{1}{2}a|\sum_{n,m} V_n^m|\tfrac{1}{2}c\rangle\right|^2 \quad (39.7)$$

the complete form of Eq. (33.6). $\boldsymbol{\eta}$ is the unit vector of $\mathbf{H}$. We must stress that the matrix elements of $\mathbf{J}$ in these equations are specifically for ethyl sulphates.

It is difficult to confirm if direct processes have ever been observed in ethyl sulphates because of the prominence of other processes, even at very low temperatures.

40. Two phonon relaxation processes. At higher temperatures the relaxation rate is determined by two phonon processes. Individually each process has a smaller chance of occurrence than the direct process but there are many more phonons available to participate in the processes. The phonons in the direct process are restricted to those of the frequency range given by the energy conservation requirements $\hbar\omega = \partial_i$. There are several variations of the two phonon mechanism.

Taking the first order term in ε_j in Eq. (38.1) twice and using it in second order perturbation theory leads to a two phonon relaxation mechanism. We will consider the levels illustrated in Fig. 8. The perturbation destroys a phonon of energy $\hbar\omega_k$ and creates another phonon of energy $\hbar\omega_j$. The quantum $\hbar\omega_k$ being absorbed by an upward spin flip $|b\rangle$ to $|c\rangle$, and the quantum $\hbar\omega_j$ being emitted by a downward spin flip $|c\rangle$ to $|a\rangle$. Energy is conserved because the decrease in energy of the spin system ∂_i equals the increase in energy of the phonon system $\hbar(\omega_j - \omega_k)$. The initial state of the system is specified by a spin being in the state $|b\rangle$ and N_j and N_k phonons in modes ω_j and ω_k respectively. The final state of the system has the spin in the state $|a\rangle$, one phonon in the mode ω_j created and one in the mode ω_k destroyed. The perturbation operator describing the orbit-lattice interaction, which causes this change of states, is

$$\mathcal{H}_{OL} = \sum_{\substack{m,n,c,j,\\ m',n'}} V_n^m(j) |c\rangle\langle c| V_{n'}^{m'}(j)\cdot \varepsilon_j \varepsilon_j', \quad (40.1)$$

where the summation is over all accessible intermediate states. It is assumed here, as before, that there are no interactions between the spins at various sites.

Using this operator the transition probabilities W_{ab} and W_{ba} are calculated and then the rate equation for the system solved, assuming that there are no direct-transitions between the two levels. This gives us the following general expression:

$$\tau_L^{-1} = \frac{9}{16\pi^3 \varrho^2 v^{10}} \cosh\left(\frac{\partial_i}{2kT}\right) \int \left| \sum_{\substack{m,n,c,\\ m',n'}} \frac{\langle a|V_n^m|c\rangle\langle c|V_{n'}^{m'}|b\rangle}{\hbar\omega_k + \tfrac{1}{2}\partial_i - A c} \right|^2 \times \\ \times \omega_k^3 \left(\omega_k - \frac{\partial_i}{\hbar}\right)^3 \exp\left(\frac{\hbar\omega_k - \partial_i}{kT}\right) \left\{\left[\exp\left(\frac{\hbar\omega_k - \partial_i}{kT}\right) - 1\right] \times \\ \times \left[\exp\left(\frac{\hbar\omega_k}{kT}\right) - 1\right]\right\}^{-1} d\omega_k. \quad (40.2)$$

α) *Raman processes in non-Kramers salts.* We now distinguish between Kramers and non-Kramers salts and consider the latter first. We assume that only state $|c\rangle$ is involved and so we have the situation shown in Fig. 8. We make the formal simplification that $\partial_i \ll kT, A_c$ which simplifies the integral in (40.2) a little. Major contributions to the integral will occur when $\hbar\omega_k \simeq kT$ and when $\hbar\omega_k = A_c$

so we differentiate between the two cases $k\Theta_D > \Delta_c$ or $k\Theta_D < \Delta_c$. The latter condition, combined with the condition $\Delta_c \gg kT$, leads to:

$$\tau_L^{-1} = \frac{9(6!)}{4\pi^3 \varrho^2 v^{10} \Delta_c^2} \left(\frac{kT}{\hbar}\right)^7 \left|\sum_{\substack{n,m,\\n',m'}} \langle a|V_n^m|c\rangle \langle c|V_{n'}^{m'}|b\rangle\right|^2 \qquad (40.3)$$

for the Raman relaxation time in the low temperature approximation, c.f. Eq. (34.11). The chief contribution to the integral is in the region $\hbar\omega_k \simeq kT$ so $\hbar\omega_k$ may be neglected with respect to Δ_c, which is the adiabatic approximation.

In the Raman process energy is conserved between the initial and the final states but not between the initial and intermediate, or the intermediate and final states. The physical reason is the very short lifetime of the intermediate state which leads to a large uncertainty in its energy. The Raman process is regarded as taking place through a virtual intermediate state, while the Orbach process, see below, takes place through a real intermediate state. In this case energy is conserved at each step.

β) *Orbach processes in non-Kramers salts.* In the case where $k\Theta_D > \Delta_c$ phonons exist which have $\hbar\omega_k = \Delta_c$, so there is a second maximum in the integral. It is not possible in this case to neglect $\hbar\omega_k$ in the denominator and when $\hbar\omega_k = \Delta_c$ the value of the integral apparently goes to infinity. When we remember that $|c\rangle$ has a finite width we realise that the integral does not diverge, but it is a much sharper maximum than that at $\hbar\omega_k \simeq kT$ so the integral may be treated as the sum of two parts, a resonant and a non-resonant part. The non-resonant part leads to the same answer as for $k\Theta_D < \Delta_c$ when the same assumptions are made. Following ORBACH [*16*], the finite width of the level $|c\rangle$ is introduced in a similar manner as in the problem of fluorescence by adding a term $\frac{1}{2}i\Gamma_c$ to the denominator. Γ is a complex quantity representing the damping caused by the reaction of the emitted quantum on the emitting atom. $\mathscr{R}e(\Gamma)$ is the linewidth due to the finite lifetime of the state and $\mathscr{I}m(\Gamma)$ is an energy term which can be ignored by a suitable choice of zero. In the fluorescence mechanism energy is conserved within Γ so the atom remembers what energy quantum it absorbed before it re-emits [*17*]. Γ_c is evaluated using the techniques described for the direct process for transitions $|c\rangle$ to $|b\rangle$ and $|c\rangle$ to $|a\rangle$, it is the sum of these two probabilities. When the integral is evaluated τ_L^{-1} is found to have an exponential dependence on temperature,

$$\tau_L^{-1} = \frac{3}{2\pi\varrho v^5 \hbar} \frac{\Delta_c^3}{\hbar^3} \left[\exp\left(\frac{\Delta_c}{kT}\right) - 1\right]^{-1} \left\{\frac{\left|\sum_{\substack{n,m,\\n',m'}} \langle a|V_n^m|c\rangle \langle c|V_{n'}^{m'}|b\rangle\right|^2}{\left|\sum_{n,m} \langle a|V_n^m|c\rangle\right|^2 + \left|\sum_{n',m'} \langle c|V_{n'}^{m'}|b\rangle\right|^2}\right\}. \qquad (40.4)$$

Even though there are only limited numbers of phonons available with $\hbar\omega_k \simeq \Delta_c$ this Orbach relaxation mechanism is of major importance in salts with a similar configuration of energy levels because it is a resonance effect. McCUMBER[1] has given the intermediate state detailed consideration.

Γ_c is the width of level $|c\rangle$ due to the finite lifetime of the level and any extra broadening, due to dipolar interactions for instance, does not alter the relaxation time. This extra broadening does destroy the coherence of the state and changes the mechanism. It can now be considered as two direct processes consecutively. However, a calculation based on this two step process[2] leads to the same result for τ_L.

γ) *Kramers salts.* Passing on to Kramers salts we have the situation of Fig. 7 with the ground states and the excited states Kramers doublets, each pair being

---
[1] D. E. McCUMBER: Phys. Rev. **130**, 2271 (1963).
[2] C. B. P. FINN, R. ORBACH, and W. P. WOLF: Proc. Phys. Soc. (London) **77**, 261 (1961).

a time conjugate pair in the absence of a magnetic field. The perturbation operator now consists of two terms, one to each level of the excited doublet. Calculating the matrix elements for this operator between the ground state levels we obtain:

$$\left\langle -\frac{a}{2}\left|\mathcal{H}_{OL}\right|\frac{a}{2}\right\rangle = \sum_{\substack{n,m \\ n',m'}} \left\{ \frac{\left\langle -\frac{a}{2}\left|V_n^m\right|\frac{c}{2}\right\rangle\left\langle \frac{c}{2}\left|V_{n'}^{m'}\right|\frac{a}{2}\right\rangle}{E_{initial} - E_{\frac{c}{2}}} + \right. \tag{40.5}$$
$$\left. + \frac{\left\langle -\frac{a}{2}\left|V_{n'}^{m'}\right|-\frac{c}{2}\right\rangle\left\langle -\frac{c}{2}\left|V_n^m\right|\frac{a}{2}\right\rangle}{E_{initial} - E_{-\frac{c}{2}}} \right\} \varepsilon_j \varepsilon_j'.$$

Taking the time conjugate of the second term and then the Hermitean conjugate and constraining the result to be Hermitean, (40.5) becomes:

$$\left\langle -\frac{a}{2}\left|\mathcal{H}_{OL}\right|\frac{a}{2}\right\rangle = \sum_{\substack{n,m, \\ n',m'}} \left\langle -\frac{a}{2}\left|V_n^m\right|\frac{c}{2}\right\rangle\left\langle \frac{c}{2}\left|V_{n'}^{m'}\right|\frac{a}{2}\right\rangle \times$$
$$\times \left\{ \frac{1}{\hbar\omega_k - \Delta_c} - \frac{1}{-\hbar\omega_j - \Delta_c} \right\} \varepsilon_j \varepsilon_j'. \tag{40.6}$$

$\left\langle -\frac{a}{2}\left|\mathcal{H}_{OL}\right|\frac{a}{2}\right\rangle$ does not go to zero because in the first term $V_{n'}^{m'}$ acts on $\left|\frac{a}{2}\right\rangle$ and because of the requirements of energy conservation a phonon is destroyed while in the second term V_n^m acting on $\left|\frac{a}{2}\right\rangle$ creates a phonon. This is an example of the non-adiabatic approximation when it is not possible to neglect $\hbar\omega_k$ because then the element $\left\langle -\frac{a}{2}\left|\mathcal{H}_{OL}\right|\frac{a}{2}\right\rangle$ goes to zero. When the time degeneracy of the wave functions acts to produce a matrix element which has two terms equal in magnitude but opposite in sign we refer to the occurrence as a Van Vleck cancellation because he was the first to discuss it in relaxation theory [13]. Eq. (40.6) is an example of a "partial" Van Vleck cancellation.

The expression for the relaxation time in Kramers salts in the case of $k\Theta_D < \Delta_c$, using similar approximations to those above, is:

$$\tau_L^{-1} = \frac{9!\,\hbar^2}{\pi^3 \varrho^2 v^{10} \Delta_c^4}\left(\frac{kT}{\hbar}\right)^9 \left|\sum_{\substack{n,m, \\ n',m'}} \left\langle -\frac{a}{2}\left|V_n^m\right|\frac{c}{2}\right\rangle\left\langle \frac{c}{2}\left|V_{n'}^{m'}\right|\frac{a}{2}\right\rangle\right|^2 \tag{40.7}$$

in the low temperature limit.

When $k\Theta_D > \Delta_c$ the resonant contribution to τ_L is given by Eq. (40.4). In contrast to the consideration of direct processes in Kramers salts we have not had to introduce a magnetic field to overcome the effects of the time conjugate degeneracy of the pairs of levels. This is because the perturbation Hamiltonian used includes elements of the type $\left\langle -\frac{a}{2}\left|V\right|\frac{c}{2}\right\rangle$ where $\left|-\frac{a}{2}\right\rangle$ and $\left|\frac{c}{2}\right\rangle$ are not time conjugate with respect to each other. It was once considered that, in analogy with the case of direct relaxation, a magnetic field was required to lift the degeneracy and allow the electrostatic operator to connect the previously time conjugate states. The magnetic field reduces the Van Vleck cancellation in two ways; firstly because the energy splittings are altered and secondly because mixing of excited states into the ground state removes the time degeneracy. The relaxation time is given by:

$$\tau_L^{-1} = \frac{9 \times 6!}{4\varrho^2 v^{10} \pi^3}\left(\frac{2\Lambda\beta H}{\Delta_c^2}\right)^2\left(\frac{kT}{\hbar}\right)^7 \left|\boldsymbol{\eta}\cdot\left\langle -\frac{c}{2}\left|\boldsymbol{J}\right|-\frac{a}{2}\right\rangle\right|^2\left|\left\langle -\frac{a}{2}\left|\sum_{n,m}V_n^m\right|\frac{c}{2}\right\rangle\right|^2 \tag{40.8}$$

in the low temperature limit. $\boldsymbol{\eta}$ is the unit vector of $\boldsymbol{H}$.

δ) *An alternative Raman process.* Second order processes also occur when the second order term in Eq. (38.1) is used in perturbation theory. The operator for this perturbation is:

$$\mathscr{H}_{OL} = \frac{1}{2} \sum_{n,m} \frac{\partial^2 V_n^m(j)}{\partial Q \, \partial Q'}\bigg|_0 \varepsilon_j Q \, \varepsilon_j Q'. \quad (40.9)$$

When this acts on the ground states of a non-Kramers salt it leads to the following expression for τ_L^{-1}:

$$\tau_L^{-1} = \frac{9 \times 6!}{8 \varrho^2 \pi^3 v^{10}} \left(\frac{kT}{\hbar}\right)^7 \times \\ \times \left|\sum_{n,m} \langle a | \frac{\partial^2 V_n^m(j)}{\partial Q \, \partial Q'}\bigg|_0 |b\rangle\right|^2 \quad (40.10)$$

in the low temperature limit. This perturbation cannot act on a Kramers doublet unless a magnetic field lifts the time degeneracy. Then τ_L is given by:

$$\tau_L^{-1} = \frac{9 \times 6!}{8 \varrho^2 \pi^3 v^{10}} \left(\frac{2\Lambda \beta H}{\Delta_c}\right)^2 \left(\frac{kT}{\hbar}\right)^7 \times \\ \times \left|\sum_{n,m} \langle -\frac{a}{2} | \frac{\partial^2 V_n^m(j)}{\partial Q \, \partial Q'}\bigg|_0 |\frac{c}{2}\rangle\right|^2. \quad (40.11)$$

The times given by Eqs. (40.8) and (40.11) are usually longer than that of Eq. (40.7). This "alternative" Raman process does not require a real intermediate state for its occurrence but, as stated above, it may be considered as taking place through a virtual intermediate state. An alternative description is that a phonon is scattered from one mode to another by a spin flipping over.

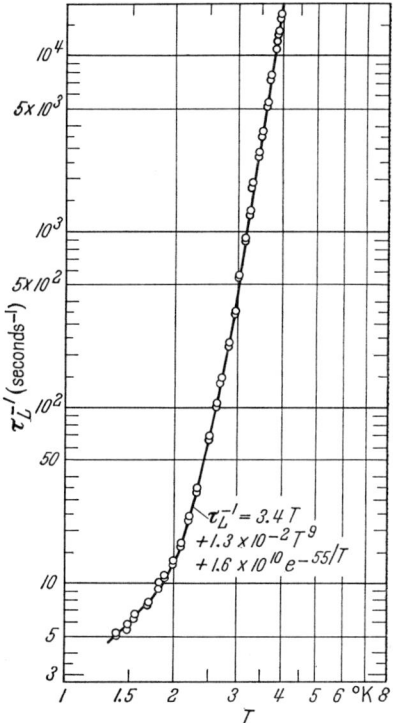

Fig. 9. Experimental measurements from ~1% Sm³⁺ in Lanthanum magnesium nitrate, $H\|z$-axis, $H = 9150$ oersted, $\nu = 9.38$ GHz. Taken from G. H. LARSON, and C. D. JEFFRIES: Phys. Rev. **141**, 461 (1966).

ORBACH discussed the situation in rare-earth ethyl sulphates [16], and the theory has been extended, with some modifications, to double nitrates by SCOTT and JEFFRIES [18], to CaF$_2$ by HUANG[1] and BIERIG et al.[2] and to garnets by

Table 2. *A comparison of theoretical predictions and experimental results for some rare-earth ions.*

Entry	Salt	Orientation	T or E	Direct τ_L^{-1} sec^{-1}	ORBACH τ_L^{-1} sec^{-1}	RAMAN τ_L^{-1} sec^{-1}	Reference
1	1% Ce³⁺: LaES	$H \perp z$ $H = 3{,}080$ oe.	E	—	2.2×10^6 $\exp(-5.6/T)$ *	—	a
			T	—	6.9×10^7 $\exp(-5.7/T)$	—	a
2	0.2% Ce³⁺: LaMN	$H \perp z$ $H = 3{,}750$ oe.	E	**	2.7×10^9 $\exp(-34/T)$	—	b
			T	**	3.5×10^9 $\exp(-34/T)$	—	b and c

[1] C. Y. HUANG: Phys. Rev. **139**, A 241 (1965).
[2] M. J. WEBER, and R. W. BIERIG: Phys. Rev. **134**, A 1492 (1964).

Table 2. (Continued.)

Entry	Salt	Orientation	T or E	Direct τ_L^{-1} sec^{-1}	ORBACH τ_L^{-1} sec^{-1}	RAMAN τ_L^{-1} sec^{-1}	Reference
3	1% Nd^{3+}: LaES	$H \perp z$ $H = 3{,}240$ oe.	E	$1.7\,T$	—	$3.6 \times 10^{-4}\,T^9$	a
			T	$1.4\,T$	—	$1.3 \times 10^{-4}\,T^9$	a
4	1% Nd^{3+}: LaMN	$H \perp z$ $H = 2{,}480$ oe.	E	$1.7\,T$	$6.3 \times 10^9 \exp(-47.6/T)$	—	a
			T	$2.6\,T$	$2.2 \times 10^{10} \exp(-47.6/T)$	$7.8 \times 10^{-4}\,T^9$	a
5	1% Nd^{3+}: YGaG		E	$17\,T$	$9 \times 10^{10} \exp(-120/T)$	—	d
			T	$10\,T$	$10^{12} \exp(-120/T)$	—	d
6	1% Nd^{3+}: YAlG		E	$34\,T$	$4.5 \times 10^{10} \exp(-110/T)$	—	d
			T	$10\,T$	$10^{11} \exp(-110/T)$	—	d
7	0.1 and 1% Nd^{3+}: YES	$H \perp z$ $H = 3{,}240$ oe.	E	$1.2\,T$	—	$1.64 \times 10^{-4}\,T^9$	e
			T	$0.74\,T$	—	$3.4 \times 10^{-5}\,T^9$	e
8a	0.1% Sm^{3+}: LaES	$H \| z$	E	$16\,T$	—	$2.6 \times 10^{-2}\,T^9$	d
			T	$0.32\,T$	—	$4.4 \times 10^{-3}\,T^9$	d
		$H \perp z$	E	$100\,T$	—	$2.6 \times 10^{-2}\,T^9$	d
			T	$1.8\,T$	—	$4.4 \times 10^{-3}\,T^9$	d
8b	0.1 and 1% Sm^{3+}: LaES	$H \| z$ $H = 11{,}240$ oe.	E	$1.0\,T$	$5.8 \times 10^8 \exp(-46/T) + 6.1 \times 10^{10} \exp(-72/T)$	$3.1 \times 10^{-3}\,T^9$	e
8c	1% Sm^{3+}: LaES	$H \perp z$ $H = 11{,}100$ oe.	E	$3.3\,T$	$5.8 \times 10^8 \exp(-46/T) + 6.1 \times 10^{10} \exp(-72/T)$	$3.1 \times 10^{-3}\,T^9$	e
9a	0.05% Sm^{3+}: LaMN	$H \| z$ $H = 9040$ oe.	E	$8\,T$	—	$4 \times 10^{-3}\,T^9$	a
			T	$80\,T$	—	$2.5 \times 10^{-2}\,T^9$	a
9b	1% Sm^{3+}: LaMN	$H \| z$ $H = 9{,}150$ oe.	E	$3.4\,T$	$1.6 \times 10^{10} \exp(-55/T)$	$1.3 \times 10^{-2}\,T^9$	e
			T	$3.4\,T$	$4.4 \times 10^{10} \exp(-55/T)$	$1.5 \times 10^{-3}\,T^9$	e
10	SmMN	$H \perp z$ $H = 17{,}300$ oe.	E	**	—	$5 \times 10^{-2}\,T^9$	a
			T	**	—	$4.6 \times 10^{-2}\,T^9$	a
11	SmES	$H \| z$ $H = 11{,}015$ oe.	E	—	$9.5 \times 10^8 \exp(-51/T)$	$5.8 \times 10^{-4}\,T^9$	e
12	1% Sm^{3+}: YES	$H \perp z$ $H = 11{,}100$ oe.	E	$1.3\,T$	$8 \times 10^8 \exp(-51/T)$	$4 \times 10^{-4}\,T^9$	e
13	$\lesssim 0.2\%$ Eu^{2+}: CaF$_2$	cubic site	E	$12\,T$	—	$5.3 \times 10^{-4}\,T^5$	d
14	Gd^{3+}: CaF$_2$	cubic site	E	**	—	$2.5 \times 10^{-4}\,T^5$	f
			T	—	—	$2.6 \times 10^{-4}\,T^5$	d

Sect. 40. Two phonon relaxation processes.

Table 2. (Continued.)

En-try	Salt	Orientation	T or E	Direct τ_L^{-1} sec^{-1}	ORBACH τ_L^{-1} sec^{-1}	RAMAN τ_L^{-1} sec^{-1}	Refer-ence
15	1% Tb^{3+}: YES	$H \parallel z$ $H = 0$ oe.	E $\nu = 12.0$ and 14.9 GHz	$30\,T$	—	$10^{-2}\,T^7$	e
			T $\nu = 12.0$ GHz	$17.4\,T$	—	$8.9 \times 10^{-3}\,T^7$	e
			T $\nu = 14.9$ GHz	$37\,T$	—	$6.1 \times 10^{-3}\,T^7$	e
		$H = 630$ oe. $\nu = 17.0$ GHz	E	$59\,T$	—	$9.2 \times 10^{-3}\,T^7$	e
			T	$67.4\,T$	—	$4.8 \times 10^{-3}\,T^7$	e
		$H = 980$ oe. $\nu = 17.0$ GHz	E	$95\,T$	—	$1.1 \times 10^{-2}\,T^7$	e
16	1% Dy^{3+}: LaMN	$H \parallel z$ $H = 1{,}558$ oe.	E	—	$7 \times 10^9 \exp(-22/T)$	—	e
			T	$< 2\,T$	$3.1 \times 10^9 \exp(-22/T)$	$\sim 0.2\,T^9$	e
17	DyES	$\widehat{Hz} = 45°$	E	$4.2\,T$	$1.1 \times 10^7 \exp(-23/T)$	—	g
			T	$6.2\,T$	$1.2 \times 10^7 \exp(-23/T)$	$1.3 \times 10^{-5}\,T^9$	h
18	0.02% Ho^{2+}: CaF$_2$	cubic site	E	$42\,T$	$8 \times 10^9 \exp(-48/T)$	—	d
			T	$37\,T$	$8.5 \times 10^9 \exp(-48.7/T)$	—	d
19	0.2% Tm^{2+}: CaF$_2$	cubic site	E	$13\,T$	—	$7.7 \times 10^{-8}\,T^9$	d
			T	$4.5\,T$	—	$1.9 \times 10^{-6}\,T^9$	d
20a	0.1% Yb^{3+}: YGaG	cubic site	E	$33\,T$	—	$1.8 \times 10^{-7}\,T^9$	d
			T	$4.2\,T$	—	$5.7 \times 10^{-7}\,T^9$	d
20b	1% Yb^{3+}: YGaG	cubic site	E	$30\,T^{1.4}$ ***	—	$1.8 \times 10^{-7}\,T^9$	d
20c	10% Yb^{3+}: YGaG	cubic site	E	$410\,T$	—	$1.8 \times 10^{-7}\,T^9$	d
21	0.1% Yb^{3+}: YAlG	cubic site	E	$15\,T$	—	$6.3 \times 10^{-7}\,T^9$	d
			T	$5.3\,T$	—	$9 \times 10^{-7}\,T^9$	d
22	1% Yb^{3+}: YAlG	cubic site	E	$11\,T^{2.3}$ ***	—	$6.3 \times 10^{-7}\,T^9$	d

* The authors suggest that the Orbach process may be bottlenecked.
** All bottlenecks in the direct region omitted.
*** The deviation from a direct dependence on T explained by the presence of exchange coupled ion pairs.

T Theoretical prediction. E Experimental result.
XMN is $X_2Mg_3(NO_3)_{12} \cdot 24\,H_2O$.
XES is $X_3(C_2H_5SO_4)_3 \cdot 9\,H_2O$.
XGaG and XAlG are gallium garnet and aluminium garnet respectively.

[a] P. L. Scott, and C. D. Jeffries: Phys. Rev. **127**, 32 (1962).
[b] R. H. Ruby, H. Benoit, and C. D. Jeffries: Phys. Rev. **127**, 51 (1962).
[c] R. P. Hudson, and B. W. Mangum: Magnetic and Electric Resonance and Relaxation, ed. J. Smidt, publ. by North Holland, Amsterdam 1963, p. 135.
[d] C. Y. Huang: Phys. Rev. **139**, A 241 (1965).
[e] G. H. Larson, and C. D. Jeffries: Phys. Rev. **141**, 461 (1966).
[f] R. W. Bierig, M. J. Weber, and S. I. Warshaw: Phys. Rev. **134**, A 1504 (1964).
[g] C. B. P. Finn, R. Orbach, and W. P. Wolf: Proc. Phys. Soc. (London) **77**, 261 (1961).
[h] R. Orbach: Proc. Roy. Soc. (London) A **264**, 458 (1961).

Huang[1]. Aminov[2] among others, has also considered the Orbach process. Table 2 is a comparison between theoretical and experimental results in various rare-earth salts. Fig. 9 shows some typical results for τ_L as a function of T.

b) Spin-lattice relaxation of the iron group transition elements.

41. Van Vleck's theory. Relaxation in this group of ions was the first to be considered. Van Vleck [13] performed detailed calculations on chromium and titanium alum and it is the technique developed then, as amended by Mattuck and Strandberg [19], that we will consider here. A detailed description will not be attempted; we prefer to indicate only the differences between the approach of these authors and that of Orbach. Because the phonons are still described by the Debije approximation in Van Vleck's theory no new temperature dependences

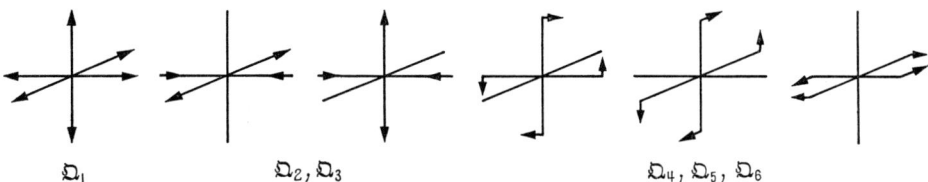

Fig. 10. Normal modes of vibration of an XY_6 complex.

are predicted, and in the case of non-Kramers salts no new field dependences either. This is also true in the Kramers salts if the state functions are calculated with the Zeeman perturbation to the same order as in Orbach's theory.

The electron energy levels and states are determined by applying the crystalline electric field, the spin-orbit coupling and the Zeeman effect as perturbations to the states of the free ion. In this group the crystalline field is stronger than the spin-orbit coupling which is in direct contrast to the case of the rare earth ions.

Van Vleck [13] used the normal coordinates of a paramagnetic cluster of ions XY_6, where X is a paramagnetic ion close to the centre of an octahedron of six water molecules. The octahedron may be slightly distorted from true cubic symmetry. Assuming each water molecule to be tightly bound and to form one unit of the structure the whole cluster has fifteen normal coordinates $\mathfrak{Q}_\gamma$, Sect. 38, but only five symmetric modes will produce the linear variations in the crystalline electric field which produce the orbit-lattice perturbation. These modes are shown in Fig. 10. The $\mathfrak{Q}_\gamma$ are related to the displacements of Eq. (31.5) by:

$$\mathfrak{Q}_\gamma = \sum_{Q\alpha} B_{\gamma Q\alpha}\, \Delta r(t)_{Q\alpha}, \tag{41.1}$$

where $\alpha = x, y, z$. Only relative displacements of the ions with respect to one another are of importance. In certain cases the mode $\mathfrak{Q}_1$ also contributes to the relaxation, see Sect. 52. Van Vleck[3] calculated the matrix elements relating the modes of vibration to the orbital component of the angular momentum. The transition probabilities are calculated using (26.1) and (26.2).

Spin state functions derived using the spin-Hamiltonian formalism are more easily used than the full state functions so Mattuck and Strandberg have derived an expression for an effective spin-lattice perturbation operator V_{eff} which

---
[1] See footnote 1, p. 53.
[2] K. Aminov: Soviet Phys. JETP **15**, 547 (1962).
[3] J. H. van Vleck: J. Chem. Phys. **7**, 72 (1939).

can be used to calculate the transition probabilities. The spin-Hamiltonian state functions are simpler to use because the influence of the excited energy levels and the orbital operators is now in the tensor operators contained in V_{eff}. During the reduction of the complete matrix elements to those of V_{eff} the Van Vleck cancellation occurs and reduces the magnitude of the first order term in S, the spin quantum number. This is the cause of the occurrence of quadrupolar selection rules in some cases.

42. Orbach's theory. The Orbach method and the Van Vleck method of calculation can be interchanged. Thus, Huang[1] could perform a complete Van Vleck calculation for the case of rare earth ions in almost cubic environments like those in CaF_2. Orbach[2] discussed the extension of his phenomenological theory to 3d ions. By considering that $V + \lambda \mathbf{L} \cdot \mathbf{S}$ is diagonalised, V is the crystalline field and $\lambda \mathbf{L} \cdot \mathbf{S}$ is the spin-orbit interaction, the formalism can be extended to both the rare earth and the 3d groups where the magnitudes of the two terms are considerably different. The Zeeman interaction and the orbit-lattice interaction can be treated as perturbations, the latter being the cause of the spin-lattice interaction. The spin-Hamiltonian formalism, as already described, adequately describes the states produced by the action of $V + \lambda \mathbf{L} \cdot \mathbf{S}$ and the Zeeman interaction so these states may be used with the orbit-lattice interaction to evaluate the transition probabilities. However, in contrast to the V_{eff} of Mattuck and Strandberg which was derived from detailed considerations of the motion of the XY_6 cluster, the orbit-lattice interaction is taken to be the dynamic crystalline field and is given by the series in Eq. (38.1). Huang adopted the Van Vleck theory to the case of rare earth ions in cubic or almost cubic sites because of the difficulty of obtaining values for the constants $A_n^m \langle r^n \rangle$, especially $A_n^m \langle r^2 \rangle$, from experimental data and from the schemes of analysis given in the next sections. Table 3 contains a comparison between theory and experiment for several salts with ions of the 3d group.

Table 3. *A comparison of theoretical predictions and experimental results for some 3d ions.*

Entry	Salt	Experimental conditions	Relaxation time τ_L sec		References
			E	T	
1	$Ti^{3+}: Al_2O_3$	Measurements at X-band 1.55 °K 4.2 °K 9.0 °K	10^{-1} 10^{-4} 5×10^{-8}	2×10^{-2} 4×10^{-5} 10^{-7}	a and b
2	$Cr^{3+}: Al_2O_3$	Transition 23*, $\widehat{Hz} = 54°$, 9.3 GHz	2×10^{-1}	2.26×10^{-1}	c and d
		Transition 23*, $\widehat{Hz} = 80°$, 7.2 GHz	5×10^{-1}	5.39×10^{-1}	c and e
		Transition 12*, $\widehat{Hz} = 60°$, 2.9 GHz	5×10^{-1}	7.5×10^{-1}	c and f
		Transition 13*, $\widehat{Hz} = 90°$, 34.6 GHz	5.4×10^{-2}	8.0×10^{-2}	c and g
3	$Mn^{2+}: MgO$	Cubic site, measurements at X-band, 4.0 °K	4×10^{-1}	4.5×10^{-1}	h

[1] See footnote 1, p. 53.
[2] R. Orbach: Proc. Phys. Soc. (London) **77**, 821 (1961).

Table 3. (Continued.)

Entry	Salt	Experimental conditions	Relaxation time τ_L sec — E	Relaxation time τ_L sec — T	References
4a	0.1 to 3% Fe^{3+}: $K_3Co(CN)_6$	H in $a-b$ plane, 8.75 GHz	$\tau_L^{-1} = 5.4\,T + 5.4 \times 10^{-3}\,T^9$	$\tau_L^{-1} = 15\,T + BT^9$ **	i
4b	Fe^{3+}: $K_3Co(CN)_6$	$H \parallel a$-axis 0.24%, 1.8 GHz	$\tau_L^{-1} = 4.3 \times 10^{-3}\,T^9$	—	j
		0.24%, 8.5 GHz	$\tau_L^{-1} = 3.1\,T + 4.3 \times 10^{-3}\,T^9$	—	
		1.7%, 1.8 GHz	$\tau_L^{-1} = 3.4 \times 10^4 \exp(-19/T) + 6.5 + 4.3 \times 10^{-3}\,T^9$	—	
		1.7%, 8.5 GHz	$\tau_L^{-1} = 3.1\,T + 2 + 3.4 \times 10^4 \exp(-19/T) + 4.3 \times 10^{-3}\,T^9$	—	
5	Co^{2+}: MgO	Cubic site	The presence of Fe spoils experimental work	$\tau_L^{-1} = 10^{-1}\,T + 2.16 \times 10^{-8}\,T^9 + 4.3 \times 10^3 \exp(-438/T)$	k
6	Co^{2+}: $La_2Zn_3(NO_3)_{12} \cdot 24\,H_2O$	1.18 °K, 13.6 GHz $H \perp z$ Electron flips only, $M = \tfrac{7}{2}$ $M = \tfrac{5}{2}$ $M = \tfrac{3}{2}$ $M = \tfrac{1}{2}$ $M = -\tfrac{1}{2}$ $M = -\tfrac{3}{2}$ $M = -\tfrac{5}{2}$ $M = -\tfrac{7}{2}$	9×10^{-3} 6×10^{-3} 1.12×10^{-2} 1.3×10^{-2} 1.45×10^{-2} 1.7×10^{-2} 1.72×10^{-2} 2.19×10^{-2}	9.57×10^{-3} 1.05×10^{-2} 1.14×10^{-2} 1.22×10^{-2} 1.39×10^{-2} 1.5×10^{-2} 1.7×10^{-2} 1.89×10^{-2}	l***
7	Cu^{2+}: $ZnK_2(SO_4)_2 \cdot 6\,H_2O$ Conc. 10^{-4}	$H \parallel [001]$, 9.445 GHz	$\tau_L^{-1} = 2.1 \times 10^{-2}\,T + 1.9 \times 10^{-6}\,T^9$	$\tau_L^{-1} = 8.4 \times 10^{-1}\,T + 1.9 \times 10^{-6}\,T^9$ †	m and n

* The energy levels are labelled 1 to 4 with level 1 the highest level and level 4 the lowest level.

** B is in the range 10^{-1} to 10^{-2}.

*** This reference contains a similar comparison of results with theory for $H \parallel z$.

† The energy level splittings involved in the calculation were used as a variable parameter to obtain this result.

T Theoretical prediction. E Experimental results. M Nuclear quantum number.

a L. S. Kornienko, i A. M. Prokhorov: Soviet Phys. JETP **11**, 1189 (1960).
b S. A. Al'tshuler, Sh. Sh. Bashkirov i M. M. Zaripov: Soviet Phys. Solid State **4**, 2465 (1963).
c P. L. Donoho: Phys. Rev. **133**, A 1080 (1964).
d Y. Nisida: J. Phys. Soc. Japan **17**, 1519 (1962).
e W. B. Mims, and J. D. McGee: Phys. Rev. **119**, 1233 (1960).
f R. A. Armstrong, and A. Szabo: Can. J. Phys. **38**, 1304 (1960).
g J. H. Pace, D. F. Sampson, and J. S. Thorp: Proc. Phys. Soc. (Lond.) **76**, 697 (1960).
h M. Blume, and R. Orbach: Phys. Rev. **127**, 1587 (1962).
i T. Bray, G. C. Brown jr., and A. Kiel: Phys. Rev. **127**, 730 (1962).
j A. Rannestad, and P. E. Wagner: Phys. Rev. **131**, 1953 (1963).
k M. H. L. Pryce: Proc. Roy. Soc. (London) A **283**, 433 (1965).
l W. P. Unruh, and J. W. Culvahouse: Phys. Rev. **129**, 2441 (1963).
m J. C. Gill: Proc. Phys. Soc. (London) **85**, 119 (1965).
n A. M. Stoneham: Proc. Phys. Soc. (London) **85**, 107 (1965).

43. Group theory and spin-lattice relaxation. An alternative approach to spin-lattice relaxation theory is to develop a general expression for the spin-lattice relaxation times by using group theoretical methods and the symmetry of the phonon structure and the state functions. The use of group theory stresses the unity which lies behind the various approaches which have been described above.

Let the symmetry group of a paramagnetic ion site be G and take the simple case where every site is equivalent. The symmetry is assumed to be produced by the nearest neighbours only and the lattice vibrations are assumed not to distort this greatly so that the expression for the Hamiltonian representing the interaction between the lattice and the spin is invariant under G. The energy V of the paramagnetic ions is a function of the interionic spacing and the spins of the ions and is obviously modified by the lattice vibrations. The spin-lattice Hamiltonian can be written as:

$$\mathcal{H}_{SL} = \tfrac{1}{2} \sum_j \left(\sum_{\mathfrak{a},\mathfrak{f}} V^{\mathfrak{a}}_{\mathfrak{f}j} q^{\mathfrak{a}}_{\mathfrak{f}j} + \sum_{\mathfrak{a},\mathfrak{b},\mathfrak{f},\mathfrak{g}} W^{\mathfrak{a}\mathfrak{b}}_{\mathfrak{f}\mathfrak{g}j} q^{\mathfrak{a}}_{\mathfrak{g}j} q^{\mathfrak{b}}_{\mathfrak{f}j} \right); \tag{43.1}$$

where

$$V^{\mathfrak{a}}_{\mathfrak{f}j} = \frac{\partial V_j}{\partial q^{\mathfrak{a}}_{\mathfrak{f}j}} \quad \text{and} \quad W^{\mathfrak{a}\mathfrak{b}}_{\mathfrak{f}\mathfrak{g}j} = \frac{\partial^2 V}{\partial q^{\mathfrak{a}}_{\mathfrak{g}j} \cdot \partial q^{\mathfrak{b}}_{\mathfrak{f}j}}.$$

The $q_{\mathfrak{f}j}$ are the normal coordinates of the nearest neighbours which can be expressed in terms of the normal coordinates of the lattice phonons, the subscript j refers to a paramagnetic ion, $\mathfrak{a}$, $\mathfrak{b}$ are irreducible representations of G and $\mathfrak{f}$, $\mathfrak{g}$ are terms in the $\mathfrak{a}$-th and $\mathfrak{b}$-th irreducible representations. If the sample is very large the effects of the boundaries can be neglected and the summation over the ions replaced by multiplying by N the total number of ions. The state functions of the electrons are chosen as the basis of the irreducible representation of G and then the matrix elements $\mathcal{H}^{\mathfrak{a}\mathfrak{b}}_{ab}$ relating the states $|a\rangle$ and $|b\rangle$ can be written in terms of constants which depend on the crystal field and the spin-lattice Hamiltonian.

$$\mathcal{H}^{\mathfrak{a}\mathfrak{b}}_{ab} = \langle a | \mathcal{H}_{SL} | b \rangle. \tag{43.2}$$

The imposition of a field H removes any degeneracy which may have been present and the new wave functions are $|a\rangle'$ and $|b\rangle'$, related to the old functions by:

$$|a\rangle' = \sum_b R_{ba'} |a\rangle; \quad \langle b|' = \langle a| \sum_a R^*_{a'b'}. \tag{43.3}$$

Then the elements for the transitions between the magnetic sublevels which are the transitions of interest are given by:

$$\mathcal{H}^{\mathfrak{a}\mathfrak{b}}_{a'b'} = \sum_{a,b} R^*_{a'b'} \mathcal{H}^{\mathfrak{a}\mathfrak{b}}_{ab} R_{ba'}. \tag{43.4}$$

The coefficients R are found by diagonalising the matrix $\mathcal{H}(H)$ in the total Hamiltonian:

$$\mathcal{H} = \mathcal{H}(0) + \mathcal{H}(H). \tag{43.5}$$

The first term does not depend on the magnetic field.

MOROCHA[1] has considered calculations of this type for ions in the 3d group. In this case the local symmetry may be represented by a cubic group and consideration of how $\mathcal{H}(H)$ transforms with the irreducible representations of the cubic group leads to the elimination of many matrix elements. He demonstrated that the com-

---

[1] A. K. MOROCHA: Soviet Phys. JETP **16**, 1275 (1963).

pletely symmetric vibration (neglected by VAN VLECK) leads to relaxations between the magnetic sublevels, this is especially true when lower symmetry fields must be considered.

RAY et al.[1] have performed a similar derivation of $\mathcal{H}_{SL}$.

c) Numerical parameters.

44. Crystal field parameters. The evaluation of the theoretical estimates for the relaxation time requires knowledge of several parameters. Perhaps the most difficult to give a value to are the V_n^m which occur in the summations. They are functions of the spherical harmonics, $Y_n^m(\vartheta, \varphi)$, the mean of the n-th power of the radius of the electron orbit $\langle r^n \rangle$, and a parameter describing the crystalline field, A_n^m. The calculation of the A_n^m from first principles requires knowledge of the ligand fields, screening effects and ionic motion.

In practice these parameters are evaluated as follows; from optical experiments, electron or ultrasonic paramagnetic resonance experiments, or stress measurements, values of A_n^m for those m and n which occur in the static case can be calculated. Then some relation is found between the static and dynamic values in order to extend them to the cases where the m values apply in the dynamic case but not in the static case. Various schemes to estimate the product $A_n^m \langle r^n \rangle$ from the static values obtained by experiment have been proposed. ORBACH [16] used an approximation which he found suitable in other problems. He assumed that $A_n^m(j) = Q(\partial A_n^m(j)/\partial Q)$, that is the first order term of the dynamic electrostatic potential equals the static term. Therefore, the perturbation operator is the simple product of the average strain and the static crystalline field potential. To obtain values of A_n^m for all m and n from the known A_n^m he used the following relations:

$$A_{n=2,4}^m \langle r^n \rangle = |A_n^0 \langle r^n \rangle|; \quad A_6^m \langle r^6 \rangle = [|A_6^0 \langle r^6 \rangle|^{6-|m|} |A_6^6 \langle r^6 \rangle|^{|m|}]^{\frac{1}{6}}. \tag{44.1}$$

HUANG[2] derived, assuming a point charge model in which $A_n^m \propto Q^{-(n+1)}$:

$$\left\| Q\left(\frac{\partial A_n^m(j)}{\partial Q}\right) \right\| = p |A_n^m(j)| \tag{44.2}$$

where $p = 3, 5, 7$ when $n = 2, 4, 6$, respectively. SCOTT and JEFFRIES [18] used $|A_n^m| \simeq A_n^0$, which was supported by experiment in rare earth double nitrates. Combined with this they used:

$$A_n^m \langle r^n \rangle = g_n^{|m|} |A_n^0 \langle r^n \rangle|, \tag{44.3}$$

when $m \neq 0$. The $g_n^{|m|}$ are normalisation factors from the crystal field potential which we quote in Table 4. In carrying out the summations they assumed the

Table 4. *Values of the normalising factor $g_n^{|m|}$.*

m	1	2	3	4	5	6	F
$n=2$	4.90	2.45	—	—	—	—	$\frac{1}{2}\sqrt{\left(\frac{5}{4}\right)}$
$n=4$	8.95	6.32	23.6	8.37	—	—	$\frac{1}{8}\sqrt{\left(\frac{9}{4}\right)}$
$n=6$	12.9	10.2	20.2	11.2	52.7	15.2	$\frac{1}{16}\sqrt{\left(\frac{13}{4}\right)}$

F is a multiplying factor for all the values in the row.

References: P. L. SCOTT, and C. D. JEFFRIES: Phys. Rev. **127**, 32 (1962). — B. W. MANGUM, and R. P. HUDSON: J. Chem. Phys. **44**, 704 (1966).

---

[1] D. K. RAY, T. RAY, and P. RUDRA: Proc. Phys. Soc. (London) **87**, 485 (1966).
[2] See footnote 1, p. 53.

terms to be incoherent, so:

$$\left|\langle a|\sum_{n,m} V_n^m|b\rangle\right|^2 = \sum_{n,m}|\langle a|V_n^m|b\rangle|^2 = \sum_{\substack{n=2,4,6 \\ -n\leq m\leq n}} |\gamma_n g_n^{|m|} A_n^0 \langle r^n\rangle \langle a|O_n^m|b\rangle|^2 \qquad (44.4)$$

where the O_n^m and the γ_n are parameters from the operator equivalent theory, see Tables 5 and 6. They also found that for several rare earth ions in double nitrates terms of the type $|\langle a|\sum V_n^m|c\rangle|^2$, $|a\rangle$ is the ground state and $|c\rangle$ is an excited state, were approximately constant.

Table 5. *Form of the operators O_n^m.*

$O_2^0 = 3J_z^2 - J(J+1)$

$O_2^1 = \frac{1}{4}[J_z(J_+ + J_-) + (J_+ + J_-)J_z]$

$O_2^2 = \frac{1}{2}(J_+^2 + J_-^2)$

$O_4^0 = 35J_z^4 - [30J(J+1) - 25]J_z^2 - 6J(J+1) + 3J^2(J+1)^2$

$O_4^1 = \frac{1}{4}\{[7J_z^3 - 3J(J+1)J_z - J_z](J_+ + J_-) + (J_+ + J_-)[7J_z^3 - 3J(J+1)J_z - J_z]\}$

$O_4^2 = \frac{1}{4}\{[7J_z^2 - J(J+1) - 5](J_+^2 + J_-^2) + (J_+^2 + J_-^2)[7J_z^2 - J(J+1) - 5]\}$

$O_4^3 = \frac{1}{4}[J_z(J_+^3 + J_-^3) + (J_+^3 + J_-^3)J_z]$

$O_4^4 = \frac{1}{2}(J_+^4 + J_-^4)$

$O_6^0 = 231J_z^6 - 105[3J(J+1) - 7]J_z^4 + [105J^2(J+1)^2 - 525J(J+1) + 294]J_z^2 -$
$\quad - 5J^3(J+1)^3 + 40J^2(J+1)^2 - 60J(J+1)$

$O_6^1 = \frac{1}{4}\{[33J_z^5 - \{30J(J+1) - 15\}J_z^3 + \{5J^2(J+1)^2 - 10J(J+1) + 12\}]$
$\quad J_z(J_+ + J_-) + (J_+ + J_-)J_z[33J_z^5 - \{30J(J+1) - \text{etc.}\}]\}$

$O_6^2 = \frac{1}{4}\{[33J_z^4 - \{18J(J+1) + 123\}J_z^2 + J^2(J+1)^2 + 10J(J+1) + 102]$
$\quad (J_+^2 + J_-^2) + (J_+^2 + J_-^2)[33J_z^4 - \{18J(J+1) + \text{etc.}\}]\}$

$O_6^3 = \frac{1}{4}\{[11J_z^3 - 3J(J+1)J_z - 59J_z](J_+^3 + J_-^3) + (J_+^3 + J_-^3)[11J_z^3 - 3J(J+1)J_z - 59J_z]\}$

$O_6^4 = \frac{1}{4}\{[11J_z^2 - J(J+1) - 38](J_+^4 - J_-^4) + (J_+^4 + J_-^4)[11J_z^2 - J(J+1) - 38]\}$

$O_6^5 = \frac{1}{4}[J_z(J_+^5 + J_-^5) + (J_+^5 + J_-^5)J_z]$

$O_6^6 = \frac{1}{2}(J_+^6 + J_-^6)$

Reference: R. Orbach: Proc. Roy. Soc. (London) A **264**, 458 (1961).

The matrix elements of all these operators except O_2^1, O_4^1, O_6^1 and O_6^5 are tabulated by M. T. Hutchings, Solid State Physics, F. Seitz and D. Turnbull, Editors (Academic Press, New York, 1964), Volume 16, p. 227. H. A. Buckmaster, Canad. J. Phys., **40**, 1670 (1962) tabulates the matrix elements for $o_2^{\pm 1}$, $o_4^{\pm 1}$, $o_6^{\pm 1}$ and $o_6^{\pm 5}$ where $O_n^m = \frac{1}{2}(o_n^m + o_n^{-m})$.

Larson and Jeffries[1] have made comparisons of these schemes in double nitrates and ethyl sulphates and they state that Scott and Jeffries' scheme (44.3), fits those cases where the A_6^m terms dominate while in the other cases they considered Orbach's simple scheme was a better fit. Huang's scheme does not give significant differences except where the matrix elements of V_6^6, V_4^3, and V_6^3 dominate.

Hüfner[2] calculated values of $A_n^m \langle r^n\rangle$ for rare earth ions in ethyl sulphates from optical data. He wrote out the energy level splittings in terms of the $A_n^m \langle r^n\rangle$ and used the method of least squares to fit these splittings to the measured spectrum.

The crystal field parameters, as so far commented upon, fit more naturally into crystal field theory developed with spherical harmonics in the solution. It is of course possible to describe the electrostatic potentials in terms of Cartesian coordinates and this allows us to consider the potential as being produced by a distribution of point charges or by a distribution of dipoles each with a definite

---

[1] G. H. Larson, and C. D. Jeffries: Phys. Rev. **141**, 461 (1966).
[2] S. Hüfner: Z. Physik **169**, 417 (1962).

moment. VAN VLECK [13] has shown that in principle there is no difference between the results from the two models. He made the assumption in the dipolar model that the dipole motion was such that at all times the dipole was radially directed from the origin of the system of coordinates. Finally the potential is

Table 6. *Multiplicative factors.*

$V_2^0 = \langle J\|\alpha\|J\rangle A_2^0 \langle r^2\rangle O_2^0$ where $\langle J\|\alpha\|J\rangle \equiv \gamma_2$
$V_4^0 = \langle J\|\beta\|J\rangle A_4^0 \langle r^4\rangle O_4^0$ where $\langle J\|\beta\|J\rangle \equiv \gamma_4$
$V_6^0 = \langle J\|\gamma\|J\rangle A_6^0 \langle r^6\rangle O_6^0$ where $\langle J\|\gamma\|J\rangle \equiv \gamma_6$

Ion	$Ce^{3+}, 4f^1\ ^2F_{\frac{5}{2}}$	$Pr^{3+}, 4f^2\ ^3H_4$	$Nd^{3+}, 4f^3\ ^4I_{\frac{9}{2}}$
$\langle J\|\alpha\|J\rangle$	$\dfrac{-2}{5\cdot 7}$	$\dfrac{-4\cdot 13}{9\cdot 11\cdot 25}$	$\dfrac{-7}{9\cdot 121}$
$\langle J\|\beta\|J\rangle$	$\dfrac{2}{9\cdot 5\cdot 7}$	$\dfrac{-4}{9\cdot 5\cdot 121}$	$\dfrac{-8\cdot 17}{9\cdot 5\cdot 121\cdot 13}$
$\langle J\|\gamma\|J\rangle$	0	$\dfrac{16\cdot 17}{81\cdot 5\cdot 7\cdot 121\cdot 169}$	$\dfrac{-5\cdot 17\cdot 19}{27\cdot 7\cdot 1321\cdot 169}$

Ion	$Pm^{3+}, 4f^4\ ^5I_4$	$Sm^{3+}, 4f^5\ ^6H_{\frac{5}{2}}$	$Tb^{3+}, 4f^8\ ^7F_6$
$\langle J\|\alpha\|J\rangle$	$\dfrac{2\cdot 7}{3\cdot 5\cdot 121}$	$\dfrac{13}{9\cdot 5\cdot 7}$	$\dfrac{-1}{9\cdot 11}$
$\langle J\|\beta\|J\rangle$	$\dfrac{8\cdot 7\cdot 17}{27\cdot 5\cdot 1321\cdot 13}$	$\dfrac{2\cdot 13}{9\cdot 5\cdot 7\cdot 11}$	$\dfrac{2}{9\cdot 5\cdot 121}$
$\langle J\|\gamma\|J\rangle$	$\dfrac{8\cdot 17\cdot 19}{27\cdot 7\cdot 121\cdot 169}$	0	$\dfrac{-1}{81\cdot 7\cdot 121\cdot 13}$

Ion	$Dy^{3+}, 4f^9\ ^6H_{\frac{15}{2}}$	$Ho^{3+}, 4f^{10}\ ^5I_8$	$Er^{3+}, 4f^{11}\ ^4I_{\frac{15}{2}}$
$\langle J\|\alpha\|J\rangle$	$\dfrac{-2}{9\cdot 5\cdot 7}$	$\dfrac{-1}{2\cdot 9\cdot 25}$	$\dfrac{4}{9\cdot 25\cdot 7}$
$\langle J\|\beta\|J\rangle$	$\dfrac{-8}{27\cdot 5\cdot 7\cdot 11\cdot 13}$	$\dfrac{-1}{2\cdot 3\cdot 5\cdot 7\cdot 11\cdot 13}$	$\dfrac{2}{9\cdot 5\cdot 7\cdot 11\cdot 13}$
$\langle J\|\gamma\|J\rangle$	$\dfrac{4}{27\cdot 7\cdot 121\cdot 169}$	$\dfrac{-5}{27\cdot 7\cdot 121\cdot 169}$	$\dfrac{8}{27\cdot 7\cdot 121\cdot 169}$

Ion	$Tm^{3+}, 4f^{12}\ ^3H_6$	$Yb^{3+}, 4f^{13}\ ^2F_{\frac{7}{2}}$
$\langle J\|\alpha\|J\rangle$	$\dfrac{1}{9\cdot 11}$	$\dfrac{2}{9\cdot 7}$
$\langle J\|\beta\|J\rangle$	$\dfrac{8}{81\cdot 5\cdot 121}$	$\dfrac{-2}{3\cdot 5\cdot 7\cdot 11}$
$\langle J\|\gamma\|J\rangle$	$\dfrac{-5}{81\cdot 7\cdot 121\cdot 13}$	$\dfrac{4}{9\cdot 7\cdot 11\cdot 13}$

The ions Eu^{3+}, $4f^6\ ^7F_0$ and Gd^{3+}, $4f^7\ ^8S$ have zero for the values of each of their factors.

Reference: R. J. ELLIOTT, and K. W. H. STEVENS: Proc. Roy. Soc. (London) A **218**, 553 (1953).

written out in terms of certain parameters which can be evaluated by comparison with experiment in a similar way to the evaluation of the A_n^m.

The crystal field parameters which we have considered are very important in making correct numerical estimates of τ_L. It is unfortunate that our only means of estimating them all requires such devious means. The values of the parameters from a certain type of experiment are average values weighted in the manner in which the experiment depends on them. In particular optical values may differ from the values which are appropriate to E.P.R. experiments because of shielding and distortion effects. In rare earth ions the 4f shell is shielded by a completely closed outer shell of electrons. Also in some materials the charge distributions of the closed and unfilled shells are distorted which causes variations in the energy level structure calculated from simple crystal field theory. Unfortunately, analysis of E.P.R. results does not give enough information to allow a whole group of crystal field parameters to be evaluated accurately[1].

The separation between the various energy levels in the system enters into the calculation, sometimes to a high power, so an accurate value of these splittings is required, but fortunately this can be obtained from E.P.R. and optical spectroscopy.

45. Phonon parameters. Making the simplest assumptions about the phonon spectrum only two parameters are required, the velocity of sound in the crystal and the density of the crystal. The former parameter enters to the tenth power in the Raman process so it should be known with reasonable accuracy in any specific calculation. A 20% change in the velocity produces an order of magnitude change in the value of τ_L. In the equations in this article no distinction is made between the transverse and longitudinal velocities of the phonons. A simple extension of the calculations allows us to include them both if required. Many workers have used a single approximate value for the velocity of sound in their predictions, but perhaps this is valid when it is realised what other approximations are made in the calculations. In many cases the theoretical predictions are not much better than order of magnitude approximations and attempts to use the formulae given in this article will, in general, only give the same sort of accuracy. Every specific calculation must be performed in detail on a more exact model than the general models considered here.

d) Anisotropy in relaxation times.

46. Anisotropy in state functions. In Kramers salts, in order to avoid non-zero matrix elements, the time degeneracy of the states has to be lifted by applying a magnetic field and mixing excited states into the ground states. This mixing may

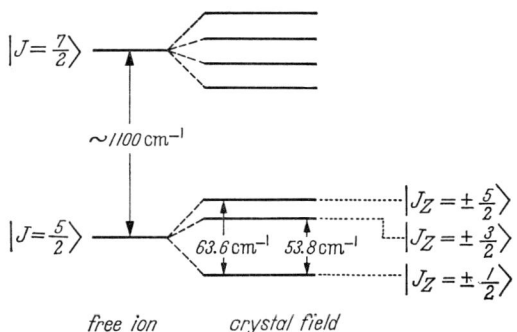

Fig. 11. Energy levels of Sm^{3+} in Lanthanum ethyl sulphate. Taken from R. ORBACH: Phys. Rev. **126**, 1349 (1962).

be anisotropic because the magnitude of the matrix elements between the ground and excited states depends on the angle between the magnetic field and the crystalline axes. Indeed, in some directions the matrix element with a certain excited state may be zero and the lifting of the degeneracy has to be brought about by another excited level. This is the case in samarium ethyl sulphate

---
[1] R. E. WATSON, and A. J. FREEMAN: Phys. Rev. **113**, A 1571 (1964).

considered by Orbach[1]. The energy levels are shown in Fig. 11. When the magnetic field is perpendicular to the z-axis the states $J_z=|\pm\tfrac{3}{2}\rangle$ mix with the $J_z=|\pm\tfrac{1}{2}\rangle$ states. When the field is parallel to the z-axis the lowest excited level which can mix with the $|\pm\tfrac{1}{2}\rangle$ level is from the $J=\tfrac{7}{2}$ multiplet, so a considerable anisotropy is predicted. In fact, the anisotropy is not so marked as may be thought because the value of the matrix elements in the perpendicular direction is smaller than those in the parallel direction. This is a case where a higher multiplet than the ground multiplet has to be considered but a straight forward application of the normal methods suffices[2]. This anisotropy is in the direct process but Mangum and Hudson[3] have observed anisotropy in the Raman region in Nd^{3+} in rare earth trichlorides. This is a Kramers salt and the anisotropy can be explained if the relaxation mechanisms which lead to the results given in Eqs. (40.8) and (40.11) dominate over the mechanism which gives rise to Eq. (40.7).

47. Anisotropy in spin-orbit coupling. Experiment shows[4] anisotropy in the relaxation times in $CuSO_4 \cdot 5\,H_2O$. This can be explained by considering the anisotropy of the spin-orbit coupling factor which Abe and Ono[5] showed was due to the effects of covalency between the Cu^{2+} and the neighbouring oxygen ions. The theoretical predictions agreed with experiment about the order of the anisotropy but not as to the absolute magnitude of the relaxation. Anisotropy in τ_L produced by anisotropy in the spin-orbit coupling can occur in the direct and the Raman regions.

The state functions of non-Kramers salts may also vary with changes in orientation of the crystal with respect to the magnetic field and so anisotropy in τ_L may occur.

e) Relaxation in S-state ions.

48. Possible spin-lattice interactions. S-state ions have no orbital angular momentum in the ground state so it may be expected that they cannot relax by any mechanism involving the influence of the crystalline field on the orbital angular momentum of an ion. However, admixtures of excited states with non-zero angular momentum do occur and produce small zero field splittings in S-state ions.

Pryce[6] proposed that the crystalline field distorted the spherical symmetry of the electronic charge distribution by introducing a contribution from an excited state and that the spin-spin interactions of the electrons within the ion produced the zero field splitting. This required second order perturbation theory. Modulation of this interaction would produce spin-lattice relaxation and has been considered by Leushin[7]. His results give a value for τ_L of Mn^{2+} in a cubic symmetry which is two orders of magnitude too long at helium temperatures. However, the theory gives adequate agreement with experiments on static deformation and acoustic paramagnetic resonance. Also there is better agreement between theory and experiment at higher temperatures where the Raman processes dominate. To simplify the calculation, the mixing between the state functions of the ground quartet was ignored and the perturbation operator was only expanded to the second power in S, whereas terms up to the fifth power

---

[1] R. Orbach: Phys. Rev. **126**, 1349 (1962).
[2] J. P. Elliott, B. R. Judd, and W. A. Runciman: Proc. Roy. Soc. (London) A **240**, 509 (1957).
[3] B. W. Mangum, and R. P. Hudson: J. Chem. Phys. **44**, 704 (1966).
[4] T. M. Volokhova: Soviet Phys. JETP **6**, 661 (1958).
[5] H. Abe, and K. Ono: J. Phys. Soc. Japan **11**, 947 (1956).
[6] M. H. L. Pryce: Phys. Rev. **80**, 1107 (1950).
[7] A. M. Leushin: Soviet Phys. Solid State **4**, 1148 (1962); **5**, 440, 623 (1963).

exist. The modulation of the spin-spin mechanism within the ion is important in the Raman processes involving the second order term in (26.3). On the other hand, the normal modulation of the spin-orbit interaction, as already discussed above, is important when the first order term in Eq. (26.3) acts twice.

Other ways of producing the zero field splitting have been considered. Van Vleck and Penney[1] considered the mixing of excited states from different configurations by the action of the crystalline field and the spin-orbit coupling taken to several orders in perturbation theory. When the specific case of third order theory, with the spin-orbit interaction considered as acting twice and the crystalline field as acting once, was considered by Blume and Orbach[2] their predictions for τ_L agreed quite well with experiment as far as magnitude was concerned but the signs of the parameters calculated for the static strain experiments were wrong. The failure seems to be connected with the use of a simple point charge crystal field model. Taking into account the influence of the lattice vibrations on the overlap and covalency of the electrons on the Mn^{2+} ion with those on the surrounding ions and assuming the overlap is the dominant term, the correct signs are found for the parameters in the Blume and Orbach theory[3].

In concentrated crystals where the paramagnetic ions are close together the mechanism considered by Waller [1] may be important in causing relaxation in S-state ions. This mechanism is the modulation of the dipolar coupling between neighbouring ions and it will be considered further in Sect. 53. It is more important than the normal orbit-lattice mechanism if:

$$\partial_i^2 \lesssim \frac{g^2 \beta^2 H^2}{r^6} (2S+1)(S+1), \tag{48.1}$$

where ∂_i is the splitting of the spin levels in the crystalline field, r is the equilibrium distance between nearest neighbouring magnetic ions and g, β and S have their usual significances.

49. Multilevel systems. By multilevel systems we mean systems with several levels which are within a few times kT from the ground levels and therefore are populated. Most S-state ions fall into this category because their zero field splitting is very small. They have been considered separately to stress the different perturbations which may be important in S-state ions. In a general discussion of multilevel systems it is not necessary to single out any one spin-lattice perturbation as important. The major difference between relaxation in multilevel systems and the other systems considered above is that the rate equations become more complex. The more complete solutions indicated in Sect. 25 are required, rather than the simple solutions which apply to two level systems only.

Dy^{3+} in CaF_2 has a quartet lowest in the cubic crystalline field and above this at 8.5 ± 1 cm^{-1} the first excited level. At 4 °K Orbach relaxation mechanisms take place via this excited level, while at 2 °K these two step processes can occur in the lowest quartet[4].

These two step processes need not necessarily take place via an upper level or set of levels as was considered in Sects. 35 and 40; they may also go via a lower level as indicated in Fig. 12. This is the inverted-Orbach process. The solution

---

[1] J. H. van Vleck, and W. G. Penney: Phil. Mag. **17**, 961 (1934).
[2] M. Blume, and R. Orbach: Phys. Rev. **127**, 1587 (1962).
[3] J. Kondo: Progr. Theoret. Phys. (Kyoto) **28**, 1026 (1962).
[4] R. W. Bierig, and M. J. Weber: Phys. Rev. **132**, 164 (1963).

of the rate equations in this case gives as the temperature dependence for τ_L:

$$\tau_L = \text{constant} \times \left(\frac{\exp\left[\frac{\Delta_c}{kT}\right]}{\exp\left[\frac{\Delta_c}{kT}\right] - 1} \right). \tag{49.1}$$

If temperatures are attained so that $kT \ll \Delta_c$, where Δ_c is the splitting between the first and second levels, then this process becomes temperature independent, see Table 1. The temperature must remain high enough for the pair of levels which are relaxing to be populated or, of course, no relaxation can occur.

The state function of a level in a multilevel system is modified by mixing with the other levels and this has interesting effects on the time-reversal properties of the functions. In the case of chromium alum, where the crystal field leaves two Kramers doublets about 0.1 cm^{-1} apart, the Van Vleck cancellation does not occur and so $\tau_L \propto T^{-7}$ instead of $\tau_L \propto T^{-9}$ as might have been expected in a Kramers salt, VAN VLECK [13].

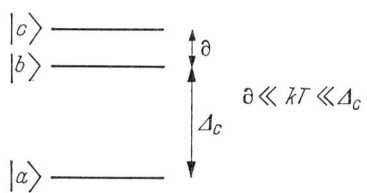

Fig. 12. Energy level scheme to demonstrate the "inverted" Orbach process.

In a multilevel system formed from Kramers doublets the Raman process shows another temperature dependence when $kT \gg \Delta_c$ where Δ_c is the energy of the excited state involved in the process. In such a multilevel system Δ_c is the splitting between doublets in the system. Following the development of Sect. 40 we obtain Eq. (40.6). This has been used in the limit $\hbar\omega \ll \Delta_c$, but now we have $\hbar\omega \gg \Delta_c$ because $\hbar\omega \simeq kT$ for a large number of the phonons. In this limit the integral over the phonons becomes:

$$I_4 = \int_0^{\omega_m} \frac{\omega^4 \exp\left(\frac{\hbar\omega}{kT}\right) d\omega}{\left[\exp\left(\frac{\hbar\omega}{kT}\right) - 1\right]^2} \tag{49.2}$$

and in the low temperature limit, $T \ll \Theta_D$, but $kT \gg \Delta_c$, this gives:

$$\tau_L^{-1} = \frac{9 \times 4!}{\hbar^2 \pi^3 \varrho^2 v^{10}} \left(\frac{kT}{\hbar}\right)^5 \left| \sum_{\substack{n,m,\\n',m'}} \left\langle -\frac{a}{2} \left| V_n^m \right| \frac{c}{2} \right\rangle \left\langle \frac{c}{2} \left| V_{n'}^{m'} \right| \frac{a}{2} \right\rangle \right|^2. \tag{49.3}$$

ORBACH and BLUME[1] have considered the relative magnitude of this process with respect to the $\tau_L \propto T^{-7}$ process which is also possible in such systems. In rare earth salts the T^{-5} term is $(\Delta_c/kT)^{-2}$ times the T^{-7} term so it is the faster process when $\Delta_c > kT$. In the 3d group salts the T^{-5} mechanism is faster when $\lambda^2/\Delta_c > kT$. λ is the spin-orbit coupling. These expressions are valid when the linear term in the elastic strain is used in second order perturbation theory. The second order term in elastic strain used in first order perturbation theory also gives a term in T^{-7} equivalent in magnitude to the term from the second order perturbation. The term in T^{-5} from the second order elastic strain may be neglected. When $kT \simeq \Delta_c$ it is necessary to solve the rate equations for the system in full in order to obtain values for the relaxation times of the system.

---

[1] R. ORBACH, and M. BLUME: Phys. Rev. Letters **8**, 478 (1962).

f) Crystal bonding.

Up to now all the crystals considered have been ionic. The other types of bonding which we will consider are covalent and molecular bonding.

50. Covalent bonding. The 4d and 5d transition groups display covalent bonding and so do ions of the 3d group in certain salts, e.g. cyanides. This bonding has several effects on the solution of the problem of spin-lattice relaxation. Firstly, the unpaired electrons responsible for the magnetism must now be considered as moving in the combined field of the whole complex and they cannot be considered as belonging to any one ion. Therefore, the electron energy levels must be calculated using the molecular orbital techniques[1]. The levels can be constructed from linear combinations of atomic orbitals, LCAO. There is no need to abandon the form of the spin-lattice perturbation in Sect. 38. The second major difference between the ionic and covalent bonding problems is the form of the three important sets of matrix elements, those of the orbital angular momentum, the spin-orbit coupling and the spin-lattice interaction.

A suitable Hamiltonian for the system formed of an electron moving in the field of the paramagnetic complex is:

$$\mathcal{H} = \mathcal{H}_x + \mathcal{H}_{LS} + \mathcal{H}_c + \mathcal{H}_v \tag{50.1}$$

where $\mathcal{H}_x$ is the Hartree operator, $\mathcal{H}_{LS}$ is the spin-orbit coupling term from $\mathcal{H}_b$, Eq. (27.1b), $\mathcal{H}_c$ is the Zeeman interaction, Eq. (27.1c) and $\mathcal{H}_v$ represents the influence of low symmetry fields produced by the neighbours of the complex. The perturbation given by Eq. (26.4) is to a first approximation equal to $\mathcal{H}_x$. For the μ-th electron we have:

$$\mathcal{H}_x^\mu = \mathcal{H}_k^\mu + \sum_j \mathcal{H}_j^\mu + \sum_l I_l^\mu \tag{50.2}$$

where the first term is the kinetic energy of the electron, the second term is the Coulomb energy between the electron and the j-th nucleus, which is screened by closed electron shells which do not form part of the molecular orbital system, and the last term is the energy of the μ-th electron in the average field of electron ν in the state $|\varphi_l\rangle$. The $|\varphi_l\rangle$ can be evaluated in terms of LCAO and then the coefficients of the LCAO are expanded in terms of the displacements of the nuclei from their equilibrium positions. The LCAO are based on the basic charge distribution of the atomic orbitals but we must neglect the charge distribution of the magnetic ion because it is assumed to be at rest in order to avoid translational motion of the whole complex. The expression for the orbit-lattice perturbation can be most conveniently rewritten as:

$$\mathcal{H}_{OL} = \sum_j q_j \{\nabla(\mathcal{H}_j^{(1)} + \mathcal{H}_j^{(2)} + \mathcal{H}_j^{(3)} + \mathcal{H}_j^{(4)})\}_{q=0}. \tag{50.3}$$

The summation is over the nuclei and q_j is the displacement of the nucleus j from its equilibrium position. The four terms in Eq. (50.3) represent respectively (1) the usual Coulomb operator with the effective charge depending on the amount of covalency, (2) the deviation from the spherical symmetry of the electron cloud whereby the nucleus is not completely shielded, (3) is the interaction with the potential field due to the quadrupole moment of the electronic shell of the ligand when the σ and π bonds in the complex are inequivalent, and (4) is similar to (2) except that the charge distribution is basically axial.

---

[1] G. G. HALL: Repts. Progr. Phys. **22**, 1 (1959).

OVCHINNIKOV[1] performed a calculation for an XY_6 complex with only one electron outside the closed shells. He presented his results as relations between the transition probabilities for the same complex regarded first as an ionic complex and then as a covalent complex. The relaxation time in the direct region for covalent bonding is about an order of magnitude longer than in ionic bonding and in the Raman region it is almost two orders of magnitude longer. The reason for this increase is the reduction in the effective charges due to covalency.

51. Molecular crystals. In molecular crystals the main perturbation of the spins is not produced by linear atomic motions but by rotational motions. ALEKSANDROV and ZHIDORIMOV[2] demonstrated this. It has interesting results as the rotational frequency spectrum is somewhat different from the translational spectrum. The linear term in the spin-lattice perturbation should produce direct processes but the rotational frequency spectrum does not extend from zero frequency, if there are no interactions with translational oscillations, that is the centres of the rotational oscillations of the individual molecules are fixed. The lower limit of the rotational frequency spectrum corresponds to phonons of energy much higher than the Zeeman energy of the system so the direct process cannot occur. The second order term produces Raman processes and in the high temperature limit the transition probability $W_{ab} \propto (T/I)^2$, I being the moment of inertia of the molecule. This temperature dependence is obtained because the density of states of the rotational spectrum is described by an ω_R^2 distribution similar to (31.6), ω_R is the rotational frequency. The insertion of this distribution into the second order perturbation theory leads to the integral I_6 (34.4) which in the high temperature limit gives a T^2 dependence (34.8).

Of course the rotational and vibrational oscillations do interact to a certain extent and so a second branch of the phonon spectrum is introduced which goes to zero frequency, but as it tends to zero the amplitude also goes to zero. In this case a weak direct process can occur and in the high temperature limit $W_{ab} \propto T$.

The presence of rotational vibrations is also important in polymers where they induce electron transitions in the free radicals associated with defects in the polymer chain. If the free radicals are not associated with defects then the unpaired electron is delocalised along some portion of the chain. ALEKSANDROV and KESSENIKH[3] pointed out that in this case the modulation by phonons of the hyperfine interaction between the electron and the protons in the chain can cause spin-lattice relaxation. The phonons required for this mechanism are not from the rotational spectrum but from the usual acoustic modes which, to a good approximation in a long chain, are described as if they are propagating in one dimension.

52. Stereochemistry. We will now make some comments about the influence of the local structure around the sites of the paramagnetic ions. We have already discussed the XY_6 coordination in Sect. 28 but here we will comment on possible sources of deviations from the true octahedral symmetry.

The two most common forms of distortion are: (1) a distortion along the (100) direction with axial symmetry about (100) which leaves the cluster with tetragonal symmetry, and (2) a distortion along (111) with axial symmetry about (111) which gives trigonal symmetry. The arrangement of the more distant ions is usually neglected in paramagnetic relaxation theories. The direct influence on X is ignored but there is obviously an influence on the Y ions. This is usually allowed

---

[1] I. V. OVCHINNIKOV: Soviet Phys. Solid State **5**, 1378 (1964).
[2] I. V. ALEKSANDROV, i G. M. ZHIDORIMOV: Soviet Phys. JETP **13**, 1211 (1961).
[3] I. V. ALEKSANDROV, i A. V. KESSENIKH: Soviet Phys. Solid State **6**, 777 (1964).

for by considering the phonon structure as being of a Debije form. The direct influence on the Y ions does play a part in distorting the octahedron and so has an indirect effect on the magnetic ion X. The other cause of distortion is the Jahn-Teller effect. This effect eliminates terms in the crystal field potential which are linear in displacement to achieve stability. If degeneracy exists, except for Kramers degeneracy, the linear terms cannot all be eliminated together in the true octahedron so it distorts and the degeneracy is reduced. Alternatively the Jahn-Teller effect is explained by the suppression of linear terms which occur in the difference between the equilibrium configurations of the complex when the spin is in the ground state and in an excited state, then the wholly symmetric mode $\mathfrak{Q}_1$ also contributes to the relaxation, Fig. 10[1]. These comments on the distortions apply to the other stereochemical arrangements to be considered below.

The eightfold coordination XY_8 also occurs. Experiments are usually performed on artificial salts by taking a diamagnetic salt and replacing some of the X ions by paramagnetic ions. Examples are rare earth ions in[2] CaF_2 and Mn^{2+} in MgO or[3] SrS. The perturbation between the lattice and the orbital component of the angular momentum is expanded in terms of the normal components of the cluster XY_8, in direct analogy to the XY_6 coordination. The normal coordinates in the two cases are different and in the XY_8 case they have been calculated by Huang[4]. The orbit-lattice perturbation is smaller in the XY_8 cluster than in the XY_6 cluster for the same interionic distance in the clusters.

The XY_9 complex with nine-fold coordination is the complex which occurs in the rare-earth ethyl sulphates whose relaxation has been considered by Orbach [16] and discussed by us in Sects. 37 to 40. The ions from the transition groups with 3d, 4d or 5d electron configurations do not seem to form XY_9 complexes but ions from these groups and from the 4f group do form XY_6 complexes naturally and they may all be substituted into XY_8 complexes. Paramagnetic relaxation of 4d and 5d ions does not seem to have been studied as intensively as in the other groups. These ions are known to form complexes with covalent bonding playing a major role.

Other possible configurations include the tetrahedral XY_4 complex which is represented by $CoCl_4$ and $ReCl_4$, the irregular octahedron in which the magnetic ion is surrounded by an octahedron consisting of five neighbours of one element and one neighbour of another element, and a second type of distorted octahedron in which the magnetic ion is surrounded by four ions of one element in a plane and two ions of another element forming the other vertices of the octahedron. An example of the former type is the $Co(NH_3)_5Cl^{2+}$ ion and of the latter is the leonite group of salts, e.g. $MnK_2(SO_4)_2 4H_2O$. The symmetry of the paramagnetic clusters with mixed ions is lower than that of the octahedron and also the thermal motion of the ions about their equilibrium positions will show anisotropies depending on the positions and masses of the ions concerned.

In tetrahedral compounds of $3d^1$ ions, the ground state levels form a nonmagnetic orbital doublet with a separation of up to 10^3 cm^{-1} and the next highest level is 10^4 cm^{-1} higher. Matrix elements of the spin-orbit coupling do not occur between the ground levels until the second order. Van Reijen et al.[5] have demonstrated that $3d^1$ ions in an orthorhombically distorted octahedron, where they have an orbital triplet ground state, relax faster than the non-magnetic states

---

[1] L. K. Aminov, i B. I. Kochelaev: Soviet Phys. JETP **15**, 903 (1962).
[2] See footnote 1 on p. 53.
[3] A. A. Manenkov, i V. A. Milyaev: Soviet Phys. JETP **14**, 75 (1962).
[4] C. Y. Huang, and M. Inoue: J. Phys. Chem. Solids **25**, 889 (1964).
[5] L. L. van Reijen, P. Cossee, and H. J. van Haren: J. Chem. Phys. **38**, 572 (1963).

in the octahedron. The relaxation time of the singlet ground state from a $3\,d^1$ ion in an orthorhombically distorted square pyramid is slower than for both the triplet and the doublet.

Rare-earth ions in double nitrates, $X_2Y_3(NO_3)_{12}\,24\,H_2O$ where X is the rare-earth ion and Y is a divalent ion, usually Mg^{2+} or Zn^{2+}, are in a C_{3v} site which for the heavier ions becomes stable in the C_s symmetry. In this case the crystal field terms A_6^m dominate as they are not screened like the lower order terms are. Therefore, the crystal field appears to be produced by an icosahedron which is somewhat irregular[1]. Theoretical estimates of τ_L for an ion in this site are likely to be difficult because of the complexity of estimating the crystal field parameters in this symmetry, even in the static case.

The Jahn-Teller effect, mentioned above, has been observed in Pt^{3+} in Al_2O_3 and Ni^{3+} in $SrTiO_3$ during the analysis of the Orbach relaxation mechanism[2]. They measured the spin-lattice relaxation between 77 °K and 500 °K and observed a temperature dependence which could be described by an Orbach relaxation mechanism. From their results they estimated the splitting Δ_c of the excited level participating in the process. The ions have a $^2\Gamma_3$ ground state in a sixfold coordinated site. This state is split only by the Jahn-Teller effect so the splittings obtained from the experimental results must be the Jahn-Teller splittings.

IV. Relaxation due to alternative perturbations.

53. Waller's mechanism. The spin-lattice perturbation described in Sect. 38 which has formed the basis of the theory up to the present section was not the first perturbation considered. WALLER [1] treated the modulation of the dipolar interaction between paramagnetic ions as the cause of the relaxation. This mechanism did not give a satisfactory explanation of the early results on paramagnetic relaxation because the predicted values of τ_L were too long, so the perturbation already described was introduced. However, the results of the dipolar modulation may apply to specimens with a high concentration of paramagnetic spins or with a large magnetic moment for each spin. It may also be of importance in S-state ions, Sect. 48, where the spin-orbit coupling is very small.

The dipolar interaction energy between two ions is given by Eq. (27.1h). For simplicity each pair is assumed to be independent of the other ions in the lattice and only pairs of ions formed from nearest neighbours are considered. This interaction lifts any time degeneracy of the electron states if it is present. The influence of all the nearest neighbours is accounted for by summing over all the pairs of neighbours, each one assumed to be independent of the others. The phonons modulate the spacing of the pairs which in turn modulates the energy and causes spin flips. Eq. (26.4) allows us to calculate the perturbation operator to be used in the time-dependent perturbation theory. The two spins may flip simultaneously, or only one may flip, and these processes may be caused by one or two phonon processes. The single-spin-flip and the double-spin-flip one-phonon processes lead to the following relation for τ_L:

$$\tau_L^{-1} \propto T H^2 (\beta/r)^6 \tag{53.1}$$

where β is the Bohr magneton and r the interionic spacing. For the double-spin-flip process the proportionality parameter is an order of magnitude larger so it is more probable.

---

[1] B. R. JUDD, and E. WONG: J. Chem. Phys. **28**, 1097 (1958).
[2] U. HÖCHLI, and K. A. MÜLLER: Phys. Rev. Letters **12**, 730 (1964).

The second order term in (26.3) leads to the Raman type process with two phonons participating. In this case τ_L is given by:

$$\tau_L^{-1} \propto T^7 (\beta/r)^6. \tag{53.2}$$

The double-spin-flip process is more probable in this case also [20]. The dipolar mechanism becomes more important in S-state ions than the modulation of the orbital component if the condition in Eq. (48.1) is satisfied.

54. Exchange interaction. Another mechanism which can cause relaxation is modulation of the exchange interaction between neighbouring ions. If the exchange interaction is so large that an exchange coupled pair is formed which has its own spectrum of energy levels then deviations in the relaxation may occur because of these extra energy levels. On the other hand when the exchange coupling is weak, modulation of the exchange may cause spin flips. The exchange interaction is given by $\mathcal{H}_g$, Eq. (27.1g), and can be used in Eq. (26.4) to obtain the perturbation operator. If the spins coupled in this way are otherwise completely free the modulation of the exchange would not cause spin flips but the introduction of crystalline fields, dipolar interactions or a slight anisotropy in the exchange effect allows spin flips to occur, ALTSHULER[1] and GILL[2]. If the anisotropy in J is much smaller than the magnitude of the isotropic J and the crystalline field splitting is small, then the exchange operator couples states whose spin quantum numbers differ by 2, but adjacent spin multiplets are not linked. Thus there are no Orbach processes using levels outside the lowest spin multiplet. The largest matrix elements are proportional to D/J, where D is the zero field crystal field splitting, and they are approximately field independent.

The distinction between weak and strong exchange interaction is usually made by considering the ratio between the exchange energy and the energy of the crystalline field splitting in the specimen. For ions which occur in pairs so close together that their exchange energy is higher than the crystalline field energy we get a new energy level spectrum. As the exchange energy varies rapidly with distance it can be seen that both strong and weak exchange may occur in the same material and the one which predominates depends on the exact conditions of the case.

The previous comments apply to the situation where there are exchange coupled pairs. This may occur in materials like copper acetate[3], where the copper ions occur in pairs, or in materials where the paramagnetic ion has been diluted and owing to deviations from a completely uniform distribution of ions some pairs occur. On the other hand some materials' paramagnetic properties are determined by the exchange interaction between neighbouring paramagnetic centres. An example is the free radical D.P.P.H. Relaxation in materials of this type will be discussed in Sect. 57 which is the section where multiple ion effects are discussed. When more than two ions are influenced by isotropic exchange it is not necessary to introduce small anisotropies to produce relaxations when the exchange is modulated.

55. Further mechanisms. Another mechanism which has been considered is modulation of the intra-electronic spin-spin interactions. It was introduced by LEUSHIN to describe the relaxation of S-state ions, see Sect. 48. LEUSHIN also based some calculations on the following perturbation scheme: the lattice perturbs the orbital motion of the electrons which perturbs the spin-orbit coupling between

---

[1] S. A. AL'TSHULER: Soviet Phys. JETP **16**, 1637 (1963).
[2] J. C. GILL: Proc. Phys. Soc. (London) **79**, 58 (1962).
[3] B. BLEANEY, and K. D. BOWERS: Proc. Roy. Soc. (London) A **214**, 451 (1952).

the spin of one electron and the orbit of one of the other electrons in the unfilled shells and this causes the spin to reorientate itself with respect to the external magnetic field. At the moment there does not appear to be any direct application of this perturbation.

Modulation of the hyperfine interaction can also produce spin flips. The interaction between the magnetic moment of the nucleus and the orbital and spin moments of the electrons is given by $\mathscr{H}_d$, Eq. (27.1d). The second term is only non-zero when there are s-electrons present and in this case the other term is zero. This is the Fermi contact potential which plays a part in the relaxation of colour centers, see Sect. 59. The hyperfine interaction lifts the Kramers degeneracy so the relaxation can occur in both Kramers and non-Kramers salts. The thermal vibrations modulate the hyperfine interaction which in turn modulates the wave functions and causes spin flips. This mechanism, in the presence of the Zeeman effect, is anisotropic and depends on the orientation of the magnetic field with respect to the crystal axes. The hyperfine interaction mixes into the ground state excited levels with a different value of m_I the nuclear quantum number so that transitions occur which would normally be forbidden. BAKER and FORD[1] have considered this mechanism in the Kramers salts Nd^{3+} in LaF_3 and LaMg nitrate.

The nuclear quadrupole interaction, $\mathscr{H}_e$ in Eq. (27.1e), can cause spin-lattice relaxation through the following perturbation: lattice motions produce variations in the electric field gradient at the site of a nucleus with a quadrupole moment and by means of the quadrupole interaction the electron wave functions are perturbed and spin flips produced.

56. Relaxation with interactions between the spins. Up to now, we have, except in Sects 53 and 54, considered only the interaction between the lattice and isolated spins. However, this is rarely the only interaction which affects the spins, except in highly diluted systems, and even then there may be a chance of pairs or isolated groups of ions occurring. Interactions between spins lead to further relaxation mechanisms and modifications of the mechanisms already described.

In spin systems with interionic interactions it is easier to visualise a temperature for the system than in the case considered at length above where each spin is assumed to be isolated. ARGYRES and KELLEY[2] amongst others, have considered spin-lattice relaxation in a spin system with interactions. The spin system will have its own highly degenerate energy level system with the spins distributed over it in a way which satisfies the Pauli exclusion principle. When it is in equilibrium it can be characterised by a spin temperature. The difficulty in the theory of isolated ions considered above is imagining a uniform spin temperature in a system where there are no interactions to maintain it. In the theory of the spin-lattice relaxation times given above a spin temperature is imposed on the system by the following device. To make up the total spin system the individual ions are distributed over the energy levels which belong to an isolated ion according to a Boltzmann distribution in which the temperature represents the temperature of the spin system.

DE VRIES et al.[3] have performed measurements on copper Tutton and cobalt Tutton salts under conditions where the spin-spin relaxation mechanism was slower than the spin-lattice relaxation mechanism. Under these conditions the

---

[1] J. M. BAKER, and N. C. FORD jr.: Phys. Rev. **136**, A 1692 (1964).
[2] P. N. ARGYRES, and P. R. KELLEY: Phys. Rev. **134**, A 98 (1964).
[3] A. J. DE VRIES, D. A. CURTIS, J W M LIVIUS, A J VAN DUIJNEVELDT, and C. J. GORTER: Physica **36**, 91 (1967).

spin-spin absorption was observed to disappear because the spin-lattice mechanism is more efficient in maintaining equilibrium. Thus, even in the case of isolated spins, that is with infinitely long spin-spin relaxation times, it is possible to think of equilibrium within the spin system which is maintained by contact with the lattice.

Ionic interactions may occur by means of exchange, dipolar interaction and through the phonons. Exchange is a contact mechanism either directly between ions or through intermediate ions, the dipolar interaction is an electromagnetic interaction through a photon field with zero retardation time while the interaction through the phonon field can take place in several ways: (1) spin-phonon interaction followed by a phonon-spin interaction which if the temperature is low enough is a retardation phenomenon, (2) at higher temperatures when the phonon mean free path is short the interaction is by means of the potential energy of the pair of ions in the phonon field, (3) the orbital moments of the ions may interact with each other through the phonon field. These interactions are usually of the same order or less than the dipolar interaction so in general they do not play a large part in spin-lattice relaxation[1].

Samples which are in a large magnetic field will behave as if their ions are independent of each other because the spins will be coupled to the magnetic field rather than to each other. This decrease in the influence of the spin-spin interaction with increasing field shows itself in experiments as an increase in the spin-spin relaxation time, DE VRIES et al.[2]. However, at fields which are of the order of the internal fields the coupling between spins will play a role in the relaxation. The dipolar coupling has two effects on the relaxation mechanism; (1) the off-diagonal terms of the dipolar interaction help the normal relaxation mechanism to flip over the spins, and (2) those spins which are close together with respect to the wavelength of the phonons relax coherently. The process of the calculation is to evaluate the state functions of the ions in the presence of the crystalline field and spin-orbit coupling and to apply perturbation theory to allow for the external magnetic field and the internal fields produced by the dipolar interactions. Then the spin-lattice perturbation is applied but because of the complexity of the state functions the summations over the ions sites become more complex. In this case it is usual to express the results in the terms of internal fields brought about by the dipolar interactions. These mean square dipolar fields depend on the energy levels encountered. The matrix elements for the dipolar coupling in the ground states of an ion will differ from those between the ground and excited states and a different mean square field is required in each case.

Another partial simplification is obtained by dividing the problem into cases where $g_\perp = 0$ and $g_\perp \neq 0$. In the former case the summations become easier because this eliminates from the perturbed state functions those admixtures from higher states caused by the component of the magnetic field perpendicular to the crystalline axis of the specimen and the components of the dipole moment which are perpendicular to this axis. HUBER[3] calculated the direct process in a Kramers salt with $g_\perp = 0$ and in this case he included two mean square dipolar fields. Even at zero field there is a relaxation time because the dipolar interaction lifts the time degeneracy of the levels. The final result is in the form:

$$\tau_L^{-1} = \tau_{L\infty}^{-1} \frac{b(H_{i\text{dip}}^2, H_{i\text{dip}}'^2, H, \vartheta)}{[H^4 \cos^2\vartheta \sin^2\vartheta (H^2 \cos^2\vartheta + \tfrac{1}{2} H_{i\text{dip}}^2)]}, \qquad (56.1)$$

---

[1] See footnote 1, p. 69.
[2] See footnote 3, p. 72.
[3] D. HUBER: Phys. Rev. **131**, 190 (1963).

where the $H^2_{i\,\text{dip}}$ are the mean square dipolar fields, ϑ is the angle between the magnetic field and the crystalline axis and $\tau_{L\infty}$ is the relaxation time when $H \gg H_{i\,\text{dip}}$.

The calculation for the Raman process leads to the Brons-Van Vleck form of the dependence of the relaxation time on magnetic field [13][1,2].

$$\tau_L^{-1} = \tau_{L\infty}^{-1} \frac{(H^2 + H^2_{i\,\text{dip}})}{(H^2 + \tfrac{1}{2} H^2_{i\,\text{dip}})} = \tau_{L0}^{-1} \frac{(b_{\text{dip}} + pCH^2)}{(b_{\text{dip}} + CH^2)} \qquad (56.2)$$

where $H^2_{i\,\text{dip}} = 2 b_{\text{dip}}/C$, C is Curie's constant and b_{dip} is the contribution of the dipolar interaction to the low temperature spin specific heat, $p \simeq \tfrac{1}{2}$. The factor $\tfrac{1}{2}$ was shown by Orbach to be exact when $g_\perp = 0$. τ_{L0} is the value of the relaxation time at zero field. Eq. (56.2) is useful in a comparison between theory and experiment because $\tau_{L\infty}$, the value of the relaxation in an infinite magnetic field, is more easily calculated than τ_{L0}. In an infinite field the spins can be regarded as isolated and the theories of **C.III** applied rather than the more complex theories indicated in this section.

In the more general case when $g_\perp \neq 0$ and when hyperfine structure is present, Eq. (56.2) becomes:

$$\tau_L^{-1} = \tau_{L\infty}^{-1} \frac{(H^2 + \mu H^2_{hfs} + \tfrac{1}{2} \mu' H^2_{i\,\text{dip}})}{(H^2 + H^2_{hfs} + \tfrac{1}{2} H^2_{i\,\text{dip}})}. \qquad (56.3)$$

Orbach gives detailed expressions for μ and μ' which occur in this equation. μ depends solely on the orbit-lattice parameters while μ' depends on the orbit-lattice parameters, the temperature and the concentration of the paramagnetic ions: as a result μ' deviates from the value 2 obtained in simple considerations which neglect correlations between spins and use the concept of an effective internal field when $g_\perp \neq 0$, where it is less suitable.

Relaxation induced by variations in the dipolar interaction has already been considered in Sect. 53. It is the Waller mechanism.

At very low temperatures the relaxation rate is controlled by the number of phonons available to carry away the energy from the spin system. The dipolar interaction mechanism which can cause more than one spin to flip at any one time is a route round this bottleneck. The $\mathscr{H}_{+2}$ and $\mathscr{H}_{-2}$ terms of the dipolar interaction, Eq. (10.8), cause two spins to flip together and these may combine to produce a phonon of twice the energy obtained from a single spin flip. Multi-spin flips are also possible and these may produce higher energy phonons which are much more numerous and so provide a convenient route for the energy to flow from the spins to the lattice and vice versa. At lower temperatures the number of phonons begins to fall so a temperature effect can be expected.

For a direct process in a non-Kramers salt four spins may be involved because the spin-Hamiltonian, with dipolar interactions, describing the lowest levels of the system is a quadratic function of the spin quantum numbers and the spin-lattice Hamiltonian to the first order in lattice strain is also quadratic in spin quantum number. The dipolar interaction then links sites with different spins and modifies the state functions so that the spin-lattice interaction is able to cause transitions between states in which four spins are involved. In Kramers salts, where it is necessary to introduce admixtures of higher states into the ground state wave functions, six spin processes may occur. The probability of more spins being involved is much smaller because it requires higher order per-

---

[1] F. Brons: Thesis University of Groningen 1938.
[2] R. Orbach: Proc. Roy. Soc. (London) A **264**, 485 (1961).

turbations to bring it about. These high number spin processes may occur, however, and have been discussed by TEMPERLEY[1].

Processes involving the turning over of more than one spin at once in a magnetic field may explain the deviations from the theoretical predictions for τ_L as a function of the magnetic field in the direct process region. The simple theory predicts a decrease in τ_L as the magnetic field increases while quite often the reverse is observed. If the relaxation is controlled by multi-spin flip processes then increasing the field will slow them down because the interaction between the spins will effectively become weaker as they couple themselves more strongly with the field. There will be a maximum relaxation time as the field increases until the point where all the ions can be regarded as isolated and the predictions of the original theory become valid.

The influence of exchange interactions makes the calculation of multi-spin processes more complex. BÖLGER[2] has made some calculations and demonstrated that the processes are not normally of major importance. The non-secular terms of the dipolar interaction, possibly modified by the exchange interaction, introduce weak satellite lines at frequencies related to the basic transition frequency. Thus, the spin system is linked to a larger number of phonons than appears to be the case when only isolated ions are considered.

57. Cross-relaxation. The dipolar interaction between ions also influences the passage of energy from the spins to the lattice by the process of cross-relaxation which rearranges the energy among the spins without reference to the lattice. This is a spin-spin process.

A much simplified, and probably rather doubtful model when compared with the considerations of part **B**, is to consider the ions as being linked by dipolar interactions which are so weak that the energy levels of an ion are those of an isolated ion. The non-secular terms of the dipolar interaction cause spin flips between these levels and the balance of the energy is taken up by the secular terms in the interaction. In the calculation of the probability for the cross-relaxation process the non-secular terms are treated by perturbation theory while the secular terms are included in the expression for the lineshape which can be described by moment theory[3]. Then for the process where one spin flips from $|i\rangle$ to $|j\rangle$ and a second from $|k\rangle$ to $|l\rangle$, the probability is:

$$W_{i,k,j,l} = \hbar^2 |\mathcal{H}^{jl,ik}_{\text{non-sec.}}|^2 g(\vartheta). \qquad (57.1)$$

$g(\vartheta)$ is the cross-relaxation lineshape. The simple process when spins flip from $|i\rangle$ to $|j\rangle$ and from $|j\rangle$ to $|i\rangle$ is a diffusion process. This process depends on interionic distances and it is slower when the spins are farther apart. The levels $|i\rangle$, $|j\rangle$, $|k\rangle$ and $|l\rangle$ may be regarded as the contributions of two ions to the highly degenerate energy level scheme of the spin system as a whole. The important parameter in the cross-relaxation process is the separation in the energy levels not the absolute values of the energy of the levels. Similar assumptions are usually made in the theory of electron spin resonance spectra.

The cross-relaxation process, which is really a spin-spin process and does not transfer energy to the lattice, affects the spin-lattice relaxation mechanism by altering the populations of the energy levels involved in the spin-lattice process. In a resonant experiment, where only one transition is monitored, this becomes very important and the relaxation time measured does not depend solely on the

---

[1] H. N. V. TEMPERLEY: Proc. Cambridge Phil. Soc. **35**, 256 (1939); **43**, 118 (1947).
[2] See footnote 4, p. 46.
[3] J. H. VAN VLECK: Phys. Rev. **74**, 1168 (1948).

transition probability for the transition but also on the populations of the other levels connected to the first by cross-relaxation. Cross-relaxation can show itself in measurements by resonant and non-resonant techniques by by-passing a slow spin-lattice process. This is shown in Fig. 13 where the energy level systems A and B are produced by two different types of ions or by different clusters of ions. If the populations of levels A are altered by some external means so that the thermal equilibrium is disturbed, cross-relaxation processes transfer energy from spin system A to spin system B which is maintained in thermal equilibrium by Orbach processes.

This process is one of several ways in which dipolar interactions between the main ions and another type of ion in a material may control the relaxation of the

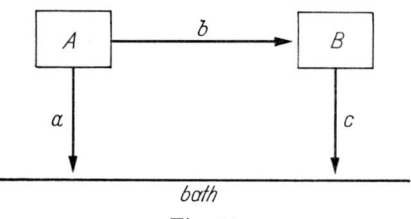

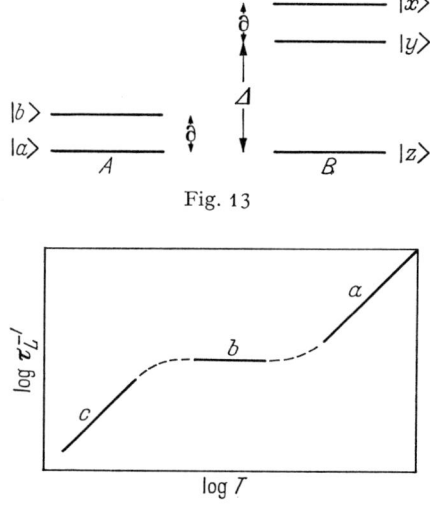

Fig. 13. Energy level scheme to demonstrate cross-relaxation processes.

Fig. 14. A block diagram showing the possible routes for relaxation among the energy levels of Fig. 13. a, b and c are impedances to the relaxation mechanisms.

Fig. 15. Schematic graph of the different regions of relaxation possible in a sample which can be represented by the block diagram of Fig. 14.

main ions. The second type of ion need not be the most numerous in the material because spin diffusion may bring energy from more distant spins to the neighbourhood of the relaxation centre where cross-relaxation can occur. If this occurs another temperature dependence for τ_L may be observed. Assuming that the two types of energy levels may be associated with two systems of spins which we label A and B, see Fig. 14, and that they are in contact with each other and the lattice, then the time constant for the relaxation from A to the lattice is

$$\tau_L^{-1} \propto \frac{a+b+c}{a(b+c)} \tag{57.2}$$

by analogy with circuit theory. Fig. 15 shows the different regions of relaxation possible when it is assumed that (1) $a \ll (b+c)$, (2) $a \gg (b+c)$ and (2a) $b \gg c$ and (2b) $b \ll c$. If b is the cross-relaxation mechanism which is temperature independent the shape of Fig. 15 is obtained.

An alternative description of the combined cross-relaxation and spin-lattice relaxation process is the following. A spin may flip from $|b\rangle$ to $|a\rangle$ simultaneously with a spin flip from $|y\rangle$ to $|z\rangle$, with the production of a phonon of energy $\partial + \Delta$. Within this simple system there are seven other possible simultaneous modes of flipping and they must all be included in the rate equations for the system when an expression is derived for the relaxation time. The cause of the simultaneous flips is the dipolar coupling between the ions of system A and system B. The modulation of the interionic spacing by the lattice vibrations is not sufficient to explain fully the rate at which the transitions occur; this is just the Waller mechanism between unlike ions. However, if the system B is in equilibrium with the lattice so that many lattice induced spin flips occur, the spin operator of the ions in B has a time variation, then the dipolar interaction also varies with time. This can be expressed as the lattice causing spin flips in system B which produce variations in the local magnetic field around B and those ions of A in the vicinity are then induced to flip over.

In materials with a strong exchange interaction the spin energy may also be divided into two systems. System A represents the Zeeman levels while system B represents the energy levels produced by the exchange interaction. The non-secular part of the dipolar interaction and the anisotropic exchange interaction link these two systems. This type of situation arises in free radicals where the g-value of the electrons is close to the free spin value so that relaxation processes involving the spin-orbit coupling, which here is very small, are very slow. Therefore, the Waller mechanism has to be considered as causing the direct relaxation from the Zeeman system to the lattice. The exchange system may relax rapidly because the energy spectrum is continuous so high energy phonons may be created by spin flips within the exchange system. Direct processes predominate at all temperatures since the spectrum is continuous. The perturbation causing the spin flips from B to the lattice is the Waller mechanism.

In some materials, particularly diluted materials, groups of spins may occur in clusters which have a distinctive set of energy levels and which may participate in cross-relaxation processes. Within the clusters there may be strong exchange while the remaining spins in the diluted material may be connected by weak interactions. These clusters can be very effective as sinks in the spin diffusion process.

Apparent anomalies in the relation between τ_L and the magnetic field have been explained by cross-relaxation processes. The theory given for isolated ions predicts monotonic relations between τ_L and H in several different cases. Hellwege et al.[1] observed local minima in the relation between τ_L and H when making measurements on $CeCl_3 \cdot 7H_2O$. These minima occurred at definite field values and were explained by the presence of a second relaxation mechanism with a second time constant τ_X and F-value F_X. Then Eqs. (4.25) and (4.26) become:

$$\begin{aligned} \frac{\chi'}{\chi_0} &= 1 - F_L - F_X + \frac{F_L}{1+\tau_L^2\omega^2} + \frac{F_X}{1+\tau_X^2\omega^2}, \\ \frac{\chi''}{\chi_0} &= \frac{F_L \tau_L \omega}{1+\tau_L^2\omega^2} + \frac{F_X \tau_X \omega}{1+\tau_X^2\omega^2}, \end{aligned} \qquad (57.3)$$

where $1 - F_L - F_X = \chi_{ad}/\chi_0$. These expressions are a natural extension of the Casimir and Du Pré formalism[2]. The second relaxation mechanism was identified as a cross-relaxation process which only became important at certain definite field values which agreed with the values of H at which the local minima occurred.

Van den Broek et al.[3] observed similar local minima when plotting τ_L against H for $Co(NH_4)_2(SO_4)_2 \cdot 6H_2O$. A similar explanation involving double

---

[1] K. H. Hellwege, R. von Klot, and G. Weber: Phys. kondens. Materie **2**, 397 (1964).
[2] J. van den Broek, L. C. van der Marel, and C. J. Gorter: Physica **27**, 661 (1961).
[3] J. van den Broek, L. C. van der Marel, and C. J. Gorter: Physica **25**, 371 (1959).

relaxations is thought to apply to this salt but a detailed description of the origin of the second relaxation has not yet been given.

58. Concentration effects. One of the discrepancies between theoretical and experimental work in spin-lattice relaxation is the influence of variations in concentration of the magnetic ions on the relaxation time. In the basic theory of Sects. 39 et seq. no influence of concentration is revealed. The state functions of the system do not depend on concentration for isolated ions, nor do the spin-lattice Hamiltonian or the lattice state. However, modification of these parameters by concentration effects is possible. Entries 20a to 22 in Table 2 show typical concentration effects in the direct relaxation region in a rare earth salt. The state functions may depend on concentration in the simple theory if the crystal field parameters vary with concentration; this has been observed in some materials[1]. The occurrence of ions in clusters also modifies the state functions because of the dipolar and exchange interactions inside the clusters. The size and local density of the cluster depend to some extent on the macroscopic concentration in the crystal. It is difficult to establish a concentration dependence from this type of effect. Calculations cannot be made by assuming that the paramagnetic ions in a crystal become further apart as it is diluted, because in many cases it is the presence of ions on neighbouring sites which is important. Therefore, the state functions of pairs of ions are not altered even though the concentration is. The questions then become, at what concentration do the pairs fail to provide an adequate path for the energy flow and what other mechanisms are available? The work of Thorp et al.[2] indicates how difficult it is in highly diluted materials to talk of a uniform distribution of ions.

The relaxation mechanisms which are concentration dependent are those whose Hamiltonian contains the interionic spacing as a parameter. These include the Waller dipolar mechanism and the exchange mechanism, either in strong exchange or weak exchange. From these Hamiltonians more definite relations between τ_L and concentration can be established. Another mechanism which is concentration dependent and influences spin-lattice relaxation in certain cases is the cross-relaxation process which has various concentration dependences, each a function of the order of the cross-relaxation process involved.

The phonon structure may be influenced by concentration effects if the paramagnetic ion is of a considerably different mass from the rest of the lattice, or if it is differently bonded. Then we have phenomena similar to those discussed in Sect. 63. The amount of scattering of lattice phonons by impurity ions obviously depends on the number of such ions present.

59. Relaxation in local electron centres. Local electron centres have properties similar to those of paramagnetic ions but their detailed structure and the relaxation processes involved are different from those of normal ions. Even the simplest model for a colour centre, say an F-centre in an alkali halide, demonstrates that the state function of the electron has a large radius and includes several neighbouring ionic sites. Thus, the interaction with these sites is of considerable importance in F-centre properties and when these sites are vibrating they have a direct influence on the relaxation processes. The metal ions in an alkali halide have a nuclear moment and this of course influences the state function of the electron which encompasses them. Let us consider an electron bound to the site of a negative ion vacancy and enclosing in its orbit several ions with nuclear spin $\frac{1}{2}$. The energy levels and the possible relaxation paths for a system like this

---

[1] K. D. Bowers, and J. Owen: Repts. Progr. Phys. **18**, 304 (1955).
[2] J. S. Thorp, D. A. Curtis, and D. R. Mason: Brit. J. Appl. Phys. **15**, 775 (1964).

Sect. 59. Relaxation in local electron centres.

are shown in Fig. 16 and Table 7. The Hamiltonian for the local electron is:

$$\mathcal{H} = \mu_{\text{elec}} \mathbf{H} \cdot \mathbf{S} + \mu_I \mathbf{H} \cdot \mathbf{I} + A \mathbf{I} \cdot \mathbf{S} + b\{3(\mathbf{I} \cdot \mathbf{r})(\mathbf{S} \cdot \mathbf{r}) - \mathbf{I} \cdot \mathbf{S}\} + \\ + \frac{8\pi}{3} \mu_{\text{elec}} \mu_I \mathbf{S} \cdot \mathbf{I} \delta(r) + \lambda \mathbf{L} \cdot \mathbf{S}, \qquad (59.1)$$

where the first term is the electronic Zeeman effect, the second term is the nuclear Zeeman effect, the third and fourth terms describe the hyperfine interactions with the nuclei, the former is isotropic while the latter is not, the fifth term is the Fermi interaction and the last term is the spin-orbit coupling. The influence of

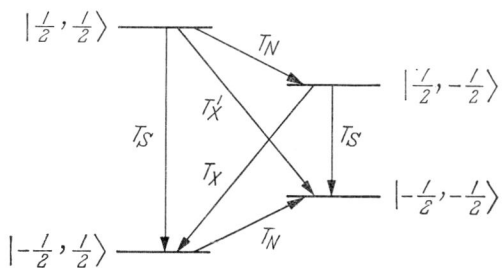

Fig. 16. Possible spin-lattice relaxation processes in a local electron centre with nuclei of spin $\frac{1}{2}$ present.

Table 7. *A list of the modes of relaxation shown in Fig. 16.*

Transition	Selection rules		Relaxation time	Comments
	Δm_S	Δm_I		
1	±1	0	T_s	Electron flips only
2	±1	∓1	T_x	Electron and nuclear flips in opposing directions
3	±1	±1	T'_x	Electron and nuclear flips in the same direction
4	0	±1	T_N	Nuclear flips only

the crystal field on the s-state functions of the local electrons has not been observed so it is omitted from Eq. (59.1), as has any interaction between neighbouring centres.

The constant A depends on the geometry and the dimensions of the site, so the passage of phonons modifies it and in first order perturbation theory it leads to the mechanisms T_x and T'_x. The b term is also time dependent when phonons disturb the lattice but this has two separate effects; firstly, in a similar way to the variation of A, b produces the relaxations described by T_x and T'_x and also T_N and T_S in a first order calculation; secondly, if the electron is being flipped rapidly by another mechanism so that S_z is a function of time, then b produces uncorrelated nuclear and electron flips. This is analogous to the mechanism discussed in Sect. 57. The Fermi contact term is also susceptible to the passage of phonons and in first order gives all four relaxation mechanisms. Even though the colour centre electrons only have s-states some orbital angular momentum is mixed in. This can be accounted for by making the F-centre state function orthogonal with respect to the state functions of the electrons on the neighbouring ions. Thus, the F-centre has the spin-orbit coupling parameter of the neighbouring electrons reduced by the square of the mutual overlap of the electron functions. This is modulated by the passage of phonons, but there is no direct process as the first order theory has a zero answer. The second order, however, leads to a rapid relaxation for all the mechanisms and indeed is now regarded as the chief Raman relaxation process, T_S, in colour centres. The main direct process seems to be the modulation of the Fermi contact potential. This mechanism taken in the second order perturbation theory has no influence on the Raman processes. The nuclear relaxation mechanism which is fastest is the variation in the hyperfine interaction between a nucleus and its electron, produced by the electron being rapidly flipped by its interaction with the lattice.

In numerical predictions of these mechanisms a model has to be used for the wave functions and the actual time variation of the parameter which depends on lattice vibrations. In alkali halides GOURARY and ADRIAN[1] have calculated certain electronic wave functions so these may be used in calculations. For those cases in which the Fermi contact is important then the detailed structure of the wave function at and close to the point of contact must be known.

It is usual to assume a Debije spectrum for the phonon structure in this case as in all others. The detailed amplitude and phase structure of the phonons about the F-centre site is neglected in most calculations, but this is not serious at temperatures of the order of Θ_D or larger, where the nearest neighbours will be out of phase with each other. The inclusion of optical phonons has no special effects unless a nuclear quadrupole moment is present. In that case another relaxation mechanism occurs, which has been described in Sect. 55. Another mistake which may be introduced by using a Debije spectrum is the omission of distortions of the local lattice structure produced by variations in the electron wave function. This is the Franck-Condon effect in optics[2]. For simple spin flips this is unlikely to have any effect but for an Orbach type relaxation involving higher energy phonons there may be some influence.

If the local electron centres do have some mutual interaction, its modulation by the phonons can cause relaxation. The phonons vary the mixing of the states and so cause the relaxation. It has received some consideration in the problem of electron donor centres[3]. If two centres are close together the electron orbits will encompass them both and this leads to a non-spherical charge distribution, a modified spin-orbit coupling and correlation effects which induce relaxation. These effects are usually stronger than the exchange effects which also occur in such a situation.

V. Relaxation in the phonon system.

60. The Debije spectrum. Up to now only the simplest considerations have been given to the phonon system. It has been treated as a Debije spectrum, but in many cases more realism is required. The important parameter in the Debije spectrum is the Debije temperature which characterises the cut-off frequency introduced to maintain a fixed and finite number of oscillators. This temperature may be measured in several ways, e.g. specific heat, elasticity, X-ray measurements, etc., and each type of measurement gives a different answer weighted according to the importance of various parts of the frequency spectrum in the measured parameter. The Debije spectrum deviates least from the true spectrum

---

[1] B. S. GOURARY, and F. J. ADRIAN: Phys. Rev. **105**, 1180 (1957).
[2] W. B. FOWLER: Phys. Rev. **135**, A 1725 (1964).
[3] A. HONIG, and E. STUPP: Phys. Rev. **117**, 69 (1960).

at the lowest frequencies, so values of Θ_D from elastic measurements which give proper weight to this region are perhaps the best to use[1]. As shown in Sect. 34, Θ_D is a convenient empirical parameter to use to describe the integral properties of the phonon spectrum.

61. Vibrations of an XY_6 cluster in a lattice. Many workers have attempted to add more realism to their considerations of the phonon spectrum. KOCHELAEV[2] extended VAN VLECK's calculations on an XY_6 cluster to include the difference in bonding between the X ion and its six Y neighbours and between the six Y groups and their neighbours. The cluster usually vibrates as a unit with respect to its neighbours with different frequencies from its internal vibrations. Spin flips give energy to the vibrations within the cluster. The problem is to calculate the effect of the neighbours of the cluster on the vibrations within the cluster. To carry out this calculation it has to be assumed that the cluster is in its ground state and that its vibrations with respect to its neighbours are of much lower frequency than the internal vibrations. The first term in the perturbation series for the interaction between the cluster and the rest of the lattice is the quadrupole term describing the interaction between the quadrupole moment of the cluster and the electric field gradient produced by the surroundings. Numerical predictions made using this perturbation, and including optical phonons, lead to agreement with experiment. The direct electrical influence of the more distant ions on X is neglected. This theory also gives reasonable values of the p-parameter in the Brons-Van Vleck equation (56.2), and it leads to p values which are structure dependent and differ from $\frac{1}{2}$.

62. Optical phonons. At higher temperatures optical mode phonons become important and these are not included at all in the Debije assumption. The optical phonons cause vibrations of neighbouring atoms in a unit cell in opposing directions so that the centre of mass of the unit cell is not displaced. The optical branches of the phonon spectrum have a finite frequency when the wave vector is zero. For optical phonons only the nearest neighbours are important because the more distant neighbours move approximately in phase with the paramagnetic ion. For acoustic phonons the more distant neighbours contribute to the relaxation as has been described above.

Numerical calculations give some support to the use of:

$$\omega = \omega_0 - \Delta\omega \frac{k}{|k_0|} \quad (62.1)$$

as the relation between the frequency and the wave vector in the optical branches. $\omega = \omega_0$ when the wave vector $k=0$, $\Delta\omega$ is the frequency interval and $k_0 = (3\pi^2/2)^{\frac{1}{3}}$ the maximum value of k. The density of the optical phonons in the energy interval dE and the solid angle interval $d\Omega$ is:

$$\varrho\, dE\, d\Omega = (N/2\pi^3)\, k^2\, dk\, d\Omega. \quad (62.2)$$

This is used with Eq. (62.1) in calculating transition probabilities involving optical phonons.

KOCHELEAV[3] calculated the nuclear quadrupole relaxation time of nuclei in NaCl lattices using an ionic model with no dipolar interactions. The result for the

---

[1] F. H. HERBSTEIN: Advances in Phys. **10**, 313 (1961).
[2] B. I. KOCHELAEV: Soviet Phys. Solid State **2**, 1294 (1960).
[3] B. I. KOCHELAEV: Soviet Phys. JETP **10**, 171 (1960).

transition probability is:

$$W = A \left(\frac{1}{\omega_\parallel^2 \Delta\omega_\parallel \sinh^2 \frac{\hbar\omega_\parallel}{2kT}} + \frac{2}{\omega_\perp^2 \Delta\omega_\perp \sinh^2 \frac{\hbar\omega_\perp}{2kT}} \right) + BT^2, \quad (62.3)$$

which reduces to:

$$W \propto \exp(-\hbar\omega/kT) \quad (62.4)$$

if the differences between the longitudinal phonons $\omega_\parallel$ and the transverse phonons $\omega_\perp$ are neglected and the energy of the optical phonons is much larger than kT. In electron spin systems several people have suggested the inclusion of optical phonons but few people have performed detailed calculations.

Nuclear quadrupole relaxation is susceptible to optical phonons. The basic theory just treats the interaction between the quadrupole moment and the electric field gradient at the site of the ion. The gradient is varied by the passage of the phonons. This theory can be modified in two ways; firstly, by taking into account the asymmetry of the charge distribution introduced by covalency and secondly, by taking account of the induced electric dipole moments produced by the passage of the phonons. The relaxation of nuclei through their magnetic dipole interaction with rapidly relaxing paramagnetic ions takes place by a mechanism similar to that considered briefly in Sect. 57 above. HEBEL [21] has discussed relaxation in nuclear spin systems and also the question of spin temperature in these systems.

63. Local phonon modes. Localised defects in a lattice introduce deviations in the phonon spectrum which may influence the relaxation rates. In artificial materials such as rare earth ions in MgO, or in highly diluted materials, the site of the paramagnetic ion may be regarded as a defect site, especially if the ion's mass is considerably different from that of the other ions in the lattice.

The defects have two main effects, firstly they scatter the lattice phonons which are usually described by the Debije spectrum and secondly, they introduce more phonon modes. The scattering mechanism affects the magnitude of the direct process. KOCHELAEV[1] considered this modification to the direct process and derived an inequality describing the range in a crystal over which a defect site affects the relaxation. Thus, his calculations include scattering from defects which are not paramagnetic ion sites. CASTLE et al.[2] referred to this as the deformation of the motions of the surrounding ions by the defect site. However, they consider the introduction of extra phonons as the most important contribution of defects to relaxation. The extra modes of phonons introduced do not necessarily provide major changes in the magnitude of the relaxation time but they can modify the temperature dependence of the relaxation time. There are two different cases; (1) the extra frequencies are above the Debije cut-off frequency, in which case they can not propagate through the lattice and so are highly localised, (2) the extra phonons are below the cut-off frequency, in this case they can propagate through the lattice.

α) *High frequency local phonons.* The effects of the high frequency local phonons become more concentrated around the defect site as the frequency increases. The chances of excitation of a high frequency become less as the temperature decreases. The local mode by itself can not cause spin flips because energy is not conserved. There are interactions with the lattice waves which broaden

---

[1] B. I. KOCHELAEV: Soviet Phys. "Doklady" **5**, 349 (1960).
[2] J. G. CASTLE jr., D. W. FELDMAN, and P. G. KLEMENS: Phys. Rev. **130**, 577 (1963).

the local modes and which allow Raman type processes to occur between the spins and the local modes. These are then followed by the decay of the local phonon into two lattice modes. There are many ways in which this interaction involving three phonons may occur and some of these are considered by FELDMAN et al.[1]. Assuming the energy of the mode is concentrated around the defect site the square of the strain at the site can be written as:

$$\varepsilon_\lambda^2 \simeq \frac{\hbar \omega_\lambda}{M v^2}, \tag{63.1}$$

where M is the mass of the ion and v the velocity of sound in the crystal. ω_λ, the local mode frequency, can be evaluated by the methods of MONTROLL and POTTS[2]. Using (63.1) the matrix elements of the dynamic spin-lattice Hamiltonian can be evaluated and the expression for τ_L at temperatures below $\Theta_\lambda/2$ is:

$$\tau_L^{-1} \propto \omega_\lambda \left(\frac{\hbar \omega_\lambda}{M v}\right)^3 \frac{1}{\partial^2} \exp\left(-\frac{\Theta_\lambda}{T}\right). \tag{63.2}$$

Θ_λ is the temperature associated with the local mode frequency in the same way that the Debije temperature is associated with the cut-off frequency. ∂ is the energy of the spin flip. At higher temperatures the relaxation time is inversely proportional to T^2, as it is for the normal Raman mechanism. The high frequencies required for this mechanism occur at the sites of light ions or when ions are more strongly bonded than usual. FELDMAN et al. observed the exponential temperature dependence predicted by (63.2) during measurements on H⁻ and D⁻ in CaF_2.

β) *Low frequency local modes.* When the defect mode has a lower frequency than the cut-off frequency a different variation with temperature is found for τ_L. The frequencies involved can be estimated simply by using the theory of a forced, damped oscillator to describe the motions of the ion at the defect site. As during the calculation of the relaxation time there is an integration over the frequencies involved, the nature of the resonance in the solution for the oscillator is lost. Then the values of the lattice strain to be used in evaluating the matrix elements are approximated as:

$$\varepsilon \simeq \frac{\omega}{\omega_m} \frac{u}{a}; \quad \varepsilon_\lambda \simeq \left(\frac{\omega}{\omega_\lambda}\right)^2 \frac{u}{a} \ (\omega < \omega_\lambda); \quad \varepsilon_\lambda \simeq \frac{u}{a} \ (\omega > \omega_\lambda), \tag{63.3}$$

where the first expression is the one for the normal lattice phonons. a is the lattice parameter, u the phonon amplitude and ω_λ is the local mode frequency. The direct process is only affected when the local mode frequencies are very low and in this case the only modification is in the dependency on magnetic field. In some cases, where the site has high symmetry, the defect modes make no contribution to the direct process at all. The Raman process in a non-Kramers salt shows the following temperature dependency:

$$\tau_L^{-1} \propto \left\{\left(\frac{T}{\Theta_D}\right)^7 J_6\left(\frac{\Theta_D}{T}\right) + \varXi\left[\left(\frac{T}{\Theta_D}\right)^3 J_2\left(\frac{\Theta_D}{T}\right) - \left(\frac{T}{\Theta_\lambda}\right)^3 J_2\left(\frac{\Theta_\lambda}{T}\right) + \frac{T^{11}}{\Theta_\lambda^8 \Theta_D^3} J_{10}\left(\frac{\Theta_\lambda}{T}\right)\right]\right\}, \tag{63.4}$$

where

$$J_n = \left(\frac{\hbar}{kT}\right)^{n+1} I_n.$$

The first term is the normal Raman process, while the other terms are due to the local modes, the final term is usually smaller than the others. $\varXi$ is a constant relating the sizes of the normal Raman process and Raman processes due to the

---

[1] D. W. FELDMAN, J. G. CASTLE jr., and J. MURPHY: Phys. Rev. **138**, A 1208 (1965).
[2] E. W. MONTROLL, and R. B. POTTS: Phys. Rev. **100**, 525 (1955).

local modes. The contribution of the frequencies above the local mode frequency is enhanced by the presence of the local mode while the frequencies below this have their contributions reduced. A similar expression to (63.4) is obtained for the case of a Kramers salt except that it is multiplied by $(T/\Theta_D)^2$ and the integrals J_n are replaced by J_{n+2}, Eq. (34.4). If the local mode frequency is close to the cut-off frequency, then τ_L varies more rapidly with temperature than the normal prediction states at temperatures very much lower than the Debije temperature. This has been observed in experiments by CASTLE et al.[1] on Cr^{3+} in MgO. When the local mode frequency is very low there may be a wide range of temperature in which $\Theta_\lambda \lesssim T \lesssim \Theta_D/3$ is satisfied. Then the T^3 dependency predominates unless the normal Raman process is strong. If $T \ll \Theta_\lambda$ then the T^{11} term may predominate, but in this region also the direct process is becoming important. The T^3 dependency has been observed by KLEMENS et al.[2] in paramagnetic centres produced in quartz by x-irradiation.

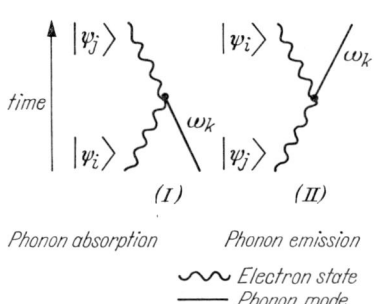

Fig. 17. A spin-phonon interaction linear in strain.

64. Diagram techniques. The techniques of quantum field theory have been applied to quantum mechanical problems.

In field theory the interest is in the interaction between a particle and a field, or between two fields, and this can be calculated by using creation and annihilation operators; the statistical mechanical problem of interactions between two particles can also be treated by creation and annihilation operators, so formally the two problems are similar.

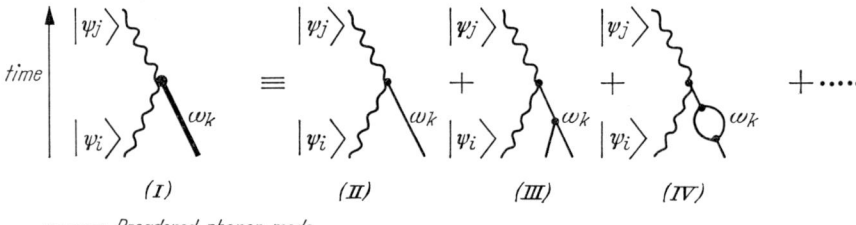

Fig. 18. A spin-phonon interaction with a broadened phonon mode participating. The interaction is linear in strain and shown expanded to cubic terms in anharmonicity.

In quantum field notation the spin-lattice interaction is given by:

$$\mathcal{H}_{SL} = \sum_{i,j} \sum_k c_{ij} S_i^* S_j b_k a_k + \sum_{i,j} \sum_k c_{ij}^* S_j^* S_i b_k a_k, \qquad (64.1)$$

where the $c_{ij} b_k$ are numbers, the a_k are the creation and annihilation operators for the phonon system, S_j^* and S_j are the creation and annihilation operators for the

---
[1] J. G. CASTLE jr., D. W. FELDMAN, and P. G. KLEMENS: Proceedings of the Second Internat. Conference on Quantum Electronics, Berkeley, p. 414. New York: Columbia University Press 1961.
[2] P. G. KLEMENS, J. G. CASTLE jr., and D. W. FELDMAN: Proceedings of the Eight Internat. Conference on Low Temperature Physics, London, p. 292. London: Butterworth & Co. 1963.

electrons in state $|\psi_j\rangle$. The sum over k is over all the phonon modes. In any calculation a definite model is required in order to evaluate $c_{ij}b_k$. The diagram technique is very useful for indicating which terms are important in the calculation. Consider a simple spin $\frac{1}{2}$ system with the states $|\psi_j\rangle$ and $|\psi_i\rangle$ $|+\frac{1}{2}\rangle$ and $|-\frac{1}{2}\rangle$ respectively separated by energy ∂. Then the direct process can be represented by Fig. 17 where $\omega_k = \partial/\hbar$. The effects of anharmonicities between the phonons can be represented by the diagrams in Figs. 18 and 19. Second order processes are shown in Fig. 20. In Fig. 17 both absorption and emission are shown but in Figs. 18 to 20 the reciprocal process is not shown. These diagrams help us to eliminate those interactions which do not conserve energy, for example in

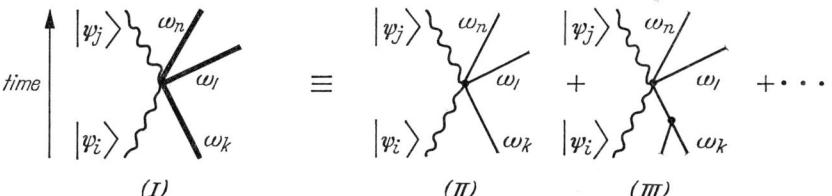

Fig. 19. A spin-phonon interaction with broadened phonon modes cubic in strain and shown expanded to the first important terms.

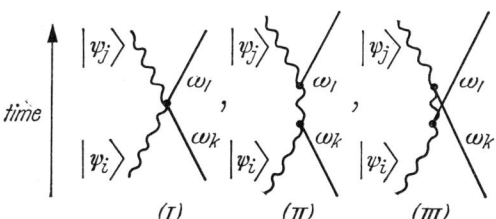

Fig. 20. Three second order processes. (i) is a spin-phonon interaction quadratic in strain, (ii) and (iii) are spin-phonon interactions linear in strain but taken twice in perturbation theory.

Figs. 18 (ii) and 19 (ii) if ω_k is a high frequency local mode [in Fig. 19 (ii) $\omega_k > 2\omega_m$]. They also indicate other possible types of interaction which ought to be considered. This theory has been used by FELDMAN et al.[1] to study the effects of deviations from the Debije spectrum of phonons, see Sect. 63.

65. Real phonon spectrums. It is possible to allow for anisotropy in the strain in the crystals by using the classical definition relating the strain to the displacement produced by the passage by the phonons, Eq. (31.4). The relation is:

$$\varepsilon_{12} = \frac{1}{2}\left(\frac{\partial q_2}{\partial x_1} + \frac{\partial q_1}{\partial x_2}\right), \quad \text{etc.,} \tag{65.1}$$

where x_1, x_2, x_3 represent x, y, z. In the evaluation of τ_L in a material with anisotropy a summation of the contributions from each direction must be made.

A few attempts have been made to calculate τ_L using a genuine frequency distribution rather than the Debije assumption. These are limited to cases where the real frequency distribution has been measured or calculated. JOSHI et al.[2] performed the calculations for relaxation of nuclei in alkali halides and obtained a slightly better agreement with theory than in the case of a Debije distribution.

---

[1] See footnote 1, p. 83.
[2] S. K. JOSHI, R. GUPTA, and T. P. DAS: Phys. Rev. **134**, A 693 (1964).

Their result gave a slightly faster temperature variation than predicted from the Debije theory which they attributed to the emphasis which the Debije distribution gives to the low frequency phonons.

66. The phonon bottleneck. Up to this point we have assumed that the energy which passes from the spins to the phonons goes immediately to the lattice and the bath but this is not always the case. In the direct and Orbach processes where only phonons of certain fixed frequencies are involved in the relaxation, we have to take account of the possibility of an inadequate heat contact between the heated phonons and the rest of the phonon system.

Consider the case of an isolated pair of levels forming the ground state of a system and relaxing by direct processes between these levels. If the phonons have a finite lifetime the relaxation may be limited by this lifetime. We are especially interested in the phonons whose energy is suitable to allow them to participate in the relaxation process. Allowing for spin transitions between the levels in both directions and labelling the higher level by i, the rate of change of population difference $n_i - n_j$ is given by:

$$\frac{d}{dt}(n_i - n_j) = \frac{-2}{\tau_L}\tanh(\partial/2kT)\{N(n_i - n_j) - n_j\}, \qquad (66.1)$$

where ∂ is the energy difference between the levels and N is the number of phonons of energy ∂. In order to write a rate equation for the phonons we must consider two points; firstly, the increase in the number of phonons in the finite frequency width $\Delta\omega$, which is in contact with the spins, and secondly, the finite lifetime of the phonons τ_{ph}, which describes the time required for the phonons to leave the range $\Delta\omega$. Apart from the interactions with spins the decay of the phonons from the interval $\Delta\omega$ is assumed to be exponential. Then we have:

$$\frac{d}{dt}(N(\partial) - N^0(\partial)) = \frac{1}{2\hbar\varrho(\partial)\Delta\omega}\frac{d}{dt}(n_i - n_j) - \frac{N(\partial) - N^0(\partial)}{\tau_{ph}}, \qquad (66.2)$$

where the superscript ^{0} refers to the equilibrium phonon occupation number, $\varrho(\partial)$ is the phonon density.

These equations are non-linear and in general can only be solved numerically. However, if the deviation of $n_i - n_j$ from equilibrium is small they can be linearised and solved in terms of two time constants T_b and T_b'. The first describes the relaxation of the spins towards the bath and the second the relaxation of the phonons. $T_b \simeq \tau_L + \tau'$ where $(\tau')^{-1}$ is proportional to $\coth^2(\partial/2kT)$, which in the high temperature limit goes to T^2. The proportionality constant contains as a parameter the spin concentration to the first power. τ_L is the spin-lattice relaxation time [18]. T_b' is proportional to T^{-1}. Usually T_b' is much shorter than T_b, so a short time after the spin temperature is altered by some external means the phonons reach the same temperature as the spins and the two systems, spins and phonons, then relax together towards equilibrium.

The solution of Eqs. (66.1) and (66.2) is more easily performed by using normalised values for the population differences. Then we have:

$$\dot{x} = \frac{1}{\tau_L}(1 - x - xy) \quad \text{and} \quad \dot{y} = b\dot{x} - \frac{y}{\tau_{ph}}, \qquad (66.3)$$

where

$$x = \frac{n_i - n_j}{n_i^0 - n_j^0}, \quad y = \frac{N(\partial) - N^0(\partial)}{N^0(\partial) + \frac{1}{2}} \quad \text{and} \quad b = \frac{n_i^0 - n_j^0}{\{N^0(\partial) + \frac{1}{2}\}2\varrho(\partial)\Delta\omega}.$$

This allows us to rewrite the equations in terms of a saturation parameter s and the bottleneck parameter σ. $s = 1 - x$ and $\sigma = \dfrac{b\tau_{ph}}{\tau_L} = \dfrac{E_s}{E_{ph}} \times \dfrac{\tau_{ph}}{\tau_L}$. The bottleneck parameter describes the increased impedance to the flow of energy from the spins to the lattice brought about by the finite lifetime of the phonons. E_s is the energy of the spin system and E_{ph} the energy of the phonon system so we see that σ is the maximum flow of energy from the spins to the phonons divided by the maximum flow of energy from the phonons to the bath. Using the definitions for s and σ and using τ_L as the unit of time Eqs. (66.3) can be written as:

$$\dot{s} = -s + \sigma z(1-s) \quad \text{and} \quad \dot{z} = \frac{\tau_L}{\tau_{ph}}[(s-z) - \sigma z(1-s)], \qquad (66.4)$$

where $z = y/\sigma$. The first equation has an exponential solution as long as the bottleneck parameter is negligible. When it is much smaller than one, but not completely negligible, Eqs. (66.4) have as approximate solutions for s and z:

$$s = \exp[-t/(1+\sigma)] \exp\left[(1-s)\left(\frac{\sigma}{1+\sigma}\right)\right], \quad z = s/[1 + \sigma(1-s)]. \qquad (66.5)$$

It is obvious that the presence of a bottleneck leads to deviations from a simple exponential recovery for the spins after they have been disturbed. In the limit of a very large bottleneck, $\sigma \gg 1$, Eqs. (66.4) have other solutions demonstrating the exponential decay of the spins and heated phonons together.

Under steady state conditions when the spin temperature is maintained at a constant value by some external means, it can be seen from Eq. (66.5) that the maximum value of z which is possible is one. Therefore, the maximum value of y, the normalised population difference for the phonons is σ. Using this maximum value, substituting into the expression for y the Bose-Einstein factors for the phonon occupation numbers $N(\partial)$ and $N^0(\partial)$, and expanding the exponentials for the case of $\partial \ll kT$ we obtain:

$$\sigma = \left(\frac{T_{ph} - T_{bath}}{T_{bath}}\right)_{max} \qquad (66.6)$$

as the relation between the bottleneck parameter, the temperature of the bath and the temperature of the excited phonons[1]. This relation must be used with some care in the low temperature region where the phonon bottleneck becomes important because in these conditions it may not be possible to expand the exponentials in the occupation numbers. From (66.6) we see that the phonon temperature does not go to infinity unless σ goes to infinity. If the spin temperature goes negative there can be no bottleneck because the production of new phonons by spins relaxing induces more spins to give out their energy. This has been demonstrated by BRYA and WAGNER[2].

There are many possible mechanisms which limit the life-time of the phonons but calculation and experiment (VAN VLECK[3], SCOTT and JEFFRIES [18], NASH[4] and MIEDEMA and MESS[5]) seem to indicate that at the temperatures at which the bottleneck occurs, the limitation on the lifetime is the inelastic scattering from the crystal surfaces or other major imperfections. Any anharmonicity such as phonon-phonon interactions, phonon-point defects and phonon-line defects can limit the lifetime but they do not appear to be of importance here. MILLS[6] has demonstrated how, in crystals containing water of hydration, the penetration of helium into the pores in the surface of the crystal, where some of the waters of hydration have been lost, can lead to absorption of phonons by the surface layers of the crystal.

---

[1] B. W. FAUGHAN, and M. W. P. STRANDBERG: J. Phys. Chem. Solids **19**, 155 (1961).
[2] W. J. BRYA, and P. E. WAGNER: Phys. Rev. Letters **14**, 431 (1965).
[3] J. H. VAN VLECK: Phys. Rev. **59**, 724 (1940).
[4] F. R. NASH: Phys. Rev. Letters **7**, 59 (1961).
[5] See footnote 3, p. 46.
[6] D. L. MILLS: Phys. Rev. **133**, A 876 (1964).

Table 8. *Some theoretical predictions for τ_L as a function of T in the presence of phonon relaxations.*

Entry	Energy levels	Phonon spectrum	Expression for τ_L^{-1}	Simplified Expression for τ_L^{-1}	Conditions	Equation no. in text	Reference	Comments
1	∧ s ∂ ∨ s	Debije + mode ω_λ, $\omega_\lambda > 1.1\,\omega_m$		$\text{const} \times \dfrac{1}{\partial^2} \exp\left(-\dfrac{\Theta_\lambda}{T}\right)$	$T \ll \dfrac{\Theta_\lambda}{2}$	(63.2)	a	Raman process. If $T \gtrsim \Theta_\lambda$ then $\tau_L^{-1} \propto T^2$ as in the "normal" Raman processes.
2a	——— s ——— s	Debije + mode ω_λ, $\omega_\lambda < \omega_m$	$\text{const}\left[\left(\dfrac{T}{\Theta_D}\right)^7 J_6\left(\dfrac{\Theta_D}{T}\right) + \right.$ $+ \text{const}\left\{\left(\dfrac{T}{\Theta_D}\right)^3 J_2\left(\dfrac{\Theta_D}{T}\right) - \right.$ $\left. - \left(\dfrac{T}{\Theta_\lambda}\right)^3 J_2\left(\dfrac{\Theta_\lambda}{T}\right)\right\}$ $\left. + \dfrac{T^{11}}{\Theta_\lambda^8 \Theta_D^3} J_{10}\left(\dfrac{\Theta_\lambda}{T}\right)\right]$	$\text{const}\left\{\left(\dfrac{T}{\Theta_D}\right)^3 J_2\left(\dfrac{\Theta_D}{T}\right) - \right.$ $\left. - \left(\dfrac{T}{\Theta_\lambda}\right)^3 J_2\left(\dfrac{\Theta_\lambda}{T}\right)\right\}$	$\Theta_\lambda \lesssim T \lesssim \dfrac{\Theta_D}{3}$	(63.4)	b	Raman processes. Entry 2a, the T^3 terms dominate only if the T^7 process is weak. Entry 2b, the T^{11} term dominates only if the direct process is weak. If ω_λ is very low then the field dependency of the direct process is altered. Entry 3 can be simplified in analogy with entry 2.
2b	——— s ——— s			$\text{const} \times \dfrac{T^{11}}{\Theta_\lambda^8 \Theta_D^3} J_{10}\left(\dfrac{\Theta_\lambda}{T}\right)$	$T \ll \Theta_\lambda$			
3	K.d. ⟨ s s ⟩ s K.d. ⟨ s s ⟩ s	Debije + mode ω_λ, $\omega_\lambda < \omega_m$	$\text{const}\left[\left(\dfrac{T}{\Theta_D}\right)^9 J_8\left(\dfrac{\Theta_D}{T}\right) + \right.$ $+ \text{const}\left\{\left(\dfrac{T}{\Theta_D}\right)^5 J_4\left(\dfrac{\Theta_D}{T}\right) - \right.$ $\left. - \left(\dfrac{T}{\Theta_\lambda}\right)^5 J_4\left(\dfrac{\Theta_\lambda}{T}\right)\right\}$ $\left. + \dfrac{T^{13}}{\Theta_\lambda^8 \Theta_D^5} J_{12}\left(\dfrac{\Theta_\lambda}{T}\right)\right]$				b	
4	∧ s ∂ ∨ s	Debije + excited mode at $\dfrac{\partial}{\hbar}$	Expression for T_b $T_b = \tau_L + \tau'$ $\tau_L^{-1} = \text{const.} \times T$ $(\tau')^{-1} = \text{const.} \times$ $\times \dfrac{\partial^2}{\tau_{ph}} \coth^2\left(\dfrac{\partial}{2kT}\right)$	Simplified expression for τ' $\text{const.} \times T^2$	$\tau_L \ll \tau'$ $\partial \ll kT$	Solution of the Eqs. (66.1) and (66.2)	c	A bottleneck in the direct process. The totalspin system energy is considered to be much larger than the phonon system energy.

The phonon bottleneck.

	Debije + excited modes at $\frac{\Delta}{\hbar}$ and $\frac{\Delta+\partial}{\hbar}$	$T_b = \tau_L + \tau'$	$\tau_L^{-1} = \text{const.} \times \Delta^3 \times$ $\times \exp\left(-\frac{\Delta}{kT}\right)$ $(\tau')^{-1} = \text{const.} \times \frac{\Delta^2}{\tau_{ph}} \times$ $\times \exp\left(-\frac{\Delta}{kT}\right)$	$\tau_L \ll \tau'$ $\partial \ll \Delta$ $kT \ll \Delta$	A bottleneck in the Orbach process.
5	Δ $\stackrel{s}{\gtrless} \partial$ $\stackrel{s}{\lessgtr}$				c

T_b is the spin-bath relaxation time.
τ_{ph} is the phonon lifetime.
The other symbols have their normal meanings.

$\Theta_\lambda = \hbar\omega_\lambda/k$.
$J_n = (\hbar/kT)^{n+1} I_n$, where I_n is given by Eq. (34.4).
s in the second column indicates a non-degenerate state.
K.d. in the second column indicates a Kramers doublet.

References: [a] P. G. KLEMENS: Phys. Rev. **125**, 1795 (1962).
[b] J. G. CASTLE, D. W. FELDMAN, P. G. KLEMENS, and R. A. WEEKS: Phys. Rev. **130**, 577 (1963).
[c] P. L. SCOTT, and C. D. JEFFRIES: Phys. Rev. **127**, 32 (1962).

This occurs because the liquid retains its viscosity even below the normal λ-point since the λ-point is depressed when the liquid is in narrow channels. Presumably a mechanism of this sort can occur in surface defects in all types of crystals.

VAN DEN BROEK et al.[1] observed deviations from a pure Casimir-Du Pré relaxation form above the λ-point in a $CuK_2Cl_4 \cdot 2H_2O$ single crystal. This they attributed to the poor contact between the phonons and the bath above the λ-point while below the λ-point penetration of liquid helium into cracks improves this contact. The cracks involved are macroscopic when compared to the pores of atomic dimensions considered by MILLS so the two explanations are not inconsistent.

The magnitude of $\Delta\omega$ in Eq. (66.2) is difficult to establish. It has been usual to assume that the electron spin resonance linewidth was a suitable estimate of the order of magnitude of this parameter. However, this does not give the full range of phonons available to take part in the relaxation. BÖLGER[2] performed calculations allowing for the interaction between the spins in the direct process. He showed that there is a possibility of other phonons taking part in the relaxation because of the occurrence of double spin flips. So, in some cases the expected bottleneck may not occur because of the larger number of phonons available.

The Orbach process also uses phonons of definite frequencies and may be bottlenecked, but this is rarely observed (see entry 1 Table 2) because the phonons involved are usually higher energy phonons which are better able to interact with phonons of neighbouring frequencies due to anharmonicities and so lose their energy. Also, by the nature of the limitations on the energy level splittings and the temperatures required for the Orbach process to occur, it occurs at higher temperatures than the direct processes, there is more rapid dissipation of the phonon energy throughout the phonon spectrum.

---
[1] See footnote 2, p. 77.
[2] See footnote 4, p. 46.

The phonon bottleneck can be by-passed by any mechanism which shortens the lifetime of the phonons in direct contact with the spins. At the moment it appears that the surface of a sample and its size, i.e. the distance from any one spin to the surface, play the most important role in determining the phonon lifetime at the very low temperatures where the direct process occurs. However, other mechanisms may put the spins in touch with more phonons. These mechanisms include the dipolar interactions which bring in many more phonons and diffusion or cross-relaxation to centres whose energy levels are more advantageous-

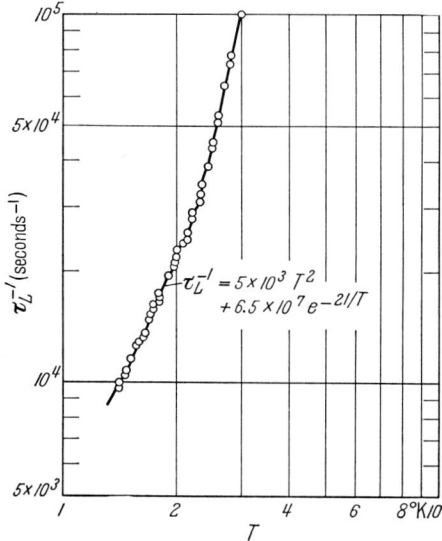

Fig. 21. Experimental results on 5% Pr^{3+} in Lanthanum ethyl sulphate which demonstrate a phonon bottleneck. $H||z$-axis, $H = 2394$ oersted, $\nu = 9.27$ GHz. Taken from G. H. LARSON, and C. D. JEFFRIES: Phys. Rev. **141**, 461 (1966).

ly situated for rapid relaxation. The centres may include clusters of one sort of ion or they may be other ions which are present as impurities. Fig. 21 shows a typical bottleneck process. Table 8 contains a list of relaxation mechanisms influenced by deviations from the Debije spectrum.

67. Relaxation without phonons. In liquids and gases where the phonon theory is inapplicable the random motions of the ions have to be taken into account in order to calculate the relaxation time. This is done by describing the parameters in the Hamiltonian of the dynamic perturbation by means of their probability distributions. The Fourier spectrum of the time dependent contributions to the Hamiltonian may contain the frequency which corresponds to a transition between the spin energy levels of the system. It is fortunate that the results of calculations based on the Fourier spectrum do not depend on the detailed shape of the spectrum as this is often not known. If the motions of the ions are very rapid then the Fourier spectrum will have a wide range. This rapid motion is characterised by a short correlation time which is the time required for the system to pass from one configuration with a certain probability to another configuration which has another certain probability.

Normal time-dependent perturbation theory gives the following expression for the transition probability per unit time between states $|\alpha\rangle$ and $|\beta\rangle$ under the influence of a perturbation $\mathcal{H}(t)$ which is a random function of time, ABRAGAM [22]:

$$W_{\alpha\beta} = \int_0^t \langle\beta|\mathcal{H}(t_1)|\alpha\rangle \langle\alpha|\mathcal{H}(t_2)|\beta\rangle \exp\{i\omega(t_2-t_1)\} dt_2 +$$
$$+ \int_0^t \langle\beta|\mathcal{H}(t_1)|\alpha\rangle \langle\alpha|\mathcal{H}(t_2)|\beta\rangle \exp\{-i\omega(t_2-t_1)\} dt_2. \quad (67.1)$$

As the perturbation is random the observed value of the transition probability is an average over an ensemble, so:

$$\overline{W}_{\alpha\beta} = \int_0^t \overline{\langle\alpha|\mathcal{H}(t_2)|\beta\rangle \langle\beta|\mathcal{H}(t_1)|\alpha\rangle} \exp\{i\omega(t_2-t_1)\} dt_2 +$$
$$+ \int_0^t \overline{\langle\alpha|\mathcal{H}(t_2)|\beta\rangle \langle\beta|\mathcal{H}(t_1)|\alpha\rangle} \exp\{-i\omega(t_2-t_1)\} dt_2, \quad (67.2)$$

where the average in the integral is the autocorrelation function of $\langle\alpha|\mathcal{H}(t)|\beta\rangle$, $G_{\alpha\beta}$, and if this is constant the integral depends only on $t_2-t_1=\tau$. Then Eq. (67.2) becomes:

$$\overline{W}_{\alpha\beta} = \int_0^t G_{\alpha\beta}(\tau) \exp(-i\omega\tau) d\tau + \int_0^t G_{\alpha\beta}(\tau) \exp(i\omega\tau) d\tau$$
$$= \int_{-\infty}^{+\infty} G_{\alpha\beta}(\tau) \exp(-i\omega\tau) d\tau, \quad (67.3)$$

where the extension of the limits of the integration to infinity is justified as we are only interested in times much longer than the correlation time $\tau_{\alpha\beta}$. Therefore, the transition probability can be written as:

$$\overline{W}_{\alpha\beta} = F_{\alpha\beta}(\omega), \quad (67.4)$$

where $F_{\alpha\beta}(\omega)$ is the Fourier transform of the correlation function. If the perturbation is separable as shown in Eq. (26.4), the part depending on the spin variables may usually be regarded as constant and only the correlation function of the other part need be calculated. However, transition probabilities for the combined cross-relaxation-spin-lattice relaxation mechanism discussed in Sect. 57 can be calculated using Eq. (67.4) when the correlation function for the spin variables is known. This has been applied to both electron and nuclear spin systems.

The theory of this section has been extended to solids by CULVAHOUSE et al.[1] amongst others.

68. The spin-phonon coupling in other work. The dynamic Hamiltonian describing the spin-lattice interaction also occurs in other work. A direct comparison may be made with acoustic paramagnetic resonance, A.P.R., where phonons are introduced into a paramagnetic system and the electron transitions produced are observed. Alternatively, but with less sensitivity, the absorption coefficient for the phonons is measured. Theoretically this technique has many advantages over the more normal relaxation methods because these latter require for their solution an average over the phases, directions and frequencies of the phonons (which as we have seen can usually only be done by putting in a Debije spectrum) while the A.P.R. work uses only phonons of one frequency, phase and polarisation, and these are induced in the crystal in one direction. This simplifies the theory considerably.

---
[1] J. W. CULVAHOUSE, W. P. UNRUH, and D. K. BRICE: Phys. Rev. **129**, 2430 (1963).

Experimentally joint A.P.R.-E.S.R. techniques are the most sensitive[1] but there are difficulties in introducing very high frequency phonons in large numbers so only direct processes are observed. This again simplifies the theory. For a general introduction to this theory see TUCKER [23].

A similar theory may be used for describing static deformation experiments. Here the influence of the lattice deformations on the spin levels of the ions is static in time so once more it is possible to leave out the complex averaging over the phonons. Indeed, this method has been called the best way of examining the spin-lattice interaction by LEUSHIN[2]. Here too, only the direct interaction is readily observable because the equivalent of the Raman process in a static strain experiment is a non-linear displacement of the E.S.R. line.

The orbit-lattice Hamiltonian (part of the spin-lattice interaction described in Sect. 38) can also mix excited levels into the ground state of a system and in this way influence all those properties of a system which depend on the exact form of the state functions. These properties include the static susceptibility, g-factors, the specific heat anomaly, etc.[3]

Acknowledgements. We express our thanks to Professor C. J. GORTER for reading the manuscript so critically and to R. H. TERWIEL, A. J. VAN DUYNEVELDT, K. VAN DER MOLEN and P. W. VERBEEK for their interest shown in many discussions about the manuscript.

General references.

[1] WALLER, I.: Z. Physik **79**, 370 (1932). The first theoretical paper on relaxation phenomena in paramagnetic materials which clearly indicated the division between spin-spin and spin-lattice phenomena and the existence of direct and Raman regions in the spin-lattice relaxation.

[2] GORTER, C. J.: Paramagnetic Relaxation. Amsterdam: Elsevier Publ. Co. 1947. A detailed account of theoretical and experimental results as known by the author at the end of the second world war.

[3] VAN VLECK, J. H.: Electric and Magnetic Susceptibilities. Oxford: Clarendon Press 1939.

[4] KUBO, R., and K. TOMITA: J. Phys. Soc. Japan **9**, 888 (1954). On the theory of paramagnetic resonance. The relaxation function as derived by the authors forms the starting point of several modern theories on spin-spin relaxation.

[5] CASPERS, W. J.: Theory of Spin Relaxation. New York: Interscience Publ. 1964. Much disputed monograph on clearly stated fundamental problems connected with paramagnetic relaxation.

[6] PAULI, W.: Handbuch der Physik, Bd. 5/1, S. 1. Berlin-Göttingen-Heidelberg: Springer 1958. An article on the technique of quantum mechanics including time-dependent perturbation theory, p. 75.

[7] BLEANEY, B., and K. W. H. STEVENS: Repts. Progr. Phys. **16**, 108 (1953). A review article on crystal field theory.

[8] BOWERS, K. D., and J. OWEN: Repts. Progr. Phys. **18**, 304 (1955). A supplementary article to reference [7] containing also a list of experimental data on crystal field parameters.

[9] GORDY, W.: Handbuch der Physik, Bd. 28, S. 1. Berlin-Göttingen-Heidelberg: Springer 1957. A description of microwave spectroscopy techniques accompanied by theoretical considerations.

[10] PRATHER, J. L.: Atomic Energy Levels in Crystals, National Bureau of Standards Monograph No. 19, Washington D.C. 1961. A description of the theoretical methods required to discuss atomic energy levels in crystals.

[11] GRIFFITH, J. S.: The Theory of Transition-Metal Ions. Cambridge: Cambridge University Press 1961. A comprehensive account of the theory of transition-metal ions.

[12] HUTCHINGS, M. T.: Solid State Phys. **16**, 227 (1964). An account of crystal field theory beginning from a more elementary level than the other references.

[13] VAN VLECK, J. H.: Phys. Rev. **57**, 426 (1940). The first detailed calculation of the spin-lattice relaxation time of actual materials.

---

[1] N. S. SHIREN, and E. B. TUCKER: Phys. Rev. Letters **6**, 105 (1961).
[2] See footnote 7, p. 64.
[3] M. INOUE: Phys. Rev. Letters **11**, 196 (1963).

[14] ZIMAN, J. M.: Electrons and Phonons. London: Oxford University Press 1960. A detailed account of the behaviour of electrons and phonons in solids.
[15] KLEMENS, P. G.: Handbuch der Physik, Bd. 14/1, S. 198. Berlin-Göttingen-Heidelberg: Springer 1956. An article on low temperature thermal conductivity including a description of the behaviour of phonons at these temperatures.
[16] ORBACH, R.: Proc. Roy. Soc. (London) A **264**, 458 (1961). Detailed calculations of the spin-lattice relaxation in rare earth ethyl sulphates with a discussion of the relaxation mechanism since called the Orbach process.
[17] HEITLER, W.: The Quantum Theory of Radiation. London: Oxford University Press 1954. A comprehensive account of quantum processes in radiation phenomena including on page 196 a description of the fluorescence process.
[18] SCOTT, P. L., and C. D. JEFFRIES: Phys. Rev. **127**, 32 (1962). This article contains a good summary of the relaxation processes possible in rare earth salts and gives a detailed discussion of phonon bottleneck processes.
[19] MATTUCK, R. D., and M. W. P. STRANDBERG: Phys. Rev. **119**, 1204 (1960). A derivation of generalised spin-lattice interaction operators for ions of the 3d group.
[20] AL'TSHULER, S. A., and B. M. KOZYREV: Electron Paramagnetic Resonance. English translation published by the Academic Press 1964. A detailed account of theory and practice in paramagnetic resonance including lists of experimental results and discussions of phenomena in colour centres, metals, liquids, irradiated organic materials and semiconductors.
[21] HEBEL, L. C.: Solid State Phys. **15**, 409 (1963). A discussion of spin-temperatures and relaxation phenomena in nuclear spin systems.
[22] ABRAGAM, A.: The Principles of Nuclear Magnetism. London: Oxford University Press 1961. A book covering the whole field of nuclear magnetism, mainly theory but it does include an account of experimental methods.
[23] TUCKER, E. B.: Proc. IEEE **53**, 1547 (1965). An account of the spin-phonon interaction in ions of the 3d group intended to describe acoustic paramagnetic resonance phenomena.

Electron Spin Resonance.

By

D. J. E. INGRAM.

With 29 Figures.

I. General principles.

The basic theory and technique of electron spin resonance is very similar to that of nuclear magnetic resonance (see Vol. 38, Part 1); the essential difference is that in the case of electron spin resonance the applied external magnetic fields are acting on the magnetic moment associated with the electron spin, and the incoming electromagnetic radiation induces transitions between energy states of this electron, whereas in the case of nuclear resonance the interaction is via the nuclear magnetic moment. Both techniques depend on the fact that an externally applied field can produce quantised energy states, and well defined spectral lines can thus be obtained from transitions between these levels. The basic theory of nuclear resonance was first developed via a classical approach in which BLOCH[1] considered the precessing motion of the magnetic moment associated with the nuclei in the applied magnetic field and determined under what conditions resonance absorption of electromagnetic radiation would occur. In this way he was able to produce a complete phenomenological account of the process, including the relaxation processes which allow the absorbed energy to be returned to the body as a whole. It is probably easier, however, to obtain an initial picture of the essential principles of the two techniques if a quantum approach is adopted first, and the general principles will, therefore, initially be considered from this point of view, and the more detailed theory then considered later. Since this article itself is concerned purely with electron spin resonance, the following description of the energy states and the transitions between them will be described in terms of the interaction between the magnetic field and the electron magnetic moment, but it may be borne in mind that exactly the same theory will apply in principle to the corresponding nuclear case.

For a substance to exhibit electron resonance, it must possess unpaired electrons, since otherwise the paired spins will cancel and there will be no net magnetic moment to interact with the applied magnetic field. It might appear at first sight, that this is a very great limitation on the technique since the vast majority of substances have paired electrons, either in the outer orbitals of closed shells, which are formed by ionic bonds, or in the shared covalent bonds existing between atoms. In fact, however, quite a large number of different types of compound do possess unpaired electrons, and these include not only the transition group complexes in which the unpaired electrons are inherent in the lower electronic shells, but also a large range of free radical type complexes in which the unpaired electrons move in a molecular orbital over a large fraction of the molecule. As

---

[1] F. BLOCH, W. W. HANSEN, and M. PACKARD: Phys. Rev. **69**, 127 (1946).

well as these two main groups of substances containing unpaired electrons there are also a large variety of others, such as those produced by irradiation damage; the unpaired electrons associated with conduction in metals and in semiconductors; the impurity atoms in semiconductor material; and a large variety of compounds of biochemical interest such as enzymes and other catalysts. It will be seen that the very property which makes these compounds of particular interest is often linked with the fact that they possess unpaired electrons, and hence one often finds that electron spin resonance can, in fact, be applied to just those compounds which have the particular chemical or physical properties of interest.

To a very rough approximation all such compounds can be divided into two main groups initially. First, those in which the unpaired electron is closely bound to one atom, as in the transition group complexes, and secondly, those in which the unpaired electron is fairly mobile and moves in a highly delocalised molecular orbital, as in the case of most free radicals. The great difference between these two, so far as electron resonance itself is concerned, is that in the first case the unpaired electron will have quite a large contribution from the orbital angular momentum, whereas in the latter case the angular momentum and hence the magnetic moment of the electron is determined entirely by its spin. The latter case is, therefore, the easiest to consider initially, although as will be seen later, the additional interaction with the orbital momentum, which changes the resonance condition in the case of the transition group complexes, can also be used to probe the physical and chemical environment of these atoms very effectively.

The case of the unpaired electrons present in the free radical specimen may, therefore, first be considered, and when such a substance is placed on the laboratory bench in the absence of any applied magnetic field, the orientation of these spins and their associated magnetic moments can take any position, and all the unpaired electrons will have the same energy whatever the direction of their spins and magnetic moments. If, however, an external field is now applied across this specimen the electron spins will have to align themselves so that their resolved components are either parallel or antiparallel to the applied field. Since the spin quantum number of the electron is one half, it follows that no other intermediate orientation is possible between these two extremes which differ by a quantum number of one. The electrons are, therefore, divided into two groups and those with their spins aligned parallel to the magnetic field will have an energy less than those lined up antiparallel, as is represented by the diverging dotted lines on the diagram in Fig. 1. The actual energy of a bar magnet, μ, in a magnetic field, H, is given by $-\mu H$. The spins lined up parallel to the field will, therefore, have their energy lowered by $\frac{1}{2}g\beta H$; and those antiparallel will be raised by $\frac{1}{2}g\beta H$. The difference in energy between these two groups of states of electrons will thus be determined by the value of the external magnetic field, and will be equal to $g\beta H$ where the constant β is the Bohr magneton and converts the units of spin angular momentum into magnetic moment. The small g parameter measures the amount of orbital motion of the electron which is admixed with its spin state, and for a free radical would be very close to the spin-only value of 2.0.

If electromagnetic radiation of frequency ν, such that $h\nu$ is equal to $g\beta H$ is now fed to this specimen, whilst it is in the strong external magnetic field, it is possible for the electrons to absorb quanta of this value, $h\nu$, and jump from the bottom to the top state, or in other words, to reverse the direction of their spins. We therefore have a resonance condition which can be written

$$h\nu = g\beta H. \tag{1}$$

This resonance effect can be detected by a decrease in the microwave power which is passing through the apparatus. If the quantitative values are inserted into this equation, and g is taken as 2.0, one obtains the relation

$$\nu = 2.8 \times 10^6\, H \text{ c/s}. \tag{2}$$

It can be seen from this that a magnetic field of about 3,300 gauss will require radiation of about 10,000 Mc/s or 3 cm wavelength for resonance to occur.

It is, of course, possible in principle to satisfy this resonance condition at any value of magnetic field and frequency provided the relation itself is obeyed, and it has often proved possible to detect electron resonance absorption at field of the

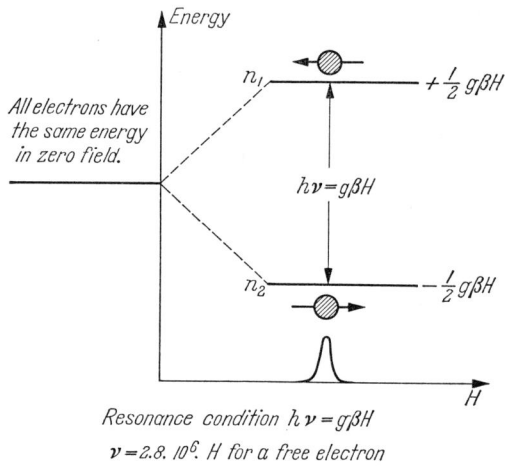

Fig. 1. Basic electron resonance condition.

order of only 10 gauss with a corresponding frequency of about 28 Mc/s. There is one very important consideration, however, which very much favours the use of as high a magnetic field strength and electromagnetic frequency as possible. This is concerned with the sensitivity of the spectrometer. The incoming electromagnetic radiation will not only cause the electrons to jump from the bottom level to the top, but will also stimulate those in the top level to fall to the bottom. The coefficients of absorption and stimulated emission are, in fact, equal and the total power absorbed or emitted therefore depends entirely on the number of electrons in these two states. The two energy levels will not contain an exactly equal number of electrons, there will be slightly fewer in the upper level due to the Boltzmann factor, i.e.

$$N_1/N_2 = \exp(-h\nu/kT). \tag{3}$$

It is only this difference in population which results in any absorption at all. It follows that there will be a greater net absorption the larger the difference in population between these two energy levels and hence the larger the value of the magnetic field which produces this splitting. It is for this reason that most electron resonance spectrometers work with magnetic fields of the order of 3,000 gauss and with microwave frequencies of the order of 9,000 Mc/s. These particular values are chosen because the 9,000 Mc/s region, known as the X band region, is used extensively for marine radar. Another spot microwave frequency, which is also often employed in electron resonance spectrometers, is the Q band wavelength region of airport control radar corresponding to wavelengths of 8 mm or frequencies of 36,000 Mc/s, and thus to magnetic fields of about 13,000 gauss.

It will be evident, therefore, that the two main requirements from the experimental side for any electron resonance experiment will be on the one hand, a large uniform magnetic field, and on the other hand, a stabilised microwave frequency source in the right wavelength region. The general way in which these are related in an actual spectrometer can then be represented schematically as in

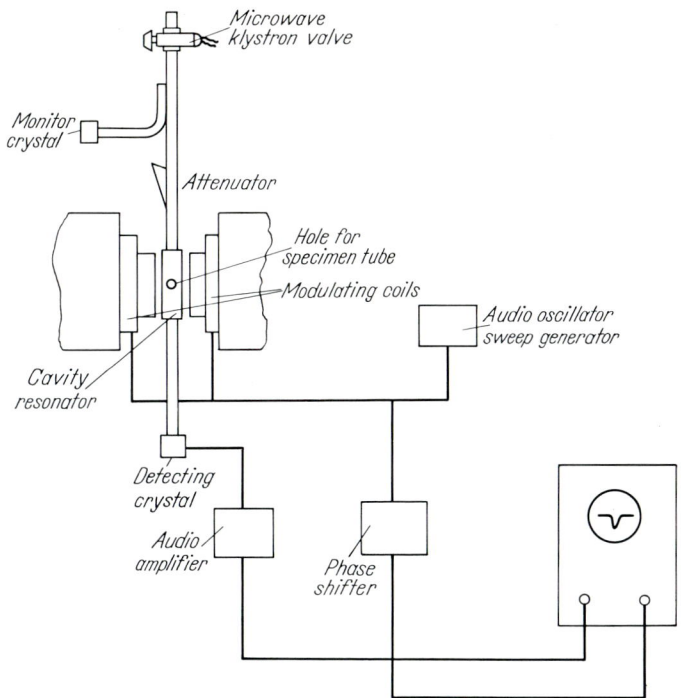

Fig. 2. Block diagram of basic electron resonance spectrometer.

Fig. 2. The basic items of any electron spin resonance spectrometer are, in fact, first the microwave radiation source; secondly the absorption cell, in which the specimen is inserted; thirdly the large uniform magnetic field into which the absorption cell containing the specimen is itself inserted; and then finally, some means of detecting the level of the microwave radiation in the absorption cell cavity and determining whether any absorption has, in fact, taken place. These items are, of course, the essential features of any spectrometer working at any wavelength, although it is noticeable that in this particular case the dispersive element, corresponding to a prism or grating can, in fact, be dispensed with, because the microwave valve, which produces the electromagnetic radiation, is so precisely built and designed that extremely monochromatic radiation can be obtained from it.

This single frequency is then fed down a waveguide to the absorption cell which takes the form of a resonant cavity, as shown in Fig. 2. The cavity resonator, which is adjusted to be of such a length that it is on tune for the incoming microwaves, serves to concentrate the radiation in its standing wave pattern and the sample is placed in it, in a position of maximum microwave magnetic field. The cavity itself is held between the pole faces of a strong electromagnet and provision is often made to cool the cavity and specimen by immersing them in a dewar containing liquid nitrogen, hydrogen or helium. The level of power in the cavity can then be measured by tapping a certain fraction of this out along the output waveguide to a crystal detector where the microwave radiation is converted into a d.c. signal. If a small a.c. magnetic field modulation is also applied at the same time, the field value can be swept through resonance twice in each cycle, and hence the output signal from the detecting crystal will be modulated at twice the frequency of the field modulation. In this way a.c. techniques can be used to amplify and display the absorption line on the screen of an oscilloscope as indicated in the block diagram. A simple spectrometer of this form is often referred to as a "crystal-video spectrometer" in that it employs simple a.c. modulation of the magnetic field and displays the absorption directly on an oscilloscope screen. It contains all the basic essentials of an electron resonance spectrometer but has a rather low sensitivity, due to the very high additional noise that is present in silicon crystal rectifiers at the low frequencies used for the modulation.

II. Parameters associated with electron resonance spectra.

The simplest form of spectrum that can be observed in electron resonance studies is, of course, a single symmetrical absorption line. Even this single line has three definite parameters associated with it, however, all of which can often give information of considerable interest.

1. Integrated intensity. In the first place there is the intensity of the line, or the integrated area under the absorption curve. It can be shown that under certain specified conditions which avoid saturation, as will be explained later, the integrated intensity of the absorption line is directly proportional to the number of unpaired electrons existing in the sample. This value may, in fact, already be known or may not be of particular interest in certain applications, but on the other hand, in other fields of study this particular information may be highly relevant especially in investigations of irradiation damage or free radical intermediates. In such cases as these, electron resonance can be used to monitor transient intermediates, or to count the amount of damage that is being done by incoming radiation as the damage actually takes place. In principle, it is possible to calibrate the electronic amplifier gains of the spectrometer right through, and make an absolute measurement each time of the absorbed microwave power and hence of the actual number of unpaired electrons present in the specimen. In practice, however, it is far more convenient to calibrate the signal strength by standards containing known numbers of unpaired electrons and make the estimation by interpolation.

2. Line width. The second parameter associated with this single absorption line is its width. Thus, a line with a given integrated area could be either sharp and narrow, or broad and shallow, and it will be evident that this second parameter reflects the spread in energy across the levels of the unpaired electron itself. In other words, the measurement of the width of the absorption line will give a measurement of the interactions between the unpaired electron and its surroundings, and this particular topic will be developed in more detail in later sections. As a general statement it may be commented that such interactions can broadly be divided into two groups, those in which the interaction is directly with the lattice or rest of the molecule or specimen as a whole, termed "spin lattice interaction", and secondly, interactions between the spins themselves, known as "spin-spin" interaction. Both of these interactions can broaden the absorption line, but it will be seen that they can be differentiated in a variety of ways. Detailed measurements on the width of the absorption lines can thus give information not only on the coupling of the electrons to the lattice as a whole, but also on interactions between unpaired electrons on various atoms, and this can be related back to the actual structure or bond nature of the specimens concerned.

3. Resonance condition and g-value. The third parameter which can be associated with this single absorption line, is the actual resonance position at which it occurs. Since the applied microwave frequency is normally held constant in all these studies, the only variable on the experimental side, is the value of the applied magnetic field, H. Reference to the basic resonance condition of Eq. (1) will show that since all the other parameters, such as β and h are constants the only other variable in the expression is the value of g. If the unpaired electron were entirely free, and did not interact at all with the orbital momentum of any atom with which it was associated, the value of g would, in fact, be the free spin value of 2.0023 (the additional 0.0023 coming from the interaction of the electron with its own radiation field). In a large number of free radical studies the g-value is found to be exceedingly close to the free spin value, indicating that the electron is moving in a highly delocalised orbital. However, in transition group complex studies, and others in which there is fairly strong bonding between the one electron and a single atom, this g-value can shift very considerably from the free spin value, and this shift reflects the strength and nature of the chemical bonding in which the electron is taking part. It therefore follows that a determination of this g-value, and in particular a determination of its angular variation, can often give very precise information on the nature of the chemical bond around the atom in question, and also on the details of the higher energy levels of the particular atomic or molecular configuration.

4. Hyperfine splitting. As well as these three parameters associated with a single absorption line there are two additional effects which will cause the single line to be split, and these splittings can in turn give very interesting additional information. In fact it is probably fair to say that the existence of these splittings of the electron resonance spectra, has opened up a very wide variety of applications for the technique which would not otherwise have existed. These splittings arise from internal magnetic fields which exist within the specimen and act on the unpaired electron in addition to the applied external magnetic field. Hence, if any internal field has the possibility of more than one steady value it is possible for the actual resonance position to be shifted from one value to another as the internal field adds or subtracts from the applied field. One of the most obvious sources for such an internal magnetic field is the nucleus of the atom itself. A large number of nuclei do in fact possess nuclear spins and magnetic moments, and although these magnetic moments are very small, being about 2,000 times smaller than that of the electron, the field which they can produce at the site of their electrons is nevertheless quite large, owing to the small distances involved. In fact a rough classical calculation will indicate that the field produced by one nuclear magneton at the position of a typical electron orbiting that atom is of the order of 100 gauss, and a splitting of the spectra of this magnitude can be observed extremely easily.

It will probably help if a definite example is taken of such an electron-nuclear interaction, and the particular case of the unpaired electron in a $3d$ orbit of a copper atom will serve to illustrate this point. This situation is represented in Fig. 3. The copper nucleus itself has a nuclear spin of $\frac{3}{2}$ and hence it can take up four possible orientations in the applied external field H_0. These four possible orientations arise from the four possible resolved magnetic moments in the direc-

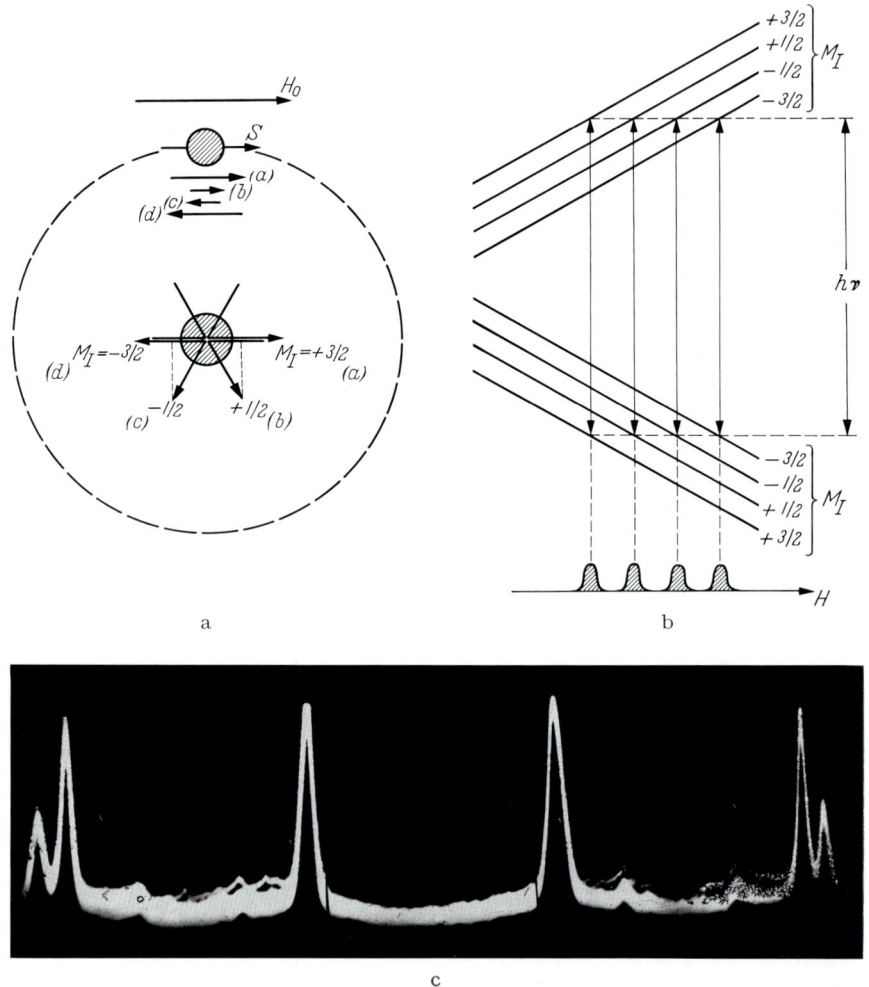

Fig. 3a—c. Hyperfine interaction with copper nucleus. (a) Incremental magnetic fields produced by the four different orientations of the copper nucleus. (b) Effect on energy level diagram. (c) Observed electron resonance spectrum from Cu^{++} salt.

tion of the field of $+\frac{3}{2}$; $+\frac{1}{2}$; $-\frac{1}{2}$; and $-\frac{3}{2}$ as indicated in Fig. 3a. The unpaired electron as it moves around the copper nucleus will now not only see the applied magnetic field H_0, but also small incremental fields arising from the copper nucleus. The magnetic magnetic moment of the copper nucleus may, in fact, be in any of the four possible orientations we have considered, and therefore the electron may have any of the four additional increments as represented by the small arrows at the top of Fig. 3a. Their effect on the energy level diagram is shown in Fig. 3b, where the two original diverging levels are now replaced by two sets of four levels, these four levels corresponding to the four possible orientations of the copper nucleus. The constant energy quantum, represented by $h\nu$, will now cause transitions between these levels at four different values of the magnetic field, as indicated by the four shaded lines below. We therefore expect our electron resonance spectrum from copper to consist not of a single line, but of four hyperfine components of equal intensity. The spectrum actually obtained is shown in Fig. 3c, and the four component lines can be clearly seen. These are actually all doubled, although in the two central lines the separate doublets are not resolved. This doubling arises from the fact that copper contains two abundant isotopes ^{63}Cu and ^{65}Cu; both of these have a nuclear spin of $\frac{3}{2}$ but they have slightly different magnetic moments and hence produce slightly different splittings. The sensitivity of this technique, in not only producing well-resolved hyperfine patterns, but also enabling isotope analysis to be carried out even when the isotopes have the same nuclear spin, is demonstrated very clearly in this figure. It follows, in general, that for a nucleus with nuclear spin I the hyperfine pattern will consist of $(2I+1)$ lines of equal intensity and equal spacing. It will be seen a little later that this can be used as a very straight forward method of identification, the known hyperfine pattern being used as a "characteristic fingerprint" of the unknown atom.

The above paragraphs describe in detail what happens when the unpaired electron experiences the magnetic field due to the one nucleus, but in a large number of cases, and especially in the study of free radicals, the unpaired electron is, in fact, moving in a delocalised molecular orbital and its orbit and wave function may well embrace several different nuclei. We therefore have to consider what type of hyperfine structure will be produced if the unpaired electron interacts with several nuclei at once. There are two particular cases which we might consider in detail, and intermediate cases can then be interpolated. The first case is where the electron interacts equally with several different nuclei of the same species. As a specific example we might take the interaction of the electron with hydrogen nuclei or protons, and take three particular cases of first an interaction with a CH group, secondly with a CH_2, and thirdly with a CH_3 group. The normal isotope of carbon does not possess a nuclear magnetic moment and therefore no splitting will be obtained from this source, and only the interaction with the protons need, therefore, be considered. In the first case, the unpaired electron will experience the additional field produced by the single proton of the CH group and this proton will either be aligned with or against the applied field since it has a spin of a half and can only take up one or other of these two orientations. It therefore follows that a doublet structure will be observed as the hyperfine pattern, and this is in fact one particular case of the $(2I+1)$ rule which we have noticed previously. The situation can, therefore, be represented as in Fig. 4a where the original electronic levels are split into two, corresponding to two different orientations of the proton, and a doublet splitting is observed in the actual spectrum.

If, however, the electron is interacting with two protons as in the CH_2 group, then the interaction with one of these will produce a doublet splitting of the original electronic levels, as discussed above for the CH group, but the interaction with the second proton of the CH_2 group will produce a further splitting in each of these levels, and if the couplings to the two protons are equal the second splitting thus produced will be equal in magnitude to the first. This is indicated in Fig. 4b and it is seen that the net result of these two splittings is to bring the central pair of resultant energy levels together so that there are now only three distinct energy levels for each electronic state, but the centre one of these is

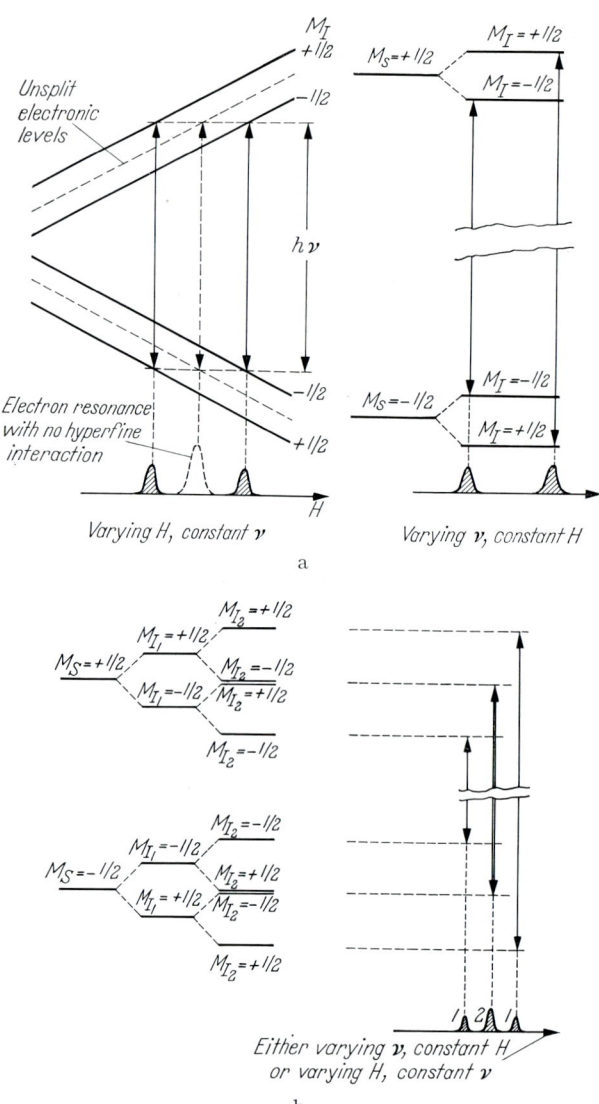

Fig. 4a—c. Hyperfine interaction with several equivalent nuclei. (a) Interactions with one proton in CH group to produce doublet. (b) Interaction with two protons in CH_2 group to produce triplet. (c) Interaction with three protons in CH_3 group to produce quartet.

formed from two which have come together, and thus will be twice as populated as either of the others. The net result of this is that the incoming microwave frequency will now find three values of applied magnetic field to satisfy the resonance condition, and a triplet hyperfine structure will thus be produced but with an intensity ratio of 1:2:1 for this pattern. In a similar way the interaction with the three protons of the CH_3 group can be considered. In this case the interaction with the two protons will produce a triplet splitting of the electronic state as discussed for the case of the CH_2 group, and each of these three will now be split again by the interaction with the third proton and again if the interaction is equal, as it will be for the CH_3 group, this last splitting will be of the same magnitude as the previous two. The net result fo this is shown in Fig. 4c and it is seen that the pattern of energy levels for each electronic state now consists of four distinct levels but the centre two are composed of three individual components which have all come together. The net result of this is that a resultant hyperfine pattern of four lines will be produced, but with an intensity ratio of $1:3:3:1$ as indicated in the bottom figure.

It will be evident that this type of reasoning can be extended indefinitely, and a simple mathematical formula built up for the hyperfine pattern to be expected for n equally coupled protons. In fact it can be rapidly shown that $(n+1)$ hyperfine lines will be produced, and that the intensity ratios of the components of this pattern will form a binomial series. This particular case of equal coupling to protons around a group has been taken as a specific example since it is one that often occurs in practice in free radical studies. The same argument can be extended, however, to any nuclei possessing magnetic moments, and which are equally coupled to the orbit of the unpaired electron.

The other extreme case of interaction with more than one nuclei is when the interaction with the different nuclei is not equal but of a very different order of

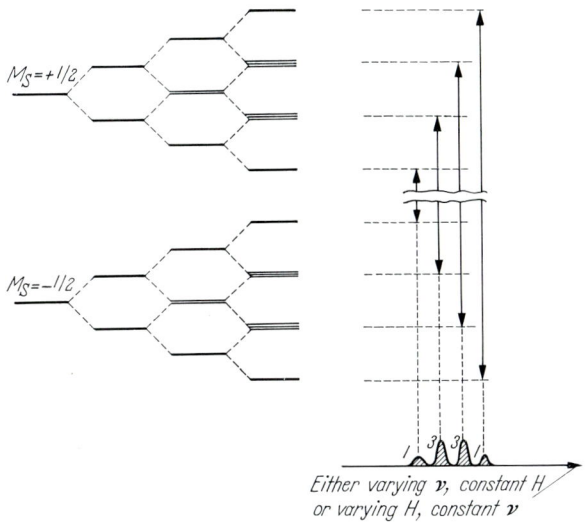

Fig. 4c.

magnitude. It can then be seen that the strongest interaction may be considered first, and $(2I_1+1)$ hyperfine lines will then be produced. The interaction with the second nucleus, will then produce a smaller splitting of each of these lines into a further $(2I_2+1)$ components. If the second interaction is very small compared with the first, it is probable that no overlapping of these different sets of super-hyperfine lines will in fact occur, and many examples will be found of such cases, where each original hyperfine line is split into a number of additional components by further interactions with a ligand nucleus. In the intermediate cases where the couplings are not equal but have the same order of magnitude, it is normally wise to consider the interactions specifically and plot out the predicted hyperfine pattern and compare this with that observed, paying particular attention to overlap between different sets of lines. The detailed analysis and theory of hyperfine patterns will be considered in a later section, but the general principles have been summarised here so that the different features which are to be met in electron resonance spectra can be appreciated.

It may help to illustrate this kind of multi-nuclear interaction, however, if two simple specific examples are given. The work of Fraenkel and his collaborators[2] on semiquinone radicals can be taken as a very good example both of equal coupling and of a case of two couplings of very different orders of magnitude. In the first case the study of the spectra obtained from benzo-semiquinone may be quoted. This molecule consists of a single aromatic ring with two oxygen atoms at either end and four protons remaining attached to the ring as illustrated in Fig. 5a. It is evident from the pure symmetry of the molecule that an unpaired electron moving in a molecular orbital round this ring must interact equally with all four protons, and hence from the argument in the above paragraphs, a five line hyperfine structure is to be expected with a binomial distribution in the intensity of the lines. The actual hyperfine pattern observed is shown in Fig. 5b and is seen to agree very well with that predicted.

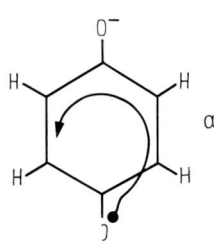

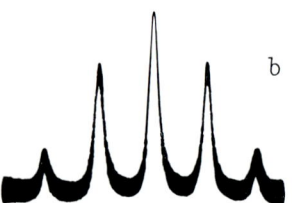

Fig. 5a and b. Hyperfine splitting obtained from benzosemiquinone. (a) Structural formula of benzosemiquinone. (b) Five line hyperfine pattern observed.

The way in which this argument can be followed through is illustrated in Fig. 6 which summarises results obtained by Wertz and Vivo[3]; on the chlorinated semiquinone in which each proton is replaced in turn by one chlorine atom. The first of these figures shows the first derivative tracing for the unchlorinated benzosemiquinone and is thus identical to that of Fig. 5, as measured by Fraenkel and his collaborators. It may be pointed out here, however, that in the Fig. 6 the spectra are presented as first derivative recordings, and it will be seen in the next section that this is the normal way in which high sensitivity electron resonance spectrometers display the spectra. Each absorption line thus appears as a swing from one side across the axis to the far side and back, and the centre of the absorption line is, in fact, defined by the crossover point on the axis. Thus Fig. 6a is seen to consist of the five absorption lines previously mentioned with the binomial distribution in their intensities. The successive tracings 'b', 'c', 'd' and 'e' represent the cases of successive steps of chlorination, thus 'b' is the spectra obtained from the monochloro derivative, and hence a quartet with an intensity ratio of 1:3:3:1 is observed from the three remaining protons on the ring. In the same way Fig. 6c is from duo-chlorinated semiquinone in which a triplet splitting with intensity ratio 1:2:1 is obtained from the two remaining protons, whilst 'd' and 'e' give the doublet splitting, and the single line spectrum, corresponding to the trichlor and tetrachlor derivatives respectively. It is evident from these examples that the equal coupling between the unpaired electron and the various protons can be readily identified from the nature of the pattern produced, and it will also be clear that this can be used as a method of analysis for proton groups if required.

The case of two hyperfine interactions of very different orders of magnitude can also be the illustrated by early work on the benzosemiquinone derivatives.

---

[2] G. K. Fraenkel: Ann. N. Y. Acad. Sci. **67**, 553 (1957); — J. Chem. Phys. **23**, 588 (1955).
[3] J. E. Wertz, and J. L. Vivo: J. Chem. Phys. **23**, 2441 (1955).

Sect. 4. Hyperfine splitting. 105

In particular, the case of the dibutyl derivative can be taken, and the structural formula of this is given in Fig. 7a. It is seen there that two protons remain attached to the central ring, while the other two positions have been taken by two butyl groups. The unpaired electron thus moves around the central ring and a

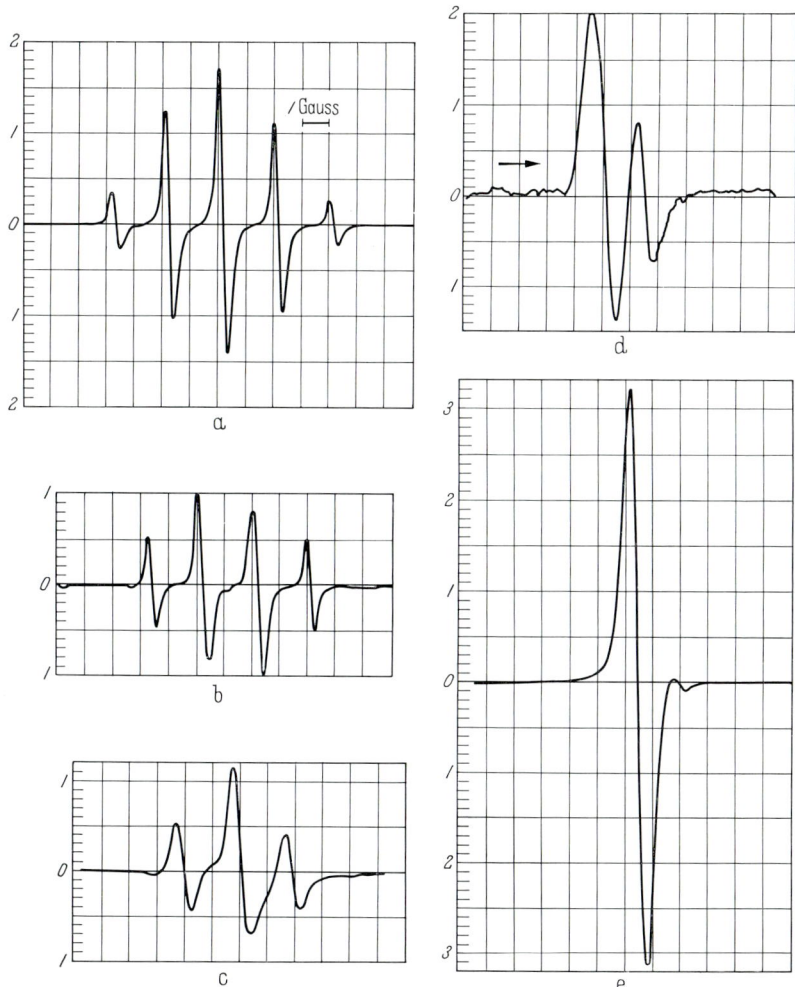

Fig. 6a—e. Hyperfine patterns obtained from chlorobenzosemiquinones. (a) Unsubstituted benzosemiquinone. (b) Monochlorsemiquinone. (c) Dichlorsemiquinone. (d) Trichlorsemiquinone. (e) Tetrachlorosemiquinone.

strong interaction with the two ring protons is to be expected, but the wave function may extend to the edges of the molecule, and a small interaction with the methyl groups around the edges might also be possible. The electron resonance spectra observed under medium resolution is shown in Fig. 7b, where the triplet splitting associated with the two ring protons is clearly seen as the dominant feature of the spectrum. If the central line is studied under higher resolution, and low dilution, further structure is resolved, however, as can be seen from Fig. 7c. This further structure does not show so well on the actual absorption line, but when plotted as a first derivative the extra splittings within the main line can be clearly resolved. These extra splittings are in fact due to the interaction of the unpaired electron with the H protons of the methyl groups, and the splitting between successive lines is of the order of 25 milligauss, or in other words, about a thousand times less than the splitting of the triplet spectra itself. This indicates that the actual interaction between the electron and these outer protons is about a thousand times weaker than with the central ring protons, and indicates the way in which electron resonance can probe out to the extremity of the wave function distribution, and obtain information on interactions which are of a very much smaller order of magnitude than those produced in the main spectrum.

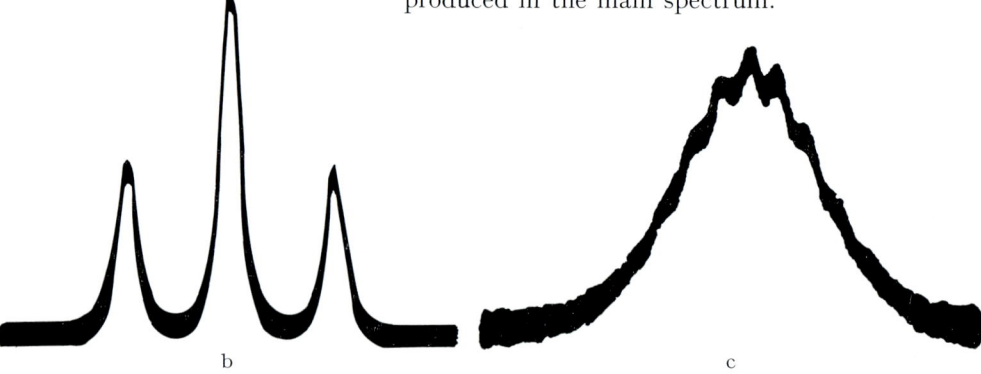

Fig. 7a—c. Hyperfine pattern obtained from dibutyl-benzosemiquinone. (a) Structural formula. (b) Hyperfine triplet under medium resolution. (c) Central line under high resolution showing structure due to edge protons.

Fig. 8. Superhyperfine interaction in copper phthalocyanine. The nine component lines on each copper hyperfine line arise from the surrounding ligand nitrogen atoms. The four groups on the left are from one copper atom.

Another example of such superhyperfine interaction, this time taken from a transition group complex, is shown in Fig. 8. This represents the spectrum obtained from a diluted copper phthalocyanine crystal in which the main interaction is between the unpaired electron and the copper nucleus, thus producing a four line spectrum, as explained earlier in this section. Each of these copper hyperfine lines is also seen to be split into further component lines, however, and these arise from the interaction with the four nitrogen atoms around the copper nucleus. The copper atom, is in fact, surrounded by square of four nitrogen atoms, and the unpaired electron associated with the orbitals of the copper interacts equally with each of these four nitrogens. Each of these has a nuclear spin of 1 and as a result a nine line hyperfine pattern is produced from the four equally coupled atoms. The magnitude of these couplings is, however, much smaller than that with the central copper nucleus, and hence the nine-line superhyperfine pattern on each copper hyperfine line can be resolved separately[4].

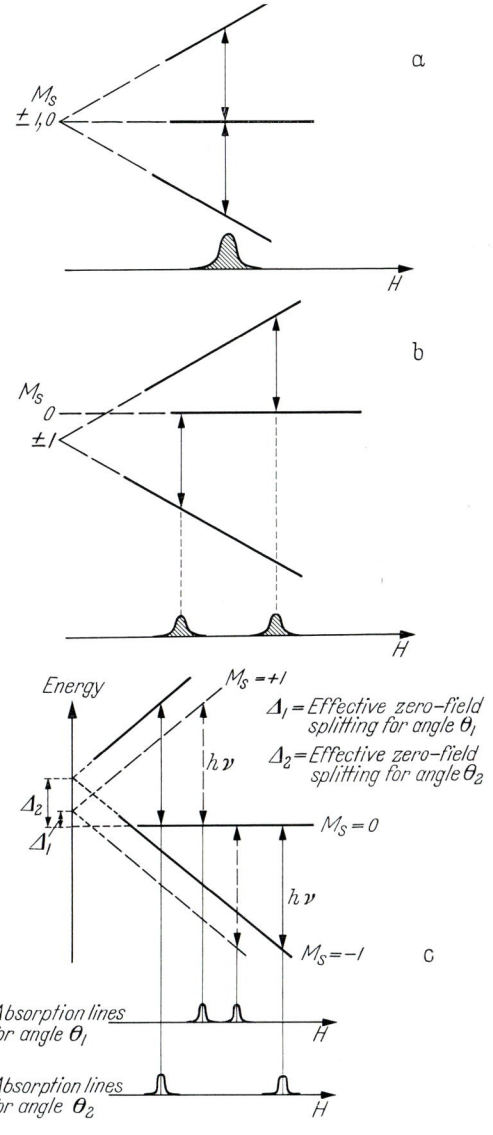

Fig. 9a—c. Electronic splitting and its angular variation. (a) Energy level separations for atom in free space. (b) Effect of internal crystalline field and production of "electronic splitting". (c) Variation of electronic splitting with angle, showing how spectrum becomes smeared out if average over all orientations is taken.

[4] R. M. DEAL, D. J. E. INGRAM, and R. SRINIVASAN: Proceedings of 12th Colloque Ampere, p. 239—246. North-Holland Publ. Co. 1963.

5. Electronic splitting. The fifth and final parameter to be associated with electron resonance spectra is the electronic splitting, which can occur if more than one unpaired electron is associated with a single atom. The particular case of an atom with two unpaired electrons associated with it may be taken as an example. In this case the two electrons will couple to give a resultant total spin quantum number of $S=1$. This overall spin quantum number can now take up three different possible orientations in the applied magnetic field, with components $M_s = +1, 0$ or -1. If the atom is in free space and has no electric or magnetic fields acting on it, these three orientations will all have the same energy, but an applied magnetic field will separate them as shown in Fig. 9a. It will be seen that incoming microwave quanta would always produce two transitions at exactly the same resonance field value in such a case, and thus only one single absorption line would be observed. If, however, the atom is located within a crystal, it will have strong internal electric field acting on it. These internal electric fields will produce a Stark effect and separate the $M_s = 0$ from the $M_s = \pm 1$ levels, even in the absence of any applied magnetic field. The situation, therefore, changes to that shown in Fig. 9b. When the magnetic field is applied in this case the levels will diverge as shown, and it is now evident that incoming microwave quanta will produce resonance absorption at two different values of the applied magnetic field strength. Two absorption lines are thus obtained, and the splitting between these, in fact, reflects the splitting between the $M_S = 0$ and $M_S = \mp 1$ levels, in zero magnetic field. This type of splitting is termed an "electronic splitting", since it arises from the different orientations of the total electronic magnetic moment. In general it is orders of magnitude greater than the hyperfine splitting discussed previously and thus can be very clearly distinguished from it.

It is this electronic splitting which produces the energy level systems of such ions as Cr^{+++} and which are now put to so much practical use in such devices as the "maser". Details of the actual mechanisms which produce such splitting and the way in which they can be represented in the general Spin Hamiltonian describing the spectra are given later in Sect. VI.

III. Experimental technique.

It will now be evident from the consideration of the various parameters discussed in the last section that an efficient electron resonance spectrometer must meet certain design specifications. Thus the very fine splittings associated with the dibutyl semiquinone will only be observable if the homogeneity of the magnetic field is considerably better than 10 milligauss (i.e. 1 part in 10^6). High sensitivity of detection, which is often required in free-radical studies, demands high stability in time for both the magnetic field strength and the klystron frequency, since large time-constants with narrow bandwidths will have to be employed in the detecting circuits, and hence all the parameters associated with the resonance condition will have to remain constant over this recording time. Highly stabilised current supplies to the magnets are therefore required, together with automatic frequency control of the klystron. The sensitivity of a spectrometer is usually one of its most important specifications, and the ways in which high sensitivity can be achieved will, therefore, first be considered.

The case of a spectrometer which has an ideal detector for the microwave radiation, and thus adds no additional noise will be considered initially. The minimum signal which can be detected by such a spectrometer can be found by calculating the root mean square noise voltage which will exist across the matching resistor fixed at the end of the waveguide run. The value of the magnetic susceptibility, and hence of the number of unpaired electrons in the sample, which will produce a signal voltage of this same amount can then be calculated, and by equating these two expressions the minimum number of unpaired electrons that can be detected can be determined. It can be shown in this way that the minimum susceptibility is given by the equation[5, 6]

$$\chi''_{min} = \frac{1}{Q_0 \eta \pi} \left(\frac{kT \cdot \Delta v}{P_0} \right)^{\frac{1}{2}} \quad (4)$$

the various parameters in this equation have the following significance; Q_0 represents the Q, or quality factor, of the resonance cavity; η is the filling factor of the cavity, and measures how much of the effective volume inside the cavity, (i.e. that containing the microwave magnetic field) is, in fact, filled by the sample; Δv is the bandwidth of the recording system, and P_0 is the microwave power inside the cavity itself.

The expression for the minimum number of unpaired electrons is related directly to the minimum susceptibility by a factor of $\Delta H/H$ this expressing the fact that a sharp resonance line, with a narrow width, is easier to detect than a broad shallow line of the same integrated area. Typical experimental values may now be substituted into these expressions, such as $Q_0 = 5,000$, $P_0 = 10$ milliwatts $\Delta v = 0.1$ cycles per sec and $\Delta H = 2$ gauss, and such figures will then give, for the minimum detectable number of unpaired spins at room temperature, a value of 10^{10}. It should be stressed that this is a theoretical minimum and has assumed no noise in the detecting system.

This does indicate immediately, however, which factors are important in the design of an electron resonance spectrometer. To obtain a high sensitivity one must obviously work with cavities with as high a Q factor as possible, and also design the sample holder so that the filling factor is as near to unity as possible. The microwave power in the cavity also affects the ultimate sensitivity although it must be mentioned here that an effect known as "saturation" can also occur if this power level becomes too large. Hence the sensitivity of the equipment cannot be indefinitely improved by raising the magnitude of P_0. In fact the parameter which is easiest to alter is the bandwidth, Δv. The smaller this can be made the higher will be the sensitivity, and in principle there is no reason why this should not be made as small as one likes. A very small bandwidth for the detecting system implies, however, that the whole of the rest of the equipment must be extremely stable, since if a long time is to be taken in tracing out the absorption curve, and a long time-constant for the detecting system is implied by a small bandwidth, then of necessity all the other parameters in the system, such as the microwave power level, the value of the magnetic field, and the frequency of the microwaves, must remain constant during this period. Before considering in detail the way in which frequency and field stability can be attained, the practical sensitivities that can be obtained, when allowance is made for the noise which is introduced in the detecting system, will be briefly considered.

The expression used in the last paragraph assumed that no noise was introduced in the actual detector. In practice, however, this detector often introduces more noise than any other element in the spectrometer. If one turns to the simple spectrometer shown in Fig. 2, this will certainly be the case, since the silicon diode detectors which are used to convert the modulated microwave power into an audio-signal have, in fact, very bad flicker noise at low audio-frequencies.

---

[5] G. FEHER: Bell System Tech. J. **36**, 449 (1957).
[6] D. J. E. INGRAM: Free Radicals as Studied by E.S.R., p. 66—76, chapt. 3. London: Butterworths & Co. 1958.

This excess flicker noise varies inversely with the frequency over a very large range, as can be seen from Fig. 10.

In this figure, the excess noise produced by the detecting system is plotted against the frequency of modulation used in the detection. It will be seen that there is a straight line graph on the left-hand side, which falls from a large value at the low frequency end, towards the zero axis at higher frequencies. This represents the excess flicker noise due to the detecting crystal itself, and this graph explains why a simple crystal video spectrometer of the type shown in Fig. 2 will be inherently very noisy. It follows that high sensitivity, in which negligible noise is added by the detecting system, will only be obtained if modula-

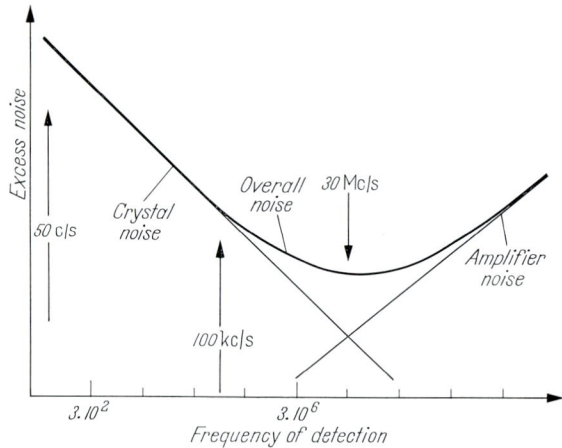

Fig. 10. Frequency variation of excess crystal and amplifier noise.

tion frequencies of much higher value than the audio range are employed. This fact by itself would suggest that it is only necessary to work at as high a frequency of modulation as possible. This is not true, however, since above about 50 Mc/s the noise of the I.F. amplifier becomes appreciable, and this is represented by the curve which rises on the right-hand side of Fig. 10. It follows that if these two curves are added together to give the total excess noise contributed by the detecting system, then a broad minimum is obtained centred on about 30 Mc/s as shown in the figure. This suggests that the maximum sensitivity in any microwave detecting system will, therefore, be obtained by employing an intermediate frequency of about 30 Mc/s, and it is of course, for this reason that radar sets use intermediate frequences of about this value. One of the possible methods of obtaining high sensitivity is to use a superheterodyne system similar to that employed in radar, and with it an I.F. frequency of about 30 Mc/s. This produces a very complex piece of equipment, however, and if similar sensitivity can be obtained without the addition of the local oscillator klystron and other ancillary apparatus a large number of difficulties can be avoided.

It is found in practice that the excess noise contributed by the crystal at 100 Kc/s has fallen to a very small amount and hence modulation and detection at these frequencies is almost as good as that at 30 Mc/s. It is relatively simple to design a spectrometer which employs 100 Kc/s magnetic field modulation, and this is, in fact, the system employed by most of the commercial spectrometers now on the market. The 100 Kc/s magnetic field modulation can be applied to the sample either by a simple loop of wire, wound around the sample and placed inside the cavity resonator, or, alternatively, by embedding a small pair of modulating coils in the wall of the resonant cavity. In the second case the walls of the cavity must be sufficiently thin to let the 100 Kc/s modulation through without attenuation, but sufficiently thick to act as a short circuit for the microwave frequencies. This can be readily achieved in practice by either coating the inner walls of a ceramic cavity with a silver lining, or casting an araldite mould

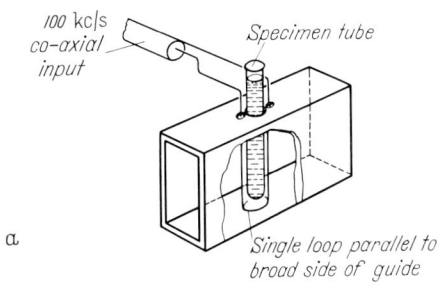

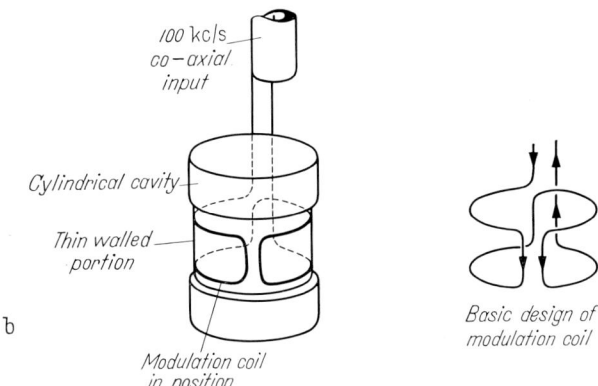

Fig. 11a and b. Methods of applying 100 kc/s modulation to specimen in cavity. (a) Single loop in rectangular cavity. (b) Helmholtz pair for cylindrical cavity.

around an electro-formed cavity wall. In both cases the thickness of the wall is adjusted to be greater than the skin depth at the microwave frequency but less than the skin depth at a 100 Kc/s. Examples of both of these forms of applying the high frequency modulation are shown in Fig. 11. In the one case the simple single loop is inserted inside the rectangular cavity and is shown located in the centre; while, in the other case, the Helmholtz coil system, which is placed around the thin walled cylindrical cavity, is indicated in an expanded form.

6. High frequency field modulation spectrometer. It may now be helpful if the particular features of electron resonance spectrometers which are currently used are described in somewhat more detail. It was pointed out in the last section that an electron resonance spectrometer employing high frequency magnetic field modulation at 100 Kc/s has almost as high a sensitivity as the super heterodyne system and far less of its complexity. It is not surprising therefore, that this type of spectrometer is the one most commonly used in practice and the one which is normally supplied by the commercial manufacturers. A block diagram of such a spectrometer system is in fact shown in Fig. 12 and there are various features in this which differ from the simple spectrometer previously discussed and illustrated in Fig. 2.

In the first case it may be noted that this spectrometer employs a microwave bridge system instead of the simple transmission type cavity indicated in Fig. 2. The great advantage of employing a microwave bridge in an electron resonance spectrometer is that it allows the level of the microwave power falling on the

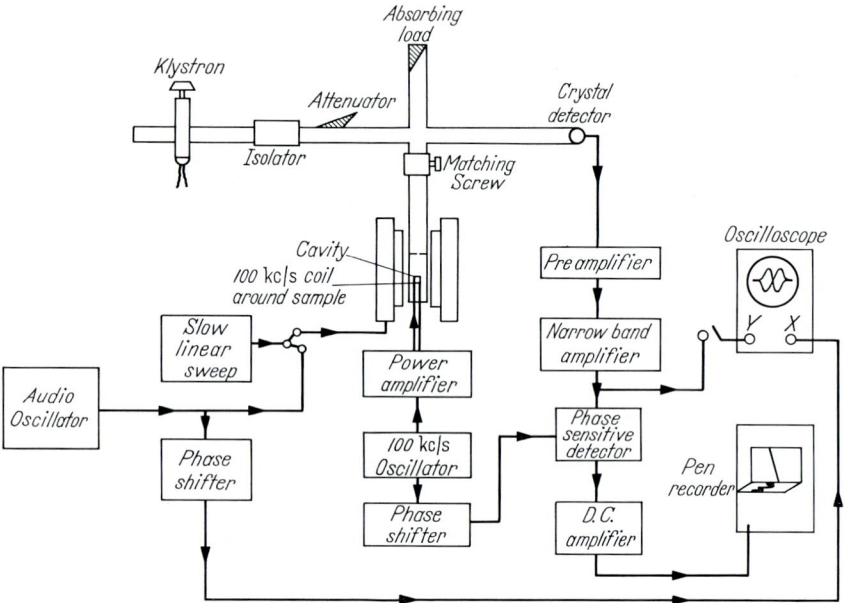

Fig. 12. Block diagram of E.S.R. spectrometer employing 100 kc/s magnetic field modulation.

crystal detector to be adjusted, quite independently of the actual power level which exists inside the cavity resonator. This is not true of the simple transmission system of Fig. 2, where optimum matching of the waveguides to the cavity would ensure that a certain proportion of the microwave power in the cavity was actually coupled out to the detecting system. Thus, if high powers were being employed to produce high microwave field levels at the specimen in the cavity, the power coupled out and applied to the crystal detector would be noticeably higher than the optimum value at which such crystals should operate. In the bridge system, however, it is possible to match the third arm of the bridge to the one containing the cavity resonator so that when the specimen is inserted in the cavity, but before the magnetic field is applied, these two arms will balance, and hence no microwave power is fed on to the fourth arm containing the detecting crystal. Application of magnetic field will then cause absorption in the cavity resonator, which unbalances the microwave bridge, and causes a small amount of microwave power to be fed into the fourth arm and on to the crystal detector. In this way the power actually reaching the crystal can be made extremely small when no resonance absorption is taking place. The presence of the absorption is then detected by a large percentage increase in the power arriving at the crystal detector itself. In practice it is found that these detecting crystals operate best with a certain bias power present such that the detected current is of the order of half a milliamp. This optimum can be readily achieved by unbalancing the microwave bridge by a certain specified amount, after balance itself has been obtained, and in this way ensuring that the detecting crystal is always operating under its optimum conditions. Since the detecting crystals are the main source of the ultimate noise in this system, and hence the final limit on its sensitivity, it is rather important to operate these under optimum conditions whenever possible.

The other advantage of using a microwave bridge system is that only one waveguide is then connected to the cavity resonator. If this has to be cooled to low temperatures, as is often the practice with electron resonance work, it follows that the heat conductivity into the liquid refrigerant is thus reduced by a factor of two from a transmission system, in which two waveguides are required to lead to and from the cavity itself.

The only other basic difference between this spectrometer and the simpler one considered previously is that the magnetic field modulation is now applied via one of the high frequency coil systems discussed at the end of the last section instead of via two modulating coils mounted on the magnetic pole pieces themselves. The 100 Kc/s signal in fact originates from a master oscillator as shown, and is fed through a power amplifier to the modulating coils themselves. The impedance of the output stage of this power amplifier needs to be matched to the impedance of the modulation coils, which is normally very low, hence large high frequency currents of low voltage are produced in the modulating system. Even so the strength of the actual magnetic field produced at this frequency, at the site of the specimen, is normally relatively small and hence the magnetic field modulation is now used to sample the gradient of the absorption line, rather than sweep right through it, as in the previous case. The way in which this 100 Kc/s modulation detects and reproduces electron resonance absorption is in fact indicated graphically in Fig. 13.

In Fig. 13a the magnitude of the 100 Kc/s magnetic field modulation is shown equal to half the width of the absorption line itself. The main value of the applied D.C. magnetic field is then slowly swept through the resonance condition, so that this high frequency modulation is swept over the profile of the absorption line, and in this way samples the gradient of the line as shown. In other words the actual depth of modulation, passed on to the microwave signal itself, will be proportional to the first derivative of the absorption line contour, going through a zero value as the modulation reaches the point of zero slope at the top of the absorption line, and swinging into the negative region as the gradient of the line changes in the second part of the absorption curve. This 100 Kc/s modulation of the microwave signal is of course detected by the crystal at the end of the microwave run, and passed on to a narrow band amplifier. This amplifies the 100 Kc/s signal specifically and then, after further amplification, it is mixed in a phase sensitive detector which has a reference signal fed from the original master oscillator. This system then has all the normal advantages to be associated with a phase sensitive detector since the noise of previous amplifier stages can be effectively eliminated, and the sensitivity of the detection and display system depends only on the bandwidth of the actual recording equipment.

Alternatively the 100 Kc/s signal, as it comes from the output of the narrow band amplifiers can be fed straight to an oscilloscope screen, before being fed to the phase sensitive detector, and the signal then observed will be as shown at the top of Fig. 13c. No phase discrimination is apparent here and the signal is seen

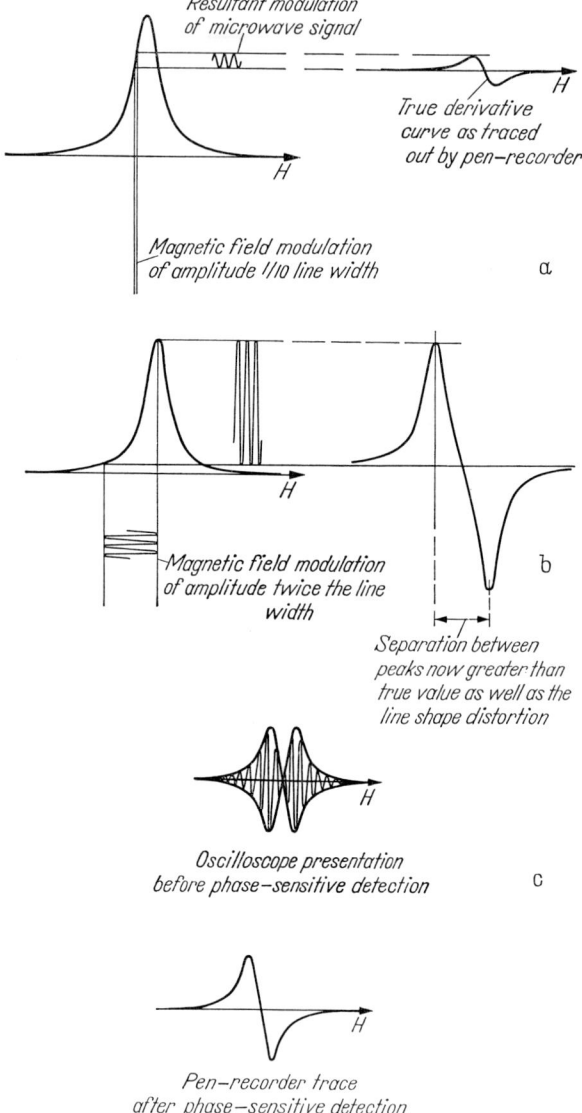

Fig. 13a—c. Detection and display of first derivative of absorption line with 100 kc/s modulation. (a) With modulation equal to half the line width. (b) With modulation equal to twice the line width. (c) Display on oscilloscope above, and by pen recorder below.

to be symmetrical about a zero point which corresponds to the centre of the absorption line as indicated before. The phase sensitive detector not only has the advantage in respect to noise, as discussed above, but also determines whether the incoming signal has changed sign, and hence the phase is retained and displayed as indicated in the bottom of Fig. 13. It is in fact, such a display as this that is normally obtained on most electron resonance output systems and these first derivative curves are normally traced out by a pen recorder on a continuously moving sheet of paper.

One other feature that should be noted when applying 100 Kc/s magnetic field modulation in this way is that a compromise must be made between high sensitivity detection, which will require as large a magnitude of the modulating field as possible, and the requirement of high resolution, or faithful reproduction of the absorption line shape. This point is indicated in Fig. 13b where the effect of employing a magnetic field modulation of twice the absorption line-width is indicated. It is clear from this example that the magnitude of the amplitude modulation produced on the microwave signal no longer follows the true first derivative of the absorption line and can now give a distorted picture of the true shape and structure of the electron resonance absorption itself. It is for this reason that normal spectrometers do not employ magnitudes of the magnetic field modulation which are likely to be greater than half the maximum line width expected.

As well as the detection and display system discussed above most spectrometers also employ some kind of automatic frequency locking circuits to maintain the frequency of the klystron at the correct value. There are essentially two different methods whereby this may be done. The frequency of the klystron can either be locked to an absolute standard of frequency, and in this case comparison with harmonics of a quartz crystal standard, or a high Q cavity resonator, can be made. It is then relatively simple to devise a discriminator circuit so that any drift of the klystron frequency from these external references will produce an error voltage which can be fed back to the klystron reflector itself and made to counteract the drift that has been produced. The great disadvantage of such an automatic frequency locking system is that it takes no account of the fact that the actual frequency of the cavity containing the specimen may also have changed. If this does happen, and the klystron is kept rigidly locked to an external frequency, then the cavity will drift away from the initial resonance condition and all the observed lines will be noticeably reduced or distorted.

The alternative automatic frequency locking system is to lock the frequency of the klystron to the cavity which forms the microwave absorption cell, itself. In this way it is possible to ensure that the incoming microwave signal to the cavity resonator is always at the correct resonance frequency and hence ensure that the microwave magnetic field strength is concentrated at the site of the sample. On the other hand, if the resonance conditions of the cavity itself have changed at all, as is likely to happen in cooling to low temperatures, the change of frequency associated with this change in cavity dimension, which the klystron has now been forced to follow, will alter the resonance condition itself somewhat. The absorption line will thus now occur at a slightly different value of the main applied magnetic field and unless this change of resonance is noted, and allowed for, an unexpected shift in the position of the resonance spectra may well be obtained and an anomalous g factor derived from it.

The choice between the two methods of automatic frequency control therefore very often depends on the main purpose to which the equipment is devoted. If accurate measurements of g values are required then it would be unwise to use a frequency locking system which kept the klystron locked to the resonance frequency of the cavity, and paid no attention to the absolute frequency of the klystron, or cavity, themselves. If on the other hand the main interest is focused on obtaining high sensitivity, and faithful reproduction of the shape and splittings of the spectra, then it would be better to use a frequency locking system which does lock the klystron frequency to that of the resonance cavity, so that no change in signal shape is produced by the cavity drifting away from the centre of the klystron mode. In fact most spectrometers now in use, including those produced commercially, normally offer as a standard item of equipment a frequency locking device which locks the klystron to the resonant cavity rather than to any external resonant device. The most common way of locking the klystron itself to the cavity is by simply applying a frequency modulation to the klystron output which can be easily produced by a voltage modulation on the reflector of the klystron valve. The frequency modulation is then converted into amplitude modulation, by the resonance curve of the absorption cavity, and if the klystron is tuned to the centre of the cavity mode then a zero net output will be obtained, since the amplitude modulation will swing equally about the positive and negative values. On the other hand, if a frequency drift does occur, the frequency modulated klystron power will then be applied, on one side or the other of the cavity mode, and a predominantly positive or negative output will thus be obtained. This can then be used to produce a d.c. correcting voltage to be fed back to the klystron reflector electrode as before.

The basic features of such an automatic frequency control unit can be added to the block diagram of figure 12 and it is evident that it can be fairly readily incorporated into the spectrometer itself. Most modern commercial electron resonance spectrometers are in fact based on some such system as that indicated in Fig. 12, and such high frequency modulation spectrometers have good sensitivity and often sufficient resolution for most of the physical or chemical systems to be studied. The particular exception to this is the study of free radicals in dilute solution at low concentrations. In this case some very small splittings in the hyperfine patterns may be expected and the broadening due to the magnetic field modulation might prevent the finest structure from being observed. In such cases as this, where both high resolution and high sensitivity are required at the same time, it will be necessary to employ some kind of superheterodyne system and such systems are now considered in somewhat more detail.

7. Superheterodyne spectrometers. It was pointed out earlier in this section that one of the ways of obtaining high sensitivity for electron resonance detection is to employ the same system as has been developed for radar detection, i.e., superheterodyne detection. This comprises a microwave circuit in which a local oscillator klystron produces a microwave signal at a frequency about 50 Mc/s higher, or lower, than that of the signal klystron so that an intermediate beat frequency is produced in the crystal detectors at this 50 Mc/s value. An intermediate frequency of this magnitude will be away from the low frequency noise of the crystal detectors, and the whole system will be working under the minimum noise conditions as indicated in Fig. 10.

A block diagram of such a superheterodyne system is shown in Fig. 14. This incorporates a microwave bridge system in just the same way as the 100 Kc/s modulation system did, and the cavity is matched to the other arm as before, so that only when resonance absorption takes place in the sample is any unbalance signal fed through to the fourth arm of the bridge. This signal, which now carries the information concerning the absorption, is not passed directly to the crystal detector, but mixed with microwave power coming from the local oscillator klystron at the right of the figure and the two frequencies together are passed on to a balanced mixer, formed by two detecting crystals in opposite arms of a second balanced bridge. The output from these two crystals are then mixed in a balancing transformer as shown, and this signal passed on to the first stage of an amplifier working at the intermediate difference frequency. The use of a balanced bridge system with two crystals detectors in this way allows the noise of the local oscillator klystron to be cancelled out in the detecting system and hence this potential additional source of noise is removed at the first detector stage. The modulation on the microwave signal is still passed on however, as an envelope on the intermediate frequency now, and after further amplification at the intermediate frequency value, this is then passed on to a phase-sensitive detector. This phase-sensitive detector operates in exactly the same way as that used in the previous spectrometer, but the frequency of magnetic field modulation can now be reduced to the audio range and hence all possibility of modulation broadening is removed. The output from the phase-sensitive detector can then be fed to a pen recorder and the first derivative of the signal plotted as before, or

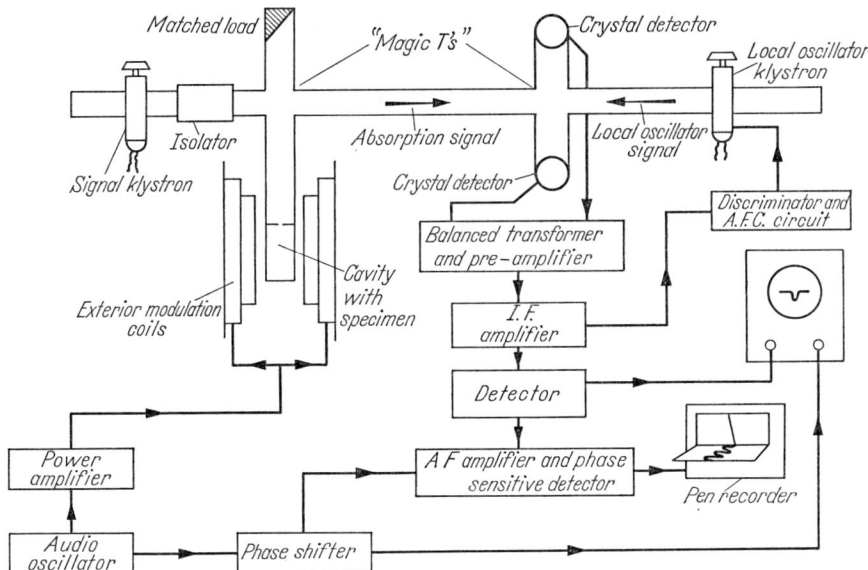

Fig. 14. Block diagram of superheterodyne spectrometer employing two klystrons.

if preferred, much larger values of magnetic field modulation can be employed and the audio frequency used to sweep right through the absorption line and the complete absorption spectra can then be traced out on the oscilloscope screen as was initially done in the case of the simple spectrometer shown in Fig. 2. If the highest sensitivity, which would require the longest time constant of the pen recorder system, is not required this rapid display of the absorption signal on the oscilloscope can be quite an advantage, and the extra sensitivity obtained by working at the high intermediate frequency is often sufficient for such a display to be made.

The basic electronic circuits which are necessary for the intermediate frequency amplification and the phase-sensitive detection network are shown in Fig. 14 and follow very closely along the lines already explained for Fig. 13. There is one additional automatic frequency control circuit however which is usually required. This is a circuit which locks the frequency of the local oscillator klystron to a value which is held at a constant difference frequency from the signal klystron, this difference being equal to the intermediate frequency value itself. The automatic frequency control circuit can take the normal form of a discriminator, as used in radio frequency applications, and this feeds a control voltage of varying polarity to the reflector of the local oscillator klystron to keep it on tune at the correct difference frequency. In some cases it has been found that klystrons can be stabilised so well by the power packs used that they will remain steady at fixed frequency values without the need of the A.F.C. circuits. It is however wise to incorporate these unless they have been shown to be unnecessary.

It will be appreciated that the main disadvantage of this superheterodyne system is the extra complexity which is produced by the insertion of the second klystron into the microwave system and the need to keep its frequency at a

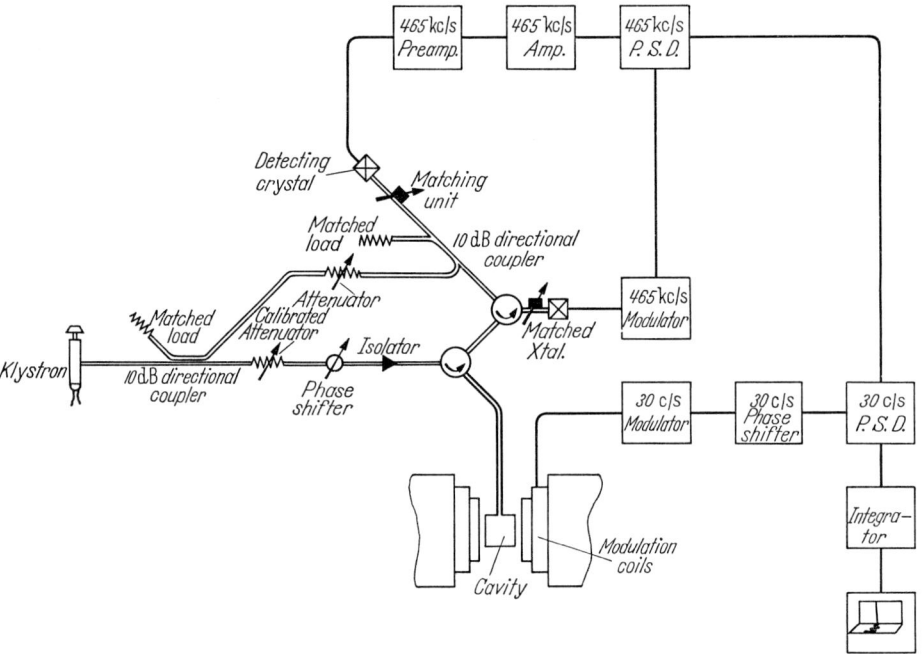

Fig. 15. Block diagram of homodyne superhet spectrometer employing only one klystron.

constant difference value from that of the signal klystron. It is possible in fact to overcome this inherent complexity and difficulty by the use of a homodyne system in which the one klystron is used to provide both the signal frequency and also the beat frequency. This may be done in principle by taking some of the power from this single klystron and passing it over a crystal mounted in a waveguide, the crystal itself being fed with high power from a modulating source. The non-linear characteristics of the crystal will then modulate the microwaves as they are reflected from it, and hence a microwave signal may be passed back into the waveguide system which has side-bands at the sum and difference frequency of the original microwave signal and the modulation frequency applied to the crystal. These side-bands can then be used as the local oscillator signal and can be made to beat with the microwave power coming from the klystron, which has not passed by the modulating crystal, and hence an intermediate frequency can be obtained in the same way as in the more normal superheterodyne system. A block diagram, showing how such a homodyne superheterodyne system can be designed, is shown in Fig. 15.

It can be seen in this figure that two circulators are employed to separate the power flow, and then mix the resultant signals as required. This circuit is based on a design due to Faulkner[7], and it can be seen how the initial frequency is fed via the first circulator to the absorption cavity. After it has been modulated by the absorption, its carrier frequency is shifted by the crystal attached to the second circulator, and finally passes on to mix with some of the initial unchanged frequency, fed directly from the klystron via the two directional couplers. It will be evident that this type of circuit includes all the advantages of superheterodyne detection, with the simplicity normally associated with the field-modulation spectrometers, and also does not waste any microwave power in the balancing arms of magic-T bridges.

8. Magnet requirements. It has already been mentioned that there are two wavelengths normally used for electron resonance in the microwave region, these corresponding to the two radar bands centred on 3.2 cm and 8 mm respectively. It follows that the magnetic fields required to observe electron resonance will either be centred around 3,300 gauss in the first case or around 13,000 gauss in the second case. The magnets which are employed in electron resonance spectrometers must therefore be capable of producing these fields without any difficulty and if transition group complexes are to be studied, which may have g-values which drop considerably below 2.0, then much higher values of magnetic field will be required. The other main feature of the magnetic field is its uniformity or homogeneity since any lack of homogeneity in the field over the volume of the specimen itself, will produce artificial broadening of the observed resonance lines. One of the examples of electron resonance spectra already quoted, i.e. that of the dibutyl semiquinone shows that quite often the fine splitting in these spectra can be of the order of 25 milligauss, or less, and thus it follows that the homogeneity of the magnetic fields employed to observe the spectra must be considerably better than this order of magnitude. It can be seen from this that the homogeneity of the fields over the volume of the specimen must in fact be better than 1 part in 10^6 or 1 part in 10^7 and it will be appreciated that at field values the order of 10,000 gauss this can be a serious engineering problem.

It is, in fact, the requirement for high field homogeneity, rather than high field magnitude, which produces the rather high cost of magnets used in both electron resonance and nuclear magnetic resonance studies. The homogeneity of the fields used for electron resonance are in fact not quite as great as those required for nuclear resonance, since the splittings are on the whole about 1,000 times greater. Nevertheless, if free radicals are to be studied in solution, or defects in solids where no broadening due to protons is present, field homogeneity of at least 1 part in 10^6 will be required.

Although permanent magnets are often employed in nuclear magnetic resonance studies they are not normally used in electron resonance spectrometers. In nuclear resonance all the spectra associated with a given nucleus can be expected to appear very close to a single magnetic field value, provided constant frequency is employed. This is certainly not the case in electron resonance, unless the equipment is only to be used for free radical studies, when all the spectra will appear very close to the position given by a g-value of 2.0. If, on the other hand, the spectrometer is to be used in a more general way for studying transition group complexes, or other such systems, the g-values and hence the fields for resonance may shift quite noticeably. In such cases as these it is imperative to employ an electromagnet so that the field can be changed easily and rapidly. Electromagnets which

---

[7] E. A. Faulkner: J. Sci. Instr. **39**, 135 (1962).

will produce fields of this kind of magnitude and homogeneity normally require pole face diameters of 20 cm or so, and the whole magnet as well as the pole gap itself has to be carefully engineered in every way. Symmetry and uniformity are required not only in the geometry of the design, but also in the magnetic material employed for the pole faces. Shimming of these is also necessary either in the form of raised shims around the circumference of the pole face itself, or in the form of flat coils which can be affixed to the pole faces and used to produce correcting fields.

As well as homogeneity in space, uniformity of the magnetic field in time is also required, since high sensitivity spectrometers will require a long time for the recording of a given spectra, and hence it is essential that there are no random changes in the magnetic field during this operation. Carefully designed stabilising systems must therefore be employed to keep the current feeding the magnet coils at a constant value, and this may be done either by passing the whole current through valves or transistors, which are themselves biased by a field-sensing device, or alternatively by taking the current from a d.c. generator, the control coils of which are automatically corrected by a feed-back system.

The exact details of the design of the magnets themselves, and of the stabilising systems for them, will not be given here since this forms a fairly detailed and technical subject on its own, but magnets and associated power supplies, suitable for electron resonance spectrometers working at either 3 cm or 8 cm wave length, can now be purchased from several commercial manufacturers and, although the cost is still somewhat high, they are preferable to any that can normally be made in a research laboratory workshop.

Superconducting magnets are now being employed with some electron resonance spectrometers, and these have particular advantages if very high values of magnetic field are required. The constant supply of liquid helium, which is needed for their operation is a noticeable drawback for spectrometers which are likely to be in constant use, however, and it is likely that they will only be used for special studies over the next few years.

IV. Basic absorption processes.

9. Relaxation phenomena. Before discussing the theory of the different parameters which are associated with electron resonance spectra, such as the g-values and the hyperfine splittings, a brief summary of the background theory of the general absorption process will first be given; and in particular the ways in which the different relaxation phenomena affect electron resonance spectra will be discussed. The fact that there are two basically different ways in which electron spins can share their energy has already been briefly mentioned in Sect. 2. It was pointed out there that, once an electron has been excited to the top energy level, it can lose this energy of excitation either by giving it to the lattice or molecule as a whole, or by sharing it with other electron spins. In the first case this interaction is termed a "spin-lattice relaxation" and has a "spin-lattice relaxation time" associated with it, whereas in the second case the terms "spin-spin interaction" and "spin-spin relaxation time" are employed.

10. Spin-lattice interaction. The interaction between the spin system and the lattice is responsible for maintaining a constant absorption signal. Thus, if no such interaction occurred, the number of electrons in the upper level would rapidly become equal to that in the lower level as absorption of the incoming microwave radiation raised them to the higher level. Moreover, if such equality were produced, it would follow that no further absorption would take place, since the stimulated emission would then be exactly equal to the absorption and hence the signal itself would disappear. This situation can, of course, only be approached asymptotically, but the onset of this effect is termed "saturation" and may produce serious line broadening, and it is discussed in more detail in Sect. 12. This saturation, where a reduced signal and broadened line are produced by the inability of the spin-lattice relaxation to remove the excited electrons fast enough, will occur when the spin lattice interaction itself is weak and the corresponding relaxation time is long. At the other end of the scale, however, a very short spin-lattice relaxation time can also produce a broadening of the absorption lines, since the lifetime of the excited state will then be reduced, and the Uncertainty Principle itself will cause a spread in the energy level. Study of spin-lattice relaxation times is thus not only of interest for its own sake, but also as one of the major possible causes for line broadening.

The theory of the actual interaction between the unpaired spins and the thermal vibrations of the lattice, or molecular array, has developed through various stages over recent years. Two types of coupling were initially suggested, the first by WALLER[8] which acted via the magnetic dipole coupling between the spins themselves, and the spatial vibrations of the paramagnetic ions. This magnetic coupling is essentially very weak, however, and the second mechanism suggested by KRONIG[9] acted via the internal crystalline electric field, which is altered by the thermal motion of the atoms, and this varying electric field then couples directly to the orbital motion of the negatively charged electrons. The final coupling to the energy of the spins then takes place via the spin-orbit coupling, which is thus the parameter which comes into all expressions deduced from this basic theory. This mechanism envisages two processes whereby the coupling can actually take place. The spins can either exchange a whole quantum directly with a lattice vibration of the appropriate frequency and this is known as the Direct Process. Alternatively the electron can scatter a quantum of the lattice and change its value in a Raman type process and, since all the quanta can take part in such a process, this is the one that is likely to predominate at the higher temperatures. KRONIG[9] in fact made an estimate for the order of magnitude of relaxation times which might be expected from these two processes and obtained expressions of the form

$$\text{Direct Process} = \frac{10^4 \cdot \Delta^4}{\lambda^2 \cdot H^4 \cdot T} \text{ secs}$$

$$\text{Raman Process} = \frac{10^4 \cdot \Delta^6}{\lambda^2 \cdot H^2 \cdot T^7} \text{ secs}$$

where Δ is the height of the next orbital state above the ground state measured in cm^{-1}, λ is the spin orbit coupling coefficient and T is the absolute temperature.

The main predictions of these two expressions are confirmed by experiment, since paramagnetic ions with small values of Δ generally have broad line widths which vary markedly with temperature, indicating a strong spin-lattice relaxation. Moreover, when detailed quantitative measurements are made of the variation of the spin-lattice relaxation time with temperature, two clearly defined parts of the curve are generally observed, one corresponding to an inverse temperature relation of the Direct Process at low temperatures, while at higher temperatures a much higher rate of change with temperature occurs.

The detailed study and measurement of spin lattice relaxation times has taken on considerable importance of recent years with the advent of the micro-

---

[8] I. WALLER: Z. Physik **79**, 370 (1932).
[9] R. DE L. KRONIG: Physica **6**, 33 (1939).

wave masers. These devices employ saturation techniques to produce an inverted energy level population, and the actual conditions under which such saturation can be achieved depend crucially on the relaxation times which exist between the different energy levels in the system. In this connection considerable study has also been made of cross-relaxation effects, but these are somewhat too complex and detailed to be included in a review article of this nature.

11. Spin-spin interaction. The other basic relaxation mechanism is that which exists between the unpaired electron spins themselves. Each such unpaired spin, whether it be attached to a molecule in a free radical, or to an ion of a single atom, may be regarded as a magnetic dipole, which will be precessing in the applied magnetic field. Its component in the direction of the field will have a steady value and this will produce an additional magnetic field at the site of the neighbouring unpaired electrons. The total value of the magnetic field seen by them is thus shifted slightly, and the value of this shift will vary markedly with the angle between the applied field and the line joining the two electron spins being considered. The angular dependence takes the general form associated with the magnetic field produced by a small dipole and varies as $(1 - 3\cos^2\vartheta)$. Different neighbouring unpaired electrons will thus experience different additional fields, and hence their energy levels will be shifted slightly and a general broadening of the observed absorption will occur.

As well as the simple dipole-dipole effect there will also be an additional broadening, which is produced by a rotating component of the precessing electron. If this rate of precession is the same as that of neighbouring unpaired electrons (i.e. they have the same g-value) the oscillating field which is set up will induce transitions in the neighbouring electrons and thus decrease the normal lifetime of their excited energy state. This increases the natural line width and hence a further broadening is produced. These two effects combine to give an expression for the mean square width of the line of the form

$$(\Delta H)^2 = \frac{3}{4} S(S+1) g^2 \cdot \beta^2 \sum_k \left\{ \frac{1 - 3\cos^2\vartheta}{r^3} \right\}^2_{jk}. \tag{5}$$

To a first approximation this broadening and interaction is independent of both the temperature and the magnitude of the applied field and the only way in which it can be reduced is by increasing the distance between the spins (i.e. the value of "r" in the formula above). In practice this means diluting the specimen with an isomorphous diamagnetic compound. When such dilution has been carried out so that the distance between the unpaired electrons is considerable, a dipolar broadening effect may then be found from the magnetic moments of the surrounding nuclei. This is particularly noticeable in hydrated salts where the nuclear magnetic moment of the protons surrounding a paramagnetic ion can pro- a line width of about 6 gauss. The only way in which such broadening can then be reduced in the solid state is by replacing the protons with deuterons and growing the crystals out of heavy water[10].

It should be noted here, however, that the dipole broadening can be removed completely if rapid motion of the molecules takes place as normally occurs in the liquid state. The value of the $(3\cos^2\vartheta - 1)$ variation is then averaged to zero in a time short compared with the inverse of the line-width frequency and as a result the microwave absorption only registers the average field value for the resonance absorption line. This phenomenon of motional narrowing is, of course, one of the

---
[10] B. BLEANEY, K. D. BOWERS, and D. J. E. INGRAM: Proc. Phys. Soc. (London) A **64**, 758 (1951).

crucial factors necessary for the observation of high resolution electron resonance spectra in solutions.

There are other similar interactions which can produce a narrowing of the absorption lines by rapidly interchanging the unpaired electrons. The most striking of these is called "Exchange Narrowing" and it occurs when electrons can be exchanged rapidly between the orbitals of different molecules. Provided

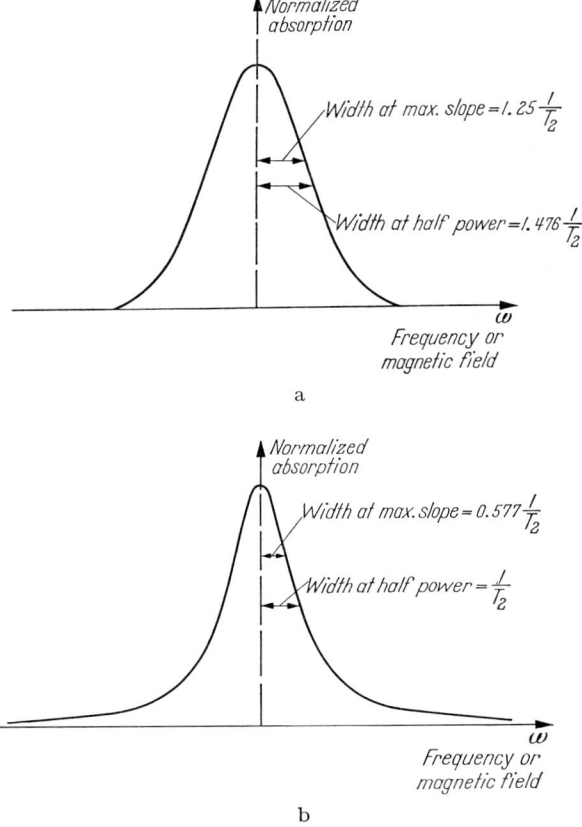

Fig. 16a and b. Gaussian and Lorentzian line shapes. (a) Gaussian shape due to dipolar broadening. (b) Lorentzian shape due to exchange or motional narrowing.

this is fast enough, an averaging of the magnetic field which they experience will then take place. The detailed theory of such exchange interaction has been treated by several authors and it has been shown in general that if exchange is between similar ions or molecules then it will narrow the absorption line in the centre and broaden it in the wings, leaving the second moment unchanged. The onset of such exchange or motional narrowing can therefore be seen from a change in shape of the absorption line. If only the normal dipole-dipole interaction is present the absorption line will have a Gaussian shape, as indicated in Fig. 16a, while the onset of Exchange Narrowing will produce a line with a Lorentzian shape with the features as shown in Fig. 16b. These two line shapes can in fact be characterised by the properties summarised in the following table

| Gaussian Line | Lorentzian Line |

Normalized Equation

$$g(\omega-\omega_0) = \frac{T_2}{\pi} e^{-(\omega-\omega_0)^2 T_2^2/\pi} \qquad g(\omega-\omega_0) = \frac{T_2/\pi}{1+(\omega-\omega_0)^2 T_2^2}$$

Width at half-height

$$(\pi \cdot \log_e 2 \cdot)^{\frac{1}{2}} \frac{1}{T_2} \qquad\qquad \frac{1}{T_2}$$

Width at point of maximum slope

$$\frac{1}{T_2}\sqrt{\frac{\pi}{2}} \qquad\qquad \frac{1}{T_2\sqrt{3}}$$

Note. In these equations T_2 is defined as the inverse of the line width parameter and is equal to π times the maximum value of $g(\omega-\omega_0)$.

12. Saturation effects. The possibility of saturation occurring if the spin lattice interaction is weak has already been briefly discussed at the beginning of the last section. The effect will now be considered in somewhat more detail, since it is basic to an understanding of quite a number of other effects in electron resonance studies.

Let us consider the case of electrons in an applied magnetic field, and with the two energy levels A and B of Fig. 1 the number of electrons in the two levels being N_1 and N_2 respectively. In thermal equilibrium:

$$N_1/N_2 = \exp \cdot (-\hbar \omega_0/k T_L) \tag{6}$$

where T_L is the temperature of the lattice. Let $n = N_2 - N_1$. When the microwave resonance radiation is applied we can write for the rate of change of n:

$$dn/dt = (dn/dt)_{\text{r.f.}} + (dn/dt)_{\text{s.l.}} . \tag{7}$$

The first term on the right is the rate of transitions produced by the microwave field and is given by radiation theory as:

$$(dn/dt)_{\text{r.f.}} = -\tfrac{1}{4} \pi \cdot \gamma^2 \cdot H_1^2 \cdot g(\omega-\omega_0) \cdot n \tag{8}$$

where γ is the gyromagnetic ratio of electron, H_1 is the strength of the microwave magnetic field and $g(\omega-\omega_0)$ is the shape function of the absorption line. This is normalized so that

$$\int_0^\infty g(\omega-\omega_0) \cdot d\omega = 1 . \tag{9}$$

The second term on the right in Eq. (7) is the rate of transitions produced by the spin-lattice interaction, and this is given by

$$(dn/dt)_{\text{s.l.}} = (n_0 - n)/T_1 \tag{10}$$

where n_0 is the value of n at thermal equilibrium and is given from equation (6) as

$$n_0 = N_{2_0} - N_{1_0} \approx (\hbar \cdot \omega_0 \cdot N_{1_0}/k T_L) . \tag{11}$$

When equilibrium conditions have set in (dn/dt) for the whole must be zero, and hence:

$$-\tfrac{1}{4} \pi \cdot \gamma^2 \cdot H_1^2 \cdot g(\omega-\omega_0) \cdot n + \frac{n_0 - n}{T_1} = 0 \tag{12}$$

therefore
$$n = n_0 \left[1 + \tfrac{1}{4}\pi \cdot \gamma^2 \cdot H_1^2 \cdot g(\omega - \omega_0) \cdot T_1\right]^{-1}. \tag{13}$$

The rate at which energy is absorbed from the magnetic field may now be calculated. Thus:
$$P_a = -\hbar\omega \, (dn/dt)_{\text{r.f.}} \tag{14}$$

and substituting (8) and then (13) into this gives:

$$\begin{aligned}P_a &= \hbar \cdot \omega \cdot \tfrac{1}{4}\pi \cdot \gamma^2 \cdot H_1^2 \cdot g(\omega - \omega_0) \cdot n_0 \left[1 + \tfrac{1}{4}\pi \cdot \gamma^2 \cdot H_1^2 \cdot g(\omega - \omega_0) \, T_1\right]^{-1} \\ &= \tfrac{1}{2}\omega \cdot \omega_0 \cdot \left(\frac{\gamma^2 \cdot \hbar \cdot n_0}{2\omega_0}\right) \cdot H_1^2 \left[\frac{\pi \cdot g(\omega - \omega_0)}{1 + \tfrac{1}{4}\pi \cdot \gamma^2 \cdot H_1^2 \cdot T_1 \cdot g(\omega - \omega_0)}\right].\end{aligned} \tag{15}$$

This rate of energy absorption can also be related to the total number of the electrons present in the sample, and to the microwave field strength. Thus it can be shown that P_a is given by

$$P_a = \tfrac{1}{2}\left(\frac{\omega^2}{\Delta\omega}\right)\chi_0 H_1^2 \tag{16}$$

where χ_0 is the d.c. susceptibility due to the electrons and is thus directly proportional to the number of electrons in the system, and $\Delta\omega$ is the linewidth of the absorption line as expressed in units of angular frequency. Equating this last equation with Eq. (15) above, and noticing that χ_0 is equal to $\gamma^2 \hbar n_0/2\omega$ it can be seen that an expression for the ratio of n_s/n_0 can be obtained in the form:

$$Z = \frac{n_s}{n_0} = \frac{1}{1 + \tfrac{1}{4}\gamma^2 \cdot H_1^2 \cdot T_1 \cdot T_2}. \tag{17}$$

In this expression n_s represents the difference in population between the two levels under the influence of the saturating microwave field, and it will be seen that the ratio n_s/n_0 effectively measures how much reduction in absorption is produced by the presence of the saturating microwave field. The equation implies that the absorption of power will be lower the larger the value of T_1, and that the fall-off in absorption will increase with increasing levels of microwave input field H_1. Moreover, this decrease in expected absorption will occur first in the centre of the lines where the greatest power is absorbed and only affect the wings as the value of H_1 rises still further. Hence, this saturation effect, produced by large values of T_1 and H_1, will not only reduce the actual power absorption but also alter the line shape, reducing it in the centre before the wings, and thus increasing the apparent width.

In considering the effect of saturation on the width of the absorption line care has to be taken to distinguish between what is known as homogeneously broadened lines and inhomogeneously broadened lines. "Homogeneously broadened" lines are those in which the initial broadening is by interactions within the spin system, or from an external interaction which is fluctuating rapidly compared with the time taken for a spin transition. It therefore includes normal dipole-dipole interaction, spin-lattice interaction and motional or exchange narrowing. Lines are said to be "inhomogeneously broadened" however, when the interactions come from outside the spin system and vary slowly in time compared with the spin transition. Typical interactions of this sort would be unresolved hyperfine components and the inhomogeneities associated with the magnetic field itself.

When saturation of homogeneously broadened lines occurs all the energy will be rapidly shared and the line shape will change in the manner discussed above, the centre of the line being reduced in intensity first, and thus an overall broadening occurs. In the case of inhomogeneously broadened lines, however, where the line shape is the envelope of several real absorption lines, saturation will occur for each of these separately and all the peak heights will be reduced by the same factor. The envelope of the lines will therefore retain the same shape and the power absorption will fall in the same proportion across the whole width.

It follows that a careful study of change in absorbed power, and any change in width on saturation, allows the mechanism of broadening to be determined, and the presence of any unresolved hyperfine structure to be detected.

V. g-values and their theoretical interpretation.

13. Free radical g-values. It has already been pointed out that the g-value of a completely free electron spin is 2.0023 but that any admixture of orbital motion may cause a considerable shift from this value. If the unpaired electron is moving in a delocalised molecular orbital, however, there will be very little coupling to any orbital motion around any particular atom, and as a result most free radicals have g-values which are very close to 2.0023. The two exceptions, in the case of free radicals, are those containing oxygen or sulphur. In these cases the electron can become highly localised close to the oxygen or sulphur atom and considerable spin-orbit coupling then occurs. This produces g-values of magnitude about 2.02, and this large shift from the free spin value can be used to identify the presence of oxygen or sulphur in a free radical.

In general, however, very little information can be deduced from the magnitude of the g-value obtained for a free radical. Some significant variations have been noted and identified with particular chemical groupings or properties[11], but in the great majority of cases information on free radicals is deduced from measurements on their hyperfine splittings rather than their g-values.

The situation is very different for the case of transition group complexes, however, since considerable spin-orbit coupling may occur for an electron moving in an atomic $3d$ or similar orbital, and hence very large shifts in the g-value may be obtained. It also follows that a careful analysis of the magnitude of the observed g-values and, in particular, their variation with angle, can give very useful and detailed information on the nature of the electron's orbital and the bonding associated with the particular atom in question. The general theoretical interpretation of the g-values associated with transition group compounds will, therefore, now be outlined.

14. Orbital level splitting in crystalline electric fields. It will be evident from Fig. 1 that, in essence, the g-value is a measure of the rate at which the two components of the ground level diverge in energy when an external magnetic field is applied. The theoretical calculation of the g-value to be associated with any particular case thus involves an accurate quantum designation of the ground level, and then the application of perturbation theory to calculate the effect of any spin-orbit coupling and then the effect of an applied magnetic field.

If the atom being studied is completely free, the ground state can be accurately specified by a certain J value, and the splitting in energy produced by an external magnetic field is given by the direct theory of the Zeeman effect. In this case the g-value is, in fact, identical with the Landé splitting factor, and electron resonance measurements on free atoms produced by photolysis of various gases has shown that this is indeed true[12].

---

[11] A. T. STONE: Molecular Physics **7**, 311 (1964).
[12] R. BERINGER, and M. A. HEALD: Phys. Rev. **95**, 1474 (1954).

If the atom under investigation is situated in a crystalline lattice, however, as is the case in most solid state studies, there will be strong crystalline electric fields acting on it, produced by the dipoles on the surrounding ligand atoms. These electric fields can produce a strong Stark effect splitting, and the first problem in any theoretical g-value analysis is to decide which orbital level will be lowest for a given electron configuration in a crystalline field of given symmetry.

The theoretical method employed to solve this particular problem has now become known as "ligand field" theory, and it is a treatment which combines the direct effects of the internal electric fields, as dealt with in the earlier "crystal field theory", with the "molecular orbital" approach of admixing the π and σ

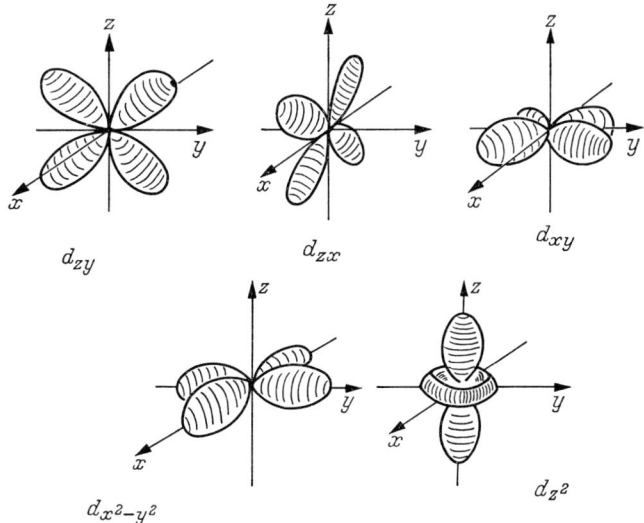

Fig. 17. Spatial distribution of 3 d orbitals.

orbitals of the atom itself with those of its surrounding ligands. The way in which this dual approach works out in practice may be considered by taking the particular case of the "d" orbitals of the first transition group atoms, and determining what energy level systems will be expected for these when placed in different crystalline electric fields.

The first step in solving this problem may be taken by considering the form of the "d" orbitals for a completely free paramagnetic atom, with no external fields to influence it. The physical distribution of these orbitals in space is shown in Fig. 17 and it can be seen that five different spatial distributions are, in fact, possible. The first corresponds to the four electron density lobes pointing along the x and y axes respectively, and this is labelled the $d_{x^2-y^2}$ orbital. The next is obtained by rotating this through 45° so that the electron density lobes are still in the xy plane but bisect the x and y axis; this is termed the d_{xy} orbital. Two more similar orbitals can be obtained when the electron density lobes bisect the axes in the yz and zx planes, these being labelled the d_{yz} and d_{zx} orbitals respectively. The only remaining configuration that is then possible is that in which the electron density is mainly concentrated along the z axis itself with a small annulus of probability density around this axis in the xy plane. It will be noticed that the first four out of these five orbitals are very similar, and consist of four lobes at right angles to each other. Only this last orbital, which is designated the d_{z^2} orbital, is significantly different in shape from the others. This difference arises from the fact that the Oz axis has been chosen as the axis of quantisation.

In the absence of any perturbing electric or magnetic field these orbitals will all have the same energy, since there is no direction to differentiate between them. It might be noticed in this connection that if all five orbital distributions are added together, a spherically symmetric distribution of charge will be produced similar to an "s" orbital distribution, and this is, of course, a fact that should be expected for all closed shells. The particular energy level splittings which are to be expected from the presence of surrounding ligand atoms may now be considered.

The simplest case to consider initially is one central paramagnetic atom surrounded by six ligand atoms, with the two members of each pair placed at

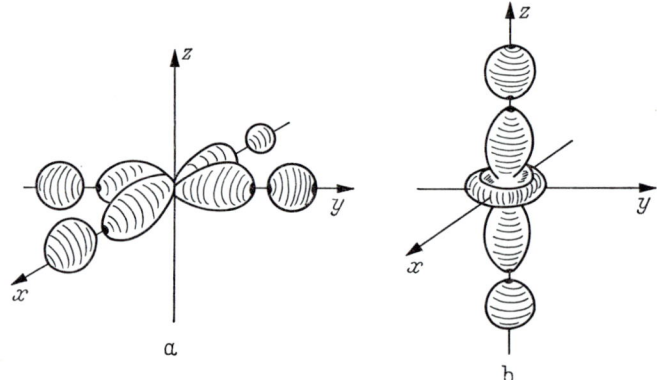

Fig. 18. Effect of octahedral crystalline field on $3d$ orbitals.

equal distances away from the central paramagnetic atom along the x, y and z axes respectively. This is indicated diagramatically in Fig. 18. It becomes clear that the $d_{x^2-y^2}$ and the d_{z^2} orbitals are now in quite a different configuration from the other remaining three. Thus, in the case of $d_{x^2-y^2}$ orbital, the electron spin density of the d orbital is concentrated so that it points along the x and y axes. There will, therefore, be considerable repulsion between this orbital and any containing an electron spin density which resides on a point along the x or y axis themselves. It may be noticed that the d_{xy} orbital does not suffer this form of repulsion in the same way, since its lobes are pointing out along the bisector of the 'xy' axis. There will, therefore, not be any very strong interaction between it and the orbitals of the surrounding ligand atoms located on the axes. It would, therefore, be expected that, as a first order effect, the atoms surrounding the ion complex will split the energy level system so that the $d_{x^2-y^2}$ level is noticeably above the d_{xy}, d_{zx} and d_{yz} levels. The same arguments apply to these last two orbitals as for the d_{xy} orbital.

The d_{z^2} orbital will also be affected in the same way as the $d_{x^2-y^2}$ orbital, since its two lobes are also pointing towards positions of high electron density, located on the ligand atoms along the z axis. The first effect of a presence of these six surrounding ligand atoms will, therefore, be to split the energy level pattern of the "d" orbitals into two groups. The d_{xy}, d_{yz} and d_{zx} will be together in the lower energy group, while the $d_{x^2-y^2}$ and d_{z^2} are both raised in energy owing to the repulsion associated with the electrons on the ligand atoms. These two groupings are also often referred to as the t_{2g} and the e_g groups of orbitals, from the nomenclature of group theory, which can also be applied to determine these sets of levels in a very general fashion.

It should be pointed out that this case, which we have considered in detail, was for the particular symmetry produced by six ligand atoms equally spaced around the central ion. This kind of field is normally said to have octahedral symmetry, but is only one of several types of crystalline field symmetry that are met in practice. Another that often occurs is produced by four ligand atoms at opposite corners of a cube with the ion at the centre of the cube. In this case a tetrahedral field symmetry is produced and this acts in exactly the opposite way on the two sets of "d" orbitals which have just been considered. Thus, in a

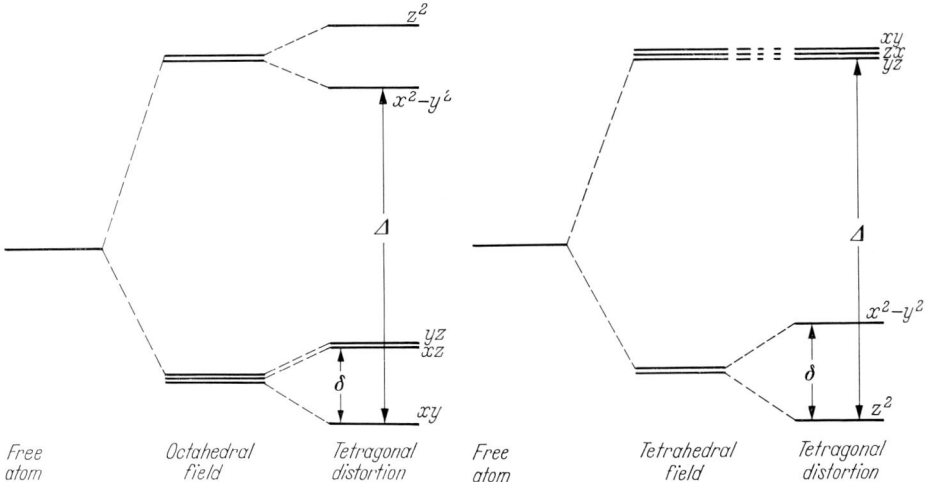

Fig. 19. Orbital energy level splitting produced by octahedral field with tetragonal distortion.

Fig. 20. Orbital energy level splitting produced by tetrahedral field with tetragonal distortion.

tetrahedral field the two orbitals belonging to the e_g group now have the lowest energy, while the t_{2g} group are raised to the higher energy position. Furthermore, it is often found that distortions of the perfect octahedral, or perfect cube, around the central ion produce lower symmetry fields as well. These lower symmetry field distortions will often split the degeneracy of the e_g and t_{2g} groups, so that only a single orbital level is left in the ground state. Examples of these further splittings are shown in Fig. 19 and 20 for the particular cases of an octahedral field with a tetragonal distortion, and also for a tetrahedral field with a tetragonal distortion. It is seen that in the first case the lowest orbital corresponds to the d_{xy}, whereas in the second case the lowest orbital is the d_{z^2}. The ways in which these particular orbital splittings can be calculated for different field symmetries have been discussed at great length in recent years and those interested in further details are referred to such books as those by ORGEL[13] or GRIFFITH[14].

Although the consideration of the symmetry of the particular crystalline field allows the details of the orbital level splitting to be determined in this way, the ground state orbital that is left is usually not a pure energy state. The spin-orbit coupling interaction will admix some of the character of the higher orbital

---

[13] L. E. ORGEL: Ligand Field Theory. London: Methuen & Co. 1960.
[14] J. S. GRIFFITHS: Theory of Transition Metal Ions, chapters 8 and 9. Cambridge University Press 1961.

states into this ground level and before the effect of the externally applied field can be considered, the result of the spin-orbit interaction must first be assessed.

15. Effect of spin orbit coupling. The spin orbit coupling interaction is essentially one of a magnetic nature, since it couples the magnetic dipole moment of the electron spin itself with the magnetic field produced by the orbital motion. The interaction can therefore be represented by a vector product of the orbital and spin quantum numbers L and S multiplied by the interaction constant which is normally denoted by λ. This energy of interaction, which can thus be represented as $\lambda \cdot \boldsymbol{L} \cdot \boldsymbol{S}$ will now act as a perturbation on the energy splittings that have already been produced by the effect of the crystalline electric field. The best way in which the effect of the perturbation can be illustrated is probably to take one of the specific examples of orbital level splittings that have already been considered.

The particular example chosen is that of the tetrahedral field with a tetragonal distortion, which has the d_{z^2} orbital level lowest. This orbital level will, in fact, need to be specified by a pair of numbers since the quantum number designating the electron spin has not yet been allocated. For this particular orbital level the ground state would therefore be twofold degenerate consisting of the two levels $|0; \tfrac{1}{2}\rangle$ and $|0; -\tfrac{1}{2}\rangle$ respectively, if there was no spin-orbit interaction.

The effect of the spin-orbit interaction will now admix small amounts of higher states with different orbital and spin quantum designations. The amount of this admixture can be determined by treating $\lambda \cdot \boldsymbol{L} \cdot \boldsymbol{S}$ as a perturbation operator, and the general application of first order perturbation theory will then give the designation of the ground state as

$$|0,\tfrac{1}{2}\rangle + \sum_{M_L}\sum_{M_S} |M_L, M_S\rangle \cdot \frac{\langle M_L, M_S|\lambda \cdot \boldsymbol{L}\cdot\boldsymbol{S}|0,\tfrac{1}{2}\rangle}{E_{(0,\tfrac{1}{2})} - E_{(M_L, M_S)}} \tag{18}$$

where the summation is to be taken over all other states, and the denominator is always the energy difference between each state being considered, and the ground state.

This summation is evaluated by expanding $\lambda \cdot \boldsymbol{L} \cdot \boldsymbol{S}$ in the standard form

$$\lambda(L_x S_x + L_y S_y + L_z S_z) = \tfrac{1}{2}\lambda \cdot (L_+ S_- + L_- S_+) + \lambda \cdot L_z S_z \tag{19}$$

where $L_\pm = L_x \pm i L_y$ and $S_\pm = S_x \pm i S_y$.

The standard results for matrix elements involving L_+, L_-, S_+ and S_- operators can then be applied. It is then found that, in the particular case being considered, the only excited state which will have a non-zero coefficient for its admixture parameter is the state $|1, -\tfrac{1}{2}\rangle$. The actual magnitude of this admixture coefficient also follows from the standard expressions for $\langle 1|L_+|0\rangle$ and $\langle -\tfrac{1}{2}|S_-|\tfrac{1}{2}\rangle$, and it is given by $-\sqrt{\tfrac{3}{2}}\cdot\dfrac{\lambda}{\varDelta}$.

In the same way it can be shown that the other of the two degenerate ground states has some of the $|-1, \tfrac{1}{2}\rangle$ higher orbital level admixed into it with the same magnitude for the admixture coefficient.

The effect of the spin-orbit coupling has therefore been to produce a degenerate ground state consisting of the two levels

$$|a\rangle \equiv |0,\tfrac{1}{2}\rangle - \frac{\lambda}{\varDelta}\sqrt{\tfrac{3}{2}}|1,-\tfrac{1}{2}\rangle \tag{20}$$

$$|b\rangle \equiv |0,-\tfrac{1}{2}\rangle - \frac{\lambda}{\varDelta}\sqrt{\tfrac{3}{2}}|-1,\tfrac{1}{2}\rangle. \tag{21}$$

Other admixtures will, of course, be obtained for other orbital splittings produced by fields of different symmetry, but this example may afford to illustrate the way in which this particular part of the calculation is effected.

16. Effect of external magnetic field and g-value derivation.

The effect that the externally-applied magnetic field has on these admixed levels is now considered. There are, in fact, two effects to be considered. Firstly, how the two degenerate states themselves may be mixed by its application, and, secondly, how the two states separate in energy as H increases. This will differ for different directions of H compared with the axis of crystalline symmetry, and a specific example may again serve to illustrate the method of calculation.

If the magnetic field is applied along the $0x$ direction — i.e. at right angles to the direction of the crystalline electric field $0z$ — then the perturbation it produces can be written in the form

$$\beta H_x [L_x + 2 S_x] \qquad (22)$$

which can be re-written in the form

$$\beta H_x (\tfrac{1}{2} L_+ + \tfrac{1}{2} L_- + S_+ + S_-). \qquad (23)$$

This operator has no diagonal elements within the pair of states $|a\rangle$ and $|b\rangle$ since each member of the KRAMER's doublet comprises two terms which differ in both M_L and M_S. The off-diagonal elements do not vanish, however, since each term in $|a\rangle$ is connected with one, or both, in $|b\rangle$ through one of the operators L_+, L_-, S_+ or S_-. The two off-diagonal terms are both equal to $\beta H_x [1 + k \sqrt{6}]$ where k is the admixture coefficient.

The two off-diagonal terms thus become $\beta H_x \left(1 - \dfrac{3\lambda}{\Delta}\right)$ and the two levels, under the influence of the externally applied field, H_x, then split into the sum and difference of these two off-diagonal terms to give an energy separation of

$$2\beta H_x \left[1 - \frac{3\lambda}{\Delta}\right] \qquad (24)$$

giving a value

$$g_\perp = 2 \left[1 - \frac{3\lambda}{\Delta}\right]. \qquad (25)$$

This particular example has been worked through in detail to show how an expression for the g-value can be derived from the orbital splitting and spin-orbit coupling constant. Different paramagnetic ions in different crystalline fields will, of course, have different admixture coefficients and hence different values for $g_\parallel$ and $g_\perp$. In general, however, the g-value is related to both Δ and λ and, a knowledge of these two parameters is necessary before a theoretical calculation of its value can be attempted.

So far in this theoretical treatment it has been assumed that the unpaired electron has been localised on one paramagnetic atom and we have been only concerned with its orbital level splitting. In a large number of cases, however, the unpaired electron becomes somewhat delocalised, even in transition group complexes, and the first-order perturbation theory summarised above then has to be modified. Thus, not only may hyperfine structure be observed from the ligand nuclei, but the actual g-values obtained will be influenced by the percentage of covalent binding present. The four general effects that may always be expected when a significant amount of covalent bonding occurs in a complex can be summarised as:

(i) a reduction in the orbital contribution to the g-value as indicated above;

(ii) a change in the magnitude of the hyperfine structure from the central ion;

(iii) an increase in the spin-lattice relaxation time;

(iv) the appearance of superhyperfine splitting from the magnetic moments of the ligand nuclei.

These effects were explained in a general way by STEVENS[15], who extended VAN VLECK's[16] theory, by demonstrating that, in suitable circumstances, the mixing of the ligand wave functions had the effect of lowering all the matrix elements of the angular momentum L by a small factor which could be related to the amount of the admixture. These reductions lead to a more complete quenching of the orbital momentum and thus produce the first three of the effects listed above. The magnitude of the superhyperfine structure can be calculated by assigning the electrons to the molecular orbital levels of the complex, when the magnetic electrons will be found to spread out into the antibonding orbitals and produce the observed transferred hyperfine interaction.

It will be evident that the basic steps that must be undertaken when applying ligand field theory to any particular complex are, therefore

(i) To tabulate the energy level diagrams of the central metal ion and of the ligand atoms separately on either side of a diagram;

(ii) To construct the energy level diagram of the complex by admixing the d-wave functions and orbitals of the central ion with those of the ligand atoms which have the appropriate symmetry. This will produce a set of bonding orbitals, with the main component from the ligand atoms, and a set of antibonding orbitals, with the main component from the d-wave functions;

(iii) The total number of electrons belonging to the $3d$-shell of the ion, and to the outer orbits of the ligands, are now fed into the molecular orbital pattern, from the bottom up, in pairs. It will be found that the bonding orbitals then become filled with paired electrons, and do not contribute to the magnetic effects, while the unpaired electron or electrons reside in the antibonding orbitals.

(iv) Attention is then concentrated on the molecular orbital containing the unpaired electron, and on the orbitals immediately above it. Detailed expressions can then be derived for the g-values and hyperfine interaction parameters in terms of the splittings of these orbital levels, the spin-orbit coupling constant of the paramagnetic ion, and the admixture coefficients.

It can be seen from this summary on the theoretical calculation of g-values that considerable information on the bond structure and state of the paramagnetic ion and its ligands can often be obtained from a detailed analysis of its g-value and the variation of this with angle.

VI. The Spin Hamiltonian.

17. The concept of the Spin Hamiltonian. One of the most useful concepts that has been introduced into the basic theory of electron resonance is that of the "Spin Hamiltonian". This idea was introduced by ABRAGAM and PRYCE[17], and its sucess depends on the fact that the energy changes, which are calculated on first and second order perturbation theory, are precisely the same if the effect of the

---

[15] K. W. H. STEVENS: Proc. Roy. Soc. (London) **219**A, 542 (1953).
[16] J. H. VAN VLECK: J. Chem. Phys. **3**, 807 (1935).
[17] A. ABRAGAM, and M. H. L. PRYCE: Proc. Roy. Soc. (London) **205**A, 135 (1951).

orbital angular momentum is ignored and is replaced by an *anisotropic* coupling between the electron spin and the magnetic field.

Thus, in the derivation of $g_\perp$ in the last section a certain energy splitting between the levels was obtained, and if $g_\parallel$ had been calculated for a magnetic field applied along the $0z$ axis, another energy splitting would have been obtained. In deriving these splittings full account of the orbital momentum was taken, but the same results would have been achieved if the coupling between the electron spin and the external field had been represented by an anisotropic expression of the form

$$\beta \cdot \mathbf{H} \cdot g \cdot \mathbf{S} = \beta \cdot g_\parallel \cdot H_z S_z + \beta \cdot g_\perp \cdot (H_x S_x + H_y S_y). \tag{26}$$

This then predicts a g-value angular variation with axial symmetry, and the actual physical effect may be explained by realising that the orbital angular momentum induced by the spinning electron makes the spin easier to orient in some directions than others.

The g-value is not the only parameter of electron resonance spectra that can be represented in this way. Thus, both the electronic splitting, and the hyperfine splitting, which have already been briefly discussed, can also be represented formally in a similar manner.

18. Formal treatment of electronic splitting. It has already been seen in Sect. II.5 that the internal crystalline electric fields will normally lift the degeneracy of the three possible M_s values associated with an $S=1$ of two unpaired electrons, and that this "electronic splitting" between the $M_s=0$ and $M_s=\pm 1$ levels will vary markedly with angle.

This interaction energy can be taken into account formally in the Spin Hamiltonian by adding three second-order terms of the form

$$D_x S_x^2 + D_y S_y^2 + D_z S_z^2. \tag{27}$$

The three D parameters represent the magnitudes of the electronic splitting when the external magnetic field is applied along different axes. They have dimensions of energy and a sum equal to zero. Hence, for an isotropic g-value the Spin Hamiltonian becomes

$$\mathcal{H} = \beta \cdot g \cdot (H_x S_x + H_y S_y + H_z S_z) + D S_x^2 + D S_y^2 + D S_z^2. \tag{28}$$

The allowed spin states and energies of the system will then be the eigenstates and eigenvalues of the Spin Hamiltonian. We therefore need to determine the matrix elements of $\mathcal{H}$ between the three levels $|+1\rangle$, $|0\rangle$ and $|-1\rangle$, which are eigenstates of S_z with eigenvalues $|1\rangle$, $|0\rangle$, and $|-1\rangle$. The standard forms of the matrix elements of the operators S_x, S_y, S_z, S_x^2, S_y^2 and S_z^2 are therefore used, and the matrix elements of the Spin Hamiltonian can then be found. For example

$$\begin{aligned}\langle 1|\mathcal{H}|1\rangle &= g\cdot\beta\cdot[H_x\langle 1|S_x|1\rangle + H_y\langle 1|S_y|1\rangle + H_z\langle 1|S_z|1\rangle]\\ &\quad + D_x\langle 1|S_x^2|1\rangle + D_y\langle 1|S_y^2|1\rangle + D_z\langle 1|S_z^2|1\rangle\\ &= g\cdot\beta\cdot[H_x\cdot 0 + H_y\cdot 0 + H_z\cdot 1] + D_x\cdot\tfrac{1}{2} + D_y\cdot\tfrac{1}{2} + D_z\cdot 1\\ &= g\cdot\beta\cdot H_z + \tfrac{1}{2}(D_x+D_y) + D_z.\end{aligned} \tag{29}$$

Other matrix elements can be determined in precisely the same manner. The general equation giving the eigenvalues E and the eigenstates

$$C_1|1\rangle + C_0|0\rangle + C_{-1}|-1\rangle$$

then becomes

$$0 = \begin{bmatrix} g\beta H_z + \tfrac{1}{2}(D_x+D_y)+D_z-E & \sqrt{\tfrac{1}{2}}g\cdot\beta\cdot(H_x-iH_y) & \tfrac{1}{2}(D_x-D_y) \\ \sqrt{\tfrac{1}{2}}g\beta(H_x+iH_y) & D_x+D_y-E & \sqrt{\tfrac{1}{2}}g\beta(H_x-iH_y) \\ \tfrac{1}{2}(D_x-D_y) & \sqrt{\tfrac{1}{2}}\cdot g\beta\cdot(H_x+i\cdot H_y) & -g\beta H_z+\tfrac{1}{2}(D_x+D_y)+D_z-E \end{bmatrix} \begin{bmatrix} C_1 \\ C_o \\ C_{-1} \end{bmatrix}. \quad (30)$$

This can in principle be solved for any conditions, but the extreme case of zero field can be considered first. Thus we put $H_x = H_y = H_z = 0$ and expand the secular determinent as a cubic equation in E.

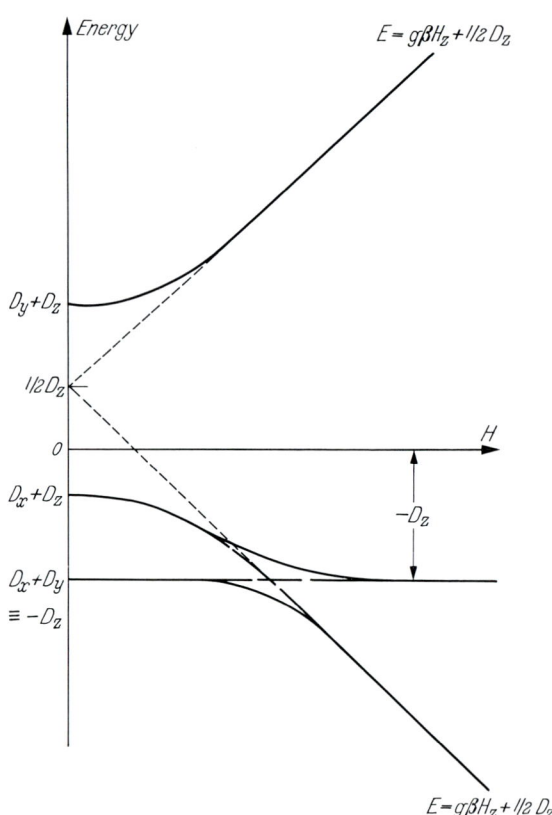

Fig. 21. Energy level splitting of $S=1$ components in zero field and with external field applied.

The three energies and their associated spin states are then found to be

$$\begin{array}{ll} \text{Energies} & \text{Spin states} \\ D_y+D_x & 0 \\ D_z+D_x & \sqrt{\tfrac{1}{2}}(|1\rangle+|-1\rangle) \\ D_z+D_y & \sqrt{\tfrac{1}{2}}(|1\rangle-|-1\rangle). \end{array} \quad (31)$$

The splitting of these three states in zero magnetic field is shown in Fig. 21 on the left-hand side.

The effect of applying an external magnetic field along the $0z$ axis can now be considered. i.e. Only
$$H_x = H_y = 0.$$
The secular equation then shows that the energy of $|0\rangle$ is unaffected by H_z. The other two energies are modified and are obtained by solving a quadratic equation in E
$$E = \tfrac{1}{2}[2D_z + D_x + D_y \pm \sqrt{\{4(g \cdot \beta \cdot H_z)^2 + (D_x - D_y)^2\}}]. \tag{32}$$
Assuming $4(g \cdot \beta \cdot H_z)^2 \gg (D_x - D_y)^2$ this gives for the energies and eigenstates

$$\left. \begin{array}{cc} \text{Energies} & \text{Spin States} \\ g\beta H_z + \tfrac{1}{2} D_z & |1\rangle \\ -g\beta H_z + \tfrac{1}{2} D_z & |-1\rangle \\ -D_z & |0\rangle. \end{array} \right\} \tag{33}$$

The allowed transitions between these three levels thus have magnitudes

and
$$\left. \begin{array}{c} g\beta H_z + \tfrac{3}{2} D_z \\ g\beta H_z - \tfrac{3}{2} D_z. \end{array} \right\} \tag{34}$$

This behaviour of the levels in large values of applied magnetic field is shown on the right-hand side in Fig. 21, and the dashed lines cross the intermediate region over the field values for which the above approximation is not valid.

It is evident that again in this case the concept and use of a Spin Hamiltonian enables the experimental data to be summarized in a very direct and convenient form. It should be stressed that the Spin Hamiltonian itself does not produce any theoretical reasons for the magnitudes of the observed g-values, electronic or hyper-fine splittings — but it does enable their angular variations to be summarized in a precise and succint way in the form of one or two parameters which can then be compared with the theoretical predictions.

VII. Hyperfine splitting and the Spin Hamiltonian.

19. Magnetic interaction. The production of a hyperfine structure on the electron resonance spectrum by the magnetic interaction between the magnetic moments of the electron and nuclear spins has already been discussed briefly in Sect. II.4. This coupling can be represented by a term $A \, \mathbf{S} \cdot \mathbf{I}$ and the Spin Hamiltonian for an axially-symmetric system in the absence of any electronic splitting then becomes

$$\left. \begin{array}{c} \mathscr{H} = g_\| \cdot \beta \cdot H_z \cdot S_z + g_\perp \cdot \beta \cdot (H_x \cdot S_x + H_y \cdot S_y) \\ + A_\| \cdot S_z \cdot I_z + A_\perp (S_x I_x + S_y I_y). \end{array} \right\} \tag{35}$$

The eigenvalues and eigenstates can then be found in exactly the same way as before, the states now being characterised by two quantum numbers $|M_S, M_I\rangle$, and the rules for operating with I_x, I_y and I_z are the same as those for S_x, S_y and S_z.

20. Electric quadrupole interaction. As well as the magnetic interaction discussed above, it is also possible for an interaction to take place between the internal electric crystalline fields and any electric quadrupole moment that the nucleus may have. This interaction will effectively try and align the nucleus up so that its electric symmetry axis is along the axis of the crystalline field, and, as such, it may be in competition with the effect of the externally applied magnetic field. When such competition occurs for an effective axis of quantisation the normal selection rules are likely to be broken down and forbidden transitions appear. Such extra lines are indeed often observed when the external magnetic field is applied nearly perpendicular to the axis of the internal crystalline field.

The effect of such a breakdown in the normal selection rules is shown in Fig. 22. This shows the hyperfine structure observed in a diluted copper salt as the direction of the applied magnetic field is moved round away from the axis of the internal crystal field towards a perpendicular direction. The four equally spaced lines, already discussed in Fig. 3, are now found to have extra components growing between them, and a detailed analysis of the intensities and spacing of these components allows the quadrupole moments of the copper nuclei to be measured.

The electric quadrupole interaction can be represented formally in the Spin Hamiltonian by a term similar in form to the electronic splitting, and, for axial symmetry, takes the form $Q[I_z^2 - \tfrac{1}{3}I(I+1)]$.

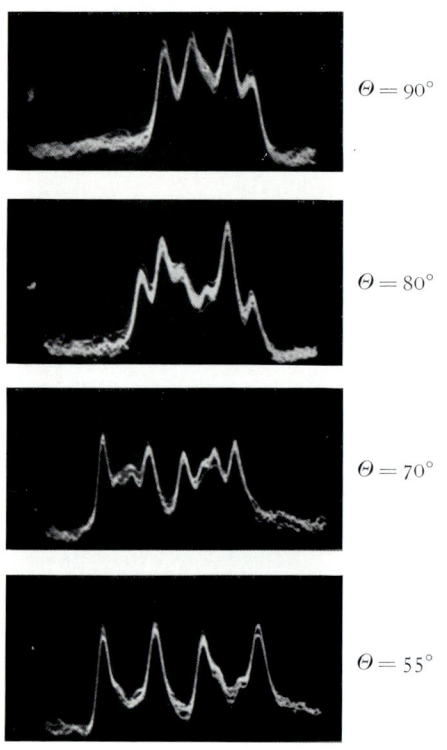

Fig. 22. Effect of quadrupolar interaction on Cu^{++} spectrum as magnetic field approaches perpendicular position.

21. The complete Spin Hamiltonian. The various interactions which have been discussed above can now be assembled in a complete Spin Hamiltonian. The one given below is, in fact, for the case of axial symmetry but other symmetries, such as rhombic only serve to introduce extra parameters to allow for the lower symmetry. The full Spin Hamiltonian for axial symmetry can therefore be written as

$$\mathcal{H} = \beta \cdot g_{||} \cdot H_z S_z + \beta \cdot g_\perp [H_x S_x + H_y S_y] \qquad \text{g-value}$$
$$+ D[S_z^2 - \tfrac{1}{3}S(S+1)] \qquad \text{electronic splitting}$$
$$+ A_{||} I_z S_z + A_\perp (I_x S_x + I_y S_y) \qquad \text{magnetic h.f.s.}$$
$$+ Q[I_z^2 - \tfrac{1}{3}I(I+1)] \qquad \text{electronic quadrupole}$$
$$- \gamma \cdot \beta_N \cdot H \cdot I. \qquad \text{direct effect of } H \text{ on nucleus.}$$

This equation is effectively stating that all the experimental results obtained on an electron resonance spectrum can be summarized by the sets of parameters $g_{||}, g_\perp; D; A_{||}, A_\perp$ and Q. The actual assignment of values to these parameters is often helped by comparing the experimentally measured angular variations with those predicted by the Spin Hamiltonian.

In this connection Bleaney[18] derived some general expressions for angular variation in terms of the principal g-values and other parameters. For example, the resonance field value, and its variation with angle Θ, when an electronic splitting D exists, is given by

$$H = H_0 - \frac{D}{g\beta}\left(M_S - \frac{1}{2}\right)\left(3\frac{g_\parallel^2}{g_\perp^2}\cos^2\Theta - 1\right) + $$
$$+ (D/g\beta)^2 \cdot \frac{1}{2H_0} \cdot \left(\frac{g_\parallel \cdot g_\perp \cdot \cos\Theta \cdot \sin\Theta}{g^2}\right)^2 (4S(S+1) - 24M_S(M_S-1) - 9) - \qquad (36)$$
$$- (D/g\beta)^2 \cdot \frac{1}{8H_0}(g_\perp \sin\Theta/g)^4 (2S(S+1) - 6M_S(M_S-1) - 3).$$

This is a complicated expression but it has been checked in detail on several transition group ions, and precise fitting of the experimental data in this way does allow very accurate values to be obtained for D etc.

It is also evident from expressions such as this that far greater information is always available from E.S.R. studies of single crystals than from liquids or polycrystalline specimens.

In fact, quite a number of the angularly dependent interactions may be averaged to zero by the rapid tumbling motion of molecules in the liquid state, and, although this does result in very narrow lines with dipolar broadening averaged to zero, it has the corresponding disadvantage that all the information associated with the angular variation has been lost.

VIII. Examples of spectrum analysis.

The way in which the different terms of the Spin Hamiltonian can be related to an experimentally measured electron resonance spectrum can probably best be illustrated by taking a specific example. The case of the Mn^{++} ion is probably the most useful in this respect, since it has a pronounced electronic splitting as well as a hyperfine splitting from the Mn^{55} nucleus.

The divalent manganese ion is in a ^6S state, having five unpaired electron spins. In an ionic type of bonding, as found in most hydrated simple inorganic salts, none of the magnetic $3d$ orbitals are required for bonding to the ligand atoms, and the five unpaired electrons therefore enter the five $3d$ orbitals singly to give a total electronic spin for the ion of $S = \frac{5}{2}$. This total spin vector can then take up different orientations with respect to the internal crystalline field, and the Stark splitting so produced splits the levels into three doublets corresponding to $M_s = \pm\frac{1}{2}$, $\pm\frac{3}{2}$ and $\pm\frac{5}{2}$. In this case the $M_s = \pm\frac{1}{2}$ components lie lowest, with the $\pm\frac{3}{2}$ and $\pm\frac{5}{2}$ some 0.05 and 0.15 cm^{-1} above them. This splitting is, in fact, the electronic splitting referred to in the Spin Hamiltonian derivation, and is denoted by the parameter D.

The three doublets can therefore be represented in zero-magnetic field to a first approximation as shown on the left-hand side of Fig. 23, and the way in which they separate as an external magnetic field is applied is shown by movig across to the right-hand side. The selection rule for transitions between these different electronic levels is $\Delta M_S = \pm 1$ and hence five transitions between the six levels are to be expected, and the field separations between these will enable the value of the splitting parameter, D, to be determined[19].

---

[18] B. Bleaney: Phil. Mag. **42**, 441 (1951).
[19] B. Bleaney, and D. J. E. Ingram: Proc. Roy. Soc. (London) **205 A**, 336 (1951).

Each of these six electronic levels is further split into $(2I+1)$ components by the magnetic interaction with the manganese nucleus, however. Mn^{55} has a nuclear spin of $I=\frac{5}{2}$ and hence six hyperfine components are produced in each transition, as shown on the right-hand side of Fig. 23. The selection rule for transitions between the levels is $\Delta M_I=0$ and hence six hyperfine components are now

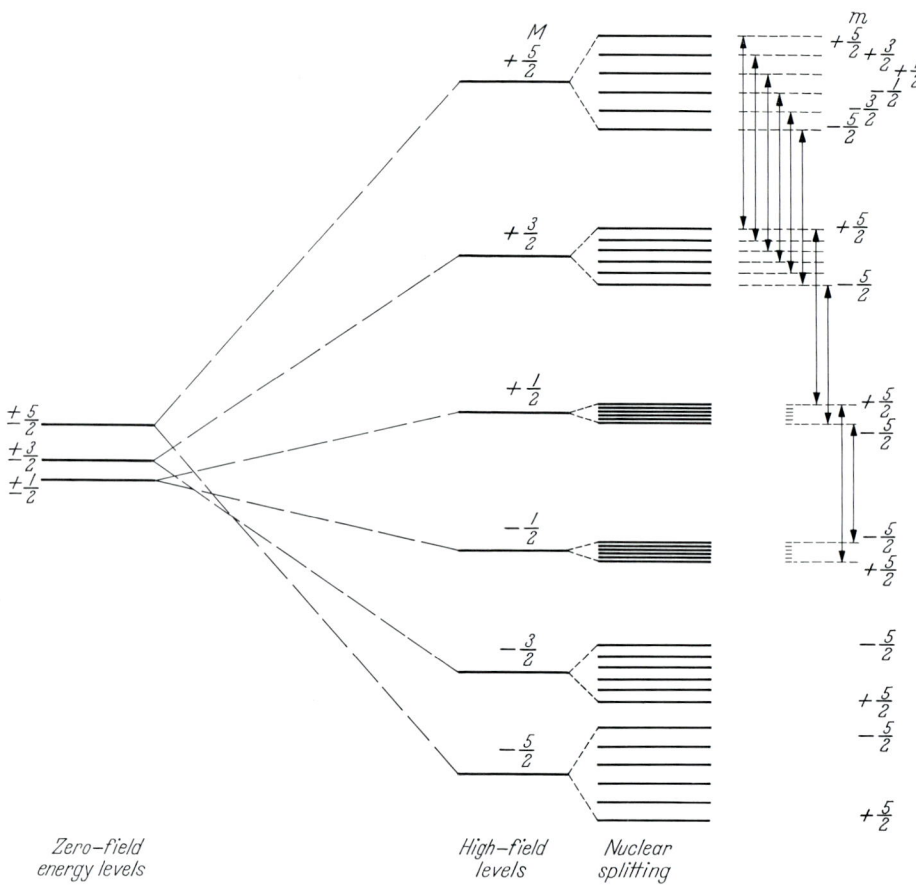

Fig. 23. Energy level splitting of six electronic levels associated with Mn^{++} ion.

expected on each electronic transition, as indicate by the vertical arrows between the top two sets of levels in Fig. 23. It follows that five sets of six hyperfine lines are to be expected for the Mn^{++} e.s.r. spectrum.

The actual spectrum[19] observed in the case of manganese fluosilicate is shown in Fig. 24. The fluosilicate lattice has the advantage that there is only one ion per unit cell and hence all the Mn^{++} ions are orientated in the same direction, and the applied magnetic field makes the same angle with the axis of crystalline electric field for all of them. In Fig. 24 the magnetic field is applied along the trigonal axis which corresponds to the direction of maximum electronic splitting. Even so, there is still some overlap between the hyperfine components of the different electronic splittings, but the five sets of six equally-intense hyperfine components can be readily sorted out, as indicated by the groupings below. It is seen that the hyperfine splitting beetween successive lines is about 100 gauss and the variation across the groups can be explained by the second-order terms in the Spin-Hamiltonian. The centre of each hyperfine group is picked out at the bottom of the diagram, and these splittings give a direct determination of D. Thus, when the magnetic field is applied along the crystalline axis, terms in D^2/H_0 disappear from the expression for the line position, and the separation of the two outer groups should be $2D-3a$, while that of the inner groups is $2D+\frac{5}{3}\cdot a$. Here 'a' is a constant in the Spin Hamiltonian which represents the direct effect of the

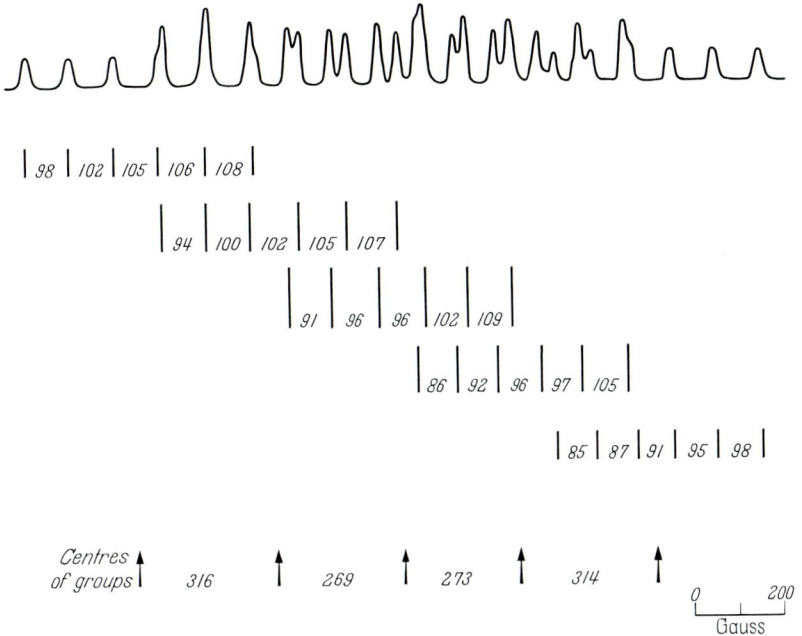

Fig. 24. Electron resonance spectrum observed from Mn^{++} in manganese fluosilicate.

cubic field, while the main electronic splitting D is produced by the trigonal disortion. The values of D and 'a' are thus determined directly as 143 and 10 gauss or 134 and 9×10^{-4} cm^{-1} respectively.

The magnitude of the electronic splitting will vary markedly with angle, however, as already pointed out in Sect. II.5, and this variation is shown quite clearly in Fig. 25. In this figure just the centre points of the five sets of hyperfine components are plotted for different angles that the applied magnetic field makes with the crystalline field axis. This, therefore, represents the angular variation of the electronic splitting and should correspond to that predicted in Eq. (36). The main feature predicted by that equation, i.e. a collapse to zero at an angle of 54° 44', which makes $(3\cos^2\Theta-1)$ equal to zero, can be clearly seen; and the fact that the splitting is not exactly zero at this angle is accounted for by the second-order terms in that equation. This practical illustration of how the observed spectrum varies with angle may help to show the actual meaning of the formal representation of the terms representing the electronic splitting in the Spin Hamiltonian.

In the case of Mn^{++} both the g-value and the hyperfine splitting are fairly isotropic, and hence $g_{||}$ is approximately equal to $g_\perp$ and $A_{||}$ to $A_\perp$. Manganese in

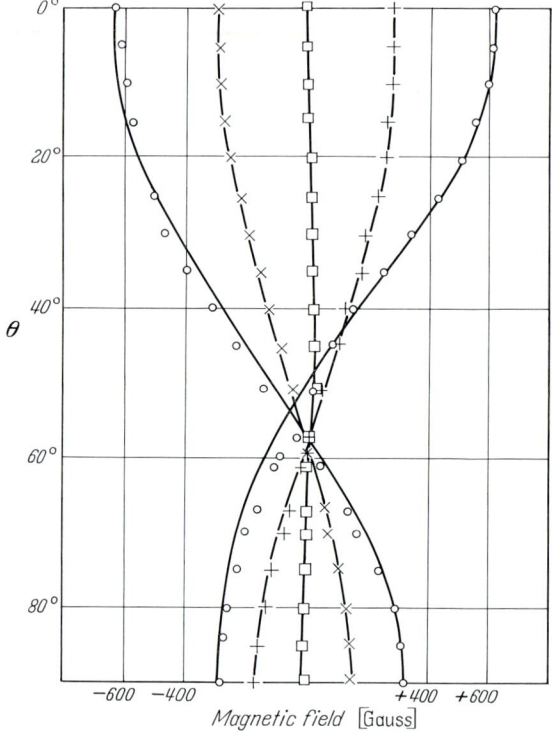

Fig. 25. Variation of electronic splitting with angle. The full lines are the calculated curves from equation 36 and the open circles are the experimental points.

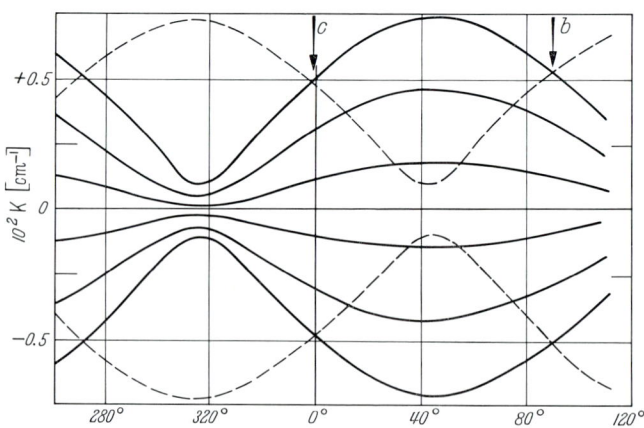

Fig. 26. Angular variation of manganate hyperfine splitting in bc plane, after the g-value variation has been removed.

another valency state has quite anisotropic g-values and hyperfine structure, however, and this case may be worth considering briefly, since it is a useful reminder that the same transition group atom can exist in different valency states which may have very different electron resonance spectra, and it also illustrates the effect of the nuclear electric quadrupole interaction experimentally.

Manganese can in fact be obtained as a manganate ion, if it is grown in a potassium chromate lattice, and, as such, it only has one unpaired electron, with $S=\frac{1}{2}$. There will, therefore, only be one electronic transition for each manganese ion, but this will be split into six hyperfine components by the magnetic interaction with the $I=\frac{5}{2}$ of Mn^{55} as before [20]. There is thus no electronic splitting in this case, but the hyperfine splitting is now very anisotropic as can be seen from Fig. 26. This plots the angular variation of the hyperfine splitting, after the effect of the anisotropic g-value has itself been removed, and the dashed curve is the envelope of a similar set of lines from the second differently-orientated ion in the lattice. The very low symmetry of the crystalline field in this particular

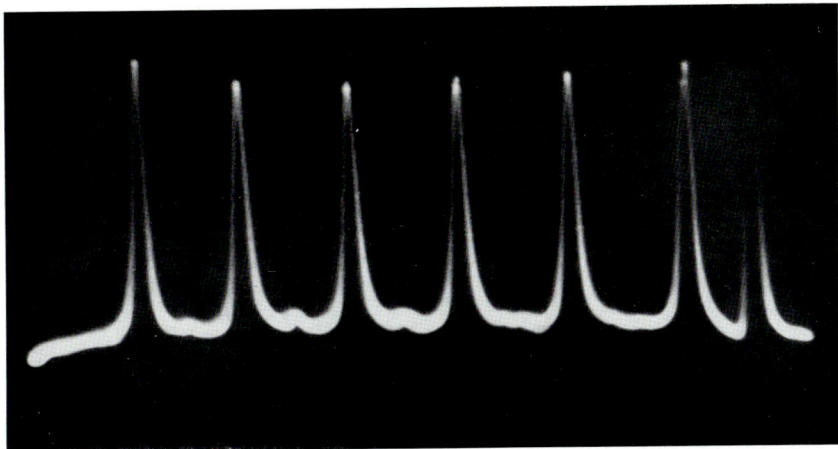

Fig. 27. Part of manganate hyperfine pattern showing weak extra transitions due do quadrupolar interaction.

lattice results in different values for the hyperfine splitting parameter A along the $0x$ and $0y$ axes, and there are thus three parameters required to specify the splitting. A_x, A_y and A_z instead of $A_\perp$ and $A_\parallel$.

The other item of interest in this particular hyperfine pattern is that it is one in which the effect of the nuclear quadrupole interaction can be seen [20]. Fig. 27 shows a portion of the manganate spectrum photographed with the magnetic field along the b axis. Some of the main hyperfine lines are clearly seen, together with a free-radical marker on the extreme right, but weak "quadrupole" lines can also be seen between the two sets of main lines on the left. These arise from the normally-forbidden $\Delta M_I = \pm 1$ transition, as explained before. A detailed analysis of their spacing and intensity allows the Q parameter of the Spin Hamiltonian to be determined, and thus an estimate made of the electric quadrupole moment of the manganese nucleus itself.

These particular examples of electron resonance spectra of transition group ions have been given as the best way of practically illustrating the effect of the different terms in the Spin Hamiltonian. It will be evident that the main power of the Spin Hamiltonian is its ability to summarise the angular variation of the spectra, as well as denote the main sources of interaction or splitting.

In the case of most free-radical spectra, which are normally studied in solution, where the molecular motion averages all the anisotropic interactions to zero, the

---

[20] A. CARRINGTON, D. J. E. INGRAM, K. A. K. LOTT, D. S. SCHONLAND, and M. C. R. SYMONS: Proc. Roy. Soc. (London) **254** A, 101 (1960).

detailed analysis of the spectrum via a Spin Hamiltonian is not so necessary. The more general lines of analysis, as already illustrated in Sect. II.4 are often then sufficient. An increasing number of studies are now being made however, of free-radicals made and trapped in single crystals, and, in these cases, a detailed analysis of the corresponding Spin Hamiltonian is necessary.

IX. General conclusion.

In this article an attempt has been made to present a precise and coherent picture of both the general principles of electron resonance and the experimental techniques that are employed, these introductory sections have then been followed by others which present the basic theoretical analysis of the spectra in somewhat more detail. No attempt has been made, however, to list or consider the many different fields of study in which this technique is now being applied. The number of different applications is continuously increasing, since the number of systems which contain unpaired electrons as an essential part of their mechanism is extremely large and the technique of electron resonance is now being applied very actively not only in physics and chemistry, but also in such fields as biochemistry, to follow active free-radical transients and valency changes in enzyme reactions for example. Some of the present applications of electron resonance are in fact considered in an accompanying article by Professor WERTZ, and no further details will be given here. It might be appropriate, however, if a brief mention is made of the two techniques of *double* resonance that have recently been introduced into this field of study, since they do increase the versatility of the technique very considerably.

The possible implications and applications of carrying out both nuclear resonance and electron resonance at the same time on the same specimen were first considered by OVERHAUSER[21] for the case of the observation of nuclear resonance in lithium metal. He showed that if the metal was irradiated with high power microwave radiation at the electron resonance frequency, at the same time, then the normal population distribution of the nuclear energy levels could be considerably modified, and, as a result, enhanced nuclear resonance signals could be obtained. This prediction was quickly verified by CARVER and SLICHTER[22] for the lithium metal, and has since been developed by a large number of workers to increase the sensitivity of ordinary nuclear resonance measurements by factors of well over one hundred.

It should be noted that the essential feature of this Overhauser technique is that the magnetic field and microwave frequency are held constant at the electron resonance condition so that the electrons are continuously pumped to the higher energy level. Their coupling to the nuclear spins then produces the much larger population difference between the nuclear spin component levels and hence the apparatus is designed to detect and analyse the nuclear resonance signal which is now available at much greater signal strength.

A few years later it was pointed out by FEHER[23] that the inverse of this kind of double resonance could also be carried out. This technique is known as ENDOR (*E*lectron *N*uclear *DO*uble *R*esonance) and again uses high power microwave radiation to saturate the electron resonance transition, but this transition and the effect of applying a simultaneous nuclear resonance frequency to it is now monitored by the equipment. As a result it is found that the high

---

[21] OVERHAUSER, A. W.: Phys. Rev. **92**, 411 (1953).
[22] T. R. CARVER, and C. P. SLICHTER: Phys. Rev. **92**, 212 (1953).
[23] G. FEHER: Phys. Rev. **103**, 500, 834 (1956).

resolution of the nuclear resonance technique is effectively fed into the electron resonance spectrum, whereas in the Overhauser effect the high sensitivity of the electron resonance technique was fed into the nuclear resonance spectrum.

The main application of the ENDOR technique is, in fact, to make measurements on electron resonance lines in which unresolved hyperfine structure is suspected. The nuclear resonance frequency is then used to pull out these unresolved hyperfine splittings. The principle of the method is probably best illustrated by a simple example in which a hyperfine interaction with a nucleus of spin $\frac{1}{2}$, such as a proton, is considered as illustrated in Fig. 28.

Each of the electronic levels will then be split into two and the energies of the four resulting level scan be written down as

Level	Energy
A	$\frac{1}{2}g\beta H + \frac{1}{4}A - \frac{1}{2}g_N\beta_N H$
B	$\frac{1}{2}g\beta H - \frac{1}{4}A + \frac{1}{2}g_N\beta_N H$
C	$-\frac{1}{2}g\beta H + \frac{1}{4}A + \frac{1}{2}g_N\beta_N H$
D	$-\frac{1}{2}g\beta H - \frac{1}{4}A - \frac{1}{2}g_N\beta_N H$

It may be noted that the second term in these expressions for the energy represents the normal hyperfine interaction, whereas the third term arises from the direct interaction of the applied magnetic field with the magnetic moment of the proton. It can be seen from these detailed expressions that the energy difference between levels A and B is not quite the same as that between C and D.

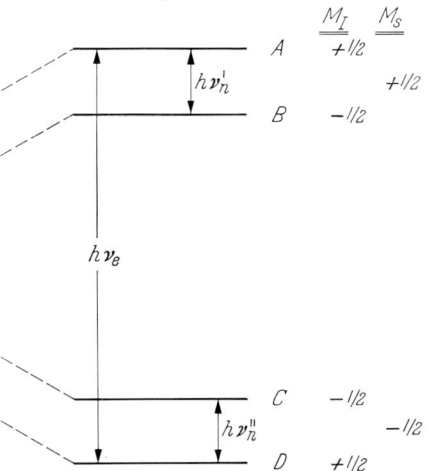

Fig. 28. Energy levels to illustrate ENDOR effect.

In the ENDOR technique a high power electron resonance frequency is first applied to saturate one of the electron resonance transitions, such as that between A and D in the figure. The result of this saturation will be to increase the energy level population of level A and hence to give level A a higher population than level B. Whilst this saturation is taking place, a radio frequency is also applied to the sample with a frequency such that $h\nu_{r.f.}$ is equal to a splitting between A and B: this then stimulates transitions from A to B and the populations of the two levels will return to their normal equilibrium values. As a result, the saturation of the electron transition will be removed, and a strong electron resonance line will then be suddenly obtained, in place of the weakened saturation condition. The net result is that if the detecting system is kept set on the electron resonance signal a sudden increase in this will be obtained when the nuclear resonance signal sweeps through the condition

$$h\nu_{r.f.} = \tfrac{1}{2}A - g_N\beta_N H.$$

It can also be seen that a similar situation arises when the radio frequency sweeps through the resonance value corresponding to the nuclear transition between levels C and D. The saturation of the electron resonance will have reduced the number of atoms in the level D, but when the nuclear resonance transition is induced by the radio frequency field the population of level C and D will be more or less equalised, and as a result the electron resonance signal being observed will suddenly become desaturated and a large signal will be obtained. It follows from this that if the radio frequency signal is slowly swept through a range of values centred on $\nu = A/2h$ a large increase in the electron resonance signal will be obtained when the frequency satisfies either of the conditions

$$h\nu_{r.f.} = \tfrac{1}{2}A \pm g_N \beta_N H.$$

From these two values of the resonance frequency of the r. f. signal the values of both A and g_N can be deduced very accurately.

These ENDOR type measurements were initially applied by FEHER[23] to study resonance from doped atoms in silicon but were also rapidly extended to other systems[24], such as those formed by F centres in irradiated KCl. Normal electron resonance techniques had not been able to resolve any hyperfine structure from such crystals but FEHER was able to resolve out not only the hyperfine splitting from the two chlorine isotopes, but also from the small quantity of potassium 41 by employing the double resonance technique. The spectrum he obtained is

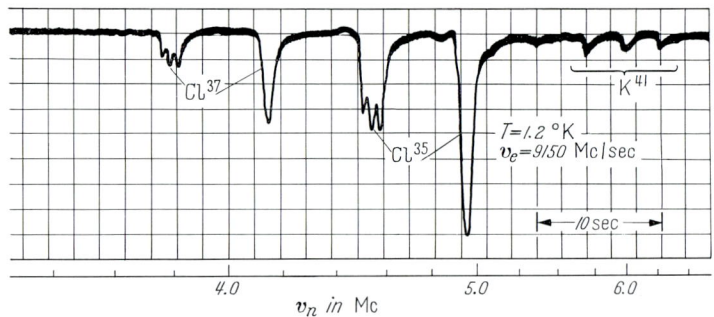

Fig. 29. Double resonance spectrum of F-centres in irradiation KCl.

shown in Fig. 29 and is typical of double resonance tracing of this type. It is to be noted that the ordinate represents the actual absorption produced by the electron resonance transition, and that both the microwave frequency and the value of the applied magnetic field are held constant throughout the experiment at this resonance condition. The abscissa corresponds to the changing value of the applied radio frequency field, and absorption lines will be obtained whenever this frequency corresponds to an actual hyperfine splitting present in the overal energy level pattern. It should be noticed in this connection that the position of the lines correspond to *splittings* and hence no symmetry is to be expected in the pattern as it is traced out. Since the factors which determine whether the desaturation is to occur or not are those which determine the width of the nuclear resonance line, it can be seen that the effective resolution has been increased by a factor of about 1,000, and this spectrum is a very good example of the very great increase in resolution that can be obtained by the ENDOR method.

It may be seen from this example that the ENDOR technique greatly enhances the possible applications of electron resonance, especially where complex hyperfine patterns are to be expected. It should possibly be mentioned, however, that the exact experimental conditions for the optimum operation of ENDOR can take some time to find in any given system since the relaxation times, and their variation with temperature, are rather crucial parameters if effective saturation is to be established.

It affords a very good example, however, of the way in which the techniques of electron resonance are being adapted and developed to apply with increasing sensitivity and resolution in an ever-widening field of research activity.

[24] G. FEHER: Phys. Rev. **105**, 1122 (1957).

Structural Information from Paramagnetic Resonance.

By

JOHN E. WERTZ.

With 34 Figures.

A. Introduction.

I. Types of systems studied.

Perhaps no other technique has given the extremely detailed descriptions of molecular structure or atomic and molecular environment which has been forthcoming from paramagnetic resonance. We follow the practice[1] of using the term to include both electron spin resonance (ESR) and electron-nuclear double resonance (ENDOR). Notwithstanding the very restrictive requirement that the system under study have one or more unpaired electrons, the great number of published data on a wide range of atoms, ions, crystalline defects or molecular radicals is such as to preclude encyclopedic coverage in one chapter. One compilation lists 57 references on Mn^{2+} alone up to 1959![2] Some applications adjudged representative of the multitude of fascinating studies are briefly summarized. Topical headings display the range of applicability of ESR methods, ranging from organic chemistry, to inorganic and physical chemistry, to solid state physics. Some of the most profitable studies have combined paramagnetic resonance techniques with optical or other techniques. Perhaps as much as in any other area, the fascination of theoreticians with experimental data and the concern of experimentalists for theoretical interpretation has contributed to the phenomenal productivity.

Most of the systems studied by paramagnetic resonance (principally ESR) techniques have a single unpaired electron and therefore a doublet ground state. Literally hundreds of free radicals (mostly organic) fall in this category (Sect. B. I.), together with numerous radiation-induced defects in solids (Sect. D.). The $S=\frac{1}{2}$ systems form a continuously graded set ranging from nearly pure spin transitions of dipoles isolated from one another and from the environment to appreciably interacting systems which may also be strongly affected by orbital contributions. Of systems with spin $S \geq \frac{1}{2}$, the transition metal ions — including the rare earth ions — form a special category, since in solids their interaction with a crystalline electric field and their spin-orbit coupling must be taken into account even in the crudest approximations. Other than such ions, there is a growing number of systems for which $S=1$, but for which the spin-orbit coupling is small. These systems may be subdivided into two categories:

1. Singlet ground state and triplet excited state accessible thermally or by optical irradiation.
2. Triplet ground state systems.

---

[1] H. SEIDEL, and H. C. WOLF: Phys. stat. sol. **11**, 3 (1965).
[2] S. A. AL'TSCHULER, and B. M. KOZYREV: Electron Paramagnetic Resonance, p. 161. Translation editor, C. P. POOLE jr. New York: Academic Press 1964.

Excited triplet systems have been extensively studied by optical means, but ESR studies have added greatly to our detailed knowledge of them and have considerably increased the number of known examples. A most important example from the solid state area is the excited state of the M-center in the alkali halides. The number of known systems with triplet ground states is growing rapidly; these include "divalent" carbon in the form of substituted methylenes R_1-C-R_2, interacting dimeric radicals, dinegative ions and charge transfer complexes (Sect. C. III.).

One of the most interesting $S=\frac{3}{2}$ systems is the excited state of the R-center (Sect. C. 26β). Many transition metal ions have $S \geq \frac{1}{2}$; there has been a tremendous output of ESR data about such ions either in diamagnetic hosts or surrounded by covalently-bonded ligands. Considering the number of presently known hosts for transition metal ions, it is unlikely that interest in them will abate. Such ions will receive scant treatment in this chapter since each involves a fairly detailed exposition.

We shall consider examples which are illustrative of composition, configuration and environment of a paramagnetic center, valence state, distribution and sign of spin polarization, molecular motion, interconversion of isomeric species, dipolar interactions, dependence upon host and effects due to temperature variation. Since the basic theory is covered in the previous chapter by INGRAM, we make only a few remarks about g- and hyperfine tensors and their principal components, from which so large a fraction of the structural information is obtained. For systems with $S \geq 1$, zero-field parameters will be an additional source of structural information. This topic, as well as phenomena involving line width, will be considered in connection with specific applications.

II. The g-tensor.

The isotropic g-factor or the components of the g-tensor are often important clues to the identification of an unknown paramagnetic center. One is frequently unable to calculate the magnitude of an expected g-value precisely enough to make a direct positive identification of an unknown species. However, in trying to choose among several possibilities one may be able to reject some because the g-discrepancy is greater than the error in calculation. One would usually be able to choose between a substitutional and an interstitial defect, for example. Lack of accuracy in calculated g-component values is often associated with uncertainty both in appropriate spin-orbit coupling constant and in knowledge of energy separations between contributing levels. For iso-electronic ions one expects g-components to be generally the same; however, the spread may be large enough so that accidental similarity of values with those of another valence state may lead to error in identification. The isotropic g-values of free radicals containing only carbon, hydrogen, oxygen and nitrogen usually deviate by less than 0.1 per cent from the free-electron value g_e; hence g-values are usually of minor value in identification unless lines are narrow and g-values have been determined with high accuracy previously. When molecules containing sulfur are irradiated, one may be able to tell whether a resultant radical contains sulfur, since the g-values of sulfur-containing radicals are about one per cent higher than g_e. Rather few generalizations applicable to groups of radicals have been made[1,2]. It is suggested that the g-values of aromatic hydrocarbon anions and cations is given by the

---

[1] A. J. STONE: Mol. Phys. **6**, 509 (1963); **7**, 311 (1964).
[2] M. S. BLOIS, H. W. BROWN, and J. E. MALING: Arch. sci. (Geneva) **13**, 243 (1960).

relation:

$$g = 2.00257 - 0.00019\,\lambda,$$

where $\alpha + \lambda\beta$ is the energy level occupied by the odd electron in the Hückel molecular orbital approximation (Sect. B. I1)[1]. A plot of $g - g_e$ vs. λ for twenty aromatic radicals whose g values have very recently been measured with unequalled precision[2] ($\pm 7 \times 10^{-6}$ maximum standard deviation) yields an excellent straight line. The experimental data, excluding several radicals, yield the expression: $\Delta g = [31.9 \pm 0.4 - (16.6 \pm 1.0)] \times 10^{-5}$. This should be regarded as excellent agreement with prediction.

Expressions have been given for the principal g-components of a diatomic π radical[3].

The tedious procedure of determining the components of g-, hyperfine- and zero-field splitting tensors for paramagnetic centers of low symmetry is greatly simplified by application of several efficient procedures which have been described[4-6]. Having data on line extrema from rotation about mutually perpendicular axes, one may establish the principal axes, the principal g-components and their direction cosines. If there is ambiguity because two sets of principal g-components satisfy the data on rotation about the original three axes, one may resolve it by rotation in yet another plane. One may also estimate the probable errors both in g-values and in direction cosines[5]. Proper attention must be given to signs of tensor components.

In powders or glassy samples one may be able to deduce g-components (and in some cases hyperfine components) if the system is isotropic[7], has axial symmetry[8], is anisotropic[9] or shows simple zero field splitting[10]. Computer programs of varying complexity have been written to synthesize spectra for isotropic or anisotropic g- and hyperfine tensors[11,12] as well as zero field splitting[13,14].

If one is determining g-components from spectra showing large hyperfine splitting, it will be necessary to make corrections because of second-order hyperfine

---

[1] See footnote 1, p. 146.
[2] B. G. SEGAL, M. KAPLAN, and G. K. FRAENKEL: J. Chem. Phys. **43**, 4191 (1965).
[3] W. KÄNZIG, and M. H. COHEN: Phys. Rev. Letters **3**, 509 (1959).
[4] M. B. PALMA VITTORELLI, M. U. PALMA, D. PALUMBO, and M. SANTANGELO: Nuovo cimento **2**, 811 (1955). — J. E. GEUSIC, and L. C. BROWN: Phys. Rev. **112**, 64 (1958). — J. A. WEIL, and J. H. ANDERSON: J. Chem. Phys. **28**, 864 (1958). — H. M. MCCONNELL, C. HELLER, T. COLE, and R. W. FESSENDEN: J. Am. Chem. Soc. **82**, 766 (1960).
[5] D. SCHONLAND: Proc. Phys. Soc. (London) **73**, 788 (1959).
[6] J. R. MORTON: Chem. Revs. **64**, 453 (1964).
[7] S. M. BLINDER: J. Chem. Phys. **33**, 748 (1960). — H. STERNLICHT: J. Chem. Phys. **33**, 1128 (1960).
[8] H. R. GERSMAN, and J. D. SWALEN: J. Chem. Phys. **36**, 3221 (1962). — R. NEIMAN, and D. KIVELSON: J. Chem. Phys. **35**, 156 (1961). — B. BLEANEY: Proc. Phys. Soc. (London) A **63**, 407 (1950). — Phil. Mag. **42**, 441 (1960); — Proc. Phys. Soc. (London) A **75**, 621 (1960). — R. H. SANDS: Phys. Rev. **99**, 1222 (1955). — J. W. SEARL, R. C. SMITH, and S. J. WYARD: Proc. Phys. Soc. (London) A **74**, 491 (1959); A **78**, 1174 (1961). — J. A. WEIL, and H. G. HECHT: J. Chem. Phys. **38**, 281 (1963). — J. A. IBERS, and J. D. SWALEN: Phys. Rev. **127**, 1914 (1962).
[9] A. K. CHIRKOV, and A. A. KOKIN: Zhur. Eksptl. i. Teoret. Fiz. **39**, 1381 (1960); — Soviet Phys. JETP **12**, 964 (1961). — F. K. KNEUBÜHL: J. Chem. Phys. **33**, 1074 (1960). — E. L. COCHRAN, F. J. ADRIAN, and V. A. BOWERS: J. Chem. Phys. **34**, 1161 (1961).
[10] L. S. SINGER: J. Chem. Phys. **23**, 379 (1955). — A. LUND, and T. VÄNNGÅRD: J. Chem. Phys. **12**, 2979 (1965). — G. BURNS: J. Appl. Phys. **32**, 2048 (1961).
[11] J. D. SWALEN, and H. M. GLADNEY: IBM J. Res. **8**, 515 (1964).
[12] R. LEFEVRE, and J. MARUANI: J. Chem. Phys. **42**, 1480 (1965).
[13] P. KOTTIS, and R. LEFEVRE: J. Chem. Phys. **41**, 379 (1964).
[14] E. WASSERMANN, L. C. SNYDER, and W. A. YAGER: J. Chem. Phys. **41**, 1763 (1964).

interaction (Sect. A. III)[1,2]. For a single nucleus with isotropic coupling constant a (in MHz), the correction $\Delta g = g \frac{a^2}{2\nu^2}[I(I+1) - M_I^2]$. For spectra with unresolved splitting of degenerate lines due to groups of n_r equivalent nuclei the g-value is given by

$$g = g_{app}\left[1 - (1/4 H_0^2)\sum_r n_r a_r^2\right].$$

Here g_{app} refers to a g-value appropriate to the center line of the spectrum (or midway between a central pair: a_r is the splitting constant (in gauss) of the r-th group[3].

Some caution must be exercised in making interferences from sign of a g-shift. While in numerous cases a positive g-shift ($g > 2.0023$) for crystal defects indicates a resonance of positive hole character, and a negative shift an electron character, there are notable exceptions. Although $\Delta g \approx +0.06$ for the R-center ESR line, the defect is unquestionably electron-like. The unusual aspect is that the electron moves about three centers, and the net spin orbit coupling is affected by electron circulation about individual positive ions as well as by electron circulation about the symmetry axis. The contribution to the g-shift of the two circulations is presumably opposite in sign[4]. A consideration of g-components of the V_K centers in the alkali halides shows that cautious interpretation is required here also[5].

III. Hyperfine splitting.

No other characteristic of ESR spectra has provided as much information about paramagnetic species as hyperfine tensor components and a specification of the principal axes. Not only may these enable a defect to be identified in many cases, but the environment may also often be described in detail. Even the qualitative isotropic hyperfine patterns of free radicals may serve to identify an odd-electron species; we shall find that spin densities derived from coupling constants describe the spin distribution.

The first-order energies for a system with $S = \frac{1}{2}$ and $I = \frac{1}{2}$ is given by a sum of ZEEMAN, isotropic and anisotropic hyperfine terms[6-8]:

$$W_{M_S M_I} = g\beta H_0 M_S + A M_I M_S + \frac{g\beta\gamma M_I M_S}{2}\left\langle\frac{1-3\cos^2\alpha}{r^3}\right\rangle(1 - 3\cos^2\theta).$$

Here $A = 8\pi\beta\gamma\psi_{ns}^2(0)/3h$ for a nucleus of magnetogyric ratio γ is obtained by taking one-third the trace of the hyperfine coupling tensor. (For free radicals in non-oriented systems it is customary to use a instead of A.) Since only s-orbitals have a non-zero value of the wave function at the nucleus, A will be a direct measure of the s-character of the unpaired electron. One compares it with the value A_0 appropriate to an electron in a pure s-orbital. In the third (anisotropic) term, r represents the unpaired electron-nucleus distance, and α is the angle between r and one of the principal axes of the (traceless) anisotropic tensor;

---

[1] R. W. FESSENDEN: J. Chem. Phys. **37**, 747 (1962).
[2] B. BLEANEY: Phil. Mag. **42**, 441 (1951).
[3] B. G. SEGAL, M. KAPLAN, and G. K. FRAENKEL: J. Chem. Phys. **43**, 4191 (1965).
[4] D. C. KRUPKA, and R. H. SILSBEE: Phys. Rev. Letters **12**, 193 (1964).
[5] C. P. SLICHTER: Principles of Magnetic Resonance, Chap. 7. New York: Harper and Row 1963.
[6] H. ZELDES, G. T. TRAMMELL, R. LIVINGSTON, and R. W. HOLMBERG: J. Chem. Phys. **32**, 618 (1960).
[7] R. A. FROSCH, and H. M. FOLEY: Phys. Rev. **88**, 1337 (1952).
[8] J. R. MORTON: Chem. Revs. **64**, 453 (1964).

θ is the angle between this same axis and the magnetic field. The average value of $(1-3\cos^2\alpha)/r^3$ is zero for s-electrons; this property allows one to use the measured value of the anisotropic splitting B to estimate the p-character of the unpaired electron orbital by reference to $B_0 = (2g\beta\gamma/5h)\langle r^{-3}\rangle_{np}$ for an electron in a pure p-orbital. Values of $\psi^2_{ns}(0)$, A_0, $\langle r^{-3}\rangle_{np}$ and B_0 for the appropriate s- and p-orbitals are given in Table 1.

Table 1 *. Atomic parameters $\psi^2_{ns}(0)$ and $\langle r^{-3}\rangle_{np}$ (A. U.) and one-electron hyperfine constants (MHz).

Nucleus	n	$\psi^2_{ns}(0)$	A_0	$\langle r^{-3}\rangle_{np}$	B_0	Reference
H^1	1		1,420			1–3
B^{11}	2	1.408	2,020	0.775	53	1, 2, 4
C^{13}		2.767	3,110	1.692	91	1, 2, 4
N^{14}		4.770	1,540	3.101	48	1, 2, 4
O^{17}		7.638	4,628	4.974	144	1, 3
F^{19}		11.966	47,910	7.546	1,515	1, 5
Si^{29}	3	3.807	3,381	2.041	87	3, 6
P^{31}		5.625	10,178	3.319	287	4, 6
S^{33}		7.919	2,715	4.814	78	4, 6
Cl^{35}		10.643	4,664	6.710	137	4, 6
As^{75}	4	12.460	9,582	7.111	255	3, 7, 8
Xe^{129}	5	26.71	33,030	17.825	1,052	2, 9

* Taken from J. R. MORTON: Chem. Revs. **64**, 453 (1964).
[1] E. CLEMENTI, C. C. J. ROOTHAAN, and M. YOSHIMINE: Phys. Rev. **127**, 1618 (1962).
[2] D. F. MAYERS: Private communication.
[3] J. R. MORTON, J. R. ROWLANDS, and D. H. WHIFFEN: National Physical Laboratory Report No. BPR 13, extracted with permission of Director.
[4] G. W. CHANTRY, A. HORSFIELD, J. R. MORTON, J. R. ROWLANDS, and D. H. WHIFFEN: Mol. Phys. **5**, 233 (1962).
[5] R. J. COOK, J. R. ROWLANDS, and D. H. WHIFFEN: Proc. Chem. Soc. **1962**, 252.
[6] R. E. WATSON, and A. J. FREEMAN: Phys. Rev. **123**, 521 (1961).
[7] W. C. LIN, and C. A. McDOWELL: Mol. Phys. **7**, 223 (1964).
[8] R. E. WATSON, and A. J. FREEMAN: Phys. Rev. **124**, 1117 (1961).
[9] J. R. MORTON, and W. E. FALCONER: J. Chem. Phys. **39**, 427 (1963).

In first order, by virtue of the selection rule $\Delta M_s = 1$, $\Delta M_I = 0$, and neglect of a possible contribution $\gamma H \cdot I$, one expects isotropic hyperfine lines from a single nucleus to have uniform spacing and intensity. With equivalent nuclei experiencing small hyperfine interactions, one expects line intensities in the ratio of binomial coefficients. Regardless of the number of nuclei involved, if the hyperfine splitting is large (> 10 G), the separations in the hyperfine multiplet will increase from low to high field due to a term in M_I^2. All lines will also be shifted downfield, as is apparent from the last section. Further, if g departs appreciably from the free electron value g_e, the separation a' of successive hyperfine lines in gauss is dependent upon the frequency used. There is no such variability in the hyperfine *coupling* parameter: $a\,(\text{MHz}) = 2.8026 \frac{g\,a'}{g_e}$ (G); hence a may properly be regarded as a parameter in the spin Hamiltonian.

The term "coupling constant" is often used loosely as a measured hyperfine splitting, usually expressed in gauss. In the remainder of this review, except where it is obvious from references cited that correction for g-value and for second-order hyperfine effects (this section) have been made, the values quoted will be referred to as splitting constants.

Even in the simplest example of hyperfine splitting, viz., the hydrogen atom, the observed separation in field of the two lines (the hyperfine *splitting*) is *not* exactly given by the expression: $H_b = H_a = h a / g \beta$, where a (MHz) is the hyperfine coupling constant and H_b and H_a are shown in Fig. 1. The four energy levels are given (in frequency) by[1]:

$$E_1 h^{-1} = -\tfrac{1}{2}(g^2 \beta^2 H_z^2 h^{-2} + a^2)^{\frac{1}{2}} - \tfrac{1}{4}a,$$
$$E_2 h^{-1} = -\tfrac{1}{2} g \beta H_z h^{-1} + \tfrac{1}{4}a,$$
$$E_3 h^{-1} = \tfrac{1}{2}(g^2 \beta^2 H_z^2 h^{-2} + a^2)^{\frac{1}{2}} - \tfrac{1}{4}a,$$
$$E_4 h^{-1} = \tfrac{1}{2} g \beta H_z h^{-1} + \tfrac{1}{4}a.$$

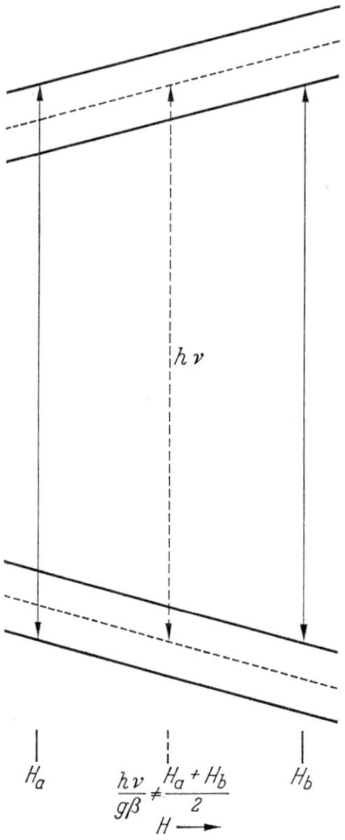

The allowed transitions *at constant field* H_z are the $\Delta E(4-1)$ and $\Delta E(3-2)$; their difference $\Delta E(4-1) - \Delta E(3-2)$ is just a. If $a^2 \ll g^2 \beta^2 H^2 h^{-2}$, and one retains the first two terms in the binomial expansion of the square root, one obtains the following expressions for the resonant field values *at constant frequency* ν:

$$H(4-1) = (h/4g\beta)[2\nu + a + (4\nu^2 + 4a\nu - 3a^2)^{\frac{1}{2}}],$$
$$H(3-2) = (h/4g\beta)[2\nu - a + (4\nu^2 - 4a\nu - 3a^2)^{\frac{1}{2}}].$$

The splitting ΔH is obviously not exactly $h a / g \beta$, the discrepancy for $\nu = 9.5$ GHz being about 3 G. More serious is the fact that the mean position of the two lines is 18 G lower than the first order value of $h\nu/g\beta$. Failure to account for such second-order hyperfine effects has sometimes resulted in erroneous reported g-values.

Fig. 1. Hyperfine levels and transitions of the hydrogen atom in large fields. Due to second order contributions, the mean position $(H_a + H_b)/2$ is not exactly equal to $h\nu/g\beta$, and $H_b - H_a$ is not equal to the hyperfine coupling constant of the spin Hamiltonian.

When hyperfine lines are very narrow, one is able to discern splitting of degenerate hyperfine components due to second-order interaction[2-6]. These are clearly discerned from Fig. 2 for PF_4, in which the 1:4:6:4:1 first order quintets (Fig. 2) are replaced by line groups: 1:1;3:1;3,2:1;3:1[5]. One notes a conservation of the total subgroup intensity. Numerous linear[4,6] or cyclic[3] saturated hydrocarbon (alkyl) radicals also show the second-order lines. In fact, some radicals require third- or fourth-order hyperfine corrections[6]. An expression (to second order) for transitions corresponding to $\Delta M_s = 1$, $\Delta I = 0$, and $\Delta M_I = 0$ is:

$$\Delta \nu = \frac{g \beta H}{h} + a M_I + \frac{a^2 h}{2 g \beta H}[I(I+1) - M^2].\,^3$$

---

[1] G. BREIT, and I. I. RABI: Phys. Rev. 38, 2082 (1931) (adapted by J. R. BOLTON), private communication).
[2] R. W. FESSENDEN: J. Chem. Phys. 37, 747 (1962).
[3] E. DE BOER, and E. L. MACKOR: Mol. Phys. 5, 493 (1962).
[4] R. W. FESSENDEN, and R. H. SCHULER: J. Chem. Phys. 39, 2147 (1963).
[5] J. R. MORTON: Can. J. Phys. 41, 706 (1963).
[6] R. W. FESSENDEN, and R. H. SCHULER: J. Chem. Phys. 43, 2704 (1965).

Paramagnetic centers in solids may show a variation in hyperfine splitting with field, except when the latter is oriented along principal hyperfine axes[1]. The invariance in the latter case is implicit in the concept of principal axes.

Numerous proton hyperfine spectra show satellite lines displaced from the central one by the proton resonant field due to simultaneous flip of proton and electron spin[2]. These therefore represent $\Delta M = 1$, $\Delta M_I = 1$ transitions. If there are two neighboring protons, the spin states of both may change concurrently with that of the electron spin[3]. Eight lines are expected at high microwave power but only some of these have been seen, due to everlap[4].

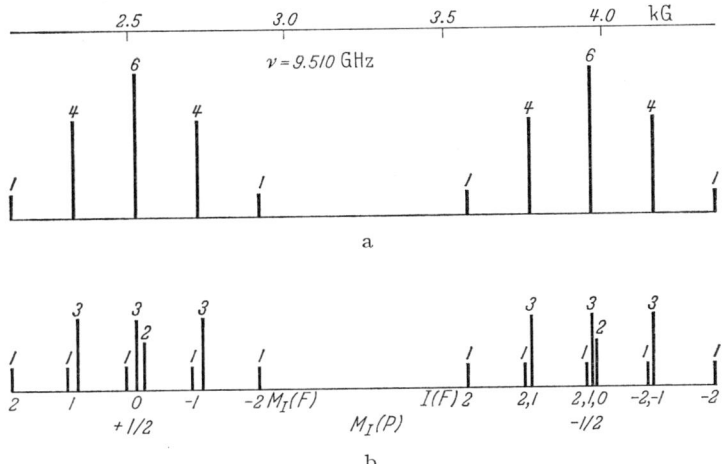

Fig. 2a and b. a) Idealized first-order spectrum of PF_4 radical. b) Observed spectrum of PF_4 in NH_4PF_6, γ-irradiated at 295 °K, showing second-order splittings [J. R. MORTON: Can J. Phys. 41, 706 (1963)].

B. Free radicals.

I. π-radicals.

1. Hückel molecular orbitals; energy levels. The large body of hyperfine splitting data for π-type organic free-radicals which we survey here would remain an undigested mass if one's only guide in interpretation were either a chemist's intuition or detailed perturbation calculations for each individual radical. The most generally useful basis of organization and understanding of hyperfine data and the prediction of now results has been the simple one-electron Hückel molecular orbital (HMO) approach[5-7]. Despite its apparently naive assumptions, drastic approximations, and its dismal failures in other areas, the molecular orbitals obtained as linear combinations of atomic orbitals (LCAO) and energy levels for

---

[1] R. J. COOK J. R. ROWLANDS, and D. H. WHIFFEN: Mol. Phys. 7, 31 (1963).
[2] H. ZELDES, and R. LIVINGSTON: Phys. Rev. 110, 630 (1958).
[3] G. T. TRAMMELL, H. ZELDES, and R. LIVINGSTON: Phys. Rev. 110, 630 (1958).
[4] W. SNIPES, and W. BERNHARD: J. Chem. Phys. 43, 2921 (1965).
[5] E. HÜCKEL: Z. Physik 70, 204 (1931); 76, 628 (1932); 83, 632 (1933).
[6] A. STREITWIESER: Molecular Orbital Calculations, p. 33. New York: John Wiley & Sons 1961.
[7] F. A. COTTON: Chemical Applications of Group Theory, p. 117. New York: Interscience Publ. 1963.

π-type free radicals often yield quantitative predictions of hyperfine splittings in amazing agreement with experiment, though some details are not accounted for. The HMO theory has given poorest agreement with data at some positions in alternant molecules (this section). Although there are many calculations involving elaborate procedures which give better agreement in specific instances, there are also numerous instances in which the HMO prediction is at least as good as that of the more elaborate calculations [1].

One may summarize the principal assumptions of the HMO approach in its simplest form as follows:

1. Molecules are regarded as having a planar skeletal arrangement of atoms held together by σ-bonds formed by hybridization of $2s$, $2p_x$ and $2p_y$ orbitals.

2. Each π-electron, occupying a $2p_z$ orbital, is distributed over the σ-framework. Examples of molecules having π-electrons are linear or cyclic hydrocarbons, conventionally written with alternating single and double bonds ("conjugated" systems), e.g., butadiene, $H_2C=\overset{H}{C}-\overset{H}{C}=CH_2$, benzene ($C_6H_6$), naphthalene ($C_{10}H_8$) or systems including "hetero" atoms (O, N, S) with unshared electron pairs, e.g., formaldehyde, $H_2C=O$.

3. Interactions between π electrons or between the σ and π systems are ignored.

4. For a system of n atoms in the σ skeleton, one writes n linear combinations of atomic orbitals:

$$\Psi_j = \sum_{i=1}^{n} c_{ij} \Phi_i. \tag{1.1}$$

One uses the variational principle to minimize the energy with respect to the LCAO coefficients and gets a secular equation:

$$\begin{vmatrix} H_{11}-ES_{11} & H_{12}-ES_{12} & \cdots & H_{1n}-ES_{1n} \\ H_{12}-ES_{12} & H_{22}-ES_{22} & \cdots & \cdots \\ \cdots & \cdots & \cdots & \cdots \\ \cdots & \cdots & \cdots & \cdots \\ H_{1n}-ES_{1n} & \cdots & \cdots & H_{nn}-ES_{nn} \end{vmatrix} = 0. \tag{1.2}$$

The H_{ii} are Coulomb integrals $\int \Phi_i \mathcal{H} \Phi_i d\tau$, H_{ij} are exchange integrals $\int \Phi_i \mathcal{H} \Phi_j d\tau$ and S_{ij} are overlap integrals $\int \Phi_i \Phi_j d\tau$. In its simplest form, one usually makes the following assumptions:

1. All H_{ii} are equal to a quantity α.
2. H_{ij} for all directly bonded atoms i and j are equal to β and zero otherwise.
3. All S_{ii} are equal to unity.
4. All S_{ij} ($i \neq j$) are zero ("neglect of overlap" assumption). Thus for linear systems, the one-off-diagonal terms are β, the diagonal terms are all $(\alpha-E)$ and other terms are zero. For linear systems, or for cyclic systems with no rings with an odd number of carbon atoms, the roots are of the form $W_i = \alpha \pm \lambda_i \beta$.

For a linear system with n odd, one of the λ_i values will be zero. Since β is essentially negative, for $\lambda_i > 0$ the $\alpha + \lambda_i \beta$ levels will lie below α. One proceeds to fill these levels with electrons, starting with the lowest-lying. In case the highest-lying occupied orbitals are degenerate and there are two electrons to distribute, one will be placed in each level, in accord with HUND's rule.

---
[1] A. CARRINGTON, and I. C. P. SMITH: Mol. Phys. 9, 137 (1965).

Sect. 1. Hückel molecular orbitals; energy levels. 153

One need not set all $H_{ii}=\alpha$, but may instead use different values of h for each distinguishable bond in the expression $H_{ii}=\alpha_0+h\beta_0$, where α_0 and β_0 are reference values, e.g., for bonds in benzene. Likewise, one may assume different values of exchange integrals by writing $H_{ij}=k\beta_0$. Finally, one may take $S_{ij}\neq 0$. For any hetero atom X (N, O, S, Cl, etc.) one generally assumes the Coulomb integral α_x as $\alpha_0+h_x\beta_0$, while the exchange integral may be assigned some nonzero value $\beta_{cx}=k_{cx}\beta_0$. One will use values of the adjustable parameters appropriate to the manner of bonding. One now has a collection of HMO parameters which

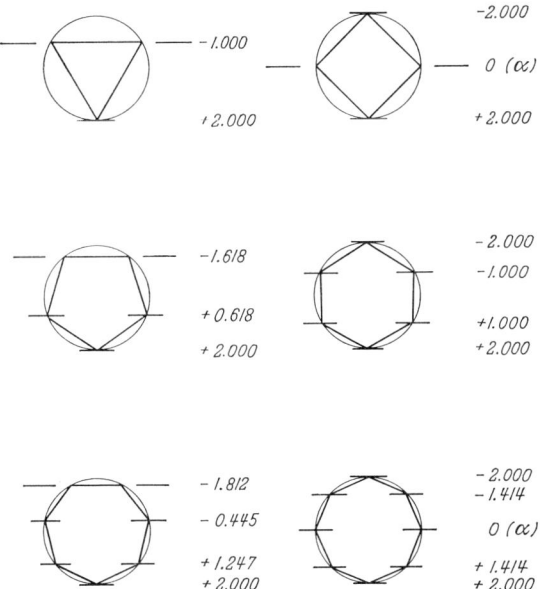

Fig. 3. Geometric scheme for determination of HMO levels for monocyclic π-conjugated systems without resort to MO calculations. [A. A. FROST, and B. MUSULIN: J. Chem. Phys. 21, 572 (1953)]. Numerical values are units of β to be added to α.

give predictions that often agree reasonably well with ESR data. Parameters are given for alkyl carbon and hydrogen[1,2], ketones[2,3], semiquinones[4,5], amino (NH_2), nitrile (CN) and nitro (NO_2) groups[2], and nitrogen in ring compounds[5,6].

A useful mnemonic device for determining the HMO levels for a cyclic *one-ring* system involves only a geometrical sketch plus simple calculation[7]. One draws a circle of radius 2β and inscribes the appropriate regular polygon with a vertex at the lowest point of the circle. Through each vertex one draws a horizontal line to represent an energy level and reckons the vertical separations m_i from the center of the circle. Taking the energy of the center level to be α, one obtains a set of levels $\alpha+\lambda_i\beta$. Thus for benzene, C_6H_6 (Fig. 3) there are two vertices at a distance β below the center and two an equal distance above, while the other

---
[1] W. T. DIXON: Mol. Phys. 9, 201 (1965).
[2] P. H. RIEGER, and G. K. FRAENKEL: J. Chem. Phys. 39, 609 (1963).
[3] P. H. RIEGER, and G. K. FRAENKEL: J. Chem. Phys. 37, 2811 (1962).
[4] G. A. VINCOW, and G. K. FRAENKEL: J. Chem. Phys. 34, 1333 (1961).
[5] J. R. BOLTON, A. CARRINGTON, and J. DOS SANTOS VEIGA: Mol. Phys. 5, 465 (1962).
[6] A. CARRINGTON, and J. DOS SANTOS VEIGA: Mol. Phys. 5, 21 (1962).
[7] A. A. FROST, and B. MUSULIN: J. Chem. Phys. 21, 572 (1953).

vertices occur at $\pm 2\beta$. The existence of energy level degeneracies is immediately apparent. Other rings which are of direct ESR interest are those with 3, 4, 5, 7 and 8 atoms. The odd numbered rings correspond to radicals in the neutral form, while the even-numbered are radicals in the negative ion form. Except in the simplest cases, one uses the symmetry properties of the molecule and group theory to factorize the secular determinant and to simplify determination of MO coefficients. In some cases, the latter can be expressed in simple form.

For linear chains with n carbon atoms, the coefficient of the r-th atom in the j-th molecular orbital will be:

$$c_{jr} = [2/(n+1)]^{\frac{1}{2}} \sin r j \, \pi/(n+1), \tag{1.3}$$

while for rings with n carbon atoms[1],

$$c_{jr} = n^{-\frac{1}{2}} \exp[2\pi i r(j-1)/n]. \tag{1.4}$$

If energy levels are symmetrically disposed about a central value, their molecular orbitals will have coefficients with the same absolute value. This "pairing property" is possessed by "alternant" hydrocarbons, i.e., those with structure such that upon identifying alternate σ skeleton atoms with an asterisk, there are no asterisks on adjacent atoms. These systems will be termed *even* alternant if the number of "starred" and "unstarred" atoms is the same, and *odd* alternant otherwise. For the latter, by convention one marks in such a way as to have more starred than unstarred atoms. Some of the even alternant hydrocarbons intensively studied include benzene, naphthalene, anthracene and a large number of other fused ring systems. Butadiene anion $CH_2=CH-CH=CH_2^-$ is an example of an even alternant linear hydrocarbon ion. Examples of simple odd-alternant systems are the allyl (1), benzyl (2) and triphenylmethyl (3) radicals.

(1)   (2)   (3)

In the HMO approximation, these molecules have a set of energy levels, equal in number to the number of carbon atoms, symmetrically disposed about a non-bonding level. The latter is so-designated because an equal number of positive and of negative terms of the same absolute magnitude occur in this molecular orbital (NBMO). The unpaired electron of a radical occupies this orbital, for which one notes that alternate atomic orbitals have zero coefficients. Further, about any starred position in the odd-alternant hydrocarbon, the sum of the coefficients of atomic orbital is zero. This rule enables one by inspection to write the coefficients in allyl (1) as $\pm \frac{1}{\sqrt{2}}$ for carbons 1 and 3. In benzyl (2), coefficients are $\pm \frac{1}{\sqrt{7}}$ for carbons 2, 4 and 6 and $\frac{2}{\sqrt{7}}$ at carbon 7. These are precisely the

---

[1] E. Hückel: Z. Physik **76**, 628 (1932). — C. A. Coulson, and H. C. Longuet-Higgins: Proc. Roy. Soc. (London) A **192**, 16 (1947). — C. A. Coulson: Proc. Roy. Soc. (London) A **164**, 383 (1938).

coefficients of the several atoms in the non-binding molecular orbital:

$$\Psi_4 = -0.378\,\Phi_2 + 0.378\,\Phi_4 - 0.378\,\Phi_6 + 0.756\,\Phi_7. \tag{1.5}$$

We shall return to this radical shortly.

2. Unpaired electron density; anion and cation radicals. We turn now to a consideration of distribution of an unpaired electron and its correlation with observed hyperfine splitting constants. That proton hyperfine splitting should

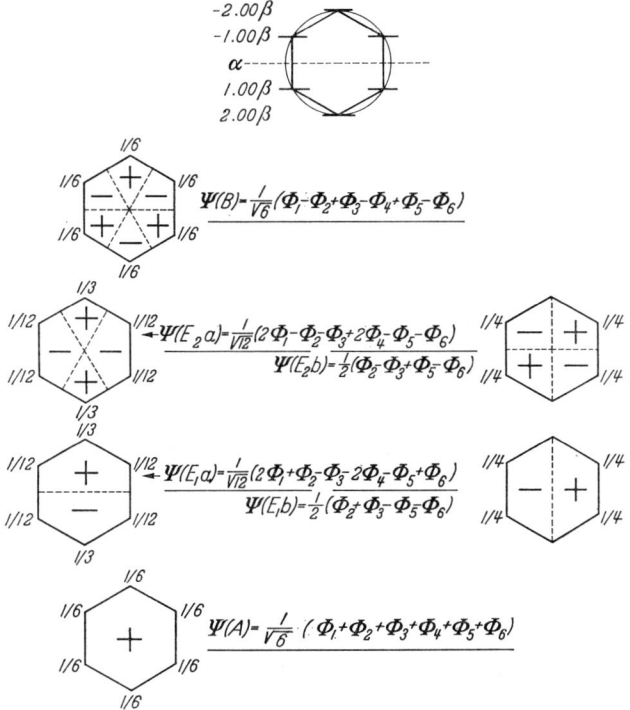

Fig. 4. Energy levels, nodal planes and unpaired electron densities in each molecular orbital of benzene.

occur at all with the unpaired electron in a $2p_z$ orbital is remarkable, since the hydrogen atoms lie in the nodal plane. We shall return to this point in the next section.

From the HMO point of view, the determination of unpaired electron density at each of a set of coupled π-centers is straightforward. Having one unpaired electron to be delocalized, one takes the square of the coefficient of each atomic orbital in the highest occupied molecular orbital to give the fractional unpaired electron density on the corresponding atom. In case the "highest occupied level" is a degenerate pair of levels, as with benzene (Fig. 4), these levels will be equally occupied, and the density at each atom will be the average of that in the two orbitals. The average unpaired electron density at each position in the benzene anion is $\frac{1}{6}$, are required by symmetry. The measured splitting constant is 3.75 G for the 1:6:15:20:15:6:1 benzene anion ESR spectrum, so the total extent of the spectrum is $6 \times 3.75 = 22.5$ G[1]. (It is important to note that not only in this especially simple case, but in general as well, the total extent of the spectrum is

---

[1] T. R. Tuttle, and S. I. Weissman: J. Am. Chem. Soc. **80**, 5342 (1958).

the sum of the hyperfine splitting constants of all nuclei.) The single-carbon π-radical CH_3 has a proton splitting constant of 23.04 G[1]. The nearly identical values for both CH_3 and $C_6H_6^-$ of ~23 G proton splitting per unit unpaired electron density suggest that individual proton splittings a^H might be directly related to the $2p_z$ unpaired electron density ϱ_C on the atom to which it is attached.

The relationship proposed is a simple proportionality[2]:

$$a^H = Q \varrho_C \quad \text{where} \quad \varrho_C = c_i^2. \tag{2.1}$$

Here Q represents the proton splitting which would result if an unpaired electron were localized upon a single carbon atom, as, e.g., the CH_3 radical. The value of Q was initially assumed to be about 28 G. Fig. 5, which shows a plot of hyperfine splitting constant vs. HMO unpaired electron density, shows a remarkable proportionality for a large number of anion and cation radicals. If one is willing to concede that some modification of the Q value is in order when one changes greatly the nature of the radical under consideration, the description seems reasonably satisfactory. Q values found for carbon have the extreme range 17 to 34 G.

We take as a simple example the butadiene anion $CH_2=CH-CH=CH_2^-$, which when produced by electrolytic reduction in liquid NH_3[3] gives a 1:4:6:4:1 spectrum of 1:2:1 triplets, with splitting constants $a_1 = 7.62$ and $a_2 = 2.79$ G respectively. The unpaired electron will be in the lowest antibonding orbital:

$$\Psi_3 = 0.600\, \Phi_1 - 0.371\, \Phi_2 - 0.371\, \Phi_3 + 0.600\, \Phi_4. \tag{2.2}$$

The ratio of the squares of coefficients is 2.62 for end vs. central atoms, while the splitting constant ratio is 2.73. The measure of agreement is reasonably representative for even-alternant hydrocarbon radicals. The values of Q which would reproduce the observed splitting constants are 21.2 G and 20.2 G (average 20.8 G) for a_1 and a_2, appreciably lower than the $Q = 23.0$ G value for the neutral CH_3 radical. It is important to note that splitting values may be dependent upon charge; we shall return to this apparently low Q value.

Table 2.

	Anthracene splitting constants[1]			HMO unpaired electron densities		
	a_1	a_2	a_9	ϱ_1	ϱ_2	ϱ_9
Cation	3.061	1.379	6.533	0.096	0.047	0.192
Anion	2.726	1.513	5.337	0.096	0.047	0.192

[1] J. R. BOLTON, and G. K. FRAENKEL: J. Chem. Phys. **40**, 3307 (1964).

The following data in Table 2 for anthracene cations (4) and anions[4] apparently show such a charge effect:

(4)

---

[1] R. W. FESSENDEN, and R. H. SCHULER: J. Chem. Phys. **39**, 2147 (1963).
[2] H. M. McCONNELL: J. Chem. Phys. **24**, 764 (1965); — Ann. Rev. Phys. Chem. **8**, 105 (1957).
[3] D. H. LEVY, and R. J. MYERS: J. Chem. Phys. **41**, 1063 (1964).
[4] J. R. BOLTON, and G. K. FRAENKEL: J. Chem. Phys. **40**, 3307 (1964).

Sect. 2. Unpaired electron density; anion and cation radicals.

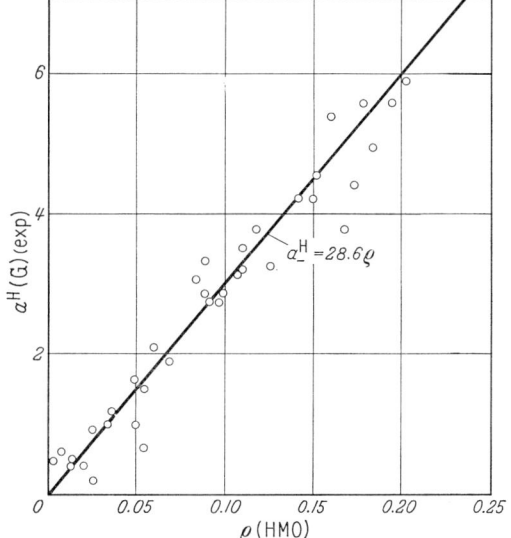

Fig. 5. Hyperfine splitting constants of aromatic hydrocarbon anions vs. HMO unpaired electron densities.

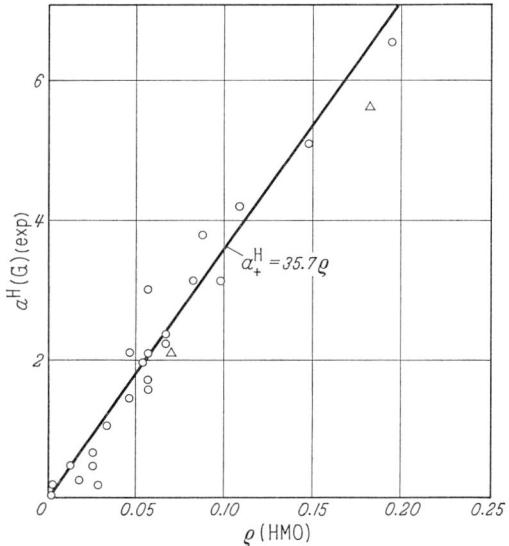

Fig. 6. Hfs constants of aromatic hydrocarbon cations vs. HMO unpaired electron densities.

A mean value of $Q_{\text{cation}} = 31.8$ G gives deviations of ± 3.2 G, while $\langle Q_{\text{anion}} \rangle = 29.5 \pm 2.6$ G. Figs. 5 and 6 show the remarkable proportionality between the measured coupling constants of a large number of negative and positive ion radicals and the squares of HMO coefficients. Some non-alternant systems are included. There is a separate point plotted for each different splitting constant with a given radical. The cations consistently show a larger value of splitting constant than the anions. It has been shown that to a high degree of approximation the cation and anion of a given alternant aromatic hydrocarbon should have identical π-electron spin distributions (pairing theorem)[1]. This has been confirmed by observation of essentially identical ^{13}C splittings in corresponding anthracene cation and anion radicals[2]. The appreciably different slopes of 28.6 and 35.7 of Figs. 5 and 6 suggest an effect of charge which may be written as[3,4]:

$$a_{i\pm}^H = -[Q_{CH}^H(0) + K\,\varepsilon_i]\varrho_i^\pi \quad (\varepsilon_i = \text{charge on carbon } i), \tag{2.3}$$

$$a_{i\pm}^H = -(Q\varrho_i \pm K\varrho_i^2). \tag{2.4}$$

Here $Q_{CH}^H(0)$ and K have variously been given the values 31.2 and 17 (27 and 12)[3] or 32.2 and 16[5]. The figures in parentheses refer to spin densities (next section),

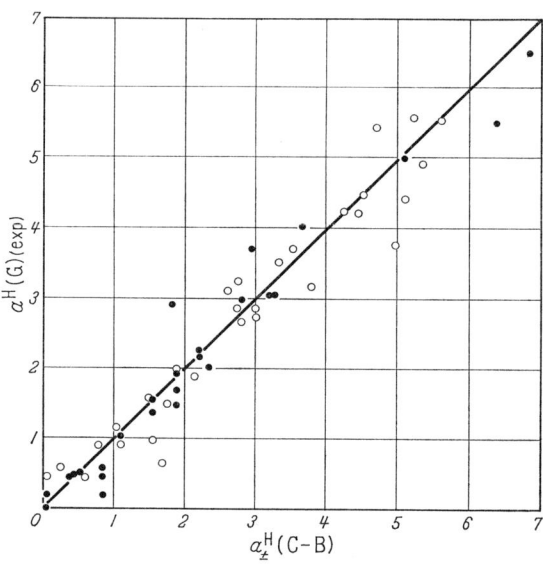

Fig. 7. Experimental splitting constants of the above anions and cations vs. calculated values from the Colpa-Bolton relation.

while other values are HMO unpaired electron densities. All points from Figs. 5 and 6 are combined in Fig. 7, which uses the last set of values. A theoretical consideration of the effect of excess charge upon the difference of exchange integrals which determines the value of Q_{CH}^H is in accord with the observation that cations have larger splitting constants than anions[4]. However, it is possible to fit the data equally well[6,5] without explicit inclusion of charge by an expression

$$a_{i\pm}^H = Q_1\varrho_i \pm Q_2 \left| \sum_j c_i\,c_j \right|, \tag{2.5}$$

where one takes products of HMO coefficients of adjacent atoms[7]. This equation is obtained by a perturbation calculation of interaction of hydrogen and next-

---

[1] A. D. McLachlan: Mol. Phys. 2, 271 (1959); 3, 233 (1960).
[2] See footnote 4, p. 156.
[3] J. P. Colpa, and J. R. Bolton: Mol. Phys. 6, 273 (1963).
[4] J. R. Bolton: J. Chem. Phys. 43, 309 (1965).
[5] I. C. Lewis, and L. S. Singer: J. Chem. Phys. 43, 2712 (1965).
[6] L. C. Snyder, and T. Amos: J. Chem. Phys. 42, 3670 (1965).
[7] G. Giacometti, P. L. Nordio, and M. V. Pavan: Theor. Chim. Acta (Berlin) 1, 404 (1963).

nearest-neighbor carbon p-orbitals, using the HMO approach[1]. For the same data given in Fig. 7, the values $Q_1=32.2$ and $Q_2=7.0$, plotted in Fig. 8 fit the data equally well[2]; one presently has no bases for preference of explanations as far as results are concerned. Eq. (2.5) requires use of calculated HMO coefficients.

A very comprehensive study of radicals derived either from linear $[\mathrm{CH_3(CH_2)}_n\mathrm{CH_3}]$ or cyclic $[\overline{\mathrm{H_2C-(CH_2)}_n\mathrm{CH_2}}]$ hydrocarbons has given a good description of proton splittings in radicals with only one carbon atom having a π-orbital[3]. This atom will be denoted with a prime and referred to as the alpha carbon atom; nearest neighbors (either side) are referred to as β-carbon atoms, etc.

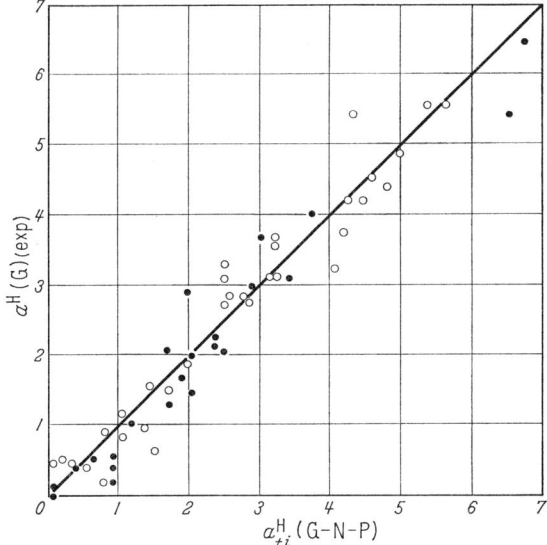

Fig. 8. Experimental vs. calculated (Giacometti-Nordio-Pavan relation) splitting constants for anions and cations [I. C. Lewis, and L. S. Singer: J. Chem. Phys. 43, 2712 (1965)].

A compilation of proton coupling constants (Table 3)[3] shows only a small decrease in a_α on substitution of one or more hydrogen atoms of $\mathrm{CH_3}$ by $\mathrm{CH_3}$ groups. Molecular orbital studies of such a group suggests that one should use slightly different Q_α values according as the carbon atom has 0, 1 or 2 carbon atoms attached to it[4,3].

In Table 4[3] there is a compilation of Q_α values which best fit the experimental data. The $a_\beta(\mathrm{CH_3})$ values are seen to decrease significantly with chain length, though $Q_\beta(\mathrm{CH_3})$ remains constant. In the section on Oriented Radicals, we shall make use of the relation $a_\beta = B + A\cos^2\theta$, where θ is the angle between the C—C—H plane and the axis of the π-orbital on the α-carbon atom. The value of B is only a few gauss, while $A \approx 50$ G. For the isotropic systems we are considering here, one takes $\langle \cos^2\theta \rangle = 0.5$, giving $a_\beta \approx 25$ G, in satisfactory accord with values in Table 4. The mechanism of hyperfine interaction of β-protons with an electron in a π-orbital is considered in the section on Hyperconjugation.

---

[1] See footnote 7, p. 158.
[2] See footnote 5, p. 158.
[3] See footnote 1, p. 156.
[4] D. B. Chesnut: J. Chem. Phys. 29, 43 (1958). — D. Lazdins, and M. Karplus: J. Chem. Phys. 44, 1600 (1966).

Table 3. *Hyperfine coupling constants[1] for alkyl and cycloalkyl radicals**.

Radical	a_α (gauss)	a_β (gauss)	a_γ (gauss)
$C'H_3$	23.04	—	—
$CH_3C'H_2$	22.38	26.87	—
$C_2H_5C'H_2$	22.08	33.2 [2]	0.38 (128 °K)
$(CH_3)_2C'H$	22.11	24.68	—
$CH_2=CHCH_2C'H_2$	22.23	29.7 [2]	0.63} 0.35} γ, δ
$C_2H_5C'HCH_3$	21.8	24.5 (CH_3) 27.9 (CH_2)	[3]
$(CH_3)_2CHC'H_2$	22.0	35.1 [2]	[3]
$(CH_3)_3C'$	—	22.72	—
$C_2H_5C'HC_2H_5$	21.8	28.8	uncertain
$(CH_3)_2C'CH_2CH_3$	—	22.8 (CH_3) 17.6 [2] (CH_2)	[3]
$(CH_3)_3CC'H_2$	22.7	—	[3]
$(C_2H_5)_3C'$	—	17.3	[3]
$(C_2H_5)_3CC'HCH_3$	21.7 ± 0.3	24.9 ± 0.3	[3]
$R_1CH_2C'HCH_2R_2$	21.0 [4]	24.8 [4]	[3]
Cyclo-C_4H_7	21.20	36.66	1.12
Cyclo-C_5H_9	21.48	35.16	0.53
Cyclo-C_6H_{11}	21.15	41} 5}	0.71 [5]
Cyclo-C_7H_{13}	21.8	24.7	[3]

* Taken from R. W. FESSENDEN, and R. H. SCHULER: J. Chem. Phys. **39**, 2147 (1963).
[1] Uncertainty 0.1 if listed to 0.1; if listed to 0.01, uncertainty 0.01—0.02.
[2] Temperature-dependent.
[3] Not resolved.
[4] Uncertainty because of line width.
[5] Two equivalent protons.

Table 4. *Empirical values of ϱ, Q_α and Q_β**.

Radical	ϱ^1	a_α	Q_α	$a_\beta(CH_3)$	$Q_\beta(CH_3)$
$C'H_3$	1	23.04	23.04	—	—
$CH_3C'H_2$	0.919	22.38	24.35	26.87	29.25
$(CH_3)_2C'H$	0.844	22.11	26.20	24.68	29.25
$(CH_3)_3C'$	0.776	—	—	22.72	29.30

* Taken from R. W. FESSENDEN, and R. H. SCHULER: J. Chem. Phys. **39**, 2147 (1963).
[1] $\varrho = (1-0.081)^n$ where n is the number of methyl groups adjacent to the carbon atom having the unpaired electron.

Returning to the apparently low value of Q for the butadiene anion, one may allow both for differences in binding and for charge. Noting the increase in Q upon going from $C'H_3$ to $C'H_2CH_3$ to $C'H(CH_3)_2$[1], one may take an adjusted value of 24.5 G as appropriate to a carbon atom bonded to one other in a neutral radical and write for CH_2 protons[2]:

$$a_{1,4} = 24.5\, \varrho_1 - 12\, \varrho_1^2.$$

For CH protons in $CH_2=CH-CH=CH_2^-$, one uses another adjusted value of $Q = 26.5$ G, appropriate to a carbon atom bonded to two others:

$$a_{2,3} = 26.5\, \varrho_2 - 12\, \varrho_2^2.$$

[1] See footnote 1, p. 156.
[2] J. R. BOLTON: Private communication.

Using again the experimental values $a_1 = 7.62$ G and $a_2 = 2.79$ G, the total spin density $\varrho = \varrho_1 + \varrho_2 = 2(0.383 + 0.111) = $[1]. 0.988 Thus there is no anomaly in Q value for this ion. Spin densities in butadiene anions, cis and trans (as well as numerous other charged and neutral radicals) have been calculated by Hartree-Fock procedures[2].

There exist paramagnetic dinegative ions, but spin-spin interaction usually leads to distinctive behavior; these are discussed in Sect. C (Triplet State). In a system having degenerate lowest-lying antibonding orbitals and a high electron affinity, one might hope to observe the ESR spectrum of a tri-negative ion. This has been observed for decacyclene[3] (5). Reduction with sodium leads first to the paramagnetic mononegative ion, but later the ESR spectrum at room temperature disappears. One may then see the triplet state spectrum of the dinegative ion at 77 °K. On still further reduction at room temperature, the trinegative ion ESR spectrum appears. Remarkably, for both the mono- and the trinegative ions the extent of the spectrum is about 23 G, though only half of the carbon atoms hold hydrogen atoms. This implies a high degree of spin polarization in the system.

(5)

The relative proton splitting for even alternant systems have shown no marked systematic departures from HMO unpaired electron densities. Discrepancies begin appear with odd alternant systems.

The squares of the coefficients in Eq. (1.5) for benzyl radical are taken as the unpaired electron density at the respective carbon atoms. Since the hyperfine splitting constants for the benzyl radical are known[4] one may readily test the

Table 5. *Benzyl radical parameters.*

Atom	a^H (G) (exp)[1]	HMO calculated value
2, 6	− 5.14	− 4.1
4	− 6.14	− 4.1
7	− 16.35	− 16.35 (reference)
3, 5	+ 1.75	0.0
Total width of spectrum	37.27 G	24.55

[1] A. CARRINGTON, and I. C. P. SMITH: Mol. Phys. **9**, 137 (1965).

validity of the HMO approach for odd-alternant radicals. In the last column of Table 5 one takes the measured value for position 7 as a reference point. The measure of agreement here for positions 2, 4 and 6 is fair. Satisfyingly, one observes a several-fold larger splitting from methylene (7) protons than from the 2, 4 or 6 protons. The 3, 5 protons, which from HMO considerations are not expected to

---

[1] See footnote 2, p. 160.
[2] L. C. SNYDER, and T. AMOS: J. Chem. Phys. **42**, 3670 (1965).
[3] P. BRASSEM, R. E. JESSE, and G. J. HOIJTINK: Mol. Phys. **7**, 587 (1964).
[4] A. CARRINGTON, and I. C. P. SMITH: Mol. Phys. **9**, 137 (1965).

Handbuch der Physik, Bd. XVIII/1.

show a splitting, show a small one, here indicated with a positive sign. The total extent of the spectrum of this *neutral* radical is somewhat greater than that of most *cation* radicals; we note this as an anomaly which is typical of odd-alternant systems.

(6a)   (6b)

The neutral odd-alternant perinaphthenyl radical (6b)[1] is remarkable because the total extent of its proton spectrum is 43 G, whereas even with cation radicals it is $\sim$35 G at most. The two splitting constants are $a_1 = -7.3$ G for protons 1, 3, 4, 6, 7, 9 and $a_2 = +2.2$ G for protons 2, 5, 8; for the latter, HMO unpaired electron densities would predict zero splitting. The marked deviation of its proton splitting behavior from HMO predictions confirmed the need of a more precise interpretation[2] of ϱ in the McConnell equation $a^H = Q\varrho$. Since the approximate magnitude of Q as $\sim$30 G is confirmed by computation for a C—H fragment, or a conjugated system[3-5], the interpretation of ϱ values as HMO unpaired electron densities would appear to limit the total extent of any free radical spectrum to $\sim$30 G. A reinterpretation of ϱ as a spin density will allow one to retain Q values applicable to even alternant systems and yet explain the perinaphthenyl results.

3. Mechanism of hyperfine splitting. Spin density.

If the assumption of pure σ-bonding of hydrogen atoms lying in the nodal plane of $2p_z$ orbitals were strictly correct, one would observe no proton hyperfine splitting in liquid solutions. It was suggested that out-of-plane vibrations of the hydrogen atoms is the mechanism leading to contact interaction[6]; however, the vibrational effect is much too small[7]. It now appears certain that a configuration interaction which admixes a small fraction of π-character into the C—H σ bond is responsible[8-12]. Both molecular orbital[12,13] and valence bond[10,12] calculations on a C—H fragment have been shown to yield the McConnell equation:

$$a = Q\varrho, \quad \text{or more precisely,} \quad a^H = Q^H_{CH}\, \varrho^\pi_C.$$

Q is predicted to have a negative sign. An alternative model of the hypothetical $C_2H_4^+$ ion leads to similar results[11].

---

[1] P. B. SOGO, M. NAKAZAKI, and M. CALVIN: J. Chem. Phys. **26**, 1343 (1957). — S. H. GLARUM, and J. H. MARSHALL: To be published. — H. M. MCCONNELL, and H. H. DEARMAN: J. Chem. Phys. **28**, 51 (1958). — L. C. SNYDER, and T. AMOS: J. Chem. Phys. **42**, 3670 (1965).
[2] H. M. MCCONNELL, and D. B. CHESNUT: J. Chem. Phys. **27**, 984 (1957).
[3] H. M. MCCONNELL, and D. B. CHESNUT: J. Chem. Phys. **28**, 107 (1958).
[4] H. M. MCCONNELL: J. Chem. Phys. **28**, 1188 (1958).
[5] A. MCLACHLAN, H. DEARMAN, and R. LEFEBVRE: J. Chem. Phys. **33**, 65 (1960).
[6] S. I. WEISSMAN, J. TOWNSEND, D. E. PAUL, and G. E. PAKE: J. Chem. Phys. **21**, 2227 (1953).
[7] B. VENKATARAMAN, and G. K. FRAENKEL: J. Chem. Phys. **24**, 737 (1956); — J. Am. Chem. Soc. **77**, 2707 (1955).
[8] H. M. MCCONNELL: J. Chem. Phys. **24**, 764 (1956).
[9] S. I. WEISSMAN: J. Chem. Phys. **25**, 890 (1956).
[10] H. S. JARRETT: J. Chem. Phys. **25**, 1289 (1956).
[11] R. BERSOHN: J. Chem. Phys. **24**, 1066 (1956).
[12] H. M. MCCONNELL, and D. B. CHESNUT: J. Chem. Phys. **28**, 107 (1958).
[13] S. AONO: Progr. Theoret. Phys. (Kyoto) **21**, 779 (1959).

A numerical evaluation of $Q = 28$ G is given for the C—H fragment[1]. The treatment has been extended to include coupled π-systems such as aromatic ions (e.g. $C_6H_6^-$)[2,3–5], with an estimate of 22.2 G for Q^4.

A pictorial representation of polarization of the C—H bond resulting from π—σ interaction is qualitatively useful. Fig. 9 indicates a (small) net polarization induced in one $sp^2 \sigma$-bond by interaction with the unpaired electron in a π-orbital. The 1s electron on hydrogen then is polarized in the sense opposite to that of the π-electron. If the direction of polarization at this carbon atom corresponds to that in the rest of the molecule (ϱ positive), then a^H is negative.

It is clear that ϱ interpreted as an unpaired electron density can only have a positive value. However, a redefinition of ϱ is desirable to allow for the possibility

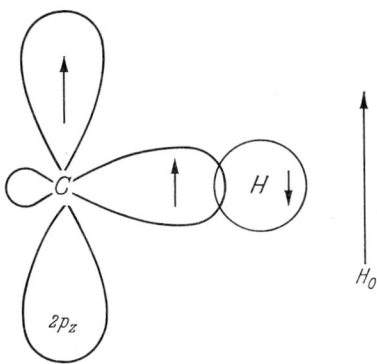

Fig. 9. Polarization of σ-bonding electrons by an unpaired electron in a π-orbital.

that the π-electron polarization at a carbon atom may be generally opposite to that of the carbon skeleton. Such an effect had been predicted[6,7]. Taking the spin of the $2p_z$ electron to be α, and $P(\alpha) - P(\beta)$ as the net probability of spin α at a given carbon atom, the spin density and charge density are respectively:

$$\varrho_k^\pi = P(\alpha) - P(\beta); \quad q_k^\pi = P(\alpha) + P(\beta).$$

With this redefinition of ϱ as a spin density, one may sketch the derivation of the McConnell relation in a direct way[8] and simultaneously discern the nature of the configuration interaction which introduces non-zero spin at the proton[3]. We start with one π and two σ electrons, described by a wave function Φ_1 corresponding to perfect pairing, with a π node at the hydrogen atom:

$$\Phi_1 = \hat{a} \, \pi(1) \, \sigma(2) \, s(3) \, (\alpha\beta\alpha - \alpha\alpha\beta)/\sqrt{2}.$$

Here
$\pi \equiv 2p_z$ orbital on carbon,
$\sigma \equiv sp^2$ hybrid on carbon,
$s \equiv 1s$ orbital on hydrogen,

---
[1] See footnote 10, p. 162.
[2] See footnote 9, p. 162.
[3] See footnote 12, p. 162.
[4] See footnote 13, p. 162.
[5] A. McLachlan, H. H. Dearman, and R. Lefebvre: J. Chem. Phys. **33**, 65 (1960).
[6] H. M. McConnell, and D. B. Chesnut: J. Chem. Phys. **27**, 984 (1957).
[7] P. Brovetto, and S. Ferroni: Nuovo cimento **5**, 142 (1957).
[8] J. R. Bolton: Private communication.

and $\hat{a}$ is the antisymmetrization operator. The spin density at the proton is zero. In order to get a non-zero value, one mixes an excited (doublet) wave function Φ_2 representing antibonding between σ and s;

$$\Phi_2 = \hat{a}\,\pi(1)\,\sigma(2)\,s(3)\,(\alpha\alpha\beta + \alpha\beta\alpha - \beta\alpha\alpha)/\sqrt{6},$$

leading to non-zero spin density at the proton. The resultant wave function is:

$$\Psi = \Phi_1 + \lambda\Phi_2, \quad \text{where} \quad 1 \gg \lambda = H_{21}/E_{21},$$

$$H_{21} = \langle \Phi_2 | r_{12}^{-1} | \Phi_1 \rangle, \quad \text{and} \quad \Delta E_{21} = H_{22} - H_{11}.$$

Using $[\pi\sigma|\pi\sigma] \equiv \langle \pi(1)\sigma(2) | r_{12}^{-1} | \pi(2)\sigma(1) \rangle$ and defining $[\pi s|\pi s]$ similarly, one writes:

$$H_{21} = -\sqrt{3}/2\{[\pi\sigma|\pi\sigma] - [\pi s|\pi s]\}.$$

The proton hyperfine splitting is given by:

$$a^H = \frac{8}{3}\frac{\pi\mu_H}{I}\langle\Psi|\sum_k \delta(r_{kH}) S_{kz}|\Psi\rangle = \frac{8}{3}\frac{\pi\mu_H}{I}\varrho^H(0),$$

where

$$\varrho^H(0) = \frac{-2\lambda}{\sqrt{3}}|S(0)|^2.$$

Using $a(H)$ to represent the splitting of a free hydrogen atom,

$$a^H \text{ (for unit } \pi\text{-spin density on atom } k) = -\frac{([\pi\sigma|\pi\sigma] - [\pi s|\pi s])}{\Delta E_{21}}\,a(H) = -Q_{CH}^H.$$

We have assumed unit atomic orbital spin density ϱ_k^π on atom k. For an electron which is delocalized over a π system:

$$a_k^H = -Q_{CH}^H\,\varrho_k^\pi.$$

The calculated sign of Q is negative, suggesting that the spin density at a proton is negative when ϱ is positive and vice versa.

It is apparent that spin density rather than unpaired electron density governs hyperfine splitting; these quantities are essentially equivalent for systems with all spin densities positive; when there are regions of negative spin density (spin polarization on some atoms antiparallel to that in the rest of the molecule) the requirement that $\sum \varrho_k^\pi = 1$ allows surprisingly large positive spin densities at some positions. In general, where HMO theory would predict zero unpaired electron density on a carbon atom, spin density on that carbon is negative. Since there is a $\pi - \sigma$ correlation, the induced σ spin density on that carbon is negative. Therefore, the spin density at the proton is positive, as given for the 3, 5 positions in benzyl (2).

The prediction that the spin polarization on some atoms of a radical may be antiparallel to the direction of the total electron spin[1,2] is now fully confirmed. Convincing evidence as to the sign of hyperfine couplings comes from ESR spectra of oriented radicals[3], from shifted proton NMR frequencies in dilute free radical systems[4-8], and from other techniques considered in Sect. 7.

---

[1] See footnote 7, p. 163.
[2] See footnote 6, p. 163.
[3] H. M. McConnell, C. Heller, T. Cole, and R. W. Fessenden: J. Am. Chem. Soc. **82**, 766 (1960).
[4] M. E. Anderson, G. E. Pake, and T. R. Tuttle: J. Chem. Phys. **33**, 1581 (1960).
[5] M. E. Anderson, P. J. Zandstra, and T. R. Tuttle: J. Chem. Phys. **33**, 1591 (1960).
[6] H. S. Gutowsky, H. Kusumoto, T. H. Brown, and D. H. Anderson: J. Chem. Phys. **30**, 860 (1959).
[7] A. Forman, J. N. Murrell, and L. E. Orgel: J. Chem. Phys. **31**, 1129 (1959).
[8] W. D. Phillips, and R. E. Benson: J. Chem. Phys. **33**, 607 (1960).

In a radical such as benzyl, one may raise the question as to whether a single value of Q is to be applied to all protons or whether separate values be used for CH_2 and ring protons[1]. We have noted that it is useful to select particular Q values for different positions in an alkyl radical. Another complication in the use of the McConnell relation is the question of whether the splitting also depends on spin densities of next-nearest carbon atoms. Such seems to be the case for protons of free radicals in crystals[2]; for ^{13}C splittings, the neighboring carbon contribution has been fully established.

An estimate has been made of the McConnell parameter $Q_{OH}^H \approx 7$ G for proton splitting from an OH group, both experimentally and from a configuration interaction calculation[3]. Other examples of $-OH$ proton splittings are found in $C'H_2OH$ [4], $HC \equiv C - C'HOH$ [5] semiquinone cations[6] and salt hydrates[7].

4. ^{13}C splitting constants. For ^{13}C splittings the simple McConnell relation does not have even approximate validity for polynuclear radicals. If the ^{13}C atom is bonded to other carbon atoms, its splitting is markedly affected by the spin density upon adjacent carbon atoms[8-10]. For a carbon atom C connected with three other atoms X_i by sp^2 bonds and having π spin density ϱ^π, the net splitting by the C nucleus may be written[10]:

$$a^C = (S^C + \sum_{i=1}^{3} Q_{CX_i}^C) \varrho^\pi + \sum_{i=1}^{3} Q_{X_iC}^C \varrho_i^\pi. \tag{4.1}$$

Here S^C represents the contribution from $1s$ electrons. The $2s$ contributions are determined by the Q-parameters and the π-electron densities on carbon (ϱ^π) and on X atoms (ϱ_i^π). The parameter Q_{BC}^A signifies the contribution to nucleus A per unit spin density on atom B from the B—C bond. If one of the X_i is H, these parameters have been assigned the values: $S^C = -12.7$ G, $Q_{CH}^C = 19.5$ G, $Q_{CC'}^C = 14.4$ G and $Q_{C'C}^C = -13.9$ G [10]. In CH_3, $Q_{CH}^C = 18.85$ G. ϱ_i^π values have been taken from the expression $\varrho_i^\pi = a_i^H/Q_{CH}^H$, using proton splitting data. While this is often satisfactory, it has led to errors in some cases, as is indicated shortly. Applied to CH_3, one writes: $a^C = S^C + 3Q_{CH}^H = +43.8$ G, which is somewhat larger than the most recent value of $a^C = 38.56$ G in CH_3 [11]. The malonic acid radical $HC'(COOH)_2$ gives $a^H = -21.17$ G, from which one may determine $\varrho_{C'}^\pi$ to be 0.89 if Q_{CH}^H is taken as -23.7 G. Neglecting the last term of Eq. (4.1), since one expects localization on oxygen rather than on the end carbon atoms, one obtains:

$$a^{C'} = (S^C + Q_{CH}^C + 2Q_{CC'}^C) 0.89 = +31.7 \text{ G}.$$

The agreement with the experimental value of $+33.0$ G is excellent[12].

The procedure has been extended to a number of aromatic systems, again yielding satisfactory agreement with available experimental values[10]. Two separate studies have predicted a positive value for the ^{13}C spin density[9,13].

---

[1] H. M. McConnell: J. Chem. Phys. **28**, 1188 (1958).
[2] A. Horsfield, J. R. Morton, and D. H. Whiffen: Trans. Faraday Soc. **57**, 1657 (1961).
[3] G. P. Rabold, K. H. Bar-Eli, E. Reid, and K. Weiss: J. Chem. Phys. **42**, 2438 (1965).
[4] H. Fischer: Mol. Phys. **9**, 149 (1965).
[5] W. T. Dixon, and R. O. C. Norman: J. Chem. Soc. **1963**, 3119.
[6] J. R. Bolton, A. Carrington, and J. dos Santos Veiga: Mol. Phys. **5**, 465 (1962).
[7] P. E. Wigen, and J. A. Cowen: J. Phys. Chem. Solids **17**, 26 (1960).
[8] F. C. Adam, and S. I. Weissman: J. Am. Chem. Soc. **80**, 2057 (1958).
[9] A. D. McLachlan, H. H. Dearman, and R. Lefebvre: J. Chem. Phys. **33**, 65 (1960).
[10] M. Karplus, and G. K. Fraenkel: J. Chem. Phys. **35**, 1312 (1961).
[11] R. W. Fessenden, and R. H. Schuler: J. Chem. Phys. **43**, 2704 (1965). — T. Cole, H. O. Pritchard, N. R. Davidson, and H. M. McConnell: Mol. Phys. **1**, 406 (1958).
[12] T. Cole, and C. Heller: J. Chem. Phys. **34**, 1085 (1961).
[13] H. M. McConnell, and D. B. Chesnut: J. Chem. Phys. **28**, 107 (1958).

compounds[1,2] suggest that although smaller, there may be an appreciable contribution to ring nitrogen splitting from adjacent carbon atoms. Allowing for a contribution from spin density on adjacent carbon atoms[3-5], one writes[3-5]

$$a^N = Q^N_{N(C_2H)} \varrho^\pi_N + \sum_i Q^N_{CN}(\varrho^\pi_{C_1} + \varrho^\pi_{C_2}). \tag{5.1}$$

From a study of dihydropyrazine cation [(10) with a hydrogen atom on each of the nitrogen atoms] and methyl derivatives, the two proportionality factors are found to be: $Q^N_{N(C_2H)} = 28.45$ G, $Q^N_{CN} = 2.62$ G[3]. These values for the most part fit excellently the data from which they were derived.

A study[5] of proton splitting constants in a large number of nitro-($-NO_2$) substituted compounds shows that they are well fitted by the simple McConnell equation with $Q = -23.7$ G and spin densities obtained from a configuration interaction calculation[5,6]. For the nitrogen splitting constants one may make use of an expression analogous to Eq. (5.1) and calculated spin densities to reproduce experimental results very well except for dinitro compounds at meta (next-nearest) orientation to one another. Similar agreement is observed for various nitriles (CN-substituted compounds)[7]. However, on applying Eq. (5.1) to a wider variety of nitrogen compounds, the range of the several Q values becomes distressingly large[8]. In view of the small value of Q^N_{CN}, (which have variously been assigned values from -1.5[9] to 9 G[2,1]) it seems likely that one may use the simple McConnell equation to fit data on very similar compounds, while an equation like (5.1) is more generally required. One principal difficulty is that of knowing the appropriate value of ϱ^π_N to use. Although HMO values of ϱ have been used with a respectable measure of success[10,11,3], one must be very cautious in use of such values for assignment of Q values to hetero atoms.

6. ^{19}F fluorine splitting constants. It has sometimes been assumed[12,13] that fluorine splittings in free radicals are given by the simple relation:

$$a^F = Q_F \varrho_C.$$

Taking $Q^H_{CH} = -23.7$ G for protons, one typically obtains Q_F values in the range 40—60 G[12-15]; a value as high as 118 G has been given[16]. For example, using the contact hyperfine shifts of fluorine NMR lines in substituted aminotroponeimineates (14), one may determine spin densities at the several positions of the variously substituted benzene rings. From the spin densities thus obtained, one gets the apparent Q_{CF} values[17,18] given in Table 6.

---

[1] J. C. M. HENNING, and C. DE WAARD: Phys. Letters **3**, 139 (1962).
[2] R. L. WARD: J. Am. Chem. Soc. **84**, 332 (1962).
[3] B. L. BARTON, and G. K. FRAENKEL: J. Chem. Phys. **41**, 1455 (1964).
[4] E. W. STONE, and A. H. MAKI: J. Chem. Phys. **39**, 1635 (1962).
[5] P. H. RIEGER, and G. K. FRAENKEL: J. Chem. Phys. **39**, 609 (1963).
[6] A. D. MCLACHLAN: Mol. Phys. **3**, 233 (1960).
[7] P. H. RIEGER, and G. K. FRAENKEL: J. Chem. Phys. **37**, 2795 (1962).
[8] E. T. STROM, G. A. RUSSELL, and R. KONAKA: J. Chem. Phys. **42**, 2033 (1965).
[9] See footnote 8, p. 166.
[10] See footnote 5, p. 166.
[11] See footnote 6, p. 166.
[12] D. H. ANDERSON, P. J. FRANK, and H. S. GUTOWSKY: J. Chem. Phys. **32**, 196 (1960).
[13] A. H. MAKI, and D. H. GESKE: J. Am. Chem. Soc. **83**, 1853 (1961).
[14] R. J. COOK, J. R. ROWLANDS, and D. H. WHIFFEN: Mol. Phys. **7**, 31 (1963).
[15] M. T. ROGERS, and D. H. WHIFFEN: J. Chem. Phys. **40**, 2662 (1964).
[16] R. J. COOK, J. R. ROWLANDS, and D. H. WHIFFEN: Proc. Chem. Soc. (London) **1962**, 252.
[17] D. R. EATON, A. D. JOSEY, W. D. PHILLIPS, and R. E. BENSON: Mol. Phys. **5**, 407 (1962).
[18] D. R. EATON, A. D. JOSEY, R. E. BENSON, W. D. PHILLIPS, and T. L. CAIRING: J. Am. Chem. Soc. **84**, 4100 (1962).

Table 6. *Fluorine splitting parameters in aminotroponeimineates*[1]

	a^F (G)	ϱ_C	Q_{CF}
p-fluorophenyl	+0.506	+0.0107	+47.2
m-fluorophenyl	−0.0325	−0.00813	+ 4.0
o-fluorophenyl	+0.270	+0.00657	+41.1

[1] D. R. Eaton, A. D. Josey, W. D. Phillips, and R. E. Benson: Mol. Phys. **5**, 407 (1962). — D. R. Eaton, A. D. Josey, R. E. Benson, W. D. Phillips, and T. L. Cairing: J. Am. Chem. Soc. **84**, 4100 (1962).

(14)

o, m, p, refer to positions of fluorine substitution for Table 6.

One must conclude that the fluorine hyperfine splitting is not simply proportional to the spin density on the adjacent carbon atom. The fluorine appears to participate in π-bonding to the ring; thus to describe the splitting one needs to consider not only the spin density on the fluorine as well as on the carbon, but the fractional double bond character of the C—F bond[1].

It therefore appears expedient to use coupling expressions analogous to those used for ^{13}C or ^{14}N, where there are contributions from atoms adjoining the carbon to which the fluorine is attached, viz.:

$$a^F = Q_{F(FC)}{}^F \varrho_F^\pi + Q_{CF}^F \varrho_C^\pi, \quad \text{where} \quad Q_{F(FC)}{}^F = P^F + Q_{FC}^F.$$

P^F includes effects from 1s electrons and the unshared electron pair on the fluorine atom. Q_{CF}^F and Q_{FC}^F are respectively the σ—π parameters denoting splitting contributions by 2s electrons caused by π-electron spin density on carbon and fluorine atoms[2]. For the 3,5-dinitrofluorobenzene anion, the value of ϱ_F^π was determined from line broadening to be approximately 1—2% of ϱ_N^π. The broadening here concerned is that resulting from electron-nuclear anisotropic intramolecular magnetic dipole interactions[2]. The procedure would also be applicable to ^{13}C or ^{14}N (but not H) or other nuclei directly conjugated in the π-system. Q parameters in the equation for a^F have not as yet been evaluated since one presently is ignorant of the spin density on the fluorine.

7. Sign of the hyperfine splitting constant. We have noted that the σ—π proton hyperfine splitting parameter $Q_{CH}{}^H$ in the relation $a_{CH}{}^H = Q_{CH}{}^H \varrho_C^\pi$ is found by calculation to have a negative sign. Hence, taking the spin density ϱ_C^π at a given carbon as positive if it has generally the same sign as the rest of the molecule, one assigns a negative value to the splitting constant for a proton on such a carbon atom. If a proton on a nitrogen atom behaves similarly one writes $a_{NH}{}^H = Q_{NH}{}^H \varrho_N^\pi$, implying an interaction between the σ-bonding framework including the nitrogen and the π-orbital on nitrogen. The parameter Q_{NH}^H is then found by molecular orbital[3] and valence bond calculations also to be negative[4]. We have noted that in a number of molecules there are carbon atoms for which it was deemed necessary to

---

[1] See footnote 17, p. 167.
[2] M. Kaplan, J. R. Bolton, and G. K. Fraenkel: J. Chem. Phys. **42**, 955 (1965).
[3] B. L. Barton, and G. K. Fraenkel. J. Chem. Phys. **41**, 695 (1964).
[4] J. C. Schug, T. H. Brown, and M. Karplus: J. Chem. Phys. **37**, 330 (1962).

assign a spin polarization opposite to that of most carbon atoms in the molecule. We now consider combinations of experimental and theoretical procedures for verifying signs of splitting constants as well as their magnitudes in some instances.

From line positions in ESR spectra of radicals (which in liquid solution show no significant departure from second-order theory), one obtains directly only the absolute magnitude of the hyperfine splitting constant. The hyperfine spectrum in such cases is independent of its sign, since to second order, change of sign alters the order of levels without affecting their separations. If splittings are large enough to discern third order effects, one may be able to determine relative signs, e.g., for ^{19}F and ^{13}C (they are identical)[1]. To determine the sign, one must usually have additional information. We shall outline several procedures, of which the most generally useful is measurement of NMR chemical shifts[2-6]. Contact interaction between electron spin density on a carbon atom and an attached proton shifts the NMR resonant field of the latter by: $\Delta H = -a_H(\gamma_e/\gamma_H)g\beta H_0/4kT$, relative to a diamagnetic reference[2]. If the proton splitting constant is negative, its resonant field will be raised and vice versa. Therefore, an NMR line displacement to low field is indicative of a positive proton splitting constant and a negative spin density on the adjacent carbon atom.

(15)

The pyrene anion (15) is a simple example in which one proton NMR line is shifted to low field and two to high field[2]. The former corresponds to protons 2 and 7 and the latter to the remaining protons. Thus the spin density at carbon atoms 2 and 7 is negative, while that at the remainder is positive, as predicted.

In some divalent nickel chelates (aminotroponeimineates) (14) the proton NMR spectrum shows such large contact hyperfine shifts (and sufficiently long proton relaxation times) that proton spin-spin interactions allow the numerous lines to be definitely identified[7,8]. Reference lines of the diamagnetic zinc compounds allow one to assign positive and negative spin density values. The large proton shifts (up to 20 kHz at 60 MHz) are attributed to spin density delocalization from metal to ligands by $d\pi-p\pi$ bonding. The spin densities at several fluorine-substituted positions are given in Table 6.

In complexes $[(C_4H_9)_4N][(C_6H_5)_3PMI_3]$, where $M =$ Co or Ni, one sees isotropic magnetic resonance shifts not only for the C_6H_5 protons but also for those of the

---

[1] R. W. FESSENDEN, and R. H. SCHULER: J. Chem. Phys. **43**, 2704 (1965).

[2] H. S. GUTOWSKY, H. KUSUMOTO, T. H. BROWN, and D. H. ANDERSON: J. Chem. Phys. **30**, 860 (1959). — T. H. BROWN, D. H. ANDERSON, and H. S. GUTOWSKY: J. Chem. Phys. **33**, 720 (1960).

[3] M. E. ANDERSON, P. J. ZANDSTRA, and T. R. TUTTLE jr.: J. Chem. Phys. **33**, 1591 (1960). — M. E. ANDERSON, G. E. PAKE, and T. R. TUTTLE jr.: J. Chem. Phys. **33**, 1581 (1960).

[4] A. FORMAN, J. N. MURRELL, and L. E. ORGEL: J. Chem. Phys. **31**, 1129 (1959).

[5] D. R. EATON, and W. D. PHILLIPS: Advances in Magnetic Resonance, p. 103. (J. S. WAUGH, ed.). New York: Acad. Press 1965.

[6] H. M. McCONNELL, and D. B. CHESNUT: J. Chem. Phys. **28**, 107 (1958).

[7] W. D. PHILLIPS, and R. E. BENSON: J. Chem. Phys. **33**, 607 (1960).

[8] D. R. EATON, and W. D. PHILLIPS: J. Chem. Phys. **43**, 392 (1965) (Numerous additional references are given here).

cation[1]. Shifts of the latter are attributed to partial ion pairing in the CDCl$_3$ solution, leading to a pseudo-contact interaction.

In a mixture of an alkyl-substituted naphthalene such as (16) with its negative ion, one notes relative displacements of NMR alkyl proton lines having different signs of the splitting constants, as the concentration of anion is increased. This results from the time-dependent contact interaction. Since the exchange rate is large, the splitting constants are readily obtained from the NMR shifts and the fraction of negative ions. When three or more carbon atoms have protons contributing to the splitting constant, the signs alternate from $+$ to $-$ as one goes from the α to β to γ atoms[2]. In (16) the δ proton coupling constants have the same sign as the γ protons.

$$\text{naphthalene}-\underset{\alpha\ \ \beta\ \ \gamma\ \ \delta}{CH_2CH_2CH_2CH_3}\ \ominus$$

(16)

NMR studies of solid WÜRSTER's blue perchlorate $(CH_3)_2NC_6H_4N(CH_3)_2^+\ ClO_4^-$ at room temperature shows CH$_3$ and ring proton lines separated by roughly 4.6 G for $H_0 = 8000$ G or $\sim 6 \times 10^{-4}$ of the field[3]. By contrast, $-N(CH_3)_2$ and ring protons of dissolved diamagnetic substances have a fractional separation of about 4×10^{-6}.

For obtaining the *relative* signs of two splitting constants in a radical, it has been suggested that the spectrum in zero external field[4] may allow the assignment to be made[5]. The spin Hamiltonian is:

$$\mathcal{H} = \sum_i a^i S_z I_z^i + \tfrac{1}{2} \sum_i a^i (S_+ I_-^i + S_- I_+^i),$$

where a^i is the splitting constant of nucleus i. Plotting levels for alternative sign choices on one nucleus, keeping the sign of the other constant, one notes which choice gives transitions in agreement with experiment.

Oriented radicals permit determination of the sign of isotropic coupling constants, e.g., ^{13}C. A basic example is the radical HOOC$-\dot{C}$(H)COOH derived from malonic acid HOOCC(H$_2$)COOH by irradiation[6]. Using the C$-$H bond as the z direction and the $2p$ orbital as x, one finds the principal values of the hyperfine tensor are $A_{xx}^C = 212.7$, $A_{yy} = 22.8$ and $A_{zz} = 42.2$ MHz, with the isotropic component one-third the trace or 93 MHz. An independent expression for ^{13}C interaction with a $2p\pi$ electron is given by 64 $(3\cos^2\theta - 1)\varrho^C$, where θ is the angle between the π orbital and the field direction. Taking $\varrho^C = 0.8$, one gets anisotropic contributions of 102 MHz for x, and -51 MHz for y and z. If the isotropic interaction is positive, $A_{xx} = 93 + 102 = 195$ MHz; $A_{yy} = 93 - 51 = 42$ MHz and $A_{zz} = +42$ MHz; the -93 MHz value would give respectively 9, -144 and -144 MHz for the three principal components. It is clear that the isotropic contribution is positive.

An ingenious procedure for aligning some radicals in solution used liquid crystals which are known to have the long axes of solvent clusters aligned in a magnetic field[7,8]. Unfortunately, only DPPH (17), tetracyanoethylene anion

---

[1] G. N. LaMar: J. Chem. Phys. **41**, 2992 (1964).
[2] E. de Boer, and C. MacLean: Mol. Phys. **9**, 191 (1965).
[3] A. Kawamori, and K. Suzuki: Mol. Phys. **8**, 95 (1964).
[4] T. Cole, T. Kushida, and H. C. Heller: J. Chem. Phys. **38**, 2915 (1963).
[5] H. C. Heller: J. Chem. Phys. **42**, 2611 (1965).
[6] T. Cole, and C. Heller: J. Chem. Phys. **34**, 1085 (1961).
[7] A. Carrington, and G. R. Luckhurst: Mol. Phys. **8**, 401 (1964).
[8] S. H. Glarum, and J. H. Marshall: J. Chem. Phys. **44**, 2884 (1966).

(TCNE$^-$), (18) and the perinaphthenyl radical (6) have thus for shown the requisite stability at elevated temperatures of the liquid crystal range. If the shape of the radical is such that extensive alignment occurs, one calculates the expected magnitude of the nonvanishing isotropic nitrogen hyperfine component. For TCNE$^-$, this (in-plane) component is calculated to be -1.06 G [1]. In the liquid-crystal region the measured nitrogen splitting is 0.62 G, while in the isotropic region it is ± 1.58 G. Thus the positive sign for isotropic coupling gives the value 0.52 G expected for complete alignment. For perinaphthenyl radical (6) one finds that the experimental values of ^{13}C splitting constants around the periphery are alternately 9.79 and -7.92 G, the signs being taken to agree best with the calculated values of 9.44 and -8.37 G.

(17) (18) (19)

These values reproduce reasonably well the proton splittings with $Q_{CH}^H = -27$ G [2]. In the liquid crystal the g value is displaced from that in the isotropic phase by $\langle 1 - 3\cos^2\beta \rangle (g_x + g_y - 2g_z)/6$. The first factor is 1 for perfect alignment and 0 for random orientation. The coupling constant is changed by the orientation factor times an anisotropic term. The hyperfine anisotropies observed are in generally good accord with calculated values. From the g-shift one finds $g_x = g_y = 2.00278$, $g_z = 2.00226$ [2]. Thus one is able to ascertain for some radicals in liquid systems the hyperfine and g-tensor data normally requiring oriented solids.

An ENDOR procedure may also be used to determine the absolute sign of hyperfine coupling constants [3].

While hyperfine line positions are (to second order) independent of signs of coupling constants, relative line widths and saturation behavior are determined by their signs. One may write expressions for the width of each hyperfine line in terms of the width of the central line [4]. Assuming $a_{CH}^H < 0$ in dihydropyrazine cation (19), one infers from the fact that high-field half of the spectrum is broader than the low field half that a^N is positive while a_{NH}^H and Q_{NH}^H are negative [5]. In this molecule the theoretical prediction [6] that for aromatic molecules the perpendicular component of the g-tensor g_3 is less than the mean of the in-plane components $(g_1 + g_2)/2$ is verified. Since high-field lines of other nitrogen heterocyclic anions and cations similarly have greater width than the low-field lines, it is predicted that in all these the nitrogen splitting is positive and the proton splitting is negative [5].

A procedure based upon anticipated g-components is useful for determining the sign of a for ^{13}C or ^{14}N splittings [7]. It is not helpful for determining the signs of proton splittings.

Separate experiments [8,9] have shown that in a radical with fluorine substituted on a carbon atom which is primarily the site of an unpaired electron, the fluorine

---

[1] See footnote 7, p. 170.
[2] See footnote 8, p. 170.
[3] R. J. Cook, and D. H. Whiffen: J. Chem. Phys. **43**, 2908 (1965).
[4] J. H. Freed, and G. K. Fraenkel: J. Chem. Phys. **39**, 326 (1963).
[5] B. L. Barton, and G. K. Fraenkel: J. Chem. Phys. **41**, 695 (1964).
[6] A. J. Stone: Mol. Phys. **6**, 509 (1963); — Proc. Roy. Soc. (London) A **271**, 424 (1963). — H. M. McConnell, and J. M. Robertson: J. Phys. Chem. **61**, 1018 (1957).
[7] E. de Boer, and E. L. Mackor: J. Chem. Phys. **38**, 1450 (1963).
[8] R. J. Cook, J. R. Rowlands, and D. H. Whiffen: Mol. Phys. **7**, 31 (1963).
[9] D. R. Eaton, A. D. Josey, W. D. Phillips, and R. E. Benson: Mol. Phys. **5**, 407 (1962).

coupling constant is positive, while that of a corresponding proton is negative. In the radical FĊHCONH$_2$ the principal values of hyperfine couplings of α-substituents (i.e., on the C' atom) are as follows[1]:

$$^{19}F: \quad +530, -11, -45 \text{ MHz},$$
$$^{1}H: \quad -96, -63, -31 \text{ MHz},$$
$$^{13}C: \quad +238 \text{ MHz}.$$

The large positive principal value for ^{19}F (in a direction perpendicular to the radical plane) insures that the isotropic coupling (one-third the trace) is positive (+158 MHz). The fluorine $2p\sigma$ spin density is -0.016, while the $2p\pi$ density is 0.119, a remarkably large value.

8. Hyperconjugation. Protons of CH$_3$ groups attached to π-electron systems typically have coupling constants comparable with or exceeding those on an atom having a π-orbital. For example, in the ethyl radical CH$_3$CH$_2$, the CH$_3$ and CH$_2$ proton splitting constants are 26.87 and 22.38 G respectively[2]. In the m-xylene anion (20) the CH$_3$ and ring proton constants are 2.26 G (CH$_3$), 1.46 (H$_4$), 6.85 (H$_2$) and 7.72 G (H$_5$)[3]. (The subscripts identify proton positions in the ring.)

In cyclohexadienyl radical (21) or in pyracene cation (22), the CH$_2$ group is outside the normal π-system, having approximately tetrahedral coordination. (21) is produced as a result of hydrogen addition to benzene upon UV, γ- or electron irradiation[4-7] or by electron irradiation of cyclohexadiene[3]. The CH$_2$ proton couplings are large, especially the 47.7 G value for the cyclohexadienyl radical[3-7].

A linear combination of methyl hydrogen atom orbitals $\psi(H_a) - [\psi(H_b) + \psi(H_c)]$ has π-type symmetry and may thus interact with a normally-conjugated π-system. This "hyperconjugation" was postulated earlier to account for dipole moments, delocalization energies, bond lengths, ionization potentials and kinetic effects[8]. It has been invoked as a mechanism for CH$_3$ (or >CH$_2$) hyperfine splitting[9-12]. For example, the CH$_3$CH$_2$ radical is treated as a three-orbital system, the

---

[1] See footnote 8, p. 171.
[2] R. W. FESSENDEN, and R. H. SCHULER: J. Chem. Phys. **38**, 773 (1963); **39**, 2147 (1963).
[3] J. R. BOLTON, and A. CARRINGTON: Mol. Phys. **4**, 497 (1961).
[4] V. A. TOLKACHEV, YU. N. MOLIN, I. I. CHKEIDZE, N. J. BUBIN, and V. V. VOEVODSKII: Doklady Akad. Nauk. S.S.S.R. **141**, 911 (1961).
[5] H. FISCHER: Z. Naturforsch. A **17**, 693 (1962); — Kolloid-Z. **180**, 64 (1962); — J. Chem. Phys. **37**, 1094 (1962).
[6] D. H. WHIFFEN: Mol. Phys. **6**, 223 (1963).
[7] S. OHNISHI, T. TANEI, and J. NITTA: J. Chem. Phys. **37**, 2402 (1962).
[8] C. A. COULSON: Valence, Oxford Univ. Press, second ed., 1961, p. 356; Series of articles in Tetrahedron **5**, 105—274 (1959).
[9] R. BERSOHN: J. Chem. Phys. **24**, 1066 (1956).
[10] D. B. CHESNUT: J. Chem. Phys. **29**, 43 (1958).
[11] M. C. R. SYMONS, in: V. GOLD, Advances in Physical Org. Chem., vol. I, p. 318. New York: Academic Press 1963.
[12] P. NORDIO, M. V. PAVAN, and G. GIACOMETTI: Theoret. Chim. Acta **1**, 302 (1963).

pseudo-π orbital of the methyl protons being coupled to $2p_z$ orbitals on the two carbons[1]. The direct participation of protons in the coupled π-system requires that the coupling constant and spin density be positive. This is in agreement with observation on pyracene cation[2] (22) and methyl-substituted organometallic compounds[3]. Also consistent with hyperconjugation as a mechanism for CH_3 splitting is the twofold larger proton splitting in methyl-substituted cations of anthracene as compared with the anions[4]. Closely allied to this is a similar ratio of CH_2 proton splittings in the cation and anion of pyracene[5,2] (22). In the anion, an electron-releasing substituent such as CH_3 would be hindered in such release by hyperconjugation; in the cation the positive charge would tend to be delocalized over the CH_3 group as a part of the π-system.

One approach to hyperconjugation is to consider the whole CH_3 group as the equivalent of a hetero-atom[6,4] (oxygen, nitrogen, sulfur and halogens are considered hetero atoms) with one bonding orbital. However, this has led to erroneous quantitative predictions as a result of neglecting the antibonding π-orbital[5].

Cyclohexadienyl (21) or pyracene cation radicals (22) because of their geometry and large coupling constants are ideal molecules for testing the validity of the hyperconjugative model for spin transfer to CH_2 protons attached to a π-system. The symmetric and antisymmetric combinations of the two atomic orbitals give respectively orbitals of σ- (σ_H) and π-symmetry (π_H). The σ_H orbitals of a CH_2 carbon atom are assumed to point toward the two other carbon atoms to which it is linked and to the π_H orbital. The procedure for computation is analogous to that used for toluene $(C_6H_5CH_3)$[6]. The largest contribution to the spin density at the CH_2 protons is given by the square of the coefficient of the π_H orbital in the singly-occupied molecular orbital. A second term arises from the non-zero value of the $2p_z$-orbital of the $C\begin{smallmatrix}H\\H\end{smallmatrix}$ carbon atom at the protons which do not lie in its nodal plane. A third term involves the product of coefficients of the $2p_z$ and π_H coefficients to take overlap into account, (unlike the HMO approach in which $S_{i,i+1}$ is set equal to zero). The total spin density at a CH_2 proton is multiplied by 508 (instead of Q) to get the coupling constant. The same procedures have been applied to acenaphthene (23) ions and the cyclohexadienyl radical[5].

(23)

The calculated CH_2 proton splittings for pyracene cation and anion are 13.50 and 7.72 G, while the experimental values are 12.80 and 6.58 G. For cyclohexadienyl, the calculated splitting is 45.2 G, as compared with the 47.71 G experimental value. The calculated spin density in the cyclohexadienyl π_H orbital is 0.178, while that at the adjacent (CH_2) carbon is given as only 0.006. The calculated charge densities at these respective positions is 0.890 and 1.052[5].

---

[1] See footnote 12, p. 172.
[2] E. DE BOER, and E. L. MACKOR: Mol. Phys. **5**, 493 (1962).
[3] A. FORMAN, J. N. MURRELL, and L. E. ORGEL: J. Chem. Phys. **31**, 1129 (1960).
[4] J. R. BOLTON, A. CARRINGTON, and A. D. McLACHLAN: Mol. Phys. **5**, 31 (1962).
[5] J. P. COLPA, and E. DE BOER: Mol. Phys. **7**, 333 (1964).
[6] C. A. COULSON, and V. A. CRAWFORD: J. Chem. Soc. **1953**, 2052.

An alternative to the delocalization of methyl group electrons into an attached π-system[1] is to invoke an "inductive" effect by which the coulomb integral α of the carbon atom at the point of substitution is altered. One may also make a valence-bond calculation of spin polarization (i.e., interaction between π and σ electrons) in the system[2]. An alternative is to change the scale of the orbital exponent of the $2p\pi$ orbital of this carbon atom[3]. This procedure does indeed give satisfactorily the ring proton splitting constants in methyl-substituted benzenes[3], and one may invoke spin polarization to get the numerical values of CH_3 proton splittings. However, it appears very improbable that in cations of methyl substituted hydrocarbons there would be a twofold larger spin density on the CH_3 carbon than in the anion to explain the factor of two in splitting constants[4]. A polarization calculation has been made for a C'—C—H fragment where only C' is part of a coupled π system. Proton coupling arises from direct polarization by the π electron on atom C' of the C—H bond. The coupling constant for this is found to be -1.76 G[5]. A second mechanism involves the polarization of the C'—C bond, with subsequent C—H polarization, and the coupling contribution for this is $+0.65$ G. The net coupling constant is then only -1.1 G, i.e., $a_H = -1.1\,\varrho_{C'}^\pi$. This constant is taken to be B_0 in the expression: $Q(\theta) = B_0 + B_1 \cos^2\theta$, where θ is the angle between the C'—C—H plane and the π orbital on C'. This equation[6-9] appears to represent satisfactorily the isotropic β-proton splittings in oriented free radicals. Extension to a C'—CH_2 fragment again gives a coupling constant of about -1.1 G for CH_2 protons, far lower than observed splittings[5]. A valence bond treatment has given $a^H = 28$ G for CH_3 protons in the $CH_3\dot{C}H_2$ radical but this treatment gives a mixture of spin polarization and hyperconjugation effects[10].

It is fair to conclude that spin polarization is important in determining coupling constants of ring protons, but to explain those of attached CH_2 or CH_3 protons, hyperconjugative transfer of spin density from the π system to a pseudo-π-orbital of protons is required. Very recent calculations indicate that for the ethyl or allyl radicals the spin delocalization is attributed 40 per cent to electron transfer (hyperconjugation) and 60 per cent to exchange polarization[11]. In $CH_3\dot{C}HOH$, the 15.0 G and 22.5 G splittings from protons on the α and β carbon atoms are typical. When the β atom is oxygen, as in $\dot{C}H_2OH$, the β proton splitting is much less, e.g., 0,96 G in this case[12]. The splitting due to protons on the γ carbon atom are far smaller than for β protons on carbon, e.g., 0.95 G for the CH_3 protons in $\dot{C}H_2\underset{\underset{O}{\|}}{C}CH_3$[13].

One may have a contribution from protons on δ atoms, provided there is a significant measure of π-bonding. This is found in radicals either of type (24) or (25)

$$\overset{\delta}{H}-\overset{\gamma}{C}-\underset{\underset{O}{\|}}{\overset{\beta}{C}}-\overset{\alpha}{O}-C' \qquad\qquad \overset{\alpha}{C'}-\underset{\underset{O}{\|}}{\overset{\beta}{C}}-\overset{\gamma}{O}-\overset{\delta}{C}-H$$

(24) \qquad\qquad\qquad (25)

---

[1] R. S. MULLIKEN: J. Chem. Phys. **7**, 339 (1939).
[2] A. D. MCLACHLAN: Mol. Phys. **1**, 233 (1958).
[3] R. L. FLURRY jr., and P. G. LYKOS: Mol. Phys. **6**, 283 (1963).
[4] See footnote 4, p. 173.
[5] See footnote 5, p. 173.
[6] M. C. R. SYMONS: J. Chem. Soc. **1959**, 277.
[7] C. HELLER, and H. M. MCCONNELL: J. Chem. Phys. **32**, 1535 (1960).
[8] E. W. STONE, and A. H. MAKI: J. Chem. Phys. **37**, 1326 (1962).
[9] A. HORSFIELD, J. R. MORTON, and D. H. WHIFFEN: Mol. Phys. **5**, 115 (1962).
[10] A. D. MCLACHLAN: Mol. Phys. **1**, 233 (1958).
[11] D. LAZDINS, and M. KARPLUS: J. Chem. Phys. **44**, 1600 (1966).
[12] H. FISCHER: Mol. Phys. **9**, 149 (1965).
[13] P. SMITH, J. T. PEARSON, P. B. WOOD, and T. C. SMITH: J. Chem. Phys. **43**, 1535 (1965).

9. Removal of orbital degeneracy by substitution.

where the δ proton splitting is 1.3 to 1.5 G. Since both C—O—C bonds in esters have double-bond character, one places a π-orbital on each of the α, β and γ atoms of either radical. Then one assumes hyperconjugation of protons with spin density on the doubly-linked oxygen of the second radical or on the γ carbon of the first radical.

9. Removal of orbital degeneracy by substitution. Rationalization of the effects of substituents upon the hyperfine splitting constants of a parent substance in terms of removal of degeneracy and occupation of either the symmetric or the

Table 7. *Splitting constants of substituted benzenes.*

Substituent(s)	Substituent position	Splitting constants (G)						Substituent proton splitting
		1	2	3	4	5	6	
CH_3	1 [1, 2]	—	5.12	5.45	0.59	5.45	5.12	0.79
CH_3	1, 2 [3]	—	—	6.93	1.81	1.81	6.93	2.00
CH_3	1, 3 [2]	6.85	—	—	1.46	7.72	1.46	2.26
CH_3	1, 4 [2]	5.34	5.34	—	5.34	5.34	0.10
—CH_2CH_3	1 [4]	—	5.06	5.06	0.92	5.06	5.06	0.92
—$CH(CH_3)_2$	1 [4]	—	5.10	5.10	1.12	5.10	5.10	0.61
—$C(CH_3)_3$	1 [4]	—	4.66	4.66	1.74	4.66	4.66	<0.10
—$Si(CH_3)_3$	1 [5]	—	2.66	1.06	8.13	1.06	2.66	0.40
—$Ge(CH_3)_3$	1 [5]	—	2.33	1.46	7.61	1.46	2.33	<0.10
CN	1 [6]	—	3.63	0.30	8.42	0.30	3.63	2.15 (N)
CN	1, 2 [6]	—	—	0.42	4.13	4.13	0.42	1.75 (N)
CN	1, 3 [6]	1.44	—	—	8.28	∼0.08	8.29	1.02 (N)
CN	1, 4 [6]	1.59	1.59	—	1.59	1.59	1.81 (N)
CN	1, 2, 4, 5 [6]	—	—	0.04	—	—	0.04	1.15 (N)

[1] T. R. TUTTLE jr., and S. I. WEISSMAN: J. Am. Chem. Soc. **80**, 5342 (1958).
[2] J. R. BOLTON, and A. CARRINGTON: Mol. Phys. **4**, 497 (1961).
[3] T. R. TUTTLE jr.: J. Am. Chem. Soc. **84**, 2839 (1962). — J. R. BOLTON: J. Chem. Phys. **41**, 2455 (1964).
[4] J. R. BOLTON, A. CARRINGTON, A. FORMAN, and L. E. ORGEL: Mol. Phys. **5**, 43 (1962).
[5] J. A. BEDFORD, J. R. BOLTON, A. CARRINGTON, and R. H. PRINCE: Trans. Faraday Soc. **59**, 53 (1962).
[6] P. RIEGER, and G. K. FRAENKEL: J. Chem. Phys. **37**, 2795 (1962). — A. CARRINGTON and P. F. TODD: Mol. Phys. **6**, 161 (1963).

antisymmetric orbital has been one of the most valuable contributions of the HMO method [1]. In the benzene anion $C_6H_6^-$, the unpaired electron occupies one of the pair of degenerate (E_2) antibonding orbitals in Fig. 4. Substitution at any position destroys the sixfold symmetry and distorts the energy level to remove the degeneracy. We shall assume that the ordering of levels is not greatly altered and that only the degeneracy of the two E_2 orbitals is removed by making one or the other the more stable.

Examination of Table 7 shows marked variations in the values of proton hyperfine splitting constants in substituted benzene anions. We note that proper assignment of the unpaired electron to one or the other of the E_2 orbitals gives HMO unpaired electron densities in rough accord with the experimental splitting constants. For methylbenzene (toluene) and 1,4-dimethylbenzene (*p*-xylene) anions, one expects that a CH_3 group (which from other types of experiments show a tendency toward electron repulsion) will lead to occupation of an antisymmetrical orbital. Such occupation would allow CH_3 groups to be placed in positions of minimum (zero predicted) unpaired electron density. Assuming a Q value of 22.5 G as

[1] A. CARRINGTON: Quart. Revs. **17**, 67 (1963).

for benzene, a value of 0.25 for ϱ' would give $a_H = 5.6$ G at the 2, 3, 5 and 6 positions of toluene with zero at position 4. The alternative orbital assignment could not place the CH_3 carbon at a node, and would predict 1.9 G for four protons and 7.5 G for the proton at position 4. Experience leads one to expect a large splitting from CH_3 protons by virtue of a hyperconjugation mechanism (Sect. 8). Agreement with predictions for the antisymmetric orbital lowest are well fulfilled, and a similarly good accord is seen for 1,4-dimethylbenzene. For the 1,3-dimethylbenzene (m-xylene), the minimum unpaired electron density available in symmetry-equivalent positions is given by the symmetrical orbital, where with a Q-value of 22.5 G, one expects splittings of 7.5 G at 2 and 5 positions (if only the antibonding levels are shifted) and 1.9 G at the 4 and 6 positions. Again, the experimental agreement is good. One may treat the other dimethyl benzenes and the mono- and dicyanobenzenes $C_6H_{6-n}(CN)_n$ in similar fashion.

If one examines monoalkyl-substituted benzenes involving progressive replacement of CH_3 hydrogens by CH_3 groups, there is a progressive increase in the splitting by the proton in position 4 from 0.59 in $C_6H_5CH_3$ to 1.74 G in $C_6H_5C(CH_3)_3$. In the latter compound, the 2, 3, 5, 6 proton splittings have dropped to 4.66 from 5.12 G, i.e., Q is effectively constant. It is assumed that due to vibrations, there is vibronic coupling of the symmetrical with the lower-lying antisymmetrical orbital, the admixture increasing as the energies of the two orbitals become more nearly equal[1].

Inverting the procedure of assignment of orbitals on the basis of known substituent properties, one may gain information regarding the electron-repelling or attracting properties, of a substituent group by examination of the splitting constants in monosubstituted benzenes. For example, the $-Si(CH_3)_3$ and

$$\begin{array}{ccc}
C(CH_3)_3^{\ominus} & Si(CH_3)_3^{\ominus} & Ge(CH_3)_3^{\ominus} \\
4.66 & 2.66 & 2.33 \\
4.66 & 1.06 & 1.46 \\
1.74 & 8.13 & 7.61 \\
(26) & (27) & (28)
\end{array}$$

$-Ge(CH_3)_3$ (28) groups are analogous to $-C(CH_3)_3$. However, for both the silicon and the germanium compounds, one notes that assignment to the symmetrical orbital is called for; hence, both $Si(CH_3)_3$ and $Ge(CH_3)_3$ should be considered as electron attracting groups[2]. For 1,4-disubstitution of the benzene anion by two groups with "electron affinity" of opposite sign, the occupation of symmetrical or antisymmetrical levels will depend upon a balance of the effects of the two groups.

The concept of removal of orbital degeneracy by making the antisymmetric or symmetric orbitals the more stable by substitution may be extended to cyclooctatetrene anion[3]. Here it is degeneracy of non-bonding levels which is involved, as can be seen by inspection of Fig. 3. Here the HMO orbital coefficients in the NBMO sum to zero about alternate positions, just as for odd-alternant hydrocarbons. Hence nodal planes may be drawn in two ways, as shown in Fig. 10, with the squares of MO coefficients 1/4 at the four other positions. Another point of difference from $C_6H_6^-$ is that one NBMO is doubly occupied. Since that orbital will be the more stable one, the odd electron will be in the orbital of higher energy. Substituted cyclooctatetrene anions appear to be planar also. If the substituent

---

[1] J. R. BOLTON, A. CARRINGTON, A. FORMAN, and L. E. ORGEL: Mol. Phys. **5**, 43 (1962).
[2] J. A. BEDFORD, J. R. BOLTON, A. CARRINGTON, and R. H. PRINCE: Trans. Faraday Soc. **59**, 53 (1962).
[3] A. CARRINGTON, and P. F. TODD: Mol. Phys. **7**, 533 (1964).

is an electron-repelling group (CH_3, C_2H_5, etc.), the odd-electron density at the substituent and at alternate carbons is expected to be 1/4, with zero at the four other positions. In unsubstituted cyclooctatetrene anion, $Q = 26.0$, and hence one expects a splitting of 6.5 G. Upon CH_3 substitution at position 1, Fig. 10, $a_{CH_3}^H = 5.1$ G and $a^H = 4.8$ for positions 3, 5 and 7. The other four protons have splitting constants of 1.6 G. Again the concept of vibronic mixing of levels is invoked to justify the observed deviations from prediction[1]. Although the accord with prediction is rough, one must nevertheless regard it as remarkably good, considering that *from symmetry arguments one should expect three sets of two equivalent protons, plus a unique proton.*

While the Hückel theory has given a remarkable semiquantitative description of the splittings of various substituted benzene or cyclooctatetrene anions, one

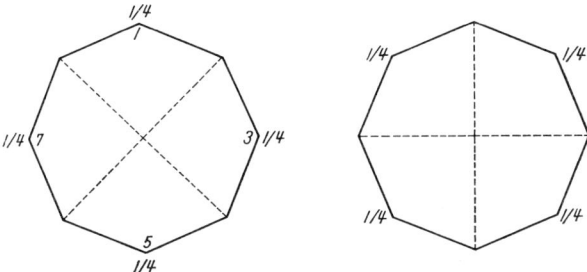

Fig. 10. MO nodes and unpaired electron distribution in the degenerate nonbonding orbitals of cyclooctatetrene anion.

Table 8.

	HMO	Config. interaction[1]
ψantisym	$\varrho_1 = \varrho_4 = 0$ $\varrho_2 = \varrho_3 = \varrho_5 = \varrho_6 = 0.2500$	$\varrho_1 = \varrho_4 = -0.0616$ $\varrho_2 = \varrho_3 = \varrho_5 = \varrho_6 = 0.2808$
ψsym	$\varrho_1 = \varrho_4 = 0.3333$ $\varrho_2 = \varrho_3 = \varrho_5 = \varrho_6 = 0.0833$	$\varrho_1 = \varrho_4 = 0.3949$ $\varrho_2 = \varrho_3 = \varrho_5 = \varrho_6 = 0.0525$

[1] W. D. Hobey: Mol. Phys. **7**, 325 (1964).

must go to more elaborate calculations for improved numerical agreement. A configuration interaction calculation[2] for hydrocarbon anions with degenerate ground states applied to $C_6H_6^-$ alters spin density distribution as shown in Table 8.

Thus one obtains negative spin densities at those positions for which the HMO method gives zero values. Some success had been achieved in calculating ESR spectra for alkyl substituents by consideration of vibronic coupling of the nearly degenerate states[3].

10. Monocyclic radicals. Of the monocyclic radicals suggested by Fig. 3, one observes the neutral cyclopentadienyl[4,5], benzene anion[6], neutral cyclohepta-

---
[1] See footnote 3, p. 176.
[2] W. D. Hobey: Mol. Phys. **7**, 325 (1964).
[3] W. D. Hobey: J. Chem. Phys. **43**, 2187 (1965).
[4] S. Ohnishi, and I. Nitta: J. Chem. Phys. **39**, 2848 (1963).
[5] P. Zandstra: J. Chem. Phys. **40**, 612 (1964).
[6] T. R. Tuttle, and S. I. Weissman: J. Am. Chem. Soc. **80**, 5342 (1958).

trienyl[1-3] and cyclooctatetrene anion[4]. Each of these in solution gives a spectrum indicative of fully-equivalent protons, and thus of a planar structure in rapid motion to average out anisotropic hyperfine contributions. One may see the effect of intramolecular distortion from non-uniform spin distribution by observing the cyclopentadienyl radical in solid bicyclopentadiene between 70 and 120 °K[5]; however, the radical still appears to be rotating in its molecular plane at 25 °K. Above 120 °K or in solution there is no evidence of distortion resulting from an orbitally non-degenerate ground state[5]. The spin densities appear to be in accord with calculated values[6].

Substitution of one hydrogen atom by deuterium causes a significant change in the spin distribution in benzene anion[7]. However, in cyclo-octatetrene-1 d ($C_8H_7D^-$) anion the proton splittings are unaltered[8].

The seven protons in the latter appear "equivalent", each of the eight line groups a triplet from the deuterium splitting[8]. The marked changes in splitting constants of C_6H_5D as compared with C_6H_6 have been quantitatively explained in terms of changes in the exchange integral β between carbon neighbors. In C_6H_6, out-of-plane vibrations of hydrogen or deuterium are presumed to involve orbital following of the $2p_z$ orbital. Substitution of D lowers the amplitude of vibration and hence of orbital following on the substituted carbon. The value of the exchange integral β between this carbon and its neighbors is then diminished in the symmetric orbital, while the anti-symmetric orbital through the substituted position is unaffected[9]. This effect provides a convincing explanation of the failure of monodeuteration of cyclo-octatetrene to destroy equivalence of the remaining protons. Both in $C_8H_8^-$ and $C_8H_7D^-$, the electron is in a non-bonding (rather than anti-bonding) orbital. Since the exchange integral between adjacent carbon atoms is zero in non-bonding orbitals in the HMO approximation, one would expect no introduction of non-equivalence by D substitution, as observed. Further, this model is in accord with results in the various deuterated naphthalenes, where D-substitution in the 1-position causes the other protons 4, 5 and 8 (which would have been symmetry-equivalent in the unsubstituted molecule) to be non-equivalent[10]. It is therefore more convincing than an alternative explanation in terms of a Jahn-Teller effect[8].

On substitution of deuterium for hydrogen one expects a reduction in hyperfine splitting by 6.514, the ratio of moments. However, marked deviations (positive and negative) have been observed[7, 11-13], the total spread in ratios being about 8 per cent. Such large deviations would seem surely to involve vibrational effects. Radical pairs such as $C_6H_6^-$, $C_6D_6^-$ or C_7H_7, C_7D_7 which have threefold or higher symmetry are expected to have strong vibronic interactions and therefore to be

---

[1] G. Vincow, M. L. Morrell, W. V. Volland, H. J. Dauben jr., and F. R. Hunter: J. Am. Chem. Soc. 87, 3527 (1965).

[2] A. Carrington, and I. C. Smith: Mol. Phys. 7, 99 (1963).

[3] D. E. Wood, and H. M. McConnell: J. Chem. Phys. 37, 1150 (1962).

[4] T. J. Katz, and H. L. Strauss: J. Chem. Phys. 32, 1873 (1960).

[5] G. R. Liebling, and H. M. McConnell: J. Chem. Phys. 42, 3931 (1965).

[6] L. C. Snyder: J. Chem. Phys. 33, 619 (1960).

[7] R. G. Lawler, J. R. Bolton, G. K. Fraenkel, and T. H. Brown: J. Am. Chem. Soc. 86, 520 (1964)

[8] A. Carrington, H. C. Longuet-Higgins, R. E. Moss, and P. F. Todd: Mol. Phys. 9, 187 (1965).

[9] M. Karplus, R. G. Lawler, and G. K. Fraenkel: J. Am. Chem. Soc. 87, 5260 (1965).

[10] R. G. Lawler, J. R. Bolton, M. Karplus, and G. K. Fraenkel: J. Chem. Phys. 41, 2149 (1967).

[11] M. T. Jones, A. Cairncross, and D. W. Wiley: J. Chem. Phys. 43, 3403 (1965).

[12] M. R. Das, and G. K. Fraenkel: J. Chem. Phys. 42, 792 (1965).

[13] R. W. Fessenden, and R. H. Shuler: J. Chem. Phys. 39, 2147 (1963).

capable of showing large deviations from expected splitting ratios. They do indeed[1] but dihydropyrazine (19) cation[2] does also, though one can hardly invoke similar reasoning for its anomalous behavior.

A prediction has been made[3] that increases in bond angle in going from six- to seven- to eight-membered rings should result in decrease in the value of Q. However, since there is also a charge effect on Q, one cannot yet say whether this prediction is contradicted in the series cyclopentadienyl, benzene⁻, cycloheptatrienyl, cyclo-octatetrene⁻, with Q values 29.9 G[4,5] 22.5[6], 27.4[7] and 25.7 G[8] respectively.

Closely related to the neutral cycloheptatrienyl radical is its dinegative ion[9], which in the HMO approximation has three electrons in antibonding orbitals (Fig. 3). The reduction in hyperfine splitting in the ion (3.48 G) as compared with the neutral radical is just that expected from the addition of two negative charges[10]. A splitting of each proton line by two Na⁺ ions is observed ($a_{2\,Na^+} = 1.74$ G).

The cyclohexadienyl radical C_6H_7 (21) produced by addition of a hydrogen atom to benzene during irradiation[11-13] or by electron irradiation of cyclohexadiene[14] is of especial interest because of the large CH_2 splitting. From the splitting constants given[14], it is obvious that no nodal plane passes through the CH_2 carbon.

The splittings in this radical are discussed under Hyperconjugation (Sect. 8).

In contrast to the planar neutral cycloheptatrienyl radical C_7H_7 which has all protons equivalent, the cycloheptatriene *anion* $C_7H_8^-$ (29) (produced from cycloheptatriene by electrolysis in liquid NH_3) has four pairs of equivalent protons. Splitting constants which are shown can be justified if the anion is non-planar

and has a "boat" (30) conformation[15]. In this example the methylene carbon doubtless lies in a nodal plane. The Q value associated with spin polarization on adjacent carbons is assigned a value of either 4.33 or -3.09 G. The positive sign is in agreement with experimental data on radicals from organic acids[16] while calculations predict a negative value[17].

Though not monocyclic, we consider briefly two cyclooctatetrene derivatives. The anion of sym-dibenzcyclo-octatetrene (31) shows splittings from three sets of

---

[1] See footnote 11, p. 178.
[2] See footnote 12, p. 178.
[3] M. Karplus, and G. K. Fraenkel: J. Chem. Phys. **35**, 1312 (1961).
[4] P. J. Zandstra: J. Chem. Phys. **40**, 612 (1964).
[5] G. R. Liebling, and H. M. McConnell: J. Chem. Phys. **42**, 3931 (1965).
[6] T. R. Tuttle, and S. I. Weissman: J. Am. Chem. Soc. **80**, 5342 (1958).
[7] A. Carrington, and I. C. P. Smith: Mol. Phys. **7**, 99 (1963).
[8] T. J. Katz, and H. L. Strauss: J. Chem. Phys. **32**, 1873 (1960).
[9] N. L. Bauld, and M. S. Brown: J. Am. Chem. Soc. **87**, 4390 (1965).
[10] J. R. Bolton: Private communication.
[11] V. A. Tolkachev, Yu. N. Molin, I. I. Chkeidze, N. J. Bubin, and V. V. Voevodskii: Doklady Akad. Nauk S.S.S.R. **141**, 911 (1961).
[12] H. Fischer: Z. Naturforsch. A **17**, 693 (1962); — Kolloid-Z. **180**, 64 (1962); — J. Chem. Phys. **37**, 1094 (1962).
[13] D. H. Whiffen: Mol. Phys. **6**, 223 (1963).
[14] R. W. Fessenden, and R. H. Schuler: J. Chem. Phys. **38**, 773 (1963); **39**, 2147 (1963).
[15] D. H. Levy, and R. J. Myers: J. Chem. Phys. **43**, 3063 (1965).
[16] J. R. Morton: J. Chem. Phys. **41**, 2956 (1964).
[17] J. P. Colpa, and E. de Boer: Mol. Phys. **7**, 333 (1964).

four equivalent protons[1,2]. This indicates appreciable coupling of the π-systems of the two benzene rings, and the considerable central ring currents required by NMR data reinforce this conclusion[1]. HMO unpaired electron densities appear insensitive to values of the exchange integral β for bonds such as 5—14[1].

Tetraphenylene $(C_6H_4)_4$ (32) has four benzene rings joined at adjacent positions so as to give a central cyclooctatetrene ring. The four outer rings have their

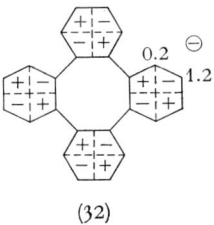

(32)

centers approximately at the vertices of a regular tetrahedron[3]. The lowest-lying antibonding orbital of tetraphenylene would be eight-fold degenerate if there were no coupling between rings; with interaction, the lowest-lying would be the one with maximum in-phase component orbital overlap[4]. This is achieved by combination of antisymmetric orbitals as shown. In this HMO approximation, the spin density of the anion should be 1/16 at the innermost and the outermost positions and zero at the others. Use of $Q=22.5$ G gives splittings of 1.41 G and 0.0 G for the ring protons.

11. Semiquinones. One of the earliest free radical intermediates observed by ESR spectra was the benzosemiquinone anion (34)[5] formed either by one-electron reduction of the quinone (33) or by one-electron oxidation of hydroquinone (37) in alkaline solution. The existence of the semiquinone anion as an intermediate was suggested early[6]; verification was demonstrated for substituted semiquinone anions by magnetic susceptibility studies[6]. Even the benzosemiquinone anion is stable enough in the absence of oxygen so that it persists for days in solvents such as dimethylsulfoxide (DMSO), $(CH_3)_2SO$. The very stable cation radical (36) formed from (37) in concentrated H_2SO_4, was first established by its optical absorption[7]. The lifetime of the hydroxyl protons under these conditions is great enough so that a marked triplet splitting is observed without significant change in the ring proton splitting[8,9]. Cations of the naphthosemiquinone (38) and of the anthra-

---

[1] T. J. KATZ, M. YOSHIDA, and L. C. SIEW: J. Am. Chem. Soc. **87**, 4516 (1965).
[2] A. CARRINGTON, H. C. LONGUET-HIGGINS, and P. F. TODD: Mol. Phys. **8**, 45 (1964).
[3] I. L. KARLE, and L. O. BROCKWAY: J. Am. Chem. Soc. **66**, 1974 (1944).
[4] A. CARRINGTON, H. C. LONGUET-HIGGINS, and P. F. TODD: Mol. Phys. **8**, 45 (1964).
[5] B. VENKATARAMAN, and G. K. FRAENKEL: J. Chem. Phys. **23**, 588 (1955).
[6] L. MICHAELIS, M. P. SCHUBERT, R. K. REBER, J. A. KUCK, and S. GRANICK: J. Am. Chem. Soc. **60**, 1678 (1938).
[7] E. J. LAND, and G. PORTER: Proc. Chem. Soc. **1960**, 84.
[8] J. R. BOLTON, and A. CARRINGTON: Proc. Chem. Soc. **1961**, 385.
[9] J. R. BOLTON, A. CARRINGTON, and J. DOS SANTOS VEIGA: Mol. Phys. **5**, 465 (1962).

Sect. 11.   Semiquinones.   181

semiquinone (39) show a hydroxyl proton splitting decreasing with increasing size of ion.

The benzosemiquinone anion or cation may be regarded formally as a benzene cation modified either by two O^- or by two OH substituents. Both have a spin distribution appropriate to a symmetric *bonding* orbital[1,2]. However, the distri-

(38)   (39)   (40)

bution in the orthosemiquinone anion (40) more nearly approximates the antisymmetric orbital. More refined MO calculations taking the oxygen explicitly into account reproduce satisfactorily the experimental observations[3].

The neutral semiquinone (35), first studied in flash photolysis[4], has a much shorter lifetime than either the anion or the cation. Successful ESR detection of this species was accomplished by in-cavity mixing of solutions containing Ti^{3+} and H_2O_2 + hydroquinone such that the pH is in the range 2—4[5,6]. Of particular interest is the very low spin density in (35) at positions 3 and 5.

To a first approximation, addition of H^+ to the O^- at position 4 largely removes it from the π-system. Thus this radical is the analog of the benzyl radical (2) which also has low spin density at the corresponding (unstarred) positions. The simple doubling of 2,6 splitting constants over those in (34) indicates no redistribution of spin density between the oxygen at position 1 and the ring carbons.

(41)   (42)   (43)

The neutral semiquinone (41) of catechol (43) further illustrates the controlling influence of the oxygen at position 1 and the essentially negligible effect of the OH oxygen[6]. The splitting constant of the protons meta (two-removed) from the oxygen at position 1 have identical values, while the ortho (adjacent) proton has a constant similar to that of the adjacent position of the neutral *p*-benzosemiquinone. The protons at the meta (3,5) and the para (4) position have values in the usual range when the symmetrical antibonding orbital of benzene is lower, but clearly the ortho value precludes such a simple description. One cannot discern an OH proton splitting from either neutral semiquinone, the lines being broader than for the anion.

Marked changes in ESR spectra with pH are also observed with other polyhydroxy compounds. One interesting example is pyrogallol[6], $C_6H_3(OH)_3$, 1, 2, 3. Be-

---

[1] See footnote 8, p. 180.
[2] See footnote 9, p. 180.
[3] G. VINCOW, and G. K. FRAENKEL: J. Chem. Phys. **34**, 1333 (1961).
[4] N. K. BRIDGE, and G. F. PORTER: Proc. Roy. Soc. (London) A **244**, 259, 276 (1958).
[5] I. YAMAZAKI, and L. PIETTE: J. Am. Chem. Soc. **87**, 986 (1965).
[6] I. C. P. SMITH, and A. CARRINGTON: Mol. Phys. **12**, 439 (1967).

ginning with pH=14 where the structure is given by (44) and going through the neutral to the acid range one observes in succession the splittings:

1. Doublet-triplet (44),
2. Doublet-doublet-doublet-doublet (pH 6.5—8 (45)),
3. Doublet-triplet-triplet (46),
4. Doublet-triplet.

Assignment of the splitting of (44) is largely based upon the splittings in (47). It appears that the bridged protons in (45) or (46) are much less rapidly exchanged than unbridged OH protons.

Analogs of the benzosemiquinones are obtained by replacement of OH with NH_2 or $N(CH_3)_2$ groups, with one-electron oxidation to form the cations in (48) and (49). The tetramethyl cation forms the stable salt WÜRSTER's Blue with the ClO_4^- anion. The appreciable nitrogen splitting both for the paraphenylene diamine (48)[1] and for the tetramethyl cation (49)[2] again is consistent with occupation of the symmetric orbital of benzene.

The dihydropyrazine (19) is analogous to hydroquinone (37) in that removal of one electron in concentrated H_2SO_4 leads to a stable cation. Analogous to the anthrasemiquinone cation (39) one has the phenazine cation (50). In these two examples it is noted that both the N and the N—H proton splitting is proportional to the spin density on the nitrogen[3].

12. Stable free radicals. Free radicals so stable that they can be isolated in solid form were a favorite subject of early ESR investigations. Owing to the close spacing in a typical solid, the exchange interaction narrows the observable portion of the line to a few gauss, whereas dipolar width might be a hundred gauss. They have thus been convenient standards for g-value, and if care is used, as intensity standards. The most familiar is diphenylpicrylhydrazyl, DPPH, (17) with $g = 2.0036$ for a polycrystalline sample[4]. There is a small anisotropy, increasing as

---

[1] M. T. MELCHIOR, and A. H. MAKI: J. Chem. Phys. **34**, 471 (1961).
[2] J. R. BOLTON, A. CARRINGTON, and J. D. S. VEIGA: Mol. Phys. **5**, 615 (1962). This paper corrects earlier erroneous assignments of a very small splitting constant to the nitrogen atoms of WÜRSTER's Blue cation.
[3] J. R. BOLTON, A. CARRINGTON, and J. DOS SANTOS VEIGA: Mol. Phys. **5**, 465 (1962).
[4] A. N. HOLDEN, C. KITTEL, F. R. MERRITT, and W. A. YAGER: Phys. Rev. **77**, 147 (1950). — C. H. TOWNES, and J. TURKEVICH: Phys. Rev. **77**, 148 (1950). — V. W. COHEN, C. KIKUCHI, and J. TURKEVICH: Phys. Rev. **85**, 379 (1952).

the temperature is lowered[1]. Inasmuch as it crystallizes with different numbers of solvent molecules in different solvents, its line width and absolute signal intensity are widely variable[2]. It was found that the exchange interaction governs relaxation processes[3,4]. The exchange was made a variable by dissolving the DPPH in polystyrene in various proportions[3]. It has also been used to examine line shapes in small magnetic fields[5]. In solution[6,7] it ordinarily shows only five broad hyperfine lines. However, if care is taken to purify the solvent, including the removal of dissolved oxygen from the solution, one is able to discern a very small proton splitting[8].

Somewhat parallel studies have been done on substances closely related to DPPH[3,7,9]. In another class of stable free radical one may have varying degrees of electron transfer between two diamagnetic materials. These donor-acceptor complexes may show a paramagnetism which indicates an excitation energy from a diamagnetic ground state to a spin-1/2 excited state[10,11]. An example is (51).

(51)

The ESR line shape at 9 and at 35 GHz[12] is just that expected for axial symmetry[13] and does not represent two different g-values for anion and cation[14]. Here $g_{\parallel} = 2.0023$, $g_{\perp} = 2.0055$ [12].

Though there are numerous radicals which may be kept for years in the solid form in contact with air, most of these react with oxygen with varying degrees of rapidity in liquid solutions. Several derivatives of NO, viz., $[(CH_3)_3C]_2NO$ [15] and $(CF_3)_2NO$ [16,17] are outstanding exceptions which appear indefinitely stable to oxygen at room temperature, where the first is a gas and the second a liquid. Both of these ought to be very useful reference substances, especially for inten-

---

[1] C. Kikuchi, and V. W. Cohen: Phys. Rev. 93, 394 (1954). — L. S. Singer, and C. Kikuchi: J. Chem. Phys. 23, 1738 (1955). — R. Livingston, and H. Zeldes: J. Chem. Phys. 24, 170 (1956).
[2] H. Ueda, Z. Kuri, and S. Shida: J. Chem. Phys. 33, 1676 (1962). — J. A. Lyons, and W. F. Watson: J. Polymer Sci. 18, 141 (1955). — E. Müller, and I. Müller-Rodloff, and W. Bunge: Liebigs Ann. Chem. 520, 244 (1935). — J. J. Lothe, and G. Eia: Acta Chem. Scand. 12, 1535 (1958).
[3] J. P. Goldsborough, M. Mandel, and G. E. Pake: Phys. Rev. Letters 4, 13 (1960) (polystyrene solutions). — Krishnaji, and B. N. Misra: J. Chem. Phys. 41, 1027 (1964). — N. Bloembergen, and S. Wang: Phys. Rev. 93, 72 (1954).
[4] J. P. Lloyd, and G. E. Pake: Phys. Rev. 92, 1576 (1953).
[5] M. A. Garstens: Phys. Rev. 93, 1228 (1954). — M. A. Garstens, L. S. Singer, and A. H. Ryan: Phys. Rev. 96, 53 (1954). — S. Becker: Phys. Rev. 99, 1681 (1955).
[6] C. A. Hutchinson, R. C. Pastor, and A. G. Kowalsky: J. Chem. Phys. 20, 534 (1952).
[7] H. S. Jarrett: J. Chem. Phys. 21, 761 (1953).
[8] Y. Deguchi: J. Chem. Phys. 32, 1584 (1960).
[9] C. Kikuchi, and V. W. Cohen: Phys. Rev. 93, 394 (1954).
[10] D. Bijl, H. Kainer, and A. C. Rose-Innes: J. Chem. Phys. 30, 765 (1959).
[11] A. Ottenberg, C. J. Hoffmann, and J. Osiecki: J. Chem. Phys. 38, 1898 (1963).
[12] M. E. Browne, A. Ottenberg, and R. L. Brandon: J. Chem. Phys. 41, 3265 (1964).
[13] J. W. Searl, R. C. Smith, and S. J. Wyard: Proc. Phys. Soc. (London) A 74, 491 (1959).
[14] Y. Matsunaga, and C. A. McDowell: Nature 185, 916 (1960).
[15] H. LeMaire, A. Rassat, and P. Servoz-Garin: J. chim. phys. 59, 1247 (1962).
[16] W. D. Blackley, and R. R. Reinhard: J. Am. Chem. Soc. 87, 802 (1965).
[17] P. J. Scheidler, and J. R. Bolton: J. Am. Chem. Soc. 88, 371 (1966).

sity comparisons. For other stable radicals, reference is made to several tabulations [1-3].

13. Ion-pairs. Formation of anions of hydrocarbons or of ketyls e.g., the hexamethylacetone anion of (52) necessarily implies the presence of charge-compensating cations (counterions), whether reduction takes place with an alkali metal or electrolytically at a cathode. In the latter case the counterion is commonly the tetramethylammonium ion $(CH_3)_4N^+$. Although the solutions are typically very dilute ($\sim 10^{-4}$ M), the dielectric constant is typically of the order of 10, and hence ion-pairing is possible. It was first clearly established for the ketyl of hexamethylacetone in tetrahydrofuran (53) at room temperature. The ESR spectrum indicates 18 equivalent protons and two equivalent Na^+ ions [4]. These are presumed to be arrayed as an ion quartet: (54).

(52)   (53)   (54)

Upon electrolytic reduction, only the spectrum of 18 equivalent protons is seen. (It will be noted later that at temperatures low enough to form a rigid glass the two separated radicals show interaction to form a ground triplet state [5]. More commonly, one observes that each proton line of an anion becomes a quartet in certain solvents when the cation is ^7Li, ^{23}Na, ^{39}K or ^{87}Rb, all of which have $I=3/2$. We shall refer to splitting by a second type of nucleus as superhyperfine (shf) splitting. For the biphenyl anion, $(C_6H_5)_2^-$ (55):

(55)   (56)

ion-pairing is observed with ^6Li, ^7Li, ^{23}Na, ^{39}K, ^{85}Rb, ^{87}Rb, ^{133}Cs [6]. Resolved octets with coupling constant 1.16 G due to ^{133}Cs ($I=7/2$) are observed for $^{133}Cs^+(C_6H_5)_2^-$. This splitting is three times that of one of the groups of ring protons. This implies a spin density of 1.42×10^{-3} at the Cs nucleus since the Cs hyperfine splitting in the free atom is 820.08 G [6].

A compound (56) of structure similar to (52) shows no detectable splitting from Na^+ ions when in the anion form. This has been attributed [7,8] to the spreading

---

[1] D. J. E. INGRAM: Free Radicals, p. 166. London: Butterworths 1958.
[2] S. A. AL'TSCHULER, and B. M. KOZYREV: Electron Paramagnetic Resonance. New York: Academic Press 1964; transl. ed. C. P. POOLE jr., p. 301.
[3] J. E. WERTZ: Chem. Revs. **55**, 829 (1955).
[4] N. HIROTA, and S. I. WEISSMAN: J. Am. Chem. Soc. **82**, 4424 (1960). — N. M. ATHERTON, and S. I. WEISSMAN: J. Am. Chem. Soc. **83**, 1330 (1961).
[5] N. HIROTA, and S. I. WEISSMAN: J. Am. Chem. Soc. **86**, 2538 (1964).
[6] H. NISHIGUCHI, Y. NAKAI, K. NAKAMURA, K. ISHIZU, Y. DEGUCHI, and H. TAKAI: Mol. Phys. **9**, 153 (1965).
[7] N. HIROTA, and S. I. WEISSMAN: J. Am. Chem. Soc. **82**, 4424 (1960).
[8] G. R. LUCKHURST, and L. E. ORGEL: Mol. Phys. **7**, 297 (1964).

of negative charge over two oxygen atoms or to the assumption of the *trans* configuration shown, which is sterically less favorable for pairing.

Ion pairing may result in large variations of ^{13}C coupling constants in the anion as the metal cation is changed, though proton couplings are only slightly affected. For example, the coupling constant of ^{13}C in $C_6H_5{}^{13}C(O)C_6H_5^-$ changes from 9.3 G with K^+ to 15.8 G with Mg^{2+}.[1]

The degree of ion pairing appears to increase with decreasing dielectric constant and increase in radius of the counterion[2]. Absence of detectable splitting by the counterion does not necessarily indicate lack of cation-anion association. Studies of proton splitting in anthracene and in azulene anions (57) show a dependence upon the alkali counterion, solvent, concentration, and temperature[2].

(57)

Dependence on the nature of the alkali ion doubtless shows a small degree of ion association. There has been speculation as to the structure of specific single ion-pairs, such as Na^+-pyrazine$^-$ (10). The sodium ion has been assumed to be directly associated with a nitrogen atom at a site of high electron density[3]. At 205 °K there is a marked variation in line width which is ascribed to a hopping of a sodium ion from one nitrogen to the other[4]. An alternative proposal[5] has the Na^+ ion above the center of the ring.

An example of a time-dependent process in cation-radical association is seen in the Na^+-benzophenone$^-$ ($C_6H_5-C(O)-C_6H_5$) system. An initially asymmetric ESR spectrum[6] eventually gives a spectrum appropriate to interaction with two cations. The g-value differs by 0.00017 from that of the one-cation system. For naphthalene$^-$, the difference in g is 0.00002[7]. The proton splittings (Table 9) show

Table 9. *Hyperfine parameters of ion pairs (or quartets?).*

Anion	Cation		a^H (G)		a(metal) (G)
Naphthalene$^-$	Na^+	4.816	1.823		1.593
Naphthalene$^-$	$(Na^+)_2$	4.936	1.858		0.488
Benzophenone$^-$	$^7Li^+$	3.536	2.712	0.938	0.673
Benzophenone$^-$	$(^7Li^+)_2$	3.567	2.855	1.017	1.125

A. H. REDDOCH: J. Chem. Phys. **43**, 3411 (1965).

significant increase for the two-cation system, while the metal splittings change even more markedly. In the case of naphthalene anion, it is supposed that repulsion of the two cations causes both to locate at sites of lower spin density than for the single cation. It is not known whether two anions are involved then there are two cations.

---

[1] N. HIROTA: J. Chem. Phys. **37**, 1884 (1962).
[2] A. H. REDDOCH: J. Chem. Phys. **43**, 225 (1965). — A. C. ATEN, J. DIELMAN, and G. J. HOIJTINK: Discussions Faraday Soc. **29**, 182 (1960). — J. R. BOLTON, and G. K. FRAENKEL: J. Chem. Phys. **40**, 3307 (1964).
[3] N. M. ATHERTON, and A. E. GOGGINS: Trans. Faraday Soc. **61**, 1399 (1965).
[4] N. M. ATHERTON, and A. E. GOGGINS: Mol. Phys. **8**, 99 (1964).
[5] C. MCDOWELL, and K. F. G. PAULUS: Can. J. Chem. **43**, 224 (1965).
[6] P. B. AYSCOUGH, and R. WILSON: J. Chem. Soc. **1963**, 5412.
[7] A. H. REDDOCH: J. Chem. Phys. **43**, 3411 (1965).

While large counterions, e.g., tetra-n-butylammonium $[CH_3(CH_2)_3]_4 N^+$ ion, do not show a demonstrable hyperfine splitting of the paramagnetic anion, one may nevertheless detect its involvement in ion-pairing due to shifts in its NMR lines when complexed with Ni^{2+}[1] or Co^{2+}[2]. The interaction is dipolar[2].

$$O_2N \langle \bigcirc \rangle NO_2 \quad {}^{\ominus}$$

(58)

Some nitro-substituted benzene anions [e.g. (58)] show drastic differences in proton and nitrogen hyperfine splittings when prepared in the presence of alkali ions[3] or by electrolytic reduction[4], seemingly by virtue of pairing with alkali ions. In the presence of Na^+ there appears to be an interaction with only two (rather than four) protons, and with only one nitrogen but with double the hyperfine splitting of the system prepared electrolytically. The ESR spectrum of this system in methanol indicates interaction of the unpaired electron with both nitrogen atoms but with slightly differing hyperfine splitting and apparently fluctuating at a frequency near the hyperfine frequency[5]. The normal 1:2:3:2:1 quintet from two equivalent nitrogens has the central three lines greatly broadened. If one sees only the outermost lines and a portion of the central one, this spectrum appears to be a 1:1:1 triplet from one nitrogen and with twice the actual splitting constant. Such line broadening phenomena from modulation of hyperfine interactions is described in the section on line width effects[6]. Thus, the seemingly large changes in proton and nitrogen coupling in the presence of alkali cations may be deceptive.

14. Radicals in oriented systems. Irradiation of rigid crystals usually leads to formation of radicals which are oriented along a limited number of directions governed by crystal symmetry, whether the trapped species arises from the host crystal or has been introduced by doping. Salts are obvious choices both from the rigidity requirement and because of ease of growing large crystals. Much of the work on oriented organic radicals has been concentrated on organic acids, largely the dibasic acids $HOOC-(CH_2)_n-COOH$, the amino acids $R-C(H)(NH_3^+)-COO^-$, as well as a few simple acids RCH_2COOH; such crystals are also easy to grow and are rigid even at room temperature. The extra information derived from oriented-, as compared with non-oriented systems are g-tensor and hyperfine tensor components instead of their averaged values. Another marked advantage is that spectra may be simplified in appearance because lines from symmetry-related radicals may be made to coincide instead of confusing the spectrum from overlapping. Especially for those radicals with simple ESR spectra it is essential to have the maximum information in order to avoid mis-identification. With inorganic radicals, one makes comparison with other iso-electronic radicals if possible. Naturally, if one is to understand the ESR spectrum in an oriented system, one ought to have extensive information about the crystal structure of the host. At a minimum, one must know the host symmetry, preferably the space group, number of molecules per unit cell and the unit cell dimensions. One is greatly aided by a knowledge

---

[1] F. A. COTTON, O. D. FANT, D. M. L. GOODGAME, and R. H. HOLM: J. Am. Chem. Soc. **83**, 1780 (1961).
[2] G. N. LAMAR: J. Chem. Phys. **41**, 2992 (1964); **43**, 235 (1965).
[3] R. WARD: J. Am. Chem. Soc. **83**, 1296 (1961); — J. Chem. Phys. **36**, 1405 (1962).
[4] A. H. MAKI, and D. H. GESKE: J. Chem. Phys. **33**, 825 (1960).
[5] J. M. GROSS, and M. C. R. SYMONS: Mol. Phys. **9**, 287 (1965).
[6] J. H. FREED, and G. K. FRAENKEL: J. Chem. Phys. **39**, 326 (1963).

of atomic positions. This rapidly-growing field has been excellently reviewed recently[1]. Our attention here is focussed on some of the salient aspects and some recent examples.

π-Radicals in Irradiated Organic Crystals. — Radicals of concern to us here have an unpaired electron primarily located in a π-orbital upon carbon (or occasionally one nitrogen) atom which we have identified as the α-carbon. We assume that is of the type $H-C'{\overset{R}{\underset{R}{\diagdown}}}$ formed by loss of a hydrogen atom from its precursor, as e.g., $H_2C(COOH)_2$ [2]. A second radical formed from malonic acid is $H_2C'-COOH$. We assume that one has either selectively bleached the less stable radical or has managed to obtain angular dependence plots in spite of its presence. Having obtained a hyperfine tensor for the α-hydrogen atom (with due regard to proper signs)[2,3], one diagonalizes it and obtains A_α (values in MHz):

$$A_\alpha = \begin{bmatrix} -63 & 0 & 0 \\ 0 & -93 & 0 \\ 0 & 0 & -33 \end{bmatrix},$$

$$T_\alpha = \begin{bmatrix} 0 & 0 & 0 \\ 0 & -30 & 0 \\ 0 & 0 & 30 \end{bmatrix}.$$

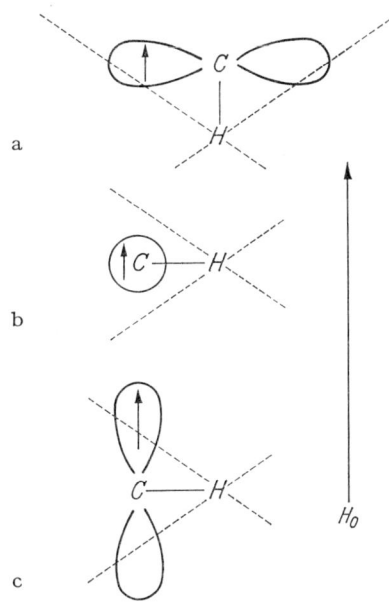

Fig. 11 a—c. Anisotropic hyperfine interaction of α-protons in oriented radicals. a) H_0 parallel to $C-H_\alpha$ bond; b) H_0 perpendicular to $C-H_\alpha$ axis and to p-orbital; c) H_0 perpendicular to radical plane [from J. R. MORTON: Chem. Rev. 64, 453 (1964)].

The A_α tensor is very typical of a $H-CR_1R_2$ radical. Since the anisotropic tensor T_α is traceless, one removes the isotropic portion by subtraction of one-third the trace from each element. This isotropic component is just the quantity $Q \approx -63$ MHz we have applied to a hydrogen atom in the nodal pane of a π-radical. We desire to assign the T_α tensor components to particular directions for our model $H-C'{\overset{R}{\underset{R}{\diagdown}}}$ radical, which by symmetry must have the H—C axis as a principal axis. Another lies in the molecular plane perpendicular to it, with a third perpendicular to the plane. By reference to Fig. 11 showing the static field successively parallel to the three axes, one may directly assign the principal tensor components[4]. In Fig. 11 a, with H_0 parallel to the H—C axis, there are indicated lines for which $(1-3\cos^2\theta)$ is zero, and one notes that most of the electron density lies in a region in which this quantity is negative. In this region the hyperfine interaction is positive, and hence the +30 MHz value is associated with the H—C axis. In

---

[1] J. R. MORTON: Chem. Revs. 64, 453 (1964). Much useful information is given in tabular form.
[2] H. M. McCONNELL, C. HELLER, T. COLE, and R. W. FESSENDEN: J. Am. Chem. Soc. 82, 766 (1960).
[3] J. A. WEIL, and J. H. ANDERSON: J. Chem. Phys. 28, 864 (1958).
[4] D. K. GHOSH, and D. H. WHIFFEN: Mol. Phys. 2, 285 (1959).

Fig. 11b, the spin density is primarily in the region of positive $(1-3\cos^2\theta)$ and hence the hyperfine interaction is negative. This justifies the association of the -30 MHz value with the axis lying in the molecular plane (y axis). Finally, one observes from Fig. 11c that there is some spin density both in negative and in positive regions of $(1-3\cos^2\theta)$ and hence one may associate the approximately zero value of the tensor with the $2p$ orbital direction, i.e., perpendicular to the plane of the radical. By analysis of the second radical produced in the irradiation of malonic acid, viz. $H_2C'COOH$ it has been established that both α-protons lie in the nodal plane of the π-orbital, with an HCH angle of $\sim 116°$[1]. The tensor components for the two atoms differ by as much as 3 MHz.

For β protons, i.e., protons on the carbon atom adjacent to the one on which the π orbital is located, the hyperfine interaction is largely isotropic, but it varies from about 14 to 130 MHz. We have already considered the mechanism of β proton interaction in Sect. 8 on hyperconjugation as interaction of a pseudo-π orbital of the β protons with the π orbital on the α carbon atom. We have likewise encountered the relation for β coupling α_β: $\alpha_\beta = B_0 + B_2 \cos^2 \theta$ [2, 3-5], where θ is the dihedral angle between the $C_\alpha - C_\beta - H$ plane and the axis of the π orbital. The range of B_0 appears to be 0 to 10 MHz[4], unless an oxygen atom is directly attached to the α-carbon, in which case B_0 may be negative (-7 MHz in HOC'HCOO$-$)[6] due to delocalization upon oxygen[7]. If one takes the average of a_β, i.e., $\frac{1}{2\pi}\int_0^{2\pi} a_\beta d\theta = B_0 + B_2/2$ and sets this equal to the coupling constant for protons in a rapidly rotating methyl group, viz., ~ 70 MHz, one may estimate B_2 as about 140 MHz if B_0 is ignored[8]. In almost all cases known, B_2 is somewhat smaller than this value[8]. In the radical $N(CH_3)_3$, the value of B_2 is 150 MHz. It is of interest to note that in the radical $^+H_3NC(H)CO_2^-$ where the NH_3 group is in rotation, the proton coupling has a somewhat lower value than for CH_3 protons, viz. 53 MHz[9].

The simplest amino acid, glycine, $^+H_3NC(H_2)CO_2^-$ upon irradiation gives not only $^+H_3NC(H)CO_2^-$ but also $CH_2CO_2^-$.[10] The latter was finally identified correctly from its ^{13}C splitting after having erroneously been identified as $CH_2NH_3^+$,[11] NH_2[12] and NH_4.[13] Some of these misinterpretations were attributable to the fact that both glycine and the partially-deuterated form $^+H_3NCD_2CO_2^-$ give essentially identical spectra. The source of this curious behaviour was shown in alanine, $CH_3CH(NH_3^+)CO_2^-$ to be an exchange of deuterium for either α or β hydrogen atoms[14].

Both ^{13}C and ^{14}N isotropic hyperfine interactions are found to be positive when the unpaired electron is primarily located upon these atoms. The ^{13}C values are often of the order of 125 MHz, while the ^{14}N couplings are about 40—50 MHz. From the form of the anisotropic hyperfine interaction[8] $B_0(3\cos^2\theta - 1)|\varrho|$, one expects the tensor T to have axial symmetry with principal value $+2B_0$ parallel

---

[1] A. Horsfield, J. R. Morton, and D. H. Whiffen: Mol. Phys. 4, 327 (1961).
[2] See footnote 2, p. 187.
[3] A. D. McLachlan: Mol. Phys. 1, 233 (1958).
[4] A. Horsfield, J. R. Morton, and D. H. Whiffen: Mol. Phys. 4, 425 (1961).
[5] M. C. R. Symons: J. Chem. Soc. 1959, 277.
[6] D. E. Henn, and D. H. Whiffen: Mol. Phys. 8, 407 (1964).
[7] W. Derbyshire: Mol. Phys. 5, 225 (1962).
[8] See footnote 1, p. 187.
[9] D. K. Ghosh, and D. H. Whiffen: Mol. Phys. 2, 285 (1959).
[10] J. R. Morton: J. Am. Chem. Soc. 86, 2325 (1964).
[11] W. Gordy, W. B. Ard, and H. Shields: Proc. Natl. Acad. Sci. U.S. 41, 983 (1955).
[12] D. K. Ghosh, and D. H. Whiffen: J. Chem. Soc. 1960, 1869.
[13] R. F. Weiner, and W. S. Kuski: J. Am. Chem. Soc. 85, 873 (1963).
[14] I. Miyagawa, and K. Itoh: J. Chem. Phys. 40, 3328 (1964).

to the $2p$ orbital and $-B_0$ and $-B_0$ at right angles. As noted in Table 1, the B_0 values for ^{13}C and ^{14}N are respectively 91 and 48 MHz. For the radicals thus far observed, the largest anisotropic component for ^{13}C is about 134 MHz and that for ^{14}N about 69 MHz[1]: both are perpendicular to the C—H or N—H directions. Thus one would assign a $2p$ spin density of the order of 0.67 to the central carbon atom. While the two smaller tensor components are approximately equal in some cases, they differ very greatly in others. Hindered rotation has been ascribed as the cause in several cases[2]. The radical $(CH_3)_2CO_2^-$ represents an interesting case of hindered rotation of one CH_3 group, while the other appears to be rotating freely at 40 °K. Coupling constants for the hindered group are 122, 30 and 52 MHz, while the coupling constant for the rotating group is 68 MHz[3].

Some radicals derived from unsaturated compounds by irradiation have an especial interest because the electron is delocalized as is usual in coupled π systems. One with great interest is HOOCC(H)=C(H)C(H)COOH, containing the (substituted) allyl radical (1), our first example of an odd alternant radical. The prediction that there is negative spin density upon the central carbon atom is borne out by its hyperfine tensor components $+17, +12$, and $+7$ MHz, while those for the two adjacent carbon atoms are $-53, -36$ and -18 MHz[4]. Several other examples of unsaturated radicals with marked delocalization have been found[5], while in other cases different kinds of radicals are formed[6], some by addition of protons[7].

"Inclusion compounds" consisting of organic acids, alcohols, halides, ethers, esters or ketones fitted into tubular cavities formed by spirals of hydrogen-bonded urea molecules $[H_2N-C(=O)NH_2]$ provide ordered structures of guest molecules which may form highly oriented radicals by x-irradiation[8]. Esters $[RCH_2CH_2C(=O)OR]$ give radicals $RCH_2C'HCOR_2$ in which hyperfine splittings
$\qquad\qquad\qquad\qquad\qquad\qquad\qquad\qquad\qquad\quad\;\;\overset{\|}{O}$
of 3—6 MHz are found for protons on R_2 groups[9]. Both α and β protons show considerable motion. No urea-derived radicals were detected. The radicals derived from oriented ketones $R_1CH_2CH_2-C-R_2'$ in urea are $R_1CH_2C'H-CR_2$. The three-
$\qquad\qquad\qquad\qquad\qquad\qquad\qquad\qquad\quad\overset{\|}{O}\qquad\qquad\qquad\qquad\qquad\;\;\overset{\|}{O}$
atom fragment C—C=O may be considered analogous to the allyl fragment C'—C=C in which each atom also contributes one electron to the system[10]. Coupling constants for α protons are $\sim$75 MHz, while those of the two β-protons are respectively 64—71 and 47—65 MHz. Those for protons on the opposite side of the C=O group range from 5.0 to 9.5 MHz. MO calculations give a spin density of 0.81 for the $2p_z$ orbital adjacent to the C=O group. When ethers $(R_1CH_2OR_2)$ are incorporated into urea and x-irradiated, one gets $R_1C'HOR_2$ radicals[11]. The C'H carbon has a spin density of about 0.70, with most of the remainder on the oxygen. α and β couplings are both about 63 MHz, while the CH_2 protons of R_2 have a coupling constant of about 8.5 MHz.

---

[1] A. Horsfield, J. R. Morton, J. R. Rowlands, and D. H. Whiffen: Mol. Phys. 5, 241 (1962).
[2] J. R. Rowlands: Mol. Phys. 5, 565 (1962). — J. R. Rowlands, and D. H. Whiffen: Nature 193, 61 (1962).
[3] J. R. Morton: J. Chem. Phys. 41, 2956 (1964).
[4] C. Heller, and T. Cole: J. Chem. Phys. 37, 243 (1962).
[5] H. C. Heller, and T. Cole: J. Am. Chem. Soc. 84, 4448 (1962); — R. J. Cook, J. R. Rowlands, and D. H. Whiffen: Mol. Phys. 7, 57 (1963).
[6] R. J. Cook, J. R. Rowlands, and D. H. Whiffen: J. Chem. Soc. (London) 1963, 3520.
[7] M. Fujimoto: J. Chem. Phys. 39, 846 (1963).
[8] O. H. Griffith, and H. M. McConnell: Proc. Natl. Acad. Sci. U.S. 48, 1877 (1962).
[9] O. H. Griffith: J. Chem. Phys. 41, 1093 (1964).
[10] O. H. Griffith: J. Chem. Phys. 42, 2644 (1965).
[11] O. H. Griffith: J. Chem. Phys. 42, 2651 (1965).

Inorganic π-Radicals. — Diatomic π-radicals which have been observed in oriented solids include N_2^- [1,2], O_2^+ [3], O_2^- [4,5], OH [6], and $PF^{\pm}$ [7], the last-named undergoing rapid reorientation so that only isotropic parameters are observed. Of these, only O_2^- has the g_{zz} component markedly different from g_e, owing to the large value of effective orbital angular momentum $l=1.04$ about the internuclear axis in the expression [5]:

$$g_{zz}=g_e+2[\lambda^2/(\lambda^2+\Delta^2)]^{\frac{1}{2}}l.$$

Here Δ represents the crystal field splitting of the π_g (antibonding) orbitals, the order of magnitude of which can be inferred from the value $\lambda/\Delta=0.23$ for O_2^-. From the g_{zz} values of the other radicals, one notes that orbital angular momentum is effectively quenched. From expressions for g_{xx} and g_{yy} [5] one may infer the levels between which spin-orbit interaction is occurring.

Of the triatomic radicals studied, the 17-electron pair CO_2^- [8,9], NO_2 [10-14] show most nearly identical g-components, with both having $g_{xx}=2.0032$; $g_{yy}=1.9975$ for CO_2^- and 1.9973 for NO_2; $g_{zz}=2.0014$ for CO_2^- and 2.0016 for NO_2. The spectrum of NO_2 is given in Fig. 12. From the CO_2^- data, including ^{13}C couplings, a $2p_z$ spin density of 0.50 and a $2s$ density of 0.14 on carbon, a bond angle of $128°$ is inferred [15]. A similar calculation for NO_2 gives a bond angle of $135°$ [15]. The 19-electron radicals ClO_2 [16,17] and NO_2^{-2} [3] (the latter misidentified as NO) have similar g_{xx} and g_{zz} values, but the g_{yy} values are respectively 2.0099 and 2.0183. No calculations have yet been made for the solid radical bond angle. The ^{35}Cl hyperfine tensor components indicate appreciable spin density in the oxygen $2p_x$ orbitals, since the $3p_x$ density on chlorine is only 0.41 and the $3p_y$ density is small [14]. The F_3^{-2} radical [18] in LiF is of interest because it is slightly bent [20,19] and has a large positive isotropic hyperfine interaction, the ratio for the center and the outer atoms being 2.6. Other triatomic ions reported are O_3^- [21], SO_2^+ [22], SO_2^- [22-24] and SeO_2^- [17].

---

[1] R. B. Horst, J. H. Anderson, and D. E. Milligan: J. Phys. Chem. Solids 23, 157 (1962).

[2] D. W. Wylie, A. J. Shuskus, C. G. Young, O. R. Gilliam, and P. W. Levy: Phys. Rev. 125, 451 (1962).

[3] C. Jaccard: Phys. Rev. 124, 60 (1961).

[4] W. Känzig: J. Phys. Chem. Solids 23, 479 (1962).

[5] W. Känzig, and M. H. Cohen: Phys. Rev. Letters 3, 509 (1959).

[6] J. A. McMillan, M. S. Matheson, and B. Smaller: J. Chem. Phys. 33, 609 (1960).

[7] J. R. Morton: Can. J. Phys. 41, 706 (1963).

[8] D. W. Ovenall, and D. H. Whiffen: Mol. Phys. 4, 135 (1961).

[9] S. A. Marshall, A. R. Reinberg, R. Serway, and J. A. Hodges: Mol. Phys. 8, 3 (1964).

[10] H. Zeldes, and R. Livingston: J. Chem. Phys. 35, 563 (1961); 41, 4011 (1964).

[11] J. Cunningham, J. A. McMillan, B. Smaller, and E. Yasaitis: J. Phys. Chem. Solids 23, 167 (1962).

[12] D. Mergerian, and S. A. Marshall: Phys. Rev. 127, 2015 (1962).

[13] H. Zeldes: Paramagnetic Resonance (W. Low, ed.), vol. II, p. 764. New York: Academic Press 1963.

[14] R. M. Golding, and M. Henchman: J. Chem. Phys. 40, 1554 (1964).

[15] J. R. Morton: Chem. Revs. 64, 453 (1964).

[16] T. Cole: Proc. Natl. Acad. Sci. U.S. 46, 506 (1960).

[17] R. J. Cook, J. R. Rowlands, and D. H. Whiffen: Mol. Phys. 8, 195 (1964).

[18] M. H. Cohen, W. Känzig, and T. O. Woodruff: J. Phys. Chem. Solids 11, 120 (1959).

[19] A. D. Walsh: J. Chem. Soc. 1953, 2266.

[20] M. C. R. Symons: J. Chem. Soc. 1963, 570.

[21] P. W. Atkins, J. A. Brivati, N. Keen, M. C. R. Symons, and P. A. Trevalion: J. Chem. Soc. 1962, 4785.

[22] J. M. de Lisle, and R. M. Golding: J. Chem. Phys. 43, 3298 (1965).

[23] H. C. Clark, A. Horsfield, and M. C. R. Symons: J. Chem. Soc. 1961, 7.

[24] R. L. Eager, and D. S. Mahadevappa: Can. J. Chem. 41, 2106 (1963).

Sect. 14. Radicals in oriented systems. 191

The tetratomic radicals known may be subdivided into planar (CO_3^- [1] and NO_3 [1,2]) and pyramidal (NO_3^{-2} [3-5], PO_3^{-2} [6,7], SO_3^- [8], ClO_3 [9] and AsO_3^{-2} [10]) structures [11]. Lines attributed to NO_3 and to NO_3^{-2} are shown in Fig. 12. There is expected to be no central-atom hyperfine splitting in CO_3^- or NO_3 since its $2s$ and $2p$ orbitals do not contribute to the unpaired electron orbital formed by combination of the oxygen lone-pair orbitals; a small negative ^{13}C coupling is found in CO_3, perhaps due to bond polarization. Non-axiality of CO_3^- and NO_3 g-tensors is attributed to a distortion to C_{2v} symmetry [1]. The large positive central-atom couplings in the pyramidal radicals indicate a spin density of about 0.1, increasing with decreasing

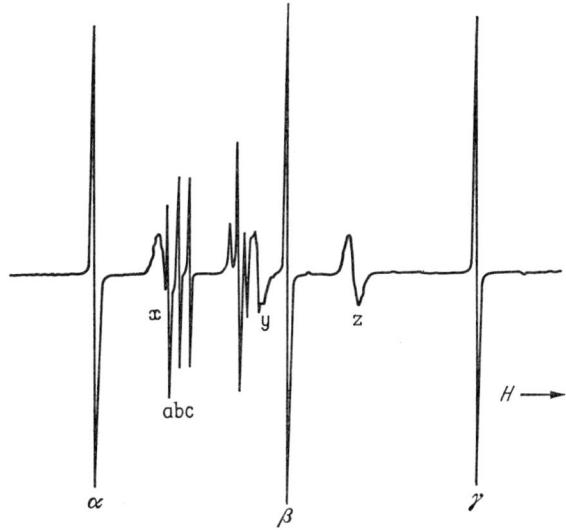

Fig. 12. ESR spectrum of $NO_2(\alpha\beta\gamma)$, $NO_3(abc)$ and $NO_3^{-2}(xyz)$ in γ-irradiated KNO_3 at 77 °K [from R. LIVINGSTON, and H. ZELDES: J. Chem. Phys. 41, 4011 (1964)].

electronegativity [12]. The $3p$ spin density increases similarly. The N_4 radical in KN_3 has a very small isotropic coupling ($\sim$14 MHz); it has variously been assumed to have a linear [13] or a square planar structure [14]. The NH_3 radical (as well as ClO_3) produced by x-irradiaton of NH_4ClO_4 is found to be planar and undergoes restricted rotation at room temperature [9]. The PF_4 radical, which reorients rapidly in irra-

---

[1] G. W. CHANTRY, A. HORSFIELD, J. R. MORTON, and D. H. WHIFFEN: Mol. Phys. 5, 589 (1962).
[2] See footnote 14, p. 190.
[3] See footnote 3, p. 190.
[4] See footnote 13, p. 190.
[5] See footnote 11, p. 190.
[6] M. W. HANNA, and L. J. ALTMAN: J. Chem. Phys. 36, 1788 (1962).
[7] A. HORSFIELD, J. R. MORTON, and D. H. WHIFFEN: Mol. Phys. 4, 475 (1961).
[8] G. W. CHANTRY, A. HORSFIELD, J. R. MORTON, J. R. ROWLANDS, and D. H. WHIFFEN: Mol. Phys. 5, 233 (1962).
[9] T. COLE: J. Chem. Phys. 35, 1169 (1961).
[10] W. C. LIN, and C. A. McDOWELL: Mol. Phys. 7, 223 (1964).
[11] A. D. WALSH: J. Chem. Soc. 1953, 2301.
[12] See footnote 15, p. 190.
[13] A. J. SHUSKUS, C. G. YOUNG, O. R. GILLIAM, and P. R. LEVY: J. Chem. Phys. 33, 622 (1960).
[14] See footnote 21, p. 190.

diated NH_4PF_6 is remarkable for its very large ^{31}P splitting, presumably due to occupation of a $4s$ orbital[1].

Radicals in Non-Oriented Irradiated Solids. — A large number of such studies have been published; references to some of these are distributed in various parts of Sect. B. Several reviews deal especially with these topics[2,3]. We take note of a few recent observations of especial interest.

Electron irradiation of CF_4 or C_2F_6 in the liquid state or of CHF_3 in krypton or xenon matrices give rise to CF_3 radicals with isotropic F coupling of 405.7 MHz[4]. This value is much larger than that for other α-fluorine radicals[5]. With such a large splitting, six lines are observed for CF_3 and one requires third-order corrections (ranging from $+2.3$ to -2.7 MHz) to reproduce line positions within experimental accuracy. These corrections have enabled the determination of identical signs for carbon and fluorine couplings. From CH_2F_2 and CH_3F one obtains the radicals CHF_2 ($a_F = 236$ and $a_H = 62.2$ MHz) and CH_2F ($a_F = 180$ and $a_H = 59.1$ MHz) respectively. The ^{13}C coupling in $^{13}CF_3$ is 761.2 MHz, as compared with that in $^{13}CH_3$ of 108 MHz. The large ^{13}C coupling of 417 MHz in $^{13}CHF_2$ also is indicative of considerable s-character in both CHF_2 and CF_3, estimated as 10% for CHF_2 and 21% for CF_3. Hence neither is planar.

The H_2CN radical has a large positive proton coupling[6]. It is presumed that the electron is primarily located upon the nitrogen atom and that one has a direct coupling to the two protons which are most favorably located in the molecular plane.

Line widths may sometimes be narrow enough by proper choice of matrix that g- and hyperfine tensor components may be extracted[7]. Computer programs have now been written for simulation of line shapes for free radicals, including effects of g- and hyperfine tensors for nuclei with $I = 3/2$ [8], one may hope that in more cases will obtain tensor components essential for full interpretation from observations on non-oriented solids.

15. Line width effects. In this section we shall be concerned with systems in which the amplitudes of hyperfine components depart appreciably from those expected on the basis of line degeneracies. Even crude measurements will usually indicate that in such cases there are significant line width variations. We assume that individual line components are symmetric, and therefore that g- or hyperfine anisotropies are negligible because to rapid reorientation. However, we shall conclude shortly that even though these magnetic anisotropies are seemingly averaged out, nevertheless they may still exert a dominant effect on relaxation processes and therefore affect line width. From a study of many spectra, one may classify line width anomalies in the following categories:

1. Line width varies continually across the spectrum (m_I effect).
2. Central line groups are narrower than groups at either side (m_I^2 effect).
3. Resolution — and hence line width — alternates in adjacent line groups. An example is shown in Fig. 13.

---

[1] See footnote 7, p. 190.
[2] C. K. JEN, in: Formation and Trapping of Free Radicals, Chap. 7. (A. M. BASS, and H. P. BROIDA, eds.). New York: Academic Press 1960.
[3] M. C. R. SYMONS, in: Advances in Physical Organic Chemistry (V. GOLD, ed.), vol. I, p. 283. New York: Academic Press 1963.
[4] R. W. FESSENDEN, and R. H. SCHULER: J. Chem. Phys. **43**, 2704 (1965).
[5] M. T. ROGERS, and D. H. WHIFFEN: J. Chem. Phys. **40**, 2662 (1964).
[6] E. L. COCHRAN, F. J. ADRIAN, and V. A. BOWERS: J. Chem. Phys. **36**, 1938 (1962).
[7] F. J. ADRIAN, E. L. COCHRAN, and V. A. BOWERS. J. Chem. Phys. **36**, 1661 (1962).
[8] R. LEFEBVRE, and J. MARUANI: J. Chem. Phys. **42**, 1480 (1965).

The appropriate starting point for study of hyperfine line width effects is a single nucleus. One observes a systematic variation of line width of the members of a hyperfine multiplet in the spectrum of Cu^{+2} [1,2] and VO^{+2} [3-5] ions. For Cu^{+2} the narrowest line may occur either at low field or at high field in liquids or in solids. Assuming that these ions in solution rigidly orient solvent molecules about them as "microcrystallites" [6], one expects anisotropies of both g- and hyperfine tensors, since both ions are welll known to organize their ligands with axial symmetry. Random reorientation of the ions plus their solvation shell has been treated in a manner analogous to the Bloembergen-Purcell-Pound approach [7] to

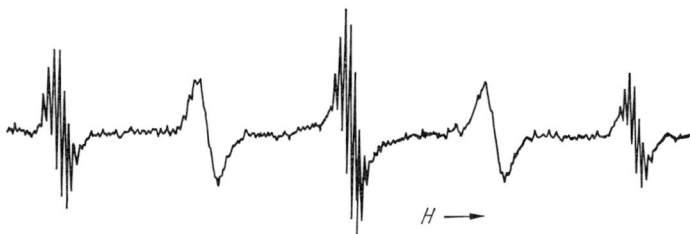

Fig. 13. Alternating line width in the ESR spectrum of 1,4-dinitro, tetramethyl benzene anion $[(CH_3)_4C_6(NO_2)_2]^-$ in dimethylformamide $[HC(O)N(CH_3)_2]$ at room temperature [from J. H. FREED, and G. K. FRAENKEL: J. Chem. Phys. 37, 1156 (1962)].

nuclear relaxation. It was concluded that both T_1 and T_2 depend upon the nuclear spin quantum number I_z, e.g.:

$$1/T_2' \propto (\Delta g\beta H + bI_z)^2 \tan^{-1}(2\tau_c/T_2'). \tag{15.1}$$

where $\Delta g = g_\| - g_\perp$, $b = 2(A-B)/3$ and τ_c is the correlation time [6]. However, one must regard the BPP approach less than satisfactory for dealing with this problem, for there is no means of taking into account relaxation process among the nuclear spin levels which lead to transitions among them. The Kubo-Tomita approach [8] is more appropriate; it has been applied to systems with small magnetic anisotropy that may have g-anisotropy, intramolecular electron-nuclear dipolar interaction, plus exchange interaction and are undergoing reorientation [9]. One conclusion of relevance here is the relation:

$$1/T_2 = \alpha + \beta m_I + \gamma m_I^2 + \varkappa m_I^4, \tag{15.2}$$

The predictions of this theory [9], and the "microcrystallite" model were though to fit the VO^{+2} behavior satisfactorily [9,10]. However, more accurate line width measure-

---

[1] B. R. McGARVEY: J. Phys. Chem. 60, 71 (1956).
[2] J. W. ORTON, P. AUZINS, J. H. E. GRIFFITHS, and J. E. WERTZ: Proc. Phys. Soc. (London) 78, 554 (1961).
[3] N. F. GARIF'YANOV, and B. M. KOZYREV: Doklady Akad. Nauk S.S.S.R. 98, 929 (1954).
[4] G. E. PAKE, and R. H. SANDS: Phys. Rev. 98, 266 (A) (1955).
[5] D. E. O'REILLY: J. Chem. Phys. 29, 1188 (1958). — H. M. McCONNELL, W. W. PORTERFIELD, and R. E. ROBERTSON: J. Chem. Phys. 30, 442 (1959). — W. A. ANDERSON, and L. H. PIETTE: J. Chem. Phys. 30, 591 (1959).
[6] H. M. McCONNELL: J. Chem. Phys. 25, 709 (1956).
[7] N. BLOEMBERGEN, E. M. PURCELL, and R. V. POUND: Phys. Rev. 73, 679 (1948).
[8] R. KUBO, and K. TOMITA: J. Phys. Soc. Japan 9, 888 (1954).
[9] D. KIVELSON: J. Chem. Phys. 33, 1094 (1960). — R. WILSON, and D. KIVELSON: J. Chem. Phys. 44, 154 (1966).
[10] R. N. ROGERS, and G. E. PAKE: J. Chem. Phys. 33, 1107 (1960).

ments are interpreted to require a term proportional to m_I^3 and to show a residual line width due to spin-rotational interaction[1,2]. For a system with g- and hyperfine anisotropy undergoing reorientation, there is a contribution to line width both from the electron spin reorientations as well as from variation in instantaneous field position of each hyperfine component. Consideration has been given to the case in which the axes of g- and hyperfine tensors are not coincident[3].

A variety of processes may give rise to fluctuations in hyperfine interactions (or "modulations" of hyperfine interactions). Intramolecular motions (rotation, vibration, torsional oscillations), solvent-radical interaction, ion-pairing, or electron transfer are some of the important modulation processes in addition to random reorientation of the whole radical. It has been shown that the narrower line width of central- as compared with outer line groups[5] arises from modulation of an anisotropic intramolecular electron-nuclear dipolar interaction[6,7]. This effect gives rise to the γm_I^2 term of Eq. 15.2. One also gets an end-to-end linewidth variation from a cross term between the motional modulation of an anisotropic g-tensor and the intramolecular dipolar interaction, viz., the βm_I term in Eq. 15.2.

The dependence of line width upon m_I and m_I^2 is well illustrated in the spectrum of dicyanotetrazine anion[8]: (59)

$$NC-C \underset{N-N}{\overset{N-N}{\diagup\diagdown}} C-CN \quad \ominus$$

(59)

The central three members of the spectrum of nine line groups (from four ring nitrogen atoms), show a quintet splitting from the two CN nitrogens. The central quintet, having m_I (ring)$=0$, shows the best resolution. The next group on the low-field side shows greater resolution than the corresponding group on the high-field side; this is an m_I contribution in agreement with prediction[3,7,9,10] if the isotropic nitrogen coupling constant has a positive sign[4]. Other inferences about determinations of sign from line-width effects (location of narrower components) have been made for ^{13}C in naphthalene anion[11] and ^{14}N in various nitro-compounds[12].

One may get very detailed structural information about radicals from hyperfine splittings when a structure persists for a time long compared with the separation of hyperfine components ($\tau \geq 10^{-7}$ sec.). While the time scale for resolution of discrete structures may be 10^4 times as short with ESR as with NMR observations, a number of instances are known in which a system is rapidly interconverting on the ESR time scale. Such behavior may be suspected when there is a significant variation in amplitude of hyperfine components which would be expected to be equal. Interconversion of isomers causes significant fluctuations in the ring proton hyperfine couplings. If the rate of interconversion is insufficient to give the

---

[1] R. WILSON, and D. KIVELSON: J. Chem. Phys. 44, 154 (1966).
[2] P. W. ATKINS, and D. KIVELSON: J. Chem. Phys. 44, 169 (1966).
[3] A. CARRINGTON, and H. C. LONGUET-HIGGINS: Mol. Phys. 5, 447 (1962); this treatment has been criticized for neglect of some terms (ref. [4]).
[4] J. GENDELL, J. H. FREED, and G. K. FRAENKEL: J. Chem. Phys. 41, 949 (1964).
[5] J. W. H. SCHREURS, G. E. BLOMGREN, and G. K. FRAENKEL: J. Chem. Phys. 32, 1861 (1960). — J. W. H. SCHREURS, and G. K. FRAENKEL: J. Chem. Phys. 34, 756 (1961).
[6] See footnote 10, p. 193.
[7] M. J. STEPHEN, and G. K. FRAENKEL: J. Chem. Phys. 32, 1435 (1960).
[8] A. CARRINGTON, and J. DOS SANTOS-VEIGA: Mol. Phys. 6, 101 (1963).
[9] See footnote 6, p. 193.
[10] See footnote 9, p. 193.
[11] E. DE BOER, and E. L. MACKOR: J. Chem. Phys. 38, 1450 (1963).
[12] J. H. FREED, and G. K. FRAENKEL: J. Chem. Phys. 40, 1815 (1964).

averaged spectrum there will be a superposition of spectra of the two forms. This intermediate case is exemplified by the terephthalaldehyde anion system, which exists as the cis (60) and trans (61) isomers[1,2].

(60)   (61)

In the NMR spectrum of the neutral molecule one sees only an averaged line, indicating that there is an interconversion of the two isomers in a time short compared with 10^{-2} sec. To see resolved NMR spectra of the two forms with lines separated by $\Delta\omega$, one requires that the mean lifetime τ be such that $\sqrt{2}\tau < \Delta\omega$. The same criterion applies to ESR spectra, but $\Delta\omega$ represents the angular frequency displacement of hyperfine lines upon converting the one isomer into the other. Assuming the corresponding isomer shift in magnetic field is ~ 2 G, one requires $\tau > 4 \times 10^{-8}$ sec for resolution of ESR lines. Each of the isomers is expected to give 27 proton hyperfine lines; the ESR spectrum observed obviously has far more, and hence must be a superposition of the two spectra. Deuterium substitution permits identification of some of the lines which are well separated[2]. A ratio of concentrations of (61):(60) is found to be 1.4 at 20 °C.

In the spectra of rapidly interconverting isomers, one observes that some line groups are well resolved, while the rest are variously broadened. The naphthazarin cation undergoes interconversion among four forms, of which two are shown as (62) and (63). At low temperatures one may distinguish the two types of hydrogen-bonded protons[3]. The ESR spectrum then shows marked alternations in resolution of hyperfine components of line groups. This effect is also shown by the tetra-

(62)   (63)   (64)

methylsemiquinone cation[4,5] (64), for which the central line and alternate lines due to the methyl protons are narrow, but the remainder are broad. Specifically, if one designates lines for the twelve methyl protons according to the total z-component of nuclear spin, those for which this sum is 0, ± 2, ± 4 and ± 6 are sharp, while the lines with ± 1, ± 3 and ± 5 are broad. In the spectrum of the analogous compound 1,4-dinitro, tetramethylbenzene anion, Fig. 13, as well as other dinitro compounds, those line groups for which the total z-component of spin for the two equivalent nitrogen atoms is 0 or ± 2 are sharp, while the ± 1 lines are broad[6]. The effect in the nitro compounds is particularly marked at low temperatures or in those compounds in which adjacent substituents hinder motion of the nitro groups.

In each of these instances there exist discrete configurations in which two or more presumably equivalent nuclei instantaneously have different hyperfine inter-

---
[1] A. H. MAKI: J. Chem. Phys. **35**, 761 (1961).
[2] E. W. STONE, and A. H. MAKI: J. Chem. Phys. **38**, 1999 (1963). — A. H. MAKI, and D. H. GESKE: J. Am. Chem. Soc. **83**, 1852 (1961).
[3] J. R. BOLTON, A. CARRINGTON, and P. F. TODD: Mol. Phys. **6**, 169 (1963).
[4] J. R. BOLTON, and A. CARRINGTON: Mol. Phys. **5**, 161 (1962).
[5] A. CARRINGTON: Mol. Phys. **5**, 425 (1962).
[6] J. H. FREED, and G. K. FRAENKEL: J. Chem. Phys. **37**, 1156 (1962); **41**, 699 (1964).

actions. There are some hyperfine levels whose separations are unaltered by these fluctuations; that one sees linewidth alternation instead of a general broadening is due to the fact that the total extent of the spectrum is unaltered. In the case of the tetramethylsemiquinone and naphthazarin cations, there are proton jumps between two different positions, and the change in their coupling constants is assignable to a different environment. Such jump modulations in hyperfine splitting have been treated by a modification of the Bloch equations[1,2]. An analogous case of broadening of some lines in the spectrum of the cyclohexyl (C_6H_{11}) radical in the interval 193 to 273 °K is ascribed to inversion between two "chair" forms[3]. The jumping of a cation between two positions in the pyracene anion has been described under Ion Pairing.

In the more general case typified by the dinitrobenzene anions, there is an overall fluctuation in distribution of spin density in the molecule. Qualitatively, one may explain these effects as being due to an out-of-phase modulation of the hyperfine interaction of the two presumably equivalent atoms, of a magnitude large compared with other line width contributions. Such fluctuations, which are different at the two atoms, may occur because of interactions with cations, with solvent or out-of-plane motions, the coupling constant of one atom increasing while the other decreases. The hindered nitro groups are rotated out of the molecular plane and have a smaller coupling constant than for groups in the plane[4]. They undergo a larger fluctuation in coupling constant for a given change in angle. The presence of a polar group such as NH_2 or O^- between two nitro groups apparently causes fluctuations to occur in phase, for no alternations in line width are observed for such compounds.

The various effects which we have just mentioned all involve sets of equivalent nuclei and therefore of degenerate spin states. The modified Kubo-Tomita theory which accounts for relaxation behavior of a single nucleus is unable to deal with this more complex problem. If one writes the Hamiltonian for such systems in the form $\hbar \mathcal{H} = \hbar \mathcal{H}_0 + \hbar \mathcal{H}_1(t)$, the first term describes a spectrum of sharp lines, while the second contains all the effects of relaxation and line broadening[5]. One is concerned with the evaluation of the time-dependent term for nuclei which are equivalent with respect to the first term, but because of instantaneously different interactions with their environment or states of motion are non-equivalent with respect to the second. We have mentioned above various interactions which instantaneously are different for two or more symmetry-equivalent nuclei. The tumbling of the molecule modulates the instantaneous g-component parallel to the magnetic field, the anisotropic intramolecular dipolar interaction and the quadrupolar interaction, if any. Internal motions and complex formation may distort the molecule, changing the spin distribution and leading to modulation of the g-tensor, the isotropic hyperfine interaction, the intramolecular dipolar interactions and the quadrupolar interaction. Dipolar interaction of the solvent plus counter ions is permitted, as is chemical exchange. Time-dependence is formulated in terms of a relaxation matrix technique[6]. The widths of individual com-

---

[1] F. Bloch: Phys. Rev. 70, 460 (1946).
[2] A. Carrington: Mol. Phys. 5, 425 (1962).
[3] S. Ogawa, and R. W. Fessenden: J. Chem. Phys. 41, 994 (1964).
[4] D. H. Geske, and J. L. Ragle: J. Am. Chem. Soc. 83, 3532 (1961). — D. H. Geske, J. L. Ragle, M. A. Bambenek, and A. L. Balch: J. Am. Chem. Soc. 86, 987 (1964). — P. H. Rieger, and G. K. Fraenkel: J. Chem. Phys. 39, 609 (1963).
[5] J. H. Freed, and G. K. Fraenkel: J. Chem. Phys. 39, 326 (1963).
[6] F. Bloch: Phys. Rev. 102, 104 (1956). — Y. Ayant: J. Physique Radium 16, 411 (1955). — A. Abragam: The Principles of Nuclear Magnetism. Chap. 8: Oxford University Press 1961. — P. S. Hubbard: Revs. Mod. Phys. 33, 249 (1961). — A. G. Redfield: I.B.M. J. Res. Develop 1, 19 (1957).

ponents of a composite line are given by the negative of the eigenvalues of the relaxation matrix. Each of the components is found to have a Lorentzian shape, and in general these will have different widths; hence the observable shape will not be Lorentzian (as had been predicted[1]) unless the variation in width is small. The difference is marked for those cases in which one can readily see variations in hyperfine line widths in ESR spectra. Our examples of molecules with two CHO, OH or NO_2 groups whose ESR spectra show alternating line widths are just such cases. A "mean T_2^{-1}" for each degenerate set of states is suggested as a useful concept[2]. The ESR spectrum of the anion of tetracyanoethylene, $(NC)_2C=C(CN)_2^-$, in an ethanol-glycerol mixture clearly shows that the line shape is not Lorentzian in this case[3]. The wide sparation of lines and appreciable degeneracies in the 1:4:10:16:19:16:10:4:1 pattern due to four symmetry-equivalent nitrogen atoms and intramolecular dipolar interactions make this radical a good choice. (However, lines in the benzosemiquinone anion are Lorentzian.) The relaxation matrix will have non-diagonal elements because of nuclear spin transitions whenever there are two or more symmetry-equivalent atoms which are not at all instants completely equivalent in their interactions. For the general case one cannot give line width expressions in closed form because of the non-diagonal form of the relaxation matrix except in the simplest cases. A serious restriction of the relaxation matrix approach applied to chemical exchange or jump problems is that it is limited to exchange or jump rates fast compared with frequency separation of hyperfine components arising from the modulations. For slow, intermediate or fast rates of exchange the modified Bloch equations[4] or a simple form of motional narrowing theory have been used. These may in favorable circumstances give satisfactory results, but the occurrence of cross-relaxation or of isotropic and anisotropic dipolar relaxation contributions are not accounted for [5]. Indeed, it has been shown that the effect of cross-relaxation upon an inhomogeneously-broadened line is to provide structure (sidebands) to the individual components; hence one is not justified in taking a weighted sum of Lorentzians or Gaussians under these conditions[6].

Under rather restrictive conditions (similar nuclei completely equivalent and exchange effects small) it is shown that under microwave saturation conditions, composite lines are superpositions of "saturated" Lorentzians. The theory allows some lines to be saturated to a greater extent than others, as is often observed[7]. It should be noted that in a system of more than two levels, spin-lattice relaxation is not in general exponential, and it is appropriate to use a relaxation probability in place of T_1^{-1} [8]. A further effect predicted when modulation of isotropic hyperfine splittings leads to marked alternation in line width is a dynamic (anomalous second order) shift from time-dependent interactions which lead to relaxation and to line broadening[9]. It is suggested that in favorable cases one might be able

---

[1] D. KIVELSON: J. Chem. Phys. **33**, 1094 (1960).
[2] D. KIVELSON: J. Chem. Phys. **41**, 1904 (1964).
[3] J. GENDELL, J. H. FREED, and G. K. FRAENKEL: J. Chem. Phys. **41**, 949 (1964).
[4] F. BLOCH: Phys. Rev. **70**, 460 (1946). — R. A. SACK: Mol. Phys. **1**, 163 (1958). — J. A. POPLE: Mol. Phys. **1**, 168 (1958). — A. CARRINGTON: Mol. Phys. **5**, 425 (1962).
[5] J. I. KAPLAN: J. Chem. Phys. **28**, 278 (1958); **29**, 462 (1958). — I. SOLOMON, and N. BLOEMBERGEN: J. Chem. Phys. **25**, 261 (1956).
[6] N. BLOEMBERGEN, S. SHAPERO, P. S. PERSHAN, and J. O. ANDERSON: Phys. Rev. **114**, 445 (1959). — J. R. KLAUDER, and P. W. ANDERSON: Phys. Rev. **125**, 912 (1962). — A. KIEL: Phys. Rev. **125**, 1451 (1962). — J. I. KAPLAN: J. Chem. Phys. **42**, 3789 (1965).
[7] J. FREED: J. Chem. Phys. **43**, 2312 (1965).
[8] J. P. LLOYD, and G. E. PAKE: Phys. Rev. **94**, 579 (1954).
[9] G. K. FRAENKEL: J. Chem. Phys. **42**, 4275 (1965).

(independent of line width effects) to measure mean-square fluctuations in hyperfine splitting as well as to estimate correlation time for fluctuating motions.

A study of ^{13}C hyperfine splittings provides ample verification of the assumption that intramolecular dipolar interaction between the unpaired electron and nuclei is an important source of line broadening. The major contribution to width of a line associated with a particular carbon atom in a π-system comes from the spin density on that carbon atom. The broadening is roughly proportional to the square of the local spin density. If one has a knowledge of spin density distribution in a molecule, one may be able to make an assignment of ^{13}C lines from their widths. Such assignment is often a difficult task both because many lines may have the same statistical weight and because of limited signal-to noise ratio. Spectra are taken at a sufficiently low temperature so that broadening is apparent. However, at high spin density positions such as the 9, 10 positions of anthracene, the ^{13}C lines in the cation must be observed well above room temperature to diminish the anisotropic line width contribution. For this cation, the approximate ratios of squares of spin densities at positions 1, 2, 9 and 11 are respectively 24:7:110:1 [1]. The method is also applicable to ^{14}N but not to H.

16. Temperature effects.

α) *Temperature-Dependent Hyperfine Splitting.* Marked variations of hyperfine constants with temperature may occur in free radical systems. Such variations may seriously complicate studies of ion pairing, exchange, solvent effects, assignment of the McConnell Q parameter, etc. An example of an unusually large variation is that in the radical (65). The NH$_2$ proton splitting constant changes from

(65)

4 G at 200 °K to 1 G at 500 °K without appreciable change in nitrogen splitting [2,3]. In this case the change is attributed to torsional oscillation of the NH$_2$ group. Upon moving out of the molecular plane, there will be a positive coupling contribution which partially compensates the normally negative one.

A more usual variation of proton splitting constants with temperature is that found for the CH$_3$ radical in solution, for which the temperature coefficient near 25 °C is -2.1×10^{-3} G °C^{-1} [4]. Using $A = 22.674$ G at 25.00 °C, this gives a value of 23.10 G at 77 °K, in better agreement with the measured value of 23.04 G (in liquid CH$_4$) than could be expected [5]. (Some variations due to the solid matrix are shown by the 4.2 °K values of 22.97, 23.06 and 23.21 G in CH$_4$, Ar and H$_2$ matrices respectively.) [6] A direct interaction resulting from vibrations would give a positive spin interaction contribution increasing with temperature [7,8] while the normal indirect interaction is negative [9].

---

[1] J. R. Bolton, and G. K. Fraenkel: J. Chem. Phys. **41**, 944 (1964).
[2] K. Scheffler, and H. B. Stegmann: Tetrahedron Letters **1964**, 3035.
[3] A. J. Stone, and A. Carrington: Trans. Faraday Soc. **61**, 2593 (1965).
[4] I. A. Zlochower, W. R. Miller jr., and G. K. Fraenkel: J. Chem. Phys. **42**, 3339 (1965).
[5] R. W. Fessenden, and R. H. Schuler: J. Chem. Phys. **39**, 2147 (1963).
[6] C. K. Jen, S. N. Foner, E. L. Cochran, and V. A. Bowers: Phys. Rev. **112**, 1169 (1958).
[7] B. Venkataraman, and G. K. Fraenkel: J. Chem. Phys. **24**, 737 (1956).
[8] D. M. Schrader, and M. Karplus: J. Chem. Phys. **40**, 1593 (1964).
[9] H. M. McConnell: J. Chem. Phys. **24**, 633, 764 (1956).

The variation with temperature of proton hyperfine splitting in the benzene anion is somewhat larger, the splitting decreasing from 22.9 to 22.4 G between 140 and 210 °K [1]. For the C_7H_7 radical the proton splitting decreases from 3.86 G at 20 °C to 3.75 G at 96 °C to 3.62 G at 176 °C. In both cases it is presumed that there is increasing occupation of a low-lying vibronic state with increasing temperature [2]; the splitting is presumably smaller in the upper state due to increased vibrational amplitude. The ^{13}C splitting increases with increasing temperature in C_7H_7 [2].

Several studies of the temperature variation of nitrogen hyperfine splitting agree on a positive coefficient. Radicals studied include the cations of dihydro- and dideuteropyrazine (19) [3], diphenyl nitroxide $(C_6H_5)_2NO$ [4] and bis(trifluoromethyl)nitroxide $(CF_3)_2NO$ [5]. For the latter, the nitrogen coupling constant is given as $a^N = (8.776 + 0.0023\,T)$ G in the range 163 to 297 °K. In the same range the fluorine coupling constant $a^F = (9.327 - 0.0036\,T)$ G.

The splitting constant of the alkali cation in an ion pair may vary considerably with temperature in some solvents. For Na^+ naphthalene$^-$ (66) in tetrahydrofuran, (53) the Na^+ splitting increases about sixfold between 210 and 310 °K [6].

(66)

One approach to the variation of hfs with temperature is to consider an averaging of hfs constants in discrete states. For two nearly-degenerate electronic levels E_1 and E_2, with coupling constants a_1 and a_2, changes in population with temperature should give an observed coupling constant $\langle a \rangle$:

$$\langle a \rangle = \frac{a_1 \exp(E_1/kT) + a_2 \exp(E_2/kt)}{\exp(E_1/kT) + \exp(E_2/kT)}.$$

For a large number of accessible vibrational levels of energy E_n and coupling constant a_n:

$$\langle a \rangle = \frac{\sum_{n=0}^{\infty} a_n \exp(-E_n/kT)}{Z},$$

where Z is the partition function.

Temperature effects on line widths are considered in Sect. 15.

A dramatic increase in resolution of hyperfine structure is shown by solutions of various radicals in a variety of solvents on increasing the temperature to an optimum extent or by removal of oxygen [7]. Further temperature increase serves to blur the lines. The initially increased resolution has been attributed to averaging of anisotropic hyperfine interaction by reorientation [8-10]. The loss of resolution on reduction of viscosity (or addition of oxygen) is attributed to spin exchange just as for broadening due to increased concentration [8,9]. This interpretation is con-

---

[1] R. W. FESSENDEN, and S. OGAWA: J. Am. Chem. Soc. **86**, 3591 (1964).
[2] G. VINCOW, M. L. MORRELL, W. V. VOLLAND, H. J. DAUBEN jr., and F. R. HUNTER: J. Am. Chem. Soc. **87**, 3527 (1965).
[3] M. R. DAS, and G. K. FRAENKEL: J. Chem. Phys. **42**, 792 (1965).
[4] D. BIJL, and A. C. ROSE-INNES: Phil. Mag. **1953**, 1187.
[5] P. J. SCHEIDLER, and J. R. BOLTON: J. Am. Chem. Soc. **88**, 371 (1966).
[6] N. M. ATHERTON, and S. I. WEISSMAN: J. Am. Chem. Soc. **83**, 1330 (1961).
[7] K. H. HAUSSER: Z. Naturforsch. **14A**, 425 (1959).
[8] G. E. PAKE, and T. R. TUTTLE: Phys. Rev. Letters **3**, 423 (1959).
[9] H. M. McCONNELL: J. Chem. Phys. **25**, 709 (1956).
[10] D. KIVELSON: J. Chem. Phys. **33**, 1094 (1960).

firmed by a study of exchange relaxation[1]. It has been pointed out that this spin exchange, as well as electron exchange[2] may both lead to error in determining the temperature coefficient of hyperfine splitting; both cause a coalescence of lines at high temperatures[3]. This would give an apparently negative coefficient.

β) *Temperature-Dependent Conformations.* Anions of diphenylethylene $C_6H_5-C(H)=C(H)-C_6H_5^-$ and azobenzene $(C_6H_5-N=N-C_6H_5^-)$ (67) show asymmetry in spin density in the ring due to restricted rotation since, free rotation would have made atoms 2 and 6 equivalent. However, the tolan anion $(C_6H_5-C\equiv C-C_6H_5^-)$ gives no evidence of hindrance to rotation[4].

(67)

Distinctly different splitting constants of the two ring protons adjacent to the substituent in benzaldehyde (C_6H_5CHO) and acetophenone $(C_6H_5C(O)CH_3)$ show that at room temperature the $C=O$ group tends to keep the substituent coplanar with the ring[5]. The nitrosobenzene (C_6H_5NO) molecule in liquid NH_3 is similarly rigid (the CNO bond is non-linear)[6]. In Sect. 15 on line widths, we noted that terephthalaldehyde, which has two CHO groups opposite one another, showed the presence of distinct isomers[7]. The same applies if the two CHO groups are adjacent to one another, or if there is one CHO and one NO_2 substituent on benzene. One should expect each substituent at sufficiently high temperature to undergo the rotation which is so common in most radicals.

17. Solvent effects.
Rather marked variations in hyperfine splitting constants with solvent have been shown for numerous nitrogen- and oxygen-containing radicals. Unless these solvent effects are taken into account, one may have difficulty in accounting for spin densities. Increases in nitrogen splitting constants in water as compared with other solvents are found for substituted nitrobenzenes, the a^N values sometimes increasing by a factor of two in water[8-11]. Marked effects of solvent are also observed on proton splittings[12-15], even in some hydrocarbons[16,17]; these effects are attributed to dipolar interaction between solvent and substituent groups, to hydrogen bonding, or to ion-pair formation[13]. Alterations both in the π-electron charge distribution and the spin density distributions will result from such interactions. Redistribution of spin density may be largely local, as among C—N—O bonds, or it may involve the entire π-system.

---

[1] J. D. Currin: Phys. Rev. 126, 1995 (1952).
[2] R. L. Ward, and S. I. Weissman: J. Am. Chem. Soc. 76, 3612 (1954); 79, 2086 (1957). — P. J. Zandstra, and S. I. Weissman: J. Chem. Phys. 35, 757 (1961).
[3] M. R. Das, and G. K. Fraenkel: J. Chem. Phys. 42, 792 (1965).
[4] C. S. Johnson jr., and R. Chang: J. Chem. Phys. 41, 3272 (1964); 43, 3183 (1965).
[5] N. Steinberger, and G. K. Fraenkel: J. Chem. Phys. 40, 723 (1964).
[6] D. H. Levy, and R. J. Myers: J. Chem. Phys. 42, 3731 (1965).
[7] E. W. Stone, and A. H. Maki: J. Chem. Phys. 38, 1999 (1963).
[8] L. H. Piette, P. Ludwig, and R. N. Adams: J. Am. Chem. Soc. 84, 4212 (1962).
[9] P. L. Kolker, T. J. Stone, and W. A. Waters: Proc. Chem. Soc. 1963, 55.
[10] J. Pannell: Mol. Phys. 7, 317 (1964).
[11] J. Q. Chambers, T. Layloff, and R. N. Adams: J. Phys. Chem. 68, 661 (1964).
[12] M. Bruin, F. W. Heinecken, and F. Bruin: J. Chem. Phys. 37, 452 (1962).
[13] J. Gendell, J. H. Freed, and G. K. Fraenkel: J. Chem. Phys. 37, 2832 (1962).
[14] E. W. Stone, and A. H. Maki: J. Chem. Phys. 36, 1944 (1962).
[15] P. B. Ayscough, and R. Wilson: J. Chem. Soc. 1963, 5412.
[16] A H Reddoch: J. Chem. Phys. 43, 225 (1965).
[17] E. de Boer, and E. L. Mackor: Proc. Chem. Soc. 1963, 23.

One can understand in a general way significant changes in ^{13}C or ^{14}N splitting constants when a radical complexes with solvent. The region of attachment of the solvent is likely to be a hetero atom, (nitrogen or oxygen), whether the interaction is dipolar or hydrogen bonding. It is just ^{13}C or ^{14}N atoms which have splitting constants strongly dependent not only upon the density on that atom but upon the densities at its neighbors. Here one has splitting contributions of nearly equal magnitude and of opposite sign. One finds that the ^{13}C splitting constant for a C—O carbon atom in a semiquinone radical changes from 0.4 G in alkaline alcoholic solution to 2.13 G in dimethylsulfoxide (DMSO)[1]. The proton splittings or ^{13}C splittings of C—H carbons are changed little. Hence there is redistribution of spin density primarily between O and C atoms of the O—C group.

While the *percentage variation* of proton splitting constants may be considerable (e.g., a 45% difference for 1, 4, 5, 8 protons of anthrasemiquinone (7) between ethanol water and DMSO), the *absolute values* are only slightly different (0.550 vs 0.303 G)[1]. It is just at positions of low spin density that one may expect large variations. In the simple case of a 1:1 radical-solvent complex in pure solvents A or B and their mixtures, one may have a displacement equilibrium:

$$RXS_A + S_B \rightleftharpoons RXS_B + S_A,$$

where S_A and S_B represent two pure solvents. However, one does not observe separate spectra of the two complexes, by virtue of exchange. With rapid exchange, the observed splitting constant is:

$$\langle a \rangle = a_A\, p_A + a_B(1 - p_A)$$

where p_A is the fraction of the radical in the form RXS_A. In terms of the equilibrium constant K and the ratio α of the mole fractions of S_B and S_A, one may write[1]:

$$\langle a \rangle = \tfrac{1}{2}(a_A + a_B) + \tfrac{1}{2}[(K\alpha - 1)/(K\alpha + 1)](a_B - a_A).$$

This relation has been shown to describe quantitatively the changes in proton splitting constants of semiquinones with solvent[1].

In some instances such as p-nitrobenzaldehyde anion (68)

(68)   (69)

in CH_3CN vs. H_2O) an additional change in hyperfine splitting on change of solvent may reflect changes in molecular symmetry, viz. acquiring a C_2 axis in water due to CHO group rotation, while being locked in planar form in CH_3CN [2-4].

Dependence of alkali superhyperfine splitting upon solvent and temperature is seen in the behavior of hexamethylacetone (52) (HMA)$^-$ ketyl in tetrahydrofuran THF, (53) or methyl tetrahydrofuran MTHF, (69) with Na$^+$. At 293 °K, HMA in THF has $a^H = 1.58$ G, whereas it is 0.65 G in MTHF. On lowering the temperature to 173 °K, there is little change in THF, but $a = 1.90$ G in MTHF[5]. The behavior in

---

[1] See footnote 13, p. 200.
[2] See footnote 10, p. 200.
[3] D. H. Geske, and A. H. Maki: J. Am. Chem. Soc. **83**, 1852 (1961).
[4] See footnote 9, p. 200.
[5] G. R. Luckhurst: Mol. Phys. **9**, 179 (1965).

MTHF has been attributed to compaction of the solvent sheath about the anion on lowering the temperature, leading to increased $Na-3s$ and radical-π interaction.

18. Transfer and exchange effects. We shall examine briefly the effects of several types of processes:

1. Chemical transfer ("exchange") of an atom or ion to or from a paramagnetic species or even between two positions in the same molecule.
2. Electron transfer within a molecule.
3. Electron transfer between an ion and a neutral molecule.
4. Heisenberg (spin) exchange.

Chemical Transfer. — In early NMR experiments on ethanol[1], it was observed that the spin-spin splitting by OH protons was observed in neutral solutions, but disappeared in acidic or basic solutions since the lifetime of one proton in an OH group became comparable with its splitting contribution. An analogous effect is seen in the CH_2OH radical formed in aqueous solution from methanol[2,3]. Above pH 1.5, the OH proton splitting is constant at 0.96 G, falling to zero at pH 1.1[3]. The exchange is attributed to the process: $CH_2OH + H_3O^+ \rightleftharpoons CH_2OH_2^+ + H_2O$. Proton transfer effects also occur in semiquinone systems (Sect. 11).

The system pyracene$^-$, M^+ (70) clearly shows the range of behavior from "unassociated" radical ions to ion pairs of a single conformation, with an intermediate range in which one observes effects due to cation jumping[4]. One may

(70)   (71)

traverse the range with appropriate choice of cation, solvent and temperature. With a long lifetime τ of an ion pair conformation, the two sets β and β' of aliphatic protons have different splitting constants a and a'. For Na^+ as cation, $a-a' \approx 0.56$ G in one case ($-80°$, MTHF (69) as solvent). "Long" lifetime is then defined by $\tau > [\gamma(a-a')]^{-1}$. All lines will then be sharp, and the alkali cation splittings will be readily discerned under favorable conditions. If the cation moves rapidly between such positions as A and B, i.e., if $\tau < [\gamma(a-a')]^{-1}$ the observed spectral lines will again be sharp and the single apparent splitting constant for the β and β' protons will be the average of a and a'. When $\tau[\gamma(a-a')]^{-1} \approx 1$, line groups with even values of quantum number m_β will show sharp lines, while the others will be broad. Such behavior is found for K^+ at -30 °C in tetrahydrofuran. It seems clear that cation motion induces fluctuations in spin density which lead to shorter spin-spin relaxation times for some of the line groups than for others.

---

[1] J. T. ARNOLD: Phys. Rev. **102**, 136 (1956).
[2] W. T. DIXON, and R. O. C. NORMAN: J. Chem. Soc. **1963**, 3119.
[3] H. FISCHER. Mol. Phys. **9**, 149 (1965).
[4] E. DE BOER, and E. L. MACKOR: J. Am. Chem. Soc. **86**, 1513 (1964).

Intramolecular Electron Transfer. — The hyperfine spectrum for a molecule consisting of two conjugated portions of a radical joined by σ-bonded $-(CH_2)_n-$[1-3], $-O-$[4,5], or $-S-$[5] groups is markedly dependent upon the bridging group and the solvent. When $n \geq 2$, the lifetime of an electron on one half of the molecule is so great that hyperfine splitting is seen only from that half. For the various other groups one sees splitting from the entire molecule, with markedly different coupling constants. For the molecule (71) also, transfer is rapid across cyclobutane anion rings[6]. In the latter, π interaction between atoms 1 and 3 is presumed to be the mechanism of transfer[7].

Transfer rates may be obtained from expressions relating observed and resolved line widths[8]. A reasonable check on correctness of transfer rate is a correspondence of calculated with observed spectra.

It has been suggested that for dipolar solvents the electron is strongly bound by oriented solvent dipoles, and that random motions of solvent molecules govern the rate of transfer between halves of the molecule. For $O_2NC_6H_4-S-C_6H_4NO_2$ (4,4') at room temperature, the transfer rate in CH_3CN is given as 2×10^6 sec^{-1}, while in $(CH_3)_2SO$ it is 9×10^6 sec^{-1} [5].

Exchange of two electrons within the same molecule is considered in Sect. C. IV. on Biradicals.

Electron transfer between charged (paramagnetic) and neutral (diamagnetic) molecules is exemplified by the process $A^- + A \rightleftharpoons A + A^-$ [9,10]. The ESR lines will be broadened if the rate of electron transfer between A^- and A is comparable with hyperfine separations. It has been shown that the rate of electron exchange is independent of nuclear spin state[9]. The line widths will increase with concentration of A. If $A^- \equiv$ naphthalene anion, the equilibrium constant ranges from 10^7 to 10^9 l mole^{-1} sec^{-1} depending on solvent and counter ion. An exchange study may involve two valence states of the same ion.

HEISENBERG or spin exchange, represented by a term $JS_1 \cdot S_2$ in the spin Hamiltonian, is responsible for the narrowing of the observable central portion of an ESR line (or line envelope) at high concentrations of radicals in solution or in the solid state[11]. At the same time the wings of the line are broadened so as to conserve the second moment.

The $(SO_3)_2NO^{-2}$ ion radical with its well-separated lines (13 G) affords an excellent opportunity for studying spin exchange frequency as a function of contration[12]. In this case this frequency may be obtained equally well from line widths or from line separations, using an expression derived from the Bloch equations for slow exchange. It is found possible to subtract the effect of oxygen by choosing

---

[1] S. I. WEISSMAN: J. Am. Chem. Soc. **80**, 6462 (1958).
[2] V. V. VOEVODSKII, S. P. SOLODOVNIKOV, and V. M. CHIBRIKIN: Doklady Akad. Nauk S.S.S.R. **129**, 1083 (1959).
[3] H. M. McCONNELL: J. Chem. Phys. **35**, 508 (1961).
[4] D. H. EARGLE, and S. I. WEISSMAN: J. Chem. Phys. **34**, 1840 (1961).
[5] J. E. HARRIMAN, and A. H. MAKI: J. Chem. Phys. **39**, 778 (1963).
[6] See footnote 2, p. 202.
[7] M. T. JONES, E. A. LALANCETTE, and R. E. BENSON: J. Chem. Phys. **41**, 401 (1964).
[8] P. W. ANDERSON, and P. R. WEISS: Revs. Mod. Phys. **25**, 269 (1953). — P. W. ANDERSON: J. Phys. Soc. Japan **9**, 316 (1954).
[9] P. J. ZANDSTRA, and S. I. WEISSMAN: J. Chem. Phys. **35**, 757 (1961).
[10] R. M. WARD, and S. I. WEISSMAN: J. Am. Chem. Soc. **79**, 2086 (1957).
[11] M. McMILLAN, and W. OPECHOWSKI: Can. J. Phys. **38**, 1168 (1960). — R. KUBO, and K. TOMITA: J. Phys. Soc. Japan **9**, 888 (1954). — J. H. VAN VLECK: Phys. Rev. **74**, 1168 (1948). — M. H. L. PRYCE, and K. W. H. STEVENS: Proc. Phys. Soc. (London) A **63**, 36 (1950). — P. W. ANDERSON, and P. R. WEISS: Revs. Mod. Phys. **25**, 269 (1953). — D. KIVELSON: J. Chem. Phys. **27**, 1087 (1957). — P. W. ANDERSON: J. Phys. Soc. Japan **9**, 316 (1954).
[12] M. T. JONES: J. Chem. Phys. **38**, 2892 (1963).

the appropriate limiting linewidths. In Sect. 16 we noted that line broadening on reduction in viscosity η, increase in temperature or on addition of oxygen may be attributable to the spin exchange[1,2], just as for broadening due to increasing concentration. In the simplest cases of exchange broadening, line width is proportional to T/η. It has been assumed[1] that the proportionality factor between spin-exchange frequency ν_{ex} and encounter frequency ν_{enc}, is simply the probability p of spin exchange during encounter. Both ν_{enc} and p are dependent on temperature as well as viscosity η. The form of p for one radical $[(CH_3)_3C]_2NO$ in propane, n-pentane and methylcyclohexane is given[3] as $p = 1 - \exp(-a\eta/T)$. Variations of viscosity independent of temperature were achieved by increased hydrostatic pressure. At moderate pressures, the three nitrogen hyperfine lines are narrowed equally because of increasing rate of spin exchange. Beyond 6500 kg cm^{-2}, the outer lines broaden much more rapidly than the inner one because of anisotropic hyperfine interactions.

II. σ-radicals.

In the preponderant majority of carbon radicals thus far known, the unpaired electron is in a $2p\pi$-orbital, either on one atom or delocalized over several atoms joined with σ-bonds. However, there are a few known radicals in which the unpaired electron occupies an sp^2 or sp^3 orbital on a carbon or phosphorus atom. For such radicals one observes extraordinarily large values of coupling constants; in extreme cases these are as large as 384 MHz for H in HCO [4,5]. Other σ-radicals include FCO [6], HPO_2^- [7], ethynyl (C≡CH) [8], and vinyl $H_2C=C'H$ [8,9] radicals.

HPO_2^- is generated as an oriented radical in $NH_4H_2PO_2^-$, with principal g-components 2.0019, 2.0035 and 2.0037; the principal phosphorus hyperfine components are 1698, 1228 and 1228 MHz. The proton coupling is 230 MHz and is almost isotropic. This appears to be the only one of the σ-radicals enumerated which has the electron occupying one of four essentially sp^3 orbitals, giving rise to a distorted tetrahedron about the phosphorus atom. Both the g- and hyperfine tensors show axial symmetry about the electron-phosphorus direction, with the $e^- - P\diagup^H$ angle about 117° [7]. In all of the cases there is hardly any doubt that the proton coupling is positive, due to a substantial s-component of the P—H, C—H or C—F bonds. In the cases of HCO and FCO radicals, it has been proposed that while the ground state structure is approximately represented as $\overset{X}{\underset{X}{\diagdown}}C'=O$, there is a low-lying excited state which might be represented as $\overset{\cdot\cdot}{C}=O$ which may be admixed by configuration interaction[8]. The exceptional stability of CO makes this configuration favorable. The H—C bond is said to be unusually weak. The computed spin density in the hydrogen 1s orbital of HCO is 0.268, while for FCO the spin density is 0.187 for a fluorine bonding orbital with 10.6 per cent s

---

[1] G. E. PAKE, and T. R. TUTTLE jr.: Phys. Rev. Letters **3**, 423 (1959).
[2] D. KIVELSON: J. Chem. Phys. **33**, 1094 (1960).
[3] N. EDELSTEIN, A. KWOK, and A. H. MAKI: J. Chem. Phys. **41**, 3473 (1964).
[4] F. J. ADRIAN, E. L. COCHRAN, and V. A. BOWERS: J. Chem. Phys. **36**, 1661 (1962). — W. GORDY, W. B. ARD, and H. SHIELDS: Proc. Natl. Acad. Sci. U.S. **41**, 996 (1955).
[5] M. C. R. SYMONS, in: Advances in Physical Organic Chemistry (V. GOLD, ed.), p. 283. New York: Academic Press 1963.
[6] F. J. ADRIAN, E. L. COCHRAN, and V. A. BOWERS: J. Chem. Phys. **43**, 462 (1965).
[7] J. R. MORTON: Mol. Phys. **5**, 217 (1962).
[8] E. L. COCHRAN, F. J. ADRIAN, and V. A. BOWERS: J. Chem. Phys. **40**, 213 (1964).
[9] R. W. FESSENDEN, and R. H. SCHULER: J. Chem. Phys. **39**, 2147 (1964).

character. For FCO the principal hyperfine tensor components are 1437.5, 708.2 and 662.0 MHz[1].

The isotropic proton splitting of 16.1 G in ethynyl radical is presumed to result from the consecutive processes: 1. Polarization of the two π-bonds by exchange interaction of the unpaired electron in the σ-orbital with the π-bond electrons. 2. Polarization of electrons in the C—H bond by exchange interaction with the π-electrons. Existence of the two π-orbitals on each carbon atom increases the effectiveness of such an indirect mechanism, but the 75 per cent s-character required in the orbital of the unpaired electron appears excessive[2]. The radical appears to be rotating at 4.2 °K. The vinyl radical $\begin{matrix}H_3\\H_2\end{matrix}C-C'\begin{matrix}\\H_1\end{matrix}$, which is isoelectronic with the formyl radical, has coupling constants: $a(H_1)=15.6$ G; $a(H_2, \text{cis})=34$ G and $a(H_3, \text{trans})=68$ G[2]. Generated in liquid ethane, this radical gives apparent coupling constants of 13.39 and 102.44 G. The H_1 proton is inverting rapidly compared with the frequency separation of H_2 and H_3, i.e., shifting from one sp^2 lobe to another on the same carbon atom. This effectively interchanges the latter two protons and smears out lines orginating from antiparallel spin states of H_2 and H_3. One thus observes lines with a separation equal to the sum of the two β proton couplings, or about 102 G[2,3]. The very large difference in α-proton coupling constants in the formyl and vinyl radicals is in part ascribed to a rapidly-varying coupling constant with bond angle[2]. Calculations of the coupling constant for HCO as a function of angle give best agreement for the known angle of 120°[2,4], if an extra 2.2 eV stabilization energy is invoked. For vinyl radical, assuming $a(H_1)$ is positive, the bond angle would appear to be 140 to 150°.

Inorganic σ-Radicals. — Known examples of σ-radicals are the X_2^- ions described in Sect. 29 as V_K centers, KrF[5] and XeF[6,7], in which the unpaired electron occupies an antibonding σ-orbital. In the series F_2^-, KrF, and XeF, the isotropic fluorine couplings are large, being respectively 883, 1683 and 1243 MHz, while the anisotropy parameter B for fluorine in these compounds is 718, 924 and 703 MHz respectively. Since B_0 for fluorine is 1515 MHz, the spin density in the fluorine $2p_z$ orbital is 0.47 for F_2^-, 0.61 for KrF and 0.47 for XeF. From ^{129}Xe coupling in XeF one obtains 0.36 for the spin density in the $5p$ orbital of xenon.

C. Triplet state systems and biradicals.

I. Introduction.

Besides atoms and atomic ions, there are numerous systems with $S=1$ (triplet state), such as metastable excited states of diamagnetic molecules; there are as well molecules with a triplet ground state. To observe systems with $S=1$, one requires an energy level array such that two electrons in degenerate states $\alpha\alpha$, $\beta\beta$, and $(\alpha\beta+\beta\alpha)/\sqrt{2}$ may exist for an adequately long time. Stable molecules such as O_2 or S_2 which fulfill this condition in their ground state are few in number. However, a respectable number of systems which may be prepared in rigid matri-

---

[1] See footnote 6, p. 204.
[2] See footnote 8, p. 204.
[3] See footnote 9, p. 204.
[4] G. HERZBERG, and D. A. RAMSAY: Proc. Roy. Soc. (London) A **233**, 34 (1955).
[5] W. E. FALCONER, J. R. MORTON, and A. G. STRENG: J. Chem. Phys. **41**, 902 (1964).
[6] W. E. FALCONER, and J. R. MORTON: Proc. Chem. Soc. **1963**, 95.
[7] J. R. MORTON, and W. E. FALCONER: J. Chem. Phys. **39**, 427 (1963).

The hazards of calculating spin densities ϱ_i^π in a very simple molecule with hetero atoms from proton splitting constants *alone* are shown by the fact that earlier calculations [1-3] gave a positive ^{13}C splitting for C_2 in p-benzosemiquinone anion (34). Using all data, including ^{13}C splittings [4], a negative value is obtained. The molecular orbital calculations of proton spin densities are too insensitive to densities at carbon or oxygen positions to be accurate for ^{13}C splittings. To get the $\sigma-\pi$ parameters $Q_{CO}^C = 17.7$ G and $Q_{OC}^C = -27.1$ G, a value of $Q_{CH}^H = -27.0$ G was used [4]. As further warning against extrapolation from one compound, it was found in anthrasemiquinone anion (7) in ethanol the splitting constant at a carbonyl oxygen is $+0.70$ G in ethanol, but -0.47 G in dimethy-oxyethane (DME). In the former, the high-field ^{13}C lines are broader, while in DME the reverse is true. We shall note later that one may sometimes determine signs of hyperfine splitting constants from line width variations. The revised $\sigma-\pi$ parameters allow quantitative description of spin density changes with solvent composition.

(7)   (8)

5. ^{14}N splitting constants. Replacement of one or more carbon atoms in anions of aromatic hydrocarbons by nitrogen atoms affords an opportunity of studying their effect on carbon spin density and on the parameter Q in the McConnell equation $a_i^H = Q_{CH}^H \varrho_i^\pi$. Unfortunately, the anion of the mono-nitrogen-substituted benzene (pyridine) has not yet been prepared because of the ease with which the bipyridyl (8) radical is formed, but the anions of the following systems have been studied [5]:

The ratios of HMO unpaired electron distributions for the nitrogen-substituted as compared with the parent hydrocarbon anion e.g. (12) or (13) vs. anthracene) are generally in good agreement with proton splitting constant ratios. The largest

(9)   (10)   (11)

(12)   (13)

deviations occur for positions of low density [5]. The nitrogen splitting constants appear to be given by the simple equation $a^N = Q^N \varrho^N$, with $Q^N = 25.3$ G for the compounds $[(8), (10)$ to $(13)]$ [6], as well as other similar ones [6]. They appear not to be sensitive to the spin density on the adjacent carbon atoms, as in the case of ^{13}C splittings [7]. However, studies on 3,5-dimethylpyridine (9) [8] and on other

---

[1] J. GENDELL, J. H. FREED, and G. K. FRAENKEL: J. Chem. Phys. **36**, 1944 (1962).
[2] G. VINCOW, and G. K. FRAENKEL: J. Chem. Phys. **34**, 1333 (1961).
[3] R. W. BRANDON, and E. A. C. LÜCKEN: J. Chem. Soc. **1961**, 4273.
[4] M. R. DAS, and G. K. FRAENKEL: J. Chem. Phys. **42**, 1350 (1965).
[5] A. CARRINGTON, and J. DOS SANTOS-VEIGA: Mol. Phys. **5**, 21 (1962).
[6] C. A. McDOWELL, and K. F. G. PAULUS: Mol. Phys. **7**, 541 (1964).
[7] M. KARPUS, and G. K. FRAENKEL: J. Chem. Phys. **35**, 1312 (1961).
[8] N. M. ATHERTON, F. GERSON, and J. N. MURRELL: Mol. Phys. **5**, 509 (1962).

ces have $S=1$ by virtue of doubly-degenerate ground states. In isolation from reactive species (e.g., oxygen) they have an indefinite lifetime. These ground-state triplet systems are discussed in Sect. C. III. Perhaps more familiar because of optical studies are systems with a metastable level (or levels) of adequate lifetime. Numerous organic compounds (such as naphthalene) upon ultraviolet excitation ($\pi \to \pi^*$) in rigid amorphous ("glassy") solution at $77\,°K$ or lower show a phosphorescence with a lifetime of the order of seconds. The suggestion that the phosphorescent state is an excited triplet state[1] was a major contribution to understanding both aspects[1]. Direct excitation to such a state has negligible probability; however excitation may be transferred by a radiationless mechanism from higher-lying singlet levels to a triplet level (Fig. 14). Magnetic susceptibility experiments gave results in qualitative accord with this assignment[2-4]. Extensive phosphorescence studies on aromatic molecules (benzene, naphthalene, anthracene, their homologs and nitrogen derivatives) are satisfactorily interpreted on the basis of triplet states. Yet numerous ESR experiments over much of a decade failed to detect a spectrum assignable to molecules in the triplet state. The first successful observation was made on naphthalene — not in the glassy solutions usually used for optical observations but oriented in single crystals of durene [1, 2, 4, 5-tetramethylbenzene] [$C_6H_2(CH_3)_4$][5,6]. Such oriented samples had earlier provided important optical polarization data[7].

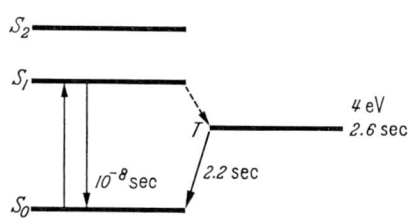

Fig. 14. Energy level diagram showing lowest excited singlet (S_1) and triplet (T_1) states of naphthalene (Jablonski diagram).

Failure of earlier attempts was recognized to be due to the considerable anisotropy of the ESR spectrum as a consequence of zero-field splitting due to spin-spin interaction of the two unpaired electrons[6]. As we shall note shortly, the form of the Hamiltonian for zero-field splitting in systems of spin $S \geq 1$ is exactly the same as for splitting due to spin-spin interactions when $S \geq 1$. Thus an energy level diagram such as Fig. 15 will be appropriate for H parallel to a molecular axis of the system. An additional cause fo failure to observe a triplet ESR signal was inadequate sensitivity, for one may observe weak ESR lines in powders from those molecules or ions which have a principal axis nearly along the magnetic field direction[8-10]. The lines detected in single crystals were later observed in glassy solutions[11-13], after improvements in sensitivity of equipment. A "half-field" ESR line of a number of triplet-state entities which was detected both in single crystals and in glassy solutions greatly stimulated interest in

---

[1] G. N. Lewis, D. Lipkin, and T. T. Magel: J. Am. Chem. Soc. 63, 3005 (1941). — A. Terenin: Acta Physicochim. U.R.S.S. 18, 210 (1943).
[2] G. N. Lewis, and M. Calvin: J. Am. Chem. Soc. 67, 1232 (1945).
[3] G. N. Lewis, M. Calvin, and M. Kasha: J. Chem. Phys. 17, 804 (1949).
[4] D. F. Evans: Nature 176, 777 (1955).
[5] C. A. Hutchinson, and B. W. Mangum jr.: J. Chem. Phys. 29, 952 (1958); 34, 908 (1961).
[6] S. I. Weissman: Cited in Ref. [5].
[7] D. S. McClure: J. Chem. Phys. 22, 1668 (1954); 24, 1 (1956).
[8] J. A. Weil, and H. G. Hecht: J. Chem. Phys. 38, 261 (1963).
[9] J. A. Ibers, and J. D. Swalen: Phys. Rev. 127, 1914 (1962).
[10] G. Burns: J. Appl. Phys. 32, 2048 (1961).
[11] W. A. Yager, E. Wasserman, and R. M. R. Cramer: J. Chem. Phys. 37, 1148 (1962).
[12] N. Hirota, and S. I. Weissman: Mol. Phys. 5, 537 (1962).
[13] P. E. Jesse, P. Biloen, R. Prins, J. D. W. van Voorst, and G. J. Hoijtink: Mol. Phys. 6, 633 (1963).

these spectra. Such a line was explained[1,2] as a "$\Delta M=2$" transition, i.e., a transition between the levels which at high fields are described as $|-1\rangle$ and $|+1\rangle$ states (Fig. 15). The intensity of these low-field transitions for non-oriented triplet molecules exceeds that of the $\Delta M=1$ lines, for essentially all molecules contribute to the former; but only the small fraction of molecules with a principal axis

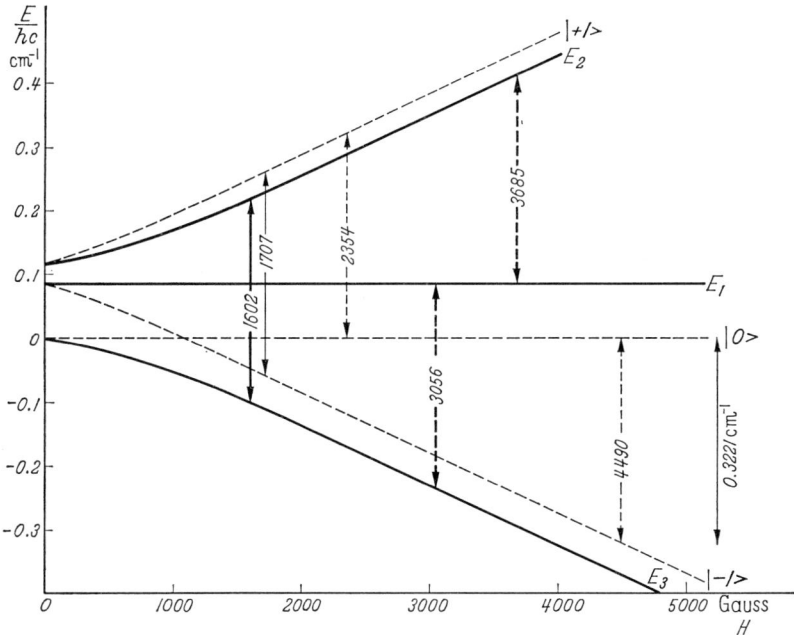

Fig. 15. ESR levels and transitions for naphthalene in the triplet excited state. Zero-field splitting parameters are shown. Solid lines, $H_0 \| y$ axis; dotted lines, $H_0 \| z$-axis. Molecular axes are shown in Fig. 16. [J. H. VAN DER WAALS, and M. S. DE GROOT: Mol. Phys. 2, 333 (1959)].

aligned along the field direction contribute to the latter. The ability to observe both of the $\Delta M=2$ and the $\Delta M=1$ lines in glassy solutions has tremendously increased the productivity of ESR triplet studies as well as increasing their number.

II. Resonant behavior of triplet systems.

It has long been recognized that when $S \geq 1$ for transition metals there is a zero-field splitting due to spin-orbit coupling. In most of the triplet-state systems of interest to us (orbitally non-degenerate systems) the first-order spin-orbit coupling vanishes and the second-order spin-orbit coupling will be small; some estimates of this effect will be given later. We are concerned with the dipole-dipole operator[3]:

$$\mathcal{H}_D = g^2 \beta^2 \left\{ \frac{\mathbf{S}_1 \cdot \mathbf{S}_2}{r^3} - \frac{3(\mathbf{S}_1 \cdot \mathbf{r})(\mathbf{S}_2 \cdot \mathbf{r})}{r^5} \right\}. \tag{1}$$

---

[1] J. H. VAN DER WAALS, and M. S. DE GROOT: Mol. Phys. 2, 333 (1959).
[2] M. S. DE GROOT, and J. H. VAN DER WAALS: Mol. Phys. 3, 190 (1960).
[3] W. HEISENBERG: Z. Physik 39, 499 (1926).

This has been shown in detail to be equivalent to the usual zero-field splitting operator[1]:
$$\mathcal{H}_D = DS_z^2 + E(S_x^2 - S_y^2) - 2D/3. \quad (2)$$

We shall be concerned with a molecule of symmetry D_2 or D_{2h}, having in mind naphthalene, the most-investigated compound. The z axis is chosen perpendicular to the plane of the molecule, the y axis is taken as the shorter-, and the x axis as the longer twofold axis. One takes the triplet spin functions to be $\alpha\alpha$, $(\alpha\beta+\beta\alpha)/\sqrt{2}$ and $\beta\beta$, as usual. Adding the Zeeman term to the spin Hamiltonian (2), one obtains as the matrix for this system[2,3].

$$\begin{pmatrix} D/3 + g\beta Hn & g\beta H(l-im)/\sqrt{2} & E \\ g\beta H(l+im)/\sqrt{2} & -2D/3 & g\beta H(l-im)/\sqrt{2} \\ E & g\beta H(l+im)/\sqrt{2} & D/3 - g\beta Hn \end{pmatrix}. \quad (3)$$

Here l, m, n are direction cosines of the applied magnetic field with respect to the symmetry axes of the triplet molecule. That is, $H(l+im) = H_x + iH_y$, etc. The axes are assigned such that $D \geq 3E$; this is always possible. By taking $H=0$, and solving the determinantal equation, the roots are found to be $D/3 + E, D/3 - E$, and $-2D/3$. These zero-field splittings are shown at the left side of Fig. 15. For $H\|z$, one obtains the following energies:

$$W_{1,-1}^z = (D/3) \pm g\beta H \pm \frac{E^2}{g\beta H + (E^2 + g^2\beta^2 H^2)^{\frac{1}{2}}} \quad (4)$$
$$W_0^z = -2D/3.$$

Here the negative signs apply to the $|-1\rangle$ state. These energies are plotted as a function of field in Fig. 15 for naphthalene, with $D = +0.1003$, $E = -0.0137$ and $g = 2.003$.

With a particular view toward observations of triplet states of organic molecules in oriented or non-oriented solids, there is some convenience in rewriting Eq. 2 (with a Zeeman term) such that all three axes assume equal importance[4]:
$$\mathcal{H}_D = -(XS_x^2 + YS_y^2 + ZS_z^2), \text{ with } X + Y + Z = 0. \quad (5)$$
Also,
$$\mathcal{H}_D = -[(D/3) + E]S_x^2 - [(D/3) - E]S_y^2 + 2DS_z^2/3. \quad (6)$$

One notes the useful relations: $D = \frac{X+Y}{2} - Z$, $E = \frac{X-Y}{2}$.

The Hamiltonian matrix assumes an especially simple form if one chooses as basis functions the spin eigenfunctions in zero magnetic field. These are[4,5]:
$$\begin{aligned} T_x &= (\beta_1\beta_2 - \alpha_1\alpha_2)/\sqrt{2}, \\ T_y &= i(\beta_1\beta_2 + \alpha_1\alpha_2)/\sqrt{2}, \\ T_z &= (\alpha_1\beta_2 + \beta_1\alpha_2)/\sqrt{2}. \end{aligned} \quad (7)$$

α and β are defined by quantization along the z axis of the molecule.

Adding a Zeeman term to Eq. 5 the Hamiltonian matrix is now[4]:
$$\begin{pmatrix} X & -ig\beta H_z & ig\beta H_y \\ ig\beta H_z & Y & -ig\beta H_x \\ -ig\beta H_y & ig\beta H_x & Z \end{pmatrix}. \quad (8)$$

---
[1] A. D. McLachlan: Mol. Phys. 6, 441 (1963).
[2] K. W. H. Stevens: Proc. Roy. Soc. (London) A 214, 237 (1952).
[3] E. Wassermann, L. C. Snyder, and W. A. Yager: J. Chem. Phys. 41, 1763 (1964).
[4] J. H. van der Waals, and M. S. de Groot: Mol. Phys. 2, 333 (1959).
[5] H. F. Hameka, and L. J. Oosterhoff: Mol. Phys. 1, 358 (1958).

The quantity X is the following integral over spatial coordinates: $\dfrac{g^2 \beta^2}{2} \times$
$\times \left\langle \psi(1,2) \left| \dfrac{3x^2}{r^5} - \dfrac{1}{r^3} \right| \psi(1,2) \right\rangle$, with Y and Z given by similar expressions. When $H=0$, X, Y and Z are just the eigenvalues appropriate to the eigenfunctions of Eq. (7).

If now a magnetic field is applied along the Z axis, $H=H_z$, $H_x=H_y=0$ and the roots of the secular equation corresponding to (8) are:

$$W_1 = Z; \quad W_{2,3} = \frac{X+Y}{2} \pm \left[\frac{(X-Y)^2}{4} + (g\beta H)^2 \right]^{\frac{1}{2}}. \tag{9}$$

For the microwave field parallel to the static field, the transition between outermost levels of Fig. 15 ("$\Delta M = 2$") is allowed; the two usual ("$\Delta M = 1$") transi-

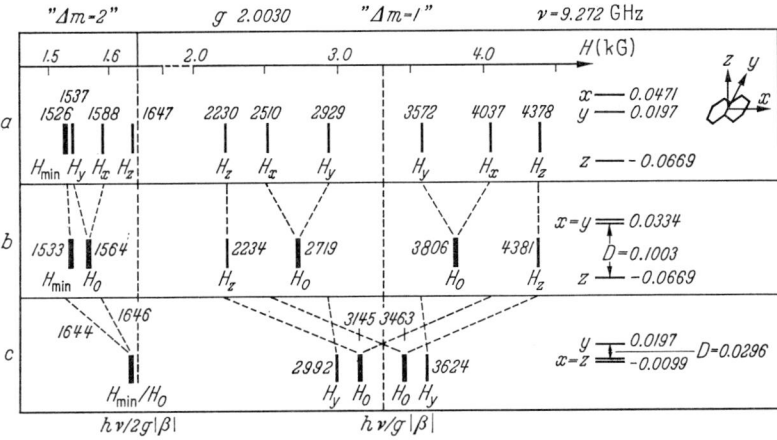

Fig. 16a—c. Stationary field values and zero-field levels for the ESR spectrum of triplet state naphthalene in dilute rigid solution. b) Stationary field values for a triplet molecule with trigonal symmetry; naphthalene rotating rapidly in its plane would give this spectrum. c) Stationary fields for tribenzotriptycene [M. S. DE GROOT, and J. H. VAN DER WAALS: Mol. Phys. **6**, 545 (1963)].

tions between adjacent levels are also shown in Fig. 15. It is usual to refer to all three as "z" transitions. From Eq. (9) one obtains at zero field either X or Y; the relative positions of levels for triplet naphthalene are shown at the right of Fig. 16a. With the static field respectively parallel to the Y or X axes, there will be one corresponding level invariant to magnetic field change; another pair of levels is obtained from Eq. (9) by cyclic permutation of axes. The levels and transitions for $H \| y$ are also shown in Fig. 15 ("y" transitions). In Fig. 16a are also displayed the "x" transitions[1] for naphthalene. It is clear from the relative positions of lines in Fig. 16a that the anisotropy is very great for the normal lines; for the "$\Delta M = 2$" transitions it is very small. The particular usefulness of Eq. (5) and matrix (8) is that a particular set of lines may be associated with the magnetic field along a specified symmetry axis. We shall note shortly that these lines may also be observed for triplet molecules in non-oriented solids. However, we shall find later that for some orientations and for low frequencies, less than nine lines are observed.

---

[1] M. S. DE GROOT, and J. H. VAN DER WAALS: Mol. Phys. **3**, 190 (1960).

19. Single crystal studies. Careful single-crystal studies with oriented guest molecules have established D, E and the principal g-tensor components for: naphthalene (72) in durene (73) or biphenyl (55); phenanthrene (74) in biphenyl; several nitrogen-substituted naphthalenes (75), (76), (11) in durene. These values are given in Table 10.

(72)　　　(73)　　　(74)

(75)　　　(76)

The sign of the parameter D for naphthalene was predicted as positive[1,2], and the prediction has been confirmed[3]. Semi-empirical calculations of D for a number of compounds in Table 10 show good agreement with experiment; the accord is much poorer for E values[4].

Phenanthrene in biphenyl, like naphthalene in durene is an oriented guest molecule of roughly the same shape as the host. An interesting aspect of the study of this system is the measurement of zero-field parameters both by studies at

Table 10. *Parameters for excited triplet systems.*

Triplet	Host	D/hc (cm^{-1})	E/hc (cm^{-1})	g_{zz}	g_{yy}	g_{xx}	Ref.
Naphthalene	Durene	+0.1003	−0.0137	2.0030	2.0030	2.0030	1–3
Anthracene		0.072	±0.007				4
Phenanthrene	Biphenyl	±0.10043	∓0.046576	2.00209	2.00279	2.0041	5
1,3,5-Triphenyl-benzene	Alphanol (glass)	0.111	0.0	2.0023			3
Triphenylene	Alphanol (glass)	0.134	0				3, 8
Coronene	Alphanol (glass)	0.096	0				3
Phenalenylium		0.053	0				2, 4
Quinoline	Durene	±0.1030	∓0.0162	2.0019	2.0029	2.0040	6
iso-Quinoline	Durene	±0.1004	∓0.0117	2.003	2.003	2.003	6
Quinoxaline	Durene	±0.1007	∓0.0182	2.0019	2.0030	2.0047	7
M-Center	KCl	−0.0172	+0.0058	1.998	1.998	1.998	9

[1] C. A. Hutchison jr., and B. W. Mangum: J. Chem. Phys. **34**, 908 (1961).
[2] A. W. Hornig, and J. S. Hyde: Mol. Phys. **6**, 33 (1963).
[3] M. S. de Groot, and J. H. van der Waals: Mol. Phys. **3**, 190 (1960).
[4] J. H. van der Waals, and G. ter Maten: Mol. Phys. **8**, 301 (1964).
[5] R. W. Brandon, R. E. Gerkin, and C. A. Hutchison jr.: J. Chem. Phys. **41**, 3717 (1964).
[6] J. S. Vincent, and A. H. Maki: J. Chem. Phys. **42**, 865 (1965).
[7] J. S. Vincent, and A. H. Maki: J. Chem. Phys. **39**, 3088 (1963).
[8] J. H. van der Waals, and M. S. de Groot: Mol. Phys. **2**, 333 (1959).
[9] H. Seidel: Phys. Letters **7**, 27 (1963). — H. Seidel, M. Schwoerer u. D. Schmid: Z. Physik **182**, 398 (1965).

[1] M. Gouterman: J. Chem. Phys. **30**, 1369 (1959).
[2] R. McWeeny: J. Chem. Phys. **34**, 399 (1961).
[3] A. W. Hornig, and J. S. Hyde. Mol. Phys. **6**, 33 (1963).
[4] J. H. van der Waals, and J. ter Maten: Mol. Phys. **8**, 301 (1964).

high frequency (23 GHz) and at frequencies corresponding to very low fields (3 G to 76 G)[1]. In the usual $S=1$ spin Hamiltonian, upon setting the external field equal to zero one gets transitions $D+E$ (or $Y-Z$), $D-E$ (or $X-Z$) and zero (having added $2D/3$ to our previous solutions). The corresponding resonant frequencies are $(D+E)/hc$, $(D-E)/hc$ and $2E/hc$. For a field of $\sim$10 G, small corrections suffice to give a better D value and an E value with at least as good precision as the high-frequency measurements. The parameters are given in Table 10.

Hyperfine structure has been observed in a number of triplet spectra; a five-line spectrum of naphthalene at 77 °K was the first example[2,3]. The splitting is due to the equivalent 1, 4, 5, 8 protons. At 1.65 °K some additional splitting is discernible[4]. By substitution of deuterons for some of the protons, it has been possible to evaluate spin densities at the several non-equivalent positions as follows[5]:

$$\varrho_\alpha(1, 4, 5, 8 \text{ positions}) = +0.219,$$
$$\varrho_\beta(2, 3, 6, 7 \text{ positions}) = +0.062,$$
$$\varrho_\gamma(9, 10 \text{ positions}) = -0.063.$$

For the α positions, the proton spin densities in the excited triplet state are essentially identical with those in the anion. This had been predicted both by extended Hartree-Fock and configurational mixing theories for the naphthalene, pyrene (15) and chrysene (77) molecules[6]. However, the ratio $\varrho_\alpha/\varrho_\beta$ for the triplet

(77)   (78)

is 4/3 that of the negative ion. Hyperfine splitting is also observed for phenanthrene in biphenyl[7]. The interpretation is consistent with theoretical predictions[8,9] as well as with spin densities from phenanthrene anion studies[10]. Hyperfine splitting has also been observed in some of the nitrogen-substituted naphthalenes of Table 10[11].

In such rigid systems one is concerned with understanding the anisotropy of hyperfine splitting. It is found that the same hyperfine tensor components which were deduced for a C—H fragment by studies on the malonic acid radical[12] also apply to proton hyperfine splittings in excited triplet state molecules.

Several defects in solids are important examples of species with singlet ground and triplet excited states. One is the M-center in the alkali halides, in which two

---

[1] R. W. Brandon, R. E. Gerkin, and C. A. Hutchison jr.: J. Chem. Phys. 41, 3717 (1964).
[2] C. A. Hutchison jr., and B. W. Mangum: J. Chem. Phys. 34, 908 (1961).
[3] A. Schmillen, u. G. v. Foerster: Z. Naturforsch. A 16, 320 (1961).
[4] See footnote 3, p. 210.
[5] N. Hirota, C. A. Hutchinson jr., and P. Palmer: J. Chem. Phys. 40, 3717 (1964).
[6] A. D. McLachlan: Mol. Phys. 5, 51 (1962).
[7] R. W. Brandon, R. E. Gerkin, and C. A. Hutchinson jr.: J. Chem. Phys. 41, 3717 (1964).
[8] A. T. Amos: Mol. Phys. 5, 91 (1962).
[9] A. D. McLachlan: Mol. Phys. 3, 233 (1960).
[10] S. H. Glarum, and L. C. Snyder: J. Chem. Phys. 36, 2989 (1962).
[11] J. S. Vincent, and A. H. Maki: J. Chem. Phys. 42, 865 (1965).
[12] T. Cole, C. Heller, and H. M. McConnell: Proc. Natl. Acad. Sci. U.S. 45, 525 (1959).

adjacent vacancies each trap one electron[1]. A second is the F_t center in CaO or MgO, also involving pairs of electrons in vacancies[2,3]. In alkali halides, the R center consists of three electrons trapped in three adjacent sites. It has a doublet ground state but an excited quartet state which is described by the spin-spin Hamiltonian of Eq. (1) or (2), with $S=3/2$[4]. These systems are described in Sect. 26 on Defects.

We now proceed to consider another representation which graphically shows how one may obtain less than nine lines even if $E=0$[5]. It also induces caution in the use of the terms "$\Delta M=1$" or "$\Delta M=2$" applied to these systems. From the secular determinant corresponding to (8), one may write[5]:

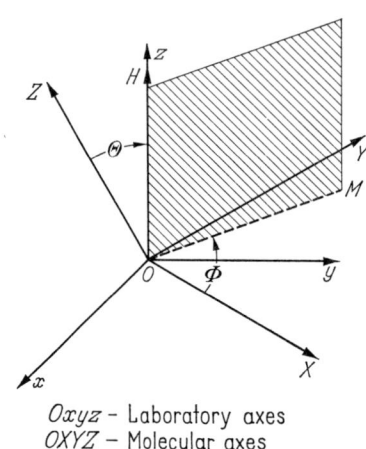

$Oxyz$ – Laboratory axes
$OXYZ$ – Molecular axes

Fig. 17. Euler angles θ and φ for Eq. (19.1). OM is the projection of Oz on the XY plane.

$$\left.\begin{aligned}E^3 - E[(g\beta H)^2 - \\ -(XY+XZ+YZ)]+ \\ +g^2\beta^2 H^2[X\sin^2\theta\cos^2\varphi+ \\ +Y\sin^2\theta\sin^2\varphi+ \\ +Z\cos^2\theta - XYZ]=0.\end{aligned}\right\} \quad (19.1)$$

Here θ and φ represent the Euler angles of Fig. 17. Using a microwave quantum of energy $h\nu$, the separation between any pair of levels may be written in the form:

$$\left.\begin{aligned}X\sin^2\theta\cos^2\varphi + Y\sin^2\theta\sin^2\varphi + Z\cos^2\theta \\ = XYZ(g\beta H)^{-2} + 3^{-\frac{3}{2}}[(g\beta H)^{-2}(h^2\nu^2+XY+XZ+YZ)-1] \\ \times[(4g\beta H)^2 - h^2\nu^2 - 4(XY+XZ+YZ)]^{\frac{1}{2}} = F(H,\nu).\end{aligned}\right\} \quad (19.2)$$

If one selects a molecule with known parameters X, Y and Z and specifies a frequency ν, then $F(H,\nu)$ may be plotted as a function of H, as in Fig. 18 for naphthalene. Here $\nu=9.279$ GHz, $X=0.0197$, $Y=0.0471$ and $Z=-0.0669$ cm^{-1}, with an average g value of 2.0030. (Here the x-axis is taken is the shorter axis of the molecule.) Then for $\theta=\pi/2$, $\varphi=0$, $F(H,\nu)=X$. By drawing a dotted line at this ordinate, one obtains three intersections corresponding to the three X transitions in Fig. 18; one proceeds similarly for Y and Z. From Figs. 19 and 20 one notes that at 3 GHz one has two transitions at every orientation, while at 750 MHz one has two transitions only for a range of orientations. Returning to Fig. 18, it is apparent that there is a minimum possible resonance field H_{min} at this frequency, occurring when the square root factor of Eq. (19.2) becomes non-real, in this case at 1527 G. H_{min} is also shown in Fig. 16. The practical significance of H_{min} at high frequencies is that for lines of reasonable width there are a large number of orientations which would contribute intensity to a line at this position. That there are field maxima also attained may be seen by writing Eq. (19.2) as $f(\theta,\varphi)=F(H,\nu)$ and differentiating: $df=F'(H,\nu)dH$, or $dH=df/F'$.

---

[1] H. SEIDEL: Phys. Letters **7**, 27 (1963).
[2] D. H. TANIMOTO, W. M. ZINIKER, and J. C. KEMP: Phys. Rev. Letters **14**, 645 (1965).
[3] B. HENDERSON: Brit. J. Appl. Phys. **17**, 851 (1966).
[4] H. SEIDEL, M. SCHWOERER u. D. SCHMID: Z. Physik **182**, 398 (1965).
[5] P. KOTTIS, and R. LEFEBVRE: J. Chem. Phys. **39**, 393 (1963).

A stationary field is attained when $df=0$ with F' finite, and this occurs whenever H is along a principal axis of the molecule. For each of these special orientations there will be three stationary fields. These are also seen in Fig. 18. The existence

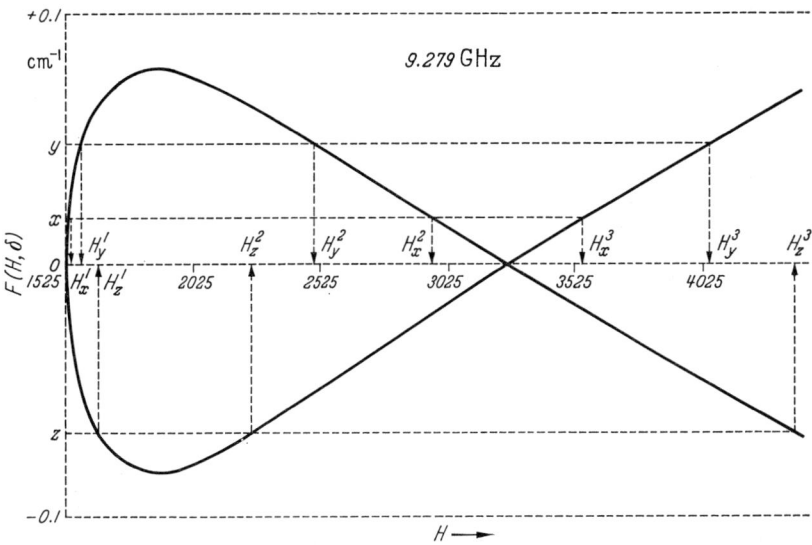

Fig. 18. Plot of $F(H, \delta)$ vs. H for naphthalene at 9.279 GHz. The three X, Y and Z transitions are shown. (Note that X and Y are interchanged as compared with Fig. 16).

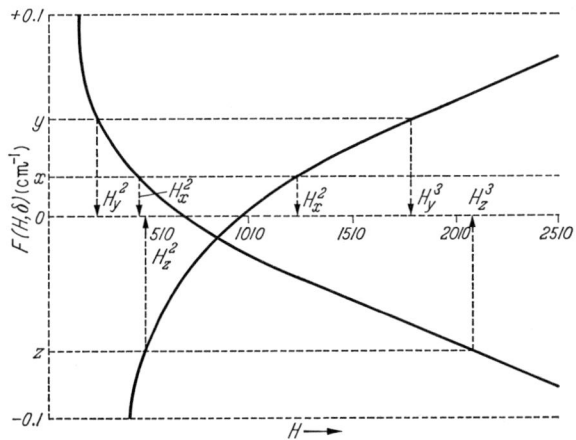

Fig. 19. The same function for $\nu = 3$ GHz.

of stationary fields both in the "$\Delta M=2$" and "$\Delta M=1$" regions thus makes it reasonable to hope for structure in the triplet spectrum of a randomly oriented sample.

The maximum intensity for $\Delta M=2$ lines is found when the static field is parallel to the y axis, being 25 percent of that for the allowed transitions[1]. When

---
[1] J. H. VAN DER WAALS, and M. S. DE GROOT: Mol. Phys. 2, 333 (1959).

the static field is along the z axis, the intensity of the forbidden line is only 2 per cent of that for the allowed lines. It is clear that one may hope to see the $\Delta M = 2$ transition even in a glass such as is typically used for phosphorescence studies. However, when $E \neq 0$, from such lines one is able to evaluate only $(D^2 + 3E^2)^{\frac{1}{2}}$ instead of the separate parameters. The intensity of the $\Delta M = 1$ transitions in a glassy solid is usually considerably less than for the strongest $\Delta M = 2$ line.

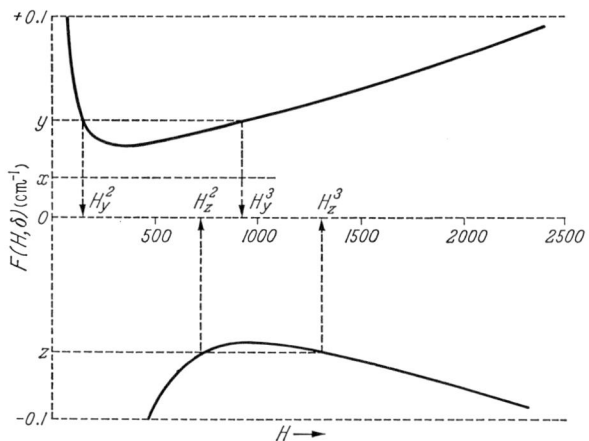

Fig. 20. The same function for $\nu = 750$ MHz [from P. KOTTIS, and R. LEFEBVRE: J. Chem. Phys. **39**, 393 (1963)].

20. Line shapes for non-oriented systems. We have noted that such ESR structure is observed in glassy solids at 77 °K or lower, using cyclohexane or other inert solvent which does not quench phosphorescence. Some measurements have been made at room temperature using a methylmethacrylate plastic[1]. However, at such temperatures diffusion of oxygen into the plastic limits the triplet lifetime.

One may proceed in various ways to obtain line shapes for triplet-state molecules in random orientation[2-4]. Approximate line shapes may be obtained by elementary considerations[3]. For exacting comparisons with spectra, one may use computer simulations[3,4].

The line shape for a random array of parallel interacting dipoles is sharply peaked. The neighbors in the "equatorial belt" around a specific dipole aligned with H_0 will provide field contributions in a sense opposite to it; those more nearly along it will contribute an aiding field, but the relative number of such is much smaller than that of the former group. We next consider a dipole opposed to the field and again get similar contribution from neighboring dipoles, but the larger contribution now comes at low field. The superposition of the local field contributions gives the absorption line shape shown in Fig. 21 or the derivative shape of Fig. 22b. One notes that this shape is closely analogous to that for the proton NMR line of a powdered hydrate, in which nuclear dipoles are closely spaced in pairs while contributions from other pairs of protons is small[5].

---

[1] C. THOMSON: J. Chem. Phys. **41**, 1 (1964).
[2] M. S. DE GROOT, and J. H. VAN DER WAALS: Mol. Phys. **3**, 190 (1960).
[3] E. WASSERMAN, L. C. SNYDER, and W. A. YAGER: J. Chem. Phys. **41**, 1763 (1964).
[4] See footnote 5, p. 212.
[5] G. E. PAKE: J. Chem. Phys. **16**, 327 (1948).

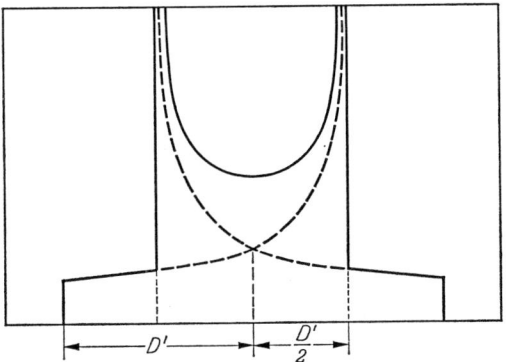

Fig. 21. Absorption line shapes for a $\Delta M = 1$ transition of a randomly-oriented triplet state molecule with $E = 0$ and for zero line width.

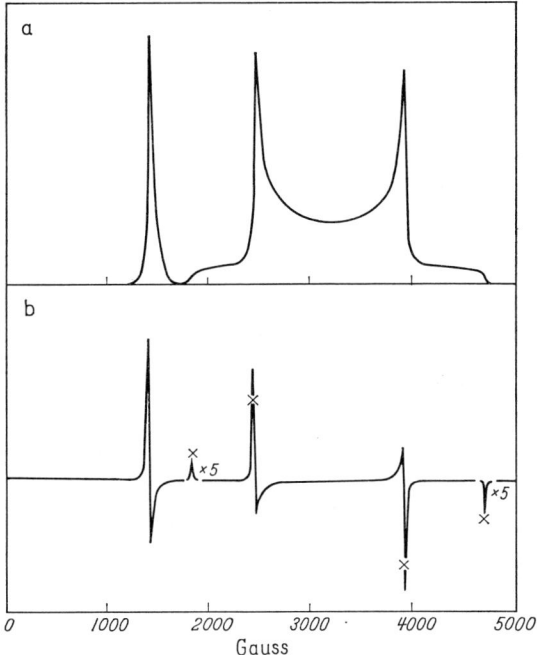

Fig. 22a and b. a) ESR absorption lines for a randomly-oriented triplet state molecule of axial symmetry [dianion of triphenylene, (79)] with $D = 0.1341$ cm^{-1}, $g = g_e$ and a line width of 15 G. b) Derivative line shape corresponding to 22a. [Figs. 22—25 from E. WASSERMAN, L. C. SNYDER, and W. A. YAGER jr.: Chem. Phys. **41**, 1763 (1964)].

If the triplet molecule has trigonal symmetry, (e.g., triphenylene (79), or triphenylbenzene (80), then four peaks are observed as in Fig. 22b, corresponding to the positions of the vertical lines of Fig. 21. When the triplet molecule lacks cylindrical symmetry, as in the case of naphthalene, one no longer has coincidence of lines from centers with axes in x and y directions: therefore, one requires the non-axial parameter E as well to describe the spectrum. We proceed to a somewhat more detailed consideration of line shape.

The allowed $0 \to 1$ and $-1 \to 0$ transitions occur approximately at fields:

$$H_r = H_0 \left[\frac{l^2}{2}(D' - 3E') + \frac{m^2}{2}(D' + 3E') - n^2 D'^2 \right]. \tag{20.1}$$

Here $H_0 = h\nu/g_e\beta$, $D' = D/g_e\beta$ and $E' = E/g_e\beta$. Thus for H respectively parallel to the x, y and z axes, one would get pairs of lines separated by $(D' - 3E')$, $(D' + 3E')$ and $2D'$. The lines are usually referred to by specifying the axes. More exactly, the x, y and z lines are given by:

$$H_x^2 = \frac{g_e^2}{g_x^2}(H_0 \mp D' \pm E')(H_0 \pm 2E'), \tag{20.2}$$

$$H_y^2 = \frac{g_e^2}{g_x^2}(H_0 \mp D' \mp E')(H_0 \mp E'), \tag{20.3}$$

$$H_z^2 = \left(\frac{g_e}{g_x}\right)^2 [(H_0 \mp D')^2 - E'^2]. \tag{20.4}$$

Here it is assumed that the g-tensor axes coincide with those of the zero-field splitting tensor.

We consider first the case $E = 0$, implying equivalent x and y axes, and make use of the approximate resonant fields Eq. (20.1) for $\Delta M = 1$ transitions. These may be written as:

$$H_r = H_0 \pm [D'/2)\sin^2\theta - D'\cos^2\theta]. \tag{20.5}$$

Changing to polar coordinates, one expects the line intensity to be directly proportional to $\sin\theta$ and inversely proportional to $\partial H_r/\partial\theta$:

$$I(H) \propto \frac{\sin\theta}{\partial H_r/\partial\theta} \tag{20.6}$$

and

$$I(H) \propto [(D'/2) \mp (H - H_0)]^{-\frac{1}{2}}. \tag{20.7}$$

Since the z lines occur at $H - H_0 = \pm D'$, the ESR intensity will be zero at greater or lesser fields. We have noted that $\partial H_r/\partial\theta = 0$ when the field is along a symmetry axis such as z. At $H = H_0 + D'/2$ the field will lie in the xy plane and all triplet molecules so oriented will contribute to the intensity. The resultant superposition of curves is shown in Fig. 21. Upon introducing broadening and investigating the behavior of the derivative curves, one notes in Fig. 22 the absorption-curve-like appearance of lines for those molecules for which H lies along the z axis; the double-cusped appearance of lines satisfying the resonance condition when H lies in any orientation in a plane is likewise expected[1]. One may see this clearly in Fig. 22 for a randomly oriented paramagnetic center with axial symmetry.

When $E \neq 0$, one would expect six field values for which $\partial H_r/\partial\theta = 0$. When D' and E' are assumed to have the same sign, in addition to the outer steps at

[1] J. A. WEIL, and H. G. HECHT: J. Chem. Phys. **38**, 281 (1963).

$\pm D'$, a second pair of steps Fig. 23 occurs at $\pm(D'+3E')/2$ from molecules for which $y\|H$. The unbounded increase in absorption occurs at $\pm(D'-3E')/2$ for molecules with $x\|H$. Computed absorption and derivative curves are given in Fig. 24.

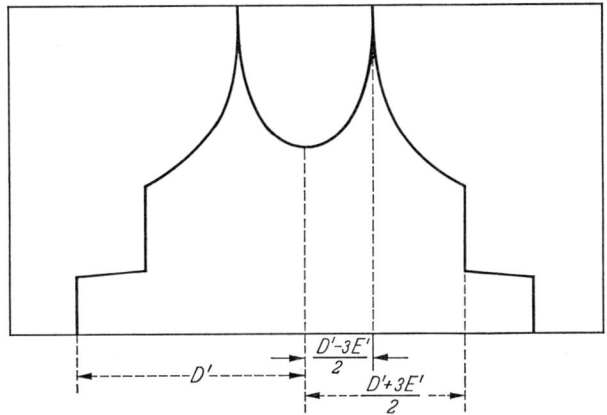

Fig. 23. Absorption line shape for randomly-oriented triplet state molecules with $E \neq 0$ and zero line width.

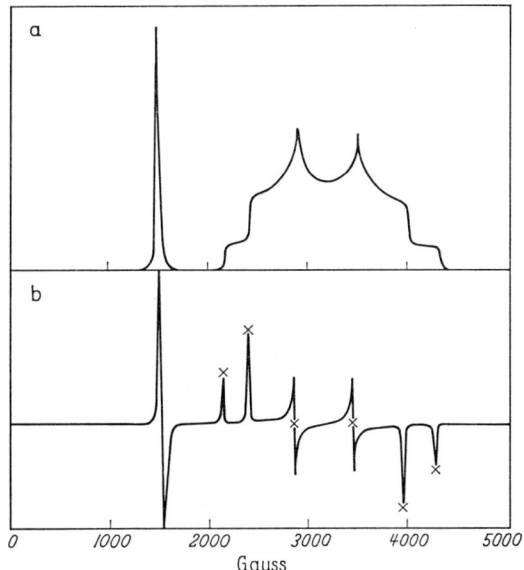

Fig. 24 a and b. a) ESR absorption lines for naphthalene ($D = 0.10046$, $E = 0.01536$ cm^{-1}), taking $g = g_e$ and a line width of 8 G. b) Derivative line shape corresponding to Fig. 24a.

Both Figs. 22 and 24 represent computer simulations[1], a most valuable technique for checking assignments, especially when D and E parameters are large.

Fig. 25 gives an indication of the change in number of lines observed as D is altered at constant E[1]. One notes that the low-field x and y lines are no longer

---

[1] See footnote 3, p. 214.

seen after D exceeds about 0.33. Note the reappearance of a low-field z line after vanishing for an interval. The diphenylmethylene indicated represents a ground-state triplet which will be discussed shortly.

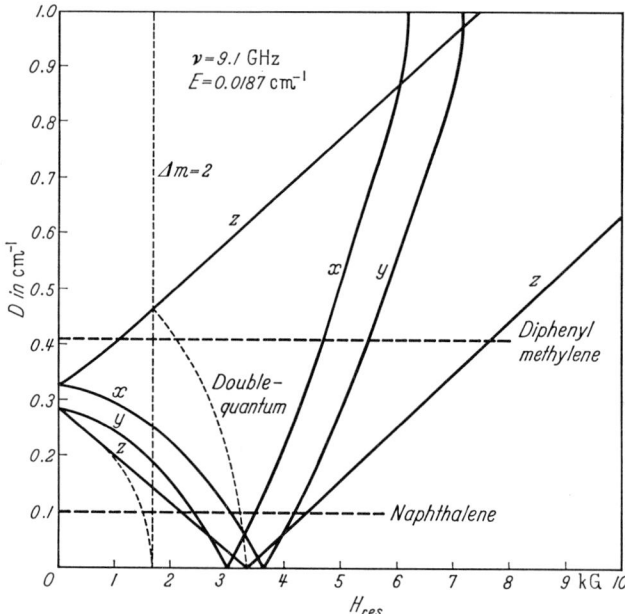

Fig. 25. "Stationary" fields as a function of D for $E = 0.0187$ cm^{-1}. Diphenylmethylene and naphthalene D values are shown.

21. Transfer of triplet excitation. Studies of excited triplet ESR spectra have contributed details about the fate of triplet excitation energy. Some of the possible processes investigated by ESR techniques follow.

1. Intramolecular intersystem conversion. In the case of naphthalene, the transition is $^3B_{2u} \rightarrow {}^1A_g$. It was shown early that the decay rate of the ESR signal is also that of the phosphorescence[1]. When both the naphthalene and the host matrix durene are completely deuterated, the mean life of the triplet state increases from 2.1 sec to 16.9 sec. Although the hyperfine anisotropy is marked, the lifetime is independent of field orientation; this indicates that the lifetime change is not a nuclear hyperfine effect[1]. Halogen atom substituents in the matrix increase the radiative decay constant more than the non-radiative[2]. Increase in singlet character of the triplet due to increased spin-orbit interaction is presumably responsible.

2. Excitation to a higher triplet state. Decrease in intensity of the naphthalene triplet spectrum on steady illumination is ascribed to such excitation followed by energy transfer to solvent and to free radical production[3,4].

3. Transfer of excitation. A systematic study of the single crystal system phenanthrene (fully deuterated)-diphenyl-naphthalene, has shown the transfer of

---
[1] C. A. HUTCHINSON, and B. W. MANGUM: J. Chem. Phys. **32**, 1261 (1960).
[2] S. SIEGEL, and H. S. JUDEIKIS: J. Chem. Phys. **42**, 3060 (1965).
[3] S. SIEGEL, and K. EISENTHAL: J. Chem. Phys. **42**, 2494 (1965).
[4] S. SIEGEL, and H. JUDEIKIS: J. Chem. Phys. **43**, 343 (1965).

triplet excitation from the phenanthrene to the naphthalene via the diphenyl[1,2]. The three molecules have ^3B states respectively 21,410, 23,010 and 21,100 cm^{-1} above their ^1A ground states. Thus transfer from phenanthrene to a diphenyl molecule in its singlet ground state (D_S) requires a vibrationally excited triplet level (P_{T*}):

$$P_{T*}+D_S \rightarrow P_S+D_T.$$

Careful measurements on separate phenanthrene-diphenyl and naphthalene-diphenyl crystals showed that the zero-field parameters in the three-component system were the same as in the two-component system, notwithstanding their sensitivity to change in environment. This demonstrates the absence of phenanthrene-naphthalene complexes. Below 85 °K the transfer of triplet excitation from phenanthrene (mean life 10 sec) to naphthalene was so fast that within 0.05 sec of cessation of illumination the naphthalene triplet decay rate was not measurably different from that in the absence of phenanthrene.

The excitation transfer in systems of very high viscosity ($>10^7$ P) appears to involve a direct electron exchange over a radius of up to 2 nm; at lower viscosities the rate is diffusion-controlled and inversely proportional to viscosity[3].

(81) (82)

In pure benzene, triplet excitation is transferred between molecules at a rate of the order of 10^{12} sec^{-1} [4]. The behavior of benzene itself is complicated by elongation of the molecule in the first triplet excited state, since both three- and sixfold symmetry are lost[5]. Distortion involving elongation of two opposite bonds had been predicted earlier[6,7]. We shall return to benzene itself after considering molecules which retain threefold symmetry in the triplet state. It has proved very interesting to study intramolecular transfer between benzene or naphthalene units rigidly held together in trigonal symmetry (D_{3h}) by two approximately tetrahedral carbons[5]. The molecules studied are triptycene (81) and tribenzotriptycene (82) respectively.

The triplet ESR spectrum of a rigid solution of a molecule not having a symmetry axis of order 3 or higher (such as naphthalene) is shown in Fig. 16a[5]. Turning to a molecule with a trigonal axis and for which the D parameter is nearly the same (Fig. 16b), one notes that the H_x and H_y lines have merged into a single degenerate line for each of the three groups. The same spectrum would also be observed for naphthalene if it were rotating about its z-axis (i.e., in the molecular plane) at a sufficiently high rate.

The first ultraviolet bands of triptycene have been shown to involve transitions in the separate π-subsystems[8]. It is believed that triplet excitation in both

---

[1] C. A. HUTCHINSON jr., and B. W. MANGUM: J. Chem. Phys. **37**, 447 (1962).
[2] N. HIROTA, and C. A. HUTCHINSON jr.: J. Chem. Phys. **42**, 2869 (1965).
[3] B. SMALLER, E. C. AVERY, and J. R. REMKO: J. Chem. Phys. **43**, 922 (1965).
[4] G. C. NIEMAN, and G. W. ROBINSON: J. Chem. Phys. **37**, 2150 (1962). — H. STERNLICHT, G. C. NIEMAN, and G. W. ROBINSON: J. Chem. Phys. **38**, 1326 (1963).
[5] M. S. DE GROOT, and J. H. VAN DER WAALS: Mol. Phys. **6**, 545 (1963).
[6] G. N. LEWIS, and M. KASHA: J. Am. Chem. Soc. **66**, 2100 (1944).
[7] O. REDLICH, and E. K. HOLT: J. Am. Chem. Soc. **67**, 1228 (1945).
[8] C. F. WILCOX: J. Chem. Phys. **33**, 1874 (1960).

triptycene and tribenzotriptycene likewise occurs within a single subsystem. If such excitation remains localized in one of the naphthalene sub-units of tribenzotriptycene for a time of the order of 10^{-9} sec, one would expect the ESR spectrum to resemble that of ordinary naphthalene. However, rapid excitation transfer among the three subunits will cause a merging of the H_x and H_y components, as well as of the low-field lines (Fig. 16c). An analogous behavior might be hoped for in the triptycene case. The ESR spectrum of the latter at 77 °K is just that expected for trigonal symmetry, with $D=0.135$ cm^{-1} and $E=0$[1]. However, at 20 °K the spectrum broadens and shifts to low field, presumably due to localization of excitation. At such temperatures there is apparently insufficient vibrational energy to effect the distortion accompanying excitation of another sub-unit. The triplet excitation transfer may be described in terms of excitons[2]. A calculation of line shape for oriented triptycene[3] shows agreement with experiments done on the glassy solution. (The calculation procedure is applicable also to a system of spin $\frac{1}{2}$ in which an electron jumps between sites with different g tensors.) Tribenzotriptycene behaves in rather analogous fashion to triptycene, but a vestigal line at 20 °K indicates that excitation transfer is still detectable[4].

The configuration of the first excited triplet state of benzene has been assumed[5,6] to be non-planar and to have D_{2h} symmetry. The low-field ESR spectrum of triplet-state benzene should show two lines if it has trigonal symmetry but only the H_{min} peak is observed. Its spectrum bears a resemblance to that of naphthalene. Thus triplet benzene cannot have trigonal symmetry. If it remained in one distorted conformation for a time long compared with the frequency separation of the H_x and H_y components of Fig. 16a one would expect these to appear, but they do not. Yet the interconversion rate is sufficient to give the averaged line H_0 of Fig. 16b. At 20 °K there still appears to be some motional process which is presumed to be tunnelling between conformational isomers; these recall an earlier suggestion of a "quinoid" triplet benzene structure involving two long bonds.

A somewhat different system showing migration of triplet excitation would be an infinite linear parallel array of spin $\frac{1}{2}$ systems with marked alternation in spacing. A model Hamiltonian would be that for such an alternating array of hydrogen atoms. For this system one would write[2,7],

$$\mathcal{H} = J_i \sum_{(odd)} \mathbf{S}_i \cdot \mathbf{S}_{i+1} + J'_j \sum_{(even)} \mathbf{S}_j \cdot \mathbf{S}_{j+1} + \hbar\omega \sum_{k=1,2,3} S_k +$$
$$+ \sum_{i(odd)} \mathbf{S}_i \cdot T \cdot \mathbf{S}_{i+1} + \sum_{j(even)} \mathbf{S}_j \cdot T' \cdot \mathbf{S}_{j+1} + \hbar a \sum_{k=1,2,3} \mathbf{S}_k \cdot \mathbf{I}_k.$$

The first two terms are exchange interactions between hydrogen atoms separated by distances l_1 and l_2 respectively, while the fourth and fifth are the spin-spin interactions for the corresponding atoms. The third is the Zeeman term, while the last is the hyperfine term which may be ignored because the excitation extends over many nuclei. If $J \gg J' \ll kT$, the zero-field splitting will be determined by the interaction between the more closely spaced atom pairs. The Hamiltonian predicts a pair of lines separated by $2D$ (Fig. 26).

---

[1] See footnote 5, p. 219.
[2] C. H. STERNLICHT, and H. M. McCONNELL: J. Chem. Phys. 35, 1793 (1961).
[3] A. HUDSON, and A. D. McLACHLAN: J. Chem. Phys. 43, 1518 (1965).
[4] See footnote 5, p. 219.
[5] G. N. LEWIS, and M. KASHA: J. Am. Chem. Soc. 66, 2100 (1944).
[6] O. REDLICH, and E. K. HOLT: J. Am. Chem. Soc. 67, 1228 (1945).
[7] H. M. McCONNELL: Molecular Biophysics (B. PULLMAN, ed.), p. 311. New York: Academic Press 1965.

Several free radical salts at low temperatures form linear arrays with alternating spacings and they show the two-line spectrum predicted for triplet excitons. These include Würster's blue cation $[(H_3C)_2-N-C_6H_4-N-(CH_3)_2]^+$ and also various salts involving tetracyanoquinodimethane anion

$$(TCNQ^-, [(NC)_2-C=C_6H_4=C-(CN)_2]^-)^{1-3}.$$

In such salts the coupling of excitons to lattice phonons is expected to be so large that an extensive nuclear displacement follows the exciton as the latter moves along ali near chain[4]. Phase transitions in such salts have been attributed to exciton interactions[4,5].

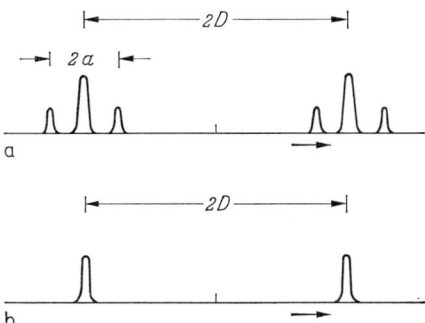

Fig. 26 a and b. a) High-field ESR spectrum expected for oriented hydrogen molecules in their triplet states. b) Spectrum of triplet excitons moving along an infinite alternating linear array of hydrogen atoms. [H. M. McConnell: Molecular Biophysics, p. 311. (B. Pullman, ed). New York: Academic Press 1965.

III. Ground state triplet species.

The preponderance of molecules having $^1\Sigma$ ground states makes it natural to regard a triplet state as an excited state. However, a respectable number of molecules are known to have a ground triplet state, with a higher-lying associated singlet state. One of the simplest examples in principle is methylene (CH_2), which has a linear ground state $^3\Sigma_g^-$ [6]. From experience with excited triplet states one expects and observes ESR absorption in rigid ground triplet systems at 77 °K or lower. Low temperatures not only avoid excitation to the singlet state but also help to prevent loss of such highly reactive entities by interaction with oxygen, solvent, or free radicals which may be present[7]. For excited triplet molecules such as naphthalene, the triplet ESR signal disappears, after cessation of excitation, with the decay constant of the phosphorescence. However, ground state triplet molecules are very stable at low temperatures (in the absence of oxygen) after formation. The usual limitation on the steady-state concentration of excited triplets imposed by the limited transmission of exciting light in some matrices does not apply to ground-triplet molecules. The methylene molecule itself may be made by low-temperature irradiation of diazomethane, CH_2N_2, with

---

[1] D. B. Chesnut, and W. D. Phillips: J. Chem. Phys. **35**, 1002 (1961).
[2] D. B. Chesnut, and P. Arthur jr.: J. Chem. Phys. **36**, 2969 (1962).
[3] D. D. Thomas, H. Keller, and H. M. McConnell: J. Chem. Phys. **39**, 2321 (1963).
[4] H. M. McConnell: Molecular Biophysics (B. Pullman, ed.), p. 311. New York: Academic Press 1965.
[5] D. B. Chesnut: J. Chem. Phys. **40**, 405 (1964).
[6] G. Herzberg: Proc. Roy. Soc. (London) A **262** 291 (1961).
[7] R. W. Murray, A. M. Trozzolo, E. Wasserman, and W. A. Yager: J. Am. Chem. Soc. **84**, 3213 (1962).

loss of nitrogen. Analogues with substituents in place of the hydrogen in CH_2N_2 may be made similarly. These "divalent carbon" compounds are referred to as carbenes. A triplet ESR spectrum of CH_2 itself at this writing has not been observed; the closest analog for which it has been is $H-C-C\equiv N$[1]. Most of the ground state triplet molecules observed have substituents with π- (or pseudo π-) electron systems and therefore have opportunities for electron delocalization. Table 11 lists parameters for some ground triplet state molecules in random orientation. The same spin Hamiltonian used for excited triplet systems is equally applicable to ground triplet systems. Likewise, the interpretation of single-crystal and of powder spectra is essentially identical with that of excited triplet systems.

A class of ground state triplet molecules analogous to the methylenes is the set of substituted nitrenes, (made from azides RN_3, such as phenyl nitrene) (83):

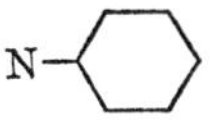

(83)

Of considerable interest are also the dicarbenes (84) and dinitrenes (85)[2], of which the following are the simplest examples:

(84) (85)

$|D|=0.0521, E<0.002$ $|D|=0.0675, E\sim 0$

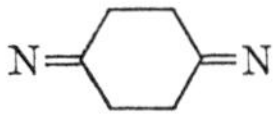

Another type of system which permits a triplet ground state is the dinegative ion with a trigonal or higher symmetry axis (group D_{nh}). Such molecules, in the HMO approximation, have degenerate lowest antibonding orbitals, each of which accommodates one electron. Examples are triphenylbenzene (80) or triphenylene (79)[1]. We have already noted that the excited triplet state has been observed for both neutral molecules. A smaller value of $D/hc=0.042$ cm^{-1} for the triphenylbenzene dinegative ion (ground triplet) as compared with $D/hc=0.111$ cm^{-1} for the neutral excited triplet state arises from charge correlation which keeps the electrons farther apart in the dinegative ion[3]. A cation with one electron in each of the two degenerate bonding orbitals is pentaphenylcyclopentadiene cation[4] (86):

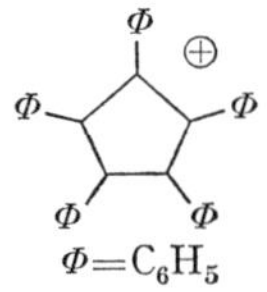

$\Phi=C_6H_5$

(86)

A triplet ground state may also result from interaction of two separated $S=\frac{1}{2}$ systems. At 77 °K a rigid glass containing ketyl anions plus alkali ions shows a typical triplet spectrum suggestive of a well-defined structure[5]: (54).

---

[1] E. WASSERMAN, L. BARASH, and W. A. YAGER: J. Am. Chem. Soc. **87**, 2075 (1965).
[2] A. M. TROZZOLO, R. W. MURRAY, G. SMOLINSKY, W. A. YAGER, and E. WASSERMAN: J. Am. Chem. Soc. **85**, 2526 (1963).
[3] R. E. JESSE, P. BILOEN, R. PRINS, J. D. W. VAN VOORST, and G. J. HOIJTINK: Mol. Phys. **6**, 633 (1963).
[4] R. BRESLOW, H. W. CHANG, and W. A. YAGER: J. Am. Chem. Soc. **85**, 2033 (1963).
[5] N. HIROTA, and S. I. WEISSMAN: J. Am. Chem. Soc. **86**, 2538 (1964).

The D/hc values for such triplet double-ion-pair combinations are small, ranging from 0.007 to 0.015 cm^{-1} [1]. One alkaline earth ion can replace two alkali ions, either for ketyls or for two radical anions, each containing two nitrogen atoms, e.g., bipyridine anion (87). In the latter example, the nearly-identical D and E

(87)

values for Be and for Mg chelates suggests that repulsion between ligands rather than size of cations governs the closeness of approach. In this same case the low value of E (0.0000 to 0.0012 cm^{-1}) leads to the suggestion that the metal ion occupies the center of a tetrahedron of nitrogen atoms[2].

Table 11.

Molecule	D*** (cm^{-1})	E (cm^{-1})	Reference
HCH	0.9055 (calc)	0	1
	1.1 (calc)	0	2
H—C—C≡N	0.8629	0	3
H—C—CF$_3$	0.72	0.021	4
H—C—C$_6$H$_5$	0.5150	0.0251	5
H—C—C≡CH	0.6276	0	3
H—C—C≡C—CH$_3$	0.6263	0	3
*H—C—C≡C—C$_6$H$_5$	0.5413	0.0035	3
H—C—C≡C—C≡C—CH$_3$	0.6087	0	3
**H—C—C≡C—C≡C—C$_6$H$_5$	0.5530	0	3
H—C—C≡C—C≡C—C(CH$_3$)$_3$	0.6055	0	3
N≡C—C—C≡N	1.002	<0.002	6
:N—C≡N	1.544	<0.002	6
:C=N$^+$=N$^-$	1.153	<0.002	6
C$_6$H$_5$—C—C$_6$H$_5$	0.4055	0.0194	5
(C$_6$H$_5$)$_3$C$_6$H$_3$(1,3,5)$^{2-}$	0.042	0	7

* Can be non-linear and hence have non-zero value of E.
** Note reduction in D value when electrons can be delocalized into ring (also * molecules).
*** As with excited triplets, these values are usually obtained by assuming $g = 2.0023$.

[1] J. HIGUCHI: J. Chem. Phys. **38**, 1237 (1963).
[2] R. D. SHARMA: J. Chem. Phys. **38**, 2350 (1963); **41**, 3259 (1963).
[3] R. A. BERNHEIM, R. J. KEMPF, J. V. GRAMAS, and P. S. SKELL: J. Chem. Phys. **43**, 196 (1965).
[4] E. WASSERMAN: J. Chem. Phys. (to be published) [cited in Ref. [3]].
[5] E. WASSERMAN, A. M. TROZZOLO, W. A. YAGER, and R. W. MURRAY: J. Chem. Phys. **40**, 2408 (1964).
[6] E. WASSERMAN, L. BARASH, and W. A. YAGER: J. Am. Chem. Soc. **87**, 2075 (1965).
[7] R. E. JESSE, P. BILOEN, R. PRINS, J. D. W. VAN VOORST, and G. H. HOIJTINK: Mol. Phys. **6**, 633 (1963).

Just as for excited triplet state species, one may study ground state triplet molecules oriented in single crystals. Diphenylmethylene (C$_6$H$_5$—C—C$_6$H$_5$) formed by irradiation (525 nm) of diphenyldiazomethane (C$_6$H$_5$—C(N$_2$)—C$_6$H$_5$) in a benzophene crystal gives ESR lines with anisotropy corresponding to that of the orthorhombic host[3]. The g tensor components, D and E values are given in Table 12. If the two phenyl rings were perpendicular, then E would be zero.

---
[1] See footnote 5, p. 222.
[2] I. M. BROWN, S. I. WEISSMAN, and L. C. SNYDER: J. Chem. Phys. **42**, 1105 (1965).
[3] R. W. BRANDON, G. L. CLOSS, C. E. DAVOUST, C. A. HUTCHISON jr., B. E. KOHLER, and R. SILBEY: J. Chem. Phys. **43**, 2006 (1965).

Inference about the relative ring orientation will be given shortly. By appropriate choice of matrix one may see many of the ESR details observed with an oriented single crystal. All the substances in Table 11 were randomly oriented in matrices. The triplet ESR line width for an organic molecule in a matrix is largely determined by the environment of the solute molecule. If the solvent molecule geometry is similar to that of the solute, the number of probable configurations will be small. Hence the dispersion of spin-spin interactions and thus of zero-field splitting parameters will be small. When the triplet molecule is $C_6H_5-C-C_6H_5$, one observes ESR line widths respectively of 94, 67, 53, 19 17 gauss when the solvents are n-pentane $CH_3(CH_2)_3CH_3$, $C_6H_5-CH_2-C_6H_5$, n-heptane $CH_3(CH_2)_5CH_3$, $C_6H_5-\underset{\underset{O}{\|}}{C}-C_6H_5$ and $C_6H_5-\underset{\underset{N_2}{}}{C}-C_6H_5$[1]. (The second compound is non-planar, in contrast to the forth and fifth.) The heptane approximates in length the $C_6H_5-C-C_6H_5$ triplet molecule. The line narrowing,

Table 12.

	g_{xx}	g_{yy}	g_{zz}	D	E	Ref.
Phenylmethylene				0.5150	0.0231	1
Diphenylmethylene	2.00251	2.00451	2.00432	± 0.40505	∓ 0.01918	2
Biphenylenemethylene (fluorenylidene) (78)	2.00512	2.00234	2.00272	± 0.40923	∓ 0.02828	2

[1] E. Wasserman, A. M. Trozzolo, W. A. Yager, and R. W. Murray: J. Chem. Phys. **40**, 2408 (1964).

[2] R. W. Brandon, G. L. Closs, C. E. Davoust, C. A. Hutchison jr., B. E. Kohler, and R. Silbey: J. Chem. Phys. **43**, 2006 (1965).

observed here when solute and solvent have similar geometry, is analogous to the appearance of narrow luminescence lines for a particular length of aliphatic hydrocarbon molecule as matrix, while longer or shorter host molecules give bands[2]. The minimum luminescence band width for anthracene occurs in hexane. Some additional effective narrowing results from rotation of the powdered sample, smearing out weak lines from small crystals with an orientation close to a principal axis. Under these conditions, lines are narrow enough to observe ^{13}C splitting of the triplet ESR lines. The line width for the methylenes and nitrenes appears to be governed by a distribution of D and E values rather than by anisotropic hyperfine interactions. Similar conclusions are reached for alkaline earth chelate triplets[3].

In excited triplet molecules such as naphthalene, the Pauli principle will not allow two electrons with identical spin to be closer than adjacent carbon atoms of the π-system. At such separation, the dipolar interaction would be about 0.1 cm^{-1}. The D values in Table 11 are not only much larger than this, but are actually larger than $h\nu$ when $\nu = 9$ GHz. Such values require considerable localization of the two electrons on one carbon or nitrogen atom. For complete localization on the central carbon of a methylene, estimates range from 0.9055 to 1.35 cm^{-1} [4,5] while on the nitrogen atom of a nitrene the corresponding values are 1.63[6] to

[1] A. M. Trozzolo, E. Wasserman, and W. A. Yager: J. chim. phys. **1964**, 1663.

[2] E. V. Shpolskii: Soviet Phys. Uspekhi **2**, 378 (1959); **3**, 327 (1960).

[3] See footnote 2, p. 223.

[4] R. W. Brandon, G. L. Closs, and C. A. Hutchison jr.: J. Chem. Phys. **37**, 1878 (1962); (D value 3 × too small).

[5] J. Higuchi: J. Chem. Phys. **39**, 1237 (1963).

[6] J. B. Lounsbury: J. Chem. Phys. **42**, 1549 (1965).

1.9 cm^{-1} [1]. Substituents permitting delocalization lead to a reduced value of D, as seen in Table 11, if the linear geometry is preserved. However, reduction in D on substitution for one or more hydrogen atoms in HCH may be indicative not only of delocalization but also of non-linearity since D diminishes with reduction of angle from 180°. For H—C—C$_6$H$_5$ the ^{1}H and ^{13}C couplings also indicate a non-linear molecule, as described later. In triplet ESR spectra it is usually assumed that a negligible contribution to D is made by spin-orbit coupling. Estimates of the spin-orbit contribution to D for CH$_2$ range from 0.12 for the linear molecule down to 0.03 for an angle of 150° [2,3].

A linear triplet molecule necessarily has cylindrical symmetry and $E=0$. From the magnitude of E, one can thus obtain some information about geometry. One electron of the triplet pair is doubtless in a $p\pi$ orbital. The other is assumed to be in a σ orbital of the central carbon atom, and this will be pure p for a linear

Table 13.

	(MHz)	A (isotropic)	A_{xx}	A_{yy}	A_{zz}	Ref.
Diphenylmethylene	(single crystal in benzophenone)	173.3	16.3	41.5	−57.8	1
Diphenylmethylene	(glass)	175	39	21	−60	2
Biphenylenemethylene		263.6	16.4	43.8	−60.2	1

[1] R. W. Brandon, G. L. Closs, C. E. Davoust, C. A. Hutchison jr., B. E. Kohler, and R. Silbey: J. Chem. Phys. 43, 2006 (1965).
[2] E. Wasserman, A. M. Trozzolo, W. A. Yager, and R. W. Murray: J. Chem. Phys. 40, 2408 (1964).

molecule. In a bent molecule the σ orbital will be a hybrid which may be written as sp^4. Table 12 gives D and E values for a few ground state triplet molecules. The proton hyperfine splitting of the triplet ESR spectrum of H—C—C$_6$H$_5$ is very small ($<$15 MHz), and it is inferred that the methylene—C—H bond is not linear, but the angle is probably closer to 140—150 degrees.

The isotropic proton hyperfine coupling in linear HCH has been estimated at −100 MHz[5], and at least 0.8 of this value would be expected in H—C—C$_6$H$_5$ if it were linear[6]. Since the coupling is less than 15 MHz, a marked positive hyperfine coupling in a non-linear configuration appears nearly to cancel the negative coupling from $\pi-\sigma$ interaction[6]. In diphenylmethylene the dihedral angle is given as 154° if the two rings are coplanar[7].

The ^{13}C splitting in contrast to that of hydrogen in the phenylmethylene (H—C—C$_5$H$_6$) is increased as the molecule is bent. Values (in MHz) of isotropic and anisotropic splittings for two examples are given in Table 13.
ENDOR measurements have shown the spin density at the methylene carbon in biphenylenemethylene (78) to be positive[8].

Remarkably, two distinct ground state triplet molecules corresponding to each of the geometrical isomers of 1- and 2-naphthylmethylene configurations

[1] E. Wasserman, G. Smolinsky, and W. A. Yager: J. Am. Chem. Soc. 86, 3166 (1964).
[3] S. J. Fogel, and H. F. Hameka: J. Chem. Phys. 42, 132 (1965).
[2] S. H. Glarum: J. Chem. Phys. 39, 3141 (1963).
[4] See footnote 6, p. 224.
[5] M. Karplus, and G. K. Fraenkel: J. Chem. Phys. 35, 1312 (1961).
[6] E. Wasserman, A. M. Trozzolo, W. A. Yager, and R. W. Murray: J. Chem. Phys. 40, 2408 (1964).
[7] R. W. Brandon, G. L. Closs, C. E. Davoust, C. A. Hutchison jr., B. E. Kohler, and R. Silbey: J. Chem. Phys. 43, 2006 (1965).
[8] C. A. Hutchison jr., and G. A. Pearson: J. Chem. Phys. 43, 2545 (1965).

below can arise because the C—C and the C—H bonds are not collinear and therefore the spin-spin interactions will be different for the two conformational isomers[1]. (88) and (89).

(88a)   (88b)   (89a)   (89b)

Just as the parent substance for the methylenes is CH_2, that of the nitrenes is the fragment NH. Predictions of the parameter D range from 1.63[2] to 1.9 cm^{-1} [3]. A value of $D = 1.86$ cm^{-1} for NH is obtained by taking $D = 2\lambda''$, where λ'' is taken from the $^3\pi \to {^3\Sigma}$ electronic spectrum[4,5]. Assuming that a σ-unpaired electron remains on the nitrogen while the π-unpaired electron is delocalized, one may use this parameter as a one-center integral to get D values for nitrenes in reasonable agreement with experiment[5]. However, the present uncertainties about spin densities on nitrogen atoms leave a considerable measure of uncertainty on any parameters derived from them.

Compared with the nitrenes, the dinitrenes have D values so low (e.g., 0.0675 cm^{-1} for (85) as assuredly to indicate large mean separation of the two degenerate electrons.

Very recently a general review of the triplet state has appeared[6].

IV. Biradicals.

We have considered molecules which in the rigid state can clearly be described as having triplet ground or excited states, with spin-spin interaction being described by D and E parameters. However, there are molecules which apparently have two unpaired electrons with such a small interaction that in liquid solution they may show narrow hyperfine lines which suggest restriction of each electron to "its" portion of the molecule. These molecules will be termed biradicals, but the definition will have to be extended to systems showing more complex behavior. The actual molecule giving rise to the ESR spectrum observed may be more complex than the apparent structural formula indicates because of "polymerization" of radicals. An example of an apparent biradical is compound (90). It is seen to be two triphenylmethyl radicals joined by oxygen.

(90)

---

[1] A. M. Trozzolo, E. Wasserman, and W. A. Yager: J. Am. Chem. Soc. **87**, 129 (1965).
[2] J. B. Lounsbury: J. Chem. Phys. **42**, 1549 (1965).
[3] See footnote 1, p. 225.
[4] R. N. Dixon: Can. J. Chem. **37**, 1171 (1959).
[5] J. A. R. Coope, J. B. Farmer, C. L. Gardner, and C. A. McDowell: J. Chem. Phys. **42**, 54 (1965).
[6] S. K. Lower, and M. A. El-Sayed: Chem. Rev. **66**, 199 (1966).

Perhaps a conceptually simpler procedure for forming a system which assuredly is a biradical is to couple weakly two independent radicals in solution. By forming the mononegative ion of 4,4' diphenyl, 2,2' bipyridine and adding the divalent ions of Be, Mg, Ca, Sr or Zn, one forms a biradical which probably has the structure[1]: (91)

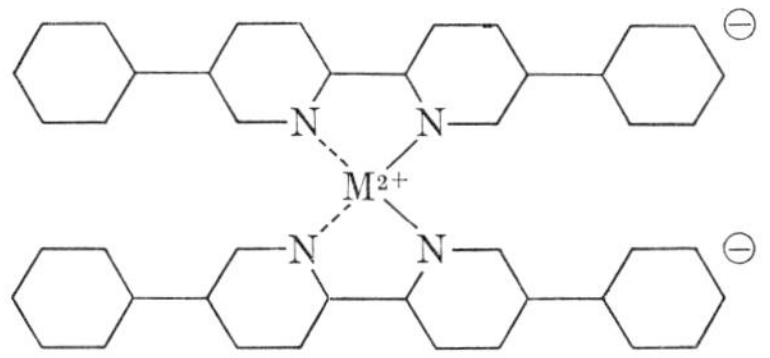

(91)

Alternatively, one Mg^{2+} or two Na^+ ions will serve to bind two ketyl radicals such as: $(CH_3)_3C-\underset{O}{C}-C(CH_3)_3^-$ to form a biradical. In liquid solutions of ketyls bound by an alkaline earth ion, the hyperfine splitting is well resolved. We have noted before that in rigid solution these systems have a ground triplet state because of dipolar interaction between two electron spins[2].

The spin Hamiltonian for a biradical system is given by[2]:

$$\mathcal{H} = g\beta H(S_1+S_2) + \sum A_{1i} S_1 \cdot I_i + \sum A_{2j} S_2 \cdot I_j + J S_1 \cdot S_2,$$

where subscripts 1 and 2 refer to electrons of the two different parts of the biradical interacting with nuclei belonging to groups i and j respectively. The appearance of the ESR spectrum is governed by the relative magnitudes of the A's and J. If $A_{1i}, A_{2j} \gg J$, then one observes the hyperfine pattern resulting from the interaction of an electron only with nuclei in its own part of the molecule. However, if $J \gg A_{1i}, A_{2j}$, then all the nuclei in the molecule will interact. Clear examples of the first case are common: in extreme examples of the latter case only a single line is observed.

An interesting transition from monoradical (92) to biradical (93) to ground state triplet molecule is afforded in the following sequence:

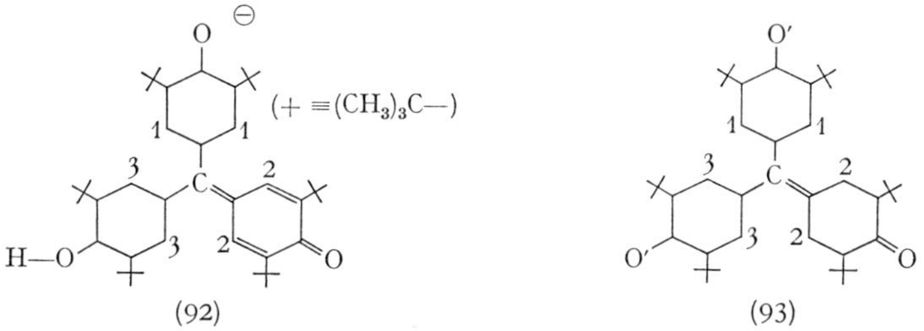

(92) (93)

$(+ \equiv (CH_3)_3C-)$

The monoradical shows a quintet from the essentially equivalent protons labeled 1,1,2,2. Conversion by oxidation to compound (93) gives interaction with six nearly equivalent protons 1,1,2,2,3,3. Lowering the temperature of (93) to give a rigid glass such that dipolar contributions are not averaged out leads to a 4-line spectrum characteristic of a triplet-state molecule with parameter $E=0$. From the separation of the inner pair of lines one calculates 9 Å as the spacing

---

[1] I. M. Brown, and S. I. Weissman: J. Am. Chem. Soc. **85**, 2528 (1963).
[2] N. Hirota, and S. I. Weissman: J. Am. Chem. Soc. **86**, 2538 (1964).

of the two unpaired electrons[1]. Another example of a molecule which in liquid solution is a biradical but in which at low temperatures spin-spin interaction leads to the ground triplet state is the dinegative ion of triphenylene cited earlier[2]. The two electrons occupy degenerate orbitals in this ion.

Evidence from quenching and from irradiation experiments has been adduced that some biradical monomers associate in solution to form either biradical or diamagnetic dimeric (or polymeric) molecules[3]. If the biradical is a long molecule (94), it is plausible that the spin exchange rate should be only in the 10^6 sec^{-1}

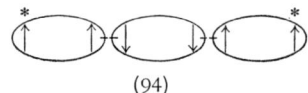

(94)

range[4] and that extensive hyperfine splitting should be observed. A large discrepancy between the singlet-triplet separation calculated from the observed exchange rate and that calculated for the monomer has been termed the "biradical paradox"[5]. However, the two ΔE values or exchange frequencies apparently refer to different molecules[1], the calculated values referring to the monomer and the observed to a dimer or polymer. One must thus beware of assuming that because one writes a plausible biradical structure and observes resonance that one is dealing with a single molecule.

D. Defects in insulating solids.

I. Electron excess centers.

22. F-centers in alkali halides. Coloration of the alkali halides by UV-, x-, γ-, or neutron irradiation, by heating in alkali metal vapors, or by high-temperature electrolysis with a pointed cathode was early ascribed to a "Farbzentrum" or F-center. The de Boer model of the F-center as an electron in a negative ion vacancy[6] had been accepted well before the advent of ESR techniques. Naively, one may regard the F-center as resembling a hydrogen atom, with the characteristic optical absorption representing the transition to the first excited state. Association of a broad ESR line in KCl with the F-band gave a satisfying unambiguous additional confirmation of the de Boer model[7]. As shown in Table 14[8] the line widths may be very great, even when the F-centers are present at very low concentrations. In fact, up to the limiting concentration (10^{18}/cm^3) which one may achieve with uniform distribution, one does not detect any effect of concentration or method of preparation or temperature upon ESR spectral properties except relaxation behavior. (An exception must be made for systems which have been irradiated in the F-band, as detailed later.) With the exception only of LiF, NaH, NaF, RbCl and CsCl, the lines are Gaussian, without appearance of resolved hyperfine structure. The structureless F-lines were shown to be inhomogeneously broadened[9] and they are the envelope resulting from overlapping of a very large

---

[1] R. Kreilick: J. Chem. Phys. **43**, 308 (1965).
[2] R. E. Jesse, P. Biloen, R. Prins, J. D. W. van Voorst, and G. J. Hoijtink: Mol. Phys. **6**, 633 (1963).
[3] R. K. Waring jr., and G. J. Sloan: J. Chem. Phys. **40**, 772 (1964).
[4] D. C. Reitz, and S. I. Weissman: J. Chem. Phys. **33**, 700 (1960).
[5] H. M. McConnell: J. Chem. Phys. **33**, 1868 (1960).
[6] J. H. de Boer: Rec. trav. chim. **56**, 301 (1937).
[7] C. A. Hutchison jr.: Phys. Rev. **75**, 1769 (1949).
[8] H. Seidel, and H. C. Wolf: Phys. stat. sol. **11**, 3 (1965).
[9] A. F. Kip, C. Kittel, R. A. Levy, and A. M. Portis: Phys. Rev. **91**, 1066 (1953).

number of unresolved hyperfine components. Taking KBr as our example, one expects a set of 19 lines from interaction with six nearest-neighbor ^{39}K$^+$ ions ($I=\frac{3}{2}$). The second-shell neighbors (Fig. 27) number twelve, and these will be ^{79}Br or ^{81}Br, each with $I=\frac{3}{2}$. Even if the moments of both of the Br isotopes were essentially identical (they are 1.4039 vs. 1.5131) one would have *each* of the original 19 lines split into 37 lines of much smaller splitting constant. As shown in Table 14, hyperfine contributions from the *eighth* shell (Fig. 27) of neighbors

Table 14. *F-center ESR spectrum characteristics**.

Crystal	$\Delta H_{1/2}$ (G)	ESR HFS data from shells	g-factor	Ref.	ENDOR data from shells	Refs.
LiH	60	0	2.004	1	—	—
LiF	150	(I + II)	2.000	2—4	I—VIII	2, 3
LiCl	57 ± 8	0	1.997	2	I—VI	2
NaH	150	I	1.998	5	I + II	14
NaF	220 ± 20	(I + II)	2.000	2, 6	I—IX (XVI)	2, 6
NaCl	140	0	1.987	2, 7	I—VI	2, 5, 9
NaBr	300	0	1.98	8	—	—
KF	100	0	2	9	I—VI	9
KCl	46	0	1.996	7	I—VI	15
KBr	125	0	1.985	7, 10	I—VIII	5, 9
KI	210	0	1.97	10	I—VI	9
RbCl	420 ± 20	(I)	2	11, 12	I—V	12
RbBr	380 ± 10	0	1.95	8, 11, 12	I—VI	12
RbI	640	0	2	11	—	—
CsCl	700	I	1.984	13	—	—

* Taken from H. Seidel, and H. C. Wolf: Phys. stat. sol. **11**, 3 (1965).
[1] W. B. Lewis, and F. E. Pretzel: J. Phys. Chem. Solids **19**, 139 (1961).
[2] W. C. Holton, and H. Blum: Phys. Rev. **125**, 89 (1962).
[3] W. C. Holton, H. Blum, and C. P. Slichter: Phys. Rev. Letters **5**, 197 (1960).
[4] R. Kaplan, and P. J. Bray: Phys. Rev. **129**, 1919 (1963).
[5] W. T. Doyle, and W. L. Williams: Phys. Rev. Letters **6**, 537 (1961).
[6] W. T. Doyle: Phys. Rev. **131**, 555 (1963).
[7] A. F. Kip, C. Kittel, R. A. Levy, and A. M. Portis: Phys. Rev. **91**, 1066 (1953).
[8] H. Blum: Thesis, Urbana (Illinois) 1961.
[9] H. Seidel: Z. Physik **165**, 218, 239 (1961).
[10] G. A. Noble: J. Chem. Phys. **31**, 931 (1959).
[11] H. C. Wolf u. K. H. Hausser: Naturwissenschaften **46**, 646 (1959).
[12] H. Seidel, u. H. C. Wolf: Z. Physik **173**, 455 (1962).
[13] F. Hughes, and J. G. Allard: Phys. Rev. **125**, 173 (1962).
[14] W. T. Doyle: Phys. Rev. **126**, 1421 (1962).
[15] V. Ya. Zevin: Soviet Phys. Solid State **3**, 662 (1961).

can be discerned by ENDOR techniques. Hence the number of hyperfine components is so great that even with a linewidth less than 1 G for each component, there is hopeless overlapping from the ESR standpoint. We shall shortly consider the ENDOR data which allow us to resolve hyperfine contributions of cations and anions, shell by shell.

First, one notes the exceptional behavior of LiF, NaH, NaF, RbCl and CsCl, for which one is able to resolve a number of hyperfine lines, as shown in the third column of Table 14. In order for partial resolution of components to be observed, it is necessary either that the hyperfine splitting of one shell appreciable to exceed that of all other shells or that the splitting constants of the first two shells are in the ratio of small integers. The first condition more precisely requires that the total extent of subsidiary splitting due to a single nucleus be less than the primary splitting. A 19-line spectrum representing only resolved first shell contributions, is observed only for NaH[1]. Although NaF was reported also to give 19 lines, careful observation shows 23, and the ENDOR data show that the splitting from first and second shells is nearly equal[2]. The low intensity of the outermost components has prevented detection of the remainder of the 31 expected lines. Next in order of complication is CsCl; with its body-centered structure, the 8 first-shell Cs neighbors of spin $\frac{7}{2}$ should give 57 lines, of which 35 are observed[3]. RbCl displays a resolved splitting due to first-shell ^{85}Rb$^+$ ions ($I=\frac{5}{2}$), which has an abundance of 72.2 per cent, the balance being ^{87}Rb$^+$ ($I=\frac{3}{2}$). The latter has a μ/I ratio 5.6 times as great as that of ^{85}Rb[4,5]. The LiF ESR spectrum represents the most complicated case, since there is strong overlapping due to splittings of the

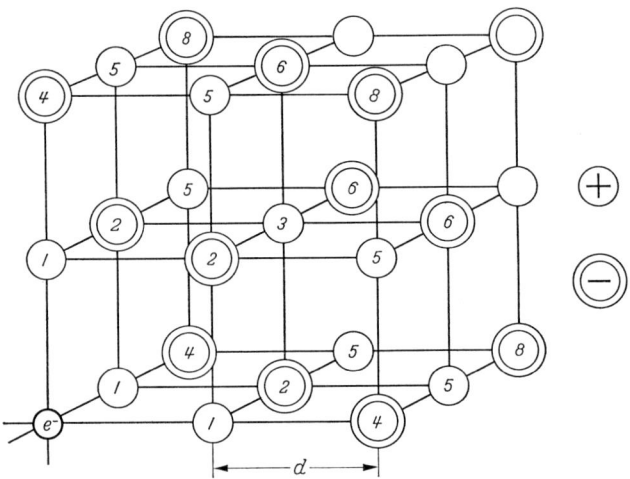

Fig. 27. Successive shells of F-center neighbors. Shell numbers represent sums of squares of Miller indices, taking d as the unit distance. [Figs. 22, 25, 26 and 33 adapted from H. SEIDEL, and H. C. WOLF: Phys. stat. sol. **11**, 3 (1965)].

first two shells and considerable anisotropy[6,7]. Investigation of ^{6}LiF allows one to observe that the principal splitting is due to the second shell F^- ions, which superimpose a marked anisotropic contribution upon the isotropic splitting of ^{6}Li[8].

The brilliant conception of the ENDOR technique for effectively resolving the numerous badly-overlapping components of inhomogeneously broadened lines has resolved many vexing problems[9], and provided an unparalleled wealth of detailed information. A modification of the technique has proved especially suitable for F-center studies[10]; in favorable cases it has permitted ENDOR observations at room temperature. Whereas the unresolved, inhomogeneously broadened F-center line gives very little information, the ENDOR data provide almost a surfeit!

[1] W. T. DOYLE, and W. L. WILLIAMS: Phys. Rev. Letters **6**, 537 (1961).
[2] W. T. DOYLE: Phys. Rev. **121**, 555 (1963).
[3] F. HUGHES, and J. G. ALLARD: Phys. Rev. **125**, 173 (1962).
[4] H. SEIDEL u. H. C. WOLF: Z. Physik **173**, 455 (1963).
[5] H. C. WOLF u. K. H. HAUSSER: Naturwissenschaften **46**, 646 (1959).
[6] W. C. HOLTON, H. BLUM, and C. P. SLICHTER: Phys. Rev. Letters **5**, 197 (1960).
[7] W. C. HOLTON, and H. BLUM: Phys. Rev. **125**, 89 (1962).
[8] R. KAPLAN, and P. J. BRAY: Phys. Rev. **129**, 1919 (1963).
[9] G. FEHER: Phys. Rev. **114**, 1219, 1245 (1959).
[10] H. SEIDEL: Z. Physik **165**, 218, 239 (1961).

ENDOR results which provide unique or extraordinarily precise information about F-centers include the following:

1. Where ESR spectra give unresolved hyperfine contributions to line width, ENDOR spectra provide unambiguous identification of responsible nuclei.

2. The great extent of the F-center wave function-out to the eight-nearest-neighbors-could hardly have been detected by any other existent method. Precise splitting constants for such a number of shells provide a theoretical challenge yet unmet. The sign of both isotropic and anisotropic splitting parameters are found to be positive for all shells, in accord with theory[1].

3. Where some resolution of ESR hyperfine structure is observed, ENDOR data provide unambiguous interpretation of the former, which is subject to misinterpretation because of inability to see all hfs components, e.g., LiF[2-4].

4. Not only the precise values of anisotropic hyperfine tensor components, but their principal axes are observed with enough precision to detect non-axiality of hyperfine interaction in some shells. Deviations of hyperfine axes in some shells from crystal axes indicate a small fraction of non-s-character.

5. The ENDOR spectra show splittings from quadrupole interaction, whose tensor components may be specified. The absolute sign of this interaction may be determined, as well as the relative signs of hyperfine and quadrupole interactions.

6. The fact that observable F-centers occur at positions isolated from other point or line defects is indicated by the narrowness of the ENDOR lines. ESR lines in solids are considered narrow when their width is 1 G or less ($\sim$2.3 MHz). ENDOR lines for the F-centers may have widths of the order of 10—100 kHz.

23. F-center interactions. Prolonged room temperature irradiation in the F-bands of KCl or KBr leads to a narrowing of the ESR F-line, for KCl by a factor of 35/46. The central portion of the residual line, attenuated by a factor of 6, is Lorentzian, whereas the line initially is Gaussian. The saturation parameter $\sqrt{T_1 T_2}$ is decreased by a factor of 8 for KCl, implying a considerably reduced T_1 value. These changes in F-center characteristics are attributed to exchange interaction between groups of near-neighbor centers[5-10]. It is not certain whether a cluster of F-centers more distant than two lattice constants or a pair of closely-spaced F-centers is responsible. The modified centers have been designated as $[F]$ centers. For the latter case, with a very high exchange frequency, the reduction in line width should be given by the factor $1/\sqrt{2}$, which is approximately what is observed. Neither the ENDOR spectrum (apart from saturation behavior) nor the optical absorption spectrum of the $[F]$ centers is observably different from the F-centers[11]. $[F]$ clusters appear to have a much shortened relaxation time, and when present, they may be involved in cross-relaxation of isolated F-centers[12].

---

[1] W. E. BLUMBERG: Bull. Am. Phys. Soc. **5**, 183 (1960).
[2] See footnote 6, p. 230.
[3] See footnote 7, p. 230.
[4] See footnote 8, p. 230.
[5] F. J. ADRIAN: Phys. Rev. **107**, 488 (1957).
[6] H. GROSS: Z. Physik **164**, 341 (1961).
[7] P. R. MORAN, S. H. CHRISTENSEN, and R. H. SILSBEE: Phys. Rev. **124**, 442 (1961).
[8] W. E. BRON: Phys. Rev. **125**, 509 (1962).
[9] M. SCHWOERER u. H. C. WOLF: Z. Physik **175**, 457 (1963).
[10] G. A. NOBLE, and J. J. MARKHAM: J. Chem. Phys. **36**, 1340 (1962).
[11] H. SEIDEL, and H. C. WOLF: Phys. stat. sol. **11**, 3 (1965).
[12] R. W. WARREN, D. W. FELDMAN, and J. G. CASTLE jr.: Phys. Rev. **136**, A 1437 (1964).

Aggregation of F-centers with the formation of colloidal metal is presumed to be responsible for the appearance of new optical bands arising either during irradiation of F-centers at elevated temperatures or upon heating of irradiated crystals. These bands are variously called R' [1] or X bands [2]. It is only in LiH that one is able unambiguously to assign colloidal particles to an ESR spectrum. A conduction electron resonance [3], an Overhauser enhancement of Li nuclear resonance and a displacement of the electron resonance frequency by polarization of Li nuclei [4] are all attributable to the colloidal particles.

It would be a difficult task to determine an F-center yield curve for neutron irradiation of LiF, for the ESR line width diminishes by more than an order of magnitude from the initial 150 G value [5]. While for low doses the F-centers are doubtless isolated, they will undergo progressive aggregation, finally to yield colloidal lithium. At low doses ($<10^{17}$) one finds the line width to *decrease* on increasing microwave power, thus revealing an underlying narrower component which can be made still narrower and more intense by heating between 250 and 400 °C. Larger doses produce an effect comparable to heating after a small dose. This narrower component shows the usual characteristics of an exchange-narrowed line, such as Lorentz shape. An extreme range of behavior — from isolated F-centers to colloidal metal — is also observed for LiH [6].

24. F-centers in other systems. The ESR spectrum of the F-center in MgO (NaCl structure) shows unambiguously that the center is an electron located at a negative ion vacancy. For $H_0 \| [100]$, a single isotropic line with $g=2.0023$ is observed from electrons having only ^{24}Mg or ^{26}Mg neighbors. It is surrounded by a many-line pattern described by the spin Hamiltonian:

$$\mathcal{H}=g\beta H \cdot S + A I_z S_z + B [I_x S_x + I_y S_y],$$

where $g=2.0023$, $I=\frac{5}{2}$ corresponding to ^{25}Mg, $A/g\beta=12.0$ G and $B/g\beta=10.6$ G. For $H \| [111]$, the anisotropic spectrum becomes greatly simplified, as shown in Fig. 28. A hyperfine sextet is the principal feature about the isotropic line, with a weaker group of ten observable lines implying an eleventh coincident with the isotropic line [7]. For a natural abundance of 10.11% of ^{25}Mg, one expects that 36 per cent of all electrons trapped at negative ion vacancies will have one ^{25}Mg neighbor, with 10 per cent having two such neighbors. The relative intensities of the isotropic line, the sextet and the other ten observable lines of the 11-line spectrum are close to those predicted for electrons with 0, 1, or 2 ^{25}Mg neighbors. This gives a very simple and unambiguous demonstration of F-center geometry. The electron is localized to a far greater extent in the vacancy in MgO than in the alkali halides.

F-centers have likewise been observed in single crystals of CaO [8] and BaO [9]; in the latter, one has marked splitting from ^{137}Ba and ^{137}Ba. F-centers are also observed in powders of SrO as well as the sulfides and most of the selenides of the alkaline earth ions after grinding to create negative ion vacancies and to populate

---

[1] A. B. Scott, and L. B. Bupp: Phys. Rev. **79**, 341 (1950).
[2] H. Schulman, and W. D. Compton: Color Centers in Solids, p. 255.: Pergamon Press 1963.
[3] W. T. Doyle, D. J. E. Ingram, and M. J. A. Smith: Phys. Rev. Letters **2**, 497 (1959).
[4] M. Gueron, and Ch. Ryter: Phys. Rev. Letters **3**, 338 (1959).
[5] R. Kaplan, and P. J. Bray: Phys. Rev. **129**, 1919 (1963).
[6] W. T. Doyle, and W. L. Williams: Phys. Rev. Letters **6**, 537 (1961). — W. L. Williams: Phys. Rev. **125**, 82 (1962).
[7] J. E. Wertz, P. Auzins, R. A. Weeks, and R. H. Silsbee: Phys. Rev. **107**, 1535 (1957).
[8] J. C. Kemp, and V. I. Neeley: J. Phys. Chem. Solids **24**, 332 (1963).
[9] J. W. Carson, D. F. Holcomb, and H. Rüchardt: J. Phys. Chem. Solids **12**, 66 (1959).

them[1]. In polycrystalline GeO_2, a center with $g=1.9957$ and associated with an optical band at 4.5 eV is produced by γ-irradiation; it is presumed to be an electron trapped at an oxygen vacancy[2].

For F-centers in CaF_2 the splitting due to second-shell F^- ions is resolved readily because the cation nuclei are non-magnetic. In BaF_2, there is also splitting of the F-center lines from first-shell ^{135}Ba and ^{137}Ba[3]. In NaN_3 irradiated with UV light at low temperatures, one obtains an ESR F-center spectrum with

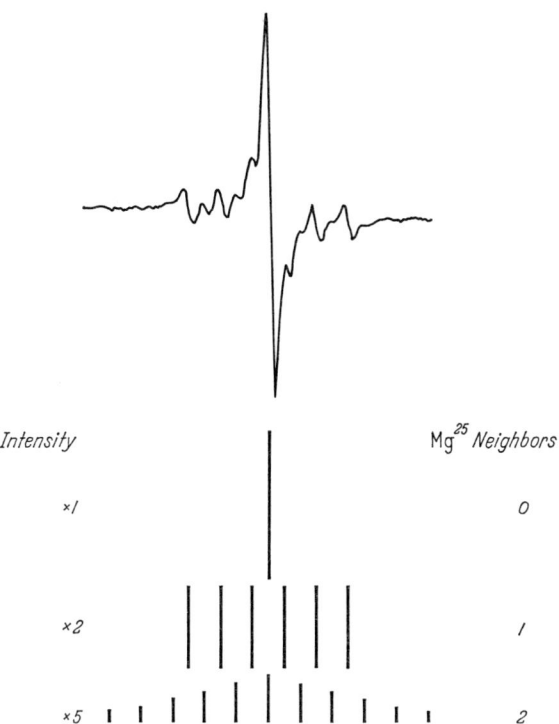

Fig. 28. ESR spectrum of the F-center in MgO for $H \parallel [111]$. The central line is due to F centers with only ^{24}Mg or ^{26}Mg nuclei.

splitting from first shell neighbors[4]. An optical band has been established for this F-center[5].

An isotropic F-center which thus far has been observed only in MgO powders neutron-irradiated in the absence of oxygen is the so-called S-center[6]. This center appears to be located very close to the surface, for its ESR spectrum and an associated blue color both vanish rapidly upon admission of oxygen.

25. Modified F-centers. It is readily possible to detect changes in F-center optical properties when one neighboring alkali ions is foreign to the lattice alkali

---

[1] P. AUZINS, J. W. ORTON, and J. E. WERTZ, in: Low, Paramagnetic Resonance, p. 90. New York: Academic Press 1963.
[2] R. A. WEEKS, and T. PURCELL: J. Chem. Phys. **43**, 483 (1965).
[3] J. ARENDS: Phys. stat. sol. **7**, 805 (1965).
[4] G. J. KING, F. F. CARLSON, B. S. MILLER, and R. C. McMILLAN: J. Chem. Phys. **34**, 1499 (1961); **35**, 1441 (1961).
[5] B. S. MILLER: J. Chem. Phys. **40**, 2371 (1964).
[6] R. L. NELSON, and A. J. TENCH: J. Chem. Phys. **40**, 2736 (1964), and private communication.

halide[1,2]. However, almost the only discernible change in the ESR spectrum is an increase in line width[3]. Two new bands arise, one showing marked dichroism along $\langle 100 \rangle$-type axes. These F_A centers (formerly called A centers) are F centers with one foreign alkali ion in the nearest-neighbor position. They are produced by irradiation in the F-band below room temperature, presumably with F' center and anion vacancy intermediates[4]. The correctness of the F_A model[1] was established unambiguously with ENDOR spectra, both with normal distribution of principal axes and with axes oriented principally along one direction[5].

Doping of alkali halides with an alkaline earth impurity leads to Z_1 centers having a distinct absorption band; the Z_2 and Z_3 centers into which they may be converted upon heating and illumination also have distinct bands[6,7]. The Z_1 ESR spectrum in KCl:Ca or KCl:Sr consists of a single line which was assumed to correspond to a monopositive Ca or Sr ion[8,9]. However, the insensitivity of the ENDOR spectrum to the nature of the alkaline earth ion requires it to be remote from the site of electron trapping[10]. The similar insensitivity of the optical Z_1 band to the identity of the divalent ion is strong supporting evidence. The Z_1 and F_A-center ENDOR spectra appear somewhat similar. The primary perturbing entity is taken to be a cation vacancy adjacent to an F-center, with an M^{+2} ion as next-nearest-neighbor to the former, as in (95):

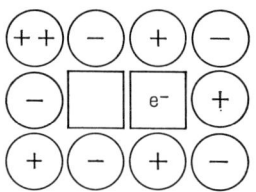

Fig. 29. The Z_1 center in the alkali halides.

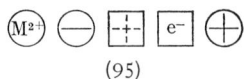

(95)

The simultaneous occurrence of another ENDOR spectrum is presumed to be due to a center with the divalent ion as nearest neighbor to the cation vacancy, as in Fig. 29.

The Z_2 center is presumably diamagnetic[11-13], while the nature of the Z_3 center is not settled.

A center for which no direct parallel has been observed in the alkali halides is an F-center with an associated nearest-neighbor cation vacancy[14], i.e., as in (96):

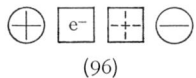

(96)

(The Z_1 center in the alkali halides has an additional divalent cation.) The anisotropy of this center is so small that the line is discernible in powders as well as in

---

[1] F. LÜTY: Z. Physik **165**, 17 (1961).
[2] K. KOJIMA, N. NISHIMAKI, and T. KOJIMA: J. Phys. Soc. Japan **16**, 2033 (1961).
[3] H. OKHURA, K. MURASE, and H. SUGIMOTO: J. Phys. Soc. Japan **17**, 708 (1962).
[4] H. HÄRTEL u. F. LÜTY: Z. Physik **177**, 369 (1964); **182**, 111 (1964).
[5] R. L. MIEHER: Phys. Rev. Letters **8**, 362 (1962).
[6] H. PICK: Ann. Physik **35**, 73 (1939); — Z. Physik **114**, 127 (1939).
[7] F. SEITZ: Phys. Rev. **83**, 134 (1951).
[8] H. KAWAMURA, and K. ISHIWATARI: J. Phys. Soc. Japan **13**, 574 (1958).
[9] G. E. CONKLIN, and R. J. FRIAUF: Phys. Rev. **132**, 189 (1963).
[10] J. C. BUSHNELL: Ph. D. Thesis, University of Illinois 1964; cited by H. SEIDEL, and H. C. WOLF: Phys. stat. solid **11**, 3 (1965).
[11] H. OKHURA, and K. MURASE: J. Phys. Soc. Japan **18**, Suppl. II, 255 (1963).
[12] N. TAKEUCHI, Y. MIZUNO, H. SASAKURA, and M. ISHIGURO: J. Phys. Soc. Japan **18**, 743 (1963).
[13] H. OKHURA: Phys. Rev. **136**, A 446 (1964).
[14] J. E. WERTZ, J. W. ORTON, and P. AUZINS: Discussions Faraday Soc. **31**, 140 (1961).

single crystals. It is partially overlapped by the simple F-center lines until the latter are preferentially bleached. This center has been observed in most of the alkaline earth oxides, sulfides and selenides[1].

26. F-aggregate centers.

α) *M-Centers.* The bleaching of the F-center optical band by irradiation to yield an M-band of considerable intensity suggests the conversion of F-centers to some successor center. (Existence of an additional M band underlying the F-band long confused the interpretation.) Painstaking ESR experiments on samples showing strong optical M bands showed that there is no measurable ESR absorption attributable to the M center[2-4]. Therefore one presumes it to be diamagnetic, in the absence of reasonable explanations as to why it should have an extremely short relaxation time. The diamagnetism is consistent with expectation of a ground singlet state for the van Doorn-Haven-Pick model of an M center as two nearest-neighbor F-centers[5,6] (Fig. 30). That the M-center axis is along one of the six $\langle 110 \rangle$-type axes had received strong support from the demonstration of M-band optical dichroism with such axes, both in selective excitation and in selective bleaching experiments. The M-bands show the phenomenon of "temporary bleaching", i.e., on irradiation in the 365—580 nm region, the band bleaches with appearance of another nearby[7,8]. Upon cessation of irradiation, the new band decays with a time constant of 50 sec[9] in KCl at 90 °K, with regrowth of the original M band. The recognition that the irradiation leads to formation of the excited triplet state with its own optical bands was a very significant step in the confirmation of the model[9]. The final verification of the excited triplet state by ESR and ENDOR spectra appears to leave no doubt of the validity of the van Doorn-Haven-Pick model. In both of types spectra, upon irradiation one observes extra lines which decay with the 50 sec time constant of the induced optical bands. The M-center ESR spectrum of KCl at 90 °K shows the number and angular dependence of lines appropriate to the Hamiltonian:

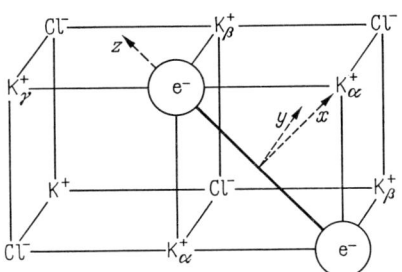

Fig. 30. The M-center in the alkali halides.

$$\mathscr{H} = g\beta\mathbf{H}\cdot\mathbf{S} - [(D/3)-E]S_x^2 - [(D/3)+E]S_y^2 + 2DS_z^2/3,$$

with $S=1$, $g=1.998$, $D/g\beta=-161$ G and $E/g\beta=+54$ G[9,10]. The principal axes are $z=\langle 110 \rangle$, $x=\langle 1\bar{1}0 \rangle$ and $y=\langle 001 \rangle$. As indicated in Sect. C. II on the Triplet State, such a Hamiltonian is appropriate to interaction of two dipoles of spin $\frac{1}{2}$ to give an $S=1$ state. Within experimental error, $(D/3)+E=0$, so the ESR spectrum is particularly simple.

---

[1] P. AUZINS, J. W. ORTON, and J. E. WERTZ, in: W. Low, Paramagnetic Resonance in Solids, p. 90. New York: Academic Press 1963.
[2] H. GROSS, and H. C. WOLF: Naturwissenschaften **48**, 299 (1961).
[3] H. GROSS: Z. Physik **164**, 341 (1961).
[4] H. BLUM: Phys. Rev. **128**, 627 (1962).
[5] C. Z. VAN DOORN, and Y. HAVEN: Philips Research Repts. **11**, 479 (1956); **12**, 309 (1957).
[6] H. PICK: Z. Physik **159**, 69 (1960).
[7] M. IKEZAWA, and M. UETA: J. Phys. Soc. Japan **18**, 145 (1963).
[8] I. SCHNEIDER, and M. E. CASPARI: Solid State Comm. **1**, 9 (1963).
[9] H. SEIDEL: Phys. Letters **7**, 27 (1963).
[10] H. SEIDEL, M. SCHWOERER, and D. SCHMID: Z. Physik **182**, 398 (1965).

The ENDOR spectrum of the KCl M-center at 90 °K shows three distinct sets of "first shell" lines appearing on irradiation corresponding to the $^{39}K^+$ ions labelled α, β, and γ in Fig. 30. While the frequency separation of β and γ lines is small, the α lines appear at nearly twice the frequency of the other two. The electron density at β and γ ions is close to that in the F-center, while that of α ions (neighboring both M electrons) is nearly twice as large. This suggests that it is a good approximation to take the M center wave function as a linear combination of two F-center functions along a [110] direction.

In neutron-irradiated CaO one has also detected the excited triplet state of a two-electron center formed by association along a ⟨100⟩ axis of two anion vacancies with one trapped electron each, joined by a cation vacancy[1], as in (97):

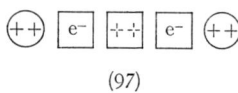

(97)

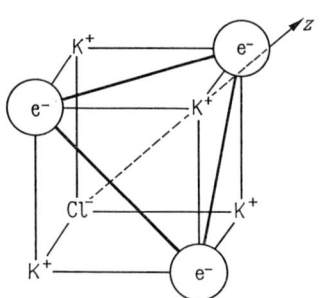

Fig. 31. The R-center in the alkali halides.

A similar system is found in MgO[2]. In both cases the singlet-triplet separation is so small that thermal excitation permits one to observe the ESR triplet spectrum at 300 °K, with a nearly isotropic value of $g \sim 2.002$.

β) *R centers.* A pair of long-wavelength bands (R_1 and R_2) which grow upon optical bleaching of F-centers were thought to arise from two distinct F-aggregate centers. It is now believed that the two bands are separate transitions of one defect, viz., the R-center, Fig. 31 having three trapped electrons in an equilateral triangle in a (111) plane[3,4]. Since the two M-center electrons interact to give a diamagnetic ground state, one expects the three R-center electrons to have a doublet ground state. One would normally assume that Kramers degeneracy would assure detection of an ESR spectrum, but early attempts were fruitless[5-7]. The behavior of the R_2 zero phonon line under uniaxial stress suggests an orbitally doubly-degenerate ground state[8]. A dynamic Jahn-Teller splitting, plus splitting due to residual strains would leave two KRAMERS' doublets lowest[9]. To explain the inability to observe an ESR signal, it is presumed that direct relaxation among these levels results in a very short value of T_1 so that the ESR signal is broadened beyond observation. The relaxation time T_1 varies as $\exp(-\Delta/kT)$, where Δ is the separation of KRAMERS' doublets[10]. Successful ESR observation was achieved on increasing Δ by application of a uniaxial stress and by working at a low temperature[9]. At a stress (3 kg/mm²) which gave reasonable R-line intensity at 2.2 °K, no signal could be seen at 4 °K, consistent with the marked variation of relaxation time assumed. While $g_{\parallel} = 2.06$, $g_{\perp} \approx 2.0$, and hence there

---

[1] D. H. Tanimoto, W. M. Ziniker, and J. C. Kemp: Phys. Rev. Letters **14**, 645 (1965).
[2] B. Henderson: Brit. J. Appl. Phys. **17**, 851 (1966).
[3] C. Z. van Doorn, and Y. Haven: Philips Research Rept. **11**, 479 (1956); **12**, 309 (1957).
[4] H. Pick: Z. Physik **159**, 69 (1960).
[5] P. R. Moran, S. H. Christensen, and R. H. Silsbee: Phys. Rev. **124**, 442 (1961).
[6] W. E. Bron: Phys. Rev. **125**, 509 (1962).
[7] M. Schwoerer, and H. C. Wolf: Z. Physik **175**, 457 (1963).
[8] R. H. Silsbee: Bull. Am. Phys. Soc. **9**, 88 (1964).
[9] D. C. Krupka, and R. H. Silsbee: Phys. Rev. Letters **12**, 193 (1964).
[10] R. Orbach: Proc. Phys. Soc. (London) A **77**, 821 (1961).

is significant overlapping by the F-center line. Additional confirmation of the model was obtained by polarized bleaching experiments.

The successful ESR and ENDOR observation of the excited triplet state of the M-center was followed by an equally impressive study of the excited quartet state of the R center[1]. During the M-center investigation, weak lines had been observed, and it was guessed that these might be quartet-R lines. The optical bleaching of F-centers to convert them into M- and R-centers leaves a considerable number of the latter in the quartet state.

Dipolar interaction in this excited quartet state of the R center is given by the same Hamiltonian for zero field splitting as for the triplet state of the M center, but with $S = \frac{3}{2}$. However, for the trigonal symmetry of the van Doorn-Haven-Pick model one expects the E parameter to vanish. The ESR spectra fully confirm the expected [111]-type symmetry axis, and one obtains $g_\| = g_\perp = 1.996$, with $D/g\beta = +168.5$ G[1]. Even the forbidden $\Delta M_s = 2$ transitions which one so frequently seens in triplet spectra are seen here.

ENDOR observations give a purely isotropic coupling constant of 17.85 MHz for the three ^{39}K nuclei which are nearest neighbors to all three trapped electrons of the R complex. The electron density at these ions is just slightly less than three times that of the F center. Again the linear combination of F-center wave functions is a good first approximation. That the LCAO approximation for the R center has considerably greater validity than for the H_2 molecule is due to the small overlap (~ 0.1) in the former case[1]. Both ESR and ENDOR lines decay with a time constant of 14.5 sec.

27. Trapped electrons in other systems. For some decades it has been known that electrons may be stably trapped in anion vacancies in alkali halides or other ionic solids and that such F-centers have an intense optical absorption band. Stable trapping of electrons in cavities in dilute liquid ammonia or amine (substituted ammonia, RNH_2) solutions of alkali metals has been proposed[2-5] to explain the volume changes, transport, and magnetic properties[6]. The success of this cavity model is impressive in the quantitative correspondence with experiment of density[7], static and radiofrequency magnetic susceptibility[8], and paramagnetic relaxation time data[9] for dilute K—NH_3 solutions. There is also reasonable correspondence of optical and thermodynamic properties with those calculated from the wave function of an unpaired electron in a cavity in a continuum[10]. By measurement of the shift in ^{14}N NMR frequency, the ^{14}N spin density is found to be nearly independent of metal, of temperature, or of concentration up to 0.6 mole/l, the spin density being 0.88 a_0^{-3} at 300 °K[11]. This further supports the existence of a trapped electron species not closely bound to an alkali ion.

---

[1] H. Seidel, M. Schwoerer, and D. Schmid: Z. Physik **182**, 398 (1965).
[2] R. A. Ogg jr.: J. Chem. Phys. **14**, 114, 295 (1946).
[3] J. Kaplan, and C. Kittel: J. Chem. Phys. **21**, 1429 (1953).
[4] C. A. Kraus: J. Am. Chem. Soc. **30**, 1197 (1908).
[5] S. Freed, and N. Sugarman: J. Chem. Phys. **11**, 354 (1943).
[6] See reviews by C. A. Kraus: J. Chem. Educ. **30**, 86 (1953). — M. C. R. Symons: Quart. Rev. (London) **13**, 99 (1959). — W. L. Jolly: Progr. Inorg. Chem. **1**, 235 (1960). — T. P. Das: Advances in Chem. Phys. **4**, 303 (1961).
[7] C. A. Hutchison jr., and D. E. O'Reilly: J. Chem. Phys. **34**, 163 (1961).
[8] C. A. Hutchison jr., and R. C. Pastor: J. Chem. Phys. **21**, 1959 (1953).
[9] D. E. O'Reilly: J. Chem. Phys. **35**, 1856 (1961); — Phys. Rev. Letters **11**, 545 (1963). — V. L. Pollak: J. Chem. Phys. **34**, 864 (1961).
[10] J. Jortner: J. Chem. Phys. **30**, 839 (1959).
[11] D. E. O'Reilly: J. Chem. Phys. **41**, 3729 (1964).

A serious obstacle to acceptance of the cavity model of a trapped electron in NH_3 was the observation of an appreciable shift of the alkali nuclear magnetic resonance field in such solutions[1,2]. It was proposed that species called a "monomer" consists of an alkali ion with an electron trapped in the vicinity[3]. Detailed NMR studies over a range of concentrations of alkali metal in NH_3 indicate that both species are present, with the "monomer" occurring even at low concentrations. Good agreement is obtained with the calculated spin density at the alkali nucleus in the monomer species and also that at a nitrogen nucleus in the cavity species[4]. When one studies amines (derivatives of ammonia in which one or more of the hydrogen atoms have been replaced by alkyl groups —CH_3, C_2H_5, etc.), the ESR behavior depends upon the alkali metal used. For ethylamine ($C_2H_5NH_2$) with potassium, one observes ^{39}K and ^{14}N but no proton splittings. With Li^+, only ^{14}N splittings are seen, but there is a sharp line present, presumably from solvated electrons[5].

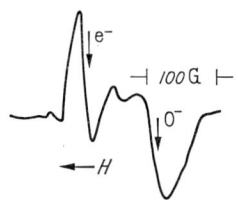

Fig. 32. ESR spectrum of the trapped electron and trapped O^- ion in an aqueous-NaOH glass [P. B. AYSCOUGH, R. G. COLLINS, and F. S. DAINTON: Nature **205**, 965 (1965)].

Eventually it became necessary to postulate a significant lifetime for hydrated electrons produced in irradiated liquid water, and a much longer lifetime for electrons trapped in ice or other rigid polar media at 77 °K. The suggestion that electrons trapped in irradiated water might be the primary reducing species[6] has stimulated efforts to detect their existence. However, it has been only very recently that the ESR spectrum of an electron trapped in rigid glassy solutions has been established[7-9]. One of the clearest experiments demonstrating the nature of the trapped species involves a deposition of alternate layers of Na atoms and H_2O at 77 °K. An ESR line with $g=2.0008$ and a deep blue color fade together in visible light. At 140 °K there are five hyperfine lines with a 5.6 G splitting, while in D_2O none such are seen. Interaction of a trapped electron (e_t^-) with four equivalent protons of water molecules thus appears responsible[10]. Lack of alkali hyperfine interaction in this and other[9] experiments are contrary to a suggestion[11] of trapping in an expanded orbital of the alkali ion. A single trapped-electron ESR line is seen in 8 M NaOH-ice containing 10^{-4} M $K_4Fe(CN)_6$ irradiated at 77 °K at 254 nm[12], the electron coming from the reaction:

$$Fe(CN)_6^{4-} + h\nu \rightarrow Fe(CN)_6^{3-} + e_t^-.$$

In Fig. 32 are shown ESR lines of the trapped electron (e_t) and of trapped O^- ions (O_t^-) at 77 °K after γ-irradiation of 8 N NaOH aqueous glass and after

---

[1] H. M. McCONNELL, and C. H. HOLM: J. Chem. Phys. **26**, 1517 (1957).
[2] D. E. O'REILLY: J. Chem. Phys. **41**, 3729 (1959).
[3] E. BECKER, R. H. LUNDQUIST, and B. J. ALDER: J. Chem. Phys. **25**, 971 (1956).
[4] D. E. O'REILLY: J. Chem. Phys. **41**, 3736 (1964).
[5] K. BAR-ELI, and T. R. TUTTLE jr.: J. Chem. Phys. **41**, 3729 (1965).
[6] N. F. BARR, and A. O. ALLEN: J. Phys. Chem. **63**, 928 (1959). — J. WEISS: Nature **186**, 751 (1960). — J. T. ALLAN, and G. SCHOLES: Nature **187**, 218 (1960). — C. CZAPSKI, and H. A. SCHWARZ: J. Phys. Chem. **66**, 471 (1962).
[7] D. SCHULTE-FROLINDE u. K. EIBEN: Z. Naturforsch. **17** A, 445 (1962); **18** A, 99 (1963).
[8] B. G. ERBOV, A. K. PIKAEV, P. Y. GLAZUNOV, and V. I. SPITSYN: Doklady Akad. Nauk. S.S.S.R. **149**, 363 (1963).
[9] M. J. BLANDAMER, L. SHIELDS, and M. C. R. SYMONS: Nature **199**, 902 (1963); — J. Chem. Soc. **1964**, 4352; **1965**, 1127.
[10] J. E. BENNETT, B. MILE, and A. THOMAS: Nature **201**, 919 (1964).
[11] J. JORTNER, and D. SHARF: J. Chem. Phys. **37**, 2506 (1962).
[12] P. B. AYSCOUGH, R. G. COLLINS, and F. S. DAINTON: Nature **205**, 965 (1965).

subsequent irradiation with visible light. The g-values are respectively 2.0023 and 2.04[1]. Decay of O_t^- and growth of OH is obvious, the latter presumably resulting from the reaction: $O_t^- + H_2O \rightarrow OH + OH^-$. Production of O_t^- during irradiation presumably occurs via the reverse reaction. Enhancement of the O_t^- concentration occurs if N_2O is added:

$$e^- + N_2O \rightarrow N_2 + O_t^-.$$

Introduction of "electron scavengers" such as CH_3Cl or $ClCH_2COOH$ before irradiation leads additionally to formation of CH_3 and of CH_2COOH radicals whose ESR spectra are observed; the concentration of the latter is increased on raising the temperature. Glassy solutions of α-methyltetrahydrofuran also give trapped electrons on γ-irradiation [1,2].

The trapped electron in rigid aqueous systems shows an intense blue coloration. In 8 N NaOH glass, $\lambda_{max} = 586$ nm [3-5]. The absorption maximum of e_t^- in glassy methanol is at 533 nm [6,7], while that of the solvated electron in liquid methanol is 650 nm. A similar shift in optical absorption to longer wavelength is noted for CH_3OH in liquid solution, compared with that in the glass [7].

By contrast, in frozen H_2SO_4, no trapped-electron spectra are observed, but only trapped hydrogen atoms (H_t) and a free radical near $g = 2$ are observed, presumably due to the reaction [1]:

$$-(SO_3)_n SO_2OH + e^{-*} \rightarrow (SO_3)_n SO_3^- + H_t'.$$

A further splitting of the proton doublet in γ-irradiated 6 N H_2SO_4 into quintets or septets with 7 G splitting and 2 G line width similar to those for ice [8] suggests that hydrogen atoms interact with four or six equivalent protons. A center produced in the sodium-compensated zeolite by γ-irradiation in vacuum shows hyperfine interaction with four sodium atoms with $A = 90.4$ MHz and $g = 1.999$ [9]. An absorption at ~500 nm appears to be correlated with it. The site appears to be a void surrounded by four Na^+ ions, and hence is an "electron in a box", whose dimensions calculated for a 500 nm transition are 0.67 nm, close to the value 0.62 nm calculated from the known structure [9].

It now appears well established that electrons in suitable hosts may have long lifetimes in a solvated or trapped state, and doubtless numerous additional stabilized states will be observed.

II. Electron-deficient centers.

28. V_1-centers. Since F-centers and the corresponding optical F-bands are produced by heating in an alkali metal, it was presumed that the V-bands induced by heating in a halogen should correspond to an "antimorph" of the F-center [10].

---

[1] See footnote 12, p. 238.
[2] M. R. RONAYNE, J. P. GUARINO, and W. H. HAMILL: J. Am. Chem. Soc. **84**, 4230 (1962).
[3] See footnote 11, p. 238.
[4] See footnote 7, p. 238.
[5] See footnote 8, p. 238.
[6] I. A. TAUB, M. C. SAUER, and L. M. DORFMAN: Discussions Faraday Soc. **36**, 206 (1963).
[7] F. S. DAINTON, J. P. KEENE, T. J. KEMP, G. A. SALMON, and J. TEPLY: Proc. Chem. Soc. **1964**, 265.
[8] J. E. BENNETT, B. MILE, and A. THOMAS: Nature **201**, 919 (1964).
[9] P. H. KASAI: J. Chem. Phys. **43**, 3322 (1965).
[10] F. SEITZ: Revs. Mod. Phys. **26**, 7 (1954).

The precise antimorph of an F-center would be a positive hole "in" a positive ion vacancy, i.e. uniformly distributed over the six halogen ions. No such isotropic V_1 center has yet been reported for the alkali halides, or, for any other compounds. A slightly modified form of the V_1 center-having tetragonal rather than octahedral symmetry has been studied in MgO[1] Fig. 33 and in CaO[2]. The observed ESR spectrum is described by a Hamiltonian with tetragonal symmetry:

$$\mathcal{H} = g_{\parallel} \beta H_z S_z + g_{\perp} \beta (H_x S_x + H_y S_y),$$

with $g_{\parallel} = 2.0032$ and $g_{\perp} = 2.0385$ for MgO. In CaO, $g_{\parallel} = 2.0011$, $g_{\perp} = 2.0170$[2]. For such symmetry, one requires only that the hole density upon the two atoms

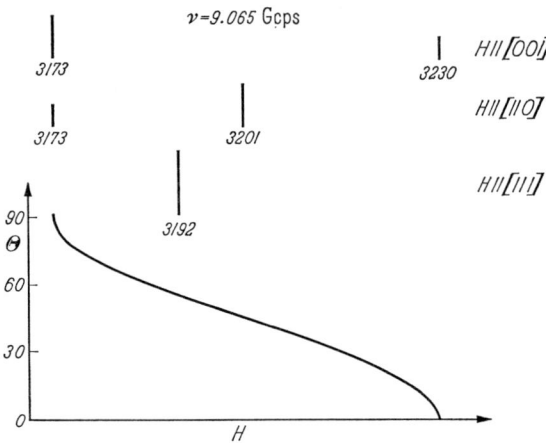

Fig. 33. ESR spectrum of the V_1 center in MgO as a function of orientation; the center has [001], [010] or [100] axes of tetragonal symmetry.

along the axis be different from that of the four atoms in the equatorial plane. This tetragonal distortion is presumed to arise from lattice relaxation taking place when the hole is trapped. The narrow line width (0.5 gauss) does require that the reorientation of the symmetry axis take place in a time long compared with 10^{-7} sec. An optical band at 2.3 eV has been shown to correspond to this V_1 center[3].

29. V_K-centers. The electron-deficient or V-centers in the alkali halides are of four principal types, each of which involves one hole which is localized respectively upon two, three or four nuclei as halogen molecule ions: These are:

1. The V_K center or self-trapped hole (A, Fig. 34).
2. The H-center (B, Fig. 34).
3. The V_F center (C, Fig. 34) (Observed only in LiF).
4. The V_t center (D, Fig. 34) (Observed only in LiF).

---

[1] J. E. WERTZ, P. AUZINS, J. H. E. GRIFFITHS, and J. W. ORTON: Discussions Faraday Soc. **28**, 136 (1959).
[2] A. SHUSKUS: J. Chem. Phys. **39**, 849 (1963).
[3] J. E. WERTZ, G. SAVILLE, P. AUZINS, and J. W. ORTON: J. Phys. Soc. Japan **18**, Suppl. II, 305 (1963).

Sect. 29. V_K-centers.   241

All V centers may be generated by x-irradiation, preferably at 77 °K or lower, since they recombine with other centers well below room temperature. The yield of hole centers may be greatly increased by doping with such electron traps as Tl⁺, Ag⁺ or Pb²⁺ [1]. The V_K center is predominantly formed at 77 °K, with H and V_t centers formed in much lower concentrations. V_F centers result from thermal bleaching of V_K centers or by x-irradiation. The ESR spectra, not only of V_F and V_t but V_K and H centers as well are simplest in the alkali fluorides, especially in LiF. All the V-centers have marked anisotropy of the g-factor, with values exceeding 2.0023. The V_K center has one hole shared by two X atoms displaced toward one another along a [110] direction to form an X_2^- molecule-ion[1-6]. In

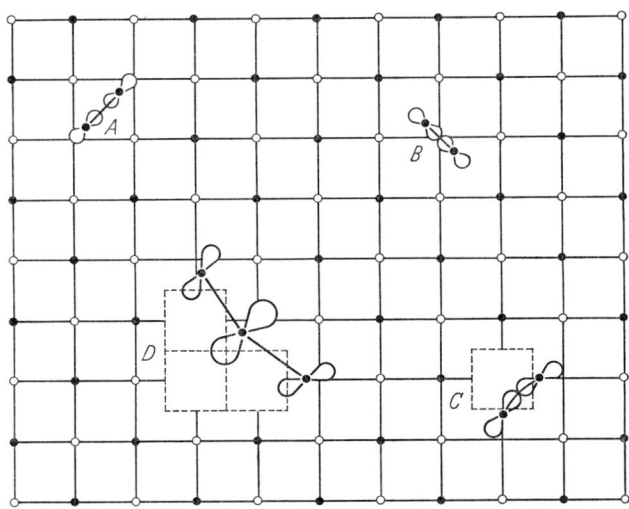

Fig. 34. V-centers in the alkali halides [W. KÄNZIG: J. Phys. Chem. Solids 17, 88 (1960)].

LiF, the F_2^- spectrum is a 1:2:1 triplet, each line of which is inhomogeneously broadened by neighbors[3,5,7]. ENDOR studies resolve the rather small hyperfine contributions from near neighbors[8] as has been done also for V_K centers in other halides. Though more complex in KCl, the interpretation of the Cl_2^- spectrum illustrates the wealth of detail which repeatedly verifies the correctness of the model. The most prominent feature of this spectrum (Fig. 35) is a seven-line group of intensity ratio 1:2:3:4:3:2:1, appropriate to two nuclei of spin $\frac{3}{2}$[3]. While both ³⁵Cl and ³⁷Cl isotopes have this value of nuclear spin, these strong lines arise from ³⁵Cl_2^-. To account for the weaker lines, one considers the ³⁷Cl_2^- and ³⁵Cl—³⁷Cl⁻ centers. Since the relative abundances of ³⁵Cl and ³⁷Cl are approximately 3:1, the relative abundances of the 35—35, 35—37 and 37—37 isotopes are 9:6:1. For the ³⁷Cl⁻ one merely expects another septet with the same intensity ratio but with a reduction in the splitting by the ratio 1.2011 of the magnetic moments

---

[1] C. J. DELBECQ, B. SMALLER, and P. H. YUSTER: Phys. Rev. 111, 1235 (1958).
[2] W. KÄNZIG: Phys. Rev. 99, 1890 (1955).
[3] T. G. CASTNER, and W. KÄNZIG: J. Phys. Chem. Solids 3, 178 (1957).
[4] T. G. CASTNER, W. KÄNZIG, and T. O. WOODRUFF: Nuovo cimento, Suppl. 7, 612 (1958).
[5] T. WOODRUFF, and W. KÄNZIG: J. Phys. Chem. Solids 5, 268 (1958).
[6] W. KÄNZIG: J. Phys. Chem. Solids 17, 80 (1960).
[7] C. E. BAILEY: Phys. Rev. 136 A 1311 (1964).
[8] R. MIEHER, and R. GAZINELLI: Phys. Rev. Letters 12, 644 (1964).

Handbuch der Physik, Bd. XVIII/1.

of ^{35}Cl vs. ^{37}Cl. The total number of lines corresponding to the ^{35}Cl—^{37}Cl center is 16, occurring as sets: 1:1,1:1,1,1:1,1,1,1:1,1,1:1,1:1. The total relative intensities remain the same as before, but since the states $[m(^{35}\text{Cl})=\frac{1}{2}, m(^{37}\text{Cl})=-\frac{3}{2}]$ and $[m(^{35}\text{Cl})=-\frac{3}{2}, m(^{37}\text{Cl})=\frac{1}{2}]$ differ in energy, there will be no energy level degeneracies and all of the sixteen lines of ^{35}Cl—^{37}Cl$^-$ will be of equal intensity. Surprisingly, the Cl$_2^-$ ion has been reported in γ-irradiated NH$_3$OH Cl[1].

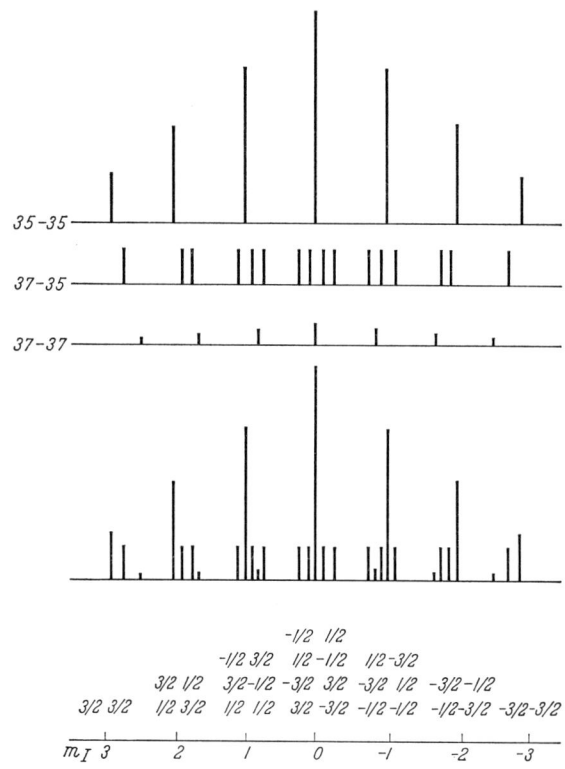

Fig. 35. ESR spectrum of the Cl$_2^-$ centers in KCl [T. G. CASTNER, and W. KÄNZIG: J. Phys. Chem. Solids 3, 178 (1957)].

The V_K spin Hamiltonian may be written in the form:

$$\frac{\mathcal{H}}{g_e\beta} = \frac{g}{g_e} H_0 S_z + a^* \mathbf{I} \cdot \mathbf{S} + b^* I_{z'} S_{z'},$$

where z and z' refer respectively to static field direction and to the molecular ion axis. Further:

$$\beta g_e(a^* + b^*) = A_{zz} = a + 2b$$

and

$$g_e \beta a^* = A_{xx} = A_{yy} = a - b,$$

where a and b are isotropic and axial coupling constants. A summary of the hyperfine and g-parameters for V_K centers in some of the alkali halides is given in Table 15.

[1] H. UEDA: J. Chem. Phys. 41, 285 (1965).

Table 15. V_K center parameters.

Center	Crystal	$a^* + b^*$ (G)	a^* (G)	g_x	g_y	g_z	Ref.
$^{19}F_2^-$	LiF	887	59	2.0227	2.0234	2.0031	1, 2
$^{35}Cl_2^-$	NaCl	101	9	2.0489	2.0425	2.0010	1
$^{35}Cl_2^-$	KCl	98	9	2.0428	2.0447	2.0010	1
$^{81}Br_2^-$	KBr	455	80	2.179	2.175	1.980	1
$^{19}F-^{35}Cl$	KF—KCl	806 (F) 126 (Cl)	81 (F) 21 (Cl)	2.030	2.030	2.0023	3
$^{35}Cl-^{81}Br$	KCl—KBr	(along $\langle 110 \rangle$) (12° to $\langle 110 \rangle$ in (100) plane)		2.136 2.193	1.9836 1.9697		4

[1] T. G. CASTNER, and W. KÄNZIG: J. Phys. Chem. Solids **3**, 178 (1957).
[2] T. WOODRUFF, and W. KÄNZIG: J. Phys. Chem. Solids **5**, 268 (1958).
[3] J. W. WILKINGS, and J. R. GABRIEL: Phys. Rev. **132**, 1950 (1963).
[4] D. SCHOEMAKER, C. J. DELBECQ, and P. H. YUSTER: Bull. Am. Phys. Soc. **9**, 629 (1964).

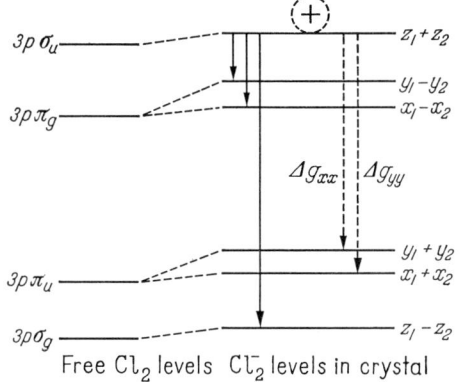

Fig. 36. Energy levels and allowed transitions for the V_K optical absorption bands [from M. H. COHEN: Phys. Rev. **101**, 1432 (1956)].

The presumptive presence of some kinds of V-centers in alkali halides was inferred from the early discovery of the V-bands. Subsequent to the elucidation of the V_K geometry by ESR studies of KCl, two optical bands were correlated with it. The optical bands for all the V-centers described here are indistinguishable from the V_K bands[1, 2]. These occur at 365 and 750 nm. They correspond to two transitions shown as solid lines in Fig. 36 for an X_2^- ion[3]. Dotted lines are forbidden transitions which are responsible for g-shifts. One electron is missing from the σ_u orbital. Accurate wave functions have been calculated for V_K systems, since one has available a wealth of detail from ESR experiments[3-5]. It has been shown from hyperfine data that the ground state, though assumed to be largely of p_z character, has a large fractional s-character (about 0.5!). Calculation of the g-factors in this system is made more complicated by its two-center nature. The

---

[1] K. TEEGARDEN u. R. MAURER: Z. Physik, **138**, 284 (1954).
[2] W. D. COMPTON, and C. C. KLICK: Phys. Rev. **110**, 349 (1958).
[3] T. G. CASTNER, and W. KÄNZIG: J. Phys. Chem. Solids **3**, 178 (1957).
[4] T. WOODRUFF, and W. KÄNZIG: J. Phys. Chem. Solids **9**, 70 (1958).
[5] M. H. COHEN: Phys. Rev. **101**, 1432 (1956).

magnetic resonance aspects of the V_K center have been discussed in considerable detail with admirable clarity[1].

Since the V_K centers are anisotropic, they should also be dichroic. Detailed experiments on selective bleaching have correlated the ESR spectrum with optical bands, as well as establishing the energies of excited states[2,3]. By following the ESR intensity, one may, e.g., see decay of intensity along the [011] direction when light polarized in this direction is employed. Simultaneously one sees growth in intensity along [0$\bar{1}$1] and the other face diagonals. This indicates that at low temperatures there is primarily a freeing and retrapping of holes rather than their destruction.

The ClF center in KCl:KF has tetragonal hyperfine symmetry with a $\langle 111 \rangle$ axis rather than $\langle 110 \rangle$. The fluorine appears to be interstitial, with the chloride ion at a normal site. Both the g-tensor and hyperfine structure parameters so closely resemble the Cl_2^- and F_2^- centers that the ClF^- may be regarded as half of each center oriented along a $\langle 111 \rangle$ axis[4].

The mixed $ClBr^-$ ions may be observed in LiCl, NaCl, KCl or RbCl which has been doped with Br^-, x-irradiated and then thermally bleached to remove Cl_2^- ions[5].

30. Other V-centers. The H-center (Fig. 34) ESR spectrum is obtained for LiF, KCl or KBr by x-irradiation at 20 °K. It disappears rapidly above 130 °K for LiF and above 60° K for KCl. The principal hyperfine splitting arises from the two central X nuclei; a tenfold smaller splitting results from the two outer nuclei on the same $\langle 110 \rangle$ axis. Thus one might refer to it as a linear X_4^{3-} center. However, with the a spin density of only 4 to 10 per cent on the end atoms, it is perhaps preferable to regard the H center as primarily an X_2^- unit occupying one anion site. The s-character of the wave function is rather smaller than for the V_K center. The H-center is considered as complementary to the F-center, for at low temperature the two are formed in pairs; recombination of a pair restores the perfect lattice[6,7]. Double occupation of one lattice site is effectively the same as a neutral atom occupying an interstitial site.

The V_F center (Fig. 34), observed only in LiF, formally fulfills one definition of an antimorph of the F center — it involves a hole localized at a positive ion vacancy. However, the sharing of the hole between two atoms, and the bent axis lead one to describe it as a perturbed V_K center. Indeed, the g-tensors and h/s constants are close to those for the F_2^- center[7,8] though the axes depart from the [110]-direction. The atomic p-functions of the two F atoms will make different angles with the applied magnetic field, and their non-equivalence will lead to extra anisotropic hyperfine structure.

The V_t center in LiF (Fig. 34) may be described as an F_3^{2-} molecule localized at a cluster of two anion and one cation vacancies. Like V_F, it has a bent configuration such that a line joining its ends lies along a [110] direction. Understanding of this center rests principally upon interpretation of hyperfine tensor data[7,9].

---

[1] C. P. SLICHTER: Principles of Magnetic Resonance, p. 195. New York: Harper and Row 1963.
[2] C. J. DELBECQ, B. SMALLER, and P. H. YUSTER: Phys. Rev. **111**, 1235 (1958).
[3] C. J. DELBECQ, W. HAYES, and P. H. YUSTER: Phys. Rev. **121**, 1043 (1961).
[4] J. W. WILKINS, and J. R. GABRIEL: Phys. Rev. **132**, 1950 (1963).
[5] D. SCHOEMAKER, C. J. DELBECQ, and P. H. YUSTER: Bull. Am. Phys. Soc. **9**, 629 (1964).
[6] See footnote 4, p. 243.
[7] W. KÄNZIG: J. Phys. Chem. Solids **17**, 88 (1960).
[8] See footnote 6, p. 241.
[9] M. H. COHEN, W. KÄNZIG, and T. O. WOODRUFF: J. Phys. Chem. Solids **11**, 120 (1959).

A defect analogous to the H-center in the alkali halides is a hole trapped on a linear array of four fluorine atoms, but oriented along a cube edge in CaF_2. If the atoms are labeled 3—1—2—4, the 1—2 bond is much shorter than the other two, as in the usual H-center. The hole is located 98% of the time on atoms 1 and 2, and 2% of the time on atoms 3 and 4, as judged by the magnitudes of (isotropic and axial) hyperfine splitting constants[1].

Evidence about one hydrogen impurity site in MgO has been obtained by ESR, ENDOR and infrared studies of a trapped hole center (V_{OH} center)[2]. The spin Hamiltonian for the V_{OH} spectrum which has a direct nuclear interaction exceeding the hyperfine interaction is[3]:

$$\mathcal{H} = \beta[g_\parallel S_z H_z + g_\perp (S_x H_x + S_y H_y)] - g_n \beta_n \mathbf{H} \cdot \mathbf{I} + A S_z I_z + B(S_x I_x + S_y I_y).$$

Here $g_\parallel = 2.0033$, $g_\perp = 2.0396$ and $g_n \approx 5.58$ refers to the proton. ENDOR measurements show unambiguously that hydrogen is the nucleus producing the hyperfine splitting and give $A/hc = 4.795$ MHz, and $B/hc = 2.331$ MHz. From the anisotropic splitting, one estimates the positive hole-hydrogen distance as 0.32 nm, appropriate to the orientation (98)

$$\overset{++}{(Mg)} \quad (O) \quad [\cdot\cdot] \quad (HO) \quad \overset{++}{(Mg)}$$

(98)

assuming the hole is localized on the opposite side of the cation vacancy. An OH-stretching band at 3323 cm^{-1} is assigned to OH in the V_{OH} center. On loss of the hole, this line shifts to 3296 cm^{-1} and thus corresponds to an OH oriented with H toward a positive ion vacancy. A spectrum very similar to that of the V_{OH} center is that due to a hole center in which a fluoride ion substitutes for an oxide ion opposite a trapped hole in MgO[4].

Upon γ-irradiation, a high alkali content borate glass gives a four-line spectrum presumably due to interaction of a positive hole with a $^{11}BO_4$ unit. Enrichment with ^{10}B gives a seven-line spectrum to support this view[5].

A hole center in x-irradiated quartz[6] shows a hyperfine sextet from a nearby aluminum atom.

III. Other defects in insulators.

31. Interstitial centers. Atoms in solids. — Observation in various hydrogen-containing solids after x- or γ-irradiation of a hyperfine doublet with a separation near the value of 506.8 G for gaseous hydrogen atoms[7] is good evidence that such atoms are trapped in sites of low reactivity. Further verification is given by replacement of H by D, which yields a triplet with appropriately reduced splitting. Some of the hosts studied include solid H_2[8]; ice[9-11], HI, H_2O and NH_3 in rare-gas

---

[1] W. Hayes, and J. W. Twidell: Proc. Phys. Soc. (London) **79**, 1295 (1962).
[2] P. W. Kirklin, P. Auzins, and J. E. Wertz: J. Phys. Chem. Solids **26**, 1067 (1965).
[3] F. S. Ham, in: H. H. Woodbury, and G. W. Ludwig: Phys. Rev. **124**, 1083 (1961).
[4] J. E. Wertz, and P. V. Auzins: Phys. Rev. **139**, A 1645 (1965).
[5] S. Lee, and P. J. Bray: J. Chem. Phys. **40**, 2982 (1964).
[6] J. H. E. Griffiths, J. Owen, and I. M. Ward: Repts. Conf. Defects in Cryst. Solids, Bristol, Phys. Soc. (London) **1954**, 81. — M. C. M. O'Brien, and M. H. L. Pryce: ibid., p. 88.
[7] J. M. B. Kellogg, I. I. Rabi, N. F. Ramsey, and J. R. Zacharias: Phys. Rev. **57**, 677 (1950).
[8] C. K. Jen, S. Foner, E. L. Cochran, and V. A. Bowers: Phys. Rev. **112**, 1169 (1958).
[9] L. Piette, R. Rempel, H. Weaver, and J. Fluornoy: J. Chem. Phys. **30**, 1623 (1959).
[10] M. S. Matheson, and B. Smaller: J. Chem. Phys. **23**, 521 (1955).
[11] S. Siegel, J. M. Fluornoy, and L. H. Baum: J. Chem. Phys. **34**, 1782 (1961).

matrices[1,2]; $HClO_4$, H_2SO_4, H_3PO_4[3,4]; KCl with U_2-centers[5-7]; quartz or silica glass[8]; CaF_2[9]; $CaSO_4 \cdot \tfrac{1}{2} H_2O$[10]; various inorganic phosphates[11,12], and in frozen products of discharges[1,2,13]. Additionally one finds hydrogen (or deuterium) atoms during electron irradiation of liquid CH_4 (or CD_4)[14]. The stability is a function both of trapping site and of host. In ice, some H atoms react rapidly at 15 °K while others disappear only slowly at 60 °K[15]. Hosts providing the other extreme of stability are CaF_2[9] and $CaSO_4 \cdot \tfrac{1}{2} H_2O$[10], in which H atoms show stability at room temperature. For the oxygen acids, much greater yields are obtained in glasses than in single crystals[4]. Line widths range from 0.1 G in quartz[8] to 68 G in KCl[5]. With such a large value of hyperfine *splitting* (measured separation in field of

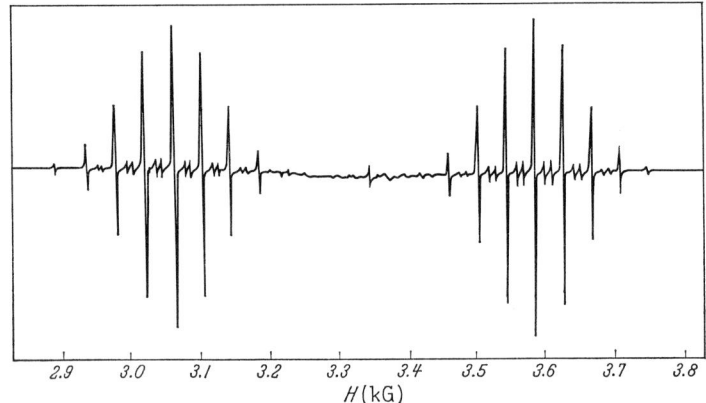

Fig. 37. ESR spectrum of interstitial hydrogen atoms in CaF_2 [J. L. HALL, and R. T. SCHUMACHER: Phys. Rev. 127, 1892 (1962)].

lines) it is essential to use the Breit-Rabi expressions to obtain accurate *coupling constants* (frequency separation at constant H) or of *g*-values. The coupling constants range from less than 1391.4[10] to 1453.1 MHz[8]. Values in excess of that of the free-atom have been attributed to "compression" of the wave function of the electron (in rare gases)[2] due to overlap interactions with the matrix. Small values are attributed to VAN DER WAALS' interaction which spreads the wave function[2]. Deuterium atoms likewise show a range of coupling constants[5], but the *g*-values are very little different from those of hydrogen atoms in phos-

---

[1] S. N. FONER, E. L. COCHRAN, V. A. BOWERS, and C. K. JEN: J. Chem. Phys. 32, 963 (1960); — Phys. Rev. 104, 846 (1956).
[2] F. J. ADRIAN: J. Chem. Phys. 32, 972 (1960).
[3] R. LIVINGSTON, H. ZELDES, and E. H. TAYLOR: Discussions Faraday Soc. 19, 166 (1955)
[4] R. LIVINGSTON, and A. J. WEINBERGER: J. Chem. Phys. 33, 499 (1960).
[5] C. J. DELBECQ, B. SMALLER, and P. H. YUSTER: Phys. Rev. 104, 599 (1956).
[6] F. KERKHOFF, W. MARTIENSSEN, u. W. SANDER: Z. Physik 173, 184 (1963).
[7] J. SPAETH, quoted in H. SEIDEL, and H. C. WOLF: Phys. stat. sol. 11, 3 (1965).
[8] R. A. WEEKS, and M. ABRAHAM: J. Chem. Phys. 42, 68 (1965).
[9] J. L. HALL, and R. T. SCHUMACHER: Phys. Rev. 127, 1892 (1962).
[10] H. KON: J. Chem. Phys. 41, 573 (1964).
[11] P. W. ATKINS, N. KEEN, M. C. R. SYMONS, and H. W. WARDALE: J. Chem. Soc. 1963, 5594. Errors in *g*-values are corrected in Ref. 12.
[12] S. OGAWA, and R. W. FESSENDEN: J. Chem. Phys. 41, 1516 (1964).
[13] T. COLE, and J. T. HARDING: J. Chem. Phys. 28, 993 (1958).
[14] R. W. FESSENDEN, and R. H. SCHULER: J. Chem. Phys. 39, 2147 (1963).
[15] See footnote 11, p. 245.

phates[1,2], viz., 2.0016 to 2.0024. Only one site in krypton ($g=1.9997$) and one in xenon ($g=2.0006$) give values outside these limits. Overlap of H atom orbitals with uncompensated p-orbitals of neighbors also leads to spin-orbit interaction, with a shift (lowering) in g-value.

The double set of nine fluorine superhyperfine lines (separation 522 G) from H atoms in CaF_2 (Fig. 37) (further corroborated by ENDOR spectra of more distant neighbors) clearly indicates occupation of some of the alternate sites not occupied by the calcium ion[3]. Hydride ions at fluorine sites are the precursors of the interstitial hydrogen atoms[4] and apparently also the product of thermal bleaching[5].

In this case and others, (e.g., frozen H_2SO_4)[6], weak "forbidden" lines provide redundant identification of the atoms giving the doublet. Between the two allowed ($\Delta M_s=1$, $\Delta M_I=0$) lines for hydrogen atoms in γ-irradiated frozen H_2SO_4, one observes $\Delta M_s=1$, $\Delta M_I=1$ lines separated by twice the proton NMR frequency at the applied field. These are in precise accord with the expressions:

$$E_4 = g\beta H/2 + A/4 + g_n\beta_n H/2,$$
$$E_3 = g\beta H/2 - A/4 - g_n\beta_n H/2,$$
$$E_2 = -g\beta H/2 + A/4 - g_n\beta_n H/2,$$
$$E_1 = -g\beta H/2 - A/4 + g_n\beta_n H/2.$$

The forbidden transitions E_3-E_1 and E_4-E_2 respectively give $g\beta H - g_n\beta_n H$ and $g\beta H + g_n\beta_n H$. Forbidden lines are seen as weak "satellites" between each pair of allowed lines.

Hydride ions (U-centers) may be incorporated substitutionally in alkali halides by doping or by heating additively-colored samples in hydrogen. Irradiation in the U bands causes the H^- ions to move to interstitial sites as U_1 centers. Both U and U_1 centers are diamagnetic. However, irradiation at 77 °K also leads to U_2 centers, which show an ESR spectrum of two line groups separated by 500 G[7] (Fig. 38). Replacement by deuterium leads to the expected triplet of line groups, confirming the assignment. However, the much greater intensity of U_2 spectrum in OH^--containing alkali halides suggests that the interstitial hydrogen comes from hydroxyl impurities[8]. With $H\|[100]$, each group contains 13 lines separated by 15.9 G in NaCl, 8.9 G in KCl and 47.5 G in KBr[8]. The resolved hyperfine structure of the center appears to be due to interaction of a hydrogen atom with four halogen ions tetrahedrally surrounding it. Hyperfine interaction by the four cations also tetrahedrally surrounding it is not sufficient to give resolved lines, but these have been studied by ENDOR[9]. The magnitudes of the U_2 splittings are approximately reproduced by calculation[10]. Although in other cases the evidence is not as compelling as for H in CaF_2 and U_2 centers, the hydrogen atoms in most solids are assumed to occupy interstitial sites[11,12]. Shfs

---

[1] See footnote 11, p. 246.
[2] See footnote 12, p. 246.
[3] See footnote 9, p. 246.
[4] W. Hayes: Private communication.
[5] B. Welber: J. Chem. Phys. **43**, 3015 (1965).
[6] See footnote 3, p. 246.
[7] C. J. Delbecq, B. Smaller, and P. H. Yuster: Phys. Rev. **104**, 599 (1956).
[8] F. Kerkhoff, W. Martienssen u. W. Sander: Z. Physik **173**, 184 (1963).
[9] J. M. Spaeth, cited by H. Seidel, and H. C. Wolf: Phys. stat. sol. **11**, 3 (1965).
[10] H. Mimura, and Y. Uemura: J. Phys. Soc. Japan **14**, 1011 (1959).
[11] See footnote 8, p. 246.
[12] See footnote 10, p. 246.

from ^{129}Xe ($I=\frac{1}{2}$) and ^{131}Xe ($I=\frac{3}{2}$) neighbors is consistent with interstitial sites for H atoms produced by photolysis of HI, H_2O or NH_3 in xenon[1,2]. However, hydrogen atoms incorporated into a rare gas matrix appear to enter substitutionally, with no superhyperfine interaction observed.

Nitrogen atoms have been produced in a variety of matrices, including H_2, N_2, and CH_4[3] and in NaN_3[4]. The g-factor is close to 2.0020 in all; the hyperfine splittings ranges from 4.1 G in H_2 to 6.2 G in NaN_3. In KN_3 upon x-irradiation at 77 °K, the initially-produced N atoms and N_2^- ions produce N_4^-, all three defects being identified by their ESR spectra[5,6].

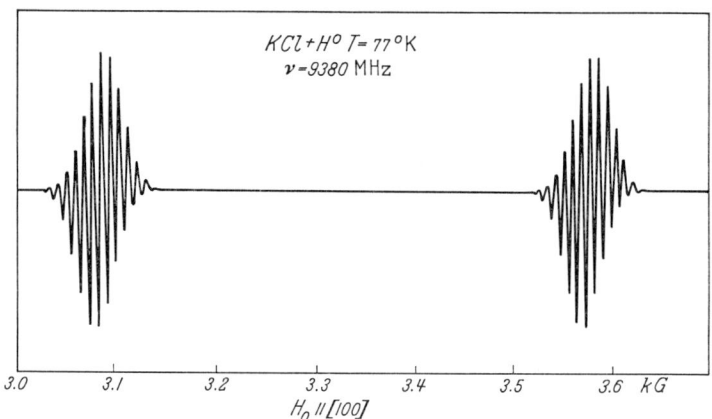

Fig. 38. ESR spectrum of interstitial hydrogen atoms in KCl (U_2 centers).

32. Anion substitutional centers. The O_2^- ion was first observed in alkali superoxides (e.g., KO_2) where it shows axial symmetry with $g_{\parallel}=2.165$ and $g_{\perp}=2.00$[7,8]. It is also formed by irradiation of alkali halides heated in oxygen[9,10]. It substitutes for a halide ion and its molecular axis lies along a [110]-type direction (Fig. 39). Its principal g-values in KBr are 1.9268, 1.9314 and 2.5203. The low value of $\lambda/\Delta=0.23$ found for O_2^- is interpreted in terms of a λ value several times lower than that of atomic oxygen, viz. 226 cm^{-1}. By application of uniaxial stress, one removes the orientational degeneracy of the six $\langle 110 \rangle$-type axes, and reorientation of the defect axes is seen in the ESR spectrum. The phenomenon has been termed "paraelasticity"[10]. The analogous S_2^- ion in KBr (formed by exposure to sulfur vapor) substitutes similarly but shows axial sym-

---

[1] See footnote 1, p. 246.
[2] See footnote 2, p. 246.
[3] C. K. JEN, S. FONER, E. L. COCHRAN, and V. A. BOWERS: Phys. Rev. **112**, 1169 (1958).
[4] G. J. KING, F. F. CARLSON, B. S. MILLER, and R. C. MCMILLAN: J. Chem. Phys. **34**, 1499 (1961).
[5] A. J. SHUSKUS, C. G. YOUNG, O. R. GILLIAM, and P. W. LEVY: J. Chem. Phys. **33**, 622 (1960).
[6] D. W. WYLIE, A. J. SHUSKUS, C. G. YOUNG, O. R. GILLIAM, and P. W. LEVY: Phys. Rev. **125**, 451 (1962).
[7] J. E. BENNETT, D. J. E. INGRAM, M. C. R. SYMONS, P. GEORGE, and J. S. GRIFFITH: Phil. Mag. **46**, 443 (1955).
[8] J. E. BENNETT, D. J. E. INGRAM, and D. SCHONLAND: Proc. Phys. Soc. (London) **69 A**, 556 (1956).
[9] W. KÄNZIG, and M. H. COHEN: Phys. Rev. Letters **3**, 509 (1959).
[10] W. KÄNZIG: J. Phys. Chem. Solids **23**, 479 (1962).

metry with $g_\| = 3.5010$ and $g_\perp = 1.05$ [1]. The larger deviations of $g_\|$ and $g_\perp$ from the free electron value than for O_2^- are consistent with the larger value of $\lambda/\Delta = 1.45$. S_2^- in KBr shows a series of luminescence bands separated by vibrational quanta [2], just as does O_2^- [3]. Yet another host for O_2^- is a zeolite (with either Ba^{2+} or Na^+ for charge compensation) which has been γ-irradiated in oxygen [4,5].

The O^- ion as a substitutional impurity in alkali halides is said to arise from O_2^- centers by reduction with electrons and irradiation with near-UV light [6]. This center has the [100] direction as a symmetry axis, with $g_\| = 1.980$, $g_\perp = 2.257$ in KCl. A Jahn-Teller displacement toward a neighboring $^{39}K^+$ ion appears to be responsible for the four-line hyperfine pattern with $A_\| = 3.9$ G, $A_\perp = 1.8$ G [7].

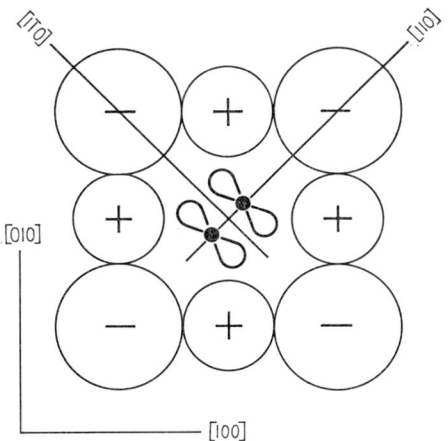

Fig. 39. The O_2^- center in an alkali halide [from W. Känzig, and M. H. Cohen: Phys. Rev. Letters 3, 509 (1959)].

An anisotropic center, presumed to be O_2^+ has been reported [8] in KCl or KBr which has either been doped in the melt with NO_2^- or NO_3^-, or which has been γ-irradiated after heating in NO.

Adsorption of oxygen on MgO which has been irradiated with ultraviolet light or with γ-rays, or upon ZnO gives rise to surface O_2^- centers [9-11].

33. Transition metal ions in insulating solids. The area covered by this subheading under Impurity Centers, is so vast that a full chapter would be required to give an adequate survey. Not only is each d- or f-electron configuration in each site symmetry a fit subject for separate consideration, but there is much to be said about each of several isoelectronic ions in similar environments (e.g.,

---

[1] J. R. Morton: J. Chem. Phys. **43**, 3418 (1965).
[2] J. H. Schulman, and R. D. Kirk: Solid State Comm. **2**, 105 (1964).
[3] J. Rolfe, F. R. Lipsett, and W. J. King: Phys. Rev. **123**, 447 (1961).
[4] P. H. Kasai: J. Chem. Phys. **43**, 3322 (1965).
[5] D. N. Stamires, and J. Turkevich: J. Am. Chem. Soc. **86**, 757 (1964).
[6] W. Sander: Z. Physik **169**, 353 (1962).
[7] W. Sander: Naturwissenschaften **51**, 404 (1964).
[8] C. Jaccard: Phys. Rev. **124**, 60 (1961).
[9] R. L. Nelson, and A. J. Tench: J. Chem. Phys. **40**, 2736 (1964).
[10] J. H. Lunsford, and J. P. Jayne: J. Phys. Chem. **69**, 2182 (1965); — J. Chem. Phys. **44**, 1487 (1966).
[11] A. J. Tench, and R. L. Nelson: J. Chem. Phys. **44**, 1714 (1966).

Table 16. *Transition metal ions in simple structures.*

Atom	Valence State	Host			
		MgO	Al$_2$O$_3$	ZnO	CaF$_2$
Ti	3	1, 2	3		
V	2	4–6	7	8	
	3		9, 10		
	4		11		
Cr	1	12			
	3	5, 6, 13–15			
			16–18		
	4		19		
Mn	2	20–22	23	24, 25	
	4	26, 27	7, 28		
Fe	1	29			
	2	30–32			
	3	5, 15, 33, 47	34, 35	36	
Co	1	29, 37			
	2	38–40	41	8	
Ni	2	29, 42, 43	44, 45		
	3	27	46–48	49	
Cu	2	27, 29, 50, 51		52–54	
	3	46, 55			
Ce	3				56
Nd	3				57
Eu	2	4			58
Gd	3		59		60, 61
Tb	3				62, 63
Ho	2				64
Er	3	32			65
Tm	2				62, 64
Yb	3				62
Ru	1	50			
	3		46		
Rh	2	50			
	3		46		
Pd	3	50	47		
Ir	3		46		
Pt	3		48		
U	3				57, 66
	4				66, 67

* Revised assignments. The ions Ru^{+1}, Rh^{+2}, and Pd^{+3} are thus analogous to Fe^{+1}, Co^{+2} and Ni^{+3}.

[1] B. BLEANEY: Proc. Phys. Soc. (London) A **63**, 407 (1950).
[2] J. E. WERTZ, G. S. SAVILLE, L. HALL, and P. AUZINS: Proc. Brit. Ceram. Soc. **1**, 59 (1964).
[3] L. S. KORNIENKO, and A. M. PROKHOROV: PR-Low**, p. 126. Soviet Phys. JETP **11**, 1189 (1960).
[4] W. LOW: Phys. Rev. **101**, 1827 (1956).
[5] W. LOW: Ann. N. Y. Acad. Sci. **72**, 69 (1958).
[6] J. S. VAN WIERINGEN, and J. G. RENSEN: PR-Low**, p. 105.
[7] N. LAURANCE, and J. LAMBE: Phys. Rev. **132**, 1029 (1963).
[8] T. L. ESTLE, and M. DE WIT: Bull. Am. Phys. Soc. **6**, 445 (1961).
[9] A. ABRAGAM, and M. H. L. PRYCE: Proc. Roy. Soc. (London) A **205**, 135 (1951).
[10] G. M. ZVEREV, and A. M. PROKHOROV: J. Exptl. Theor. Phys. (U.S.S.R.) **34**, 1023 (1958).
[11] J. LAMBE, and C. KIKUCHI: Phys. Rev. **118**, 71 (1960).
[12] J. WERTZ, and P. AUZINS: To be published.
[13] W. LOW: Phys. Rev. **105**, 801 (1957).
[14] J. E. WERTZ, and P. AUZINS: Phys. Rev. **106**, 484 (1957); — J. Chem. Phys. **43**, 1229 (1965).
[15] W. M. WALSH jr.: Phys. Rev. **122**, 762 (1961).
[16] J. E. GEUSIC: Phys. Rev. **102**, 1252 (1956).

References of Table 16 (continued).

[17] R. W. Terhune, J. Lambe, C. Kikuchi, and J. Baker: Phys. Rev. **123**, 1265 (1961).
[18] N. Laurance, E. C. McIrvine, and J. Lambe: J. Phys. Chem. Solids **23**, 515 (1962).
[19] R. H. Hoskins, and B. H. Soffer: Phys. Rev. **133**, A 490 (1964).
[20] W. Low: Phys. Rev. **105**, 793 (1957).
[21] J. E. Drumheller, and R. S. Rubins: Phys. Rev. **133**, A 1099 (1964).
[22] E. R. Feher: Phys. Rev. **136**, A 145 (1964).
[23] W. Low, and J. T. Suss: Phys. Rev. **119**, 132 (1960).
[24] P. B. Dorain: Phys. Rev. **112**, 1058 (1958).
[25] J. Schneider, u. J. R. Sircar: Z. Naturforsch. **17 a**, 570 (1962).
[26] B. Henderson, and T. P. P. Hall: Proc. Phys. Soc. (London) **90**, 511 (1967).
[27] U. Höchli, K. A. Müller, and P. Wysling: Phys. Letters **15**, 5 (1965).
[28] S. Geschwind, P. Kisliuk, M. P. Klein, J. P. Remeika, and D. L. Wood: PR-Low**, p. 113.
[29] J. W. Orton, P. Auzins, J. H. E. Griffiths, and J. E. Wertz: Proc. Phys. Soc. (London) **78**, 554 (1961).
[30] W. Low: Phys. Rev. **101**, 1827 (1956).
[31] W. Low, and M. Weger: Phys. Rev. **118**, 1130 (1960).
[32] W. Low, and E. Offenbacher: Solid State Physics, vol. 17, p. 135, (F. Seitz, and D. Turnbull, eds.). New York: Academic Press 1965.
[33] W. Low: Proc. Phys. Soc. (London) B **69**, 1169 (1956).
[34] L. S. Kornienko, and A. M. Prokhorov: J. Exptl. Theoret. Phys. (U.S.S.R.) **33**, 805 (1957).
[35] G. S. Bogle, and H. F. Symmons: Proc. Phys. Soc. (London) B **73**, 531 (1959).
[36] W. M. Walsh jr., and L. W. Rupp jr.: Phys. Rev. **126**, 952 (1962).
[37] J. W. Orton, P. Auzins, and J. E. Wertz: Phys. Rev. **119**, 1691 (1960).
[38] B. Bleaney, and W. Hayes: Proc. Phys. Soc. (London) B **70**, 626 (1957).
[39] W. Low: Phys. Rev. **109**, 256 (1958).
[40] D. J. I. Fry, and P. M. Llewellyn: Proc. Roy. Soc. (London) A **266**, 84 (1962).
[41] G. M. Zverev, and A. M. Prokhorov: Soviet Phys. JETP **9**, 451 (1959); **12**, 41 (1961).
[42] W. Low: Phys. Rev. **109**, 247 (1958).
[43] J. W. Orton, P. Auzins, and J. E. Wertz: Phys. Rev. Letters **4**, 128 (1960).
[44] S. A. Marshall, and A. R. Reinberg: J. Appl. Phys. **31**, 336 S (1960).
[45] S. A. Marshall, T. T. Kikuchi, and A. R. Reinberg: Phys. Rev. **125**, 453 (1962).
[46] S. Geschwind, and J. P. Remeika: J. Appl. Phys. **33**, 370 S (1962).
[47] R. Lacroix, U. Höchli, and K. A. Müller: Helv. Phys. Acta **37**, 627 (1964).
[48] U. Höchli, and K. A. Müller: Phys. Rev. Letters **12**, 730 (1964).
[49] W. C. Holton, J. Schneider, and T. L. Estle: Phys. Rev. **133**, A 1638 (1964).
[50] P. Auzins, J. W. Orton, and J. E. Wertz: PR-Low**, p. 90.
[51] R. Coffman: Phys. Letters **19**, 425 (1965).
[52] H. Kamimura, and A. Yariv: Bull. Am. Phys. Soc. **8**, 23 (1963).
[53] R. E. Dietz, H. Kamimura, M. D. Sturge, and A. Yariv: Phys. Rev. **132**, 1559 (1963).
[54] M. de Wit, and T. L. Estle: Bull. Am. Phys. Soc. **8**, 24 (1963).
[55] W. E. Blumberg, J. Eisinger, S. Geschwind, and J. P. Remeika: PR-Low**, p. 125; — Phys. Rev. **130**, 900 (1963).
[56] J. M. Baker, W. Hayes, and M. C. M. O'Brien: Proc. Phys. Soc. (London) A **254**, 273 (1958).
[57] B. Bleaney, P. M. Llewellyn, and D. A. Jones: Proc. Phys. Soc. (London) B **69**, 858 (1956).
[58] J. N. Baker, J. P. Hurrell, and F. I. B. Williams, PR-Low**, p. 202.
[59] S. Geschwind, and J. P. Remeika: Phys. Rev. **122**, 757 (1961).
[60] J. Sierro: J. Chem. Phys. **34**, 2183 (1961).
[61] B. Bleaney: J. Appl. Phys. **33**, 358 S (1962).
[62] W. Low, and U. Ranon: PR-Low**, p. 167.
[63] P. A. Forrester, and C. E. Hempstead: Phys. Rev. **126**, 923 (1962).
[64] W. Hayes, and J. W. Twidell: PR-Low**, p. 163.
[65] J. M. Baker, W. Hayes, and D. A. Jones: Proc. Phys. Soc. (London) **73**, 942 (1959).
[66] R. S. Title: PR-Low**, p. 178.
[67] A. Yariv: PR-Low**, p. 189.

** Paramagnetic Resonance, vol. I. (W. Low, ed.) New York: Academic Press 1963.

V^{2+}, Cr^{3+}, Mn^{4+}). Even restricting consideration to Cr^{3+}, one finds e.g., that the extra trigonal crystal field component in Al_2O_3 as compared with an octahedral environment has profound effects upon the ESR spectrum. The iron-group (3 d) elements have been the most extensively investigated. Next in order are the rare earth (4f), the 4d, the 5d and 5f (actinide) groups. There have appeared several general treatments[1-3] and a number of very useful tabulations of spin Hamiltonian parameters of transition metal ions[3-5].

We shall be content to cite some examples of transition metal ions for a limited number of hosts of rather simple structure. Ions in alkali halides and in simple oxides, and CaF_2 will be mentioned here. Transition metals in semiconductors are considered in Sect. E. Transition metals in solution are described in Sect. F.II.

The most frequently investigated ion in the alkali halides is Mn^{2+}, which in NaCl shows a wide range of behavior with temperature[6,7]. At 25 °C exchange interaction between the Mn^{2+} ions in precipitates leads to a single broad line. Upon heating at 175 °C the broad line disappears and one observes spectra showing Mn^{2+} ions in isolated sites, with a cation vacancy in the nearest or next-nearest neighbor positions[7,8], or associated with a charge-compensating impurity[7,9]. As the temperature is raised, the vacancies dissociate from the ion-vacancy pairs; their binding energy is found to be 0.4 eV. Similar associated vacancy centers have been studied in LiCl and KCl. At intermediate temperatures, vacancy migration among equivalent positions about the Mn^{2+} ion leads to line broadening. Dielectric loss measurements have confirmed this interpretation[7]. In LiF[10], NaF[11] and KF[12], the fluorine superhyperfine structure is observed; chlorine shfs has been observed with Mn^{2+} in NaCl[13]. Spectra of Mn^{2+} associated with F^- and a cation vacancy are reported[9].

Other divalent ions observed in NaF (after UV irradiation) include Fe^{+2}, Co^{+2}, and Ni^{+2} [14]. Ag^{+2} is reported in KCl[15], while Ag^0 has been studied in KCl both by ESR[16] and by ENDOR[17]. Its spectrum is akin to that of the U_2 center, with a primary doublet split by six Cl^- neighbors. The monovalent ions Cr^+, Fe^+, Co^+ and Ni^+ are produced by x-irradiation of NaF at room temperature, while Fe^+ and Ni^+ are observed in LiF[14]. Cr^+ has been observed in NaCl[18]. For Ni^+ in NaF and Ag^{2+} in KCl, the symmetry is tetragonal rather than cubic.

[1] B. BLEANEY, and K. W. H. STEVENS: Repts. Progr. in Phys. **16**, 108 (1953). — J. S. GRIFFITH: The Theory of Transition Ions, Chap. 12.: Cambridge Univ. Press 1961.
[2] G. E. PAKE: Paramagnetic Resonance. New York: Benjamin 1962. — W. LOW: Paramagnetic Resonance in Solids, Suppl. 2 of Solid State Physics (F. SEITZ and D. TURNBULL, eds). New York: Academic Press 1960. — W. LOW, and E. L. OFFENBACHER: Same series, vol. 17, p. 135, 1965 (Complex oxides).
[3] S. A. AL'TSCHULER, and B. M. KOZYREV: Electron Paramagnetic Resonance. Translation editor, C. P. POOLE jr. New York: Academic Press 1964.
[4] K. D. BOWERS, and J. OWEN: Repts. Progr. in Phys. **18**, 304 (1955).
[5] J. W. ORTON: Repts. Progr. in Phys. **22**, 204 (1959).
[6] E. E. SCHNEIDER, and J. E. CAFFYN: Repts. Conf. on Defects in Crystalline Solids, Bristol. Phys. Soc. (London) **1955**, 74. — P. A. FORRESTER, and E. E. SCHNEIDER: Proc. Phys. Soc. (London) **69**, 833 (1956).
[7] G. D. WATKINS: Phys. Rev. **113**, 79, 91 (1959).
[8] K. MORIGAKI, M. FUJIMOTO, and J. ITOH: J. Phys. Soc. Japan **13**, 1174 (1958).
[9] Y. YOKOZAWA, and K. KAZUMATA: J. Phys. Soc. Japan **16** 694 (1961).
[10] T. T. CHANG W. H. TANTILLA and J. S. WELLS: J. Chem. Phys. **39** 2453 (1963).
[11] W. HAYES and D. A. JONES: Proc. Phys. Soc. (London) **71** 503 (1958).
[12] W. J. VEIGELE and W. H. TANTILLA: J. Chem. Phys. **41**, 274 (1964).
[13] H. SEIDEL, and H. C. WOLF: Phys. stat. sol. **11**, 3 (1965).
[14] W. HAYES: J. Appl. Phys. **33**, 329 S (1962); — Discussions Faraday Soc. **26**, 58 (1958).
[15] W. HAYES: J. Phys. Soc. Japan **17**, Suppl. I, 456 (1962).
[16] C. J. DELBECQ, W. HAYES, M. C. M. O'BRIEN, and P. H. YUSTER: Proc. Roy. Soc. (London) A **271**, 243 (1963).
[17] H. SEIDEL: Phys. Letters **6**, 150 (1963).
[18] B. WELBER: Phys. Rev. **138**, A 1481 (1965).

This section is concluded with references to transition metals in MgO, Al_2O_3, ZnO and CaF_2 (Table 16) as representative of important classes of solids. A compilation of work on magnetic ions in rutile, perovskites, spinels and garnet structures has appeared recently[1].

E. Defects in semiconductors.

Electron spin resonance studies of semiconductors cover a very wide range of material properties. On the metallic side, one may be able to see only a conduction electron resonance as in graphite[2] or in InSb with $g = -50.7$ [3]. On the other hand, in ZnS or other II—VI compounds one may observe substitutional impurities with g-values hardly different from those in insulators. Semiconductors commonly have tetrahedral coordination about one type of atom in the II—VI compounds, while silicon or germanium of group IV have the diamond structure. Covalent character is predominant in this case, with increase in ionic character as one goes to II—VI compounds. In ionic compounds, a slight measure of covalence may manifest itself in such ways as reduction in spin-orbit coupling constant of an impurity ion, abnormally small hyperfine splitting of a substitutional ion or by superhyperfine splitting. Semiconductors (at least some of them) have the advantage that one is able to control the charge state of a particular type of atom by controlling the total impurity content.

Paramagnetic centers in silicon have been among the most extensively investigated. These include the following:

34. α) *Shallow Donor Impurities (P, As, Sb, Bi).* These are Group V elements whose extra (donor) electron in the neutral atom has only a small binding energy (~ 0.05 eV) compared to the band gap of 1.2 eV. The donor electron wave function is hydrogenic. In low concentrations one may observe hyperfine splitting by donor nuclei[4]. One of the earliest applications of ENDOR was to the mapping of hyperfine interaction of the donor electron at various ^{29}Si sites. The remarkable observation was made that even at the twenty-third-nearest-neighbor positions[4] the wave function amplitude is greater than that at the nearest-neighbor positions. At low donor concentrations, the donor spin-lattice relaxation time is very long — of the order of 10 sec at 4.2 °K[5-8]. By application of uniaxial stresses, it has been possible to obtain detailed information about the band structure.

In silicon simultaneously doped with phosphorus and arsenic, one is able to observe progressively increasing exchange coupling between impurity atoms as the phosphorus concentration is increased from 10^{16} to 10^{18} cm^{-3} while the arsenic concentration is kept at 10^{17} cm^{-3} [9,10]. Pair line positions are given by $\gamma_e h H_0 + \frac{1}{2}(a_1 m_1 + a_2 m_2)$, where a_1 and a_2 are coupling constants and m_1 and m_2 are quantum numbers for the impurity nuclei[11]. γ_e is the electron magnetogyric ratio.

---

[1] W. Low, and E. L. Offenbacher: Solid State Physics, Vol. 17 p. 135. (F. Seitz, and D. Turnbull, eds.). New York: Academic Press 1965.
[2] G. Wagoner: Phys. Rev. **118**, 647 (1960).
[3] G. Bemski: Phys. Rev. Letters **4**, 62 (1960).
[4] G. Feher: Phys. Rev. **114**, 1219 (1959).
[5] A. Honig: Phys. Rev. **96**, 234 (1954).
[6] A. Honig, and J. Combrisson: Phys. Rev. **102**, 917 (1956).
[7] D. Pines, J. Bardeen, and C. P. Slichter: Phys. Rev. **106**, 489 (1957).
[8] G. Feher, R. C. Fletcher, and E. A. Gere: Phys. Rev. **100**, 1784 (1955).
[9] Y. Sugiura: J. Phys. Soc. Japan **20**, 1272 (1965).
[10] R. C. Fletcher, W. A. Yager, G. L. Pearson, and F. R. Merritt: Phys. Rev. **95**, 844 (1954).
[11] C. P. Slichter: Phys. Rev. **99**, 479 (1955).

β) *Deep Donor Impurities (S)*. Sulfur as S^+ is reported to be a deep donor. ENDOR studies show hyperfine interaction with eight different ^{29}Si sites. A pair center, presumed to be S_2^+, and an Fe—S center are also observed[1].

35. Shallow acceptor impurities (B, Al, Ga, In). Observation has required lifting the degeneracy of valence bands by uniaxial stress[2].

36. Transition metal ions. The d^3 and the d^5 to d^9 states occur interstitially, while d^2 and d^5 state ions also occur substitutionally[3-5]. Depending upon the Fermi level, these ions may donate or accept an electron (in some cases, several). Unlike the hydrogenic states of the shallow donors, these states lie deep within the valence band. Ions of the 4d and 5d groups as well as the 3d have been studied[3].

37. Impurity pairs. These are of the type: [Interstitial transition metal ion]$^+$ [Substitutional acceptor]$^-$. The cation may be Cr, Mn or Fe, and the acceptor B, Al, Ga, In, Zn, Cu, Au or Pt[3]. An unusual pair is Li—O[6].

A Mn_4 cluster is also reported[7].

38. Radiation-induced centers. Neutron, electron, or ionizing radiation leads to vacancies, either isolated, associated as divacancies or with some impurity[8,9]. The Si—A center (in pulled *n*-type silicon) has an interstitial oxygen-vacancy pair, with the electron located primarily in the antibonding orbital of a pair of Si atoms adjoining the vacancy[10-12]. The Si—E center observed in P-doped vacuum floating zone silicon involves a substitutional phosphorus-vacancy pair. Two of the adjoining Si atoms bond to one another, and the unpaired electron is in an orbital pointing from the third Si atom toward the vacancy[8,12]. A number of other centers are known. Reorientation of Si bonds about the vacancy occurs by low temperature stress. High temperature stress leads to reorientation of the phosphorus-vacancy direction. Both singly-positive and singly-negative divacancies may be produced by electron irradiation. The number of "broken" bonds is reduced from six to two by formation of two "bent" bonds between adjacent atoms. This leaves an extended orbital passing through the divacancy from two remote Si atoms. One or three electrons are in this orbital. This orbital is reoriented by low-temperature stress. High temperature stress reorients the divacancy axis, inducing migration[9].

Germanium is a much less favorable host than silicon for ESR detection of paramagnetic centers, owing to a considerably larger spin-orbit coupling, resulting in short spin-lattice relaxation times. Shallow donor centers observed include (as neutral atoms) phosphorus, arsenic[13], antimony[14] Mn^{-2}[15] Ni^{-2}[16], both of the latter being substitutional.

---

[1] G. W. Ludwig: Phys. Rev. **137**, A 1520 (1965).

[2] G. Feher, J. C. Hensel, and E. A. Gere: Phys. Rev. Letters **5**, 309 (1960).

[3] G. W. Ludwig, and H. H. Woodbury, in: F. Seitz, and D. Turnbull: Solid State Physics, vol. 13, p. 223. New York: Academic Press 1962.

[4] H. H. Woodbury, and G. W. Ludwig: Phys. Rev. Letters **5**, 96 (1960).

[5] H. H. Woodbury, and G. W. Ludwig: Phys. Rev. **117**, 102 (1960).

[6] See footnote 4, p. 253.

[7] G. W. Ludwig, H. H. Woodbury, and R. O. Carlson: J. Phys. Chem. Solids **8**, 490 (1959).

[8] G. W. Watkins, and J. W. Corbett: Discussions Faraday Soc. **31**, 86 (1961).

[9] G. W. Watkins, and J. W. Corbett: Phys. Rev. **138**, A 543 (1965).

[10] G. Bemski, G. Feher, and E. Gere: Bull. Am. Phys. Soc. **3**, 135 (1958).

[11] G. D. Watkins, and J. W. Corbett: Phys. Rev. **121**, 1001 (1961).

[12] G. D. Watkins, J. W. Corbett, and R. M. Walker: J. Appl. Phys. **30**, 1198 (1959).

[13] G. Feher, D. K. Wilson, and E. A. Gere: Phys. Rev. Letters **3**, 25 (1959).

[14] R. E. Pontinen, and T. M. Sanders jr.: Phys. Rev. Letters **5**, 311 (1960).

[15] G. D. Watkins: Bull. Am. Phys. Soc. **2**, 345 (1957).

[16] G. W. Ludwig, and H. H. Woodbury: Phys. Rev. **113**, 1014 (1959).

In Type I diamond, substitutional nitrogen is a deep donor which interacts with one nearest-neighbor carbon atom, as shown by hyperfine splitting from one ^{13}C atom, the electron occupying an antibonding orbital[1].

In Type II diamonds, a seemingly single ESR line is shown to be a superposition of three lines with g values 2.00230, 2.00234 and 2.00238 due to centers labelled A, B and C, respectively. The A center is believed to be the positively ionized state of the isolated vacancy[2].

In graphite only conduction electron resonance has been observed[3]. The $g_\parallel$ value increases markedly as temperature is lowered, being 2.0495 at room temperature and 2.127 at 77 °K; $g_\perp = 2.0026$.

In the hexagonal form of silicon carbide, nitrogen appears to substitute for carbon[4-7] giving a large isotropic ^{14}N hyperfine splitting consistent with a hydrogenic model for the donor electron[5-7]. Boron as an acceptor also substitutes for carbon, as inferred from ^{29}Si superhyperfine splitting of boron lines[5]. An unusual aspect of the boron splitting is that at one position, the values of hyperfine parameters A and B are of opposite sign. In contrast with nitrogen, the wave function has primarily p-character.

A single ESR line in commercial boron is enhanced by addition of carbon or silicon. It is surmised that the center may be related to defects in a $B_{12}C_3$-type structure. The $B_{12}C_3$ also shows a single line with $g = 2.003$ [8].

Few ESR data are available on III—V compounds. In InSb one observes conduction electron resonance[9]. In GaP, Fe^{3+} substitutes for Ga; Mn has also been observed[10].

I. Compounds.

Like the previous semiconductors considered, there is a tetrahedral coordination about the Zn or Cd atoms in ZnSe, ZnTe or CdTe (cubic) or CdS and CdSe (hexagonal). ZnS has both cubic and hexagonal forms. Most of the ESR work has been concerned with iron ground or rare earth ions substituting for Zn or Cd and retaining full site symmetry. The earlier work is summarized in Ref. [11]. An unusual shallow donor example is Cl in CdS[12], and Cl or Br in ZnS[13]. Upon irradiation at 365 nm, one observes an anisotropic line attributed to halogen atoms trapped next to Zn vacancies. Fe^{3+} has been reported in CdS[14].

A long-known blue self-activated luminescence in ZnS had been assigned to a center involving association between a zinc vacancy and a substitutional im-

---

[1] W. V. Smith, P. P. Sorokin, I. L. Gelles, and G. J. Lasher: Phys. Rev. **115**, 1546 (1959).
[2] J. A. Baldwin: Phys. Rev. Letters **10**, 220 (1963).
[3] G. Wagoner: Phys. Rev. **118**, 647 (1960).
[4] J. S. van Wieringen, in: M. Schön, and H. Welker (eds.), Semiconductors and Phosphors. New York: Interscience 1958.
[5] H. H. Woodbury, and G. W. Ludwig: Phys. Rev. **124**, 1083 (1961).
[6] J. A. Lely: Ber. deut. keram. Ges. **32**, 229 (1955).
[7] J. A. Lely, and F. Kroger, in: M. Schön, and H. Welker (eds.), Semiconductors and Phosphors, p. 525. New York: Interscience 1958.
[8] D. Geist, and H. J. Gläser: J. Phys. Chem. Solids **26**, 57 (1965); — Phys. Letters **17**, 186 (1965).
[9] G. Bemski: Phys. Rev. Letters **4**, 62 (1960).
[10] H. H. Woodbury, and G. W. Ludwig: Bull. Am. Phys. Soc. **6**, 118 (1961).
[11] G. W. Ludwig, and H. H. Woodbury, in: F. Seitz, and D. Turnbull (eds.). Solid State Physics, vol. 13, p. 223. New York: Academic Press 1962.
[12] J. Lambe, and C. Kikuchi: J. Phys. Chem. Solids **8**, 492 (1959).
[13] P. H. Kasai, and Y. Otomo: Phys. Rev. Letters **7**, 17 (1961).
[14] K. Morigaki, and T. Hoshina: Phys. Letters **17**, 85 (1965).

purity from group III or group VII elements[1]. A series of ESR experiments have shown the presence of various ESR spectra when doping was done with such impurities[2-6].

Hyperfine structure found for Ga, Br and I shows direct involvement of these impurities[6].

In cubic ZnS the A center at 77 °K is described as a hole shared among three equivalent S atoms about a Zn vacancy associated with a substitutional halogen atom, the system showing axial symmetry[4]. At 1.3 °K the center has orthorhombic symmetry, suggesting localization at one S atom not equivalent to the other two[7].

In hexagonal ZnS: Cu, Ga, photoexcitation leads to a multiple-line ESR spectrum presumably due to a triplet state of separated Cu—Ga donor-acceptor pairs[8].

F. Miscellaneous topics.

I. ESR of gases.

Paramagnetic resonance spectra have been observed for numerous atoms in the gas phase. These include hydrogen[9]; nitrogen[10-13], including ^{15}N in natural abundance[14]; oxygen[13, 15] fluorine[16]; phosphorus[17]; chlorine[18]; bromine[19]; and iodine[20]. For atoms the g-factor is the Lande value $g_J = 1 + \dfrac{J(J+1)+S(S+1)-L(L+1)}{2J(J+1)}$. Theoretical intensity relationships have been derived for the determination of H, N and O atoms[11], Cl, Br and I atoms, as well as electrons, using O_2 as the reference gas[21]. The spectra of gaseous di- or triatomic molecules are considerably more complex, due to the molecular rotational angular momentum and appreciable spin-orbit interaction.

Oxygen was the first diatomic molecule whose ESR spectrum was observed[22]. The detailed interpretation of the many-line spectrum has been worked out[23]

---

[1] J. S. Prener, and F. E. Williams: J. Chem. Phys. 25, 361 (1956).
[2] P. H. Kasai, and Y. Otomo: Phys. Rev. Letters 7, 17 (1961); — J. Chem. Phys. 37, 1263 (1962); — J. Phys. Soc. Japan Suppl. 2, 295 (1963).
[3] A. Räuber, and J. Schneider: Phys. Letters 3, 230 (1963).
[4] J. Schneider, W. C. Holton, T. L. Estle, and A. Räuber: Phys. Letters 5, 312 (1963).
[5] B. Dischler, A. Räuber, and J. Schneider: Phys. stat. sol. 6, 507 (1964).
[6] J. Schneider, A. Räuber, B. Dischler, T. Estle, and W. C. Holton: J. Chem. Phys. 42, 1839 (1965).
[7] K. Narita, and H. Kusumoto: Phys. Letters 10, 267 (1964).
[8] H. D. Fair, R. D. Ewing, and F. E. Williams: Phys. Rev. Letters 15, 355 (1965).
[9] R. Beringer, and E. B. Rawson: Phys. Rev. 87, 228 (1952). — R. Beringer, and M. A. Heald: Phys. Rev. 95, 1474 (1954).
[10] M. A. Heald, and R. Beringer: Phys. Rev. 96, 645 (1954).
[11] A. A. Westenberg, and N. de Haas: J. Chem. Phys. 40, 3087 (1964).
[12] C. J. Ultee: J. Phys. Chem. 64, 1873 (1960).
[13] T. C. Marshall: Phys. Fluids 5, 743 (1962).
[14] C. J. Ultee: J. Chem. Phys. 43, 1080 (1965).
[15] E. B. Rawson, and R. Beringer: Phys. Rev. 88, 677 (1952).
[16] H. E. Radford, V. W. Hughes, and V. Beltran-Lopez: Phys. Rev. 123, 153 (1961). — S. Aditya, and J. E. Willard: J. Chem. Phys. 44, 833 (1966).
[17] H. Dehmelt: Phys. Rev. 99, 527 (1955).
[18] V. Beltran-Lopez, and H. G. Robinson: Phys. Rev. 123, 161 (1961).
[19] N. Vanderkooi, and J. S. MacKenzie: Advances in Chem. Ser. 36, 98 (1962).
[20] K. D. Bowers, R. A. Kamper, and G. D. Lustig: Proc. Phys. Soc. (London) B 70, 1176 (1957).
[21] S. Krongelb, and M. W. B. Strandberg: J. Chem. Phys. 31, 1196 (1959).
[22] R. Beringer, and J. C. Castle jr.: Phys. Rev. 75, 1963 (1949).
[23] M. Tinkham, and M. W. P. Strandberg: Phys. Rev. 97, 951 (1955).

thus paving the way for use of gaseous O_2, including $^{16}O^{18}O$ [1,2] as an intensity standard [3,4]. Resonance of O_2 in the $^1\Delta_g$ excited state gives four lines as expected, with $\Delta M_J = 1$ transitions between levels for which the leading term is $-g_J H M_J$. Here g_J is observed to be -0.66662 [5]. NO was the second molecule observed [6], followed by OH [7]. In OH, the actual transitions observed are Zeeman splittings of the (electric dipole) Λ-doubling transitions in each of the $J=\frac{3}{2}$, $J=\frac{5}{2}$ and $J=\frac{7}{2}$, $^2\pi_{\frac{3}{2}}$ rotational levels [8]. Intensity expressions relating OH to NO concentrations allow an independent check since the latter also undergoes an electric dipole transition [9]. Recently SO [10,11], SH [10], SeH [12], TeH [12], ClO, BrO and NS [13] have all been reported in the gas phase. The list of triatomic gaseous molecules is confined to NO_2 [9,14] and NF_2 [15]. Recently, NO_2 has been used to "titrate" H atoms quantitatively [9]:

$$NO_2 + H \rightarrow NO + OH.$$

This reaction has allowed verification of an intensity expression for H atoms.

II. Transition metal ions in liquid solution.

ESR spectra are observed for relatively few hydrated transition metal ions in aqueous solution. Reasonably long relaxation times and small spin-orbit couplings are characteristic of the ions observed. These include $TiF_2(H_2O)_4^+$ [16,17] or $Ti^{3+}:CH_3O^-$ [16] complexes, $VO(H_2O)_5^{+2}$ [18], $Cr(H_2O)_6^{+3}$ [18,19], $Mn(H_2O)_6^{+2}$ [20-24], Cu^{+2} [19,25] and Mo^{+5} [26,27] complexes. The Fe^{+3} ion in some form gives resonance in an aqueous solution of FeF_3, but it is very doubtful that it is the hexahydrated ion. It is much more likely to be a fluoride complex FeF_2^+ or FeF_3 [18]. The spectra of numerous complexes of Cr^{+3} and Cu^{+2} which have relatively small spin-orbit coupling and long spin-lattice relaxation times have been observed in liquid solutions.

---

[1] G. Filipovich, and T. M. Sanders: Rev. Sci. Instr. **30**, 293 (1959).
[2] K. D. Bowers, R. A. Kamper, and R. B. D. Knight: J. Sci. Instr. **34**, 49 (1957).
[3] See footnote 21, p. 256.
[4] See footnote 11, p. 256.
[5] A. M. Falick, B. Mahan, and R. J. Myers: J. Chem. Phys. **42**, 1837 (1965).
[6] R. Beringer, and J. G. Castle jr.: Phys. Rev. **76**, 868 (1949).
[7] See footnote 13, p. 256.
[8] H. E. Radford: Phys. Rev. **122**, 114 (1961).
[9] A. A. Westenberg, and N. de Haas: J. Chem. Phys. **43**, 1551 (1965).
[10] C. C. McDonald: J. Chem. Phys. **39**, 2587 (1963).
[11] J. M. Daniels, and P. B. Dorain: J. Chem. Phys. **40**, 1160 (1964).
[12] H. E. Radford: J. Chem. Phys. **40**, 2732 (1964).
[13] A. Carrington, and D. H. Levy: J. Chem. Phys. **44**, 1298 (1966).
[14] J. G. Castle jr., and R. Beringer: Phys. Rev. **80**, 114 (1950).
[15] L. H. Piette, F. A. Johnson, K. A. Booman, and C. B. Colburn: J. Chem. Phys. **35**, 1481 (1961).
[16] E. L. Waters, and A. H. Maki: Phys. Rev. **125**, 233 (1962).
[17] A. Carrington, and G. R. Luckhurst: Mol. Phys. **8**, 125 (1964).
[18] B. R. McGarvey: J. Phys. Chem. **61**, 1232 (1957).
[19] S. A. Al'tschuler, and B. M. Kozyrev: Electron Paramagnetic Resonance, p. 170. C. P. Poole jr., translation editor. New York: Academic Press 1961.
[20] N. S. Garif'yanov, and B. M. Kozyrev: Doklady Akad. Nauk U.S.S.R. **98**, 929 (1954).
[21] M. Tinkham, R. Weinstein, and A. F. Kip: Phys. Rev. **84**, 848 (1951).
[22] E. E. Schneider, and T. S. England: Physica **17**, 221 (1951).
[23] M. A. Garstens, and S. H. Liebson: J. Chem. Phys. **20**, 1647 (1952).
[24] B. B. Garrett, and L. O. Morgan: J. Chem. Phys. **44**, 890 (1966). — C. C. Hinckley, and L. O. Morgan: ibid. **44**, 898 (1966).
[25] S. Fujiwara, and H. Hayashi: J. Chem. Phys. **43**, 23 (1965).
[26] N. S. Garif'yanov, and V. N. Fedotov: Soviet Phys. JETP **16**, 269 (1963).
[27] S. I. Weissman, and M. Cohn: J. Chem. Phys. **27**, 1440 (1957).

In the $Ti^{3+}(d^1)$ complexes thus far observed in solution, viz. $TiF_2(H_2O)_4^+$ or $Ti^{3+}:OCH_3^-(CH_3OH)_n$, the structures presumably provide an axial distortion to remove the orbital degeneracy which prevents observation of Ti^{3+} in octahedral fields even in solids[1,2].

The VO^{2+} ion is also an example of a $3d^1$ system, which in aqueous solutions has an octahedral configuration of five water molecules and an oxygen atom, the latter occupying the sixth position. For $VOCl_2$ in aqueous solution, $g = 1.965$ and $a = 113$ G [3]. Alternatively, ligands may occupy four of the five available positions, the remaining one being occupied by solvent[3-9]. The ESR problem is thus analogous to Cu^{2+}, in that both ions tend toward axial symmetry.

Two other d^1 ions observed in sulfuric acid solution are V^{4+} and Cr^{5+} [10]. In the ions $CrOCl_4^{-1}$ and $CrOCl_5^{-2}$, observable in acetic acid at $77\ °K$, the pentavalent chromium has a single d-electron which is presumed to occupy an antibonding orbital[11,12]. While crystal field considerations predict $g_\parallel < g_\perp$, experimental values are 2.008 and 1.977 respectively. MO calculations allowing for contributions from excited configurations gives the experimental ordering of g-values[13].

Complexes with trivalent chromium (d^3) studied in solution probably outnumber those of any other transition metal ion. Their ESR parameters have been treated[14-17] at length[18]. In some of these complexes, e.g., $Cr(C_6H_6)_2^+$ [19] and $[Cr(CN)_5NO]^{3-}$, one observes hyperfine splitting from the ligand atoms at room temperature. A second d^3 ion observable in $H_2SO_4-SO_3$ (oleum) solutions at room temperature is Mn^{4+} [10]. The most frequently observed ion in aqueous solution is Mn^{2+}, dilute solutions (10^{-6} or 10^{-7} M) of which may serve to test the sensitivity of an ESR spectrometer. The hyperfine sextet is well resolved, with splitting constant 95.6 G [20-24]. Upon adding a complex-forming substance with measurable association constant, the observed amplitude diminishes, since the very broad line of the covalently-bonded Mn^{2+} is not detected. Association constants have been measured in this way, e.g., with malonic acid $[HOOCCH_2COOH]$ [25] or pyridine $C_5(H_5)N$ [26].

---

[1] See footnote 16, p. 257.
[2] See footnote 17, p. 257.
[3] D. KIVELSON, and S. K. LEE: J. Chem. Phys. **41**, 1896 (1964).
[4] D. E. O'REILLY: J. Chem. Phys. **29**, 1188 (1958); **30**, 591 (1959).
[5] E. M. ROBERTS, and W. S. KOSKI: J. Chem. Phys. **34**, 591 (1961).
[6] I. BERNAL, and P. RIEGER: Inorg. Chem. **2**, 256 (1963).
[7] C. J. BALLHAUSEN, and H. B. GRAY: Inorg. Chem. **1**, 111 (1962).
[8] H. R. GERSMANN, and J. D. SWALEN: J. Chem. Phys. **36**, 3221 (1962).
[9] N. S. GARIF'YANOV, B. M. KOZYREV, R. KH. TIMEROV, and N. F. USACHEVA: Soviet Phys. JETP **16**, 269 (1963).
[10] H. C. MISHRA, and M. C. R. SYMONS: J. Chem. Soc. **1963**, 4490.
[11] H. B. GRAY, and C. R. HARE: Inorg. Chem. **1**, 363 (1962).
[12] C. R. HARE, I. BERNAL, and H. B. GRAY: Inorg. Chem. **1**, 831 (1962).
[13] H. KON, and N. E. SHARPLESS: J. Chem. Phys. **42**, 906 (1965).
[14] H. A. KUSKA, and M. T. ROGERS: J. Chem. Phys. **40**, 910 (1964).
[15] I. BERNAL, and S. E. HARRISON: J. Chem. Phys. **34**, 102 (1961); **38**, 2581 (1963).
[16] R. G. HAYES: J. Chem. Phys. **38**, 2580 (1963).
[17] J. B. SPENCER, and R. J. MYERS: J. Am. Chem. Soc. **86**, 522 (1964).
[18] H. S. JARRETT: Solid State Physics, vol. 14, p. 215. F. SEITZ, and D. TURNBULL, eds. New York: Academic Press 1963.
[19] R. FELTHAM: J. Inorg. & Nuclear Chem. **16**, 197 (1961).
[20] See footnote 20, p. 257.
[21] See footnote 21, p. 257.
[22] See footnote 22, p. 257.
[23] See footnote 23, p. 257.
[24] See footnote 24, p. 257.
[25] M. COHN, and J. TOWNSEND: Nature **173**, 1090 (1954).
[26] See footnote 18, p. 257.

Frozen aqueous solutions of Mn^{2+} show hyperfine splitting if methanol[1,2] or an excess of an anion such as chloride[2] or perchlorate[3], is present. To minimize effects of spin-spin interactions incident upon segregation of a salt upon freezing, it is advantageous to use quick-freezing techniques found suitable for biological samples[4]. Under favorable conditions one may see the forbidden $\Delta m_I = \pm 1$ transitions[5] which are seen in such non-oriented systems as modeling clay.

For Cu^{2+} in aqueous solutions, the nature of the anion makes little difference in the g-value which ranges from 2.182 for the chloride to 2.187 for the sulfate, with 2.185 for the nitrate[6,7]. Values of the spin orbit coupling parameter for the same three salts are 593, 609 and 603 cm^{-1} respectively, taking $\Delta = 12,500$ cm^{-1}. The line width decrease with increasing concentration and with reduction in temperature is interpreted as showing relaxation within the hydration sheath[7]. Hyperfine structure of Cu^{2+} has been reported for alkaline solutions (pH > 10) (T. Takeshita, cited in [7]). A large number of Cu^{2+} complexes in solution has been studied. One of the earliest was the acetylacetonate in which each of the four hyperfine lines has a different width[8].

At room temperature, relative ESR line intensities for $TiF_2(H_2O)_4^+$ $(S=\frac{1}{2})$, $Cr(NH_3)_6^{3+}$ $(S=\frac{3}{2})$ and $Mn(H_2O)_6^{2+}$ $(S=\frac{5}{2})$ in liquid solution are in the ratio of $S(S+1)$, indicating that all of the transitions of each contribute to line intensity[9] However, for Mn^{4+} in oleum $(H_2SO_4-SO_3)$ only the $\frac{1}{2} \leftrightarrow -\frac{1}{2}$ transition is said to contribute to the observed line intensity in viscous solution[10].

III. Generation of odd-electron centers.

Formation of free radicals and defects in solids by irradiation with ultraviolet light, electrical discharges, x-rays, γ-rays, neutrons or electrons is standard procedure when the temperature and the matrix are such that the center has a long life. Careful selection of materials to be irradiated may lead to greatly enhanced yields of an odd-electron product. For example, if one is seeking to generate alkyl radicals by γ-irradiation, then the appropriate halide RX in 3-methylpentane at 77 °K is a good choice if the electron affinity of X exceeds the bond dissociation energy of R—X[11]. The mechanism appears to be: $RX + e \rightarrow R' + X^-$. A variant of this procedure involves deposition alternately of thin layers of alkali metal and of the halide[12]. Bombardment of organic solids with hydrogen atoms[13,14] has been used in a few instances; it usually produces one radical (by hydrogen atom abstraction), whereas x- or γ-irradiation often produce two or more. Malonic acid [HOOCCH$_2$COOH] or its derivatives, as well as polystyrene have been successfully used, while a number of other acids give no radicals. Hydrogen

---

[1] B. M. Kozyrev: Discussions Faraday Soc. **19**, 135 (1955). — N. S. Garif'yanov: Proc. Acad. Sci. U.S.S.R. **103**, 41 (1955).
[2] B. T. Allen, and D. W. Nebert: J. Chem. Phys. **41**, 1983 (1964).
[3] R. T. Ross: J. Chem. Phys. **42**, 3919 (1965).
[4] G. Palmer, R. C. Bray, and H. Beinert: J. Biol. Chem. **239**, 2657 (1964). — R. C. Bray: Biochem. J. **81**, 189 (1961).
[5] B. Bleaney, and R. S. Rubins: Proc. Phys. Soc. (London) **77**, 103 (1961).
[6] See footnote 19, p. 257.
[7] See footnote 25, p. 257.
[8] B. R. McGarvey: J. Phys. Chem. **60**, 71 (1956).
[9] See footnote 17, p. 257.
[10] See footnote 15, p. 257.
[11] D. W. Skelly, R. G. Hayes, and W. H. Hamill: J. Chem. Phys. **43**, 2795 (1965).
[12] J. E. Bennett, and A. Thomas: Nature **195**, 995 (1962).
[13] R. B. Ingalls, and L. A. Wall: J. Chem. Phys. **35**, 370 (1961). — L. A. Wall, and R. B. Ingalls: J. Chem. Phys. **41**, 1112 (1964).
[14] T. Cole, and H. C. Heller: J. Chem. Phys. **42**, 1668 (1965).

atoms penetrate to distances of the order of ten layers, judged in part by the narrowness of the lines and the persistence of signal [1].

Short-lived radicals may be generated continuously in the microwave cavity by ultraviolet light or even with x-rays. An elegant approach using continuous 2.8 MeV electron irradiation has furnished unique data on a large number of alkyl (saturated hydrocarbon) radicals [2]. The trapping of continuously produced reactive species in an inert gas matrix has been widely applied [3]. A more generally available procedure for producing short-lived radicals in solution uses "OH" radicals generated in a continuous flow system to abstract a hydrogen atom from molecules of interest [4-6]. (The quotation marks are intended to convey uncertainty as to the structure of the solvated species. The spectrum originally attributed to "OH" has been reassigned to a radical containing titanium, since Ti hfs is observed.) [6] If solutions containing Ti^{3+} (or Ce^{4+}) [5] and H_2O_2 are efficiently mixed and caused to flow through the microwave cavity, one may observe two independent ESR lines if the pH is not too low [4,6]. If an oxidizable substance is added to the H_2O_2 solution, hydrogen abstraction will often occurs. A typical reaction product from CH_3OH is $C'H_2OH$, identified by splittings from both types of protons (unless the OH protons are being too rapidly exchanged [4,7]. This technique has permitted convenient generation of many radicals which had not been produced in any other way; in addition it produces radicals in liquid solution which previously have only been observed in rigid media at low temperature, e.g., the CH_3 radical, produced from dimethylsulfoxide, $(CH_3)SO$. It is only required that the radical lifetime be of the order of milliseconds at the temperature of observation to permit adequate steady-state radical concentrations in the cavity. Liquid solutions permit observation of very small splittings from distant atoms; if one wishes to observe anisotropic interactions, he need merely cool the solutions or increase their viscosity by addition of inert substances.

For the production of charged radical species of short (or long) life, one may conveniently use an electrolytic generation process [8]. Most frequently it is the anions which are the more readily generated. It has long been known that during reduction of some systems at dropping mercury electrodes one sees two steps or "waves". These are attributed to two successive one-electron processes. The technique is readily adaptable to ESR use, either as a static system with the electrode within the cavity or as a flow system with the electrode immediately outside. As electrodes one may also use gold or platinum, especially for cation formation. Anions of aromatic hydrocarbons, semiquinones (Sect. B. I. 11), ketyls (from ketones), nitro compounds RNO_2, azo compounds $R-N=N-R'$ and azoxy-compounds $R-N-N(O)-R'$ are readily produced. As noted in Sect. B. I. 13

[1] See footnote 14, p. 259.
[2] R. W. FESSENDEN, and R. H. SCHULER: J. Chem. Phys. **39**, 2147 (1963).
[3] A. M. BASS, and H. P. BROIDA, ed.: Formation and Trapping of Free Radicals. New York: Academic Press 1960.
[4] W. T. DIXON, and R. O. C. NORMAN: Nature **196**, 891 (1962); — Proc. Chem. Soc. **1963**, 97; — J. Chem. Soc. **1963**, 3119; ibid., **1964**, 4850, 4857.
[5] E. SAITO, and B. H. J. BIELSKE: J. Am. Chem. Soc. **83**, 4467 (1961).
[6] H. FISCHER: Private communication.
[7] H. FISCHER: Mol. Phys. **9**, 149 (1965).
[8] A. H. MAKI, and D. H. GESKE: J. Chem. Phys. **30**, 1356 (1959); **33**, 825 (1960); — J. Am. Chem. Soc. **82**, 2671 (1960); **83**, 1852 (1961). — L. H. PIETTE, P. LUDWIG, and R. N. ADAMS: Anal. Chem. **34**, 916 (1962); — J. Am. Chem. Soc. **83**, 3909 (1961), — P. H. RIEGER, I. BERNAL, W. H. REINMUTH, and G. K. FRAENKEL: J. Am. Chem. Soc. **85**, 683 (1963). — J. R. BOLTON, and G. K. FRAENKEL: J. Chem. Phys. **40**, 3307 (1964). — J. E. HARRIMAN, and A. H. MAKI: J. Chem. Phys. **39**, 778 (1963). — D. H. LEVY, and R. J. MYERS: J. Chem. Phys. **41**, 1062 (1964).

on Ion Pairing, the bulky cations used as counter-ions in the electrolysis do not produce an additional splitting of the spectra of anion radicals which is often observed when alkali metals are used for reduction. A few inorganic radicals of short life have been generated electrolytically.

IV. Sources of error.

Sources of confusion in the interpretation of paramagnetic resonance spectra are numerous enough to merit enumeration, since some of them have led to published errors. References cited discuss sources of error or make corrections of erroneous interpretations. Sources of confusion include:

1. Presence of unsuspected impurities. In the case of hexene-1 [$CH_2=CH(CH_2)_3CH_3$], the apparent formation of a free radical upon irradiation of the hydrocarbon was shown to be due to photolysis of a peroxide such as may be formed in highly purified hydrocarbons by exposure to air[1].

2. Production of two or more paramagnetic centers. Spectral lines may appear unrelated, in which case one may observe the effects of changed conditions of irradiation, selective thermal or optical bleaching, or change of microwave frequency; if g-values are different, one will observe a relative shift of the several spectra. In some cases, a high degree of coincidence of isotropic lines may occur; a missing member of a hydrogen atom doublet has been shown to underlie a line from another center[2].

3. Low intensity of outer members of a spectrum from two or more groups of non-equivalent nuclei. One may be able to achieve an apparent fit of the spectrum with very erroneous splitting constants which assume an incorrect number of spectral lines[3].

4. Close correspondence of one hyperfine coupling constant with an integral multiple of another or with the sum of two others. Such effects have led to conclusions that some nuclei do not contribute a measurable splitting.

5. Low intensity of lines from ^{13}C in natural abundance. Failure to seek these out led to three erroneous assignments of one radical[4].

6. Unresolved shoulders on lines from non-oriented solids. This has led in some cases to assumption of the presence of two centers when one center with an anisotropic g-tensor was present. It has also led to erroneous assignment of the origin of hyperfine splitting, e.g., assignment of lines from $CrOCl_5^{-2}$ to ^{53}Cr instead of ^{35}Cl[5].

7. Formation of a paramagnetic byproduct, when the main product is diamagnetic. One may be misled in interpretation of a process by failure carefully to consider intensity of ESR spectra observed and to make measurements using other techniques. This has at times caused other types of spectroscopists to heap scorn on some whose sole preoccupation was an ESR spectrum.

8. Consecutive reactions of the primary product of irradiation. One is amply warned of the possibility of fast reactions of primary products by the fact that irradiation of neutral molecules are commonly neutral instead of ionic. Reactions may be slow enough that one may observe two, three or even up to five successive

---

[1] E. B. BEASLEY, and R. S. ANDERSEN: J. Chem. Phys. **40**, 2565 (1964).
[2] S. FONER, E. L. COCHRAN, V. A. BOWERS, and C. K. JEN: J. Chem. Phys. **32**, 963 (1960).
[3] J. R. BOLTON, A. CARRINGTON, and J. D. S. VEIGA: Mol. Phys. **5**, 615 (1962).
[4] J. R. MORTON: J. Am. Chem. Soc. **86**, 2325 (1964).
[5] H. KON, and N. E. SHARPLESS: J. Chem. Phys. **43**, 1081 (1965).

paramagnetic entities. The last figure applies to some semiquinones observed by the author in alkaline solution. A few examples of successive radical formation follow.

(a) $\quad C'H_2OH \to HC'O \quad$ 1,2

(b) $\quad O'CH_2CH_3 \to C'H_3 + ? \quad$ 3

(c) $\quad (CH_3)_2C'OH \to (CH_3)_2C'HO \quad$ 4

(d)

(99)

In some sulfur compounds one may observe both consecutive and parallel reactions. For $CH_3S(O_2)CH(CH_3)_2$ it is reported that radical products include CH_3, $(CH_3)_2CH$, and $CH_3SO_2C(CH_3)_2$, which decay, leaving an RSO_2 radical. Inferences may be drawn about relative ease of breaking C—S and C—H bonds from the preponderance of CH_3 and $(CH_3)_2CH$ radicals[7]. One may also observe effects of interaction of primary radicals with added gases such as NO, CO, O_2, SO_2, or H_2S[8].

9. Deuterium substitution, which is intended to confirm an interpretation based on a hydrogen-containing center, may lead to a remarkable degree of deuterium-hydrogen exchange. While stable guest molecules may be shown not to exchange with the host, free radicals may exchange with surprising rapidity[9]. Correct interpretation of spectra from irradiated amino acids was doubtless delayed by the isotopic substitution experiments. Another example of deuterium-hydrogen exchange occurs with the irradiation products of benzene dissolved in boric acid glass[10].

V. Special techniques.

ENDOR in Solids and Liquids. — Examples of the use of ENDOR techniques for resolution of unresolved hyperfine splittings are given in Sects. D. I and D. II and include the F, M and R centers in the alkali halides, the V_{OH} center in MgO, and defects in silicon. The technique also allows the determination of additional hyperfine coupling constants for oriented free radicals derived from succinic acid $[HOOC(CH_2)COOH]$[11] or adipic acid $[HOOC(CH_2)_3COOH]$[12]. Adaptation of the technique to radicals in liquid solutions requires far higher microwave fields than for solids; pulsed operation is used to avoid excessive heating[13,14]. The several

---

[1] R. S. Alger, T. H. Anderson, and L. A. Webb: J. Chem. Phys. 30, 695 (1959).
[2] R. H. Johnsen: J. Phys. Chem. 65, 2144 (1961).
[3] Boubnov, et al., cited by T. Omae, et al.: J. Chem. Phys. 42, 4053 (1965).
[4] T. Omae, S. I. Ohnishi, H. Sakurai, and I. Nitta: J. Chem. Phys. 42, 4053 (1965).
[5] P. B. Ayscough, and C. Thomson: Trans. Faraday Soc. 58, 1477 (1962).
[6] J. F. Gibson, D. J. E. Ingram, M. C. R. Symons, and M. G. Townsend: Trans. Faraday Soc. 53, 914 (1957).
[7] P. B. Ayscough, K. J. Ivin, and J. H. O'Donnell: Trans. Faraday Soc. 61, 1110 (1965).
[8] H. N. Rexroad, and W. Gordy: J. Chem. Phys. 30, 399 (1959). — Z. Kuri, H. Ueda, and S. Shida: J. Chem. Phys. 32, 371 (1960).
[9] K. Itoh, and I. Miyagawa: J. Chem. Phys. 40, 3328 (1964).
[10] F. Hughes, R. D. Kirk, and F. W. Patten: J. Chem. Phys. 40, 872 (1964).
[11] T. Cole, C. Heller, and J. Lambe: J. Chem. Phys. 34, 1447 (1961).
[12] A. L. Quiram, and J. S. Hyde: J. Chem. Phys. 42, 791 (1965).
[13] J. S. Hyde, and A. H. Maki: J. Chem. Phys. 40, 3117 (1964).
[14] J. S. Hyde: J. Chem. Phys. 43, 1806 (1965).

necessary conditions for successful ENDOR observation on liquids are as follows: $T_e \geq (T_1 T_2)^{\frac{1}{2}} \geq (\gamma_e H_1)^{-1} \approx (\gamma_n H_{rf})^{-1}$. Here T_e is the spin-spin exchange time for electrons, T_1 and T_2 are nuclear relaxation times, γ_e and γ_n are electron and nuclear magnetogyric ratios; H_1 is the microwave- and H_{rf} the ENDOR field. Since electron exchange increases with temperature, one must use progressively more dilute solutions at higher temperatures[1]. Three advantages are claimed for the ENDOR technique over ESR in the study of radicals in solution: 1. Spectra are simplified because one sees one *pair* of lines for each *set* of equivalent nuclei. 2. Resolution is improved, enabling determination of small coupling constants. 3. Coupling constants are measured with increased precision. It is also possible to observe an ESR spectrum induced by nuclear transitions if the normal ESR spectrum is subtracted[2].

Spin Echo Studies. — In nuclear magnetic resonance the spin echo technique[3] is widely used for measurement of relaxation times. With appropriate changes in circuitry[4], the technique has been used for studies of spin diffusion of Ce^{3+} and Er^{3+} in $CaWO_4$[5], and for measurement of electric field shift in ESR studies[6]. Modulation of the electron spin echo envelope by dipolar-coupled nuclei adjoining an unpaired electron may in some cases be interpreted to give information normally sought from ENDOR experiments. The echo technique is applicable to systems with such a rapid spin diffusion rate that it would be difficult to maintain saturation for an ENDOR experiment[7]. The technique appears to be limited to systems with phase memory times of the order of five times the pulse duration[4].

Acknowledgments. Support from the United States Air Force Grant 200—66 during the writing of this chapter is gratefully acknowledged. It is also a pleasure to acknowledge helpful discussions with my colleague, Dr. JAMES R. BOLTON, who also kindly criticized portions of the manuscript.

General references.

General treatments of paramagnetic resonance.

AL'TSCHULER, S. A., and B. M. KOZYREV: Electron Paramagnetic Resonance, Transl. ed. C. P. POOLE jr. New York: Academic Press 1964.

PAKE, G. E.: Paramagnetic Resonance. New York: Benjamin 1962.

CARRINGTON, A., and A. D. MCLACHLAN: Introduction to Magnetic Resonance. New York: Harper & Row 1967.

Transition metal ions.

LOW, W.: Paramagnetic Resonance in Solids, Suppl. 2 of Solid State Physics (F. SEITZ, and D. TURNBULL, eds). New York: Academic Press 1960.

BLEANEY, B., and K. W. H. STEVENS: Repts. Progr. in Phys. **16**, 108 (1953).

BOWERS, K. D., and J. OWEN: Repts. Progr. in Phys. **18**, 304 (1955).

ORTON, J. W.: Repts. Progr. in Phys. **22**, 204 (1959).

CARRINGTON, A., and H. C. LONGUET-HIGGINS: Quart. Revs. (London) **14**, 427 (1960).

GRIFFITH, J. S.: The Theory of Transition Metal Ions, chap. 12. Cambridge Univ. Press 1961.

LOW, W., and E. L. OFFENBACHER, in: Solid State Physics, vol. 17, p. 135. (F. SEITZ and D. TURNBULL, eds.). New York: Academic Press 1965.

---

[1] See footnote 13, p. 262.
[2] See footnote 14, p. 262.
[3] E. L. HAHN: Phys. Rev. **80**, 580 (1950). — H. Y. CARR, and E. M. PURCELL: Phys. Rev. **94**, 630 (1954).
[4] W. B. MIMS: Rev. Sci. Instr. **36**, 1472 (1965).
[5] W. B. MIMS, K. NASSAU, and J. D. MCGEE: Phys. Rev. **123**, 2059 (1961).
[6] W. B. MIMS: Proc. Roy. Soc. (London) A **283**, 452 (1965); — Phys. Rev. **133**, A 835 (1964).
[7] L. G. ROWAN, E. L. HAHN, and W. B. MIMS: Phys. Rev. **137**, A 61 (1965).

Semiconductors.

LUDWIG, G. W., and H. H. WOODBURY: Solid State Physics, vol. 13, p. 223. (F. SEITZ, and D. TURNBULL, eds.). New York: Academic Press 1962.

Alkali halides.

SEIDEL, H., and H. C. WOLF: Phys. stat. sol. **11**, 3 (1965).
SLICHTER, C. P.: Principles of Magnetic Resonance, chap. 7 (V_K centers). New York: Harper and Row 1963.

Organometallic compounds (chelates).

JARRETT, H. S.: Solid State Physics, vol. 14, p. 215. (F. SEITZ, and D. TURNBULL, eds.). New York: Academic Press 1963.
ROBERTSON, R. E., in: Determination of Organic Structures by Physical Methods, p. 617. (F. C. NACHOD, and W. D. PHILLIPS, eds.). New York: Academic Press 1962.

Free radicals.

INGRAM, D. J. E.: Free Radicals. London: Butterworths 1958.
JEN, C. K., in: Formation and Trapping of Free Radicals, chap. 7. (A. M. BASS, and H. P. BROIDA, eds.). New York: Academic Press 1960.
SYMONS, M. C. R.: Advances in Physical Organic Chemistry, vol. I, p. 283. (V. GOLD, ed.). New York: Academic Press 1963.
CARRINGTON, A.: Quart. Revs. **17**, 67 (1963).
MORTON, J. R.: Chem. Revs. **64**, 453 (1964) [Oriented radicals].
FESSENDEN, R. W., and R. H. SCHULER: J. Chem. Phys. **39**, 2147 (1963) [Alkyl radicals].
MCCLELLAND, B. J.: Chem. Rev. **64**, 301 (1964) [Anionic Free Radicals].
FISCHER, H.: Magnetic Properties of Free Radials Landolt-Bornstein Tabellen, Group II, Vol. 1 (1965). This is a comprehensive listing of g-values and hyperfine splitting constants.
FRAENKEL, G. K.: J. Phys. Chem. **71**, 139 (1967) [Line width- and line shift effects].
JOHNSON, JR., C. S.: Advances in Magnetic Resonance (J. S. WAUGH, ed.), v. 1, 33 (1965) [NMR and ESR spectra applied to chemical rate processes].

Austauschwechselwirkung in Isolatoren[1].

Von

S. V. VONSOVSKY und B. V. KARPENKO[2].

Mit 9 Figuren.

Einführung.

In Stoffen, die aus Atomen oder Ionen der Übergangselemente (mit nicht abgeschlossenen inneren d- oder f-Elektronenschalen, die in Übereinstimmung mit der Hundschen Regel [1] nichtkompensierte Spin- und Bahnmomente besitzen) aufgebaut sind, können unter bestimmten Bedingungen magnetisch geordnete Atomzustände vom Typ des Ferro- oder Antiferromagnetismus auftreten. Diese Zustände beobachtet man im Temperaturintervall von 0 °K bis zu einer bestimmten kritischen Temperatur, der Curie-Temperatur Θ_C (bei Ferromagnetika) oder der Néel-Temperatur Θ_N (bei Antiferro- oder Ferrimagnetika). Offensichtlich wird die Größe dieser Temperaturen durch den Betrag des Energieparameters der Wechselwirkung zwischen den Elektronen bestimmt, die das Auftreten geordneter Verteilungen für die magnetischen Momente der Elektronen im Kristallgitter ermöglicht.

Die physikalische Ursache dieser Wechselwirkung wurde im Rahmen der Quantenmechanik geklärt, nachdem das Pauli-Prinzip aufgestellt wurde, d.h. nachdem die Symmetrieeigenschaften der Wellenfunktionen des Vielelektronensystems und ihre Antisymmetrie in bezug auf Vertauschungen der Koordinaten einzelner Teilchen erkannt wurden [2] bis [4].

Die Abhängigkeit der Energie eines n-Teilchensystems von der Größe des Gesamtspins ist eine Folge der Coulombschen Wechselwirkung zwischen den Teilchen, der Antisymmetrie der Gesamtwellenfunktion hinsichtlich der Koordinatenvertauschung (sowohl hinsichtlich der Orts- als auch der Spinkoordinaten) und der Überlappung verschiedener Bestandteile der Gesamtwellenfunktion, die von den Ortskoordinaten abhängen und sich untereinander nur durch Koordinatenvertauschungen unterscheiden, im $3n$-dimensionalen Phasenraum. Der letztgenannte Punkt wird in der Einelektronendarstellung als Forderung nach von Null verschiedener Überlappung der Einelektronenwellenfunktionen formuliert.

Für isolierte Atome und Moleküle wurde diese Wechselwirkung erstmalig quantitativ behandelt in den bekannten Arbeiten von HEISENBERG [2], [3] über die Theorie des Heliumatoms, des einfachsten Vielelektronenatoms, und von HEITLER und LONDON [5] über die Theorie des Wasserstoffmoleküls, des einfachsten Zweiatommoleküls. Für diese *Zwei*elektronensysteme wurde die Existenz einer Energiedifferenz (Aufspaltung) zwischen den Niveaus des *Singulett*- (mit dem resultierenden Elektronenspin $S = 0$) und *Triplett*zustandes (mit dem

---

[1] Das Manuskript wurde durch freundliche Vermittlung von Izdatel'stvo APN (Verlag der Presseagentur Nowosti) zur Verfügung gestellt.

[2] Das originale russische Manuskript wurde übersetzt von den Herren U. HOFMANN und H. HEMSCHIK, Dresden.

Spin $S = 1$) bewiesen:

$$\Delta E = E_s - E_t. \tag{1}$$

Die Differenz ΔE wird durch die Abhängigkeit der elektrostatischen Energie des Elektronensystems von der Größe des resultierenden Spins S bestimmt und trägt die Bezeichnung *Austausch*energie. Gewöhnlich nennt man die Hälfte der Differenz (1) das Austauschintegral J:

$$J = \tfrac{1}{2}\Delta E = \tfrac{1}{2}(E_s - E_t). \tag{2}$$

In Abhängigkeit vom Vorzeichen der Energie J ist der Grundzustand des Zweielektronensystems entweder ein *Singulett* ($J < 0$) oder ein *Triplett* ($J > 0$). Hieraus folgt, daß der unmagnetische Zustand (analog dem Antiferromagnetismus) einer negativen Austauschenergie und der magnetische Zustand (analog dem Ferromagnetismus) einer positiven Austauschenergie entspricht. Es läßt sich zeigen, daß die Beziehung (2) formal die Eigenwerte für E_s und E_t liefert, wenn man vom Diracschen Spin-Hamilton-Operator

$$\hat{H}_{sp} = E_0 - 2\vec{s}_1\vec{s}_2 J \tag{3}$$

ausgeht, wobei $\vec{s}_1$ und $\vec{s}_2$ die Spinoperatoren zweier Elektronen und E_0 die mittlere Energie aller Spinzustände

$$E_0 = \tfrac{1}{4}(E_s + 3E_t)$$

bedeuten. In Übereinstimmung mit der Formel (3) verhalten sich die Elektronen wie quasimagnetisch stark gekoppelte Teilchen, obwohl der Energieparameter J rein elektrostatischer Herkunft ist.

FRENKEL [7] und unabhängig davon HEISENBERG [8] verallgemeinerten diese Ergebnisse für zwei Elektronen auf die Vielelektronensysteme in Kristallen. Dabei ging FRENKEL vom Modell frei beweglicher Leitungselektronen (Fermi-Gas) für Metalle aus und klärte die Bedingungen, unter welchen der magnetisierte Zustand eines stark entarteten Fermi-Gases energetisch günstiger ist als der unmagnetische. Hierfür ist erforderlich, daß die Energieverringerung des Fermi-Gases auf Grund der Austauschwechselwirkung größer ausfällt als sein Zuwachs an kinetischer Energie. Später wurde dieses Verfahren in den Arbeiten von BLOCH [9], STONER [10] bis [12], WIGNER [13], MOTT [14] bis [15], SLATER [16], WOHLFARTH [17] u. a. weiterentwickelt[3]. Diese Richtung führt natürlich zur Theorie des Ferro- und Antiferromagnetismus von $3d$-Übergangsmetallen und -legierungen, in denen die ursprünglich magnetisch aktiven $3d$-Elektronen der isolierten Atome in starkem Maße kollektiviert werden und in der Gesamtheit der Leitungselektronen metallischer Kristalle aufgehen.

In der Arbeit von HEISENBERG [8] wurde für die Ferromagnetika ein anderes Modell entwickelt, das eine Verallgemeinerung des Heitler-London-Modells für Wasserstoffmoleküle darstellt. VAN VLECK [26] verallgemeinerte den Diracschen Spin-Hamilton-Operator (3) für willkürliche Spins ($S > \tfrac{1}{2}$)

$$H_{allg.} = -2\sum_{i>j} J_{ij}\vec{S}_i\vec{S}_j - g\mu_B H \sum_i S_i^z. \tag{4}$$

---

[3] Außerordentlich interessant sind die Arbeiten von DZYALOSHINSKY [18] und KONDRATENKO [19], in denen die Kollektivelektronentheorie der ferromagnetischen Metalle auf der Basis der Theorie der Fermi-Flüssigkeit nach LANDAU [20—22] entwickelt wurde. Aufmerksamkeit verdienen auch die Übersichtsartikel von MOTT [23], THOMPSON [24] und HERRING [25].

Hierin bedeuten $\vec{S}_i$ und $\vec{S}_j$ die Vektoroperatoren der Gesamtspins der Atome i und j, J_{ij} das Austauschintegral zwischen den d- (oder f-) Schalen des Atompaares (ij), g den Landé-Faktor, $\mu_B = \frac{e\hbar}{2mc}$ das Bohrsche Magneton, H die Stärke eines homogenen magnetischen Gleichfeldes in Richtung der Achse z und S_i^z den Operator der z-Komponente des Spinvektors eines Atomes i. Die Summation erfolgt über alle Atome i oder über alle Atompaare $(i > j)$ des Kristalls.

Es läßt sich zeigen (s. z.B. FAULKNER [27] oder in Ziff. 11), daß im allgemeinen der „quadratische" Spin-Hamilton-Operator (4), den man gewöhnlich als *Heisenberg- oder Dirac-van Vleck-Hamilton*-Operator bezeichnet, nicht exakt ist, d.h. nicht die gesamte Abhängigkeit der elektrostatischen Wechselwirkung vom Gesamtspin des Systems berücksichtigt. Tatsächlich wird der *exakte* Spin-Hamilton-Operator neben einer Summe vom Typ (4) noch eine Reihe von Summen höherer Ordnung in den Spinoperatoren enthalten (biquadratische usw.). Jedoch bedient man sich in erster Näherung gewöhnlich eines Spin-Hamilton-Operators vom Typ (4).

Es ist ferner zu beachten, daß im Hamilton-Operator (4) die Möglichkeit einer Änderung der Multiplizität der d- oder f-Schale von magnetisch aktiven Atomen in Kristallen [28] sowie die Teilnahme der Bahnmomente an der Austauschwechselwirkung nicht berücksichtigt werden [29c].

Verständlicherweise bestimmt die Größenordnung der Parameter J_{ij} auch die Größe der kritischen Temperaturen Θ_C oder Θ_N: $|J_{ij}| \approx k\Theta_C$, $k\Theta_N$ (k ist die Boltzmann-Konstante).

Das Heisenbergsche Verfahren ist für die Theorie des Ferro- und Antiferromagnetismus von *Isolatoren* und *Halbleitern* am besten geeignet. Letztere sind niemals Kristalle reiner Elemente (reine Ferromagnetika wie Fe, Co und Ni und reine Antiferromagnetika wie Cr und Mn sind Metalle), sondern Ionenverbindungen der d-Übergangselemente oder $4f$-Seltenerdmetalle. Die am besten untersuchte Klasse dieser Stoffe sind verschiedene Oxide, darunter die Ferrite mit Spinell- und Granatstruktur. In den meisten Fällen liegt hier eine antiferromagnetische Kopplung vor, d.h. entweder kompensierter Antiferromagnetismus oder öfter nichtkompensierter Antiferromagnetismus oder Ferrimagnetismus (z.B. bei Spinellen und Granaten). In der letzten Zeit wurden jedoch auch nichtmetallische Ionenverbindungen mit ferromagnetischer Kopplung beobachtet. So sind verschiedene Europiumverbindungen (EuO [30], EuS, EuSe, EuI_2, Eu_2SiO_4 [31]) sowie $CrBr_3$ [32] und einige andere Verbindungen [33] ferromagnetische Dielektrika.

Da man es bei allen nichtmetallischen Ferro-, Antiferro- und Ferrimagnetika mit Kristallen zu tun hat, die aus Ionen sowohl magnetisch aktiver (d- oder f-Elemente) als auch magnetisch neutraler Komponenten aufgebaut sind, entsteht hier eine spezifische Situation. Betrachten wir der Anschaulichkeit halber ein typisches Beispiel einer antiferromagnetischen Ionenverbindung, nämlich MnO. In diesem Stoff bilden die magnetisch aktiven zweiwertigen Kationen des d-Übergangsmetalles Mn^{++} und die magnetisch neutralen Sauerstoffanionen O^{--} eine Schachbrettordnung. Somit ist der Abstand zwischen den nächsten Kationennachbarn Mn^{++} entweder zweimal (für Wechselwirkungen vom 180°-Typ: $Mn^{++}-O^{--}-Mn^{++}$ oder vom 90°-Typ: $Mn^{++}-O^{--}$) oder $\sqrt{2}$-mal (bei unmittel-
$\qquad\qquad\qquad\qquad\qquad\qquad\qquad\qquad\qquad\quad\;\; |$
$\qquad\qquad\qquad\qquad\qquad\qquad\qquad\qquad\quad\;\; Mn^{++}$
baren Wechselwirkungen — direkter Austausch) größer als bei Kristallen aus reinen Elementen, z.B. aus Eisen oder Nickel[4]. Andererseits ist aus Berechnun-

---

[4] Folglich können die Abstände zwischen den Kationen der d-Metalle in ihren Monoxiden ~ 4 Å erreichen, wo sich die tatsächlichen Wellenfunktionen der d-Schale nicht mehr überlappen (s. z.B. SHULL u.a. [34]).

gen des Austauschparameters J_{ij} bekannt, daß er sehr stark vom Abstand R_{ij} zwischen den Gitterpunkten i und j abhängt. Seine Größe fällt mit zunehmendem R_{ij} exponentiell ab. Daher muß die *direkte* Austauschwechselwirkung zwischen den nächsten Nachbarn und um so mehr zwischen den auf die nächsten Nachbarn folgenden Ionen Mn^{++} im Kristall MnO um einige Größenordnungen kleiner sein als z.B. in ferromagnetischen Metallen. Das Experiment zeigt jedoch, daß in solch typischen antiferromagnetischen Isolatoren wie MnO, NiO und α-Fe_2O_3 die Néel-Temperaturen immerhin 120, 515 bzw. 950 °K und in den Ferriten Fe_3O_4, $Fe[Li_{0,5}Fe_{1,5}]O_4$ und $Y_3[Fe_5O_{12}]$ entsprechend 858, 943 bzw. 560 °K erreichen. Folglich treten unter diesen Verbindungen in vielen Fällen Néel-Temperaturen von der Größenordnung der Curie-Temperaturen der ferromagnetischen Elemente Eisen, Kobalt und Nickel auf. Im Zusammenhang damit sprach F. BLOCH[5] schon im Jahre 1933 die Idee aus, daß sich die Austauschkopplung in magnetisch geordneten Kristallen von Verbindungen unter aktiver Teilnahme der Elektronen in den spin-gesättigten p-Schalen der magnetisch neutralen Anionen (O^{--}, S^{--}, F^- usw.) vollzieht, die zwischen den magnetischen Kationen der d- oder f-Metalle (Fe^{++}, Ni^{++}, Co^{++} usw.) angeordnet sind. Die Berechnung dieser *indirekten* Austauschwechselwirkung wurde erstmals von KRAMERS [35] durchgeführt und von ANDERSON [36] und VAN VLECK [37], [38] weiterentwickelt. Das Schema der Berechnung läßt sich am Beispiel der kollinearen (180°-)Anordnung zweier Kationen Mn^{++} mit einem Zwischenanion O^{--} demonstrieren. Im einfachsten Modell wird angenommen, daß sich in der Nähe von O^{--} zwei p-Elektronen aufhalten, die eine Schale mit abgesättigten Spins bilden, und bei den Kationen Mn^{++} auf der äußeren Bahn je ein d-Elektron (das sog. Problem der drei Zentren und vier Elektronen). Im Grundzustand befinden sich die d-Elektronen der Kationen in den Zuständen φ_{d_1} und φ_{d_2}. Die beiden p-Elektronen des Anions sind auf einer Bahn im Zustand φ_p untergebracht (mit entgegengesetzten Spinprojektionen $s_z = \pm \frac{1}{2}$). Die Verteilung der p-Elektronen des Anions φ_p besitzt die Form einer Hantel, die entlang der alle drei Zentren verbindenden Achse orientiert ist. Wegen der Wechselwirkung zwischen den Ionen Mn^{++} und O^{--} besteht eine gewisse Wahrscheinlichkeit, daß eines der p-Elektronen vom Ion O^{--} zu einem der Kationen Mn^{++} (z.B. zum linken) übergeht. In diesem virtuellen angeregten Zustand besteht eine Kopplung zwischen den Elektronen in den Zuständen φ_{d_1} und $\varphi_{d'_1}$ des linken Kations Mn^{++} und eine Kopplung zwischen dem beim Anion O^{--} verbliebenen p-Elektron und dem d_2-Elektron des rechten Kations Mn^{++}. Selbstverständlich hängt diese Kopplung von der gegenseitigen Orientierung der Spins jedes Elektronenpaares ab. Weiterhin ist ein Austausch zwischen dem am Anion O^{--} verbliebenen p-Elektron und dem d_2-Elektron des rechten Kations Mn^{++} möglich. Als Ergebnis erhält man einen Grundzustand mit antiferromagnetischer Kopplung. Aus dieser schematischen Betrachtung wird deutlich, warum die indirekte Austauschkopplung für die auf die nächsten Nachbarn folgenden Nachbarn Mn^{++} (bei kollinearer 180°-Anordnung) stärker ist als eine ebensolche Kopplung für nächste Nachbarn Mn^{++} bei 90°-Anordnung mit dem Zwischenanion O^{--} am Scheitelpunkt. Bei dieser Kopplung bilden die Verbindungslinien der beiden benachbarten Kationen mit dem Anion einen rechten Winkel. Jede der beiden hantelförmigen p-Wellenfunktionen des Anions O^{--} überlappt sich merklich nur mit einem der nächsten Nachbarkationen Mn^{++}. Aus diesem Grunde ist die indirekte Kopplung der Kationen bei 90°-Anordnung wesentlich schwächer als bei der 180°-Konfiguration. Offensichtlich erfordert diese Kopplung die Beteiligung des hybridisierten Zustandes

---

[5] Siehe Hinweis darüber in der Arbeit von KRAMERS [35].

des Anions O^{--} mit einer Superposition von 2s- und 2p-Funktionen oder die Einbeziehung des angeregten 3s-Zustandes in die Betrachtung. Möglicherweise tritt eine solche Situation bei der Ausbildung der indirekten Austauschkopplung in den Verbindungen vom Typ MnF$_2$ auf.

Die weitere Entwicklung und Vervollständigung der Theorie des indirekten Austausches nach KRAMERS[6] erfolgte in den Arbeiten vieler anderer Autoren.

Ein etwas anderes Verfahren schlug SLATER [46] vor, der die Spinpolarisation des Zwischenanions berücksichtigte. Es zeigt sich, daß die Konfiguration mit zwei antiparallelen Spins, z.B. zweier Manganionen, stabiler wird, wenn die beiden äußeren p-Elektronen des Anions O^{--} leicht unterschiedliche Bahnfunktionen besitzen. Auf der einen davon wird der (+)-Spin zum rechten Kation Mn^{++} verschoben, auf der anderen verlagert sich der (−)-Spin zum linken Kation Mn^{++}. Bei parallelen Spins der d-Schalen beider Nachbarkationen würde dieser Effekt nicht entstehen. Die Energiedifferenz kann hierbei mittels eines effektiven Austausch-Spin-Hamilton-Operators vom Heisenberg-Typ mit indirektem Austauschintegral ausgedrückt werden.

Eine wichtige Etappe in der Entwicklung der Theorie der Austauschkopplung in Isolatoren stellte die Arbeit von ANDERSON [47] dar, wo dieser folgende, nach seiner Meinung wichtige, allen vorhergehenden Untersuchungen anhaftende Unzulänglichkeiten vermerkt. Erstens tritt in all diesen Berechnungen der indirekte Austauscheffekt erst in höheren Näherungen der Störungstheorie (in der dritten oder vierten) auf, ungeachtet seiner relativ bedeutenden Größe ($\sim 10^2$ °K). Nach ANDERSON deutet dies darauf hin, daß die Konvergenz der Reihen der Störungstheorie bei diesen Berechnungen schlecht ist. Zweitens wurden neben dem ursprünglichen Bloch-Kramers-Mechanismus noch andere Mechanismen vorgeschlagen, in denen zwei- oder einfache Übergangsintegrale verwendet werden. Ungeachtet dessen führen alle diese *nicht-Kramersschen* Austauschmechanismen etwa zu den gleichen Folgerungen über das Vorzeichen und die Größenordnung der indirekten Austauschkopplung. Ferner muß man beachten, daß geringfügige Änderungen der bei den Berechnungen verwendeten Wellenfunktionen oftmals den einen Austauschmechanismus in einen anderen überführen können. Der letzte Umstand deutet darauf hin, daß die zu untersuchende Erscheinung der indirekten Austauschkopplung ihrem physikalischen Inhalt nach wesentlich universeller ist, als das in den von ANDERSON kritisierten Auslegungen angenommen wurde. Schließlich gibt es noch die komplizierte Frage, die das Problem der Orthogonalität sowie der Wahl der Wellenfunktionen bei diesen Berechnungen betrifft. In der Regel verwendet man die Wellenfunktionen eines freien Ions. Bei realen Kristalldichten überlappen sich jedoch diese Funktionen, die die Zustände der äußeren Elektronen beschreiben, für benachbarte Gitterpunkte sehr stark. Eben in dieser Überlappung ist auch die Ursache für das Auftreten des einen oder anderen Typs der magnetischen Atomordnung im Kristall zu suchen. Zugleich führt die merkliche Überlappung der Funktionen zu einer erheblichen Komplikation aller Berechnungen. Insbesondere treten sofort Schwierigkeiten bei der Auswahl des richtigen Satzes von Basisfunktionen auf. Die Benutzung der Wellenfunktionen des freien Ions wird überdies unzulässig, da sich diese Funktionen im Kristall tatsächlich grundlegend ändern. Bei Versuchen,

---

[6] Für den Fall der 3d- oder 4f-Metalle wird der Zustand der magnetischen Ordnung auf der Basis des $s-d(f)$-Austauschmodelles beschrieben, das zuerst von SCHUBIN und VONSOVSKY [39] bis [41] vorgeschlagen wurde. Nach diesem Modell kann die Austauschkopplung zwischen den magnetisch aktiven 4f-Schalen der Nachbaratome mit Hilfe des Rudermann-Kittel-Yosida-Kasuya-Mechanismus [42] bis [45] des indirekten Austausches über das System der Leitungselektronen erfolgen.

diese Änderungen in der Theorie in irgendeiner Form zu berücksichtigen, müssen unbedingt sehr viele willkürliche Parameter eingeführt werden, was praktisch die Möglichkeit quantitativer theoretischer Abschätzungen ausschließt.

Mit dem Ziel, die erwähnten Schwierigkeiten irgendwie zu umgehen, schlug ANDERSON [47] sein „neues Verfahren" in der Theorie des indirekten Austausches in dielektrischen Kristallen vor und zerlegt die Aufgabe in zwei Etappen. Im *ersten* Schritt muß die Wellenfunktion eines magnetischen Ions, das in einem diamagnetischen Medium unmagnetischer Kristallionen gelöst ist, gefunden werden. Dabei wird nicht berücksichtigt, wie die Wellenfunktionen, die nach der Methode des Ligandenfeldes [48], [49] zu berechnen sind, durch Austauschkopplung mit anderen magnetischen Ionen beeinflußt werden. Der *zweite* Schritt besteht in der Berechnung der Wechselwirkung der magnetischen Ionen, deren Zustände durch die vorher berechneten Wellenfunktionen beschrieben werden.

ANDERSON verweist auch auf experimentelle Begründungen für die Berechtigung einer solchen Unterteilung der Aufgabe in die oben angegebenen zwei Schritte. Da ist zunächst die beobachtete Unveränderlichkeit der Hyperfeinwechselwirkungen an den Kernen der Liganden in verdünnten und konzentrierten Modifikationen derselben Verbindungen (CLOGSTON u. a. [50]) sowie in para- und antiferromagnetischen Bereichen ein und derselben Verbindung (SHULMAN und JACCARINO [51] bis [53]). Zweitens ist dies die Übereinstimmung der Größen der Parameter des Ligandenfeldes und der Austauschintegrale sowohl in konzentrierten als auch in verdünnten Systemen (GRIFFITHS u. a. [54], STOUT [55]). Somit darf man vom experimentellen Standpunkt aus annehmen, daß durchaus definierte Wellenfunktionen und Energieniveaus der magnetischen Ionen existieren, die nur unwesentlich von der magnetischen Umgebung abhängen, und deren Berechnung daher praktisch vollständig in das Gebiet der Ligandenfeldtheorie gehört[7].

Es sollen noch zwei Vorteile der neuen Methode von ANDERSON Erwähnung finden [47]: 1. Im zweiten Schritt hat man es bei der Berechnung nach der Störungstheorie nur mit Wechselwirkungen zwischen magnetischen Ionen zu tun (da der Einfluß der dazwischen liegenden Liganden bereits im ersten Schritt berücksichtigt wurde), d. h. tatsächlich mit schwacher Wechselwirkung, was die Schwierigkeiten mit der schlechten Konvergenz der Reihen beseitigt. 2. Da bei der abschließenden Berechnung der Austauschwechselwirkung die unmagnetischen (Liganden-) Zwischenionen explizit ausgeschaltet sind, verschwindet formal grundsätzlich der Unterschied zwischen den Bezeichnungen „direkter" und „indirekter" Austausch. Statt dessen sollte man, wie in Ziff. 15 und 26 noch gezeigt wird, von „*kinetischem*" und „*potentiellem*" Austausch sprechen. Der erste ist verantwortlich für die antiferromagnetische Kopplung, der zweite für die ferromagnetische.

Im Zusammenhang mit der neuen Methode von ANDERSON [47] für den indirekten Austausch in Isolatoren erlangen die Berechnungen der Quantenzustände der magnetisch aktiven Ionen in einem Kristallfeld mit vorgegebener Symmetrie (Ligandenfeldtheorie) große Bedeutung.

Die vorliegende Übersichtsarbeit besteht aus vier Teilen: Im Teil A wird als erläuternde Einführung hinreichend ausführlich die Frage der Austauschkopplung im einfachsten Zweielektronensystem, d. h. im Wasserstoffmolekül H_2, betrachtet. Im Teil B wird die Theorie der direkten Austauschwechselwirkung in Isolatorkristallen dargelegt.

---

[7] Das Verfahren von ANDERSON mit den angegebenen zwei Etappen ist in gewissem Sinne analog der Methode der effektiven Masse in der Einelektronentheorie metallischer und halbleitender Kristalle.

Teil C ist dem zentralen Problem der indirekten Austauschwechselwirkung bei drei Zentren und vier Elektronen (IV, a und b) sowie in Kristallen (IVc, V) gewidmet.

Schließlich befaßt sich Teil D mit der Anwendung der Ligandenfeldtheorie auf das Problem des indirekten Austausches in Isolatoren.

In der Übersicht fand der gegenwärtige Stand der Theorie der Austauschwechselwirkung in nichtleitenden Kristallen seinen Niederschlag. Es wird eine weitgehend vollständige Bibliographie von Übersichts- und Originalarbeiten zu diesem Problem gegeben.

A. Die Theorie des Wasserstoffmoleküls in der Heitler-London-Näherung

1. Das Auffinden der Wellenfunktionen nullter Näherung und der Energie erster Näherung für das Molekül H_2. Die Definition des Austauschintegrals. Wie bereits in der Einführung erwähnt, entstand die Theorie der Austauschwechselwirkung in Kristallen historisch gesehen als Verallgemeinerung der einfachsten Probleme der Quantenchemie und der Theorie der Vielelektronenatome, nämlich des zweiatomigen Wasserstoffmoleküls H_2 [5] und des Heliumatoms He [2], [3]. Im Zusammenhang damit erscheint es zweckmäßig, zur Veranschaulichung die Theorie des Wasserstoffmoleküls H_2 kurz zu behandeln.

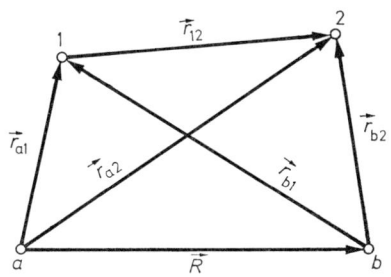

Fig. 1. Zur Geometrie des Wasserstoffmoleküls.

Das H_2-System besteht aus den beiden Kernen (Protonen) a und b mit den Radiusvektoren $\vec{R}_a$ und $\vec{R}_b$ und den beiden Elektronen 1 und 2 mit den Radiusvektoren $\vec{r}_1$ und $\vec{r}_2$. Fig. 1 zeigt seine Geometrie. Es bedeuten $|\vec{r}_{12}|=|\vec{r}_1-\vec{r}_2|$ den Abstand zwischen den Elektronen, $|\vec{R}|=|\vec{R}_a-\vec{R}_b|$ den Abstand zwischen den Kernen sowie $|\vec{r}_{a1}|=|\vec{R}_a-\vec{r}_1|$, $|\vec{r}_{a2}|=|\vec{R}_a-\vec{r}_2|$, $|\vec{r}_{b1}|=|\vec{R}_b-\vec{r}_1|$ und $|\vec{r}_{b2}|=|\vec{R}_b-\vec{r}_2|$ die Abstände zwischen den entsprechenden Kernen und Elektronen. Der Hamilton-Operator $\hat{H}$ lautet (ohne Berücksichtigung relativistisch-magnetischer Wechselwirkungen)

$$\hat{H} = -\frac{\hbar^2}{2m}(\Delta_1+\Delta_2) - \frac{e^2}{r_{a1}} - \frac{e^2}{r_{b1}} - \frac{e^2}{r_{a2}} - \frac{e^2}{r_{b2}} + \frac{e^2}{r_{12}} + \frac{e^2}{R} \qquad (1.1)$$

oder

$$\hat{H} = \hat{H}_a(\vec{r}_1) + \hat{H}_b(\vec{r}_2) + V_a(\vec{r}_2) + V_b(\vec{r}_1) + V(\vec{r}_1,\vec{r}_2) + V(\vec{R}), \qquad (1.2)$$

wobei folgende Bezeichnungen eingeführt wurden:

$$\left.\begin{aligned}
\hat{H}_a(\vec{r}_1) &= -\frac{\hbar^2}{2m}\Delta_1 - \frac{e^2}{r_{a1}}, \quad \hat{H}_b(\vec{r}_2) = -\frac{\hbar^2}{2m}\Delta_2 - \frac{e^2}{r_{b2}}, \\
V_a(\vec{r}_2) &= -\frac{e^2}{r_{a2}}, \quad V_b(\vec{r}_1) = -\frac{e^2}{r_{b1}}, \quad V(\vec{r}_1,\vec{r}_2) = \frac{e^2}{r_{12}}, \\
V(\vec{R}) &= \frac{e^2}{R}.
\end{aligned}\right\} \qquad (1.3)$$

Hierin bedeuten $2\pi\hbar = h$ die Plancksche Konstante, m die Elektronenruhemasse, e die Elektronenladung und Δ_i den Laplace-Operator.

In der Quantenmechanik ist die Behandlung des Wasserstoffmoleküls gleichbedeutend mit der Lösung der Schrödinger-Gleichung

$$\hat{H}\Psi(q_1, q_2) = E\Psi(q_1, q_2),\qquad(1.4)$$

wobei $\Psi(q_1, q_2)$ die Wellenfunktionen und E die Energie des Moleküls sowie q_i die Gesamtheit der Orts- ($\vec{r}_i$) und Spinkoordinaten (σ_i) des i-ten Elektrons bedeuten ($i=1, 2$).

Die allgemeinste Definition der *Austausch*energie eines *Zwei*elektronensystems, wie es der Hamilton-Operator vom Typ (1.1) beschreibt, beruht darauf, daß es zwei Typen von Zuständen geben kann: *Singulett*zustände mit der Quantenzahl des Gesamtspins $S=0$ und der Energie E_s und *Triplett*zustände mit $S=1$ und E_t. Die Austauschenergie oder das Austauschintegral J wird als halbe Differenz der jeweils kleinsten Energiewerte beider Zustandstypen E_s^0 und E_t^0 definiert [56]:

$$J = \tfrac{1}{2}(E_s^0 - E_t^0).\qquad(1.5)$$

Aus der Formel (1.5) ist ersichtlich, daß bei $J<0$, $E_s^0 < E_t^0$, der Grundzustand ein *Singulett*zustand ist und die Elektronenspins antiparallel gerichtet sind ($\uparrow\downarrow$ — Prototyp des Antiferromagnetismus), während bei $J>0$, $E_t^0 < E_s^0$, der Grundzustand ein *Triplett*zustand ist und beide Spins „parallel" liegen ($\uparrow\uparrow$ — Prototyp des Ferromagnetismus)[8].

Unter Verwendung des Vektormodells für die Summation der Spinmomente der einzelnen Elektronen läßt sich eine geometrische Interpretation des Singulett- und Triplettzustandes

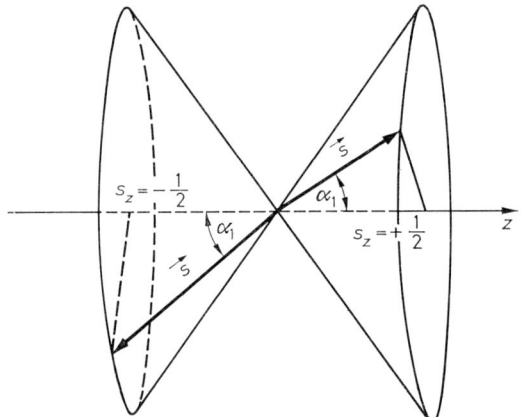

Fig. 2. Präzessionskegel des Spins eines Elektrons in den Zuständen $s_z = +\tfrac{1}{2}$ und $s_z = -\tfrac{1}{2}$.

(mit drei Projektionen) geben. Das Spinmoment $\vec{s}$ eines Elektrons besitzt auf der Quantisierungsachse z zwei Projektionen: $s_z = \pm\tfrac{1}{2}$ (in $\hbar$-Einheiten). Die Länge des Vektors $\vec{s}$ wird durch die Beziehung

$$|\vec{M}| = \sqrt{M(M+1)}$$

bestimmt, die für beliebige Drehimpulse $\vec{M}$ gültig ist. Hierin bedeutet M die maximal mögliche Projektion des Vektors $\vec{M}$ auf die Quantisierungsachse. Es folgt daraus $|\vec{s}| = \sqrt{\tfrac{1}{2}(\tfrac{1}{2}+1)}$ $\sqrt{3}/2$. Aus der Beziehung

$$(\vec{s})^2 = s_x^2 + s_y^2 + s_z^2$$

findet man bei Kenntnis von s_z und $|\vec{s}|$

$$s_x^2 + s_y^2 = \tfrac{1}{2}.$$

---

[8] Siehe hierzu nachfolgenden Kleindruck.

Ziff. 1. Die nullte Näherung beim H_2-Molekül. 273

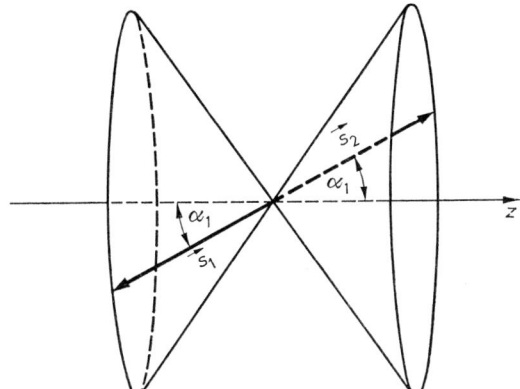

Fig. 3. Präzessionskegel der Elektronenspins im Singulettzustand eines Zweielektronensystems.

Folglich liegt der Vektor $\vec{s}$ so auf einem Kegel, daß seine Projektion auf die Achse z gleich $+\frac{1}{2}$ oder $-\frac{1}{2}$ ist und seine Spitze in Übereinstimmung mit der letzten Gleichung auf der Fläche (x, y) einen Kreis mit dem Radius $1/\sqrt{2}$ beschreibt. Fig. 2 zeigt die zeichnerische Darstellung dieses Verhaltens.

Der Neigungswinkel α_1 wird durch die Gleichung $\cos \alpha_1 = 1/\sqrt{3}$ bestimmt. Bei der Addition der Spins zweier Elektronen kann sich jeder der Spins auf irgendeinem der beiden Kegel in Fig. 2 befinden.

Im Singulettzustand $(S = 0)$ ist die Länge des Gesamtmomentes gleich Null. Folglich ordnen sich die Elektronenspins auf verschie-

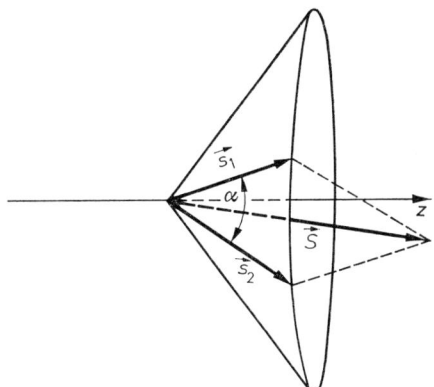
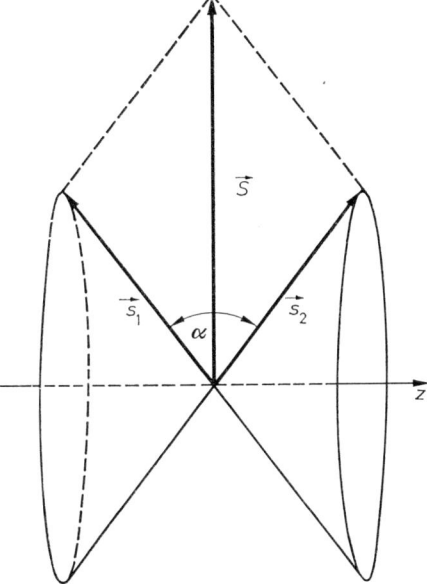

Fig. 4. Präzessionskegel der Elektronenspins im Triplettzustand eines Zweielektronensystems mit $S_z = 1$.

Fig. 5. Präzessionskegel der Elektronenspins im Triplettzustand eines Zweielektronensystems mit $S_z = 0$.

denen Kegeln an und sind antiparallel zueinander entlang den Kegelmantellinien gerichtet, wie in Fig. 3 dargestellt ist.

Man kann sagen, daß im Singulettzustand die Spins der beiden Elektronen um die Achse z „präzedieren" und sich stets in Gegenphase befinden.

Im Triplettzustand $(S = 1)$ muß man drei Fälle unterscheiden: $S_z = +1, 0, -1$. Der Betrag des Gesamtmomentes ist dann $|\vec{S}| = \sqrt{S(S+1)} = \sqrt{2}$. Um den Summenvektor mit der Länge $\sqrt{2}$ zu erhalten, müssen die Spins der beiden Elektronen, wie aus elementaren Betrachtungen folgt, untereinander den Winkel α bilden mit der Bedingung $\cos \alpha = \frac{1}{3}$. So sind bei

Handbuch der Physik, Bd. XVIII/1. 18

$S_z = +1$ beide Spins auf dem rechten Kegel angeordnet (was die erforderliche Gesamtprojektion $+\frac{1}{2}+\frac{1}{2}=+1$ ergibt), und der Winkel zwischen ihnen beträgt α, wie Fig. 4 zeigt.

In diesem Falle erfolgt die „Präzession" um z mit konstanter Phasendifferenz.

Für den Fall $S_z = -1$ erhält man ein zu Fig. 4 für $S_z = +1$ analoges Bild. Beide Spins sind jetzt lediglich auf dem linken Kegel angeordnet. Den Fall $S_z = 0$ kann man sich so vorstellen, daß beide Spins sich zwar auf verschiedenen Kegeln bewegen ($+\frac{1}{2} - \frac{1}{2} = 0$), aber im Gegensatz zum Singulettzustand den Winkel α einschließen, was zur richtigen Länge $\sqrt{2}$ des Gesamtvektors führt. Wie elementare Überlegungen zeigen, liegen die beiden Spins und die Quantisierungsachse in einer Ebene (in Fig. 5 sind sie in der Zeichenebene wiedergegeben).

Die „Präzession" der Spins erfolgt in Phase auf verschiedenen Kegeln, und der Gesamtvektor $\vec{S}$ „rotiert" um die Achse z, indem er stets senkrecht zu ihr bleibt ($S_z = 0$).

Es sei bemerkt, daß die Definitionsgleichung (1.5) aus den allgemeinsten Symmetrieeigenschaften der Wellenfunktion $\Psi(q_1, q_2)$ hinsichtlich der Vertauschung zweier Elektronen folgt. Tatsächlich ergibt sich aus der allgemeinen Forderung der Quantenmechanik über die Antisymmetrie der Wellenfunktion des Elektronensystems gegenüber der Vertauschung der Teilchenkoordinaten q_1 und q_2 speziell für das *Zwei*elektronensystem

$$\Psi(q_1, q_2) = -\Psi(q_2, q_1). \tag{1.6}$$

Bei Vernachlässigung der magnetischen (Spin-)Kräfte läßt sich die Zweielektronenwellenfunktion hinsichtlich ihrer Abhängigkeit von den Orts- und Spinkoordinaten wie folgt separieren[9]:

$$\Psi(q_1, q_2) = \psi(\vec{r}_1, \vec{r}_2)\chi(\sigma_1, \sigma_2). \tag{1.7}$$

Wie im vorliegenden Falle aus der Symmetriebedingung (1.6) sofort folgt, können Orts- und Spinanteil der Wellenfunktion sowohl symmetrisch

$$\psi_{\text{sym.}}(\vec{r}_1, \vec{r}_2) = \psi_{\text{sym.}}(\vec{r}_2, \vec{r}_1), \quad \chi_{\text{sym.}}(\sigma_1, \sigma_2) = \chi_{\text{sym.}}(\sigma_2, \sigma_1)$$

als auch antisymmetrisch

$$\psi_a(\vec{r}_1, \vec{r}_2) = -\psi_a(\vec{r}_2, \vec{r}_1), \quad \chi_a(\sigma_1, \sigma_2) = -\chi_a(\sigma_2, \sigma_1)$$

sein.

Hieraus ergibt sich auch die Möglichkeit der Existenz zweier Typen von Zuständen, des Singulettzustandes mit einer Wellenfunktion

$$\Psi_s(q_1, q_2) = \psi_{\text{sym.}}(\vec{r}_1, \vec{r}_2)\chi_a(\sigma_1, \sigma_2) \tag{1.8a}$$

und des Triplettzustandes mit einer Funktion

$$\Psi_t(q_1, q_2) = \psi_a(\vec{r}_1, \vec{r}_2)\chi_{\text{sym.}}(\sigma_1, \sigma_2). \tag{1.8b}$$

Die Definition (1.5) gestattet es, ohne Schwierigkeit den Austauschanteil der elektrostatischen Energie des *Zwei*elektronensystems, das von der gegenseitigen Orientierung der Spins abhängt, in der Schreibweise des einfachen Vektormodells anzugeben. Man schreibt die Energie des zu betrachtenden Systems in der Form

$$E = \tfrac{1}{2}(E_s^0 + E_t^0) - \tfrac{1}{2}\varkappa(E_s^0 - E_t^0) \quad (\varkappa = \pm 1). \tag{1.9}$$

Aus (1.9) folgt, daß bei $\varkappa = -1$ $E = E_s^0$ und bei $\varkappa = 1$ $E = E_t^0$ ist. Das Quadrat des Gesamtspins des Systems $\vec{S} = \vec{s}_1 + \vec{s}_2$ (in $\hbar$-Einheiten) beträgt

$$\vec{S}^2 = \tfrac{3}{2} + 2(\vec{s}_1 \vec{s}_2) = S(S+1), \tag{1.10}$$

---

[9] Im Falle von Vielelektronensystemen mit mehr als zwei Teilchen wird die Situation kompliziert, da die Wellenfunktion jetzt nicht mehr als Produkt zweier Faktoren dargestellt werden kann, wobei der eine lediglich von den Orts- und der andere von den Spinkoordinaten abhängt (s. z.B. [57], [58]).

da der Operator des Quadrates eines Elektronenspins $\vec{s}_i^2$ den Eigenwert $\frac{1}{2}(\frac{1}{2}+1)=\frac{3}{4}$ besitzt und der Eigenwert des Operators (1.10) gleich $S(S+1)$ ist. Entsprechend (1.10) hat dann der Operator

$$\tfrac{1}{2}+2(\vec{s}_1\vec{s}_2) \tag{1.11}$$

die Eigenwerte $S(S+1)-1$, d.h. -1 bei $S=0$ und $+1$ bei $S=1$, und diese stimmen mit den Werten $\varkappa$ in (1.9) überein. Ersetzt man $\varkappa$ in (1.9) durch den Operator (1.11), so erhält man für den Energieoperator des Systems

$$\widehat{W}=E_0-2J(\vec{s}_1\vec{s}_2). \tag{1.12}$$

Der Operator (1.12) besitzt die richtigen Spineigenfunktionen sowie die richtigen Energiewerte. Hierin ist $E_0=\frac{1}{4}(E_s^0+3E_t^0)$ die mittlere Energie aller vier Spinzustände unter Berücksichtigung ihrer statistischen Gewichte. Die Beziehung (1.12) ist der exakte Ausdruck für die Energie im Vektormodell. Sie hängt von keinerlei Annahmen ab, wie man sie üblicherweise bei der Behandlung des Wasserstoffmoleküls verwendet.

Aus (1.2) und (1.3) ist ersichtlich, daß der Operator der Teilchenvertauschung mit dem Hamilton-Operator $\widehat{H}$ vertauschbar (d.h. eine Erhaltungsgröße) ist. Wir führen eine Probefunktion $\Phi(\vec{r}_1,\vec{r}_2)$ der räumlichen Elektronenkoordinaten ein und zerlegen sie nach ihren symmetrischen und antisymmetrischen Anteilen, die zueinander orthogonal sind und unter dem Einfluß des Operators (1.1) nicht miteinander wechselwirken:

$$\left.\begin{array}{l}\Phi(\vec{r}_1,\vec{r}_2)=\tfrac{1}{2}[\Phi(\vec{r}_1,\vec{r}_2)+\Phi(\vec{r}_2,\vec{r}_1)]+\tfrac{1}{2}[\Phi(\vec{r}_1,\vec{r}_2)-\Phi(\vec{r}_2,\vec{r}_1)]\\=\tfrac{1}{2}[1+\widehat{P}_{12}]\Phi(\vec{r}_1,\vec{r}_2)+\tfrac{1}{2}[1-\widehat{P}_{12}]\Phi(\vec{r}_1,\vec{r}_2).\end{array}\right\} \tag{1.13}$$

Aus der oben erwähnten Forderung der Antisymmetrie der Gesamtwellenfunktion (1.6) folgt, daß der erste Summand in (1.13) dem Singulett- und der zweite dem Triplettzustand entspricht. Nach der Normierung der einzelnen Komponenten in (1.13) erhält man

$$E_{s,t}^0=\frac{\langle\Phi|\widehat{H}|\Phi\rangle\pm\langle\Phi|\widehat{H}\widehat{P}_{12}|\Phi\rangle}{1\pm\langle\Phi|\widehat{P}_{12}|\Phi\rangle}, \tag{1.14}$$

wo das obere Vorzeichen zum Singulett- und das untere zum Triplettzustand gehört. Hier fanden die Diracschen Bezeichnungen der Matrixelemente Anwendung:

$$\langle\Phi|\widehat{H}|\Phi\rangle=\int\Phi^*(\vec{r}_1,\vec{r}_2)\widehat{H}\Phi(\vec{r}_1,\vec{r}_2)d\vec{r}_1 d\vec{r}_2\quad\text{und}\quad\langle\Phi|\Phi\rangle=1$$

(Normierungsintegral).

(1.14) in (1.5) eingesetzt ergibt

$$J=\frac{\langle\Phi|\widehat{H}\widehat{P}_{12}|\Phi\rangle-\langle\Phi|\widehat{P}_{12}|\Phi\rangle\langle\Phi|\widehat{H}|\Phi\rangle}{1-\langle\Phi|\widehat{P}_{12}|\Phi\rangle^2}. \tag{1.15}$$

Falls die Funktion Φ eine „Mischung" zweier exakter Eigenfunktionen des Systems darstellt, so ist (1.15) der exakte Ausdruck für die Austauschenergie J — ein recht brauchbarer Ausgangspunkt für die Diskussion der Ausdrücke für J, die man aus Näherungstheorien erhält. Nimmt man jedoch an, daß die Probefunktion Φ selbst entweder symmetrisch oder antisymmetrisch ist, so liefert (1.15) keine hinreichende Information hinsichtlich des anderen Zustandes, und deshalb verliert die Beziehung (1.15) für J jeden Sinn.

Betrachten wir jetzt das *konkrete* Näherungsmodell nach HEITLER und LONDON [5], in dem die Elektronen zwei Atombahnzustände mit den normierten Funktionen $\varphi_a(\vec{r}_1)$ und $\varphi_b(\vec{r}_2)$ einnehmen, so daß

$$\Phi(\vec{r}_1, \vec{r}_2) = \varphi_a(\vec{r}_1)\,\varphi_b(\vec{r}_2) \tag{1.16}$$

ist. Für das Überlappungsintegral oder die Nichtorthogonalität wird die Bezeichnung

$$\langle \Phi|\hat{P}_{12}|\Phi\rangle = \int \varphi_a^*(\vec{r}_1)\,\varphi_b^*(\vec{r}_2)\,\varphi_a(\vec{r}_2)\,\varphi_b(\vec{r}_1)\,d\vec{r}_1\,d\vec{r}_2 = |S_{ab}|^2 \tag{1.17}$$

eingeführt. Dann erhält man wegen (1.15) schon einen *Näherungs*ausdruck für das Austauschintegral:

$$J = \frac{\langle ab|\hat{H}|ba\rangle - |S_{ab}|^2 \langle ab|\hat{H}|ab\rangle}{1 - |S_{ab}|^4}. \tag{1.18}$$

Hier sind in den Matrixelementen zur Abkürzung φ_a und φ_b durch die Symbole a und b ersetzt. Zur Vereinfachung wird oftmals das Überlappungsintegral (1.17) vernachlässigt. Anstelle von (1.18) erhält man dann für das Austauschintegral die Näherung[10]

$$J(S_{ab}=0) = \langle ab|\hat{H}|ba\rangle. \tag{1.19}$$

Setzt man (1.2) unter Berücksichtigung von (1.3) in (1.18) bzw. (1.19) ein, dann folgt

$$J = (1 - |S_{ab}|^4)^{-1}\{\langle a|\hat{H}_a+V_b|b\rangle S_{ba} + \langle b|\hat{H}_b+V_a|a\rangle S_{ab} + \\ + \langle ab|V_{12}|ba\rangle - |S_{ab}|^2(\langle a|\hat{H}_a+V_b|a\rangle + \langle b|\hat{H}_b+V_a|b\rangle + \langle ab|V_{12}|ab\rangle)\} \tag{1.20}$$

bzw.

$$J(S_{ab}=0) = \langle ab|V_{12}|ba\rangle = e^2 \int \frac{\varphi_a^*(\vec{r}_1)\,\varphi_b^*(\vec{r}_2)\,\varphi_a(\vec{r}_2)\,\varphi_b(\vec{r}_1)}{r_{12}}\,d\vec{r}_1\,d\vec{r}_2. \tag{1.21}$$

Wie (1.21) zeigt, ist das Integral $J(S_{ab}=0)$ die Selbstenergie einer Elektronenwolke mit der Dichte $e \cdot \varphi_a^*(\vec{r})\,\varphi_b(\vec{r})$ und damit positiv definit,

$$J(S_{ab}=0) > 0. \tag{1.22}$$

Aus (1.22) folgt, daß bei Vernachlässigung des Nichtorthogonalitätsintegrales (1.17) ($S_{ab}=0$) das auf der Verallgemeinerung des Wasserstoffmolekülproblems beruhende Modell für die Beschreibung der magnetischen Ordnung im Kristall nicht imstande ist, die Existenz von Antiferromagnetismus zu erklären, denn das Austauschintegral ist nach (1.22) *stets* positiv und kann sein Vorzeichen nicht ändern. Das hängt damit zusammen, daß bei *orthogonalisierten* Bahnen die Verwendung von Probefunktionen der Form (1.16) stets einen *Triplett*zustand als Grundzustand zur Folge hat ($E_t^0 < E_s^0$) und demnach ein Singulettzustand (im Rahmen eines solchen Modells) niemals eine chemische Bindung ermöglichen kann. Mit Hilfe der orthogonalisierten Bahnen läßt sich eine adäquate Beschreibung des Wasserstoffmoleküls lediglich unter Berücksichtigung der „polaren" Zustände[11] $\varphi_a(\vec{r}_1)\,\varphi_a(\vec{r}_2)$ und $\varphi_b(\vec{r}_1)\,\varphi_b(\vec{r}_2)$ erhalten, welche die Energie des Singulettzustandes verändern.

Im Falle nichtorthogonaler Bahnfunktionen φ_a und φ_b (wenn $S_{ab} \neq 0$ ist) enthält das Austauschintegral (1.20) die Operatoren $\hat{H}_a+V_b$ oder $\hat{H}_b+V_a$ und somit die Operatoren der kinetischen Energie. Die Einelektronenfunktionen φ_a und φ_b werden gewöhnlich als Lösungen der Schrödinger-Gleichung für isolierte

---

[10] Diese Vereinfachung ist nicht folgerichtig, da der Ausdruck (1.19) selbst von der Größenordnung $|S_{ab}|^2$ ist.

[11] Polare Zustände dieses Typs wurden erstmals von SLATER [59] und dann von SCHUBIN und VONSOVSKY [39], [60] bis [62] eingeführt; s. auch Abschnitt III der Arbeit von SOMMERFELD und BETHE [63].

Die nullte Näherung beim H$_2$-Molekül.

Atome angesehen:
$$\hat{H}_a \varphi_a = E_0 \varphi_a, \quad \hat{H}_b \varphi_b = E_0 \varphi_b. \tag{1.23}$$

Unter Verwendung von (1.23) folgt aus (1.20)

$$\begin{aligned} J = (1-|S_{ab}|^4)^{-1} \{ &\int \varphi_a^*(\vec{r}_1) \varphi_b^*(\vec{r}_2) [V_a(\vec{r}_2) + V_b(\vec{r}_1) + \\ &+ V(\vec{r}_1 \vec{r}_2)] \varphi_b(\vec{r}_1) \varphi_a(\vec{r}_2) d\vec{r}_1 d\vec{r}_2 - \\ -|S_{ab}|^2 &\int \varphi_a^*(\vec{r}_1) \varphi_b^*(\vec{r}_2) [V_a(\vec{r}_2) + V_b(\vec{r}_1) + V(\vec{r}_1 \vec{r}_2)] \varphi_a(\vec{r}_1) \varphi_b(\vec{r}_2) d\vec{r}_1 d\vec{r}_2 \}. \end{aligned} \tag{1.24}$$

Wenn man in (1.24) die Glieder mit dem Faktor $|S_{ab}|$ wegläßt, dann erhält man den Ausdruck

$$J = \int \varphi_a^*(\vec{r}_1) \varphi_b^*(\vec{r}_2) \left(-\frac{e^2}{r_{a2}} - \frac{e^2}{r_{b1}} + \frac{e^2}{r_{12}} \right) \varphi_b(\vec{r}_1) \varphi_a(\vec{r}_2) d\vec{r}_1 d\vec{r}_2. \tag{1.25}$$

Es sei bemerkt, daß in die Ausdrücke für das Austauschintegral (1.24) und (1.25) das durch die Kern-Kern-Wechselwirkung bedingte Glied $V(R)$ nicht eingeht. Wenn man jedoch im Zähler des Ausdruckes (1.24) dieses Glied $V(R)$, das zu den zwei Ausdrücken unter dem Integral die von den Elektronenkoordinaten unabhängige Größe e^2/R beisteuert, hinzufügt bzw. abzieht, so stellt das Integral mit dem Faktor $|S_{ab}|^2$ die Energie der Coulomb-Wechselwirkung zweier neutraler Atome dar:

$$e^2 \int |\varphi_a(\vec{r}_1)|^2 |\varphi_b(\vec{r}_2)|^2 \left(-\frac{1}{r_{a2}} - \frac{1}{r_{b1}} + \frac{1}{r_{12}} + \frac{1}{R} \right) d\vec{r}_1 d\vec{r}_2.$$

Dieser Ausdruck ist der Exponentialform e^{-2R} proportional [67]. Dadurch wird das Integral bei großem R klein. Diese Tatsache dient zugleich zur Rechtfertigung der weit verbreiteten Näherung, in der dieses mit dem Quadrat des Überlappungsintegrals $|S_{ab}|^2$ multiplizierte Glied vernachlässigt wird. In diesem Falle lautet das Austauschintegral

$$J_H = \int \varphi_a^*(\vec{r}_1) \varphi_b^*(\vec{r}_2) \left(-\frac{e^2}{r_{a2}} - \frac{e^2}{r_{b1}} + \frac{e^2}{r_{12}} + \frac{e^2}{R} \right) \varphi_b(\vec{r}_1) \varphi_a(\vec{r}_2) d\vec{r}_1 d\vec{r}_2. \tag{1.25a}$$

Die Größe (1.25a) ist in der Literatur unter der Bezeichnung *Heisenbergsches* Austauschintegral bekannt.

Beim Zweielektronenproblem macht es keine Schwierigkeiten, die Nichtorthogonalität der Funktionen $\varphi_a(\vec{r})$ und $\varphi_b(\vec{r})$ zu berücksichtigen und die Rechnung nach der genaueren Formel (1.24) durchzuführen. Tatsächlich erfolgt jedoch bei Annäherung der Atome und Bildung eines Moleküls eine Deformation der Elektronenhüllen, weshalb die Funktionen φ_a und φ_b schon nicht mehr den Gleichungen (1.23) unterliegen und daher z.B. mit Hilfe des Virialsatzes bestimmt werden müssen [64].

Aus dieser kurzen Darstellung des quantenmechanischen H$_2$-Problems sehen wir, daß die eigentliche Bestimmung der Austauschenergie nach (1.15) keine besonderen Schwierigkeiten macht. Alle Komplikationen entstehen im Zusammenhang mit der Wahl einer konkreten Probefunktion $\Phi(\vec{r}_1, \vec{r}_2)$. Es ist klar, daß beispielsweise ein Ansatz der Form (1.16), wobei die Funktionen φ_a und φ_b den Einelektronengleichungen der Wasserstoffatome (1.23) genügen, nur eine äußerst grobe Näherung darstellt. Das äußert sich vor allem darin, daß man die exakte Lösung nicht in Form einer Entwicklung nach einem *vollständigen* System bekannter orthonormierter Funktionen ansetzt, sondern als eine plötzlich *abbrechende* Entwicklung.

Wenden wir uns noch einmal einer ausführlichen Diskussion der Heitler-London-Näherung zu, um die oben erwähnten Schwierigkeiten im einzelnen zu klären. Wie bereits bemerkt, sind in der Heitler-London-Näherung die Wellenfunktionen eines Zweielektronenmoleküls aus den Einelektronenwellenfunktionen der isolierten Atome aufgebaut, d.h. aus den Wellenfunktionen des Wasserstoffatoms. Wie aus dem Pauli-Prinzip in Übereinstimmung mit der Bedingung (1.6) folgt, wird die Wellenfunktion für das Molekül in nullter Näherung als antisymmetrische Linearkombination von Produkten der Orts- und Spinfunktionen der einzelnen Elektronen angesetzt. Dabei ist es zweckmäßig, eine Entwicklung der Vielelektronenwellenfunktion nach Slater-Determinanten [65] anzuwenden, die bereits antisymmetrisch hinsichtlich der Koordinatenvertauschung sind. Berücksichtigt man lediglich die Funktionen des Grundzustandes des Wasserstoffatoms und vernachlässigt die ionisierten (polaren) Zustände mit zwei Elektronen an einem Atom, so lauten die entsprechenden Slater-Determinanten:

$$\left.\begin{aligned}
u_1 &= \frac{1}{\sqrt{2}} \begin{vmatrix} \varphi_a(\vec{r}_1)\,\alpha(\sigma_1) & \varphi_a(\vec{r}_2)\,\alpha(\sigma_2) \\ \varphi_b(\vec{r}_1)\,\beta(\sigma_1) & \varphi_b(\vec{r}_2)\,\beta(\sigma_2) \end{vmatrix}, \\
u_2 &= \frac{1}{\sqrt{2}} \begin{vmatrix} \varphi_a(\vec{r}_1)\,\beta(\sigma_1) & \varphi_a(\vec{r}_2)\,\beta(\sigma_2) \\ \varphi_b(\vec{r}_1)\,\alpha(\sigma_1) & \varphi_b(\vec{r}_2)\,\alpha(\sigma_2) \end{vmatrix}, \\
v_1 &= \frac{1}{\sqrt{2(1-S_{ab}^2)}} \begin{vmatrix} \varphi_a(\vec{r}_1)\,\alpha(\sigma_1) & \varphi_a(\vec{r}_2)\,\alpha(\sigma_2) \\ \varphi_b(\vec{r}_1)\,\alpha(\sigma_1) & \varphi_b(\vec{r}_2)\,\alpha(\sigma_2) \end{vmatrix}, \\
v_2 &= \frac{1}{\sqrt{2(1-S_{ab}^2)}} \begin{vmatrix} \varphi_a(\vec{r}_1)\,\beta(\sigma_1) & \varphi_a(\vec{r}_2)\,\beta(\sigma_2) \\ \varphi_b(\vec{r}_1)\,\beta(\sigma_1) & \varphi_b(\vec{r}_2)\,\beta(\sigma_2) \end{vmatrix},
\end{aligned}\right\} \quad (1.26)$$

wobei $\alpha(\sigma)$ die Spinfunktion mit $s_z = +\tfrac{1}{2}$ als Projektion des Elektronenspins und $\beta(\sigma)$ diejenige mit $s_z = -\tfrac{1}{2}$ bedeuten. (Im folgenden werden der Kürze halber die Bezeichnungen „Rechts-" und „Links-"Spin verwendet.) Wir nehmen an, daß die Funktionen φ_a und φ_b den Gleichungen (1.23) genügen. Die Molekülfunktion $\Psi(q_1, q_2)$ stellen wir in Form einer Entwicklung nach den Funktionen u_i und v_i aus (1.26) dar. Aus der elementaren Störungstheorie wissen wir, daß zur Bestimmung der Entwicklungskoeffizienten für ein entartetes Niveau die Säkulargleichung

$$|H_{ij} - E S_{ij}| = 0 \qquad (1.27)$$

gelöst werden muß, worin E die Energie des Vielelektronensystems in erster Näherung, H_{ij} das Matrixelement des Hamilton-Operators $\hat{H}$, berechnet für Determinanten aus Basisfunktionen,

$$H_{ij} = \int \Psi_i^* \hat{H} \Psi_j \, d\tau, \qquad (1.28)$$

und S_{ij} das Nichtorthogonalitätsintegral [s. (1.17)]

$$S_{ij} = \int \Psi_i^* \Psi_j \, d\tau$$

bedeuten. Hier kann man unter Ψ_i und Ψ_j die Funktionen u_i und v_i aus (1.26) verstehen. Das Symbol $\int \ldots d\tau$ in (1.28) bedeutet Integration über Orts- und Summation über Spinveränderliche. Um die Determinante in (1.27) zu vereinfachen, d.h. um ihr eine diagonale oder quasidiagonale (Block-)Form zu geben und somit den Grad der Gleichung (1.27) bezüglich E zu erniedrigen, ist es zweckmäßig, die Vielelektronenfunktion als eine Entwicklung nach Funktionen anzusetzen, die gleichzeitig Eigenfunktionen der Operatoren des Quadrates des Gesamtspins $\hat{S}^2$ und seiner z-Komponente $\hat{S}_z$ sind. Die Matrixelemente zwischen Zuständen, die zu verschiedenen Eigenwerten der Operatoren $\hat{S}^2$ und $\hat{S}_z$ gehören, ver-

Die nullte Näherung beim H$_2$-Molekül.

schwinden dann, wenn $\hat{H}$ selbst nicht von den Spins abhängt. Man kann weiter bei Kenntnis der Eigenfunktionen dieser Operatoren sofort die Energien für Zustände unterschiedlicher Multiplizität angeben. Folglich muß man zuerst aus den Determinanten u_i und v_i die Eigenfunktion der Triplett- und Singulettzustände Ψ_t und Ψ_s aufbauen. Das läßt sich leicht bewerkstelligen, indem man den gesuchten Funktionen folgende Forderungen auferlegt:

$$\hat{S}^2\, {}^{S_z}\Psi_t = S(S+1)\, {}^{S_z}\Psi_t = 2\cdot {}^{S_z}\Psi_t, \quad \hat{S}_z\, {}^{S_z}\Psi_t = S_z \cdot {}^{S_z}\Psi_t, \\ \hat{S}^2\Psi_s = S(S+1)\Psi_s = 0, \qquad \hat{S}_z\Psi_s = 0, \tag{1.29}$$

wobei $S_z = -1, 0, 1$ ist. Allgemein können bei Problemen mit mehr als zwei Elektronen mehrere Multipletts mit gleichem S auftreten. Deshalb muß dann außer S_z und S noch ein weiterer Index eingeführt werden.

Man sieht leicht, daß u_1 und u_2 selbst Eigenfunktionen von $\hat{S}_z$ mit dem Eigenwert $S_z = 0$ sowie v_1 und v_2 solche mit $S_z = +1$ bzw. $S_z = -1$ darstellen. Für die Anwendung auf die Determinanten ist die günstigste Schreibweise für den Operator $\hat{S}^2$

$$\hat{S}^2 = \hat{S}^- \hat{S}^+ + \hat{S}_z + \hat{S}_z^2, \tag{1.30}$$

wobei

$$\hat{S}^+ = \hat{S}_x + i\hat{S}_y, \\ \hat{S}^- = \hat{S}_x - i\hat{S}_y \tag{1.31}$$

ist. Die Operatoren S^+ und S^- wirken auf eine beliebige willkürliche Funktion $\Phi(S, S_z)$ entsprechend den bekannten Gleichungen

$$\hat{S}^+ \Phi(S, S_z) = (S-S_z)^{\frac{1}{2}}(S+S_z+1)^{\frac{1}{2}}\Phi(S, S_z+1), \\ \hat{S}^- \Phi(S, S_z) = (S+S_z)^{\frac{1}{2}}(S-S_z+1)^{\frac{1}{2}}\Phi(S, S_z-1). \tag{1.32}$$

Man erhält dann unter Verwendung von (1.29), (1.30) und (1.32) leicht

$$\begin{aligned}{}^{-1}\Psi_t = v_2 &= \frac{1}{\sqrt{2(1-S_{ab}^2)}}\{\varphi_a(\vec{r}_1)\varphi_b(\vec{r}_2) - \varphi_a(\vec{r}_2)\varphi_b(\vec{r}_1)\}\beta(\sigma_1)\beta(\sigma_2), \\ {}^{0}\Psi_t &= \frac{1}{\sqrt{2(1-S_{ab}^2)}}(u_1+u_2) = \frac{1}{2\sqrt{1-S_{ab}^2}}\{\varphi_a(\vec{r}_1)\varphi_b(\vec{r}_2) - \\ &\quad - \varphi_b(\vec{r}_1)\varphi_a(\vec{r}_2)\}\{\alpha(\sigma_1)\beta(\sigma_2) + \beta(\sigma_1)\alpha(\sigma_2)\}, \\ {}^{+1}\Psi_t = v_1 &= \frac{1}{\sqrt{2(1-S_{ab}^2)}}\{\varphi_a(\vec{r}_1)\varphi_b(\vec{r}_2) - \varphi_b(\vec{r}_1)\varphi_a(\vec{r}_2)\}\alpha(\sigma_1)\alpha(\sigma_2), \\ \Psi_s &= \frac{1}{\sqrt{2(1+S_{ab}^2)}}(u_1-u_2) = \frac{1}{2\sqrt{1+S_{ab}^2}}\{\varphi_a(\vec{r}_1)\varphi_b(\vec{r}_2) + \\ &\quad + \varphi_a(\vec{r}_2)\varphi_b(\vec{r}_1)\}\{\alpha(\sigma_1)\beta(\sigma_2) - \beta(\sigma_1)\alpha(\sigma_2)\}.\end{aligned} \tag{1.33}$$

Alle Funktionen (1.33) sind normiert und orthogonal zueinander. Die Faktoren $1/\sqrt{2(1 \mp S_{ab}^2)}$ und $1/2\sqrt{1 \mp S_{ab}^2}$ gewährleisten die Normierung der entsprechenden Funktionen. Die auf den Funktionen (1.33) aufgebaute Energiematrix besitzt eine diagonale Form, d.h. alle Funktionen unterscheiden sich entweder durch S oder S_z. Folglich ist die Säkulargleichung vom ersten Grad. Alle Triplettzustände haben die gleiche Energie. Dies ist die Folge eines allgemeineren Theorems, nach dem die Austauschwechselwirkung die Terme mit unterschiedlichem S aufspaltet, hinsichtlich der Quantenzahl S_z innerhalb eines Multipletts aber keine Aufspaltung bewirkt. Somit ist:

$$E_t = \langle {}^{-1}\Psi_t | \hat{H} | {}^{-1}\Psi_t \rangle = \langle {}^{0}\Psi_t | \hat{H} | {}^{0}\Psi_t \rangle = \langle {}^{1}\Psi_t | \hat{H} | {}^{1}\Psi_t \rangle, \\ E_s = \langle \Psi_s | \hat{H} | \Psi_s \rangle.$$
(1.34)

Durch Einsetzen der Wellenfunktionen (1.33) in den Ausdruck für die Energie (1.34) erhält man für den Singulett- bzw. Triplettzustand

$$E_s = 2E_0 + \frac{C+A}{1+S_{ab}^2}, \\ E_t = 2E_0 + \frac{C-A}{1-S_{ab}^2}$$
(1.35)

mit den Bezeichnungen

$$C = e^2 \left\langle \varphi_a(\vec{r}_1)\varphi_b(\vec{r}_2) \left| \frac{1}{R} - \frac{1}{r_{a2}} - \frac{1}{r_{b1}} + \frac{1}{r_{12}} \right| \varphi_a(\vec{r}_1)\varphi_b(\vec{r}_2) \right\rangle, \\ A = e^2 \left\langle \varphi_a(\vec{r}_1)\varphi_b(\vec{r}_2) \left| \frac{1}{R} - \frac{1}{r_{a2}} - \frac{1}{r_{b1}} + \frac{1}{r_{12}} \right| \varphi_b(\vec{r}_1)\varphi_a(\vec{r}_2) \right\rangle.$$
(1.36)

Wie man leicht sieht, ergeben sich die Formeln (1.35) und (1.36) aus der allgemeinen Formel (1.14), falls die Probefunktion $\Phi(\vec{r}_1, \vec{r}_2)$ in der Form (1.16) gewählt wird.

2. Das Problem der Orthogonalisierung der Wellenfunktionen für H$_2$. Am Beispiel des unter Ziff. 1 betrachteten H$_2$-Problems sehen wir bereits, daß man zur Bestimmung der Molekülwellenfunktion Atomfunktionen verwenden muß. Dabei ergibt sich automatisch die Frage nach Berücksichtigung (oder Nichtberücksichtigung) ihrer Nichtorthogonalität im Verlaufe aller Rechnungen. Insbesondere könnte man versuchen, von den nichtorthogonalen Atomwellenfunktionen zu deren orthogonalisierten Linearkombinationen überzugehen. Beim Übergang zum Kristall wird das ganze Problem wesentlich komplizierter. Deshalb scheint es zweckmäßig, das Problem wiederum am einfachsten Beispiel, dem Wasserstoffmolekül, ausführlicher zu betrachten, um alle seine realen Schwierigkeiten zu klären [66].

Gehen wir nun von den nichtorthogonalen Atomfunktionen $\varphi(\vec{r})$ beim Problem des Wasserstoffmoleküls über zu deren orthogonalen Linearkombinationen (Molekülbahnen), und zwar mit Hilfe der Formeln

$$f_1(\vec{r}) = \frac{1}{\sqrt{2(1-S_{ab})}} \{\varphi_a(\vec{r}) - \varphi_b(\vec{r})\}, \\ f_2(\vec{r}) = \frac{1}{\sqrt{2(1+S_{ab})}} \{\varphi_a(\vec{r}) + \varphi_b(\vec{r})\}.$$
(2.1)

Die Rücktransformation erfolgt dann nach folgenden Beziehungen:

$$\varphi_a(\vec{r}) = \frac{1}{2} \{f_1(\vec{r})\sqrt{2(1-S_{ab})} + f_2(\vec{r})\sqrt{2(1+S_{ab})}\}, \\ \varphi_b(\vec{r}) = \frac{1}{2} \{-f_1(\vec{r})\sqrt{2(1-S_{ab})} + f_2(\vec{r})\sqrt{2(1+S_{ab})}\}.$$
(2.2)

Nun bilden wir aus Slater-Determinanten vom Typ (1.26), die jedoch aus den Funktionen f_1 und f_2 aufgebaut sind, Triplett- und Singulettzustände, bei denen die Bahnbewegung der Elektronen durch die in (2.1) abgeleiteten Molekülorbitale f_1 und f_2 beschrieben wird.

Für das Triplett werden wir der Kürze halber nur die eine Funktion, die der Projektion $S_z = 1$ entspricht, betrachten. Die Slater-Determinanten nehmen jetzt

Ziff. 2. Das Problem der Orthogonalisierung der Wellenfunktionen für H_2.

in Analogie zu (1.26) folgende Formen an:

$$\omega_1 = \frac{1}{\sqrt{2}} \begin{vmatrix} f_1(\vec{r}_1)\alpha(\sigma_1) & f_1(\vec{r}_2)\alpha(\sigma_2) \\ f_2(\vec{r}_1)\beta(\sigma_1) & f_2(\vec{r}_2)\beta(\sigma_2) \end{vmatrix}, \quad \omega_2 = \frac{1}{\sqrt{2}} \begin{vmatrix} f_1(\vec{r}_1)\beta(\sigma_1) & f_1(\vec{r}_2)\beta(\sigma_2) \\ f_2(\vec{r}_1)\alpha(\sigma_1) & f_2(\vec{r}_2)\alpha(\sigma_2) \end{vmatrix},$$
$$w_1 = \frac{1}{\sqrt{2}} \begin{vmatrix} f_1(\vec{r}_1)\alpha(\sigma_1) & f_1(\vec{r}_2)\alpha(\sigma_2) \\ f_2(\vec{r}_1)\alpha(\sigma_1) & f_2(\vec{r}_2)\alpha(\sigma_2) \end{vmatrix}, \quad w_2 = \frac{1}{\sqrt{2}} \begin{vmatrix} f_1(\vec{r}_1)\beta(\sigma_1) & f_1(\vec{r}_2)\beta(\sigma_2) \\ f_2(\vec{r}_1)\beta(\sigma_1) & f_2(\vec{r}_2)\beta(\sigma_2) \end{vmatrix}, \quad (2.3)$$

und die Wellenfunktionen des Triplettzustandes $(S=1, S_z=1)$ Φ_t und des Singulettzustandes Φ_s lauten dann entsprechend[12]:

$$\Phi_t = w_1 = \frac{1}{\sqrt{2}} \{f_1(\vec{r}_1) f_2(\vec{r}_2) - f_2(\vec{r}_1) f_1(\vec{r}_2)\} \alpha(\sigma_1)\alpha(\sigma_2),$$
$$\Phi_{1s} = \frac{1}{\sqrt{2}}\{\omega_1 - \omega_2\} = \frac{1}{2}\{f_1(\vec{r}_1) f_2(\vec{r}_2) + f_2(\vec{r}_1) f_1(\vec{r}_2)\}\{\alpha(\sigma_1)\beta(\sigma_2) - \beta(\sigma_1)\alpha(\sigma_2)\}. \quad (2.4)$$

Berechnet man mit Hilfe der Funktionen (2.4) nach Formeln, die (1.34) entsprechen, die Energien der Triplett- und Singulettzustände, so erhält man für sie die Differenz

$$E_s - E_t = 2 J_{12}, \quad (2.5)$$

wobei

$$J_{12} = \int f_1^*(\vec{r}_1) f_2^*(\vec{r}_2) V(\vec{r}_1, \vec{r}_2) f_2(\vec{r}_1) f_1(\vec{r}_2) d\vec{r}_1 d\vec{r}_2 \quad (2.6)$$

ist. Aus (2.6) sieht man, daß die Größe $J_{12} > 0$ wird; dies ist eine allgemeine Eigenschaft der Integrale dieses Typs. Das Triplett liegt also stets unterhalb des Singuletts, und es bleibt keine Möglichkeit einer umgekehrten Anordnung der E_s- und E_t-Niveaus. Dieses Ergebnis erhielten wir bereits in Ziff. 1 bei Vernachlässigung des Nichtorthogonalitätsintegrals S_{ab} in (1.20). Diese Schlußfolgerung ist selbstverständlich nicht richtig und widerspricht sowohl dem Experiment für H_2, wo eben $E_s < E_t$ ist, als auch dem Sturm-Liouville-Theorem (s. Ziff. 3).

Um die Ursache dieses falschen Ergebnisses zu erkennen, wird in den Ausdrücken (2.4) mit Hilfe der Transformationsformeln (2.2) zu den alten Atomfunktionen übergegangen. Man erhält dann anstelle von (2.4):

$$\Phi_t = {}^1\Psi_t, \quad (v_1 = w_1), \quad (2.7)$$

$$\Phi_{1s} = \frac{1}{\sqrt{2(1-S_{ab}^2)}} \{u_3 - u_4\} = \frac{1}{2\sqrt{1-S_{ab}^2}} \{\varphi_a(\vec{r}_1)\varphi_a(\vec{r}_2) - \varphi_b(\vec{r}_1)\varphi_b(\vec{r}_2)\} \times \\ \times \{\alpha(\sigma_1)\beta(\sigma_2) - \beta(\sigma_1)\alpha(\sigma_2)\}, \quad (2.8)$$

wobei für die neuen Slater-Determinanten die Bezeichnungen

$$u_3 = \frac{1}{\sqrt{2}} \begin{vmatrix} \varphi_a(\vec{r}_1)\alpha(\sigma_1) & \varphi_a(\vec{r}_2)\alpha(\sigma_2) \\ \varphi_a(\vec{r}_1)\beta(\sigma_1) & \varphi_a(\vec{r}_2)\beta(\sigma_2) \end{vmatrix}, \quad u_4 = \frac{1}{\sqrt{2}} \begin{vmatrix} \varphi_b(\vec{r}_1)\alpha(\sigma_1) & \varphi_b(\vec{r}_2)\alpha(\sigma_2) \\ \varphi_b(\vec{r}_1)\beta(\sigma_1) & \varphi_b(\vec{r}_2)\beta(\sigma_2) \end{vmatrix} \quad (2.9)$$

eingeführt wurden und ${}^1\Psi_t$ mit Hilfe der Formel (1.33) bestimmt wird. Somit bleibt beim Übergang zu den orthogonalen Funktionen f_1 und f_2 die Triplett-

---

[12] Wir machen gleich den Vorbehalt, daß hier die Beschreibung des Systems mit Hilfe einfach besetzter Bahnen f_1 und f_2 rein formalen Charakter trägt und lediglich in der Absicht, das Orthogonalisierungsproblem zu untersuchen, begründet ist. In der Methode der Molekülbahnen werden diese Zustände unterschiedliche Energien besitzen, und deshalb ist der tiefste Singulettzustand naturgemäß durch eine zweifache Besetzung der Molekülbahn mit der niedrigsten Energie darzustellen.

wellenfunktion ungeändert ($\Phi_t = {}^1\Psi_t$) erhalten. Dagegen stimmt die Singulettfunktion Φ_{1s} nicht mehr mit der alten Funktion Ψ_s [s. (1.33)] überein, sondern stellt eine Linearkombination von Slater-Determinanten dar, die die Superposition von polaren oder Ionenzuständen beschreiben, in denen sich an irgendeinem Atom zwei Elektronen mit entgegengesetzten Spins befinden.

Natürlich wird auch die Energie des Zustandes Φ_{1s} vollkommen verschieden sein von der Energie des Zustandes Ψ_s. Somit haben wir uns unmittelbar davon überzeugt, daß beim Orthogonalisierungsprozeß neue angeregte (Ionen-)Zustände mit eingeführt werden, und wir haben es nun, genau genommen, schon nicht mehr mit dem Heitler-London-Modell zu tun, bei dem ja angenommen wurde, daß sich an jedem Atom lediglich ein Elektron befindet. Wir zeigen jetzt, daß man das Heitler-London-Modell von Anbeginn an durch die Einführung polarer Zustände „verwässern" muß, damit beim Orthogonalisierungsprozeß keine neuen Zustände eingeführt werden, die ursprünglich unter den nichtorthogonalen Bahnen nicht vorhanden waren. Tatsächlich kann man, wenn man das System der Slater-Determinanten u_1 und u_2 aus (1.26) durch die Determinanten u_3 und u_4 aus (2.9), welche die polaren Zustände beschreiben, ergänzt, zwei neue Singulettfunktionen aufbauen:

$$\Psi_s' = u_3, \quad \Psi_s'' = u_4. \tag{2.10}$$

Indem man bei der Beschreibung des betrachteten Systems mit den orthogonalen Funktionen f_1 und f_2 polare Zustände einführt, erhält man ebenfalls zwei zusätzliche Singulettfunktionen

$$\left.\begin{aligned}
\Phi_{2s} &= \frac{1}{\sqrt{2}} \begin{vmatrix} f_1(\vec{r}_1)\alpha(\sigma_1) & f_1(\vec{r}_2)\alpha(\sigma_2) \\ f_1(\vec{r}_1)\beta(\sigma_1) & f_1(\vec{r}_2)\beta(\sigma_2) \end{vmatrix} = \omega_3, \\
\Phi_{3s} &= \frac{1}{\sqrt{2}} \begin{vmatrix} f_2(\vec{r}_1)\alpha(\sigma_1) & f_2(\vec{r}_2)\alpha(\sigma_2) \\ f_2(\vec{r}_1)\beta(\sigma_1) & f_2(\vec{r}_2)\beta(\sigma_2) \end{vmatrix} = \omega_4.
\end{aligned}\right\} \tag{2.11}$$

Drückt man in Φ_{2s} und Φ_{3s} aus (2.11) die Funktionen f_1 und f_2 entsprechend (2.1) durch φ_a und φ_b aus, so erhält man

$$\Phi_{1s} = \frac{1}{\sqrt{2(1-S_{ab}^2)}} \{\Psi_s' - \Psi_s''\}, \tag{2.12}$$

$$\Phi_{2s} = \frac{1}{2(1-S_{ab})} \{\Psi_s' + \Psi_s''\} - \frac{\sqrt{1-S_{ab}^2}}{\sqrt{2}(1-S_{ab})} \Psi_s, \tag{2.13}$$

$$\Phi_{3s} = \frac{1}{2(1+S_{ab})} \{\Psi_s' + \Psi_s''\} + \frac{\sqrt{1+S_{ab}^2}}{\sqrt{2}(1+S_{ab})} \Psi_s. \tag{2.14}$$

Wir bemerken, daß die Funktionen Ψ_s, Ψ_s' und Ψ_s'' normiert, jedoch nicht orthogonal, die Funktionen Φ_{1s}, Φ_{2s} und Φ_{3s} dagegen orthonormiert sind.

Bildet man hingegen aus den mittels der nichtorthogonalen Atomfunktionen φ_a und φ_b aufgebauten Funktionen (2.10) und ψ_s [aus (1.33)] Linearkombinationen, die den rechten Seiten der Gleichungen (2.12) bis (2.14) ähneln, so erhält man folgende drei orthonormierte Singulettfunktionen:

$$\left.\begin{aligned}
\Psi_{1s} &= \frac{1}{\sqrt{2(1-S_{ab}^2)}} \{u_3 - u_4\}, \\
\Psi_{2s} &= \frac{1}{2(1-S_{ab})} \{-u_1 + u_2 + u_3 + u_4\}, \\
\Psi_{3s} &= \frac{1}{2(1+S_{ab})} \{u_1 - u_2 + u_3 + u_4\}.
\end{aligned}\right\} \tag{2.15}$$

Wie aus (2.9), (1.26), (2.8) und (2.11) zu erkennen ist, lassen sich die Singulettfunktionen stets so wählen, daß die beiden Sätze für orthogonale und nichtorthogonale Orbitale identisch werden. In unserem Falle gilt

$$\left.\begin{array}{l}\Psi_{1s}\equiv\Phi_{1s},\\ \Psi_{2s}\equiv\Phi_{2s},\\ \Psi_{3s}\equiv\Phi_{3s}.\end{array}\right\} \qquad (2.16)$$

Auf diese Weise gewährleistet die Einführung angeregter Konfigurationen in den Ansatz die Erhaltung des ursprünglichen Raumes der Wellenfunktionen während des Orthogonalisierungsprozesses. Das ermöglicht die Verwendung orthogonaler Bahnen, und das ist für die Berechnung der Matrixelemente vorteilhaft. Dabei müssen jedoch zusätzliche Zustände eingeführt und die Wechselwirkung zwischen den Konfigurationen berechnet werden, was die Aufgabe außerordentlich kompliziert. Die Säkulargleichung für die Energie des Singuletts E_s ist in diesem Falle vom dritten Grade[13]. Löst man sie, so kann man die Ungleichung $E_s < E_t$ erhalten, die bei der Orthogonalisierung ohne Einführung zusätzlicher Konfigurationen niemals erfüllt werden kann. Ausführlicher wird diese Frage nochmals unter Ziff. 16 betrachtet.

3. Die asymptotisch exakte Lösung für das Wasserstoffmolekül. Alle Integrale, die in den Ausdruck (1.24) für das Austauschintegral eingehen, wurden im Jahre 1927 von SUGIURA [67] unter Verwendung der $1s$-Wellenfunktionen des Wasserstoffatoms

$$\varphi(\vec{r}) = \frac{1}{\sqrt{\pi}} e^{-r} \qquad (3.1)$$

berechnet. [(3.1) wird in atomaren Einheiten wiedergegeben.] Die Ergebnisse der Berechnungen mit Hilfe von (3.1) befinden sich in recht guter Übereinstimmung mit dem Experiment, indem sie das Vorzeichen der Differenz $(E_s - E_t)$ bei vernünftigen interatomaren Abständen richtig wiedergeben. Jedoch ändert bei Abständen $R \approx 50\, a_H$ (a_H ist der Bohrsche Radius) die Singulett-Triplett-Aufspaltung $E_s - E_t$ ihr Vorzeichen [68]. Für diese Abstände läßt sich das Austauschintegral J mit großer Genauigkeit durch den Ausdruck

$$J \approx -\left[\frac{28}{45} - \frac{2}{15}\gamma - \frac{2}{15}\ln R\right] R^3 e^{-2R} \qquad (3.2)$$

annähern, wobei $\gamma = 0{,}5772$ die Euler-Konstante bedeutet. Aus (3.2) folgt, daß bei großen Abständen R die Größe J ihr Vorzeichen wechselt wegen des Gliedes mit $\ln R$, das bei $R \to \infty$ unbegrenzt wächst. Da bei großen R das Austauschintegral positiv wird, d.h. als Grundzustand ein Triplett auftritt, steht fest, daß die Heitler-London-Methode asymptotisch nicht richtig sein kann. Nach dem Sturm-Liouville-Theorem [69] darf die Wellenfunktion des Grundzustandes nämlich keine Knoten besitzen. Folglich kann im Falle zweier Elektronen das Triplett niemals unterhalb des Singuletts liegen, da die Koordinatenwellenfunktion des Tripletts $\Phi_t(\vec{r}_1, \vec{r}_2)$ antisymmetrisch ist $\left(\Phi_t(\vec{r}_1, \vec{r}_2) = -\Phi_t(\vec{r}_2, \vec{r}_1)\right)$ und bei $\vec{r}_1 = \vec{r}_2$

---

[13] Der Grad der Säkulargleichung für den Singulettzustand kann herabgesetzt werden (zwei Gleichungen: eine vom zweiten und die andere vom ersten Grade an Stelle einer Gleichung dritten Grades), wenn man aus den Funktionen (2.16) andere Funktionen konstruiert, die die erforderliche Symmetrie hinsichtlich der Inversionsoperation besitzen (gerade und ungerade Funktionen).

(oder) gleich Null wird, d. h. Knoten[14] besitzt. Aus diesem Grunde stellt die Wahl der Probefunktion $\Phi(\vec{r}_1, \vec{r}_2)$ in der Form $\varphi_a(\vec{r}_1)\,\varphi_b(\vec{r}_2)$, wobei φ_a und φ_b die 1s-Wasserstoffunktionen (3.1) bedeuten, eine, wenigstens für große R, viel zu grobe Näherung dar, bei der die räumliche Korrelation der wechselwirkenden Elektronen nicht berücksichtigt wird. Die Größe $\ln R$ in (3.2) rührt von der Elektron-Elektron-Wechselwirkung $\left\langle \varphi_a(\vec{r}_1)\,\varphi_b(\vec{r}_2)\,\frac{1}{r_{12}}\,\varphi_b(\vec{r}_1)\,\varphi_b(\vec{r}_2)\right\rangle$ her. Somit wird deutlich, daß die Heitler-London-Methode die Elektron-Elektron-Wechselwirkung stark überschätzt, d.h. nicht berücksichtigt, daß die Elektronen einander ausweichen.

Die Berücksichtigung der Elektronenkorrelation und die Herleitung der asymptotisch exakten Lösung für das Wasserstoffmolekül erfolgten in Arbeiten von GORKOV und PITAEVSKI [70] sowie HERRING und FLICKER [71]. Die von den Autoren dieser Arbeiten gestellte Grundaufgabe besteht im Auffinden einer „guten" Probefunktion $\Phi(\vec{r}_1, \vec{r}_2)$. Es wurde bereits gezeigt [s. (1.13)], daß die Funktion $\Phi(\vec{r}_1, \vec{r}_2)$ mit den exakten Singulett- und Triplettfunktionen durch die Beziehung

$$\Phi = \frac{1}{\sqrt{2}}(\Phi_s + \Phi_t) \tag{3.3}$$

zusammenhängt, wobei Φ_s und Φ_t Lösungen der Schrödinger-Gleichung

$$\hat{H}\,\Phi_{s,t} = E_{s,t}\,\Phi_{s,t} \tag{3.4}$$

darstellen. Die Umkehrung von (3.3) lautet in Übereinstimmung mit (1.13)

$$\Phi_{s,t} = \frac{1}{\sqrt{2}}(\Phi \pm \hat{P}_{12}\Phi). \tag{3.5}$$

Betrachten wir die Größe

$$\Phi_t \hat{H} \Phi_s - \Phi_s \hat{H} \Phi_t, \tag{3.6}$$

die mit Hilfe von (3.5) in die Form

$$\Phi \hat{H} \hat{P}_{12} \Phi - \hat{P}_{12} \Phi \hat{H} \Phi \tag{3.7}$$

überführt wird. Setzt man in (3.7) den Hamilton-Operator (1.1) ein, so erhält man den Ausdruck

$$-\tfrac{1}{2}\vec{V}[\Phi\vec{V}\hat{P}_{12}\Phi - \hat{P}_{12}\Phi\vec{V}\Phi], \tag{3.8}$$

worin $\vec{V}$ einen 6-dimensionalen Gradienten bedeutet.

Integriert man nun (3.6) über ein bestimmtes Volumen V, so erhält man unter Verwendung der Beziehungen (3.7) und (3.8) sowie des Gaußschen Satzes

$$\int_V (\Phi_t \hat{H} \Phi_s - \Phi_s \hat{H} \Phi_t)\,dV = \tfrac{1}{2}\int_\Sigma d\vec{S}\,[(\hat{P}_{12}\Phi)\vec{V}\Phi - \Phi\vec{V}(\hat{P}_{12}\Phi)]. \tag{3.9}$$

Erinnert man sich der allgemeinen Definition (1.5) des Austauschintegrals

$$J = \tfrac{1}{2}(E_s^0 - E_t^0), \tag{3.10}$$

---

[14] Das erwähnte Theorem ist im allgemeinen nur bei höchstens zwei Elektronen anwendbar. Bei Systemen aus drei und mehr Elektronen ist eine hinsichtlich der Vertauschung der Koordinaten $r_1, r_2, r_3, \ldots$ vollkommen symmetrische Ortsfunktion überhaupt nicht zulässig, und deshalb besitzt die Wellenfunktion unbedingt Knoten. Demnach wird auch die Wellenfunktion des Grundzustandes Knoten aufweisen, und der Grundzustand des Systems wird nicht dem kleinsten der Eigenwerte des Hamilton-Operators entsprechen.

so ergibt sich mit Hilfe der Gleichung (3.4) dafür folgende Beziehung:

$$J \int\int_V [\Phi^2 - (\hat{P}_{12}\Phi)^2] dV = \int_\Sigma d\vec{S} (\hat{P}_{12}\Phi) \vec{\nabla}\Phi. \tag{3.11}$$

[In (3.9) sind die Integrale auf der rechten Seite einander gleich, was man durch partielle Integration und unter Verwendung der Eigenschaft $\Phi \underset{r\to\infty}{\to} 0$ beweisen kann.]
Wie aus (3.11) folgt, besteht die Aufgabe jetzt darin, ein Gebiet im Konfigurationsraum auszuwählen mit einer Grenzfläche Σ, auf der sich Φ und $\hat{P}_{12}\Phi$ leicht bestimmen lassen, wobei das Volumenintegral auf der linken Seite von (3.11) etwa gleich 1 sein sollte. Als eine solche Fläche wird die durch die Gleichung

$$r_{a1}^2 + r_{b2}^2 = r_{b1}^2 + r_{a2}^2 \tag{3.12}$$

bestimmte Hyperfläche gewählt. Führt man die Achse z ein, die mit der Verbindungslinie der beiden Kerne a und b zusammenfällt, so nimmt (3.12) die Form

$$z_1 = z_2 \tag{3.13}$$

an.
Die Funktion Φ hat große Werte, wenn sich das Elektron 1 neben dem Atom a und das Elektron 2 neben dem Atom b befindet. Bezeichnen wir dieses Gebiet als die „nahe" Seite der Fläche Σ. Die Funktion $\hat{P}_{12}\Phi$ ist dagegen hauptsächlich auf der „fernen" Seite der Fläche Σ lokalisiert. Versteht man unter dem Volumen V in (3.11) die „nahe" Seite, so kann man das Volumenintegral auf der linken Seite von (3.11) über Φ^2 durch 1 annähern, und das Integral über $(\hat{P}\Phi)^2$ ist von der Größenordnung e^{-2R}. Da das Austauschintegral selbst proportional e^{-2R} ist, läßt sich die Größe J mit einer Genauigkeit e^{-4R} durch den Ausdruck

$$J = \int_\Sigma d\vec{S} (\hat{P}_{12}\Phi) \vec{\nabla}\Phi \tag{3.14}$$

annähern, wobei das Vektorelement der Fläche vom „nahen" zum „fernen" Gebiet gerichtet ist.
Nunmehr besteht die Aufgabe in der Bestimmung der Funktion Φ. Schreiben wir den Hamilton-Operator des Systems in der Form

$$\hat{H} = \hat{H}_0 + \hat{H}' \tag{3.15}$$

mit

$$\hat{H}_0 = -\frac{\hbar^2}{2m}(\Delta_1 + \Delta_2) - \frac{e^2}{r_{a1}} - \frac{e^2}{r_{b2}}, \tag{3.16}$$

$$\hat{H}' = \frac{e^2}{r_{12}} - \frac{e^2}{r_{a2}} - \frac{e^2}{r_{b1}} + \frac{e^2}{R}. \tag{3.17}$$

Die Funktion

$$\Phi_0 = \varphi_a(\vec{r}_1)\varphi_b(\vec{r}_2) = \pi^{-1} \exp[-r_{a1} - r_{b2}] \tag{3.18}$$

befriedigt die Gleichung

$$\hat{H}_0 \Phi_0 = 2E_0 \Phi_0, \tag{3.19}$$

worin E_0 die Energie eines isolierten Wasserstoffatoms bedeutet. Mit Hilfe der Formeln (3.3), (3.4), (3.5) und (3.10) kann leicht bewiesen werden, daß die Funktion Φ der Gleichung

$$(\hat{H} - 2E_0 - E)\Phi = J\hat{P}_{12}\Phi \tag{3.20}$$

genügt, wo die Bezeichnung
$$E = \tfrac{1}{2}(E_s + E_t) - 2E_0 \qquad (3.21)$$
eingeführt wurde.

Die Größe E (verallgemeinerte van-der-Waals-Energie) hat die Größenordnung R^{-6} und kann im Vergleich zu den anderen Gliedern auf der linken Seite von (3.20) vernachlässigt werden. Im gesamten Gebiet zwischen den Kernen a und b ist die Größe $J\hat{P}_{12}\Phi$ auf der rechten Seite von (3.20) exponentiell klein im Vergleich zu jedem der Glieder auf der linken Seite und kann ebenfalls vernachlässigt werden. Schließlich lautet die Näherungsgleichung für Φ

$$(\hat{H} - 2E_0)\Phi = 0. \qquad (3.22)$$

Die Gleichung (3.22) wird mit Hilfe der Störungstheorie gelöst, und als nullte Näherung für Φ wählt man die Heitler-London-Probefunktion (3.18), die der Gleichung (3.19) genügt. Die Funktion Φ setzt man dabei in der Form

$$\Phi = \chi \Phi_0 \qquad (3.23)$$

an, wobei χ eine zunächst unbekannte Funktion ist.

Die Lösung der Gleichung (3.22) mit der Funktion Φ in der Form (3.23) wurde ausführlich in den oben erwähnten Arbeiten [70], [71] untersucht, und wir wollen sie hier nicht wiederholen.

Der endgültige Ausdruck für das Austauschintegral hat die Form

$$J = -0{,}821 \cdot R^{5/2} e^{-2R} + O(R^2 e^{-2R}). \qquad (3.24)$$

Die Beziehung (3.24) ist asymptotisch exakt für $R \to \infty$ und bleibt in Übereinstimmung mit dem Sturm-Liouville-Theorem stets negativ.

Die Heitler-London-Methode liefert bei großen Molekülabständen kein richtiges Ergebnis. Hat man jedoch die Anwendung dieser Methode auf den Kristall im Auge, so ist die Heitler-London-Methode annehmbar, da der Austausch über solch große Abstände kaum eine wesentliche Rolle spielen wird. Allerdings entsteht beim Übergang zu Vielelektronenatomen, die einen großen Spin besitzen, die Gefahr, daß die erwähnte Methode bereits bei kleineren Abständen versagt. Wie Carr und Ashkin [72] zeigten, ergeben die Heitler-London-Methode sowie die Gorkov-Pitaevski- und Herring-Flicker-Verfahren merklich unterschiedliche Ergebnisse erst in einem Abstandsbereich, in dem die relativistische Spin-Spin-Wechselwirkung (proportional R^{-3}) dominiert und der entsprechende „magnetische" Anteil der Energie des Moleküls das Austauschintegral J ohnedies übersteigt. Bei Atomen mit großem Spin nehmen diese relativistischen Wechselwirkungen in entsprechendem Ausmaß zu. Hieraus folgt für die praktische Behandlung des Problems, daß in einem Abstandsbereich, der durch die Größe der Spin-Spin-Wechselwirkung begrenzt wird, die Heitler-London-Methode sowie die Gorkov-Pitaevski- und die Herring-Flicker-Methode fast die gleichen Ergebnisse liefern.

B. Die direkte Austauschwechselwirkung.

I. Die Verallgemeinerung der Heitler-London-Methode auf den Kristall.

Nachdem das Wasserstoffmolekülproblem im Rahmen der Heitler-London-Methode behandelt worden ist, kommen wir jetzt zu deren Verallgemeinerung auf Kristalle mit einer atomaren magnetischen Struktur (Ferro- oder Antiferromagnetismus). Eine solche Verallgemeinerung wurde erstmalig von Heisenberg

[8] vorgeschlagen, um die Erscheinung des Ferromagnetismus quantenmechanisch zu deuten. Später wurden diese Überlegungen von VAN VLECK [26] im Rahmen des Vektormodells und von BLOCH [73], MÖLLER [74] sowie HOLSTEIN und PRIMAKOFF [75] mittels der zweiten Quantelung formuliert. Als Grundlage dieser Verallgemeinerung wurde ein relativ einfaches Modell vorgeschlagen, in welchem der Kristall aus N elektrisch neutralen Atomen besteht, wobei jedes Atom außerhalb der abgeschlossenen, magnetisch neutralen Elektronenschalen je ein s-Valenzelektron[15] mit nichtkompensiertem Spin aufweist. Diese Elektronen sind, so wird angenommen, an den Gitterpunkten lokalisiert; ein Landungstransport ist in diesem Modell vollständig ausgeschlossen (homöopolares Modell). In diesem Sinne haben wir es streng genommen nicht mit einem Metall, sondern mit einem Dielektrikum zu tun. In Arbeiten von SLATER [59], SCHUBIN und VONSOVSKY [39], [60], [61], VONSOVSKY und AGAFONOVA [76] wurde versucht, diese Methode auf Metalle und Halbleiter zu verallgemeinern, indem im Rahmen des Heisenberg-Heitler-London-Modells auch polare und Exzitonenzustände Berücksichtigung fanden.

4. Die Säkulargleichung für den Kristall. Betrachten wir einen Kristall, der aus N Kernen und n Elektronen besteht. Der Hamilton-Operator dieses Systems hat (wenn man lediglich die kinetische Energie der Elektronen sowie elektrostatische Wechselwirkungen berücksichtigt) die Form

$$\hat{H} = \sum_{i=1}^{n} \hat{H}_1(\vec{r}_i) + \frac{1}{2} \sum_{i,j=1}^{n}{'} V(\vec{r}_i, \vec{r}_j) + C, \tag{4.1}$$

wobei $\hat{H}_1(\vec{r}_i)$ die Summe der kinetischen Energie des i-ten Elektrons und seiner Coulomb-Energie im Feld sämtlicher Kerne des Kristalls, $\hat{V}(\vec{r}_i, \vec{r}_j)$ die Coulomb-Wechselwirkung des i-ten und j-ten Elektrons sowie C die Coulomb-Wechselwirkung der Kerne bedeuten; der Strich beim Summenzeichen besagt, daß die Glieder mit $i=j$ wegzulassen sind. Wie schon im Falle des Wasserstoffmoleküls kann man die Wellenfunktion des Systems der wechselwirkenden Elektronen im Kernfeld in Form einer Entwicklung nach antisymmetrisierten Produkten der Einelektronenfunktionen $u(\vec{r}, \sigma)$ ansetzen. Diese Einelektronenfunktionen erscheinen als Produkt aus einer Koordinatenfunktion φ und einer Spinfunktion χ:

$$u(\vec{r}, \sigma) = \varphi(\vec{r})\chi(\sigma). \tag{4.2}$$

Die Funktionen (4.2) bezeichnen wir als Spin-Bahn-Funktionen. Bei der Heitler-London-Methode gehören die Spin-Bahn-Funktionen (4.2) zu bestimmten Kernen, also Gitterpunkten des Kristalls, und sind lokalisierte Atomfunktionen, die durch Indices zu kennzeichnen sind, welche die Nummer des Atoms, den Bahnzustand des Elektrons im Atom und seine Spinprojektion bezeichnen. Die Gesamtfunktion ist dann als Entwicklung in den Basis-Slater-Determinanten

$$\Psi_i = (D_{ii})^{-\frac{1}{2}} (n!)^{-\frac{1}{2}} \det\{u_i\} \tag{4.3}$$

anzusetzen. Der Faktor vor $\det\{u_i\}$ gewährleistet die Normierung der Funktion Ψ_i. Es läßt sich zeigen [77], daß für Ψ_i und eine andere Funktion dieses Typs

$$\Psi_j' = (D_{jj})^{-\frac{1}{2}} (n!)^{-\frac{1}{2}} \det\{v_j\} \tag{4.4}$$

---

[15] Die Wahl des s-Zustandes für dieses Elektron ist gleichbedeutend mit der vollständigen Vernachlässigung des Bahnbeitrages zum Magnetismus des Kristalles. Im Falle der ferromagnetischen d-Metalle (Fe, Ni, Co) ist in Übereinstimmung mit den beobachteten Werten des g-Faktors (der ungefähr gleich dem g-Faktor des Elektronenspins ist) eine solche Vernachlässigung des praktisch vollkommen ausgelöschten Bahnmagnetismus durchaus gerechtfertigt.

eine Beziehung

$$\int \Psi_i^* \Psi_j \, d\tau = (D_{ii} D_{jj})^{-\frac{1}{2}} D_{ij} = (D_{ii} D_{jj})^{-\frac{1}{2}} \det \{S'_{kl}\} \qquad (4.5)$$

gilt, worin

$$S'_{kl} = \int u_k^* v_l \, d\tau \qquad (4.6)$$

das Nichtorthogonalitätsintegral der beiden Spin-Bahn-Funktionen u_k und v_l bedeutet und

$$D_{ii} = \det \{\int u_k^* u_l d\tau\} = \det S_{kl} \qquad (4.7)$$

mit

$$S_{kl} = \int u_k^* u_l d\tau \qquad (4.8)$$

ist. Das Integral $\int \ldots d\tau$ bezeichnet, wie schon oben, eine Integration über die Orts- und eine Summation über die Spinkoordinaten.

Zur Bestimmung der Wellenfunktion und der Energie des durch den Hamilton-Operator (4.1) beschriebenen Systems muß man die Säkulargleichung

$$\det \{\langle \Psi_i \hat{H} \Psi_j \rangle - E \langle \Psi_i \Psi_j \rangle \} = 0 \qquad (4.9)$$

lösen, wo anstelle von $\int \ldots d\tau$ das Symbol $\langle \ldots \rangle$ benutzt wurde und worin Ψ_i und Ψ_j beliebige antisymmetrische Funktionen darstellen können, also nicht unbedingt die Slater-Determinanten (4.3) und (4.4) sein müssen.

Eine vorteilhafte Art, die mittels der Slater-Determinanten (4.3) und (4.4) berechneten Matrixelemente des Hamilton-Operators $\hat{H}$ zu schreiben, wird in der Arbeit [77] angegeben:

$$\left. \begin{array}{l} \langle \Psi_i \hat{H} \Psi_j \rangle = (D_{ii} D_{jj})^{-\frac{1}{2}} \sum_{k,l} \langle u_k \hat{H}_1 v_l \rangle D_{ij}(k/l) + \\ + \frac{1}{2} (D_{ii} D_{jj})^{-\frac{1}{2}} \sum_{k_1, k_2, l_1, l_2} \langle u_{k_1} u_{k_2} V v_{l_1} v_{l_2} \rangle D_{ij}(k_1 k_2 / l_1 l_2). \end{array} \right\} \qquad (4.10)$$

Darin bedeuten $D_{ij}(k/l)$ und $D_{ij}(k_1 k_2/l_1 l_2)$ die Minoren ersten und zweiten Grades der Determinanten D_{ij}.

Das idealisierte Kristallmodell, in dem angenommen wird, daß sich bei jedem Atom je ein Elektron im 1s-Zustand befindet ($n=N$; Heisenberg-Modell des Ferromagnetismus), spielte in der Entwicklung der Theorie des Magnetismus eine große Rolle. Man kann in diesem Falle das vollständige System der Basis-Slater-Determinanten lediglich durch eine stark eingeschränkte Gesamtheit von Determinanten annähern. Da nur eine „Bahnkonfiguration" betrachtet wird (eine festgelegte Numerierung der Atome und pro Atom ein bestimmter Bahnzustand), können sich die verschiedenen Slater-Determinanten offenbar nur durch unterschiedliche Zahlen von „Rechts-" und „Linksspins" sowie durch deren Verteilung auf bestimmte Atome voneinander unterscheiden. Man überlegt sich leicht, daß die Zahl der Möglichkeiten, n Plätze irgendwie mit „Rechts-" oder „Linksspins" zu besetzen,

$$v = 2^n \qquad (4.11)$$

ist und der Gesamtspin S zwischen 0 (bei geradem n) oder $\frac{1}{2}$ (bei ungeradem n) und $n/2$ variiert.

Somit ist in diesem Falle die Zahl der Basisdeterminanten ebenfalls 2^n und dementsprechend die Säkulardeterminante (4.9) 2^n-reihig. Wie man zeigen kann [9], [78], wird die Zahl der Multipletts mit einem vorgegebenen Wert S für den

Spin des Systems durch die Beziehung[16]

$$K(S) = \frac{n!\,(2S+1)}{(n/2+S+1)!\,(n/2-S)!} \qquad (4.12)$$

bestimmt.

Das Aufsuchen der Energien aller Multipletts des Systems ist auch im Grunde genommen das Ziel der Theorie des Magnetismus. Leider ist diese Aufgabe selbst im idealisierten Heisenberg-Modell des Magnetismus nicht gelöst.

In Wirklichkeit muß man ein Modell betrachten, in dem auf jedes Atom mehrere Elektronen entfallen, d.h. $n > N$. Nimmt man wiederum an, daß das Heitler-London-Schema adäquat ist, so werden bei der Beschreibung der Kristallatome die gleichen Begriffe (Atomquantenzahlen, Elektronenschalen) verwendet wie beim freien Atom (oder damit zusammenhängende). Dabei führt man dann gewöhnlich neue Näherungen ein. So nimmt man an, daß eine abgeschlossene Schale mit verschwindendem Spin sich nicht wesentlich auf das magnetische Verhalten des Systems auswirken kann und explizit nur jene Elektronen berücksichtigt werden müssen, die ein Spinmoment des Atoms oder Ions erzeugen. Wenn sich außerhalb einer abgeschlossenen Schale nur ein Elektron befindet, kann man formal den Kristall als ein System von Einelektronenatomen betrachten, indem man im Einelektronenterm von (4.1) die Wechselwirkungen des i-ten Elektrons mit den Kernen durch den Operator seiner Wechselwirkung mit den Ionenrümpfen ersetzt, denen eine effektive Ladung zugeschrieben, deren Elektronenstruktur aber explizit nicht berücksichtigt wird. Zugleich verbleiben im Ausdruck für die Elektron-Elektron-Wechselwirkung nur Glieder $V(\vec{r}_i, \vec{r}_j)$, für die beide Elektronen i und j sich außerhalb der abgeschlossenen Schalen aufhalten. Wird das Spinmoment des Atoms von mehreren außerhalb der abgeschlossenen Schalen befindlichen Elektronen erzeugt, so wird die Coulomb-Wechselwirkung explizit wiederum nur zwischen diesen berücksichtigt, und die abgeschlossene Schale führt lediglich zu einer Umnormierung der Kernladung. Hier muß auf einige Besonderheiten bei der Bestimmung des Grades der Säkulargleichung und der Anzahl der Spinzustände des Kristalls sowie bei der Auswahl der antisymmetrisierten Basisfunktionen hingewiesen werden. Die Elektronen außerhalb abgeschlossener Atomschalen ergeben für ein Atom den Gesamtspin S_a. Dieser Spin kann $(2S_a+1)$ verschiedene Projektionen auf der Quantisierungsachse besitzen. Läßt man Anregungen unter Änderung der atomaren Multiplizität unberücksichtigt (der Atomspin ist also „starr"), dann beträgt die Zahl aller Spinzustände des Kristalls offensichtlich

$$v' = \prod_{i=1}^{N}(2S_{a\,i}+1) = (2S_a+1)^N. \qquad (4.13)$$

(Es wurde angenommen, daß alle Atome gleich sind.) Falls jedoch der Spin jedes einzelnen Elektrons einer nichtabgeschlossenen Schale unabhängig von den anderen Elektronenspins eine von den zwei Orientierungen annähme, dann wäre

---

[16] Im Prinzip läßt sich der Grad (2^n) der Ausgangs-Säkulargleichung herabsetzen, wenn man von den Basisfunktionen (4.3) zu deren Linearkombinationen $\Phi_i(S, S_z)$ übergeht, die Eigenfunktionen der Operatoren des Quadrates des Gesamtspins und seiner Projektion $\hat{S}_z$ darstellen. Da die Matrixelemente des Operators (4.1) zwischen Funktionen mit verschiedenen S oder S_z gleich Null werden, erhält die Säkulardeterminante (4.9) eine Block- oder Stufenform. Die Stufen der Energiematrix sind jeweils $K(S)$-reihig. Da keine Aufspaltung im Hinblick auf S_z erfolgt, genügt es, wenn man sich bei jedem Multiplett lediglich auf eine Projektion S_z beschränkt. Folglich hat man es nun an Stelle einer Säkulargleichung 2^n-ten Grades mit $n/2+1$ Säkulargleichungen (die den $n/2+1$-Werten des Gesamtspins S entsprechen) zu tun, wobei jede vom Grade $K(S)$ ist. Ein Schema für den Aufbau der Funktionen $\Phi(S, S_z)$ aus den Funktionen (4.3) ist in den Arbeiten [79], [80] und [81] angegeben.

die Zahl der möglichen Spinzustände

$$v = 2^n, \qquad (4.14)$$

wobei wir unter n die Zahl aller Elektronen verstehen, die zu den resultierenden Atomspins beitragen (die Elektronen der abgeschlossenen Schalen werden nicht berücksichtigt). Ist die Zahl der magnetisch aktiven Elektronen pro Atom $n_a > 1$, gilt offenbar die Ungleichung

$$v > v'. \qquad (4.15)$$

Wenn man als Basisfunktionen Slater-Determinanten (4.3) mit den Einelektronenfunktionen (4.2) wählt, ergäbe sich im letzten Falle für die Säkulargleichung (4.9) wiederum der Grad 2^n. Jedoch würden hierbei offenbar auch Spinzustände des Systems berücksichtigt werden, bei denen sich die Multiplizität einzelner Atome geändert hat. Um diese „überzähligen" Zustände auszuschließen und damit entsprechend der Ungleichung (4.15) den Grad der Säkulargleichung herabzusetzen, muß man einen anderen Satz von Basisfunktionen benutzen. Als natürliche Verallgemeinerung der Heitler-London-Methode ist die Verwendung antisymmetrisierter Produkte der lokalisierten Vielelektronenatomfunktion als Basisfunktion anzusehen. Dabei (in der Einkonfigurations-Näherung) wird angenommen, daß sich die einzelnen Atome des Kristalls im Grundzustand befinden und daß sie ein festes Spinmoment und eine unveränderliche Elektronenzahl besitzen. Diese Näherung entspricht dann auch tatsächlich der Heitler-London-Methode; der Atomzustand wird in erster Linie durch die intraatomaren Wechselwirkungen bestimmt, die interatomare Wechselwirkung wird als Störung angesehen. Die antisymmetrisierten Basisfunktionen lassen sich leicht aufstellen. Eine solche Funktion hat die Form

$$\sum_P (-1)^P \hat{P} \prod_{i=1}^N U_i,$$

wobei $\hat{P}$ das Vertauschungssymbol für die Viererkoordinaten $[(-1)^P = \pm 1$ in Abhängigkeit von der Parität der Vertauschung] bedeutet und U_i die Vielelektronen-Atomfunktion der Koordinaten und Spins von n_a Elektronen.

5. Die „Nichtorthogonalitätskatastrophe". Wie wir bereits weiter oben sahen, führt die Verwendung nichtorthogonaler Bahnen bei der Konstruktion der Vielelektronenwellenfunktion des Kristalles zu wesentlichen Komplikationen bei der Energieberechnung; der Übergang zu einer orthogonalisierten Basis ist jedoch auch ein schwieriges Problem.

SLATER [82], [83] und INGLIS [84] wiesen darauf hin, daß allgemein die Verwendung nichtorthogonaler Bahnen gefährlich sein kann, da sie zu physikalisch sinnlosen Ergebnissen führt. So wird, wenn man die Normierungsfaktoren für antisymmetrisierte Vielelektronen-Basisfunktionen (z.B. für die Slater-Determinanten), die „Nichtorthogonalitätsintegrale" verschiedener solcher Funktionen und die Matrixelemente des Energieoperators berechnet, der durch Permutationen höherer Ordnung bedingte Beitrag anomal groß. Tatsächlich beträgt die Zahl der Vertauschungen zweier Elemente (der „Transpositionen")

$$\frac{N!}{2!(N-2)!} = \frac{N(N-1)}{2} \sim N^2.$$

Nimmt man an, daß nur die Wechselwirkungen und die Überlappungsintegrale mit nächsten Nachbarn wesentlich sind, so beträgt die Zahl der Transpositionen, die einen merklichen Beitrag liefern, größenordnungsmäßig

$$Nz,$$

wenn z die Zahl der nächsten Nachbarn ist. Bei der Berechnung der Normierungskoeffizienten werden also diese Transpositionen den Beitrag

$$NzS^2$$

ergeben, wo mit S das Nichtorthogonalitätsintegral bezeichnet ist. Der Beitrag der Vertauschungen dritter und vierter Ordnung, der proportional S^3 und S^4 ist, wird jedoch Zahlenfaktoren N^2 und N^3 enthalten. Daher wird die Reihe ungeachtet der Tatsache, daß $S>S^2>S^3>\ldots$ ist, wegen der Größe von N divergieren. Man darf deshalb offensichtlich, unabhängig davon, wie klein das Integral S ausfällt, die Beiträge höherer Vertauschungen nicht vernachlässigen, und es besteht die Gefahr, daß für sehr große N ein Normierungsfaktor gar nicht mehr definiert werden kann. Eine ähnliche Schwierigkeit tritt bei der Berechnung der Matrixelemente der Energie auf.

Im Zusammenhang damit versuchten einige Autoren zu beweisen, 1. daß bei korrekter Behandlung der Aufgabe bei physikalischen Größen keine Divergenz eintritt, 2. daß die Vertauschungen höheren Grades nicht berücksichtigt zu werden brauchen und folglich der effektive Spin-Hamilton-Operator nach DIRAC und VAN VLECK begründet ist und schließlich 3., daß ein konkreter Ausdruck für die Energie auch bei Verwendung nichtorthogonaler Bahnen angegeben werden kann.

Im folgenden werden wir diese Versuche kurz behandeln. Leser, die ausführlich mit dem Problem vertraut werden wollen, werden auf den Übersichtsartikel von HERRING [85, Ch. 7] und die Originalarbeiten von ARAI [86], [87] verwiesen.

Als erster versuchte VAN VLECK [88] zu beweisen, daß bei Verwendung nichtorthogonaler Einelektronenfunktionen keinerlei Divergenzen bei physikalischen Größen auftreten. Man kann eine wesentliche Vereinfachung aller Ausdrücke erzielen, wenn man zwei Typen von Vertauschungen unterscheidet: faktorisierbare und nichtfaktorisierbare. Bekanntlich läßt sich eine beliebige Vertauschung als Produkt cyclischer Vertauschungen darstellen (die cyclische Vertauschung $(ijk\ldots l)$ bedeutet, daß $i\to j, j\to k, \ldots, l\to i$). Wenn sich eine Vertauschung in kürzere elementfremde Cyclen zerlegen läßt, nennt man sie nach VAN VLECK faktorisierbar, im anderen Falle nichtfaktorisierbar. Betrachten wir Vertauschungen dieser beiden Typen, die die gleiche Zahl n von Koordinaten betreffen. Wenigstens solange $n\ll N$ ist, ist die Zahl der Vertauschungen beider Typen eine Größe von der Ordnung N^n. Tatsächlich beträgt die Zahl der nichtfaktorisierbaren Vertauschungen $(a_1\ldots a_n)$

$$\frac{N!}{n!(N-n)!} = \frac{(N-n+1)(N-n-2)\ldots N}{n!} \sim N^n.$$

Die Zahl der faktorisierbaren Vertauschungen $n!(a_1\ldots a_l)(a_{l+1}\ldots a_n)$ ist

$$\frac{N!}{l!(N-l)!} \cdot \frac{N!}{(n-l)!(N-n-l)!} \sim N^n.$$

Wenn also jedes Atom mit jedem beliebigen anderen, sogar mit entfernten Atomen wechselwirkt, so haben die Beiträge von Vertauschungen der verschiedenen Typen ein und dieselbe Größenordnung. Eine wesentliche Änderung tritt dann ein, wenn lediglich die Wechselwirkungen zwischen nächsten Nachbarn berücksichtigt werden. In diesem Falle ist die Anzahl der wirksamen nichtfaktorisierbaren Vertauschungen n-ter Ordnung gleich N, die entsprechende Zahl der faktorisierbaren Vertauschungen derselben Ordnung, soweit sie sich als Produkt zweier Cyclen darstellen lassen, ist N^2, bei drei Cyclen N^3 usw. Hieraus folgt, daß den Hauptbeitrag zur Energie Glieder liefern, die von den faktorisierbaren Vertauschungen herrühren. Wie van Vleck zeigt, gilt eine Gleichung

$$H_{P_1\ldots P_n} = H_{P_1} S_{P_2} \ldots S_{P_n} + H_{P_2} S_{P_1} S_{P_3} \ldots S_{P_n} + \ldots$$

in guter Näherung. Dabei ist $H_P = \langle \Psi H \hat{P} \Psi \rangle$, $S_P = \langle \Psi \hat{P} \Psi \rangle$ und Ψ das Produkt von N Einelektronenfunktionen. Werden alle Vertauschungen, die sich nicht als Produkt einfacher Transpositionen P_{ij} darstellen lassen, vernachlässigt, dann hat man die Eigenwerte eines effektiven Hamilton-Operators

$$\hat{H} = \sum_{i>j} H_{P_{ij}} \hat{P}_{ij}$$

aufzusuchen. Van Vleck beschäftigte sich nun mit der Bestimmung der Energie W eines Systems, in dem alle Spins parallel liegen (vollkommen antisymmetrischer Ortsanteil der Funktion), sowie mit dem Zustand mit vollkommen symmetrischem Ortsanteil (einer Aufgabe, die bei einem System von mehr als zwei Elektronen keinen Sinn hätte). Für den dreidimensionalen Fall wird das Ergebnis durch den Ausdruck

$$W = -\tfrac{1}{2} N z J [1 - (2z-1) S^2]^{-1}$$

beschrieben, wobei J das Austauschintegral zwischen nächsten Nachbarn bedeutet. Man sieht so, daß glücklicherweise die durch die Nichtorthogonalität bedingte Korrektur proportional ist der Zahl z der nächsten Nachbarn und nicht der Gesamtzahl der Atome N.

Die Schlußfolgerung von van Vleck gilt jedoch lediglich für den *Grund*-*zustand* ($S=N/2$) im einfachsten Fall, beim kollinearen Ferromagnetismus. Was alle übrigen Multipletts anbetrifft, so ist der erbrachte Beweis auf sie nicht übertragbar.

Als nächster Schritt bei der Lösung des Nichtorthogonalitätsproblems ist eine Arbeit von Takano [89] anzusehen. Er berechnete darin die Energie des Systems nicht nur für die Zustände maximaler Multiplizität $\left(S = \dfrac{N}{2}\right)$, sondern auch für die Zustände mit dem Spin $S = \dfrac{N}{2} - 1$, indem er die Methode der Entwicklungskoeffizienten des Energieoperators benutzte. Das Wesen dieser Methode besteht im folgenden: Ein System von Funktionen Φ_i sei vollständig, jedoch nicht orthonormiert. Dann gilt für den Hamilton-Operator $\hat{H}$ (wie auch für jeden anderen Operator) die Beziehung

$$\hat{H} \Phi_i = \sum_j h_{ji} \Phi_j. \tag{5.1}$$

Setzt man die Eigenfunktion Ψ_i des Operators $\hat{H}$ als Entwicklung nach den Funktionen Φ_i an und benützt in der Schrödinger-Gleichung die Beziehung (5.1), so erhält man (bei Beachtung der Tatsache, daß alle Φ_i linear unabhängig voneinander sind) eine Säkulargleichung

$$|h_{ij} - E \delta_{ij}| = 0 \tag{5.2}$$

anstelle der gewöhnlich verwendeten

$$|H_{ij} - E S_{ij}| = 0. \tag{5.3}$$

Dabei ist $H_{ij} = \int \Phi_i^* \hat{H} \Phi_j d\tau$ und $S_{ij} = \int \Phi_i^* \Phi_j d\tau$.

Zwischen den Matrixelementen H_{ij} und den Entwicklungskoeffizienten h_{ij} läßt sich leicht ein Zusammenhang herstellen:

$$H_{ij} = \sum_k S_{ik} h_{kj}. \tag{5.4}$$

Die Funktionen Φ_i können verschieden gewählt werden, z.B. auch als Produkte von Einelektronenfunktionen

$$\Phi_E = \varphi_1(\vec{r}_1)\, \varphi_2(\vec{r}_2) \ldots \varphi_N(\vec{r}_N) \tag{5.5}$$

sowie als Funktionen, wie man sie aus (5.5) mit Hilfe von Vertauschungen erhält,

$$\hat{P}\, \Phi_E. \tag{5.6}$$

Allerdings wird hierbei die Säkulargleichung von sehr hohem Grade ($\sim N!$) sein, was zu großen Schwierigkeiten führt. Eine zweite Möglichkeit besteht darin, daß man die Einelektronenfunktionen φ_i selbst und ihre Produkte vom Typ $\varphi_i \varphi_j$ für den Funktionensatz Φ_i auswählt. Der Hamilton-Operator des Systems ist gleich der Summe eines Einelektronen- und eines Zweielektronenanteils

$$\hat{H} = \sum_{i=1}^{N} \hat{f}(\vec{r}_i) + \sum_{i>j} \hat{g}(\vec{r}_1, \vec{r}_2). \tag{5.7}$$

In Übereinstimmung mit (5.1) können dann die Funktionen f und g in der Form

$$\hat{f}(\vec{r})\, \varphi_i(\vec{r}) = \sum_{j=1}^{N} \varphi_j(\vec{r})\, f_{ji}, \tag{5.8}$$

$$\hat{g}(\vec{r}_1, \vec{r}_2)\, \varphi_i(\vec{r}_1)\, \varphi_j(\vec{r}_2) = \sum_{k=1}^{N} \sum_{l=1}^{N} \varphi_k(\vec{r}_1)\, \varphi_l(\vec{r}_2)\, g_{kl,\, ij} \tag{5.9}$$

als Reihen geschrieben werden. Führt man für die gewöhnlichen Matrixelemente der Operatoren $\hat{f}$ und $\hat{g}$ die Bezeichnungen

$$F_{ij} = \langle \varphi_i | \hat{f} | \varphi_j \rangle, \quad G_{kl,\, ij} = \langle \varphi_k \varphi_l | \hat{g} | \varphi_i \varphi_j \rangle \tag{5.10}$$

ein, so finden wir zwischen diesen und den Entwicklungskoeffizienten in (5.7) und (5.8) den Zusammenhang

$$\left.\begin{aligned} f_{ji} &= \sum_k t_{jk} F_{ki}, \\ g_{kl,\, ij} &= \sum_m \sum_n t_{km} t_{ln} G_{mn,\, ij}. \end{aligned}\right\} \tag{5.11}$$

Die Größen t_{kl} bedeuten Elemente der Matrix S^{-1}, die der Überlappungsmatrix S ($S_{ij} = \langle \varphi_i \varphi_j \rangle$) reziprok ist. Diese Matrizen sind vom Range N. Wie aus (5.9) zu erkennen ist, enthält die Größe

$$\hat{H}\, \varphi_1(\vec{r}_1) \ldots \varphi_N(\vec{r}_N)$$

Ionenzustände, d.h. Zustände, in denen sich an einem Atom zwei Elektronen befinden und an einem anderen überhaupt keins. Es ist nicht möglich, sich von Anfang an dieser Polarzustände zu entledigen. TAKANO schlug ein Verfahren zur Bestimmung der Energie des Kristalls vor, das darin besteht, daß die Ionenzustände zunächst bewußt berücksichtigt werden, daß anschließend aber die für die Ionisation des Atoms erforderliche Energie gegen Unendlich geführt wird. Dieser Grenzübergang führt dazu, daß alle Polarzustände aus der Betrachtung ausgeschaltet werden, und wir zum gewöhnlichen homöopolaren Heisenberg-

Modell gelangen. Ein solcher Schritt, d. h. die Vernachlässigung der Polarzustände, ist jedoch nur in einem ganz bestimmten Stadium der Rechnungen gestattet. Zur Bestimmung der Energie des Grundzustandes $\left(S=\dfrac{N}{2}\right)$ und des angeregten Zustandes $\left(S=\dfrac{N}{2}-1\right)$ verwendet TAKANO den vorteilhaften Formalismus der zweiten Quantelung. Es läßt sich zeigen, daß in der Darstellung der zweiten Quantelung der Hamilton-Operator die Form

$$\hat{H} = \sum_{ij}\sum_{\sigma} f_{ji}\, a_{j\sigma}^{+} a_{i\sigma} + \tfrac{1}{2}\sum_{kl,ij}\sum_{\sigma\sigma'} g_{kl,ij}\, a_{k\sigma}^{+} a_{l\sigma'}^{+} a_{j\sigma'} a_{i\sigma} \qquad (5.12)$$

erhält, wobei f_{ij} und $g_{kl,ij}$ durch (5.11) definiert sind. Der Operator (5.12) besitzt die gleiche Form wie der Operator, den man bei Verwendung orthogonaler Bahnen erhält. Nur stehen hier die Größen f und g anstelle der gewöhnlichen Integrale F und G.

Die Wellenfunktion des Grundzustandes des Kristalles (alle Spins in positiver Richtung parallel zur Achse z) besitzt offensichtlich die Eigenschaft

$$a_{i+}^{+}\,\Phi_0 = 0, \qquad a_{i-}\,\Phi_0 = 0. \qquad (5.13)$$

Unter Verwendung von (5.13) finden wir sofort

$$\hat{H}\,\Phi_0 = W_0\,\Phi_0$$

mit

$$W_0 = \sum_{i=1}^{N} f_{ii} + \tfrac{1}{2}\sum_{i,j=1}^{N}(g_{ij,ij}-g_{ij,ji}). \qquad (5.14)$$

In der gewöhnlichen Darstellung mit Hilfe der Formel (5.11) erhält man

$$W_0 = \sum_{ij} t_{ij}\,F_{ij} + \tfrac{1}{2}\sum_{ijkl}(t_{ik}\,t_{jl} - t_{il}\,t_{jk})\,G_{kl,ij}. \qquad (5.15)$$

Dieses Ergebnis wurde zum ersten Male von LÖWDIN [90] sowie von CARR [91] gefunden.

Zur Behandlung der angeregten Zustände führen wir den Operator $B_{lj}^{+} = a_{l-}^{+}\,a_{j+}$ ein. Dann beschreibt die Funktion

$$\Phi_{lj} = B_{lj}^{+}\,\Phi_0$$

einen Zustand, bei dem am Gitterpunkt l Spins beider Richtungen vorliegen und am Gitterpunkt j ein Loch ($l \neq j$). Φ_{lj} genügt der Gleichung

$$\hat{H}\,\Phi_{lj} = W_0\,\Phi_{lj} + [\hat{H},\,B_{lj}^{+}]\,\Phi_0. \qquad (5.16)$$

Der Kommutator auf der rechten Seite von (5.16) hat die Form

$$[\hat{H},\,B_{lj}^{+}]\,\Phi_0 = \Bigg\{\sum_{m}\!\Big(f_{ml}+\sum_{n}g_{mn,ln}\Big)B_{mj}^{+} - \\ -\sum_{i}\!\Big(f_{ji}+\sum_{n}(g_{nj,ni}-g_{nj,in})\Big)B_{li}^{+} - \sum_{m}\sum_{i}g_{mj,li}B_{mi}^{+}\Bigg\}\Phi_0. \qquad (5.17)$$

Trennt man in (5.17) jene Glieder ab, die proportional $G_{ii,ii}=G_0$ sind und der Ionisationsenergie entsprechen, und geht mit $G_0\to\infty$, so erhält man

$$(\hat{H}-W_0)\,\Phi_{lj} \approx G_0\sum_{m}(t_{ml}\,t_{ll}\,B_{mj}^{+} - t_{ml}\,t_{jl}\,B_{ml}^{+})\,\Phi_0. \qquad (5.18)$$

Für die nichtpolaren Zustände $l=j$ strebt (5.18) gegen Null, und man muß auf den exakten Ausdruck (5.17) zurückgreifen. Unter Verwendung der Fourier-

Transformationen für den Operator B_{lj} und die Funktion Φ_{lj} erhalten wir anstelle von (5.16) und (5.18)

$$\begin{aligned}(\hat{H}-W_0)\Phi_{k0} &= \sum_n g_{0n,n0}(1-e^{ikn})\Phi_{k0} + \sum_{j\neq 0}\Big\{f_{j0}(e^{-ikj}-1) - \\ &\quad -\sum_n(g_{jn,0n}-g_{jn,n0}) + e^{-ikj}\sum_n(g_{jn,0n}-g_{nj,0n}e^{-ikn})\Big\}\Phi_{kj} \\ &\equiv \varepsilon'(k)\Phi_{k0} + \sum_{j\neq 0}\beta_k(j)\Phi_{kj};\end{aligned} \qquad (5.19)$$

$$\begin{aligned}(\hat{H}-W_0)\Phi_{kj} &= G_0\Big\{t_0 t_j(e^{ikj}-1)\Phi_{k0} + \\ &\quad +\sum_{j'\neq 0}(t_0 t_{j-j'}e^{ik(j-j')}-t_j t_{-j'}e^{-ikj'})\Phi_{kj'}\Big\}, \quad (j\neq 0).\end{aligned} \qquad (5.20)$$

Die Lösung der Gleichungen (5.19) und (5.20) bei $G_0\to\infty$ lautet

$$(H_0-W)\Phi_{k0} = \varepsilon\Phi_{k0},$$

wobei

$$\varepsilon(k) = \varepsilon'(k) + \sum_{j\neq 0}\alpha_k(j)\beta_k(j) \qquad (5.21)$$

und

$$\alpha_k(j) = \Big(\sum_{j_1}S_{j_1}e^{ikj_1}t_{-j_1}\Big)^{-1}\sum_{j_1}S_{j_1}e^{ikj_1}t_{j_1 j} \qquad (5.22)$$

mit

$$S_j \equiv S_{0j} = \langle\varphi_0\varphi_j\rangle$$

bedeuten. Es ist sehr schwierig, die Energie (5.21) in den gewöhnlichen Matrixelementen auszudrücken. Aus diesem Grunde werden folgende Näherungen eingeführt: man berücksichtigt die Wechselwirkung lediglich zwischen nächsten Nachbarn und nimmt an, daß die Größe

$$zS < 1 \qquad (5.23)$$

ist. Dann wird aus (5.21)

$$\varepsilon(k) = (\gamma(0)-\gamma(k))\{J+2\alpha F_1-2\alpha^2 F_0+\alpha^2(1-2\gamma(0))K - \\ -\alpha^2[2-\gamma(0)-2\gamma(k)]J\}. \qquad (5.24)$$

Dabei bedeuten $F_0 = \langle\varphi_i|\hat{f}|\varphi_i\rangle$, $F_1 = \langle\varphi_i|\hat{f}|\varphi_j\rangle$ (i und j bezeichnen nächste Nachbarn), $K = \langle\varphi_i\varphi_j|\hat{g}|\varphi_i\varphi_j\rangle$, $J = \langle\varphi_i\varphi_j|\hat{g}|\varphi_j\varphi_i\rangle$ und $\gamma(k) = \sum_r e^{ikr}$ (Summation über die nächsten Nachbarn).

Auf diese Weise gelang TAKANO der Beweis, daß es im Falle nichtorthogonaler Einelektronenfunktionen nicht nur für Zustände mit maximaler Multiplizität, sondern auch für das erste angeregte Multiplett $\left(S=\frac{N}{2}-1\right)$ keine Divergenzen gibt. Außerdem erhielt TAKANO eine konkrete Formel für die Energie des letzten Zustandes.

Die konsequenteste Analyse der Nichtorthogonalitätskatastrophe beim Problem des direkten Austausches wurde in zwei Arbeiten von ARAI durchgeführt [86], [87]. In der ersten Arbeit wurde gezeigt, daß in dem betrachteten System aus N Spins für die Energie eine Entwicklung

$$E = \sum_P J_P \tilde{U}(P) \qquad (5.25)$$

existiert, wobei die Summation über alle $N!$ Elemente der Vertauschungsgruppe zu erfolgen hat und $\tilde{U}(P)$ die Matrix der irreduziblen Darstellung der Vertauschungsgruppe bedeutet. Da die Entwicklungskoeffizienten J_P endlich sind, läßt sich sofort sagen, daß im wörtlichen Sinne keine Nichtorthogonalitätskatastrophe auftreten kann. Jedoch ist die Bestimmung aller Koeffizienten J_P eine praktisch nicht lösbare Aufgabe, und man muß sich auf die Bestimmung eines Teiles dieser Koeffizienten beschränken. Nach ARAI liefern den Hauptbeitrag gerade die Glieder, die den Vertauschungen niederer Ordnung entsprechen. Folglich darf man annehmen, daß das Heisenberg-Modell mit dem Dirac-van Vleck-Hamilton-Operator wohl begründet ist. Wie HERRING [85, Ch. 7] zeigte, gibt es aber in den komplizierten mathematischen Berechnungen von ARAI einige Ungenauigkeiten, und die von ihm erhaltenen zahlreichen Ungleichungen besitzen nicht immer genügend enge Grenzen, um die erforderliche Genauigkeit zu garantieren. Ähnliche schwache Stellen bemerkte HERRING auch in einer Arbeit von MIZUNO und IZUYAMA [92], die ebenfalls das Problem der Nichtorthogonalität behandelt.

Trotz einiger Erfolge bei der Überwindung der Nichtorthogonalitätskatastrophe und bei der strengen Ableitung des Spin-Hamilton-Operators darf man dieses Problem im streng mathematischen Sinne nicht als gelöst ansehen, mit Ausnahme der Einzelfälle $S = \dfrac{N}{2}$ und $S = \dfrac{N}{2} - 1$. Deshalb entbehrt auch das Heisenberg-Modell des Ferromagnetismus bis heute einer solchen strengen Begründung[17].

II. Das Vektormodell von Dirac und van Vleck.

6. Die Ableitung des Vektormodells für *ein* s-Elektron pro Atom. DIRAC [6] hat erstmals gezeigt, daß man annäherungsweise anstatt die Energieeigenwerte des Vielelektronensystems aufzusuchen, auch die Energie eines Systems bestimmen kann, das durch einen nur von den Spinoperatoren der einzelnen Elektronen abhängigen effektiven Hamilton-Operator beschrieben wird. Dieser effektive Hamilton-Operator lautet[18]

$$\hat{H} = 2 \sum_{i>j} J_{ij} \vec{s}_i \vec{s}_j. \qquad (6.1)$$

Bei der Ableitung von (6.1) wurden die Einelektronenbahnen als streng orthogonal angenommen und lediglich *eine* Elektronenkonfiguration berücksichtigt. Die Verallgemeinerung von (6.1) für den Fall vieler Konfigurationen ist von SERBER [93], [94] durchgeführt worden.

Die Diracsche Schreibweise der Austauschwechselwirkung verwendete VAN VLECK [26] bei der Behandlung des Verhaltens der magnetischen Systeme in Kristallen. Als Austauschintegral J_{ij} benützte er den Heisenbergschen Ausdruck (1.25a) aus der Wasserstoffmolekültheorie. Dies war der Ansatzpunkt für die Kritik des Vektormodells durch SLATER, der bemerkte, daß das Austauschintegral J_{ij} unbedingt die Form (1.21) annimmt und folglich der Hamilton-Operator (6.1) stets zu einem Grundzustand mit maximaler Multiplizität führt; denn (6.1) gilt ja exakt nur bei orthogonalen Funktionen [95]. Beschreibt man also das System mit Hilfe von (6.1), so kann man niemals Antiferromagnetismus erhalten. Die Einführung negativer Glieder in (1.21) (zur Berücksichtigung der Nichtorthogonalität der Bahnen) ist aber nicht erlaubt, da hierdurch eine grundlegende Bedingung bei der Ableitung des Hamilton-Operators (6.1) verletzt wird.

---

[17] In seinem Übersichtsartikel bedauert HERRING den großen Aufwand an Mühe zur Beseitigung der Nichtorthogonalitätskatastrophe beim Problem des direkten Heisenberg-Austausches, da dieses Problem selbst ein nur außerordentlich enges Anwendungsgebiet besitzt, während für das Problem der *indirekten* Austauschkopplung, die wesentlich weiter verbreitet ist (sowohl in Isolatoren als auch in Elektronenleitern mit magnetischer Ordnung der Atome), die Frage der Nichtorthogonalität überhaupt noch nicht gestellt worden ist.

[18] Wir erinnern, daß (6.1) für zwei Elektronen immer bestimmt werden kann [s. (1.12)].

Die Ableitung des Vektormodells für *ein* s-Elektron pro Atom.

Wie jedoch konsequentere Überlegungen verschiedener Autoren (BOGOLYUBOV [96], CARR [91], [97], [98] und in allgemeinerer Form HERRING [85]) zeigten, läßt sich (6.1) dennoch im Rahmen bestimmter Näherungen begründen.

Wir werden jetzt durch ein Näherungsverfahren die Beziehung (6.1) für das Heisenberg-Modell gewinnen, in dem der Kristall aus Einelektronenatomen bestehen soll und der Elektronenzustand durch lokalisierte nichtentartete Atomfunktionen beschrieben wird. Vielelektronenatome und entartete Bahnzustände werden in den folgenden Ziff. 7 und 8 behandelt.

Wir gehen von der Säkulargleichung (4.9) aus. Die Funktionen aus (4.3) stellen antisymmetrisierte Produkte der orts- und spinabhängigen Atomfunktionen (4.2) dar[19]:

$$\Psi_\alpha = \sum_P (-1)^P \hat{P} \{\varphi_1(\vec{r}_1)\chi_1(\sigma_1) \ldots \varphi_n(\vec{r}_n)\chi_n(\sigma_n)\}$$
$$= \sum_P (-1)^P \hat{P} \prod_{k=1}^n \{\varphi_k(\vec{r}_k)\chi_k(\sigma_k)\}. \qquad (6.2)$$

Für alle k ist $\sigma_k = \pm \frac{1}{2}$, also gleich der z-Komponente eines einzelnen Spins. Da in unserem Modell die Elektronen an jedem Atom in einem festen Zustand lokalisiert sind, bezeichnet der Index α bei der Funktion Ψ_α eine bestimmte Zuordnung der Spinindices zu den einzelnen Atomen: $\alpha = \{\alpha_1, \ldots, \alpha_n\}$, wobei jedes α_i gleich $\pm \frac{1}{2}$ ist. Weil in der Säkulargleichung die Matrixelemente zwischen den Zuständen mit unterschiedlichen z-Projektionen des Gesamtspins $S_z = \sum_{i=1}^n s_{iz}$ verschwinden, brauchen wir nur jene Ψ_α zu betrachten, die in den Projektionen S_z übereinstimmen. Zwei Funktionen Ψ_α und Ψ'_α mit gleichem S_z können sich aber lediglich durch die Reihenfolge der „Plus" und „Minus" in den Gesamtheiten

$$\{\alpha_1, \alpha_2, \ldots, \alpha_n\} \quad \text{und} \quad \{\alpha'_1, \alpha'_2, \ldots, \alpha'_n\}$$

voneinander unterscheiden.

Zur Berechnung der Matrixelemente in (4.9) verwendet man zweckmäßigerweise die Beziehung

$$\langle \Psi_\alpha (\hat{H}-E) \Psi_{\alpha'} \rangle = n! \langle \varphi_1(\vec{r}_1)\chi(\sigma_1) \ldots \varphi_n(\vec{r}_n)\chi_n(\sigma_n)(\hat{H}-E)\Psi_{\alpha'}\rangle. \qquad (6.3)$$

Da $\Psi_{\alpha'}$ in (6.3) die Summe aller möglichen Permutationen des Produkts $\varphi_1(\vec{r}_1)\chi'_1(\sigma_1) \ldots \varphi_n(\vec{r}_n)\chi'_n(\sigma_n)$ darstellt und $\hat{H}$ eine Summe von Gliedern ist, die höchstens von zwei Koordinaten $\vec{r}_i$ und $\vec{r}_j$ abhängen, so erhält man (6.3) als Summe von Matrixelementen. Das erste Glied ist „diagonal" in bezug auf die Anordnungen $\{\vec{r}_1, \sigma_1\}, \ldots, \{\vec{r}_n, \sigma_n\}$, d. h. innerhalb dieses Matrixelements gibt es keine Vertauschungen von Koordinaten; das zweite Glied wird unter dem Integral rechts ein Produkt enthalten, in dem ein Koordinatenpaar $\{\vec{r}_i, \sigma_i\}$ und $\{\vec{r}_j, \sigma_j\}$ vertauscht wird usw. Von den Funktionen ohne Vertauschungen werden dabei in (6.3) Glieder vom Typ

$$\langle \varphi_1(\vec{r}_1)\chi_1(\sigma_1) \ldots \varphi_n(\vec{r}_n)\chi_n(\sigma_n)\hat{H}\,\varphi_1(\vec{r}_1)\chi'_1(\sigma_1) \ldots \varphi_n(\vec{r}_n)\chi'_n(\sigma_n)\rangle$$

oder

$$\langle \varphi_1(\vec{r}_1)\chi_1(\sigma_1) \ldots \varphi_n(\vec{r}_n)\chi_n(\sigma_n)E\,\varphi_1(\vec{r}_1)\chi'_1(\sigma_1) \ldots \varphi_n(\vec{r}_n)\chi'_n(\sigma_n)\rangle$$

beigesteuert, die keine Überlappungen von Bahnfunktionen mit verschiedenen i und j enthalten. Der einer (einfachen) Koordinatenvertauschung entsprechende Beitrag enthält Glieder, die proportional dem Quadrat der Überlappung

---
[19] Hier sind der Kürze halber die in (4.3) eingeführten Normierungskoeffizienten weggelassen.

$|\varphi_i^*(\vec{r})\, \varphi_j(\vec{r})|^2$ sind, wobei das Überlappungsintegral sowohl explizit als $S_{ij} = \int \varphi_i^*(\vec{r})\, \varphi_j(\vec{r})\, d\vec{r}$ als auch in der Form $\langle \varphi_i^*(\vec{r}_1)\, \varphi_j^*(\vec{r}_2) \hat{H}\, \varphi_j(\vec{r}_1)\, \varphi_i(\vec{r}_2) \rangle$ eingehen kann. Der letzte Ausdruck sollte von der Ordnung $|S_{ij}|^2$ sein.

Kompliziertere Vertauschungen geben entsprechend höhere Ordnungen in S_{ij}. In der Annahme, daß die Integrale S_{ij} genügend klein sind, vernachlässigen wir in (6.3) und folglich auch in (4.9) alle Glieder, die in S_{ij} von höherer Ordnung als der zweiten sind. Wir können allerdings den damit verbundenen Fehler nicht abschätzen.

Auf diese Weise erhalten wir in der angenommenen Näherung anstelle von (6.3)

$$\begin{aligned}
\langle \Psi_\alpha (\hat{H}-E) \Psi_{\alpha'} \rangle \cong\ & n! \Big\langle \prod_{k=1}^{n} \{\varphi_k(\vec{r}_k)\chi_k(\sigma_k)\} (\hat{H}-E) \Big[\prod_{k=1}^{n} \{\varphi_k(\vec{r}_k)\chi'_k(\sigma_k)\} - \\
& - \tfrac{1}{2} \sum_{i,j=1}^{n}{}' \hat{P}_{ij}^{r} \hat{P}_{ij}^{\sigma} \prod_{k=1}^{n} \{\varphi_k(\vec{r}_k)\chi'_k(\sigma_k)\} \Big] \Big\rangle \\
=\ & n! \Big\langle \prod_{k=1}^{n} \{\varphi_k(\vec{r}_k)\chi_k(\sigma_k)\} (\hat{H}-E) \prod_{k=1}^{n} \{\varphi_k(\vec{r}_k)\chi'_k(\sigma_k)\} \Big\rangle - \\
& - \tfrac{1}{2} n! \Big\langle \prod_{k=1}^{n} \{\varphi_k(\vec{r}_k)\chi_k(\sigma_k)\} (\hat{H}-E) \sum_{i,j=1}^{n}{}' \hat{P}_{ij}^{r} \hat{P}_{ij}^{\sigma} \prod_{k=1}^{n} \{\varphi_k(r_k)\chi'_k(\sigma_k)\} \Big\rangle,
\end{aligned} \tag{6.4}$$

wobei $\hat{P}_{ij}$ einen Operator bezeichnet, der i und j vertauscht, und wo ein Vertauschungsoperator $\hat{P}$ als Produkt der Operatoren der Koordinatenvertauschung $\hat{P}^r$ und der Spinvertauschung $\hat{P}^\sigma$ dargestellt ist:

$$\hat{P} = \hat{P}^r \hat{P}^\sigma. \tag{6.5}$$

Weiterhin formen wir das erste Produkt auf der rechten Seite von (6.4) um in

$$\begin{aligned}
& \Big\langle \prod_{k=1}^{n} \{\varphi_k(\vec{r}_k)\chi_k(\sigma_k)\} (\hat{H}-E) \prod_{k=1}^{n} \{\varphi_k(\vec{r}_k)\chi'_k(\sigma_k)\} \Big\rangle \\
& = \Big\langle \prod_{k=1}^{n} \varphi_k(\vec{r}_k) (\hat{H}-E) \prod_{k=1}^{n} \varphi_k(\vec{r}_k) \Big\rangle_r \Big\langle \prod_{k=1}^{n} \chi_k(\sigma_k) \Big| \prod_{k=1}^{n} \chi'_k(\sigma_k) \Big\rangle_\sigma = (\overline{E}-E)\, \delta_{\alpha\alpha'},
\end{aligned} \tag{6.6}$$

wobei

$$\begin{aligned}
\overline{E} = & \Big\langle \prod_{k=1}^{n} \varphi_k(\vec{r}_k) \hat{H} \prod_{k=1}^{n} \varphi_k(\vec{r}_k) \Big\rangle_r = \sum_{i=1}^{n} \langle \varphi_i(\vec{r})\hat{H}_1(\vec{r})\varphi_i(\vec{r}) \rangle_r + \\
& + \tfrac{1}{2} \sum_{i,j=1}^{n}{}' \langle \varphi_i(\vec{r}_1)\varphi_j(\vec{r}_2) V(\vec{r}_1,\vec{r}_2) \varphi_i(\vec{r}_1)\varphi_j(\vec{r}_2) \rangle_r
\end{aligned} \tag{6.7}$$

ist und der Faktor

$$\delta_{\alpha\alpha'} = \delta_{\{\alpha_1 \ldots \alpha_n\},\{\alpha'_1 \ldots \alpha'_n\}}$$

wegen der strengen Orthogonalität der Spinfunktionen erscheint. Das zweite Glied in (6.4) läßt sich wie folgt umformen:

$$\begin{aligned}
& \Big\langle \prod_{k=1}^{n} \{\varphi_k(\vec{r}_k)\chi(\sigma_k)\} (\hat{H}-E) \sum_{i,j=1}^{n}{}' \hat{P}_{ij}^{r} \hat{P}_{ij}^{\sigma} \prod_{k=1}^{n} \{\varphi_k(\vec{r}_k)\chi'_k(\sigma_k)\} \Big\rangle \\
& = \Big\langle \prod_{k=1}^{n} \{\varphi_k(\vec{r}_k)\} (\hat{H}-E) \sum_{i,j=1}^{n}{}' \hat{P}_{ij}^{r} \prod_{k=1}^{n} \{\varphi_k(\vec{r}_k)\} \Big\rangle_r \Big\langle \prod_{k=1}^{n} \chi_k(\sigma_k) \hat{P}_{ij}^{\sigma} \prod_{k=1}^{n} \chi'_k(\sigma_k) \Big\rangle_\sigma \\
& = \sum_{i,j=1}^{n}{}'' \Big\langle \prod_{k=1}^{n} \{\varphi_k(\vec{r}_k)\} (\hat{H}-E) \hat{P}_{ij}^{r} \prod_{k=1}^{n} \{\varphi_k(\vec{r}_k)\} \Big\rangle_r \Big\langle \prod_{k=1}^{n} \chi_k(\sigma_k) \hat{P}_{ij}^{\sigma} \prod_{k=1}^{n} \chi'_k(\sigma_k) \Big\rangle_\sigma.
\end{aligned} \tag{6.8}$$

Die Ableitung des Vektormodells für *ein* s-Elektron pro Atom.

Es läßt sich leicht zeigen, daß der Ausdruck $\langle \ldots \rangle_r$ in (6.8) gleich

$$\langle \varphi_i(\vec{r}_1)\,\varphi_j(\vec{r}_2)\,(H_{ij}(\vec{r}_1,\vec{r}_2)-\tfrac{1}{2}E)\,\varphi_j(\vec{r}_1)\,\varphi_i(\vec{r}_2)\rangle_r, \tag{6.9}$$

ist. Hierin bedeutet

$$\left.\begin{aligned}H_{ij}(\vec{r}_1,\vec{r}_2) &= H_1(\vec{r}_1) + \tfrac{1}{2}\sum_{k(\neq i,j)=1}^{n}\langle \varphi_k(\vec{r})\,\hat{H}_1(\vec{r})\,\varphi_k(\vec{r})\rangle_r + \\ &\quad + \tfrac{1}{2}V(\vec{r}_1,\vec{r}_2) + \sum_{k(\neq i,j)=1}^{n}\langle \varphi_k(\vec{r})\,V(\vec{r}_1,\vec{r})\,\varphi_k(\vec{r})\rangle_r + \\ &\quad + \tfrac{1}{4}\sum_{k,l(\neq i,j)=1}^{n}{}'\langle \varphi_k(\vec{r})\,\varphi_l(\vec{r}')\,V(\vec{r},\vec{r}')\,\varphi_k(\vec{r})\,\varphi_l(\vec{r}')\rangle_{r,r'}.\end{aligned}\right\} \tag{6.10}$$

Somit nimmt die Säkulargleichung wegen (6.6), (6.8), (6.9) folgende Form an:

$$\det\left\{(\overline{E}-E)\delta_{\alpha\alpha'} - \sum_{i,j=1}^{n}{}'\langle \varphi_i(\vec{r}_1)\,\varphi_j(\vec{r}_2)\,(H_{ij}(\vec{r}_1,\vec{r}_2)-\tfrac{1}{2}E)\times \right.\\ \left. \times \varphi_j(\vec{r}_1)\,\varphi_i(\vec{r}_2)\rangle_r \Big\langle \prod_{k=1}^{n}\chi_k(\sigma_k)\,\hat{P}_{ij}^{\sigma}\prod_{k=1}^{n}\chi_k'(\sigma_k)\Big\rangle_\sigma\right\}. \tag{6.11}$$

Unter Verwendung der Diracschen Beziehung, die den Vertauschungsoperator $\hat{P}_{ij}^{\sigma}$ mit den Spinoperatoren verknüpft,

$$\hat{P}_{ij}^{\sigma} = \tfrac{1}{2} + 2\vec{s}_i\vec{s}_j \tag{6.12}$$

schreiben wir die Gleichung (6.11) um in

$$\left.\begin{aligned}\det\Big\{\Big(\overline{E}-E-\tfrac{1}{2}\sum_{i,j=1}^{n}{}'\langle \varphi_i(\vec{r}_1)\,\varphi_j(\vec{r}_2)\,(H_{ij}(\vec{r}_1,\vec{r}_2)-\tfrac{1}{2}E)\,\varphi_j(\vec{r}_1)\,\varphi_i(\vec{r}_2)\rangle_r\Big)\times \\ \times \delta_{\alpha\alpha'} - \sum_{i,j=1}^{n}{}'\langle \varphi_i(\vec{r}_1)\,\varphi_j(\vec{r}_2)|2H_{ij}(\vec{r}_1,\vec{r}_2)-E|\varphi_j(\vec{r}_1)\,\varphi_i(\vec{r}_2)\rangle_r\times \\ \times \Big\langle \prod_{k=1}^{n}\chi_k(\sigma_k)|\vec{s}_i\vec{s}_j|\prod_{k=1}^{n}\chi_k'(\sigma_k)\Big\rangle_\sigma\Big\}=0.\end{aligned}\right\} \tag{6.13}$$

Innerhalb der verwendeten Näherung können wir unter dem Summenzeichen in (6.13) die Größe E durch $\overline{E}$ ersetzen und den Beitrag der ersten Summe mit $\Delta\overline{E}$ bezeichnen, da er nicht von den Spinindices α,α' abhängt und nur eine additive Korrektur zur Energie beisteuert. Die Säkulargleichung erhält so die endgültige Form

$$\det\Big\{(E+\Delta\overline{E}-E)\delta_{\alpha\alpha'} - \sum_{i,j=1}^{n}{}'J_{ij}\Big\langle \prod_{k=1}^{n}\chi_k(\sigma_k)|\vec{s}_i\vec{s}_j|\prod_{k=1}^{n}\chi_k'(\sigma_k)\Big\rangle_\sigma\Big\}=0, \tag{6.14}$$

wobei das Austauschintegral durch

$$\left.\begin{aligned}J_{ij} &= \langle \varphi_i(\vec{r}_1)\,\varphi_j(\vec{r}_2)|\hat{h}(\vec{r}_1,\vec{r}_2) - \langle \varphi_i(\vec{r})\,\varphi_j(\vec{r}')|\hat{h}(\vec{r},\vec{r}')|\varphi_i(\vec{r})\,\varphi_j(\vec{r}')\rangle_{r,r'}|\times \\ &\quad \times \varphi_j(\vec{r}_1)\,\varphi_i(\vec{r}_2)\rangle_{r_1,r_2} = \langle \varphi_i(\vec{r}_1)\,\varphi_j(\vec{r}_2)|\hat{h}(\vec{r}_1,\vec{r}_2)|\varphi_j(\vec{r}_1)\,\varphi_i(\vec{r}_2)\rangle - \\ &\quad - |S_{ij}|^2\langle \varphi_i(\vec{r}_1)\,\varphi_j(\vec{r}_2)|\hat{h}(\vec{r}_1,\vec{r}_2)|\varphi_i(\vec{r}_1)\,\varphi_j(\vec{r}_2)\rangle\end{aligned}\right\} \tag{6.15}$$

bestimmt wird und wo an Stelle von (6.10) die symmetrischen Ausdrücke verwendet wurden, die CARR [91] beim Aufsuchen der Vektorform des effektiven

Hamilton-Operators in der Eindeterminantennäherung erhielt:

$$h(\vec{r}_1,\vec{r}_2) = \hat{H}_1(\vec{r}_1) + \hat{H}_1(\vec{r}_2) + V(\vec{r}_1,\vec{r}_2) +$$
$$+ \sum_{k(\pm i,j)=1}^{n} \{\langle \varphi_k(\vec{r}) V(\vec{r},\vec{r}_1) \varphi_k(\vec{r})\rangle_r + \langle \varphi_k(\vec{r}) V(\vec{r},\vec{r}_2) \varphi_k(\vec{r})\rangle\}. \tag{6.16}$$

Auf diese Weise ist gezeigt worden, daß bis zu Gliedern proportional zu $|S_{ij}|^2$ die Säkulargleichung (4.9) äquivalent ist der Gleichung (6.14) in den Spinveränderlichen; das heißt weiterhin, daß in der angegebenen Näherung das System durch den Hamilton-Operator von DIRAC und VAN VLECK (6.1) mit einem durch (6.15) und (6.16) bestimmten Austauschintegral beschrieben wird.

Vergleichen wir das Austauschintegral (6.15) mit dem von uns für das Wasserstoffmolekül erhaltenen Ausdruck (1.18). Wir betrachten das Integral (1.18) ebenso wie (6.15) nur bis zu den quadratischen Gliedern in S_{ij} und vernachlässigen deshalb in (1.18) oder (1.20) im Nenner die Glieder mit S_{ab}^4. Dann unterscheidet sich (1.18) oder (1.20) von (6.15) dadurch, daß im Ausdruck (6.15) auch die Wechselwirkung unseres Elektronenpaares, dessen Elektronen an den Atomen i und j sitzen, mit allen anderen Kernen sowie mit der gemittelten Ladung aller übrigen Elektronen des Kristalls auftritt.

Wenn das Vorzeichen des Austauschintegrals für das Wasserstoffmolekül (1.18) negativ ist, ist schwerlich zu erwarten, daß der Ausdruck (6.15) wegen der [gegenüber (1.18)] zusätzlichen Glieder positiv sein kann. Die Glieder der Elektron-Kern-Wechselwirkung müssen auch hier dem Betrage nach größer sein als der von der Wechselwirkung mit der gemittelten Elektronenladung herrührende zusätzliche Beitrag.

7. Die Situation bei Atomen mit mehreren Elektronen ohne Bahnentartung.
Wir wollen jetzt den effektiven Hamilton-Operator des Vektormodelles vom Typ (6.1) näherungsweise für den Fall ableiten, daß der Kristall aus Vielelektronenatomen oder -ionen besteht.

Zunächst einige einleitende Bemerkungen. In Übereinstimmung mit den Ergebnissen früherer Überlegungen (s. Ende von Ziff. 4) nehmen wir an, daß die Wellenfunktion des Kristalls am besten durch ein Produkt von Vielelektronen-Atomfunktionen U_i angenähert werden kann und nicht durch ein Produkt von Einelektronen-Atomfunktionen φ_i. Ferner sollen sich die Atome im Grundzustand befinden und alle angeregten Atomzustände vernachlässigt werden. Folglich arbeiten wir in einem atomaren „Einkonfigurationsraum". Hier wird zunächst der Fall behandelt, daß es unter den Atomfunktionen $keine$ Bahnentartung gibt, d. h., daß diese Vielelektronenfunktionen vom S-Typ darstellen (der Fall der Entartung wird in Ziff. 8 behandelt). Jedes Atom i besitzt also einen bestimmten Gesamtspin $\vec{S}_i$, und die einzige Quantenzahl, die mehrere Werte annehmen kann, ist die Projektion S_z des Atomspins $\vec{S}$: $S_z = -S, \ldots, +S$. Wir bezeichnen die z-Komponente des Spins $\vec{S}_i$ des i-ten Atoms mit k_i.

Die antisymmetrisierte Kristallfunktion lautet

$$\Psi_k = \sum_P (-1)^P \hat{P} \prod_{i=1}^{N} U_{k_i}, \tag{7.1}$$

wobei $k = \{k_1, \ldots, k_N\}$ die Gesamtheit der z-Komponenten der Spins aller Atome und N die Anzahl der Atome bedeuten. Die Argumente (Orts- und Spinkoordinaten) der Atomfunktionen U_{k_i} sind der Kürze halber weggelassen. Im folgenden bezeichnen wir mit dem Symbol n_i die Zahl der Elektronen des i-ten Atoms. Im

Ziff. 7. Die Situation bei Atomen mit mehreren Elektronen ohne Bahnentartung.

oben behandelten Falle eines Elektrons pro Atom wurde angenommen, daß den Hauptbeitrag zur Säkulargleichung (4.9) jene Matrixelemente liefern, die Produkte von Einelektronenfunktionen mit nichtvertauschten Argumenten enthalten oder Produkte, bei denen sich der rechte Faktor im Matrixelement vom linken lediglich durch die Vertauschung zweier Argumente unterscheidet. Im vorliegenden Falle aber darf man offenbar nicht glauben, daß nur diese Vertauschungen (die identische Vertauschung und Transpositionen) einen wesentlichen Beitrag liefern werden. Man kann vielmehr von vornherein folgendes annehmen: Die Vertauschungen von Elektronen, die *ein und demselben* Atom angehören, ergeben einen ebenso großen Beitrag wie die identische Vertauschung. Die Beiträge, die Vertauschungen zweier Elektronen an verschiedenen Atomen entsprechen, sind unabhängig davon, ob *innerhalb* des einen oder anderen Atoms noch weitere Elektronen vertauscht worden sind oder nicht.

Wir bezeichnen die Gruppe der intraatomaren Vertauschungen mit $\{G_0\}$. Offensichtlich ist die Ordnung von $\{G_0\}$

$$\prod_{i=1}^{N}(n_i!). \tag{7.2}$$

Da alle U_{k_i} antisymmetrisierte Atomfunktionen darstellen, wird folgende Beziehung erfüllt

$$G_0 U_{k_i} = (-1)^{G_0} U_{k_i}, \tag{7.3}$$

wobei G_0 ein beliebiges Element der Gruppe $\{G_0\}$ bedeutet, das auf die Argumente der vorgegebenen Funktion U_{k_i} einwirkt.

Zur Vereinfachung der Schreibweise benutzen wir die bereits bekannte Beziehung (6.3), die im vorliegenden Falle die Form

$$\langle \Psi_k(\hat{H}-E)\Psi_{k'}\rangle = n!\left\langle \prod_{i=1}^{N} U_{k_i}(\hat{H}-E)\sum_{P}(-1)^P \prod_{i=1}^{N} U_{k'_i}\right\rangle \tag{7.4}$$

annimmt, wobei $n=\sum_{i=1}^{N} n_i$ die Gesamtzahl der Elektronen im Kristall außerhalb der abgeschlossenen Schalen der Ionenrümpfe bedeutet.

In Übereinstimmung mit obigen Ausführungen machen wir für die Funktion folgenden Näherungsansatz:

$$\Psi_{k'} = \sum_P (-1)^P \hat{P} \prod_{i=1}^{N} U_{k'_i} \approx \sum_{G_0} (-1)^{G_0} G_0 \prod_{i=1}^{N} U_{k'_i} + \sum_{P_1, G_0}(-1)^{P_1 G_0} \hat{P}_1 G_0 \prod_{i=1}^{N} U_{k'_i}. \tag{7.5}$$

Darin ist P_1 eine Vertauschung zweier Elektronen, die zwei verschiedenen Atomen angehören. Mit Hilfe von (7.3) und den trivialen Beziehungen

$$(-1)^{RQ} = (-1)^R (-1)^Q$$

(für beliebige Vertauschungen R und Q) und

$$(-1)^{P_1} = -1$$

läßt sich (7.5) leicht umformen:

$$\Psi_{k'} \cong \prod_{i=1}^{N}(n_i!)\left[\prod_{i=1}^{N} U_{k'_i} - \sum_{P_1}\prod_{i=1}^{N} \hat{P}_1 U_{k'_i}\right]. \tag{7.6}$$

Das erste Glied in der eckigen Klammer in (7.6) rührt her von den Vertauschungen der Gruppe $\{G_0\}$, das zweite Glied von Vertauschungen $\{P_1 G_0\}$. Setzt man die Entwicklung (7.6) in die Säkulargleichung (4.9) ein und kürzt den gemeinsamen

Faktor $n!\prod_{i=1}^{N}(n_i!)$, so erhält man unter Beachtung von (6.5)

$$\left\langle \prod_{i=1}^{N} U_{k_i}(\hat{H}-E) \prod_{i=1}^{N} U_{k_i} \right\rangle \delta_{kk'} - \sum_{P_1} \left\langle \prod_{i=1}^{N} U_{k_i}(\hat{H}-E) \hat{P}_1^r \hat{P}_1^\sigma \prod_{i=1}^{N} U_{k'_i} \right\rangle = 0. \qquad (7.7)$$

Da alle Vertauschungen von Elektronen zwischen zwei ganz bestimmten Atomen i und j den gleichen Beitrag liefern, werden wir im folgenden an Stelle von $\sum_{P_1} \hat{P}_1 \ldots$ den Ausdruck $\sum_{P_{m_i m_j}} \hat{P}_{m_i m_j} n_i n_j$ verwenden, wobei $\hat{P}_{m_i m_j}$ die Vertauschung der Koordinaten und Spins jener beiden Elektronen bezeichnet, die von allen Elektronen der Atome i bzw. j in der vorher gewählten Anordnung der Orts- und Spinkoordinaten die erste Stelle einnehmen.

Führen wir wieder für $\hat{P}^\sigma_{m_i m_j}$ die Darstellung (6.12) ein,

$$\hat{P}^\sigma_{m_i m_j} = \tfrac{1}{2} + 2\vec{s}_{m_i}\vec{s}_{m_j},$$

so erhalten wir die Säkulargleichung (7.7) in der Form

$$(\bar{E}-E)\delta_{kk'} - \sum_{i>j} n_i n_j \left\langle \prod_{i=1}^{N} U_{k_i}(\hat{H}-E)\hat{P}^r_{m_i m_j}(\tfrac{1}{2}+2\vec{s}_{m_i}\vec{s}_{m_j}) \prod_{i=1}^{N} U_{k'_i} \right\rangle = 0 \qquad (7.8)$$

mit

$$\bar{E} = \left\langle \prod_{i=1}^{N} U_{k_i} | \hat{H} | \prod_{i=1}^{N} U_{k_i} \right\rangle. \qquad (7.9)$$

Ersetzt man in (7.8) unter dem Summenzeichen E durch $\bar{E}$ (s. oben) und bezeichnet das von dem Summanden $\tfrac{1}{2}$ in der runden Klammer herrührende Glied durch $\Delta \bar{E}\, \delta_{kk'}$ [s. (6.14)], so folgt

$$(\bar{E}+\Delta\bar{E}-E)\delta_{kk'} - 2\sum_{i>j} n_i n_j \left\langle \prod_{i=1}^{N} U_{k_i}(\hat{H}-E)\hat{P}^r_{m_i m_j}\vec{s}_{m_i}\vec{s}_{m_j} \prod_{i=1}^{N} U_{k'_i} \right\rangle. \qquad (7.10)$$

Da $\bar{E}+\Delta\bar{E}$ von k unabhängig ist, kann man es im ersten Glied von (7.10) einfach weglassen; dadurch wird nur der Energienullpunkt verschoben. Man erhält

$$-E\delta_{kk'} - 2\sum_{i>j} n_i n_j \left\langle \prod_{i=1}^{N} U_{k_i}(\hat{H}-\bar{E})\hat{P}^r_{m_i m_j}\vec{s}_{m_i}\vec{s}_{m_j} \prod_{i=1}^{N} U_{k'_i} \right\rangle. \qquad (7.11)$$

Jetzt müssen wir noch von den Spins $\vec{s}_{m_i}$ und $\vec{s}_{m_j}$ der einzelnen Elekronen zu den Atomspins $\vec{S}_i$ und $\vec{S}_j$ übergehen. Dabei unterscheidet man zweckmäßigerweise mehrere Fälle:

1. Der einfachste Fall ist ein Elektron pro Atom, $n_i = 1$. Dann ist $\vec{s}_{m_i} = \vec{S}_i$, wenn $\vec{S}_i$ den Atomspin bezeichnet. Indem man $U_{k_i} = \varphi_k(\vec{r}_i)\chi_k(\sigma_i)$ setzt, erhält man sofort ein Ergebnis, das mit (6.14) übereinstimmt. Der effektive Hamilton-Operator erhält die Form (6.1) mit dem Austauschintegral

$$J_{ij} = \left\langle \prod_{k} \varphi_k(\vec{r}_k)(\hat{H}-\bar{E})\hat{P}^r_{ij} \prod_{k} \varphi_k(\vec{r}_k) \right\rangle_r. \qquad (7.12)$$

2. Den nächstkomplizierten Fall haben wir, wenn alle Elektronen eines Atoms parallel gerichtete Spins besitzen (z.B. eine d-Atomschale mit fünf Elektronen[20]). Dann ist die Funktion U_{k_i} symmetrisch hinsichtlich der Vertauschungen der Spinveränderlichen und antisymmetrisch hinsichtlich der Koordinatenvertauschungen. Die bezüglich der gleichzeitigen Vertauschung der Elektronenkoordinaten und

---

[20] Im Falle einer d-Schale, die weniger als halb gefüllt ist, wird bei Parallelstellung der Spins aller Elektronen ein S-Zustand nicht realisiert.

Ziff. 7. Die Situation bei Atomen mit mehreren Elektronen ohne Bahnentartung.

-spins antisymmetrische Funktion U_{k_i} muß dann selbst als einfaches Produkt eines Koordinaten- und Spinteiles darstellbar sein:

$$U_{k_i}(\vec{r}_1, \ldots, \vec{r}_n, \sigma_1, \ldots, \sigma_n) = \varphi_1(\vec{r}_1, \ldots, \vec{r}_n)\chi_k(\sigma_1, \ldots, \sigma_n). \tag{7.13}$$

Durch Einsetzen von (7.13) in (7.11) und unter Verwendung der Beziehung

$$\left.\begin{array}{l}\langle \prod U_{k_i}(\hat{H}-\bar{E})\hat{P}'_{m_i m_j}\vec{s}_{m_i}\vec{s}_{m_j} \prod U_{k_i}\rangle = n_i^{-1}n_j^{-1}\langle \prod U_{k_i}(\hat{H}-\bar{E})\hat{P}'_{m_i m_j}\vec{S}_i\vec{S}_j \prod U_{k_i}\rangle, \\ \left(\vec{S}_i = \sum_{m_i=1}^{n_i}\vec{s}_{m_i}\right),\end{array}\right\} \tag{7.14}$$

die wegen der Symmetrie der Produkte $\prod_{i=1}^{N} U_{k_i}$ hinsichtlich der Spinargumente gilt, erhalten wir wiederum den effektiven Hamilton-Operator (6.1) mit dem Austauschintegral

$$J_{ij} = \langle \prod \varphi_k(\hat{H}-\bar{E})\hat{P}'_{m_i m_j}\prod \varphi_k\rangle, \tag{7.15}$$

wobei unter den Spinoperatoren jetzt die resultierenden Atomspins zu verstehen sind.

3. Ein noch komplizierterer Fall liegt vor, wenn die Elektronenspins eines Atoms nicht alle parallel zueinander sind (als Beispiel könnten die Atome der Übergangsmetalle mit sechs und mehr d-Elektronen angesehen werden, wo ein Vielelektronen-S-Zustand realisiert werden kann, obwohl er nicht Grundzustand ist). In diesem Falle kann die Wellenfunktion des Atoms entsprechend der allgemeinen Theorie der Vertauschungsgruppen dargestellt werden als

$$U_{k_1}(\vec{r}_1, \ldots, \vec{r}_{n_1}, \sigma_1, \ldots, \sigma_{n_1}) = M_1^{-\frac{1}{2}}\sum_{\xi_1=1}^{M_1}\varphi_{\xi_1}(\vec{r}_1, \ldots, \vec{r}_{n_1})\chi_{\xi_1 k_1}(\sigma_1, \ldots, \sigma_{n_1}), \tag{7.16}$$

wobei die Funktionen φ_{ξ_1} und $\chi_{\xi_1 k_1}$ die Basis einer irreduziblen Darstellung der Gruppe der Vertauschungen von n_i Koordinaten bzw. Spins bilden. M_1 ist die Dimension dieser Darstellungen.

Bezeichnet man die Darstellung der Gruppe der Koordinatenvertauschungen mit G, dann besitzt die Gruppe der Spinvertauschungen eine Darstellung $\bar{G}^*$, deren Elemente zu den Elementen einer Darstellung $\bar{G}$ konjugiert komplex sind. $\bar{G}$ bedeutet eine mit G verknüpfte Darstellung. Für jedes Element von $\bar{G}$ gilt die Beziehung

$$\bar{G}(P) = (-1)^P G(P), \tag{7.17}$$

worin P ein beliebiges Element der Vertauschungsgruppe sein kann. Durch die Angabe der Darstellung G wird der Gesamtspin des Systems S bestimmt, und der Index „k" der Funktion $\chi_{\xi k}$ durchläuft $2S+1$ Werte: $-S, -S+1, \ldots, S-1, S$, die $2S+1$ möglichen Projektionen des Gesamtspins S entsprechen. Führt man für jede Funktion U_{k_i} die Entwicklung (7.16) ein, so erhält man

$$\prod_{i=1}^{N}U_{k_i} = \left(\prod_{i=1}^{N}M_i\right)^{-\frac{1}{2}}\sum_{\xi_1=1}^{M_1}\cdots\sum_{\xi_N=1}^{M_N}\varphi_{\xi_1}\cdots\varphi_{\xi_N}\chi_{\xi_1 k_1}\cdots\chi_{\xi_N k_N}. \tag{7.18}$$

Für die weiteren Berechnungen benützen wir eine Beziehung, die die Matrixelemente des Gesamtspins mit dem Spin der einzelnen Elektronen verknüpft:

$$\langle \chi_{\xi_i k_i}\vec{s}_{m_i}\chi_{\xi_i' k_i'}\rangle = f_i(\xi_i \xi_i')\langle \chi_{\xi_i k_i}\vec{S}_i \chi_{\xi_i' k_i'}\rangle. \tag{7.19}$$

Die Koeffizienten f genügen den Beziehungen

$$f_i(\xi_i, \xi_i') = f_i^*(\xi_i', \xi_i), \quad \sum_{\xi_i=1}^{M_i} f_i(\xi_i, \xi_i') = \frac{M_i}{n_i}. \tag{7.20}$$

Indem man (7.18) in das zweite Glied von (7.11) einsetzt und (7.19) sowie die Orthogonalität der Funktionen $\chi_{\xi_i k_i}$ verwendet, erhält man leicht

$$\left. \begin{aligned} &-2 \sum_{i>j} n_i n_j \langle \prod U_{k_i}(\hat{H}-E) \hat{P}^r_{m_i m_j} \vec{s}_{m_i} \vec{s}_{m_j} \prod U_{k_i'} \rangle \\ &= -2 \sum_{i>j} J_{ij} \langle \prod_{\lambda=1}^{N} \chi_{\xi_\lambda k_\lambda} \vec{S}_i \vec{S}_j \prod_{\lambda=1}^{N} \chi_{\xi_\lambda k_\lambda'} \rangle_\sigma, \end{aligned} \right\} \tag{7.21}$$

worin das Austauschintegral J_{ij} durch

$$\left. \begin{aligned} J_{ij} = &\frac{n_i n_j}{M_i M_j} \sum_{\xi_i, \xi_i'=1}^{M_i} \sum_{\xi_j, \xi_j'=1}^{M_j} f_i(\xi_i, \xi_i') f_j(\xi_j, \xi_j') \times \\ &\times \sum_{\lambda(\neq i,j)}^{N} \sum_{\xi_\lambda=1}^{M_\lambda} \left(\prod_{\varkappa(\neq i,j)=1}^{N} M_\varkappa \right)^{-1} \langle \varphi_{\xi_1} \cdots \varphi_{\xi_{i-1}} \varphi_{\xi_i} \varphi_{\xi_{i+1}} \cdots \varphi_{\xi_{j-1}} \varphi_{\xi_j} \varphi_{\xi_{j+1}} \cdots \varphi_{\xi_N} \times \\ &\times (\hat{H}-E) \hat{P}^r_{m_i m_j} \varphi_{\xi_1} \cdots \varphi_{\xi_{i-1}} \varphi_{\xi_i'} \varphi_{\xi_{i+1}} \cdots \varphi_{\xi_{j-1}} \varphi_{\xi_j'} \varphi_{\xi_{j+1}} \cdots \varphi_{\xi_N} \rangle \end{aligned} \right\} \tag{7.22}$$

bestimmt wird. Somit erhielten wir für den Fall mehrerer Elektronen pro Atom (ohne Bahnentartung) eine Dirac-van Vleck-Formel (6.1) mit dem Austauschintegral (7.22).

8. Der Fall mehrerer Elektronen pro Atom unter Berücksichtigung der Bahnentartung. Betrachten wir die Säkulargleichung des Kristalls in der Heitler-London-Näherung für den Fall, daß sich die isolierten Atome in einem gewissen entarteten Bahnzustand (mit einem Bahnmoment $L \neq 0$) befinden. Hierbei wird die Wellenfunktion des isolierten Atoms in der Näherung der L-S-Kopplung durch die vier Quantenzahlen L, S, L_z und S_z charakterisiert. In der Annahme, daß die Größe des Bahn- und Spinmomentes an jedem i-ten Atom konstant (gleich L_i bzw. S_i) bleibt, bezeichnen wir die Vielelektronenatomfunktion mit dem **Symbol**

$$U_{k_i l_i},$$

wobei k_i die z-Komponente des Spins $\vec{S}_i$ bedeutet und $(2S_i+1)$ Werte durchläuft sowie l_i die z-Komponente des Bahnmoments $\vec{L}_i$ bezeichnet und $(2L_i+1)$ Werte annehmen kann. (Den Atomzustand kann man nur unter der Bedingung, daß das Atom isoliert ist, mit der Quantenzahl L charakterisieren. Will man die Heitler-London-Methode vervollständigen und dabei nicht von den Wellenfunktionen des freien Atoms, sondern von den Wellenfunktionen des im Kristallfeld befindlichen Atoms ausgehen, dann ist L keine gute Quantenzahl mehr, und man kann lediglich von gewissen effektiven Bahnquantenzahlen sprechen.)

Wie schon in Ziff. 7 besitzen die antisymmetrisierten Basisfunktionen die Form:

$$\Psi_{kl} = \sum_P (-1)^P \hat{P} \prod_{i=1}^{N} U_{k_i l_i},$$

worin die Symbole k und l die Gesamtheit der Spinquantenzahlen bzw. Bahnquantenzahlen bedeuten, also

$$k = \{k_1, \ldots, k_N\}, \quad l = \{l_1, \ldots, l_N\}.$$

Ziff. 8. Mehrere Elektronen pro Atom und Bahnentartung.

Bei der Berechnung der Matrixelemente der Säkulargleichung berücksichtigen wir wie zuvor nur Vertauschungen der Gruppe $\{G_0\}$, welche Koordinaten innerhalb der Atome austauschen, sowie die Vertauschungen, die zum Austausch von maximal zwei Elektronen zwischen verschiedenen Atomen führen. Für die Funktion $\prod U_{k_i l_i}$ benutzen wir eine zu (7.18) analoge Darstellung

$$\prod_{i=1}^{N} U_{k_i l_i} = \left(\prod_{i=1}^{N} M_i\right)^{-\frac{1}{2}} \sum_{\xi_1=1}^{M_1} \cdots \sum_{\xi_N=1}^{M_N} \varphi_{\xi_1 l_1} \cdots \varphi_{\xi_N l_N} \chi_{\xi_1 k_1} \cdots \chi_{\xi_N k_N}; \quad (8.1)$$

alle hier vorkommenden Bezeichnungen sind bereits in Ziff. 7 erklärt worden. Berechnet man die Matrixelemente in der oben angegebenen Näherung und verwendet wieder die Darstellung (6.12), so erhält man die Säkulargleichung in folgender Form

$$\left. \begin{array}{l} (E_{l\,l'} - E\delta_{l\,l'})\delta_{k\,k'} - \displaystyle\sum_{i>j} \frac{n_i n_j}{M_i M_j} \sum_{\xi_i \xi_j} \sum_{\xi_1 \ldots \xi_N (\neq \xi_i, \xi_j)} \left(\prod_{\varkappa \neq i,j} M_\varkappa\right)^{-1} \times \\ \times \langle \varphi_{\xi_1 l_1} \cdots \varphi_{\xi_N l_N} |(\hat{H} - E)\hat{P}^r_{m_i m_j}| \varphi_{\xi_1 l'_1} \cdots \varphi_{\xi_N l'_N}\rangle \times \left\{\frac{1}{2}\delta_{k\,k'} + \right. \\ \left. + 2 f_i(\xi_i, \xi_i) f_j(\xi_j, \xi_j) \langle \chi_{\xi_1 k_1} \cdots \chi_{\xi_N k_N} |\vec{S}_i \vec{S}_j| \chi_{\xi_1 k'_1} \cdots \chi_{\xi_N k'_N}\rangle\right\} = 0 \end{array} \right\} \quad (8.2)$$

mit

$$E_{l\,l'} = \left\langle \prod_{i=1}^{N} U_{k_i l_i} \middle| \hat{H} \middle| \prod_{i=1}^{N} U_{k_i l'_i} \right\rangle. \quad (8.3)$$

Das Glied außerhalb der Summe in (8.2) stammt von den Vertauschungen der Gruppe $\{G_0\}$, und die Summe ist durch die Paarvertauschungen bedingt.

Im Gegensatz zum nichtentarteten Fall (Ziff. 7) ist im Ausdruck (8.2) die Abweichung der Größe $E_{l\,l'}$ (ihr entspricht $\overline{E}$ in Ziff. 7) von $E\delta_{l\,l'}$ nicht exponentiell klein, sondern verringert sich mit zunehmendem Abstand nach einem Potenzgesetz. Ersetzt man daher unter der Summe in (8.2) die Energie E durch $E_{l\,l'}$, dann führt das zu einem Fehler von der Größenordnung des Produkts aus dem Quadrat des Überlappungsintegrals und einer negativen Potenz des Abstandes. Dagegen war in Ziff. 7 der Fehler beim Ersetzen von E durch $\overline{E}$ von der vierten Ordnung im Überlappungsintegral.

Die Gleichung (8.2) läßt sich prinzipiell vereinfachen, indem man Grenzfälle betrachtet, nämlich daß die Multipolwechselwirkung groß ist gegenüber der Austauschwechselwirkung (die von den Spinveränderlichen abhängt) oder daß die umgekehrte Situation verwirklicht ist. Im ersten Fall muß man zunächst solche Kombinationen der Atomfunktionen bestimmen, die die Größe $E_{l\,l'}$ diagonalisieren. Sodann wird, indem man diese Funktionen als gegeben betrachtet, der Spin-Hamilton-Operator für die in diesen Zuständen befindlichen Atome ermittelt.

Im Allgemeinfall muß man den Ausdruck (8.2) gleichzeitig sowohl in bezug auf $\{l\}$ als auch hinsichtlich $\{k\}$ diagonalisieren. Auf alle Fälle ist klar, daß die Diracsche Schreibweise einer Säkulargleichung nur für den Spinraum hier nicht mehr adäquat ist.

Für die Behandlung der Vielelektronenatome mit einem Spin $S > \frac{1}{2}$ dürfte bei der Berechnung der konkreten Konfigurationen eine auf der modernen Theorie der Addition der Drehimpulse beruhende Methode sehr aussichtsreich sein. Ihre Anwendung gestattet es, leicht von den Einelektronen- zu den Vielelektronenquantenzahlen äußerst komplizierter Konfigurationen überzugehen, wie sie in realen Magnetika auftreten, und die verschiedenen Matrixelemente der Coulomb-, Austausch- und Spin-Bahn-Wechselwirkung zu berechnen [29a, b].

Besonders vorteilhaft ist die Anwendung dieser Methode in der Darstellung der zweiten Quantelung, wo eine Verallgemeinerung der für den Fall $S=\frac{1}{2}$ bekannten Analogie zwischen den Spinoperatoren und den Fermi-Operatoren der Elektronen auf den Fall $S>\frac{1}{2}$ möglich ist [29c, d].

9. Das Vorzeichen des Austauschintegrals.
Die Frage des Vorzeichens des Austauschintegrals wurde bereits von HEISENBERG [8] gestellt, als er das Heitler-London-Modell zur Beschreibung des Ferromagnetismus von Kristallen verwendete. Da das Vorzeichen der Singulett-Triplett-Aufspaltung für das Wasserstoffmolekül demjenigen entgegengesetzt ist, das für das Auftreten des Ferromagnetismus erforderlich wäre, nahm HEISENBERG an, daß das Austauschintegral sein Vorzeichen ins Positive umkehren kann, wenn man Elektronen betrachtet, die Schalen mit einer großen Hauptquantenzahl n (z.B. $n=3$) besetzen. Diese Annahme ist jedoch keineswegs befriedigend, schon deshalb nicht, weil sich in der Palladium-Platinreihe die Elektronen in nicht spin-abgeschlossenen Schalen mit den Quantenzahlen 4 und 5 befinden, die entsprechenden Stoffe aber nicht ferromagnetisch sind.

Eine qualitative Abschätzung für das Vorzeichen des Austauschintegrals führten SLATER [99], [100] und BETHE [63] durch. Sie versuchten dabei, das Vorzeichen und die Abhängigkeit des Austauschintegrals vom Abstand festzulegen, indem sie von der Formel

$$J=\int \varphi_a^*(\vec{r}_1)\,\varphi_b^*(\vec{r}_2)\left[\frac{e^2}{R}+\frac{e^2}{r_{12}}-\frac{e^2}{r_{1b}}-\frac{e^2}{r_{2a}}\right]\varphi_b(\vec{r}_1)\,\varphi_a(\vec{r}_2)\,d\vec{r}_1\,d\vec{r}_2 \qquad (9.1)$$

ausgingen. Wie bereits in Ziff. 1 erwähnt, stellt die Größe (9.1) (Heisenbergsches Austauschintegral) nicht einmal für das Wasserstoffmolekül einen exakten Austauschparameter dar. Ungeachtet dessen wurde in den erwähnten Arbeiten angenommen, daß der Ausdruck (9.1) auf reale ferromagnetische Stoffe angewendet werden kann, wenn man unter φ_a und φ_b die $3d$-Wellenfunktionen des Atoms versteht.

Wenn man zunächst von der Winkelabhängigkeit der Atomfunktionen φ_a und φ_b absieht, läßt sich die Abhängigkeit von J vom interatomaren Abstand R qualitativ wie folgt erklären: Bei kleinen R wird die Austauschdichte der Ladung $\varphi_a^*(\vec{r})\,\varphi_b(\vec{r})$ etwa in gleicher Weise im Raum zwischen den Kernen und in der Nähe der Kerne konzentriert. In diesem Falle ist der durch die Elektron-Kern-Wechselwirkung bedingte Beitrag zu (9.1), der ein negatives Vorzeichen aufweist, groß. Bei Zunahme von R wird die Austauschdichte vornehmlich im Raum zwischen den Kernen konzentriert, sobald der Kernabstand R den Radius der Elektronenhülle des Atoms wesentlich übersteigt. Dann wächst die Bedeutung des Gliedes der Elektron-Elektron-Wechselwirkung e^2/r_{12}, das zu (9.1) einen positiven Beitrag liefert. An dieser Stelle soll der qualitative Charakter dieser Überlegungen nochmals unterstrichen werden. Betrachtet man nämlich nur ein Zweielektronensystem, so kann das Austauschintegral bei keinem noch so großen Abstand zwischen den Kernen positiv werden, und zwar auf Grund des Postulats einer knotenfreien Wellenfunktion des Grundzustandes (s. Ziff. 3). Der Umstand, daß das Glied e^2/r_{12} im Heitler-London-Verfahren bei großen Abständen zu einem positiven Austauschintegral führt, ist gerade ein Beweis dafür, daß diese Methode und folglich auch der Ausdruck für die Austauschenergie für $R\to\infty$ versagen. Der Mangel der qualitativen Analyse von BETHE und SLATER besteht daher in der Unterschätzung der Bedeutung des Korrelationseffektes, d.h. des gegenseitigen Ausweichens der beiden Elektronen. Das gilt auch für die folgenden Überlegungen. Das Integral (9.1) wird mit noch größerer Wahrscheinlichkeit positiv, wenn die Austauschdichte in unmittelbarer Nähe des Kerns klein ist. Diese Bedingung wird um so besser erfüllt, je größer die Bahnquantenzahl l ausfällt; denn der Radialteil der Funktion $\varphi_a(\vec{r})$ ist in der Nähe des Kerns a proportional zu r^l. Dementsprechend wird auch die Austauschdichte $\varphi_a^*(\vec{r})\,\varphi_b(\vec{r})$ in der Nähe der Kerne um so kleiner, je größer die Quantenzahl l der am Austausch beteiligten Elektronen ist.

Aus diesen einfachen Überlegungen folgt der Schluß, daß ein Stoff mit einer um so größeren Wahrscheinlichkeit ferromagnetisch wird, je größer das Verhältnis des interatomaren Abstandes zum Radius der nichtabgeschlossenen Schale und je größer das Bahnmoment der Elektronen in dieser Schale ist. Nach der zweiten Bedingung entsteht Ferromagnetismus am wahrscheinlichsten bei den Elementen mit nichtaufgefüllter d- oder f-Schale. Die von SLATER durchgeführten Rechnungen zeigen, daß das Verhältnis des Abstandes zwischen den Atomen zum Radius der Schale groß ist bei den $3\,d$-Übergangselementen und maximale Werte bei den Ferromagnetika Fe, Co und Ni (3,26; 3,64 bzw. 3,96) erreicht. Bei den nicht ferromagnetischen Metallen Pd ($4\,d$-Schale) und Pt ($5\,d$-Schale) ist dieses Verhältnis kleiner (2,82 bzw. 2,49) und erreicht bei den seltenen Erdmetallen ($4f$-Schale) ebenfalls hohe Werte, höhere als in der Eisengruppe. Unter der Annahme, daß der Ferromagnetismus der Metalle durch direkten Austausch zwischen den nichtaufgefüllten Schalen bedingt wird, muß der qualitative Verlauf des Austauschparameters in Abhängigkeit vom Verhältnis des interatomaren Abstandes R zum Schalenradius r_0 folgender sein: Bei kleinen R/r_0 ist das Integral J negativ. Mit zunehmendem R/r_0 wird J positiv und erreicht ein Maximum im ferromagnetischen Gebiet der $3\,d$-Metalle, um dann bis zu sehr kleinen Werten im Gebiet der seltenen Erden abzufallen. Die geringe Größe von J bei großen Werten von R/r_0 ist vollkommen verständlich, denn bei großen Abständen fällt J exponentiell ab. In Übereinstimmung mit diesen Überlegungen ist bei den $4f$-Metallen das Vorzeichen von J mit größter Wahrscheinlichkeit positiv, aber J selbst betragsmäßig klein. Die graphische Darstellung des soeben beschriebenen Verhaltens von J erhielt in der Literatur die Bezeichnung Bethe-Slater-Kurve.

Die Winkelabhängigkeit der Atomfunktionen wurde jedoch bei den oben angeführten Überlegungen vollkommen ignoriert.

Eine qualitative Analyse von BETHE [63] hat ergeben, daß diese Abhängigkeit wichtig und sogar bestimmend sein kann für das Vorzeichen des Austauschintegrals. Betrachten wir das Austauschintegral für Funktionen, die keine Kugelsymmetrie besitzen. Zunächst untersuchen wir den Zustand mit $m=l$. In diesem Falle nimmt der Winkelteil der Atomfunktion die Form

$$\sin^l\Theta \, e^{il\varphi}$$

an, wobei Θ und φ räumliche Polarkoordinaten bedeuten. Als gemeinsame Quantisierungsachse wählen wir die Verbindungslinie zwischen den Kernen a und b. Dann erhält man das Koordinatensystem für das Atom „b", indem man das Koordinatensystem des Atoms „a" entlang der Quantisierungsachse z parallel verschiebt. Die Austauschdichte, die proportional $\sin^l\Theta_a \sin^l\Theta_b$ ist, wird auf der z-Achse verschwinden und in deren Nähe klein sein. Dagegen wird die Dichte relativ groß in den Ebenen, die senkrecht zur Verbindungslinie liegen und die Kerne enthalten. Demnach konzentriert sich die Austauschdichte nicht zwischen den Kernen. Hier ist schwerlich ein positives Austauschintegral zu erwarten. Die Situation ändert sich, wenn die Elektronen Bahnen mit $m<l$ einnehmen. Als Beispiel betrachten wir den Fall $l=1$ und $m=0$. Der Winkelanteil wird dann

cos Θ. Diese Funktion besitzt eine Knotenebene bei $\Theta = 90°$, die durch den Kern geht und senkrecht zur Achse liegt. Die Austauschdichte ändert ihr Vorzeichen beim Übergang über die Knotenebenen beider Atome, während das Vorzeichen zwischen den Ebenen überall gleich ist. Die Wechselwirkung der Elektronen mit den Kernen verringert sich durch das Vorhandensein der Knotenebenen wegen der fast vollständigen Kompensation der unterschiedlichen Austauschladungen. Im Gegensatz dazu ist die Elektron-Elektron-Wechselwirkung zwischen den Kernen groß ($\varphi_a^* \varphi_b \sim \cos \Theta_a \cos \Theta_b$) und klein auf der Außenseite der Kerne, da in diesem Gebiet eine der Funktionen klein ist. Folglich kann man in dieser Situation ein positives Vorzeichen für das Austauschintegral erwarten.

Solche qualitativen Überlegungen riefen jedoch Einwände hervor. So postulierte z.B. ZENER [101] bis [104], daß das Vorzeichen des Austauschparameters stets negativ sei und sich der Parameter mit dem Abstand R zwischen den Kernen monoton verringere. Er wies besonders darauf hin, daß der Kern des Atoms oder Ions bei weitem nicht vollständig durch die inneren Elektronen abgeschirmt wird, was zu einer Zunahme des negativen Gliedes im Austauschintegral führen muß. Der Ferromagnetismus wird, gemäß der Theorie von ZENER, in den Übergangsmetallen auf Grund der indirekten Austauschwechselwirkung der lokalisierten Elektronen über die Leitungselektronen verwirklicht.

So weit die qualitativen Untersuchungen zum Vorzeichen des Austauschintegrals.

10. Zahlenmäßige Abschätzungen für das Austauschintegral. Das Austauschintegral (9.1) hat BARTLETT [105] für die $2p$-Funktionen mit $m = 0$ berechnet. Wie sich herausstellte, besitzt das Integral in Übereinstimmung mit der qualitativen Herleitung durch BETHE ein positives Vorzeichen. Mit dem positiven Vorzeichen von J erklärte BARTLETT den Paramagnetismus des molekularen Sauerstoffs.

Als erster Versuch, das Integral (9.1) unter Verwendung der $3d$-Funktionen eines wasserstoffähnlichen Atoms zu berechnen, ist eine Arbeit [106] von WOHLFARTH anzusehen. WOHLFARTH vernachlässigte jedoch vollständig die Winkelabhängigkeit der Funktionen, indem er für sie den kugelsymmetrischen Ausdruck $r^2 e^{-\alpha r}$ (also lediglich den Radialanteil der Wellenfunktionen) benutzte. Die Berechnungen erfolgten in einem breiten Intervall von Kernabständen, und das Vorzeichen von J erwies sich überall als negativ. Das Ergebnis von WOHLFARTH beinhaltet aber nicht viel nützliche Information, da der Ersatz der wahren $3d$-Funktionen durch kugelsymmetrische Funktionen eine überaus grobe Vereinfachung darstellt.

KAPLAN [107] berechnete (9.1) für die $3d$-Funktionen mit $m = 0$. Für die Anwendung auf reale Stoffe besaß der Exponentialkoeffizient die Form $e^{-r/2r_0}$, wobei r_0 gleich $0,25\,a_H$ gesetzt wurde (eine für Eisen vernünftige Größe). Das Vorzeichen des Integrals erwies sich als positiv, und der Betrag war $J = 0,6$ eV, d.h. zehnmal mehr als experimentell ermittelt. Dieses Ergebnis ist ebenfalls unbefriedigend, da der entsprechende Kernabstand zu gering ist, nämlich 0,75 Å. KAPLAN schätzte auch für einen doppelt so großen Abstand das Integral ab, das nach wie vor positiv bleibt, aber auch dieser Abstand ist klein gegenüber den realen Abständen im Kristall.

Eine Arbeit von STUART und MARSHALL [108] brachte eine weitere Präzisierung in der Berechnung des Austauschparameters. Wesentlich dabei war, daß die Autoren darauf verzichteten, den Radialteil der Atomfunktion in Form einer wasserstoffähnlichen Funktion anzusetzen. Die Werte für den Radialteil entnahmen STUART und MARSHALL den Rechnungen von WOOD und PRATT [109] für die $3d$-Einelektronenfunktionen des neutralen Fe-Atoms, das die Konfiguration $3d^6 4s^2$ besitzt. Diese Funktionen wurden nach der Hartee-Fock-Methode berechnet. Das Austauschintegral wurde nicht für alle 25 Paare der $3d$-Funktionen berechnet, sondern lediglich für zwei Funktionen mit $m=0$ (mit der Kern-Verbindungslinie als Quantisierungsachse), deren Winkelanteil die Form

$$3 \cos^2 \Theta - 1$$

besitzt. Es wurde der Austauschparameter nach der Formel (9.1) sowie nach der Formel (1.20) berechnet. Wie sich zeigte, unterscheiden sich die nach diesen beiden Formeln bestimmten Werte lediglich bei sehr kleinen Kernabständen ($<1 a_H$) merklich voneinander. Im gesamten Abstandsbereich ist J positiv, die Abhängigkeit $J(R)$ folgt also nicht der Bethe-Slater-Kurve. Bei dem beobachteten Wert von $4,7 a_H$ für den Abstand zwischen den nächsten Nachbarn im Eisenkristall nimmt das Integral J den Wert $J = 0,0068$ eV an. Ferner versucht der Autor in der Annahme, daß beim metallischen Nickel der Wert des Austauschparameters etwa gleich dem von Eisen ist und die Nickelatome im Kristall teilweise die Konfiguration $3d^9$ besitzen [die Beschreibung der Austauschwechselwirkung ist daher mit Hilfe der Formel (1.20) möglich], den theoretischen Wert mit den experimentellen Angaben zu vergleichen. Die experimentellen Werte für J betragen für Nickel nach Spinwellen-Messungen 0,020 eV und nach Messungen der spezifischen Wärme 0,010 eV. STUART und MARSHALL empfehlen nicht, den theoretischen Wert für J unmittelbar mit dem Experiment zu vergleichen, sondern vorher eine Mittelung durchzuführen. Unter Hinweis auf Neutronenbeugungsversuche [110], [111] nahmen sie an, daß die am Austausch beteiligten Elektronen mit gleicher Wahrscheinlichkeit verschiedene Atombahnen besetzen. Da die Besetzungswahrscheinlichkeit für jede der fünf $3d$-Bahnen 0,2 beträgt, ist die Wahrscheinlichkeit der Paarbildung mit $m=0$ an jedem der Atome gleich 0,04. Aus diesem Grunde muß man mit dem Experiment die Größe $0,04 J$ vergleichen, und zwar unter der Bedingung, daß alle übrigen Austauschintegrale vernachlässigbar klein sind. (Zur Kontrolle berechneten die Autoren das Austauschintegral für $R = 4,7 a_H$ zwischen den Bahnen mit $m=0$ und einer „kubischen" Bahn zx mit dem Winkelteil $\cos\Theta \sin\Theta \cos\varphi$. Dieses Integral war 50mal kleiner als dasjenige zwischen zwei Bahnen mit $m=0$.)

Die Mittelung erfolgte ebenfalls für Eisen unter der Annahme, daß dort die Atomspins gleich 1 sind. In diesem Falle nimmt der Austauschparameter wiederum den Wert $0,04 J$ an. Die experimentellen Werte betragen nach Spinwellen-Messungen 0,018 eV und nach Messungen der spezifischen Wärme 0,011 eV [112]. Wie die Autoren bemerken, ergeben sich die kleinen theoretischen Werte des Austauschparameters nicht wegen der Kleinheit des Austauschintegrals zwischen Bahnen mit $m=0$, sondern wegen der geringen Besetzungswahrscheinlichkeit dieser Bahnen.

Als nächste wichtige Arbeit zur Berechnung des Austauschparameters ist eine Veröffentlichung von FREEMAN und WATSON [113] anzusehen, in der neben den zahlenmäßigen Ergebnissen eine umfangreiche Diskussion des Begriffes Austauschintegral enthalten ist und wo versucht wird, den Rahmen der groben Näherungen zu sprengen, die in den früheren Arbeiten üblich waren.

In erster Linie berechneten FREEMAN und WATSON die Austauschintegrale zwischen den $3d$-Funktionen des Wasserstoffatoms nach der Formel (1.20), um endgültig Klarheit zu schaffen über das Vorzeichen und die Abhängigkeit des Austauschintegrals für wasserstoffähnliche Funktionen, was historisch gesehen von großer Wichtigkeit ist. Berechnet wurden die Integrale für Paare von σ- ($m=0$), π- ($m=1$) und δ-Funktionen ($m=2$): $J_{\sigma\sigma}$, $J_{\pi\pi}$ und $J_{\delta\delta}$. Das Integral

$J_{\sigma\sigma}$ hat überall positives Vorzeichen, $J_{\pi\pi}$ ist bei kleinen Abständen der Atome positiv und wird negativ bei großen, und $J_{\delta\delta}$ hat überall negatives Vorzeichen. Für sehr kleine Abstände wurden die Integrale nicht berechnet. Da, wie FREEMAN und WATSON bemerken, der Radialteil der $3d$-Funktionen des Wasserstoffs ein verwaschenes Maximum bei $\approx 9 a_H$ aufweist, die $3d$-Funktionen der Übergangsmetalle dagegen bei $\approx 1 a_H$, muß man für einen groben Vergleich mit dem Experiment wenigstens den Maßstab so verändern, daß ein Kernabstand von 40 bis 50 a_H für Wasserstoffunktionen den im Gitter beobachteten Abständen entspricht. Bei $R \approx 40$ bis 50 a_H sind die berechneten Integrale sehr klein.

Nach FREEMAN und WATSON folgt nicht ein einziges der berechneten Integrale der Bethe-Slater-Kurve, es sind auch nicht alle negativ — im Gegensatz zur Hypothese von ZENER. Geht man jedoch von diesen Berechnungen aus, wäre ein Urteil über die Richtigkeit der Überlegungen von BETHE und SLATER sowie ZENER verfrüht, um so mehr, da BETHE selbst bemerkte, daß die $3d$-Wasserstofffunktionen für die Berechnungen kaum geeignet sind. Auch wies ZENER darauf hin, daß die Annahme $Z_{Kern}=1$ eine schlechte Näherung darstellt, da die Kernladung von den inneren Elektronen nicht vollkommen abgeschirmt wird. Andererseits wird die von BETHE gezogene Schlußfolgerung, nach der der Parameter $J_{\sigma\sigma}$ vorwiegend positiv und $J_{\delta\delta}$ negativ ist, durch die vorliegenden Rechnungen bestätigt [streng genommen basiert BETHEs Schlußfolgerung auf (9.1) und nicht auf (1.20)].

Als weiteren Schritt führten FREEMAN und WATSON eine Berechnung der Austauschintegrale für alle möglichen 25 Paare durch unter Verwendung der $3d$-Wellenfunktionen des Kobalt-Ions Co^{+2} und des neutralen Kobalts Co, die bereits früher von WATSON [114] nach der Hartree-Fock-Methode berechnet wurden. Die Rechnung erfolgte sowohl nach (1.20) als auch nach (1.18). Die Autoren betonen besonders, daß die Formel (1.20) — die nur gilt, wenn die Elektronenfunktionen Wasserstoffunktionen darstellen — für die Berechnung der Austauschparameter realer Atome von vornherein vollkommen ungeeignet ist. Die Hartree-Fock-Funktionen haben nämlich eine andere radiale Abhängigkeit als die Wasserstoffunktionen. Wie sich bei der Berechnung nach (1.20) herausstellte, hat $J_{\sigma\sigma}$ sowohl im Falle Co^{+2} (für $R=4{,}75$ bzw. 2,25 a_H) als auch im Falle Co (für $R=4{,}75$ a_H) positives Vorzeichen. Folglich verhält sich das Vorzeichen des Austauschparameters ebenso wie bei STUART und MARSHALL, die auch die Beziehung (1.20) benutzten, jedoch unter Verwendung der Wellenfunktionen von Eisen.

Der Ausdruck (1.18), in dem die Funktionen nicht vom Wasserstofftyp zu sein brauchen (wo aber das Modell einer Punktladung mit $Z=1$ benutzt wird), ist genauer als (1.20). Die Berechnungen nach (1.18) ergaben ein negatives Vorzeichen für alle Integrale: $J_{\sigma\sigma}$, $J_{\pi\pi}$ und $J_{\delta\delta}$. Weiterhin wurden die Austauschintegrale für gewisse effektive Kernladungen $Z \neq 1$ ermittelt, für die die Ausdrücke (1.20) und (1.18) (in denen jetzt die Ladung 1 durch Z zu ersetzen ist) identisch werden. In Abhängigkeit von R und m liegt Z etwa zwischen 9 und 17. In diesem Falle sind alle diagonalen Austauschintegrale negativ.

FREEMAN und WATSON berechneten den Austauschparameter auch, indem sie in (1.18) und (1.20) auf die Annäherung des Ionenrumpfes durch eine Punktladung $Z=1$ oder die effektiven Ladungen Z_a bzw. Z_b verzichteten. Zu diesem Zwecke werden im Hamilton-Operator (1.1) die Glieder der Elektron-Kern-Wechselwirkung durch die Größen

$$V_{1a} = -\frac{Z}{r_{1a}} + \sum_{ia} \int \varphi_{ia}^*(q_2) \frac{1-\widehat{P}_{12}}{r_{12}} \varphi_{ia}(q_2) dq_2 \qquad (10.1)$$

ersetzt, wobei Z die Kernladung und $\hat{P}_{12}$ den Vertauschungsoperator für die Koordinaten q_1 und q_2 bedeuten; die Summation erfolgt über die „gepaarten" Elektronen. Das zweite Glied in (10.1) ist das Hartree-Fock-Potential, mit dem die Elektronen in den spin-abgeschlossenen Schalen auf das am Austausch beteiligte Elektron einwirken. Ein analoger Ausdruck muß auch für die übrigen Elektron-Ionen-Wechselwirkungen Anwendung finden. Die Funktionen φ_{ia} sind Lösungen der Hartree-Fock-Gleichungen

$$\left(-\frac{\hbar^2}{2m}\Delta + V_{1a}\right)\varphi_{ia} = \varepsilon_i \varphi_{ia}. \tag{10.2}$$

Im Falle einer Schale mit nicht abgeschlossenem Spin ergeben die Gleichungen (10.2) wegen des Spinpolarisationseffektes im allgemeinen für verschiedene Bahnen auch unterschiedliche Radialteile der Funktion. FREEMAN und WATSON vernachlässigten jedoch diesen Unterschied in der Annahme, daß dies zu keinem wesentlichen Fehler führt. Die Berücksichtigung der Struktur der Elektronenrümpfe beschränkte sich auf die $3s$-, $3p$- und $3d$-Elektronen. Die übrigen, tiefer liegenden Elektronen wurden nur im Rahmen der Methode der effektiven Ladung berücksichtigt; die Kernladung Z wurde also durch den Wert $Z-10$ ersetzt. Die Berechnung des Austauschparameters erfolgte nach der allgemeinen Formel (1.18). Dabei nahmen die Austauschintegrale $J_{\sigma\sigma}$, $J_{\pi\pi}$ und $J_{\delta\delta}$ (für Co und Co^{+2}) alle negatives Vorzeichen an. Die nichtdiagonalen positiven Austauschintegrale sind jedoch so groß, daß sie das Resultat der Mittelung wesentlich beeinflussen, ja den Mittelwert von J (wenn auch nur sehr wenig) positiv machen können. Ein weiterer, von FREEMAN und WATSON unternommener Schritt war die explizite Berücksichtigung der Überlappung der Wellenfunktionen der Elektronen in den spinabgeschlossenen Bahnen. Im Falle eines ungepaarten Elektrons in jedem Atom a und b haben die Slater-Determinanten für $S_z = 0$ die Form

$$\left.\begin{array}{l}\Psi_1 = \det\{(\varphi_a\alpha)(\varphi_b\beta)(\varphi_{1a}\alpha)(\varphi_{1a}\beta)(\varphi_{1b}\alpha)(\varphi_{1b}\beta)\ldots\}, \\ \Psi_2 = \det\{(\varphi_a\beta)(\varphi_b\alpha)(\varphi_{1a}\alpha)(\varphi_{1a}\beta)(\varphi_{1b}\alpha)(\varphi_{1b}\beta)\ldots\},\end{array}\right\} \tag{10.3}$$

wo φ_a und φ_b die Atomfunktionen der ungepaarten Elektronen und φ_{ia} und φ_{ib} die zweifach besetzten Bahnzustände bezeichnen; die Argumente sind der Kürze halber weggelassen. Die Singulett- und Triplettfunktionen lauten

$$\Psi_s = \frac{1}{\sqrt{2(S_{11}-S_{12})}}(\Psi_1 - \Psi_2), \quad \Psi_t = \frac{1}{\sqrt{2(S_{11}+S_{12})}}(\Psi_1 + \Psi_2) \tag{10.4}$$

mit

$$S_{ij} = \langle \Psi_i \Psi_j \rangle, \quad i,j = 1, 2. \tag{10.5}$$

Der Austauschparameter wird wie folgt bestimmt

$$J = \frac{S_{12}H_{11} - S_{11}H_{12}}{S_{11}^2 - S_{12}^2}, \tag{10.6}$$

wo H_{ij} durch

$$H_{ij} = \langle \Psi_i \hat{H} \Psi_j \rangle; \quad i,j = 1, 2$$

definiert ist.

Da die Überlappungsintegrale für die Funktionen verschiedener Atome klein sind, kann man sich in (10.6) auf Glieder zweiter Ordnung in der Überlappung beschränken. Nach dem Ergebnis der Rechnungen unterscheiden sich die diagonalen Austauschintegrale nicht wesentlich von den früher ohne Berücksichtigung der Überlappung der Rumpffunktionen erhaltenen. Bei den nichtdiagonalen Austauschparametern sind die Änderungen merklicher. Die gemittelten Werte des Austauschintegrals sind negativ. Die Rechungen von FREEMAN und WATSON haben also gezeigt, daß im vorgegebenen Modell der direkten Austauschwechselwirkung das Vorzeichen des Austauschparameters nicht den Ferromagnetismus, sondern den Antiferromagnetismus begünstigt.

Mit den Schlußfolgerungen von FREEMAN und WATSON stimmt das Ergebnis der Untersuchungen von CARR [97] überein. CARR wählte als Einelektronenfunktionen ebenfalls nicht Eigenfunktionen eines wasserstoffähnlichen Atoms, sondern Hartree-Funktionen. Das Austauschintegral war für Eisen negativ. FREEMAN, NESBETT und WATSON [115] berechneten das Austauschintegral unter Verwendung orthogonalisierter Bahnen und unter Berücksichtigung der Konfigurationswechselwirkung. Falls φ_a und φ_b Atombahnen gleicher Symmetrie sind, so lassen sich daraus die orthogonalen Bahnen

$$\psi_a = c_1 \varphi_a + c_2 \varphi_b, \quad \psi_b = c_2 \varphi_a + c_1 \varphi_b \qquad (10.7)$$

konstruieren, wobei die Transformationskoeffizienten den Beziehungen

$$c_1 + c_2 = \frac{1}{\sqrt{1 + S_{ab}}}, \quad c_1 - c_2 = \frac{1}{\sqrt{1 - S_{ab}}} \qquad (10.8)$$

gehorchen. Unter Berücksichtigung der Konfigurationen, in denen einer der Bahnzustände ψ_a oder ψ_b doppelt besetzt ist, haben die Basisdeterminanten die Form (2.3) und (2.11), die Triplett- und Singulettfunktionen die Form (2.4) bzw. (2.11). Stellt man die Säkulargleichung dritten Grades für das Singulett auf und löst sie mit Hilfe der Störungstheorie, erhält man leicht folgenden Ausdruck für den Austauschparameter:

$$J = -\frac{2H_{ab}^2}{\varDelta} + C \qquad (10.9)$$

mit den Bezeichnungen

$$H_{ab} = \langle \psi_a | \hat{H}_1 | \psi_b \rangle + \langle \psi_a \psi_b | V | \psi_a \psi_a \rangle,$$
$$\varDelta = \langle \psi_a \psi_a | V | \psi_a \psi_a \rangle - \langle \psi_a \psi_b | V | \psi_a \psi_b \rangle,$$
$$C = \langle \psi_a \psi_b | V | \psi_a \psi_b \rangle.$$

(H_1 ist der Einelektronenteil des effektiven Hamilton-Operators und V die Elektron-Elektron-Wechselwirkung.) Das erste Glied in (10.9) ist negativ, das zweite positiv. Der Parameter (10.9) wurde bestimmt unter Verwendung der Integrale mit Hartree-Fock-Funktionen, die von WATSON [116] für Co und Co^{+2} für die Kernladungen $Z=1$ und $Z=10$ berechnet wurden. In allen den Fällen, wo die Heitler-London-Näherung annehmbar ist, waren die Austauschintegrale negativ.

11. Die biquadratische Austauschwechselwirkung. In Ziff. 6—7 wurde der effektive Hamilton-Operator (6.1) des Vektormodells unter einer wesentlichen Annahme abgeleitet: Bei der Berechnung der Matrixelemente des Energieoperators wurden in der Säkulargleichung (4.9) alle Beiträge vernachlässigt, die höheren Vertauschungen entsprechen als der identischen Vertauschung oder einfachen Transpositionen (zwischen verschiedenen Atomen). Die Näherung (7.6) führte dann dazu, daß der effektive Hamilton-Operator nur quadratisch in den Elektronenspinoperatoren wird. Bei Berücksichtigung der Vertauschungen höheren Grades würde der effektive Spin-Hamilton-Operator offensichtlich neben den in den Spinoperatoren quadratischen Gliedern auch Glieder höherer Ordnung in den Spins enthalten. Man kann ([27]) eine allgemeine Form des effektiven Spin-

Ziff. 11. Die biquadratische Austauschwechselwirkung. 313

Hamilton-Operators finden und die Anwendungsgrenzen des Diracschen bilinearen Hamilton-Operators festlegen. Betrachten wir den Fall ohne Bahnwechselwirkung. Die Aufgabe wird folgendermaßen formuliert. Es ist ein Ausdruck für einen Hamilton-Operator aufzusuchen, der lediglich im Spinraum wirkt und dessen Eigenwerte mit den Eigenwerten des im gewöhnlichen Raum definierten Ausgangs-Hamilton-Operators des Systems zusammenfallen. Das betrachtete n-Elektronensystem besitzt λ verschiedene Energieniveaus, und zwar $n!\left(\frac{n}{2}!\right)^{-2}$ für gerade n und $n!\left(\frac{n-1}{2}!\right)^{-1}\left(\frac{n+1}{2}!\right)^{-1}$ für ungerade. Diese λ-Werte stimmen überein mit der Gesamtzahl der verschiedenen Multipletts des Systems, und man kann sie durch Summation der Größe $K(S)$ aus (4.12) über alle möglichen Spinwerte des Systems erhalten:

$$\lambda = \sum_S K(S) = \sum_S \frac{n!\,(2S+1)}{\left(\frac{n}{2}+S+1\right)!\left(\frac{n}{2}-S\right)!}. \tag{11.1}$$

Wir bezeichnen die exakten Energieeigenwerte, d.h. die Energien der verschiedenen Multipletts des Systems mit E_{iS} ($i=1, 2, \ldots, K(S)$ bedeutet die Nummer des Multipletts S) und führen folgenden Operator ein

$$\widehat{H}_{\text{eff}} = \frac{1}{n!}\sum_S \sum_{i=1}^{K(S)} \sum_P K(S)\, E_{iS}\, D_{ii}^{S*}(P)\, \widehat{P}. \tag{11.2}$$

Dabei ist $\widehat{P}$ ein Vertauschungsoperator, der lediglich auf die Spinveränderlichen wirkt, und $D_{ii}^S(P)$ ein Diagonalelement der Matrix der irreduziblen Darstellung $[D_{ij}^S(P)]$ der Vertauschungsgruppe mit der Dimension $K(S)$. Als Basisfunktionen der Transformationsmatrix $[D_{ij}^S(P)]$ dienen willkürliche Multiplettfunktionen, d.h. Eigenfunktionen der Operatoren $\widehat{S}^2$. Beim Matrixelement der irreduziblen Darstellung ist der Index S_z weggelassen, da die Form (11.2) beim vorgegebenen S für beliebiges S_z die gleichen Eigenwerte besitzt.

Man kann sich unmittelbar davon überzeugen, daß der Operator (11.2) die Energien E_{jS} der Multipletts des Systems als Eigenwerte besitzt, indem man diesen Operator auf eine beliebige Eigenfunktion $\Theta_j(S)$ des Operators $\widehat{S}^2$ wirken läßt. Dabei muß man die Identität

$$\widehat{P}\,\Theta_j(S) = \sum_{i=1}^{K(S)} D_{ij}^S(P)\,\Theta_i(S) \tag{11.3}$$

und die Orthogonalitätsbedingung

$$\sum_P D_{\alpha\gamma}^{S*}(P)\, D_{\beta\delta}^{S'}(P) = \delta_{SS'}\,\delta_{\alpha\beta}\,\delta_{\gamma\delta}\,\frac{n!}{K(S)} \tag{11.4}$$

beachten. [$n!$ bedeutet die Zahl der Elemente der Vertauschungsgruppe, $\Theta_j(S)$ in (11.3) und $D_{ii}^S(P)$ in (11.2) müssen zum gleichen S_z gehören.] Somit ist bewiesen, daß man den Operator (11.2) als effektiven Spin-Hamilton-Operator des untersuchten Systems ansprechen kann.

Jede beliebige Vertauschung P kann dargestellt werden als Produkt von Transpositionen $\widehat{P}_{ij}$, die mit den Spinoperatoren unmittelbar durch die Diracsche Beziehung (6.12) verknüpft sind. Folglich enthält der effektive Spin-Hamilton-Operator (11.2) lediglich *gerade Potenzen* der Spinoperatoren. Die Matrizen $[D_{ij}^S(P)]$ sind bis auf unitäre Transformationen bestimmt, womit die Form (11.2) nicht eindeutig ist. FAULKNER [27] untersuchte die Grenzen der Anwendbarkeit für einen Diracschen Spin-Hamilton-Operator der Form

$$\hat{H}_D = C - \sum_{i>j} J_{ij}(\tfrac{1}{2} + 2\vec{s}_i\vec{s}_j). \tag{11.5}$$

Außerdem behandelte er das damit zusammenhängende Problem der Bestimmung des „Austauschintegrals" J_{ij}. Die Größen J_{ij} sind unbekannt und müssen ebenso bestimmt werden, wie auch der Austauschparameter für zwei Elektronen in Ziff. 1 bestimmt wurde. Mit Hilfe von (6.12) läßt sich der Operator (11.5) darstellen als

$$\hat{H}_D = C - \sum_{i<j} J_{ij}\hat{P}_{ij}. \tag{11.6}$$

Die mit Hilfe willkürlicher Eigenfunktionen $\Theta_i(S)$ des Operators $\hat{S}^2$ berechneten Matrixelemente des Hamilton-Operators (11.6) lauten

$$\langle \Theta_i(S)|\hat{H}_D|\Theta_j(S')\rangle = \delta_{SS'}\left[C\delta_{ij} - \sum_{k>l} J_{kl} D^S_{ji}(P_{kl})\right]. \tag{11.7}$$

[Bei der Ableitung von (11.7) muß man die Gl. (11.3) und die Orthonormierungsbedingung für die Funktion $\Theta_i(S)$ verwenden.] Wenn die Funktionen $\Theta_i(S)$ Eigenfunktionen des Hamilton-Operators (11.6) darstellten, hätte die Matrix (11.7) Diagonalform, und ihre Elemente wären die Eigenwerte E_{iS}. Durch Vergleich der Matrixelemente (11.7) mit der Diagonalmatrix erhält man ein System von Bestimmungsgleichungen für die Parameter J_{ij} und C. Die Zahl dieser Unbekannten beträgt $\tfrac{1}{2}n(n-1)+1$. Die Zahl der Gleichungen, die einen Matrixblock der Ordnung $K(S)$ ergibt, beträgt $(K(S))^2$. Lediglich $\tfrac{1}{2}K(S)(K(S)+1)$ Gleichungen davon sind unabhängig, denn die Matrizen der irreduziblen Darstellung $[D^{(S)}_{ij}(P_{kl})]$ sind symmetrisch wegen $\hat{P}^2_{kl} = 1$ und weil die Eigenfunktionen $\Theta_i(S)$ reell sind. Die Gesamtzahl der unabhängigen Gleichungen, die durch die Summe $\sum_S \tfrac{1}{2}K(S)(K(S)+1)$ bestimmt wird, übersteigt die Zahl der unbekannten Parameter J_{ij} und C ab $n=4$ (davon überzeugt man sich leicht mit Hilfe von (4.12)). Das bedeutet, daß bei Elektronenzahlen, die größer als drei sind, das Gleichungssystem in den Parametern J_{ij} und C unverträglich wird. Folglich ist für $n>3$ der Diracsche Hamilton-Operator (11.5) wegen der Unmöglichkeit der Bestimmung der Parameter C und J_{ij} nicht mehr adäquat. Betrachten wir ausführlicher den Fall dreier Elektronen, wo es vier Parameter J_{12}, J_{13}, J_{23}, C gibt, über die verfügt werden kann. Zur Aufstellung des Gleichungssystems suchen wir die Matrizen der irreduziblen Darstellungen $[D^S_{ij}(P)]$, wobei $S=\tfrac{1}{2}$ oder $\tfrac{3}{2}$ ist und $\hat{P}$ 3! Werte: 1, $\hat{P}_{12}$, $\hat{P}_{13}$, $\hat{P}_{23}$, $\hat{P}_{13,12}$, $\hat{P}_{23,12}$ durchläuft. Als Spinfunktionen wählen wir folgende Ausdrücke

$$\Theta^{\tfrac{3}{2}}_1 = \alpha(\sigma_1)\alpha(\sigma_2)\alpha(\sigma_3), \tag{11.8}$$

$$\Theta^{\tfrac{1}{2}}_1 = \frac{1}{2}[\alpha(\sigma_1)\alpha(\sigma_2)\beta(\sigma_3) - \alpha(\sigma_1)\beta(\sigma_2)\alpha(\sigma_3)], \tag{11.9}$$

$$\Theta^{\tfrac{1}{2}}_2 = \frac{1}{\sqrt{6}}[\alpha(\sigma_1)\alpha(\sigma_2)\beta(\sigma_3) + \alpha(\sigma_1)\beta(\sigma_2)\alpha(\sigma_3) - 2\beta(\sigma_1)\alpha(\sigma_2)\alpha(\sigma_3)], \tag{11.10}$$

wobei $\Theta^{\tfrac{3}{2}}_1$ die Eigenfunktion des Quartettzustandes mit $S_z = \tfrac{3}{2}$, $\Theta^{\tfrac{1}{2}}_1$ und $\Theta^{\tfrac{1}{2}}_2$ zwei Dublettfunktionen mit $S_z = \tfrac{1}{2}$ bedeuten. Die Funktion $\Theta^{\tfrac{3}{2}}_1$ stellt die Basis der singulären Darstellung $D^{\tfrac{3}{2}}_{11}(P) = 1$ dar. Die Funktionen $\Theta^{\tfrac{1}{2}}_1$ und $\Theta^{\tfrac{1}{2}}_2$ bilden die Basis einer zweidimensionalen irreduziblen Darstellung. Die Matrizen dieser irreduziblen Darstellung haben die Form:

$$D^{\frac{1}{2}}(P=1)=\begin{pmatrix}1 & 0\\ 0 & 1\end{pmatrix}, \qquad D^{\frac{1}{2}}(P_{12})=\begin{pmatrix}\frac{1}{2} & \frac{1}{2}\sqrt{3}\\ \frac{1}{2}\sqrt{3} & -\frac{1}{2}\end{pmatrix},$$

$$D^{\frac{1}{2}}(P_{13})=\begin{pmatrix}\frac{1}{2} & -\frac{1}{2}\sqrt{3}\\ -\frac{1}{2}\sqrt{3} & -\frac{1}{2}\end{pmatrix}, \qquad D^{\frac{1}{2}}(P_{23})=\begin{pmatrix}-1 & 0\\ 0 & 1\end{pmatrix}, \qquad (11.11)$$

$$D^{\frac{1}{2}}(P_{13,12})=\begin{pmatrix}-\frac{1}{2} & \frac{1}{2}\sqrt{3}\\ \frac{1}{2}\sqrt{3} & -\frac{1}{2}\end{pmatrix}, \qquad D^{\frac{1}{2}}(P_{23,12})=\begin{pmatrix}-\frac{1}{2} & \frac{1}{2}\sqrt{3}\\ \frac{1}{2}\sqrt{3} & -\frac{1}{2}\end{pmatrix}.$$

Verwendet man (11.7) und (11.11) und vergleicht die Energiematrix mit der Diagonalmatrix, dann erhält man ein System von vier Gleichungen, dessen Lösungen die folgenden Ausdrücke sind:

$$J_{12}=\tfrac{1}{3}(-E_q+E_{2d}), \qquad J_{13}=\tfrac{1}{3}(-E_q+E_{2d}),$$
$$J_{23}=-\tfrac{1}{3}E_q+\tfrac{1}{2}E_{1d}-\tfrac{1}{6}E_{2d}, \qquad C=\tfrac{1}{2}(E_{1d}+E_{2d}). \qquad (11.12)$$

E_q ist hier die Energie des Quartetts, und E_{1d} bzw. E_{2d} stellen die Energien der beiden Dubletts dar. Dabei sind die Funktionen (11.8), (11.9) und (11.10) Eigenfunktionen des Diracschen Hamilton-Operators. Die Ausdrücke (11.12) erscheinen als Definitionen für den Begriff der Austauschparameter im Diracschen Hamilton-Operator bei $n=3$ und bedeuten eine „Verallgemeinerung" der Definition (1.5) auf den Fall von drei Elektronen. FAULKNER [27] machte darauf aufmerksam, daß der Diracsche Hamilton-Operator (11.5) richtige Energiewerte für den Zustand maximaler Multiplizität $\left(S=\frac{n}{2}\right)$ und für die $(n-1)$ Multipletts mit $S=\frac{n}{2}-1$ liefern kann. Der Funktion des Zustandes maximaler Multiplizität entspricht eine singuläre Darstellung, und die Multipletts mit $S=\frac{n}{2}-1$ bedingen eine $(n-1)$-dimensionale irreduzible Darstellung. Folglich ergeben die Zustände mit $S=\frac{n}{2}$ und $S=\frac{n}{2}-1$ insgesamt $\frac{1}{2}n(n-1)+1$ unabhängige Gleichungen für die Bestimmung ebenso vieler unbekannter Parameter J_{ij} und C.

Wie man sieht, ist man berechtigt, in den Fällen, wo der Grundzustand das Spinsystem die maximale Multiplizität $\left(S=\frac{n}{2}\right)$ besitzt, für die Berechnung der niedrigsten angeregten Zustände den Dirac-Spin-Hamilton-Operator (11.5) zu verwenden.

Alle bisherigen Überlegungen dieses Abschnittes über ein auf den Kristall anwendbares n-Elektronensystem gehen davon aus, daß der Kristall aus n Einelektronenatomen besteht. Die Anwendbarkeit des bilinearen Operators wird noch weiter eingeschränkt, wenn die Atome einen starren Spin $S_i>\tfrac{1}{2}$ besitzen. Sogar für Systeme, die lediglich aus zwei Atomen mit den Spins $S_a, S_b>\tfrac{1}{2}$ bestehen, kann die Diracsche Form

$$\hat{H}_D=C-2J\vec{S}_a\vec{S}_b \qquad (11.13)$$

keine richtigen Werte für die Energie der Multipletts mehr liefern. Der Gesamtspin des Systems kann nämlich $2S_a+1$-Werte: $S_b-S_a, S_b-S_a+1, \ldots, S_b+S_a$ annehmen (wir setzen hier $S_a<S_b$). Somit erhalten wir durch Vergleich der Eigenwerte des Hamilton-Operators (11.13) mit den richtigen Energien der Multipletts $(2S_a+1)>2$ Gleichungen für die Bestimmung der beiden Parameter C und J.

Wir gehen nun über zur Ableitung des Spin-Hamilton-Operators unter Berücksichtigung der Vertauschungen höherer Ordnung [85, Ch. 8]. Unter den Vertauschungen zwischen zwei Atomen ist von nächsthöherer Ordnung eine Vertauschung, in der zwei Elektronenpaare ihre Plätze wechseln. Soweit drei Atome daran beteiligt sind, liegt eine cyclische Vertauschung der Elektronen vor, und zwar geht ein Elektron vom Atom i zum Atom j, ein Elektron vom Atom j zum Atom l und schließlich ein Elektron vom Atom l zum Atom i. Bei vier Atomen handelt es sich dabei um unabhängige Transpositionen innerhalb zweier getrennter Atompaare. Außerdem gibt es auch bei vier Atomen eine cyclische Vertauschung:
$i \to j \to l \to m \to i$.

Betrachten wir den Beitrag der cyclischen Vertauschung zwischen drei Atomen. Unter Verwendung von (6.12) schreiben wir:

$$\hat{P}^\sigma_{ijl} = \hat{P}^\sigma_{lj} \hat{P}^\sigma_{ji} = \tfrac{1}{4} + \vec{s}_i \vec{s}_j + \vec{s}_j \vec{s}_l + \vec{s}_l \vec{s}_i + 2i\,\vec{s}_i [\vec{s}_j \times \vec{s}_l]. \tag{11.14}$$

Wiederholt man hier die Rechnungen aus Ziff. 7, so führen die Vertauschungen vom Typ (11.14) zum effektiven Spin-Hamilton-Operator nach Dirac. Das Austauschintegral nimmt dabei folgende Form an:

$$\begin{aligned} J = -\tfrac{1}{2} \left(\prod_{\lambda=1}^N M_\lambda \right)^{-1} \sum_{\xi_i \xi'_i}^{M_i} \sum_{\xi_j \xi'_j}^{M_j} f_i(\xi_i, \xi'_i) f_j(\xi_j, \xi'_j) \times \\ \times \sum_{\xi_1 \ldots \xi_N (\neq \xi_i, \xi_j)} \langle \varphi_{\xi_1} \ldots \varphi_{\xi_{i-1}} \varphi_{\xi'_i} \varphi_{\xi_{i+1}} \ldots \varphi_{\xi_{j-1}} \varphi_{\xi'_j} \varphi_{\xi_{j+1}} \ldots \varphi_{\xi_N} (\hat{H} - E) \times \\ \times \left(\sum_{l(l>i,j)} v_{ijl} \hat{P}_{ijl} + \sum_{l(l<i,j)} v_{lij} \hat{P}_{lij} + \sum_{l(i<l<j)} v_{ilj} \hat{P}_{ilj} \right) \times \\ \times \varphi_{\xi_1} \varphi_{\xi_{i-1}} \varphi_{\xi_i} \varphi_{\xi_{i+1}} \ldots \varphi_{\xi_{j-1}} \varphi_{\xi_j} \varphi_{\xi_{j+1}} \ldots \varphi_{\xi_N} \rangle_r, \end{aligned} \tag{11.15}$$

wobei $v_{lij} = n_l n_i n_j$ bedeutet. Das Glied mit den drei Spinoperatoren [das letzte in der Formel (11.14)] liefert in Übereinstimmung mit allgemeinen Überlegungen keinen Beitrag. Die Beiträge der übrigen erwähnten Vertauschungen lauten

$$\sum_{i>j} K_{ij} (\vec{S}_i \vec{S}_j)^2, \quad \sum_{i>j>l} A_{ijl} (\vec{S}_i \vec{S}_j)(\vec{S}_j \vec{S}_l), \quad \sum_{i>j>l>m} B_{ijlm} (\vec{S}_i \vec{S}_j)(\vec{S}_l \vec{S}_m), \tag{11.16}$$

wobei die Koeffizienten K, A und B eine recht komplizierte Struktur besitzen. Der Koeffizient K_{ij} wird beispielsweise durch den Ausdruck

$$\begin{aligned} K_{ij} = \frac{n_i(n_i-1)\,n_j(n_j-1)}{M_i M_j} \sum_{\xi_i \xi'_i} \sum_{\xi_j \xi'_j} h_i(\xi_i \xi'_i)\, h_j(\xi_j \xi'_j) \times \\ \times \langle \varphi_{\xi_1} \ldots \varphi_{\xi_{i-1}} \varphi_{\xi'_i} \varphi_{\xi_{i+1}} \ldots \varphi_{\xi_{j-1}} \varphi_{\xi'_j} \varphi_{\xi_{j+1}} \ldots \varphi_{\xi_N} (\hat{H} - E) \hat{P}' \times \\ \times \varphi_{\xi_1} \ldots \varphi_{\xi_{i-1}} \varphi_{\xi'_i} \varphi_{\xi_{i+1}} \ldots \varphi_{\xi_{j-1}} \varphi_{\xi'_j} \varphi_{\xi_{j+1}} \ldots \varphi_{\xi_N} \rangle_r \end{aligned} \tag{11.17}$$

wiedergegeben, worin die $h_i(\xi_i \xi'_i)$ die Analoga zu den Wigner-Eckart-Koeffizienten aus (7.19) und (7.20) darstellen. Bei $S = \tfrac{1}{2}$ tritt ein biquadratisches Glied im engeren Sinne nicht auf: $K_{ij} = 0$.

Wir ersehen also aus (11.16), daß die Berücksichtigung höherer Vertauschungen bei der Berechnung der Matrixelemente für die Säkulargleichung im Spin-Hamilton-Operator zu Gliedern führt, welche die allgemeinste biquadratische Form in den Spinoperatoren enthalten.

Theoretische Untersuchungen zur biquadratischen Wechselwirkung sowie deren Berücksichtigung innerhalb des Problems der indirekten Wechselwirkung und bei der Erklärung der magnetischen Eigenschaften realer Stoffe findet man in Originalarbeiten von ANDERSON [47], KITTEL [117], BEAN und RODBELL [118], HARRIS und OWEN [119], RODBELL u.a. [120], WILL u.a. [121], HUANG und ORBACH [122].

12. Einige Bemerkungen. In Ziff. 4—11 ist der Spin-Hamilton-Operator nach DIRAC und VAN VLECK abgeleitet sowie das Austauschintegral zahlenmäßig abgeschätzt worden. Grundlage war das Heisenberg-Modell des Ferromagnetismus, eine Verallgemeinerung der Heitler-London-Methode für den Kristall. Kennzeichnend für die Heitler-London-Methode ist, daß die Basisfunktionen der Säkulargleichung in der Form antisymmetrisierter Produkte der Wellenfunktionen isolierter Atome eingesetzt werden [s. z.B. (7.1)].

In Ziff. 3 wurde bereits auf eine Schwäche der Heitler-London-Methode bei der Anwendung auf das Wasserstoffmolekül hingewiesen. Da nämlich diese Methode die Korrelation der Elektronenbewegung nicht berücksichtigt, also das gegenseitige Ausweichen der Elektronen im Raum als Folge der elektrostatischen Abstoßung vernachlässigt, wird der Beitrag der Elektron-Elektron-Wechselwirkung zur Singulett-Triplett-Aufspaltung viel zu groß, wenn der Kernabstand $R \to \infty$ geht. Obwohl beim „Verdünnen" die Energie des Systems dem richtigen Wert zustrebt (der Energie zweier isolierter Atome), wächst der Fehler bei der Bestimmung der Singulett-Triplett-Aufspaltung und überschreitet bei einem gewissen Wert des Abstandes R die wahre Größe der Aufspaltung. Die Versuche verschiedener Autoren, die Heitler-London-Methode zu verbessern, wurden von HERRING in seiner Übersichtsarbeit eingehend behandelt [85, Ch. 5].

Ein Weg zur Verbesserung der Heitler-London-Methode besteht in der Anwendung der Methode der Konfigurationswechselwirkung, bei der neue Konfigurationen entstehen durch den virtuellen Übergang eines Elektrons von *einem* Atom zu einer einfach besetzten Bahn eines *anderen* Atoms [59], [66] (s. Ziff. 2). Jedoch wird dieses Heitler-London-Schema, das durch polare Zustände ergänzt ist, auch keine asymptotisch exakte Lösung liefern. Beim Wasserstoffmolekül kann man sich unmittelbar davon überzeugen, indem man im Ausdruck für die Energie die bekannten Integrale mit 1s-Wasserstofffunktionen verwendet [67]. Für Abstände in der Nähe des Gleichgewichts-Kernabstandes im Wasserstoffmolekül werden durch die Berücksichtigung polarer Zustände die Ergebnisse etwas verbessert [123], [66].

MATTHEISS [124] führte zahlenmäßige Berechnungen für ein gleichseitiges Sechseck aus Wasserstoffatomen durch, wobei er 1s-Funktionen verwendete und alle möglichen Ionisationsgrade berücksichtigte. Die Rechnungen zeigten, daß das gewöhnliche Heitler-London-Verfahren und die Methode, die polare Zustände einbezieht, für Kernabstände oberhalb $\sim 4\,a_H$ gut miteinander übereinstimmen.

Eine Verbesserung der Heitler-London-Methode durch die Berücksichtigung angeregter Zustände bringt die Gefahr mit sich, daß man im Ausdruck für das Austauschintegral einen falschen Exponentialfaktor erhält, falls man nur eine begrenzte Zahl von Konfigurationen in Betracht zieht. Man muß aber die Unzahl der angeregten Eigenzustände des Atoms vollständig berücksichtigen, einschließlich der Zustände des kontinuierlichen Spektrums [125].

HERRING [126], [85] schlug ein Verfahren zur Verbesserung der Heitler-London-Methode vor, das zu einer asymptotisch richtigen Lösung führen sollte. Die Idee hierbei besteht darin, das wahre Kristallpotential so genau wie möglich zu berücksichtigen. Der Gedankengang ist dabei folgender: Zum Auffinden der $\nu = \prod_i (2S_i + 1)$ Wurzeln einer Säkulargleichung vom Typ (4.9), die die ν niedrigsten exakten Eigenwerte eines Kristalls aus n Elektronen und N Atomen annähern, verwendet man in der Heitler-London-Methode Basisfunktionen der Form

$$\Psi_k = \sum_P (-1)^P \hat{P} \prod_{i=1}^N U_{k_i}, \tag{12.1}$$

wobei die U_{k_i} Wellenfunktionen der isolierten Atome bedeuten. Wären die exakten Eigenfunktionen bekannt, die zu den ν niedrigsten Eigenwerten gehören, dann wäre das Analogon der Basisfunktionen (12.1) ein Ausdruck der Form

$$\sum_P (-1)^P \hat{P} \Phi_k. \tag{12.2}$$

Hierin stellt Φ_k im Gegensatz zum Heitler-London-Produkt $U_k = \prod_{i=1} U_{k_i}$ eine exakte Funktion dar. Die Funktionen Φ_k selbst entsprechen keinem der Zustände des Systems, da sie nicht antisymmetrisch hinsichtlich der Vertauschungen der Orts- und Spinkoordinaten sind. Die exakten Eigenfunktionen werden erst durch die Φ_k und ihnen entsprechende Funktionen mit vertauschten Argumenten ausgedrückt. Bei großen Kernabständen werden die Funktionen Φ_k und U_k einander ähnlich. Bei den Funktionen Φ_k sind, ähnlich wie bei den U_k, die Elektronen 1 bis n_1 am Kern mit der Nummer 1, die Elektronen n_1+1 bis n_1+n_2 am Kern mit der Nummer 2 usw. weitgehend lokalisiert. Da eine Berechnung der Funktionen Φ_k ebenso kompliziert ist wie die der Eigenfunktionen selbst, ist man für ihre praktische Ermittlung auf Näherungen angewiesen. Die Funktionen Φ_k sind am genauesten in dem Gebiet des Konfigurationsraumes zu bestimmen, das den größten Beitrag zur van-der-Waalsschen Energie liefert, d.h. die gegenseitige Polarisation der Atome muß sehr genau erfaßt werden. Zwar ist auch in einem begrenzten Gebiet des Konfigurationsraumes eine exakte Bestimmung von Φ_k ebenso kompliziert wie das Auffinden der Gesamtlösung, doch kann man hier ein Näherungspotential dergestalt bestimmen, daß der Unterschied gegenüber der Gleichung für U_k als kleine Störung behandelt werden kann. Das Heitler-London-Produkt U_k in (12.1) stellt die Eigenfunktion eines Hamilton-Operators mit einem Potentialglied V_0 dar, das die Coulomb-Wechselwirkungen der Elektronen, die anfangs bestimmten Atomen zugeschrieben werden, mit ihrem Kern und die Coulomb-Wechselwirkungen der Elektronen untereinander innerhalb ihres Atoms enthält. In V_0 gehen nicht die Glieder der Coulomb-Wechselwirkung der Elektronen mit fremden Kernen und mit den Elektronen fremder Atome ein. Eine genauere Funktion als U_k kann man erhalten als Eigenfunktion eines Hamilton-Operators mit einem künstlichen Potential, das jedes Elektron in der Nähe seines eigenen Atoms festhalten und zugleich dem wahren Potential der wechselwirkenden Atome weitestgehend ähneln sollte. HERRING, der dieses Potential „verstümmelt" (truncated) nennt, konstruiert es so, daß es im größten Teil des $3n$-dimensionalen Konfigurationsraumes mit dem wahren Kristallpotential übereinstimmt, in gewissen Gebieten des Raumes aber so verändert wird, daß die gesuchte Näherungsfunktion für Φ_k, ähnlich wie die Produkte U_k, die Lokalisierung der Elektronen an ihren Atomen gut beschreibt.

Konkret wird dieses Potential wie folgt bestimmt. Wir bezeichnen als Ausgangs-Potentialmulde das Gebiet des Phasenraumes, in dem sich die Elektronen in der Nähe der eigenen Atome befinden (die Elektronen 1 bis n_1 am Atom 1, die Elektronen n_1+1 bis n_1+n_2 am Atom 2 usw.). Diese Ausgangs-Potentialmulde geht unter Wirkung der Vertauschungsoperatoren der Gruppe G_0, welche die Elektronenkoordinaten innerhalb der Atome vertauschen, in sich selbst über. P nennen wir die Potentialmulde im Konfigurationsraum, die man aus der

Ausgangs-Potentialmulde unter der Wirkung eines Operators $\hat{P}$ erhält, der nicht zur Gruppe G_0 gehört. Da G_0 von der Ordnung $\prod\limits_{i=1}^{N} n_i!$ ist, gibt es $n! \Big/ \prod\limits_{i=1}^{N} n_i!$ verschiedene Potentialmulden. Wir bestimmen $(3n-1)$-dimensionale Hyperflächen Σ_P im Konfigurationsraum mit Hilfe der Gleichungen

$$\max_m |\vec{r}_m - \vec{R}_m^P| = L. \tag{12.3}$$

Dabei bedeutet $\vec{r}_m$ die Ortskoordinate des m-ten Elektrons und $\vec{R}_m^P$ die Lage des Kerns, in dessen Nähe sich das m-te Elektron in der Potentialmulde P befindet. Der Abstand L muß so klein wie möglich gewählt werden und der Forderung genügen, daß der niedrigste Eigenwert für ein System aus n Elektronen, von denen eines in einem Abstand $\geq L$ vom nächsten Kern fixiert ist, eine Größe $E_0 + \eta\, \varepsilon_1$ übersteigt. Dabei ist E_0 die Energie des Grundzustandes des verdünnten Systems der Atome, ε_1 das kleinste Ionisationspotential eines Atoms und η eine Größe von der Ordnung 1. Dann kann ein Potential V' konstruiert werden gemäß

$$\left. \begin{array}{ll} V' = V & \text{innerhalb } \Sigma_1 \\ V' = V & \text{außerhalb aller } \Sigma_P \\ V' = E_0 + \varepsilon_1 & \text{innerhalb } \Sigma_P \text{ bei } P \neq 1, \end{array} \right\} \tag{12.4}$$

wobei V das wahre Potential der wechselwirkenden Atome darstellt. Die Tatsache, daß die Potentialmulden innerhalb der Flächen Σ_P abgeschnitten sind [dritte Zeile in (12.4)], führt zu einer Lokalisierung der Elektronen an den Atomen. Die Eigenfunktionen χ_k eines Hamilton-Operators mit dem Potential V' aus (12.4) beschreiben diese Lokalisierung und nähern die exakten Lösungen Φ_k an. HERRING untersucht anschließend ausführlicher ihre Eigenschaften und verwendet sie zur Abschätzung des Fehlers bei der Bestimmung der Eigenwerte des untersuchten Systems. Die Funktionen χ_k sind antisymmetrisch hinsichtlich der Vertauschungen der Gruppe G_0. Wie auch im Heitler-London-Produkt U_k kann man in χ_k die Spinindizes k_1 der Gesamtheit der Elektronen mit den Nummern 1 bis n_1 zuschreiben, die Indizes k_2 den Elektronen mit den Nummern n_1+1 bis n_1+n_2 usw. Der Index k bezeichnet dann die Gesamtheit $\{k_1, \ldots, k_N\}$. Die χ_k entsprechenden Eigenwerte E_χ hängen nicht von k ab.

Da das Hauptanliegen in der Gewinnung einer asymptotisch exakten Lösung besteht, ist es wesentlich, die Form des Abfallens der Funktionen χ_k bei großen Werten des Abstandes im Konfigurationsraum zu bestimmen. Obwohl dies keine leichte Aufgabe ist, gelangt HERRING durch detaillierte, allerdings mathematisch nicht vollkommen strenge Untersuchungen zu dem Schluß, daß der Abfall von χ_k mit Hilfe einer Exponentialform $\exp(-F)$ beschrieben werden kann. Für F gilt dabei

$$\sum_m \alpha_i(m) |\vec{r}_m - \vec{R}_i(m)| \geq F \geq \alpha_{\min} \sum_m |\vec{r}_m - \vec{R}_i(m)|, \tag{12.5}$$

worin $\vec{R}_i(m)$ die Lage des Kerns bezeichnet, zu dem das m-te Elektron gehört. Die Parameter $\alpha_i(m)$ hängen mit dem Ionisationspotential des Atoms zusammen und beschreiben den Abfall der Funktion des freien Atoms. Die Bestimmung von χ_k als Eigenfunktion des Hamilton-Operators mit dem Potential (12.4) und die Aussage (12.5) über das asymptotische Verhalten gestatten, mit Hilfe einer leicht modifizierten Formel von KATO [127] den Fehler bei der Berechnung der Energieniveaus des Systems für große Abstände zwischen den Atomen zu ermitteln. Dieser Fehler wird durch eine Formel $\Omega e^{-4\alpha R}$ bestimmt, worin Ω eine algebraische Funktion des Kernabstandes bedeutet und αR gewöhnlich das kleinste Produkt eines α_i mit dem Abstand eines Atoms i vom Nachbaratom darstellt.

Die Ableitung eines Spin-Hamilton-Operators der Form (6.1) läuft bei Verwendung der χ_k an Stelle der Heitler-London-Funktionen U_k vollkommen analog der Darstellung in Ziff. 7. Die von HERRING erhaltene Form des Austauschintegrals stimmt mit dem Ausdruck (7.22) überein bis auf *einen* Unterschied, daß nämlich die Produkte der Funktionen φ_{ξ_i} (der Koordinatenanteile der Wellenfunktionen der einzelnen Atome) durch Funktionen $\varphi_{\xi_1\ldots\xi_N}(\vec{r}_1, \ldots, \vec{r}_N)$ ersetzt sind, die von den Ortskoordinaten aller n Elektronen der N Atome abhängen.

Da bei der Ableitung des Spin-Hamilton-Operators Vertauschungen von Elektronen verschiedener Atome bis zur höchsten Ordnung berücksichtigt werden, hat im stark aufgeweiteten Kristall das Austauschintegral die Größenordnung $\exp(-2\alpha R)$. Da ferner der Fehler bei der Bestimmung der Niveaudifferenz nach HERRING proportional $\exp(-4\alpha R)$ war, folgt hieraus, daß die Eigenwerte des Spin-Hamilton-Operators asymptotisch exakt sind.

Wie am Ende von Ziff. 6 im Rahmen der Heitler-London-Methode gezeigt wurde, ist bei großen Aufweitungen der Parameter der Austauschkopplung zwischen zwei neutralen Atomen eine Funktion des Abstandes zwischen diesen beiden Atomen und ist unabhängig von der Anwesenheit der anderen Atome des Kristalls. Diese Additivität der Austauschkopplung zwischen den einzelnen Paaren ist durch die Neutralität der Atome sowie durch die Kleinheit der Überlappung ihrer Wellenfunktionen bedingt. Bei der Herring-Methode ist die Additivität der Austauschkopplungen zwischen Paaren wegen der Berücksichtigung der van-der-Waalsschen Polarisationseffekte asymptotisch nur bis auf Glieder gewährleistet, die proportional zu negativen Potenzen des Kernabstandes sind, aber nicht exponentiell klein wie bei der Heitler-London-Methode mit nichtpolarisierten Atomen. Wie bereits in Ziff. 3 vermerkt wurde [s. z.B. (3.9) und (3.14)], schlug HERRING ein vorteilhaftes Verfahren zur Darstellung des Austauschparameters mit Hilfe der Integrale über die $(3n-1)$-dimensionale Hyperfläche Σ vor.

Es sei betont, daß nach HERRING eine Darstellung der Austauschparameter durch Flächenintegrale innerhalb des Heitler-London-Schemas nicht möglich ist.

Durch die Untersuchung der in den Ausdruck für das Austauschintegral eingehenden Funktionen Φ, $P\Phi$, $\nabla\Phi$ in jenen Gebieten der Hyperfläche Σ, die den Hauptbeitrag zum Flächenintegral liefern, konnte HERRING nachweisen, daß vorwiegend eine antiferromagnetische Kopplung realisiert wird. Dabei wird auf die drei wichtigsten Umstände hingewiesen, die das Vorzeichen des Austauschintegrals bestimmen: 1. das Gebiet, das bei der Integration den Hauptbeitrag liefert, 2. das Verhältnis der Vorzeichen von Φ und $P\Phi$, 3. der Einfluß des Vorzeichenunterschiedes bei den Wigner-Eckart-Koeffizienten f [s. (7.19)]. Die Betrachtung gestattet den allgemeinen Schluß, daß in einem beliebigen System identischer Atome oder Moleküle ohne Bahnentartung der Austauschparameter J_{ij} für nächste Nachbarn im Bereich großer Kernabstände mit großer Wahrscheinlichkeit negativ ist, also den Antiferromagnetismus begünstigt.

Die Schlußfolgerung, daß hauptsächlich eine antiferromagnetische Kopplung besteht, braucht im Prinzip bei Bahnentartung und bei realen Kernabständen nicht zu gelten.

Insgesamt gesehen, ist sowohl innerhalb der Heitler-London-Methode als auch innerhalb der Herring-Methode die Realisierung einer ferromagnetischen Kopplung relativ schwierig. LIEB und MATTIS [*128*] haben streng bewiesen, daß in einem eindimensionalen System von Teilchen stets die Beziehung $E(S) \leq E(S+1)$ erfüllt wird, wobei S den Spin des Systems bedeutet. Obwohl der eindimensionale Fall keinerlei Beziehung zu den realen Kristallen besitzt, beweist dieses Beispiel ein weiteres Mal die Bevorzugung einer antiferromagnetischen Kopplung.

III. Das Vektormodell bei orthogonalisierten Wellenfunktionen.

13. Die Orthogonalisierung der Einelektronen-Atomfunktionen und die „angeregten" Konfigurationen. Da die Verwendung orthogonaler Bahnen für praktische Berechnungen sehr vorteilhaft ist, wurden verschiedene Verfahren zur Konstruktion solcher Bahnen entwickelt. In der Quantenchemie werden in großem Umfang orthogonale Molekülbahnen benutzt, die als lineare Kombinationen von Atombahnen aufgebaut werden (MO-LCAO-Methode); bei den Untersuchungen der Elektronenstruktur fester Körper verwendet man die Methode der orthogonalisierten ebenen Wellen (OPW-Methode), die Methoden der Wannier-Funktionen und der Bloch-Wellen sowie eine spezielle Methode von BOGOLJUBOV und LÖWDIN zum Auffinden orthogonaler lokalisierter Funktionen durch eine Entwicklung nach dem Nichtorthogonalitätsintegral. Obwohl die Methoden mit orthogonalen Bahnen wenigstens in einem gewissen Stadium der Berechnungen von Vorteil sind, so ist doch bei ihrer Anwendung Vorsicht geboten. So führt im einfachsten Fall des Wasserstoffmoleküls die Beschreibung des Systems mit Hilfe lediglich zweier einfach besetzter orthogonaler Molekülbahnen (2.1) zu einem falschen Vorzeichen für die Singulett-Triplett-Aufspaltung und ergibt keine chemische Bindung. Die Verwendung orthogonaler Bahnen führt in der Einkonfigurationsnäherung auch bei der Behandlung des Magnetismus von Kristallen zu einem absurden Ergebnis, da wegen des positiv definiten Austauschintegrals alle Stoffe Ferromagnetika sein müßten. Daß die Einkonfigurationsnäherung bei der Methode der orthogonalen Bahnen außerordentlich nachteilig ist und daß deshalb unbedingt zusätzliche Konfigurationen in die Betrachtung einbezogen und die Wechselwirkungen zwischen ihnen berücksichtigt werden müssen, wurde bereits unter Ziff. 2 besprochen. Insbesondere wurde dort darauf hingewiesen, daß bei Einführung polarer Zustände die Methode der orthogonalen Molekülbahnen qualitativ richtige Resultate liefert, die mit den unter Verwendung nichtorthogonaler Bahnen (Heitler-London-Methode) gewonnenen Ergebnissen übereinstimmen, wenn beim letztgenannten Verfahren auch die polaren Zustände berücksichtigt werden. Unten wird unter Ziff. 14, 15 und 16 eine spezielle Methode von BOGOLYUBOV in der Störungstheorie für orthogonale Bahnen dargestellt und der Zusammenhang mit den Ergebnissen der Heitler-London-Methode aufgezeigt werden.

Hier soll als Einleitung zu der unter Ziff. 14 bis 16 folgenden Darlegung der Beweis für einen Satz erbracht werden, nach dem die Energie eines Systems sich beim Übergang von der Beschreibung durch ein vollständiges System nichtorthogonaler Funktionen zur Beschreibung durch ein vollständiges orthonormiertes Funktionensystem nicht ändert [96]. Das orthonormierte Funktionensystem $\{\Theta_\alpha(\vec{r}, \sigma)\}$ und das System der nichtorthogonalen Funktionen $\{\varphi_\alpha(\vec{r}, \sigma)\}$ sind äquivalent, wenn jede Funktion Θ_α eine Linearkombination von Funktionen φ_α darstellt und umgekehrt, wenn also eine Transformationsmatrix $\|A\|$ und die zugehörige reziproke Matrix existieren:

$$\left.\begin{aligned}\Theta_\alpha(\vec{r}, \sigma) &= \sum_{\alpha'} A_{\alpha\alpha'} \varphi_{\alpha'}(\vec{r}, \sigma), \\ \varphi_\alpha(\vec{r}, \sigma) &= \sum_{\alpha'} A^{-1}_{\alpha\alpha'} \Theta_{\alpha'}(\vec{r}, \sigma).\end{aligned}\right\} \quad (13.1)$$

Es sei $\Psi(\vec{r}_1, \sigma_1, \ldots, \vec{r}_n, \sigma_n)$ eine exakte Vielelektronenfunktion des Systems. Wir schreiben sie als Entwicklung nach den Slater-Determinanten, die aus den nichtorthogonalen Funktionen φ_α aufgebaut sind:

$$\Psi(\vec{r}_1, \sigma_1, \ldots, \vec{r}_n, \sigma_n) = \sum_{\alpha_1 \ldots \alpha_n} K(\alpha_1 \ldots \alpha_n) \Psi_{\alpha_1 \ldots \alpha_n}(\vec{r}_1 \sigma_1, \ldots, \vec{r}_n \sigma_n), \quad (13.2)$$

wobei die $K(\alpha_1 \ldots \alpha_n)$ unbekannte Entwicklungskoeffizienten und die $\Psi_{\alpha_1\ldots\alpha_n}$ Slater-Determinanten

$$\Psi_{\alpha_1\ldots\alpha_n} = \sum_P (-1)^P \hat{P}\{\varphi_{\alpha_1}(\vec{r}_1, \sigma_1) \ldots \varphi_{\alpha_n}(\vec{r}_n, \sigma_n)\} \tag{13.3}$$

darstellen. Die Entwicklungskoeffizienten $K(\alpha_1 \ldots \alpha_n)$ in (13.2) werden durch ein Gleichungssystem

$$\langle \Psi_{\alpha_1\ldots\alpha_n}(\hat{H}-E)\Psi\rangle = 0 \tag{13.4}$$

bestimmt.

Der Beweis des Theorems besteht nun in folgendem: Jede Funktion (13.2), deren Entwicklungskoeffizienten die Gl. (13.4) befriedigen, kann in der Form

$$\Psi = \sum_{\alpha_1\ldots\alpha_n} Q(\alpha_1 \ldots \alpha_n) \Theta_{\alpha_1\ldots\alpha_n} \tag{13.5}$$

dargestellt werden, wobei

$$\Theta_{\alpha_1\ldots\alpha_n} = \sum_P (-1)^P \hat{P} \Theta_{\alpha_1}(\vec{r}_1, \sigma_1) \ldots \Theta_{\alpha_n}(\vec{r}_n, \sigma_n) \tag{13.6}$$

ist und die Koeffizienten Q einer Gleichung

$$\langle \Theta_{\alpha_1\ldots\alpha_n}(\hat{H}-E)\Psi\rangle = 0 \tag{13.7}$$

genügen. Unter Verwendung der Beziehung (13.1) schreiben wir

$$\Psi_{\alpha_1\ldots\alpha_n} = \sum_{\alpha'_1\ldots\alpha'_n} A^{-1}_{\alpha_1\alpha'_1} \ldots A^{-1}_{\alpha_n\alpha'_n} \Theta_{\alpha'_1\ldots\alpha'_n}. \tag{13.8}$$

Hieraus folgt, daß die Funktion (13.2) tatsächlich die Form (13.5) annimmt mit den Koeffizienten

$$Q(\alpha_1 \ldots \alpha_n) = \sum_{\alpha'_1\ldots\alpha'_n} K(\alpha'_1 \ldots \alpha'_n) A^{-1}_{\alpha'_1\alpha_1} \ldots A^{-1}_{\alpha'_n\alpha_n}. \tag{13.9}$$

Um uns davon zu überzeugen, daß Ψ der Gl. (13.7) gehorcht, kehren wir die Beziehung (13.8) um, was

$$\Theta_{\alpha_1\ldots\alpha_n} = \sum_{\alpha'_1\ldots\alpha'_n} A_{\alpha_1\alpha'_1} \ldots A_{\alpha_n\alpha'_n} \Psi_{\alpha'_1\ldots\alpha'_n} \tag{13.10}$$

ergibt. Mit Hilfe von (13.10) und (13.4) erhalten wir

$$\langle \Theta_{\alpha_1\ldots\alpha_n}(\hat{H}-E)\Psi\rangle = \sum_{\alpha'_1\ldots\alpha'_n} A^*_{\alpha_1\alpha'_1} \ldots A^*_{\alpha_n\alpha'_n} \langle \Psi_{\alpha'_1\ldots\alpha'_n}(\hat{H}-E)\Psi\rangle = 0,$$

und damit ist das Theorem bewiesen.

14. Die Entwicklung nach dem Nichtorthogonalitätsintegral und die Operatorform der Störungstheorie nach Bogolyubov. Bekanntlich ([96]) ist die Säkulargleichung, die man bei der Entwicklung der Vielelektronenwellenfunktion des Systems nach den antisymmetrisierten Produkten der Einelektronenfunktionen erhält, in der Darstellung der zweiten Quantelung folgendem Hamilton-Operator äquivalent:

$$\hat{H} = \sum_{\alpha\alpha'} L(\alpha, \alpha') a^+_\alpha a_{\alpha'} + \frac{1}{2} \sum_{\alpha_1\alpha_2\alpha'_1\alpha'_2} F(\alpha_1, \alpha_2, \alpha'_1, \alpha'_2) a^+_{\alpha_1} a^+_{\alpha_2} a_{\alpha'_2} a_{\alpha'_1}. \tag{14.1}$$

Dabei bezeichnen die Indizes α und α' Einelektronenzustände (d.h. „Bahn" und z-Projektion des Spins); die Größen L und F stellen die Matrixelemente der Ein-bzw. Zweiteilchenterme im Hamilton-Operator dar:

$$L(\alpha, \alpha') = \int \Theta_\alpha^* \hat{H}_1(\vec{r}) \Theta_{\alpha'} d\tau, \tag{14.2}$$

$$F(\alpha_1, \alpha_2, \alpha_1', \alpha_2') = \int \Theta_{\alpha_1}^* \Theta_{\alpha_2}^* V(\vec{r}_1, \vec{r}_2) \Theta_{\alpha_1'} \Theta_{\alpha_2'} d\tau_1 d\tau_2. \tag{14.3}$$

Die Einelektronenfunktionen Θ_α bilden einen vollständigen Satz orthonormierter Wellenfunktionen. Die a_α^+ und a_α sind Fermi-Operatoren, die der Erzeugung bzw. Vernichtung eines Elektrons im Zustand α entsprechen und folgenden Vertauschungsbeziehungen gehorchen:

$$a_\alpha^+ a_{\alpha'} + a_{\alpha'} a_\alpha^+ = \delta_{\alpha\alpha'}, \quad a_\alpha^+ a_{\alpha'}^+ + a_{\alpha'}^+ a_\alpha^+ = 0, \quad a_\alpha a_{\alpha'} + a_{\alpha'} a_\alpha = 0. \tag{14.4}$$

$$a_\alpha^+ a_\alpha = N_\alpha \tag{14.5}$$

ist der Operator der Zahl der Teilchen im Zustand α und kann lediglich zwei Werte annehmen: 0 und 1.

Indem wir in den Zuständen α den Spinindex σ und den „Bahnindex" f explizit unterscheiden und uns den Umstand zunutze machen, daß der Ausgangs-Hamilton-Operator des Systems selbst keine Spinoperatoren enthält, schreiben wir (14.1) um in die Form

$$\hat{H} = \sum_{f f' \sigma} L(f f') a_{f\sigma}^+ a_{f'\sigma} + \frac{1}{2} \sum_{f_1 f_2 f_1' f_2' \sigma_1 \sigma_2} F(f_1 f_2 f_1' f_2') a_{f_1\sigma_1}^+ a_{f_2\sigma_2}^+ a_{f_2'\sigma_2} a_{f_1'\sigma_1}, \tag{14.6}$$

wo der Index σ zwei Werte, $\frac{1}{2}$ und $-\frac{1}{2}$, annimmt. Für die Anwendung des Hamilton-Operators (14.6) auf den Kristall wird angenommen, daß die Indices f die Gesamtheit der Atomnummern (Gitterpunkte) und der „intraatomaren" Quantenzahlen darstellen. In diesem Abschnitt behandeln wir den Fall eines Elektrons pro Atom, wobei jedes der Elektronen sich stets in einem Atomquantenzustand $1s$ befindet und lediglich ein Typ virtueller Übergänge möglich ist, nämlich die Bildung von $1s$-Elektronenpaaren mit entgegengesetztem Spin an dem einen oder anderen Gitterpunkt. Aus diesem Grunde wird der Index f hier nur die Atomnummer bezeichnen.

Die Einelektronenatomfunktionen $\varphi_f(\vec{r})$ sind nicht orthogonal, es existiert vielmehr das Überlappungsintegral[21]

$$S_{ff'} = \int \varphi_f^*(\vec{r}) \varphi_{f'}(\vec{r}) d\vec{r} - \delta_{ff'}. \tag{14.7}$$

Es ist in Ziff. 13 gezeigt worden, daß die Energie des Systems sich gegenüber einem Übergang von der Beschreibung des Einelektronenzustandes mit Hilfe nichtorthogonaler Funktionen zur Beschreibung mittels eines vollständigen Systemes orthogonaler Funktionen invariant verhält.

Wenn $\{\Theta_\alpha(\vec{r}, \sigma) = \Theta_f(\vec{r}) \chi(\sigma)\}$ einen vollständigen Satz zueinander *orthogonaler* Spin-Bahn-Funktionen darstellt und $\{u_\alpha(\vec{r}, \sigma) = \varphi_f(\vec{r}) \chi(\sigma)\}$ ein System *nicht-orthogonaler* spin- und ortsabhängiger Atomfunktionen, so existiert also eine lineare Transformation

$$\Theta_\alpha = \sum_\beta A_{\alpha\beta} u_\beta, \tag{14.8}$$

die die Energie des Systems unverändert läßt.

Die Transformationsmatrix $\|A_{\alpha\beta}\|$ gestattet eine Reihenentwicklung nach zunehmenden Potenzen des Nichtorthogonalitätsintegrals (14.7). So erhalten wir

---

[21] Anders als unter Ziff. 1 und 4 ist eine solche Definition des Überlappungsintegrals hier für die Rechnungen zweckmäßiger. Auf die exponentielle Abnahme des Integrals $S_{ff'}$ bei zunehmendem Abstand $|\vec{R}_f - \vec{R}_{f'}|$ sei hier ebenfalls hingewiesen.

für die Funktionen $\Theta_f(\vec{r})$, mit denen die Integrale $L(ff')$ und $F(f_1 f_2 f'_1 f'_2)$ in (14.6) berechnet werden, als erste Glieder der Entwicklung

$$\Theta_f(\vec{r}) = \varphi_f(\vec{r}) - \tfrac{1}{2} \sum_{f' \neq f} S_{f f'} \varphi_{f'}(\vec{r}) + \tfrac{3}{8} \sum_{f'} \sum_{f'' \neq f, f'} S_{f f''} S_{f'' f'} \varphi_{f'}(\vec{r}) + \cdots \quad (14.9)$$

Dabei wurde folgendermaßen vorgegangen: Die Matrixelemente $\|A_{\alpha\beta}\|$ in (14.8) stellt man als Reihe nach zunehmenden Potenzen des kleinen Parameters (14.7) dar. Die Koeffizienten der Reihenentwicklung werden dann aus der Orthogonalitätsbedingung für die Θ_α

$$\langle \Theta_\alpha \Theta_{\alpha'} \rangle = \delta_{\alpha \alpha'}$$

bestimmt.

Im Hamilton-Operator (14.6) wird der Zustand des Systems durch einen Satz orthogonaler Einelektronenfunktionen $\Theta_f(\vec{r})$ und nicht durch die nichtorthogonalen Atomfunktionen $\varphi_f(\vec{r})$ beschrieben. Wenn die Funktionen $\Theta_f(\vec{r})$ auch keine Atomfunktionen darstellen, wie in der Heitler-London-Näherung, so sind sie doch einem bestimmten Gitterpunkt zugeordnet und dort lokalisiert. Ihr Lokalisierungsgrad ist schwächer als bei den φ_f, da um jeden Gitterpunkt Oscillationen auftreten. Wir können aber nach wie vor von „atomaren" Einelektronenzuständen sprechen, ohne dabei zu vergessen, daß eine „Deformation" der wahren Atomfunktionen in die orthogonalen Funktionen $\Theta_f(\vec{r})$ erfolgt ist, die stärker über den Kristall „verschmiert" sind und eine Superposition aller Atomfunktionen φ_f darstellen.

Wenn man die Entwicklung (14.9) in die Ausdrücke für die Matrixelemente (14.2) und (14.3) einsetzt, erhält man eine Reihe, in der alle möglichen Potenzen des Nichtorthogonalitätsintegrals (14.7) vertreten sind, sowohl explizit als auch in der Form von Integralen, welche die nichtorthogonalen Atomfunktionen $\varphi_f(\vec{r})$ enthalten und dadurch von verschiedener Ordnung in (14.7) sind.

Es ist wesentlich zu bemerken, daß die Entwicklung der Größen $L(ff)$ und $F(ffff)$ mit den Gliedern nullter Ordnung in (14.7) beginnt. Die Elemente $L(ff')$ und $F(fgf'g)$ beginnen für $f \neq f'$ mit den Gliedern erster Ordnung, während $F(f_1 f_2 f'_1 f'_2)$ für $f_1 \neq f'_1$ und $f_2 \neq f'_2$ bei der zweiten Ordnung anfängt. Selbstverständlich handelt es sich hierbei nur um die jeweils *niedrigste* nichtverschwindende Ordnung, und alle angeführten Größen enthalten auch Glieder höherer Ordnung im Parameter (14.7).

Die Zerlegung der Ausdrücke $L(ff')$ und $F(f_1 f_2 f'_1 f'_2)$ nach Größen verschiedener Ordnung ermöglicht es, auch den Hamilton-Operator (14.6) nach dem kleinen Parameter zu entwickeln und eine Operatormethode der Störungstheorie zu entwickeln. Dabei wird der Operator (14.6) in der Form

$$\hat{H} = \hat{H}_0 + \varepsilon \hat{H}_1 + \varepsilon^2 \hat{H}_2 \quad (14.10)$$

dargestellt mit

$$\hat{H}_0 = \sum_f [L(ff) - \tfrac{1}{2} F(ffff)] N_f + \tfrac{1}{2} \sum_{f_1 f_2} F(f_1 f_2 f_1 f_2) N_{f_1} N_{f_2}, \quad (14.11)$$

$$\varepsilon \hat{H}_1 = \sum_{ff'\sigma} a^+_{f\sigma} [L(ff') + \sum_{f''} F(ff'' f' f'') N_{f''}] a_{f'\sigma}, \quad (14.12)$$

$$\varepsilon^2 \hat{H}_2 = \tfrac{1}{2} \sum_{\substack{f_1 f_2 f'_1 f'_2 \sigma_1 \sigma_2 \\ (f_1 \neq f'_1, f_2 \neq f'_2)}} F(f_1 f_2 f'_1 f'_2) a^+_{f_1 \sigma_1} a^+_{f_2 \sigma_2} a_{f'_2 \sigma_2} a_{f'_1 \sigma_1} \quad (14.13)$$

und

$$N_f = \sum_\sigma N_{f\sigma} = \sum_\sigma a^+_{f\sigma} a_{f\sigma}. \quad (14.14)$$

Der Exponent von ε bezeichnet die Ordnung des Gliedes, vor dem die entsprechende Potenz steht. Die Ausdrücke $\hat{H}_0$, $\hat{H}_1$ und $\hat{H}_2$ werden natürlich jeweils auch Glieder aller höheren Ordnungen im Parameter (14.7) enthalten. ε muß am Schluß der Rechnungen durch 1 ersetzt werden.

Wir bezeichnen die Eigenfunktion des Hamilton-Operators nullter Ordnung $\hat{H}_0$ mit C'. Es gilt

$$\hat{H}_0 C' = E' C'. \tag{14.15}$$

Aus (14.11) und (14.14) folgt, daß die Energieniveaus in nullter Näherung durch die Gesamtheit der Werte $\ldots N_f^0 \ldots$ der Besetzungszahlen $\ldots N_f \ldots$ bestimmt werden,

$$E' = E_{\ldots N_f^0 \ldots} = \sum_f [L(ff) - \tfrac{1}{2} F(ffff)] N_f^0 + \tfrac{1}{2} \sum_{f_1 f_2} F(f_1 f_2 f_1 f_2) N_{f_1}^0 N_{f_2}^0. \tag{14.16}$$

Im allgemeinen enthalten diese Gesamtheiten der Besetzungszahlen $\ldots N_f^0 \ldots$ die Zahlen $N_f^0 = 0$, 1 und 2 und genügen der Bedingung $\sum_f N_f^0 = n$, wenn n die Gesamtzahl der Elektronen im Kristall ist. Das Energieniveau E_0, für das alle Zahlen $N_f^0 = 1$ sind, ist selbstverständlich (in nullter Näherung) das niedrigste. Lösungen dieses Typs kann man „quasihomöopolar" nennen, da sie (im Heitler-London-Sinne) nicht exakt homöopolar sind, und zwar deshalb nicht, weil sie wegen der Verwendung der Funktion Θ_f an Stelle von φ_f eine „Beimengung" polarer Zustände enthalten (ausführlicher s. Ziff. 16). Zustände, bei denen neben $N_f^0 = 1$ auch die Besetzungszahlen $N_f^0 = 2$ auftreten sowie in gleicher Anzahl $N_f^0 = 0$, sind angeregt.

Zur Berücksichtigung des Einflusses der Glieder erster und zweiter Ordnung (14.12) und (14.13) muß man auf die Störungstheorie zurückgreifen. Es seien C_0 die Funktionen, die dem niedrigsten Energieniveau E_0 des Hamilton-Operators $\hat{H}_0$ entsprechen:

$$\hat{H}_0 C_0 = E_0 C_0. \tag{14.17}$$

Wir führen einen Operator $\hat{P}$ ein, der eine beliebige Eigenfunktion C des Operators $\hat{H}$ auf den linearen Raum der Funktionen C_0 projiziert:

$$\hat{P} C = C_0. \tag{14.18}$$

Folglich gilt für eine willkürliche Funktion C

$$\hat{H}_0 \hat{P} C = E_0 \hat{P} C. \tag{14.17'}$$

Jede beliebige Funktion C läßt sich darstellen als eine Summe

$$C = \hat{P} C + (1 - \hat{P}) C = C_0 + C_1. \tag{14.19}$$

Man kann dann die Wellengleichung mit dem Energieoperator (14.10) wie folgt schreiben:

$$(E - \hat{H}_0 - \varepsilon \hat{H}_1 - \varepsilon^2 \hat{H}_2) \hat{P} C + (E - \hat{H}_0 - \varepsilon \hat{H}_1 - \varepsilon^2 \hat{H}_2) C_1 = 0. \tag{14.20}$$

Multipliziert man weiterhin (14.20) von links mit dem Operator $\hat{P}$ und beachtet die folgenden Beziehungen für den Projektionsoperator

$$\hat{P}^2 = \hat{P}, \quad \hat{P} \hat{H}_0 = \hat{H}_0 \hat{P}, \quad \hat{P}(E - \hat{H}_0) C_1 = (E - \hat{H}_0) \hat{P}(1 - \hat{P}) C = 0, \tag{14.21}$$

so erhält man

$$(E - \hat{H}_0 - \varepsilon \hat{P} \hat{H}_1 \hat{P} - \varepsilon^2 \hat{P} \hat{H}_2 \hat{P}) \hat{P} C - \varepsilon \hat{P} \hat{H}_1 C_1 - \varepsilon^2 \hat{P} \hat{H}_2 C_1 = 0. \tag{14.22}$$

Zieht man (14.22) von (14.20) ab, dann gilt

$$(E - \hat{H}_0 - \varepsilon \hat{H}_1 - \varepsilon^2 \hat{H}_2 + \varepsilon \hat{P} \hat{H}_1 + \varepsilon^2 \hat{P} \hat{H}_2) C_1 + \\ + \varepsilon (\hat{P} \hat{H}_1 \hat{P} - \hat{H}_1) \hat{P} C + \varepsilon^2 (\hat{P} \hat{H}_2 \hat{P} - \hat{H}_2) \hat{P} C = 0. \quad (14.23)$$

Die Energie und die Korrektur zur Wellenfunktion setzen wir als Reihe nach Potenzen des kleinen Parameters ε an:

$$E = E_0 + \varepsilon \varDelta_0 + \cdots, \qquad C_1 = \varepsilon K + \varepsilon^2 L. \quad (14.24)$$

Unter Verwendung der Entwicklungen (14.24) finden wir auf der Basis von (14.23)

$$(E_0 - \hat{H}_0) K = (\hat{H}_1 - \hat{P} \hat{H}_1 \hat{P}) \hat{P} C, \\ (E_0 - \hat{H}_0) L = (\hat{H}_2 - \hat{P} \hat{H}_2 \hat{P}) \hat{P} C + (\hat{H}_1 - \hat{P} \hat{H}_1 - \varDelta_0) K. \quad (14.25)$$

Hieraus folgt

$$K = (\hat{H}_0 - E_0)^{-1} (\hat{P} \hat{H}_1 \hat{P} - \hat{H}_1) \hat{P} C, \quad (14.26)$$

$$L = (\hat{H}_0 - E_0)^{-1} (-\hat{H}_2 + \hat{P} \hat{H}_2 \hat{P}) \hat{P} C + \\ + (\hat{H}_0 - E_0)^{-1} (-\hat{H}_1 + \hat{P} \hat{H}_1 + \varDelta_0) (\hat{H}_0 - E_0)^{-1} (\hat{P} \hat{H}_1 \hat{P} - \hat{H}_1) \hat{P} C. \quad (14.27)$$

Wegen (14.24) und (14.17) erhält (14.22) jetzt die Form

$$(E - E_0 - \varepsilon \hat{P} \hat{H}_1 \hat{P} - \varepsilon^2 \hat{P} \hat{H}_2 \hat{P}) \hat{P} C - \varepsilon^2 \hat{P} \hat{H}_1 K - \varepsilon^2 \hat{P} \hat{H}_1 L - \\ - \varepsilon^3 \hat{P} \hat{H}_2 K + O(\varepsilon^4) = 0. \quad (14.28)$$

Wir bemerken weiterhin, daß wegen (14.21), (14.26) und (14.27) $\hat{P} K = \hat{P} L = 0$ gilt. Deshalb kann man $\hat{P} \hat{H}_1 K = \hat{P}(\hat{H}_1 - \hat{P} \hat{H}_1 \hat{P}) K$, $\hat{P} \hat{H}_1 L = \hat{P}(\hat{H}_1 - \hat{P} \hat{H}_1 \hat{P}) L$, $\hat{P} \hat{H}_2 K = \hat{P}(\hat{H}_2 - \hat{P} \hat{H}_2 \hat{P}) K$ schreiben und die Gl. (14.28) unter Berücksichtigung der Ausdrücke (14.26) und (14.27) wie folgt darstellen:

$$(E - E_0) \hat{P} C = \hat{P} \{ \varepsilon \hat{H}_1 + \varepsilon^2 \hat{H}_2 - \varepsilon^2 (\hat{H}_1 - \hat{P} \hat{H}_1 \hat{P}) (\hat{H}_0 - E_0)^{-1} (\hat{H}_1 - \hat{P} \hat{H}_1 \hat{P}) - \\ - \varepsilon^3 (\hat{H}_2 - \hat{P} \hat{H}_2 \hat{P}) (\hat{H}_0 - E_0)^{-1} (\hat{H}_1 - \hat{P} \hat{H}_1 \hat{P}) - \\ - \varepsilon^3 (\hat{H}_1 - \hat{P} \hat{H}_1 P) (\hat{H}_0 - E_0)^{-1} (\hat{H}_2 - \hat{P} \hat{H}_2 \hat{P}) + \\ + \varepsilon^3 (\hat{H}_1 - \hat{P} \hat{H}_1 \hat{P}) (\hat{H}_0 - E_0)^{-1} (\hat{H}_1 - \hat{P} \hat{H}_1 - \varDelta_0) \times \\ \times (\hat{H}_0 - E_0)^{-1} (\hat{H}_1 - \hat{P} \hat{H}_1 \hat{P}) + O(\varepsilon^4) \} \hat{P} C. \quad (14.29)$$

Hieraus erhalten wir für die Bestimmung der Wellenfunktion $C_0 = \hat{P} C$ als Gleichungen erster, zweiter und dritter Näherung:

$$(E - E_0) C_0 = \varepsilon \hat{P} \hat{H}_1 \hat{P} C_0, \quad (14.30)$$

$$(E - E_0) C_0 = \hat{P} \{ \varepsilon \hat{H}_1 + \varepsilon^2 \hat{H}_2 - \varepsilon^2 (\hat{H}_1 - \hat{P} \hat{H}_1 \hat{P}) (\hat{H}_0 - E_0)^{-1} \times \\ \times (\hat{H}_1 - \hat{P} \hat{H}_1 \hat{P}) \} \hat{P} C_0, \quad (14.31)$$

$$(E - E_0) C_0 = \hat{P} \{ \varepsilon \hat{H}_1 + \varepsilon^2 \hat{H}_2 - \varepsilon^2 (\hat{H}_1 - \hat{P} \hat{H}_1 \hat{P}) (\hat{H}_0 - E_0)^{-1} \times \\ \times (\hat{H}_1 - \hat{P} \hat{H}_1 \hat{P}) - \varepsilon^3 (\hat{H}_2 - \hat{P} \hat{H}_2 \hat{P}) (\hat{H}_0 - E_0)^{-1} (\hat{H}_1 - \hat{P} \hat{H}_1 \hat{P}) - \\ - \varepsilon^3 (\hat{H}_1 - \hat{P} \hat{H}_1 \hat{P}) (\hat{H}_0 - E_0)^{-1} (\hat{H}_2 - \hat{P} \hat{H}_2 \hat{P}) + \\ + \varepsilon^3 (\hat{H}_1 - \hat{P} \hat{H}_1 \hat{P}) (\hat{H}_0 - E_0)^{-1} (\hat{H}_1 - \hat{P} \hat{H}_1 - \varDelta_0) (\hat{H}_0 - E_0)^{-1} \times \\ \times (\hat{H}_1 - \hat{P} \hat{H}_1 P) \} \hat{P} C_0. \quad (14.32)$$

In allen Näherungen handelt es sich um eine Gleichung, in der die Wellenfunktion zum Raum der Wellenfunktionen C_0 gehört. Daher kann man diese Gleichungen als „Projektionen" von (14.20) in dem Raum der Funktionen C_0 betrachten.

Die oben dargestellte Variante der Störungstheorie kann man jetzt zur Lösung der Gl. (14.20) anwenden, in welcher der Energieoperator durch die Ausdrücke (14.11) bis (14.13) gegeben ist. In diesem Falle treten als nullte Näherung die Funktionen

$$C_0(\ldots N_{f\sigma}^0 \ldots) = \Psi(\ldots N_{f\sigma} \ldots) \prod_f \delta_{N_f^0, 1} \tag{14.33}$$

auf, wobei $\Psi(\ldots N_{f\sigma} \ldots)$ eine willkürliche Funktion von $N_{f\sigma}$ bedeutet. Die Elemente des linearen Raumes der Funktionen nullter Ordnung sind also durch die Bedingung gekennzeichnet, daß alle Besetzungszahlen N_f gleich 1 sind. Wie aus den Formeln (14.30) bis (14.32) und (14.12), (14.13) folgt, sind zuerst die Ausdrücke

$$\hat{P} a_{f_1 \sigma_1}^+ a_{f_1' \sigma_1'} \hat{P}, \quad \hat{P} a_{f_1 \sigma_1}^+ a_{f_2 \sigma_2}^+ a_{f_2' \sigma_2'} a_{f_1' \sigma_1'} \hat{P} \tag{14.34}$$

zu bestimmen. Zur Berechnung der ersten Form in (14.34) berücksichtigen wir, daß

$$a_{f_1 \sigma_1}^+ a_{f_1' \sigma_1'} N_f - N_f a_{f_1 \sigma_1}^+ a_{f_1' \sigma_1'} = a_{f_1 \sigma_1}^+ a_{f_1' \sigma_1'} \sum_\sigma a_{f\sigma}^+ a_{f\sigma} -$$
$$- \sum_\sigma a_{f\sigma}^+ a_{f\sigma} a_{f_1 \sigma_1}^+ a_{f_1' \sigma_1'} = \sum_\sigma \{ a_{f_1 \sigma_1}^+ a_{f_1' \sigma_1'} a_{f\sigma}^+ a_{f\sigma} - a_{f\sigma}^+ a_{f\sigma} a_{f_1 \sigma_1}^+ a_{f_1' \sigma_1'} \}$$

ist. Aus den Vertauschungsrelationen (14.4) erhalten wir

$$a_{f_1 \sigma_1}^+ a_{f_1' \sigma_1'} a_{f\sigma}^+ a_{f\sigma} - a_{f\sigma}^+ a_{f\sigma} a_{f_1 \sigma_1}^+ a_{f_1' \sigma_1'} =$$
$$= a_{f_1 \sigma_1}^+ a_{f_1' \sigma_1'} a_{f\sigma}^+ a_{f\sigma} + a_{f_1 \sigma_1}^+ a_{f\sigma}^+ a_{f_1' \sigma_1'} a_{f\sigma} + a_{f\sigma}^+ a_{f_1 \sigma_1}^+ a_{f_1' \sigma_1'} a_{f\sigma} +$$
$$+ a_{f\sigma}^+ a_{f_1 \sigma_1}^+ a_{f\sigma} a_{f_1' \sigma_1'} + a_{f\sigma}^+ (a_{f\sigma} a_{f_1 \sigma_1}^+ - \delta_{f f_1} \delta_{\sigma \sigma_1}) a_{f_1' \sigma_1'} - a_{f\sigma}^+ a_{f\sigma} a_{f_1 \sigma_1}^+ a_{f_1' \sigma_1'} =$$
$$= \delta_{f f_1'} \delta_{\sigma \sigma_1'} a_{f_1 \sigma_1}^+ a_{f\sigma} - \delta_{f f_1} \delta_{\sigma \sigma_1} a_{f\sigma}^+ a_{f_1' \sigma_1'}.$$

Folglich kann man schreiben:

$$a_{f_1 \sigma_1}^+ a_{f_1' \sigma_1'} N_f - N_f a_{f_1 \sigma_1}^+ a_{f_1' \sigma_1'} = [\delta_{f f_1'} - \delta_{f f_1}] a_{f_1 \sigma_1}^+ a_{f_1' \sigma_1'}. \tag{14.35}$$

Betrachten wir nun die willkürliche Funktion C. Wegen (14.35) gilt

$$\langle C_0 a_{f_1 \sigma_1}^+ a_{f_1' \sigma_1'} N_f C_0 \rangle - \langle C_0 N_f a_{f_1 \sigma_1}^+ a_{f_1' \sigma_1'} C_0 \rangle =$$
$$= [\delta_{f f_1'} - \delta_{f f_1}] \langle C_0 a_{f_1 \sigma_1}^+ a_{f_1' \sigma_1'} C_0 \rangle =$$
$$= [\delta_{f f_1'} - \delta_{f f_1}] \langle C \hat{P} a_{f_1 \sigma_1}^+ a_{f_1' \sigma_1'} \hat{P} C \rangle.$$

Nun ist
$$N_f C_0 = C_0, \quad C_0^* N_f = C_0^*$$
und daher
$$[\delta_{f f_1'} - \delta_{f f_1}] \langle C \hat{P} a_{f_1 \sigma_1}^+ a_{f_1' \sigma_1'} \hat{P} C \rangle = 0.$$

Setzt man hier $f = f_1$, so findet man

$$\langle C \hat{P} a_{f_1 \sigma_1}^+ a_{f_1' \sigma_1'} \hat{P} C \rangle = 0, \quad f_1 \neq f_1',$$

und wegen der Willkürlichkeit der Funktion C erhält man

$$\hat{P} a_{f_1 \sigma_1}^+ a_{f_1' \sigma_1'} \hat{P} = 0, \tag{14.36}$$

wenn $f_1 \neq f_1'$ ist. Im Falle $f_1 = f_1'$ ändert der Operator $a_{f_1 \sigma_1}^+ a_{f_1' \sigma_1'}$ nicht die Zahl N_{f_1}, da er gleichzeitig die Erzeugung eines Spins σ_1 im Zustand f_1 und die Vernichtung eines Spins σ_1' im selben Zustand beschreibt. Deshalb kommutiert auch der

Operator $a^+_{f_1\sigma_1} a_{f'_1\sigma'_1}$ mit dem Projektionsoperator $\hat{P}$, führt er doch jede quasihomöopolare Funktion wiederum in eine solche über. Folglich erhalten wir

$$\hat{P} a^+_{f_1\sigma_1} a_{f_1\sigma'_1} \hat{P} = a^+_{f_1\sigma_1} a_{f_1\sigma'_1} \hat{P}. \qquad (14.37)$$

Der zweite der Ausdrücke (14.34) wird auf ähnliche Weise berechnet. Das Ergebnis lautet:

$$\hat{P} a^+_{f_1\sigma_1} a^+_{f_2\sigma_2} a_{f'_2\sigma'_2} a_{f'_1\sigma'_1} \hat{P} = 0, \quad (f_1 f_2) \neq (f'_1 f'_2). \qquad (14.38)$$

Die symbolische Ungleichung $(f_1 f_2) \neq (f'_1 f'_2)$ bedeutet, daß das Punktepaar $(f'_1 f'_2)$ ohne Beachtung der Reihenfolge verschieden ist vom Paar $(f_1 f_2)$. Im Falle $(f_1 f_2) = (f'_1 f'_2)$ erhalten wir

$$\hat{P} a^+_{f_1\sigma_1} a^+_{f_2\sigma_2} a_{f'_2\sigma'_2} a_{f'_1\sigma'_1} \hat{P} = a^+_{f_1\sigma_1} a^+_{f_2\sigma_2} a_{f'_2\sigma'_2} a_{f'_1\sigma'_1} \hat{P}. \qquad (14.39)$$

Vor der Anwendung der Störungstheorie muß noch die Bedeutung des Ausdruckes

$$(\hat{H}_0 - E_0)^{-1} a^+_{f_1\sigma_1} a^+_{f_2\sigma_2} a_{f'_2\sigma'_2} a_{f'_1\sigma'_1} \hat{P} \qquad (14.40)$$

für

$$(f_1 f_2) \neq (f'_1 f'_2)$$

geklärt werden. Der Operator $a^+_{f_1\sigma_1} a^+_{f_2\sigma_2} a_{f'_2\sigma'_2} a_{f'_1\sigma'_1}$ beschreibt die Vernichtung von Elektronen mit den Spins σ'_1, σ'_2 in den Zuständen f'_1, f'_2 unter gleichzeitiger Erzeugung von Elektronen mit den Spins σ_1, σ_2 in den Zuständen f_1, f_2. Im Ausdruck (14.40) wird der erwähnte Operator dank dem Faktor $\hat{P}$ immer auf eine quasihomöopolare Wellenfunktion angewendet, weshalb er „Löcher" in den Zuständen f'_1, f'_2 und zugleich „Doppelbesetzungen" in den Zuständen f_1, f_2 erzeugt. Der Zuwachs, den die Energie des Grundzustands beim Auftreten des Komplexes aus Löchern f'_1, f'_2 und Doppelbesetzungen f_1, f_2 im quasihomöopolaren Zustand erfährt, beträgt nach (14.16) für ein Paar

$$\Delta(f_1, f'_1) = \tfrac{1}{2} C(f_1 f_1) + \tfrac{1}{2} C(f'_1 f'_1) - C(f_1 f'_1) - C(f'_1 f_1) \qquad (14.41)$$

und für zwei Paare

$$\Delta(f_1 f_2, f'_1 f'_2) = \Delta(f_1, f'_1) + \Delta(f_2, f'_2) - C(f_1 f'_2) - C(f_2 f'_1), \qquad (14.42)$$

wobei

$$C(f_1 f_2) = F(f_1 f_2 f_1 f_2)$$

ist.

Auf diese Weise erhält man

$$(\hat{H}_0 - E_0)^{-1} a^+_{f_1\sigma_1} a^+_{f_2\sigma_2} a_{f'_2\sigma'_2} a_{f'_1\sigma'_1} \hat{P} = \frac{1}{\Delta(f_1 f_2, f'_1 f'_2)} a^+_{f_1\sigma_1} a^+_{f_2\sigma_2} a_{f'_2\sigma'_2} a_{f'_1\sigma'_1} \hat{P} \qquad (14.43)$$

und analog dazu

$$\hat{P} a^+_{f_1\sigma_1} a^+_{f_2\sigma_2} a_{f'_2\sigma'_2} a_{f'_1\sigma'_1} (\hat{H}_0 - E_0)^{-1} = \frac{1}{\Delta(f_1 f_2, f'_1 f'_2)} \hat{P} a^+_{f_1\sigma_1} a^+_{f_2\sigma_2} a_{f'_2\sigma'_2} a_{f'_1\sigma'_1}. \qquad (14.44)$$

Die gefundenen Formeln gestatten es, zur expliziten Form der Gleichungen in den verschiedenen Näherungen überzugehen.

Zur Gewinnung der Gleichung erster Näherung multiplizieren wir (14.12) von rechts und links mit dem Operator $\hat{P}$. Dann finden wir unter Berücksichtigung von (14.36) und (14.38)

$$\varepsilon \hat{P} \hat{H}_1 \hat{P} = 0. \qquad (14.45)$$

Hiernach unterscheidet sich die Gleichung erster Näherung (14.30) nicht von ihrer nullten Näherung (14.17), und folglich wird die Entartung auch nicht aufgehoben.

Die Gleichung zweiter Näherung (14.31) nimmt unter Benützung von (14.45), (14.37), (14.12), (14.44) folgende Form an:

$$(E-E_0)C_0 = \frac{1}{2} \sum_{\substack{f_1 f_2 \sigma_1 \sigma_2 \\ (f_1 \neq f_2)}} F(f_1 f_2 f_2 f_1) a^+_{f_1\sigma_1} a^+_{f_2\sigma_2} a_{f_1\sigma_2} a_{f_2\sigma_1} C_0 +$$

$$+ \sum_{\substack{f_1 f_2 f'_1 f'_2 \sigma_1 \sigma_2 \\ (f_1 \neq f'_1, f_2 \neq f'_2)}} \hat{P} a^+_{f_1\sigma_1} [L(f_1 f'_1) + \sum_{f''} F(f_1 f'' f'_1 f'') N_{f''}] a_{f'_1\sigma_1} \times \quad (14.46)$$

$$\times \frac{1}{\Delta(f_2, f'_2)} a^+_{f_2\sigma_2} [L(f_2 f'_2) + \sum_{f''} F(f_2 f'' f'_2 f'') N_{f''}] a_{f'_2\sigma_2} \hat{P} C_0.$$

Im weiteren verwenden wir folgende, leicht ableitbare Beziehungen

$$N_{f''} a_{f_1\sigma_2} \hat{P} = a_{f_1\sigma_2} N_{f''} \hat{P} = a_{f_1\sigma_2} \hat{P}, \quad (f'' \neq f_1),$$

$$N_{f_1} a_{f_1\sigma_2} \hat{P} = \sum_\sigma a^+_{f_1\sigma} a_{f_1\sigma} a_{f_1\sigma_2} \hat{P} = -\sum_\sigma a^+_{f_1\sigma} a_{f_1\sigma_2} a_{f_1\sigma} \hat{P} =$$

$$= \sum_\sigma [a_{f_1\sigma_2} a^+_{f_1\sigma} - \delta_{\sigma_2\sigma}] a_{f_1\sigma} \hat{P} = a_{f_1\sigma_2}(N_{f_1}-1) \hat{P} = 0,$$

sowie
$$\hat{P} a^+_{f_1\sigma_1} N_{f''} = \hat{P} a^+_{f_1\sigma_1}, \quad (f'' \neq f_1), \quad \hat{P} a^+_{f_1\sigma_1} N_{f_1} = 0$$

$$\hat{P} a^+_{f_1\sigma_1} a_{f_2\sigma_1} a^+_{f_2\sigma_2} a_{f_1\sigma_2} \hat{P} = \hat{P} a^+_{f_1\sigma_1} [\delta_{\sigma_1\sigma_2} - a^+_{f_2\sigma_2} a_{f_2\sigma_1}] a_{f_1\sigma_2} \hat{P} =$$

$$= \hat{P} a^+_{f_1\sigma_1} a_{f_1\sigma_1} \hat{P} \delta_{\sigma_1\sigma_2} + \hat{P} a^+_{f_1\sigma_1} a^+_{f_2\sigma_2} a_{f_1\sigma_2} a_{f_2\sigma_1} \hat{P}.$$

Die endgültige Gleichung zweiter Näherung (14.31) lautet dann:

$$(E-E^{(0)})C_0 = \frac{1}{2} \sum_{\substack{f_1 f_2 \sigma_1 \sigma_2 \\ (f_1 \neq f_2)}} J(f_1 f_2) a^+_{f_1\sigma_1} a^+_{f_2\sigma_2} a_{f_1\sigma_2} a_{f_2\sigma_1} C_0, \quad (14.47)$$

wobei folgende Abkürzungen eingeführt wurden:

$$E^{(0)} = E_0 - \sum_{\substack{f_1 f_2 \\ (f_1 \neq f_2)}} \left[L(f_1 f_2) + \sum_{f'(\neq f_2)} F(f_1 f' f_2 f') \right] \times$$

$$\times \left[L(f_2 f_1) + \sum_{f'(\neq f_1)} F(f_2 f' f_1 f') \right] \frac{1}{\Delta(f_1, f_2)}, \quad (14.48)$$

$$J(f_1 f_2) = F(f_1 f_2 f_2 f_1) - \frac{2}{\Delta(f_1, f_2)} \left[L(f_1 f_2) + \sum_{f'(\neq f_1)} F(f_1 f' f_2 f') \right] \times$$

$$\times \left[L(f_2 f_1) + \sum_{f'(\neq f_1)} F(f_2 f' f_1 f') \right]. \quad (14.49)$$

Geht man in (14.47) mittels

$$s^x_f = \frac{1}{2}(a^+_{f-} a_{f+} + a^+_{f+} a_{f-}),$$

$$s^y_f = \frac{i}{2}(a^+_{f+} a_{f-} - a^+_{f-} a_{f+}), \quad (14.50)$$

$$s^z_f = \frac{1}{2}(a^+_{f-} a_{f-} - a^+_{f+} a_{f+})$$

zu den Operatoren der Elektronenspins über, so erhalten wir (bis auf eine additive Konstante) unter der Bedingung $\sum_\sigma a^+_{f\sigma} a_{f\sigma} = 1$ (Quasihomöopolarität) für den effektiven Hamilton-Operator einen Ausdruck von der Dirac-van Vleckschen

Vektorform

$$\hat{H} = -2 \sum_{f_1 > f_2} J(f_1 f_2) \vec{s}_{f_1} \vec{s}_{f_2}. \tag{14.51}$$

Nun ist die erhaltene Formel (14.51) zu diskutieren und das Austauschintegral $J(f_1 f_2)$ mit dem Austauschintegral zu vergleichen, das aus der Heitler-London-Methode in Ziff. 6 folgte. Diese Fragen werden unter Ziff. 15 und 16 behandelt.

15. Der potentielle und kinetische Austausch. Wir kommen jetzt zur Diskussion der in Ziff. 14 erhaltenen Ergebnisse. Wie die dort durchgeführten Rechnungen zeigten, wird die Spinentartung des Systems erst in der zweiten Näherung der Theorie aufgehoben [s. (14.45) und (14.47)].

Das Austauschintegral (14.49) besteht aus zwei Teilen: dem „direkten" Austauschintegral $F(f_1 f_2 f_2 f_1)$, das durch die homöopolare Grundkonfiguration bedingt ist, und dem Glied, das den Energienenner $\Delta(f_1 f_2)$ enthält und durch die Berücksichtigung der virtuellen Übergänge mit Bildung einer Doppelbesetzung und eines Loches zustande kommt. Das Glied $F(f_1 f_2 f_2 f_1)$ hängt lediglich von der Elektron-Elektron-Wechselwirkung e^2/r_{ij} ab und ist notwendig positiv. Das zweite Glied im Ausdruck (14.49) ist, wie man sieht, stets negativ (bei Berücksichtigung des Minuszeichens davor). Um also bei Benützung der orthogonalen Bahnen die Möglichkeit eines negativen Vorzeichens für das Austauschintegral zu erhalten, muß man, wie bereits mehrfach bemerkt, virtuelle angeregte Zustände einbeziehen. Diese erhöhen das „Gewicht" der Zustände niedriger Multiplizität.

Indem ANDERSON [47] seine neue Methode bei der Behandlung des Problems der indirekten Austauschwechselwirkung weiterentwickelte (s. Abschnitt V), schlug er auch eine neue Terminologie vor zur Bezeichnung der beiden grundlegenden „Elektronenprozesse", die zur Aufhebung der Spinentartung führen. Das erste Glied im Ausdruck für das Austauschintegral (14.49) $F(f_1 f_2 f_2 f_1)$, das lediglich von der unmittelbaren Überlappung der orthogonalen „Atom"-Funktion herrührt, entspricht dem *potentiellen* Austausch der Elektronen in der homöopolaren Konfiguration. Das zweite Glied in (14.49), das mit dem virtuellen Übergang eines Elektrons von einem Atom zum anderen zusammenhängt, entspricht dem *kinetischen* Austausch[22]. Somit steht also fest, daß der potentielle Austausch den Ferromagnetismus und der kinetische den Antiferromagnetismus begünstigt.

16. Der Zusammenhang mit dem Heisenbergschen Austauschintegral. Interessant ist ein Vergleich des nach der Methode der orthogonalen Bahnen erhaltenen Ausdruckes (14.49) für das Austauschintegral mit dem (auch Heitler-London- oder Heisenbergsches Austauschintegral genannten) Ausdruck (6.15) für den Fall nichtorthogonaler Bahnen. Sowohl (14.49) als auch (6.15) wurden unter der Annahme abgeleitet, daß es nur ein 1s-Elektron pro Atom gibt. (Wir erinnern daran, daß man die orthogonalen Bahnen nur bedingt als Atombahnen bezeichnen kann.) Beide Ausdrücke sind von gleicher (quadratischer) Ordnung hinsichtlich des Überlappungsintegrals (14.7). Um den Vergleich der Ausdrücke (6.15) und (14.49) durchführen zu können, muß man in der letztgenannten Beziehung mit Hilfe der Entwicklung (14.9) die Größen $L(f_1 f_1)$, $F(f_1 f_2 f_2 f_1)$, $F(f_1 f_3 f_2 f_3)$ und $\Delta(f_1, f_2)$ als Potenzreihen in (14.7) darstellen. Wir schreiben die

---

[22] In den entsprechenden Formeln der Arbeit von ANDERSON [47] steht im Zähler des Ausdruckes für den kinetischen Austausch an Stelle unseres Gliedes $L(f_1 f_2) + \sum_{f' \neq f_1} F(f_1 f' f_2 f')$
eine mit $b_{f_1 - f_2}$ bezeichnete Größe, die man *Einelektronen*-Übergangsintegral nennt und die nicht explizit angegeben ist. Dieses Übergangsintegral stellt jedoch nicht ein Matrixelement nur des Einelektronenanteils des Hamilton-Operators dar (die Größe $L(f_1 f_2)$ in unserer Bezeichnung), sondern ein Matrixelement des *vollständigen* Hamilton-Operators, der auch die Elektronenwechselwirkung enthält.

orthogonalen Funktionen Θ_{f_1}, Θ_{f_2} und Θ_{f_3} in der Form

$$\left.\begin{aligned}\Theta_{f_1} &= \varphi_{f_1} - \tfrac{1}{2}S_{f_1f_2}\varphi_{f_2} - \tfrac{1}{2}S_{f_1f_3}\varphi_{f_3} - \tfrac{1}{2}\sum_{\alpha \neq f_1,f_2,f_3} S_{f_1\alpha}\varphi_\alpha + \cdots \\ \Theta_{f_2} &= -\tfrac{1}{2}S_{f_1f_2}\varphi_{f_1} + \varphi_{f_2} - \tfrac{1}{2}S_{f_2f_3}\varphi_{f_3} - \tfrac{1}{2}\sum_{\alpha \neq f_1,f_2,f_3} S_{f_2\alpha}\varphi_\alpha + \cdots \\ \Theta_{f_3} &= -\tfrac{1}{2}S_{f_1f_3}\varphi_{f_1} - \tfrac{1}{2}S_{f_2f_3}\varphi_{f_2} + \varphi_{f_3} - \tfrac{1}{2}\sum_{\alpha \neq f_1,f_2,f_3} S_{f_3\alpha}\varphi_\alpha + \cdots\end{aligned}\right\} \quad (16.1)$$

Die im Integral (14.49) in den Ausdruck für den kinetischen Austausch eingehenden Integrale $L(f_1 f_2)$ und $F(f_1 f_3 f_2 f_3)$ berechnen wir lediglich in niedrigster (erster) Ordnung hinsichtlich des Parameters (14.7), da in (14.49) das Quadrat ihrer Summe erscheint. Setzt man die Entwicklungen (16.1) in die Formeln (14.2) und (14.3) ein, so ergibt sich

$$L(f_1 f_2) = -S_{f_1f_2}\langle\varphi_{f_1}|\varphi_{f_1}\rangle + \langle\varphi_{f_1}|\varphi_{f_2}\rangle, \quad (16.2)$$

$$F(f_1 f_3 f_2 f_3) = \langle\varphi_{f_1}\varphi_{f_3}|\varphi_{f_2}\varphi_{f_3}\rangle - \tfrac{1}{2}S_{f_1f_2}[\langle\varphi_{f_1}\varphi_{f_3}|\varphi_{f_1}\varphi_{f_3}\rangle + \langle\varphi_{f_2}\varphi_{f_3}|\varphi_{f_2}\varphi_{f_3}\rangle]. \quad (16.3)$$

Das Integral $F(f_1 f_2 f_2 f_1)$ lautet bis zu den quadratischen Gliedern in (14.7)

$$\left.\begin{aligned}F(f_1 f_2 f_2 f_1) &= \langle\varphi_{f_1}\varphi_{f_2}|\varphi_{f_2}\varphi_{f_1}\rangle - 2S_{f_1f_2}\langle\varphi_{f_1}\varphi_{f_1}|\varphi_{f_2}\varphi_{f_1}\rangle + \\ &\quad + \tfrac{1}{2}S_{f_1f_2}^2[\langle\varphi_{f_1}\varphi_{f_1}|\varphi_{f_1}\varphi_{f_1}\rangle + \langle\varphi_{f_1}\varphi_{f_2}|\varphi_{f_1}\varphi_{f_2}\rangle].\end{aligned}\right\} \quad (16.4)$$

Hier wurden der Kürze halber die Bezeichnungen

$$\langle\varphi_{f_i}|\varphi_{f_j}\rangle = \int \varphi_{f_i}^*(\vec{r})\hat{H}_1(\vec{r})\varphi_j(\vec{r})\,d\vec{r},$$
$$\langle\varphi_{f_i}\varphi_{f_j}|\varphi_{f_k}\varphi_{f_l}\rangle = \int \varphi_{f_i}^*(\vec{r}_1)\varphi_{f_j}^*(\vec{r}_2)V(\vec{r}_1,\vec{r}_2)\varphi_{f_k}(\vec{r}_1)\varphi_{f_l}(\vec{r}_2)\,d\vec{r}_1\,d\vec{r}_2$$

eingeführt. Die Größe $\Delta(f_1, f_2)$ wird natürlich nur in nullter Ordnung berechnet, da der Zähler im Integral (14.49) von zweiter Ordnung ist:

$$\Delta(f_1, f_2) = F(f_1 f_1 f_1 f_1) - F(f_1 f_2 f_1 f_2) = U. \quad (16.5)$$

Wir stellen die in den kinetischen Summanden von (14.49) eingehende Summe dar als

$$\left.\begin{aligned}\sum_{f_3 \neq f_1} F(f_1 f_3 f_2 f_3) &= \langle\varphi_{f_1}\varphi_{f_2}|\varphi_{f_2}\varphi_{f_2}\rangle - \tfrac{1}{2}S_{f_1f_2}(\langle\varphi_{f_2}\varphi_{f_2}|\varphi_{f_2}\varphi_{f_2}\rangle - \\ &\quad -\langle\varphi_1\varphi_2|\varphi_1\varphi_2\rangle) - S_{f_1f_2}\langle\varphi_{f_1}\varphi_{f_2}|\varphi_{f_1}\varphi_{f_2}\rangle + \\ &\quad + \sum_{f_3 \neq f_1,f_2}\{\langle\varphi_{f_1}\varphi_{f_3}|\varphi_{f_2}\varphi_{f_3}\rangle - \tfrac{1}{2}S_{f_1f_2}\langle\varphi_{f_2}\varphi_{f_3}|\varphi_{f_2}\varphi_{f_3}\rangle - \tfrac{1}{2}S_{f_1f_2}\langle\varphi_{f_1}\varphi_{f_3}|\varphi_{f_1}\varphi_{f_3}\rangle\}.\end{aligned}\right\} \quad (16.6)$$

Wir ziehen jetzt den Faktor $\langle\varphi_{f_2}\varphi_{f_2}|\varphi_{f_2}\varphi_{f_2}\rangle - \langle\varphi_{f_1}\varphi_{f_2}|\varphi_{f_1}\varphi_{f_2}\rangle = U$, der gleich der Größe $\Delta(f_1, f_2)$ aus (16.5) ist, vor den Zähler im kinetischen Teil des Integrals (14.49). Setzt man (16.2), (16.4), (16.5) und (16.6) in (14.49) ein, so erhält man das Austauschintegral in der endgültigen Form

$$\left.\begin{aligned}J(f_1 f_2) &= \langle\varphi_{f_1}\varphi_{f_2}|\varphi_{f_2}\varphi_{f_1}\rangle - S_{f_1f_2}^2\langle\varphi_{f_1}\varphi_{f_1}|\varphi_{f_1}\varphi_{f_1}\rangle + 2S_{f_1f_2}\langle\varphi_{f_1}|\varphi_{f_2}\rangle - \\ &\quad - 2S_{f_1f_2}^2\langle\varphi_{f_1}|\varphi_{f_1}\rangle + S_{f_1f_2}\sum_{f_3 \neq f_1,f_2}\{2\langle\varphi_{f_1}\varphi_{f_3}|\varphi_{f_2}\varphi_{f_3}\rangle - S_{f_1f_2}(\langle\varphi_{f_2}\varphi_{f_3}|\varphi_{f_2}\varphi_{f_3}\rangle + \\ &\quad + \langle\varphi_{f_1}\varphi_{f_3}|\varphi_{f_1}\varphi_{f_3}\rangle)\} - \tfrac{2}{U}\Big[-S_{f_1f_2}\langle\varphi_{f_1}|\varphi_{f_1}\rangle + \langle\varphi_{f_1}|\varphi_{f_2}\rangle + \langle\varphi_{f_1}\varphi_{f_2}|\varphi_{f_2}\varphi_{f_2}\rangle - \\ &\quad - S_{f_1f_2}\langle\varphi_{f_1}\varphi_{f_2}|\varphi_{f_1}\varphi_{f_2}\rangle + \sum_{f_3 \neq f_1,f_2}\{\langle\varphi_{f_1}\varphi_{f_3}|\varphi_{f_2}\varphi_{f_3}\rangle - \tfrac{1}{2}S_{f_1f_2}(\langle\varphi_{f_2}\varphi_{f_3}|\varphi_{f_2}\varphi_{f_3}\rangle + \\ &\quad + \langle\varphi_{f_1}\varphi_{f_3}|\varphi_{f_1}\varphi_{f_3}\rangle)\}\Big]^2.\end{aligned}\right\} \quad (16.7)$$

Wenn man den Ausdruck (6.15) für das Austauschintegral unter Benützung der hier verwendeten Bezeichnungen umschreibt, so folgt

$$\left.\begin{aligned}J_{H-L} =& \langle \varphi_{f_1}\,\varphi_{f_2} | \varphi_{f_2}\,\varphi_{f_1}\rangle - S_{f_1 f_2}^2 \langle \varphi_{f_1}\,\varphi_{f_2} | \varphi_{f_1}\,\varphi_{f_2}\rangle + \\ &+ 2 S_{f_1 f_2} \langle \varphi_{f_1} | \varphi_{f_2}\rangle - 2 S_{f_1 f_2}^2 \langle \varphi_{f_1} | \varphi_{f_1}\rangle + \\ &+ S_{f_1 f_2} \sum_{f_3 \neq f_1, f_2} \{ 2\langle \varphi_{f_1}\,\varphi_{f_3} | \varphi_{f_2}\,\varphi_{f_3}\rangle - S_{f_1 f_2}(\langle \varphi_{f_2}\,\varphi_{f_3} | \varphi_{f_2}\,\varphi_{f_3}\rangle + \langle \varphi_{f_1}\,\varphi_{f_3} | \varphi_{f_1}\,\varphi_{f_3}\rangle)\}.\end{aligned}\right\} \quad (16.8)$$

Da bei der Ableitung des Ausdruckes für J_{H-L} in Ziff. 6 die virtuellen Übergänge unter Bildung von Doppelbesetzungen und Löchern nicht berücksichtigt wurden, fehlen in (16.8) gegenüber (16.7) die Glieder mit dem Energienenner U. Ein Vergleich dieser beiden Ausdrücke für das Austauschintegral ist daher nur bei Vernachlässigung des letzten Gliedes von (16.7) durchführbar. In diesem Falle stimmt das mit Hilfe des Projektionsformalismus für orthogonale Bahnen erhaltene Integral (16.7) vollständig mit dem Heitler-London-Integral (16.8) überein.

Bei der Entwicklung der orthogonalisierten Funktionen mittels (14.9) treten im Zähler des Ausdruckes für den kinetischen Austausch zum Teil Glieder auf, die den Nenner U enthalten, was nach entsprechender Kürzung bei der Addition mit dem potentiellen Austauschterm $F(f_1\,f_2\,f_2\,f_1)$ dann zu einer partiellen Kompensation von Gliedern verschiedener Herkunft führt. Das zeigt, daß, obwohl der Ausdruck für den kinetischen Austausch in der Terminologie der gewöhnlichen Störungstheorie (wegen des Auftretens des Nenners U) von höherer (zweiter) Ordnung ist als der Ausdruck für den potentiellen Austausch, der kinetische Austausch im allgemeinen nicht kleiner ist als der potentielle. Im gegebenen Falle wird die Größe der verschiedenen Summanden im Ausdruck (14.49) für das Austauschintegral nicht durch deren Ordnung in der Störungstheorie, sondern durch ihre Ordnung im Hinblick auf die Potenz eines kleinen Parameters, des Nichtorthogonalitätsintegrales (14.7), bestimmt: alle Glieder in (14.49) und (16.7) sind nach der jeweiligen Lesart von zweiter Ordnung.

Um das Verhältnis der beiden Methoden der orthogonalen und nichtorthogonalen Bahnen besser verstehen zu können, wird im folgenden als Beispiel das Wasserstoffmolekülproblem betrachtet. Im Gegensatz zu Ziff. 2 erfolgt die Orthogonalisierung der Wellenfunktionen nicht exakt, sondern angenähert in Übereinstimmung mit der Entwicklung (14.9).

Zum Aufbau der Determinanten wählen wir wie schon in Ziff. 3 die Funktionen

$$\omega_1 = \frac{1}{\sqrt{2}} \begin{vmatrix} \Theta_1\,\alpha \\ \Theta_2\,\beta \end{vmatrix}, \quad \omega_2 = \frac{1}{\sqrt{2}} \begin{vmatrix} \Theta_1\,\beta \\ \Theta_2\,\alpha \end{vmatrix}, \quad \omega_3 = \frac{1}{\sqrt{2}} \begin{vmatrix} \Theta_1\,\alpha \\ \Theta_1\,\beta \end{vmatrix}, \quad \omega_4 = \frac{1}{\sqrt{2}} \begin{vmatrix} \Theta_2\,\alpha \\ \Theta_2\,\beta \end{vmatrix} \quad (16.9)$$

mit den orthogonalen Bahnen Θ_1 und Θ_2 sowie

$$u_1 = \frac{1}{\sqrt{2}} \begin{vmatrix} \varphi_a\,\alpha \\ \varphi_b\,\beta \end{vmatrix}, \quad u_2 = \frac{1}{\sqrt{2}} \begin{vmatrix} \varphi_a\,\beta \\ \varphi_b\,\alpha \end{vmatrix}, \quad u_3 = \frac{1}{\sqrt{2}} \begin{vmatrix} \varphi_a\,\alpha \\ \varphi_a\,\beta \end{vmatrix}, \quad u_4 = \frac{1}{\sqrt{2}} \begin{vmatrix} \varphi_b\,\alpha \\ \varphi_b\,\beta \end{vmatrix} \quad (16.10)$$

mit den nichtorthogonalen Atomfunktionen φ_a und φ_b. Die orthogonalen Funktionen werden durch die Atomfunktionen mit Hilfe der Beziehungen[23]

$$\left.\begin{aligned}\Theta_1 &= (1 + \tfrac{3}{8} S_{ab}^2)\,\varphi_a - \tfrac{1}{2} S_{ab}\,\varphi_b + \cdots \\ \Theta_2 &= -\tfrac{1}{2} S_{ab}\,\varphi_a + (1 + \tfrac{3}{8} S_{ab}^2)\,\varphi_b + \cdots\end{aligned}\right\} \quad (16.11)$$

---

[23] Wir bemerken, daß eine analoge Zerlegung auch für die Funktionen (10.7) gilt.

Ziff. 16. Der Zusammenhang mit dem Heisenbergschen Austauschintegral.

ausgedrückt. In den orthogonalen Funktionen haben die drei Singulettfunktionen [s. (2.4) und (2.11)] und die eine Triplettfunktion (für $S_z=0$) die Form

$$\left.\begin{array}{l}\Phi_{1s}=\dfrac{1}{\sqrt{2}}\{\omega_1-\omega_2\},\\ \Phi_{2s}=\omega_3,\qquad\Phi_t=\dfrac{1}{\sqrt{2}}\{\omega_1+\omega_2\}.\\ \Phi_{3s}=\omega_4,\end{array}\right\} \qquad (16.12)$$

Für die Bestimmung der Singulettenergie haben wir eine Säkulargleichung dritten Grades und für die Triplettenergie eine ersten Grades. Unmittelbare Berechnungen ergeben für die Matrixelemente $H^s_{ij}=\langle\Phi_{is}\hat{H}\Phi_{js}\rangle$ und $H^t=\langle\Phi_t\hat{H}\Phi_t\rangle$ die folgenden Ausdrücke:

$$\left.\begin{array}{l}H^s_{11}=2\langle\Theta_1|\Theta_1\rangle+\langle\Theta_1\Theta_2|\Theta_2\Theta_1\rangle+\langle\Theta_1\Theta_2|\Theta_1\Theta_2\rangle,\\ H^s_{12}=\sqrt{2}\{\langle\Theta_1|\Theta_2\rangle+\langle\Theta_1\Theta_2|\Theta_1\Theta_1\rangle\},\\ H^s_{13}=\sqrt{2}\{\langle\Theta_1|\Theta_2\rangle+\langle\Theta_1\Theta_2|\Theta_2\Theta_2\rangle\},\\ H^s_{22}=H^s_{33}=2\langle\Theta_1|\Theta_1\rangle+\langle\Theta_1\Theta_1|\Theta_1\Theta_1\rangle,\\ H^s_{23}=\langle\Theta_1\Theta_1|\Theta_2\Theta_2\rangle,\\ H^t\ =2\langle\Theta_1|\Theta_1\rangle-\langle\Theta_1\Theta_2|\Theta_2\Theta_1\rangle+\langle\Theta_1\Theta_2|\Theta_1\Theta_2\rangle.\end{array}\right\}\qquad(16.13)$$

Beim Aufsuchen der Energie des Singulettzustandes verwenden wir zur Lösung der Gleichung dritten Grades die Störungstheorie. Als „Grund"-Konfiguration wählen wir in Übereinstimmung mit dem Hauptgedanken von Ziff. 15 den durch die Funktion Φ_{1s} beschriebenen (quasihomöopolaren) Zustand, und die Zustände Φ_{2s} und Φ_{3s} sollen diesen Grundzustand stören.

Wir führen die Bezeichnung

$$\Delta_{12}=H^s_{22}-H^s_{11}=H^s_{33}-H^s_{11} \qquad (16.14)$$

ein und suchen die Lösung der Säkulargleichung in der Form einer Reihe nach reziproken Potenzen von Δ, der Energiedifferenz zwischen den „angeregten" Konfigurationen und der „Grund"-Konfiguration. Die Energie des Tripletts wird dagegen einfach durch das diagonale Matrixelement

$$E_t=H^t \qquad (16.15)$$

bestimmt. Die unmittelbare Rechnung ergibt für das Austauschintegral folgenden Ausdruck:

$$J=\frac{1}{2}(E_s-E_t)=\langle\Theta_1\Theta_2|\Theta_2\Theta_1\rangle-2\frac{\{\langle\Theta_1|\Theta_2\rangle+\langle\Theta_1\Theta_2|\Theta_1\Theta_1\rangle\}^2}{\Delta_{12}}. \qquad (16.16)$$

Um sich wiederum nur mit der zweiten Ordnung im Hinblick auf das Nichtorthogonalitätsintegral S_{ab} zu begnügen, muß man in (16.16) nur Glieder nullten Grades in Δ_{12} behalten, indem man (16.14) durch den Ausdruck

$$\Delta_{12}=\langle\Theta_1\Theta_1|\Theta_1\Theta_1\rangle-\langle\Theta_1\Theta_2|\Theta_2\Theta_2\rangle \qquad (16.17)$$

annähert, der mit (14.41) übereinstimmt.

Wie man leicht sieht, erhält man den Ausdruck (16.16) aus der allgemeinen Formel (14.49), wenn man dort die Tatsache beachtet, daß es beim vorliegenden Problem im ganzen nur zwei Zentren gibt. Setzt man die Entwicklung (16.11) in

(16.16) ein, so erhält man

$$\begin{aligned}J = \langle \varphi_a \varphi_b | \varphi_a \varphi_a \rangle - S_{ab}^2 \langle \varphi_a \varphi_b | \varphi_a \varphi_b \rangle + 2 S_{ab} \langle \varphi_a | \varphi_b \rangle - \\ - 2 S_{ab}^2 \langle \varphi_a | \varphi_a \rangle - \frac{2}{U} [- S_{ab} \langle \varphi_a | \varphi_a \rangle + \langle \varphi_a | \varphi_b \rangle + \\ + \langle \varphi_a \varphi_b | \varphi_a \varphi_a \rangle - S_{ab} \langle \varphi_a \varphi_b | \varphi_a \varphi_b \rangle]^2 \end{aligned} \qquad (16.18)$$

mit
$$U = \langle \varphi_a \varphi_a | \varphi_a \varphi_a \rangle - \langle \varphi_a \varphi_b | \varphi_a \varphi_b \rangle. \qquad (16.19)$$

Wir berechnen jetzt das Austauschintegral für nichtorthogonale Bahnen mit Hilfe der Basisfunktionen $u_1, \ldots, u_4$ aus (16.10). Die bis zur Ordnung S_{ab}^2 orthonormierten drei Singulettfunktionen und die eine Triplettfunktion ($S_z = 0$) lauten:

$$\begin{aligned} \Psi_{1s} &= \frac{1}{\sqrt{2}} \left(1 - \frac{1}{2} S_{ab}^2\right)(u_1 - u_2), \\ \Psi_{2s} &= S_{ab}(u_1 - u_2) - (1 + S_{ab}^2) u_3, \\ \Psi_{3s} &= S_{ab}(u_1 - u_2) - S_{ab}^2 u_3 - (1 + S_{ab}^2) u_4, \\ \Psi_t &= \frac{1}{\sqrt{2}} \left(1 + \frac{1}{2} S_{ab}^2\right)(u_1 + u_2). \end{aligned} \qquad (16.20)$$

In Übereinstimmung mit der Heitler-London-Methode wählen wir als „Grund"-Konfiguration den Zustand Ψ_{1s}, der streng homöopolar ist, und betrachten Ψ_{2s} und Ψ_{3s} als Funktionen angeregter Zustände. Wir bemerken, daß die Funktionen Ψ_{2s} und Ψ_{3s} neben den polaren Basisfunktionen u_3 und u_4 auch die Funktionen u_1 und u_2 der „Grund"-Konfiguration enthalten. Man überzeugt sich leicht davon, daß es, wenn man eine der Singulettfunktionen in der Form Ψ_{1s} nach (16.20) gewählt hat, nicht möglich ist, die beiden anderen Funktionen so aufzubauen, daß sie linear unabhängig voneinander und zueinander sowie zu Ψ_{1s} orthogonal sind und dabei die Determinanten u_1 und u_2 nicht enthalten.

Weiterhin werden genau wie bei orthogonalen Bahnen die Matrixelemente der Säkulargleichung berechnet und die Lösung als Reihe nach negativen Potenzen der Energiedifferenz zwischen der „Grund"-Konfiguration und den „angeregten" Konfigurationen angesetzt. Die Rechnungen ergeben bis zu Gliedern der Ordnung S_{ab}^2

$$E_s = (1 - S_{ab}^2) h_{11} - h_{12} + \frac{4(S_{ab} h_{11} - h_{13})^2}{U}, \qquad (16.21)$$

$$E_t = (1 + S_{ab}^2) h_{11} + h_{12}, \qquad (16.22)$$

wobei die Bezeichnungen

$$\begin{aligned} h_{11} &= 2 \langle \varphi_a | \varphi_a \rangle + \langle \varphi_a \varphi_b | \varphi_a \varphi_b \rangle, \\ h_{12} &= -2 S_{ab} \langle \varphi_a | \varphi_b \rangle - \langle \varphi_a \varphi_b | \varphi_b \varphi_a \rangle, \\ h_{13} &= S_{ab} \langle \varphi_a | \varphi_a \rangle + \langle \varphi_a | \varphi_b \rangle + \langle \varphi_a \varphi_b | \varphi_a \varphi_a \rangle \end{aligned} \qquad (16.23)$$

eingeführt wurden. Geht man mit (16.23) in (16.21) und (16.22), erhält man genauso das Ergebnis (16.18) wie vorher bei Benützung orthogonalisierter Bahnen. Wie unter Ziff. 2 bemerkt wurde, kann man bei exakter Orthogonalisierung die Multiplettfunktionen stets so wählen, daß sie bei den Methoden der orthogonalen und nichtorthogonalen Bahnen übereinstimmen und folglich auch die Wurzeln der Säkulargleichung. Hier ist gezeigt worden, daß die kleinsten Wurzeln auch bei nur angenäherter Orthogonalisierung übereinstimmen, wenn man die „Grund"-Konfigurationen entsprechend wählt. Die Berücksichtigung der polaren Zustände in der Heitler-London-Methode gestattet beim Wasserstoffmolekül eine vollkommenere Übereinstimmung zwischen beiden Methoden als beim Kristall [vgl. (16.7) und (16.8)].

C. Die indirekte Austauschwechselwirkung.

Die Entwicklung einer mikroskopischen Theorie der indirekten Austauschwechselwirkung in den nichtmetallischen Verbindungen der Übergangsmetalle, wo die magnetischen Kationen im Kristall durch ein nichtmagnetisches Anion oder eine diamagnetische Gruppe getrennt sind, begann früher als die systematische experimentelle Untersuchung der atomaren magnetischen Struktur dieser Verbindungen. Im Jahre 1934 zeigte KRAMERS [35] in Fortentwicklung einer Idee von BLOCH, daß in Verbindungen, in denen die direkte Austauschwechselwirkung zwischen den magnetischen Ionen vernachlässigt werden kann, eine Abhängigkeit der Energie des Kristalls von der Spinordnung dennoch möglich ist, falls der Grundzustand des Systems nicht als reine Ionenkonfiguration aufgefaßt wird, sondern „angeregte" Konfigurationen beigemischt werden, die einem virtuellen Übergang eines Valenzelektrons vom Anion zum Kation entsprechen [24].

Anregungen zu einer detaillierten Erforschung der Natur der Austauschkopplungen in Nichtleitern gaben Arbeiten zur Theorie [129] bis [131] und experimentelle Untersuchungen über den Antiferromagnetismus und Ferrimagnetismus von Isolatoren, insbesondere von Ferriten mit Spinell- und Granatstruktur [132], [133]. Eine wichtige Rolle in der Entwicklung einer Theorie der Austauschwechselwirkung in Isolatoren spielten eine Arbeit von SHULL und SMART [134] über die Neutronenbeugung in der antiferromagnetischen Verbindung MnO und die folgenden Arbeiten auf diesem Gebiet.

Für MnO, das die NaCl-Struktur besitzt, zeigt eine Analyse der magnetischen Ordnung, daß die stärkste magnetische Kopplung zwischen solchen Manganionen auftritt, die durch ein Anion O^{--} getrennt sind und sich mit diesem auf einer Geraden befinden. Dem Experiment zufolge existiert also eine starke indirekte Austauschkopplung zwischen Mn^{++}-Ionen, die Nachbarn zweiter Ordnung sind. Diese Kopplung ist größer als die Wechselwirkung zwischen nächstbenachbarten Manganionen, obwohl der Abstand zwischen diesen $\sqrt{2}$-mal kleiner ist als zwischen den Nachbarn zweiter Sphäre.

Experimentelle Beweise für die Existenz einer indirekten Wechselwirkung wurden von vielen Forschern für eine sehr große Anzahl verschiedener Verbindungen erbracht. ANDERSON [36] entwickelte die Idee von KRAMERS weiter und erhielt 1950 einen Ausdruck für das effektive Integral des indirekten Austausches im sog. Problem der drei Zentren und vier Elektronen. Er untersuchte ferner das Vorzeichen der Wechselwirkung in Abhängigkeit vom Auffüllungsgrad der d-Schale des Übergangsmetallions. Dieses Problem wurde dann später von vielen anderen Autoren aufgegriffen. In einigen dieser Arbeiten wurden angeregte Konfigurationen eingeführt, die ursprünglich bei ANDERSON keine Berücksichtigung fanden; andere verallgemeinerten dieses Problem auf den Kristall.

1959 schlug ANDERSON [47] eine neue Art der Behandlung für die indirekte Wechselwirkung in Isolatoren vor, nachdem er darauf hingewiesen hatte, daß die wesentliche Schwäche der in den erwähnten Arbeiten verwendeten Modelle in der schwachen Konvergenz der Reihen in der Störungstheorie besteht: die indirekte Austauschwechselwirkung erscheint dort erst in der dritten oder vierten Ordnung.

---

[24] Da die Wellenfunktion für den stationären Zustand des Vielelektronensystems als Überlagerung antisymmetrisierter Produkte von spin- und ortsabhängigen Einelektronenfunktionen dargestellt wird, müssen in diese Superposition auch verschiedene angeregte Einelektronenzustände eingehen. Deshalb sind die unten folgenden Bezeichnungen angeregte Konfigurationen und Elektronenübergänge lediglich im Sinne virtueller Zustände und Übergänge, aber nicht als wahre Anregung des Vielelektronensystems zu verstehen (falls dies nicht besonders vermerkt ist). Die erwähnten Bezeichnungen dürften demnach im folgenden Text nicht zu Unstimmigkeiten führen.

Außerdem ergeben sich Schwierigkeiten bei der Interpretation der Ergebnisse wegen der großen Zahl verschiedener Energieausdrücke, die durch die Einführung verschiedenartiger angeregter Konfigurationen bedingt sind. Bei dem von ANDERSON vorgeschlagenen neuen Lösungsweg tritt die Austauschkopplung schon in niedriger Störungsordnung auf, weil das magnetische Kation und seine diamagnetische Umgebung als einheitlicher Komplex behandelt werden; die Aufgabe, den exakten magnetischen Zustand der einzelnen Kationen zu bestimmen, die mit der diamagnetischen Matrix wechselwirken, wird also als *im Prinzip* von vornherein gelöst betrachtet. Wir haben es also hier im Vergleich zu den früheren Modellen mit einer eigentümlichen Umnormierung der nullten Näherung zu tun.

Im folgenden stellen wir die Theorie der Austauschwechselwirkung in den nichtmetallischen Verbindungen der Übergangsmetalle in ihrer ursprünglichen Formulierung dar und anschließend in der von ANDERSON gegebenen modernen Form. Den ersten dieser Lösungswege bezeichnen wir als Theorie des Superaustausches nach KRAMERS und ANDERSON.

IV. Der Superaustausch nach Kramers und Anderson.

a) Das Problem der drei Zentren und vier Elektronen. Orthogonale Bahnen.

Es ist zu beachten, daß das zu betrachtende Problem der drei Zentren und vier Elektronen eine möglichst genaue qualitative Erklärung der wichtigsten Charakteristika der indirekten Austauschwechselwirkung in realen Kristallen vom Typ MnO erlauben soll. Wichtig ist, wie bereits erwähnt, der Fall, daß sich zwei Ionen eines Übergangsmetalls und das dazwischenliegende nichtmagnetische Anion auf einer Geraden befinden (π- oder 180°-Kopplung, s. Fig. 6) und daß die direkte Wechselwirkung zwischen ihnen nicht merklich ist.

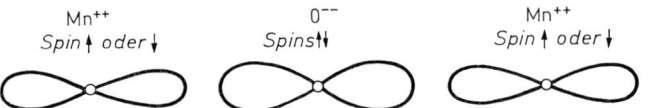

Fig. 6. Grundkonfiguration beim Problem der drei Zentren und vier Elektronen (reine Ionenkonfiguration).

Alle Einelektronen-Wellenfunktionen werden als orthogonal vorausgesetzt. Der Einfachheit halber wird angenommen, daß sich bei den Kationen je ein d-Elektron und am Anion zwei Elektronen in der p-Schale aufhalten[25].

In realen Verbindungen können sich mehrere Elektronen in der d-Schale befinden; jedoch gestattet, wie wir sehen werden, auch das vereinfachte Modell mit nur einem Elektron pro Kation Schlußfolgerungen über das Verhalten des Systems in Abhängigkeit vom Auffüllungsgrad der d-Schale. Es läßt sich in gewissem Grade auch rechtfertigen, daß wir beim Sauerstoff nur zwei Elektronen betrachten an Stelle der sechs im realen Fall. Bekanntlich haben die p-Bahnen die Form dreier Hanteln, deren Achsen senkrecht zueinander stehen. Lediglich die Hantel, deren Achse mit der Achse des Systems Me–O–Me zusammenfällt, liefert einen merklichen Beitrag zur Wechselwirkung. Die Überlappungen der anderen beiden p-Hanteln mit den d-Wellenfunktionen der Kationen werden gewöhnlich vernachlässigt[26].

---

[25] Diese Vereinfachung ist darin begründet, daß sich ein Problem mit mehreren Elektronen pro Kation und mehr als zwei Elektronen am Anion wegen der großen Schwierigkeiten bei der Rechnung praktisch nicht mehr lösen läßt.

[26] Im folgenden wird bei der Berechnung der Energie des Systems nicht unbedingt angenommen, daß die Anionenfunktionen die Symmetrie eben dieser p-Funktionen besitzen.

Wie aus den unten durchgeführten Rechnungen folgt, stimmen die Energien E_s bzw. E_t des Singulettzustandes und des Triplettzustandes des Systems überein, wenn als Grundzustand eine reine Ionenkonfiguration gewählt wird. Somit fehlt in einem Grundzustand, wie er in Fig. 6 schematisch dargestellt ist, die von der gegenseitigen Orientierung der Kationenspins abhängige Wechselwirkung.

Wenn jedoch in der Wellenfunktion des Grundzustandes neben der reinen Ionenkonfiguration auch „angeregte" Konfigurationen berücksichtigt werden, bei denen eines der p-Elektronen des Anions zu einem der Kationen übergeht (s. Fig. 7 und 8), so ändert sich das Bild, und es erfolgt eine Aufspaltung der Multipletts: $E_s \neq E_t$.

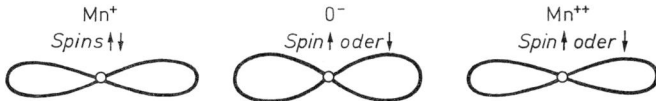

Fig. 7. Angeregte Konfiguration: ein p-Elektron des Anions ist in eine einfach besetzte d-Bahn des Kations übergegangen.

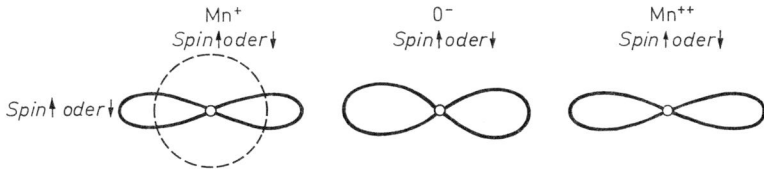

Fig. 8. Angeregte Konfiguration: ein p-Elektron des Anions ist in die erste freie d-Bahn des Kations übergegangen.

Die zusätzlichen Glieder in der Wellenfunktion des Grundzustandes, die mit den „angeregten" Konfigurationen zusammenhängen, führen innerhalb der Störungstheorie zu entsprechenden Zusatzgliedern in der Energie des Systems, die bei quantenmechanischen Rechnungen unter der Bezeichnung „Glieder der *Konfigurationswechselwirkung*" bekannt sind.

Bei der Durchführung konkreter Rechnungen ist mit der Berücksichtigung jener „angeregten" Konfigurationen zu beginnen, die hinsichtlich ihrer Energie dem reinen Ionenzustand am nächsten liegen. Im Zusammenhang damit werden unten die folgenden Konfigurationen betrachtet: 1. die reine Ionenkonfiguration, 2. der Übergang eines p-Elektrons des Anions auf die erste freie Bahn eines Kations, 3. ein entsprechender Übergang auf die einfach besetzte d-Bahn eines Kations, 4. der gleichzeitige Übergang zweier p-Elektronen auf die einfach besetzten d-Bahnen beider Kationen und schließlich 5. der Übergang eines Elektrons von einem Kation zum Anion bei gleichzeitigem Übergang eines Elektrons vom Anion zum anderen Kation.

17. Der „Übergang" in die leere Bahn. Wir kommen nun zur Berechnung der Energie des Systems [*36*], [*38*], [*135*]. Zunächst werden alle möglichen Basisfunktionen des Systems in der Form von Slater-Determinanten herausgeschrieben. Es werden folgende Bezeichnungen eingeführt: $\varphi_{d_1}(\vec{r})$ und $\varphi_{d_2}(\vec{r})$ bedeuten die d-Bahnfunktionen des ersten und zweiten Kations in der Ionenkonfiguration, $\varphi_{d_1'}(\vec{r})$ und $\varphi_{d_2'}(\vec{r})$ die Bahnfunktionen der angeregten Zustände derselben Kationen, $\varphi_p(\vec{r})$ die Bahnfunktion des Anions; die Indizes α und β bezeichnen die zwei möglichen Orientierungen des Elektronenspins. Die Betrachtung beginnen wir mit den „an-

---

(also ungerade hinsichtlich des Inversionsoperators sind). Die Anionenfunktionen können sogar gerade sein, also die Symmetrie des s-Zustandes aufweisen. Folglich ist die verwendete Bezeichnung p für die Anionenfunktionen nur bedingt richtig.

geregten" Konfigurationen vom Typ 2. In einem System mit vier Elektronen können drei Arten von Multipletts auftreten: das Singulett ($S=0$), das Triplett ($S=1$) und das Quintett ($S=2$). Im reinen Ionenzustand sind zwar lediglich Singuletts und Tripletts möglich, in der jetzt berücksichtigten „angeregten" Konfiguration aber auch das Quintett. Da aber die Wellenfunktion des Systems als Superposition der reinen Ionen- und der „angeregten" Konfiguration angesetzt wird, müssen beide Summanden gleiche Multiplizität besitzen. Folglich können die „angeregten" Quintettkonfigurationen hier überhaupt aus der Betrachtung ausgeschlossen werden. Wir bezeichnen mit u_i die Basis-Slater-Determinanten, die der z-Komponente $S_z=0$ des Gesamtspins entsprechen, und mit v_i diejenigen für $S_z=\pm 1$; sie lauten:

$$u_1 = \begin{vmatrix} d_1\alpha \\ p\alpha \\ p\beta \\ d_2\beta \end{vmatrix},\quad u_2 = \begin{vmatrix} d_1\beta \\ p\alpha \\ p\beta \\ d_2\alpha \end{vmatrix},\quad u_3 = \begin{vmatrix} d_1'\alpha \\ d_1\beta \\ p\alpha \\ d_2\beta \end{vmatrix},\quad u_4 = \begin{vmatrix} d_1'\beta \\ d_1\alpha \\ p\alpha \\ d_2\beta \end{vmatrix},\quad u_5 = \begin{vmatrix} d_1'\alpha \\ d_1\alpha \\ p\beta \\ d_2\beta \end{vmatrix},$$

$$u_6 = \begin{vmatrix} d_1'\beta \\ d_1\alpha \\ p\beta \\ d_2\alpha \end{vmatrix},\quad u_7 = \begin{vmatrix} d_1'\alpha \\ d_1\beta \\ p\beta \\ d_2\alpha \end{vmatrix},\quad u_8 = \begin{vmatrix} d_1'\beta \\ d_1\beta \\ p\alpha \\ d_2\alpha \end{vmatrix},\quad u_9 = \begin{vmatrix} d_1\beta \\ p\alpha \\ d_2'\beta \\ d_2\alpha \end{vmatrix},\quad u_{10} = \begin{vmatrix} d_1\beta \\ p\beta \\ d_2'\alpha \\ d_2\alpha \end{vmatrix},$$

$$u_{11} = \begin{vmatrix} d_1\alpha \\ p\beta \\ d_2\beta \\ d_2\alpha \end{vmatrix},\quad u_{12} = \begin{vmatrix} d_1\alpha \\ p\beta \\ d_2'\alpha \\ d_2\beta \end{vmatrix},\quad u_{13} = \begin{vmatrix} d_1\alpha \\ p\alpha \\ d_2'\beta \\ d_2\beta \end{vmatrix},\quad u_{14} = \begin{vmatrix} d_1\beta \\ p\alpha \\ d_2'\alpha \\ d_2\beta \end{vmatrix},$$

$$v_1 = \begin{vmatrix} d_1\alpha \\ p\alpha \\ p\beta \\ d_2\alpha \end{vmatrix},\quad v_2 = \begin{vmatrix} d_1'\beta \\ d_1\alpha \\ p\alpha \\ d_2\alpha \end{vmatrix},\quad v_3 = \begin{vmatrix} d_1'\alpha \\ d_1\beta \\ p\alpha \\ d_2\alpha \end{vmatrix},\quad v_4 = \begin{vmatrix} d_1'\alpha \\ d_1\alpha \\ p\beta \\ d_2\alpha \end{vmatrix},\quad v_5 = \begin{vmatrix} d_1'\alpha \\ d_1\alpha \\ p\alpha \\ d_2\beta \end{vmatrix}, \quad (17.1)$$

$$v_6 = \begin{vmatrix} d_1\alpha \\ p\alpha \\ d_2'\beta \\ d_2\alpha \end{vmatrix},\quad v_7 = \begin{vmatrix} d_1\alpha \\ p\alpha \\ d_2'\alpha \\ d_2\beta \end{vmatrix},\quad v_8 = \begin{vmatrix} d_1\alpha \\ p\beta \\ d_2'\alpha \\ d_2\alpha \end{vmatrix},\quad v_9 = \begin{vmatrix} d_1\beta \\ p\alpha \\ d_2'\alpha \\ d_2\alpha \end{vmatrix},\quad v_{10} = \begin{vmatrix} d_1\beta \\ p\alpha \\ p\beta \\ d_2\beta \end{vmatrix},$$

$$v_{11} = \begin{vmatrix} d_1'\alpha \\ d_1\beta \\ p\beta \\ d_2\beta \end{vmatrix},\quad v_{12} = \begin{vmatrix} d_1'\beta \\ d_1\alpha \\ p\beta \\ d_2\beta \end{vmatrix},\quad v_{13} = \begin{vmatrix} d_1'\beta \\ d_1\beta \\ p\alpha \\ d_2\beta \end{vmatrix},\quad v_{14} = \begin{vmatrix} d_1'\beta \\ d_1\beta \\ p\beta \\ d_2\alpha \end{vmatrix},$$

$$v_{15} = \begin{vmatrix} d_1\beta \\ p\beta \\ d_2'\alpha \\ d_2\beta \end{vmatrix},\quad v_{16} = \begin{vmatrix} d_1\beta \\ p\beta \\ d_2'\beta \\ d_2\alpha \end{vmatrix},\quad v_{17} = \begin{vmatrix} d_1\beta \\ p\alpha \\ d_2'\beta \\ d_2\beta \end{vmatrix},\quad v_{18} = \begin{vmatrix} d_1\alpha \\ p\beta \\ d_2'\beta \\ d_2\beta \end{vmatrix}.$$

Darin sind der Kürze halber Funktionen φ_k mit k bezeichnet. Von den 32 Determinanten gehören die Funktionen u_1, u_2, v_1 und v_{10} zur reinen Ionenkonfiguration und die übrigen 28 zu den „angeregten" Konfigurationen. Zur weitestgehenden Vereinfachung der Säkulargleichung benutzen wir eine bereits oben erwähnte Methode von GRAY und WILLS [136], wobei folgendermaßen verfahren wird: Beim Aufbau der richtigen Funktionen des entarteten Systems kann man diese entweder als Linearkombination der Determinanten u_i und v_i selbst darstellen oder als Überlagerung von Linearkombinationen dieser Determinanten, wobei diese Linearkombinationen dann auch Eigenfunktionen des Operators $\hat{S}^2$ sind und nicht nur von S_z. Im ersten Falle läßt sich die Säkulargleichung vom 32. Grad in eine Gleichung 14. Grades (für $S_z = 0$) und in zwei Gleichungen 9. Grades (für $S_z = \pm 1$) aufspalten. Im zweiten Falle kann die Säkulargleichung in ein Gleichungssystem noch niedrigerer Ordnung überführt werden; denn bei der Berechnung der Matrixelemente zwischen Eigenfunktionen des Operators $\hat{S}^2$ werden alle Elemente gleich Null, die Funktionen mit unterschiedlichem S enthalten, und außerdem verschwinden innerhalb jedes Multipletts (mit vorgegebenem S) alle Matrixelemente zwischen Funktionen mit unterschiedlichem S_z. Zum Aufbau der Multiplett-Basisfunktionen hat man sich folgender bekannter Beziehungen zu bedienen:

wobei
$$\hat{s}^+ \alpha = 0, \quad \hat{s}^+ \beta = \alpha, \quad \hat{s}^- \alpha = \beta, \quad \hat{s}^- \beta = 0,$$
$$\hat{s}^+ = \hat{s}_x + i\hat{s}_y, \quad \hat{s}^- = \hat{s}_x - i\hat{s}_y. \tag{17.2}$$

$\hat{s}_x$ und $\hat{s}_y$ sind die Operatoren der x- bzw. y-Komponente des Spins eines einzelnen Elektrons, und α bzw. β bezeichnet die Spinfunktion eines Elektrons für $s_z = \pm \frac{1}{2}$. Für eine Spinfunktion $\Phi(S, S_z)$ mit dem willkürlichen Spin S gilt

$$\hat{S}^+ \Phi(S, S_z) = (S - S_z)^{\frac{1}{2}}(S + S_z + 1)^{\frac{1}{2}} \Phi(S, S_z + 1),$$
$$\hat{S}^- \Phi(S, S_z) = (S + S_z)^{\frac{1}{2}}(S - S_z + 1)^{\frac{1}{2}} \Phi(S, S_z - 1). \tag{17.3}$$

Da sich die Quantenzahl S im Gegensatz zu S_z unter der Wirkung der Operatoren $\hat{S}^+$ oder $\hat{S}^-$ nicht ändert, wird verständlich, daß man die Funktionen mit vorgegebenem S und allen möglichen S_z vollständig erhält, wenn man mit dem Operator $\hat{S}^-$ auf die Funktion des Satzes einwirkt, die zum maximalen S_z gehört, und diese Prozedur unterschiedlich oft wiederholt, bis man schließlich bei der Funktion mit $S_z = -S$ anlangt. Dabei sind diese Funktionen automatisch orthogonal zueinander und normiert.

Wir bezeichnen die Singulettfunktionen mit $\Psi_{\alpha s}$ und die Triplettfunktionen mit $^{S_z}\Psi_{\alpha t}$. Die Funktionen $^{S_z}\Psi_{\alpha t}$ setzen wir als Reihen an, die nur die Funktionen v (bei $S_z = \pm 1$) bzw. lediglich die Funktionen u (bei $S_z = 0$) aus (17.1) enthalten, in Übereinstimmung mit dem wohlbekannten Umstand, daß Funktionen mit unterschiedlichen S_z miteinander nicht kombinieren. So erhalten wir für $S_z = +1$

$$^1\Psi_{\alpha t} = \sum_{i=1}^{9} c_{\alpha i} v_i. \tag{17.4}$$

Die Bestimmungsgleichung für die Koeffizienten $c_{\alpha i}$ folgt aus der Orthogonalitätsbedingung für die Funktion $^1\Psi_{\alpha t}$

$$\int {}^1\Psi_{\alpha t}^* \, {}^1\Psi_{\beta t} \, d\tau = \delta_{\alpha\beta} \tag{17.5}$$

und aus der Forderung, daß die Funktionen $^1\Psi_{\alpha t}$ Eigenfunktionen des Operators $\hat{S}^2$ mit dem Eigenwert $S(S+1)$ sein sollen, wobei hier $S = 1$ ist:

$$\hat{S}^2 \, {}^1\Psi_{\alpha t} = S(S+1) \, {}^1\Psi_{\alpha t} = 2 \, {}^1\Psi_{\alpha t}. \tag{17.6}$$

Den Operator $\hat{S}^2$ stellt man zweckmäßigerweise in der Form

$$\hat{S}^2 = \hat{S}^- \hat{S}^+ + \hat{S}_z + \hat{S}_z^2 \tag{17.7}$$

dar, da die Eigenschaften der Operatoren auf der rechten Seite von (17.7) schon bekannt sind. Die Bedingungen (17.5) und (17.6) führen zu den folgenden Gleichungen für die $c_{\alpha i}$:

$$\sum_{i=1}^{9} c_{\alpha i} c_{\beta i} = \delta_{\alpha\beta}, \quad \sum_{i=2}^{5} c_{\alpha i} = 0, \quad \sum_{i=6}^{9} c_{\alpha i} = 0. \tag{17.8}$$

Die Lösungen des Gleichungssystems (17.8) kann man leicht finden; die ihnen entsprechenden Funktionen haben die Form [27]:

$$\left.\begin{aligned}
{}^1\Psi_{1t} &= v_1, & {}^1\Psi_{4t} &= \frac{1}{\sqrt{2}}(v_2 - v_4), \\
{}^1\Psi_{2t} &= \frac{1}{2}(v_3 + v_5 - v_2 - v_4), & {}^1\Psi_{5t} &= \frac{1}{2}(v_7 + v_9 - v_6 - v_8), \\
{}^1\Psi_{3t} &= \frac{1}{\sqrt{2}}(v_5 - v_3), & {}^1\Psi_{6t} &= \frac{1}{\sqrt{2}}(v_7 - v_9), \\
& & {}^1\Psi_{7t} &= \frac{1}{\sqrt{2}}(v_6 - v_8).
\end{aligned}\right\} \tag{17.9}$$

Die Triplettfunktionen ${}^0\Psi_{\alpha t}$ erhält man durch Einwirkung des Operators $\hat{S}^-$ aus den Funktionen (17.9) und die Funktionen ${}^{-1}\Psi_{\alpha t}$ durch Anwendung desselben Operators auf die ${}^0\Psi_{\alpha t}$:

$$\left.\begin{aligned}
{}^0\Psi_{1t} &= \frac{1}{\sqrt{2}}(u_1 + u_2), & {}^0\Psi_{4t} &= \frac{1}{2}(u_4 - u_5 - u_7 + u_8), \\
{}^0\Psi_{2t} &= \frac{1}{\sqrt{2}}(u_3 - u_6), & {}^0\Psi_{5t} &= \frac{1}{\sqrt{2}}(u_{14} - u_{11}), \\
{}^0\Psi_{3t} &= \frac{1}{2}(u_4 + u_5 - u_7 - u_8), & {}^0\Psi_{6t} &= \frac{1}{2}(u_{12} + u_{13} - u_9 - u_{10}), \\
& & {}^0\Psi_{7t} &= \frac{1}{2}(u_9 - u_{10} - u_{12} + u_{13});
\end{aligned}\right\} \tag{17.10}$$

$$\left.\begin{aligned}
{}^{-1}\Psi_{1t} &= v_{10}, & {}^{-1}\Psi_{4t} &= \frac{1}{\sqrt{2}}(v_{13} - v_{11}), \\
{}^{-1}\Psi_{2t} &= \frac{1}{2}(v_{11} + v_{13} - v_{12} - v_{14}), & {}^{-1}\Psi_{5t} &= \frac{1}{2}(v_{15} + v_{17} - v_{16} - v_{18}), \\
{}^{-1}\Psi_{3t} &= \frac{1}{\sqrt{2}}(v_{12} - v_{14}), & {}^{-1}\Psi_{6t} &= \frac{1}{\sqrt{2}}(v_{18} - v_{16}), \\
& & {}^{-1}\Psi_{7t} &= \frac{1}{\sqrt{2}}(v_{17} - v_{15}).
\end{aligned}\right\} \tag{17.11}$$

Zu (17.5) und (17.6) analoge Bedingungen führen auf folgendes Gleichungssystem für die Koeffizienten $c_{\alpha i}$ in der Entwicklung der Singulettfunktion $\Psi_{\alpha s}$:

$$\left.\begin{aligned}
&\sum_{i=1}^{14} c_{\alpha i} c_{\beta i} = \delta_{\alpha\beta}, \quad c_{\alpha 1} + c_{\alpha 2} = 0, \\
&\sum_{i=3}^{8} c_{\alpha i} = 0, \quad \sum_{i=9}^{14} c_{\alpha i} = 0, \\
&c_{\alpha 3} - c_{\alpha 6} = c_{\alpha 4} - c_{\alpha 7} = c_{\alpha 5} - c_{\alpha 8} = c_{\alpha 9} - c_{\alpha 12} = c_{\alpha 10} - c_{\alpha 13} = c_{\alpha 11} - c_{\alpha 14} = 0.
\end{aligned}\right\} \tag{17.12}$$

---

[27] Selbstverständlich ist der Satz (17.9) durch (17.8) nur bis auf eine unitäre Transformation bestimmt.

Schließlich lauten die Singulettfunktionen selbst:

$$\Psi_{1s} = \frac{1}{\sqrt{2}}(u_1 - u_2), \qquad \Psi_{4s} = \frac{1}{2}(u_9 - u_{10} + u_{12} - u_{13}),$$

$$\Psi_{2s} = \frac{1}{2}(u_3 - u_4 + u_6 - u_7), \qquad \Psi_{5s} = \frac{1}{2\sqrt{3}}(u_9 + u_{10} + u_{12} + u_{13} - 2u_{11} - 2u_{14}). \quad (17.13)$$

$$\Psi_{3s} = \frac{1}{2\sqrt{3}}(u_3 + u_4 + u_6 + u_7 - 2u_5 - 2u_8),$$

Nachdem man die Multiplettfunktionen gefunden hat, muß man die Energie des Systems bestimmen, die gegeben ist durch die Wurzeln der Säkulargleichung

$$|H_{ij} - E\delta_{ij}| = 0. \qquad (17.14)$$

Dabei ist

$$H_{ij} = \int \Psi_i^* \hat{H} \Psi_j \, d\tau, \qquad (17.15)$$

$\hat{H}$ der Hamilton-Operator des Systems und Ψ_i eine Wellenfunktion $^{S_z}\Psi_{\alpha t}$ oder $\Psi_{\alpha s}$. (17.14) ergibt im vorliegenden Fall drei Gleichungen siebenten Grades für den Triplettzustand [s. (17.9) bis (17.11)] und eine Gleichung fünften Grades für den Singulettzustand [s. (17.13)]. Uns interessiert die Energiedifferenz zwischen den niedrigsten Triplett- und Singulettniveaus. Wir gehen nun zur Bestimmung der Matrixelemente (17.15) über und nehmen, wie bereits oben erwähnt, an, daß sich die Wellenfunktionen der Kationen nicht überlappen und folglich Integrale, welche die Produkte $\varphi_{d_1}(\vec{r}) \cdot \varphi_{d_2}(\vec{r})$, $\varphi_{d_1'}(\vec{r}) \cdot \varphi_{d_2'}(\vec{r})$ sowie $\varphi_{d_1'}(\vec{r}) \cdot \varphi_{d_2}(\vec{r})$ und $\varphi_{d_1}(\vec{r}) \cdot \varphi_{d_2'}(\vec{r})$ enthalten, verschwinden.

Wir schreiben den Hamilton-Operator des Systems in der Form

$$\hat{H} = \hat{H}_1 + V, \qquad (17.16)$$

$$\hat{H}_1 = \sum_{i=1}^{4} \hat{H}_1(i), \qquad (17.17)$$

$$V = \tfrac{1}{2} \sum_{i,j=1}^{4}{}' V(i,j), \qquad (17.18)$$

worin $\hat{H}_1(i)$ den Einelektronenanteil des Hamilton-Operators einschließlich des kinetischen Gliedes sowie der Wechselwirkung des i-ten Elektrons mit allen Kernen und $V(i,j)$ den Operator der Elektron-Elektron-Wechselwirkung für das Paar ij bedeuten. Zur Berechnung der Matrixelemente $\langle u_i \hat{H} u_j \rangle$ und $\langle v_i \hat{H} v_j \rangle$ muß man links vom Operator $\hat{H}$ einen „Repräsentanten" (z.B. ein Diagonalelement) aus der Determinante (17.1) einsetzen und unter den 4! Gliedern der Determinante, die rechts von $\hat{H}$ steht, jene heraussuchen, die ein von Null verschiedenes Integral liefern. Dabei ist die Orthogonalität der Bahn- und Spinfunktionen zu beachten. Wie man leicht sieht, ist das nichtdiagonale Matrixelement des Einelektronenanteils $\hat{H}_1$ des Hamilton-Operators nur dann von Null verschieden, wenn rechts und links vom Operator Funktionen stehen, die sich lediglich in einem Atombahnzustand unterscheiden, während die drei restlichen einander paarweise gleich sind, und wenn die Elektronen in diesen Bahnzuständen paarweise auch in den Spinquantenzahlen übereinstimmen. Offensichtlich können die nichtdiagonalen Einelektronen-Matrixelemente nur dann von Null verschieden sein, wenn eine der Funktionen zur reinen Ionen- und die andere zur „angeregten" Konfiguration gehört. Die Matrixelemente des Zweielektronenoperators V stellen Austauschintegrale dar. Sie sind von Null verschieden, wenn innerhalb der Wellenfunktionen rechts und links von V wenigstens zwei Spin-Bahn-Zustände gleich sind.

Nach Berechnung der Matrixelemente (17.15) in der Säkulargleichung (17.14) erhält man folgende Gleichungen zur Bestimmung der Energien der Triplett- bzw. Singulettzustände:

$$\begin{vmatrix} A & 0 & 0 & \sqrt{2}\varrho & 0 & 0 & \pm\sqrt{2}\varrho \\ 0 & H_{11} & H_{12} & H_{13} & -D & 0 & 0 \\ 0 & H_{12} & H_{22} & H_{23} & 0 & -D & 0 \\ \sqrt{2}\varrho & H_{13} & H_{23} & H_{33} & 0 & 0 & D \\ 0 & -D & 0 & 0 & H_{11} & -H_{12} & H_{13} \\ 0 & 0 & -D & 0 & -H_{12} & H_{22} & -H_{23} \\ \pm\sqrt{2}\varrho & 0 & 0 & D & H_{13} & -H_{23} & H_{33} \end{vmatrix} = 0, \quad (17.19)$$

$$\begin{vmatrix} A & 0 & \sqrt{2}\varrho & 0 & \pm\sqrt{2}\varrho \\ 0 & h_{11} & h_{12} & -D & 0 \\ \sqrt{2}\varrho & h_{12} & h_{22} & 0 & -D \\ 0 & -D & 0 & h_{11} & -h_{12} \\ \pm\sqrt{2}\varrho & 0 & -D & -h_{12} & h_{22} \end{vmatrix} = 0 \quad (17.20)$$

mit

$$\left. \begin{aligned} &A = E_0 - 2J_{pd} - E, \\ &H_{11} = E_1 - J_{pd'} - E, \\ &H_{22} = E_1 - J_{pd} - J_{pd'} - \tfrac{1}{2}J_{d'd} - E, \\ &H_{33} = E_1 - J_{pd} + J_{pd'} - \tfrac{1}{2}J_{dd'} - E, \\ &H_{12} = -\tfrac{1}{\sqrt{2}} J_{dd'}, \qquad h_{11} = E_1 + J_{pd} - J_{pd'} + \tfrac{1}{2}J_{dd'} - E, \\ &H_{13} = \tfrac{1}{\sqrt{2}}(2J_{pd} - J_{dd'}), \quad h_{22} = E_1 - J_{pd} + J_{pd'} - \tfrac{1}{2}J_{dd'} - E, \\ &H_{23} = \tfrac{1}{2} J_{dd'}, \qquad h_{12} = \tfrac{\sqrt{3}}{2} J_{dd'}. \end{aligned} \right\} \quad (17.21)$$

Hierin bedeuten

$$\left. \begin{aligned} J_{pd} &= \int \varphi_p^*(\vec{r}_1)\, \varphi_d^*(\vec{r}_2)\, V(\vec{r}_1, \vec{r}_2)\, \varphi_p(\vec{r}_2)\, \varphi_d(\vec{r}_1)\, d\vec{r}_1\, d\vec{r}_2, \\ J_{pd'} &= \int \varphi_p^*(\vec{r}_1)\, \varphi_{d'}^*(\vec{r}_2)\, V(\vec{r}_1, \vec{r}_2)\, \varphi_p(\vec{r}_2)\, \varphi_{d'}(\vec{r}_1)\, d\vec{r}_1\, d\vec{r}_2, \\ J_{dd'} &= \int \varphi_{d'}^*(\vec{r}_1)\, \varphi_d^*(\vec{r}_2)\, V(\vec{r}_1, \vec{r}_2)\, \varphi_{d'}(\vec{r}_2)\, \varphi_d(\vec{r}_1)\, d\vec{r}_1\, d\vec{r}_2, \\ D &= \int \varphi_p^*(\vec{r}_1)\, \varphi_{d_1'}^*(\vec{r}_2)\, V(\vec{r}_1, \vec{r}_2)\, \varphi_p(\vec{r}_2)\, \varphi_{d_2'}(\vec{r}_1)\, d\vec{r}_1\, d\vec{r}_2, \\ \varrho &= \int \varphi_p^*(\vec{r})\, \hat{H}_1(\vec{r})\, \varphi_{d_1'}(\vec{r})\, d\vec{r} = \pm \int \varphi_p^*(\vec{r})\, \hat{H}_1(\vec{r})\, \varphi_{d_2'}(\vec{r})\, d\vec{r}. \end{aligned} \right\} \quad (17.22)$$

E_0 und E_1 sind die Summen der kinetischen und der Coulomb-Energie im Grund- bzw. angeregten Zustand. Das Vorzeichen von ϱ richtet sich nach dem Symmetriecharakter der Anionenfunktion. Bei der Aufstellung der Säkulargleichungen (17.19) und (17.20) vernachlässigten wir ein Integral,

$$\varrho_1 = \int \varphi_p^*(\vec{r}_1)\, \varphi_d^*(\vec{r}_2)\, V(\vec{r}_1, \vec{r}_2)\, \varphi_{d'}(\vec{r}_2)\, \varphi_d(\vec{r}_1)\, d\vec{r}_1\, d\vec{r}_2, \quad (17.23)$$

das den indirekten Übergang $p \to d$, $d \to d'$ beschreibt; es wird gewöhnlich im Vergleich zu ϱ, das dem direkten Übergang $p \to d'$ entspricht, als klein angenommen. Das Integral D beschreibt die Wechselwirkung zwischen zwei angeregten Zuständen.

Wir bemerken, daß die mit $^1\Psi_{\alpha t}$, $^0\Psi_{\alpha t}$ und $^{-1}\Psi_{\alpha t}$ aufgebauten Stufen der Determinante (17.14) zu ein und derselben Gl. (17.19) führen.

Die Determinanten auf der linken Seite der Gln. (17.19) und (17.20) lassen sich in zwei Blöcke zerlegen, von denen nur einer mit der reinen Ionenkonfiguration zusammenhängt. Wir erhalten so an Stelle von (17.19) und (17.20)

$$\begin{vmatrix} A & 0 & 0 & 2\varrho \\ 0 & H_{11}-D & H_{12} & H_{13} \\ 0 & H_{12} & H_{22}\pm D & H_{23} \\ 2\varrho & H_{13} & H_{23} & H_{33}\pm D \end{vmatrix} = 0, \tag{17.24}$$

$$\begin{vmatrix} H_{11}\pm D & -H_{12} & H_{13} \\ -H_{12} & H_{22}\mp D & -H_{23} \\ H_{13} & -H_{23} & H_{33}\mp D \end{vmatrix} = 0, \tag{17.25}$$

$$\begin{vmatrix} A & 0 & 2\varrho \\ 0 & h_{11}\pm D & h_{12} \\ 2\varrho & h_{12} & h_{22}\mp D \end{vmatrix} = 0, \tag{17.26}$$

$$\begin{vmatrix} h_{11}\mp D & -h_{12} \\ -h_{12} & h_{22}\pm D \end{vmatrix} = 0, \tag{17.27}$$

wobei das obere Vorzeichen einer geraden und das untere einer ungeraden Anionenfunktion entspricht.

Die Gln. (17.24) und (17.26) enthalten die reine Ionenkonfiguration. Da uns lediglich Zustände interessieren, in denen unbedingt eine reine Ionenkonfiguration vorliegt, betrachten wir im folgenden nur diese Gleichungen. Sie lassen sich zwar exakt lösen, doch ergeben sich dabei für die Lösungen so komplizierte Ausdrücke, daß sie sehr schwer zu analysieren sind. Deshalb lösen wir diese Gleichungen nach der Störungsmethode. Wir nehmen an, daß alle Matrixelemente in (17.24) und (17.26) klein sind im Vergleich zur Energiedifferenz E_1-E_0. Dann kann man die Lösungen in Form einer Reihe nach reziproken Potenzen der Differenz E_1-E_0 ansetzen. Diese Prozedur ist elementar, jedoch recht umständlich. Nach dem Ergebnis unterscheiden sich die Energien E_t und E_s des Tripletts bzw. des Singuletts erst im vierten Glied der Reihe, also in der vierten Ordnung der Störungstheorie. Die Differenz dieser Energien ist gleich

$$E_s - E_t = -\frac{8\varrho^2}{(E_1-E_0)^3} J_{pd}(J_{dd'}-J_{pd}). \tag{17.28}$$

Das Matrixelement D tritt erst in den folgenden Ordnungen der Störungstheorie auf.

Wenn wir für das effektive Integral des indirekten Austausches die Bezeichnung

$$J = -\frac{4\varrho^2}{(E_1-E_0)^3} J_{pd}(J_{dd'}-J_{pd}) \tag{17.29}$$

einführen, erhält (17.28) die Form

$$E_s - E_t = 2J. \tag{17.30}$$

In diesem Falle läßt sich, wie bereits in Ziff. 1 gezeigt wurde, der Energieoperator des Systems in der üblichen Diracschen Form darstellen,

$$\hat{H} = -2J\vec{s}_{d_1}\vec{s}_{d_2}, \tag{17.31}$$

wobei $\vec{s}_{d_i}$ den Spinoperator des d-Elektrons am Kation i bedeutet.

Das Ergebnis (17.28) kann man auch auf eine weniger aufwendige Art erhalten. Hierfür verwenden wir eine spezielle Methode, nämlich die von LÖWDIN [*137*] vorgeschlagene quantenmechanische Störungstheorie, die dann auch weiter benutzt werden wird. Das Wesentlichste an dieser Methode besteht darin, daß mit Hilfe eines Kunstgriffes der Grad der Säkulargleichung (17.14) herabgesetzt und damit deren Lösung erleichtert wird. Der Satz der Basis-Wellenfunktionen, mit denen die Matrixelemente (17.15) berechnet werden, wird willkürlich in zwei Systeme $\{A\}$ und $\{B\}$ zerlegt. Im weiteren muß der Einfluß der Zustände $\{B\}$ auf die Zustände der Klasse $\{A\}$ untersucht werden. Wir schreiben das Gleichungssystem zur Bestimmung der Koeffizienten in den wahren Funktionen nullter Näherung in der Form

$$(E - H_{mm}) b_m = \sum_n^{\{A\}} H'_{mn} b_n + \sum_n^{\{B\}} H'_{mn} b_n, \qquad (17.32)$$

wobei die Symbole über den Summenzeichen bedeuten, daß die Summation über die Zustände der angegebenen Klasse von Basisfunktionen zu erfolgen hat. Der Strich beim Matrixelement besagt, daß alle diagonalen Elemente wegzulassen sind. Unter Verwendung der Bezeichnung

$$h_{mn} = \frac{H_{mn}}{E - H_{mm}} \qquad (17.33)$$

erhalten wir an Stelle von (17.32)

$$b_m = \sum_n^{\{A\}} h'_{mn} b_n + \sum_n^{\{B\}} h'_{mn} b_n. \qquad (17.34)$$

Ersetzt man die Koeffizienten b_n in der zweiten Summe auf der rechten Seite der Gl. (17.34) durch ihre Ausdrücke nach (17.34) (Iterationsverfahren), so erhalten wir die Entwicklung

$$b_m = \sum_n^{\{A\}} \left(h'_{mn} + \sum_\alpha^{\{B\}} h'_{m\alpha} h'_{\alpha n} + \sum_{\alpha\beta}^{\{B\}} h'_{m\alpha} h'_{\alpha\beta} h'_{\beta n} + \ldots \right) b_n. \qquad (17.35)$$

Bei Einführung der neuen Bezeichnung

$$U^A_{mn} = H_{mn} + \sum_\alpha^{\{B\}} \frac{H'_{m\alpha} H'_{\alpha n}}{E - H_{\alpha\alpha}} + \sum_{\alpha\beta}^{\{B\}} \frac{H'_{m\alpha} H'_{\alpha\beta} H'_{\beta n}}{(E - H_{\alpha\alpha})(E - H_{\beta\beta})} + \cdots \qquad (17.36)$$

folgt endgültig

$$b_m = \sum_n^{\{A\}} \frac{U^A_{mn} - H_{mn} \delta_{mn}}{E - H_{mm}}. \qquad (17.37)$$

Somit besitzen wir zur Bestimmung der Koeffizienten b_m mit m aus der Klasse $\{A\}$ oder aus der Klasse $\{B\}$ die beiden Beziehungen

$$\sum_n^{\{A\}} (U^A_{mn} - E \delta_{mn}) b_n = 0, \qquad m \in \{A\}, \qquad (17.38)$$

$$b_m = \sum_n^{\{A\}} \frac{U^A_{mn}}{E - H_{mm}} b_n, \qquad m \in \{B\}. \qquad (17.39)$$

Die Verträglichkeitsbedingung für das homogene System (17.38) lautet

$$|U^A_{mn} - E \delta_{mn}| = 0. \qquad (17.40)$$

(17.40) hat die gleiche Form wie die Ausgangs-Säkulargleichung (17.14); sie ist jedoch von niedrigerem Grad, da die Indizes m und n lediglich das System $\{A\}$

und nicht wie bei (17.14) beide Systeme durchlaufen. Somit erhielten wir eine neue Säkulargleichung für das System mit einem effektiven Hamilton-Operator U, dessen Matrixelemente durch die Beziehung (17.36) bestimmt werden. Die Gl. (17.40) ist im Gegensatz zu (17.14) nicht algebraisch, sondern transzendent, da die Matrixelemente (17.36) selbst Funktionen der unbekannten Energie E darstellen. Da sie sich nicht exakt lösen läßt, wird sie im folgenden störungstheoretisch behandelt.

Als System $\{A\}$ wählen wir naturgemäß die Funktion Ψ_{1s} mit reiner Ionenkonfiguration für den Singulettzustand und eine der Funktionen $^1\Psi_{1t}$, $^0\Psi_{1t}$ oder $^{-1}\Psi_{1t}$ für den Triplettzustand derselben Konfiguration. Das System $\{B\}$ bilden von den Singulettzuständen die $\Psi_{\alpha s}$ mit $\alpha = 2, 3, 4, 5$ sowie einer der Sätze $^1\Psi_{\alpha t}$, $^0\Psi_{\alpha t}$ oder $^{-1}\Psi_{\alpha t}$ mit $\alpha = 2, 3, 4, 5, 6, 7$ für das Triplett. Sie besitzen alle eine „angeregte" Konfiguration. Es ist gleichgültig, welcher Funktionensatz $^{S_z}\Psi_{\alpha t}$ mit $\alpha = 1, 2, \ldots, 7$ bei den Rechnungen als System $\{A\}+\{B\}$ Verwendung findet, da alle selbstverständlich zu ein und demselben Ergebnis führen. Der Einfluß der „angeregten" Zustände $\Psi_{\alpha s}$ ($\alpha = 2, \ldots, 5$) und $^{S_z}\Psi_{\alpha t}$ ($\alpha = 2, \ldots, 7$) auf die Zustände Ψ_{1s} bzw. $^{S_z}\Psi_{1t}$ wird durch die Beziehung (17.36) berücksichtigt.

Berechnen wir die Energie des Singulettzustandes. Da im System $\{A\}$ nur eine Funktion Ψ_{1s} enthalten ist, erhalten wir aus (17.40) sofort

$$E = U_{11}^A. \tag{17.41}$$

Wegen (17.36) genügt somit die Energie E einer Gleichung

$$\left.\begin{aligned}E = H_{11} + \sum_{\alpha=2}^{5} \frac{H_{1\alpha}^2}{E - H_{\alpha\alpha}} + \sum_{\alpha,\beta=2}^{5} \frac{H_{1\alpha} H'_{\alpha\beta} H_{\beta 1}}{(E - H_{\alpha\alpha})(E - H_{\beta\beta})} + \\ + \sum_{\alpha,\beta,\gamma=2}^{5} \frac{H_{1\alpha} H'_{\alpha\beta} H'_{\beta\gamma} H_{\gamma 1}}{(E - H_{\alpha\alpha})(E - H_{\beta\beta})(E - H_{\gamma\gamma})} + \cdots,\end{aligned}\right\} \tag{17.42}$$

wo die Summation über die Funktionen $\Psi_{\alpha s}$ der „angeregten" Konfigurationen erfolgt. Mit Hilfe einer Iteration erhalten wir für $(E - H_{\alpha\alpha})^{-1}$ die Reihenentwicklung

$$\frac{1}{E - H_{\alpha\alpha}} = \frac{1}{E_0 - E_1} + \frac{-\widetilde{H}_{11} + \widetilde{H}_{\alpha\alpha}}{(E_0 - E_1)^2} + \frac{(-\widetilde{H}_{11} + \widetilde{H}_{\alpha\alpha})^2 - \sum_{\beta=2}^{5} H_{1\beta}^2}{(E_0 - E_1)^3} + \cdots \tag{17.43}$$

mit den Bezeichnungen

$$\widetilde{H}_{11} = H_{11} - E_0, \quad \widetilde{H}_{\alpha\alpha} = H_{\alpha\alpha} - E_1. \tag{17.44}$$

Setzt man (17.43) in (17.42) ein, erhält man

$$\left.\begin{aligned}E = H_{11} + \frac{1}{E_0 - E_1} \sum_{\alpha=2}^{5} H_{1\alpha}^2 + \frac{1}{(E_0 - E_1)^2} \times \\ \times \left\{ \sum_{\alpha=2}^{5} H_{1\alpha}^2 (-\widetilde{H}_{11} + \widetilde{H}_{\alpha\alpha}) + \sum_{\alpha,\beta=2}^{5} H_{1\alpha} H'_{\alpha\beta} H_{\beta 1} \right\} + \\ + \frac{1}{(E_0 - E_1)^3} \left\{ \sum_{\alpha=2}^{5} H_{1\alpha}^2 \left[(-\widetilde{H}_{11} + \widetilde{H}_{\alpha\alpha})^2 - \sum_{\beta=2}^{5} H_{1\beta}^2 \right] + \right. \\ \left. + \sum_{\alpha,\beta=2}^{5} H_{1\alpha} H'_{\alpha\beta} H_{\beta 1} (-2\widetilde{H}_{11} + \widetilde{H}_{\alpha\alpha} + \widetilde{H}_{\beta\beta}) + \sum_{\alpha,\beta,\gamma=2}^{5} H_{1\alpha} H'_{\alpha\beta} H'_{\beta\gamma} H_{\gamma 1} \right\} + \cdots\end{aligned}\right\} \tag{17.45}$$

In der Annahme, daß die Differenz $|E_0 - E_1|$ größer ist als die Integrale in den Zählern der Summanden von (17.45), können wir diesen Ausdruck wie eine Reihe der Störungstheorie verwenden. Setzt man in (17.45) die Matrixelemente aus (17.21) und (17.22) ein, so erhält man für die Energie des Singulettzustandes den Ausdruck

$$\begin{aligned} E_s = E_0 - 2J_{pd} + \frac{4\varrho^2}{E_0 - E_1} + \frac{4\varrho^2}{(E_0 - E_1)^2}\left(J_{pd} + J_{pd'} - \tfrac{1}{2}J_{d'd} - D\right) + \\ + \frac{4\varrho^2}{(E_0 - E_1)^3}\{-4\varrho^2 + J_{pd}^2 + J_{pd'}^2 + J_{d'd}^2 + D^2 + 2J_{pd}J_{pd'} - \\ - J_{pd'}J_{d'd} - J_{pd}J_{d'd} - D(2J_{pd} + 2J_{pd'} - J_{d'd})\}. \end{aligned} \quad (17.46)$$

Die Energie des Tripletts, wo sich in der Klasse $\{A\}$ ebenfalls nur je eine Funktion $^{S_z}\Psi_{1t}$ befindet, erhält man nach genau demselben Verfahren. Die Energiedifferenz $E_s - E_t$ wird wieder durch die Beziehung (17.28) bestimmt.

Zum Schluß betrachten wir den interessanten Fall, daß in der angeregten Konfiguration das zum Kation übergegangene zusätzliche d'-Elektron so stark mit den dort befindlichen d-Elektronen wechselwirkt, daß das Austauschintegral $J_{d'd}$ zwischen ihnen schon nicht mehr als klein im Vergleich zur Energiedifferenz $|E_0 - E_1|$ betrachtet werden kann. In diesem Fall muß der Parameter $J_{d'd}$ unbedingt in den nichtangeregten Zustand einbezogen werden. Bei der Berechnung der Energie dürfen die Funktionen $^{S_z}\Psi_{\alpha t}$ und $\Psi_{\alpha s}$ nicht unmittelbar verwendet werden, da die mittels dieser Funktionen aufgebauten Energiematrizen hinsichtlich $J_{d'd}$ nicht diagonal sind [s. (17.19) und (17.20)]. Jedoch läßt sich eine solche Diagonalisierung leicht durchführen, indem man entsprechende Linearkombinationen der ursprünglichen Funktionen $^{S_z}\Psi_{\alpha t}$ aus (17.9) bis (17.11) und $\Psi_{\alpha s}$ aus (17.13) aufsucht. Die im Hinblick auf $J_{d'd}$ diagonale Darstellung wird in folgenden Funktionen verwirklicht:

$$\begin{aligned} ^1\Psi'_{1t} &= {}^1\Psi_{1t}, \\ ^1\Psi'_{2t} &= \tfrac{1}{2}(\sqrt{2}\,{}^1\Psi_{2t} - {}^1\Psi_{3t} - {}^1\Psi_{4t}), & ^1\Psi'_{5t} &= \tfrac{1}{2}(\sqrt{2}\,{}^1\Psi_{5t} - {}^1\Psi_{6t} + {}^1\Psi_{7t}), \\ ^1\Psi'_{3t} &= \tfrac{1}{2}(\sqrt{2}\,\Psi_{5t} + {}^1\Psi_{6t} - {}^1\Psi_{7t}), & ^1\Psi'_{6t} &= \tfrac{1}{\sqrt{2}}({}^1\Psi_{3t} - {}^1\Psi_{4t}), \\ ^1\Psi'_{4t} &= \tfrac{1}{2}(\sqrt{2}\,{}^1\Psi_{2t} + {}^1\Psi_{3t} + {}^1\Psi_{4t}), & ^1\Psi'_{7t} &= \tfrac{1}{\sqrt{2}}({}^1\Psi_{6t} + {}^1\Psi_{7t}), \\ \Psi'_{1s} &= \Psi_{1s}, \\ \Psi'_{2s} &= \tfrac{1}{2}(\sqrt{3}\,\Psi_{2s} + \Psi_{3s}), & \Psi'_{4s} &= \tfrac{1}{2}(\Psi_{2s} - \sqrt{3}\,\Psi_{3s}), \\ \Psi'_{3s} &= \tfrac{1}{2}(\sqrt{3}\,\Psi_{4s} - \Psi_{5s}), & \Psi'_{5s} &= \tfrac{1}{2}(\Psi_{4s} + \sqrt{3}\,\Psi_{5s}). \end{aligned} \quad (17.47)$$

Die mittels der Funktionen (17.47) aufgestellten Energiematrizen lauten für das Triplett bzw. Singulett

$$\begin{pmatrix} A & -\varrho/\sqrt{2} & -\varrho/\sqrt{2} & \varrho/\sqrt{2} & \varrho/\sqrt{2} & -\varrho & \varrho \\ -\varrho/\sqrt{2} & H'_{11} & H'_{12} & H'_{13} & H'_{12} & -\sqrt{2}H'_{13} & -\sqrt{2}H'_{12} \\ -\varrho/\sqrt{2} & H'_{12} & H'_{11} & J'_{12} & H'_{13} & \sqrt{2}H'_{12} & \sqrt{2}H'_{13} \\ \varrho/\sqrt{2} & H'_{13} & H'_{12} & H'_{22} & H'_{12} & B & \sqrt{2}H'_{12} \\ \varrho/\sqrt{2} & H'_{12} & H'_{13} & H'_{12} & H'_{22} & -\sqrt{2}H'_{12} & -B \\ \varrho & -\sqrt{2}H'_{13} & \sqrt{2}H'_{12} & B & -\sqrt{2}H'_{12} & H'_{33} & 0 \\ \varrho & -\sqrt{2}H'_{12} & \sqrt{2}H'_{13} & \sqrt{2}H'_{12} & -B & 0 & H'_{33} \end{pmatrix} \quad (17.48)$$

bzw.

$$\begin{pmatrix} A & \varrho/\sqrt{2} & -\varrho/\sqrt{2} & -\sqrt{\frac{3}{2}}\varrho & \sqrt{\frac{3}{2}}\varrho \\ \varrho/\sqrt{2} & h'_{11} & H'_{12} & \sqrt{3}\,H'_{13} & \sqrt{3}\,H'_{12} \\ -\varrho/\sqrt{2} & H'_{12} & h'_{11} & \sqrt{3}\,H'_{12} & \sqrt{3}\,H'_{13} \\ -\sqrt{\frac{3}{2}}\varrho & \sqrt{3}\,H'_{13} & \sqrt{3}\,H'_{12} & h'_{22} & -H'_{12} \\ \sqrt{\frac{3}{2}}\varrho & \sqrt{3}\,H'_{12} & \sqrt{3}\,H'_{13} & -H'_{12} & h'_{22} \end{pmatrix}. \qquad (17.49)$$

Dabei bedeuten

$$\left.\begin{array}{l} H'_{11}=E_1-\dfrac{3}{2}J_{pd}+J_{d'd}-\dfrac{1}{2}J_{pd'}-E, \quad H'_{12}=-\dfrac{1}{2}J_{pd_1,d_2p}, \\[4pt] B=-\dfrac{1}{\sqrt{2}}(J_{pd}+J_{pd'}), \\[4pt] H'_{22}=E_1+\dfrac{1}{2}J_{pd}-\dfrac{1}{2}J_{pd'}-J_{d'd}-E, \quad H'_{13}=\dfrac{1}{2}(J_{pd}-J_{pd'}), \\[4pt] H'_{33}=E_1-J_{pd}-J_{d'd}, \\[4pt] h_{11}=E_1+\dfrac{1}{2}J_{pd}-\dfrac{1}{2}J_{pd'}+J_{d'd}-E, \\[4pt] h_{22}=E_1-\dfrac{1}{2}J_{pd}+\dfrac{1}{2}J_{pd'}-J_{d'd}-E; \end{array}\right\} \qquad (17.50)$$

die übrigen Bezeichnungen sind bereits erklärt worden. (Der Einfachheit halber wurde die Anionenfunktion als gerade angenommen.)

$E_1+J_{d'd}$ ist die Summe der kinetischen, Coulomb- und Austauschenergien bei antiparalleler Einstellung der d'- und d-Elektronenspins und $E_1-J_{d'd}$ dieselbe Größe bei paralleler Ausrichtung der Spins. Da wir annehmen, daß die Austauschenergie $J_{d'd}$ in den nichtangeregten Zustand des Systems einzubeziehen ist, muß diese Größe unbedingt in den Energienennern der Störungsreihe verbleiben. Demzufolge lautet die der Entwicklung (17.43) entsprechende Reihe für $(E-H_{\alpha\alpha})^{-1}$ jetzt

$$\frac{1}{E-H_{\alpha\alpha}}=\frac{1}{E_0-E_\alpha}+\frac{-\widetilde{H}_{11}+\widetilde{H}_{\alpha\alpha}}{(E_0-E_\alpha)^2}+\cdots \qquad (17.51)$$

mit

$$\widetilde{H}_{\alpha\alpha}=H_{\alpha\alpha}-E_\alpha, \quad E_\alpha=\begin{cases}E_1-J_{d'd}\\ E_1+J_{d'd}\end{cases}. \qquad (17.52)$$

Unter Verwendung der Schemata (17.48) und (17.49) konstruieren wir wiederum eine Störungsreihe vom Typ (17.45). Einfache Berechnungen ergeben folgenden Wert für die Energiedifferenz zwischen Singulett und Triplett:

$$E_s-E_t=2\varrho^2 J_{pd}\left(\frac{1}{E_{\uparrow\downarrow}^2}-\frac{1}{E_{\uparrow\uparrow}^2}\right), \qquad (17.53)$$

worin

$$E_{\uparrow\downarrow}=E_1+J_{d'd}-E_0, \qquad (17.54)$$

$$E_{\uparrow\uparrow}=E_1-J_{d'd}-E_0 \qquad (17.55)$$

bedeuten. $E_{\uparrow\downarrow}$ ist die Energiezunahme beim Übergang eines Elektrons vom Anion zum Kation, wenn dann die Spins der d'- und d-Elektronen antiparallel liegen; $E_{\uparrow\uparrow}$ ist der analoge Energiezuwachs bei parallelen Spins der beiden Elektronen am Kation. Man erhält so für das der Diracschen Form (17.31) entsprechende effektive Integral des indirekten Austausches

$$J=\varrho^2 J_{pd}\left(\frac{1}{E_{\uparrow\downarrow}^2}-\frac{1}{E_{\uparrow\uparrow}^2}\right). \qquad (17.56)$$

Die Diskussion der erhaltenen Ergebnisse verschieben wir auf das Ende von Ziff. 19.

18. Der „Übergang" in die einfach besetzte Bahn. Oben betrachteten wir den Fall, daß das zum Kation übergegangene zusätzliche Elektron dort einen freien Zustand einnimmt, den wir als d'-Zustand bezeichneten. Wesentlich war dabei, daß der d'-Zustand nicht mit dem einfach besetzten d-Zustand des Kations übereinstimmte. Es sind jedoch auch solche Übergänge möglich, wo das Zusatzelektron in den bereits mit einem Elektron besetzten d-Zustand des Kations übergeht. Dabei stellt sich wegen des Pauli-Prinzips der Spin des übergegangenen Elektrons antiparallel zum Spin des dort bereits vorhandenen Elektrons ein. Wir betrachten jetzt ein Modell, das die Möglichkeit solcher Übergänge berücksichtigt. Man könnte nun die den neuen Konfigurationen entsprechenden Wellenfunktionen jenem Satz (17.1) hinzufügen, den wir bereits oben besaßen, und wieder die Energiewerte aufsuchen. Der Einfachheit halber wollen wir jedoch das Problem isoliert lösen und dabei annehmen, daß nur solche Konfigurationen möglich sind, bei denen das Zusatzelektron in einfach besetzte d-Zustände übergeht, aber nicht in die freien d'-Zustände. Somit lauten die zu den Eigenwerten $S_z = 0, 1, -1$ gehörenden Basiswellenfunktionen

$$u_1 = \begin{vmatrix} d_1 \alpha \\ p \alpha \\ p \beta \\ d_2 \beta \end{vmatrix}, \quad u_2 = \begin{vmatrix} d_1 \beta \\ p \alpha \\ p \beta \\ d_2 \alpha \end{vmatrix}, \quad u_3 = \begin{vmatrix} d_1 \alpha \\ d_1 \beta \\ p \alpha \\ d_2 \beta \end{vmatrix},$$

$$u_4 = \begin{vmatrix} d_1 \alpha \\ d_1 \beta \\ p \beta \\ d_2 \alpha \end{vmatrix}, \quad u_5 = \begin{vmatrix} d_1 \alpha \\ p \beta \\ d_2 \alpha \\ d_2 \beta \end{vmatrix}, \quad u_6 = \begin{vmatrix} d_1 \beta \\ p \alpha \\ d_2 \alpha \\ d_2 \beta \end{vmatrix},$$

$$v_1 = \begin{vmatrix} d_1 \alpha \\ p \alpha \\ p \beta \\ d_2 \alpha \end{vmatrix}, \quad v_2 = \begin{vmatrix} d_1 \beta \\ p \alpha \\ p \beta \\ d_2 \alpha \end{vmatrix}, \quad v_3 = \begin{vmatrix} d_1 \alpha \\ p \alpha \\ d_2 \alpha \\ d_2 \beta \end{vmatrix},$$

$$v_4 = \begin{vmatrix} d_1 \beta \\ p \alpha \\ p \beta \\ d_2 \beta \end{vmatrix}, \quad v_5 = \begin{vmatrix} d_1 \alpha \\ d_1 \beta \\ p \beta \\ d_2 \beta \end{vmatrix}, \quad v_6 = \begin{vmatrix} d_1 \beta \\ p \beta \\ d_2 \alpha \\ d_2 \beta \end{vmatrix}.$$

(18.1)

Mit Hilfe der unter Ziff. 17 beschriebenen Methode stellen wir die Singulettfunktionen $\Psi_{\alpha s}$ und die Triplettfunktionen $^{S_z}\Psi_{\alpha t}$ zusammen:

$$\left.\begin{aligned} \Psi_{1s} &= \tfrac{1}{\sqrt{2}}(u_1 - u_2), \\ \Psi_{2s} &= \tfrac{1}{\sqrt{2}}(u_3 - u_4), \\ \Psi_{3s} &= \tfrac{1}{\sqrt{2}}(u_5 - u_6), \end{aligned}\right\} \quad (18.2)$$

$$^{1}\Psi_{1t}=v_1, \quad {}^{0}\Psi_{1t}=\frac{1}{\sqrt{2}}(u_1+u_2), \quad {}^{-1}\Psi_{1t}=v_4,$$
$$^{1}\Psi_{2t}=v_2, \quad {}^{0}\Psi_{2t}=\frac{1}{\sqrt{2}}(u_3+u_4), \quad {}^{-1}\Psi_{2t}=v_5, \quad\quad (18.3)$$
$$^{1}\Psi_{3t}=v_3, \quad {}^{0}\Psi_{3t}=\frac{1}{\sqrt{2}}(u_5+u_6), \quad {}^{-1}\Psi_{3t}=v_6.$$

Die Matrixelemente (17.15) für diese Funktionen werden wie oben berechnet. Die Matrixelemente H_{ij}^s und H_{ij}^t der Säkulargleichung für den Singulettzustand bzw. den Triplettzustand lauten

$$\begin{aligned} H_{11}^s &= E_0 - 2J_{pd}, \\ H_{12}^s &= H_{13}^s = -\varrho - J_{d_1p,\,d_2d_1} - J_{d_1p,\,d_1d_1} - J_{pp,\,pd_1} = \varrho_1, \\ H_{22}^s &= H_{33}^s = E_1, \quad H_{23}^s = 2J_{pd_1,\,d_2p}, \\ H_{11}^t &= E_0 - 2J_{pd}, \quad H_{12}^t = H_{13}^t = \varrho_1, \\ H_{22}^t &= E_1 - 2J_{pd}, \quad H_{23}^t = 0 \end{aligned} \quad (18.4)$$

mit

$$\varrho = \int \varphi_d^*(\vec{r})\hat{H}_1(\vec{r})\varphi_p(\vec{r})d\vec{r}, \quad J_{kl,mn} = \int \varphi_k^*(\vec{r}_1)\varphi_l^*(\vec{r}_2)V(\vec{r}_1,\vec{r}_2)\varphi_m(\vec{r}_1)\varphi_n(\vec{r}_2)d\vec{r}_1 d\vec{r}_2.$$

Nach Aufteilung in ein System $\{A\}$ (die Funktionen Ψ_{1s} und $^{S_z}\Psi_{1t}$) und ein System $\{B\}$ erhalten wir eine Störungsreihe vom Typ (17.45) zur Bestimmung der Energie. Die Differenz zwischen den Singulett- und Triplettenergien beträgt in diesem Falle

$$E_s - E_t = \frac{4\varrho_1^2}{(E_1-E_0)^2}(J_{pd} \pm J_{pd_1,\,d_2p}). \quad (18.5)$$

(18.5) geht in den erstmals von ANDERSON [36] aufgestellten üblichen Ausdruck über, wenn Glieder, die Coulomb-Integrale enthalten, vernachlässigt werden:

$$E_s - E_t = \frac{4\varrho^2 J_{pd}}{(E_1-E_0)^2}. \quad (18.6)$$

Für die Diskussion der gewonnenen Ergebnisse verweisen wir wieder auf das Ende von Ziff. 19.

19. Die zweifache „Anregung". Wir klären jetzt, welchen Beitrag zur Energiedifferenz die „angeregte" Konfiguration beisteuert, die bei gleichzeitigem Übergang zweier p-Elektronen des Anions — je eins geht dabei zu jedem der beiden Kationen — auf die einfach besetzten d-Bahnen entsteht [138] bis [140].

In diesem Falle wird zu den Basiswellenfunktionen (18.1) als einzige Determinante

$$u_7 = \begin{vmatrix} d_1\alpha \\ d_1\beta \\ d_2\alpha \\ d_2\beta \end{vmatrix} \quad (19.1)$$

hinzugefügt. Die Funktion u_7 selbst stellt bereits eine Singulettfunktion $\Psi_{4s}=u_7$ dar. Folglich wirkt sich die Berücksichtigung der neuen „angeregten" Konfiguration auf die Triplettzustände nicht aus. Die Matrixelemente $H_{4i}^s = \langle \Psi_{4s}\hat{H}\Psi_{is}\rangle$ mit $i=1, 2, 3, 4$, die den zweifach ionisierten Zustand (19.1) mit dem reinen Ionenzustand, mit den einfach ionisierten Zuständen (18.2) und mit sich selbst verbinden, besitzen (bei gerader Anionenfunktion) die Werte

$$H^s_{41} = -\sqrt{2}\, J_{p\,d_1,\,d_2 p},$$
$$H^s_{42} = H^s_{43} = \sqrt{2}\{\varrho + 2 J_{d_1 p,\,d_1 d_2} + J_{p\,d_1,\,d_1 d_1}\}, \quad H^s_{44} = E_2. \qquad (19.2)$$

Bei der Durchführung der Rechnungen für die Energie des Systems nach dem in Ziff. 17 und 18 verwendetem Schema erhält man bereits in der ersten Ordnung der Störungstheorie zur gesuchten Differenz $E_s - E_t$ einen Beitrag

$$\frac{2 J^2_{p\,d_1,\,d_2 p}}{E_0 - E_2}, \qquad (19.3)$$

wobei mit E_2 die kinetische und Coulomb-Energie in der zweifach ionisierten Konfiguration bezeichnet ist.

Wir kommen jetzt zur Diskussion der in den Ziff. 17 bis 19 erhaltenen Ergebnisse, welche die Bestimmung der Austauschkopplung in einem System mit drei Zentren und vier Elektronen unter Verwendung orthogonaler Atomfunktionen betreffen. Bei einer normalen chemischen Bindung ist zu erwarten, daß das Austauschintegral J_{pd} in allen betrachteten Fällen negativ ist. Was die angeregten Konfigurationen mit Übergang eines Elektrons auf die leere Bahn eines Kations anbelangt, darf man annehmen, daß dieser Fall der realen Situation entspricht, wenn die d-Schale des Kations weniger als halb gefüllt ist. Da wegen der Hundschen Regel das Zusatzelektron seinen Spin vorzugsweise parallel zum Spin der Schale einstellt, besteht die Ungleichung $E_{\uparrow\downarrow} > E_{\uparrow\uparrow}$. Dann folgt aus (17.53) $E_s > E_t$; die ferromagnetische Ordnung ist also bevorzugt.

Der Übergang eines p-Elektrons auf die besetzte Bahn des Kations entspricht der Situation, daß die d-Schale des Kations zur Hälfte gefüllt ist oder mehr. In diesem Falle erhalten wir aus (18.6) sofort $E_s < E_t$; die Austauschkopplung begünstigt also die antiferromagnetische Atomanordnung. Somit folgt, daß bei festgehaltenem Vorzeichen von J_{pd} das Auftreten einer ferro- oder antiferromagnetischen Ordnung durch den Auffüllungsgrad der d-Schale des Kations bestimmt wird.

Bei Berücksichtigung der zweifachen Anregung, die in der Energiedifferenz $E_s - E_t$ das Glied (19.3) bedingt, wird die antiferromagnetische Kopplung weiter begünstigt. Obwohl (19.3) von zweiter Ordnung ist, ist es wegen des großen Energienenners doch klein. Weiter unten kommen wir auf die Diskussion dieser Größe noch zurück.

Es wird jedoch kaum Sinn haben, die hier erhaltenen Schlußfolgerungen auch nur qualitativ mit dem Experiment zu vergleichen, da das vorgelegte Modell doch recht grob ist, allein schon wegen der Annahme einer strengen Orthogonalität der Atomwellenfunktionen. Tatsächlich sind diese Funktionen nicht orthogonal, und dieser Umstand kann alle Schlußfolgerungen in Frage stellen.

Nach PRATT [138] kann bei Verwendung orthogonaler Bahnen unter alleiniger Berücksichtigung des einfachen Überganges in eine einfach besetzte Bahn (vgl. Ziff. 18) im allgemeinen der Singulettzustand nicht tiefer liegen als der Triplettzustand, selbst wenn das Problem der Konfigurationswechselwirkung exakt gelöst wird. Der mit Hilfe der Störungstheorie in der Methode der Konfigurationswechselwirkung erhaltene Ausdruck (18.6) stimmt ebenfalls mit dieser Schlußfolgerung überein. Es ist zwar so, daß, wie bereits oben erwähnt, die Formel (18.6) zu antiferromagnetischer Kopplung führt, wenn man annimmt, daß das Austauschintegral J_{pd} negativ ist. Aus den allgemeinen Eigenschaften der Integrale vom Typ J_{pd} zwischen den orthogonalen Bahnen φ_p und φ_d folgt jedoch, daß sie stets *positiv* sein müssen. Daher ist die oben angenommene Ungleichung $J_{pd} < 0$

künstlich und unreal. Auf Grund des positiven Vorzeichens des Integrals J_{pd} muß vielmehr in (18.6) stets die Ungleichung $E_s > E_t$ erfüllt sein; eine antiferromagnetische Kopplung ist also im vorliegenden Schema nicht möglich. Bei Berücksichtigung der zweifachen Anregung trifft diese Behauptung schon nicht mehr zu. Das durch die Einbeziehung dieser Konfiguration bedingte Glied liefert, wie aus (19.3) folgt, einen negativen Beitrag zur Differenz $E_s - E_t$ und begünstigt somit die antiferromagnetische Kopplung. Die Situation hier ist derjenigen sehr ähnlich, die wir bereits bei der Behandlung des Problems der Orthogonalisierung der Atomfunktionen beim Wasserstoffmolekül antrafen (s. Ziff. 2). Die Verwendung orthogonaler Bahnen konnte dort nicht zum richtigen Vorzeichen für die Singulett-Triplett-Aufspaltung führen und auch nicht zur Ausbildung einer chemischen Bindung [66], falls nicht zusätzlich angeregte Konfigurationen vom Typ ionisierter (polarer) Zustände berücksichtigt wurden.

Die Verwendung nichtorthogonaler Atombahnen beim Problem der drei Zentren und vier Elektronen kompliziert selbstverständlich dessen Lösung, man kann dann jedoch erwarten, daß man bereits bei Berücksichtigung einer geringeren Zahl angeregter Konfigurationen eine Austauschkopplung zwischen den Kationen erhält.

20. Der Doppelübergang. Ein weiterer möglicher Mechanismus der indirekten Austauschwechselwirkung ist durch einen virtuellen Übergang bedingt, den man als Doppelübergang bezeichnen kann. Er kann aufgefaßt werden als Übergang eines der Elektronen eines Kations zum Anion bei gleichzeitigem Übergang eines Elektrons vom Anion zum anderen Kation [141]. Den gleichen, in leitenden Medien mit gemischter Wertigkeit untersuchten Mechanismus nennt man Doppelaustausch. Wir betrachten hier einen solchen Übergang, bei dem sich im angeregten Zustand ein Elektronenpaar auf *einer* Bahn bei einem der Kationen befindet, also den Gesamtspin Null besitzt. Die Basisdeterminanten sind dann

$$u_1 = \begin{vmatrix} d_1\alpha \\ p\alpha \\ p\beta \\ d_2\beta \end{vmatrix}, \quad u_2 = \begin{vmatrix} d_1\beta \\ p\alpha \\ p\beta \\ d_2\alpha \end{vmatrix}, \quad u_3 = \begin{vmatrix} d_1\alpha \\ d_1\beta \\ p\alpha \\ p\beta \end{vmatrix}, \quad u_4 = \begin{vmatrix} p\alpha \\ p\beta \\ d_2\alpha \\ d_2\beta \end{vmatrix}, \quad (20.1)$$

wobei u_3 und u_4 der angeregten Konfiguration entsprechen. Die drei Singulettfunktionen und die eine Triplettfunktion ($S_z = 0$) lauten

$$\left.\begin{aligned} \Psi_{1s} &= \frac{1}{\sqrt{2}}(u_1 - u_2), \\ \Psi_{2s} &= u_3, \\ \Psi_{3s} &= u_4, \end{aligned}\right\} \quad (20.2)$$

$$\Psi_{1t} = \frac{1}{\sqrt{2}}(u_1 + u_2). \quad (20.3)$$

Die Funktionen u_3 und u_4 liefern selbstverständlich keinen Beitrag zum Triplettzustand. Da im vorliegenden Problem nur eine Triplettfunktion (20.3) auftritt, ist die Energie des Triplettzustandes E_t einfach gleich

$$E_t = \langle \Psi_{1t} | \hat{H} | \Psi_{1t} \rangle. \quad (20.4)$$

Die Energie des Singulettzustandes wird wieder aus der Gl. (17.45) bestimmt. Einfache Rechnungen ergeben für die Singulett-Triplett-Aufspaltung bereits in der zweiten Ordnung der Störungstheorie einen Ausdruck

$$E_s - E_t = -\frac{4 J^2_{d_1 p, p d_2}}{E_1 - E_0} \tag{20.5}$$

mit

$$\left.\begin{array}{l} E_1 - E_0 = \int \varphi^*_{d_1}(\vec{r}_1) \varphi^*_{d_2}(\vec{r}_2) V(\vec{r}_1, \vec{r}_2) \varphi_{d_1}(\vec{r}_1) \varphi_{d_2}(\vec{r}_2) d\vec{r}_1 d\vec{r}_2 - \\ - \int \varphi^*_{d_1}(\vec{r}_1) \varphi^*_{d_2}(\vec{r}_2) V(\vec{r}_1, \vec{r}_2) \varphi_{d_2}(\vec{r}_1) \varphi_{d_1}(\vec{r}_2) d\vec{r}_1 d\vec{r}_2. \end{array}\right\} \tag{20.6}$$

Das negative Vorzeichen der Differenz (20.5) ist nicht verwunderlich, wenn man sich daran erinnert, daß die Funktionen u_3 und u_4 nur für den Singulettzustand das Gewicht erhöhen.

21. Der Polarisationsmechanismus. Eine indirekte Austauschkopplung zwischen den magnetischen Kationen in den Kristallen vom Typ MnO kann man auch im Rahmen des sog. uneingeschränkten Hartree-Fock-Verfahrens erhalten. Bei der Standardableitung der Hartree-Fock-Gleichungen wird der in der Eindeterminantennäherung berechnete Mittelwert der Energie des Systems unabhängig nach jeder der Spin-Bahn-Funktionen variiert. Im Ergebnis zeigen die Radialanteile der Einelektronenfunktionen eine parametrische Abhängigkeit von der Richtung des Elektronenspins in dem Zustand, der durch die gegebene Einelektronenfunktion beschrieben wird. So entsteht im uneingeschränkten Hartree-Fock-Verfahren eine Situation, in der man nicht mehr von doppelt besetzten Einelektronenzuständen (oder Bahnen) sprechen kann, da sich die „ursprünglich doppelt besetzte Bahn" in zwei Bahnen aufspaltet, von denen jede zu einem Spin bestimmter Richtung gehört. Die Gleichungen, bei deren Ableitung Determinanten mit zweifach besetzten Bahnen als Vielelektronenfunktionen benutzt wurden und Variationen lediglich nach dem Koordinatenanteil der Einelektronenfunktion erfolgten, werden als Gleichungen des beschränkten Hartree-Fock-Verfahrens bezeichnet. Bei der Benützung der Gleichungen der uneingeschränkten Hartree-Fock-Methode gibt es jedoch Komplikationen. Da im Determinantenansatz zweifach besetzte Bahnen nicht mehr auftreten, ist diese Ausgangsdeterminante keine Eigenfunktion zum Operator $\hat{S}^2$ des Quadrates des Systemspins mehr. Wählt man beispielsweise die Funktion des Grundzustandes für das Lithiumatom ^2S in Form der Determinante

$$\begin{vmatrix} \varphi_{2s} \alpha \\ \varphi_{1s} \alpha \\ \varphi_{1s} \beta \end{vmatrix} \tag{21.1}$$

und variiert man die mit Hilfe von (21.1) berechnete mittlere Energie nach den Spin-Bahn-Funktionen $\varphi_{1s}\alpha$ und $\varphi_{1s}\beta$, so erhält man verschiedene Koordinatenanteile für die Funktionen $\varphi_{1s}\alpha$ und $\varphi_{1s}\beta$. Um diese Koordinatenfunktionen unterscheiden zu können, versehen wir eine davon mit einem Strich. Somit wird das System der drei Elektronen des Lithiumatoms nicht mit Hilfe von (21.1), sondern mittels einer Determinante

$$\begin{vmatrix} \varphi_{2s} \alpha \\ \varphi_{1s} \alpha \\ \varphi'_{1s} \beta \end{vmatrix} \tag{21.2}$$

beschrieben, worin die Koordinatenanteile für die „spin-abgeschlossene" Schale φ_{1s} und φ'_{1s} nicht mehr übereinstimmen. Dieses Ergebnis ist die Folge davon, daß Elektronen der inneren Schale je nach ihrer Spinrichtung eine unterschiedliche Austauschwechselwirkung mit der hinsichtlich des Spins unabgeschlossenen

Schale 2s erfahren (so enthält der Ausdruck für die Energie lediglich ein Austauschglied zwischen den Bahnen φ_{1s} und φ_{2s}, während zwischen den Elektronen mit entgegengesetztem Spin auf den Bahnen φ'_{1s} und φ'_{2s} ein „Austausch" entfällt). Dieses Auseinanderrücken der Bahnen in den inneren Schalen auf Grund der „Vormagnetisierung" seitens der im Spin nichtabgeschlossenen Bahnen erhielt die Bezeichnung Austauschpolarisation. Sie hat große Bedeutung insbesondere bei der Bildung anomal starker lokaler Magnetfelder an den Kernen in Kristallen mit magnetisch aktiven Ionen; diese Felder der HFS-Wechselwirkung wurden bei der Messung des Kernbeitrages zur spezifischen Wärme, beim Mößbauer-Effekt und bei der magnetischen Kernresonanz an Kristallen dieses Typs experimentell beobachtet (s. z.B. die Übersichtsarbeit [142] von WATSON und FREEMAN). Im Gegensatz zur Funktion (21.1), die einen Dublettzustand mit $S_z = +\frac{1}{2}$ beschreibt, stellt die Funktion (21.2) keine Eigenfunktion des Operators $\hat{S}^2$ dar; sie beschreibt also keinen reinen Spinzustand des Systems. Wie leicht zu zeigen ist, erscheint (21.2) als Mischung von Zuständen verschiedener Multiplizität. Im Zusammenhang damit ergibt sich die Frage, inwieweit die Funktion (21.2) überhaupt zur Berechnung des Grundzustandes des Systems verwendbar ist. Die Einführung „unterschiedlicher Bahnen für verschiedene Spins", wie sie für das uneingeschränkte Hartree-Fock-Verfahren charakteristisch ist, macht im allgemeinen eine besondere Untersuchung des gesamten Problems der Spinentartung und des Aufbaues der Multiplettzustände des Systems erforderlich. Diesem Problem ist, des öfteren im Zusammenhang mit dem Lithiumatom, eine große Anzahl von Arbeiten gewidmet [77], [109], [143] bis [150].

In unserem Falle, wo es um die Bestimmung der Singulett-Triplett-Aufspaltung im Problem der drei Zentren und vier Elektronen geht, führt eine Anwendung des uneingeschränkten Hartree-Fock-Verfahrens zu überaus umfangreichen Rechnungen.

Wir gehen jetzt zur Behandlung dieses Problems über. Die im System Me-O-Me zu erwartende Singulett-Triplett-Aufspaltung wird hierbei durch die Austauschpolarisation der abgeschlossenen p-Schale des Sauerstoffions beeinflußt, die ihre Ursache in der Wechselwirkung mit den Elektronen der Kationen hat. Auch wenn unterschiedliche p-Bahnen für das Sauerstoffion nicht eingeführt worden sind, werden die Multiplettzustände des Systems von Anfang an nicht durch *eine* Determinante (der Triplettzustand wird mit $S_z = 0$ gewählt), sondern durch eine Überlagerung von Determinanten dargestellt [s. die Formeln für $^0\Psi_{1t}$ und Ψ_{1s} in (17.10) und (17.13)][28]. Mit der Einführung unterschiedlicher Sauerstoff-p-Bahnen wächst die Zahl der Multipletts. Die Basis-Slater-Determinanten lauten dann

$$u_1 = \begin{vmatrix} d_1\alpha \\ p_1\alpha \\ p_2\beta \\ d_2\beta \end{vmatrix}, \quad u_2 = \begin{vmatrix} d_1\alpha \\ p_1\beta \\ p_2\alpha \\ d_2\beta \end{vmatrix}, \quad u_3 = \begin{vmatrix} d_1\beta \\ p_1\alpha \\ p_2\beta \\ d_2\alpha \end{vmatrix},$$

$$u_4 = \begin{vmatrix} d_1\beta \\ p_1\beta \\ p_2\alpha \\ d_2\alpha \end{vmatrix}, \quad u_5 = \begin{vmatrix} d_1\alpha \\ p_1\beta \\ p_2\beta \\ d_2\alpha \end{vmatrix}, \quad u_6 = \begin{vmatrix} d_1\beta \\ p_1\alpha \\ p_2\alpha \\ d_2\beta \end{vmatrix}. \qquad (21.3)$$

---

[28] Die Hartree-Fock-Gleichungen wurden von LÖWDIN [56], [79], [80] für den Fall, daß Spinentartung vorliegt und die Basisfunktionen als Eigenfunktionen von $\hat{S}^2$ und als Kombinationen von Determinanten erscheinen, mit Hilfe einer Projektionsoperatormethode erhalten.

Bei der Konstruktion der Determinanten (21.3) wurden auch die Funktionen u_5 und u_6 berücksichtigt, die einer hinsichtlich des Spins unabgeschlossenen Sauerstoffschale entsprechen. Die vollständige Freiheit bei der Besetzung der Bahnen p_1 und p_2 mit „Rechts"- und „Links"-Spins führte somit in unserem System sofort zu *neuen Konfigurationen*. Es ist nicht schwer, mit Hilfe von (21.3) folgende zwei Singulett- und drei Triplettfunktionen aufzustellen:

$$\left.\begin{aligned}\Psi_{1s} &= \frac{1}{2\sqrt{1+S^2}}(u_1 - u_2 - u_3 + u_4),\\ \Psi_{2s} &= \frac{1}{2\sqrt{3(1-S^2)}}(u_1 + u_2 + u_3 + u_4 - 2u_5 - 2u_6),\\ \Psi_{1t} &= \frac{1}{2\sqrt{1+S^2}}(u_1 - u_2 + u_3 - u_4),\\ \Psi_{2t} &= \frac{1}{2\sqrt{1-S^2}}(u_1 + u_2 - u_3 - u_4),\\ \Psi_{3t} &= \frac{1}{\sqrt{2(1-S^2)}}(u_5 - u_6).\end{aligned}\right\} \quad (21.4)$$

Die Größe S in (21.4) stellt das Nichtorthogonalitätsintegral für die Bahnen p_1 und p_2 dar:

$$S = \int p_1^*(\vec{r}) p_2(\vec{r}) d\vec{r}. \qquad (21.5)$$

In (21.4) sind alle Funktionen orthonormiert. Die Triplettfunktionen in (21.4) gehören zur Komponente Null des Gesamtspins.

Die Funktionen Ψ_{1s} und Ψ_{1t} entsprechen einer Spinkonfiguration, in der die Elektronen in den Sauerstoffbahnen p_1 und p_2 ein Singulett aufbauen [sie sind also den entsprechenden Funktionen der reinen Ionenkonfiguration aus (17.10) und (17.13) mit spin-abgeschlossener Sauerstoffschale analog]. Davon kann man sich überzeugen, indem man zunächst den Ausdrücken für Ψ_{1s} und Ψ_{1t} die Determinanten u_1, u_2, u_3 und u_4 nach den Elementen der ersten Zeile (d_1-Zeile), dann die erhaltenen dreireihigen Determinanten nach den Elementen ihrer dritten Zeile (d_2-Zeile) entwickelt und schließlich die zweireihigen Determinanten, die jetzt nur noch die Bahnen p_1 und p_2 enthalten, wie erforderlich gruppiert. Die zweireihigen Determinanten bilden Singulettfunktionen für die Elektronen in den Bahnen p_1 und p_2:

$$\begin{vmatrix} p_1 \alpha \\ p_2 \beta \end{vmatrix} - \begin{vmatrix} p_1 \beta \\ p_2 \alpha \end{vmatrix}. \qquad (21.6)$$

Die Funktion Ψ_{1s} lautet dann

$$\Psi_{1s} = \sum_P c_P \hat{P}\, \Psi_{s(p_1,p_2)}(q_1, q_2)\, \Psi_{d_1}(q_3)\, \Psi_{d_2}(q_4), \qquad (21.7)$$

wobei $\hat{P}$ der Vertauschungsoperator für die Koordinaten $q_1, \ldots, q_4$ ist; $\Psi_{s(p_1,p_2)}(q_i, q_j)$ bezeichnet die Singulettfunktion (21.6) in den Koordinaten der Elektronen i und j, die die Bahnen p_1 und p_2 einnehmen, $\Psi_{d_k}(q_i)$ die Dublettfunktion für ein einzelnes Elektron i auf einer d-Bahn ($k = 1$ oder 2) mit Rechts- oder Linksorientierung des Spins [d.h. eine Einelektronenfunktion $d_{1,2}(\vec{r})\chi(\sigma)$], und schließlich ist $c_P = \pm 1$. Die Funktion Ψ_{1t} besitzt eine zu (21.6) analoge Form.

Auf die gleiche Weise kann man sich davon überzeugen, daß die zweite Singulettfunktion Ψ_{2s} und die restlichen Triplettfunktionen Ψ_{2t} und Ψ_{3t} einer Konfiguration entsprechen, bei der die Elektronenspins auf den p_1- und p_2-Bahnen ein Triplett bilden.

Wir wollen jetzt feststellen, ob man den Polarisationseffekt bei ausschließlicher Benutzung der Funktionen Ψ_{1s} und Ψ_{1t} (die der reinen Ionenkonfiguration entsprechen) erhalten kann, ohne die übrigen Multiplettzustände aus (21.4) zu betrachten, die ja „angeregten" Sauerstoffbahnen entsprechen. Da die Hartree-Fock-Gleichungen bei der Minimalisierung der Singulett- und der Triplettenergie im allgemeinen eine unterschiedliche Polarisation der p-Bahnen ergeben, muß man letztere in diesen beiden Fällen unterscheiden. Die Bezeichnungen p_1 und p_2 behalten wir zur Beschreibung der Sauerstoffbahnen im Singulettzustand Ψ_{1s} bei, während wir für den Triplettzustand Ψ_{1t} die Buchstaben p_1' und p_2' benutzen.

Die Berechnung der Energiemittelwerte für den Singulettzustand, $E_s = \langle \Psi_{1s} \hat{H} \Psi_{1s} \rangle$, und für den Triplettzustand, $E_t = \langle \Psi_{1t} \hat{H} \Psi_{1t} \rangle$, ergibt:

$$E_s = \frac{1}{2(1+S^2)} (h_{11} + h_{22} - 2h_{12}), \qquad (21.8)$$

$$E_t = \frac{1}{2(1+S'^2)} (h_{11}' + h_{22}' - 2h_{12}'), \qquad (21.9)$$

worin die $h_{ij} = \langle u_i \hat{H} u_j \rangle$ sind und die gestrichenen Größen aus den analogen ungestrichenen folgen, wenn man in den u_i die Funktionen p_1 und p_2 jeweils durch p_1' und p_2' ersetzt. [Bei der Ableitung von (21.8) und (21.9) wurden die Beziehungen $h_{11} = h_{44}$, $h_{22} = h_{33}$, $h_{12} = h_{34}$, $h_{13} = h_{24} = h_{14} = h_{23} = 0$ verwendet, die aus Symmetriegründen oder wegen der „Orthogonalität der Bahnen" folgen.] Da E_s und E_t vollkommen gleich gebaut sind, führt ihre Variation nach den Einelektronenfunktionen zu identischen Gleichungen für p_1 und p_1' bzw. für p_2 und p_2'. In Anbetracht der erhaltenen Identitäten $p_1 = p_1'$ und $p_2 = p_2'$ folgt aus (21.8) und (21.9), daß E_s und E_t gleich sind. Die Gleichheit der Polarisation der Sauerstoffschale im Singulett- und Triplettzustand, Ψ_{1s} und Ψ_{1t}, gibt somit keine Möglichkeit für die Entstehung einer Austauschkopplung zwischen den magnetischen Kationen.

Folglich *muß* man, um den gesuchten Effekt zu erhalten, die Funktionen Ψ_{2s} bzw. Ψ_{2t} und Ψ_{3t} aus (21.4) verwenden. Hierbei ergibt sich allerdings sofort eine Komplikation bei der Berechnung der Energie und beim Aufstellen der Variationsgleichungen für den Triplettzustand. Da die Funktionen Ψ_{2t} und Ψ_{3t} gleichzeitig berücksichtigt werden müssen, muß man die Säkulargleichung und die Variationsgleichung gemeinsam lösen. Wir wollen das hier nicht durchführen. Löst man jedoch die Säkulargleichung nicht und berechnet die Energien E_{1t} und E_{2t} für die Zustände Ψ_{1t} und Ψ_{2t} einzeln, so sieht man sofort, daß sich die Variationsgleichungen für die beiden Sauerstoffbahnen von den Gleichungen für dieselben Bahnen im Singulettzustand unterscheiden werden. Bei Einführung der Funktionen Ψ_{2t} und Ψ_{3t} führt also die Polarisation der Sauerstoffbahnen zu unterschiedlichen Energiewerten im Singulett- und Triplettzustand, was der Existenz einer effektiven Austauschkopplung zwischen den magnetischen Kationen gleichzusetzen ist.

Das Entstehen einer solchen Kopplung läßt sich bei der Behandlung unseres Problems in der Eindeterminantennäherung veranschaulichen, die bei der Anwendung des uneingeschränkten Hartree-Fock-Verfahrens häufig benützt wird. Für die Singulettfunktion wählen wir die Determinante

$$\Psi_s = \begin{vmatrix} d_1 \alpha \\ p_1 \alpha \\ p_2 \beta \\ d_2 \beta \end{vmatrix} \qquad (21.10)$$

und für die Triplettfunktion den Ansatz

$$\Psi_t = \begin{vmatrix} d_1\alpha \\ p'_1\alpha \\ p'_2\beta \\ d_2\alpha \end{vmatrix}. \tag{21.11}$$

Die Funktionen (21.10) und (21.11) sind dem Singulett- bzw. Triplettzustand lediglich ähnlich, stellen diese Zustände selbst aber nicht dar. (21.10) ist eine Kombination des Singulettzustandes und der Quintett- und Triplettzustände mit $S_z = 0$ und (21.11) eine Kombination der Triplettzustände und des Quintettzustandes mit $S_z = +1$. Gewöhnlich rechtfertigt man bei der Berechnung des Spektrums und der Wellenfunktionen der Atome die Eindeterminantennäherung vom Typ (21.10) und (21.11) dadurch, daß der Polarisationseffekt nicht groß sei und folglich z.B. die Funktion (21.2) fast eine Dublettfunktion darstelle [d.h. dem exakten Dublett (21.1) sehr nahe komme] und daß der entstehende Fehler unbedeutend sei. Diese Begründung der Eindeterminantennäherung läßt sich in unserem Fall nicht einfach übernehmen, da der exakte Singulettzustand (bei übereinstimmenden Bahnen p_1 und p_2) ohnehin nicht als einzelne Determinante erscheint, sondern als Kombination zweier Determinanten [s. (17.13)]. Da jedoch im konkreten Fall Ψ_{1s} aus (17.3) die Energie $E_s = \langle \Psi_{1s} H \Psi_{1s} \rangle$ mit der Energie übereinstimmt, die man mit Hilfe einer der beiden Determinanten u_1 oder u_2 aus (17.1) berechnet, kann der erwähnte Umstand auch hier zur Begründung der Näherung (21.10) dienen. (Im Falle nichtorthogonaler p- und d-Bahnen unterscheidet sich der mit Hilfe der exakten Singulettfunktion berechnete effektive Austauschparameter der Spinwechselwirkung in der reinen Ionenkonfiguration von dem gleichen Parameter, den man in der Eindeterminantennäherung erhält, lediglich durch den unwesentlichen Faktor 2.)

Die Ausdrücke für die Energien E_s und E_t der Zustände (21.10) und (21.11) lauten

$$\left.\begin{aligned} E_s &= \langle d_1|d_1\rangle + \langle p_1|p_1\rangle + \langle p_2|p_2\rangle + \langle d_2|d_2\rangle + \langle p_1 d_1|p_1 d_1\rangle + \\ &\quad + \langle p_2 d_1|p_2 d_1\rangle + \langle d_1 d_2|d_1 d_2\rangle + \langle p_1 p_2|p_1 p_2\rangle + \langle p_1 d_2|p_1 d_2\rangle + \\ &\quad + \langle p_2 d_2|p_2 d_2\rangle - \langle p_1 d_1|d_1 p_1\rangle - \langle p_2 d_2|d_2 p_2\rangle, \end{aligned}\right\} \tag{21.12}$$

$$\left.\begin{aligned} E_t &= \langle d_1|d_1\rangle + \langle p'_1|p'_1\rangle + \langle p'_2|p'_2\rangle + \langle d_2|d_2\rangle + \langle p'_1 d_1|p'_1 d_1\rangle + \\ &\quad + \langle p'_2 d_1|p'_2 d_1\rangle + \langle d_1 d_2|d_1 d_2\rangle + \langle p'_1 p'_2|p'_1 p'_2\rangle + \langle p'_1 d_2|p'_1 d_2\rangle + \\ &\quad + \langle p'_2 d_2|p'_2 d_2\rangle - \langle p'_1 d_1|d_1 p'_1\rangle - \langle p'_1 d_2|d_2 p'_1\rangle, \end{aligned}\right\} \tag{21.13}$$

worin $\langle a|b\rangle = \int \varphi_a^*(\vec{r})\, \hat{H}_1(\vec{r})\, \varphi_b(\vec{r})\, d\vec{r}$ und $\langle ab|cd\rangle = \int \varphi_a^*(\vec{r}_1)\, \varphi_b^*(\vec{r}_2)\, V(\vec{r}_1 \vec{r}_2) \times \varphi_c(\vec{r}_1)\, \varphi_d(\vec{r}_2)\, d\vec{r}_1\, d\vec{r}_2$ bedeuten.

Die Variation der Ausdrücke (21.12) und (21.13) nach den Einelektronenfunktionen φ_{p_1}, φ_{p_2}, $\varphi_{p'_1}$ und $\varphi_{p'_2}$ führt zu den folgenden Hartree-Fock-Gleichungen:

$$\left.\begin{aligned} &[\hat{H}_1(\vec{r}_1) + \int |\varphi_{d_1}(\vec{r}_2)|^2 V(\vec{r}_1, \vec{r}_2)\, d\vec{r}_2 + \int |\varphi_{p_2}(\vec{r}_2)|^2 V(\vec{r}_1, \vec{r}_2)\, d\vec{r}_2 + \\ &\quad + \int |\varphi_{d_2}(\vec{r}_2)|^2 V(\vec{r}_1, \vec{r}_2)\, d\vec{r}_2]\, \varphi_{p_1}(\vec{r}_1) - [\int \varphi_{d_1}^*(\vec{r}_2)\, \varphi_{p_1}(\vec{r}_2)\, V(\vec{r}_1, \vec{r}_2)\, d\vec{r}_2]\, \varphi_{d_1}(\vec{r}_1) = \\ &= \varepsilon_{p_1}\, \varphi_{p_1}(\vec{r}_1), \end{aligned}\right\} \tag{21.14}$$

$$\left.\begin{aligned} &[\hat{H}_1(\vec{r}_1) + \int |\varphi_{d_1}(\vec{r}_2)|^2 V(\vec{r}_1, \vec{r}_2)\, d\vec{r}_2 + \int |\varphi_{p_1}(\vec{r}_2)|^2 V(\vec{r}_1, \vec{r}_2)\, d\vec{r}_2 + \\ &\quad + \int |\varphi_{d_2}(\vec{r}_2)|^2 V(\vec{r}_1, \vec{r}_2)\, d\vec{r}_2]\, \varphi_{p_2}(\vec{r}_1) - [\int \varphi_{d_2}^*(\vec{r}_2)\, \varphi_{p_2}(\vec{r}_2)\, V(\vec{r}_1, \vec{r}_2)\, d\vec{r}_2]\, \varphi_{d_2}(\vec{r}_1) = \\ &= \varepsilon_{p_2}\, \varphi_{p_2}(\vec{r}_1), \end{aligned}\right\} \tag{21.15}$$

Der Polarisationsmechanismus.

$$[\hat{H}_1(\vec{r}_1)+\int|\varphi_{d_1}(\vec{r}_2)|^2 V(\vec{r}_1,\vec{r}_2)\,d\vec{r}_2+\int|\varphi_{p_1'}(\vec{r}_2)|^2 V(\vec{r}_1,\vec{r}_2)\,d\vec{r}_2+$$
$$+\int|\varphi_{d_2}(\vec{r}_2)|^2 V(\vec{r}_1,\vec{r}_2)\,d\vec{r}_2]\varphi_{p_1'}(\vec{r}_1)-[\int\varphi_{d_1}^*(\vec{r}_2)\varphi_{p_1'}(\vec{r}_2)V(\vec{r}_1,\vec{r}_2)\,d\vec{r}_2]\varphi_{d_1}(\vec{r}_1)-$$
$$-[\int\varphi_{d_2}^*(\vec{r}_2)\varphi_{p_1'}(\vec{r}_2)V(\vec{r}_1,\vec{r}_2)\,d\vec{r}_2]\varphi_{d_2}(\vec{r}_1)=\varepsilon_{p_1'}\varphi_{p_1'}(\vec{r}_1),$$
(21.16)

$$[\hat{H}_1(\vec{r}_1)+\int|\varphi_{d_1}(\vec{r}_2)|^2 V(\vec{r}_1,\vec{r}_2)\,d\vec{r}_2+\int|\varphi_{p_1'}(\vec{r}_2)|^2 V(\vec{r}_1,\vec{r}_2)\,dr_2+$$
$$+\int|\varphi_{d_2}(\vec{r}_2)|^2 V(\vec{r}_1,\vec{r}_2)\,d\vec{r}_2]\varphi_{p_2'}(\vec{r}_1)=\varepsilon_{p_2'}\varphi_{p_2'}(\vec{r}_1).$$
(21.17)

Die Gln. (21.14) und (21.15) besitzen einen symmetrischen Aufbau und dem Augenschein nach gleiche Austauschglieder; wir setzen deshalb $\varphi_{p_1}=\varphi_{p_2}$. Die Gln. (21.16) und (21.17) unterscheiden sich voneinander dadurch, daß in (21.17) die Austauschterme fehlen. Dieser Umstand führt auch zu unterschiedlichen Ergebnissen für die Funktionen $\varphi_{p_1'}$ und $\varphi_{p_2'}$. Zur Gewinnung von Beziehungen zwischen den Funktionen φ_{p_1}, $\varphi_{p_1'}$ und $\varphi_{p_2'}$ bedienen wir uns der Störungstheorie. In erster Näherung kann man annehmen, daß die Coulomb- (nicht Austausch-) Hartree-Fock-Felder in den Gln. (21.14) bis (21.17) übereinstimmen. Wir führen die Bezeichnung

$$\hat{H}_1(\vec{r}_1)+\int|\varphi_{d_1}(\vec{r}_2)|^2 V(\vec{r}_1,\vec{r}_2)\,d\vec{r}_2+\int|\varphi_{p_1}(\vec{r}_2)|^2 V(\vec{r}_1,\vec{r}_2)\,d\vec{r}_2+$$
$$+\int|\varphi_{d_2}(\vec{r}_2)|^2 V(\vec{r}_1,\vec{r}_2)\,d\vec{r}_2=\hat{H}_0(\vec{r}_1)$$
(21.18)

ein. Weiterhin wird der austauschbedingte Teil der Gleichungen dargestellt als

$$-\int\varphi_{d_1}^*(\vec{r}_2)\varphi_{p_1'}(\vec{r}_2)V(\vec{r}_1,\vec{r}_2)\,d\vec{r}_2\,\varphi_{d_1}(\vec{r}_1)=$$
$$=-\int\varphi_{d_1}^*(\vec{r}_2)V(\vec{r}_1,\vec{r}_2)\,d\vec{r}_2\hat{P}_{12}\varphi_{d_1}(\vec{r}_2)\varphi_{p_1'}(\vec{r}_1)\equiv\hat{W}\varphi_{p_1'}(\vec{r}_1),$$

worin der Operator $\hat{W}$ durch den Ausdruck

$$\hat{W}=-\int\varphi_{d_1}^*(\vec{r}_2)V(\vec{r}_1,\vec{r}_2)\,d\vec{r}_2\hat{P}_{12}\varphi_{d_1}(\vec{r}_2)$$
(21.19)

gegeben ist. $\hat{P}_{12}$ bedeutet den Vertauschungsoperator für die Koordinaten $\vec{r}_1$ und $\vec{r}_2$. In den Bezeichnungen (21.18) und (21.19) nehmen (21.14) bis (21.17) die Form von Eigenwertgleichungen an:

$$(\hat{H}_0+\hat{W})\varphi_{p_1}=\varepsilon_{p_1}\varphi_{p_1},$$
(21.20)

$$(\hat{H}_0+2\hat{W})\varphi_{p_1'}=\varepsilon_{p_1'}\varphi_{p_1'},$$
(21.21)

$$\hat{H}_0\varphi_{p_2'}=\varepsilon_{p_2'}\varphi_{p_2'}.$$
(21.22)

(21.20) folgt aus den identischen Gleichungen (21.14) und (21.15). In (21.21) wurde die Gleichheit der beiden Austauschterme in (21.16) berücksichtigt. Die Gl. (21.22) kann im Vergleich zu (21.20) und (21.21), wo $\hat{W}$ bzw. $2\hat{W}$ die Störung des Hamilton-Operators $\hat{H}_0$ darstellen, als ungestörtes Problem aufgefaßt werden. Verwendet man die Standardtechnik der Störungstheorie zum Aufsuchen der gestörten Eigenfunktionen, so erhält man leicht

$$\varphi_{p_1}=\varphi_{p_2'}+\Delta\varphi_{p_2'},$$
(21.23)

$$\varphi_{p_1'}=\varphi_{p_2'}+2\Delta\varphi_{p_2'}$$
(21.24)

mit

$$\Delta\varphi_{p_2'}=-\sum_k{}'\frac{\langle p_{2k}'d|dp_2'\rangle}{\varepsilon_{p_2'}-\varepsilon_{p_{2k}'}}$$
(21.25)

und

$$\langle p_{2k}'d|dp_2'\rangle=\int\varphi_{p_{2k}'}^*(\vec{r}_1)\varphi_d^*(\vec{r}_2)V(\vec{r}_1,\vec{r}_2)\varphi_d(\vec{r}_1)\varphi_{p_2'}(\vec{r}_2)\,d\vec{r}_1\,d\vec{r}_2.$$
(21.26)

In (21.25) und (21.26) ist der Index bei der d-Bahn weggelassen; der Index k durchläuft die Nummern der verschiedenen p'_2-Bahnen; der Strich bei der Summe bedeutet, daß man beim Summieren die Glieder mit $p'_{2k}=p'_2$ weglassen muß. $\varepsilon_{p'_2}$ und $\varepsilon_{p'_{2k}}$ sind die Hartree-Fockschen Einelektronenenergien der p'_2- und p'_{2k}-Zustände. Setzt man (21.23) und (21.24) in die Ausdrücke für die Energie des Singulett- bzw. Triplettzustandes (21.12) und (21.13) ein und bildet die halbe Differenz dieser Energien, so erhält man folgenden Ausdruck für das effektive Austauschintegral:

$$J = \frac{1}{2}(E_s - E_t) = -\sum_k{}' \frac{|\langle p'_{2k} d | d p'_2 \rangle|^2}{\varepsilon_{p'_2} - \varepsilon_{p'_{2k}}}. \qquad (21.27)$$

[Um bei der Ableitung von (21.27) im Rahmen der gegebenen Näherungen (21.12) und (21.13) zu bleiben, muß man die Coulomb- (nicht die Austausch-) Terme einander gleichsetzen, was auch bei der Ableitung der Gln. (21.20) bis (21.22) geschehen ist.] Der Ausdruck (21.27) wird in Arbeiten von ANDERSON [151], [152] angegeben. Das Vorzeichen von (21.27) kann in Abhängigkeit vom konkreten Typ einer realen magnetischen Verbindung sowohl positiv als auch negativ sein.

Noch eine Bemerkung zum Polarisationsmechanismus. Die verbreitete Aussage darüber, daß die Elektronen mit nichtkompensiertem Spinmoment die in den Spins abgeschlossene Schale polarisieren, ist eine zweckmäßige und anschauliche Art der Beschreibung für diese Erscheinung; sie ist jedoch nicht exakt. Wie wir sahen, erfolgt im Zustand Ψ_{1t} keine Polarisation der abgeschlossenen Sauerstoffschale, obwohl im System das nichtkompensierte Spinmoment der Elektronen auf den Bahnen d_1 und d_2 existiert. Im Zustand Ψ_{1t} sind die beiden Elektronen auf der Sauerstoffbahn in einem Singulett gebunden. Analog dazu gibt es auch im Lithiumatom [146] keine Polarisation der 1 s-Bahn, falls lediglich eine Dublettfunktion in der Form

$$\begin{vmatrix} \varphi_{2s}\alpha \\ \varphi_{1s}\alpha \\ \varphi'_{1s}\beta \end{vmatrix} - \begin{vmatrix} \varphi_{2s}\alpha \\ \varphi_{1s}\beta \\ \varphi'_{1s}\alpha \end{vmatrix} \qquad (21.28)$$

angesetzt wird.

Die Dublettfunktion (21.28) enthält, wie auch die Triplettfunktion Ψ_{1t}, in den Zuständen $1s$ und $1s'$ ein Zweielektronen-Singulett ($1s$ und $1s'$ analog zu p_1 und p_2).

Eine Polarisation der Bahnen erfolgt nur dann, wenn sowohl beim Lithiumatom als auch im Problem der drei Zentren und vier Elektronen Triplettzustände für diejenigen Elektronen eingeführt werden, die die „ursprünglich" abgeschlossenen Schalen besetzen. Beim Auftreten von Triplettzuständen in den polarisierten Bahnen hat es jedoch keinen Sinn mehr, von der Polarisation einer abgeschlossenen Schale zu sprechen, da die ursprünglich abgeschlossene Schale hinsichtlich des Spins unabgeschlossen wird (ein Triplettzustand an Stelle eines Singulettzustandes). So führt das uneingeschränkte Hartree-Fock-Verfahren zu neuen Bahnkonfigurationen ($\varphi_{1s} \neq \varphi_{1s'}$ für Lithium und $\varphi_{p'_1} \neq \varphi_{p'_2}$ für unser Problem) und damit zusammenhängenden Spinkonfigurationen. In diesem Sinne ist das uneingeschränkte Hartree-Fock-Verfahren der Methode der Konfigurationswechselwirkung sehr ähnlich. Tatsächlich kann man beim Sauerstoffatom eine angeregte Konfiguration einführen (indem man von vornherein $\varphi_{p_1} \neq \varphi_{p_2}$ annimmt) und das Problem nach der Methode der Konfigurationswechselwirkung lösen.

b) Das Problem der drei Zentren und vier Elektronen. Nichtorthogonale Bahnen.

22. Die einfache und zweifache „Anregung". Wie bereits oben erwähnt (s. Ziff. 2), führt die Vernachlässigung der Nichtorthogonalitätsintegrale in der Heitler-London-Näherung zu recht ungenauen Ergebnissen. Deshalb bringen wir hier die Berechnung der Singulett-Triplett-Aufspaltung beim Problem der drei Zentren und vier Elektronen unter Verwendung nichtorthogonaler Atomfunktionen. Selbstverständlich werden dabei die Rechnungen wesentlich komplizierter als bei Verwendung orthogonaler Bahnen. Im folgenden betrachten wir lediglich die beiden Fälle, wo unter den angeregten Konfigurationen die virtuellen Übergänge von Elektronen des Anions in die halbbesetzten Bahnen eines Kations bzw. beider Kationen berücksichtigt werden, sowie den Doppelübergang (s. Ziff. 23). Wie schon bei orthogonalen Bahnen werden als antisymmetrische Basisfunktionen die folgenden Slater-Determinanten verwendet:

$$u_1 = c_1 \begin{vmatrix} d_1\alpha \\ p\alpha \\ p\beta \\ d_2\beta \end{vmatrix}, \quad u_2 = c_2 \begin{vmatrix} d_1\beta \\ p\alpha \\ p\beta \\ d_2\alpha \end{vmatrix}, \quad u_3 = c_3 \begin{vmatrix} d_1\alpha \\ d_1\beta \\ p\alpha \\ d_2\beta \end{vmatrix}, \quad u_4 = c_4 \begin{vmatrix} d_1\alpha \\ d_1\beta \\ p\beta \\ d_2\alpha \end{vmatrix},$$

$$u_5 = c_5 \begin{vmatrix} d_1\alpha \\ p\beta \\ d_2\alpha \\ d_2\beta \end{vmatrix}, \quad u_6 = c_6 \begin{vmatrix} d_1\beta \\ p\alpha \\ d_2\alpha \\ d_2\beta \end{vmatrix}, \quad u_7 = c_7 \begin{vmatrix} d_1\alpha \\ d_1\beta \\ d_2\alpha \\ d_2\beta \end{vmatrix}, \quad (22.1)$$

worin die c_i Normierungskoeffizienten bedeuten. Die Überlappung der Funktionen d_1 und d_2 wird wiederum vernachlässigt, dagegen wird die Nichtorthogonalität der p-Funktionen zu d_1 und d_2 berücksichtigt. Es erscheinen also die von Null verschiedenen Nichtorthogonalitätsintegrale

$$S = \int \varphi_{d_1}^*(\vec{r})\, \varphi_p(\vec{r})\, d\vec{r} = \pm \int \varphi_{d_2}^*(\vec{r})\, \varphi_p(\vec{r})\, d\vec{r}. \quad (22.2)$$

Das Plus entspricht dabei einer geraden Anionenfunktion $\varphi_p(\vec{r})$ und das Minus einer ungeraden. Die Funktionen (22.1) gehören zu $S_z = 0$. Mit ihnen werden Singulett-Vielelektronenfunktionen aufgebaut und Triplettfunktionen mit verschwindender z-Komponente des Spins[29]. Zur Bestimmung der Normierungskoeffizienten c_i aus (22.1) und der „Nichtorthogonalitätsintegrale" $S_{ij} = \int u_i^* u_j\, d\tau$ für diese Funktionen bedient man sich zweckmäßigerweise der Formeln von LÖWDIN (4.5) und (4.7). Die unmittelbare Rechnung ergibt für die Normierungskoeffizienten die Werte

$$c_1 = c_2 = \frac{1}{1 - S^2}, \quad c_3 = c_4 = c_5 = c_6 = \frac{1}{\sqrt{1 - S^2}}, \quad c_7 = 1. \quad (22.3)$$

---

[29] KEFFER und OGUCHI [153] verwendeten die elegante Methode der Operatordarstellung nach DIRAC, VAN VLECK und SERBER, die jedoch weniger übersichtlich ist als die Methode des Aufbaues der Eigenfunktionen zu $\hat{S}^2$ und außerdem den komplizierten Apparat der Theorie der Vertauschungsgruppe erfordert.

Für die Nichtorthogonalitätsintegrale S_{ij} erhält man

$$S_{12} = \frac{-S^4}{(1-S^2)^2}, \quad S_{13} = \frac{-S(1-S^2)}{(1-S^2)^{\frac{3}{2}}}, \quad S_{14} = \frac{S^3}{(1-S^2)^{\frac{3}{2}}},$$

$$S_{15} = \frac{\mp S(1-S^2)}{(1-S^2)^{\frac{3}{2}}}, \quad S_{16} = \frac{\pm S^3}{(1-S^2)^{\frac{3}{2}}}, \quad S_{34} = \frac{-S^2}{1-S^2},$$

$$S_{35} = \frac{\pm S^2}{1-S^2}, \quad S_{36} = \frac{\mp S^2}{1-S^2} \tag{22.4}$$

und

$$S_{23} = S_{14}, \quad S_{24} = S_{13}, \quad S_{25} = S_{16}, \quad S_{26} = S_{15},$$
$$S_{45} = S_{36}, \quad S_{46} = S_{35}, \quad S_{56} = S_{34}.$$

Die Funktion u_7 ist orthogonal zu allen übrigen. Unter Verwendung von (22.3) und (22.4) lassen sich nach der in Ziff. 17 beschriebenen Methode leicht vier orthonormierte Singulettfunktionen Ψ_s und drei orthonormierte Triplettfunktionen Ψ_t aufbauen:

$$\Psi_{1s} = \frac{1-S^2}{\sqrt{2(1-2S^2+2S^4)}} \{u_1 - u_2\},$$

$$\Psi_{2s} = \frac{S(1-S^2)}{\sqrt{2(1-2S^2+2S^4)(1-S^2)(1-2S^2)}} \times$$

$$\times \left\{ u_1 - u_2 + \frac{1-2S^2+2S^4}{S\sqrt{1-S^2}} (u_3 - u_4) \right\},$$

$$\Psi_{3s} = \frac{S(1-S^2)}{\sqrt{2[1-3S^2+2S^6(3-2S^2)]}} \left\{ u_1 - u_2 - \frac{S(1-2S^2)}{\sqrt{1-S^2}} (u_3 - u_4) \pm \right.$$

$$\left. \pm \frac{\sqrt{1-S^2}}{S} (u_5 - u_6) \right\},$$

$$\Psi_{4s} = u_7;$$

$$\Psi_{1t} = \frac{1-S^2}{\sqrt{2(1-2S^2)}} \{u_1 + u_2\},$$

$$\Psi_{2t} = \frac{S\sqrt{1-S^2}}{\sqrt{2(1-2S^2)}} \left\{ u_1 + u_2 + \frac{1}{S\sqrt{1-S^2}} (u_3 + u_4) \right\},$$

$$\Psi_{3t} = \frac{S\sqrt{1-S^2}}{\sqrt{2(1-2S^2)}} \left\{ u_1 + u_2 + \frac{S}{\sqrt{1-S^2}} (u_3 + u_4) \pm \frac{\sqrt{1-S^2}}{S} (u_5 + u_6) \right\}.$$

(22.5)

(22.6)

Die Funktion u_7 liefert keinen Beitrag zu den Triplettfunktionen. Zu beachten ist, daß im Gegensatz zu dem Fall orthogonaler Bahnen in die angeregten Zustände $\Psi_{2s}, \Psi_{3s}, \Psi_{2t}$ und Ψ_{3t} auch die Funktionen der reinen Ionenkonfiguration u_1 und u_2 eingehen. Man könnte natürlich die angeregten Funktionen auch ohne Berücksichtigung von u_1 und u_2 aufbauen (ebenso wie bei orthogonalen Bahnen), allerdings wären dann die unterschiedlichen Ψ_{is} (und Ψ_{it}) nicht mehr orthogonal zueinander. Bei der Lösung der Säkulargleichung benutzen wir die vereinfachende Zusatzannahme, daß das Überlappungsintegral (22.2) wesentlich kleiner ist als 1 und folglich die Lösung als Potenzreihe in S angesetzt werden kann. Vorläufig beschränken wir unsere Betrachtung bei den Singulettfunktionen auf $\Psi_{1s}, \Psi_{2s}, \Psi_{3s}$.

Vorgreifend sei gesagt, daß im vorliegenden Falle das effektive Austauschintegral von vierter Ordnung im Nichtorthogonalitätsintegral (22.2) ist. Deshalb

Ziff. 22. Die einfache und zweifache „Anregung".

verwenden wir an Stelle der Funktionen (22.5) und (22.6) die Näherungsausdrücke

$$\left.\begin{aligned}
\Psi_{1s} &= \frac{1}{\sqrt{2}} \left(1 - \frac{S^4}{2}\right)(u_1 - u_2), \\
\Psi_{2s} &= \frac{1}{\sqrt{2}} \left\{ S\left(1 + \frac{3}{2} S^2\right)(u_1 - u_2) + (1 + S^4)(u_3 - u_4) \right\}, \\
\Psi_{3s} &= \frac{1}{\sqrt{2}} \left\{ S\left(1 + \frac{S^2}{2}\right)(u_1 - u_2) - S^2(1 - S^2)(u_3 - u_4) \pm \left(1 + \frac{3}{2} S^4\right)(u_5 - u_6) \right\},
\end{aligned}\right\} \quad (22.7)$$

$$\left.\begin{aligned}
\Psi_{1t} &= \frac{1}{\sqrt{2}} \left(1 + \frac{S^4}{2}\right)(u_1 + u_2), \\
\Psi_{2t} &= \frac{1}{\sqrt{2}} \left\{ S\left(1 + \frac{S^2}{2}\right)(u_1 + u_2) + \left(1 + S^2 + \frac{3}{2} S^4\right)(u_3 + u_4) \right\}, \\
\Psi_{3t} &= \frac{1}{\sqrt{2}} \left\{ S\left(1 + \frac{3}{2} S^2\right)(u_1 + u_2) + S^2(1 + 2S^2)(u_3 + u_4) \pm \right. \\
&\qquad \left. \pm (1 + S^2 + 2S^4)(u_5 + u_6) \right\}.
\end{aligned}\right\} \quad (22.8)$$

Die Funktionen (22.7) und (22.8) sind nur bis zur Ordnung S^4 orthonormiert. [In den entsprechenden Orthonormierungsgleichungen müssen die Ausdrücke (22.4) nur bis zu den Gliedern S^4 reproduziert werden.]

Im Prinzip erfolgt das Aufsuchen der Singulett- und Triplettenergien genauso wie im Falle orthogonaler Bahnen. Wir verwenden wiederum die Störungstheorie von LÖWDIN zur Bestimmung der durch die Konfigurationswechselwirkung bedingten Korrekturen zu den Energien der reinen Ionenzustände, die durch die Funktionen Ψ_{1s} und Ψ_{1t} beschrieben werden.

Für die weiteren Berechnungen benötigen wir die Größen $h_{ij} = \langle u_i \hat{H} u_j \rangle$. Zunächst folgen dafür aus Symmetriebetrachtungen die Beziehungen

$$\left.\begin{aligned}
h_{11} &= h_{22}, & h_{33} &= h_{44} = h_{55} = h_{66}, & h_{13} &= h_{24} = \pm h_{15} = \pm h_{26}, \\
h_{14} &= h_{23}, & h_{16} &= h_{25}, & h_{34} &= h_{56}, & h_{35} &= h_{46}, & h_{36} &= h_{45}.
\end{aligned}\right\} \quad (22.9)$$

Dann kann man aus (22.4) die Ordnung jedes einzelnen h_{ij} bestimmen, ohne dieselben gleich explizit auszurechnen. Man erhält so

$$\left.\begin{aligned}
h_{11} &\sim S^0, & h_{12} &\sim S^4, & h_{13} &\sim S, & h_{14} &\sim S^3, & h_{15} &\sim S, & h_{16} &\sim S^3, \\
h_{33} &\sim S^0, & h_{34} &\sim S^2, & h_{35} &\sim S^2, & h_{36} &\sim S^2, & h_{55} &\sim S^0.
\end{aligned}\right\} \quad (22.10)$$

Aus (17.45) folgt, daß es für die Berechnung der Energien in der hier beabsichtigten Näherung genügt, die Diagonalelemente H_{22}^s, H_{33}^s, H_{22}^t und H_{33}^t sowie die nichtdiagonalen Matrixelemente H_{ij}^s und H_{ij}^t nur bis zu den im Parameter (22.2) quadratischen Gliedern zu berechnen. In der jeweils erforderlichen Näherung erhalten wir:

$$\left.\begin{aligned}
H_{11}^s &= (1 - S^4) h_{11} - h_{12}, & H_{12}^s &= S h_{11} + h_{13}, & H_{13}^s &= S h_{11} + h_{13}, \\
H_{22}^s &= S^2 h_{11} + h_{33} + 2S h_{13} - h_{34}, \\
H_{23}^s &= S^2 h_{11} - S^2 h_{33} + 2S h_{13} \pm (h_{35} - h_{36}), \\
H_{33}^s &= S^2 h_{11} + h_{33} + 2S h_{13} - h_{34},
\end{aligned}\right\} \quad (22.11)$$

$$\left.\begin{aligned}
H_{11}^t &= (1 + S^4) h_{11} + h_{12}, & H_{12}^t &= S h_{11} + h_{13}, & H_{13}^t &= S h_{11} + h_{13}, \\
H_{22}^t &= S^2 h_{11} + (1 + 2S^2) h_{33} + 2S h_{13} + h_{34}, \\
H_{23}^t &= S^2 h_{11} + S^2 h_{33} + 2S h_{13} \pm (h_{35} + h_{36}), \\
H_{33}^t &= S^2 h_{11} + (1 + 2S^2) h_{33} + 2S h_{13} + h_{34}.
\end{aligned}\right\} \quad (22.12)$$

Mit Hilfe von (17.45), (22.11) und (22.12) ergibt sich für die Singulett-Triplett-Aufspaltung exakt bis zu Gliedern der Ordnung S^4 der Ausdruck

$$E_s - E_t = -2 \left[(S^4 E_0 + h_{12}) + 2 \frac{(S E_0 + h_{13})^2}{(E_0 - E_1)^2} (2 S^2 E_1 + h_{34} \pm h_{36}) \right], \quad (22.13)$$

wobei E_0 und E_1 die Energien der Grund- bzw. der angeregten Konfiguration bedeuten.

Die in (22.13) eingehenden Größen sind wie folgt durch Integrale mit nichtorthogonalen Einelektronenfunktionen auszudrücken:

$$\begin{aligned}
E_0 &= 2\langle d_1|d_1\rangle + 2\langle p|p\rangle + 4\langle d_1 p|d_1 p\rangle + \langle d_1 d_2|d_1 d_2\rangle + \langle p p|p p\rangle, \\
E_1 &= 3\langle d_1|d_1\rangle + \langle p|p\rangle + \langle d_1 d_1|d_1 d_1\rangle + 2\langle d_1 d_2|d_1 d_2\rangle + 3\langle d_1 p|d_1 p\rangle, \\
h_{12} &= -2\{2 S^3 \langle d_1|p\rangle + S^2 (\pm 2 \langle d_1 d_2|p p\rangle + \langle d_1 p|p d_1\rangle)\}, \\
h_{13} &= -\{S(2\langle d_1|d_1\rangle + \langle p|p\rangle + \langle d_1 d_2|d_1 d_2\rangle + 2\langle d_1 p|d_1 p\rangle) + \langle d_1|p\rangle + \\
&\quad + \langle d_1 p|d_1 d_1\rangle + \langle d_1 p|p p\rangle \pm \langle d_1 d_2|d_1 p\rangle\}, \\
h_{34} &= -\{S(2 S \langle d_1|d_1\rangle + S \langle d_1 d_1|d_1 d_1\rangle + 2\langle d_1|p\rangle \pm \\
&\quad \pm 4\langle d_1 d_2|d_1 p\rangle) + \langle d_1 p|p d_1\rangle\}, \\
h_{36} &= -\{S(\pm 2\langle d_1|p\rangle + 2\langle d_1 d_2|d_1 p\rangle \pm 2\langle d_1 p|d_1 d_1\rangle \pm 2 S \langle d_1|d_1\rangle \pm \\
&\quad \pm S \langle d_1 d_2|d_1 d_2\rangle) + \langle d_1 d_2|p p\rangle\}.
\end{aligned} \quad (22.14)$$

Hierbei wurde die folgende abgekürzte Schreibweise angewandt:

$$\left. \begin{aligned}
\langle \varphi_1|\varphi_2\rangle &= \int \varphi_1^*(\vec{r}) \hat{H}_1(\vec{r}) \varphi_2(\vec{r}) \, d\vec{r}, \\
\langle \varphi_1 \varphi_2|\varphi_3 \varphi_4\rangle &= \int \varphi_1^*(\vec{r}_1) \varphi_2^*(\vec{r}_2) V(\vec{r}_1, \vec{r}_2) \varphi_3(\vec{r}_1) \varphi_4(\vec{r}_2) \, d\vec{r}_1 \, d\vec{r}_2.
\end{aligned} \right\} \quad (22.15)$$

Bei der Reihenentwicklung nach S wurde ganz einfach den verschiedenen Integralen je nach der Zahl der Produkte $\varphi_{d_1}(\vec{r}) \varphi_p(\vec{r})$ [oder $\varphi_{d_2}(\vec{r}) \varphi_p(\vec{r})$] unter dem Integralzeichen eine bestimmte Ordnung bezüglich S zugeschrieben. So besitzen z.B. die Integrale $\langle d_1|d_1\rangle$, $\langle p|p\rangle$, $\langle d_1 d_1|d_1 d_1\rangle$, $\langle d_1 d_2|d_1 d_2\rangle$, $\langle d_1 p|d_1 p\rangle$ den Grad S^0; die Integrale $\langle d_1|p\rangle$, $\langle d_1 d_2|d_1 p\rangle$, $\langle d_1 p|d_1 d_1\rangle$ den Grad S^1 und die Integrale $\langle d_1 p|p d_1\rangle$, $\langle d_1 d_2|p p\rangle$ den Grad S^2. Wenn man in (22.13) den Parameter S dort, wo er *explizit* auftritt, gleich Null setzt, so ergibt sich derselbe Ausdruck, den wir bereits früher für orthogonale Bahnen erhielten.

Die Begriffe „verschwindende Überlappung" und „Orthogonalität" der Funktionen sind zu unterscheiden. Die Orthogonalität zweier Funktionen φ_1 und φ_2,

$$\int \varphi_1^*(\vec{r}) \varphi_2(\vec{r}) \, d\vec{r} = 0 \quad (22.16)$$

bedeutet im allgemeinen nicht, daß diese Funktionen sich nicht überlappen, d.h. in einem Gebiet des Raumes, wo $\varphi_1(\vec{r})$ verschieden von Null ist, muß $\varphi_2(\vec{r})$ nicht Null sein und umgekehrt. Bei einander überlappenden orthogonalen Funktionen besagt die Bedingung (22.16) nur, daß das Produkt $\varphi_1^*(\vec{r}) \varphi_2(\vec{r})$ sowohl positive als auch negative Werte annimmt, deren Anteile sich kompensieren, was bei der Integration Null ergibt. Wenn zwei Funktionen sich nicht überlappen,

$$\varphi_1^*(\vec{r}) \varphi_2(\vec{r}) = 0, \quad (22.17)$$

erfüllen sie automatisch die Orthogonalitätsbedingung (22.16). Somit ist die Bedingung (22.17) strenger als (22.16). Die Existenz eines Nichtorthogonalitäts- oder

Überlappungsintegrals zwischen zwei Funktionen

$$\int \varphi_1^*(\vec{r})\,\varphi_2(\vec{r})\,d\vec{r} = S \tag{22.18}$$

mit $S \neq 0$, bedeutet, daß unbedingt ein Raumgebiet existiert, in dem das Produkt $\varphi_1^*(\vec{r})\,\varphi_2(\vec{r})$ von Null verschieden ist. Jedoch ist (22.18) im allgemeinen kein eindeutiges Maß für die Überlappung zweier Funktionen, denn die Größe S ist für einander beliebig stark überlappende, aber orthogonale Funktionen gleich Null. In Anbetracht dessen erfolgte bei den Berechnungen eigentlich nicht eine Entwicklung nach dem Überlappungsintegral (22.18), sondern nach Größen, die funktional von dem Produkt

$$\varphi_1^*(\vec{r})\,\varphi_2(\vec{r}) \tag{22.19}$$

abhängen. Aus diesem Grunde sind z.B. die Ausdrücke $\langle \varphi_1\,\varphi_2 | \varphi_2\,\varphi_1 \rangle$ und $S^2 \langle \varphi_1\,\varphi_2 | \varphi_1\,\varphi_2 \rangle$ von gleicher Größenordnung. Den Grenzübergang

$$S \to 0$$

kann man folglich in zweierlei Hinsicht verstehen: 1. als Übergang zu orthogonalen Funktionen, wo zwar (22.16), nicht jedoch (22.17) erfüllt wird; 2. als Übergang zu einander nicht überlappenden Funktionen, wo die Bedingung (22.17) und demzufolge auch (22.16) gilt. Im ersten Falle bleibt der Ausdruck $\langle \varphi_1\,\varphi_2 | \varphi_2\,\varphi_1 \rangle$ endlich, und $S^2 \langle \varphi_1\,\varphi_2 | \varphi_2\,\varphi_1 \rangle$ strebt gegen Null, genau wie alle anderen Glieder, die den Faktor S explizit enthalten. Im zweiten Falle verschwinden in gleicher Weise sowohl $\langle \varphi_1\,\varphi_2 | \varphi_2\,\varphi_1 \rangle$ als auch $S^2 \langle \varphi_1\,\varphi_2 | \varphi_2\,\varphi_1 \rangle$. So ist der Ausdruck in den ersten runden Klammern von (22.13) vollständig durch die reine Ionenkonfiguration bedingt und gleich

$$\langle \Psi_{1s} | \hat{H} | \Psi_{1s} \rangle - \langle \Psi_{1t} | \hat{H} | \Psi_{1t} \rangle. \tag{22.20}$$

Bei Verwendung orthogonaler Bahnen wird die Differenz (22.20) insgesamt Null, hier dagegen liefert sie einen Beitrag zur Größe $E_s - E_t$. In eben dieser Beziehung unterscheidet sich die Verwendung nichtorthogonaler Bahnen prinzipiell von der Methode der orthogonalen Bahnen. Im gegebenen Falle kommt es im Gegensatz zu den Ergebnissen bei Anwendung orthogonaler Bahnen bereits ohne Einführung angeregter Konfigurationen zu einer Singulett-Triplett-Aufspaltung. Diesen Unterschied in den Ergebnissen kann man verstehen, wenn man einen Zusammenhang zwischen den Wellenfunktionen im ersten und zweiten Falle herstellt. Dazu suchen wir die orthogonalen Linearkombinationen der Funktionen d_1, d_2 und p auf. Sie lauten

$$\left.\begin{array}{l} f_1 = d_1, \\ f_2 = \dfrac{1}{\sqrt{1-S^2}}(S\,d_1 \pm S\,d_2 - p), \\ f_3 = d_2. \end{array}\right\} \tag{22.21}$$

Drückt man jetzt beispielsweise in u_1 die nichtorthogonalen Funktionen d_1, d_2, p durch die orthonormierten f_1, f_2, f_3 aus,

$$\left.\begin{array}{l} d_1 = f_1, \\ p = S(f_1 \pm f_3) - \sqrt{1-S^2}\,f_2, \\ d_2 = f_2, \end{array}\right\} \tag{22.22}$$

so erhält man für die einzelne Determinante u_1 eine Darstellung in Form einer Superposition anderer Determinanten mit orthogonalen Funktionen, und unter den Gliedern dieser Reihe werden sich auch Determinanten befinden, die Zustände mit einfacher und zweifacher Anregung beschreiben. Somit wird klar, daß die Verwendung nichtorthogonaler Funktionen in der reinen Ionenkonfiguration des

Systems gleichbedeutend ist der Einführung angeregter Konfigurationen bei der Methode orthogonaler Bahnen. Diese angeregten Konfigurationen führen dann ebenfalls zur Singulett-Triplett-Aufspaltung. Wie wir sehen, ist die Situation etwa derjenigen zu vergleichen, die wir bei der Orthogonalisierung im Wasserstoffmolekülproblem antrafen (s. Ziff. 2).

Alle Glieder in (22.13) sind von der gleichen Ordnung S^4, gehören jedoch innerhalb der Störungstheorie zu unterschiedlichen Ordnungen. Die erste Zeile von (22.13) ist von erster Ordnung, und in der zweiten Zeile steht ein Glied dritter Ordnung. Die Funktion u_7, die lediglich zum Singulettzustand beiträgt, führt auf Grund der Wechselwirkung mit der reinen Ionenkonfiguration in der ersten Ordnung der Störungstheorie zu einer indirekten Kopplung. Der entsprechende Beitrag zur Größe $E_s - E_t$ wird durch den Ausdruck

$$\frac{2}{E_0 - E_2} \{2S^2 \langle d_1 | d_1 \rangle + 2S \langle d_1 | p \rangle + 2S \langle d_1 \, p | d_1 \, d_1 \rangle + S^2 \langle d_1 \, d_2 | d_1 \, d_2 \rangle \pm$$
$$\pm 2S \langle d_1 \, p | d_1 \, d_2 \rangle \pm \langle p \, p | d_2 \, d_1 \rangle \}^2$$

gegeben und besitzt ebenfalls die Ordnung S^4. Die Bestimmung der relativen Größe jedes einzelnen Gliedes ist sehr schwierig, da die Einelektronenfunktionen nicht bekannt sind. KEFFER und OGUCHI [153] versuchten eine grobe Abschätzung für die Singulett-Triplett-Aufspaltung in der Annahme, daß die p-Funktionen des Sauerstoffs durch eine Slater-Funktion vom Typ

$$\sqrt{\frac{\delta^5}{\pi}} \, r \, e^{-\delta r} \, P_1(\cos \Theta) \tag{22.23}$$

und die Kationenbahnen durch eine Funktion

$$\sqrt{\frac{\delta^5}{3\pi}} \, r \, e^{-\delta r} \tag{22.24}$$

beschrieben werden, wobei δ einen Variationsparameter und $P_1(x)$ ein Legendresches Polynom bedeuten.

Indem sie den Parameter δ so wählten, daß das Integral S einen vernünftigen Wert annimmt, und die Übergangsenergie zwischen 5 eV und 20 eV variierten, erhielten sie für das Austauschintegral $(E_s - E_t)/2$ einen Wert von $\approx 10-50$ °K. Wie sich herausstellte, ergeben alle Glieder einen negativen Beitrag, und die reine Ionenkonfiguration [erste Zeile in (22.13)] liefert dabei den niedrigsten Wert. Es lohnt sich jedoch kaum, diesen Abschätzungen besonderen Wert beizumessen, wenn man an die dabei verwendeten groben Näherungen denkt. Deshalb scheint es zweckmäßiger zu sein, einen Vergleich der Theorie mit dem Experiment erst dann anzustellen, wenn man über realistischere Funktionen für den Kristall verfügt.

23. Der Doppelübergang. Der Mechanismus des indirekten Austausches, der durch einen Doppelübergang bedingt ist und in Ziff. 20 für den Fall orthogonaler Bahnen betrachtet wurde, kann auch innerhalb der Methode nichtorthogonaler Bahnen leicht dargestellt werden.

Die Basis-Slater-Determinanten (für $S_z = 0$) lauten, wie schon bei orthogonalen Bahnen [s. (20.1)],

$$u_1 = c_1 \begin{vmatrix} d_1 \alpha \\ p \alpha \\ p \beta \\ d_2 \beta \end{vmatrix}, \quad u_2 = c_2 \begin{vmatrix} d_1 \beta \\ p \alpha \\ p \beta \\ d_2 \alpha \end{vmatrix}, \quad u_3 = c_3 \begin{vmatrix} d_1 \alpha \\ d_1 \beta \\ p \alpha \\ p \beta \end{vmatrix}, \quad u_4 = c_4 \begin{vmatrix} p \alpha \\ p \beta \\ d_2 \alpha \\ d_2 \beta \end{vmatrix}. \tag{23.1}$$

Mit Hilfe von (4.5) und (4.7) finden wir die Normierungskoeffizienten in (23.1) und die Nichtorthogonalitätsintegrale $S_{ij}=\langle u_i u_j \rangle$:

$$c_1 = c_2 = c_3 = c_4 = \frac{1}{1-S^2}, \tag{23.2}$$

$$\left.\begin{array}{l} S_{12} = -\dfrac{S^4}{(1-S^2)^2}, \quad S_{13} = S_{14} = \mp \dfrac{S^2}{1-S^2}, \\[4pt] S_{23} = -S_{14}, \quad S_{24} = -S_{13}, \quad S_{34} = -S_{12}. \end{array}\right\} \tag{23.3}$$

Offensichtlich kann man aus den vier Funktionen (23.1) eine Triplett- und drei Singulettfunktionen aufbauen. In der Näherung aus Ziff. 22 lauten diese Funktionen

$$\left.\begin{array}{l} \Psi_{1s} = \dfrac{1}{\sqrt{2}} \left(1 - \dfrac{S^4}{2}\right)(u_1 - u_2), \\[4pt] \Psi_{2s} = S^2(1+S^2)(u_1 - u_2) \pm (1+S^4) u_3, \\[4pt] \Psi_{3s} = S^2(1+S^2)(u_1 - u_2) \pm S^4 u_3 \pm (1+S^4) u_4, \end{array}\right\} \tag{23.4}$$

$$\Psi_{1t} = \frac{1}{\sqrt{2}}\left(1 + \frac{S^4}{2}\right)(u_1 + u_2). \tag{23.5}$$

Da bei der gegebenen Art eines virtuellen Übergangs (sowohl die Anionen- als auch die Kationenbahnen sind zweifach besetzt) keine angeregten Triplettzustände existieren, gehen in die Größe der Singulett-Triplett-Aufspaltung im allgemeinen Beiträge aller Ordnungen der Störungstheorien ein [in Ziff. 22 lieferte zu (22.13) die zweite Störungsordnung keinen Beitrag].

So nimmt in der ersten, durch die reine Ionenkonfiguration bedingten Ordnung der Störungstheorie der Ausdruck für die Differenz $E_s - E_t$ folgende Formen an:

mit
$$E_s - E_t = \langle \Psi_{1s} \hat{H} \Psi_{1s} \rangle - \langle \Psi_{1t} \hat{H} \Psi_{1t} \rangle = -2\{h_{12} + S^4 h_{11}\} \tag{23.6}$$

$$\left.\begin{array}{l} h_{11} = 2\langle d_1|d_1\rangle + 2\langle p|p\rangle + 4\langle d_1 p|d_1 p\rangle + \langle pp|pp\rangle + \langle d_1 d_2|d_1 d_2\rangle, \\[4pt] h_{12} = 2S^2\{2S\langle d_1|p\rangle \pm \langle d_1 p|p d_2\rangle + \langle d_1 p|p d_1\rangle \pm \langle d_1 d_2|pp\rangle. \end{array}\right\} \tag{23.7}$$

Der Ausdruck (23.6) ist von der Ordnung S^4. Den Beitrag höherer Ordnungen der Störungstheorie kann man ebenso berechnen wie in Ziff. 22.

c) Die Verallgemeinerung auf den Kristall.

24. Die Herleitung des effektiven Hamilton-Operators für orthogonale Bahnen. Eine Verallgemeinerung der Theorie der indirekten Austauschwechselwirkung auf den Kristall wurde von SHIMIZU [135] unter der Annahme orthogonaler Bahnen sowie von VONSOVSKY und SEIDOV [154], [155], YAMASHITA [156], KONDO [157] und FUKUCHI [158], [159] in der Näherung der orthogonalisierten Bahnen durchgeführt. Die vollständigsten Ausdrücke für die Spinkopplung wurden von FUKUCHI erhalten. YAMASHITA [156] und KONDO [157] verwendeten den Formalismus von LÖWDIN [160], um die Energie des Systems in der Eindeterminantennäherung in eine Reihe nach dem Nichtorthogonalitätsintegral zu entwickeln. In den Arbeiten [154], [155], [158], [159] wurde die Operatormethode der Störungstheorie benützt, wie sie von BOGOLYUBOV [96] entwickelt und in Ziff. 14 ausführlich dargestellt worden ist. Für die Behandlung von Kristallen ist die Bogolyubov-Methode am besten geeignet. Auch das hier wiedergegebene Ergebnis wurde damit erhalten. Wir betrachten lediglich die Prozesse der einfachen und zweifachen Anregung in die einfach besetzte Bahn. Wie oben nehmen wir an, daß in einem Kristall vom Typ MnO in der reinen Ionenkonfiguration je ein Elektron auf die Kationen und je zwei Elektronen auf die abgeschlossene Schale jedes Anions entfallen. Der Kristall wird in der Heitler-London-Näherung behandelt, jedem Elektron entspricht also eine lokalisierte Atomfunktion φ_f, wobei f die Nummer des Gitterpunktes angibt. Die Wellenfunktionen der verschiedenen Gitterpunkte sind nicht orthogonal zueinander, ihr Überlappungsintegral[30] ist

$$S_{f_1 f_2} = \int \varphi_{f_1}^*(\vec{r}) \, \varphi_{f_2}(\vec{r}) \, d\vec{r} - \delta_{f_1 f_2}. \tag{24.1}$$

Wir nehmen hier an, daß von den Nichtorthogonalitätsintegralen (24.1) lediglich die Integrale zwischen den Wellenfunktionen der Kationen und den Funktionen der dazu nächstgelegenen Anionen von Null verschieden sind; alle anderen Integrale werden vernachlässigt. Nach diesem Integral als einzigem Entwicklungsparameter werden auch in Übereinstimmung mit dem unter Ziff. 14 ausführlich dargelegten Schema alle Entwicklungen erfolgen. Im Gegensatz zu den Berechnungen bei der Herleitung der Ausgangsformel (14.10) für die Entwicklung des Hamilton-Operators nach einem kleinen Parameter muß man hier zwei Sorten von Gitterpunkten unterscheiden, die mit magnetischen Kationen bzw. mit nichtmagnetischen Anionen besetzt sind. Das kompliziert selbstverständlich alle Rechnungen wesentlich, obwohl sie sich im Prinzip nicht von dem in Ziff. 14 dargestellten Fall unterscheiden. Deshalb lassen wir sie hier weg und bringen lediglich das Endergebnis, das eine Verallgemeinerung der Formeln (14.49) und (14.51) darstellt. Das effektive Integral des indirekten Austausches lautet im vorliegenden Fall

$$\begin{aligned} J_{m_1 m_2} = F(m_1 m_2 m_2 m_1) + \sum_n \Big\{ &\frac{\Lambda_1(m_1 n) \Lambda_2(n m_2) F(m_1 n n m_2)}{\Delta(m_1, n)\Delta(m_2, n)} + \\ + \frac{1}{2}\sum_{n'} \Big(&\frac{\Lambda_1(m_1 n') \Lambda_2(n m_1) F(m_2 n' n m_2)}{\Delta(m_1, n)\Delta(m_1, n')} + \\ + &\frac{\Lambda_1(m_2 n') \Lambda_2(n m_2) F(m_1 n' n m_1)}{\Delta(m_2, n)\Delta(m_2, n')} \Big) - \frac{1}{2}\frac{F^2(m_1 m_2 n n)}{\Delta(m_1 m_2, n n)} \Big\} \end{aligned} \tag{24.2}$$

mit den Bezeichnungen

$$\left.\begin{aligned} \Lambda_1(f_1 f_2) &= L(f_1 f_2) + \sum_f F(f_1 f f_2 f) \{N_f - \delta_{f f_2}\}, \\ \Lambda_2(f_1 f_2) &= L(f_1 f_2) + \sum_f F(f_1 f f_2 f) \{N_f - \delta_{f f_1}\}. \end{aligned}\right\} \tag{24.3}$$

Hierin bedeuten m bzw. n die Nummern der magnetischen bzw. nichtmagnetischen Gitterpunkte. Somit bleibt die Diracsche Form (14.51) der spinabhängigen Wechselwirkung auch für den Kristall gültig.

Setzt man in die Austauschintegrale (24.2) die Entwicklung (14.9) ein, so erhält man Ausdrücke, die analog sind den vorher in Ziff. 18, 19 und 22 beim Problem der drei Zentren und vier Elektronen erhaltenen. Der Übergang zu orthogonalen Bahnen erfolgt, indem man S gleich Null setzt. Die Bedeutung der einzelnen Glieder in (24.2) ist offensichtlich. Der erste Summand entspricht der ersten Zeile in (22.13) und ist folglich durch die reine Ionenkonfiguration bedingt. Die ersten drei Glieder zwischen den geschweiften Klammern entsprechen dem Kramers-Anderson-Mechanismus, wo ein Elektron von einem Anion in die einfach besetzte Bahn eines Kations übergeht mit einer Anregungsenergie $\Delta(m, n)$. Der

---

[30] Die Definition (24.1) des Überlappungsintegrals ist für den Bogolyubov-Formalismus günstiger als die Definition $S_{f_1 f_2} = \int \varphi_{f_1}^*(\vec{r}) \, \varphi_{f_2}(\vec{r}) \, d\vec{r}$, die wir oben benützten [s. z. B. (22.12)].

letzte Term beschreibt schließlich den Anregungsmechanismus, der im gleichzeitigen Übergang zweier Elektronen eines Anions zu zwei benachbarten Kationen besteht. In (24.2) wurden Glieder vernachlässigt, die der Wechselwirkung zweier angeregter Konfigurationen entsprechen. Diese Ausdrücke sind recht umfangreich, man kann sie in der Arbeit von FUKUCHI [*158*], [*159*] finden.

FUKUCHI berücksichtigte auch Konfigurationen, die dadurch bedingt sind, daß an einem Anion ein Elektron in eine höhere Bahn desselben Anions übergeht. Dabei erhält man wiederum einen effektiven Hamilton-Operator von Diracscher Form mit einem sehr komplizierten Ausdruck für das Austauschintegral. Alle Glieder darin sind ebenfalls von der Ordnung S^4.

25. Bemerkungen zum Doppelaustausch. In Ziff. 17—24 wurden verschiedene Typen physikalischer Wechselwirkungsmechanismen zwischen den Elektronen betrachtet, die in nichtleitenden Kristallen zur Ausbildung einer indirekten Austauschkopplung zwischen den magnetisch aktiven Kationen führen können. Ein Vergleich der Methoden der orthogonalen und nichtorthogonalen Bahnen zeigt, daß es oftmals nicht möglich ist, die diversen Mechanismen voneinander zu unterscheiden, da ein ganz bestimmter Typ eines virtuellen Überganges aus dem Blickwinkel einer anderen Methode betrachtet ganz anderen Typen entsprechen kann. Die Untersuchung des Polarisationsmechanismus zeigte auch den engen Zusammenhang eines solchen Überganges mit dem Problem der Konfigurationswechselwirkung. Aus diesem Grunde muß man die Rollen vergleichen, welche die unterschiedlichen Konfigurationen spielen, wie das bereits in Ziff. 2 und 16 erfolgt ist.

Offen bleibt noch die Frage eines quantitativen Vergleichs der Theorie mit dem Experiment. Darauf kommen wir unter Ziff. 30 zurück, nachdem wir ANDERSONs neue Methode [*47*] beim Problem der indirekten Austauschwechselwirkung dargestellt und einen kurzen Abriß der Ergebnisse der Ligandenfeldtheorie gegeben haben.

Hier wollen wir uns mit der kurzen Diskussion eines der möglichen Mechanismen der Austauschwechselwirkung begnügen, nämlich des Doppelaustausches, den man in der Literatur oft als Anderson-Hasegawa-Mechanismus bezeichnet [*161*].

ANDERSON und HASEGAWA untersuchten eingehend den Mechanismus des Doppelaustausches, dessen Existenz und grundlegende Eigenschaften von ZENER [*101*] bis [*104*] beschrieben wurden. ZENER erklärte den Ferromagnetismus leitender Kristalle durch die indirekte Wechselwirkung der lokalisierten Spins über die Leitungselektronen. Eine der wichtigsten Gruppen von Stoffen, für die der Doppelaustausch nach Meinung von ZENER entscheidende Bedeutung besitzt, sind Verbindungen mit Ionen wechselnder Wertigkeit. Ein besonders lohnendes Objekt für die Anwendung dieser Theorie sind Stoffe mit Perowskitstruktur, die von JONKER und VAN SANTEN [*162*] bis [*164*] untersucht wurden. Die Verbindung $La_{1-x}Ca_xMnO_3$ enthält bei einer Calciumkonzentration $x=0$ lediglich dreiwertige Manganionen Mn^{+3}. Werden die dreiwertigen Lanthanionen durch zweiwertige Calciumionen ersetzt, so treten zur Gewährleistung der elektrischen Neutralität vierwertige Manganionen Mn^{+4} in einem Ausmaß auf, das der Menge der zugefügten Calciumionen entspricht. Dabei wurde folgende Korrelation beobachtet: Bei den Konzentrationen $x=0$ und $x=1$ ist die Verbindung ein schlechter Leiter, und die Wechselwirkung besitzt einen antiferromagnetischen Charakter. Dagegen nimmt etwa im Konzentrationsbereich $0,2 < x < 0,4$ die Leitfähigkeit um einige Größenordnungen zu, und ein Übergang zum ferromagnetischen Zustand wird beobachtet. ANDERSON und HASEGAWA [*161*] untersuchten die Natur der

Austauschwechselwirkung in Verbindungen mit Ionen wechselnder Wertigkeit, indem sie lediglich ein Paar von Kationen, deren Ladungen sich um 1 voneinander unterscheiden, betrachteten. Die Energie des Systems wird durch eine Säkulargleichung bestimmt, deren Matrixelemente Zustände verbinden, in denen sich das zusätzliche Elektron entweder am linken oder am rechten Kation befindet. Das Übergangsintegral hängt, wenn das Elektron seine Spinrichtung beibehält, vom Winkel zwischen den Gesamtspins $\vec{S}_1$ und $\vec{S}_2$ der Kationen ab. Sie untersuchten ferner das Energiespektrum des Systems in Abhängigkeit vom Größenverhältnis des intraatomaren Austauschintegrals und des Übergangsintegrals und beschrieben die thermodynamischen Eigenschaften solcher Magnetika. Weitere eingehende Untersuchungen über die Besonderheiten des Doppelaustauschs enthält eine Arbeit von DE GENNES [165]. Der oben beschriebene Mechanismus des Doppelaustauschs ist im allgemeinen für Isolatoren nicht charakteristisch. Da der Zustand mit einem zusätzlichen Elektron am linken Kation und der Zustand mit diesem Elektron am rechten Kation miteinander entartet sind, sieht das vorliegende Modell eine „freie" Elektronenbewegung vor, also eine hohe Leitfähigkeit der Substanz. Bei der Anwendung auf Dielektrika kann der Übergang eines Elektrons z.B. von Mn^{+3} zu Mn^{+4} als ein angemessener virtueller Prozeß angesehen werden, wenn die Lage des drei- und vierwertigen Manganions im Kristall fixiert wird. Demnach sind ein Zustand, bei dem sich das dreiwertige Ion links und das vierwertige rechts vom Anion befindet, und ein Zustand mit umgekehrter Anordnung energetisch nicht äquivalent. Die Bezeichnung „Doppelaustausch" hat folglich bei Anwendung auf leitende Stoffe und Dielektrika einen unterschiedlichen Sinn. Betrachtet man das Zwischenanion explizit wie im Problem der drei Zentren, so kann man den Mechanismus des Doppelaustauschs in Isolatoren beschreiben, indem man „angeregte" Konfigurationen berücksichtigt. In den angeregten Konfigurationen enthält dabei die Schale des Anions dieselbe Zahl von Elektronen wie in der reinen Ionenkonfiguration, während die Schale des einen Kations ein Elektron mehr und die des anderen ein Elektron weniger aufweist als in der Ionenkonfiguration (vgl. Ziff. 20 und 23). In der Terminologie der Einelektronenübergänge erhält man die angeregte Konfiguration aus der reinen Ionenkonfiguration, indem ein Elektron von einem Kation in die p-Schale eines Anions unter Beibehaltung seiner Spinrichtung überwechselt bei gleichzeitigem Übergang eines Elektrons (mit gleicher Spinrichtung) dieser p-Schale zu einem anderen Kation.

V. Der neue Lösungsweg Andersons für das Problem der indirekten Wechselwirkung.

26. Magnetische Quasiteilchen. Wir bereits bemerkt (s. Einführung in Teil C), unterzog ANDERSON in seinen Arbeiten [47] den in Kapitel IV dargestellten Ansatz für die Theorie der indirekten Austauschwechselwirkung in Isolatoren einer Kritik. Er verwies dabei auf den betonten Modellcharakter der Theorien, auf die schwache Konvergenz der Störungsreihen und auf die Schwierigkeit bei der Interpretation und bei der zahlenmäßigen Abschätzung der einzelnen, durch verschiedene Konfigurationen und ihre Wechselwirkung bedingten Summanden im indirekten Austauschintegral. Er versuchte auch, einige allgemeingültige Kriterien zu erhalten, also jene grundlegenden Wechselwirkungen und virtuellen Elektronenübergänge herauszustellen, die zu bestimmten Typen der Austauschwechselwirkungen in Kristallen führen.

Vor allen Dingen schlagt ANDERSON vor, die magnetischen Kationen im Kristallgitter des Isolators und ihre Anionenumgebung als einheitliches System zu betrachten und das Problem der Bestimmung der im Kristall lokalisierten Elektronenzustände mit von Null verschiedenem magnetischen Moment als im Prinzip exakt gelöst anzusehen. Folglich kann man die nichtkompensierten d-Schalen der magnetischen Kationen, die sich unter der Einwirkung desKristallfeldes ihrer diamagnetischen Umgebung befinden, etwa in Analogie zum Polaron ($\equiv$ dem Leitungselektron und der von ihm verursachten Polarisation der Umgebung im Ionenkristall) als ein gewisses Spin- (oder d-) Quasiteilchen betrachten. Die „Einelektronen-Wellenfunktionen" dieser an den Kationen lokalisierten Quasiteilchen können als Lösungen der selbstkonsistenten Hartree-Fock-Gleichungen ermittelt werden und müssen orthogonal zueinander sein. Genauer gesagt, müssen die vollständigen spin- und ortsabhängigen Funktionen orthogonal sein. Deshalb sind auch die Koordinatenanteile der Funktion nur für Zustände mit parallelen Spins orthogonal, jedoch nicht für antiparallele Spins, da die vollständigen Spin-Bahn-Funktionen in diesem Falle wegen der Spinfaktoren orthogonal zueinander sind (uneingeschränkte Hartree-Fock-Methode).

Wir betrachten nun die verschiedenen Typen der Wechselwirkung der Elektronen an den Kationen, also der Spin-Quasiteilchen, und zwar in der Annahme, daß sich an jedem Kation mehrere d-Elektronen aufhalten können. Wir schreiben den Energieoperator wie schon oben in der Darstellung der zweiten Quantelung

$$\hat{H} = \sum_{ff'\sigma} L(ff') c_{f\sigma}^+ c_{f'\sigma} + \frac{1}{2} \sum_{f_1 f_2 f_1' f_2' \sigma_1 \sigma_2} F(f_1 f_2 f_1' f_2') c_{f_1\sigma_1}^+ c_{f_2\sigma_2}^+ c_{f_2'\sigma_2} c_{f_1'\sigma_1}, \qquad (26.1)$$

wobei c^+ und c die Erzeugungs- und Vernichtungsoperatoren für die „angezogenen" Elektronen (d.h. die Quasiteilchen) bedeuten, für die die gleichen Vertauschungsrelationen (14.4) gelten wie für die analogen Operatoren a^+ und a der Elektronen. Die Indices f durchlaufen jetzt lediglich die Nummern der Gitterpunkte und Bahnen der Kationen, nicht der Kationen und Anionen wie in Ziff. 24. Die Atomnummer bezeichnen wir mit dem Index n und die Bahn innerhalb eines Atoms mit dem Index m. Folglich ist $f = (n, m)$. Die Bedeutung der übrigen Bezeichnungen in (26.1) ist die gleiche wie oben.

Verwendet man wieder die Störungstheorie aus Ziff. 14, so finden wir den effektiven Hamilton-Operator als eine Summe von Operatoren, die unterschiedliche Beiträge zur Energie des Systems liefern. Geordnet nach abnehmender Größe ihres Beitrages sind das die kinetische Energie, die Coulomb- und Austauschwechselwirkung innerhalb der Atome, die Coulombsche Wechselwirkung zwischen den Atomen und schließlich die Austauschwechselwirkung zwischen den Atomen auf Grund virtueller Elektronenübergänge von einem Kation zum anderen sowie die direkte interatomare Austauschwechselwirkung. In der Störungsreihe treten natürlich auch kompliziertere Terme auf, die der Wechselwirkung verschiedener Konfigurationen entsprechen, sowie Terme von höherer Ordnung hinsichtlich der Spinoperatoren als $\vec{s}_{n_1 m_1} \vec{s}_{n_2 m_2}$. Sie alle sind jedoch kleiner als die oben aufgezählten. Somit ergibt die in Ziff. 14 dargestellte Theorie für die von den Spinoperatoren abhängigen Glieder in der interatomaren Wechselwirkung folgende Ausdrücke[31]:

$$\hat{H}_k = \sum_{n \neq n', m, m'} \frac{\left| L(nm, n'm') + \sum_{n_1 \neq n', m_1} F(nm, n_1 m_1, n'm', n_1 m_1) \right|^2}{\varDelta(n'm', nm)} \left(-\frac{1}{2} + 2\vec{s}_{nm} \vec{s}_{n'm'} \right), \qquad (26.2)$$

$$\hat{H}_p = -\frac{1}{2} \sum_{n \neq n', m, m'} F(nm, n'm', n'm', nm) \left(\frac{1}{2} + 2\vec{s}_{nm} \vec{s}_{n'm'} \right), \qquad (26.3)$$

---

[31] Wir bemerken, daß der Ausdruck (26.2) erstmals von N. N. Bogolyubov in der Arbeit [96] für den Fall eines Elektrons pro Atom erhalten wurde [s. z.B. (14.49)].

wobei $\hat{H}_k$ den Austauschoperator auf Grund virtueller Übergänge $nm \to n'm'$ darstellt und $\hat{H}_p$ den durch direkten Austausch bedingten Operator.

Die Größe $\Delta(n'm', nm)$ in (26.2) läßt sich leicht aus dem Ausdruck für den Operator $\hat{H}_0$ herleiten [vgl. (14.11)]:

$$\hat{H}_0 = \sum_{n_1 m_1} \{\varepsilon_{n_1 m_1} - \tfrac{1}{2} F(n_1 m_1, n_1 m_1, n_1 m_1, n_1 m_1)\} N_{n_1 m_1} + \\ + \tfrac{1}{2} \sum_{n_1 m_1, n_2 m_2} F(n_1 m_1, n_2 m_2, n_1 m_1, n_2 m_2) N_{n_1 m_1} N_{n_2 m_2}. \quad (26.4)$$

Dabei ist $\varepsilon_{nm} = L(nm, nm)$ die „Bahn"-Energie des Elektrons. Der Übergang $nm \to n'm'$ bedeutet, daß die Elektronenzahl am Atom n im Bahnzustand m um eins vermindert und die Elektronenzahl am Atom n' im Bahnzustand m' entsprechend erhöht wird: $N_{nm} \to N_{nm} - 1$ und $N_{n'm'} \to N_{n'm'} + 1$. Dann folgt aus (26.4)

$$\Delta(n'm', nm) = \varepsilon_{n'm'} - \varepsilon_{nm} + F(n'm', n'm', n'm', n'm') - \\ - F(nm, n'm', nm, n'm') + \sum_{n_1 m_1 \neq nm, n'm'} F(n'm', n_1 m_1, n'm', n_1 m_1) N_{n_1 m_1} - \\ - \sum_{n_1 m_1 \neq nm, n'm'} F(nm, n_1 m_1, nm, n_1 m_1) N_{n_1 m_1}. \quad (26.5)$$

Den Hauptbeitrag in (26.5) liefert das Matrixelement $F(n'm', n'm'; n'm', n'm')$. Die Bahnenergien $\varepsilon(n, m)$ sind von m schwach abhängig, und daher ist $\varepsilon_{n'm'} - \varepsilon_{nm} \approx 0$. Das vierte Glied in (26.5) ist kleiner als das dritte, da es die Coulomb-Wechselwirkung zwischen verschiedenen Atomen beschreibt. Wegen der schwachen Abhängigkeit der Summen in (26.5) von nm und $n'm'$ (innerhalb eines Atoms unterscheiden sich die F wie die bekannten Slater-Integrale) kompensieren sich praktisch die Summanden mit unterschiedlichem Vorzeichen. Folglich wird die Anregungsenergie $\Delta(n'm', nm)$ hauptsächlich durch die Energie U der Coulomb-Wechselwirkung zweier Elektronen innerhalb eines Atoms bestimmt. Wir bezeichnen das „Einelektronenübergangsintegral" $L + \sum F$ mit $b^{nm}_{n'm'}$ und die Koeffizienten F in (26.3) mit $J(nm, n'm')$ und schreiben (26.2) und (26.3) um in

$$H_k = \sum_{nm \neq n'm'} \frac{|b^{nm}_{n'm'}|^2}{U} \left(-\frac{1}{2} + 2\vec{s}_{nm} \vec{s}_{n'm'}\right), \quad (26.6)$$

$$H_p = -\tfrac{1}{2} \sum_{nm \neq n'm'} J(nm, n'm') \left(\tfrac{1}{2} + 2\vec{s}_{nm} \vec{s}_{n'm'}\right). \quad (26.7)$$

Da beim vorliegenden Ansatz für das Problem des indirekten Austausches alle Effekte der Wechselwirkung mit der Matrix der Anionen des Kristalls schon in die Parameter „b" und „J" einbezogen sind und die Anionen gar nicht explizit betrachtet werden, hat es keinen Sinn mehr, von einem indirekten Austauscheffekt zu sprechen.

Die Beziehung (26.6) beschreibt die Austauschwechselwirkung auf Grund des virtuellen Elektronenüberganges von einem Atom zum anderen, und somit ist es, wie bereits in Ziff. 15 bemerkt, nur natürlich, diesen Typ der Kopplung *kinetischen* Austausch zu nennen. Dann ist es ebenso selbstverständlich, den Ausdruck (26.6), der die Wechselwirkung auf Grund der direkten Überlappung der Kationenfunktionen darstellt, als *potentiellen* Austausch zu bezeichnen. Die Vorzeichen der Austauschintegrale in (26.6) und (26.7) sind bestimmt: der kinetische Austausch ergibt stets eine antiferromagnetische Kopplung und der potentielle Austausch eine ferromagnetische ($J_{nm, n'm'}$ ist positiv definit). Diese beiden Typen der Wechselwirkung sind offenbar in jedem beliebigen magnetischen Dielektrikum bestimmend. Neben (26.6) und (26.7) kann man auch die durch die Spinpolarisation der

Wellenfunktion der Anionen hervorgerufenen Effekte sowie die Effekte der zweifachen Anregung heranziehen.

27. Die relative Bedeutung der verschiedenen Konfigurationen. In Ziff. 26 wurden die Coulombschen und Austauschwechselwirkungen *innerhalb* eines Atoms sowie der potentielle und der kinetische Austausch zwischen *verschiedenen* Atomen betrachtet. Der Ausdruck für den kinetischen Austausch war durch eine Konfiguration bedingt, bei der ein Elektron von einem der Atome in die einfach besetzte Bahn eines anderen Atoms übergeht. Dieser Konfigurationstyp entspricht bei der üblichen Behandlung des Problems der indirekten Wechselwirkung dem Fall einer zur Hälfte oder stärker besetzten Kationenschale.

Wir betrachten nun (bei weniger als halb besetzten Schalen) den Übergang eines Elektrons in die leere Bahn eines anderen Atoms. Der Einfachheit halber nehmen wir an, daß in der „Grund"-Konfiguration auf jedes Atom ein Elektron entfällt. Für jedes Atom n_i bezeichnen wir die Bahnfunktion des Grundzustands mit dem Index k_i und die angeregte Bahn mit l_i. In der „Grund"-Konfiguration sind überall lediglich die k-Bahnen besetzt, die l-Bahnen dagegen frei. In einer beliebigen „angeregten" Konfiguration befinden sich an einem der Atome überhaupt kein Elektron, an einem anderen dafür zwei Elektronen: eins in der k-Schale, das andere in der l-Schale. Folglich handelt es sich hier um virtuelle Übergänge $n_i k_i \rightarrow n_j l_j$. Der Teil des Hamilton-Operators des Systems, der von nullter Ordnung im Überlappungsintegral ist, wird im vorliegenden Falle lauten

$$\left.\begin{aligned}\hat{H}' = & \sum_{nmm'\sigma} L(nm, nm') c^+_{nm\sigma} c_{nm'\sigma} + \\ & + \tfrac{1}{2} \sum_{n_1 m_1 m'_1 n_2 m_2 m'_2 \sigma_1 \sigma_2} F(n_1 m_1, n_2 m_2, n_1 m'_1, n_2 m'_2) c^+_{n_1 m_1 \sigma_1} c^+_{n_2 m_2 \sigma_2} c_{n_2 m'_2 \sigma_2} c_{n_1 m'_1 \sigma_1},\end{aligned}\right\} \quad (27.1)$$

wobei der Index m nur die beiden Werte k und l annimmt. Der Ausdruck (27.1) ist das Analogon der Form (14.11) (er hat hinsichtlich der Überlappung die gleiche Ordnung), ist jedoch im Gegensatz dazu im Hinblick auf die Besetzungszahlen nicht diagonal.

Um (27.1) teilweise zu diagonalisieren, stellen wir diesen Ausdruck in Form einer Summe $\hat{H}' = \hat{H}_0 + \hat{H}'_0$ dar, wobei

$$\left.\begin{aligned}\hat{H}_0 = & \sum_{nm} L(nm, nm) N_{nm} + \tfrac{1}{2} \sum_{\substack{n_1 m_1 n_2 m_2 \\ (n_1 \neq n_2)}} C(n_1 m_1, n_2 m_2) N_{n_1 m_1} N_{n_2 m_2} + \\ & + \sum_i J(n_i k_i, n_i l_i) \sum_{\sigma_1 \sigma_2} c^+_{n_i k_i \sigma_1} c^+_{n_i l_i \sigma_2} c_{n_i k_i \sigma_2} c_{n_i l_i \sigma_1},\end{aligned}\right\} \quad (27.2)$$

$$\left.\begin{aligned}\hat{H}'_0 = & \sum_{\substack{nmm' \\ (m \neq m')}} \left\{ L(nm, nm') - \tfrac{1}{2} \sum_{m''} F(nm, nm'', nm'', nm') \right\} c^+_{nm\sigma} c_{nm'\sigma} + \\ & + \tfrac{1}{2} \sum_{\substack{nmm'\sigma_1\sigma_2 \\ (m \neq m')}} F(nm, nm, nm', nm') c^+_{nm\sigma_1} c^+_{nm\sigma_2} c_{nm'\sigma_2} c_{nm'\sigma_1} + \\ & + \tfrac{1}{2} \sum_{\substack{n_1 m_1 m'_1 n_2 m_2 m'_2 \sigma_1 \sigma_2 \\ (n_1 \neq n_2, m_1 \neq m'_1, m_2 \neq m'_2)}} F(n_1 m_1, n_2 m_2, n_1 m'_1, n_2 m'_2) \times \\ & \times c^+_{n_1 m_1 \sigma_1} c^+_{n_2 m_2 \sigma_2} c_{n_2 m'_2 \sigma_2} c_{n_1 m'_1 \sigma_1}\end{aligned}\right\} \quad (27.3)$$

mit

$$C(n_1 m_1, n_2 m_2) = F(n_1 m_1, n_1 m_1, n_2 m_2, n_2 m_2),$$
$$J(nm, nm') = F(nm, nm', nm', nm).$$

Mit Hilfe der Beziehungen (14.50) läßt sich der Ausdruck für $\hat{H}_0$ umbilden in

$$\left.\begin{array}{l}\hat{H}_0 = \sum_{nm} L(nm, nm) N_{nm} + \tfrac{1}{2} \sum_{\substack{nmm' \\ (m \neq m')}} C(nm, nm') N_{nm} N_{nm'} + \\ + \tfrac{1}{2} \sum_{\substack{n_1 m_1 n_2 m_2 \\ (n_1 \neq n_2)}} C(n_1 m_1, n_2 m_2) N_{n_1 m_1} N_{n_2 m_2} + \sum_i J(n_i k_i, n_i l_i)(-\tfrac{1}{2} - 2\vec{s}_{n_i k_i} \vec{s}_{n_i l_i}).\end{array}\right\} \quad (27.4)$$

Weil hierbei $N_{nm} = N_{nm}^2$ gesetzt worden ist, gilt (27.4) nur so weit, als im weiteren lediglich Übergänge $n_i k_i \to n_j l_j$ berücksichtigt werden. Der Austauschanteil von (27.4) [die Summe mit $J(nk, nl)$] liefert selbstverständlich einen Beitrag zur Energie nur unter der Bedingung, daß in der Funktion, auf die der Operator $\hat{H}_0$ wirkt, die Besetzungszahlen N_{nk} und N_{nl} beide gleich eins sind. Wenn eine dieser Besetzungszahlen gleich Null ist — und das ist bei den Funktionen C_0 der „Grund"-konfiguration für alle n der Fall — trägt der entsprechende Summand nicht zur Energie bei. Somit ergibt die Einwirkung des Operators $\hat{H}_0$ auf eine beliebige dieser 2^n Funktionen C_0 (n ist die Gesamtzahl der Elektronen)

$$\hat{H}_0 C_0 = E_0 C_0 \qquad (27.5)$$

mit

$$E_0 = \sum_{nk} L(nk, nk) + \tfrac{1}{2} \sum_{\substack{n_1 k_1 n_2 k_2 \\ (n_1 \neq n_2)}} C(n_1 k_1, n_2 k_2). \qquad (27.6)$$

Beim Übergang $n_i k_i \to n_j l_j$ kann das in der l-Bahn des Atoms n_j auftretende Elektron zusammen mit dem bereits in der k-Bahn desselben Atoms befindlichen entweder einen Singulett- oder einen Triplettzustand bilden. Deshalb wird der Operator (27.4) entsprechend diesen beiden Spinzuständen die Eigenwerte $-J(n_j k_j, n_j l_j)$ für das Triplett und $+J(n_j k_j, n_j l_j)$ für das Singulett annehmen. Die Anregungsenergie lautet

$$\left.\begin{array}{l}\Delta_{t,s}(n_j l_j, n_i k_i) = L(n_j l_j, n_j l_j) - L(n_i k_i, n_i k_i) + \\ + \sum_{\substack{nk \\ (n \neq n_i)}} [C(nk, n_j l_j) - C(nk, n_i k_i)] \mp J(n_j k_j, n_j l_j).\end{array}\right\} \quad (27.7)$$

Man hat nun den Ausdruck $\hat{H}_0$ aus (27.4) als ungestörten Hamilton-Operator, die Form $\hat{H}_0'$ aus (27.3) sowie alle übrigen Glieder des Hamilton-Operators, die im Überlappungsparameter von höherer Ordnung sind, als Störung zu verwenden.

ANDERSON [47], der in die Energienenner der Störungstheorie das innere Austauschintegral $J(nk, nl)$ nicht einbezog, erhielt für den Parameter der effektiven Austauschwechselwirkung zwischen den Spins der Kationen den Ausdruck

$$\frac{b^2 J(nk, nl)}{U^2}, \qquad (27.8)$$

wobei b das Übergangsintegral und U die Coulomb-Wechselwirkung zweier Elektronen am gleichen Atom bedeuten. Die Größe (27.8) ist positiv, sie begünstigt also die ferromagnetische Kopplung. Sie ist absolut genommen $J(nk, nl)/U$ mal kleiner als das Integral des kinetischen Austauschs in (26.6), allerdings in der Annahme, daß in diesen beiden Fällen alle Parameter gleich sind.

Man kann die Beiträge der beiden untersuchten Prozesse (Übergang in eine besetzte Bahn und Übergang in eine leere Bahn) noch eingehender vergleichen, wenn man die durch sie bedingten Beiträge zur effektiven Austauschkopplung in der gleichen Ordnung der Störungstheorie und nicht in verschiedenen Ordnungen [Formeln (26.6) und (27.8)] betrachtet. Der Einfachheit halber nehmen wir uns das Problem der zwei Zentren mit zwei Elektronen vor. Bei jedem der beiden gleichen Atome a und b berücksichtigen wir je zwei Bahnen, die Wellenfunktionen seien a_1 und a_2 für das Atom a sowie b_1 und b_2 für das Atom b. Für Übergänge in die besetzte Bahn $a_1 \to b_1$ und $b_1 \to a_1$ lautet das effektive Austauschintegral [s. (16.16)]

$$J_1 = -\frac{2\{\langle a_1|b_1\rangle + \langle a_1 b_1|a_1 a_1\rangle\}^2}{E_1 - E_0}. \tag{27.9}$$

Hierin bezeichnen E_0 und E_1 die Summen der kinetischen und Coulomb-Energien in der Grund- bzw. angeregten Konfiguration. Man erhält auch leicht den Ausdruck für das effektive Integral auf Grund der Übergänge in die leere Bahn $a_1 \to b_2$ und $b_1 \to a_2$:

$$\left. \begin{array}{l} J_2 = -\dfrac{\{\langle a_1|b_2\rangle + \langle a_1 b_1|a_1 a_2\rangle - \langle a_1 b_1|a_2 a_1\rangle\}^2}{E_2 + J_{12} - E_0} + \\[2mm] \qquad + \dfrac{\{\langle a_1|b_2\rangle + \langle a_1 b_1|a_1 a_2\rangle - \langle a_1 b_1|a_2 a_1\rangle\}^2}{E_2 - J_{12} - E_0}. \end{array} \right\} \tag{27.10}$$

E_2 bedeutet hier die Summe der kinetischen und Coulomb-Energien in den Konfigurationen, wo eins der Elektronen sich auf der Bahn a_2 oder b_2 befindet, und J_{12} ist das Austauschintegral zwischen den Bahnen a_1 und a_2 oder b_1 und b_2. Das erste Glied in (27.10) hat ein negatives Vorzeichen und ist bedingt durch einen virtuellen Übergang, bei dem sich die Spins der Elektronen an einem Atom antiparallel einstellen. Da ein solcher Prozeß das Gewicht des Singulettzustandes erhöht, wird das negative Vorzeichen dieses Gliedes in (27.10) verständlich. Das zweite Glied in (27.10) ist positiv und durch die Übergänge bedingt, bei denen die Spins beider Elektronen an einem Atom parallel liegen. Das positive Vorzeichen dieses Gliedes erklärt sich durch die Gewichtszunahme des Triplettzustandes bei der gegebenen Form eines virtuellen Überganges. Auf Grund der Hundschen Regel ist ein Übergang in einen Zustand mit parallelen Spins am gleichen Atom bevorzugt. Jedoch besitzen die beiden Glieder in (27.10) nicht nur verschiedene Nenner, sondern auch unterschiedliche Zähler (die den Übergangswahrscheinlichkeiten proportional sind). Demzufolge ist von einem allgemeinen Standpunkt aus das Vorzeichen von (27.10) unbestimmt. Bei gleichen Übergangsintegralen ist (27.10) positiv definit. Die Bestimmung des Vorzeichens für das resultierende Integral $J = J_1 + J_2$ ist wegen der Verschiedenheit der in die Formeln (27.9) und (27.10) eingehenden Parameter nicht leicht. Um uns aber die Tendenz zur Ausbildung des einen oder anderen Typs der magnetischen Kopplung klar zu machen, betrachten wir das relative Gewicht jedes einzelnen virtuellen Überganges, indem wir möglichst vereinfachende Annahmen treffen. So nehmen wir an, daß für alle Übergänge (sowohl in die besetzte als auch in die leere Bahn) die Übergangsintegrale gleich sind. Weiterhin werden wie in Ziff. 26 die Differenzen $E_1 - E_0$ und $E_2 - E_0$ durch die eine Größe U, die Coulomb-Wechselwirkung zweier Elektronen am gleichen Atom, angenähert. Dann stimmt das Vorzeichen des resultierenden Integrals J mit dem Vorzeichen des folgenden Ausdruckes überein:

$$-\frac{1}{U} - \frac{1}{U + J_{12}} + \frac{1}{U - J_{12}}. \tag{27.11}$$

Das letzte Glied in (27.11) ist positiv und größer als jedes der beiden anderen. Das ist vollkommen verständlich, da bei im übrigen gleichen Bedingungen der Übergang in einen Zustand mit parallelen Spins an einem Atom günstiger ist als der Übergang in eine besetzte Bahn oder in eine leere Bahn unter Bildung eines Zweielektronen-Singuletts. Das Vorzeichen des gesamten Ausdrucks hängt jedoch von dem Verhältnis der Größen U und J_{12} ab. So ist (27.11) für $J_{12} > 0,4\,U$ positiv und für $J_{12} < 0,4\,U$ negativ. Da offensichtlich stets die letzte Ungleichung erfüllt ist, ist auch das Vorzeichen des resultierenden Integrals der effektiven

Austauschwechselwirkung negativ. Die Schwäche der ferromagnetischen Wechselwirkung ist eine Folge davon, daß sie auf einem Differenzeffekt beruht. d. h., daß der Übergang in die leere Bahn das Gewicht sowohl für den Triplettzustand als auch für den Singulettzustand erhöht. Zweifellos kann in realen Fällen die allgemeine Tendenz zum antiferromagnetischen Wechselwirkungstyp wegen der Ungleichheit der Übergangsintegrale für die verschiedenen Typen von „Anregungen" verletzt werden.

D. Die Austauschwechselwirkung und die Wellenfunktionen der Elektronen im Kristall.

28. Kurze Angaben aus der Theorie des Kristallfeldes. In diesem Abschnitt werden wir einige Aspekte der Theorie der indirekten Austauschwechselwirkung unter eingehender Berücksichtigung der Eigenschaften der Wellenfunktionen für die Elektronen im Kristall behandeln. Der Diskussion des Problems der Austauschwechselwirkung seien in aller Kürze einige Angaben aus der Theorie des Kristallfeldes vorangeschickt.

Die ersten Betrachtungen über das Verhalten der Einelektronenatomfunktionen sowie der Wellenfunktionen des Atoms im Kristallverband stammen von BETHE [166] aus dem Jahre 1929. Sein Verfahren beruhte auf einer konsequenten Behandlung der Wellenfunktionen des Atoms in einem elektrischen Feld vorgegebener Symmetrie. Betrachtet wurde ein Zentralion in der Umgebung anderer Ionen oder Komplexe. Diese Umgebung erzeugt ein elektrisches Feld von ganz bestimmter Symmetrie. Läßt man beim Aufbau der Funktionen des Zentralions und bei der Berechnung seines Energiespektrums die Spin-Bahn-Wechselwirkungen außer acht, so lassen sich zwei Hauptfälle behandeln, die sog. Näherungen des schwachen und des starken Kristallfeldes.

In der Näherung des schwachen Kristallfeldes wird angenommen, daß die Wechselwirkungen der Elektronen innerhalb des zentralen Ions wesentlich stärker sind als die Wechselwirkungen dieser Elektronen mit den Ionen der nächsten Umgebung. Folglich hat man in nullter Näherung die gleichen Funktionen und das gleiche Spektrum wie beim freien Ion.

Tabelle 1.

n	Term	f	Determinante
1	^2D	10	$(\varphi_2)^1$
2	^3F	21	$\hat{A}\,(\varphi_1)^1(\varphi_2)^1$
3	^4F	28	$\hat{A}\,(\varphi_0)^1(\varphi_1)^1(\varphi_2)^1$
4	^5D	25	$\hat{A}\,(\varphi_{-1})^1(\varphi_0)^1(\varphi_1)^1(\varphi_2)^1$
5	^6S	0	$\hat{A}\,(\varphi_{-2})^1(\varphi_{-1})^1(\varphi_0)^1(\varphi_1)^1(\varphi_2)^1$
6	^5D	25	$\hat{A}\,(\varphi_{-2})^1(\varphi_{-1})^1(\varphi_0)^1(\varphi_1)^1(\varphi_2)^2$
7	^4F	28	$\hat{A}\,(\varphi_{-2})^1(\varphi_{-1})^1(\varphi_0)^1(\varphi_1)^2(\varphi_2)^2$
8	^3F	21	$\hat{A}\,(\varphi_{-2})^1(\varphi_{-1})^1(\varphi_0)^2(\varphi_1)^2(\varphi_2)^2$
9	^2D	10	$\hat{A}\,(\varphi_{-2})^1(\varphi_{-1})^2(\varphi_0)^2(\varphi_1)^2(\varphi_2)^2$

Betrachten wir Ionen der d-Metalle in einem Feld bestimmter Symmetrie. In der Tabelle 1 sind die Grundzustände für freie d-Ionen in Abhängigkeit von der Zahl n der Elektronen in der d-Schale (für $n=1, \ldots, 9$) wiedergegeben. Dort sind auch der Grad f der Entartung (Bahn- und Spinentartung) des Grundzustandes und dessen Wellenfunktion für Maximalwerte der Projektionen des Bahn- und Spinmoments in der Eindeterminantennäherung angegeben.

In der Tabelle 1 symbolisiert $\hat{A}$ den Antisymmetrisierungsoperator $(n!)^{-\frac{1}{2}}\sum_P (-1)^P \hat{P}$, φ_i ist die Einelektronen-d-Funktion mit der Bahnkomponente $i = \pm 2, \pm 1, 0$; die Ziffer rechts oben von φ_i gibt die Anzahl der Elektronen im gegebenen Zustand an. Die Funktionen der Terme des freien Ions bilden die Basis für die irreduzible Darstellung der Gruppe der kontinuierlichen Drehungen. Wenn sich dieses Ion im Kristallverband befindet, kann im allgemeinen die Darstellung der kontinuierlichen Gruppe wegen der niedrigeren Symmetrie des Feldes reduzibel werden.

Eine gruppentheoretische Analyse [166] zeigt, daß beim Anlegen eines elektrischen Feldes von beispielsweise kubischer Symmetrie eine Termaufspaltung beim Zentralion erfolgt. In der Gruppentheorie bedeutet dies, daß die einem bestimmten Term des freien Ions entsprechende Darstellung nach den irreduziblen Darstellungen der Symmetriegruppe des kubischen Kristalls zerlegt werden kann. Die entsprechenden Darstellungen haben die Bezeichnungen A_1, A_2 (eindimensionale Darstellungen), E (zweidimensionale Darstellung) sowie T_1 und T_2 (dreidimensionale Darstellungen).

So existiert für die Darstellungen der D-Terme des Systems der d-Elektronen eine Zerlegung $E_g + T_{2g}$, für die F-Terme gilt $A_{2g} + T_{1g} + T_{2g}$, für den S-Term gibt es die Darstellung A_{1g}. Demzufolge spaltet sich in einem Kristall mit kubischer Symmetrie der D-Term in zwei Niveaus auf, in ein zweifach entartetes und in ein dreifach entartetes, der F-Term in drei Niveaus, in zwei dreifach entartete und in ein nichtentartetes; beim S-Term des Ions gibt es überhaupt keine Aufspaltung. Die Angaben über den Entartungsgrad der Niveaus erfolgten hier natürlich lediglich im Hinblick auf die „Bahn"-Entartung. Die Näherung des schwachen Feldes gilt, wenn der Abstand zwischen den Termen des freien Ions groß ist und die Aufspaltung im äußeren Feld übertrifft.

Bevor wir die Näherung des starken Feldes behandeln, betrachten wir ein d-Elektron im kubischen Feld. Der ^2D-Term eines einzelnen d-Elektrons spaltet sich dabei in zwei Niveaus auf, in ein zweifach entartetes (e_g-Niveau) und in ein dreifach entartetes (t_{2g}-Niveau). Der Abstand zwischen ihnen wird gewöhnlich durch eine Größe $\Delta = 10\,Dq$ gekennzeichnet. Die Funktionen e_g und t_{2g} kann man, wenn man die Achse C_4 als Quantisierungsachse wählt, in der Form folgender Linearkombinationen der d-Einelektronenfunktionen φ_i mit $i = \pm 2, \pm 1, 0$ darstellen:

$$e_g: \quad \begin{cases} d_{x^2-y^2} = \dfrac{1}{\sqrt{2}}(\varphi_2 + \varphi_{-2}), \\ d_{z^2} = \varphi_0, \end{cases} \tag{28.1}$$

$$t_{2g}: \quad \begin{cases} d_{xy} = \dfrac{1}{i\sqrt{2}}(\varphi_2 - \varphi_{-2}), \\ d_{xz} = -\dfrac{1}{\sqrt{2}}(\varphi_1 - \varphi_{-1}), \\ d_{yz} = -\dfrac{1}{i\sqrt{2}}(\varphi_1 + \varphi_{-1}). \end{cases} \tag{28.2}$$

Die den Funktionen e_g und t_{2g} entsprechenden Elektronendichten werden durch die folgenden graphischen Darstellungen wiedergegeben (Fig. 9).

Die Lage der e_g- und t_{2g}-Niveaus zueinander wird durch die konkrete Umgebung des Zentralions bestimmt. Hat z.B. das Zentralkation eine oktaedrische Anionenumgebung, so besitzen die t_{2g}-Funktionen Ausläufer, die in den Raum zwischen den Anionen gerichtet sind, und wechselwirken daher mit letzteren nur schwach.

Die Elektronendichten der e_g-Bahnen sind dagegen in der Richtung zu den Anionen langgestreckt und erfahren eine starke Wechselwirkung. Folglich liegen bei einer oktaedrischen Anionenumgebung die t_{2g}-Niveaus der Kationen tiefer als die e_g-Niveaus. Im Falle einer tetraedrischen Anionenumgebung und wenn die benachbarten Anionen Würfelecken oder die Mitten der Würfelkanten einnehmen, ändert sich die Situation, und das e_g-Niveau kommt unter dem t_{2g}-Niveau zu liegen.

Gehen wir jetzt über zur Betrachtung der Situation bei starkem Kristallfeld. Hierbei ist die Aufspaltung im Felde so groß, daß L selbst in niedrigster Nähe-

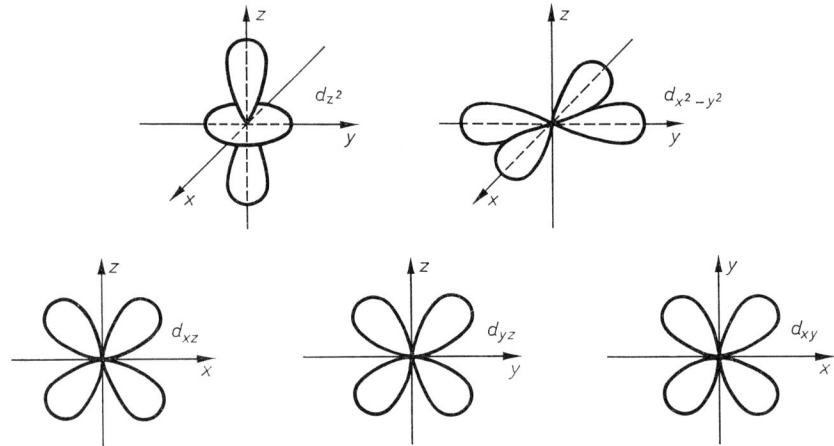

Fig. 9. Elektronendichten in den Zuständen e_g (oben) und t_{2g} (unten).

rung keine gute Quantenzahl mehr ist. Die Funktionen und das Spektrum der Elektronen des Kations im Kristallfeld muß man jetzt aufsuchen, indem man von den Einelektronen-„Kristall"-Bahnen e_g und t_{2g} ausgeht.

Betrachten wir z. B. den oktaedrischen Fall. Bei Zunahme der Elektronenzahl in der d-Schale füllen sich zunächst die tiefstgelegenen t_{2g}-Niveaus und dann die weniger stabilen e_g-Zustände. Die Antisymmetrisierung erfolgt wie üblich. Die möglichen Terme (A_{1g}, A_{2g}, E_g, T_{1g} und T_{2g}) kann man durch eine Zerlegung des unmittelbaren Produktes der t_{2g}- und e_g-Bahnen erhalten. Wenn bei einer solchen Auffüllung der Bahnen das unmittelbare antisymmetrisierte Produkt eine Summe verschiedener Darstellungen der kubischen Symmetrie ist, dann müßte sich das System in dem Zustand befinden, dem die niedrigste Energie entspricht.

Im allgemeinen läßt sich der Vielelektronenzustand durch eine „reine" Besetzung von e_g- und t_{2g}-Bahnen nicht exakt darstellen. Zum Beispiel spaltet sich im Falle des schwachen Kristallfeldes der ^3F-Zustand eines Kations mit zwei Elektronen in drei Niveaus auf mit den Wellenfunktionen

$$T_{1g}: \begin{cases} \hat{A}\,\dfrac{1}{\sqrt{20}}\,\{(d_{x^2-y^2}+\sqrt{3}\,d_{z^2})\,d_{yz}+4 d_{xy}\,d_{xz}\} \\ \hat{A}\,\dfrac{1}{\sqrt{20}}\,\{(d_{x^2-y^2}-\sqrt{3}\,d_{z^2})\,d_{xz}-4 d_{xy}\,d_{yz}\} \\ \hat{A}\,\dfrac{1}{\sqrt{5}}\,(d_{x^2-y^2}\,d_{xy}-2 d_{xz}\,d_{yz}), \end{cases} \qquad (28.3)$$

$$T_{2g}: \begin{cases} \hat{A} \frac{1}{2} (\sqrt{3}\, d_{x^2-y^2} - d_{z^2})\, d_{yz} \\ \hat{A} \frac{1}{2} (\sqrt{3}\, d_{x^2-y^2} + d_{z^2})\, d_{xz} \\ \hat{A}\, d_{z^2}\, d_{xy}, \end{cases} \qquad (28.4)$$

$$A_{2g}: \quad \hat{A}\, d_{x^2-y^2}\, d_{z^2}, \qquad (28.5)$$

worin $\hat{A}$ den Antisymmetrisierungsoperator symbolisiert.

Die Stabilisierungsenergien im Feld sind $-6Dq$, $2Dq$ bzw. $12Dq$. Symbolisch kann man die Besetzung jedes dieser drei Niveaus wie folgt schreiben:

$$T_{1g} - e_g t_{2g} + (t_{2g})^2, \qquad T_{2g} - e_g t_{2g}, \qquad A_g - (e_g)^2. \qquad (28.6)$$

Somit entsteht das niedrigste Niveau T_{1g} nicht allein durch Besetzung der am tiefsten liegenden t_{2g}-Bahnen [s. das Glied $(t_{2g})^2$ in (28.6)], es enthält vielmehr auch eine Beimischung höher liegender Zustände, nämlich $e_g t_{2g}$. Geht man dagegen von der Näherung des starken Kristallfeldes aus, so enthält nicht nur das unmittelbare Produkt $t_{2g} \times t_{2g}$ den Term T_{1g}, sondern auch das Produkt $t_{2g} \times e_g$. Das „Aneinanderfügen" der beiden Näherungen (des starken und schwachen Kristallfeldes) erfolgt in der Methode des sog. mittleren Kristallfeldes, deren Ergebnisse tatsächlich als die Ergebnisse erweiterter Methoden des starken und schwachen Kristallfeldes erscheinen. Diese beiden Methoden liefern nämlich übereinstimmende Ergebnisse, wenn man in beiden eine entsprechende Anzahl von Konfigurationen berücksichtigt. So muß man bei der Methode des schwachen Feldes neben der Konfiguration ^3F bei zwei Elektronen noch die Konfiguration ^3P berücksichtigen, die dem im kubischen Feld nicht aufgespaltenen Niveau mit der Darstellung $T_{1g}(^3P)$ entspricht. Folglich ist zur Bestimmung der Energie T_{1g} eine Säkulargleichung zweiten Grades zu lösen mit Matrixelementen, die man mittels der Funktionen $T_{1g}(^3F)$ und $T_{1g}(^3P)$ berechnet.

Analog dazu erhält man auch in der Näherung des starken Feldes eine Säkulargleichung zur Bestimmung der Energie T_{1g}, da, wie bereits erwähnt, die Darstellung T_{1g} in den beiden Produkten $t_{2g} \times t_{2g}$ und $t_{2g} \times e_g$ auftritt.

Bei der praktischen Lösung einer Reihe von Aufgaben kann man sich jedoch einer einfacheren Methode bedienen, die den Umstand ausnützt, daß die Zustände t_{2g} und e_g nacheinander besetzt werden. Eine Beschreibung dieser Methode ist von GRIFFITH und ORGEL [167] gegeben worden. Für die Reihenfolge, in der die zehn Zustände t_{2g} und e_g mit Elektronen besetzt werden, sind zwei Umstände maßgeblich. Erstens haben die Elektronen das Bestreben, möglichst den energetisch niedrigsten Bahnzustand zu besetzen. Zweitens suchen sie, wegen der Austauschwechselwirkung ihre Spins parallel zueinander auszurichten (Hundsche Regel) und folglich verschiedene Bahnzustände einzunehmen. Unter bestimmten Bedingungen können diese beiden Tendenzen miteinander in Konflikt geraten. Betrachten wir die Besetzung der Bahnen t_{2g} und e_g im oktaedrischen Feld. Da die t_{2g}-Bahnen energetisch unterhalb der e_g-Bahnen liegen, besetzt das erste Elektron eine t_{2g}-Bahn. Das zweite Elektron geht ebenfalls in eine solche Bahn, doch nehmen die beiden Elektronen wegen der Hundschen Regel unterschiedliche t_{2g}-Bahnen ein. Sind drei Elektronen vorhanden, dann sind alle drei t_{2g}-Bahnen einfach besetzt, und die Spins der Elektronen liegen parallel. Bei vier Elektronen gibt es eine neue Möglichkeit in der Reihenfolge der Niveaubesetzung. Das vierte Elektron kann entweder in eine der e_g-Bahnen gehen und damit den Spin des Kations weiter um $\frac{1}{2}$ erhöhen oder eine der einfach besetzten t_{2g}-Bahnen einnehmen und dabei den Gesamtspin um $\frac{1}{2}$ verringern. In Abhängigkeit von der Energiebilanz aus Gewinn bzw. Verlust an Austausch- und Bahnenergie wird die eine oder die andere Möglichkeit realisiert. Wenn die Energie Δ der Aufspaltung im Kristallfeld größer ist als die Energie der Austauschwechselwirkung des vierten Elektrons auf der e_g-Bahn mit den drei Elektronen in den Bahnen t_{2g} (starkes Feld), dann besetzt das vierte Elektron eine dieser Bahnen. Im umgekehrten Falle (schwaches Feld) geht es in die e_g-Bahn. Die folgende, der Arbeit von GRIFFITH und ORGEL [167] entnommene Tabelle 2 enthält für die beiden Fälle die Besetzung der d-Schale für Elektronenzahlen zwischen 1 und 9. N ist die Zahl der möglichen Kombinationen von jeweils zwei Elektronen mit parallelen Spins.

Tabelle 2.

Zahl der d-Elektronen	Verteilung im schwachen Kristallfeld		N	Verteilung im starken Kristallfeld		N	Gewinn an Bahnenergie im starken Feld
	t_{2g}	e_g		t_{2g}	e_g		
1	↑		0	↑		0	0
2	↑↑		1	↑↑		1	0
3	↑↑↑		3	↑↑↑		3	0
4	↑↑↑	↑	6	↑↓↑↑		3	Δ
5	↑↑↑	↑↑	10	↑↓↑↓↑		4	2Δ
6	↑↓↑↑	↑↑	10	↑↓↑↓↑↓		6	2Δ
7	↑↓↑↓↑	↑↑	11	↑↓↑↓↑↓	↑	9	Δ
8	↑↓↑↓↑↓	↑↑	13	↑↓↑↓↑↓	↑↑	13	0
9	↑↓↑↓↑↓	↑↓↑	16	↑↓↑↓↑↓	↑↓↑	16	0

Wie man aus Tabelle 2 ersieht, besteht ein Unterschied in der Besetzung der Bahnen im starken und im schwachen Kristallfeld nur bei vier, fünf, sechs und sieben d-Elektronen. Dieses einfache Schema gestattet es, die Verringerung des Spinmomentes eines Kations im starken Feld gegenüber dem freien oder im schwachen Feld befindlichen Kation zu erklären. So erhält man im starken Feld an Stelle der Spins 2, $\frac{5}{2}$, 2, $\frac{3}{2}$ für vier bis sieben Elektronen die Werte 1, $\frac{1}{2}$, 0, $\frac{1}{2}$.

Ein anderer Zugang zum Problem „Zentralkation-Anionenumgebung" wird in der Methode der Molekülbahnen entwickelt, wie sie erstmalig von VAN VLECK [168] vorgeschlagen wurde. Wenn innerhalb des soeben beschriebenen Verfahrens die Ionen der Umgebung ein Kristallfeld verursachten, das zu einer Aufspaltung der Terme des Zentralkations führte, so wird bei der Methode der Molekülbahnen ein Übergang der Elektronen von den Kationen zu den Anionen zugelassen. Ausgangspunkt ist der Aufbau von Einelektronen-Molekülbahnen mit den erforderlichen Transformationseigenschaften. Dazu verwendet man die LCAO-Methode, die Molekülbahnen stellen also Linearkombinationen von Atombahnen dar. Mit Hilfe der Gruppentheorie kann man die Anzahl und Form der Linearkombinationen bestimmen. So bilden sich in einer oktaedrischen Umgebung σ-Kopplungen aus (Funktionen, die symmetrisch sind hinsichtlich der Verbindungslinien zwischen den Atomen der Umgebung und dem Zentralion), die sich gemäß der singulären Darstellung a_{1g}, der zweidimensionalen e_g und der dreidimensionalen t_{1u} transformieren. Die entsprechenden Funktionen können aufgebaut werden aus den Funktionen $4s$, $d_{x^2-y^2}$, d_{z^2} und $4p$ des Zentralions sowie aus den Funktionen der umgebenden Ionen. Die Koeffizienten in der Linearkombination der Atomfunktionen geben den Grad der Kovalenz für die vorliegende Bindung an. Die Variationsgleichungen zur Bestimmung dieser Koeffizienten können für dieselben mehrere Sätze von Lösungen liefern. Die zu den niedrigsten Energiewerten gehörenden Bahnen bezeichnet man als bindende (bonding), die mit höheren Energiewerten als lockernde (antibonding) Bahnen. Die Elektronenstruktur des Komplexes erhält man durch Besetzung der am niedrigsten gelegenen Molekülbahnen. Auch in der Methode der Molekülbahnen läßt sich der Begriff des Parameters $10\,Dq$ der Aufspaltung einführen. Das ist die Größe des Energiespaltes zwischen einer lockernden t_{2g}-Bahn und einer lockernden e_g-Bahn.

Zum Schluß bemerken wir, daß die für jede Verbindung charakteristische Größe Δ die Größenordnung 10^4 cm^{-1} besitzt. Verfahren zur experimentellen Bestimmung von Δ findet man in den Arbeiten [169] bis [172] beschrieben. Die modernsten theoretischen Berechnungen der Parameter in der Ligandentheorie sowie eine Diskussion der verschiedenen Aspekte und Schwierigkeiten der Theorie sind in den Arbeiten [173] bis [176] dargestellt.

Eine Analyse der verschiedenen theoretischen Berechnungen wird in einer Übersichtsarbeit von ANDERSON [151], [152] geboten.

29. Halbempirische Regeln. Hier wird die indirekte Austauschwechselwirkung unter Berücksichtigung der Symmetrie der Wellenfunktionen im Kristall in aller Kürze behandelt. Die Zusammenhänge zwischen der Austauschwechselwirkung und der konkreten Symmetrie des Kristalls wurden in Arbeiten von GOODENOUGH u.a. [177] bis [181], KANAMORI [182], WOLLAN u.a. [183] bis [186] untersucht.

Hier werden wir uns der Darstellung von KANAMORI [182] anschließen, der mehrere Austauschmechanismen untersuchte und zeigte, daß das Vorzeichen der entsprechenden Integrale oftmals schon durch die Symmetrie der Einelektronenwellenfunktionen bestimmt wird. Die von KANAMORI untersuchten Wechselwirkungstypen waren folgende: Der Kramers-Anderson-Mechanismus, d.h. der Übergang eines Elektrons vom Anion in eine leere oder halbbesetzte Kationenbahn. Der Goodenough-Mechanismus, der die Bildung „partieller kovalenter Bindungen" des Anions mit zwei Kationen zugleich vorsieht. In der Terminologie unserer Darstellung entspricht der Goodenough-Mechanismus der in Ziff. 19 und 22 betrachteten zweifachen Anregung nach PRATT und NESBET. Weiterhin wurden der Mechanismus des Doppelaustausches und der Slatersche Polarisationsmechanismus berücksichtigt. Was die Polarisation anbetrifft, wäre zu bemerken, daß KANAMORI nicht die Darstellung benutzte, die unserer Betrachtung in Ziff. 21 zugrunde lag. Dieser Effekt erscheint bei KANAMORI nicht auf Grund einer „Vormagnetisierung" der Anionenschale durch den Gesamtspin des Systems; der mit der Polarisation verbundene Gewinn an Austauschenergie entsteht vielmehr dadurch, daß je nach dem Vorzeichen des Austauschintegrals J_{pd} zwischen den Anionen- und Kationenbahnen die letzteren einander anziehen oder abstoßen. So kann beispielsweise eine Polarisation der Anionenschale in einem Singulettzustand auftreten, im Triplettzustand aber fehlen. Im Anschluß an KANAMORI betrachten wir nun einige konkrete Fälle.

Die Verbindung $CaMnO_3$ ist ein Antiferromagnetikum mit Perowskitstruktur. Jedes Mn^{+4}-Ion besitzt eine oktaedrische Umgebung von Sauerstoffionen. Das Ion Mn^{+4} enthält drei Elektronen in der d-Schale. Im freien Zustand lautet sein Grundterm ^4F. Dieser Term spaltet sich im oktaedrischen Feld in drei auf, von denen das nichtentartete Niveau $^4A_{2g}$ mit der Elektronenkonfiguration $(t_{2g})^3$ am tiefsten liegt. Die Funktionen t_{2g} und die p-Bahn des Sauerstoffs mit einer Achse entlang der Verbindungslinie Anion-Kation ($p\sigma$-Bahn) sind wegen ihrer unterschiedlichen Winkelabhängigkeiten orthogonal zueinander. Auf Grund der unterschiedlichen Symmetrie verschwindet auch das Übergangsintegral zwischen ihnen, und damit ist der durch den Übergang eines Elektrons in eine einfach besetzte Bahn bedingte Mechanismus der indirekten Wechselwirkung nach ANDERSON ausgeschlossen. Ein Übergang von der $p\sigma$-Bahn in die leere e_g-Bahn ist nach den Symmetriegesetzen nicht verboten. Nach diesem Übergang liegt der Spin des Elektrons in der e_g-Bahn wegen der Hundschen Regel parallel zum resultierenden Spin der drei Elektronen in den t_{2g}-Bahnen. Da das in der $p\sigma$-Bahn verbliebene nichtabgesättigte Elektron mit den Elektronen des anderen Kations (zu dem kein Übergang erfolgt ist) ferromagnetisch wechselwirkt ($J_{pd} > 0$ wegen der Orthogonalität von t_{2g} und $p\sigma$), wird die endgültige Orientierung der Gesamtspins zweier Kationen antiparallel. Dieses Ergebnis entspricht vollkommen unserer Schlußfolgerung aus Ziff. 17.

Der Polarisationseffekt führt zum gleichen Vorzeichen für die effektive Austauschwechselwirkung. Wenn die Kationenspins antiparallel liegen, wird die Bahn eines $p\sigma$-Elektrons mit „Rechts"-Spin (wegen des positiven Austauschintegrals J_{pd}) von dem Kation mit „Rechts"-Spin angezogen. Eine entsprechende Anziehung tritt auch zwischen den „Links"-Spins auf. Dabei wird Austauschenergie gewonnen; ihr Betrag (das Vorzeichen ist negativ) ist umso größer, je stärker die Dichteverteilungen der am Austausch beteiligten Elektronen einander überlappen. Im Falle paralleler Kationenspins ist eine solche Polarisation (d.h. die Verschiebung der beiden $p\sigma$-Bahnen nach rechts bzw. nach links) nicht möglich, da in der $p\sigma$-Bahn die Spins antiparallel zueinander liegen müssen. Zu einer antiferromagnetischen Kopplung führen auch der Mechanismus von GOODENOUGH und der Doppelaustausch. In hierzu analoger Weise kann man das Vorzeichen der indirekten Austauschwechselwirkung auch für andere Verbindungen betrachten.

Eine indirekte Wechselwirkung kann außer über die $p\sigma$-Bahnen auch über $p\pi$-Bahnen, deren Achsen senkrecht auf den Achsen der ersteren stehen, verwirklicht werden. So kann im gegebenen Fall der Verbindung $CaMnO_3$ ein virtueller Übergang von der $p\pi$-Bahn in die einfach besetzte t_{2g}-Bahn eines Kations erfolgen. Dabei stellt sich selbstverständlich der Spin des übergehenden Elektrons antiparallel zum resultierenden Spin der drei t_{2g}-Elektronen ein. Die endgültige Orientierung der Kationenspins wird durch das Vorzeichen des Austauschintegrals zwischen der $p\pi$-Bahn und der t_{2g}-Bahn bestimmt. Da die Bahnen $p\pi$ zu einer der t_{2g}-Bahnen nicht orthogonal sind, kann das entsprechende Austauschintegral negativ sein; dadurch sollten sich der nichtabgesättigte Spin des einen $p\pi$-Elektrons am Anion und der Gesamtspin des Kations mit den drei t_{2g}-Elektronen antiparallel ausrichten. Die indirekte Wechselwirkung wird schließlich wieder antiferromagnetisch.

Der Fall, daß die d-Schale gerade halb gefüllt ist, wird z.B. in der Verbindung MnO realisiert. Die Elektronenkonfiguration im kubischen Feld ist $(t_{2g})^3 (e_g)^2$. Ein Übergang $p\sigma - e_g$ verkleinert den Spin des Kations um $\frac{1}{2}$. Das Austauschintegral zwischen dem nichtabgesättigten $p\sigma$-Elektron und den Elektronen des Kations ist im Mittel negativ, so daß die daraus resultierende Kopplung antiferromagnetisch wird. Wenn das mittlere Austauschintegral zwischen der $p\pi$-Anionbahn und den Kationenbahnen auch negativ ist, führt diese Wechselwirkung ebenfalls zum Antiferromagnetismus. Bei einem positiven Austauschintegral mit der $p\pi$-Bahn wird die resultierende Kopplung durch die Konkurrenz der Beiträge von $p\sigma$- und $p\pi$-Bahnen bedingt. Selbstverständlich ist dabei eine Kopplung über die $p\sigma$-Bahn wegen deren starker Überlappung mit den d-Funktionen der Kationen im Vorteil.

KANAMORI betrachtete auch andere Verbindungen mit Kationen in oktaedrischer Umgebung. Das Vorzeichen der Austauschwechselwirkung kann man auch hier erklären, wenn man sich auf die oben dargelegten Überlegungen stützt und relativ wenige Annahmen über das Verhältnis der Überlappungen zwischen den Wellenfunktionen sowie zur Größe der Austauschintegrale trifft.

ANDERSON [151], [152] gab, indem er diese halbempirische Argumentation von GOODENOUGH, KANAMORI, WOLLAN u.a. verallgemeinerte, die folgende kurze Zusammenstellung von Regeln, die den Kopplungstyp bei fast allen magnetischen Verbindungen erklären können: Wenn sich bei oktaedrischer Umgebung Elektronen in e_g-Bahnen befinden, die in Richtung auf die Liganden und gegeneinander langgestreckt sind (180°-Position, starke Überlappung), so liegt eine starke antiferromagnetische Kopplung vor. In der 180°-Position wechselwirken bei oktaedrischer Umgebung auch die Elektronen in t_{2g}-Bahnen antiferromagnetisch über die $p\pi$-Bahnen. In der 90°-Position gibt es bei oktaedrischer Umgebung eine antiferromagnetische Wechselwirkung zwischen den besetzten e_g-Bahnen eines Kations und den t_{2g}-Bahnen eines anderen über die $p\pi$- und $p\sigma$-Bahnen des Anions. Bei 180°-Position und oktaedrischer Umgebung ist die Austauschkopplung positiv für zwei verschiedene Kationen, deren besetzte Bahnen orthogonal zueinander sind [t_{2g}- und e_g-Bahnen, z.B. in der Konfiguration $(t_{2g})^6(e_g)^2$ für das eine Kation und $(t_{2g})^3$ für das andere].

30. Diskussion und Schlußfolgerung. Zieht man das allgemeine Fazit hinsichtlich der Entstehung der indirekten Austauschwechselwirkung in nichtleitenden Kristallen, so bemerkt man, daß wahrscheinlich die wichtigsten und am meisten charakteristischen Züge des physikalischen Mechanismus dieser Erscheinung verständlich sind. Ob man von der ursprünglichen Darstellung nach KRAMERS und ANDERSON oder von ANDERSONs späteren Ansätzen ausgeht, man kann wenigstens im Prinzip die vielfältigen, für die Austauschkopplung verantwortlichen Mechanismen angeben. Wenn die Herkunft der indirekten Wechselwirkung in Isolatoren vom Prinzip her auch klar zu sein scheint, so ist doch die moderne quantitative Theorie noch weit von der Vollkommenheit entfernt. Die zahlenmäßigen Berechnungen müßten, wenn auch nur angenähert, zeigen, welcher der verschiedenen Typen physikalisch möglicher Mechanismen bei der Austauschkopplung die entscheidende Rolle spielt und welcher vernachlässigt werden kann. Leider hat uns die Theorie bisher keine zuverlässigen Zahlenangaben bieten können. Viele zahlenmäßige Abschätzungen und Rechnungen leiden außer unter den Unzulänglichkeiten des Modells darunter, daß zu wenige Konfigurationen berücksichtigt werden, daß zu rohe Ausdrücke für die Einelektronenfunktionen verwendet werden, daß mit sehr zweifelhaften Angaben über die verschiedenen Energieparameter operiert wird usw.

Im folgenden zählen wir die grundlegenden Arbeiten auf, deren Autoren versuchten, die Austauschparameter zahlenmäßig abzuschätzen bzw. halbquantitativ oder quantitativ zu berechnen.

Eine erste grobe Abschätzung für das Integral der indirekten Austauschwechselwirkung wurde bereits von KRAMERS in seiner Pionierarbeit [35] gegeben. Indem er das Verhältnis zwischen Übergangsintegral und Anregungsenergie [s. (18.6)] mit 0,1 ansetzte, nahm er an, daß die indirekte Wechselwirkung ungefähr 100mal schwächer ist als die direkte in den Kristallen, wo sie auftritt. Ähnliche qualitative Abschätzungen stammen auch von ANDERSON [36].

Ausgehend von der Formel (17.56) bemerkte VAN VLECK [38], daß, wenn man in Anlehnung an experimentelle Daten (Néel-Temperaturen) für das indirekte Austauschintegral J einen Wert von 10^{-3} eV annimmt, man diesen Wert auch theoretisch begründen kann, indem man folgende vernünftige Abschätzungen für die in den Ausdruck für J eingehenden Parameter trifft:

$$E_{\uparrow\downarrow}-E_{\uparrow\uparrow}\sim 1\text{ eV}, \quad \varrho\sim J_{pd}\sim 0{,}1\text{ eV}.$$

Erste Versuche zu einer zahlenmäßigen Berechnung des effektiven Austauschintegrals wurden von YAMASHITA [187] durchgeführt auf der Basis bereits früher ([156]) von ihm erhaltener Formeln, die unter Verwendung nichtorthogonaler Einelektronenfunktionen abgeleitet worden waren. Einige Parameter der Theorie wurden dabei mit Hilfe von Hartree-Fock-Funktionen berechnet, andere dagegen nur abgeschätzt. So wurde z.B. für die Aktivierungsenergie ΔE ein Wert von ~ 3 eV faktisch angenommen. Eine solche halbquantitative Berechnung für die Verbindung MnO ergab für das Austauschintegral den Wert $\sim 4\cdot 10^{-3}$ eV.

Von PRATT [138] stammt eine vollständige zahlenmäßige Berechnung der Singulett-Triplett-Aufspaltung im Problem der drei Zentren und vier Elektronen, wobei der virtuelle Übergang in die einfach besetzte Bahn berücksichtigt wurde. Der Abstand zwischen Kation und Anion wurde zu 2,5 atomaren Einheiten gewählt und die Modellwellenfunktionen der beiden Kationen und des Anions in der Form $\exp[-(r/a_H)^2]$. Das Austauschintegral war gleich 0,27 eV. Selbstverständlich hat eine solche Berechnung nur rein illustrativen Wert.

KONDO [188] behandelte im Rahmen der Bändertheorie das Problem der indirekten Austauschwechselwirkung im Kristall NiO. Er berechnete die Energiedifferenz zwischen den geordneten ferromagnetischen und den geordneten antiferromagnetischen Zuständen des gesamten Kristalls. Einer der Parameter, die diese Differenz bestimmen, ist eine Größe, die dem Verhältnis b^2/U analog ist, wie es in der Theorie von ANDERSON [s. (26.6)] auftritt. Die Anregungsenergie U wurde von KONDO zu 12,55 eV abgeschätzt. Diesen Wert erhält man bei Berücksichtigung der Übergänge eines Elektrons von einem Ion Ni^{+2} zum anderen. Die entsprechende Energie kann man aus der Differenz der Ionisationspotentiale für das dritte und zweite Elektron in Nickel — diese ist ~ 15 eV — berechnen, wenn man die Wechselwirkungsenergie zweier Elektronen, die sich bei benachbarten Nickelionen befinden, abzieht. Zur Berechnung der Übergangsintegrale wurden knotenfreie Slater-Funktionen benutzt. Der Beitrag des betrachteten Prozesses zur fraglichen Differenz betrug 0,0069 eV pro Elementarzelle.

Auch YAMASHITA und KONDO [189] führten zahlenmäßige Berechnungen für das System MnO durch, und zwar für die reine Ionenkonfiguration sowie für den Kramers-Anderson-Mechanismus. Für die Einelektronen-Funktionen des Kations wurden $2s$-Wasserstofffunktionen und für die Anionenfunktion $2p$-Wasserstofffunktionen angesetzt. Die Übergangsenergie wurde wie in der früheren Arbeit von YAMASHITA [187] gleich 3 eV gesetzt. Die effektiven Ladungen der Mangan- und Sauerstoffionen wurden mit $+2$ bzw. -2 angenommen. Die Berechnungen erfolgten für ein breites Intervall interatomarer Abstände. YAMASHITA und KONDO versuchten auch, den Beitrag des Slaterschen Polarisationsmechanismus abzuschätzen, indem sie sich an den bekannten dielektrischen Eigenschaften des Kristalls MgO und der Alkalihalogenide orientierten. Wie sich zeigt, muß der untersuchte Mechanismus eine beträchtliche Rolle spielen.

NESBET [139] schätzte den Beitrag des Prozesses der zweifachen Anregung ab [s. Formel (19.3)]. Für die Anregungsenregie erhält er Werte zwischen 1,69 eV für CoO und 7,77 eV für MnO. In einer letzten Arbeit [140] berücksichtigte NESBET zusätzlich den von uns in Ziff. 20 und 23 beschriebenen Prozeß und schätzte seinen Beitrag zur Anregungsenergie ab.

KEFFER und OGUCHI [153] wiesen darauf hin, daß NESBET die Anregungsenergie beim zweifachen Übergang höchstwahrscheinlich zu klein eingeschätzt hat.

ANDERSON unterstreicht ebenfalls, daß NESBET die Bedeutung seines Mechanismus überschätzt.

KONDO [157], [190] schätzte unter Verwendung nichtorthogonaler Bahnen die Beiträge der reinen Ionenkonfiguration, des Kramers-Anderson-Mechanismus, des Polarisationseffekts und des Anderson-Hasegawa-Mechanismus ab. Wie sich herausstellte, rührt der größte Beitrag vom Polarisationsmechanismus her.

KOIDE, SINHA und TANABE [191] untersuchten den Mechanismus des indirekten Austausches innerhalb eines Schemas, in dem Elektronenübergänge über die Polarisation der Anionenschale berücksichtigt werden. Die Wellenfunktionen des Zwischenions wurden als Superposition von $2p$- und $3d\text{-}e_g$-Funktionen des Sauerstoffions bzw. der unbesetzten hybridisierten Kationenbahnen dargestellt. Die Bestimmung der Singulett-Triplett-Aufspaltung erfolgte unter Berücksichtigung der Wechselwirkung mit angeregten Konfigurationen, die infolge der Nichtabgeschlossenheit der Schale des Zwischenions hinsichtlich des Spins entstehen [s. Formel (21.4)]. Die Berechnungen erfolgten im Hinblick auf MnO. Die Einelektronenwellenfunktionen wurden als Slater- (oder modifizierte Slater-) Funktionen angesetzt. Nach einer angenäherten Berechnung der Integrale erhalten die Autoren für das gesuchte effektive Austauschintegral $\sim 3 \cdot 10^{-3}$ eV, was größenordnungsmäßig wahrscheinlich richtig ist.

ANDERSON [47] schätzte die Größe des Integrals des kinetischen Austausches b^2/U ab [s. Formel (26.6)]. Die Größe U, die in nullter Näherung gleich der Energie der Coulomb-Wechselwirkung zweier Elektronen an einem Atom gesetzt werden kann, muß eigentlich genauer berechnet werden. Sie ist näherungsweise gleich der Differenz des dritten und zweiten Ionisationspotentials der freien Ionen (für Verbindungen mit zweiwertigen Kationen, wie MnO, FeO, CoO, NiO), vermindert um die Wechselwirkungsenergie zweier Elektronen an benachbarten Gitterpunkten und um die Polarisationsenergie der Elektronen der Umgebung. Außerdem muß sie noch in bezug auf die Kovalenz korrigiert werden. Indem er für die aufgezählten Korrekturglieder vernünftige Größen wählt, erhält ANDERSON für Mn^{+2}, Fe^{+2}, Co^{+2} und Ni^{+2} die folgenden Werte für U: 9,9 eV; 5,1 eV; 5,8 eV; 6,3 eV, wobei der Fehler bei der Bestimmung dieser Größen etwa 1 eV beträgt. Er weist weiter darauf hin, daß man zur Bestimmung des Übergangsintegrals „b" Daten über die Größe $10Dq$ der Aufspaltung im Ligandenfeld verwenden kann. Die Übereinstimmung zwischen dem Integral „b" und der Größe $10Dq$ wurde von ihm mit größerer Ausführlichkeit in der Übersichtsarbeit [151], [152] abgehandelt.

STUART und MARSHALL [192] versuchten, das Integral des potentiellen Austauschs J, wie es in (26.3) und (26.7) auftritt, für die Verbindungen $KNiF_3$ und NiO zahlenmäßig zu berechnen. In seinen Arbeiten [47], [151], [152] nahm ANDERSON an, daß J gegenüber dem Integral des kinetischen Austauschs sehr klein ist, und zwar infolge des sog. McWeeny-Theorems [193]: Die Größe J ist klein, da sie der Wechselwirkungsenergie von Austauschladungen mit den Dichten $\varphi_1^*(\vec{r})\,\varphi_2(\vec{r})$ entspricht, zumal für orthogonale Funktionen φ_1 und φ_2 die gesamte Austauschladung Null ist: $\int \varphi_1^*(\vec{r})\,\varphi_2(\vec{r})\,d\vec{r}=0$. STUART und MARSHALL konstruierten die Wellenfunktionen der magnetischen Kationen für das Problem von drei Zentren: zwei Kationen und ein Anion auf einer Geraden. Die Wellenfunktion φ des magnetischen Quasiteilchens ist die Superposition einer lockernden e_g-Funktion des einen oktaedrischen Komplexes und einer d-Funktion des anderen Kations. Die beiden (nichtnormierten) lockernden Funktionen lauten

$$\left.\begin{array}{l}\psi_1=\varphi_{d_1}-A_\sigma\,\varphi_p-A_s\,\varphi_s,\\ \psi_2=\varphi_{d_2}+A_\sigma\,\varphi_p-A_s\,\varphi_s,\end{array}\right\} \tag{30.1}$$

wobei φ_{d_1} und φ_{d_2} die Wellenfunktionen des „linken" und „rechten" Kations, φ_p und φ_s die p- und s-Funktion des Anions bedeuten. Die Größen A_σ und A_s geben den Kovalenzgrad der Bindung an. Die Wellenfunktionen der mit Hilfe von (30.1) konstruierten magnetischen Quasiteilchen sind

$$\begin{aligned}\varphi_1 &= \psi_1 + B\,\varphi_{d_2},\\ \varphi_2 &= \psi_2 + B\,\varphi_{d_1},\end{aligned} \quad (30.2)$$

wobei die Größe B aus der Forderung nach Orthogonalität der Funktionen (30.2) folgt. Das Austauschintegral wurde mit Hilfe der Funktionen (30.2) unter Vernachlässigung der Überlappung der Kationenfunktionen φ_{d_1} und φ_{d_2} berechnet. Für die zahlenmäßigen Berechnungen kamen die Hartree-Fock-Funktionen der Ionen von Nickel [*194*], Sauerstoff [*195*] und Fluor [*196*] zur Anwendung. Die Berechnung der Integrale für $KNiF_3$ und NiO erfolgte in einem breiten Intervall von Werten für A_s und A_σ. Es zeigte sich, daß für die experimentellen Werte $A_s^2 = 0{,}0054$ und $A_\sigma^2 = 0{,}033$ bei der Verbindung $KNiF_3$ [*197*] das Integral des potentiellen Austausches viel zu groß wird, was mit dem Experiment und Andersons Annahmen über die Schwäche des potentiellen Austausches im Widerspruch steht. Anderson selbst kam, allerdings in einer späteren Arbeit [*198*], auf Grund zahlenmäßiger Abschätzungen zu dem Schluß, daß der potentielle Austausch in seiner Theorie eine wesentlich größere Rolle spielt als angenommen wurde.

Zum Schluß sei nochmals unterstrichen, daß die Richtigkeit der physikalischen Konzeption, die der Theorie der indirekten Austauschkopplung in Isolatoren zugrunde liegt, im allgemeinen und qualitativ vom Experiment bestätigt wird. Jedoch trifft die vorliegende Theorie im Quantitativen auf große mathematische Schwierigkeiten, die noch lange nicht überwunden sind. Diese Schwierigkeiten auf dem Teilgebiet der Theorie des indirekten Austausches in Isolatoren hängen im Grunde genommen mit der allgemeineren Schwierigkeit der Vielelektronenbehandlung des Festkörpers und in erster Linie mit der unkorrekten Anwendung von Einelektronennäherungen zusammen.

Literatur.

[1] Hund, F.: Linienspektren und periodisches System der Elemente. Berlin 1927.
[2] Heisenberg, W.: Z. Physik **38**, 411 (1926).
[3] Heisenberg, W.: Z. Physik **39**, 499 (1926).
[4] Dirac, P. A. M.: Proc. Roy. Soc. (London) A **112**, 661 (1926).
[5] Heitler, W., u. F. London: Z. Physik **44**, 455 (1927).
[6] Dirac, P. A. M.: The Principles of Quantum Mechanics, ch. IX, fourth edit. Oxford: Clarendon Press 1958.
[7] Frenkel, J.: Z. Physik **49**, 31 (1928); — Proc. Roy. Soc. (London) A **123**, 714 (1929).
[8] Heisenberg, W.: Z. Physik **49**, 619 (1928).
[9] Bloch, F.: Z. Physik **57**, 545 (1929).
[10] Stoner, E. C.: Proc. Roy. Soc. (London) A **154**, 656 (1936).
[11] Stoner, E. C.: Repts. Progr. in Phys. **11**, 43 (1948).
[12] Stoner, E. C.: Repts. Progr. in Phys. **13**, 83 (1950).
[13] Wigner, E.: Trans. Faraday Soc. **34**, 678 (1938).
[14] Mott, N. F.: Proc. Roy. Soc. (London) A **153**, 699 (1936).
[15] Mott, N. F.: Proc. Roy. Soc. (London) A **156**, 368 (1936).
[16] Slater, J. C.: Phys. Rev. **49**, 537, 931 (1936).
[17] Wohlfarth, E. P.: Revs. Mod. Phys. **25**, 211 (1953).
[18] Abrikosov, A. A., i I. E. Dzyaloshinsky: Zhur. Ekspt̃l. i Teoret. Fiz. **35**, 771 (1958).
[19] Kondratenko, P. S.: Zhur. Eksptl. i Teoret. Fiz. **46**, 1438 (1964).
[20] Landau, L. D.: Zhur. Eksptl. i Teoret. Fiz. **30**, 1058 (1956).
[21] Landau, L. D.: Zhur. Eksptl. i Teoret. Fiz. **32**, 59 (1957).
[22] Landau, L. D.: Zhur. Eksptl. i Teoret. Fiz. **35**, 97 (1958).

[23] MOTT, N. F.: Advances in Phys. **13**, 525 (1964).
[24] THOMPSON, E. D.: Advances in Phys. **14**, 213 (1965).
[25] HERRING, C.: Exchange Interactions among Itinerant Electrons. Magnetism, ed. by G. T. RADO and H. SUHL, vol. IV. New York: Academic Press 1966.
[26] VAN VLECK, J. H.: The Theory of Electric and Magnetic Susceptibilities. Oxford: Univ. Press 1932; — Phys. Rev. **45**, 405 (1934).
[27] FAULKNER, J. S.: Phys. Rev. **128**, 202 (1962).
[28] VONSOVSKY, S. V., i M. S. SVIRSKY: Zhur. Eksptl. i Teoret. Fiz. **47**, 1354 (1964).
[29a] KAPLAN, T. A., and D. H. LYONS: Phys. Rev. **129**, 2072 (1963).
[29b] KASUYA, T., and D. H. LYONS: ISSP Report Ser. ANI 67 (1965).
[29c] IRKHIN, YU. P.: Zhur. Eksptl. i Teoret. Fiz. **50**, 379 (1966).
[29d] VONSOVSKY, S. V., YU. P. IRKHIN, i M. S. SVIRSKY: Izvest. Akad. Nauk S.S.S.R., Ser. Fiz. **30**, 906 (1966).
[30] MATTIAS, B. T., R. M. BOZORTH, and J. H. VAN VLECK: Phys. Rev. Letters **7**, 160 (1961).
[31] MCGUIRE, T. R., and M. W. SHAFER: J. Appl. Phys. **35**, 984 (1964).
[32] TSUBOKAWA, I.: J. Phys. Soc. Japan **15**, 1664 (1960).
[33] WOOD, D. W., and N. W. DALTON: Proc. Phys. Soc. (London) **87**, 755 (1966).
[34] SHULL, C. G., W. A. STRAUSER, and E. O. WOLLAN: Phys. Rev. **83**, 333 (1951).
[35] KRAMERS, H. A.: Physica **1**, 182 (1934).
[36] ANDERSON, P. W.: Phys. Rev. **79**, 350 (1950).
[37] VAN VLECK, J. H.: Phys. Rev. **78**, 266 (1950).
[38] VAN VLECK, J. H.: J. phys. radium **12**, 262 (1951).
[39] SCHUBIN, S., u. S. WONSOWSKY: Phys. Z. Sowjetunion **7**, 292 (1935).
[40] VONSOVSKY, S. V.: Zhur. Eksptl. i Teoret. Fiz. **10**, 468 (1946).
[41] VONSOVSKY, S. V., i E. A. TUROV: Zhur. Eksptl. i Teoret. Fiz. **24**, 419 (1953).
[42] RUDERMAN, M. A., and C. KITTEL: Phys. Rev. **96**, 99 (1954).
[43] KASUYA, T.: Progr. Theoret. Phys. (Kyoto) **16**, 45 (1956).
[44] YOSIDA, K.: Phys. Rev. **106**, 893 (1957).
[45] YOSIDA, K.: Phys. Rev. **107**, 369 (1957).
[46] SLATER, J. C.: Quarterly Progress Report. Massachusetts Institute of Technology, July 15, Oct. 15, 1953 (unveröffentlicht).
[47] ANDERSON, P. W.: Phys. Rev. **115**, 2 (1959).
[48] STEVENS, R. W. H.: Proc. Roy. Soc. (London) A **219**, 542 (1953).
[49] OWEN, J.: Proc. Roy. Soc. (London) A **227**, 183 (1954).
[50] CLOGSTON, A. M., J. P. GORDON, V. JACCARINO, M. PETER, and L. R. WALKER: Phys. Rev. **117**, 1222 (1960).
[51] SHULMAN, R. G., and V. JACCARINO: Phys. Rev. **103**, 1126 (1956).
[52] SHULMAN, R. G., and V. JACCARINO: Phys. Rev. **108**, 1219 (1957).
[53] JACCARINO, V., and R. G. SHULMAN: Phys. Rev. **107**, 1196 (1957).
[54] GRIFFITHS, J. H. E., J. OWEN, J. G. PARK, and M. F. PARTRIDGE: Proc. Roy. Soc. (London) A **250**, 84 (1959).
[55] STOUT, J. W.: J. Chem. Phys. **31**, 709 (1959).
[56] LÖWDIN, P. O.: Rev. Mod. Phys. **34** (I), 80 (1962).
[57] LANDAU, L. D., i E. M. LIFSHITZ: Kvantovaya Mekhanika, Kap. 9. Moskau 1963.
[58] WIGNER, E.: Group Theory and Its Application to the Quantum Mechanics of Atomic Spectra, ch. 22. New York and London: Academic Press 1959.
[59] SLATER, J. C.: Phys. Rev. **35**, 509 (1930).
[60] SCHUBIN, S. P., and S. V. VONSOVSKY: Proc. Roy. Soc. (London) A **105**, 159 (1934).
[61] SCHUBIN, S., u. S. WONSOWSKY: Phys. Z. Sowjetunion **10**, 348 (1936).
[62] VONSOVSKY, S. V.: Izvest. Akad. Nauk S.S.S.R., Ser. Fiz. **12**, 337 (1948).
[63] SOMMERFELD, A., u. H. BETHE: Elektronentheorie der Metalle. Handbuch der Physik, Bd. XXIV/2. Berlin: Springer 1933.
[64] BETHE, H. A., and E. E. SALPETER: Quantum mechanics of one- and two-electron atoms. Berlin-Göttingen-Heidelberg: Springer 1957.
[65] SLATER, J. C.: Phys. Rev. **34**, 1293 (1929).
[66] SLATER, J. C.: J. Chem. Phys. **19**, 220 (1951).
[67] SUGIURA, Y.: Z. Physik **45**, 484 (1927).
[68] DELBRÜCK, M.: Ann. Phys. **5**, 36 (1930).
[69] COURANT, R., u. D. HILBERT: Methoden der Mathematischen Physik, 2. Aufl., Kap. VI. Berlin: Springer 1931.
[70] GORKOV, L. P., i L. P. PITAEVSKI: Dokl. Akad. Nauk S.S.S.R. **151**, 822 (1963) [Engl. Übersetz.: Soviet Phys.-Dokl. **8**, 788 (1964)].
[71] HERRING, C., and M. FLICKER: Phys. Rev. **134**, A 362 (1964).

[72] Carr jr., W. J., and M. Ashkin: J. Chem. Phys. **42** (8), 2796 (1964).
[73] Bloch, F.: Z. Physik **74**, 295 (1932).
[74] Möller, C.: Z. Physik **82**, 559 (1933).
[75] Holstein, T., and H. Primakoff: Phys. Rev. **58**, 1098 (1940).
[76] Vonsovsky, S. V., i E. N. Agafonova: Sbornik posvyasheny 70-letiyu akademika Joffe, A. F., 1950.
[77] Löwdin, P. O.: Phys. Rev. **97**, 1474 (1955).
[78] Bloch, F.: Handbuch der Radiologie, Bd. 6, Teil 2. Leipzig 1934.
[79] Löwdin, P. O.: Phys. Rev. **97**, 1509 (1955).
[80] Löwdin, P. O.: Advances in Phys. **5** (17), 1 (1956).
[81] Secrest, D., and L. M. Holm: J. Math. Phys. **5**, 738 (1964).
[82] Slater, J. C.: Phys. Rev. **35**, 509 (1930).
[83] Slater, J. C.: Rev. Mod. Phys. **25**, 199 (1953).
[84] Inglis, D. R.: Phys. Rev. **46**, 135 (1934).
[85] Herring, C.: Direct Exchange Between Well-Separated Atoms. Magnetism, ed. by G. T. Rado and H. Suhl, vol. II B. New York: Academic Press 1966.
[86] Arai, T.: Phys. Rev. **126**, 471 (1962).
[87] Arai, T.: Phys. Rev. **134**, A 824 (1964).
[88] Van Vleck, J. H.: Phys. Rev. **49**, 232 (1936).
[89] Takano, F.: J. Phys. Soc. Japan **14**, 348 (1959).
[90] Löwdin, P. O.: J. Chem. Phys. **18**, 365 (1950).
[91] Carr jr., W. J.: Phys. Rev. **92**, 28 (1953).
[92] Mizuno, Y., and T. Izuyama: Progr. Theoret. Phys. **22**, 344 (1959).
[93] Serber, R.: Phys. Rev. **45**, 461 (1934).
[94] Serber, R.: J. Chem. Phys. **2**, 697 (1934).
[95] Koster, G. F.: Phys. Rev. **98**, 514 (1955).
[96] Bogolyubov, N. N.: Lektzii z kvantovoi statistiki. Kiew (UdSSR) 1949.
[97] Carr jr., W. J.: J. Phys. Soc. Japan **17**, Suppl. B-I. 36 (1962).
[98] Carr jr., W. J.: Phys. Rev. **131**, 1947 (1963).
[99] Slater, J. C.: Phys. Rev. **35**, 509 (1930).
[100] Slater, J. C.: Phys. Rev. **36**, 57 (1930).
[101] Zener, C.: Phys. Rev. **81**, 440 (1951).
[102] Zener, C.: Phys. Rev. **82**, 403 (1951).
[103] Zener, C.: Phys. Rev. **83**, 299 (1951).
[104] Zener, C., and R. R. Heikes: Rev. Mod. Phys. **25**, 191 (1953).
[105] Bartlett, J. H.: Phys. Rev. **37**, 507 (1931).
[106] Wohlfarth, E. P.: Nature **163**, 57 (1949).
[107] Kaplan, H.: Phys. Rev. **85**, 1038 (1952).
[108] Stuart, R. N., and W. Marshall: Phys. Rev. **120**, 353 (1960).
[109] Wood, J. H., and G. W. Pratt: Phys. Rev. **107**, 995 (1957).
[110] Nathans, R., C. G. Shull, G. Shirane, and A. Anderson: J. Phys. Chem. Solids **10**, 138 (1959).
[111] Weiss, R. J., and A. J. Freeman: J. Phys. Chem. Solids **10**, 147 (1959).
[112] Hofman, J. A., A. Paskin, K. J. Tauer, and R. J. Weiss: J. Phys. Chem. Solids **1**, 45 (1956).
[113] Freeman, A. J., and R. E. Watson: Phys. Rev. **124**, 1439 (1961).
[114] Watson, R. E.: Phys. Rev. **118**, 1036 (1960).
[115] Freeman, A. J., R. K. Nesbet, and R. E. Watson: Phys. Rev. **125**, 1978 (1962).
[116] Watson, R. E.: Solid State and Molecular Theory Group. Massachusetts Institute of Technology, Technical Report N 12, 1959 (unveröffentlicht).
[117] Kittel, C.: Phys. Rev. **120**, 335 (1960).
[118] Bean, C. P., and D. S. Rodbell: Phys. Rev. **126**, 104 (1962).
[119] Harris, E. A., and J. Owen: Phys. Rev. Letters **11**, 9 (1963).
[120] Rodbell, D. S., I. S. Jacobs, and J. Owen: Phys. Rev. **11**, 10 (1963).
[121] Will, G., S. J. Pickart, and H. A. Alperin: J. Phys. Chem. Solids **24**, 1679 (1963).
[122] Huang, N. L., and R. Orbach: Phys. Rev. Letters **12** (11), 275 (1964).
[123] Weinbaum, S.: J. Chem. Phys. **1**, 317, 593 (1933).
[124] Mattheiss, L. F.: Phys. Rev. **123**, 1209 (1961).
[125] Eisenschitz, R., u. F. London: Z. Physik **60**, 491 (1930).
[126] Herring, C.: Rev. Mod. Phys. **34** (4), 631 (1962).
[127] Kato, T.: J. Phys. Soc. Japan **4**, 334 (1949).
[128] Lieb, E., and D. Mattis: Phys. Rev. **125**, 164 (1962).
[129] Landau, L.: Phys. Z. Sowjetunion **4**, 675 (1933).
[130] Néel, L.: Ann. Phys. (Paris) (10) **17**, 5 (1932).
[131] Néel, L.: Ann. Phys. (Paris) (11) **5**, 232 (1936).

[132] SNOEK, J. L.: New Developments in Ferromagnetic Materials. Amsterdam, Holland: Elsevier 1947.
[133] SMIT, J., and H. P. J. WIJN: Ferrites. Philips Research Laboratories, N. V. Philips' Gloeilampenfabrieken Eindhoven, The Netherlands. Philips' Technical Library, 1959.
[134] SHULL, C. G., and J. S. SMART: Phys. Rev. **76**, 125 (1949).
[135] SHIMIZU, M.: Progr. Theoret. Phys. **8**, 416 (1952).
[136] GRAY, N. M., and L. A. WILLS: Phys. Rev. **38**, 248 (1931).
[137] LÖWDIN, P. O.: J. Chem. Phys. **19** (11), 1396 (1951).
[138] PRATT, G. W.: Phys. Rev. **97**, 926 (1955).
[139] NESBET, R. K.: Ann. Phys. (N.Y.) **4**, 87 (1958).
[140] NESBET, R. K.: Phys. Rev. **119**, 658 (1960).
[141] MARSHALL, W.: Perspectives in Materials Research. Panels on Materials Research of the National Science Foundation, Washington, D.C., 1961.
[142] WATSON, R. E., and A. J. FREEMAN: Phys. Rev. **123**, 2027 (1961).
[143] SLATER, J. C.: Phys. Rev. **81**, 385 (1951).
[144] SLATER, J. C.: Phys. Rev. **82**, 538 (1951).
[145] LÖWDIN, P. O.: Phys. Rev. **97**, 1490 (1955).
[146] MARSHALL, W.: Preprint.
[147] NESBET, R. K.: Proc. Roy. Soc. (London) A **230**, 312 (1955).
[148] PRATT, G. W.: Phys. Rev. **102**, 1303 (1956).
[149] HEINE, V.: Phys. Rev. **107**, 1002 (1957).
[150] NESBET, R. K., and R. E. WATSON: Ann. Phys. (N.Y.) **9**, 260 (1960).
[151] ANDERSON, P. W.: Solid State Physics, vol. 14, p. 99—214. New York: Academic Press 1963.
[152] ANDERSON, P. W.: Exchange in Insulators. Magnetism, edit. by G. T. RADO and H. SUHL, vol. I, New York and London: Academic Press 1963.
[153] KEFFER, F., and T. OGUCHI: Phys. Rev. **115**, 1428 (1959).
[154] VONSOVSKY, S. V., i Y. M. SEIDOV: Izv. Akad. Nauk S.S.S.R., Ser. Fiz. **18**, 319 (1954).
[155] VONSOVSKY, S. V., i Y. M. SEIDOV: Dokl. Akad. Nauk S.S.S.R. **107**, 37 (1956).
[156] YAMASHITA, J.: J. Phys. Soc. Japan **9**, 339 (1954).
[157] KONDO, J.: Progr. Theoret. Phys. (Kyoto) **22**, 41 (1959).
[158] FUKUCHI, M.: Progr. Theoret. Phys. (Kyoto) **25**, 939 (1961).
[159] FUKUCHI, M.: Progr. Theoret. Phys. (Kyoto) **25**, 956 (1961).
[160] LÖWDIN, P. O.: A Theoretical Investigation of some Properties of Ionic Crystals. (Thesis), Uppsala (1948).
[161] ANDERSON, P. W., and H. HASEGAWA: Phys. Rev. **100**, 675 (1955).
[162] JONKER, G. H., and J. H. VAN SANTEN: Physica **16**, 337 (1950).
[163] JONKER, G. H., and J. H. VAN SANTEN: Physica **16**, 929 (1950).
[164] JONKER, G. H., and J. H. VAN SANTEN: Physica **19**, 120 (1953).
[165] DE GENNES, P.-G.: Phys. Rev. **118**, 141 (1960).
[166] BETHE, H. A.: Ann. Phys. (Leipzig) (5), **3**, 133 (1929).
[167] GRIFFITH, J. S., and L. E. ORGEL: Quart. Rev. (London) **11**, 381 (1957).
[168] VAN VLECK, J. H.: J. Chem. Phys. **3**, 843 (1935).
[169] HOLMES, O. G., and D. S. MCCLURE: J. Chem. Phys. **26**, 1686 (1957).
[170] MORIN, F. J.: Bell System Techn. J. **37** (4) 1047 (1958).
[171] LOW, W.: Paramagnetic Resonance in Solids. Solid State Physics, Suppl. 2. New York and London: Academic Press 1960.
[172] ORGEL, L. E.: An Introduction to Transition-Metal Chemistry. London: Methuen & Co. Ltd. 1961.
[173] BALLHAUSEN, C. J.: Introduction to Ligand Field Theory. New York-San Francisco-Toronto-London: McGraw-Hill Book Co. 1961.
[174] SHULMAN, R. G., and S. SUGANO: Phys. Rev. Letters **7**, 157 (1961).
[175] SHULMAN, R. G., and S. SUGANO: Phys. Rev. **130**, 517 (1963).
[176] SHULMAN, R. G., and S. SUGANO: Phys. Rev. **130**, 506 (1963).
[177] GOODENOUGH, J. B., and A. L. LOEB: Phys. Rev. **98**, 391 (1955).
[178] GOODENOUGH, J. B.: Phys. Rev. **100**, 564 (1955).
[179] GOODENOUGH, J. B.: Phys. and Chem. Solids **6**, 287 (1958).
[180] GOODENOUGH, J. B., A. WOLD, R. J. ARNOTT, and N. MENYUK: Phys. Rev. **124**, 373 (1961).
[181] GOODENOUGH, J. B.: Magnetism and the Chemical Bond. New York: John Wiley & Sons 1963.
[182] KANAMORI, J.: Phys. and Chem. Solids **10**, 87 (1959).
[183] WOLLAN, E. O., and W. C. KOEHLER: Phys. Rev. **100**, 545 (1955).
[184] WOLLAN, E. O.: Phys. Rev. **117**, 387 (1960).

[185] Gilleo, M. A.: Phys. Rev. **109**, 777 (1958).
[186] Osmond, W. P.: Proc. Phys. Soc. (London) **87**, 767 (1966).
[187] Yamashita, J.: Progr. Theoret. Phys. (Kyoto) **12**, 808 (1954).
[188] Kondo, J.: Progr. Theoret. Phys. (Kyoto) **18**, 541 (1957).
[189] Yamashita, J., and J. Kondo: Phys. Rev. **109**, 730 (1958).
[190] Kondo, J.: Progr. Theoret. Phys. (Kyoto) **22**, 819 (1959).
[191] Koide, S., K. P. Sinha, and Y. Tanabe: Progr. Theoret. Phys. (Kyoto) **22**, 647 (1959).
[192] Stuart, R. N., and W. Marshall: Proc. Phys. Soc. (London) **87**, 749 (1966).
[193] McWeeny, R.: Proc. Roy. Soc. (London) A **227**, 288 (1955).
[194] Watson, R. E.: Massachusetts Institute of Technology, Technical Report No. 12, 1959.
[195] Watson, R. E.: Phys. Rev. **111**, 1108 (1958).
[196] Allen, J. C.: Quarterly Progress Report, Solid State and Molecular Theory Groop, Massachusetts Institute of Technology **22**, 4 (1956).
[197] Knox, K., and R. G. Shulman: Phys. Rev. **119**, 94 (1960).
[198] Anderson, P. W.: Buhl Internat. Conference on Materials: Transition Metal Compounds. New York: Gordon & Breach 1963.

Magnetic Semiconductors.

By

Siegfried Methfessel and Daniel C. Mattis*.

With 52 Figures.

A. Preface.

I. General remarks.

1. "Magnetic" materials. All known materials show, of course, some response or susceptibility to applied magnetic fields. The classification "magnetic materials", however, is usually reserved for substances in which at least some of the atoms or ions maintain their unpaired electron spins or orbits, spontaneously, without applied magnetic fields, in spite of the chemical bonding mechanisms which hold the material together. The study of the correlations and interference of the local moments with the chemical bonding mechanism in the solid state is one of the central tasks of magnetic research.

There is an old challenge that is to find magnetic materials which contain only electrons with low angular momentum quantum numbers, such as electrons with s and p character. General experience, however, indicates that the appearance of spontaneous magnetic moments at certain atoms or ions is essentially coupled with the occurrence of incompletely filled d or f electron shells. Thus, the physics of magnetic materials is actually the major part of the physics of transition elements and their compounds. The occurence of spontaneous magnetic moments in elements and compounds appears to be restricted to atoms only with compact $3d$ and $4f$ electron shells: Fig. 1 gives the calculated radii of the outer occupied electron shells of the elements throughout the periodic system. It is interesting to note that for all "magnetic" elements, such as Mn, Fe, Co, Ni and the lanthanides, the magnetic $3d$ or $4f$ shells, respectively, have a smaller radial extension than the filled outer p-shell of the relevant noble gas core. In the solid state the more extended valence electrons with s or p character provide the chemical bonding. The resulting interatomic distances become large enough for the inner $3d$ or $4f$ orbits to retain some of their free ion properties. Some residual spin-spin or orbit-orbit interactions provide atomic magnetic moments which, however, may deviate appreciably from values expected from the multiplet structure of the free atom. The well localized $4f$ electrons in the lanthanides have magnetic moments nearly identical with the free atom moments, while the more extended $3d$ electrons show much stronger deviations due to quenching of contributions of the orbital moments by the crystalline environment. In compounds with large anions, such as halogens, the interatomic distances are large enough to admit spontaneous moments in nearly all d-ions[1].

---

* Research supported in part by a grant of the United States Air Force Office of Scientific Research No. AFOSR 1075-66.

[1] For examples see Landolt-Börnstein, II/9. Berlin-Göttingen-Heidelberg: Springer 1962.

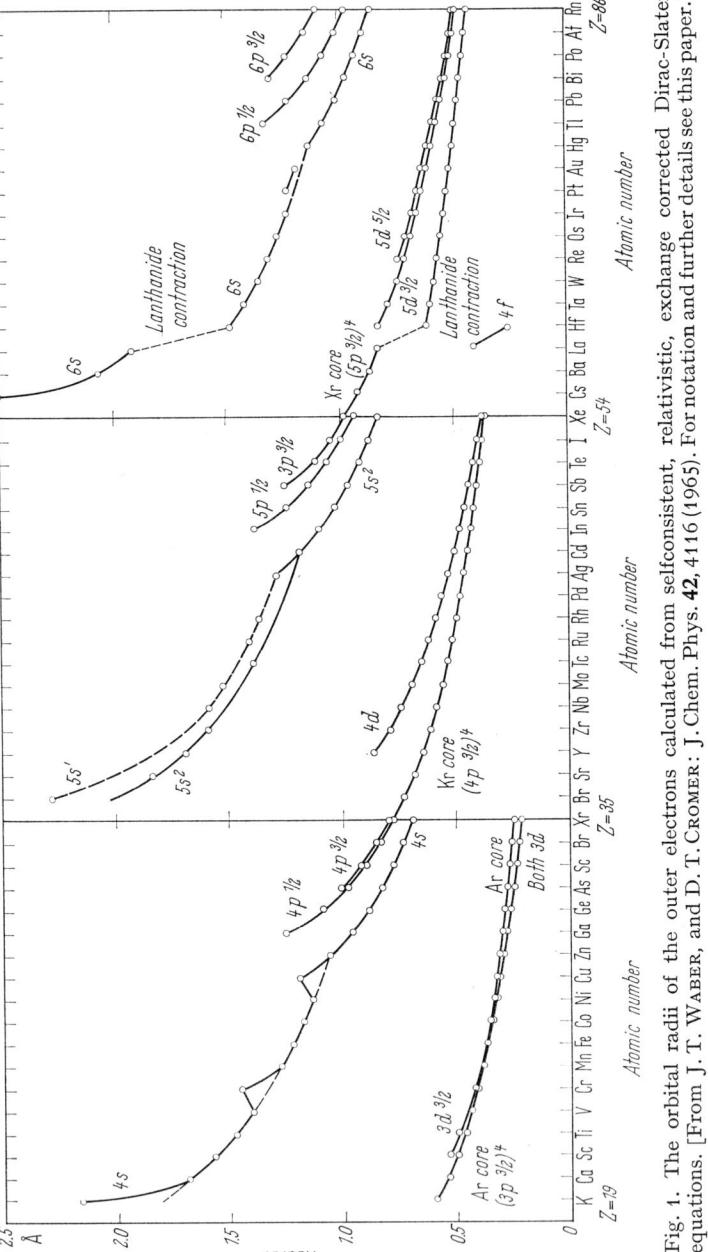

Fig. 1. The orbital radii of the outer electrons calculated from selfconsistent, relativistic, exchange corrected Dirac-Slater equations. [From J. T. WABER, and D. T. CROMER: J. Chem. Phys. **42**, 4116 (1965). For notation and further details see this paper.]

The interaction of the magnetic electron shells with their environment does not only affect the atomic magnetic moment but can also establish exchange interactions between all the magnetic atoms in the same crystal, which result at low temperatures into ferromagnetic order or more complicated spin structure. For well localized magnetic moments, such as the $4f$ electrons in the lanthanides, one expects to find mainly the weak dipole interactions between neighboring atoms.

The magnetic properties of some lanthanide compounds are, therefore, rather simple and comparable to those of a gas of noninteracting paramagnetic particles. In other lanthanide compounds and most elemental metals, however, the interaction of the $4f$ electrons with their environment is strong enough that the relative orientation of neighboring moments to one another contributes somewhat to the chemical binding energy. For $3d$ elements the connection between spin alignment and chemical bonding is much stronger than for the lanthanides, but the total contribution of magnetic interactions to the overall stability of the crystal still remains quite small. As an example, the forced alignment of magnetic moments in strong external fields results usually only in magnetostrictive distortions of one part in a million or so, but rarely in crystallographic transformations.

The strength of magnetic interactions between neighboring atoms in crystals can be measured by the Curie (or Néel) temperature at which the ferromagnetic (antiferromagnetic) order decays into the disordered paramagnetic state. It rarely reaches values above 1000 °K and is, therefore, in most magnetic materials several orders of magnitude weaker than the electrostatic lattice energy and the Coulomb or exchange energy of electrons in the free atom.

The spontaneous magnetization in crystals is caused by mechanisms which are analogous to those responsible for the spontaneous magnetic moment in a single atom. In both instances exchange corrections to the strong electrostatic forces between electrons in similar quantum states are caused by the Pauli principle (and, to a lesser degree, by spin-orbit interaction). In atoms, molecules, and most particularly in solids, tremendous mathematical difficulties arise when the necessary electron-electron interactions are added to the single-electron description of the atom or crystal. The lower symmetry of a crystal as compared to the spherical symmetry of a single atom and the complexity of the chemical bonding bring additional difficulties and characterize the theoretical treatment of electron-electron interactions as one of the most demanding of problems in solid state physics. For these reasons, the investigation of magnetic properties of crystals and their relationship to other electronic properties has greater implications than might be suggested merely by the well known importance of magnetism in modern electro-engineering and electronics.

2. Magnetic metals, insulators and semiconductors. Depending on the character of the chemical bonding mechanism we classify magnetic materials into magnetic metals, insulators, and semiconductors[2]. For metals the occurrence of conduction-electron magnetism needs special theoretical justifications which are not the concern of this article[3]. For magnetic semiconductors and insulators the bonding lies somewhere between pure ionic bonding and covalency, depending on the respective electron affinities of the anions and cations. The physical properties are strongly influenced by the contribution of magnetic electrons to the bonding process. *Qualitatively*, the exchange interactions and Curie temperature are expected to increase with increasing covalency while the atomic magnetic moments are increasingly perturbed and diminished below their free ion value. When finally, for quite small interatomic distances, the overlap between the magnetic electron

---

[2] The relation between magnetism and chemistry has been forwarded particularly by J. B. GOODENOUGH in his book "Magnetism and the Chemical Bond" (New York: Interscience Publishing Co. 1963) and several other publications, including P. W. SELWOOD, "Magnetochemistry" (New York: Interscience Publishing Co. 1956).

[3] See C. HERRING, in: Magnetism, Vol. 4 (G. F. RADO, and H. SUHL, Eds.). New York: Academic Press 1966, J. FRIEDEL: J. Phys. Radium **23**, 501, 692 (1962); N. F. MOTT: Advances in Phys. **13**, 325 (1965); D. MATTIS: The Theory of Magnetism. New York: Harper & Row 1965.

states at neighboring ions becomes comparable to the binding forces at the individual atom, the atomic moments disappear and all magnetic properties dissolve into the competing Pauli susceptibility and Landau diamagnetism of the free electron gas.

For *insulating* compounds of $3d$ elements the interatomic exchange derives mainly from the spin dependence of the covalent bond between the magnetic cation and the nonmagnetic anion via overlap of the d electrons with the filled p shell at the anion. The theory of this mechanism of *superexchange* (as it is called) is only briefly discussed here, as it is the subject of a companion article in this volume. Since superexchange is a property of true insulators with perfect stoichiometry and charge balance between all ions, any connections between this mechanism and the electrical conductivity due to electrons or holes are not obvious. For this reason, the main work on transition metal compounds has been concerned with the crystal structure and the (mostly ferri- or antiferromagnetic) spin structures of $3d$ oxides in the neighborhood of room temperature. Only a few compounds have been subjected to systematic studies of their electrical or optical properties. Considerations of the electronic band structure are very sparse[4] and large scale band structure calculations by modern computers are rarely made for nonmetallic magnetic compounds[5]. This seems to be surprising when we consider that the transition element compounds represent the largest group of semiconducting materials. The wide variety of magnetic and electrical properties found in paramagnets, ferromagnets, antiferromagnets, ferrites, metals, semiconductors, and insulators combined with nearly unlimited possibilities of forming solid solutions between elements and compounds should provide a rich field for scientific and technical exploitation.

The problems which retarded the progress in the field of transition element semiconductors are chemical-technical as well as purely scientific in nature. Since most transition metal compounds combine relatively high melting temperatures with high chemical agressivity to most crucible materials and wide flexibility in valence and stoichiometry, it is quite difficult to prepare materials with sufficient purity and homogeneity for physical investigations. The majority of available samples consists of sintered ceramics which can provide good magnetic or structural data but are quite unreliable for electrical and transport measurements. The effects of crystal boundaries, inhomogeneities, pores and deviations from stoichiometry confuse the experimental results and even causal relations between electrical transport and magnetic order can escape observation and analysis. In recent years, however, advanced physico-chemical technologies of material preparation have been devised and successfully used for growing single crystals of sufficient purity and size of several materials. The supply of reliable samples stimulated intensive investigations and reinvestigations of semiconducting properties in transition element compounds.

The results are by far not simple and their bewilderingly large variety makes the reduction of all properties to a common description very difficult. In series of isostructural compounds some members can be found to be insulators while others show typically metallic properties. The electrical properties of some magnetic compounds, such as many ferrites, show very small variation with magnetic

---

[4] J. F. MORIN: Bell System Tech. J. **37**, 1047 (1958); in: N. B. HANNAY: Semiconductors, New York: Reinhold 1959; G. H. YONKER, and S. VAN HOUTEN, in: Halbleiterprobleme, Bd. II (F. SAUTER, Ed.), p. 118 and ff. Braunschweig: F. Vieweg & Sohn 1961; and J. B. GOODENOUGH: Magnetism and the Chemical Bond. New York: Interscience Publ. 1963.

[5] Some recent calculations are: V. ERN, and A. C. SWITENDICK: Phys. Rev. **137**, A 1927 (1965): TiC, TiN, TiO; S. J. CHO: Phys. Rev. **157**, 632 (1967): EuS.

order only. Some oxides of vanadium and titanium, however, abruptly change their electrical conductivity by several orders of magnitude, from metallic to semiconducting, at the magnetic ordering temperature. Some ferromagnetic compounds show giant magneto-resistance effects in the neighborhood of their Curie temperature. The nature of electrical conductivity and other transport phenomena, and their relation to the magnetic ordering, require novel concepts and experimental techniques quite distinct from those which have been successfully applied to the wide band semiconductors of the germanium type. Present-day understanding of exchange forces which order the atomic magnetic moments over long distances is unsatisfactory. Many other aspects, such as the response of optical properties to magnetic ordering, and the inter-relationships of all the above mentioned properties on the basis of a clear microscopic picture are just at the beginning of experimental and theoretical exploitation.

The field of magnetic semiconductors, new and tentative as it must seem, receives, as modern magnetism in general, much stimulation from studies of the RE[6] atoms and compounds. VAN VLECK has commented: "The rare earths have been a proving ground for certain aspects of theoretical solid state physics ever since quantum mechanics was discovered in 1926."[7] The reason is mainly simplicity: all 14 RE atoms of equal valence are chemically indistinguishable and differ physically almost entirely in their atomic number and the number of electrons (hence, in the magnetic moment) of their tightly-bound f shell. Whereas the chemistry and solid state of d series elements is substantially complicated by the fact that d electrons generally participate in the chemical bond, there is in the simple RE atoms a clean separation of the electronic states into bound magnetic states and chemically active valence electrons. But going beyond this simple picture, one of the important magneto-electronic effects is associated with the ground state fluctuations, such as f shell electrons being virtually excited into valence states, a spin-dependent effect. Thus, the study of the ground state of magnetic rare earth semiconductors should already be essential to the understanding of dynamic effects such as the dependence of the mobility of carriers on the magnetic state of the lattice, etc., discussed later in this article.

Some twenty years ago the rare earth elements were considered as scientific curiosities and were studied only in a few laboratories, because their chemical and physical similarity made the preparation as pure elements extremely difficult. However, the need for pure thorium in nuclear technology forced the development of effective methods for the elimination of lanthanides with large neutron capture cross sections. During the last ten years an increasing quantity of rare earth materials with defined purity has become commercially available. This stimulated many research projects on the basic properties of lanthanide compounds in order to verify existing theories in solid state physics. RE chemistry has developed into one of the most active areas in the entire field of solid state chemistry, although it is still considered to be in an early state of development. The number of known rare earth compounds has approximately doubled during the last four years. Numerous compounds have been formed with the elements in the periodic table to the right of the Cr-Mo-W column[8]. The physical properties of only a few of these compounds have been investigated, but one can expect

---

[6] Henceforth we shall use the abbreviation RE for "rare earth". The name rare earth was originally assigned by chemists to the oxidic minerals of the lanthanide and actinide elements. In physics, however, the name "rare earth" became recently more specialized to the lanthanide elements and we will use it often in this way.

[7] J. H. VAN VLECK: Physical Sciences, p. 113. New York: University Press 1962.

[8] O. D. MASTERS, and K. A. GSCHNEIDNER: Nuclear Metallurgy Series 10, 93 (1964).

that most of them will show some form of magnetic ordering. Since the chemical properties of the RE are close to those of the alkaline earth elements, many compounds with the more electronegative elements from the three right-side columns of the periodic table are semiconductors or insulators. The investigation of their properties has just started with the availability of single crystalline samples of several compounds and has significance for the general understanding of magnetic semiconductors.

II. Novel features.

3. Conductivity and magnetic order. What are some of the phenomena which will determine the future of this field? Without wishing to repeat the catalogue of properties listed in the remainder of this article, *we may just concentrate here on a unique property, which so far has been manifest mainly in the Europium chalcogen compounds with small doping of a second rare earth, this property being the very great variation in electrical resistivity* (several orders of magnitude) *over a narrow range of temperatures of the order of $\pm 20\,°K$ from the Curie temperature*. The maximum in the resistance at the Curie temperature is a form of "critical fluctuations" associated with the magnetic phase transition, and shows that the *band structure* and the mechanisms which *scatter* electrons are both strong functions of the magnetic long range order.

Now, unlike crystal structure, donor and acceptor concentrations, and scattering and other properties of traps, all of which have in common that they are "frozen in" to a particular semiconductor and cannot be altered without destruction of the material, *the magnetic order parameters are easily changed by the application of magnetic force fields involving rather trivial energies*. The greatest virtue of ordinary semiconductors is their sensitive and nonlinear reactions to electronic changes; in the magnetic semiconductors we have, in addition to this, a very sensitive and nonlinear reaction to *magnetic* forces. The *spin* of the electron, a hitherto neglected "handle" on the fundamental particle of solid state physics, is as important a parameter in the magnetic semiconductors as its charge. The two values of spin do not merely double the number of possibilities, but they create entirely new areas of exploration in physics and its applications.

The ferromagnetism and antiferromagnetism of the RE metals must be explained by some form of indirect exchange, generally via the conduction band electrons, between the localized $4f$ shells. The indirect mechanisms which couple the moments of $4f$ electrons in insulating compounds are generally complex and quite sensitive to many nonmagnetic phenomena; this is their great virtue. One has a large variety of RE compounds ranging from metals to insulators, in which various and variable indirect exchange mechanisms connect with varying degrees of leverage, the electronic to the magnetic properties. It will be seen that magnetic interactions change drastically even in isostructural compound series, as the coupling electrons change from a free metallic character into bound orbitals characteristic of valence states in insulating salts.

4. Technological utility of the coupling of magnetic and electronic states. We point here briefly to the technological utility of the remarkable properties which will result from the coupling of magnetic and electronic states of the solid, as our understanding develops over the next decades. We can anticipate devices based on the electron's spin which internally couple magnetic and electric circuits without use of inductive coupling. Local variations in the conduction carrier concentration (as produced by light radiation, electric fields, carrier injection, etc., in many known semiconductor devices) can result immediately in local variations in the magnetic order[9]. One can imagine the electronic state of certain semiconducting materials strongly influencing their magneto-optic properties, with applications to laser technology. A polarized laser beam could be modulated in passing through a transparent magnetic semiconductor: the magnetic state of the crystal and the rotation of the angle of polarization of the beam could be modulated at electronic frequencies, which have been remarkably high in recent semiconductor technology. The strong variations of electrical and optical properties with magnetic ordering observed in some materials suggest their use in many semiconductor devices where the variation of known properties by applied magnetic field provides desirable new parameters. In general, the technical application of magnetic semiconductors has, at present, the typical characteristics of a new field where the imperfection of scientific knowledge leaves still room for exciting speculations.

III. Organization of this article.

In the following article we have attempted to survey present day theory and experiments concerning magnetic semiconductors, with particular reference to the lanthanide Rare Earths.

To make each Chapter (denoted by letters A, B, etc.) reasonably self-contained, we have not hesitated to repeat the footnoted references wherever appropriate. Moreover, some of the subject material will be found repeated in several chapters, wherever it would help the discussion, or wherever a new application required it, or whenever it was deemed useful to discuss it from a new point of view. The broad topical index below shows this, but the reader should turn to the detailed Table of Contents for the descriptive title of each chapter and paragraph.

Atomic properties: Chapters B, E, F
Transport properties: C, H
Optical properties: F
Magnetic interactions: B, D, G

Of the various Chapters, C and D are mostly concerned with the solution of idealized theoretical models and the detailed consequences thereof. B and E—H are devoted to the experimental data, and to our present-day qualitative interpretation thereof. A number of Tables and a larger number of Figures have been used to convey a maximum of information on our subject, although neither the data nor the explanations are offered as complete or definitive; this demurral also applies to a number of original ideas and interpretations that were incorporated, without hesitation, into the text. The large variety of referenced material in both research and review literature should be sufficient to help the interested reader arrive at a fruitful understanding of the state of this new, lively, art of magnetic semiconductors.

B. Problems related to the electron band structure of magnetic materials.

5. The ambiguous character of electrons in magnetic semiconductors.

J. C. SLATER has reviewed the electronic structure of solids and the relevant mathematical methods in volume 19 of this handbook. He examines the two basic approaches

---

[9] S. METHFESSEL: Z. angew. Phys. **18**, 414 (1965); IEEE Transactions on Magnetics **MAG 1**, 144 (1965); U.S. Patent 3,271,709 (1963): B. V. KARPENKO, and A. A. BERDYSHEV: Sovj. phys. solid state **5**, 2494 (1964).

solid state physics takes in solving the 10^{23} body problem of electrons interacting with the atomic cores in the crystal: the single electron band model and the localized, atomic or Heitler-London model. In general, the band model gives a suitable description of those solid state properties which can be approximated as the response of the whole electron gas in the periodic crystal potential to applied external perturbations such as electrical fields, etc. The localized model, however, is more successful in describing those properties in which the interaction of the electrons with anion core and other core electrons has fundamental importance, e.g., for the description of chemical bonding mechanism in molecules, some optical properties and magnetism.

The specialized assumptions of both extreme approximations seriously restrict our attempts to find quantitative descriptions of observed material properties and therefore one of the central problems in modern solid state physics is to find more refined "intermediate-coupling" theories which can give a unified description of all aspects of electron behavior in crystals. Since the "magnetic state" with its strong electron-electron interactions and local atomic moments seems to be the logical counterpoint to the "metallic state" of free conduction electrons[10], vehement discussions about the electronic band structure of magnetic transition metals pervade the literature of the last decades.

Recently, C. Herring[11] gave a survey of the historical development and the present situation of the controversy between band theory and local description of ferromagnetism in transition metals. His conclusion is that *both* approaches are now sufficiently refined to cover all properties from magnetism to conductivity. Their discrepancies have more theoretical-mathematical significance than representing a physical ambiguity in the character of the magnetic electrons.

Other scientists[12] advocate the existence of two fundamentally different thermodynamic states of the electrons: the "collective" *vs* the localized state. Band-theory and the existence of a Fermi-surface characterizes the one state, the other is described by ligand field theory and superexchange. The transition between both states is expected to occur abruptly when the orbital overlap at neighboring atoms exceeds a certain critical value. It should be accompanied by the abrupt change in some crystalline, magnetic, electrical or optical properties.

There is no doubt that the magnetic $3d$ electrons contribute to the Fermi surface and transport phenomena in the elemental metals. The description of their magnetic properties by exchange interactions between itinerant band electrons is to be preferred to localized approaches. In insulating compounds with localized magnetic moments, on the other hand, the suitable zeroth approximation is the concept of a crystal lattice of free atoms with weak interactions, for the absence of electrical conductivity makes the principal shortcomings of this approach physically acceptable.

The materials we name *"magnetic semiconductors"* have a position intermediate between the magnetic metals and insulators, where the differences in the concept of band theory and local description imply immediate physical significance. The simple attempt to classify these materials according to their electrical and magnetic properties raises several very basic and so far not conclusively answered questions such as, for instance, "How does the character of localized elec-

---

[10] P. W. Anderson, in: Transition Metal Compounds (Ed. E. R. Schatz). New York: Gordon and Breach 1964.

[11] C. Herring: Magnetism IV. (Ed. G. T. Rado, and H. Suhl). New York: Academic Press 1966.

[12] For example: J. B. Goodenough; J. Appl. Phys. **37**, 1415 (1966); **39**, 403 (1968); Phys. Rev. **164**, 785 (1967); Czech. J. Phys. B **17**, 304 (1967); J. Friedel: J. Phys. Radium **23**, 501, 692 (1962); Bull. Soc. Chim. France 1965, p. 1186.

trons vary with decreasing distance to the neighbor atom?" "Is the transition from the localized magnetic states into the metallic band states a smooth function of the interatomic distances or does it take the form of an abrupt 'Mott transition' at a certain critical distance?" "Which parameters determine this critical distance?" "What is the relation between crystalline order and magnetic properties?" etc. The possibility of describing the electron behavior in crystals by Bloch states or localized atomic states now acquires a physical meaning beyond that of selecting the most suitable basis functions for the mathematical description of a complex phenomenon. Central problems are: how to introduce electron-electron interactions into the theoretical treatment of semiconductors to account for the occurrence of magnetic moments, how to describe the characteristics of electron transport phenomena involving localizing magnetic states and their dependence on magnetic order.

In the following paragraphs we review in a very qualitative way some typical features of the single electron band model and of the localized description in order to give a general background to our topic and to connect it with related parts in this handbook and other, more recent, review articles.

I. The one-electron band model of semiconductors.

6. Description of the one-electron band model. The single electron band theory, which has proved to be an extremely powerful tool for the description of charge transport and all other collective electron effects in metals and semiconductors, assumes that the binding of the outer electrons to the individual atom core is so weak that the whole crystal can be visualized as one big molecule. The energies of the nearly free electrons are obtained as one-electron-solutions from a Schroedinger equation which includes a periodic potential made up of the averaged charge distribution of the nuclei and electrons. The symmetry of the crystal lattice is represented by the periodicity of the perturbing potential and causes electrons with certain wavelengths to undergo Bragg scattering. Therefore, forbidden energy gaps appear. The gaps separate the otherwise continuous energy spectrum of free electrons into bands of allowed energies. The filled inner atomic shells are described by deep lying bands which are occupied with an equal number of spin up and spin down electrons. Such inner electrons do not contribute to transport and magnetic phenomena as long as the external perturbations such as applied fields, temperatures, etc., are not strong enough to excite electrons across the forbidden energy gap into unoccupied energy states[13]. Materials with filled bands which do not overlap empty bands are insulators if the forbidden energy gap above the last filled band is much larger than kT and, otherwise, semiconductors with temperature-dependent free carrier concentration. In materials with partially filled bands the electron occupation of the lowest energy states in the band defines a Fermi surface in the momentum space defined by the coordinates of the electron momentum $\boldsymbol{k}$. The continuum of empty states above the Fermi level enables the conduction electrons to respond to arbitrarily weak external perturbations without violating the Pauli principle and metallic properties are observed at 0 °K. The mobility $\mu = \frac{e\tau}{m^*}$ which measures how an electron e reacts to external perturbations during the scattering time τ depends on the ef-

---

[13] A. H. WILSON: Proc. Roy. Soc. (London) A **133**, 458 (1931). Several survey articles on mathematical methods of bandstructure calculations have appeared in the series "Solid State Physics" (Ed. F. SEITZ, D. TURNBULL). New York: Academic Press. F. HERMAN: Rev. Mod. Phys. **30**, 102 (1958).

fective mass tensor $m^*\left(\dfrac{1}{m^*}=\dfrac{1}{\hbar^2}\dfrac{d^2 E_k}{dk_i\,dk_j}\right)$, hence on the curvature of the energy-wavenumber relationship in different directions in the crystals and, therefore, on the position of the Fermi level and on the bandwidth. The stronger the interaction of the electron with the periodic core potential the wider are the forbidden gaps, and the narrower the bands of allowed energies. Therefore, the energy bands are quite narrow for inner electrons or in crystals with widely separated atoms and the resultant mobility for free carriers is expected to be small. For magnetic $3d$ electrons mobilities between 10 and 10^{-5} cm^2 V^{-1} sec^{-1} are found in a wide variety of magnetic compounds. For the much stronger localized $4f$ electrons even smaller mobilities are expected.

The strength and weakness of the band model lies in the possibility of concentrating the information about 10^{23} electron states into a single determinantal wave function consisting of Bloch states only or into a single function $E(\mathbf{k})$ giving the energy of a single electron as a function of its wave vector $\mathbf{k}$. Consequently, difficulties arise when interactions between electrons must be considered. The correlations in electron motion which decrease the interelectronic repulsion energy between electrons with antiparallel spin as well as the occurrence of localized magnetic moments by spin alignment of several electrons at a lattice site are beyond the basic purview of the free electron band model. The ferromagnetism of transition metals is only qualitatively described by differences in the Fermi surfaces for electrons with different spin directions. The occurrence of non-integral values of the magnetic moment per atom is taken as evidence for the validity of the band description for itinerant d electrons in metals, while the description of spin wave effects, critical fluctuations near the magnetic ordering temperature and similar effects are easiest described assuming magnetic electrons localized at lattice sites. Since the problem of exchange interactions between itinerant electrons in magnetic transition *metals* has been treated recently in depth by C. HERRING and N. F. MOTT[14] the present article can spare further elaboration on this topic.

7. Limitations of the one-electron band model. The difficulty of describing dynamic interactions between electrons in the band model leads to particularly drastic discrepancies between theoretical expectation and experimental reality in the case of magnetic compounds for which the separation between the magnetic atoms is somewhat larger than in the elemental transition metals. When the atoms are taken apart the band picture suggests relative increase of the core potential; the forbidden energy gaps should become broader and the conductivity finally will disappear because of vanishing bandwidth. At some intermediate atomic distances, when the band width is finite, thermal excitations of the ion lattice are comparable in energy with the band width and the Born-Oppenheimer approximation of the band theory (treating electron-phonon interactions as small perturbations) loses its validity. The Fermi surface of the conduction electrons may disappear abruptly due to vanishing screening effects and increasing importance of electron-phonon or electron-magnon interactions. Strong electron-phonon interactions in polar materials lead to a new charged entity of more localized character, the "polaron". The electrostatic coupling of the band electron to the polar vibrations of the dielectric medium surrounds every electron with a region of displaced ions and the resultant Coulomb forces bind the electron to the lattice polarization. The strength of the interaction between electrons and lattice can be

---

[14] C. HERRING: Vol. 4 of Magnetism (ed. G. T. RADO, and H. SUHL). New York: Academic Press 1966. N. F. MOTT: Advances in Phys. **13**, 325 (1964).

defined by a dimensionless coupling constant[15] which contains the effective mass of the electron in the stationary lattice, the polar phonon frequency and the static and dynamic dielectric constant of the lattice. As a result one obtains for the polaron a large, temperature-dependent effective mass. The electron-magnon interaction may lead to formation of magnetic polarons in analogy to the electrostatic polarons (Sections 16 to 18).

In many magnetic semiconductors, such as NiO, the insulating properties cannot be simply explained by assuming a sufficiently narrow bandwidth and small carrier mobility in the $3d$ band of the far-separated Ni ions, *because the creation of holes in the d band by suitable chemical doping results in electrical conductivity with small but finite values of the mobility*. The suggestion of SLATER[16] that the larger periodicity of the antiferromagnetic order could split the conduction band into filled and empty parts found several objections (Sect. 14). The most obvious one is that the disappearance of the antiferromagnetic order above the Néel temperature is not found to be accompanied with the expected transition from insulating behavior to metallic conductivity. These points are elaborated in Chapter C.

Obviously, it is difficult to escape the need to provide the band model with additional features which suppress the motion of electrons by sufficiently strong electron-electron repulsion. Since the localized Heitler-London approach contains the electron-electron repulsion as its most significant element from the beginning, the problem can be turned into the question of how the Bloch electrons transform into localized electrons when the interatomic distances are continuously increased in an expanding lattice. During the last two decades, this "Mott problem" has been investigated theoretically and experimentally from quite different aspects but has not yet found a satisfactory solution (Sects. 12 and 13). Much insight was derived not only from experiments with magnetic materials but also with the hydrogenic localized states of electrons around impurity atoms in wideband semiconductors[17]. The basic qualitative arguments have been given by MOTT in a number of papers[18] since 1949 and have been summarized in Sects. 12—14 below.

MOTT has argued for a sharp breakdown of the band model at some critical separation R_c since in a narrow band the correlation can be strong enough to prevent over distances comparable to the interatomic distances the screening of the Coulomb attraction $-e/\varkappa R$ ($\varkappa$=dielectric constant) between the moving electron and the hole it leaves behind. The resulting exciton state is electrically neutral and cannot transport current (Mott trapping). For an array of hydrogen atoms Mott estimates the critical distance to be roughly two times the equilibrium distance in a hydrogen molecule.

GOODENOUGH[19] has described in great detail the experimental indications for the existence of a critical distance. From the electrical and magnetic properties of a large number of transition metals and compounds he tries to define empirical

---

[15] H. FRÖHLICH, and N. F. MOTT: Proc. Roy. Soc. (London) A **171**, 496 (1939); H. FRÖHLICH: Advances in Phys. **3**, 325 (1954); H. FRÖHLICH, and G. L. SEWELL: Proc. Phys. Soc. (London) **74**, 643 (1959); H. HAKEN: Halbleiterprobleme II. (Ed. W. SCHOTTKY), (Braunschweig: Vieweg & Sohn 1955; G. L. SEWELL: Phil. Mag. **3**, 1361 (1958); J. YAMASHITA, and T. KUROSAWA: J. Phys. Soc. Japan **15**, 802 (1960); J. Phys. Chem. Solids **5**, 34 (1958); T. HOLSTEIN: Ann. Phys. **8**, 325, 343 (1959).

[16] J. C. SLATER: Phys. Rev. **82**, 538 (1951); this Handbook, Vol. 19/1 (1961).

[17] N. F. MOTT, and W. D. TWOSE: Advances in Phys. **10**, 107 (1961).

[18] N. F. MOTT: Can. J. Phys. **34**, 1356 (1956); Nuovo cimento, Suppl. **7**, 318 (1958); Phil. Mag. **6**, 1287 (1961), Proceedings of 1968 Conference on metal-insulator transitions, to appear in Rev. Mod. Phys., Oct. 1968.

[19] J. B. GOODENOUGH: Magnetism and the Chemical Bond. New York: Interscience Publ. 1963. J. Appl. Phys. **37**, 1415 (1966); **39**, 403 (1968); Phys. Rev. **164**, 785 (1967).

values for R_c which can be used as guidelines for the classification of materials and for the prediction of their properties. Typical numbers for the d-electrons in metals are: Ni: 3.06 Å, Pd: 3.94 Å, Pt: 4.42 Å, Gd: 5.8 Å, Eu: 6.5 Å. In non-metallic transition-element compounds such as the oxides, the critical radius is found to be about 7—10% smaller than in metals. The physical meaning of these numbers is, of course, very limited and purely phenomenological since an accurate value of such a critical radius, if it exists, can be expected to depend for each material on many individual parameters, such as crystal symmetry, details of the band structure, screening by other electrons, etc.

Recently, Hubbard[20] has made a theoretical attempt to bridge the gap between the weakly interacting electrons of the band model and the very strongly correlated electrons of the localized theory by treating electron correlations in narrow energy bands with the Green function technique. Since this work has been discussed already in Herring's book[21] we have only to mention the resulting conjecture that when the Coulomb repulsion between electrons at the same lattice site becomes equal to the bandwidth, an s and d band may split into 2 or 10 separate sub-bands, respectively. Since each sub-band contains only one electron state per atom, conductivity cannot occur for integral numbers of electrons per atom. The transition occurs abruptly at a certain value of the band width to interaction ratio, or has a smooth character, depending on details in the density of band states. Kemeny[22] applied equations which describe the normal to superconducting transition in the BCS theory to the metal-insulator transition replacing the coupled electron pairs of the superconductor by electron-hole pairs. Kohn[23] has shown that under certain conditions for large lattice parameters a system with one electron per atom can be an insulator instead of a metal because the many-electron wavefunctions fall off exponentially with the overlap integral when one coordinate is moved along a circular wire. With decreasing interatomic distance, the system may finally become a metal, but the nature of this transition cannot be described easily.

II. Localized electrons and magnetic moments.

8. Orbitals and magnetic moments in isolated atoms. At the opposite extreme of band theory stands the localized electron model, which allows the consideration of strong electron-electron interactions at the same atom, as is necessary for the interpretation of optical spectra, but cannot give sufficient emphasis to the periodic regularity of the crystal lattice, as is essential for the description of transport phenomena. It is generally assumed that electron core potential and repulsion between electrons on the same core are the most significant forces, and the derivation of electron energies in a crystal starts with the Schrödinger equation for the energy of a single electron in the spherical effective potential of the atom core, screened by the other electrons at the same atom. The value of the central effective field depends on the particular quantum state under consideration because screening of the nuclear charge depends on the character of the electron wavefunctions. A suitable potential is usually derived by the self-consistent field method of Hartree and Fock[24], which ensures that the screening properties

---

[20] J. Hubbard: Proc. Roy. Soc. (London) A **276**, 238; **277**, 237; **281**, 401 (1964); and in: Transition Metal Compounds (Ed. E. R. Schatz). New York: Gordon and Breach 1964.

[21] C. Herring: Magnetism, Vol. 4 (Ed. G. T. Rado, and H. Suhl). New York: Academic Press 1966.

[22] G. Kemeny: Phys. Letters **14**, 87 (1965).

[23] W. Kohn: Phys. Rev. **133**, A 171 (1964).

[24] D. R. Hartree: The Calculation of Atomic Structures. New York: John Wiley & Sons 1957; A. J. Freeman: Phys. Rev. **91**, 1410 (1953).

of the final wave functions are consistent with the assumed characteristics of the screened spherical potential of the core. Detailed descriptions of the methods for calculating the atomic energy levels are given in volumes 35 (1957) and 36 (1956) of this handbook, in the famous book of CONDON and SHORTLEY[25] on the theory of atomic spectra as well as in many textbooks on solid state physics, spectroscopy or physical chemistry. We shall make here only a few superficial remarks so far as is suitable to connect our topic with the well known theories of free atom energy levels and spectroscopy.

The exact quantum theory of the many-electron atom is an extremely complicated many body problem. The approximate one-electron theory considers the motion of individual electrons in an effective atomic field taken from the Coulomb attraction by the nucleus and a screening field which results from the motion of all other electrons averaged over long time intervals. In the simplest assumption the wavefunction of the electron is calculated as though the atomic field were spherically symmetric. The spherical nature of the central field problem permits us to define individual electron states by individual sets of four quantum numbers. The principal or total quantum number $n = 1, 2, \ldots \infty$ is related to the radius of the electron orbits. The orbital quantum number $l\,(0 \leq l \leq n-1)$ gives the magnitude of the orbital angular momentum in units of $\hbar$, while the magnetic quantum number $m\,(-l \leq m \leq l)$ is assigned to the component of the angular momentum in a certain direction. Each of these energy states can be occupied by two and only two electrons with different spin quantum numbers $s = \pm \tfrac{1}{2}$ (Pauli principle). When no external magnetic field acts on the free atom the $2 \times (2l+1)$ states with the same quantum numbers n and l are degenerate in energy and a configuration containing q noninteracting electrons is usually denoted as nl^q, wherein $l = 0, 1, 2, 3, \ldots$ is identified with the letters $s, p, d, \ldots$, respectively. Wave functions with odd or even l values are said to have ungerade (u) or gerade (g) parity corresponding to their response to inversion at the nucleus. In cartesian coordinates, the odd wave functions change sign if all coordinates are replaced by their negative values, the even wave functions do not. Odd and even wave functions are necessarily orthogonal to one another.

Another consequence of the spherical symmetry of the problem is that each individual wavefunction can be expanded in harmonic functions written in the form of a product $\psi_{n\,l\,m} = R_{n\,l}(r)\,\theta_{l\,m}(\vartheta)\,\varphi_m(\psi)$, where each term depends on the quantum numbers given in its lower index and on one of the polar coordinates r, ϑ or ψ only. The radial part R is related to the square root of the probability of finding an individual electron somewhere in a shell of width dr at the distance r from the center of the nucleus. The radial probability function $P(r) = r^2 \psi \psi^* \, dr$ of an electron in a given quantum state shows one major peak and a number of minor peaks. The major peak represents the most probable radial position of the electron and is roughly equal to the classical orbit radius for which the attractive Coulomb force balances the centrifugal force. Examples of radial probability functions are shown later in this article. The angular parts of the wave function are usually normalized to unity or 4π and the $\varphi_m(\psi)$ have the form $\varphi_0 = (2\pi)^{-\frac{1}{2}}$ and $\varphi_{\pm m} = e^{\pm i m \psi}$ while the $\theta_{l\,m}$ part is described by Legendre polynomials. The angular distribution of the probability density depends essentially on $\theta_{l\,m}$ only. The polar graphs of $\theta_{l\,m}$ where the $(2l+1)$ electron wavefunctions with the same quantum number l appear as certain arrangements of three dimensional lobes will be quite familiar to the reader. The mathematical presentations of

---

[25] E. U. CONDON, and G. H. SHORTLEY: The Theory of Atomic Spectra. Cambridge: Cambridge University Press 1957.

the s, p, d, and f orbits with $l=0, 1, 2, 3$, respectively, are given, for example, in the appendix III of PAULING's book[26], and in another article in this volume.

These "hydrogen-like" electron wavefunctions are only concerned with a single electron in each n, l-orbital. Relativistic effects which relate the electron's mass to its velocity, magnetic spin-orbit interactions and also electrostatic exchange effects are added as corrections to the calculated wavefunctions. The central field model works well for atoms with a single electron outside a closed shell, because the screening of the nuclear potential by the inner electrons can be adapted by adjustments in the self-consistent spherical field. Difficulties arise when the electrons can be distributed in more than one way over configurations with about the same energy so that resulting "configuration interactions" invalidate the one-electron approximation.

When there is more than one electron outside the closed shells, the Coulomb and exchange interactions between these outer electrons destroy the spherical field. *In reality*, electron-electron forces keep electrons in the same orbital further apart from one another than the one-electron theory is able to admit and the resulting *correlations* in electron motion are expected to lower the actual orbital energy compared to the calculated values. The correlation energy[27] is defined, somewhat tautologically, as the difference between the theoretically predicted and experimentally determined energy levels. Since the single-electron atomic wavefunctions are also used for setting up the periodic potential for band structure calculations, the correlations between core electrons will be treated rather poorly in the band model also. (There is the additional problem of correlations among the band electrons themselves, and between the band electrons and the core electrons.)

In the treatment of the free atom one can go one step further and introduce some of the electron-electron interactions in the same nl^q state, as far as the spin and orbit interactions are concerned, by defining additional quantum numbers by suitable combinations of spins and orbital moments. The electrostatic spin-spin, orbit-orbit interactions are described simply by scalar products which means that the coupling strength is assumed to be proportional to a coupling constant times the cosine of the angle between spins and orbits. The multiplet splitting of the nl^q states which results from the spin and orbit interactions cannot be treated in a general theory including all possible values for the coupling constants but only for certain appropriate coupling schemes. For not too large atoms, the Russell-Saunders coupling[28] has proved to be a good approximation for the interpretation of optical spectra by assignment of observed lines to calculated multiplet terms. The Russell-Saunders approximation is valid for atoms in which electrostatic exchange interactions and orbit-orbit interactions between different electrons in the same state are larger than the magnetic spin-orbit interactions at the same electron. If the spin-orbit interaction is neglected additional quantum numbers can be defined by the several ways in which the individual electron spin and orbital vectors can be summed up geometrically: $S = \sum_q s_q$, $L = \sum_q l_q$. The possible

---

[26] L. PAULING: The Nature of the Chemical Bond. Ithaca, N.Y.: Cornell University Press 1960. Graphical representation of f orbits are given by: H. G. FRIEDMAN, G. R. CHOPPIN, and D. G. FEUERBACHER: J. Chem. Educ. **41**, 354 (1964); C. BECKER: J. Chem. Educ. **41**, 358 (1964). — J. T. WABER, and J. E. HOCKETT: RE-research, Vol. 4 (Ed. L. R. EYRING), p. 281. New York: Gordon and Breach 1965.

[27] A summary on correlation effects can be found in C. K. JORGENSEN: Orbitals in Atoms and Molecules. London: Academic Press 1962; Solid State Phys. **13**, 376 (1962); P. O. LÖWDIN, and H. YOSHIZUMI: Advances in Chem. Phys. **2**, 207, 323 (1959).

[28] H. N. RUSSELL, and F. A. SAUNDERS: Astrophys. J. **61**, 38 (1925).

values for q electrons are $L = 0, 1, 2, \ldots$ and $S = q/2, q/2-1, \ldots 0$. If no magnetic field is applied L and S can be combined vectorially to another quantum number, the total angular momentum $J = L + S$ with the possible values $(L+S), (L+S-1), \ldots (L-S)$. Because of the neglect of the spin orbit interaction the energy levels with the same L and S values but different J values are degenerate and form a single "multiplet term" with the number of multiplet components $2S+1$ if $L \geq S$ or $2L+1$ if $L \leq S$. Each multiplet component is characterized by one of the magnetic quantum numbers $M = -J, -J+1, \ldots, J-1, J$, which give the quantized projections of the total angular momentum J along an axis of its motion. When small spin-orbit interactions are introduced as perturbations the multiplets split into their individual components which then show a separation proportional to the strength of the L, S interaction [29] and to the larger value of J (Landé interval rule). When the L, S interaction $\xi_{S,L}$ becomes negative for a more than half filled electron shell the multiplet changes from a "regular" to an "inverted" one, in which the components with the highest J values now have the lowest energy. (In an exactly half filled shell L is zero, and $J = S$.) In spectroscopy the multiplets are usually identified by the notation ^{2S+1}L where $2S+1$ indicates the maximum multiplicity of the multiplet and L is given by the capital letters S, P, D, F, G, etc., standing for $L = 0, 1, 2, 3, 4$, etc. Individual components are labelled with the individual J value as a lower index to the multiplet notation $^{2S+1}L_J$.

In heavy atoms, or for electrons excited into outer shells, the spin-orbit interactions (which depend on the nuclear charge as Z^4) can become comparable in strength with the electrostatic spin-spin, orbit-orbit interactions which increase approximately proportionally to Z only. The total multiplet width, (i.e., the energy difference $S(2L+1)\xi_{S,L}$ for $S \leq L$ and $L(2S+1)\xi_{S,L}$ for $L \leq S$ between the components $J = L+S$ and $J = L-S$), then becomes comparable to the separation between the multiplets and components with a certain J value mix increasingly with the corresponding J components in the next multiplet which has different L and S values. The transition from Russell-Saunders coupling to intermediate coupling schemes is indicated by deviations from the Landé interval rule and in the gyromagnetic ratio g. The extreme case of $j-j$ coupling where the magnetic spin-orbit coupling $\xi_{s,l}$ is larger than the electrostatic interaction between the orbits and the HEISENBERG's exchange forces between the spins is rarely obtained in reality. The calculation of multiplet terms has been discussed by EDLEN in Vol. 27 (1964) of this encyclopedia.

While spectroscopy is concerned with distances between energy levels in order to assign observed absorption or emission lines to electron transitions between them, magnetism is mainly interested in the electron configuration of the ground state. Therefore, only the interactions between equivalent electrons in the same n, l state usually have to be considered while spectroscopy often has to worry about coupling schemes between electrons in excited configurations including more than one electron shell. But problems of coupling between different electron shells occur also in magnetism of rare earth materials, where two magnetically potent, partially filled electron shells, the $4f$ and $5d$ levels, sometimes coexist. Since the Pauli principle demands $L = S = 0$ in filled shells, only partially filled shells contribute to the paramagnetic moment of the atom and the influence of electrons in closed shells has to be considered only in the form of small corrections for the diamagnetic properties to measured values of the paramagnetic

---

[29] Methods for calculating the L, S interaction from the l, s interaction of the individual electron have been developed by S. GOUDSMIT: Phys. Rev. **31**, 946 (1928).

moment. Optical, resonance and magnetic measurements complement one another as tools for the investigation of partially filled electron states in atoms, in solids.

The decision as to which of the possible Russell-Saunders multiplets ^{2S+1}L has the lowest energy requires knowledge about the correlation energy which has to be added to the result of the central field approximation. The necessary information is furnished primitively by Hund's rules[30] which assert empirically, that in a given electron configuration the Russell-Saunders multiplets with the highest value of the spin multiplicity $2S+1$ always have the lowest energy; within comparable terms with equal S values the one with the highest L value forms the ground multiplet with lowest energy; in the ground multiplet the component with $J=|L-S|$ is the lowest one (regular multiplet) for a less than half filled shell, for a more than half filled shell, however, the component $J=L+S$ forms the ground state (inverted multiplet).

For the interpretation of optical spectra it is not sufficient to know the ground state but the distances between multiplets must be derived also. Methods for calculating multiplets have been elaborated by Slater, Condon and Shortley and particularly by Racah[31]. Magnetic investigations need only information about the electron states of lowest energies. In order to assign a constant paramagnetic moment to each atom, one has first to make sure that the next multiplet component is separated far enough from the groundstate that mixing between these components during measurement can be neglected. If this is the case, the paramagnetic moment can be considered to be a property of the ground state alone and the paramagnetic susceptibility can be described by the quantized alignment of the total angular moment J of the ground state by the applied field H. The applied field, of course, must be so weak that it cannot break up the Russell-Saunders coupling by forcing L and S to align separately. The angular momentum J can then be visualized as precessing slowly on a cone about the direction of the applied field H while the vectors L and S precess rapidly around J. Since the *spherical* symmetry [which caused the $(2J+1)$ fold degeneracy of the multiplet components] is reduced to *axial* symmetry by the applied field each component should now be characterized by its projection $M=J, J-1, J-2, \ldots -J$ on the field direction. The induced magnetic moment, $g\beta M$, in the direction of the field is related to the projection M of the total angular momentum J by the Landé splitting factor

$$g = 1 + \frac{J(J+1) + S(S+1) - L(L+1)}{2J(J+1)}. \tag{8.1}$$

For a pure spin state with $L=0$ this yields $g=2$ and for pure orbital motion with $S=0$ one finds $g=1$. The value of the Bohr magneton

$$\beta = \frac{e\hbar}{2mc} = 0.92837 \times 10^{-20} \text{ erg/gauss}, \tag{8.2}$$

---

[30] F. Hund: Z. Physik **33**, 345 (1925); Linienspektren und periodisches System der Elemente, Berlin: Springer 1927; Vol. 36, pp. 1—108 (1956) in this Handbook.

[31] J. C. Slater: Quantum Theory of Atomic Structure, New York: McGraw-Hill 1960; E. U. Condon, and G. H. Shockley: Theory of Atomic Spectra, Cambridge: Cambridge University Press 1953; G. Racah: Phys. Rev. **62**, 438 (1942); **63**, 367 (1943); **76**, 1352 (1949); J. S. Griffiths: Theory of Transition Metal Ions, Cambridge: Cambridge University Press 1961; B. R. Judd: Operator Techniques in Atomic Spectroscopy, New York: McGraw-Hill 1963; B. G. Wybourne: Spectroscopic Properties of Rare Earths, New York: Interscience Publishers 1965. E. P. Wigner: Group Theory, New York: Academic Press 1959; A. R. Edmonds: Angular Momentum in Quantum Mechanics, Princeton, New Jersey: Princeton University Press 1957; U. Fano, and G. Racah: Irreducible Tensorial Sets, New York: Academic Press 1959; C. W. Nielson, and G. F. Koster: Spectroscopic Coefficients for the p^n, d^n and f^n Configurations, Cambridge, Massachusetts: MIT Press 1963.

which represents the magnetic moment of an orbit with the smallest possible orbital moment $\hbar$, corresponding to $l=1$, is used as a unit. Placed in the field H at a temperature T, the probability of finding the atom with a certain value of M can be described by a Boltzmann distribution and the Langevin-Debye theory derives for a gram mol of a gas containing $N=6.023\times 10^{23}$ atoms the susceptibility[32]

$$\chi = N\beta gJ \frac{B(\alpha)}{H}, \tag{8.3}$$

where $B(\alpha) = \frac{1}{J}\left[\left(J+\frac{1}{2}\right)\coth\left(J+\frac{1}{2}\right)\alpha - \frac{1}{2}\coth\frac{\alpha}{2}\right]$ is the Brillouin function with the argument $\alpha = \frac{1}{kT}\beta gH$. For applied fields $\beta H \gg kT$ (i.e., $\alpha \gg 1$) $B(\alpha)$ approaches unity and the completely aligned angular moments result in the magnetic saturation moment $N\beta gJ$. For small values of α, $gB(\alpha)$ becomes a linear function in α and the susceptibility can be described by the Curie law:

$$\chi = \frac{N}{3kT} g^2 \beta^2 J(J+1) = \frac{C}{T}. \tag{8.4}$$

The Curie law is convenient for deriving the Curie constant C experimentally from susceptibilities measured as functions of temperature by plotting $\frac{1}{\chi} = f(T)$. If the assumption $\alpha \ll 1$ is satisfied a straight line results which passes through the origin and has the slope $1/C$. After suitable correction for the diamagnetic contribution of the electrons in closed shells and for the van Vleck paramagnetism due to contributions from the next multiplet components above the ground state the effective magneton number per atom $P_{\text{eff}} = g[J(J+1)]^{\frac{1}{2}}$ can be derived from the Curie constant as

$$P_{\text{eff}} = \frac{1}{\beta}\left(\frac{3kC}{N}\right)^{\frac{1}{2}} = 2.839\, C^{\frac{1}{2}} \approx (8C)^{\frac{1}{2}} \tag{8.5}$$

and the validity of the coupling scheme defining the ground state can be checked. In 1925, several months before the introduction of quantum mechanics, F. Hund[33] calculated the effective paramagnetic moments for transition elements and lanthanide ions and found remarkably good agreement with susceptibility measurements on lanthanide sulphates, while his analysis failed for the $3d$ transition elements Sc to Ni. As we discuss later, the perturbation of the crystalline environment on the groundstate is very considerable for $3d$ electrons, but the $4f$ electrons of the lanthanides are so much better screened against environmental perturbations that many lanthanide compounds can be accurately described as an arrangement of noninteracting paramagnetic atoms.

When the separation of the multiplet components with different J values in the ground multiplet is comparable to kT the derivation of the magnetic susceptibility becomes very complicated since the well defined ground state for all atoms has to be replaced by the average of the excited J values with appropriate Boltzmann weights. The susceptibility cannot be described by the Curie law except by introducing, somewhat artificially, the concept of a temperature-dependent

---

[32] J. H. van Vleck: The Theory of Electric and Magnetic Susceptibilities, London: Oxford University Press 1932 (last reprint 1965); L. F. Bates: Modern Magnetism, Cambridge: Cambridge University Press 1961; F. Seitz: The Modern Theory of Solids, New York: McGraw-Hill 1940; Ch. Kittel: Introduction to Solid State Physics, New York: John Wiley & Sons 1953; E. C. Stoner: Magnetism and Matter, London: Methuen 1934; J. S. Smart: Effective Field Theories of Magnetism, Philadelphia: W. B. Saunders Co. 1966, includes very useful numerical tables for the Brillouin and related functions.

[33] F. Hund: Z. Physik **33**, 855 (1925).

effective moment. The classical examples for this case are the trivalent Eu and Sm ions.

In the approximation that kT is much larger than the multiplet width the mathematical treatment becomes simpler again, since the Boltzmann factors for all multiplet components may be considered to be equal. The susceptibility is now a property of the complete ground multiplet. The total angular momentum J is still a constant of motion but the quantization of L and S in the applied field (which made only a temperature independent contribution to the susceptibility in the first case of a wide multiplet) now becomes the characteristic feature. The Curie law for the susceptibility has the form

$$\chi = \frac{N}{3kT}\beta^2[4S(S+1)+L(L+1)] \tag{8.6}$$

with the effective atomic moment $P_{\text{eff}} = [L(L+1)+4S(S+1)]^{\frac{1}{2}}$. This case is realized in several $3d$ transition metal compounds and applies also to paramagnetic atoms in fields strong enough to break up the L, S coupling but not the stronger electrostatic interactions.

9. Magnetic moments of atoms in a solid material.

Arrangements of noninteracting atoms are, of course, very rarely found in reality. The significance of their theoretical treatment is not so much to explain the magnetic and optical behavior of paramagnetic vapors or gases of very low density than to give the zeroth approximation for the description of interacting atoms in crystals. The fact that the $4f$ shells in lanthanide compounds sometimes approach very closely the idealization of non-interacting atoms gives them a quite unique position in solid state physics, and these properties of the lanthanides will concern us extensively in this article.

Of all the unfilled shells only the d and f electrons can be expected to maintain any properties of the free atom state in a solid materials. Electrons in the outermost s or p shells have radial dimensions substantially larger than the d or f orbital extension and keep the ions apart. Therefore, the occurrence of localized magnetic moments in materials is closely related to the delayed building up of the d and f shells in the periodic system of the elements.

The most basic perturbation of the free ion states in a crystalline environment is the superposition of the Coulomb field from the neighboring ions, denoted *ligands* by chemists, on the spherical core potential of the single atom and the covalent overlap with ligand wave functions. The electrostatic ligand field or crystalline field reflects the symmetry of the crystal which is, of course, less than spherical and produces, therefore, a splitting of the multiplets by a Stark effect. For the calculation of the Stark field one uses the very simplified but successful approximation that the ligands are represented by point charges equal to the ion valency at the neighboring lattice sites determined by the crystal symmetry. Attempts to consider more realistic charge distributions include the overlap of electron clouds at neighboring atoms [34] or the shielding and polarization produced by closed shells [35] and become, in general, very complicated. For d electrons, the

---

[34] W. H. KLEINER: J. Chem. Phys. **20**, 1784 (1952); T. MORIYA, K. MOTIZUKI, J. KANOMORI, and T. NAGAMIYA: J. Phys. Soc. Japan **11**, 211 (1956); Y. TANABE, and S. SUGANO: J. Phys. Soc. Japan **9**, 766 (1964); **11**, 864 (1956); J. OWEN: Proc. Roy. Soc. (London) A **227**, 183 (1954); S. SUGANO, and R. G. SHULMAN: Phys. Rev. **130**, 506, 517 (1963); R. E. WATSON, and A. J. FREEMAN: Phys. Rev. **134**, A 1526 (1964).

[35] C. J. LENANDER, and E. Y. WONG: J. Chem. Phys. **38**, 2750 (1963); D. K. RAY: Proc. Phys. Soc. (London) **82**, 47 (1963); R. E. WATSON, and A. J. FREEMAN: Phys. Rev. **133**, A 1571 (1964); K. RAYNAK, and B. G. WYBOURNE: J. Chem. Phys. **41**, 565 (1964).

largest contribution to the ligand field comes from covalency effects[36]. The splitting parameter 10 Dq is determined mainly by the admixture parameter of d-orbitals to ligand orbitals and much less by the electrostatic Stark-field. The mixture between d orbitals and ligand s or p functions maintains the symmetry but increases the radial extension of the d-orbitals. Therefore, the symmetry considerations are the same as in the purely electrostatic theory but the magnitude of 10 Dq is different, and conservation of energy does not hold for the level splittings. For $4f$ electrons, however, the electrostatic model appears to be a good approximation. The first successful applications of the point charge model were made by KRAMERS and VAN VLECK[37] in order to explain the magnetic properties of lanthanide compounds. Mathematical methods for calculating the crystal field splittings have been discussed in Volume 28 (1957) of this handbook and in many other places[38]. In principle, group theoretical methods[39] are used to reduce the full rotation group of the spherical symmetry in the free ion to the lower symmetry group of the ligand field which can be expanded again in spherical harmonic functions. In order to find the number of levels produced by the Stark splitting one has to investigate how often the irreducible representations of the point group describing the crystal symmetry are contained in the representation of the energy level l of the free ion. (The number of rows in the representation gives the degeneracies of the levels.) The irreducible representations of the cubic group are usually given in the notation $\Gamma_1 - \Gamma_5$ of BETHE or in the notation A_1, A_2, E, T_1, T_2 of MULLIKEN[40] for states with integral angular momentum. One finds, as an example, that a p or P state remains threefold degenerate in a cubic field but splits in tetragonal symmetry into a twofold degenerate and a single state. Many-electron D states or a single d electron in a cubic field split from five fold degeneracy into a twofold $\Gamma_3 = E$ and a threefold $\Gamma_5 = T_2$ state. The fivefold degenerate f or F states decompose into a single A_2 state, a threefold T_1 and a threefold T_2 state. (For single electrons lower case letters are used instead of upper case.) The splitting can be visualized intuitively by considering the symmetry of the free ion wavefunctions with respect to the coordination of the negatively charged ligands in the crystal lattice. One finds that degenerate states consist of wavefunctions with similar orientations in the ligand coordination. Groups of wavefunctions which are more directed toward ligands have higher energy than other groups which extend more in the space between the ligands. The crystal field splitting affects directly only the orbital part of the wavefunctions. Spins are affected only indirectly via the spin-orbit interactions.

---

[36] S. SUGANO, and R. G. SHULMAN: Phys. Rev. **115**, 2 (1959); **129**, 2481 (1963); **130**, 506, 512, 517 (1963); P. W. ANDERSON: Solid State Physics **14**, 99 (1963); J. B. GOODENOUGH: Coll. Intern. C.N.R.S. No. 157, Oct. 1965.

[37] H. B. KRAMERS: Proc. Amsterdam Acad. Sci. **32**, 1176 (1929); J. H. VAN VLECK: Phys. Rev. **41**, 208 (1932); W. G. PENNY, and R. SCHLAPP: Phys. Rev. **41**, 194 (1932); **42**, 666 (1932).

[38] J. S. GRIFFITHS: The Theory of Transition Metal Ions, London: Cambridge University Press 1961; C. J. BALLHAUSEN: Introduction to Ligand Field Theory, New York: McGraw-Hill 1962; L. E. ORGEL: Ligand Field Theory, London: Methuen 1960; T. M. DUNN, D. S. McCLURE, and R. G. PEARSON: Some Aspects of Crystal Field Theory, New York: Harper & Row 1965; D. S. McCLURE: Solid State Physics **9**, 399 (1959); C. M. HERZFELD, and P. H. E. MEIJER: Solid State Physics **12**, 1 (1961); M. T. HUTCHINGS: Solid State Physics **16**, 277 (1964); W. Low: Solid State Physics, Suppl. 2 (1960); W. MOFFITT, and C. J. BALLHAUSEN: Ann. Rev. Phys. Chem. **7**, 107 (1956); B. BLEANEY, and K. W. H. STEVENS: Repts. Progr. in Phys. **16**, 108 (1953); K. W. H. STEVENS: Proc. Phys. Soc. (London) A **65**, 209 (1962); J. M. BAKER, B. BLEANEY, and W. HEYES: Proc. Roy. Soc. (London) A **247**, 141 (1958).

[39] H. A. BETHE: Ann. Physik **3**, 133 (1929).

[40] R. J. MULLIKEN: J. Chem. Phys. **3**, 375 (1935).

The crystal field theory is reliable only with respect to the multiplicities and relative order of the Stark levels which result from symmetry considerations. The strength of the Stark field can be calculated easily for point change configurations but the results deviate from experimental data because of covalency effects which are observable even for $4f$ electrons (Sect. 40). Therefore, crystal field parameters are usually derived experimentally from observed line splittings in optical spectra of crystals, from the position of Stark effect lines in the infrared spectrum, from g-factors obtained from microwave resonance experiments, from magnetic susceptibilities, specific heats, etc. For the $3d$-electrons in transition elements one finds crystal field splittings of the order of 10,000 cm^{-1}, depending, of course, on the character of the investigated crystals. The inner f orbitals of the rare earth elements, however, are less exposed to the crystal field and the splittings are only of the order of 100 to 1000 cm^{-1}.

The strength of the crystal field relative to the free ion potentials can be used to classify the magnetic materials in three major groups:

1) *Weak crystal fields* smaller than the spin-orbit interaction. In this case (which applies to the $4f$ electrons in most lanthanide compounds), the crystal field splitting is smaller than the distance between the multiplet components. When the splitting in the subcomponents is also small compared to kT the crystal field can be neglected to a good approximation.

2) *Medium fields* weaker than the electrostatic spin-spin and orbit-orbit interactions, which are of the order of $10^4 - 10^5$ cm^{-1}, but stronger than the spin-orbit coupling which is $\sim 10^2$ cm^{-1} in $3d$ elements and about 10^3 cm^{-1} in lanthanides and actinides. The multiplet structure remains generally unperturbed but the order of the components is derived by applying first the perturbation from the crystal field and then the spin-orbit splitting. The Russell-Saunders scheme and Hund's rules still apply but J ceases to be a good quantum number. Vector L precesses now around the axis of the crystal field giving the projections $M_L = L$, $L-1 \ldots -L$ with the same energy for equal but opposite projections. The spin, however, is unaffected by the crystal field and precesses about the magnetic field made up by the external field and the spin-orbit coupling. As a consequence of the change in the spatial quantization of L by the crystal field the orbital contribution to the magnetic moment is more or less quenched depending on the strength and asymmetry of the crystalline field. The effective magnetic moment has a value between $[4S(S+1)]^{\frac{1}{2}}$ and $[L(L+1)+4S(S+1)]^{\frac{1}{2}}$ as observed in most compounds of $3d$ transition elements and the experimental g factor deviates from the theoretical free ion value. Orbital quenching can also be a consequence of covalency or of the Jahn-Teller effect[41]. The latter can lift the degeneracy in the ground level by mechanical distortions inducing an asymmetry of the crystal field which is larger than the spin-orbit coupling.

3) *Strong fields* comparable to the electrostatic couplings. The crystal field splitting confuses the multiplet structure. The Russell-Saunders coupling scheme and Hund's rules break down and the energy levels are determined essentially by the crystal field splitting to which the exchange interactions and orbit-orbit couplings are perturbations. This case applies, in principle, to compounds in which $5d$ and $4d$ electrons contribute considerably to the chemical bonding mechanism, but is also found in $3d$ compounds, where the magnetic cations occur in an anomalously high valence state. Since the exchange interactions are unable to overcome the crystal field splitting, the ions are then in a low-spin state with a magnetic moment smaller than in the free ion.

---

[41] H. A. Jahn, and E. Teller: Proc. Roy. Soc. (London) A **161**, 220 (1937); A **164**, 117 (1937).

10. Interactions between magnetic moments in a solid. Magnetic interactions between neighboring atoms manifest themselves in the occurence of a long range spin order below a certain Néel or Curie temperature. In the paramagnetic region (at temperatures high enough above the ordering temperature) the Curie law (8.4) is replaced by the Curie-Weiss law

$$\chi = \frac{C}{T-\Theta}. \tag{10.1}$$

The parameter

$$\Theta = \frac{2}{3k} \sum_i z_i J_i S(S+1) \tag{10.2}$$

is the Curie-Weiss temperature which sums the magnetic interactions J_i of each atom with its z_i neighbors.

The only magnetic interactions which are admitted in the model of a crystal discussed so far are magnetic dipole interactions which can be expected to align the atomic moments against the thermal agitation only at very low temperatures below 1 °K. In order to explain the interatomic exchange interactions responsible for the Weiss molecular field one has to leave the fiction of a pure ionic crystal and to include the overlap of electron wavefunctions and covalent binding between neighboring atoms. The orbital overlap occuring between adjacent atoms is the concern of molecular theories discussed in Volume 37 of this encyclopedia. The molecular orbital theory[42] approximates the energy states of symmetry suitable to the molecule by Linear Combination of the outer Atomic Orbitals (LCAO-method). Differences and sums of orbitals are called antibonding or bonding orbitals and correspond to electron concentrations either at the individual atoms or between them. When the number of atoms contained in the molecule becomes larger and finally equal to the whole crystal more and more orbitals become included in the bonding and antibonding states which finally merge to the valence and conduction bands of the tight binding model of the crystal. The molecular orbitals of all atoms are then filled by the outer electrons according to the Pauli principle beginning with two antiparallel spins in the lowest orbit, etc. In this procedure the electron-electron interactions are not better treated than in the band theory in zeroth order. In the Heitler-London approach, however, the electron-electron interactions of the free-ion description can be maintained and the interatomic overlap is introduced by configuration interactions (Vol. 19) which add excited states in which electrons are transferred to other atoms in the molecule. When one speaks about the localized electron picture he usually refers to the Heitler-London model. The interactions between electrons in the same quantum state results in the multiplet structure which connects readily with the observation of localized atomic moments in crystals. Therefore, the localized model has been favored for the theoretical treatment of magnetic and optical properties of non-metals. The observation of integral numbers of magnetic electrons per atom can be taken as assurance for the validity of this approach.

Typical difficulties, arise in the description of collective phenomena such as long range magnetic ordering, electron transport mechanism, etc., and one can easily run into very complicated or insoluble many-body problems. The understanding of the origin of interatomic exchange forces which produce the

---

[42] F. Hund: Z. Physik **51**, 759 (1928); **63**, 719 (1930); R. S. Mulliken: Phys. Rev. **32**, 186 (1928); **33**, 730 (1928); L. Pauling: The Nature of the Chemical Bond, New York: Cornell University Press 1960; K. W. H. Stevens: Proc. Roy. Soc. London) A **219**, 542 (1953); J. Owen: Proc. Roy. Soc. (London) A **227**, 183 (1955); S. Sugano, and R. G. Shulman: Phys. Rev. **130**, 506, 517 (1963). References to summarizing literature are given in subdivision F II.

ferromagnetic, antiferromagnetic or ferrimagnetic alignment of localized magnetic moments throughout a crystal is very unsatisfactory. The problem is usually attacked by effective field theories[43] which permit the reduction of the problem to the values of a few effective exchange parameters describing the magnetic interactions between nearest and next-nearest-neighbor ions. The simple case of direct exchange interactions between neighbor atoms by overlap of their corresponding electron wavefunctions, which has been suggested by HEISENBERG[44], is, however, very rarely found in reality and pertains only to spin coupling between atoms which are far enough separated to have almost free atom characteristics. A most recent survey on the theory of direct exchange between well separated atoms has been given by HERRING[45].

The concept of *indirect exchange*[46] which is especially applicable to the magnetic coupling between the very localized $4f$ electron spins in rare earth metals[47] uses the local spin polarization of the conduction electrons, induced by the intra-atomic exchange interaction to electrons in the $4f$ shell, to explain long range "effective" interactions between localized magnetic moments (see Chapter D).

The model of *superexchange*[48] pertains to exchange coupling between localized magnetic moments which are surrounded by nonmagnetic ligands. It derives from the variation of the covalent chemical bonding energy with relative spin orientation of the valence electrons. The ground state energy of the lattive can be lowered by mixing the magnetic antibonding electron states with excited states which transfer spin to the next magnetic cation. The transfer can be direct or indirect via ligands. It costs intra-atomic exchange at the cation and Coulomb energy (in materials where all cations have the same electron configuration). The resulting superexchange depends on the nature of the cation wave functions in the groundstate and excited state, on the coordination and distance of the neighboring ions. Consideration of these factors leads to the empirical Goodenough-Kanamori rules (Sect. 30). Superexchange does not produce electrical conductivity. However, transfer between identical cations of *different* valence produces electrical conductivity and *double exchange*.

Charge transport phenomena are realized in the localized electron model as local fluctuations in ion valency. In the case of more separated ions in a crystal lattice the overlap between wavefunctions at neighboring ions can be comparable to or smaller than the interelectronic repulsion between electrons at the same ion and the crystal will be an insulator even with partially filled electron levels. Conductivity by free carriers requires the presence of atoms with different chemical valence in the crystal and can be visualized as a diffusion process by hopping of electrons from lower valent ions to neighbors with higher valence. Thermal excitations of the crystal

---

[43] The most recent reviews have been given by J. S. SMART: Effective Field Theories of Magnetism, Philadelphia: W. B. Saunders Co. 1966, and in: Magnetism, Vol. III (Ed. G. T. RADO, and H. SUHL). New York: Academic Press 1963.

[44] W. HEISENBERG: Z. Physik 49, 619 (1928).

[45] C. HERRING: Magnetism, Vol. II B (Ed. G. T. RADO, and H. SUHL). New York: Academic Press 1966.

[46] C. ZENER: Phys. Rev. 81, 440 (1951); 83, 299 (1951); S.V. VONSOVSKI: Zhur. Eksptl. i Theoret. Fix. 16, 981 (1946); 24, 419 (1953); J. Phys. (U.S.S.R.) 10, 468 (1946); T. KASUYA: Progr. Theoret. Phys. (Kyoto) 16, 45 (1956); K. YOSHIDA: Phys. Rev. 106, 893 (1957); M. A. RUDERMAN, and C. KITTEL: Phys. Rev. 96, 99 (1954); N. BLOEMBERGEN, and T. J. ROWLAND: Phys. Rev. 97, 1679 (1955).

[47] Review article: T. KASUYA: Magnetism, Vol. II B (Ed. G. T. RADO, and H. SUHL). New York: Academic Press 1966; P. G. DEGENNES: J. Phys. Radium 23, 510 (1962).

[48] H. A. KRAMERS: Physica 1, 182 (1934); P. W. ANDERSON: Phys. Rev. 79, 350 (1950); 115, 2 (1959); Review articles: P. W. ANDERSON, in: Magnetism, Vol. I (Ed. G. T. RADO, and H. SUHL), New York: Academic Press 1963; P. W. ANDERSON: Solid State Physics 14, 99 (1963). Article by VONSOVSKY and KARPENKO in this volume.

lattice are, here, in obvious contrast to the single electron band model, very strong perturbations to electron motion. As a consequence, one observes small electron mobilities of the order of 10^{-1} and 10^{-5} cm^2/V·sec, which increase exponentially with temperature,

$$\mu = (eD_0/kT)\exp(-q/kT) \tag{10.3}$$

and depend on a "diffusion constant" D_0 which is given by the concentration and geometrical distribution of the ions with different valency and a jump frequency in the order of 10^{11} to 10^{14} sec^{-1}. The activation energy q is related to factors which stabilize the electron at the site of the lower valent ion. This stabilization has no direct connection to the insufficient orbital overlap with the higher valent neighbor cation, because the electron can always tunnel through the narrow potential barrier. It is rather a consequence of the Coulomb attraction between the electrons and holes (Mott localization) or is related to the elastic relaxation of the crystal around the cation with abnormal valence (Landau trapping[49]) or to local fluctuations in the chemical composition. Of special interest with respect to observable relations between electrical and magnetic properties is the influence of the localized ion spins' direction on the electron transfer between neighboring ions of different valency. One has now to consider also the modifications of the electron-electron repulsion at the same ion by different spin and orbital alignments which, in the free ion, lead to the multiplet structure of optical spectra. The electron prefers then to hop to such higher valent neighbors to the extent that its own spin direction agrees with the intra-atomic coupling conditions for the lowest multiplet configuration of the next lower ionization state. The electron hopping between ions of suitable spin directions induces a spin coupling by carrier motion which is always ferromagnetic and has been named *double exchange*[50]. The theory is discussed in Chapters C and D.

BLOEMBERGEN and ROWLAND[51] have suggested a hybrid model in which the atomic magnetic moments are accepted as a property of the ion cores just as the nuclear magnetic moments, but in which the coupling electrons are described by the band model. By extending the concept of indirect exchange via conduction electrons in metals to valence electrons in insulators they can explain anomalous nuclear resonance linewidths in Tl_2O_3 by indirect hyperfine interaction via the filled valence band. The Bloembergen-Rowland coupling should reduce to the superexchange mechanism when the coupled outer electrons are described by molecular orbital approach instead of Bloch functions. Indeed, DE GRAF and STRÄSSLER[52] could demonstrate that the antiferromagnetic interactions in insulating salts (such as MnF_2, CoF_2, NiO, FeO, etc.) which are usually considered to be typical examples of superexchange via intermediate anions, can be described also by indirect exchange via valence electrons if suitable assumptions are made about the nature of the electron band structure. KARPENKO and BERDYSHEV[53] superimpose on the indirect exchange through the filled valence band the exchange effects due to the free conduction carriers when intrinsic conductivity is produced by thermal excitation. The resulting exchange interactions become

---

[49] L. D. LANDAU: Phys. Z. S. Socoj. **3**, 664 (1933); N. F. MOTT, and P. W. GURNEY: Electronic Processes in Ionic Compounds. London: Oxford University Press 1940.

[50] C. ZENER, and R. R. HEIKES: Conf. Magnetism and Magnetic Materials, Boston, 1956, p. 216, AIEE, New York, Phys. Rev. **82**, 403 (1951); P. W. ANDERSON and H. HASEGAWA: Phys. Rev. **100**, 675 (1955); P. G. DE GENNES: Phys. Rev. **118**, 141 (1960).

[51] N. BLOEMBERGEN, and T. J. ROWLAND: Phys. Rev. 97, 1679 (1955).

[52] A. M. DE GRAF, and S. STRÄSSLER: Phys. Kond. Materie **1**, 13 (1963).

[53] B. V. KARPENKO i A. A. BERDYSHEV: Soviet Phys. — Solid State **5**, 2494 (1964); A. A. BERDYSHEV: Soviet Phys. — Solid State **8**, 1104 (1966).

temperature dependent by the temperature dependence of the free carrier concentration. In principle, every variation of the free carrier concentration induced by impurity atoms, optical absorption, carrier injection, strong electric fields, etc., could potentially vary the Curie temperature of a magnetic semiconductor and provide, therefore, an entirely new means of controlling the magnetization state of certain materials in technical applications [54].

C. Theory of electron transport properties.

11. Survey. This section is primarily concerned with the added electrons or holes that transform an insulator into a semiconductor. While it is always true that the band structure and the scattering mechanisms, taken together, determine the electrical and thermal transport properties of the semiconductor, in magnetic substances there is an added important ingredient absent from nonmagnetic materials: the magnitude and orientation of the atomic spins greatly influence both the band structure and the scattering of carriers through the local exchange coupling. It is this feature which must be thoroughly understood if one wishes, someday, to extract maximum advantage from the magnetic semiconductors. But to understand it one must almost immediately become involved with the statistical mechanics of magnetic systems, a subject which is not yet quantitatively well understood even in idealized systems, and which in the present context, assumes formidable difficulties. The approximations to which one then resorts are often of the crudest type: molecular field theory and the like. Hopefully, as the understanding of magnetic semiconductors grows, so will the theory of cooperative phenomena and statistical mechanics, so that perhaps the future will see one of these phenomena being used to study the other. For the present, one must assume that detailed agreement between theory and experiment, when it occurs, is partly fortuitous and one must be satisfied with a gross explanation of the observed phenomena. The principal difficulties will now be outlined.

First and foremost, *the very existence of magnetism in a solid is tantamount to a breakdown of the conventional electron energy-band picture.* The magnetism results from strong Coulomb and exchange forces such that spatial and spin correlations among the electrons cannot be neglected; the band picture, on the other hand, is essentially based on the one-electron approximation, which in turn is founded on the neglect of dynamic correlations among the particles. In the conventional band picture KRAMERS' degeneracy means that for every state of a given energy for an electron of spin up, there is an equivalent state of equal energy for an electron of spin down; this ensures that the ground state spin is negligibly small, whereas in ferromagnetic solids it is macroscopically large! Thus straightforward use of BLOCH's theorem and of band structure calculations does not result in a sensible explanation of the transport properties of magnetic semiconductors, and, among other topics, this chapter will be concerned with the modifications that are required for these ordinary concepts to make sense in the new context.

Aside from modifications of the transport coefficients, and from detailed considerations such as whether the carriers undergo wave-mechanical motion or are "hopping" from site to site, there is the fundamental problem of explaining the very semiconducting (or insulating) nature of the solids under consideration. One of the great successes of the one-electron band theory of solids is the prediction that those substances with a sufficient number of electrons to fill the last zone are insulators; with added impurities, optical or thermal excitations, or in-

---

[54] S. METHFESSEL: Z. Physik **18**, 414 (1965), U.S. Patent 3 271 709 (1963).

jected carriers, these become the semiconductors. The band theory has no such predictive success when it comes to magnetic materials. In the case of the magnetic semiconductors, the size of the zone may change the degree of magnetic order, as affected by thermal fluctuations or by external magnetic fields. In the (weakly magnetic) vanadium oxides a transition from the metallic to the insulating state *is* observed, which, because of the intrinsic interest of this phenomenon, will be discussed below. Nevertheless, this is an outstanding *exception*; most magnetic materials maintain their metallic or insulating nature over the entire range of temperatures where the zone structure undergoes basic modifications, which implies a significant breakdown of the conventional band theory. It is the electron correlations and not the periodic potential which are responsible for the state of the solid, and the conventional theory must be adapted to take this into account.

The discussion will be organized as follows: we start by reviewing the fundamental problem of metal *vs.* insulator and of the "Mott transition" from the one to the other. It will then become apparent why this transition need not happen merely because some magnetic order evanesces above T_c. If the insulating state is stable, then what modification of band theory can be used to study it? This question is studied next, followed by considerations about various modes of transport and scattering mechanisms, with some reference to critical phenomena at the Curie or Néel temperature.

I. Magnetic insulators, the Mott transition, and the two types of energy gap.

12. Qualitative explanation of the Mott transition. There are two distinct sorts of energy gaps in a magnetic insulator: the first is the ordinary gap separating valence and conduction bands, which originates in the periodic potential and ensures that weak forces accelerate only *added* carriers (such as the few electrons in the conduction bands or holes in the valence bands as may be present owing to spontaneous thermal fluctuations, or owing to optical generation or injection at a surface) but prohibits the vast number of valence electrons from participation in the transport processes. The second gap is owed mainly to the ionization energy of the d-shells (or f-shells). In order for a magnetic electron to participate in transport phenomena it must leave its "home atom" positively ionized[55], at the cost of an energy U; however, the delocalization lowers its energy, say, by an amount E_c. Thus if U exceeds E_c there is a net energy gap $E_0 \equiv U - E_c$ which, generally speaking, insures that f-shell electrons do not form ordinary one-electron energy bands, whereas sometimes for d-shell electrons U may be sufficiently small that the energy gap E_0 in fact disappears. If E_0 vanishes, the d-electrons participate in conduction phenomena and may impart metallic character to the solid. We therefore turn our attention to this second type of gap (that due to the Coulomb repulsion among the magnetic electrons) and study under what conditions it is positive and so maintains the insulating or semiconducting properties of the solid, and under what conditions it vanishes, allowing metallic conductivity, together with a lowering, or total disappearance, of the atomic magnetic moments.

Such a study has historical antecedents in the work by WIGNER[56] on the hypothetical low-density electron lattice which now bears his name. MOTT[57] then sug-

---
[55] Cf. discussion preceding Eq. (23.5).
[56] E. WIGNER: Trans. Faraday Soc. **34**, 678 (1938).
[57] N. F. MOTT: Proc. Phys. Soc. (London) A **62**, 416 (1949); Nuovo cimento, Suppl. **7**, 312 (1958), and Phil. Mag. **6**, 287 (1961).

gested a mechanism to explain the nature of transition to the insulating state. The following rather qualitative considerations may serve to explain how a substance, which is an insulator at low temperature, may become metallic as the temperature is raised, the "Mott transition". If, in a magnetic insulator, we have a bound hole-electron pair (the "hole" is the missing electron of the KRAMERS' doublet) in its state of lowest energy — an exciton — the binding energy is of order $U = e^2/2\varkappa a_H$, where

$$a_H = \hbar^2 \varkappa / m^* e^2 \tag{12.1}$$

is the radius of a hydrogenic orbit in a medium of dielectric constant $\varkappa$. This pair, being neutral, does not carry a current at $T=0$. The ground state of the crystal includes a large number of pairs of this type, but if the parameter a_H does not exceed an interatomic distance, the overlap among pairs is small and the ground state is surely of the insulating type. If, however, a considerable concentration of free carriers, n, is created it is possible for the pairs to become dissociated and an "avalanche" occurs to the metallic state. The reason is that in the presence of free carriers the Coulomb attraction (between hole and electron) becomes screened and may not, in fact, have a bound state. As a crude estimate of this effect, consider the Thomas-Fermi screening length r_{TF}:

$$r_{TF}^2 = a_H/4 n^{\frac{1}{3}} \tag{12.2}$$

in terms of which the screened Coulomb interaction is:

$$-(e^2/\varkappa r)\exp - r/r_{TF}. \tag{12.3}$$

Using (12.2) and a variational estimate for the existence of a bound state, one readily obtains the following criterion:

$$\begin{aligned} \text{insulator:} \quad & n^{\frac{1}{3}} a_H < \tfrac{1}{4} \\ \text{metal:} \quad & n^{\frac{1}{3}} a_H > \tfrac{1}{4}. \end{aligned} \tag{12.4}$$

[However the number, $\tfrac{1}{4}$, is not known with any certainty. A similar estimate may be obtained more simply by postulating that when the overlap between electrons and bound excitons exceeds a certain fraction, say p, the latter may dissociate allowing a certain form of conductivity to set in. This leads to the same criterion as above, with $p^{\frac{1}{3}}$ replacing $\tfrac{1}{4}$, on the right-hand side. In any event, this fundamental question has not yet had a quantitative solution, but the general form of MOTT's result, above, must be correct.] The determination of n, the number of carriers participating in the screening process, may be accomplished as follows[58,59]: assume that each time an exciton is dissociated the over-all correlation energy, hence the Coulomb gap, decreases on the average by a fixed fraction. We assume the following linear relation between the gap energy E_g and n:

$$E_g = E_0(1 - 2n), \tag{12.5}$$

where $E_0 = U - E_c$ is the $T=0$ energy gap introduced. U is the Coulomb binding energy of the exciton given just prior to Eq. (12.1), and E_c is related to some overlap integral and to the number of nearest neighbor atoms. The number of

---

[58] D. CALECKI: Modèle Théorique pour un Semiconducteur Ferromagnétique, Proc. 7th Intl. Conf. on Physics of Semiconductors, Braunschweig: F. Vieweg & Sohn, and Paris: Dunod Editeur 1964; p. 113.

[59] D. ADLER, and H. BROOKS: Theory of Semiconductor-to-Metal Transitions, Phys. Rev. **155**, 826 (1967); J. FEINLEIB, and W. PAUL: Semiconductor to Metal Transition in V_2O_3, Phys. Rev. **155**, 841 (1967); D. ADLER, J. FEINLEIB, H. BROOKS, and W. PAUL: Phys. Rev. **155**, 851 (1967).

electrons is determined by the Fermi function,

$$n = \frac{1}{e^{\frac{1}{2}E_g/kT}+1} \tag{12.6}$$

with the Fermi level at mid-gap. The above simultaneous equations for n and E_g may be solved, with the easiest procedure being to eliminate n so as to obtain the transcendental equation:

$$E_g = E_0 \tanh \tfrac{1}{4} E_g/kT. \tag{12.7}$$

[This is precisely the constitutive equation of molecular field theory for ferromagnets of spin $\tfrac{1}{2}$, with E_g playing the role of the magnetization[60].] A nonlinear dependence of E_g on n, or a linear dependence with a coefficient different from 2 may give somewhat different results, although the qualitative features which we now outline should persist. At high temperatures $kT > \tfrac{1}{4} E_0$ only the "trivial" solution $E_g = 0$ exists, and thus only the metallic phase is stable. Below the Mott transition temperature, which the present formula puts at $kT_c = \tfrac{1}{4} E_0$, a nontrivial solution develops which is thermodynamically more stable than the metallic one. It corresponds to a nonvanishing gap E_g which increases approximately as $(T_c - T)^{\frac{1}{2}}$ as the temperature is lowered below T_c, to a maximum value of E_0 at $T = 0$. Whether the transition is continuous (second-order) as in the present calculation or discontinuous (first-order) as postulated by MOTT, depends on many details near T which we have so far neglected. The most important of these may be related to lattice deformations and strain[61], and of course the nature of critical fluctuations must also be known before a definitive explanation can be proposed.

The following simple argument helps explain the instability of tightly-bound electrons in a half-filled band against either strain or antiferromagnetic ordering. Let us consider the sc or bcc structures, for which the Bloch energies are, respectively:

$$\mathfrak{E}(k) = -W(\cos k_x a + \cos k_y a + \cos k_z a) \quad \text{(sc)} \tag{12.8}$$

$$\mathfrak{E}(k) = -W(\cos k_x a \cos k_y a \cos k_z a) \quad \text{(bcc)} \tag{12.9}$$

For both there exists the important symmetry.

$$\mathfrak{E}(k \pm Q) = -\mathfrak{E}(k), \tag{12.10}$$

where Q is a special wavevector,

$$Q = \frac{\pi}{a}(1, 1, 1). \tag{12.11}$$

Thus for a half-filled band, any point on the Fermi surface $\mathfrak{E}(k) = 0$ is mapped by a translation of Q onto another point of the Fremi surface. (Note also that $2Q$ is a vector of the reciprocal lattice.)

The frequency- and wavevector-dependent susceptibility is given by

$$\chi(\omega, k) = \sum_{k_1, k_2} \frac{f(k_2)[1 - f(k_1)]\delta(k - k_1 + k_2)}{\omega + \mathfrak{E}(k_1) - \mathfrak{E}(k_2)}, \tag{12.12}$$

where $f(k)$ is the Fermi function. The substitution $k_1 \to k_1 + Q$ changes the sign of $\mathfrak{E}(k_1)$ according to (12.10), and changes $f_1 \to 1 - f_1$. The susceptibility then

---

[60] See, e.g., D. MATTIS: The Theory of Magnetism, p. 228. New York: Harper & Row 1965.

[61] D. ADLER: "Insulating and Metallic States in Transition Metal Oxides", a review appearing in Solid State Physics **21**, (1968); J. B. GOODENOUGH: Czech. J. Phys. B **17**, 304 (1967).

becomes:

$$\chi(\omega, k) = \sum_{k_1, k_2} \frac{f(k_1) f(k_2) \delta(k - Q - k_1 + k_2)}{\omega - \mathfrak{E}(k_1) - \mathfrak{E}(k_2)}. \tag{12.13}$$

At $k = Q$ momentum conservation requires $k_1 = k_2$. Introducing the density of states function $N(\mathfrak{E})$ we find:

$$\chi(\omega, Q) = \int d\mathfrak{E} \, N(\mathfrak{E}) [f(\mathfrak{E})]^2 [\omega - 2\mathfrak{E}]^{-1}. \tag{12.14}$$

At low temperature and frequencies, this expression diverges approximately as $\log(|\omega| + kT)^{-1}$. This means that for $kT \ll W$ the assumed band structure is unstable against lattice deformations to a new structure having two atoms in a unit cell, or against antiferromagnetism. In either case a gap develops, and we therefore have a Mott transition at arbitrarily small coupling constant. If the gap develops because of lattice strain, then the instability toward antiferromagnetism may be eliminated; explaining the absence of anitferromagnetism observed in some of the transition oxides. Conversely, if the gap develops because of antiferromagnetic ordering, then the instability against lattice deformation may disappear.

In either case, the instability disappears as the temperature is raised, because of the Fermi functions in (12.14). It may be seen that at sufficiently high temperatures the susceptibility ceases even to have a maximum at $k = Q$. Thus the metallic state may be stable at high temperature, while at low temperature some sort of insulating state with an energy gap is surely stable, with or without magnetic long-range ordering, depending on the relative magnitudes of the electron-phonon and electron-electron interactions.

13. Examples of Mott insulators. Nature provides examples which seemingly can be explained by this sort of effect, in the form of some transition metal oxides such as V_2O_3 and TiO (Sect. 79). These materials are insulators below the Néel temperature, and show a large discontinuity in the resistance at this critical temperature, becoming metallic above it. Morin[62] proposed that the $3d$ band was split by crystal fields into $d\varepsilon$ and $d\gamma$ sub-bands, and that the $d\varepsilon$ band was further split by the Coulomb repulsion energies into two collective bands separated by an energy gap of the second type. The measured critical temperatures are related to the experimental gap at $T = 0$ as[61]

$$kT_c = E_0/8.3 \quad (V_2O_3) \tag{13.1}$$

and

$$kT_c = E_0/1.2 \quad (Ti_2O_3) \tag{13.2}$$

values which bracket the hypothetical value $E_0/4$ derived above. A plot of the experimental resistivities showing the discontinuities at the critical temperatures is given in Fig. 2. Morin[63] estimates the magnitude of the jump in resistivity by using the usual semiconductor formula, which gives for the number of carriers Eq. (12.6) in the limit $kT \ll E_g$:

$$n \sim \exp(-\tfrac{1}{2} E_g/kT). \tag{13.3}$$

---

[62] F. J. Morin: Phys. Rev. Letters **3**, 34 (1959). The band structure of TiO has been calculated by V. Ergs, and A. Switendick: Phys. Rev. **137**, A 1927 (1965) and agrees with metallic behavior.

[63] F. J. Morin: Oxides of the $3d$ Transition Metals, Chap. 14 in Semiconductors (Ed. N. B. Hannay). New York: Reinhold 1959. More recent measurements indicate that VO does not show a semiconductor-metal transition: see J. B. Goodenough: J. Appl. Phys. **39**, 409 (1968).

He then obtains E_g from the slope of the resistivity curve below the critical temperature, assuming a full complement of carriers above the critical temperature and an approximately constant mobility. For V_2O_3 this formula predicts a jump of approximately 6 orders of magnitude for the resistivity, in good agreement with the observed value.

The transition oxides also provide proof that strong correlations do not necessarily result in remarkable magnetic properties; indeed, nowhere in the above discussion did the value of the atomic magnetic moments appear as an important parameter. Experimentally, even in the insulating state the upper limit for the magnetic moment *per* cation in V_2O_3 is found to be (only) $\frac{1}{5}$ Bohr magneton[64], as in pure vanadium metal; and neutron diffraction studies have so far failed to show any discernible magnetic long range order in this substance. The inference is that charge correlations are related only indirectly to spin correlations, and that, whereas all magnetic insulators can be presumed to be Mott insulators of some type, the converse does not hold — all Mott insulators do not necessarily possess well defined magnetic properties.

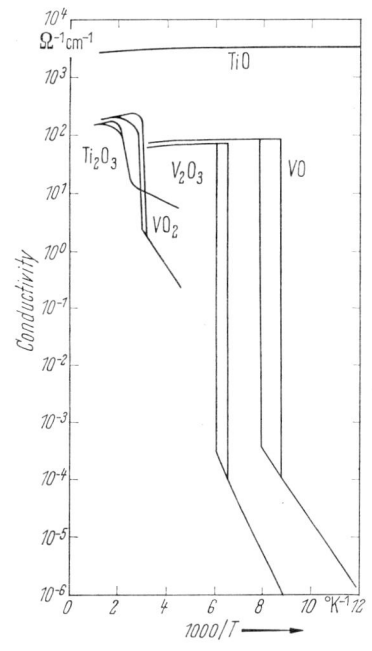

Fig. 2. Conductivity as a function of reciprocal temperature for the lower oxides of titanium and vanadium. Measurements were made along the [100] direction in VO, and along the c axis in V_2O_3 and VO_2. [F. J. MORIN: Phys. Rev. Letters **3**, 34 (1959).]

14. Slater's band model of antiferromagnetism. An opposing point of view was originally taken by SLATER[65] in his study of the band theory of antiferromagnetic insulators. If one follows that author and assumes a two-sublattice model of an antiferromagnet, and includes the exchange contributions in the periodic lattice potential with magnetic atoms (one up, the other down) in each unit cell, he finds indeed a splitting of each band into two sub-bands corresponding to Bloch states with oscillating spin polarization. The lower of each pair of sub-bands are polarized parallel to the assumed exchange potential polarization, and the upper sub-bands, antiparallel. The energy gap which separates them is the consequence not of the Coulomb interaction, but of the decrease of the size of the magnetic Brillouin zone (i.e., the doubling of the size of the elementary magnetic unit cell) and the exchange splitting. Such a theory is incapable of explaining the persistence of the energy gap above T_N in most antiferromagnetic oxides, nor can it explain the existence of ferromagnetic insulators such as EuS. Nevertheless, CALLAWAY[66] has carried out a self-consistent calculation on an idealized body-centered cubic structure and found that a gap might indeed occur when the interaction parameter was sufficiently strong for the energy of an s-like state at a zone corner to lie

---

[64] A. PAOLETTI, and S. PICKART: J. Chem. Phys. **32**, 308 (1960).
[65] J. C. SLATER: Phys. Rev. **82**, 538 (1951); **87**, 807 (1952).
[66] J. CALLAWAY: Proc. of Conf. on Semiconductors, Exeter, 1962, p. 584.

below that of a *p*-like state at the zone face, so that the Fermi surface lay in a forbidden region. Similar energy gaps also result as a consequence of spiral long range order in the theory of spin density waves[67]. The generally unsatisfactory aspect of this theory is the causal inter-relation of electronic properties and magnetic long-range order. In the following section, we shall study the very important effects that magnetic long range order *can* have on electronic band structure. However, as we have noted, the Mott transition and other similar properties may be explained without the assumption of any particular band structure and without vestige of magnetic long range order. Elementary quantum mechanical considerations dictate that if some sort of magnetic order is important in the determination of the self-consistent ground state of a solid, then it can only be the *short-range* order as the *d*-band structures (and *a fortiori f* bands) are determined mainly by nearest-neighbor type overlaps and intra-atomic forces.

II. Band structure of rare earth semiconductors.

We now turn to the interesting class of rare earth semiconductors, in which the magnitudes of the atomic magnetic moments are usually well conserved due to a strong energy gap of the second type U and strong Hund's rule exchange forces lining up the electron spins. These are related but not identical mechanisms, and it happens that the second one is the more important in determining the band structure and mean free path of added carriers responsible for the semiconducting properties (regardless of the mechanisms important for the ground state).

15. Band structure for an insulator with magnetic exchange. In the present section we examine the extent to which a band structure can be defined for an insulator with one added carrier (electron or hole) in a potential which is the sum of two terms: a periodic electrostatic part and an exchange part arising from the Hund's rule interaction of the carrier with the magnetic *f*-shell nearest to it. Exchange forces being typically short-ranged, this describes the interactions in a rare earth semiconductor fairly well. The new difficulty is a lack of parallelism of the atomic spins, except at $T=0$ in the special case of ferromagnets. Let us, therefore, start by considering this special case, where one has the added advantage of Cho's[68] recent calculation of the band structure for the ferromagnetic ground state in $Eu^{++}S^{--}$. This we now treat in some detail.

The exchange forces on the added particle are approximately derivable from a potential energy function which differs for a carrier of spin up and down (the *f*-shells are all up). Fig. 3 shows the muffin tin potential energy outside the *f*-shell of Eu^{++}, and Fig. 4 an enlargment of the potential energy *difference* for spins up and down. Fig. 5 gives the Brillouin zone for the relevant (fcc) crystal structure and the nomenclature for some principal points. Fig. 6 is the calculated band structure for up spin electrons (dashed curve) and down spin electrons (solid curve). Because of large correlation energies, the calculated position of the *f*-levels

---

[67] The most detailed treatment of the Slater-type model (antiferromagnetic case) has been given by J. DesCloizeaux: J. Phys. Radium **20**, 607, 751 (1959), and of the Overhauser-type model (spiral antiferromagnetism) in a review by C. Herring in Vol. IV of Magnetism (G. Rado and H. Suhl, Eds.), p. 298. New York: Academic Press 1966.

[68] S. J. Cho: Phys. Rev. **157**, 632 (1967). J. C. Slater: J. Appl. Phys. **39**, 761 (1968): The results are not yet made self consistent. Dr. Cho has informed us in a private communication that more recent calculations show an extremly strong variations of the $4f$ levels with the exchange term $A\varrho^{\frac{1}{3}}$. For $A=0.7$ the $4f$ levels move above the valence band to a position which is in agreement with optical absorption data discussed in Chapter F.

is not meaningful, nor is it in good agreement with optical and structural data given in Sect. F III. On the other hand, it may be assumed that the valence bands, with a maximum at $\varGamma_{15}$ and the conduction bands, with a minimum at X_3, are given with the usual great accuracy of SLATER's APW method which is used in this calculation.

At temperatures high compared to the Curie temperature of 16.5 °K, but low compared to any of the band structure parameters, the magnetic order disappears. Indeed, while the magnetic short-range order persists out to some multiples of the Curie temperature, the long range order vanishes — almost by definition — at all temperatures above T_c. What meaning, then, are we to attribute to these spin "up" and "down" bands, when the lack of magnetization makes all orientations in spin space equivalent? Obviously, the separation of these bands does *not* relate to the inter-atomic coupling which determines the Curie temperature but to the intra-atomic exchange between $4f$ electrons at the same ion.

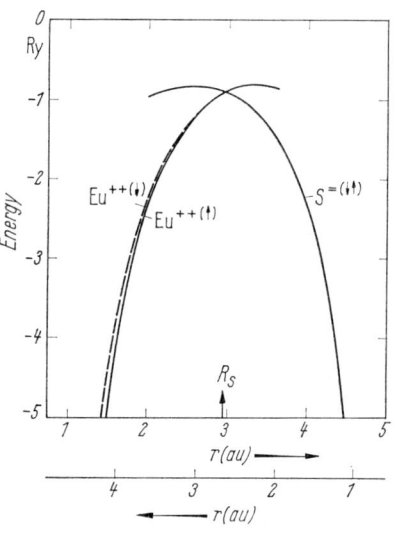

Fig. 3. Muffin-tin crystal potential energies for up- and down-spin cases for $Eu^{++}S^{=}$. The upper abscissa is the distance measured from the Eu^{++} and lower abscissa is the distance measured from the $S^{=}$. [T. CHO: Phys. Rev. **157**, 632 (1967).]

16. The effective Hamiltonian and the magnetic polaron. Let us consider a simplified case. In the effective mass approximation the energy of an electron near the conduction band minimum (or of a hole near the valence band maximum) is a parabolic function of the wave vector; let us assume $E(k) = k^2/2m^*$ (in units $\hbar = 1$). The analogue of the region near X_3 for this parabolic band structure is given in Fig. 7; dashed curve for spin up, solid curve for spin down, and dotted curve for the average. If the exchange splitting between up and down spin bands

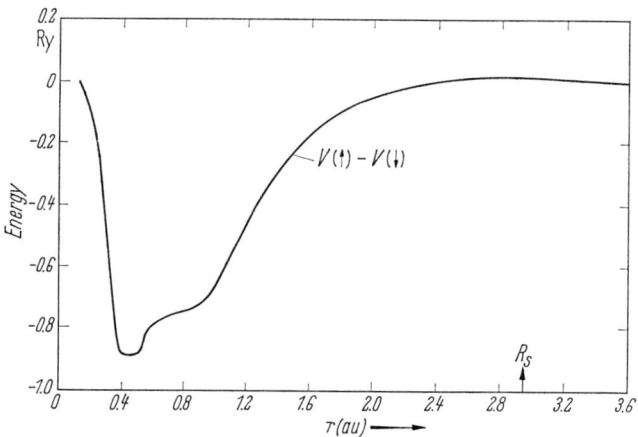

Fig. 4. Difference in the crystal potential energies for up- and down-spin in Eu^{++} in $Eu^{++}S^{=}$. Note change in scale from previous figure. [T. CHO: Phys. Rev. **157**, 632 (1967).]

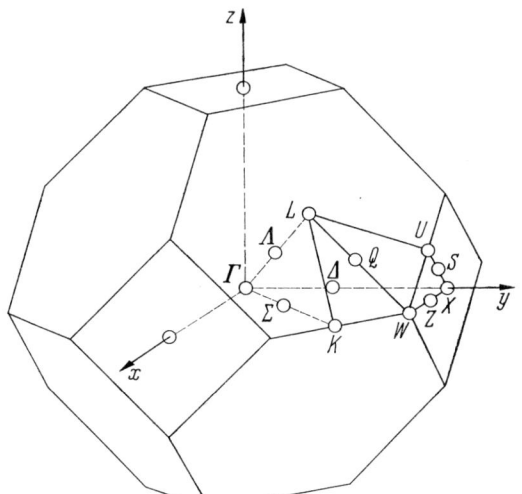

Fig. 5. Brillouin zone for fcc structure.

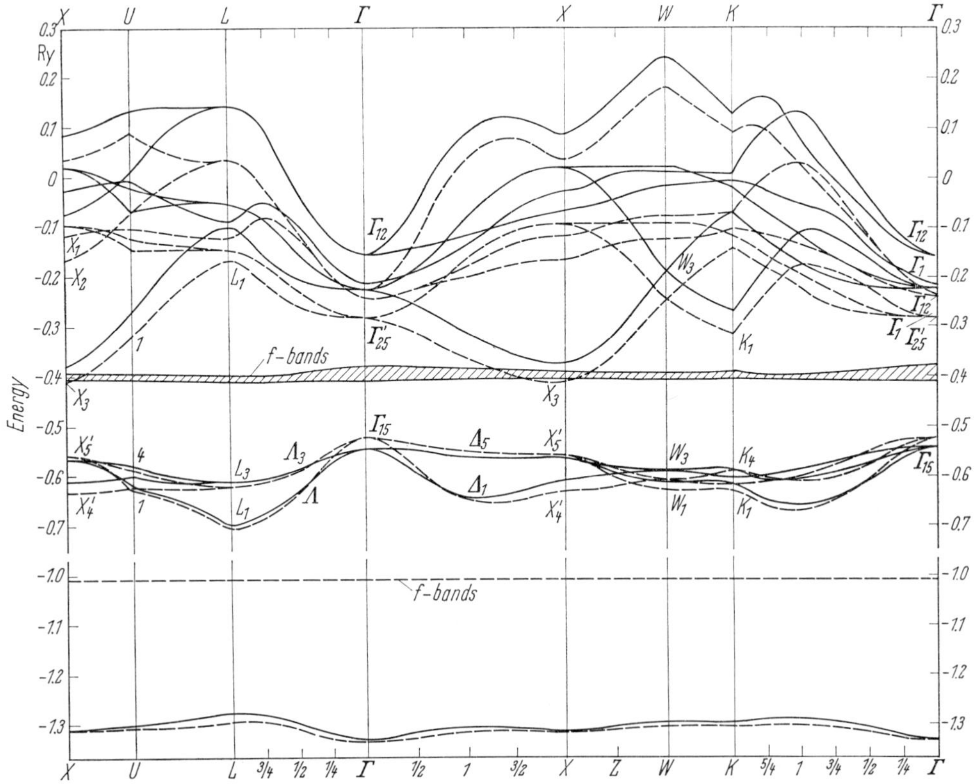

Fig. 6. Energy bands as a function of k in $Eu^{++}S^{=}$: ——— for up-spin and ——— for down-spin. Bands with s-character: Γ_1 p-bands: L_1, X_5', Γ_{15} d-bands: X_3, Γ_{25}'', Γ_{12}'. [T. Cho: Phys. Rev. **157**, 632 (1967).]

is JS^z, in which S^z (the spin of the Europium ions) is taken to be $+\frac{7}{2}$, the maximum, then the simplest effective spin Hamiltonian for an electron, the eigenvalues of which reproduce the energy levels of the carrier, is:

$$k^2/2m^* - J\tfrac{7}{2} s^z$$

where s^z, the z component of spin of the added carrier, has eigenvalues $\pm\frac{1}{2}$. This effective mass analogue to the Cho calculation suffers from the same defect as the latter. At high temperature it does not yield the dotted band structure nor the scattering effects, nor is it a rotationally invariant formulation in spin space. However, the above does suggest an acceptable effective Hamiltonian[69], for the magnetic structure plus an added carrier, which is free from these objections:

$$H = H_0 + H_{\text{mag}} \qquad (16.1)$$

where

$$H_0 = -\nabla^2/2m^* - \mathbf{s} \cdot \sum_j \mathbf{S}_j J(\mathbf{r} - \mathbf{R}_j)$$

$$\left[\text{or } E(-i\nabla) - \mathbf{s} \cdot \sum_j \mathbf{S}_j J(\mathbf{r} - \mathbf{R}_j)\right] \qquad (16.2)$$

and H_{mag} involves only the ionic spins:

$$H_{\text{mag}} = -\sum_{ij} J_{ij} \mathbf{S}_i \cdot \mathbf{S}_j. \qquad (16.3)$$

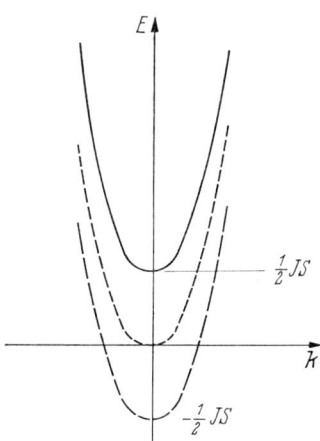

Fig. 7. Parabolic spin-band structure: (— — —) for spin up, (———) spin down, and (- - - - -) spin-averaged band structure.

In this Hamiltonian s is the (dimensionless) spin one-half vector operator of the carrier, S_i is the vector spin operator of the i-th magnetic ion. $J(\mathbf{r} - \mathbf{R}_i)$ is the exchange potential connecting the carrier at r with an ionic spin at R_i and the J_{ij} are the Heisenberg exchange constants responsible for the magnetic properties of the solid without the added carrier. If we assume $J(\mathbf{r} - \mathbf{R}_i) = J/\Omega =$ constant for r within the i-th cell, of volume Ω, and zero outside it, then the Fourier transformation of the interaction is just the constant J independent of k. This is tantamount to an assumption of constant exchange splitting for the entire band, which simplification is qualitatively supported by the calculation of Cho, Fig. 6. We obtain J from his calculation by equating $\frac{7}{2} J$ to $E_\downarrow(k=X_3) - E_\uparrow(k=X_3)$ for an electron; for a hole, we must evaluate it at $k=\Gamma_{15}$. Finally, the effective Hamiltonian above can be extended beyond the band extrema by using $E(-i\nabla)$ instead of $-\nabla^2/2m^*$ for the kinetic energy operator, where $E(k)$ is the spin-averaged calculated band structure. As for the exchange parameters J_{ij}, in such carefully studied materials as EuS they are known from low temperature spin wave analysis or from studies of their high temperature properties[70].

The term "magnetic polaron" denotes, by analogy with the ordinary polaron, the electronic carrier and its associated polarization cloud, the Hamiltonian of which is given above. Whereas the polaron interacts with polar modes of an ionic crystal, the new particle interacts with magnetic degrees of freedom. Both have

---

[69] T. WOLFRAM, and J. CALLAWAY: Phys. Rev. **127**, 1605 (1962) already considered such a Hamiltonian in connection with electrons in ferromagnetic metals. They also treat spin-orbit coupling, which we suppress in the present text for simplicity.

[70] S. CHARAP, and E. BOYD: Phys. Rev. **133**, A 811 (1964). Data on anisotropy is given in S. VON MOLNAR, and A. W. LAWSON: Phys. Rev. **139**, A 1598 (1965). See Sect. 59.

to be studied by approximate methods of varying degrees of sophistication. Actual magnetic semiconductors may be sufficiently ionic that *both* polaron and magnetic polaron effects may be simultaeneously active and interfere with one another, but neither theory nor experiment are yet sufficiently advanced to test this interference.

17. Solution of the magnetic polaron problem at $T=0$. The magnetic polaron problem can be solved at $T=0$ for a ferromagnet with one added carrier of spin up or down. If the spin is up, parallel to the ionic magnetization, the eigenstates of $H_0 + H_{mag}$ are simply

$$\psi_{\uparrow k} \sim e^{ik \cdot r} \begin{bmatrix} 1 \\ 0 \end{bmatrix} \prod_i \varphi(\xi_i) \tag{17.1}$$

in which $\varphi(\xi_i)$ is the eigenstate of S_i^z having maximum ionic spin up $S_i^z = S$ (e.g. $\tfrac{7}{2}$ for Eu^{++}) and the spinor refers to the state of electronic spin. The eigenvalue associated with each eigenfunction of this form is obtained by substitution of it into Schrödinger's equation with the H above, yielding

$$E_{mag}^0 + k^2/2m^* - \tfrac{1}{2} JS \tag{17.2}$$

in which E_{mag}^0 is the eigenvalue of H_{mag} in the ferromagnetic state, and the remainder $k^2/2m^* - \tfrac{1}{2} JS$ is the energy of the added carrier in the spin up subband.

A carrier with spin antiparallel to the magnetization is capable of emitting a spin wave, reversing its spin direction in the process, and of reabsorbing it. This "dresses" the carrier with a cloud of virtual excitations, but may also scatter it if energy is conserved in the process, as well as momentum. The eigenstates must be of the form:

$$\psi_{\downarrow k} \sim e^{ik \cdot r} \left\{ \begin{bmatrix} 0 \\ 1 \end{bmatrix} + \sum_q F_q\, e^{-iq \cdot r} \begin{bmatrix} 1 \\ 0 \end{bmatrix} \sum_j S_j^-\, e^{iq \cdot R_j} \right\} \prod_i \varphi(\xi_i). \tag{17.3}$$

Substitution of this form into Schrödinger's equation with our Hamiltonian $H_0 + H_{mag}$ yields an equation for the amplitudes F_q, and, in turn, an energy eigenvalue in the form

$$E_{mag}^0 + w_k \tag{17.4}$$

in which $w(k)$ is the solution of a transcendental equation,

$$w_k = k^2/2m^* - \tfrac{1}{2} JS + \frac{JS}{1 + \tfrac{1}{2} J \Lambda(w_k)} \tag{17.5}$$

where

$$\Lambda(w_k) = \frac{1}{N} \sum_q \left[\frac{(k-q)^2}{2m^*} + \hbar\omega_q - \tfrac{1}{2} JS - w_k \right]^{-1} \tag{17.6}$$

and $E_{mag}^0 + \hbar\omega_q$ is the energy of the ionic lattice with one spin wave above the ground state. The sums on wavevectors are restricted to the first BZ. If w_k is picked so as to make one of the terms in the summand to be infinite, i.e.,

$$w_k = (k-q)^2/2m^* + \hbar\omega_q - \tfrac{1}{2} JS$$

then the last term in (17.5) vanishes and we have, simultaneously,

$$w_k = k^2/2m^* - \tfrac{1}{2} JS.$$

Taken together, these two equations have a solution only at $q=0$, the energy of which coincides with (17.2). This, in fact, is merely the totally ferromagnetic solution (17.1) in the one-reversed spin subspace. All other solutions, therefore, belong to total spin equal to one unit less than maximum; and as the summand will not diverge for any of these states, it is permissible to replace the sum in (17.6) by an integration, the principal part of which yields the energy shift and the imaginary part of which yields the scattering lifetime. Thus, if J is small and it is legitimate to expand the denominator in (17.5) and replace w_k in the summand by the unperturbed value, we find the following results:

$$w_k = k^2/2m^* + \tfrac{1}{2}JS - \tfrac{1}{2}J^2S(2\pi)^{-3}\int_{B.Z.}d^3q[(q^2-2k\cdot q)/2m^* + \hbar\omega_q - JS]^{-1} \qquad (17.7)$$

for the spin down polaron energy and

$$\frac{1}{\tau_k} \propto J^2 S \int_{B.Z.} d^3q\,\delta[(q^2-2k\cdot q)/2m^* + \hbar\omega_q - JS] \qquad (17.8)$$

for the lifetime, in agreement with second order perturbation theory and first Born approximation. For electrons in ferromagnetic metals, similar expressions have recently been derived[71] in second-order perturbation theory. The results (17.1)—(17.6) are, of course, exact to all orders in the coupling constant J for ferromagnets at $T=0$ but fail at finite temperature.

18. Solution of the magnetic polaron problem at finite temperature. A finite temperature approach designed for the range $T<T_c$ was used by WOLFRAM and CALLAWAY[72] in their treatment of the magnetic polaron problem. Their theory can be summarized as follows: the spin operators are first transformed to the idealized spin wave representation of HOLSTEIN and PRIMAKOFF,

$$\left.\begin{aligned} S_i^z &= S - b_i^* b_i \\ S_i^+ &= (2S)^{\frac{1}{2}} b_i \\ S_i^- &= (2S)^{\frac{1}{2}} b_i^* \\ a_k &= (N)^{-\frac{1}{2}} \sum_i \exp(-ik\cdot R_i) b_i \end{aligned}\right\} \qquad (18.1)$$

H_{mag} in this representation, has the form $\sum \hbar\omega_q a_q^* a_q$ and H_0 has some formal similarity to the electron-phonon coupling Hamiltonian of the polaron model. The variational method used to approximate the solutions of SCHROEDINGER'S equation and the related transport properties than follow along the lines of what has come to be termed the "small polaron" model[73]. The carrier is localized in a more-or-less ad hoc manner, and the scattering of spin waves resulting in a "polarization cloud" about the particle is resolved. This results in a state function, involving the space and spin coordinates r, ξ of the electron and a_k^* of the magnon field:

$$\varphi(r - R_i, \xi; \ldots a_k^* \ldots)\,|0\rangle \qquad (18.2)$$

centered about the point R_i which can be a member of a discrete or continuous set. The variational Bloch function, which includes the polarization effects to a satisfactory extent, is just the Fourier transform,

$$\psi_k \propto \sum_i e^{ik\cdot R_i} \varphi(r - R_i) \qquad (18.3)$$

---

[71] L. C. DAVIS, and S. H. LIU: Phys. Rev. **163**, 503 (1967).
[72] T. WOLFRAM, and J. CALLAWAY: Phys. Rev. **127**, 1605 (1962).
[73] References in Sect. 20.

(suppressing the other coordinates for typographical clarity). This is a better variational function than the purely localized one above because of the matrix elements in the total Hamiltonian which connect $\varphi(r-R_i)$ to a function localized about a different point $\varphi(r-R_j)$. These matrix elements are responsible for the band structure, and if J is sufficiently small we must recover $k^2/2m^*$ as the energy eigenvalue associated with the Bloch function of wavevector k. At large values of J the polarization clouds at the different sites have only a very small overlap, with a correspondingly small matrix element connecting them. The effective mass must, therefore, increase with increasing J (or in the terminology of Wolfram and Callaway, the effective width of the conduction band must be reduced). The associated magnon drag effect, and scattering lifetimes are not treated in that work. The reduction in bandwidth (or increase in effective mass) is given as roughly proportional to the exponential factor,

$$\exp-\frac{4S}{N}\sum_k (2\bar{n}_k+1)(J'_{12}/\hbar\omega_k)^2 \sin^2 \frac{1}{2}kR \qquad (18.4)$$

in which R may be taken to be the distance between nearest neighbor magnetic ions, J'_{12} is essentially proportional to $\frac{1}{2}J$ *times* the overlap integral of the electronic part of a localized function (18.2) at one site with a similar function at a nearest neighbor site, and $\bar{n}_k$ is the Bose-Einstein distribution function of magnons of energy $\hbar\omega_k$. An approximate evaluation of this at $T=0$ yields

$$\exp-\{\text{const} \cdot S^{-1}(J'_{12}/kT_c)^2\}$$

which may be enormously small in a substance for which kT_c is small, as in EuS. However, in comparing this with the exact results [obtained in Eq. (17.5) for $T=0$] one fails to detect any such exponentially small factor in the effective mass or in the band width. It is, therefore, not precisely clear yet just what the applicability of the quoted zero temperature and finite temperature results might be.

19. Scattering of carriers by spin fluctuations. What is a more serious objection to the "small polaron-type model" is that, in order to apply the theory to temperatures comparable to, or higher than T_c, it is clear the spin wave representation must be entirely abandoned. From Eqs. (17.6) and (17.7) above we see that a quite different approach might be successful provided $q^2/2m^* \gg \hbar\omega_q$. The spin wave energy $\hbar\omega_q$ can then be omitted from the problem in the usual case, and one may instead consider the simpler problem of *elastic* scattering of carriers by spin fluctuations in the magnetic medium, similar in its detailed aspects to the much studied problem of *neutron* scattering[74]. One treats the interaction in (16.2) by perturbation theory so that first- and second-order self-energy corrections for carriers of spin $m=\pm\frac{1}{2}$ are, respectively,

$$\Sigma^1_m(k)=-mJM \qquad (19.1)$$

in which M is the thermal average of S^z_j, and, with neglect of $\hbar\omega_q$,

$$\Sigma^2_m(k)=-\frac{J^2}{2N}\sum_{\substack{q\neq 0 \\ \varepsilon\to 0}}\frac{|S_q|^2}{(q^2-2k\cdot q)/2m^*+i\varepsilon} \qquad (19.2)$$

in which $\vec{S}_q$ are the three-dimensional thermal averaged Fourier-transformed spin fluctuations. The principal part of this integral yields the energy shift, and the

---

[74] See review by P. G. de Gennes in Vol. III of Magnetism (Ed. G. Rado and H. Suhl). New York: Academic Press 1963.

imaginary part yields the scattering lifetime in the first Born approximation. These expressions are, however, not valid in the low temperature regime where spin wave theory holds, as we have not distinguished between the absorption and emission of spin waves, i.e., we have violated the conservation laws for total spin which permitted an exact solution at $T=0$. Nor are these formulae valid when J is comparable to the kinetic energy of the carrier[75]. But subject to these restrictions, they do permit the study of intermediate and high temperature ranges, including the neighborhood of the critical temperature T_c.

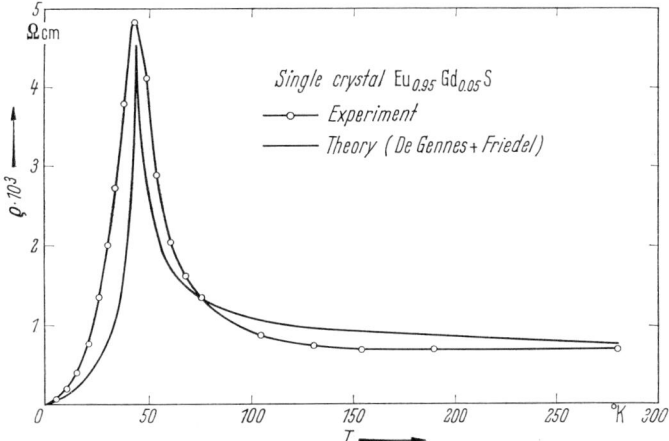

Fig. 8. Comparison of the de Gennes-Friedel formulae to experimental data obtained on a $Eu_{0.95} Gd_{0.05} S$ single crystal with about 2×10^{20} cm^{-3} carriers:

$$T > T_c : \varrho = \frac{\varrho_0}{4} \int_0^2 \frac{x^3\, dx}{1 - \frac{T_c \sin k_0 D x}{T\, k_0 D x}}$$

$$T < T_c : \varrho = \frac{\varrho_0}{4} \int_0^2 \frac{x^3\, dx}{1 - \left(3 - 2\frac{T_c}{T}\right) \frac{\sin k_0 D x}{k_0 D x}}$$

with $k_0 D = 0.702$. The resistivity at room temperature has been used as the only adjustable parameter (S. VON MOLNAR: To be published). Compare also Figs. 49 to 52.

An early detailed investigation of this type of scattering of carriers by spin fluctuations was undertaken by DE GENNES and FRIEDEL[76] who, using molecular field theory to obtain short- and long-range order parameters, calculated the temperature dependence of the scattering cross-section as a function of the energy of the carrier. Their results, and the fit to experiment are shown in Fig. 8 and discussed in Sect. 75. The problem of evaluating $|S_q|^2$ has not been settled in general, as it is tied in with a detailed thermodynamical solution of the Hamiltonian H_{mag} which is not available at this date. Generalized molecular

---

[75] This case is studied in Sect. 28 entitled "Double Exchange".
[76] P. G. DE GENNES, and J. FRIEDEL: J. Phys. Chem. Solids **4**, 71 (1958). This work has been critized recently by M. E. FISHER, and J. S. LANGER: Phys. Rev. Letters **20**, 665 (1968) on the basis that in metals the fluctuations $q \approx 2k_F$ contribute far more to the resistance than the "critical" modes $q \approx 0$.

field theory[77-79] gives the following:

$$|S_q|^2 \sim \frac{kT}{Q+Aq^2} \quad T \geq T_c \tag{19.3}$$

where $Q \sim T-T_c$ and A is constant. Below T_c one has,

$$|S_q^x|^2 = |S_q^y|^2 \sim \frac{kT}{(h/M)+Aq^2} \tag{19.4}$$

and from spin wave theory, an indication[78] that

$$|S_q^z|^2 \sim \left(\frac{kT}{AM}\right)^2 \frac{\tan^{-1}\left[\frac{q^2MA}{4h}\right]^{\frac{1}{2}}}{q}. \tag{19.5}$$

Use of these formulas also yields results of the sort shown in the Figures. However recent research into critical point phenomena[79, 80] have made detailed application of such formulas quite obsoletes near T_c. Applying an external magnetic field to line up the ionic spins has the same effect as lowering the temperature in a ferromagnet. Such an effect (giant negative magnetoresistance) has been reported on Gd-doped EuSe compounds[81] and n-type $CdCr_2Se_4$-spinels[82].

A more systematic investigation of the general aspects of the magnetic polaron problem, whether at high temperatures (where an expansion in T^{-1} is presumably possible), in the critical range of temperatures near T_c, or at low temperatures or strong magnetic fields (where spin wave theory is applicable) is not yet available but would obviously be of the greatest interest.

III. Conductivity properties of magnetic semiconductors.

20. Modes of motion of a carrier. One distinguishes three, essentially different, modes of motion of a carrier, listed in order of decreasing mobility:

(a) *Free motion*, such as that of a Bloch electron. In the case of a magnetic polaron, the carrier and associated polarization cloud should be able to move either freely through the lattice or with a very long mean free path.

(b) *Hopping motion*, as in the small polaron[83]. The mean free path as determined by interaction with the magnetization or lattice vibrational fields is not much greater than the dimensions of a unit cell.

(c) *Impurity band type hopping motion*[84]. The carriers move from one donor to the next in a partly compensated sample, so that, added to the specifications

---

[77] Also known as "Landau theory", "spherical" or "gaussian" model, "Ornstein-Zernicke" theory, etc. A fairly general discussion is given in R. BROUTS' book "Phase Transitions", New York: Benjamin 1965, pp. 56 *et seq.*

[78] P. G. DE GENNES: Theory of Neutron Scattering in Vol. III of Magnetism (RADO and SUHL, Eds.), Chap. 3, Sect. 3. New York: Academic Press 1963.

[79] L. KADANOFF et al.: Static Phenomena near Critical Points. Revs. Mod. Phys. **39**, 395 (1967).

[80] M. E. FISHER: The Theory of Equilibrium Critical Phenomena. Rpts. Progr. in Phys. **30**, 615 (1967).

[81] S. VON MOLNAR, and S. METHFESSEL: J. Appl. Phys. **38**, 959 (1967). See Sect. 82.

[82] H. W. LEHMANN: Phys. Rev. **163**, 488 (1967). See Sect. 66.

[83] Literatur is given in Sect. 76, reference 471.

[84] T. KASUYA: J. Phys. Soc. Japan **13**, 1096 (1958); T. KASUYA, and S. KOIDE: J. Phys. Soc. Japan **13**, 1287 (1958); A. MILLER, and E. ABRAHAMS: Phys. Rev. **120**, 745 (1960). Specific application to RE chalcogenides in T. KASUYA, and A. YANOSE: To be publ., J. Phys. Soc. Japan.

in (b), there is a statistical factor related to the probability of a carrier being near a suitable ionized donor.

Of the three mechanisms, only (a) results in the usual Hall effect, $R_H = 1/nec$, with (b) and (c) obeying different relations for this coefficient[85]. Formulas for the electrical conductivity, thermoelectric coefficients, optical properties of free carriers, etc., also all depend on the mode of motion which is presumed correct for a carrier in the solid in question. Some of these properties are discussed in turn in the following sections.

21. Electrical conductivity. (a) The *dc.* mobility of carriers in *ordinary* non-magnetic semiconductors such as Ge or Si has the following idealized temperature dependence[86], due to the temperature dependence of τ in $\mu \equiv e\tau/m^*$:

$$
\left.\begin{array}{l}
\text{Acoustical phonon scattering:} \quad \mu \sim T^{-\frac{3}{2}} \\
\text{Optical phonon scattering} \\
\quad (T_D = \text{Debye temperature}): \quad \mu \sim (\exp T_D/T - 1), \\
\text{Ionized impurity scattering:} \quad \mu \sim T^{\frac{3}{2}} \\
\text{Neutral impurity scattering:} \quad \mu \sim \text{temperature independent} \\
\text{Edge dislocation scattering:} \quad \mu \sim T \\
\text{Magnetic scattering:} \quad \text{given in (19.2); power law unknown}
\end{array}\right\} \quad (21.1)
$$

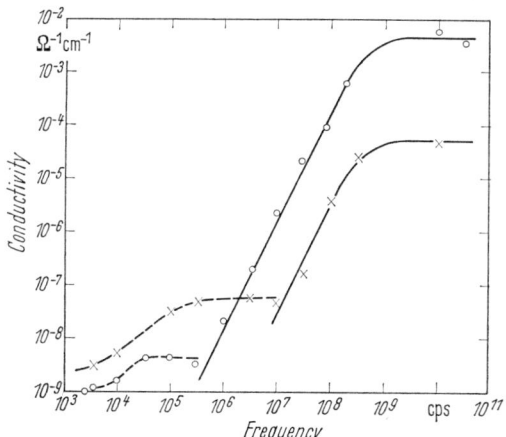

Fig. 9. Frequency-dependent conductivity of two samples of NiO. The solid curves are obtained from Eq. (21.5). [SNOWDEN and SALTSBURG: Phys. Rev. Letters **14**, 497 (1965).]

In the case of alternating fields, the frequency-dependence of the real part is approximately given by

$$\mu(\omega) = \mu(0) \frac{1}{1 + (\omega\tau)^2}. \quad (21.2)$$

(b) and (c) The mobility of carriers undergoing "hopping" type motion is a diffusion process determined by an activation energy q (possibly a function of T)

---

[85] T. HOLSTEIN: Phys. Rev. **124**, 1329 (1961); L. FRIEDMAN: Phys. Rev. **131**, 2445 (1963), YU. A. FIRSOV: Soviet Phys.-Solid State **5**, 1566 (1964), trans. from Fiz. Tverdogo Tela **5**, 2149 (1963); and additional references in Sect. 76.

[86] American Institute of Physics Handbook, pp. 9—53. New York: McGraw-Hill 1963.

and a jump frequency ν, and is of the general form

$$\mu = \frac{eD_0}{kT} \exp(-q/kT), \quad D_0 \propto a^2 \nu \qquad (21.3)$$

(except possibility at very low temperatures), with $a=$ average distance separating hopping sites and ν typical jump frequency. For antiferromagnetic ordering, the following forms have been suggested[87]:

$$\mu \sim \nu \frac{T}{T_c} \exp(-q/kT) \quad (T<T_c)$$

and $\qquad (21.4)$

$$\mu \sim \nu \frac{T_c}{T} \exp(-q/kT) \quad (T>T_c).$$

Polaron theory should be at least qualitatively applicable to the ferromagnetic semiconductors; two contributions are important, each in a different range of temperature. At low temperature (depending on the parameters, this range of temperatures may be too low to be ordinarily observable) there is a band type conductivity with the typical enhanced effective mass (or reduced bandwidth) and scattering caused by phonon and/or magnon drag and the discussion of Sect. 16 is applicable. At higher temperatures the thermally activated hopping processes dominate rather complicated formulas for both effects applicable to the small polaron are given in the literature[88]. In alternating fields, the frequency dependence of the real part of the mobility[89,90] is, schematically, (Fig. 9),

$$\mu(\omega) = \mu(0) + \delta\mu \frac{\omega^2 \tau^2}{1 + (\omega\tau)^2} \qquad (21.5)$$

POLLACK[89] has pointed out that the increase with frequency in the hopping regime, as contrasted with the decrease (21.2) of ordinary mobility at high frequency, can be a direct test of the characteristic mode of motion.

It is well known that in ordinary semiconductors the discrepancy between drift mobility and Hall mobility is related to various trapping mechanisms. In the hopping regime the carriers are almost always trapped, so that the Hall mobility bears essentially no relation to the electrical conductivity, although theoretical estimates show that it may even exceed the drift mobility by as much as two orders of magnitude[91]. Note that in the hopping regime μ is already so small that the discrepancy may be academic.

In order to explain the electronic, optical, and magnetic properties of Gd-doped EuS and EuSe, (Sects. 44, 48, 61, 62) KASUYA and collaborators[92] have recently exploited an adaptation of the theory of impurity banded hopping conduction, which can be easiest explained in the following terms: the impurities (Gd in this case) are partly compensated allowing an electron to leave a Gd^{2+}

---

[87] R. HEIKES, A. MARADUDIN et R. MILLER: Etude des Propriétés de Transport des Semiconducteurs de Valence Mixte. Ann. phys. **8**, 733 (1963).

[88] I. G. LAND, and YU. A. FIRSOV: Soviet Phys. JETP **16**, 1301 (1963); trans. from Zhur. Eksptl. i Teoret. Fiz. **43**, 1843 (1962); L. FRIEDMAN: Phys. Rev. **135**, A 233 (1964) and most recently using Green function and Kubo formalism for the conductivity, by J. SCHNAKENBERG: Phys. Letters **14**, 266 (1965).

[89] M. POLLACK in Proc. of Int. Conf. on Phys. of Semiconductors, Exeter, 1962 (Inst. of Physics and Physical Soc., London, 1962), p. 86.

[90] D. SNOWDEN, and H. SALTSBURG: Hopping Conduction in NiO. Phys. Rev. Letters **14**, 497 (1965).

[91] T. HOLSTEIN: Hall Effect in Impurity Conduction, Phys. Rev. **124**, 1329 (1961), and references in footnote 83, Sect. 20.

[92] T. KASUYA, and A. YANASE: To be publ.; see also, references in footnote 84, Sect. 20 and the discussion of experimental data in sections 61 and 62.

atom for any neighboring Gd³⁺ center. There is an activation energy q associated with this process, however, because the spin of the electron has partly aligned the spins of the neighboring Eu atoms to that of the Gd donor, whereas the new Gd³⁺ center to which it hops is initially unpolarized. We may overestimate q from the infinite wavelength differential paramagnetic susceptibility,

$$q \propto \chi = dM/dh|_{h=0} \tag{21.6}$$

a quantity which vanishes both at low and at high temperatures, and diverges at T_c. Inserting such a function, e.g., $\chi \propto |T-T_c|^{-1}$ into (21.3) gives a mobility

$$\mu \sim \frac{v}{kT} \exp -\frac{-A}{T|T-T_c|} \tag{21.7}$$

which vanishes strongly in the neighborhood of T_c due to critical fluctuations. The Born scattering theory, which yields the imaginary part of (19.2) can also be estimated, to the same crude accuracy, to yield a magnetic mobility which varies as χ^{-1}, i.e., as $|T-T_c|$. The two extremes of types of motion thus result, respectively, in a power-law-type behavior near T_c (ordinary motion) and an essential singularity at T_c (hopping motion). However, see the remarks concerning the resistivity of metals in footnote 76; here too, if we are more realistic and use χ at *finite* wavelength, the singularity at T_c is rounded off.

CALLEN has pointed out[93] that the motion (hence the activation energy q) may be limited in part by magnetoelastic effects, i.e., the local lattice parameter in the neighborhood of the electron may be somewhat altered from the value in the absence of this particle, a form of the polaron effect. This, in turn, affects the band structure (through the *deformation potential*), shifts the optical absorption spectra, etc.

22. Thermoelectric properties. The data on conductivity yields only a particular combination of constants, e.g., $\sigma = ne^2\tau/m^*$ for quasi-free motion, so it is necessary to investigate various other transport properties to untangle the dependence on temperature and on other parameters of each of the factors in this formula. The usually popular Hall effect is generally too small in the magnetic semiconductors to be measurable, but many investigators[94] have found some of the thermoelectric properties, such as the Seebeck effect, convenient to measure and interpret. We recall the definition of the thermoelectric force: when particles carry an electrical or thermal current down a potential or thermal gradient, an irreversible dissipation of energy occurs which, in the case of an electric current, is known as the Joule heat, (and has, in fact, nothing to do with the reversible effects with which we are presently concerned). But if the current passes through a solid in which there is simultaneously a thermal gradient dT/dx an extra heat is generated reversibly, i.e., dQ changes sign with dT/dx, and is proportional both to this quantity and to the current. The Seebeck coefficient α is defined as the constant of proportionality. The Peltier heat liberated or absorbed at the junction of two dissimilar solids is related to the Seebeck coefficient as follows: $\alpha = \pi/T$, in which π is the absolute Peltier coefficient (energy flux/unit current when the zero is chosen at the Fermi level, i.e., the *electrochemical potential*) so that a knowledge of either is sufficient, and determines the position of E_f relative to the band structure.

---

[93] E. R. CALLEN: Phys. Rev. Letters **20**, 1045 (1968).
[94] F. J. MORIN, in: HANNAY, Semiconductors, New York: Reinhold 1959; G. H. JONKER u. S. VAN HOUTEN, in: Halbleiterprobleme (Ed. F. SAUTER), Braunschweig: F. Vieweg & Sohn 1961; A. F. JOFFE: Semiconductor Thermoelements and Thermoelectric Cooling, London: Infosearch 1951; Physics of Semiconductors, New York: Academic Press 1960.

In ordinary semiconductors, carrier motion of type (a), there is an additional-correction effect[95] known as "phonon drag" which HERRING has described as follows[95]: in a thermal gradient the phonon distribution is not isotropic, as more phonons must move from the hot side to the cold one than the converse, so that in scattering the carriers, the phonons must impart a net momentum to them in such a direction as to push them more often toward the cold side than not. Zero current results only when the cold end acquires an excess of carriers, the electrostatic field of which counteracts phonon drag and normal carrier diffusion. This effect is expected to be strongest at temperatures such that anharmonic forces are relatively weak and cannot easily maintain the phonon distribution at thermal equilibrium. The result is an anomalous contribution to α.

A similar effect recent observed[96] in the antiferromagnet MnTe has been ascribed to "magnon drag"; here magnons have an energy proportional to their wave vector, hence a group velocity which is approximately constant. (In ferromagnets the energy varies as the square of the wave vector, and the analysis which is found appropriate to phonons cannot be so simply transcribed to the magnetic case.)

In the hopping regimes (b) and (c) the Seebeck coefficients are no longer merely an interesting curiosity but provide one of the few available measurements of the number of carriers. For if $\alpha \doteq (k/e)(E_F/kT)$ with the zero of energy chosen appropriately, and the density of carriers is

$$n = \frac{1}{e^{E_F/kT} + 1} \qquad (22.1)$$

we obtain

$$\alpha \doteq (k/e) \log \frac{1-n}{n}. \qquad (22.2)$$

Aside from magnon and phonon drag, a number of corrections (e.g., the activation energy for a hop) must also be taken into account[97].

D. Theory of indirect exchange.

The interactions among magnetic atoms through the medium of conduction-band electrons (or valence-band holes) is known variously as *indirect exchange, s—d exchange, double exchange, conduction-band superexchange, Ruderman-Kittel interaction*, etc. Not all these terms are synonymous and various mechanisms are involved, some of which we shall touch upon in the present chapter.

The experimental data and the manner in which they have been interpreted in favor of various physical mechanisms, such as "indirect exchange" in its various manifestations, are the topic of chapter G ("Magnetic Interactions via Free Carriers"). The situation in real materials can be quite complex. The theoretical analysis has to be limited to some rather idealized and tractable models.

Indirect exchange of one sort or another occurs to varying degrees in metals and insulators as well as in semiconductors[98], and generally operates in the fol-

---

[95] L. GUREVICH: J. Phys. (U.S.S.R.) **9**, 477 (1945); **10**, 67 (1946); H. FREDERIKSE: Phys. Rev. **91**, 491 (1953); **92**, 248 (1953); C. HERRING: Phys. Rev. **96**, 1163 (1954).

[96] J. D. WASSCHER, and C. HAAS: Phys. Letters **8**, 302 (1964). M. BAILYN: Phys. Rev. **126**, 2040 (1963).

[97] R. HEIKES, A. MARADUDIN, and R. MILLER: Ann. phys. **8**, 733 (1963).

[98] General references specific to magnetic semiconductors are G. H. JONKER and S. VAN HOUTEN's "Semiconducting Properties of Transition Metal Oxides", in: Halbleiterprobleme, Dd. VI (F. SAUTER, Ed.), p. 118 and ff. Braunschweig: F. VIEWEG & Sohn 1961, and F. J. MORIN: Oxides of the 3d Transition Metals, in: N. B. HANNAY's "Semiconductors", pp. 600—631, New York: Reinhold Publ. Co. 1959, or in: Bell System Tech. J. **37**, 1047 (1958).

lowing manner: the spins of mobile particles (electrons or holes) are polarized by a magnetic atom through a localized exchange interaction (which can be ferromagnetic or antiferromagnetic in nature). As a result of this interaction a net spin density cloud is formed, centered about the magnetic atom and decaying exponentially or in some oscillatory manner at large distances from it. A second magnetic atom interacts with the first by way of this induced spin density, hence the terminology: "indirect exchange". We start the analysis therefore with a study of the local exchange interaction, continue with the study of indirect exchange, and mention associated topics (electrical resistance, magnetic susceptibility, etc.) only insofar as they arise in the development.

23. Local exchange mechanism. The magnetic atom is characterized by an unfilled d- or f-shell. Let us assume it is less than half filled, and speak of "electrons"; much of the following discussion applies to more than half filled shells merely by substituting "hole" for electron. The electrons occupy orbital states $\varphi_n(r-R)$ where R is the position of the magnetic atom, and $n = 1, 2, \ldots 2l+1$ labels the orbital; if crystal fields do not destroy the orbital angular momentum, this quantum number is related to the eigenvalue of L_z. Electrons in the conduction band are characterized, on the other hand, by Bloch states $\psi_k(r)$. The Bloch states are the product of a phase factor $\exp i k \cdot r$ and a function periodic from cell to cell $u_k(r)$; for many purposes the periodic function is averaged and it is only the change in the plane-wavelike phase factor which is of importance, although this is not always strictly correct.

The potential which scatters conduction electrons in the vicinity of the magnetic atom has two main contributions: an electrostatic term, which we shall (approximately) characterize by a potential $V(r)$ with matrix elements $V_{k,k'}$, and a direct exchange correction to the electron-electron repulsion, favoring conduction electrons of spin parallel to that of the magnetic electrons. Therefore, to the zeroth-order Hamiltonian H_0 of the Bloch states,

$$H_0 = \sum_{\substack{k \\ m=\pm\frac{1}{2}}} \varepsilon_k c^*_{k,m} c_{k,m} \tag{23.1}$$

it is necessary to add a scattering term,

$$H_{\text{electrost.}} = \sum_{\substack{k,k' \\ m=\pm\frac{1}{2}}} V_{k,k'} c^*_{k,m} c_{k',m} \tag{23.2}$$

which includes the (suitably averaged) 2-body repulsions as well as the attractive nuclear potential, and to introduce the spin-dependent exchange corrections thereto:

$$H' = -\sum_{k,k'} J_{k,k'} \{ S^z (c^*_{k+} c_{k'+} - c^*_{k-} c_{k'-}) + S^+ c^*_{k-} c_{k'+} + S^- c^*_{k+} c_{k'-} \}. \tag{23.3}$$

We have assumed that the total spin S of the $2s$ electrons in the magnetic shell is a good quantum number, and S^z and $S^\pm$ are the appropriate azimuthal and raising/lowering operators of the total spin. When $J = L + S$ rather than S yields good quantum numbers, as is apt to be the case for the rare earths, we may replace S in the above operators by $(g-1)J$, where $g =$ Landé g-factor. Additional anisotropic corrections have been derived by Liu[99] and by Kaplan and Lyons[100], the

---

[99] S. H. Liu: Phys. Rev. **121**, 451 (1961).
[100] T. A. Kaplan, and D. H. Lyons: Phys. Rev. **129**, 2072 (1963). See also survey by T. Kasuya: s—d and s—f Interaction in: Magnetism IIB (Rado and Suhl, Eds.). New York: Academic Press 1966.

latter authors estimating the effects of such corrections on the spin-only formulas to range in magnitude from 0% (for Gd, in which $L=0$) to 10% for Er, Ho, Dy and Tb, to a maximum of about 100% for Tm^{3+} and Yb^{3+} ions. Here we shall not consider these corrections, and moreover for simplicity we shall assume that because of sufficiently strong Hund's rule (intra-atomic exchange) coupling all the spins of the $2s$ magnetic electrons are aligned, let us say "up", so that the exchange integral is given by

$$J_{k,k'}^{\text{exch.}} = \frac{1}{2s} \sum_{n=1}^{2s} \int d_3 r \int d_3 r' \, \varphi_n^*(r) \varphi_n(r') \frac{e^2}{|r-r'|} \psi_{k'}^*(r') \psi_k(r). \quad (23.4)$$

We assume (Hund's *second* rule) that orbitals $n=1, \ldots, 2s$ are occupied and the remainder empty. This rule is, mathematically, the consequence of exchange and Coulomb integrals such as the above, but among only localized orbitals. These lift the degeneracy of the various magnetic multiplets in such a manner as to result in maximum magnetic moment. This mechanism is not presumed to differ, in the solid, from its much studied and well understood counterpart in the free atom.

For $k \approx k'$ the above *direct* exchange, being a Coulomb integral, is basically positive, i.e., ferromagnetic. It is well known that if the Bloch states are not orthogonal to the atomic orbitals, there is an "overlap" exchange correction similar to the above, but of *opposite* sign, i.e., antiferromagnetic; so that the *total* exchange scattering may in fact be of either sign. There is, however, no need to use nonorthogonal basis functions to derive this fact. Let us calculate, using second-order perturbation theory, the effect of "interband mixing", i.e., virtual excitations of conduction electrons into unoccupied orbital states on the magnetic atom or vice versa, of magnetic electrons into the conduction band. For the purpose of this calculation, we may assume that the $\varphi_n(r)$ are *Wannier* orbitals orthogonal to the conduction band states. If in the energy denominators we may neglect the Bloch energy ε_k or $\varepsilon_{k'}$ relative to the average internal Coulomb energy U required to add an extra particle (electron or hole) in the unfilled shell, then we are describing a mechanism of order of magnitude V^2/U favoring conduction electrons of spin *anti*parallel to the magnetic shell, as the exclusion principle prohibits even virtual excitations to the occupied states. This antiferromagnetic exchange mechanism is labelled "kinetic exchange", and is given by:

$$J_{k,k'}^{\text{kin.}} = -\frac{1}{2s} \sum_{n=1}^{2s} \int d_3 r \int d_3 r' \, \varphi_n^*(r) \varphi_n(r') \frac{V(r) V(r')}{U} \psi_{k'}^*(r') \psi_k(r). \quad (23.5)$$

Thus the *net* exchange parameter which appears in H', the exchange Hamiltonian of Eq. (23.3), is given by

$$J_{k,k'} = J_{k,k'}^{\text{exch.}} + J_{k,k'}^{\text{kin.}} \quad (23.6)$$

i.e., by the same sort of integral as $J_{k,k'}^{\text{exch.}}$ but with $\left(\frac{e^2}{|r-r'|} - \frac{V(r) V(r')}{U}\right)$ replacing the pure Coulomb repulsion. We paraphrase from the work of Watson *et al.*[101]: "the effective exchange parameter, investigated with a model involving simple orthogonalized-plane wave conduction states and rare earth ion cores, may be dominated by interband mixing causing a net negative conduction-electron polarization, in agreement with a number of experiments. The interband mixing is found to be very sensitive to conduction-electron character and for ions

---

[101] R. E. Watson, S. Koide, M. Peter, and A. J. Freeman: Phys. Rev. **139**, A 167 (1965).

other than spherical Gd, to vary with conduction electron k direction, leading to a significant source of anisotropy in the induced spin density." Our formulas agree with the results of these authors. While the potential $V(r)$ can be partly eliminated by the use of atomic orbital states not orthogonalized to the Bloch states, an overlap integral must then be taken into account, which once again leads to similar results.

An additional ferromagnetic mechanism for less than half-filled shells arises from virtual excitations into *un*occupied orbitals, such as $n = 2s+1, \ldots, 2l+1$ (as well as into empty higher orbital states). This type of interband mixing favors ferromagnetic alignment because of atomic HUND's rule exchange splitting. If, as before, the excitation energy for antiparallel electrons is U, then the excitation energy for spin up electrons is $U-J$, where J is the atomic exchange parameter, which, being a Coulomb integral, is necessarily positive. The matrix elements for virtual excitations of both type electrons remain the same. Therefore, we see that a decrease in magnitude of the energy denominator favors parallel spin electrons. However, as one proceeds to higher-order perturbation corrections, other corresponding antiferromagnetic terms appear. While various such complications make it difficult to obtain $J_{k,k'}$ from "first principles" they do not affect qualitatively the properties of magnetic exchange scattering which can be obtained from the above equations.

24. Polarization cloud: perturbation theory. If the conduction band is totally empty or totally filled, as in an insulator, the spin-dependent scattering H' of Eq. (23.3) has no effect. Otherwise the magnetic atom is able to polarize the spins of the conduction particles in its vicinity and out to a certain distance. In metals or degenerate semiconductors with a definite Fermi surface, the spin polarization can be shown not only to decrease with increasing distance from the source, but also to oscillate with characteristic wavelength $\lambda_F = \pi/k_F$. In ordinary semiconductors the characteristic distance is proportional to $(\hbar^2/2m^*kT)^{\frac{1}{2}}$

The simplest derivation is based on perturbation theory, and is therefore only valid provided $J_{kk'}$ is much smaller in magnitude than the characteristic energy of a conduction particle. The strong-coupling situation is treated in the following section. We calculate the z-component of spin polarization, for example:

$$s^z(r) = \frac{1}{2N} \sum_{k,k'} \cos(\mathbf{k}-\mathbf{k}') \mathbf{r} \langle 0| c^*_{k+} c_{k'+} - c^*_{k-} c_{k'-} |0\rangle. \tag{24.1}$$

We may take $|0\rangle$ to be the state obtained in lowest order perturbation theory, assuming H' (23.3) to be the perturbation and H_0 (23.1) to be the unperturbed Hamiltonian. The above formula neglects fluctuations on an atomic scale, i.e., treats the Bloch functions as though they were plane waves. It is therefore consistent[102] to put $\varepsilon_k = \hbar^2 k^2/2m^*$ in the evaluation of the following result:

$$s^z(r) \propto S^z \int d_3 k \, d_3 k' \, f(k') \, J_{k\,k'} \frac{\cos(\mathbf{k}-\mathbf{k}') \cdot \mathbf{r}}{\varepsilon_k - \varepsilon_{k'}}. \tag{24.2}$$

Similar expressions hold for $s^x(r)$ and $s^y(r)$. Approximating $J_{k,k'}$ by a constant parameter J/N one finds the integrals can be evaluated in terms of elementary functions, both for metals and for semiconductors. In the case of metals, we take the Fermi function $f(k)$ to equal unity for $k \leq k_F$ and zero for $k > k_F$ [ignoring

---

[102] See, however, the discussion of the band structure of the lanthanides in (Sect. 41) concerning deviations from the simple theory. The validity of the assumption $J_{k,k'} = $ const has been discussed and criticized by R. E. WATSON, and A. J. FREEMAN in "Hyper fine-interactions", p. 53. New York: Academic press 1967.

thermal fluctuations as the source of small corrections $0(kT/E_F)$]. Straightforward evaluation of (24.2) results in

$$\mathbf{s}(r) \equiv [s^x(r), s^y(r), s^z(r)] \propto \mathbf{S} J F(2k_F r) \quad \text{(metals)} \tag{24.3}$$

where F is the oscillatory function of RUDERMAN and KITTEL[103],

$$F(x) = \frac{\sin x - x \cos x}{x^4} \xrightarrow[(x \gg 1)]{} \frac{-\cos x}{x^3}. \tag{24.4}$$

In semiconductors, the Boltzmann distribution must be used and one expects a nontrivial dependence on the temperature. Starting with

$$f(k) = A \exp[-\varepsilon(k)/kT] \tag{24.5}$$

one finds the total number of carriers to be

$$n = 2 \sum_k f(k) \propto A \left| \frac{2m^* kT}{\hbar^2} \right|^{\frac{3}{2}} \tag{24.6}$$

assuming parabolic bands and effective mass m^*. The parameter A will depend exponentially on $1/T$ for intrinsic semiconductors, and less sensitively on the temperature for extrinsic or degenerate semiconductors, as discussed in any general reference on semiconductors. Again assuming a constant J one can evaluate the analogue of the Ruderman-Kittel function for the semiconductor[104]:

$$\begin{aligned}\mathbf{s}(\mathbf{r}) &\propto \frac{SJ}{(2\pi)^6} \int d_3 k \, d_3 k' \, A \, e^{-\varepsilon(k')/kT} \frac{\cos(\mathbf{k}-\mathbf{k}') \cdot \mathbf{r}}{\varepsilon(k) - \varepsilon(k')} \\ &\propto \frac{SJn}{r} e^{-\left(\frac{2mkT r^2}{\hbar^2}\right)} \quad \text{(semiconductors)}\end{aligned} \tag{24.7}$$

in which we make use of expression (24.6) for the density of carriers.

If one carries the interaction H' to higher order in the perturbation expansion, he finds that the operator nature of the scattering Hamiltonian (the various components of S do not commute) gives rise to certain interference terms of the type first discovered by KONDO[105] in the study of the electrical resistance of magnetic impurity atoms. The simultaneous effect of this operator interference and of the exclusion principle on the spin polarization has been studied by FULLENBAUM and FALK[106] who have calculated a small *logarithmic* correction to the Ruderman-Kittel function for metals. In ordinary semiconductors there is no such correction, as the total number of electrons in the vicinity of a magnetic impurity will be so low that one may ignore the exclusion principle, as is often done implicitly in replacing the Fermi function by its Boltzmann analogue.

---

[103] M. A. RUDERMAN, and C. KITTEL: Phys. Rev. **96**, 99 (1954); T. KASUYA, Progr. Theoret. Phys. (Kyoto) **16**, 45 (1956); K. YOSIDA: Phys. Rev. **106**, 893 (1957). — J. H. VAN VLECK: Rev. Mod. Phys. **34**, 681 (1962) discusses the point whether this is first-, second-, or mixed first- and second-order perturbation theory.

[104] B. V. KARPENKO, and A. A. BERDYSHEV: Soviet Phys.-Solid State **5**, 2494 (1964); transl. from Fiz. Tverdogo Tela **5**, 3397 (1963). [An extraneous and incorrect statement is made in this paper that the Ruderman-Kittel interaction in metals is capable of leading only to ferromagnetism. It is well known that because of its oscillatory character, it can in fact lead to all sorts of antiferromagnetic arrangements, for sufficiently large k_F such as those discussed in D. MATTIS and W. DONATH, Phys. Rev. **128**, 1618 (1962) and in Sect. 27 of the present article. The nonoscillatory function (24.7), similar to the Ruderman-Kittel function for $k_F \to 0$, leads only to ferromagnetism.] A. J. FEDRO, and T. ARAI: Phys. Rev. **170**, 583 (1968) have given a related discussion.

[105] J. KONDO: Progr. Theoret. Phys. (Kyoto) **32**, 37 (1964).

[106] M. FULLENBAUM, and D. FALK: Phys. Rev. **157**, 452 (1967).

25. Polarization cloud: strong coupling.

In order to determine the magnitude of the error introduced by use of lowest-order perturbation theory, it is valuable to have an exactly soluble model which one can investigate when the coupling is no longer weak. Such a model is provided by $H_0 + H'$ of Eqs. (23.1) and (23.3) provided we discard the troublesome transverse terms (S^+ and S^-) and assume $J_{k,k'} = J/N$ (constant) as before. We use the formalism of ZUBAREV[107] who defines one-electron Green functions in the following manner:

$$g_{k,k'}^m (t) = -i\vartheta(t) \langle \{c_{km}(t), c_{k'm}^*(0)\} \rangle \tag{25.1}$$

where the curly brackets indicate an anticommutator, the triangular brackets a thermal average, and the time dependence is given in the Heisenberg representation using the specified Hamiltonian, i.e., for arbitrary operators A, we have: $A(t) = e^{+iHt} A e^{-iHt}$. In the calculation, one actually uses the Fourier transform of the above, denoted

$$G_{k,k'}^m (\omega) = \frac{1}{2\pi} \int_{-\infty}^{\infty} dt\, e^{i\omega t} g_{k,k'}^m (t) \tag{25.2}$$

in terms of which the following thermal averages can be calculated:

$$\langle c_{k'm}^*(0) c_{km}(t) \rangle = i \int_{-\infty}^{\infty} d\omega\, e^{-i\omega t} f(\omega) [G_{k,k'}(\omega + i0^+) - G_{k,k'}(\omega - i0^+)]. \tag{25.3}$$

This, or rather the special case $t=0$ of this expression, is precisely what one must calculate to obtain the spin polarization, cf. Eq. (24.1). To evaluate (25.3) we shall develop the equations of motion of the Green function (25.2), by noting that the time derivative of an operator, calculable as the commutator of that operator with H, is also equivalent to multiplication by $i\omega$ according to (25.2). Thus, taking $S^z = \tfrac{1}{2}$ in (23.3) and $m = \pm \tfrac{1}{2}$,

$$(\omega - \varepsilon_k) G_{k,k'}^m = \frac{1}{2\pi} \delta_{k,k'} - \frac{Jm}{N} \sum_{k''} G_{k'',k'}^m. \tag{25.4}$$

This is an integral equation which is readily solved. Start by defining the sum on the right-hand side to be $\frac{1}{2\pi} \varphi^m(k')$, thus

$$G_{k,k'}^m = \frac{1}{2\pi} \frac{\delta_{k,k'} - Jm\, \varphi^m(k')}{\omega - \varepsilon_k} \tag{25.5}$$

and

$$\frac{1}{2\pi} \varphi^m(k') = \frac{1}{N} \sum_{k''} G_{k'',k'}^m = \frac{1/2\pi}{\omega - \varepsilon_{k'}} - (Jm\, \varphi^m(k')/2\pi N) \sum_{k''} \frac{1}{\omega - \varepsilon_{k''}}$$

so that

$$\varphi^m(k') = \frac{1}{\omega - \varepsilon_{k'}} \left(1 + \frac{Jm}{N} \sum_{k''} \frac{1}{\omega - \varepsilon_{k''}} \right)^{-1} \tag{25.6}$$

and finally:

$$G_{k,k'}^m = \frac{\delta_{k,k'}/2\pi}{\omega - \varepsilon_k} - \frac{Jm/2\pi N}{(\omega - \varepsilon_k)(\omega - \varepsilon_{k'}) \left(1 + \frac{Jm}{N} \sum_{k''} \frac{1}{\omega - \varepsilon_{k''}}\right)} \tag{25.7}$$

---

[107] D. N. ZUBAREV: Uspekhi Fiz. Nauk **71**, 71 (1960). English trans.: Soviet Phys. Uspekhi **3**, 320 (1960).

To evaluate this slightly off the real axis, we use the identity

$$\frac{1}{x \pm i0^+} = P(1/x) \mp i\pi\delta(x) \tag{25.8}$$

so that, in terms of the density-of-states-per-atom function $N(\varepsilon_k)/N \equiv n(\varepsilon_k)$

$$\frac{1}{N}\sum_{k''}\frac{1}{\omega-\varepsilon''} = R(\omega) \mp i\pi n(\omega). \tag{25.9}$$

It should be noted that $Jn(\omega)$ is a dimensionless quantity which is independent of N, and is of the order of magnitude of the exchange energy multiplied by the density of states/atom in the conduction band at energy ω. Similarly, JR is

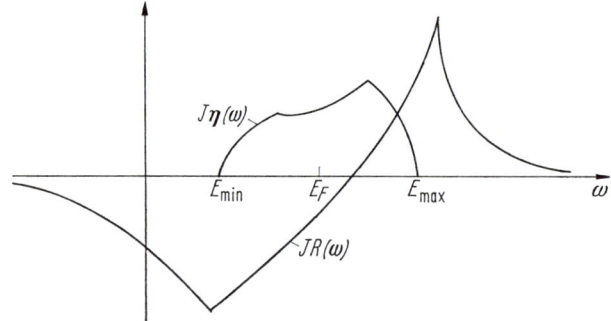

Fig. 10. Real part R of scattering function and imaginary part. The latter is proportional to band structure density of states $n(\omega)$.

a dimensionless function, of magnitude approximately on the order of the ratio of exchange energy to conduction-band-width. A schematic plot of these quantities is given in Fig. 10.

Because the poles of G determine the nature of the perturbed states, we examine the denominators in (25.7) carefully. In addition to the usual singularity at ε_k (corresponding to the unperturbed Bloch state), we see that a new pole might appear if $(1+mJR)$ vanishes. If it vanishes outside the band, where $N(\omega)=0$, this is indeed a pole corresponding to a new *bound state*. If this quantity vanishes *within* the band, the imaginary part remains finite and one speaks of a "virtual bound state", i.e., an in-band resonance. From a study of Fig. 10 we see that if there is a bound state a resonance will necessarily also occur within the band.

Now we combine the expression for G^m with the equation for thermal averages (25.3), insert into the spin polarization (24.1) to obtain:

$$s^z(r) = \frac{J}{\pi N^2}\sum_{k,k'}'\frac{\cos(k-k')\cdot r}{\varepsilon_k - \varepsilon_{k'}}\int d\omega\, f(\omega) \times \\ \times \operatorname{Im}\left\{\left[\frac{1}{\omega-\varepsilon_k-i0^+} - \frac{1}{\omega-\varepsilon_k'-i0^+}\right]Q(\omega)\right\} \tag{25.10}$$

where

$$Q(\omega) = [1 - J^2(R(\omega)+i\pi n(\omega))^2]^{-1}.$$

Near the band middle $R\sim 0$, and Q is approximately real. It is also approximately real near the band edges where $n(\omega)\to 0$, which is the usual case of interest to semiconductors. In either case after neglect of small terms involving $\mathrm{Im}Q$, we obtain:

$$s^z(r) = \frac{2J}{(2\pi)^6} \int d_3 k \int d_3 k' \frac{f(\varepsilon_k)\cos(k-k')\cdot r}{\varepsilon_k - \varepsilon_{k'}} \,\mathrm{Re}\,(Q(\varepsilon_k)) \tag{25.11}$$

which reduces to (24.2) when $Q=1$ in the limit $J \to 0$. We see that for $J \neq 0$ *the perturbation-theoretic results* (24.3) *and* (24.7) *remain essentially correct if J is replaced by* J_{eff},

$$J_{\mathrm{eff}} = J/[1 - J^2 R^2(E_0)] \quad \text{(semiconductor)} \tag{25.12}$$

provided $|JR(E_0)| < 1$, where E_0 is the typical energy (e.g., the energy at the bottom of the band for the conduction band of a semiconductor). In the case of a metal with E_F in the middle of a band, it is more nearly accurate to set $R=0$ and take

$$J_{\mathrm{eff}} = J/[1 + J^2 \pi^2 n^2(E_F)] \quad \text{(metal)} \tag{25.13}$$

where $n(E_F) = N(E_F)/N =$ density of states/atom at the Fermi level. Whereas the exchange parameter tends to be boosted near the band edges, it is reduced in the middle of a band. We emphasize that J can not be supposed large enough to produce bound states [the neglected Coulomb interaction is determining in that respect, so that in deriving (25.11) we have neglected any possible bound state contributions to (25.10)], which is the reason why we restrict J to values $|JR| < 1$.

26. Friedel's phase shift analysis.

(a) *Metals.* The delta-function potential considered in the preceding section is not often applicable to real impurity atoms, in which the Coulomb interactions play an important role. Nor can the Coulomb interactions among conduction particles be totally ignored. Motivated by considerations of electrical neutrality, FRIEDEL[108] and his collaborators have reformulated the problem for an arbitrary potential, in terms of a small number of phase shifts. When the potential is unknown these can not be directly calculated but are then obtained from experiment. This simple theory has thus been highly successful in correlating known experimental and theoretical information.

In briefly reviewing this theory for metals, we start by recalling the expansion of a plane wave in spherical Bessel functions:

$$\psi_k^0 = e^{i k \cdot r} = \sum_l i^l [2(2l+1)]^{\frac{1}{2}} j_l(kr) P_l(\cos\vartheta). \tag{26.1}$$

In the presence of a scatterer (without bound states) the above is modified by the introduction of phase shifts η_l and Neumann functions:

$$\psi_k(r) = \sum_l i^l e^{i\eta_l} [2(2l+1)]^{\frac{1}{2}} P_l(\cos\vartheta) [j_l(kr)\cos\eta_l - n_l(kr)\sin\eta_l] \tag{26.2}$$

an expansion valid outside the range of the scatterer. Let us now compare perturbed and unperturbed charge densities. Originally we have:

$$\varrho^0 = \int d_3 k\, f_k\, \varrho_k^0(r) = \int d_3 k\, f_k |e^{i k \cdot r}|^2 = \int d_3 k\, f_k \tag{26.3}$$

a result independent of r. The perturbed charge density, however, will be a function of the radial distance from the scatterer, given by

$$\delta\varrho_k(r) = |\psi_k|^2 - |e^{i k \cdot r}|^2. \tag{26.4}$$

---

[108] J. FRIEDEL: Nuovo Cimento **7**, Suppl., 287 (1958); Can. J. Phys. **34**, 1190 (1956). Applications to magnetic metals are discussed in J. FRIEDEL, J. Phys. Radium **23**, 501 (1962).

When this is averaged over orientations of k (prior to summing over magnitudes of k in the range $0 < k < k_F$) the result is the formula of BLANDIN, DANIEL and FRIEDEL[109]:

$$\langle \delta \varrho_k(r) \rangle = \sum_{l=0}^{\infty} (2l+1) \{\sin^2 \eta_l [n_l^2(kr) - j_l^2(kr)] - j_l(kr) n_l(kr) \sin 2\eta_l\}. \quad (26.5)$$

In this procedure one ignores crystalline symmetries, and assumes isotropy in coordinate and momentum spaces. These assumptions are satisfactory at large distances from the scatterer, for crystals of reasonably high symmetry and simple band structure[110]. Compatible with these assumptions, let us now examine the asymptotic form of the spherical functions valid at large distances r:

$$\left. \begin{aligned} j_l(kr) &\sim \frac{\sin(kr - \tfrac{1}{2}\pi l)}{kr}, \\ n_l(kr) &\sim \frac{\cos(kr - \tfrac{1}{2}\pi l)}{kr}. \end{aligned} \right\} \quad (26.6)$$

Evaluating (26.5) asymptotically we find:

$$\langle \delta \varrho_k(r) \rangle_{\text{angles}} = \sum_l (-1)^l (2l+1) \sin \eta_l \frac{\sin(2kr + \eta_l)}{(kr)^2}. \quad (26.7)$$

Finally, we integrate this expression over the magnitude of k up to the Fermi level and divide by ϱ^0 to normalize the result. It is assumed that the phase shifts η_l vanish at $k=0$, but that they may be taken constant at the upper limit of integration. Thus:

$$\left. \begin{aligned} \frac{\delta \varrho(r)}{\varrho_0(r)} &= \sum (-1)^l (2l+1) \frac{\int^{k_F} dk\, k^2 \sin \eta_l \sin(2kr + \eta_l)/(kr)^2}{\int k_F\, dk\, k^2} \\ &= -\frac{3}{2} \sum (-1)^l (2l+1) \sin \eta_l^F \left[\frac{\cos(2k_F r + \eta_l^F)}{(k_F r)^3} \right] \\ &= -A \frac{\cos(2k_F r + \vartheta)}{(k_F r)^3} \end{aligned} \right\} \quad (26.8)$$

a result which differs from Ruderman-Kittel theory in the asymptotic range only by the possible differences in amplitude A and in the phase angle ϑ. These two quantities are evaluated using (26.8):

$$A = \tfrac{3}{2} \left[\left| \sum_{l=0}^{\infty} (-1)^l (2l+1) \sin \eta_l^F e^{i \eta_l^F} \right|^2 \right]^{\frac{1}{2}}. \quad (26.9)$$

and

$$\tan \vartheta = \frac{\sum (-1)^l (2l+1) \sin^2 \eta_l}{\sum (-1)^l (2l+1) \sin \eta_l \cos \eta_l} \quad (26.10)$$

The phase shifts for electrons of spin up (relative to the magnetic scatterer) will differ[111] from those of spin down. Thus $A_\uparrow$, $\vartheta_\uparrow$ will differ from $A_\downarrow$, $\vartheta_\downarrow$ and the spin density will be given by

$$s^z(r) = \tfrac{1}{2}(\delta \varrho_\uparrow - \delta \varrho_\downarrow). \quad (26.11)$$

---

[109] A. BLANDIN, E. DANIEL, and J. FRIEDEL: Phil. Mag. **4**, 180 (1959).

[110] The theory of scattering taking into account the crystalline nature of the solid is developed in J. CALLAWAY, J. Math. Phys. **5**, 783 (1964).

[111] The concept of using different phase shifts for spins up and down is already in A. BLANDIN and J. FRIEDEL, J. Phys. Radium **20**, 160 (1959), and J. FRIEDEL, J. Phys. Radium **23**, 501 (1962). It is valid as long as one may ignore $S\pm$ matrix elements.

The spin up and spin down charge densities must add up to cancel any net valency difference Z between host and impurity atoms. Thus by FRIEDEL's sum rule[112]:

$$Z = \frac{1}{\pi} \sum_l (2l+1)(\eta^F_{l\uparrow} + \eta^F_{l\downarrow}). \tag{26.12}$$

For the electrical resistance we may take the currents of electrons of spin up and down to be in parallel, thus the resistivity R is given by

$$R = \left(\frac{1}{R_\uparrow} + \frac{1}{R_\downarrow}\right)^{-1} \tag{26.13}$$

where $R_\pm$ is the sum of nonmagnetic resistivity mechanisms R_0 and of the magnetic contribution $r_\pm$, assuming the validity of MATHIESSON's rule which states that various mechanisms of resistance are additive (i.e., in series). The magnetic contribution can also be calculated in terms of phase shifts[112,113]:

$$r_\pm = \frac{4\pi c_\pm}{p_\pm k_F} \sum_l l \sin^2(\eta^F_{l-1,\pm} - \eta^F_{l,\pm}), \tag{26.14}$$

where $c_\pm$ gives the concentration of impurities with spin up/down, which are generally equal in the absence of external magnetic field to line them up. The unperturbed density of electrons of spin up/spin down per unit cell is $p_\pm$.

If a sufficiently small number of phase shifts belonging to appropriate l is chosen to characterize a given impurity atom, the resistivity, valency and spin polarizations can all be adequately correlated with any a-priori assumptions about the exact properties of the scattering potential. We should note that the relation of (26.12) to (26.14) holds equally well when one allows for interactions among the conduction particles, as proved by LANGER and AMBEGAOKAR[114] subject only to the validity of infinite-order perturbation theory (specifically, the existence of a sharp Fermi surface). But this apparently innocuous limitation might be fatal to the direct application of the theory to magnetic impurities, as there are reasons to believe that the appearance of a localized moment signifies the breakdown of perturbation theory. There is an example of this in an exactly soluble 2-body problem concerning the magnetization of donors in semiconductors[115], but for metals or degenerate semiconductors this point has not been sufficiently investigated.

(b) *Semiconductors.* A modification of FRIEDEL's sum rule applicable to the Boltzmann distribution was recently proposed by STERN[116]. We now rederive his formulas and obtain some new results on the nature of the polarization cloud induced in the carriers near the impurity[117] by following the treatment in KITTEL's text "Quantum Theory of Solids", replacing the Fermi distribution by BOLTZMANN's whenever appropriate. One generalizes to the case where several species of carriers are present (e.g. electrons and holes, or electrons belonging to distinct "valleys") by labelling each species i, introducing a degeneracy index g_i ($g_i = 2$ for the simplest case of KRAMER's degeneracy, but as this degeneracy is lifted for magnetic scatterers, $g_i = 1$ for them), and a charge $q_i = +e$ for holes and $-e$

---

[112] Note that this formula can *not* be obtained from (26.8) directly, although it does follow from elementary arguments in reference 113.
[113] J. FRIEDEL: Phil. Mag. **43**, 153 (1952); Can. J. Phys. **34**, 1190 (1956).
[114] J. LANGER, and V. AMBEGAOKAR: Phys. Rev. **121**, 1090 (1961).
[115] D. MATTIS, and E. LIEB: Theory of Paramagnetic Impurities in Semiconductors. J. Math. Phys. **7**, 2045 (1966).
[116] F. STERN: Phys. Rev. **158**, 697 (1967).
[117] D. MATTIS, and O. SINHA: To be published.

for electrons. For whatever distribution, one first obtains a general expression for Δq, the net charge displaced at a large distance r from the impurity:

$$\Delta q = \frac{1}{\pi} \sum_{i,l} (2l+1) g_i q_i \times \\ \times \int_0^\infty dk\, f_k \left[\frac{\partial \eta_{i,l}}{\partial k} - k^{-1} \sin \eta_{i,l} \cos 2\left(kr + \eta_{i,l} - \frac{1}{2}\pi l\right) \right]. \tag{26.15}$$

The first term in the integrand is the systematic non-oscillatory charge deviation which must have a value ensuring charge neutrality. Thus,

$$Ze = -\frac{1}{\pi} \sum_{i,l} (2l+1) g_i q_i \int dk\, f_k \frac{\partial \eta_{i,l}}{\partial k} \tag{26.16}$$

which is FRIEDEL's sum rule as derived by STERN. The second term in the integrand above yields the oscillatory charge density $\varrho(r)$:

$$\varrho(r) = \frac{1}{4\pi r^2} \frac{\partial}{\partial r} \Delta q = \frac{1}{4\pi^2 r^2} \sum_{i,l} g_i q_i (2l+1) \times \\ \times \int_0^\infty dk\, f_k \left[\cos 2\left(kr - \frac{1}{2}l\pi\right) - \cos 2\left(kr + \eta_{i,l} - \frac{1}{2}l\pi\right) \right]. \tag{26.17}$$

The evaluation of this formula, so straightforward in the case of the Fermi distribution, depends in the case of the Boltzmann distribution, on the functional dependence of the phase shift on the energy. For simplicity, in what follows, let us consider only s-wave scattering and assume the validity of the Born approximation. In this manner we can evaluate the above expression and make contact with the perturbation-theoretic result of Eq. (24.7). Thus only $l=0$ is retained in the above sum, and we assume $\eta = ak$, where

$$a_i = -\int j_0^2(kr) \frac{2m_i V(r)}{\hbar^2} d^3r \quad \text{(Born approx.)} \tag{26.18}$$

and using distribution function as given in (24.5) we find:

$$\varrho(r) = \sum_i \frac{g_i q_i A_i}{4\pi^2 r^2} \int_0^\infty dk\, e^{-(k d_i)^2} [\cos 2kr - \cos 2k(r+a_1)] \tag{26.19}$$

in which we have defined the thermal length,

$$d_i = \hbar (2m_i kT)^{-\frac{1}{2}}. \tag{26.20}$$

Evaluation of the integral yields:

$$\varrho(r) = \frac{1}{8\pi^{\frac{3}{2}} r^2} \sum_i \frac{g_i q_i A_i}{d_i} \left[e^{-r^2/d_i^2} - e^{-(r+a_i)^2/d_i^2} \right]. \tag{26.21}$$

This reduces to the perturbation-theoretic result (24.7) at low temperatures and for moderate distances r, such that $a_i \ll d_i$ and $a_i r \ll d_i^2$, for it then becomes permissible to expand the exponentials:

$$[e^{-r^2/d^2} - e^{-(r+a)^2/d^2}] \doteq e^{-r^2/d^2} (2ar/d^2) \tag{26.22}$$

which leads to a linear dependence on the small parameter a, compatible with the use of the Born approximation. Use of this expansion in the proceding equation, and identification of $V(r)$ with the exchange parameter $J(r)$ for spin-up carriers, and $-J(r)$ for spin-down carriers, leads to an expression identical in form with the result of second-order perturbation theory, Eq. (24.7).

The new derivation suggests, however, that *no matter how weak* the coupling constant may be taken to be, the *perturbation-theoretic analysis breaks down* at large distances $r \gtrsim \frac{1}{2} d^2/a$, where the exponential terms may no longer be expanded as in (26.22) above.

For the electrical resistance, we make approximate use of the formulas used for metals. Currents associated with each species i are ordinarily in parallel. Adapting Eqs. (26.13) and (26.14), with replacement of k_F and $\eta(k_F)$ by d_i^{-1} and $\eta(d_i^{-1})$ results in:

$$R = \left[\sum_i \frac{1}{R_{i,\uparrow}} + \sum_i \frac{1}{R_{i,\downarrow}} \right]^{-1} \tag{26.23}$$

where each $R_{i,\pm}$ is the sum of a nonmagnetic contribution R_i^0 and of a magnetic scattering contribution $r_{i,\pm}$, with the latter being:

$$r_{i,\pm} = \frac{4 c_{\pm} d_i}{p_{\pm}} \sum_l l \sin^2(\eta_{i,l-1,\pm} - \eta_{i,l,\pm}). \tag{26.24}$$

The application of such formulas to semiconductors and their evaluation beyond the Born approximation, as required for low-energy carriers, appears to be as promising as for the case of metallic alloys. It is, however, an area of investigation mainly unexplored at the present (1968).

27. Indirect exchange. The exchange interaction of an impurity at a point R_i causes a spin polarization of the conduction medium at point R_j,

$$\mathbf{s}_i(R_j) = [s_i^x(R_j), s_i^y(R_j), s_i^z(R_j)]. \tag{27.1}$$

Assuming once more a constant $J_{k,k'} = J/N$, according to (23.3) the (first-order) interaction of another magnetic atom at R_j with the spin polarization caused by the atom at R_i is

$$-J S_j \cdot \mathbf{s}_i(R_j) \tag{27.2}$$

and the total *indirect exchange* magnetic interaction Hamiltonian is a sum over all magnetic atoms at points R_i:

$$H = -J \sum_{i,j} S_j \cdot \mathbf{s}_i(R_j). \tag{27.3}$$

This may be further simplified of J is sufficiently small that lowest-order perturbation theory applies in the calculation of the spin-density function itself.

Then we obtain the Ruderman-Kittel type Hamiltonian among pairs of solute spins:

$$H = -J^2 \sum_{i,j} F(R_{ij}) S_i \cdot S_j, \tag{27.4}$$

where for metals $F(R) = (\sin 2k_F R - 2k_F R \cos 2k^F R)/R^4$, as found in (24.3) and for semiconductors the exponential form (24.7) applies. If the average distance among magnetic atoms in a *semiconductor* is R, so that in terms of the lattice constant a the concentration c of impurities is

$$c = (a/R)^3, \tag{27.5}$$

we find for the average coupling constant F to be used in (27.4):

$$\langle F(R) \rangle \approx F(\langle R \rangle) = n c^{\frac{4}{3}} \exp - (2m^* kT a^2/\hbar^2 c^{\frac{2}{3}}) \quad \text{(semiconductor)}. \tag{27.6}$$

The sign favors ferromagnetism, but the magnitude effectively vanishes at high temperatures

$$kT \gg \hbar^2 c_F^{\frac{2}{3}}/2m^*a^2 \tag{27.7}$$

in which case we are left with independent noninteracting paramagnetic impurities, and the high-temperature paramagnetic susceptibility will be expected to strictly obey CURIE's law $\chi = C/T$ with *no* Weiss T^{-2} molecular field corrections. At intermediate or low temperatures, the magnetic properties can be compared to those of an ordinary Heisenberg nearest-neighbor ferromagnet, subject to the slight modifications required to take the temperature dependence of the interaction F into account.

The case of a metal or doped semiconductor is more difficult, although the Hamiltonian (27.4) can be factored into two parts, only one of which depends on lattice spacing and geometry, and electron concentration. We recall that the effective spin operator is not $\boldsymbol{S}_i$ but $(g-1)\boldsymbol{J}_i = (g-1)(\boldsymbol{S}_i + \boldsymbol{L}_i)$ so that for magnetic ions all of the same species the Hamiltonian takes the form:

$$\left.\begin{array}{r}H = -J^2(g-1)^2 j\,(j+1) \times \\ \times \sum_{ij} F(R_{ij})\hat{u}_i \cdot \hat{u}_j\end{array}\right\} \tag{27.8}$$

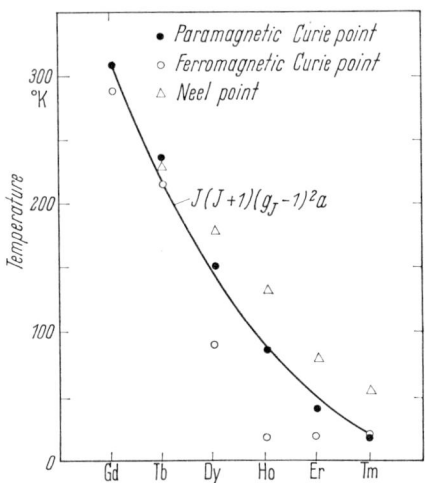

Fig. 11. Comparison of the Curie points for various heavy rare earths metals with the theoretical value based on molecular field theory. The constant a has been assumed to be independent of atomic number and has been selected so as to give agreement with the experimental paramagnetic Curie point of Gd. The figure also shows the observed Néel points below which the metal acquires antiferromagnetic ordering prior to becoming ferromagnetic at a still lower temperature. [From V. VLECK, in: Progr. in the Science and Technology of RE, Vol. 2 (L. EYRING, Ed.). Oxford: Pergamon Press 1966.]

in which j is the magnitude of $\boldsymbol{J}$ and $\hat{u}_i$ is a normalized vector operator which, for the large spin magnitudes characterizing the rare earths, may be accurately represented by an ordinary three-dimensional unit vector. The oscillatory summand has a period which depends on lattice spacing and Fermi wavevector through the parameter $k_F a$, the cancellation of various terms in the sum being dependent on the particularities of the lattice structure. Early numerical investigations[118,119] which were restricted to the first few neighbors were somewhat unreliable as the basis for comparison with experiment, but the factor $(g-1)^2 j\,(j+1)$, first derived by DE GENNES[118], could be independently checked on the heavy rare earth metals. As these all have similar crystal structures and equal numbers of conduction electrons, their magnetic properties should obey a law of corresponding states scaled by DE GENNES' factor (assuming constant J for all RE elements). A comparison between experimental critical temperatures and the theoretical factor is given in Fig. 11. The paramagnetic Curie temperatures show excellent agreement with a law of corresponding states, but other thermodynamic features (the ferromagnetic Curie point, the Néel point as well as the absence thereof in the case of

---

[118] P. G. DE GENNES: Compt. rend. **247**, 1836 (1958).
[119] E. WOLL, and S. NETTEL: Phys. Rev. **123**, 796 (1961). J. CHEVALIER, and W. BALTENSPERGER: Helv. Phys. Acta **34**, 859 (1961).

Gd) show discrepancies, suggesting the importance of Fermi surface structure and of anisotropy (small in Gd but strong to varying degrees in the other elements) as well as related effects such as magnetostriction, in the detailed interpretation of the magnetism of these substances[120]. (Additional discussion will be given in Sect. 51 below.) Thus (27.8) can provide only an approximate explanation of rare earth magnetism as long as anisotropic factors are not explicitly taken into account. However, a detailed investigation of the oscillatory factor could be expected to explain the magnetism of Gd and of isotropic or idealized magnetic metals and

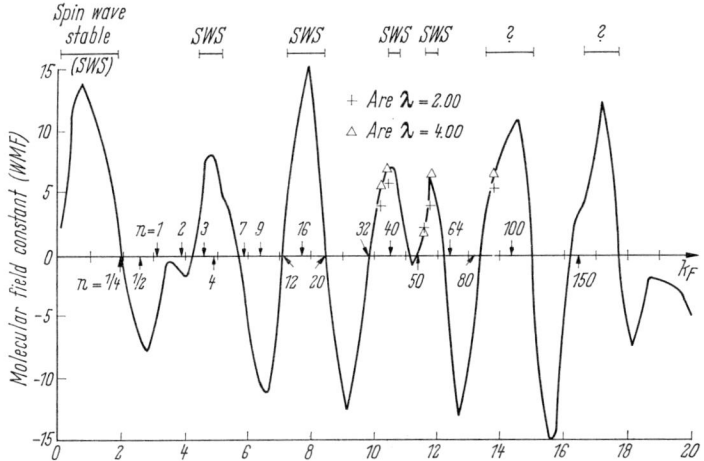

Fig. 12. Calculated WMF (proportional to paramagnetic Curie temperature) plotted for s.c. lattice as function of k_F assuming unit separation between spins. Relevant values of $n = $ No. of conduction electrons/spin $= 8\pi/3\,(k_F/2\pi)^3$ are indicated, as well as regions of spin wave stable ferromagnetism. Main curve is for free path $\lambda = 3.00$, with some points indicating values for WMF for $\lambda = 2.00$ and 4.00. [From D. MATTIS: The Theory of Magnetism. New York: Harper & Row 1965].

doped semiconductors. This hope prompted extensive computer calculations[121] of the following quantities: the (normalized) factor in the paramagnetic Curie temperature:

$$WMF \equiv \sum_{i \neq 0} F(R_i)\, e^{-\lambda|R_i|} \tag{27.9}$$

and the spin wave spectrum given by:

$$\hbar\omega(k) = \sum_i F(R_i)(1 - e^{i k \cdot R_i}), \tag{27.10}$$

where λ is a convergence factor tantamount to a mean free path. In Figs. 12—14 we plot $WMF(k_F a)$ for three cubic magnetic lattices, and in Figs. 15a and b we show the spin wave spectrum for two representative values of $k_F a$ in the simple cubic lattice. In the first of these the spin waves all have positive energies and the ferromagnetic ground state is stable at low temperature, although the ferro-

---

[120] R. J. ELLIOT: Theory of Magnetism in the Rare Earth Metals, in: Magnetism, vol. IIA (RADO and SUHL, Eds.), New York: Academic Press 1965.

[121] Earliest results given in D. MATTIS, and W. DONATH: Phys. Rev. **128**, 1618 (1962). Completed calculations in D. MATTIS, The Theory of Magnetism, New York: Harper & Row 1965 (see Appendix) and in somewhat greater detail, in: D. MATTIS et al.: Tables for Theory of Magnetism. IBM Report RC-945, May 23, 1963 (unpublished).

magnetic Curie temperature will not agree precisely with the paramagnetic one. The figure (15b) illustrates a material which has a positive paramagnetic Curie temperature but which, even at $T=0$, is unstable with respect to antiferromagnetic ordering.

The stability conditions for the spin order are here derived with the assumption of a constant $s-f$ exchange parameter J and neglecting the anisotropy of the

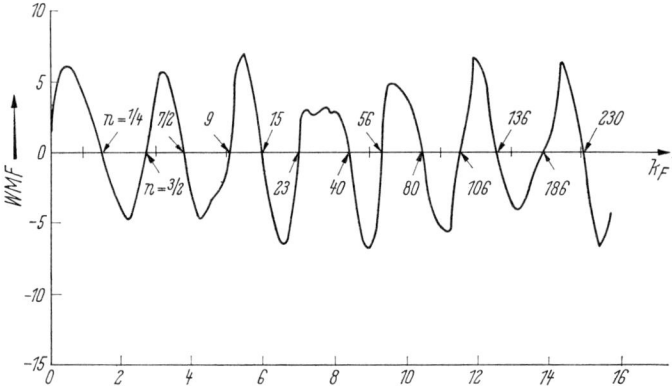

Fig. 13. WMF in f.c.c. lattice, $n = 16\pi/3\,(k_F/2\pi)^3$.

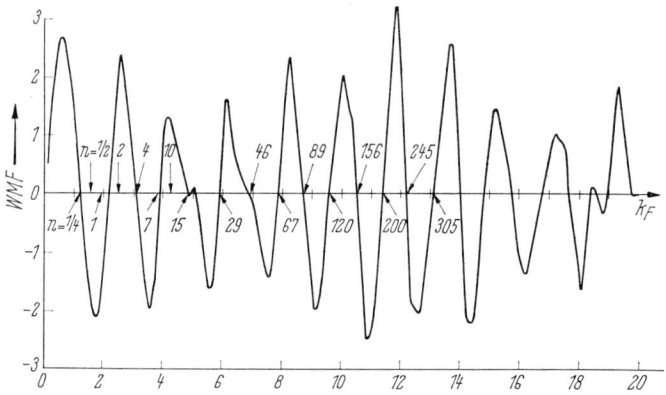

Fig. 14. WMF in b.c.c. lattice, $n = 32\pi/3\,(k_F/2\pi)^3$.

crystals. If one uses more realistic parameters[122] $J(q)$ the oscillatory behavior of having ferro- and antiferromagnetic states is found to depend critically on the functional form of $J(q)$. The $J(q)$ obtained by WATSON and FREEMAN decays too fast for large q and yields only the ferromagnetic state for all values of k_F and a. If the range of $J(q)$ exceeds a certain value of q, however, the oscillatory behavior similar to that found here appear suddenly.

In semiconductors the band minimum is often at a point $k \neq 0$ of the Brillouin zone, with the result that several minima or *valleys* are centered about the equivalent points in the zone. The evaluation of integrals such as (24.2) in the case of the multi-valley bandstructure leads to oscillation associated with intervalley

---

[122] A. J. FEDRO, and T. ARAI: Phys. Rev. **170**, 583 (1968); R. E. WATSON, and A. J. FREEMAN: Phys. Rev. **152**, 566 (1966);

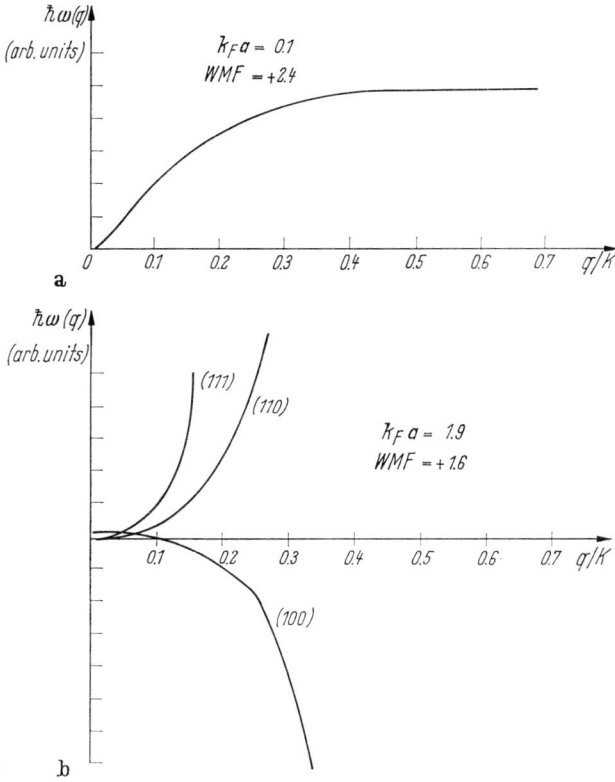

Fig. 15a and b. Spin waves in S.C. RE metals, Eq. (27.10). (a) For $k_F a \ll 1$ spin waves are nearly isotropic and constant over the entire B.Z. (except near $q=0$). This is typical of long ranged interactions. (b) Example of spin-wave instability: WMF > 0 but structure is unstable against antiferromagnetism in (100) direction. (From data in D. MATTIS: The Theory of Magnetism. New York: Harper & Row 1965.)

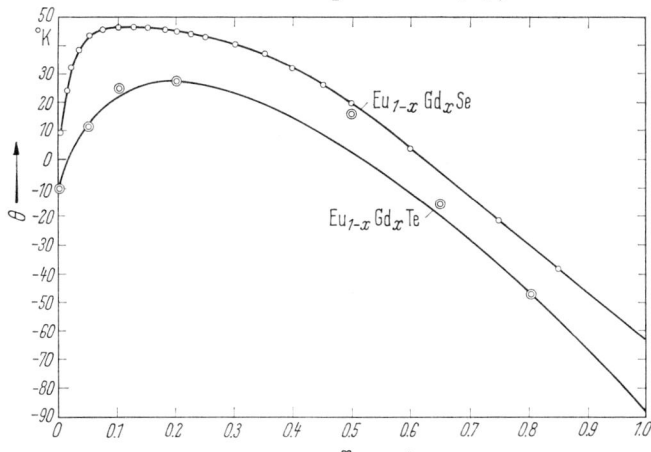

Fig. 16. Paramagnetic Curie temperature of $Eu_{1-x}Gd_xTe$ and $Eu_{1-x}Gd_xSe$. Conduction electron density increases with x, filling d equivalent valleys. The indirect exchange interaction, and thereby the Curie temperature is extremely sensitive to x, particularly at low densities. The theoretical solid lines are adjusted to the experimental points for $x \sim 0.1$ and by selecting a suitable parameter value d. It is $d \sim 16$ for taken for the telluride and $d \sim 4$ for the selenide.
[From CULLEN, CALLEN, and LUTHER: J. Appl. Phys. 39, 1105 (1968).]

scattering[123]. In Fig. 16 we show the result of such a calculation by CULLEN, CALLEN and LUTHER, as applied to semiconducting EuSe and EuTe doped with varying amounts of Gd donors (in which it is assumed that each Gd atom contributes precisely one electron to the conduction band). This assumption cannot be valid at both low donor concentrations and low temperatures, when the conduction electrons are trapped at the impurity centers. But at the high temperatures from which the paramagnetic Curie temperature is extrapolated, and at high impurity concentrations, the donors can be considered to be sufficiently ionized[124] for the simple theory to be applicable.

28. Double exchange[125]. "Double exchange" is the (somewhat inappropriate) name given to the indirect exchange mechanism when the number of carriers is small compared to the number of magnetic atoms and the coupling constant Js is large compared to kT or E_F. Suppose two similar magnetic atoms M_1 and M_2 are nearest neighbors, and that the first has an extra (conduction or d-band) electron, i.e., a configuration $M_1^- M_2$. It is degenerate with $M_1 M_2^-$ and if a finite overlap exists, there will be a matrix element to lift this degeneracy, creating in effect a molecular orbital. *The coupling is large because the extra electron fits into the unfilled, magnetic d-shell of the atoms.*

The basic Hamiltonian consists of an overlap term which results in a kinetic energy Hamiltonian similar to (23.1), and HUND's rule energy similar to (23.3) for each atom. A complete solution in the case of two atoms has been given[125] but the essential results may be reproduced here quite simply, by assuming that the atomic spins are classical vector S_1 and S_2 at some specified angle given by $\cos\vartheta = S_1 \cdot S_2/s^2$, and by making the additional assumption that states violating HUND's rule have very high energy and may be entirely discarded. Let us pick the direction of S_1 for our z-axis, so that when the electron is on atom 1 its wavefunction is

$$\psi_1 = \begin{pmatrix} 1 \\ 0 \end{pmatrix} \varphi(r - R_1). \tag{28.1}$$

Similarly when the electron is on atom 2 its wavefunction is

$$\psi_2 = \begin{pmatrix} \cos\dfrac{\vartheta}{2} \\ \sin\dfrac{\vartheta}{2} \end{pmatrix} \varphi(r - R_2) \tag{28.2}$$

in the same coordinate system. We define the Hamiltonian and overlap matrices H_{ij} and Ω_{ij} by:

$$H_{ij} = \int d\tau\, \varphi^*(r - R_i) H(r) \varphi(r - R_j)$$

and

$$\Omega_{ij} = \int d\tau\, \varphi^*(r - R_i) \varphi(r - R_j). \tag{28.3}$$

If we use normalized atomic orbitals, $\Omega_{11} = \Omega_{22} = 1$, and if the phases are chosen properly, the above are both real, symmetric matrices. Making these simplifying assumptions, we find the ground state energy as the lowest eigenvalue of the

---

[123] J. CULLEN, E. CALLEN, and A. LUTHER: J. Appl. Phys. **39**, 1105 (1968); Phys. Rev. **170**, 733 (1968).

[124] See a competing theory by KASUYA in the preceding section C which would be valid for trapping energies $\gg kT$, and the discussion of experimental data in Sects. 61 and 62.

[125] P. W. ANDERSON, and H. HASEGAWA: Phys. Rev. **100**, 675 (1955); P. G. DE GENNES: Phys. Rev. **118**, 141 (1960); P. W. ANDERSON: Theory of Exchange in Insulators. Solid State Physics (SEITZ and TURNBULL, Eds.), Vol. 14, p. 99 and ff. New York: Academic Press 1963.

following 2×2 matrix equation:

$$\begin{bmatrix} H_{11} & H_{12}\cos\frac{\vartheta}{2} \\ H_{12}\cos\frac{\vartheta}{2} & H_{22} \end{bmatrix}\psi = E \begin{bmatrix} 1 & \Omega_{12}\cos\frac{\vartheta}{2} \\ \Omega_{12}\cos\frac{\vartheta}{2} & 1 \end{bmatrix}\psi. \tag{28.4}$$

As the two atoms are equivalent, $H_{11}=H_{22}$, and the ground state energy is:

$$E_0\left(\cos\frac{\vartheta}{2}\right) = \frac{H_{11}-H_{12}\cos\frac{\vartheta}{2}}{1-\Omega_{12}\cos\frac{\vartheta}{2}} \tag{28.5a}$$

or

$$E_0'\left(\cos\frac{\theta}{2}\right) = \frac{H_{11}+H_{12}\cos\frac{\theta}{2}}{1+\Omega_{12}\cos\frac{\theta}{2}} \tag{28.5b}$$

whichever is lower. By keeping the electron on one atom or the other we obtain a variational upper bound to the ground state energy, $E_{\mathrm{var}}=H_{11}$; this also happens to be the value of E_0, Eq. (28.5a) or (28.5b), at $\vartheta=\pi$, i.e., for antiparallel spins. Thus by the variational principle *the energy for parallel spins must be lower* than for *antiparallel spins*. If we neglect overlap (Ω_{12}) it has the simple form $H_{11}-\left|H_{12}\cos\frac{\vartheta}{2}\right|$, which may be compared to the usual exchange energy

$$a-b\cos\vartheta = (a+b)-2b\cos^2\frac{\vartheta}{2}.$$

If, for example, we choose $H_{11}=H_{12}=2a=2b=1$, then in the interval $0\leq\vartheta\leq\pi$ the double-exchange energy is

$$DEx: \quad 1-\cos\frac{\vartheta}{2} = 2\sin^2\frac{\vartheta}{4} \tag{28.6a}$$

and, by way of comparison, the appropriately scaled Heisenberg exchange energy is

$$HEx: \quad \frac{1}{2}(1-\cos\vartheta) = \sin^2\frac{\vartheta}{2}. \tag{28.6b}$$

The two expressions agree at $\vartheta=0$ and $\vartheta=\pi$, but at intermediate values the expression for double exchange always lies lower (approximately 40% lower at $\pi/4$). Whether this difference in angular dependence could be experimentally established is a moot point.

The kinetic energy operator in a crystal made up of N like atoms M_i with co-planar spins for any number of double-exchanging electrons is best written in terms of the creation (c_{im}^*) and annihilation (c_{im}) operators of fermions occupying orbitals $\psi(r-R_i)$ with spin index suppressed (only the exact spin value obeying Hund's rule is allowed on any given site). It is, the basic Hamiltonian of double exchange:

$$H_{DEx} = -\sum |H_{ij}\cos\tfrac{1}{2}\vartheta_{ij}| c_i^* c_j \tag{28.7}$$

neglecting overlap Ω_{ij}. [If the spins are ferromagnetically aligned $\vartheta_{ij}=0$, and if the atoms form a Bravais lattice we may straightaway transform to Bloch functions:

$$c_i = \frac{1}{\sqrt{N}} \sum_k e^{i k \cdot R_i} c_k \tag{28.8}$$

and
$$\varepsilon(k) = -\frac{1}{N}\sum_{i,j}|H_{ij}|\cos k \cdot R_{ij} \qquad (28.9)$$

to obtain the analogue of H_0, Eq. (23.1):

Ferro:
$$H_{DEx} = -\sum_k \varepsilon(k) c_k^* c_k. \qquad (28.10)$$

Note however the lack of KRAMERS' degeneracy.]

When the spins are not all parallel the electronic energy is perforce raised. Consider, for example, the electronic energy at temperatures T high compared to the ordering temperature T_N or T_c of the material. The spins may be assumed to be completely disordered so that the angles ϑ_{ij} are random variables. This randomness restores some measure of translational invariance, and after comparison with (28.7) we have approximately the Bloch-like Hamiltonian (28.10) above (plus important scattering terms which we shall not yet discuss) but in which the Bloch energies are given by

High T:
$$\varepsilon(k) \cong -\frac{1}{N}\sum_{i,j}|H_{ij}|\gamma_{ij}\cos k \cdot R_{ij} \qquad (28.11)$$

where
$$\gamma_{ii} = 1$$
and
$$\gamma_{ij} = \tfrac{1}{2}\int_0^\pi d\vartheta \sin\vartheta \cos\tfrac{1}{2}\vartheta = \tfrac{2}{3}, \quad i \neq j. \qquad (28.12)$$

The center of gravity of the band is not affected by the disorder, but the band width is reduced by a factor $\tfrac{2}{3}$.

DE GENNES[126] has given an interesting calculation of how the ferromagnetic ordering tendency of double exchange may compete with the antiferromagnetic ordering tendency of the usual superexchange type, in materials such as $(La_{1-x}Ca_x)(Mn_{1-x}^{3+}Mn_x^{4+})O_3$, to yield: (a) antiferromagnetism; (b) canted antiferromagnetism (weak ferromagnetism); and (c) ferromagnetism, as well as the possibility of thermodynamic phase transitions from one of these states to another. There have been no other applications of this theory to date although the application to magnetic metals and rare earth chalcogenides seem to be obvious unexplored possibilities.

29. Other mechanisms in magnetism. Among the physical mechanisms we have not yet discussed in this chapter, the most important is known as "superexchange". It is the subject of a brief description in the following Sect. 30 as well as of an extensive treatment in an accompanying article in this volume. In this section we merely list three additional items which may sometimes play important roles in the magnetic properties of semiconductors, but which, for reasons of limited space and time, will not be discussed here in detail: (a) direct exchange may be significant in some europium salts and other exceptional materials (see Sect. 59) but the Curie temperatures are generally low (less than 100°K); (b) magnetostatic dipole-dipole forces between the magnetic moments of the localized spins[127]. Together with (c) $L \cdot S$ coupling, these forces are responsible for magnetic anisotropy.

---

[126] P. G. DEGENNES: Phys. Rev. **118**, 141 (1960). Magnetic properties of Mn-perovskites are discussed in Sect. 56.

[127] W. P. WOLF, M. J. M. LEASK, B. MAGNUM, and A. F. WYATT: J. Phys. Soc. Japan **17**, Suppl. BI, 487 (1962); A. H. COOKE, D. T. EDMONDS, C. B. P. FINN, and W. P. WOLF: Proc. Phys. Soc. (London) **74**, 791 (1959).

Such forces may be important in a given semiconducting magnetic material, depending upon the accidents of crystal structure, lattice parameters, band structure, number of carriers, etc. But in many insulators superexchange is the dominant factor.

30. Mechanisms of superexchange[128]. Typical magnetic salts such as MnF_2 are highly ionic. If they were *entirely* so, the entire binding energy would be the sum of individual ionic energies and the Madelung energy of the crystal; the exchange constants would be ferromagnetic and microscopically small. In fact, for MnF_2 as an example, the exchange constant *is* small but *anti*ferromagnetic, the critical temperature T_N being 68 °K. Thus, part of the binding must be due to covalent bonding. While the degree of covalency may (or may not) be a negligible factor in the lattice dynamics or cohesive energy of the solid, it is found to be the key to the magnetic properties: the crystal field splitting, the partial quenching of orbital angular momentum of d-electrons, and the exchange interaction itself; and it is also responsible for the p-type conductivity (if any), for a hole in a purely ionic lattice cannot move.

The ligand ion (oxygen, sulfur, fluorine, etc.) is negatively charged in the ground state (e.g., O^{--}) and will accept additional electrons only with difficulty; it may readily *give up* one or more of its charges, however, in a virtual excited state. Similarly, the positively charged magnetic ion may temporarily *accept* an extra electron but it will not so readily give one up. If the magnetic shell is less than half-filled, this extra electron will occupy a d-orbital with spin parallel to the total spin of the atom; the antiparallel orientation violates HUND's rule and requires the expenditure of extra energy. If the magnetic shell is half- or more than half-filled, the extra electron can only occupy a state of antiparallel spin.

The ligand bond is a p-orbital pointing directly at the magnetic atom; it interacts most strongly with the d-orbitals having lobes pointing to the ligand ion. Out of this lattice we pick out a typical pair of magnetic ions and the ligand between them, and study the eigenstates and eigenvalues of this cluster. [It is easily shown that the true ground state energy of the crystal is *higher* than the ground state of the cluster times the number of such clusters which can be picked in the crystal, and the simple calculation, therefore, provides a valuable lower bound to the energy. This can be supplemented by a variational upper bound, to bracket the true ground state energy.] The quantum mechanics becomes tractable after a few simplifying assumptions: (a) treat the magnetic atoms as classical spins of fixed length "s" but variable orientation; (b) allow the excess electron only one state of space and spin when on the magnetic atom, i.e., the state most favored by HUND's rule; (c) consider only the two opposite spin electrons which are originally in the ligand orbital connecting the two magnetic atoms; and (d) assume the latter have less than half-filled shells.

Four configurations are allowed: (1) the unperturbed ionic configuration (energy e_0); (2) the transfer of an electron to the right-hand (r—h) magnetic atom (energy e_0+U); (3) the transfer of an electron to the l—h magnetic atom (also energy e_0+U); and (4) the transfer of an electron to each of the two magnetic atoms (total energy e_0+U+V). Let us assume an angle ϑ between the spins of the two magnetic atoms; a ferromagnetic alignments is equivalent to $\vartheta=0$, an antiferromagnetic one to $\vartheta=\pi$. Intermediate values are not generally considered, but with the present model it is not difficult to calculate the energy as a function of ϑ. We may safely assume U and $V > 0$.

---

[128] See the article by VONSOVSKY and KARPENKO in this volume for a complete discussion.

The Hamiltonian may be assumed to have small matrix elements b connecting (1) to (2) or (3), and $b \sin \tfrac{1}{2}\vartheta$ connecting (2) or (3) to (4). The half-angle formula has already been discussed in connection with the double exchange, Eqs. (28.4) and ff. In the present case, the first electron has its spin parallel to one of the magnetic atoms; the second electron is antiparallel to the first, hence has a matrix element $b \cos \frac{\pi-\theta}{2} = b \sin \frac{\theta}{2}$ to transfer to the other magnetic atom. The eigenvalues E are given by the solution to the following 4×4 determinantal equation:

$$\det \begin{vmatrix} -E+e_0 & b & b & 0 \\ b & -E+e_0+U & 0 & b\sin\frac{\vartheta}{2} \\ b & 0 & -E+e_0+U & b\sin\frac{\vartheta}{2} \\ 0 & b\sin\frac{\vartheta}{2} & b\sin\frac{\vartheta}{2} & -E+e_0+U+V \end{vmatrix} = 0. \quad (30.1)$$

In seeking the ground state energy, it is advantageous first to combine the second and third rows and columns of the above determinant, obtaining:

$$\det \begin{vmatrix} -E+e_0 & b\sqrt{2} & 0 \\ b\sqrt{2} & -E+e_0+U & b\sqrt{2}\sin\frac{\vartheta}{2} \\ 0 & b\sqrt{2}\sin\frac{\vartheta}{2} & -E+e_0+U+V \end{vmatrix} = 0. \quad (30.2)$$

Expanding this cubic equation, we find the ground state energy E_0:

$$E_0 = e_0 - \frac{2b^2}{U} + 2\left(\frac{b^2}{U}\right)^2 \left(\frac{\cos\vartheta - 1}{U+V}\right) + O(b^6). \quad (30.3)$$

This leads to the Heisenberg model with an effective J_{12} having a negative (antiferromagnetic) sign and with magnitude:

$$J_{12} = -\frac{2}{s^2}\left(\frac{b^2}{U}\right)^2 \frac{1}{U+V}. \quad (30.4)$$

There are small correction terms $O(\cos^2 \vartheta)$ which correspond to *biquadratic* $(S_1 \cdot S_2)^2$ corrections to the Heisenberg exchange Hamiltonian, but they may be ignored for present purposes. While the above was derived for less than half-filled shells, entirely analogous results are obtained for more than half-filled shells.

First introduced by KRAMERS[129] in 1934, this mechanism and others have been discussed by ANDERSON[130] who feels, however, that superexchange may be viewed in yet another light. In simplest terms, one of the magnetic electrons may hop onto the ligand atom and then to the other magnetic atom. This is similar to the *double exchange* discussed in a previous section, except than the final state has higher energy than the starting configuration, and, therefore, occurs only virtually; in double exchange the starting and final configurations are degenerate.

The matrix element for a single electron transfer will be $b' \cos \tfrac{1}{2}\vartheta$ and the excitation energy of the virtual state is denoted U', introducing two constants b' and U' not related to b and U previously used. According to second-order

---

[129] H. A. KRAMERS: Physica **1**, 182 (1934).
[130] P. W. ANDERSON, in: Solid State Physics (SEITZ and TURNBULL, Eds.) **14**, 99 (1963); J. B. GOODENOUGH: J. Phys. Chem. Solids **6**, 287 (1958); J. KANAMORI: J. Phys. Chem. Solids **10**, 87 (1959).

perturbation theory, the energy is

$$E'_0 = e_0 - \frac{2(b' \cos \tfrac{1}{2} \vartheta)^2}{U'} \qquad (30.5)$$

and the effective coupling constant, aside from J_{12}, is

$$J'_{12} = -\frac{b'^2}{U'}. \qquad (30.6)$$

Going beyond these calculations, GOODENOUGH and KANAMORI have established semi-empirical rules for the sign and magnitude of superexchange. We quote ANDERSON's version of these rules:

"First, one observes which of the d orbitals contains the magnetic electrons according to ligand field theory. This is simple in the octahedral ions that are isoelectronic with Cr^{3+}, Mn^{++}, and Ni^{++}. The first has magnetic electrons in d_{xy} orbitals alone, the second has them in all, and the third has them only in d_z^2. For other ions, ligand field theory must be studied in relation to any distortion from cubic symmetry in the surroundings.

In a form slightly modified from the original Goodenough-Kanamori formulation, the rules are the following.

1. When the two ions have lobes of magnetic orbitals pointing towards each other in such a way that the orbitals would have a reasonably large overlap integral, the exchange is antiferromagnetic. There are several subcases:

a) When the lobes are d_z^2 type orbitals in the octahedral case, particularly in the "180° position" in which these lobes point directly towards a ligand and each other, one obtains particularly large superexchange.

b) When d_{xy} orbitals are in the 180° position relative to each other, so that they can interact via p_π orbitals on the ligand, one again obtains antiferromagnetism.

c) In a situation involving a 90° ligand in which one ion has a d_z^2 orbital occupied and the other a d_{xy} occupied, the p_π for one is the p_σ for the other and one expects strong overlap and thus antiferromagnetic exchange.

2. When the orbitals are arranged in such a way that they expected to be in contact but have no overlap integral, most notably, a d_z^2 and a d_{xy} state in the 180° position for which the overlap is zero by symmetry, the rule gives ferromagnetic interaction. This interaction, however, is not usually as strong as the antiferromagnetic one. Some authors put this condition in an equivalent form by saying that the effect of an empty d orbital in "contact" with a magnetically filled one is ferromagnetic.

These two rules seem to explain almost all the spin-pattern data for a wide variety of substances".

E. Electron orbitals and energies in rare earth ions.

31. General remarks. The occurrence of magnetic moments on individual atoms requires a certain localization of electrons by the Coulomb attraction of the ion cores, while electrical properties and magnetic interactions in a crystal depend on the delocalized character of the outer electrons. The position which a transition element compound has in the spectrum of covalency between a perfectly ionic, magnetic insulator and a ferromagnetic metal with itinerant magnetic electrons is determined by the character of the d or f — electron orbits in the atoms and their stereochemical situation in the crystal symmetry of this compound. Consequently, we discuss in this chapter some experimental data on electron

Table 1. *The lanthanide series.*

	Z = Element:	57 La	58 Ce	59 Pr	60 Nd	62 Sm	63 Eu
I – neutral RE	Configuration Groundstate Ioniz. pot (eV) ζ_{4f} theor. cm^{-1} exp.	$5d\ 6s^2$ $^3D_{3/2}$ 5.61 — —	$4f\ 5d\ 6s^2$ $^1G_4$ 5.6 930 623	$4f^3\ \ \ 6s^2$ $^4I_{9/2}$ 5.48	$4f^4\ \ \ 6s^2$ $^5I_4$ 5.51 1,240 777	$4f^6\ \ \ 6s^2$ $^7F_0$ 5.6 1,596 1,150	$4f^7\ \ \ 6s^2$ $^8S_{7/2}$ 5.67
II – RE+	Configuration Groundstate Ioniz. pot (eV)	$5d^2$ $^3F_2$ 11.43	$4f\ 5d^2$ $^4H_{7/2}$ 14.8	$4f^3\ \ \ 6s$ $^5I_4$ (10.55)	$4f^4\ \ \ 6s$ $^6I_{7/2}$ (10.73)	$4f^6\ \ \ 6s$ $^8F_{1/2}$ 11.4	$4f^7\ \ \ 6s$ $^9S_4$ 11.22
III – RE2+	Configuration Groundstate Ioniz. pot (eV) ζ_{4f} (cm^{-1})	$5d$ $^2D_{3/2}$ 19.17	$(4f^2)$ $(^3H_4)$ 20	$4f^3$ $^4I_{9/2}$ 23.2	$(4f^4)$ $(^5I_4)$	$4f^6$ $^7F_0$ 21.3 1,090	$4f^7$ $^8S_{7/2}$ 22
IV – RE3+	Configuration Groundstate Ioniz. pot (eV) ξ_{4f} theor. (cm^{-1}) exp. g-value $n_B = g \cdot J (\mu_B)$ $p = g[J(J+1)]^{\frac{1}{2}}$	— — 36.6 0 0 0	$4f^1$ $^2F_{5/2}$ 830 644 6/7 15/7 2.535	$4f^2$ $^3H_4$ 980 729.5 4/5 16/5 3.578	$4f^3$ $^4I_{9/2}$ 1,130 884.6 8/11 36/11 3.618	$4f^5$ $^6H_{5/2}$ 1,200 2/7 5/7 0.845	$4f^6$ $^7F_0$ 1,320 0 0 0
V – RE4+	Configuration Groundstate			$4f^1$ $^2F_{5/2}$	$(4f^2)$ $(^3H_4)$		

orbitals and their energies in free atoms and free ions. The situation of magnetic ions in crystals is derived from the optical properties of magnetic crystals in Part F. We will concentrate here mainly on lanthanides for two reasons. Firstly, the properties of d-electrons have been described already in a large number of excellent survey articles in this handbook and elsewhere to which we will give reference in suitable context. Secondly, the study of optical, magnetic and electrical properties of lanthanide ions and compounds is very illustrative for the general understanding of principles and phenomena which, in d-element compounds, are often hidden under extraneous chemical complications. The lanthanides represent the most extreme case of transition element behavior and localization of magnetic electrons, properties which are gradually lost as one goes from the lanthanides over the actinides to the $3d$, $4d$ and $5d$ transition elements.

32. Orbital radii. The rare earth (RE) elements, which include the lanthanides and actinides, have a very special position in chemistry and solid state physics because of their unique outer electron configurations including *two* unfilled shells; see Table 1. Lanthanum (atomic number 57) and lutetium (atomic number 71) the first and last member of the lanthanide series, occupy the place of the first $5d$-transition element, both having the valence electron configuration $5d^1 6s^2$ around the closed shells of the xenon core. With the elements following lanthanum, the $4f$ shell is progressively filled by 14 electrons and then the $5d$ series is continued with hafnium 72.

Sect. 32. Orbital radii.

(ζ = spin-orbit parameter).

64 Gd	65 Tb	66 Dy	67 Ho	68 Er	69 Tm	70 Yb	71 Lu
$4f^7\ 5d\ 6s^2$ $^9D_2$ 6.16 2,143	$4f^8\ 5d\ 6s^2$ $^8G_{15/2}$ 5.98 (6.7)	$4f^{10}\ 6s^2$ $^5I_8$ 6.82 2,479	$4f^{11}\ 6s^2$ $^4I_{15/2}$	$4f^{12}\ 6s^2$ $^3H_6$ 6.08 3,100	$4f^{13}\ 6s^2$ $^2F_{1/2}$ 5.81 2,506	$4f^{14}\ 6s^2$ $^1S_0$ 6.22 3,620	$4f^{15}\ 5d\ 6s^2$ $^2D_{3/2}$ 6.15
$4f^7\ 5d\ 6s$ $^{10}D_{5/2}$ 12.1	$(4f^9\ 6s)$ $(^7H_8)$ (11.52)	$4f^{10}\ 6s$ $^6I_{17/2}$ (11.67)	$(4f^{11}\ 6s)$ $(^5I_8)$ (11.8)	$4f^{12}\ 6s$ $^4H_{13/2}$ (11.93)	$4f^{13}\ 6s$ $^3F_4$ (12.05)	$4f^{14}\ 6s$ $^2S_{1/2}$ 12.18	$4f^{14}\ 6s^2$ $^1S_0$ 14.7
$4f^7\ 5d$ $^9D_2$ 1,050	$4f^9$ $^6H_{15/2}$	$4f^{10}$ $^5I_8$	$4f^{11}$ $^4I_{15/2}$	$4f^{12}$ $^3H_6$	$4f^{13}$ $^2F_{7/2}$	$4f^{14}$ $^1S_0$ 22.6	$4f^{14}\ 6s$ $^2S_{1/2}$ 21
$4f^7$ $^8S_{7/2}$ 1,583 2 7 7.937	$4f^8$ $^7F_6$ 1,705 3/2 9 9.721	$4f^9$ $^6H_{15/2}$ 2,310 1,900 4/3 10 10.646	$4f^{10}$ $^5I_8$ 2,163 5/4 10 10.607	$4f^{11}$ $^4I_{15/2}$ 2,830 2,393 6/5 9 9.581	$4f^{12}$ $^3H_6$ 2,656 7/6 7 7.561	$4f^{13}$ $^2F_{7/2}$ 3,400 2,883 8/7 4 4.536	$4f^{14}$ $^1S_0$ 0 0 0
	$4f^7$ $^8S_{7/2}$						

The special situation of the $4f$ electron shell, which has together with the $5f$ shell of the actinides the highest angular momentum quantum number l occurring in the Periodic Table, is best demonstrated by the radial density distribution of the outer electrons of the neutral Gd atom which is shown in Fig. 17. The electron wave functions have been calculated from nonrelativistic Hartree-Fock equations[131] and modified for exchange and spin-orbit interactions. A plot of the radii for which the charge densities show their outermost maxima[132] for

---

[131] F. HERMAN, and S. SKILLMAN: Atomic Structure Calculations, New Jersey: Prentice-Hall 1963; A. J. FREEMAN, and R. E. WATSON: Phys. Rev. **127**, 2058 (1962). O. J. SOUERS: J. Phys. Chem. Solids **28**, 1073 (1967). Recent calculations of the relativistic Dirac-Slater wave functions by WABER and CROMER [J. Chem. Phys. **42**, 4116 (1965)] give (within the accuracy of our drawing) the same results (compare Fig. 1). The dotted line refers to empirical values for the trivalent ion radii as derived from the packing density with different anions in ionic crystals with octahedral coordination. The ions are visualized as hard spheres and one tends to relate their radius to the tail of the theoretical charge density curve of the largest occupied electron shell. [J. C. SLATER, J. Chem. Phys. **41**, 3199 (1964)]. In covalent or metallic bonding where electrons are shared between ions, the atomic or metallic radii correspond to the radius of the maximum in the theoretical charge density distribution. The influence of several parameters on the result of atomic orbital calculations has been discussed in detail by A. J. FREEMAN, and R. E. WATSON [Phys. Rev. **127**, 2058 (1962)].

[132] The angular parts of the $4f$ wave-functions have been exhibited graphically by: H. G. FRIEDMAN, G. R. CHOPPIN, and D. G. FEUERBACHER: J. Chem. Educ. **41**, 358 (1964); J. T. WALKER, and J. E. HOCKETT: RE research IV (Ed. L. R. EYRING), p. 281. New York: Gordon and Breach 1965.

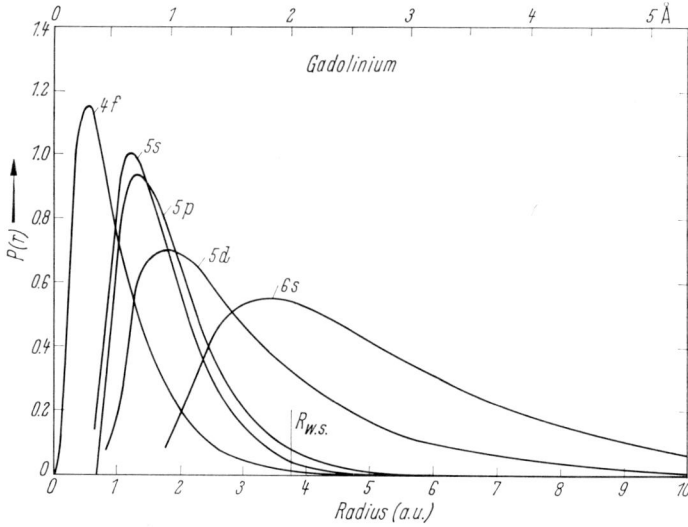

Fig. 17. Radial part of the modified electron Hartree-Fock wave functions for a neutral Gd-atom with the electron configuration $4f^7\, 5d^1\, 6s^2$. Only the outer parts of the wave functions are shown and the inner nodes omitted. R_{ws} is the Wigner-Seitz radius for Gd-metal. (1 $au = h/me^2 = 0.529172$ Å.) (From A. J. Freeman, J. O. Dimmock, and R. E. Watson, in: Quantum Theory of Atoms, Molecules, Solid State. New York: Academic press 1966.)

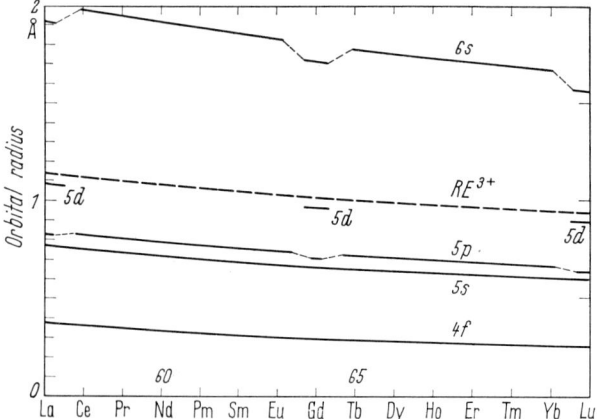

Fig. 18. The radii of the principal maxima in the radial electron charge density calculated from modified nonrelativistic Hartree-Fock equations or relativistic Dirac-Slater equations. The electron configuration for La, Gd, Lu is $4f^{n-1}\, 5d^1\, 6s^2$ ($n = 1$ for La) and $4f^n\, 6s^2$ for the other elements. The dotted line gives the empirical hard sphere radii for the trivalent ions RE^{3+} in a coordination of six neighbors.

the various lanthanides is given in Fig. 18. It is obvious that the $4f$ electrons are much more tightly bound to the atom core than the $5d$ and $6s$ valence electrons. The $4f$-orbitals extend to less than 50% of the trivalent ion radii which are mainly determined by the filled $5s^2$ and $5p^6$ electron shells. The spatial seclusion of the partially filled $4f$ shell within the xenon core and its screening by two surrounding filled shells characterizes the lanthanides and motivates the interest in their chemical and physical properties. It is common knowledge that the $4f$-electrons contribute only negligibly to the chemical properties of the lanthanide elements (see Sect. 40). Most lanthanides prefer the trivalent ionization state in which the three $5d^1 6s^2$ electrons participate. In this valence state, they can be distinguished in chemical processes only by the small, but regular variation in their ionic size (lanthanide contraction). Consequently, their mutual separation requires special physico-chemical techniques, such as the ion-exchange process[133] which was developed for industrial use after 1945. The chemist and metallurgist finds in the lanthanides a unique workshop for the investigation of ion size effects on the stability of crystal structures, chemical compounds and alloys[134]. The physicist enjoys the opportunity of studying the interaction of electrons in nearly free ions in a solid-state matrix. Optical and magnetic measurements on $4f$-electrons can be explained by the usual atomic theory on which the influence of the crystalline environment can be grafted as a small perturbations.

33. Electron configurations in the ground state. The characteristic spatial seclusion of the $4f$-electrons in the interior of the xenon core does not imply an energetic segregation. The $4f$-electrons experience very strong attraction by the positive core potential (as indicated by their small orbital radius) but the *interelectronic repulsion cancels most of the attraction*. The resulting ionization energy of a $4f$-electron is *smaller* than the binding energy of the $5s$ and $5p$ electrons. The similarity in binding energy for the last $4f$ and the first $5d$-electron characterizes the position of the lanthanides in the Periodic Table as *transition elements with two partially filled shells*. The resulting uncertainty about the electron configuration in the ground state of the free atoms and ions has not yet been fully resolved for all lanthanides. Magnetic resonance experiments on atomic beams or the interpretation of emission spectra of free atoms excited in electrical arc or spark discharges are required for this purpose. Because of the large number of $4f$-states, the lanthanides exhibit, in common with the actinides, the most complex and least known spectra of all elements[135]. Concentrated efforts to interpret the lanthanide spectra systematically have been undertaken recently in several laboratories. Table 1 shows most of the known ground states as determined from spectroscopic investigations or by atomic beam resonance experiments. Estimates appear in brackets. Only the outer electrons are considered and the 54 electrons in the closed shells of the xenon core omitted.

The data in Table 1 give some interesting qualitative indication about the energetic situation of the outer electron states, which we may sketch out before we present more quantitative results of atomic orbital calculations. *In lanthanum*, the first element of the group, the $4f$ levels lie several eV above the $5d$ states of the neutral atom. This energy difference decreases with increasing ionization due to differences in the screening of the core charge for f and d electrons but is, for the valence electron of La^{2+}, still about 1 eV[136]. With increasing core charge and decreasing ionic radius the regular state $4f^n 6s^2$ gains increasing stability. But in Ce the energy difference between the $4f$ and $5d$-levels is still smaller than the repulsion between two $4f$ electrons and the irregular configura-

---

[133] F. H. SPEDDING, and A. H. DAANE: The Rare Earth, New York: J. Wiley & Sons 1961; D. M. YOST, H. RUSSELL, and C. S. GARNER: The Rare Earth Elements and Their Compounds, New York: J. Wiley & Sons 1947, and several articles in the book series: Progress in Science and Technology of the Rare Earth (Ed. L. EYRING). New York: Macmillan 1964.

[134] K. A. GSCHNEIDNER: Rare Earth Alloys. Princeton: D. van Nostrand Co. 1961.

[135] If only transitions from $4f$ into the $5d$, $6s$ and $6p$ states are considered, one already expects for Pr with only three $4f$-electrons, as an example, 107 levels with 5780 allowed transitions between them! For 5 electrons there are 977 levels with a countless number of transitions. [G. H. DIEKE, H. M. CROSSWHITE, and B. DUNN: J. Opt. Soc. Am. **51**, 820 (1961).]

[136] J. SUGAR, and V. KAUFMANN: J. Opt. Soc. Am. **55**, 1283 (1965). C. E. MOORE: Atomic Energy Levels, Circ. Natl. Bur. Stand. 467, May 1 (1958).

tions $4f\,5d\,6s^2$ form the groundstate of Ce I and Ce II, instead of the regular configuration $4f^n 6s^2$ which is stable in Pr and the following elements.

The similarity in binding energy between $4f$ and $5f$ electrons in Ce results in an interesting phase transformation between *two isostructural phases for the elemental Ce-metal*. At room temperature γ-Ce is stable in a Cu-type face centered cubic structure with a lattice parameter of 5.16 Å and an average valency of 3.1 per Ce-atom. The localized single electron in the $4f$-shell can be identified by magnetic moment measurements. When cooled below 120 °K at 1 atm the metal collapses into the α-phase which has the same structure but a lattice parameter of only 4.85 Å. The average valency per Ce-atom is now 3.6, i.e. about half of the $4f$-electrons are ionized into the conduction band. The $4f^1$-levels which in γ-Ce lie below the Fermi energy, possibly within the conduction band, move partially above the Fermi energy in the collapsed α-phase[137].

It has been suggested[138] that the hybridization of $4f$-electrons with outer valence electron states may be responsible for the low melting points, in general, of the tri-valent lanthanides and the actinides and for the occurrence of the double hexagonal structure in α-La, β-Ce, α-Pr, α-Nd, Am and Cm as well as of the Sm-structure at lower temperatures. The possibility of forming virtual $4f$-levels which are localized but broadened in energy by resonances with nonlocalized metallic $6s$ and $5d$ states has been discussed by ROCHER[139] and others.

WABER, LIBERMAN and CROMER[140] have investigated how the energy of the Dirac-Slater wave functions of the free Ce-atom varies with atomic radius and with number of electrons in the $4f$-shell. Due to the strong repulsions at the small interelectronic distances in the $4f$-shell the total energy of the shell rises much more steeply with electron occupation than the $5d$ and $6s$ levels. For the same reason the binding energy of the $4f$ electrons increases strongly with increasing degree of ionization and only the elements at the beginning of the series and Tb can be ionized beyond their trivalent state. DIEKE and CROSSWHITE[141] derived a more quantitative demonstration of the *influence of the ionization state on the energy levels* from the free ion spectra of the four first lanthanides, by comparing the wave numbers of corresponding line groups in La I, Ce II, Pr III and Nd IV, which have an identical number of electrons outside the Xe-core but different ionization states. The configuration $4f^3$ which is the stable ground state of Pr III and Nd IV lies in the lesser ionized Ce II ion more than 4 eV above the $4f\,5d^2$ state, and in La I far beyond the near ultraviolet region. The energy difference

---

[137] K. A. GSCHNEIDNER: Rare Earth Alloys, Princeton: D. van Nostrand Co. 1961; Rare Earth Research IV (Ed. L. EYRING), p. 153. New York: Gordon and Breach 1965. — K. A. GSCHNEIDNER, and R. SMOLUCHOWSKI: Less-Common Metals **5**, 372 (1963). — K. A. GSCHNEIDNER, R. O. ELLIOTT, and R. R. McDONALD: J. Phys. Chem. Solids **23**, 555, 1191, 1201 (1962).
It has been suggested that the heavy RE-metals (other than Gd and Lu) have a partial divalent character in the liquid state, in the bcc structure at high temperatures and under high pressure. [Review by E. B. ROYCE: Phys. Rev. **164**, 929 (1967).]
[138] B. T. MATTHIAS, W. H. ZACHARIASEN, G. W. WEBB, and J. J. ENGELHARDT: Phys. Rev. Letters **18**, 781 (1967); A. R. MAKINTOSH: Rare Earth Research (Ed. J. F. NACHMAN, C. E. LUNDIN), p. 272. New York: Gordon and Breach 1962; D. B. McWHAN, and A. L. STEVENS: Phys. Rev. **139**, A 682 (1965); D. C. HAMILTON, and M. A. JENSEN: Phys. Rev. Letters **11**, 205 (1963); K. A. GSCHEIDNER, and R. M. VALETTA: Acta Metal. **16**, 1477 (1968).
[139] Y. A. ROCHER: Advances in Phys. **11**, 233 (1962); Rare Earth Research IV (Ed. L. EYRING), p. 127. New York: Gordon and Breach 1965; J. Phys. Chem. Solids **23**, 1621 (1962); A. BLANDIN, B. COQBLIN, and J. FRIEDEL: Phys. of Solids at High Pressure (Ed. C. T. TOMIZUKA, R. M. EMRICK), p. 233. New York: Academic Press 1965; A. BLANDIN: J. Appl. Phys. **39**, 1285 (1968).
[140] J. T. WABER, D. LIBERMAN, and D. T. CROMER: Rare Earth Research IV (Ed. L. EYRING), p. 187. New York: Gordon and Breach 1965.
[141] H. DIEKE, and H. M. CROSSWHITE: Appl. Optics **2**, 675 (1963).

between the $4f\,5d$ state and the $4f\,5d^2$ ground state is quite small in Ce II but increases by about 3 eV with each step of increasing core charge and is larger than 6 eV in Nd IV.

In the middle of the series the well known effect of increased binding energy for electrons in half filled shells dominates the situation. In Eu the stabilization energy of the half-filled shell[142] provides enough compensation of interelectronic repulsion that the divalent ionization state becomes remarkably stable. For most of the other lanthanides the divalent state can be obtained only in very ionic compounds, such as halides, after special treatment[143]. However, Eu^{2+}, Sm^{2+} and Yb^{2+} are stable in many solutions and compounds with suitable valency and stoichiometry of electronegative anions. Even the elemental Eu-metal has been found, in magnetic measurements[144], to be divalent in a body centered cubic structure of the W-type, while all of the other lanthanides, except Ce and Yb, crystallize as trivalent ions in hexagonal structures. Also with respect to its metallic radius (2.04 Å), ionic radius (1.12 Å), melting point (826 °C), boiling point (1430 °C) and many other physical and chemical properties Eu (and to a certain extent Sm) resembles more closely the divalent alkaline earth elements, especially Strontium (2.15 Å; 1.13 Å; 768 °C; 1380 °C) than Gd (1.79 Å; 1.02 Å; 1312 °C; 3000 °C), the next neighbor in the lanthanide series.

Burnett and Cunningham[145] derived an experimental value for the stabilization energy of the $4f^7$ configuration from heat of formation measurements for Eu^{2+} and Eu^{3+} in solutions. They find that the sum of the first three ionization potentials for Eu must be about 75 Kcal (~ 3.5 eV) larger than the value expected from comparison with the other lanthanide elements. This value agrees with Jorgensen's estimate of 79 Kcal for the additional stabilization energy gained by minimizing Coulomb repulsion in the half filled shell[146].

In Gd I, Gd II and Gd III the binding energy of the first $4f$-electrons in excess of the half filled shell drops to low values and the ground state configuration for the low ionization states of Gd and Tb again becomes $4f^{n-1}$ similar to Ce. The regular $4f^n$ configuration is, however, expected to lie quite close to the ground state and in Tb the energy difference has been found by atomic beam resonance to be not larger than 1000 cm^{-1} (0.124 eV)[147]. Tb forms also tetravalent ions with the $4f^{n-2}$ ground state in similarity to the elements at the beginning of the series. There is no experimental evidence for the position of the empty $4f^8$ level in Gd, but it must be expected to lie quite close to the ground state; the absorption lines which connect the $4f^7\,5d$ and the $4f^8$ state in Gd^{2+} may be found in the far infrared.

The similarities and differences in the properties of lanthanide ions offer the possibility of investigating the influence of individual ionic parameters on the physical and chemical properties of compounds by *replacing one lanthanide element by another*. Exchange of Gd^{3+} with Eu^{2+} or Tb^{4+}, as an example, reduces or increases the number of valence electrons by one, but maintains the electron

---

[142] C. K. Jorgensen: Orbitals in Atoms and Molecules, London: Academic Press 1962; Lanthanides and $5f$ Group Elements, London: Acad. Press 1965; Mol. Phys. **2**, 309 (1959); **5**, 271 (1962); Solid State Physics **13**, 375 (1962).

[143] Review articles: F. K. Fong: Progress in Solid State Chemistry (Ed. H. Reiss), Vol. 3, p. 135, Oxford: Pergamon Press 1967; F. K. Fong: Rare Earth Research IV (Ed. L. R. Eyring), p. 373. New York: Gordon and Breach 1965.

[144] R. M. Bozorth, and J. H. van Vleck: Phys. Rev. **118**, 1493 (1960).

[145] J. L. Burnett, and B. B. Cunningham: Rare Earth Research IV (Ed. L. Eyring), p. 585. New York: Gordon and Breach 1965.

[146] C. K. Jorgensen: Orbitals in Atoms and Molecules. London: Academic Press 1962.

[147] W. E. Albertson, H. Bruynes, and R. Hanau: Phys. Rev. **57**, 292 (1940); I. Bender, S. Penschin u. K. Schlüpmann: Z. Physik **179**, 4 (1964).

configuration and ground state of the magnetic $4f$-shell unchanged. Sr^{2+} and La^{3+} can serve as a nonmagnetic replacement for the chemically equivalent magnetic elements Eu and Gd, respectively, and so on.

34. Energy levels. HERMAN and SKILLMAN[148] calculated the ionization energies necessary to remove a single Hartree-Fock-Slater electron from a certain energy level for all neutral atoms in the periodic table. The calculations include the spherical part of Coulomb interactions and exchange effects between electrons as far as is describable by spherical potentials. The spin-orbit interactions and relativistic effects of the electron velocity on their energy are added as a correction to the result of the calculations.

In Fig. 19 we show the *theoretical energy levels* obtained for neutral lanthanide atoms. The splitting of energy levels due to spin-orbit interaction is neglected in the drawing but the value of the spin-orbit coupling for $4f$-electrons is given in Table 1. The energies in Fig. 19 represent, therefore, the centers of gravity of the multiplets, while the lines observed in optical spectra refer, in general, to excitations from the ground term which has the lowest energy. The ground state configuration $4f^n 6s^2$ has been used for all atoms with the exception of Gd and Lu, for which the configuration $4f^{n-1} 5d 6s^2$ is assumed; the energy difference between both states is about one half of a Rydberg for the $4f$-electrons and demonstrates again the strong repulsion between $4f$ electrons.

When we compare the experimental values for the first ionization energy given in Table 1 with the position of $6s$-levels in Fig. 19 we find excellent agreement. This agreement becomes quite impressive when we realize that the total energy of all electrons in a lanthanum atom is of the order of 10^4 Ry and that small errors in the assessment of the electron-core and electron-electron interactions could easily lead to uncertainties in the outer electron energies which become comparable to measurable chemical and optical energies (usually less than 0.5 Ry).

Since the higher ionization potentials of Table 1 refer to higher ionized configurations, they cannot be compared numerically with the position of the deeper energy levels of the neutral atoms in Fig. 19. The *experimental points for the energies* of the $4f$ and $5p$ electrons are derived from the position of the X-ray absorption edges in elemental metals as discussed in Vol. 30 of this handbook[149].

---

[148] F. HERMAN, and S. SKILLMAN: Atomic Structure Calculations. New Jersey: Prentice Hall 1963.

The first systematic calculations of Fermi and Thomas-Fermi potentials for all atoms in the periodic table have been made by R. LATTER [Phys. Rev. **99**, 510 (1955)]. Corrections for the self-interaction between $4f$ electrons have been discussed by C. A. COULSON, and C. S. SHARMA: Proc. Phys. Soc. (London) **79**, 920 (1962).

Detailed introductions to atomic structure calculations can be found in many textbooks such as: E. U. CONDON, and G. H. SHORTLEY: The Theory of Atomic Spectra, Cambridge: Cambridge University Press 1935; J. S. GRIFFITH: The Transition-Metal Ions, Cambridge: Cambridge University Press 1961, D. R. HARTREE: The Calculation of Atomic Structures, New York: John Wiley & Sons 1957; J. C. SLATER: Quantum Theory of Atomic Structure, 2 Volumes, New York: McGraw Hill 1960; R. K. NESBET: Rev. Mod. Phys. **33**, 28 (1961); R. S. KNOX: Solid State Physics, Vol. 4, p. 413, New York: Academic Press 1957.

The more recent literature with respect to lanthanide elements is given in A. J. FREEMAN, and R. E. WATSON: Phys. Rev. **127**, 2058 (1962), and B. G. WYBOURNE: Spectroscopic Properties of Rare Earth, New York: Interscience Publishers 1965; D. LIBERMAN, J. T. WABER, and D. T. CROMER: Phys. Rev. **137**, A 27 (1965); J. T. WABER, and A. C. LARSON: Rare Earth Research II (Ed. K. S. VORRES), p. 351, New York: Gordon and Breach 1964; Non-relativistic Hartree-calculations give about the same orbital radii but binding energies which are smaller by about 0.5; 0.24; 0.15 and 0.04 Ry than the energy levels of the $4f$; $5p$; $5d$ and $6s$ levels, respectively, in Fig. 19.

[149] A. E. SANDSTRÖM: Handbuch der Physik, Vol. 30, p. 78—333 Berlin-Göttingen-Heidelberg: Springer 1957.

Since electron excitation by X-rays ends at the Fermi energy of the conduction electrons in the metal and not at infinity, the zero point of the X-ray energies had to be adjusted by a procedure described by SLATER[150]. We note, that the theoretical energies lie well within the scattered range of the adjusted experimental data.

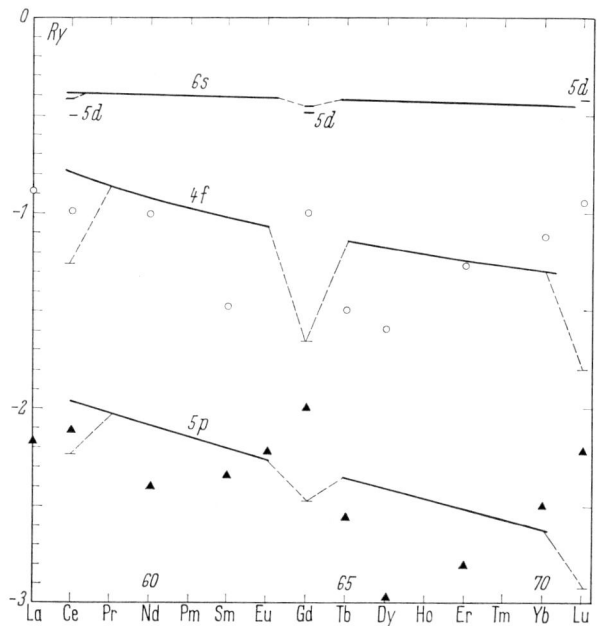

Fig. 19. The energy of the outermost Hartree-Fock-Slater electrons in neutral atoms with the configuration $4f^n 6s^2$. For La, Gd and also Ce the configuration $4f^{n-1} 5d 6s$ has been assumed. The lines give theoretical values and the points experimental data. The multiplet splitting by spin-orbit interaction has been neglected. Energy scale: 1 Ry = 109.7×10^3 cm^{-1} = 13.60 eV. (HERMAN, SKILLMAN: Atomic Structure Calculations. New Jersey: Prentice Hall 1963.)

35. The interpretation of experimental RE-spectra. The separation of energy levels can be derived experimentally from *optical or microwave spectra*. For this purpose large numbers of observed spectral lines have to be ordered in multiplets and assigned to transitions or combinations between different energy levels. The intensities of the transitions are subject to certain selection rules depending on the nature of the coupling scheme. For the Russell-Saunders coupling (see Sect. 8) the ground states are denoted by the usual spectroscopic notation in Table 1.

The *Russell-Saunders selection rules* permit electric-dipole transitions between levels where L remains equal or goes to $L \pm 1$, J must go to J or $J \pm 1$, (but $J=0$ to $J=0$ is excluded) and $\Delta S = 0$. In addition, the Laporte rule must be satisfied according to which the only possible transitions occur between terms of different parity. It is also possible to have interaction of radiation with changes in the atomic electric moment of higher order than dipole polarization. Such *electric quadropole* and *magnetic dipole* transitions obey other selection rules[151] and

---

[150] J. C. SLATER: Phys. Rev. **98**, 1039 (1955).
[151] J. H. VAN VLECK: J. Phys. Chem. **41**, 67 (1937); L. J. BROER, C. J. GORTER, and J. HOOGSCHAGEN: Physica **11**, 231 (1945).

occur only between terms of equal parity in contrast to the Laporte rule for electric dipole transitions. Since their intensities are usually *8 to 5 orders of magnitude smaller* than those of electric dipole transitions their use in the interpretation of atomic energy levels is limited.

The main problem in the interpretation of RE-spectra is the handling of very large numbers of spectral lines. The Russell-Saunders terms of free ions can be calculated with high accuracy[152]. SLATER, RACAH and JUDD have developed methods for calculating such terms without solving complicated secular equations. The term energies of the $4f$-electrons are expressed as functions of the Slater radial integrals F^2, F^4 and F^6 or of Racah-parameters which can be used as adjustable parameters in order to fit a theoretical set of terms to the observed multiplet lines. The mathematical techniques for handling a large number of states by digital computers is well developed and is very successful for the lanthanides, in contrast to most transition element spectra. Detailed discussions of the general theory of spectra and the spectral properties of the lanthanides are given in volumes 27, 28 of this Handbook or in many other books and review articles[153].

With the additional assumption that the ratios between the Slater radial integrals F^2, F^4, F^6 are the same as for hydrogenic orbitals the number of parameters can be further reduced to two, namely the parameter F^2 which is determined by the effective charge of the screened inner core, and the spin orbit coupling parameter ζ. Especially for R.E. ions with more than four $4f$-electrons or holes this simplification is necessary in order to reduce the theoretical analysis to manage-

---

[152] f^2: F. H. SPEDDING: Phys. Rev. **58**, 255 (1940); f^3: R. R. JUDD, and R. LOUDON: Proc. Roy. Soc. (London) A **251**, 127 (1959); f^4, f^{10}: M. H. CROZIER, and W. A. RUNCIMAN: J. Chem. Phys. **35**, 1392 (1959); f^5, f^6: B. G. WYBOURNE: J. Chem. Phys. **32**, 639 (1960); **34**, 279 (1961); **35**, 340 (1961); f^7: W. A. RUNCIMAN: J. Chem. Phys. **36**, 1481 (1962); f^5, f^9: W. G. WYBOURNE: J. Chem. Phys. **36**, 2295, 2309 (1962).

[153] Some books and review papers which give access to the extended literature on lanthanide spectra and their interpretation: E. U. CONDON, and G. H. SHORTLEY: The Theory of Atomic Spectra, Cambridge: Univers. Press 1957; U. FANO, and G. RACAH: Irreducible Tensorial Sets, New York: Academic Press 1959; J. S. GRIFFITHS: The Theory of Transition Metal Ions: Cambridge: Cambridge University Press 1961; C. K. JORGENSEN: Solid State Physics, Vol. 13, New York: Academic Press 1962; W. LOW: Paramagnetic Resonance in Solids, New York: Academic Press 1960; J. C. SLATER: Quantum Theory of Atomic Structure, New York: McGraw-Hill 1960; Phys. Rev. **34**, 1293 (1929); B. G. WYBOURNE: Spectroscopic Properties of Rare Earth, New York: Interscience Publ. 1965; G. RACAH: Phys. Rev. **61**, 186 (1942); **62**, 438 (1942); **63**, 367 (1942); **76**, 1352 (1949); J. Opt. Soc. Am. **50**, 408 (1960); G. RACAH, and Y. SHADMI: Phys. Rev. **119**, 156 (1960); J. P. ELLIOT, B. R. JUDD, and W. A. RUNCIMAN: Proc. Roy. Soc. (London) A **240**, 509 (1957); B. R. JUDD: Phys. Rev. **127**, 750 (1962); Proc. Roy. Soc. (London) A **227**, 552 (1955); Operator Techniques in Atomic Spectroscopy, New York: McGraw-Hill 1963; E. P. WIGNER: Group Theory, New York: Academic Press 1959; A. R. EDMONDS: Angular Momentum in Quantum Mechanics, Princeton and New York: University Press 1957; C. W. NIELSON, and G. F. KOSTER: Spectroscopic Coefficients for the p^n, d^n and f^n Configurations, Cambridge, Mass.: MIT Press 1963; B. G. WYBOURNE: J. Chem. Phys. **32**, 639 (1960); **36**, 2295, 2301 (1962); A. J. FREEMAN, and R. E. WATSON: Phys. Rev. **127**, 2058 (1962); W. F. MEGGERS: Rev. Mod. Phys. **14**, 96 (1942); F. A. KLINKENBERG: Physica **13**, 1 (1947); H. G. KUHN: Atomic Spectra: New York: Academic Press 1962; C. A. MOORE: Circular Natl. Bur.Stand. 467, May 1, 1958; J. Opt. Soc. Am. **50**, 1407 (1960); **53** (1963); Appl. Optics **2**, 665 (1963); D. S. McCLURE, and Z. KISS: Optical Lasers, p. 357, New York: Polytechn. Press 1963; J. Chem. Phys. **39**, 3251 (1963); D. S. McCLURE: Solid State Physics, Vol. 9 (1959); S. P. SINHA: Complexes of the Rare Earth, Oxford: Pergamon Press 1966; M. A. ELYASHEVICH: Spectra of the Rare Earth, Moscow: State Publ. House of Technical-Theoretical Literature 1953. Transl. by U. S. Atomic Energy Comm. AEC-tr.-4403. LANDOLT-BÖRNSTEIN: Tabellen (6) Vol. 1. Part, 1950; K. H. HELLWEGE: Ann Phys. **4**, 95, 127, 143 (1948); Naturwissenschaften **34**, 225 (1947); K. W. H. STEVENS: Proc. Phys. Soc. (London) A **65**, 209 (1952); R. J. ELLIOTT, and K. W. H. STEVENS: Proc. Roy. Soc. (London) A **215**, 437 (1952); A **218**, 553 (1953); A **219**, 387 (1953).

able proportions. DIEKE and CROSSWHITE[154] have reviewed and analyzed within the hydrogenic approximation the experimental spectra for the doubly and triply ionized free ions which are mainly derived from the crystal spectra discussed in the next paragraph. Fig. 20 gives the approximate *position of several other energy levels with respect to the* $4f^{n-1} 6s$ *state*. The levels are defined by the center points of the corresponding $4f$ configurations which can disperse over energy ranges

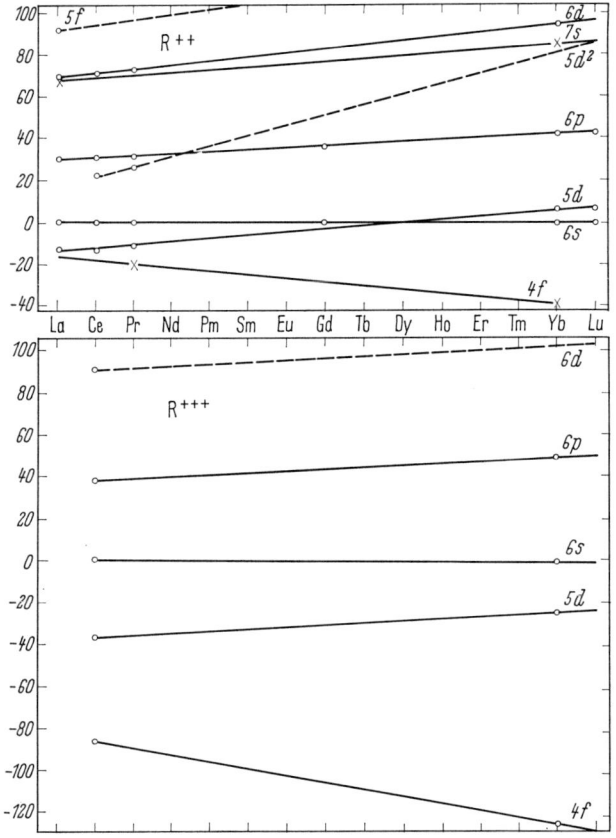

Fig. 20. Relative positions of the centers of the low configurations in the spectrum of RE^{2+} and RE^{3+} ions with respect to the $6s$ level. The nl symbols refer to the excited electrons, e.g. $6s$ means the state $4f^{n-1} 6s$. Experimental data are given by circles. [From H. DIEKE, and H. M. CROSSWHITE. Appl. Optics **2**, 673 (1963).]

wider than 10^5 cm^{-1}, especially for the elements in the middle of the lanthanide series. We note in Fig. 20 that the difference in binding energy between the $4f$ and $5d$ electrons increases by about 4,000—5,000 cm^{-1} with each step in increasing nucleus charge, and that the energy of the $4f^n$-$5d$-transition is, in general, about 50,000 cm^{-1} lower in the divalent state than for trivalent ions. The corresponding absorption bands which are about 1,000—2,000 cm^{-1} wide lie in the red and visible for divalent Sm, Eu, Yb and in the near or far ultraviolet for all trivalent ions. For La^{2+} the $4f$ and $5d$ levels must practically coincide and no

---

[154] G. H. DIEKE, H. M. CROSSWHITE, and B. DUNN: J. Opt. Soc. Am. **51**, 820 (1961); G. H. DIEKE, and H. M. CROSSWHITE: Appl. Optics **2**, 673 (1963).

transitions between them have yet been found. With increasing core charge the transitions between $4f$ and $5d$ levels which have large oscillator strength move through the visible range and group around 2,000 Å for Yb^{2+}. In the fourth spectrum all transitions between $4f$ and $5d$ levels lie in the vacuum ultraviolet between 2,000 and 1,000 Å.

The experimental energy levels lie closer together than is expected from the theoretical orbital calculations. For the Hartree-Fock electrons of the neutral Ce-atom, as an example, we read from Fig. 19 an energy of about 0.3 Ry for the transition from the $4f^2 6s^2$ in the $4f 5d 6s^2$ state. The experimentally found excitation energy in Ce^{2+} is, however, less than 10,000 cm^{-1} or 0.1 Ry. Similar deviations occur in the assignment of $4f$ multiplets to observed lines when Slater radial integrals are used which have been calculated from the Hartree-Fock, Hartree or Thomas-Fermi wave functions. FREEMAN and WATSON[155] have discussed this problem and the related questions about reliability of approximations in orbital calculations. They derived theoretical values for the Slater integrals F^k and ζ from the Hartree-Fock $4f$-wavefunctions for several trivalent ions. Their values are larger by 40—50% and 20—30%, respectively, than the parameter values suitable for fitting the experimental spectra. We include in Table 1 experimental and theoretical values for the spin-orbit coupling ζ of the $4f$-electrons. Technical details in the fitting procedure, deviation of the excited state from the Russell-Saunders coupling scheme[156] and configuration interaction with other electron configurations of the same parity[157] may be quoted as possible sources for inaccuracy of the experimental numbers. The main deviation between experiment and theory, however, has to be ascribed to principal deficiencies in the calculations. The integrals F^k do not agree with experiments because of correlation effects; discrepancies in ζ are found to derive from the inaccuracy of formula (35.1).

Because the spin-orbit interaction constant increases with the core charge Ze and with decreasing orbital radius r like

$$\zeta = \frac{2}{m^2 c^2} \frac{1}{r} \frac{d}{dr} \frac{Ze}{r} \tag{35.1}$$

it is not surprising to find ζ values about five times larger for $4f$ electrons than for the $3d$-electrons in iron group elements. The question arises if *the Russell-Saunders coupling remains a good approximation* for lanthanides. Large spin-orbit interaction can mix in higher-order ground state and excited states which have the same J but different L and S-values. Especially for the elements in the middle of the series, where the multiplet levels lie close together, the use of intermediate coupling schemes[158] appears to be more appropriate for the interpretation of

---

[155] A. J. FREEMAN, and R. E. WATSON: Phys. Rev. **127**, 2058 (1962). Magnetism II A (Ed. SUHL and RADO), New York: Academic Press 1965; Hyperfine Interactions (Ed. A. J. FREEMAN, and R. B. FREANKEL), p. 53, New York: Academic Press 1967.

[156] For excited electrons in the $5d$, $6p$ and higher states the own spin orbit interaction is expected to become comparable to or larger than the electrostatic interaction with the other electrons. Consequently, the suitable description shifts more and more from the Russell-Saunders coupling to intermediate coupling, to J_j-coupling [G. RACAH: J. Opt. Soc. Am. **50**, 408 (1960)]. Investigations of the Gd^{2+} spectrum by W. R. CALLAHAN [J. Opt. Soc. Am. **53**, 659 (1963)] shows LS-coupling to be most appropriate for the $4f^7 5d$-states while the $4f^7 6p$ levels are most closely described by J_j-coupling.

[157] B. G. WYBOURNE: Spectroscopic Properties of Rare Earth, New York: Interscience Publ. 1965; W. A. RUNCIMAN, and B. G. WYBOURNE: J. Chem. Phys. **31**, 149 (1959); J. S. MARGOLIS: J. Chem. Phys. **35**, 1367 (1961).

[158] G. RACAH: Phys. Rev. **61**, 537 (1942); B. R. JUDD: Proc. Phys. Soc. (London) A **69**, 157 (1956); J. P. ELLIOT, B. R. JUDD, and W. A. RUNCIMAN: Proc. Roy. Soc. (London) A **240**, 509 (1957); B. G. WYBOURNE: Spectroscopic Properties of Rare Earth, New York: Interscience Publ. 1965.

line positions and intensities than the Russell-Saunders coupling. However the extreme case of predominant $j-j$ coupling between $4f$-electrons is not expected to occur and has never been found experimentally in lanthanides. Deviations from LS-coupling become notable particularly in higher excitation levels of the $4f$-configuration, while the lowest multiplets of the lanthanides are usually better than 95—98% pure Russell-Saunders states. Since magnetic and solid state properties are mainly related to the lowest multiplet levels, *it is sufficient to assume that in magnetic semiconductors the $4f$-electrons of the lanthanides conform always to the Russell-Saunders coupling scheme.*

F. Optical properties of crystals.

While the interpretation of free ion spectra is quite complicated due to the large number of possible transitions and is still in its beginnings, the interpretation of the low lying energy levels of divalent and trivalent lanthanide ions diluted in transparent crystals has made remarkable progress by a fortunate combination of experimental and theoretical analysis during the last decade and is understood rather well. One has the unusual situation that most of our knowledge about the free ion spectra does not originate from excitations of gaseous ions in electric arcs and sparks but rather from the interpretation of crystal spectra.

I. Crystalline field effects on $4f$ electrons.

36. The character of crystal field spectra. In the crystalline environment the free ion spectrum becomes more or less strongly modified by the electrostatic Stark field and covalency effects. In addition, charge transfer transitions to neighbor ions[159] and also $4f-4f$ transitions become observable. Transitions within the same configuration which are strictly parity forbidden in the free ion by the Laporte rule (i.e., no electric dipole transitions between terms of equal parity!) become weakly allowed under the influence of the crystalline field. Since the $4f$ electrons are more concentrated around the ion core than $3d$ electrons, and are electrostatically screened by the outer $5s$, $5p$ electrons, the crystalline field acting on the $4f$ electrons is only a few 100 cm^{-1} as compared to about 10^4 cm^{-1} for the $3d$ electrons in iron group elements where covalency dominates. The spin-orbit interaction for $4f$ electrons is, on the other hand, nearly one order of magnitude larger than for $3d$ electrons and the Stark effect from the crystal field remains small compared to the level separation in the multiplets. In crystals where the environment of the diluted lanthanide ions has no center of symmetry the crystalline field is asymmetric enough to intermix the $4f$ levels slightly with the d levels of opposite parity so that needle sharp transitions between multiplet levels within the same $4f^n$ configuration can be observed[160], although a close relationship to the term scheme of the free ion is still preserved. The intensities of those forced electric-dipole transitions are about 10^6 times smaller than for allowed transitions, such as $4f-5d$. Their width, however, is only of the order of a few cm^{-1}. In many cubic crystals the crystal field contains even terms only

---

[159] G. K. JORGENSEN: Absorption Spectra and Chemical Bonding in Complexes, London: Pergamon Press 1961; G. K. JORGENSEN: Mol. Phys. **5**, 271 (1962); C. K. JORGENSON: Orbitals in Atoms and Molecules, London: Academic Press 1962; S. P. SINHA: Complexes of the RE, Oxford: Pergamon Press 1966.

[160] J. H. VAN VLECK: J. Phys. Chem. **41**, 67 (1937); L. J. F. BROER, C. J. GORTER, and J. HOOGSCHAGEN: Physica **11**, 231 (1945); B. R. JUDD: Phys. Rev. **127**, 750 (1962); G. S. OFELT: J. Chem. Phys. **37**, 511 (1962).

and $4f-4f$ excitations of complex nature appear due to thermal lattice vibrations[161], electric quadrupole or magnetic dipole transitions. The occurence of the needle sharp $4f-4f$ transitions in crystal spectra is the reason why the $4f$ multiplets in RE ions are now better defined in spectroscopy than any other atomic levels.

37. The $4f$ multiplets in host crystals. The crystal spectra of many divalent and trivalent RE ions in a variety of host crystals, such as ethyl sulfates, double nitrates, garnets, lanthanum trihalides and alkaline earth fluorides, have been investigated extensively[162], interpreted without ambiguity and their energy levels accounted for within a few cm^{-1}. The results made remarkable contributions to the elucidation of the term schemes of many ions and provided appreciable information about the interaction of the $4f$ electrons with their crystalline environment[163]. In Fig. 21 we reproduce from a paper of DIEKE et al.[164] the lower parts of the $4f^n$ term schemes obtained from the absorption and fluorescence spectra of the trivalent RE ions in crystalline salts. In many cases the anhydrous trichlorides have been used because their structure provides the RE ions with an environment of relatively high symmetry of the hexagonal C_{3h}-type. In those salts, the fluorescence and absorption spectra are very clear with sharp lines and relatively free of superimposed vibrational excitations. The thickness of the lines in Fig. 21 represents the total crystal field splitting and their centers of gravity are quite close to the theoretical positions of the free-ion levels. Consequently, one finds for all compounds in which the crystalline field is only a weak perturbation to the multiplet energies, nearly the same term arrangements. Comparison of Fig. 21 with the Fig. 13 on page 270 in the volume 28 of this Handbook demonstrates impressively the progress in RE spectroscopy during the last ten years[165].

---

[161] R. A. SATTEN: J. Chem. Phys. **27**, 286 (1957); **29**, 658 (1958); G. J. BALLHAUSEN: Progr. Inorg. Chem. **2**, 251 (1960); W. E. BRON, and W. R. HELLER: Phys. Rev. **136**, A 1433 (1964); J. D. AXE, and P. P. SOROKIN: Phys. Rev. **130**, 945 (1963); I. RICHMAN, R. A. SATTEN, and E. WONG: J. Chem. Phys. **39**, 1833 (1963); **40**, 1451 (1964); M. WAGNER, and W. E. BRON: Phys. Rev. **139**, A 223 (1965); **145**, 689 (1966); S. YATSIV, I. ADATO, and A. GOREN: Phys. Rev. Letters **11**, 108 (1963).

[162] Several investigations are motivated by technical aspects. The multitude of sharp energy levels in the visible and red characterizes the RE salts as promising laser materials and laser oscillations have been observed for many of these systems: P. P. SOROKIN: Proc. III. Int. Congr. on Quantum Electronics (Ed. P. GRIVET, and N. BLOEMBERGEN), Vol. 2, p. 985, New York: Columbia University Press 1966; op. cit., G. G. B. GARRET, p. 971; W. V. SMITH, and P. P. SOROKIN: Lasers. New York: McGraw-Hill 1966.

[163] Reviews: D. S. McCLURE: Solid State Physics, Vol. 9 (1959); M. A. ELYASHEVICH: Spectra of the RE (in Russian), Gosudarst. Izdatelstua Tekh. Teoret. Lit., Moscow, 1953 (trans. by U.S. Atomic Energy Commision: AEC-tr-4403); LANDOLT-BÖRNSTEIN: Tabellen, Vol. I, Berlin-Göttingen-Heidelberg: Springer 1950; B. G. WYBOURNE: Spectroscopic properties of RE, New York: Interscience Publ. 1965; S. P. SINHA: Complexes of RE, Oxford: Pergamon Press 1966; E. FICK, and G. JOOS: This handbook, Vol. 28 (1957); W. LOW: Progr. Science, Technol. RE (ed. L. R. EYRING), Vol. 2, Oxford: Pergamon Press 1966.

[164] G. H. DIEKE, H. M. CROSSWHITE, and B. DUNN: J. Opt. Soc. Am. **51**, 820 (1961); G. H. DIEKE, and H. M. CROSSWHITE: Appl. Optics **2**, 673 (1963); G. H. DIEKE: Adv. in Quantum Electronics, p. 164. New York: Columbia University Press 1961. Energy levels for several tetravalent ions are given by L. P. VARGA and L. B. ASPREY: J. Chem. Phys. **48**, 139 (1968).

[165] A numerical tabulation of averaged experimental energy levels for di- and trivalent RE ions up to 15,000 cm^{-1} has been given by D. H. DENNISON, and K. A. GSCHNEIDNER: A Tabulation of the Specific Heat Contribution due to Thermal Excitation of $4f$ Electrons of Some di- and trivalent Lanthanide Elements, U.S. AEC Research and Development Report 15-1156 available from Clearinghouse for Federal Scientific and Technical Information, National Bureau of Standards, Springfield, Virginia.

Sect. 37.   The 4f multiplets in host crystals.   465

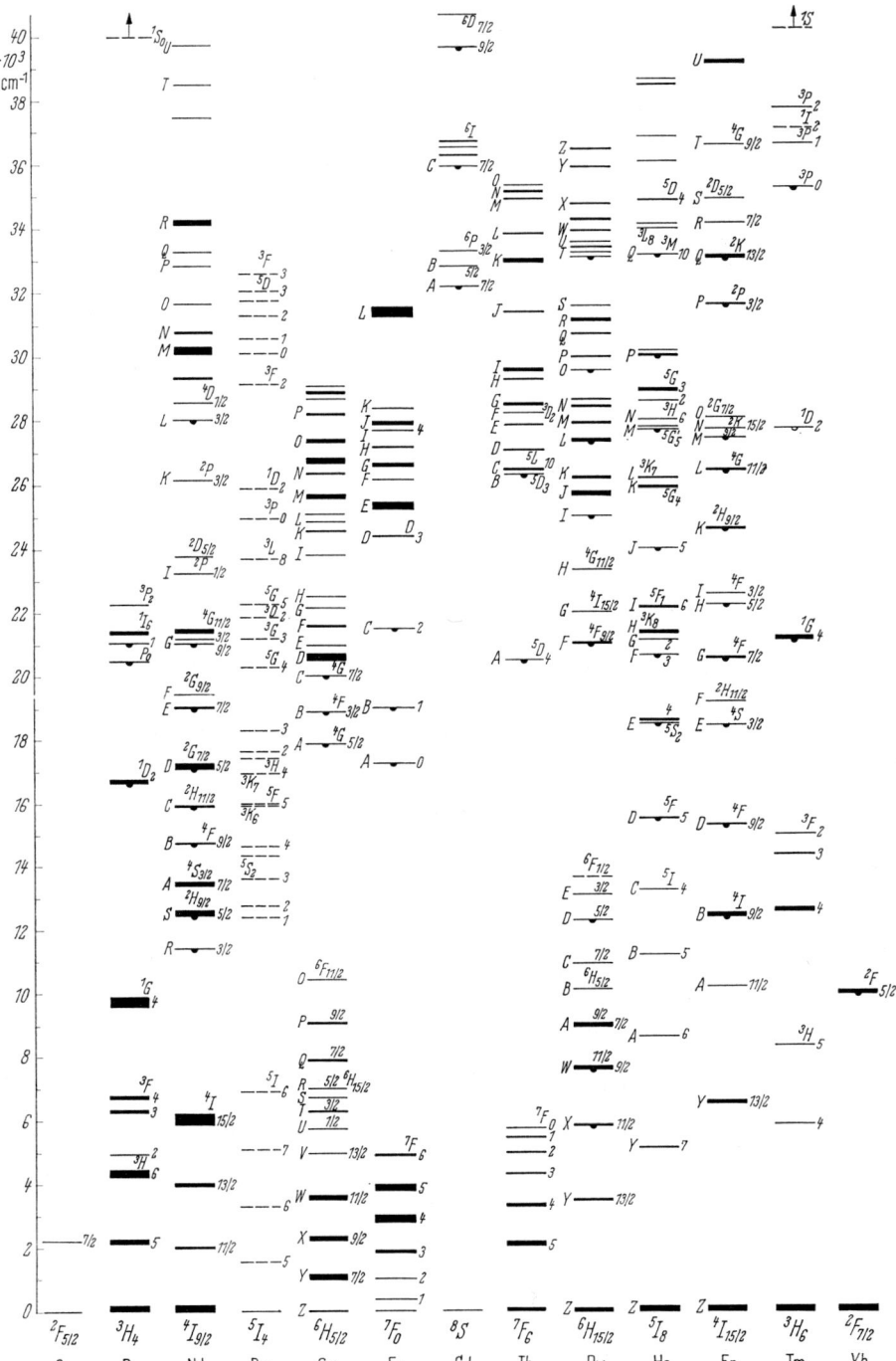

Fig. 21. The lower energy levels of the 4f^n configuration for RE^{3+} ions as obtained from crystal absorption and fluorescence spectra. The line thickness represents the total crystal field splitting. Half circles indicate levels fluorescing in the LaCl$_3$ lattice. (From G. H. Dieke, H. M. Crosswhite, and B. Dunn.)

38. Crystal fields in electrostatic approximation. The special properties of the $4f$ electrons make the lanthanides unique tools for studying the influence of the environment on atomic electrons in crystals, liquids[166] or even in the stars and sun[167] by interpreting the difference between experimentally observed spectra and the ideal free ion case. In solid state physics this is the main source for quantitative information about the crystalline field and chemical bonding mechanism. Until recently it was quite usual to assume that the $4f$ electrons, in contrast to d electrons, are so localized and well screened by the outer $5s$ and $5p$ shells of the Xe core that the electrostatic crystal field gives a sufficient approximation for the description of the perturbations by the crystalline environment.

As has been mentioned in part B, the simplest form of crystal field theory neglects all chemical bonding, ligand size, etc., and considers only the electrostatic field from point charges placed at the neighboring ligand sites. The potential of point charges which do not overlap with the magnetic ion can be derived as the solution to a Laplace equation[168]. It can be written in the general form of an expansion in spherical harmonics. The first term of this expansion is angularly independent. It contains the number, charge and distance of the ligands and is related to the binding energy of the ionic lattice. It shifts all levels of a given configuration by the same amount. The higher terms contain as an essential feature the microsymmetry of the crystal around the magnetic cation. STEVENS and others[169] have replaced the harmonics by so called "operator equivalents" which are certain functions of the total angular momentum J of the magnetic ion and transform in the same manner as the successive terms in the spherical expansion. The number of terms needed depends on the symmetry and the orbital angular momentum of the magnetic electrons. For d electrons with $l=2$ it is not necessary to go beyond terms of fourth order, and the crystal field splittings in a cubic environment can be expressed by a single parameter

$$10\,\mathrm{Dq} \sim \frac{5}{3} \frac{\langle r^4 \rangle}{R^5} \tag{38.1}$$

($r=$ orbital radius, $R=$ ligand distance). Configuration interactions with higher electron levels are, of course, neglected in this, and would require more terms. For electrons in a pure f configuration, potentials up to sixth order have to be considered. As an example, the most general Hamilton operator for f electrons in a cubic point symmetry is given by an expansion in operator equivalents O of the following form[170] (center of a cube of 8 anions as in CaF_2, garnets or center of octahedron in NaCl):

$$H = B_4(O_4^0 + 5 \times O_4^4) + B_6(O_6^0 - 21 \times O_6^4). \tag{38.2}$$

---

[166] C. V. BANKS, and D. W. KLINGMAN: Anal. Chim. Acta **15**, 356 (1956); C. K. JORGENSEN, and B. R. JUDD: Mol. Phys. **8**, 281 (1964); W. T. CARNALL, D. M. GRUEN, and R. L. MACBETH: J. Phys. Chem. **66**, 2159 (1962); G. P. SMITH, in: Molten Salt Chemistry (Ed. M. BLANDER), p. 427. New York: Interscience 1964.

[167] P. W. MERRILL: Lines of the Chemical Elements in Astronomical Spectra, Carnegie Inst. Wash. Publ. **610**, 75 (1956); D. N. DAVIS: Astrophys. J. **106**, 28 (1947); W. F. MEGGERS, and J. A. WHEELER: J. Research Natl. Bur. Standards **6**, 239 (1931); C. E. MOORE: Appl. Optics **2**, 665 (1963).

[168] Volume 28 of this Handbook.

[169] K. W. H. STEVENS: Proc. Phys. Soc. (London) A **65**, 209 (1952); B. BLEANEY, and K. W. H. STEVENS: Repts. Progr. Phys. **16**, 108 (1953); J. M. BAKER, B. BLEANEY, and W. HAYES: Proc. Roy. Soc. (London) A **247**, 141 (1958); M. T. HUTCHINGS: Solid State Physics **16**, 227 (1964); R. J. ELLIOT, and K. W. H. STEVENS: Proc. Roy. Soc. (London) A **219**, 1387 (1953); A **218**, 553 (1953).

[170] K. R. LEA, M. J. M. LEASK, and W. P. WOLF: J. Phys. Chem. Solids **23**, 1381 (1962); J. A. WHITE: J. Phys. Chem. Solids **23**, 1787 (1962).

The coefficients B_4 and B_6 determine the magnitude of the crystal field splittings. In a purely electrostatic model the coefficients are linear functions of the mean fourth and sixth power of the orbital radius r of the magnetic electrons,

$$B_4 = \beta A_4 \langle r^4 \rangle, \quad B_6 = \gamma A_6 \langle r^6 \rangle, \tag{38.3}$$

where β and γ are tabulated constants and A_4 and A_6 include lattice sums[171] over the charged ligands and depend on the distance R to the ligands as R^{-5} and R^{-7}, respectively[172]. For noncubic crystal symmetries the ligand field has about the same magnitude as in the cubic case but an increasing number of small terms is added which describe the deviation from the cubic pointsymmetry around the cation. The hexagonal C_{3h} symmetry in the trichlorides and ethylsulfates, as an example, needs for its description four parameters $A_2^0 \langle r^2 \rangle$, $A_4^0 \langle r^4 \rangle$, $A_6^0 \langle r^6 \rangle$ and $A_6^6 \langle r^6 \rangle$, while a C_{3v} symmetry requires six independent parameters, etc. The magnitude of these parameters can be calculated on purely electrostatic principles but, as we discuss later, the agreement with experimental data is often found to be quite disappointing.

39. Observed crystal field splittings. In order to avoid the pertinent questions about radial extensions of the magnetic electrons and ligand charges, etc., for the interpretation of observed crystal spectra one uses the coefficients in the harmonic expansion as adjustable "crystal field parameters". What then remains in the mathematical description are symmetry arguments which can be handled by group-theoretical means[173]. The validity of the symmetry aspects is rather independent of the character of the chemical bonding[174] and accounts for the success of the phenomenological treatment of the crystal field theory in predicting from the crystal symmetry alone, the right multiplicities of the multiplet sublevels in crystals. It may be useful to demonstrate here an example of crystal field splittings in the point charge approximation, and the variation of these splittings throughout the lanthanide series.

In Fig. 22 we show from a paper of WHITE[175] the relative splittings of the ground state of the trivalent RE ions in a cubic field calculated from the Hamiltonian (38.2). White assumes that A_4 and $\langle r^4 \rangle$ are constant in the lanthanide series and uses the ratio of the sixth order to the fourth order potential

$$\frac{A_6 \langle r^6 \rangle}{A_4 \langle r^4 \rangle} = -\frac{2}{7} \frac{\langle r^6 \rangle}{\langle r^4 \rangle R^2} \tag{39.1}$$

and a scaling constant W as parameters instead of B_4 and B_6. In reality, one can expect[176] the mean radius $\langle r^n \rangle$ to vary with the atomic number Z roughly as

$$\langle r^n \rangle \sim (Z - 55)^{-n/4}, \tag{39.2}$$

---

[171] B. BLEANEY: Proc. Roy. Soc. (London) **277**, 289 (1964).

[172] Extensive calculations of the crystal field in garnets have been made by M. T. HUTCHINS, and W. P. WOLF: J. Appl. Phys. **35**, 1060 (1964); J. Chem. Phys. **41**, 617 (1964).

[173] H. BETHE: Am. Phys. **3**, 133 (1929); W. G. PENNEY, and R. SCHLAPP: Phys. Rev. **41**, 194 (1932).

[174] J. H. VAN VLECK: J. Chem. Phys. **3**, 803, 807 (1935).

[175] J. A. WHITE: J. Phys. Chem. Solids **23**, 1787 (1962).

[176] R. J. ELLIOT, and K. W. H. STEVENS: Proc. Roy. Soc. (London) A **218**, 553 (1953); B. R. JUDD: Proc. Roy. Soc. (London) A **251**, 134 (1959); M. J. D. POWELL, and R. ORBACH: Proc. Phys. Soc. (London) **78**, 753 (1961); A. J. FREEMAN, and R. E. WATSON: Phys. Rev. **127**, 2058 (1962) discuss the validity of this assumption in comparison with $\langle r^n \rangle$ values calculated from their Hartree-Fock wavefunctions. A. J. FREEMAN, and R. F. WATSON: Magnetism 2A (Ed. G. T. RADO, and H. SUHL). New York: Academic Press 1965.

and $A_4 \langle r^4 \rangle$ may decrease by about a factor of four in going from Ce^{3+} to Yb^{3+}. In Fig. 22 we select as an example the crystal field splittings for a ratio of -0.10 which best suits some experimental observations in garnet crystals[177] with a cation-anion distance of $R=2.4$ Å. If one uses, however, the $\langle r^4 \rangle$ and $\langle r^6 \rangle$ values calculated from the Hartree-Fock wavefunctions of the $4f$ electrons the ratio should have a 2 to 3 times smaller value.

It is, at present, a major problem in crystal field theory to define the physical reasons for the large differences between theoretical crystal field parameters

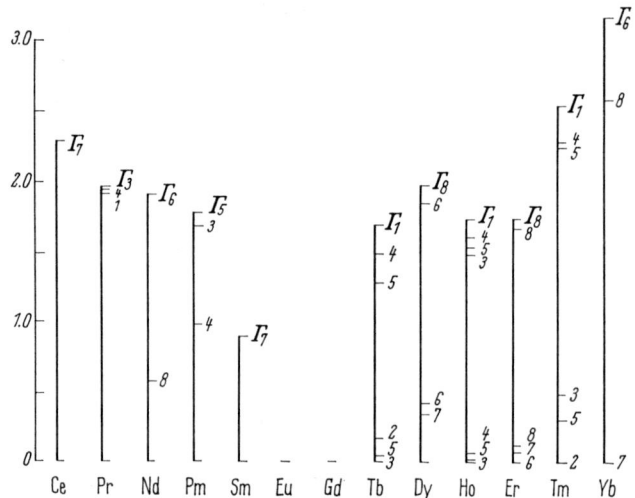

Fig. 22. Relative crystal field splittings of the ground levels of RE^{3+} given in Table 1 in cubic symmetry (garnets). The states Γ_1 and Γ_2 are nondegenerate, the levels $\Gamma_3, \Gamma_4, \Gamma_5, \Gamma_6, \Gamma_7$, and Γ_8 are 2, 3, 3, 2, 2, and 4 fold degenerate, respectively (notation of BETHE). The ordinate is arbitrary and should be multiplied by some reasonable value for $A_4 \langle r \rangle$ for each ion. For Yb^{3+}, the separation between Γ_1 and Γ_8 should be about 550 cm^{-1}. (From survey article by J. H. VAN VLECK: Progress in Science and Technology of RE, Vol. 2, p. 11. Oxford: Pergamon Press 1966.)

and their experimental values. Accurate experimental data can be obtained from a large number of optical observations, microwave resonance experiments[178], specific heat[179], magnetic susceptibility and paramagnetic relaxation[180] measurement as well as inelastic scattering of thermal neutrons[181] and Mössbauer studies[182]. One finds, in general, that the interpretation of the experimental crystal field splittings needs larger contributions of higher order potentials than the simple point charge electrostatic theory permits for given ligand distribution

---

[177] J. H. VAN VLECK: Progress in Science and Technology of R.E. (Ed. L. R. EYRING), Vol. 2, p. 1, Oxford: Pergamon Press 1966; J. Chem. Phys. Solids **27**, 1047 (1966).

[178] W. LOW: Paramagnetic resonance in solids. New York: Academic Press 1960.

[179] W. SCHOTTKY: Phys. Z. **23**, 448 (1922); H. MEYER, and P. L. SMITH: J. Phys. Chem. Solids **9**, 285, 296 (1959); K. H. HELLWEGE, U. JOHNSEN u. W. PFEFFER: Z. Physik **154**, 301 (1959).

[180] W. M. YEN, W. C. SCOTT, and P. L. SCOTT: Phys. Rev. **137**, A 1109 (1965); B. BLEANEY, and K. W. H. STEVENS: Repts. Progr. in Phys. **16**, 109 (1953); K. D. BOWERS, and J. OWEN: Repts. Progr. in Phys. **18**, 304 (1955).

[181] D. CRIBIER et B. JACROT: Compt. rend. **250**, 2871 (1960); S. ODIOT, and D. SAINT-JAMES: J. Phys. Chem. Solids **17**, 117 (1960).

[182] Several papers in Rev. Mod. Phys. **36**, 385 (1964).

and radial extensions of the Hartree-Fock wavefunctions[183]. Since numerical data for the electron wavefunctions became available during the last five years, the discrepancies between theory and experiment can be formulated more quantitatively. From their understanding, one learns about the role of localized electrons in chemical bonding. The separation of the splitting constants into contributions from the ligand charge distribution and from the radial part of the magnetic electron wave functions suggests that corrections to the point charge model should be sought in two directions, namely through a more sophisticated description of crystalline environment or from a more accurate theory of free ion wave functions and their modifications when subjected to the crystalline environment.

BURNS[184] has investigated how the uncertainty in the positions of the ligands, and polarization effects due to the finite extent of the ligand charge, affect the accuracy of the theoretical splitting parameters. Several attempts have been made to correct the electrostatic potential for mixing with configurations of opposite parity by odd crystal field terms[185] and for the shielding and polarization by electrons in closed shells[186]. In particular, the deformation of the closed $5s$ and $5p$ shell of the R.E. ion in the crystalline field has been studied by FREEMAN and WATSON. It seems to account for major parts of the crystal field in materials with noncubic symmetry. In many aspects, however, these corrections are still insufficient and more serious modifications of the $4f$ wavefunctions in crystals have to be considered.

40. Influence of covalent bonding on $4f$ electrons. JORGENSEN[187] has suggested that the crystal field splitting in cubic lanthanide compounds are dominantly due to the covalent σ-antibonding between the $4f$ electrons and the ligand electrons. For $3d$ element compounds the basic significance of the covalency for the properties of the magnetic electrons has never been doubted seriously. It has been suggested for the actinides, that the $5f$ electrons may also be used in formation of covalent bonds in compounds which have a predominantly ionic character[188], such as, for example, in the octahedral coordination of UCl_6.

---

[183] G. BURNS: Phys. Rev. **128**, 2121 (1962); J. Chem. Phys. **42**, 377 (1965); J. C. EISENSTEIN: J. Chem. Phys. **39**, 2128 (1963); Symposium on paramagnetic resonance (Ed. W. Low), Vol. 1, p. 253, New York: Academic Press 1963; R. J. ELLIOT, and K. W. H. STEVENS: Proc. Roy. Soc. (London) A **215**, 437 (1952); A **218**, 553 (1952); A **219**, 387 (1952); A. J. FREEMAN, and R. F. WATSON: Phys. Rev. **127**, 2058 (1962); M. T. HUTCHINGS, and W. P. WOLF: J. Chem. Phys. **41**, 617 (1964); M. T. HUTCHINGS, and D. K. RAY: Proc. Phys. Soc. (London) **81**, 663 (1963); C. A. HUTCHINSON, and E. Y. WONG: J. Chem. Phys. **29**, 754 (1958); K. R. LEA, M. J. M. LEASK, and W. P. WOLF: J. Phys. Chem. Solids **23**, 1381 (1962); J. S. MARGOLIS: J. Chem. Phys. **35**, 1367 (1961); J. A. WHITE: J. Phys. Chem. Solids **23**, 1787 (1962); E. Y. WONG, and J. RICHMAN: J. Chem. Phys. **36**, 1889 (1962); **37**, 2270, 2498 (1962); **39**, 793 (1963).

[184] G. BURNS: J. Chem. Phys. **42**, 377 (1965).

[185] B. R. JUDD: Phys. Rev. **127**, 750 (1962); S. OFELT: J. Chem. Phys. **38**, 2171 (1962); J. D. AXE: J. Chem. Phys. **39**, 1154 (1963).

[186] G. BURNS: Phys. Rev. **128**, 2121 (1962); C. J. LENANDER, and E. Y. WONG: J. Chem. Phys. **38**, 2750 (1963); D. K. RAY: Proc. Phys. Soc. (London) **82**, 47 (1963); R. E. WATSON, and A. J. FREEMAN: Phys. Rev. **133**, A 1577 (1964); K. RAJNAK, B. G. WYBOURNE: J. Chem. Phys. **41**, 565 (1964); A. J. FREEMAN, and R. E. WATSON: Phys. Rev. **139**, A 1066 (1965); R. M. STERNHEIMER: Phys. Rev. **146**, 140 (1966).

[187] C. K. JORGENSEN, R. PAPPALARDO, and H. SCHMIDTKE: J. Chem. Phys. **39**, 1422 (1963); C. K. JORGENSEN: Absorption Spectra and Chemical Bonding in Complexes, Reading: Addison-Wesley Publ. Comp. 1962; C. K. JORGENSEN: Orbitals in Atoms and Molecules, New York: Academic Press 1962; C. K. JORGENSEN: Phys. Stat. Solidi **2**, 1146 (1962); Progr. Inorg. Chemistry **4**, 73 (1962); Faraday Soc. Dis. **26**, 110 (1958).

[188] J. C. EISENSTEIN: J. Chem. Phys. **25**, 142 (1956); J. C. EISENSTEIN, and M. H. L. PRYCE: Proc. Roy. Soc. (London) A **255**, 181 (1960) where references to earlier literature are given.

For the lanthanides, however, the possibility of covalent bonding contributions of the $4f$ electrons became accepted increasingly only during recent years. The resulting effects are here so small, that the concept of fully localized magnetic electrons remains a good approximation for the description of most properties.

The treatment of the crystalline environment of a magnetic ion certainly has a quite different aspect in the case of extreme covalent bonding than for a weak electrostatic crystal field. The nearly free atom approximation has to be replaced by a cluster approximation in which the ion and the surrounding ligands are treated as one molecular unit. The atomic orbitals have to be replaced by the molecular bonding and antibonding orbitals which can be derived from linear combinations of the atomic orbitals. The overlap with electron wave functions at neighboring atoms depends strongly on their angular symmetry with respect to the symmetry of the surrounding ligand cluster. Therefore, the multiplicity of the bonding and antibonding levels is determined again by group-theoretical considerations. For transition element compounds, it is a well accepted procedure to describe the properties of d electrons semiquantitatively, with adjustable parameters, by molecular orbital theory[189], but quantitative calculations from first principles have been made also[190]. EISENSTEIN[191] has tabulated for f electrons in several ligand configurations the group-theoretical functions relevant to the hybridization of atomic f orbitals to molecular orbitals. First principle molecular orbital calculations, however, are not available for rare earth compounds, at present.

AXE and BURNS[192] have investigated experimentally the effect of unaxial stress on the optical transition between crystal levels in the $^2F_{7/2}$ groundstate of Tm^{2+} in CaF_2 and SrF_2. The crystal symmetry of the fluorite structure provides the RE site with the configuration of eight F ions in the shape of a simple cube. The observed frequency shift of about 1.75×10^{-10} cm^{-1} for a strain of 1 dyn cm^{-2} has been used to calculate the radial dependence of the crystal field splitting which is found to be somewhat larger than predicted by a purely electrostatic model. The effect of covalency has been treated in a semi-empirical molecular-orbital model in which the overlaps of the $4f$ orbitals with the fluorines are derived from Hartree-Fock wavefunctions and interatomic distances. The largest group overlap was found to be 3.6% and leads to a covalent contribution of about 50% to the observed crystal field splittings. The largest contribution comes from the σ bond of $4f$ orbitals of suitable symmetry with the fluorine p electrons.

---

[189] Some survey literature: J. S. GRIFFITH: Theory of Transition Metal Ions, London: Cambridge Univ. Press 1961; C. J. BALLHAUSEN: Introd. to Ligand Field Theory, New York: McGraw-Hill 1962; L. E. ORGEL: Ligand Field Theory, London: Methuen 1960; C. J. BALLHAUSEN, and H. GRAY: Molecular Orbital Theory, New York: W. A. Benjamin Inc. 1964; C. K. JORGENSEN: Orbitals in Atoms and Molecules, New York: Academic Press 1962; F. A. COTTON: Chemical Applications of Group Theory, New York: Whiley-Interscience 1963; C. A. COULSON: Valence, Oxford: University Press 1961; H. EYRING, J. WALTER, and G. E. KIMBALL, Quantum Chemistry, New York: Wiley 1960; H. B. GRAY: Electrons and Chemical Bonding, New York: W. A. Benjamin 1965; W. MOFFIT, and C. J. BALLHAUSEN: Ann. Rev. Phys. Chem. 7, 107 (1956).

[190] L. E. ORGEL: J. Chem. Phys. 23, 1004 (1955); V. TANABE, and S. SUGANO: J. Phys. Soc. Japan 9, 753 (1954); M. KOTANI: J. Phys. Soc. Japan 4, 293 (1949); R. G. SHULMAN, and S. SUGANO: Phys. Rev. 130, 506 (1962); K. KNOX, R. G. SHULMAN, and S. SUGANO: Phys. Rev. 130, 512 (1962); S. SUGANO, and R. G. SHULMAN: Phys. Rev. 130, 517 (1962); R. E. WATSON, and A. J. FREEMAN: Phys. Rev. 134, A 1526 (1964); E. SIMANEK, and Z. SROUBECK: Phys. Stat. Solidi 4, 251 (1964); S. SUGANO, and Y. TANABE: J. Phys. Soc. Japan 20, 1155 (1965); J. HUBBARD, D. E. RIMMER, and F. R. A. HOPGOOD: Proc. Phys. Soc. (London) 88, 13 (1966).

[191] J. C. EISENSTEIN: J. Chem. Phys. 25, 142 (1956).

[192] J. D. AXE, and G. BURNS: Phys. Rev. 152, 331 (1966).

Throughout the lanthanide series the various overlaps decrease with increasing number of f electrons by a factor of about 2 to 2.5. It appears that the $4f$ orbital overlap shrinks faster than the variation of the theoretical orbital radii in Fig. 18 would permit. Watson and Freeman[193] have discussed further implications of the above calculations. They conclude that another effect may be important too, namely the contribution of an indirect mechanism via the closed $5p$ shell. The covalent bonding originates here from mixing the empty f-states with filled $5p$ states which thus become available for covalent interactions with the filled p states at the neighboring anions. The strength of the "$5p$-covalency" depends on the nature of the matrix elements which provide the $4f-5p$ mixing and can seemingly account in some ions for the full amount of observed crystal field splittings.

The possible covalency of $4f$ electrons is not only based on observations of crystal field spectra but is also suggested by investigations of the Zeeman effect[194] and several other effects[195]. Jorgensen[196] has explained the shift of absorption spectra of RE ions in solutions by "nephelauxetic", i.e., cloud enlarging, effects, in which the covalent mixing with ligand orbitals is assumed to affect the radial (but not the angular) part of the $4f$ orbitals by shielding the effective core charge. As a consequence, the $4f$ orbital radius expands and the Slater integrals and spin-orbit interaction decrease compared to their free ion values. Besides several energy transfer and quenching effects in fluorescence of RE ions, strong evidence for some amount of covalency derives also from nuclear and double resonance experiments where a transfer of the RE hyperfine splitting (e.g., of Eu^{2+}) to the anion core (e.g., F^- in $Eu:CaF_2$) has been observed[197]. Of special relevance to magnetic properties of RE compounds is the question, whether and how strongly the covalency of the $4f$ orbitals can contribute to superexchange interactions via anions, analogous to the superexchange usually found in $3d$ compounds. Since the intra-orbital exchange between $4f$ electrons is very large (e.g. the spin-up and spin-down $4f$ bands in Fig. 6 are separated by about $0.6\,Ry = 8\,eV$), reasonably strong superexchange may result even from small overlap with ligand orbitals.

II. The $5d$ orbitals in lanthanides.

41. $5d$ electrons in RE-elements. Concentrating on the extraordinary properties of the f electrons, the solid-state physics of RE compounds has tended to ignore the other filled shells. Until recently, it was quite customary to assume a large radial extension for the $5d$ electrons in lanthanides. Particularly in metals where

---

[193] R. E. Watson, and A. J. Freeman: Phys. Rev. **156**, 251 (1967).

[194] J. M. Baker, and J. P. Hurrel: Proc. Phys. Soc. (London) **82**, 742 (1963); W. Low, and R. S. Rubins: Phys. Rev. **131**, 2527 (1963); M. J. M. Leask, R. Orbach, M. J. D. Powell, and W. P. Wolf: Proc. Roy. Soc. (London) A **272**, 371 (1963); B. Bleaney: Proc. Roy. Soc. (London) A **277**, 289 (1964).

[195] B. G. Wybourne: Spectroscopic Properties of R.E., New York: Wiley-Interscience 1965; L. Katzin, and L. Barnett: J. Phys. Chem. **68**, 3779 (1964); F. A. Deeney, J. A. Delaney, and V. P. Ruddy: Phys. Letters **25 A**, 370 (1967); K. A. Gschneidner, and R. M. Valetta: Acta Metall. **16**, 477 (1968).

[196] W. Marshall, and R. Stuart: Phys. Rev. **123**, 2048 (1961); C. K. Jorgensen: Progr. Inorg. Chem. **4**, 73 (1962); C. K. Jorgensen, R. Pappalardo u. E. Rittershaus: Z. Naturforsch. **20 A**, 54 (1964); C. K. Jorgensen: Orbitals in Atoms and Molecules, New York: Academic Press 1962; R. Rajnak, and B. G. Wybourne: J. Chem. Phys. **41**, 565 (1964); S. P. Sinha: Complexes of R.E., Oxford: Pergamon Press 1966.

[197] J. M. Baker, and J. P. Hurrell: Proc. Phys. Soc. (London) **82**, 742 (1963); R. G. Shulman, and B. J. Wyhida: J. Chem. Phys. **30**, 335 (1959); R. E. Watson, and A. J. Freeman: Phys. Rev. **156**, 251 (1967) with more literature, Phys. Rev. Letters **6**, 1277, 388 (1961).

the outer conduction electrons provide the magnetic interactions via indirect exchange[198], the description of the conduction electrons as s-type Bloch electrons was considered to be satisfactory. However, several results from magnetic[199], electric[200], specific heat[201], optic[202], and photoelectric[203] measurements have been found to be in strict contradiction to the free electron model for the conduction electrons in RE metals. Dimmock, Watson and Freeman[204] have calculated the electronic band structure for Gd, La, Tm and Lu metal applying the augmented plane wave method to a muffin-tin potential composed of the Hartree-Fock wave functions we have discussed earlier in this chapter. Their results indicate that the assumption of a free electron model for the conduction electrons is completely incorrect. The $5d$ band contributions provide a high density of states in the vicinity of the Fermi energy (about 3 times the free electron value). The conduction electron band width is comparable to that of the magnetic $3d$ electrons in iron (~ 6 eV). The Fermi surfaces are correspondingly complicated. The irregularities in the density of conduction electron states are in agreement with the nearly 10% increase of the magnetic moment of Gd and other RE in the metals and some metallic compounds by exchange polarization of the d electrons. They account also for experimental observations of irregularly high electronic specific heat contributions or special features in the optical properties of RE metals.

The strong qualitative resemblance between the properties of the $5d$ electrons and magnetic $3d$ electrons in metals raises a question concerning the character of $5d$ electrons in compounds, where larger interatomic distances sustain better localization of d electrons. In perfectly ionic crystals the outer $6s$ and $5d$ orbitals of the RE ions are empty, of course. However, in geometrically suitable configurations with less electronegative anions, a covalent transfer of ligand electrons into rather localized (and relatively strongly $4f$ exchange coupled) $5d$ orbitals could contribute significantly to magnetic interactions of the superexchange type between magnetic ions via intermediate anions. This f-d superexchange depends on the overlap symmetry of the d and ligand orbitals as described by the Goodenough-Kanamori rules[205] for superexchange between $3d$ ions but it is weaker since magnetic electron orbitals are not involved in covalency directly but indirectly via the intra-atomic f-d exchange.

42. Optical $4f$—$5d$ transitions in crystals. Properties of $5d$ orbitals in crystals have been studied optically for divalent lanthanide ions diluted in alkali-halides or alkaline earth halides and chalcogenides where the $4f$—$5d$ transition lies in the visible range and is unperturbed by the intrinsic absorption of the host crystal[206].

---

[198] T. Kasuya: $s-d$ and $s-f$ interaction in RE metals, Magnetism 2b (Ed. G. T. Rado, and H. Suhl), p. 215. New York: Academic Press 1966.

[199] H. Nigh, S. Legvold, and F. H. Spedding: Phys. Rev. **132**, 1092 (1963).

[200] Discussed in A. J. Freeman, J. O. Dimmock, and R. E. Watson: Quantum Theory of Atoms, Molecules and the Solid State, p. 361. New York: Academic Press 1966.

[201] O. V. Lounasma: Phys. Rev. **129**, 2460 (1963); **133**, A 219 (1964).

[202] C. C. Schüler: Phys. Letters **12**, 84 (1964); 5th R.E. Conf. Proc. Vol. 2, 35 (1965); B. R. Cooper, and R. W. Redington: Phys. Rev. Letters **14**, 1066 (1965).

[203] A. J. Blodgett, W. E. Spicer, and Y. C. Yu: Optical properties and electronic structure of metals and alloys (Ed. F. Abeles), p. 246. Amsterdam: North-Holland Publ. 1966.

[204] J. O. Dimmock, and A. J. Freeman: Phys. Rev. Letters **13**, 750 (1964); **16**, 558 (1966); Quantum Theory of Atoms, Molecules and the Solid State, p. 361. New York: Academic Press 1966.

[205] See article by Vonsovsky and Karpenko in this volume, or J. Goodenough: Magnetism and Chemical bond, New York: Wiley-Interscience 1963, and Sect. 30.

[206] Recent Reviews: B. G. Wybourne: Spectroscopic Properties of RE, New York: Wiley-Interscience 1965; D. S. McClure, and Z. Kiss: J. Chem. Phys. **39**, 3251 (1963); in: Optical Lasers, p. 357, New York: Polytechnic Press 1963; S. P. Sinha: Complexes of R.E., London: Pergamon Press 1966.

Fig. 23 shows, as an example, the absorption spectrum of Eu^{2+} in the fluorite lattice. While Eu^{2+}, Sm^{2+} and Yb^{2+} form stable ions in those crystals, the other lanthanides have a larger chemical reduction potential for the divalent state and have to be transformed from the trivalent into the divalent state by radiation or chemical means [207].

From the spectrum of the internal $4f-4f$ transitions one concludes that the ground state of most divalent ions is isoelectronic to that of the next higher trivalent ion in the series. The lower nuclear charge of the divalent ion, however, scales down the Coulomb energy and energy levels (e.g., by a factor of about 0.85 for Sm^{2+} compared to Eu^{3+}). Only Gd^{2+} does not take the excess electron back into the $4f$ shell but maintains the stable $4f^7$ configuration and forms a $4f^7\,5d$ ground state. In Ce^{2+} and Tb^{2+} the $4f^n$ and $4f^{n-1}\,5d$ configurations lie energetically close together.

The absorption spectra of divalent ions are characterized by few strong, about 1,000 to 2,000 cm^{-1} wide bands which extend from the infrared or visible to the violet and do not sharpen with decreasing temperature but rather resolve into several sublines. In addition to these strong bands one observes many weak sharp lines due to magnetic dipole $4f-4f$ and vibronic transitions [208]. In the first studies on RE^{2+} ions, the broad bands were interpreted already as electric dipole transitions between $4f$ and $5d$ or $6s$ states [209], which are expected for trivalent ions to lie much more in the ultraviolet (Fig. 20).

McClure and Kiss [210] have calculated in a first approximation the energy differences between the $4f^n$ and $4f^{n-1}\,5d$ states using the Slater-Condon theory for free ion spectra with some modifications by electrostatic crystal fields for all RE^{2+} ions in CaF_2. Their results account satisfactorily for the observed positions of the broad bands through the lanthanide series (see Fig. 20). Exceptions are Gd^{2+}, Ce^{2+} and Tb^{2+} for which we have already mentioned the preference for a $4f^{n-1}\,5d$ groundstate in the CaF_2 matrix. The majority of authors assumes that the main spectral structure of the $4f^n - 4f^{n-1}\,5d$ transitions arises from the crystal field splitting of the $5d$ orbitals in the excited state [211]. The $5d$ orbitals are more sen-

[207] W. Hayes, and J. W. Twidell: J. Chem. Phys. **35**, 1521 (1961); Z. J. Kiss, and R. C. Duncan: Proc. I.R.E. **50**, 1531, 1532 (1962); Z. J. Kiss: Phys. Rev. **127**, 718 (1962); Z. J. Kiss, in: Lasers and Applications (Ed. W. S. C. Chang), Columbus: Ohio State University Press 1963; Recent survey: F. K. Fong: Progress in Solid State Chemistry (Ed. H. Reiss), Vol. 3, p. 135, Oxford: Pergamon Press 1967; R.E. Research 4 (Ed. L. R. Eyring), p. 373, New York: Gordon and Breach 1965; J. L. Merz, and P. S. Pershan: Phys. Rev. **162**, 217, 235 (1967).

[208] J. D. Axe, and P. P. Sorokin: Phys. Rev. **130**, 945 (1963); W. Bron, and M. Wagner: Phys. Rev. **139**, A 223, A 233 (1965); **145**, 689 (1966).

[209] S. Freed, and S. Katcoff: Physica **14**, 17 (1948); F. D. S. Butement: Trans. Faraday Soc. **44**, 617 (1948); P. P. Feofilov: Optics i Spectr. **1**, 992 (1956); W. Low: Nuovo Cimento **17**, 607 (1960); W. Kaiser, C. G. B. Garett, and D. L. Wood: Phys. Rev. **123**, 766 (1961); A. Kaplyanskii i P. Feofilov: Optics i Spectr. **12**, 272 (1962); **13**, 129 (1962); P. P. Sorokin, M. J. Stevenson, J. R. Lankard, and G. D. Pettit: Phys. Rev. **127**, 503 (1962); R. Reisfeld, and A. Glasner: J. Opt. Soc. Am. **54**, 331 (1964); D. L. Wood, and W. Kaiser: Phys. Rev. **126**, 2079 (1962); Z. Kiss: Phys. Rev. **127**, 718 (1962).

[210] D. S. McClure, and Z. Kiss: J. Chem. Phys. **39**, 3251 (1963); in: Optical Lasers, p. 357. New York: Polytechnic Press 1963.

[211] Y. R. Shen, and N. Bloembergen: [Phys. Rev. **133**, A 511, A 515 (1964)] have argued from their magneto-optical measurements on Eu^{2+} diluted in CaF_2 that the excited $4f^6 5d$ configuration has the Hund's rule state ^8P which does not split in the crystal field. It is, however, observed often that in crystals with large crystalline fields the Hund's rule state of the $4f^{n-1}\,5d$ configuration in the free ions is not necessarily the lowest state in the crystal. M. J. Freiser, S. Methfessel and F. Holtzberg have shown that the magnetization dependence of the Faraday rotation can be equally well described with the assumption of a crystalline field splitted $5d$ orbital, weakly coupled to a $4f^6$ Hund rule state (J. Appl. Phys. **39**, 900 (1968)).

sitive to the crystal field than the more localized $4f$ electrons (splitting of $4f^7$ of Eu^{2+} in CaF_2 is about 0.2 cm^{-1}) [212]. They split in cubic environment by the amount 10 Dq into two branches with different symmetry, the doubly degenerate e_g and the triply degenerate t_{2g} levels. In the eightfold anion configuration of the fluorite lattice the e_g level has the lower energy while for the octahedral configuration of the NaCl-type lattices the order is reversed. RUNCIMAN and STAGER [213] have investigated the Zeeman effect and the influence of uniaxial stress on the absorption and fluorescence spectrum of Sm^{2+} and Eu^{2+} in CaF_2 single crystals. The line splittings observed in the magnetic field and under pressure with respect to different crystal directions indicate that the lower state of the d-electrons has e_g character.

43. Coupling scheme of the $4f$—$5d$ interaction. In principle, calculations of the spectra of divalent lanthanides in crystals could be made, but this needs very voluminous calculations since so many parameters have to be considered: the crystal field splittings of the f electrons in the ground and excited configurations, the exchange and spin-orbit coupling between f and d electrons, the spin-orbit coupling of the d electrons and their strong interaction with the crystal lattice. It is difficult to assign, a priori, a reliable coupling scheme to the $f-d$ interaction in the excited $4f^{n-1}5d$ configuration, since the crystal field splitting of the d electron, its spin-orbit interaction and the coupling to the $4f^{n-1}$ electron state are expected to be of comparable magnitude. Several coupling schemes have been proposed [214].

A superior approach considers the $5d$ electron in zeroth approximation as being decoupled from the $4f^{n-1}$ core in the excited state. The crystal field splits and separates the d levels by about $10,000 \text{ cm}^{-1}$, while the crystal field effect at the $4f$ electrons can be neglected. The electrostatic interactions between the $5d$ and $4f$ electrons are treated as a small perturbation to the crystal field levels. The corresponding complex calculations are valuable only if the fine structure of the absorption bands can be resolved experimentally and interpreted. The situation is simplest, in this respect, for lanthanide ions with relatively simple $4f$ configurations in the ground state. Eu^{2+} has with seven f electrons an ^8S configuration with widely spaced levels (similar to Gd^{3+} in Fig. 21, but scaled down [215] by a factor of 0.934). Yb^{2+} has a closed shell $4f$ configuration and hence no $4f-4f$ transitions. PIPER, BROWN and MCCLURE [216] have calculated the energy levels of the configurations $4f^{13}5d$ in cubic crystal fields with variable crystal field parameters for f and d electrons. The results of the calculation are used to explain successfully the spectrum of Yb^{2+} in $SrCl_2$ and support the assumption of a d electron level strongly split by the crystal field, and weakly coupled to the $4f$ core.

44. Optical absorption spectra of Eu^{2+} ions. We discuss the *absorption spectrum of Eu^{2+} in the fluorite lattice* as a relatively simple example of a crystal spectrum of divalent RE ions. Fig. 23a shows the absorption spectrum of thin evaporated films of EuF_2, which is quite similar to the spectrum of Eu^{2+} dilute in alkali

---

[212] C. RYTER: Helv. Phys. Acta **30**, 353 (1957); R. LACROIX: Helv. Phys. Acta **30**, 374, 478 (1957); Proc. Phys. Soc. (London) **77**, 550 (1961); J. OVERMEYER, and R. J. GAMBINO: Physics Letters **9**, 108 (1964).

[213] W. A. RUNCIMAN, and C. V. STAGER: J. Chem. Phys. **37**, 196 (1962); **38**, 279 (1963); A. A. KAPLYANSKII i A. K. PRZHEVUSKII: Optics i Spectr. **12**, 272 (1962); B. P. ZAKHARCHENYA i A. Y. RYSKIN: Optics i Spectr. **13**, 501 (1962).

[214] J. D. AXE, and P. P. SOROKIN: Phys. Rev. **130**, 945 (1963); W. V. SMITH, and P. P. SOROKIN: Lasers, New York: McGraw-Hill 1966; B. R. JUDD: Phys. Rev. **125**, 613 (1962); D. L. WOOD, and W. KAISER: Phys. Rev. **126**, 2079 (1962); T. S. PIPER, J. P. BROWN, and D. S. MCCLURE: J. Chem. Phys. **46**, 1353 (1967).

[215] F. D. S. BUTENENT: Trans. Faraday Soc. **44**, 617 (1948).

[216] T. S. PIPER, J. P. BROWN, and D. S. MCCLURE: J. Chem. Phys. **46**, 1353 (1967).

halides. Eu^{2+} is isoelectronic with Gd^{3+} and has a half filled $4f$ shell in the $^8S_{7/2}$ Russell-Saunders configuration. The $^8S_{7/2}$ state provides a strong atomic moment of $7\,\mu_B$ and is, as a pure spin state, insensitive to lower order crystalline field splittings. The spectrum of Eu^{2+} diluted in host crystals with fluoride lattice is dominated by two absorption bands. For Eu^{2+} in CaF_2, as an example, the bands have peaks with an oscillator strength[217] of about $f = 0.01$ at 29,000 cm^{-1} and of $f = 0.03$ at about 45,000 cm^{-1}. The pure CaF_2 host crystal is transparent up to 80,000 cm^{-1}. The high oscillator strength in the two peaks is characteristic of highly allowed $4f - 5d$ electric-dipole transitions. The separation of the band peaks gives the crystal field splitting of the $5d$ states as about 16,000 cm^{-1}. This value is larger than the Stark splitting of the $5d$ states in Ce^{3+} (10,000 cm^{-1}) and Yb^{2+} (13,000 cm^{-1}) but still much smaller than the splittings one finds in the $5d$ elements following the lanthanide series in the Periodic Table (for Ir^{3+} and Pt^{4+} one finds, as an example, values for the splittings in octahedral configurations of the order of 25,000 cm^{-1} and 30,000 cm^{-1}, respectively[218]). The crystal field splitting of $3d$ ions has its largest value at the beginning of the series (Ti^{3+}: 20,300 cm^{-1} in octahedral coordination) and decreases with increasing $3d$ localization throughout the series (Fe^{2+}, Co^{2+}: ~10,000 cm^{-1}). Consequently, we find here again indication that the $5d$ electrons of lanthanides resemble in their properties $3d$ rather than the $5d$ electrons of transition elements.

The "staircase" structure of the low energy excitation into the $5d\,e_g$ level of Eu^{2+} in Fig. 23 can be explained by considering the multiplet structure of the $^7F_0$ configuration of the six $4f$ electrons in the excited ion[219]. The sequence of the subpeaks agrees closely with the sequence of the $J = 0, 1, 2, \ldots 6$ levels in the free Eu^{3+} ion. We find that the width of the absorption band is not determined by the width of the $5d$ levels but by the total multiplet splitting of about 5,000 cm^{-1} in the ^7F term. This explains also why the band width is found to be temperature independent, in contrast to the expectation that outer electron orbitals should strongly interact with the thermal excitations of the lattice.

The actual width of the $5d$ levels is indicated by the width of the subpeaks. It is of the order of about 0.1 eV. The optical level width is, of course, not related, in a simple manner, to the width of electron bands as calculated in the electronic band structure. The reduced population of the $4f$ shell in the excited state increases locally the effective core charge. Therefore, the excited electron can be bound to the hole in the f-shell and form an exciton-like state below the edge of the $5d$ band states[220]. In the other absorption band at higher energy (which results from excitations into the $5d$-t_{2g} states) the substructure is wiped out by a spin-orbit interaction of the order of 1,000 cm^{-1} which does not apply to the e_g states.

The spectrum of Fig. 23a is taken on evaporated EuF_2 films at a temperature of about 20 °K[222]. EuF_2 is a paramagnetic material with negligibly small magnetic interactions[221]. We find that in the CaF_2 lattice the absorption spectrum (particularly the "staircase" structure of the $5d$-e_g band) is practically independent[222] of Eu^{2+} distances from small concentrations of Eu^{2+} in CaF_2 to EuF_2. Only the

---

[217] The f number is related to the absorption coefficient α cm^{-1} by $f = \dfrac{mc}{\pi e^2 N} \int \alpha\, dv$ (N: density of excitable ions in cm^{-3}, m, e: electron mass and charge v: light frequency).

[218] D. S. McClure: Solid State Phys. **9**, 399 (1959).

[219] M. J. Freiser, S. Methfessel, and F. Holtzberg: J. Appl. Phys. **39**, 900 (1968).

[220] R. S. Knox: Theory of Excitons. New York: Academic Press 1963.

[221] K. Lee, H. Muir, and E. Catalano: J. Phys. Chem. Solids **26**, 523 (1965).

[222] S. Methfessel, M. J. Freiser, and F. Holtzberg: 2nd Int. Conf. on Solid Comp. of Trans. Elements, Univ. of Twente, June 1967, to be published; M. J. Freiser, S. Methfessel, and F. Holtzberg: J. Appl. Phys. **39**, 900 (1968).

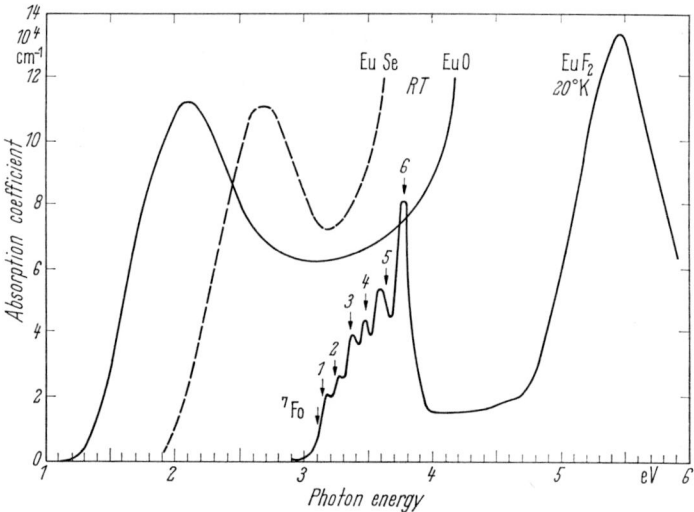

Fig. 23a. Absorption curves for 1000 Å to 5000 Å thick evaporated films of EuSe, EuO and EuF_2. The arrows indicate the relative positions of the multiplets in Eu^{3+} at intervals taken from free ion spectra and adjusted at the highest peak with the $J=6$ level. (From Ref. [225].)

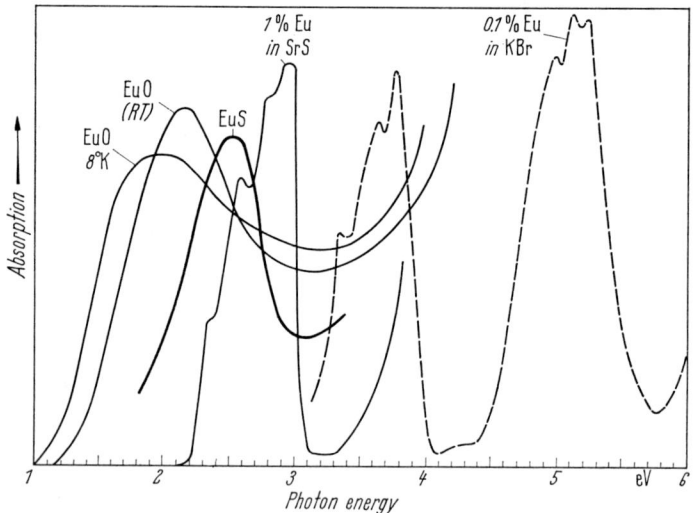

Fig. 23b. Absorption curves of EuO, EuS films compared with the absorption of Eu^{2+} ions dilute in SrS crystals and in KBr. (From Ref. [225].)

separation between the bands varies with the lattice constant as usually observed for different host crystals. This concentration independence is quite in contrast to the properties of Eu^{2+} in the NaCl lattice of the chalcogenides, which we will discuss now.

Fig. 23 includes also absorption curves obtained from *evaporated films of ferromagnetic Eu chalcogenides*. It is surprising to find that the absorption edge in the isostructural Eu chalcogenide series shifts with decreasing electron affinity of the anions from O^{2-} to Te^{2-} to higher energies and not to lower ones (as usually found in semiconductors and explained by the dependence of the forbidden gap width on the difference between cation ionization potential and anion electron affinity). In bulk Eu chalcogenide powders or single crystals the onset of absorption is observed at about 1.1 eV, 1.65 eV, 1.8 eV, and 2.0 eV for EuO, EuS, EuSe, and EuTe, respectively[223]. Attempts have been made to explain this irregularity by the variation of $5d$ crystal field splitting with the lattice parameter[224]. Since EuO has a smaller lattice constant than EuTe, as an example, a larger crystalline field on the $5d$ electron states and a smaller separation of the excited $5d\text{-}t_{2g}$ level from the $4f^7$ ground state can be expected. The observed effects, however, are too large to be readily connected with variations in electrostatic crystal field splittings. More over, the argument neglects the large covalent contribution to the crystal field splittings of d electrons.

Studies of optical absorption in solid solution systems between Eu and Sr chalcogenides show that the irregular variation of the absorption edge in the Eu chalcogenide series is to a large part a concentration effect[225]. Europium and Strontium are very similar in their ionic size and chemical properties, and their corresponding 1:1 chalcogenides are isostructural, with lattice parameters equal within 1%. For Eu^{2+} diluted to about 1% in Sr chalcogenides the peak found in evaporated thin films can be observed also in bulk single crystals (Fig. 23 b). It lies at energies around 23,000 cm^{-1} in SrS, SrSe and SrTe and at 21,000 cm^{-1} in SrO. The shape and structure of the band is very similar to that of the low energy peak at about 12,000 cm^{-1} in the absorption of Eu^{2+} dilute in alkali halides[226]. With increasing Eu^{2+} concentration the absorption peak shifts continuously to longer wave lengths and reaches energies of 17,000 cm^{-1}, 19,400 cm^{-1}, 21,500 cm^{-1} and 22,000 cm^{-1} for evaporated films of concentrated EuO, EuS, EuSe, and EuTe, respectively. The magnetic interactions between Eu ions, as represented by the magnetic Curie temperature, vary with increasing concentrations as predicted by molecular field theory[227].

45. The bandstructure of Eu chalcogenides. A qualitative *energy level scheme*[228] of the Eu chalcogenides, as shown in Fig. 24, can be derived from the spectroscopic properties of Eu^{2+} ions in CaF_2, Sr chalcogenides and Eu chalcogenide thin films. The fluorite and rock salt structure belong to the same crystal symmetry group $Fm3m - O_h^5$ and the cation sites form actually identical sublattices. The shortest $Eu^{2+} - Eu^{2+}$ distances are 4.12 Å in EuF_2, 3.92 Å in EuO and 4.21 Å in EuS.

---

[223] G. Busch, P. Junod, and P. Wachter: Phys. Letters **12**, 11 (1964); S. Methfessel: Z. angew. Phys. **18**, 414 (1965); F. Holtzberg, T. R. McGuire, and S. Methfessel: J. Appl. Phys. **37**, 976 (1966); Transparency of EuS to beyond 3 eV has been reported by G. Busch, P. Junod, M. Risi, and O. Vogt: Int. Conf. in Phys. of Semiconductors, Exeter, 1962, p. 729; and the value of 1.050 eV has been given for the absorption edge of EuTe [G. Busch, P. Junod, and P. Wachter: Phys. Letters **12**, 11 (1964)]. Those discrepancies are probably due to variations in sample purity and stoichiometry and are not fully understood at present.

[224] S. Methfessel, F. Holtzberg, and T. R. McGuire: IEEE Trans. Magn. **2**, 305 (1966).

[225] S. Methfessel, M. J. Freiser, and F. Holtzberg: The contribution of the $5d$ states to the ferromagnetic and semiconducting properties of Eu chalcogenides. 2nd Int. Conf. on Solid Comp. of Trans. Elements, Univ. of Twente, June 1967; M. Freiser, F. Holtzberg, S. Methfessel, D. Pettit, M. Shafer, and J. C. Suits: To be published in Helv. Phys. Acta, Sept. 1968.

[226] R. Reisfeld, and A. Glasner: J. Opt. Soc. Am. **54**, 331 (1964).

[227] T. R. McGuire, M. W. Shafer, and W. Palmer: Proc. Int. Conf. on Magnetism, p. 474. Nottingham 1964.

[228] S. Methfessel, F. Holtzberg, and T. R. McGuire: IEEE Trans. Magn. **2**, 305 (1966); S. Methfessel: Z. angew. Phys. **18**, 414 (1965); J. C. Suits, and B. E. Argyle: J. Appl. Phys. **36**, 1251 (1965); J. M. McClure: J. Phys. Chem. Solids **24**, 871 (1963).

This close similarity of the Eu^{2+} ion arrangement in both structures raises the question why EuO and EuS are ferromagnetic below 76 °K and 16.5 °K, respectively, while EuF_2 and TbO_2 show no measurable magnetic interactions at all (TbO_2 has a fluorite lattice with Tb^{4+} isoelectronic to Eu^{2+} and Gd^{3+}). The crucial difference lies in the anion coordination which forms a cube in the fluorite lattice and an octahedron in the rocksalt lattice. As a consequence, the crystal field split t_{2g} and e_g levels of the $5d$ configuration interchange with one another. The occurrence of ferromagnetic cation-cation exchange, the magnetic and the concentration dependent red shift (Sect. 48) may be related to the symmetry of the lowest $5d$ orbitals. Orbitals with t_{2g} symmetry extend in the NaCl type structure between the nearest Eu^{2+} neighbors and establish cation-cation interactions.

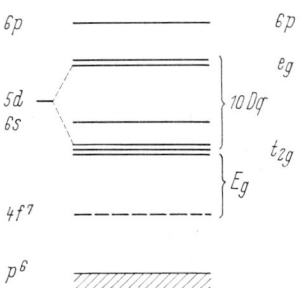

Fig. 24. Qualitative energy level scheme for Eu chalcogenides. The $5d$ levels are split by a crystal field 10 Dq=10,000 cm^{-1} into e_g and t_{2g} orbitals. The t_{2g} orbitals connect the nearest Eu neighbors, and the cation-cation overlap lowers the $5d$ level by 0.5 eV in EuO and by 0.1 eV in EuTe. The separation between the $4f^7$ and $5d\text{-}t_{2g}$ levels gives the experimental absorption edge E_g which has a smaller value for EuO (1.1 eV) than for EuTe ($\sim$2 eV). The bottom of the $6s$ band is uncertain and can lie below the t_{2g} level. Excitations from p^6 into the conduction band have similar energies and properties in Sr chalcogenides. (SrO $\sim$5.7 eV; SrS $\sim$4.6 eV; SrSe $\sim$4.1 eV, SrTe $\sim$3.5 eV.) (From Ref. [224].)

For Eu^{2+} diluted in KBr or KCl the two bands known from Eu^{2+} in CaF_2 appear [229] at about 29,000 cm^{-1} and 40,000 cm^{-1} (Fig. 23 b). For Eu^{2+} in Sr chalcogenides only one band is observable with a structure similar to that in alkali halides. The peak position at 23,000 cm^{-1} defines the separation of the $5d\text{-}t_{2g}$ levels from the $4f$ levels to about 1.85 eV for diluted, non-interacting Eu^{2+} ions in chalcogenides. The second band which is expected to lie about 10,000 cm^{-1} higher is cut off by the charge transfer absorption between valence and conduction states which are found in SrTe and EuTe films to be similar to one another in energy and shape [230].

The earlier discussed shift of the t_{2g} band to lower energies with increasing Eu concentration by about 0.5 eV, 0.45 eV, 0.18 eV and 0.1 eV in the oxide, sulfide, selenide and telluride, respectively, reflects probably the kinetic broadening and stabilization of the t_{2g} levels by overlap with the Eu neighbors. The optical exciton states are expected to lie below the empty d-band states and to shift with them. The cation-cation overlap decreases with increasing lattice constant (EuO: 5.14 Å, EuS: 5.96 Å, EuSe: 6.19 Å, EuTe: 6.60 Å) and increasing electronegativity of the anion (Fig. 45). The fact that this overlap occurs in EuS but not in EuF_2, which has about the same Eu-Eu distances, must be related to the larger dielectric constant ε or optical refraction index n of EuS ($n \approx 2.3$) [231] compared to EuF_2 ($n=1.555$, $\varepsilon=7$) [232]. The larger dielectric constant results from the smaller electronegativity of sulfur and enlarges the effective orbital radius in the excited state proportional to n^2.

---

[229] R. REISFELD, and A. GLASNER: J. Opt. Soc. Am. **54**, 331 (1964).
[230] F. HOLTZBERG, T. R. McGUIRE, and S. METHFESSEL: J. Appl. Phys. **37**, 976 (1966); The absorption properties of Sr chalcogenides have been measured by F. C. JAHODA: Phys. Rev. **107**, 1261 (1957); R. J. ZOLLWEG: Phys. Rev. **111**, 113 (1958); G. A. SAUM, and E. B. HENSLEY: Phys. Rev. **113**, 1019 (1959) and their fine structure discussed in terms of an exciton model by A. W. OVERHAUSER: Phys. Rev. **101**, 1702 (1956).
[231] J. D. AXE: Private communication.
[232] J. D. AXE, and G. D. PETTIT: J. Chem. Phys. Solids **27**, 621 (1966).

Suits et al.[233] have investigated and extensively discussed the magneto-optical effects of Eu^{2+} in chalcogenides, EuF_2, and glasses containing Eu^{2+}. The results prove that the absorption band in Eu chalcogenides resembles the $4f-5d$ transition of Eu^{2+} in other materials not only by shape and oscillator strength but also in their magneto-optical parameters[234]. Another obvious similarity is the temperature independence of the slope of the absorption curves which we have discussed earlier for Eu^{2+} in CaF_2 and which is also observed in the Eu chalcogenides, as shown in Fig. 27 for EuO. The magnetic red shift of the absorption edge (which is shown in Fig. 23b by the two EuO-film curves at different temperatures and will be discussed in Sect. 48) seems to be coupled to the concentration dependent redshift. It may result from a further increase of cation-cation interactions with increasing magnetic order.

The band structure calculations for EuS by Cho[235] were discussed in Sect. 15 and their result is shown in Fig. 6. From the proximity of the values obtained for the energy gap from electrical resistivity[236] and optical measurements it was derived that the optical absorption is due to indirect transitions between occupied valence states (essentially the p electron states of the anions) and empty conduction band states. This concept can not explain the large absorption peak of the order of 10^{+5} cm^{-1} and the large magneto-optical effects. Moreover, the meaning of the band gap determinations from electrical measurements is doubtful in chalcogen compounds which are prepared from RE materials with impurity concentrations of the order of 10^{-2} to 10^{-4}.

Band structure calculations, on the other hand, are probably good only for the valence and conduction bands, but cannot give reliable information about the position of the $4f$ levels. These lie outside the scope of the band theory due to their extreme localization. The easy reducibility of Eu^{2+} to Eu^{3+} by deviation from stoichiometry or reduction with monovalent alkali ions[237] is incompatible with the calculated band structure (Fig. 6) which shows the occupied $4f^7$ levels to lie 0.6 Ry ($\sim$8 eV) below the conduction band and even 0.3 Ry ($\sim$4 eV) below the valence band.

By application of pressure of about 25 to 40 kbar it was possible to transfer EuTe into a NaCl-type lattice with reduced lattice parameter (to 6.22 Å from 6.59 Å at normal pressure)[238]. The accompanying variations in electrical and magnetic properties indicate a partial transformation of Eu^{2+} into Eu^{3+} by promotion of electrons from the $4f^7$ level into the conduction band. In SmS the $4f^6$ levels of Sm^{2+} have higher energy than in Eu^{2+} compounds and the f electrons can be excited thermally into the conduction band[239] at temperatures below

---

[233] J. C. Suits, B. E. Argyle, and M. J. Freiser: J. Appl. Phys. **37**, 1391 (1966) with references to earlier papers of the authors.

[234] M. J. Freiser, S. Methfessel, and F. Holtzberg: J. Appl. Phys., **39**, 900 (1968).

[235] S. J. Cho: Phys. Rev. **157**, 632 (1967); (see part C).

[236] G. Busch, P. Junod, M. Risi, and O. Vogy: Conf. on Phys. of Semicon., Exeter, 1962, p. 727.

[237] M. W. Shafer: J. Appl. Phys. **36**, 1145 (1965); T. R. McGuire, M. W. Shafer, and W. Palmer: Proc. Int. Conf. on Magnetism, Nottingham, 1965, p. 474.

[238] C. J. M. Rooymans: Sol. State Comm. **3**, 421 (1956); Ber. Bunsenges. phys. Chemie **70**, 1036 (1966).

[239] V. P. Zhuze, A. V. Golubkov, E. V. Goncharova, T. I. Komarova i V. M. Sergeeva: Soviet Phys. — Solid State **6**, 205, 213 (1965); V. E. Adamyan, A. V. Galubkov i G. M. Loginov: Soviet Phys. — Solid State **7**, 239 (1966); A. V. Golubkov, E. V. Goncharova, V. P. Zhuze i I. G. Manoilova: Soviet Phys. — Solid State **7**, 1963 (1966); S. M. Blokhin, E. E. Vainshtein i V. M. Bertenev: Soviet Phys. Solid State **7**, 2870 (1966); F. J. Reid, L. K. Matson, J. F. Miller, and R. C. Himes: J. Phys. Chem. Solids **25**, 969 (1964).

1,000 °K. The gap between $4f^6$ states and the bottom of the conduction band is found to be 0.18 eV in SmS and the mobility of the conduction electrons is of the order of 10 cm^2/V.sec.

III. Spectroscopy of exchange interactions.

For the discussion of optical properties, it is usually assumed that the exchange between magnetic ions is small compared to the optical line width. Since the exchange integrals decrease quite rapidly with distance, this assumption is usually justified for concentrations of magnetic ions below some 0.1 Mol.-% in a diamagnetic host crystal. In materials with higher concentrations, exchange effects of the order 1 to 30 cm^{-1} in ion pairs and clusters become observable[240]. For even larger concentrations, the crystal field or molecular orbital treatment has to account for an increasing amount of direct or indirect exchange mechanisms in addition to the Coulomb interactions and covalency. But the strength of the exchange interactions remains of the order of 100 cm^{-1} or less even in magnetically concentrated RE compounds. Therefore, microwave resonance[241] and far infrared spectroscopy[242] are usually considered to be more suitable tools than visible optics for investigation of exchange splittings in the magnetic ground state. Optical spectroscopy, however, gives information about magnetic effects on the excited states and is, therefore, more closely related to the phenomena of semiconductor physics.

46. Interaction between light polarization and magnetized materials. *Polarization studies on light beams* interacting with magnetized materials are among to the oldest techniques of magnetic measurements. The first accurate values of the magnetization in transition metals and their alloys were actually obtained by Faraday and Kerr effect measurements. More recently, magneto-optical techniques have been used often to explore the nature of impurity states in non-magnetic semiconductors. All those effects are well understood on the basis of Maxwellian optics and dispersion theory[243]. The variations with magnetization in refraction index and absorption, for differently polarized light, are caused, in principle, by Zeeman splittings of the ground and excited states in the molecular exchange field and by polarization-dependent selection rules for optical transitions between these levels.

---

[240] P. KISLINK, A. L. SCHAWLOW, and M. D. STURGE: Proc. 3rd Int. Congr. Quantum Electronics, p. 725, New York: Columbia Univ. Press 1964; A. L. SCHAWLOW, D. L. WOOD, and A. M. CLOGSTON: Phys. Rev. Letters 3, 271 (1959); K. W. BLAZEY, and G. BURNS: Phys. Letters 15, 117 (1965); D. S. McCLURE: J. Chem. Phys. 39, 2850 (1963); Survey articles: S. SUGANO, and Y. TANABE: Magnetism (Ed. G. T. RADO, and H. SUHL), Vol. 1, p. 243, New York: Academic Press 1963; P. KISLINK, and W. F. KRUPKE: J. Appl Phys. 36, 1025 (1965).

[241] W. LOW: Paramagnetic Resonance in Solids, New York: Academic Press 1960; S. A. ALTSHULLER, and B. M. KOZYREV: Electron Paramagnetic Resonance. New York: Academic Press 1964.

[242] A. J. SIEVERS, and M. TINKHAM: Phys. Rev. 124, 321 (1961); 129, 1995 (1963); J. Appl. Phys. 34, 1235 (1963); M. TINKHAM: J. Appl. Phys. 33, 1248 S (1962); Phys. Rev. 124, 311 (1961); R. C. OHLMANN, and M. TINKHAM: Phys. Rev. 123, 425 (1961); H. KONDOH: J. Phys. Soc. Japan 15, 1970 (1960); R. L. RICHARDS: J. Appl. Phys. 34, 1237 (1963); R. G. WHEELER, F. M. REAMES, and E. J. WACHTEL: J. Appl. Phys. 39, 915 (1968).

[243] W. VOIGT: Magneto- und Elektrooptik, Leipzig: Teubner 1908; H. R. HULME: Proc. Roy. Soc. (London) A 135, 237 (1932); W. SCHÜTZ: Handbuch der Experimentalphysik, Vol. 16/1 (Ed. W. WIEN u. F. HARMS), Leipzig: Akademie-Verlag 1936; M. v. LAUE: op cit., Vol. 18, p. 185; P. N. ARGYRES: Phys. Rev. 97, 334 (1955); P. S. PERSHAN: J. Appl. Phys. 38, 1462 (1967); A. V. SOKOLOV: Optical Properties of Metals, New York: Amer. Elsevier 1967; J. C. SUITS: Magneto-Optical Properties, Handb. of Magnetic Materials, REINHOLD, to be published; D. S. SMITH: Magnetooptics, Vol. 25/29 (1967) of this Handbook; E. D. POLIK: Appl. Optics 6, 597, 603 (1967); M. J. FREISER: IEEE Trans. Magn. 4, 152 (1968).

In actual materials however, the interpretation of the magneto-optical dispersion by certain energy levels is often very difficult since most ferromagnetic or ferrimagnetic materials show rather high optical absorption. Many insulating magnets are only transparent for infrared light. The majority of transparent salts are antiferromagnetic and the occurrence of magnetization and related magneto-optical effects needs application of extremely large magnetic fields beyond the reach of most laboratories. (An applied magnetic field of 10^4 oersted splits the energy of a single electron spin by about $1\,\text{cm}^{-1}$ or $1\,°\text{K}$ only!) In recent years, the interest in the relation of magneto-optics to the electronic band structure has been revived by the availability of highly transparent crystals of optical quality from several compounds which become spontaneously ferrimagnetic or ferromagnetic at low temperatures.

Ferrimagnetic garnets are transparent in the visible and red part of the spectrum and give at room temperature Faraday rotations [244] of the order of 10^2 to 10^4 deg/cm. Comparison of the magneto-optical dispersion with optical absorption makes it possible to identify the character of the optical transitions which respond to magnetic effects. For iron garnets, CLOGSTON has assigned the magneto-optical rotation to a charge transfer of $3d^5$ electrons from Fe^{3+} into $4p$ orbitals at the anions. *The Chromium halides*, $CrBr_3$ and CrI_3, are also transparent in the visible and become truly ferromagnetic below $32.5\,°\text{K}$ and $68\,°\text{K}$, respectively [245]. At $1.5\,°\text{K}$ the Faraday rotation in $CrBr_3$ becomes comparable to that of ferromagnetic iron metal (almost 5×10^5 deg/cm for green light near the absorption edge). The structure of the magneto-optical dispersion and absorption spectrum can be analyzed in detail for effects of the crystal field and the molecular exchange field on the localized $3d$ orbitals at Cr^{3+} ion and on the molecular orbitals of the nearly octahedral halogen coordination [246]. The large Faraday rotations do not arise from the exchange splittings in the ground state but from spin-orbit interactions in the excited state where electrons are transferred from molecular orbits at the halogen octahedron into localized $3d$-e_g orbitals at the central Cr^{3+} ion. Extremely large magneto-optical effects of the order of 10^5 deg/cm have been found in *ferromagnetic Eu chalcogenides* near their absorption edge and interpreted by exciton-like $4f-5d$ transitions [247].

The ferromagnetic spinel $CdCr_2Se_4$ shows [248] large negative Faraday rotation of about 10^4 deg/cm near the absorption edge at $1.2\,\mu$. The rotation changes sign between 7 and 8 μ and approaches a constant value of about 150 deg/cm at longer wave lengths up to $18\,\mu$.

The large magneto-optical effects in ferromagnetic and ferrimagnetic semiconductors suggest the use of these materials in *magneto-optical devices*. The small

---

[244] J. F. DILLON: J. Appl. Phys. **29**, 539 (1958); J. phys. radium **20**, 374 (1959); J. T. CHANG, J. F. DILLON, and U. F. GIANOLA: J. Appl. Phys. **36**, 1110 (1965); R. C. SHERWOOD, J. P. REMEIKA, and H. J. WILLIAMS: J. Appl. Phys. **30**, 217 (1959); G. V. KRINCHIK i M. V. CHETKIN: Soviet Phys. JETP **13**, 509 (1961); **14**, 485 (1962); J. Phys. Soc. Japan **17**, B **1**, 358 (1962); Soviet Phys. — Solid State **5**, 273 (1963); G. S. KRINCHIK i G. K. TYUTNEVA: Soviet Phys. JETP **19**, 292 (1964); J. Appl. Phys. **35**, 1014 (1964); A. M. CLOGSTON: J. phys. radium **20**, 151 (1959); J. Appl. Phys. **31**, 198 S (1959).

[245] I. TSUBOKAWA: J. Phys. Soc. Japan **15**, 1664 (1960); J. F. DILLON: J. Phys. Soc. Japan **19**, 1662 (1964); W. H. HANSEN, and M. GRIFFET: J. Chem. Phys. **20**, 902 (1958); **30**, 913 (1959); J. Appl. Phys. **30**, 304 S (1959).

[246] J. F. DILLON, H. KAMIMURA, and J. P. REMEIKA: Phys. Rev. Letters **9**, 161 (1962); J. Appl. Phys. **34**, 1240 (1963); J. Phys. Chem. Solids **27**, 1531 (1966).

[247] J. C. SUITS, B. E. ARGYLE, and M. J. FREISER: J. Appl. Phys. **37**, 1391 (1966); **36**, 1251 (1965); Phys. Rev. Letters **14**, 687 (1965); **15**, 822 (1965); J. H. GREINER, and G. J. FAN: Appl. Phys. Letters 9, 27 (1966).

[248] P. F. BONGERS, and G. ZANMARCHI: Solid State Comm. **6**, 291 (1968).

absorption and good optical quality in the transparent region makes it possible to produce sufficiently large changes of light polarization without large absorption losses. Magneto-optical devices for reading magnetic computer memories or tapes, for laser beam modulations and deflections have been suggested [249].

47. Magnetic effects in absorption spectra. Information about electron states and energies involved in magnetic exchange and ordering phenomena can be derived much easier from magnetic effects in absorption spectra than from the dispersion of magneto-optical rotations, provided the relevant absorption lines are sufficiently narrow. Especially suitable are the forced electric or magnetic dipole transitions between magnetic electron orbitals, which in $3d$ ions are much stronger ($\alpha \sim 10^3$ cm^{-1}) and broader than the $4f-4f$ transitions in lanthanides, where the lines are only a few cm^{-1} wide. They produce vivid colors in some transition element salts. The crystalline field even in cubic symmetry is larger than the multiplet width. Therefore, one cannot rely on free ion approximations for the interpretation of spectra, but must derive the necessary information from considerations of possible molecular orbital combinations in the crystal symmetry.

Thermal variations of absorption lines or bands in the neighborhood of the magnetic ordering temperature are of special interest because of possible relations to the molecular field. The temperature dependence of absorption and luminescence spectra has been investigated for only a few antiferromagnetic $3d$ compounds which can be prepared with sufficient optical quality and have relatively simple crystal structures, such as the 1:1 oxides and chalcogenides [250], the halogenides [251], and perovskites [252]. The results have been reviewed and discussed by Sugano, and Tanabe, and Wickersheim [253]. One finds line splittings, abrupt line shifts to shorter wave lengths and intensity variations at the Néel temperature. Sometimes new satellite lines appear in the magnetically ordered state. The observed blue shifts are of the order of one to 100 cm^{-1} only, and it can be difficult to separate with sufficient accuracy genuine exchange effects from variations in the crystalline field splittings by magnetostrictive effects. The situation becomes even more complicated when crystal field effects and exchange interactions are of comparable magnitude and have to be treated simultaneously, such as in RE iron garnets or other lanthanide salts with high Curie temperatures.

Theoretical interpretations are usually based on crystal field theory, while magnetic interactions are described by the molecular field approximation. An ob-

---

[249] R. L. Conger, and J. L. Tomlinson: J. Appl. Phys. **33**, 1059 (1962); J. T. Chang, J. F. Dillon, and U. F. Gianola: J. Appl. Phys. **36**, 1110 (1965); G. Y. Fan, and J. H. Greiner: J. Appl. Phys. **39**, 1216 (1968); L. K. Anderson: J. Appl. Phys. **34**, 1230 (1963); J. E. Geusic, and H. E. D. Scovil: Bell System Tech. J. **41**, 1371 (1962); L. J. Aplet, and J. W. Carson: Appl. Optics **3**, 544 (1964); C. B. Rubinstein, L. C. van Uitert, and W. H. Grodkiewicz: J. Appl. Phys. **35**, 3069 (1964); C. S. Porter, E. G. Spencer, and R. C. Le Graw: J. Appl. Phys. **29**, 495 (1958); R. N. Zitter, and E. G. Spencer: J. Appl. Phys. **37**, 1089 (1966).

[250] R. Newman, and R. M. Chrenko: Phys. Rev. **114**, 1507 (1959); **115**, 882 1147 (1959); G. W. Pratt, and R. Coelho: Phys. Rev. **116**, 281 (1959); D. R. Huffman, R. L. Wild, and J. Callaway: J. Phys. Soc. Japan **21**, 623 (1966).

[251] J. W. Stout: J. Chem. Phys. **31**, 709 (1959); I. Tsujikawa, and E. Kanda: J. Phys. Soc. Japan **18**, 1382 (1963); I. Tsujikawa: J. Phys. Soc. Japan **13**, 315 (1958); V. V. Eremenko, and Y. A. Popkov: Phys. Status Solidi **12**, 627 (1965).

[252] W. W. Holloway, and M. Kestigian: Phys. Rev. Letters **15**, 17 (1965); W. W. Holloway, E. W. Prokofsky, and M. Kestigian: Phys. Rev. **129**, A 954 (1965); W. M. Yen, G. F. Imbusch, and D. L. Huber: Opt. Prop. of Ions in Crystals (Ed. H. M. Crosswhite, and H. W. Moos), p. 301, New York: Wiley-Interscience 1967; A. I. Belyaeva i V. V. Eremenko: Soviet Phys. JETP **17**, 319 (1963).

[253] S. Sugano, and Y. Tanabe: Optical Spectra in Magnetically Ordered Materials, Magnetism 1 (Ed. G. T. Rado, and H. Suhl), p. 243, New York. Academic Press 1963; K. A. Wickersheim: Optical and Infrared Properties of Magnetic Materials, op cit., p. 269.

vious weakness of this approach is the neglect of local fluctuations in the ordered magnetic state in the neighborhood and above the magnetic transition temperature. However, the observable magnetic effects in the optical spectra are usually very small and such details easily escape experimental observation.

The appearance of weak satellite lines in antiferromagnetic difluorides of transition elements has been interpreted as *exciton-magnon absorptions or spin wave side bands*[254]. Analogous to vibronic excitations, the dipole transitions which are originally forbidden, become allowed by the exchange coupling to the spin waves in the antiferromagnetic sublattices. The resulting absorption lines are stronger and broader than expected for magnetic dipole transitions, and show in antiferromagnetic crystals of suitable symmetries typical variations in applied magnetic fields. The optical transitions in antiferromagnetic crystals show DAVYDOV splitting because the magnetic sublattices have lower symmetry than the atomic lattice[254a].

Transparent RE salts are expected to offer some favorable conditions for optical investigations of magnetic effects on excited electron states. Their crystal spectra have sharper lines than those in $3d$ compounds and the closer relationship to theoretical free ion spectra simplifies their interpretation. Line splittings of $4f-4f$ transitions by exchange effects have been studied recently in iron free garnets, such as $Dy_3Al_5O_{12}$, etc., in perovskites and in trihalides at the Clarendon Laboratory in Oxford and in Darmstadt[255]. The exchange interactions derived from the observed optical effects are in good agreement with values obtained from magnetic and thermal measurements at Yale by W. P. WOLF and collaborators[256]. The physical properties of mixed cation oxides, such as garnets and perovskites, containing both RE and $3d$ ions have been reviewed recently by LOW[257].

48. The magnetic red shift. BUSCH and collaborators[258] have found a very interesting and unusual temperature dependence of the absorption edge in Eu chalcogenide powders by diffuse reflection measurements. Fig. 25 shows this effect for EuO, as an example. By cooling from room temperature to 90 °K the edge shifts about 1 to 1.4×10^{-4} eV/deg. The direction of the shift is toward higher

---

[254] Y. TANABE, T. MORIYA, and S. SUGANO: Phys. Rev. Letters **15**, 1023 (1965); R. L. GREENE, D. D. SELL, W. M. YEN, A. L. SCHAWLOW, and R. M. WHITE: Phys. Rev. Letters **15**, 656 (1965); D. D. SELL, R. L. GREENE, and R. M. WHITE: Phys. Rev. **158**, 489 (1967); and several papers in: Optical Properties of Ions in Crystals (Ed. H. M. CROSSWHITE, and H. W. MOOS), New York: Wiley-Interscience 1967.

[254a] A. S. DAVYDOV: J. Exp. Theor. Phys. (U.S.S.R.) **18**, 210 (1948); **21**, 673 (1951); H. C. WOLF: Solid State Physics **9**, 1 (1959); D. S. MCCLURE: Solid State Physics **8**, 1 (1958); J. P. VAN DER ZIEL: Phys. Rev. Letters **18**, 237 (1967).

[255] Recent publications with references to earlier literature: M. J. M. LEASK: J. Appl. Phys. **39**, 908 (1968); K. H. HELLWEGE, S. HÜFNER, M. SCHINKMANN, and H. SCHMIDT: Phys. Letters **12**, 107 (1964); A. H. COOKE, K. A. GEHRING, M. J. M. LEASK, D. SMITH, and J. H. M. THORNLEY: Phys. Rev. Letters **14**, 685 (1965); S. HÜFNER, M. SCHINKMANN, and H. SCHMIDT: Phys. Kondens. Materie **4**, 108 (1965); S. HÜFNER, and H. SCHMIDT: Phys. Kons. **4**, 262 (1965); R. FAULHABER, G. HÜFNER, E. OLRICH, H. SCHMIDT, and H. SCHUCHERT, in: Optical Properties of Ions in Crystals (Ed. H. M. CROSSWHITE, and H. W. MOOS), p. 329. New York: Wiley-Interscience 1967.

[256] W. P. WOLF, M. BALL, M. T. HUTCHINGS, M. J. M. LEASK, and A. F. G. WATT: J. Phys. Soc. Japan **17**, Suppl. B 1, 443, 487 (1962); W. P. WOLF: Proc. Nottingham Conf. Inst. of Phys., London, 1964, p. 342, 555; R. J. BIRGENAU, M. T. HUTCHINGS, and W. P. WOLF: J. Appl. Phys. **38**, 957 (1967).

[257] W. LOW, in: Progress in the Science and Technology of R.E. (Ed. L. E. EYRING) Oxford: Pergamon Press 1966.

[258] G. BUSCH, P. JUNOD, and P. WACHTER: Phys. Letters **12**, 11 (1964); B. E. ARGYLE, J. C. SUITS, and M. J. FREISER: Phys. Rev. Letters **15**, 882 (1965); G. BUSCH, and P. WACHTER: Phys. Kondens. Materie **5**, 232 (1966); G. BUSCH: J. Appl. Phys. **38**, 1386 (1967); Variations in luminescence with magnetic order have been reported also: G. BUSCH, and P. WACHTER: Phys. Letters **20**, 617 (1966); G. BUSCH: J. Appl. Phys. **38**, 1386 (1967).

energies with decreasing temperature, as it is quite usual in covalent semiconductors. Below 90 °K, however, the shift reverses and the absorption edge goes to lower energies with decreasing temperature. At 20 °K the onset of the absorption lies at about 0.95 eV, so that a total red shift of more than 0.2 eV results by cooling. BUSCH et al. have demonstrated that the red shift is closely related to the onset of ferromagnetic order in EuO around 76 °K and responds to applied magnetic fields proportionally to the induced magnetization. In EuS, which becomes ferromagnetic below 16.5 °K, the blue shift at higher temperatures is 1.7×10^{-4} eV/deg. and the magnetic red shift amounts to 0.18 eV. EuSe has below 4.6 °K a complicated spin order which switches to ferromagnetism in applied magnetic fields of the order of a few koe. Fig. 26a and 26b demonstrate how those variations in magnetic ordering result in absorption edge shifts with tem-

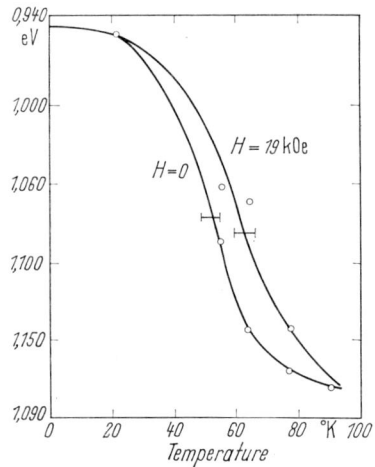

Fig. 25. The magnetic red shift of the absorption edge with magnetic order in EuO. [From G. BUSCH, and P. WACHTER: Phys. Kondens. Mat. 5, 232 (1966).]

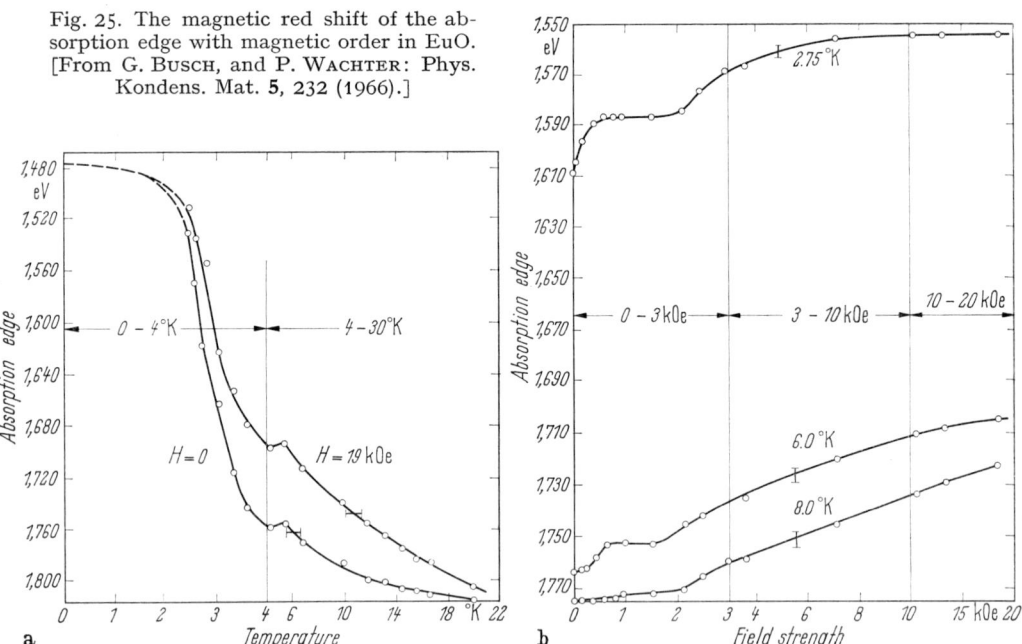

Fig. 26a and b. Position of the absorption edge in EuSe as a function of temperature (a) and applied magnetic field (b). Variations in the spin structure and spontaneous magnetization around 4 °K and in applied magnetic fields which are known from magnetic measurements show up clearly. [From G. BUSCH, and P. WACHTER: Phys. Kondens. Mat. 5, 232 (1966).]

perature and applied magnetic field. It is also interesting to note that EuTe which becomes antiferromagnetic with a spin order of second type (as in MnO) between 8 and 10 °K shows no irregular red shift. BUSCH and collaborators have suggested exploiting the red shift as a new experimental method for determining the spontaneous magnetization by optical measurements. In contrast to the Faraday and Kerr effects, it gives the spontaneous magnetization independent of domain effects and does not need polarized light or applied magnetic fields for observation. Applied magnetic fields are only effective as far as their strength is sufficient to increase the spontaneous magnetization. The red shift does not disappear at the Curie temperature, as the molecular field does, but extends to temperatures

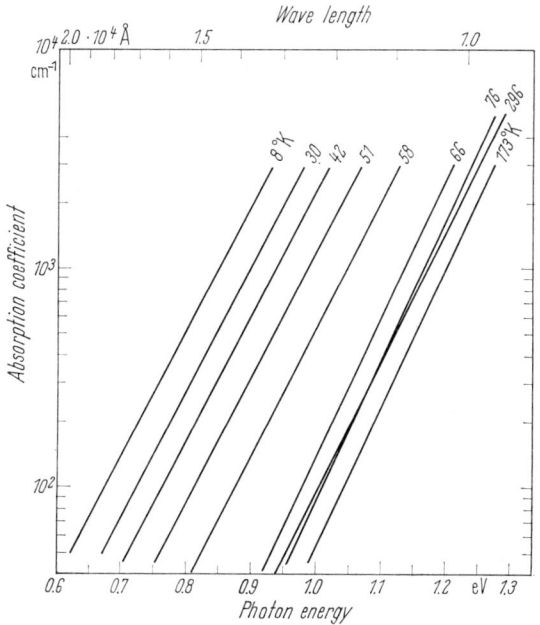

Fig. 27. EuO single crystals have an exponential absorption edge which shifts to low energy with decreasing temperature and increasing ferromagnetic order, but does not change its slope. [From S. METHFESSEL, M. J. FREISER, G. D. PETTIT, and J. C. SUITS: J. Appl. Phys. **38**, 1500 (1967).]

nearly twice as large as the Curie temperature. Proportionality to the spontaneous magnetization (as measured or calculated at low temperatures) becomes, of course, invalid near or above the Curie temperature. For these temperatures, the red shift has been found in EuO and EuS single crystals[259] to be approximately proportional to the nearest neighbor spin correlation function $\langle S_1 S_2 \rangle / S^2$. This suggests that the electron states responsible for the red shift must be sufficiently localized to experience only the spin order at neighboring atoms instead of the crystal's molecular field. This concept is supported also by the observation that the exponential absorption edge does not change its slope with temperature while shifting to low energies in single crystals (Fig. 27), because the width of the excited state is narrower than the $4f^6$ multiplet (Sect. 44).

The *interpretation of the magnetic red shift* does not depend so much on the model accepted for the absorption edge as one could expect. The anomalous red

---

[259] S. METHFESSEL, M. J. FREISER, G. D. PETTIT, and J. C. SUITS: J. Appl. Phys. **38**, 1500 (1967).

shift of 1,000 to 2,000 cm^{-1} is too large to be connected with the exchange splittings of the $4f$ ground state which can be only of the order of 100 cm^{-1} in EuO as inferred from the magnetic Curie temperature. The observed shift must, therefore, be the property of the excited state. A similar conclusion is obtained from the assumption of indirect transitions between valence and conduction bands since the valence electrons experience very weak exchange splittings. If the electron in the excited state is a Bloch electron in a good conduction band, the band gap shift relates to the polarization of free conduction electrons by intra-atomic $s-f$ exchange of the Ruderman-Kittel type (part D) and to magnetic disorder resistivity (see part C). In ferromagnetically ordered crystals the conduction band splits by an amount proportional to the magnetization and the optical transitions which conserve spin reduce their energy correspondingly. Near to and above the Curie temperature the correlation between two ion spins *modulates the conduction band and quasi bound states* can occur in regions with parallel spin. BALTENSBERGER and collaborators[260] gave the theoretical treatment of the interaction of conduction electron states with ion spin correlations. This is in agreement with experiment when an $s-f$ exchange integral of the order of 0.2 eV, as derived from magnetic interactions in RE metals, is used.

If the concept of a localized, exciton-like $4f-5d$ transition is used for the explanation of the absorption edge, the magnetic shift appears to be related to the dependence of the cation-cation overlap of the excited t_{2g} electrons on the relative ion spin orientation through the intra-atomic $f-d$ exchange mechanism at the neighboring Eu ions[261]. KASUYA[262] has introduced the name "magnetic exciton" for this situation, which he discussed in connection with the effect of conduction electrons on the Curie temperature and transport properties in Eu chalcogenides (Sects. 82—84). He points out that there can be several types of excitons depending on the possible combinations of wave functions at the excited ion and its neighbors. The magnetic exciton of lowest energy is probably made up with $6s$ wave functions only, but is optically inactive. The largest oscillator strength is assigned to the $5d-5d$ excitons. When the $f-d$ exchange derived from free ion spectra[263] is used for the coupling of the excited electron to the neighbor spins the electron must be shared by some 45% by the 12 Eu neighbors in order to explain the experimentally observed magnetic red shift. In luminescence[258] the emitted light has longer wave length than the absorbed light since the lattice can relax around the excited state which leads to a delocalization of 85% of the $5d$ orbitals to neighboring Eu ions.

The shift of the lowest empty conduction states with magnetic order in Eu-chalcogenides has been observed also by electron tunneling through these materials. The barrier voltage between EuS, EuSe and Al-contacts reduces drastically below the magnetic Curie temperature or in applied magnetic fields. The effect of "magneto-internal field emission" has been interpreted by electron tunneling from the metal electrodes into the empty conduction band of the

---

[260] F. RYS, J. HELMAN, and W. BALTENSPERGER: Helv. Phys. Acta **39**, 197 (1966); Phys. Kondens. Materie **6**, 105 (1967); G. BUSCH: J. Appl. Phys. **38**, 1386 (1967).

[261] J. C. SUITS, B. E. ARGYLE, and M. J. FREISER: Phys. Rev. Letters **15**, 882 (1965); J. Appl. Phys. **37**, 1391 (1967).

[262] T. KASUYA, and A. YANASE: Mechanism for the anomalous properties of Eu chalcogenide alloys. I; to be publ. J. Phys. Soc. Japan; A. YANASE, and T. KASUYA: J. Appl. Phys. **39**, 430 (1968).

[263] The exchange interaction in cm^{-1} for Eu$^+$ are: $I_{sf}=209$, $I_{pf}=114$, $I_{df}=787$ and the spin orbit couplings $\zeta_p - 1610$ cm^{-1}, $\zeta_d = 648$ cm^{-1} [H. N. RUSSEL, W. ALBERTSON, and D. N. DAVIS: Phys. Rev. **60**, 641 (1941)]; and for Gd^{2+}: $I_{sf}=294$, $I_{pf}=246$, $I_{df}-1013$; $\zeta_p - 3050$, $\zeta_d = 1050$ [W. R. CALLAHAN: J. Opt. Soc. Am. **53**, 695 (1963)].

magnetic semiconductor which shifts to lower energies with increasing spin order[264].

The red shift of absorption bands with ferromagnetic order is not restricted to Eu chalcogenides but has been found also in *ferromagnetic $CdCr_2Se_4$*. As we will discuss in later parts of this chapter, the spinel compounds $CdCr_2S_4$ and $CdCr_2Se_4$ are semiconductors and become ferromagnetic below about 84—97 °K and 130—140 °K, respectively (see Table 5). In contrast to the Eu chalcogenides magnetic exchange interactions can be described by superexchange interactions via anions. The magnetic Curie temperature and optical absorption edge ($CdCr_2S_4$: 1.57 eV; $CdCr_2Se_4$: 1.32 eV) depend on the anion character in the same manner as usually found in transition metal compounds. The temperature dependence of the absorption edge in $CdCr_2Se_4$, however, is very similar to that found in Eu chalcogenides. By cooling between 300 and 150 °K, the edge shifts to blue by about 10^{-4} eV/deg, reverses its motion at about double the Curie temperature and shows a red shift of about 0.18 eV with increasing ferromagnetic order at lower temperatures or in applied magnetic fields[265]. Very surprising is the fact that the related compound $CdCr_2S_4$ behaves just oppositely in shifting its absorption edge with increasing magnetization by about 0.1 eV to higher energies.

Several interpretation have been suggested for the magnetic red shift in Cr spinels[266]. GOODENOUGH assigns the optical absorption edge to excitations from the valence band into empty Cr^{2+} states (Fig. 35). He assumes, that in the sulfides the high-spin state ^5Eg of Cr^{2+} lies below the low spin state $^3T_{1g}$. Magnetic order of the Cr spins polarizes the valence band, i.e. parallel-spin states gain energy. Optical transitions have to conserve the electron spin. Therefore, the stabilization of the parallel-spin valence electrons is observable as a blue shift of the absorption edge with increasing magnetic order. In the selenide, however, the larger covalency and crystal field stabilizes the low-spin state of Cr^{2+}. Optical excitations at the absorption edge transfer antiparallel-spin electrons from the valence band into the Cr^{2+} state. The antiparallel-spin electrons become destabilized with increasing magnetic order, and the absorption edge shifts to longer wavelengths. LEHMANN describes the magnetic shift of the absorption edge by excitations from donor levels to the bottom of the conduction band which is supposed to shift with increasing magnetic order. CALLEN relates the magnetic shift of absorption in Cr-spinels and Eu-chalcogenides to magneto-elastic effects.

G. Experimental evidence for indirect exchange.

49. General remarks. A complete review of the present state of knowledge of the magnetic properties of nonmetallic magnets is certainly out of the scope of this paragraph. Several review articles have been published about 3 d-compounds[267]

---

[264] L. ESAKI, P. J. STILES, and S. VON MOLNAR: Phys. Rev. Letters **19**, 852 (1967).

[265] G. BUSCH, B. MAGYAR, and P. WACHTER: Phys. Letters **23**, 438 (1966); G. HARBEKE, and H. PINCH: Phys. Rev. Letters **17**, 1090 (1966); H. W. LEHMANN, and G. HARBEKE: J. Appl. Phys. **38**, 946 (1967).

[266] J. B. GOODENOUGH: Private communication; B. F. BONGERS, and G. ZANMARCHI: Solid State Comm. **6**, 291 (1968); W. H. LEHMANN: Phys. Rev. **163**, 488 (1967); E. CALLEN: Phys. Rev. Letters **20**, 1045 (1968).

[267] For examples: J. B. GOODENOUGH: Magnetism and the Chemical Bond, New York: Wiley-Interscience 1963, and several articles in Magnetism (Ed. G. T. RADO, and H. SUHL), New York: Academic Press 1963—1966; J. H. VAN VLECK: Theory of Electric and Magnetic Susceptibilities, London: Oxford Univ. Press 1961; Physical Sciences 1962, p. 113; R.E. Research, Vol. 4 (Ed. L. R. EYRING), p. 3, New York: Gordon and Breach 1964; Progr. in Science and Techn. of RE (Ed. L. R. EYRING), Vol. 2, p. 1, Oxford: Pergamon Press 1966; J. Phys. Soc. Japan **17**, B1, 352 (1962).

and RE-materials[268]. We intend to concentrate here on magnetic ordering phenomena which appear to be related to the existence of free charge carriers in semiconductors. As a most simple model of a magnetic semiconductor one conceives of a material which has localized atomic moments at a large portion of lattice sites and free carriers. The concentration of the free carriers can be varied as usually in semiconductors, by optical or thermal excitations, carrier injection etc. The carriers are supposed to be spinpolarized to a certain extent by intra-atomic exchange and produce magnetic interactions between the local moments and magnetic order, as described in part D. The question is, how does the magnetic order depend on free carrier concentration? In an ideal experiment one would like to vary the free carrier concentration over a wide range, let us say orders of magnitude, without changing other material parameters, such as the size of the local moments, crystal symmetry, electronic band structure etc. Experience has shown that RE-compounds can be expected to come most closely to this ideal. For $3d$-compounds, it is more difficult to obtain a large variation in free carrier concentration without changes in the valence state of the magnetic ions.

The existence of free carriers should produce always some indirect exchange. In reality, however, such interactions are very weak in most semiconductors and can be detected only by small variations in microwave spectra. Most materials in the multitude of semiconducting transition element and RE-compounds have been found to show only very weak changes in semiconducting properties with magnetic order. The present indirect-exchange theories are not refined enough to account reliably for all parameters which can, in reality, influence the strength of indirect exchange mechanisms and the intra-atomic exchange integrals. Such parameters include the density of states of the coupling electrons at the Fermi surfaces and the width of the conduction band. Materials with strong interactions via conduction electrons have usually much more complicated band structures than alkali metals or high mobility semiconductors, as examples. Even the conduction electrons in heavy RE-elements, usually considered to be prototypes of the conduction electron exchange mechanism, are not free particles but rather narrow band d-electrons, as discussed in F III.

Complications come from the fact also, that measurable contributions to magnetic interactions need, in general, quite large carrier concentrations, of the order of 10^{18} cm^{-3} or more. Techniques which provide variations in such large carrier concentrations (such as controlled valency doping by chemical substitution of certain ions by other ions with different valency) induce inevitably serious perturbations in other material properties and the effects of carrier concentration are difficult to separate out. Considering the difficulties of preparing sufficiently pure transition and lanthanide elements and compounds (which have been mentioned in part A) one concludes that magnetic semiconductors are rather dirty materials measured by standards of modern semiconductor technology even when single crystal preparation has been increasingly successful in recent years.

Another oversimplified conclusion often drawn from idealized $4f$-localization and indirect exchange mechanism implies that the only possible interaction

---

[268] K. Yosida: Progr. in Low Temp. Phys. **4**, 265 (1964); S. Methfessel: IEEE Trans. Magn. **1**, 144 (1965); Z. angew. Phys. **18**, 414 (1965); W. C. Koehler: J. Appl. Phys. **36**, 1078 (1965); RE Research, Vol. 2, p. 199, New York: Gordon and Breach 1964; K. P. Belov, M. A. Beljanchikova, P. Z. Levitin, and S. A. Nikitin: RE-Ferro- and Antiferromagnetics [in Russ.], Moscow: Science publ. house 1965; K. P. Below, R. Z. Levitin i S. A. Nikitin: Uspekhi Fiz. Nauk **82**, 449 (1964); T. R. McGuire, and M. W. Shafer: J. Appl. Phys. **35**, 984 (1964); and many articles in the book series: RE-research, Proceedings of RE conferences, published by Gordon and Breach, New York, and in Progress in Science and Technology of RE (Ed. L. R. Eyring), Publ. by Pergamon Press, Oxford.

between neighboring $4f$-moments goes via Bloch electrons in the conduction or valence band. In many compounds certain exchange effects cannot be disregarded *a priori*, such as those which may result from the covalency of the $4f$-electrons and $5d$-orbitals, from the polarization of filled inner shells and, in general, from effects which we have discussed in part F as the possible sources for increased interactions of $4f$-electrons with their crystalline environment. Such contributions to exchange are not too well understood theoretically, at present, and their magnitude cannot be determined reliably in actual materials. As a matter of fact, some Eu-chalcogenides are ferromagnetic with remarkably high Curie temperatures without free conduction electrons. Later we will give some arguments in favor of cation-cation nearest neighbor exchange without involvement of anion valence electrons. And the exchange interactions between lanthanide and $3d$-ions in garnets seem to be closer to a description by superexchange via anions than by indirect exchange.

Because of all those uncertainties in the interpretation of exchange interaction in RE-compounds and because the possibility of conduction electron exchange is so suggestive for materials with $4f$-moments, we will discuss in part I some general magnetic properties of metallic and insulating RE-compounds. The special topic of conduction electron exchange in semiconducting RE-chalcogenides and some $3d$-compounds will then be investigated in part II.

I. Metallic and insulating rare earth materials.

50. The magnitude of the atomic magnetic moment of RE-materials. The magnitude of the atomic magnetic moment and its response to external perturbations, such as exchange fields, applied fields, temperature etc. can be predicted when the ground state and the low lying levels are known from spectroscopic investigations and vice versa. The relationship between paramagnetic susceptibility and multiplet states has been sketched out in Sect. 8. In Gd^{3+} and Eu^{2+} the level separations are quite large compared to energies available in most laboratories (about 30,000 cm^{-1} are indicated in Fig. 21 compared to 1 cm^{-1} ∼ 10 Koe ∼ 1 °K) while the *crystalline field splittings* of the ^8S state are too small to interfere with measurements in usual temperature ranges. Therefore, both ions can be expected to have temperature-independent atomic moments close to the free ion value of 7 μ_B under usual conditions, in most materials. The magnetic susceptibility should conform well to the Curie law (8.4) in paramagnetic materials or to the Curie-Weiss law (10.1) in ferromagnets sufficiently above the Curie temperature. The stability of the magnetic moment can be used even to complement or to replace chemical analysis for unknown Gd^{3+} or Eu^{2+} concentrations.

For the heavier and lighter lanthanides *crystal field splittings of the order of some 100 cm^{-1}* have to be considered for their temperature dependent effect on the magnetic moment. The classical example is the different temperature dependence of magnetic susceptibility in Pr^{3+} and Nd^{3+} in sulphates below 100 °K. This was explained as an early application of crystal field theory[269] by the fact that Pr^{3+} has two $4f$-electrons while Nd^{3+} has an odd electron number of electrons (3). The Nd electrons form Kramer doublets in the ground state which cannot further split in the electrostatic crystalline field but only in magnetic fields. In many materials the RE ions crystallize in rather complicated coordinations of low symmetry and reliable crystal field calculations become very difficult. However, in cases where magnetic and spectroscopic measurements have been made on the

---

[269] W. G. PENNEY, and R. SCHLAPP: Phys. Rev. **41**. 194 (1932).

same material [270] the expected relations between the results have generally been observed within the quantitative limitations of the crystal field theory. In Eu^{3+} and Sm^{2+} the *separations between low lying levels in the free ion multiplet* are already comparable to laboratory energies (Fig. 21) and temperature dependent moments occur already in the free ion state.

The quantitative calculations of magnetic moments and their dependence on environmental conditions have been given in the classic book of VAN VLECK on electric and magnetic susceptibilities [271]. In Table 1 we have listed the magnetic moments and g-factors expected for the ground state configurations of the $4f$-electrons in trivalent free RE-ions. The value of $n_B = gJ$ gives the saturation moment per RE-ion measured at 0 °K with perfectly parallel aligned moments and $P_{\text{eff}} = g[J(J+1)]^{\frac{1}{2}}$ is the effective magneton number per ion as derived from the Curie law (8.4) in (8.5).

In RE-materials the free ion moments give a good zeroth approximation to the magnetic properties to be expected for compounds, in contrast to $3d$-element compounds where the crystal field quenches the orbital contribution and makes predictions of magnetic moments much more complicated. The unquenched orbital contributions to the $4f$-moments result in large magnetization compared to transition elements; the largest saturation magnetization is found in Dy and Ho metal (37,000 Gauss as compared to ~22,000 Gauss in iron metal). The lighter RE-ions with less than seven $4f$-electrons have smaller moments which result from orbital moments L oppositely aligned to the spin contributions S. In the f^6 state L and S cancel one another and the ground state of Eu^{3+} and Sm^{2+} has no magnetic moment at 0 °K. Magnetic moments of RE-ions with higher or lower valency are equal to that of a trivalent ion with the same $4f$-ground state configuration. As an obvious consequence of the $4f$-localization the magnetic moments of lanthanides do not mix and average out with one another in alloys or solid solutions, as found in $3d$-elements.

For ferromagnetic RE-materials the *magnetic moment n_B* in Bohr magnetons per molecule can be derived most directly from the absolute saturation magnetization M at 0 °K:

$$M = \frac{1}{4\pi}\pi(B-H) = N\beta d\, n_B/W = 5585\, n_B\, d/W \text{ (Gauss)} \tag{50.1}$$

[d = material density, W = molecular weight, N = Avogadro number, β = Bohr magneton (8.2)]. In the RE-elements and several metallic compounds the atomic magnetic moment is found to be larger than predicted from the HUND's rule ground state of the $4f$-electrons. This phenomenon has not yet been fully explained but the polarization of the conduction electrons by intra-atomic exchange gives a convincing explanation (compare Sect. 41). Another possibility for deviations in magnetic moments could result from partial filling of virtual $4f^{n+1}$ states by conduction electrons. We have discussed this situation in Sect. 33 for Ce-metal where large effects have been observed [272]. The small separation of the $4f^{n+1}$ states from the $5d$ and $6s$ free ion levels (at least at the ends and in the middle of the

---

[270] See, for instance, review by J. H. VAN VLECK, in: Interaction of Radiation with Solids (Ed. A. BISHAY), New York: Plenum Press 1967; E. F. WESTRUM, in: Progr. in Science Techn. of RE (Ed. L. R. EYRING), Oxford: Pergamon Press, K. H. HELLWEGE, W. SCHEMBS u. B. SCHNEIDER: Z. Physik **167**, 477 (1962); A. H. CROOKE, R. LAZENBY, and M. J. LEASK: Proc. Phys. Soc. (London) **85**, 767 (1965); Y. AYANT et J. THOMAS: Compt. rend. **248**, 387, 1955 (1959); W. H. BRUMAGE, C. C. LIN, and J. H. VAN VLECK: Phys. Rev. **132**, 608 (1963).

[271] J. H. VAN VLECK: The Theory of Electric and Magnetic Susceptibilities. London: Oxford Univ. Press 1934, last reprint 1965.

[272] Recent discussion of virtual $4f$ levels by A. BLANDIN: J. Appl. Phys. **39**, 1285 (1968).

lanthanide series) suggests to consider the existence of empty $4f$-levels near the Fermi energy in metals.

In many cases chemical impurities or deviations from stoichiometry have simulated magnetic irregularities. The present supply of RE-elements contains often large amounts of oxygen, nitrogen, hydrogen and carbon as impurities. These impurities can form inclusions in compounds with deviating magnetic properties. An historic example is the discovery of ferromagnetic EuO at the Bell-Laboratories from irregularly high Curie temperatures in Laves phase compounds of Eu with iridium [273].

51. Magnetic interactions in RE-elements. The magnetic interactions in RE-elements and metallic compounds are usually assigned to *Ruderman-Kittel*-type interactions via conduction electrons as discussed in part D. The conditions are, however, more complicated than described in part D by the model of indirect exchange via Bloch-electrons. The electron band structure can deviate seriously from the free electron case. The crystal field splittings in metals are smaller relativ to the band width than in insulators and weaker than the exchange fields in RE-elements. This is indicated by the close proximity of the observed moments to the values for free RE^{3+} ions [274]. The crystal field then induces very large anisotropies of the magnetization and magneto-elastic effects [275]. The crystal field splittings are smallest for the ^8S state in close packed hexagonal Gd-metal (which is ferromagnetic below 290 °K) and produce here a crystal anisotropy of about 10^6 erg/cm^3 at 80 °K which is of the same order of magnitude as that of cobalt and shows complicated temperature dependence [276]. For the heavier RE-elements with $L \neq 0$ the crystal anisotropy is one to two orders of magnitude larger than in Gd. The sixfold c-axis of the hexagonal structure becomes an extremely hard axis for the magnetization in Tb, Dy and Ho metal and an easy axis in Er, Tm. The combination of temperature-dependent crystal anisotropy and long range exchange interactions produces complicated spin structures in the temperature range between Néel and Curie temperature [277] in a variety not equaled in any $3d$-material; neutron diffraction is the only reliable means to disentangle those complex magnetic properties. For the lighter RE-elements, from Ce to Eu, which do not become ferromagnetic at low temperatures, the atomic moments at low temperatures and the character of the ground state can be obtained (with low accuracy) only from neutron diffraction. For these RE, the magnetic ordering phenomena are by far not as well understood as for the heavy RE elements, from Gd to Tm. There is, for instance, no convincing explanation in simple *Ruderman-Kittel* theory for the fact that Eu and Gd-metal which have the same $4f$ ground state configuration show such different magnetic properties.

The proportionality of magnetic transition temperatures to the de Gennes-factor $(g-1)^2 J(J+1)$ has been investigated for solid solutions of heavy RE-elements with one another and with La, Y, Sc, but applies approximately only to the paramagnetic Curie-Weiss temperature. The Néel-temperature varies rather

---

[273] B. T. MATTHIAS, R. M. BOZORTH, and J. H. VAN VLECK: Phys. Rev. Letters **7**, 160 (1961).

[274] The crystal field effects on the magnetic properties of Ce, Pr, Nd have been discussed by B. BLEANEY: RE-Research, Vol. 2, p. 417 (Ed. K. S. VORRES). New York: Gordon and Breach 1964.

[275] Survey article by J. KANAMORI: Magnetism Vol. 1, p. 127 (ed. G. T. RADO, and H. SUHL). New York: Academic Press 1963.

[276] C. D. GRAHAM: J. Appl. Phys. **34**, 1341 (1963).

[277] W. C. KOEHLER: J. Appl. Phys. **36**, 1078 (1965); K. YOSIDA: Progr. in Low Temp. Phys. **4**, 265 (1964); T. KASUYA: Magnetism, Vol. 2b, p. 215 (Ed. G. T. RADO, and H. SUHL), New York: Academic Press 1966.

with the $^2/_3$ power of the de Gennes-factor and the Curie temperatures behave even more irregularly[278]. The reason for this is not understood, some discussion is given in Sect. 27.

52. Magnetic interactions in RE-compounds with elements of the first transition series. With transition elements the RE tend to form a large number of intermetallic compounds but show relatively small solubility because of their larger size and smaller electronegativity[279]. The phase diagrams and magnetic properties of most systems are insufficiently known. RE-compounds including magnetic $3d$-elements promise, in principle, the technically interesting possibility of new magnetic materials which combine the high Curie temperatures of the $3d$-compounds with the large magnetization of RE-materials. This expectation, however, has been generally unfullfilled. The very complex compound Gd_6Mn_{23} is so far the only known case in which RE and $3d$-spins couple ferromagnetically with one another[280]. Details in the temperature dependence of the magnetization, however, indicate that the Curie temperature of 478 °K is probably related only to the ferromagnetism of the Mn-spins and that the Gd-spins need lower temperatures for ordering.

The most comprehensive studies[281] have been made on intermetallic compounds of the composition AB_2 and AB_5 ($A=$RE, $B=3d$-element) which crystallize in the *Laves Phase structures* C 15 (cubic $MgCu_2$-type) and D $2d$ (hexagonal $CaZn_5$-type), respectively. The Laves phases represent the most abundant structure of intermetallic compounds. They provide a very effective packing of different size ions and are stable for certain numbers of conduction electrons per atom[282] (e.g. 1.5 to 1.8 and 2.5 to 2.8 for the C 15-structure). The magnetic moments of the $3d$-ions as well as of the RE-ions have been found to be lower than in the metallic elements. The $3d$-levels are progressively filled, from Mn to Ni, by $5d$ or $6s$ electrons of the RE-ions. Ni has very often, and Co sometimes, no magnetic moment at all in these compounds, whereas Mn has almost normal moments. The lower moments of the RE^{3+} ions, however, have been explained by BLEANEY[283] as partial quenching of the orbital contributions by crystal fields comparable to or larger than the exchange fields. In a primitive theory using molecular field approximation and electrostatic crystal fields, he has derived approximate formulae for Curie-temperature and saturation magnetization in magnetic substances where the exchange interaction is comparable with crystal field effects. Application to ANi_2 compounds gives a satisfactory explanation of the magnetic properties of the RE-ions.

In Laves phase compounds with magnetic RE and $3d$-ions magnetic ordering temperatures up to and above 1,000 °K can be found. However, this Curie-tem-

---

[278] W. C. KOEHLER, E. O. WOLLAN, H. R. CHILD, and J. W. CABLE: R.E. Research, Vol. 2, p. 199 (Ed. K. S. VORRES), New York: Gordon and Breach 1964; R. M. BOZORTH, and R. J. GAMBINO: Phys. Rev. **147**, 487 (1966); R. M. BOZORTH: J. Appl. Phys. **38**, 1366 (1967). It is interesting to note that the experimental Néel points could be interpolated also with an $[S(S+1)]^{\frac{3}{8}}$ curve instead with the $\frac{2}{3}$ power of the De Gennes-factor.

[279] For example, some RE with Mn, Fe form up to 4, with Co 6 and with Ni 9 intermetallic compounds. See K. A. GSCHNEIDNER: R.E. Alloys, New York: D. van Nostrand Co. 1961; O. D. McMASTERS, and K. A. GSCHNEIDNER: Nuclear Met. **10**, 92 (1964).

[280] B. F. DE SAVAGE, R. M. BOZORTH, and F. E. WANG: J. Appl. Phys. **36**, 992 (1965).

[281] Survey and references of magnetic properties in: H. WEIK u. K. STRNAT: LANDOLT-BÖRNSTEIN, IV/2c, Berlin-Göttingen-Heidelberg-New York: Springer 1964; R. M. BOZORTH, and C. D. GRAHAM, in: Handb. of Magn. Mater. (Ed. P. ALBERT, and F. LUBORSKY), to be publ. New York: Reinhold Press.

[282] W. E. WALLACE: Ann. Rev. Phys. Chem. **15**, 109 (1964).

[283] B. BLEANEY: J Phys. Soc. Japan **17**, B1, 435 (1962); Proc. Phys. Soc. (London) A **276**, 19, 28 (1963); A **277**, 289 (1964); RE-Research, Vol. 2, p. 499 (Ed. K. S. VORRES), New York: Gordon and Breach 1964.

perature refers only to the alignment of $3d$-spins, which is ferromagnetic or more complex ferrimagnetic (when sublattices at different lattice sites can be formed). The RE-spins remain paramagnetic and need lower temperatures for ordering into a spin sublattice which is coupled antiparallel to the net magnetization of the $3d$-spins. Obviously, the situation has some phenomenological similarity with the well investigated magnetic behavior of insulating garnets. Depending on the net magnetization of the $3d$ and RE sublattice, maxima also occur here in the temperature dependence of the sample magnetization. Some compounds have compensation points with disappearing magnetization at temperatures where both sublattices have equal but opposite magnetization. Since the lighter RE-ions, from Ce to Sm, have their atomic magnetic moments opposite to the spins in $J = L - S$ ground states, "pseudo"ferromagnetism is observed in compounds with those ions, i.e. the antiferromagnetic exchange interaction between spins produces parallel alignment of magnetic moments. A simple rule[284] is to replace S_i by $(g_i - 1) J_i$ so that the magnetic Hamiltonian becomes

$$H = -J^2 \Sigma F(R_{ij}) (g_i - 1)(g_j - 1) J_i J_j \qquad (52.1)$$

replacing Eq. (27.4). One observes that $(g_i - 1)$ can be positive (heavy RE) or negative (light RE), and this degree of freedom, in addition to the sign of $F(R_{ij})$, will determine the nature of the magnetic bond.

The mechanism of exchange interactions between RE and $3d$-spins in metallic compounds and its almost always antiferromagnetic nature is not understood, at present, but is often attributed to some form of indirect exchange via conduction electrons. The electronic band structure of materials including more or less localized orbitals of two transition elements is expected to be very complicated. Extended magnetic investigations on larger numbers of isostructural compounds, resonance and Mössbauer measurements in compounds and solutions will contribute to the further understanding of this very interesting field.

53. General properties of III—V-compounds of RE^{3+} with N, P, As, Sb, Bi

(pnictides). The compounds with 1:1 composition have the advantage of a quite simple crystal structure (cubic Fm3m-O_h^5 of the NaCl-type) but their physical properties are of a rather ambigous nature. From comparison with III-V-compounds of other elements which are semiconductors, SCLAR[285] has suggested that the RE-pnictides should be also semiconductors or insulators. In evaporated films of DyN, ErN and HoN he has observed, in agreement with his expectation, absorption edges of about 2 eV which are interpreted as direct transitions between the valence and conduction band[285]. On the other hand, electrical measurements[285a] give for the RE-nitrides (and most other III-V-compounds with RE^{3+}) metallic resistivities of the order of 10^{-4} ohm. cm with positive temperature coefficients of about 10^{-3} deg^{-1}. These values have the same order of magnitude as found for the electrical properties of the pure rare earth metals, and LaN even becomes superconducting at low temperatures[286]. The origin and nature of the electrical conduction is complex and not yet explained. The nitrogen defect which can amount to as much as 5% deviations from stoichiometry in many RE-nitride samples is a sufficient but probably not necessary condition for conductivity. Irregularities in the conduction band structure such as partial localization of

---

[284] A more complete discussion will be found in D. MATTIS: The Theory of Magnetism, p. 207. New York: Harper and Row 1965.

[285] N. SCLAR: J. Appl. Phys. **33**, 2999 (1962); **35**, 1534 (1964).

[285a] R. DIDCHENKO, and F. P. GORTSEMA: J. Phys. Chem. Solids **24**, 863 (1963); F. J. REID, L. K. MATSON, J. F. MILLER, and R. C. HIMES: J. Electrochem., Soc. **111**, 943 (1964).

[286] J. J. VEYSSIC, D. BROCHIER, A. NEMOZ, and J. BLANC: Phys. Letters **14**, 261 (1965).

Table 2. *Physical and magnetic properties of RE-pnictides*

			Ce	Pr	Nd	Sm	Eu
Number of 4f electrons			1	2	3	5	6
Ground state of Ln^{3+}			$^2F_{5/2}$	$^3H_4$	$^4J_{9/2}$	$^6H_{5/2}$	$^7F_0$
$g[J(J+1)]^{\frac{1}{2}}$			2.56	3.62	3.68	0.83	0
Anions	N	Order Θ_p T_c n_s (0 °K)	anti ...	n.o. 0	ferri (?) +24 32 3.1	anti (?) ... Van Vleck paramagnetism	n.o. −200
	P	Order Θ_p T_C, T_N	anti −8 +8	n.o. −2	ferri (?) +11 +11		
	As	Order Θ_p T_C, T_N	anti −5 +7	n.o. −6	anti +4 +11		
	Sb	Order Θ_p T_C, T_N	anti +8 +18		anti −3 +16		

Θ_p = paramagnetic Curie point.
T_C, T_N = critical point for order.

5 d-electrons at the RE-ions could lead to irregular values of the effective valencies and excess carrier concentration in samples with accurate 1:1 stoichiometry.

In this context it is interesting to note that Gd and Dy have been found to form with Sb and Bi (the least electronegative members of the fifth group) not only 1:1, but also metallic 4:3 compounds [287] which have an inverted $I\bar{4}3d$-T_d^6-structure of the Th$_3$P$_4$-type. The RE ions are located at the sixfold coordinated lattice sites which are normally occupied by the larger anions [288] whereas Sb and Bi have the cation positions. The electrical properties of those compounds are also quite similar to those of the pure RE-metals. Gd$_4$Bi$_3$ and Gd$_4$Sb$_3$ are ferromagnets with Curie temperatures of 340 °K and 260 °K, respectively, compared to 290 °K for pure Gd-metal. Dy$_4$Sb$_3$ has a paramagnetic Curie temperature $\Theta = 70$ °K, becomes antiferromagnetic with complicated spin structure below 67 °K and ferromagnetic below 21 °K. The coercive force in the ferromagnetic state at 4.2 °K is about 3,000 oersted and the first anisotropy constant amounts to $10^7 - 10^8$ erg/cm^3.

It is quite surprising that the dilution of the magnetic Gd-ions in Gd$_4$Bi$_3$ by diamagnetic Bi to almost 50% results in a Curie temperature increase of 50 °K compared to pure Gd-metal. On the other hand, the Gd-Gd distances are about 2% shorter than the nearest-neighbor distance in Gd-metal. The ionic Gd-radius derived from cation-anion distances is in the 4:3 compounds 1.02 Å compared to 0.94 Å in the NaCl-type compounds GdBi and GdSb. Carter [289] has discussed the covalent bonding aspects in Th$_3$P$_4$-compounds using bidirectional hybridized orbitals of p and d-character which link ions together in chains. The conducting electrons might be localized to a chain instead to individual atoms and produce the strong magnetic interaction. The transfer of 0.75 electrons from the anions back to each Gd-ion has been suggested.

---

[287] S. Methfessel, and E. Kneller: Appl. Phys. Letters **2**, 115 (1963); F. Holtzberg, S. Methfessel, and J. C. Suits: R.E. Research, Vol. 2, p. 385, New York: Gordon and Breach 1963; F. Holtzberg, T. R. McGuire, S. Methfessel, and J. G. Suits: J. Appl. Phys. **35**, 1033 (1964); R. J. Gambino: J. Less Common Metals **12**, 344 (1967).
[288] F. Holtzberg, and S. Methfessel: J. Appl. Phys. **37**, 1433 (1966).
[289] F. L. Carter: R.E. Research, Vol. 4, p. 495, New York: Gordon and Breach 1965.

[G. Busch: J. appl. Phys. **38**, 1386 (1967)].

Gd	Tb	Dy	Ho	Er	Tm	Yb
7	8	9	10	11	12	13
$^8S_{7/2}$	$^7F_6$	$^6H_{15/2}$	$^5I_8$	$^4I_{15/2}$	$^3H_6$	$^2F_{7/2}$
7.94	9.7	10.6	10.6	9.6	7.57	4.5
ferro +69	ferri +34	ferri +20	ferri +12	ferri +4	n.o. −18	anti (?) −116
72 6.6	40 6.3	21 6.3	13 9.2	6 5.5		
anti +2 +15	anti +3 +8	ferri (?) +8 +8	ferri +6 +6	anti +2 +4	n.o. −2 ...	n.o. −55 ...
anti −12 +25	anti −4 +10.5	ferri (?) +2 +8.5	anti +1 +4.8	anti −1.5 +3.5	n.o. −2 ...	n.o. −25 ...
anti −42 +28	anti −14 +16.5	anti −4 +9.5	anti −2.5 +5.5	anti −3 +3.5	n.o. −1 ...	n.o. −60 (?)

anti, ferri, ferro = (antiferro-, ferri, ferro-)magnetic.
n_s = Spontaneous magnetization in μ_B/Ion.

54. Magnetic interactions in the 1:1 RE-pnictides. The magnetic properties of the 1:1 RE-pnictides have been reviewed recently by Busch[290] and Table 2 is reproduced from his paper. Only GdN is a simple ferromagnet. All the other compounds have complex ferrimagnetic or antiferromagnetic MnO-type structures. Extensive neutron diffraction studies have been made at the Oak Ridge National Laboratory[291]. Darby and Taylor[292] have attempted to relate the variation of magnetic transition temperatures with anion character to the indirect exchange via conduction electrons using the lattice constant as parameter. Very recently, the increasing line width of electron spin resonance in GdN and GdBi with decreasing temperature has been interpreted as indication of a temperature dependent exchange mechanism due to the temperature dependent carrier concentration of semiconductors[293]. However, the role of the conduction electrons in the magnetic properties of the pnictides is by far not clear.

The more important parameters determining the magnetic properties appear to be connected with *crystal field effects* which are here, in contrast to the RE-metals, often larger than exchange interactions. As a consequence, the free ion values of the atomic moments are obtained only at *high* temperatures in paramagnetic measurements, but the *low temperature* saturation moments in strong applied fields are lower than expected because of partial orbital quenching. Trammel[294] has interpreted some of the experimental properties of the pnictides, such as moments and anisotropies, by considering the decomposition of the magnetic ground state in the point charge field of the octahedral ligand coordination under influence of the exchange fields (described by molecular fields at high temperature and spin-wave or exciton treatment at low temperature). In sixfold coordination the crystal field levels have a different order than for the

---

[290] G. Busch: J. Appl. Phys. **38**, 1386 (1967); D. P. Schumacher, and W. E. Wallace: J. Appl. Phys. **36**, 984 (1965).
[291] H. R. Child, M. K. Wilkinson, J. W. Cable, W. C. Koehler, and E. O. Wollan: Phys. Rev. **131**, 922 (1963).
[292] M. I. Darby, and K. N. R. Taylor: Phys. Letters **14**, 179 (1965).
[293] M. B. Allenson, and K. N. R. Taylor: J. Appl. Phys. **39**, 1094 (1968).
[294] G. T. Trammel: J. Appl. Phys. **31**, 362 S (1960); Phys. Rev. **131**, 932 (1963).

eightfold coordination shown in Fig. 22, as example, but the lowest state is also a nondegenerate level for most ions with even number of $4f$-electrons. In this case magnetic moments disappear at low temperature if the exchange interactions, applied fields and temperatures are smaller than the crystal field splittings of some 100 cm^{-1} and can not mix higher levels into the nonmagnetic ground level. Therefore, Pr and Tm compounds are nonmagnetic at low temperatures and show only temperature independent Van Vleck-susceptibility by contributions from the next level. For Ho^{3+} and Tb^{3+} the exchange field is sufficiently strong compared to the crystal field splitting to produce magnetic order, but the moments are still lower than in free ions and elemental metals where all levels contribute to the magnetic properties. For RE-ions with odd numbers of $4f$-electrons Kramer's theorem asserts that the crystal field cannot remove degeneracy to less than doublet states and there will be always magnetic moments at low temperatures. Irregularities in magnetic susceptibility and lattice parameter of Ce-pnictides have been interpreted [295] as partial valence changes from Ce^{3+} to Ce^{4+}. Details in the magnetic properties of pnictides with exchange and crystal fields of competing magnitude have been discussed in several papers of Cooper et al. [296] and Tsuya and Ebina [297].

All these experiments and theories do not give any information about the nature of the ferromagnetic and antiferromagnetic interactions between the RE-ions and no simple relationship to the electrical conductivity is indicated. In a formal way one can derive by molecular field analysis of the magnetic transition temperatures values of a ferromagnetic exchange integral J_1 (which connects each RE-ion with its twelve nearest neighbors) and an antiferromagnetic exchange integral J_2 (which connects each RE-ion with its six next nearest neighbors straight over the ligands) [298]. J_1 decreases with increasing lattice distances as is typical for cation-cation interactions, whereas J_2 increases in the line N, P, As, Sb as expected for 180° superexchange. In this respect, the conducting RE-pnictides have characteristics qualitatively quite similar to insulating NaCl-type compounds such as Mn^{2+} and Eu^{2+} chalcogenides.

55. Chemistry of compounds of RE-elements with elements of group VI. The elements of group VI form with the RE-elements quite stable compounds with high melting temperatures around 2,000 °C. The anions have sufficient electronegativity to maintain their divalency in combination with many-valent RE-elements such as Ce, Pr, Sm, Eu, Tb, Yb which adjust then their valency to the stoichiometry. Compounds in which cation valency varies continuously (e.g. from divalent to trivalent) are known to occur in crystal structures which are insensitive to ionic size, or which have a complex cell structure. Examples are the systems $Ce_2O_3-CeO_2$, $Tb_2O_3-TbO_2$, $Pr_2O_3-PrO_2$, $Eu_2O_3-Eu_3O_4$, $Eu_2S_3-Eu_3S_4$, etc.[299]. However, it has not been observed that chalcogens can force trivalent RE other than Sm, Eu, Yb

---

[295] R. Didchenko, and F. P. Gortsema: J. Phys. Chem. Solids **24**, 863 (1962); T. Tsuchida, and W. E. Wallace: J. Chem. Phys. **43**, 2885 (1965); G. Busch, and O. Vogt: Phys. Letters **20**, 152 (1966); T. Tsuchida, and Y. Nakamura: J. Phys. Soc. Japan **22**, 942 (1967).

[296] B. R. Cooper: Phys. Rev. **163**, 444 (1967) with references to earlier papers; B. R. Cooper, and R. C. Fedder: Phys. Rev. **163**, 506 (1967); P. Junod, and A. Menth: Phys. Letters **25**A, 602 (1967).

[297] N. Tsuya, and Y. Ebina: J. Appl. Phys. **35**, 800 (1964).

[298] G. Busch: J. Appl. Phys. **38**, 1386 (1967).

[299] References to more details can be found in K. A. Gschneidner: RE-Alloys, New York: D. van Nostrand Co. 1961 and in several more recent papers by D. J. M. Bevan, G. Brauer, L. R. Eyring, and J. Flahaut in the book series R.E. Research, proceedings of RE-conferences published by Gordon and Breach, New York. E. F. Westrum gave review articles on thermodynamic properties of RE-chalcogenides in progress in Science and Technology of RE (Ed. L. R. Eyring), Oxford: Pergamon Press 1964, 1966.

in a state with a higher f-electron number, as was mentioned earlier in this article for halogens (Sects. 33 and 44). Instead, compounds with metallic conductivity occur in which the excess electron of the RE^{3+} ions goes into conducting states such as the $5d$ or $6s$ levels. We mention here only examples such as GdS, TbS, $Gd_{2+x}S_3$ etc. and postpone the discussion of more details to later (Sect. 59).

The stepwise decrease in electronegativity and increase in anion size between O^{2-} and the other chalcogens S^{2-}, Se^{2-}, Te^{2-} separates the oxides in their chemical properties somewhat from the other chalcogenides, which have been studied much less extensively. The following four sections discuss examples of magnetic interactions in oxides and chalcogenides.

56. Magnetic interactions in $3d$ garnets and perovskites containing RE. The most frequently investigated and best understood materials of all RE as well as $3d$ oxides are, of course, the *garnets*. Their sharp stoichiometry distinguishes them favorably from other $3d$-oxides, such as ferrites and perovskites, which easily form defect structures with non-uniform cation valency, electrical conductivity and optical opacity. The garnets include a wide variety of $A_3B_2C_3O_{12}$-compounds with three cation-sublattices. Trivalent RE-ions can be placed on the A sites which are coordinated with 8 oxygen ions in form of a distorted cube. In ferrogarnets the B and C sites are both occupied by Fe^{3+}-ions in an S-state with octahedral and tetrahedral oxygen coordination, respectively. Since all cation sites are filled, the garnets are insulators with resistivities of the order of 10^{13} ohm. cm at room temperature, are transparent with intense colors and have elastic properties comparable to quartz. The small damping at high frequencies permits extremely sharp microwave resonance lines (line width ~ 1 oe compared to $100-1,000$ oe in ferrites) useful for technical applications [300]. The magnetic properties [301] have been studied very successfully from the point of view of molecular field theory and ligand field theory but without any significant contribution to the understanding of the nature of exchange forces. The crystal structure is too complicated for detailed studies of wave functions and their overlap. Below Curie temperatures of near 550 °K, which are almost independent of the nature of the RE-ions, the Fe-spins on the B and C sites order into two ferromagnetic sublattices which couple antiparallel to one another. Since there are more C than B sites a net magnetization results which couples antiferromagnetically again, but much more weakly, to the RE-spins on the A-sites. The interaction between RE-spins is usually neglected as small, compared to the $3d$-interactions. Below room temperature clear compensation points of zero magnetization can be observed when the ordering of the $4f$-spins in the exchange field from the iron sublattices yields an equal but opposite magnetization to the $3d$-magnetization. For more details we have to refer to some of the survey articles [302] which review the numerous literature accumulated in this field during recent years.

Another large group of mixed oxides containing RE together with $3d$ ions crystallizes in the *perovskite structure* with the general composition ABO_3. The ideal perovskite structure O_h^1-Pm3m is cubic with two simple cubic cation sub-

---

[300] See, for instance, A. F. HARVEY: Microwave Engineering. New York: Academic Press 1963.
[301] F. BERTAUT, and R. PAUTHENET: Proc. I.E.E. (London), Suppl. B **104**, 261 (1957); R. PAUTHENET: J. Appl. Phys. **29**, 253 (1958); R. V. JONES, in: LANDOLT-BÖRNSTEIN, Vol. II/9, 2. Berlin-Göttingen-Heidelberg: Springer 1962.
[302] J. H. VAN VLECK: Physical Sciences, New York: University Press 1962; Progr. in Science, Techn. of R.E. (Ed. L. R. EYRING), Vol. 2, 1, Oxford: Pergamon Press 1966; J. Appl. Phys. **35**, 882 (1964); W. Low: Progr. in Science, Techn. of R.E. (Ed. L. R. EYRING), Vol. 2, 123, Oxford: Pergamon Press 1966.

lattices of A and B-ions, respectively (Fig. 28). The structural symmetry can be visualized as a cube with the A-cation in its center, the B-cations on the 8 corners and the oxygen anions in the middle of 12 edges. Each B-cation is surrounded by a cube of 8 A-cations and an octahedron of 6 oxygen ions (with the symmetry axes pointing parallel to the cube edges). GOLDSCHMIDT[303] has introduced as critical stability parameter the ratio of the cube edges (determined by the sum of ionic radii of oxygen and cation A) to the octahedron size, determined by the size of the B-cations. When those two dimensions are compatible with one another a perfectly cubic structure is obtained. However, this case is rare. The usual perovskites are either orthorhombic or rhombohedral, depending on the relative ion sizes, and phase transitions are quite abundant.

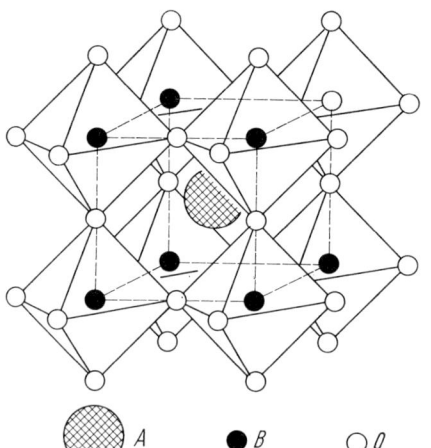

Fig. 28. The undistorted cubic perovskite structure.

The magnetic perovskite materials have played a significant role in experimental studies of the mechanism of 180°-superexchange via anions and its relationship to the character of the involved orbitals of magnetic $3d$-ions on B-sites. The structural symmetry provides the magnetic B-ions only with nonmagnetic nearest neighbors when the large A-cation is nonmagnetic. Direct cation-cation interactions are excluded in this case as well as 90° superexchange. Most $3d$-perovskites are antiferromagnetic. In *orthoferrites*, with Fe in the B-sites and RE-ions at the A-sites, the Néel-temperature is around 700 °K and decreases from La to Lu. Neutron diffraction shows four distinct iron sublattices with each ion having six antiparallel neighbor spins. At low temperatures weak ferromagnetism occurs with low magnetization values due to canting of spins out of the antiparallel alignment. The spin canting results from single ion anisotropies with different axis for inequivalent ions or from DZYALOSHINSKY'S antisymmetric exchange interactions[304] which tend to align spins perpendicular to one another. Magnetic measurements and Mössbauer studies indicate predominance of the antisymmetric exchange[305].

JONKER and VAN SANTEN[306] found interesting magnetic properties in *Mn and Co perovskites*. Early in their studies they found that the strongly distorted orthorombic perovskite $LaMnO_3$ became weakly ferromagnetic at liquid nitrogen temperature whereas $LaCrO_3$ and $LaFeO_3$ did not. The ferromagnetic Curie temperature increased up to 210 °K by heat treatment in oxygen atmosphere. Chemical analysis ascertained that the ferromagnetism was accompanied by the presence of about 3% of the Mn-ions in the Mn^{4+} state due to the excess oxygen in the lattice.

---

[303] V. M. GOLDSCHMIDT: Geochemische Verteilungsgesetze der Elemente **7**, 8 (1927—1928).

[304] I. DZYALOSHINSKY: J. Phys. Chem. Solids **4**, 241 (1958); T. MORIYA: Phys. Rev. **120**, 91 (1960).

[305] Survey article with literature references: D. TREVES: J. Appl. Phys. **36**, 1033 (1965); W. LOW: Progr. Science, Techn. of RE (ed. L. R. EYRING), vol. 2, Oxford: Pergamon Press 1966.

[306] G. H. JONKER, and J. H. VAN SANTEN: Physica **16**, 337, 599 (1950); **19**, 120 (1953); **22**, 707 (1956); in: Halbleiterprobleme **6**, 118 (1961).

Larger concentrations of Mn^{4+} ions were introduced by solid solution with manganites containing divalent alkaline earth ions instead of La^{3+}. Materials such as $CaMnO_3$, $SrMnO_3$, $BaMnO_3$ contain Mn^{4+} ions only and are antiferromagnetic below about 100 °K. Pure materials which have *only* Mn^{3+} or Mn^{4+} ions are semiconductors with resistivities between 10 and 10^4 ohm. cm at room temperature, depending on the preparation conditions. The solid solutions of the type

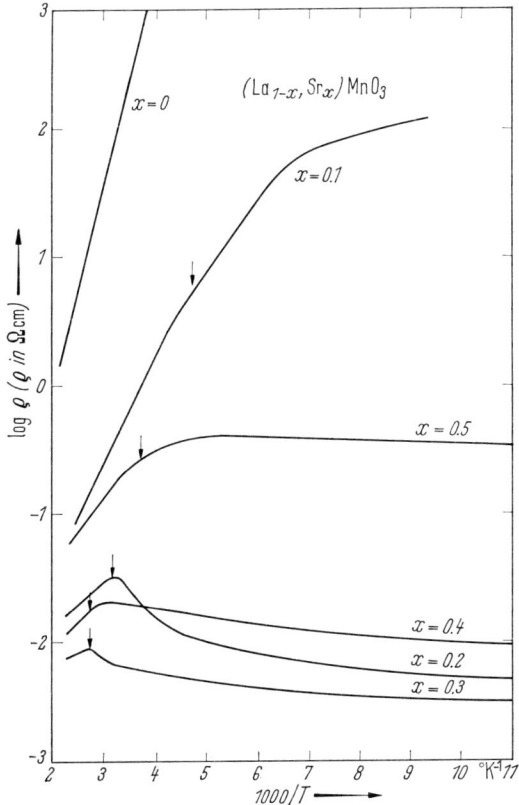

Fig. 29. The electrical resistivity of $La_{1-x}Sr_xMnO_3$ as function of reciprocal temperature. Similar curves are obtained for manganites with Ca or Ba. [J. H. VAN SANTEN and G. H. JONKER: Physica **16**, 599 (1950).] Arrows mark the Curie temperatures.

$La_{1-x}Ca_xMnO_3$ become, with increasing x, first p-type conductors, then, in the range from $x = 20\%$ to 40%, metals with resistivities of the order of 10^{-2} ohm. cm and finally p-type semiconductors at higher concentrations (Fig. 29). Simultaneously a large variety of crystalline and magnetic structures appears with variation of x.

The most remarkable fact is the occurrence of ferromagnetism together with metallic conductivity in the narrow concentration region containing 25 to 35% Mn^{4+}-ions. The Curie temperatures rise here to about 300 °K and the observed saturation magnetization equals the sum of all available Mn^{3+} and Mn^{4+} moments. In the beginning, these results were explained by the assumption of very weak positive interactions between Mn^{3+}-ions, strong antiferromagnetic Mn^{4+}-coupling and a strongly ferromagnetic interaction between ions of different valency. ZENER has related the ferromagnetic Mn^{3+}-Mn^{4+} interaction to the occurrence of metallic conductivity in this concentration region and suggested double exchange as a new type of magnetic interaction, due to the tendency of travelling electrons to keep their spin orientation, while jumping from lower to higher valent ions. The theory of double exchange and its refinements by ANDERSON, HASEGAWA and DE GENNES had been discussed in part D.

The resistivity anomalies in the neighborhood of the Curie temperatures of the ferromagnetic metallic samples have been investigated by VOLGER[307]. The

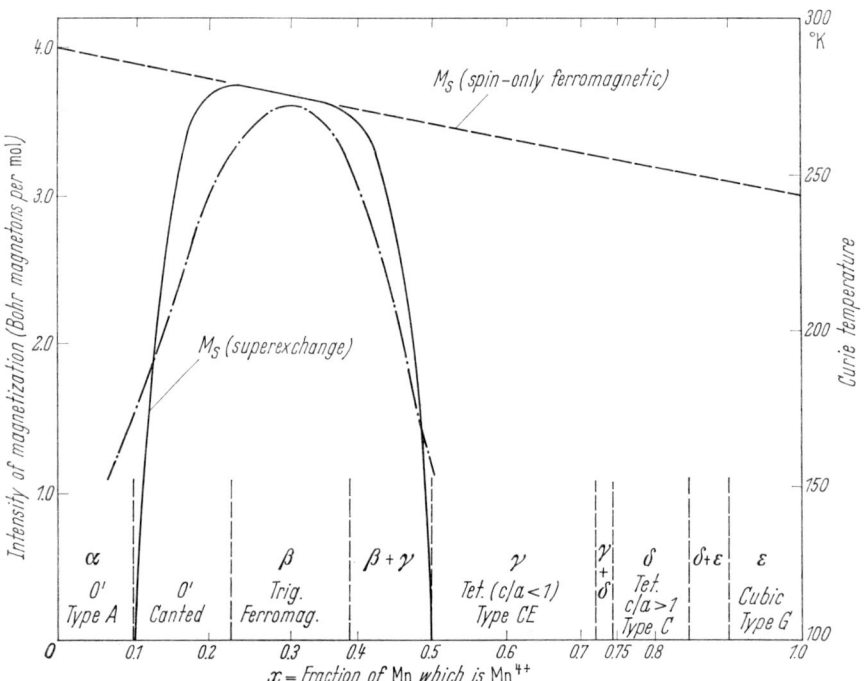

Fig. 30. Concentration dependence of magnetic Curie temperature (·—·—·—·) and saturation magnetization (———) of the system $La_{1-x}Ca_xMnO_3$ in a semiempirical phase diagram given by GOODENOUGH. [Phys. Rev. **100**, 564 (1955).]

analysis is complicated by the ceramic nature of the samples where conducting grains are separated by insulating layers.

The magnetic, electric and crystallographic properties of perovskites have suggested extensive studies of spin structures and superexchange mechanism. WOLLAN and KOEHLER[308] have made neutron diffraction investigations of spin-structures with varying Mn^{4+} concentration. A few semiempirical rules for the sign of the 180°-superexchange and its variation with interatomic distance are sufficient for explanation of the whole variety of observed ferromagnetic and

---

[307] J. VOLGER: Physica **20**, 49 (1954); more recently the transport properties have been investigated extensively and interpreted by small polaron hopping: P. GERTHSEN u. K. H. HÄRDTL: Z. Naturforsch. **17**a, 514 (1962); H. G. REIK, E. KAUER, and P. GERTHSEN: Physics Letters **8**, 29 (1964); K. H. HÄRDTL, and P. GERTHSEN: Solid State Comm. **3**, 283 (1965); R. R. HEIKES, R. C. MILLER, and R. MAZELSKY: Physica **30**, 1600 (1964).

[308] E. O. WOLLAN, and W. C. KOEHLER: Phys. Rev. **100**, 545 (1955); J. Phys. Chem. Solids **6**, 287 (1958).

antiferromagnetic spin structures indicated in Fig. 30. GOODENOUGH[309] has given a qualitative theory on magnetic exchange via covalent bonds which depends strongly on distances. This type of 180°-superexchange is ferromagnetic for large lattice parameters and becomes negative for small dimensions. Generalization has led to the Goodenough-Kanamori rules, (see Sect. 30) and to the realization that perovskite compounds are ideal materials for investigation of details in the superexchange mechanism. Ferromagnetic exchange interactions, such as the Mn^{3+}—Mn^{3+}, Mn^{3+}—Mn^{4+}, Cr^{3+}—Mn^{3+}, Co^{4+}—Co^{3+}, Co^{3+}—Mn^{4+} interactions, are found to be in accordance with these semiempirical rules of superexchange.

JONKER[310] has given experimental evidence that ferromagnetism in Mn- and Cr-perovskites can occur also between ions of equal valency in nonconducting compounds. Ferromagnetic double exchange makes small contributions only to interactions between ions of different valency and produces spin cantings not explainable by superexchange alone. Moreover, the occurrence of ferromagnetism at certain concentrations in solid solutions is found to be related to the removal of the orthorhombic lattice distortion in $LaMnO_3$ when divalent ions, of different size than La, adapt the Goldschmidt factor to values mandatory for undistorted cubic symmetry. Magnetic susceptibility measurements at high temperatures actually indicate that pure $LaMnO_3$ behaves like a ferromagnet with a Curie temperature of about 200 °K when the Jahn-Teller distortion is removed in a phase transformation around 700 °K.

In recent years extensive investigations have been made on numerous perovskite materials with a large variety of magnetic and nonmagnetic cations in solid solutions. Values for critical distances in the transition from localized magnetic $3d$-electrons to conducting band states have been derived and discussed, especially by GOODENOUGH. His recent publications[311] give good introductions to this complicated but very interesting field where, at present, a large amount of experimental data is interpreted by qualitative and semiquantitative concepts which have not yet yielded to more accurate mathematical treatment.

57. Magnetic interactions in RE-sesquioxides. The magnetic properties of mixed oxides which contain both, $3d$- and $4f$-elements, are so dominated by the strong exchange interactions of $3d$-ions that contributions of RE-ions to the coupling mechanism are negligible. Such contributions should be studied in *pure RE-sesquioxides*, but these materials appear to be rather trivial with respect to their magnetic interactions. Their moments have essentially free ion values and the magnetization temperature dependence follows the Curie-Weiss law with small, usually negative intersections with the temperature axis indicating the possibility of weak antiferromagnetic interactions[312]. Spin orderings, however, are restricted to very low temperatures[313] of the order of a few K°. Deviations of the

---

[309] J. B. GOODENOUGH: Phys. Rev. **100**, 564 (1955); J. B. GOODENOUGH, A. WORLD, A. J. ARNOTT, and N. MENYUK: Phys. Rev. **124**, 373 (1961); J. B. GOODENOUGH, and P. M. RACAH: J. Appl. Phys. **36**, 1031 (1965).

[310] G. H. JONKER: Physica **22**, 707 (1956); J. Appl. Phys. **37**, 1424 (1966).

[311] J. B. GOODENOUGH: Magnetism and the Chemical Bond, New York: Wiley-Interscience 1963; J. B. GOODENOUGH, and P. M. RACCAH: J. Appl. Phys. **36**, 1031 (1965); J. B. GOODENOUGH: J. Appl. Phys. **37**, 1415 (1966); **39**, 403 (1968), in Landolt-Börnstein, Vol. 9/2, Berlin: Springer 1962; Czech. J. Phys. B **17**, 304 (1967); Phys. Rev. **164**, 785 (1961).

[312] For references see several recent papers in RE-Research (Proc. of R.E. conf. publ. by Gordon and Breach, New York) and in Progr. in Science, Techn. of R.E. (book series publ. by Pergamon Press, Oxford).

[313] Spin structure investigations by neutron diffraction: W. C. KOEHLER, and E. O. WOLLAN: Acta Cryst. **6**, 741 (1953); Phys. Rev. **100**, 545 (1955); **110**, 37 (1958); J. phys. radium **20**, 180 (1959).

paramagnetic susceptibility from the Curie-Weiss law occur often and can be explained by crystal field splittings of the magnetic ground state, van Vleck paramagnetism etc. At temperatures below 5 °K metamagnetism has been found in Eu_3O_4 which contains Eu^{2+} and Eu^{3+} ions on the Ca- and Fe-sites of the $CaFe_2O_4$-structure, respectively. The magnetic properties have been interpreted by a model of ferromagnetic chains of Eu^{2+}-spins which are antiferromagnetically coupled with one another by dipole-interactions [314].

Most RE-sesquioxides have cubic Mn_2O_3-type structure (C-type) at room temperature with polymorphic transformations to hexagonal (A-type) and monoclinic structures (B-type) at higher temperatures for Gd and the lighter RE. They are good, optically transparent insulators with extremely high resistivities at room temperature. Electrical conductivity at high temperatures occurs by ion diffusion or electron hopping between different valent cations if those are available. The melting points of the sesquioxides lie around 2,500 °C.

The stability of the *sesqui compounds of the remaining chalcogens S, Se, Te* is lower corresponding to melting temperatures between 1,500 °C and 2,000 °C. The phase diagrams for chalcogen rich concentrations are rather complicated with several compounds of unknown structure and many polymorphic transformations [315]. The methods of preparation are not very reliable since the reaction with crucible materials or residual gases are difficult to avoid for such very reactive materials with inconveniently high melting temperatures. Electrical resistivities vary over many orders of magnitude depending on preparation and annealing conditions. Magnetic measurements are in most cases applied only in a small range around room temperature. Their results are close enough to expected free ion values to support chemical analysis of RE concentrations.

58. Magnetic interactions of 2:3 compounds with Th_3P_4-structure containing RE.

Of special interest are 2:3 compounds which crystallize in the cubic Th_3P_4-structure [316], because their electrical resistivity can be varied by many orders of magnitude, from insulating to metallic, within the same crystal structure [317]. In some cases the lattice parameter even remains constant.

The Th_3P_4-structure ($I\bar{4}3d$-T_d^6) occurs in 3:4 compounds of tetravalent actinides, with anions of Group V and for small trivalent RE-ions with chalcogens. The interesting properties result from the unfilled cation sites in the 2:3 composition. The number $4/3$ of vacant sites per unit cell is required in order to provide charge compensation between anions and cations in an insulating 2:3 compound. The vacant sites have a formal negative charge of three and must be distributed statistically over all cation sites, since no superstructure has been found in X-ray investigations. Each unit cell contains four molecules. The 12 cations have fixed

---

[314] L. Holmes, and M. Schieber: J. Appl. Phys. **37**, 968 (1965).

[315] Reviews: A. Gschneidner: R.E. Alloys, Princeton: D. van Nostrand Co. 1961; E. F. Westrum, in: Science and Techn. of R.E., Vol. 1, 2. London: Pergamon Press 1964, 1966; H. Weik, u. K. Strnat: Seltene Erden, in: Landolt-Börnstein, Vol. 4/2c, Berlin-Göttingen-Heidelberg-New York: Springer 1964.

[316] A. Benacerraf, and M. Guittard: Compt. Rend. **248**, 1672, 2012 (1959); M. Picon, L. Domange, J. Flahaut, M. Guittard et M. Patrie: Bull. Soc. Chim. France **2**, 221 (1960) with earlier literature; Compt. Rend. **257**, 1530 (1963).

[317] M. C. Picon et J. Flahaut: Compt. Rend. **242**, 1321 (1956); **243**, 1769, 2074 (1956); J. F. Miller, F. J. Reid, and R. C. Himes: J. Electrochem. Soc. **106**, 1043 (1959); **111**, 943 (1964), and several papers at RE-conferences; M. Cutler, R. L. Fitzpatrick, and J. F. Leary: J. Phys. Chem. Solids **24**, 2, 319 (1963); D. J. Haase, and H. Steinfink: J. Appl. Phys. **37**, 2246 (1966); **38**, 3490 (1967); R.E. Research, Vol. 4, p. 535, New York: Gordon and Breach 1964; J. R. Henderson, M. Nuramoto, E. Loh, and J. B. Gruber: J. Chem. Phys. **47**, 3347 (1967); R. C. Vickerey, and H. M. Muir: Nature **190**, 336 (1961).

positions, whereas the 16 anion sites can be shifted by a free parameter along cube diagonals[318]. The parameter varies the shape of the octaverticon formed by the 8 anions surrounding each cation, and shifts the anions through the octahedra formed by the cation coordination. It is interesting to note that for the experimental parameter values found in insulating Gd_2Se_3, the Se-anions have the same position relative to Gd-triangle in the octahedra as in the metallic NaCl-compound GdSe, which we will discuss later.

As a consequence of the large vacancy concentration a large number of additional cations (such as divalent alkaline earth ions, divalent or trivalent RE-ions or transition and nontransition metal ions) can be introduced into the lattice[319]. A very special property is that for trivalent RE-ions charge balance does not have to be maintained. The excess of cations supplies the material with donors for conduction electrons, resulting in a variation from semiconducting to metallic character. As a consequence, the Th_3P_4-structure has an unusually large homogenity range which extends for La and the large RE-ions from the metallic 3:4 compound[320] to the semiconducting 2:3 material, without observable variation in lattice constant. With decreasing RE-ion size and increasing anion size the stability region of the Th_3P_4-structure narrows. Dy_2S_3 has comparably small solubility for trivalent ions in the Th_3P_4-structure. Gd_2Te_3 is orthorhombic. The chemical bonding in the Th_3P_4-structure and its significance for the large resistivity variations is not too well understood[321]. Relations may exist to the conditions in metallic NaCl-type compounds of lanthanides and actinides, where excess electrons are promoted into a conduction band, consisting primarily of d-wave functions. The crystal structure provides in both cases the necessary symmetry and distances for overlap of $5d$ wave functions between RE-neighbors[322].

Special measures are required (such as excess chalcogen pressure in closed containers, annealing in chalcogen vapor etc.) for the preparation of nearly stoichiometric 2:3 compounds. Samples of Gd_2Se_3 with an electrical resistivity of the order of 10^6 ohm. cm and an optical absorption edge of about 2.2 eV have been prepared. Since $4f$-$5d$ transitions in Gd^{3+} ions are expected at much shorter wavelengths this edge is probably due to transitions from the valence into the conduction band. Compositions near Dy_2S_3 have the absorption edge near 3 eV and internal $4f$-transitions of Dy^{3+} span the red part of the spectrum in a ladderlike fashion[323]. The optical properties of Ce_2S_3 have been investigated by KURNICK and MEYER[324]. The gap energy is about 2.6 eV, the refraction index 2.73 and the static dielectric constant 19. Absorption edges in the neighborhood of 2 eV are

---

[318] K. MEISEL: Z. anorg. u. allgem. Chem. **240**, 300 (1939); W. H. ZACHARIASEN: Acta Cryst. **2**, 57 (1949); A. BENACERRAF et M. GUITTARD: Compt. Rend. **248**, 1672, 2012 (1959); P. I. KRIPYAKEVICH: Sov. Phys. Cryst. **7**, 556 (1963); S. METHFESSEL: Z. angew. Phys. **18**, 414 (1964); F. HOLTZBERG, and S. METHFESSEL: J. Appl. Phys. **37**, 1433 (1966).

[319] J. FLAHAUT, L. DOMANGE, and M. PATRIE: Bull. Soc. Chim. France (1962) 2048; S. M. GALABI, J. FLAHAUT, and L. DOMANGE: Compt. Rend. **259**, 820, 4039 (1964).

[320] La_3Se_4 and La_3S_4 are even superconductive below about 10 °K [R. M. BOZORTH, F. HOLTZBERG, and S. METHFESSEL: Phys. Rev. Letters **14**, 952 (1965); G. L. GUTHRIE, and R. L. PALMER: Phys. Rev. **141**, 346 (1966)]. F. HOLZBERG, P. E. SEIDEN, and S. VON MOLNAR: Phys. Rev. **168**, 408 (1968).

[321] F. L. CARTER: R.E. Research, Vol. 4, p. 495, New York: Gordon and Breach 1965; S. METHFESSEL: Z. angew. Phys. **18**, 414 (1965).

[322] J. W. McCLURE: Phys. Chem. Solids **24**, 871 (1963); S. METHFESSEL: Z. angew. Phys. **18**, 414 (1965).

[323] J. R. HENDERSON, M. MURAMOTO, E. LOH, and J. B. GRUBER: J. Chem. Phys. **47**, 3347 (1967).

[324] S. W. KURNICK, and C. MEYER: J. Phys. Chem. Solids **25**, 115 (1964).

in approximate agreement with values of the forbidden band gap derived from electrical measurements [325]. The nephelauxetic effects (Sect. 40) in RE-sesquisulfide powders have been discussed by JORGENSEN et al. [326]. No magnetic ordering was observed in most insulating 2:3 compounds [327]. Paramagnetic susceptibilities extrapolate to negative temperatures of the order of 10 °K. Gd_2Se_3 becomes antiferromagnetic [328] below a Néel temperature of 6 °K. Insulating 3:4 compounds with Eu^{2+} ions filling the cation vacancies of the Th_3P_4-structure or with orthorhombic $CaFe_2O_4$-type structure show indications of weakly ferromagnetic Eu^{2+}-Eu^{2+} interactions [329].

Metallic compounds, such as $Gd_{2.5}Se_4$, have electrical resistivities of the order of 10^{-3} to 10^{-4} ohm. cm, increasing with increasing temperature. Conduction electron concentrations as large as 10^{22} cm^{-3} with mobilities between 1 and 100 cm^2/V. sec are observed in materials close to the 3:4 composition [330]. The electrical, thermoelectrical and optical properties of Ce-sesquisulfides and their variation with Ce-concentration has been investigated in more detail [331]. In Ce_3S_4 the 6×10^{21} cm^{-3} conduction eletrons with rather large mobility have semimetallic character and their resistivity and Seebeck effect between 10 and 1,000 °K can be described by the equations of conventional transport theory. Deviation from the 3:4 stoichiometry introduces large effective charges at the vacant cation sites. The resulting fluctuation in the lattice potential influence increasingly the transport properties as screening decreases with decreasing electron concentration. For compositions close to the 2:3 ratio large cross sections per vacancy and large screening distances of several lattice constants result in anomalous transport properties, which can be described by hopping processes with activation energies of the order of some 10^{-2} eV, and mean free paths shorter than interatomic distances. However, the quantitative interpretation is seriously restricted not only by the complicated electron band structure [332] but also by the possibility of impurity effects. The interesting fact, that electrical conductivity in Th_3P_4-compounds of Gd can produce ferromagnetic interactions of magnitude up to 100 °K will be discussed in Sect. 63.

59. 1:1 RE-chalcogen compounds. The 1:1 chalcogen-compounds have the advantage of simplicity in the NaCl-type structure. The chalcogenides with

---

[325] V. I. MARCHENKO, and G. V. SAMSONOV: Chemical Bonds in Semicond. and Solids (Ed. N. N. SIROTA), New York: Consultants Bureau 1967.

[326] C. K. JORGENSEN, R. PAPPALARDO, and J. FLAHAUT: H. chim. Phys. **62**, 444 (1965), see also Sect. 40.

[327] R. C. VICKERY, and H. M. MUIR: R.E. Research (Ed. E. V. KLEBER), p. 223, New York: Macmillan 1961; G. BUSCH, P. JUNOD, M. RISI, and O. VOGT: Conf. on Phys. of Semicond., Exeter 1962, p. 727; W. KLEMM u. A. KOCZY: Z. anorg. Chem. **233**, 84 (1937).

[328] F. HOLTZBERG, T. R. McGUIRE, S. METHFESSEL, and J. C. SUITS: J. Appl. Phys. **35**, 1033 (1964).

[329] F. HULLIGER, and O. VOGT: Phys. Letters **17**, 238 (1965); **21**, 138 (1966); Helv. Phys. Acta **34**, 199 (1966).

[330] J. F. MILLER, F. J. REID, and R. C. HIMES: J. Electrochem. Soc. **106**, 1043 (1959); **111**, 943 (1964); M. C. PICON et J. FLAHAUT: Compt. rend. **243**, 1210 (1956); M. C. PICON, L. DOMANGE, J. FLAHAUT, M. GUITTARD et M. PATRIE: Bull. Soc. Chim. **2**, 221 (1960). F. L. CARTER, R. C. MILLER, and F. M. RYAN: Adv. Energy Conv. **1**, 165 (1961); D. J. HAASE, and H. STEINFINK: J. Appl. Phys. **36**, 3490 (1965); **37**, 2246 (1966).

[331] S. W. KURNICK, R. L. FITZPATRICK, and M. F. MERRIAM: R.E. Research (Ed. J. F. NACHMAN, and C. E. LUNDIN), p. 249, New York: Gordon and Breach 1962; J. APPEL, and S. W. KURNICK: J. Appl. Phys. **32**, 2206 (1961); M. CUTLER, J. F. LEAVY, and R. L. FITZPATRICK: Phys. Rev. **133**, A 1143, A 1153 (1964); J. Phys. Chem. Solids **24**, 319 (1963); S. W. KURNICK, and C. MEYER: J. Phys. Chem. Solids **25**, 115 (1964).

[332] G. F. KARAVAEV, N. Y. KUDRYAVTSEVA 1 V. A. CHALDYSHEV. Soviet Phys. Solid State **4**, 2540 (1963).

RE^{3+} ions form metallic compounds with electrical resistivities[333] comparable to pure RE-metals ($\sim 10^{-4}$ ohm. cm), with mobilities of the order of 1 cm^2/V. sec. The one valence electron in excess of the divalent ionic bonding is promoted to the conduction band and contributes there to the total stability by metallic bonding. The addition of metallic bonding increases the melting points to quite high values in the neighborhood of 2,500 °C. NaCl-compounds with excess conduction electrons are also found with 3d-elements, such as TiO, VO, and with 1:1 actinides-chalcogenides (including 1:1 oxides). It is questionable, if metallic 1:1 *oxides* with RE^{3+} ions exist as homogeneous materials.

It is tempting to make the partial occupation of d-orbitals by conduction electrons responsible for the stability of metallic NaCl-type compounds. The low energy crystal field levels with t_{2g}-symmetry overlap well with the 12 next cation neighbors in the NaCl-type structure and can form a metallic bond superimposed on the covalent bond in the directions of the 6 neighboring anions. On the other hand, one has to realize that the kinetic widening of the large s-orbitals by overlap is probably larger (~ 10 eV) than the energy gain of the $5d$-t_{2g} orbitals in the crystal field and that, therefore, the bottom of the conduction band may be expected to have essentially s-character. The golden color of many metallic 1:1 RE^{3+}-chalcogenides results from a gap of small optical absorption for blue light between the intrinsic absorption edge for valence-conduction electron excitations at high energies and the conduction electron absorption in the red part of the spectrum. Magnetic susceptibility measurements at room temperature[334] give the normal ground state of the free RE^{3+} ions. At low temperatures antiferromagnetic order occurs as qualitatively expected for indirect exchange via an appropriate number of free carriers[335]. GdSe, GdS, as examples, have Néel temperatures of about 50 °K and a paramagnetic Curie-Weiss temperature of -60 °K. The magnetic properties of the lighter monochalcogenides are determined by the competition between exchange and crystal field effects. Only RE-ions with odd numbers of $4f$-electrons exhibit antiferromagnetic order below about 10 °K[336].

Insulating monochalcogenides are formed with RE^{2+} ions, such as Sm^{2+}, Eu^{2+}, Yb^{2+}. The discovery of ferromagnetism of EuO below 70 °K, several years ago[337], came as quite a surprise. Insulating compounds with ferromagnetic order are unusual materials, in any case. Whereas ferromagnetic superexchange interactions occur often in 3d-compounds, following the Goodenough-Kanamori-rules, their effect on magnetic order is in most materials annihilated by stronger antiferro-

---

[333] E. D. Eastman, L. Brewer, L. A. Bromley, P. W. Gilles, and N. L. Lofgreen: J. Am. Chem. Soc. **72**, 4019 (1950); R. Didchenko, and F. P. Gortsema: J. Phys. Chem. Solids **24**, 863 (1963); J. F. Miller, F. J. Reid, and R. C. Himes: J. Electrochem. Soc. **106**, 1043 (1959); J. Phys. Chem. Solids **25**, 969 (1964); V. P. Zhuze, A. V. Golubkov, E. V. Goncharova i V. M. Sergeeva: Soviet Phys. — Solid State **6**, 205, 213, 343 (1965).

[334] W. Klemm u. H. Senff: Z. anorg. Chem. **241**, 259 (1939); M. C. Picon et M. Patric: Compt. rend. **242**, 1321 (1956); A. Benacerraf, L. Domange et J. Flahaut: Compt. rend. **248**, 1672 (1959); L. Domange, J. Flahaut et M. Guittard: Compt. rend. **249**, 697 (1959); A. Iandelli: R.E. Research, p. 135, New York: Macmillan 1961; J. W. McClure: Phys. Chem. Solids **24**, 871 (1963).

[335] See part D for theory, experimental details are found in: F. Holtzberg, T. R. McGuire, S. Methfessel, and J. C. Suits: Phys. Rev. Letters **13**, 18 (1964); Proc. Internat. Conf. Magnetism Nottingham 1964; S. Methfessel: Z. angew. Phys. **18**, 414 (1965); Relationship between antiferromagnetic ordering and electron-concentration in metallic CsCl-type compounds of Tb with Pd, Ag, In have been investigated by J. W. Cable, W. C. Koehler, and H. R. Child: J. Appl. Phys. **36**, 1096 (1965).

[336] G. A. Smolenskii, V. E. Adamyan, and G. M. Loginov: Phys. Letters **23**, 16 (1966); J. Appl. Phys. **39**, 786 (1968); Phys. Stat. Solidi **18**, 873 (1966).

[337] B. T. Matthias, R. M. Bozorth, and J. H. van Vleck: Phys. Rev. Letters **7**, 160 (1961).

magnetic 180° interactions. An extrapolation from properties of other RE-materials would not lead one to expect the possibility of relatively strong ferromagnetic interactions in an insulating RE-compound. Application of rules of superexchange to half-filled 4f-orbitals gives antiferromagnetic cation-anion-cation and cation-cation interactions.

The remaining NaCl-type chalcogenides EuS, EuSe and EuTe had also been in literature[338] as paramagnetic materials, but then, after discovery of ferromagnetism in EuO, they were investigated immediately in several laboratories[339] for the possible existence of magnetic order at low temperatures. The results have been discussed in a review article by McGuire and Shafer[340], and, more

Table 3. *Some properties of Eu-chalcogenides.*

	Lattice constant (Å)	Curie-Weiss Temp. (°K)	Magnetic order	Ordering Temp. (°K)	Exchange param. J_1/k (°K)	Exchange param. J_2/k (°K)	Moment at $T=0$ in μ_B per Eu^{2+}	Sat. magnetization $4\pi M$ (gauss)	Opt. absorption edge (eV) at R.T.	Absorption peak (Å) at R.T.	Faraday-rotation (deg/cm) in 20 Koe at 4.2 °K
EuO	5.15	76	ferro	69.4	0.58 to 0.67	−0.07	6.8	∼24,000	1.12	6,000	520,000 at $\lambda=7,000$Å
EuS	5.96	19	ferro	16.5	0.2	−0.06 to −0.14	6.87	∼21,000	1.64	5,200	?
EuSe	6.19	9	anti-ferro ferro	4.6 2.8	$J_1 \sim J_2$	$J_1 \sim J_2$	6.7	∼17,000	1.85	4,800	95,000 at $\lambda=8,000$Å
EuTe	6.60	−6	anti-ferro	7.8 to 11	∼0.03	−0.12 to −0.17	—	—	2.0	4,650	?

recently, by Busch[341]. Table 3 compiles some properties of Eu-chalcogenides. The ferromagnetism of EuO has been confirmed by neutron diffraction[342]. The ferromagnetic Curie temperature of EuS has been determined by specific heat measurements[343]. EuSe has a quite complex spin structure immediately below the ordering temperature in zero field, but becomes ferromagnetic in applied fields[344]. As a consequence the magnetization-field dependence shows a pronounced step between 1 to 3 koe. Magnetization measurements on single crystals

---

[338] W. Klemm u. G. Winkelman: Z. anorg. Chem. **288**, 87 (1956).

[339] T. R. McGuire, B. E. Argyle, M. W. Shafer, and J. S. Smart: Appl. Phys. Letters **1**, 17 (1962); J. Appl. Phys. **34**, 1345 (1963); G. Busch, P. Junod, M. Risi, and O. Vogt: Proc. Int. Conf. on Semicond., Exeter 1962, p. 727; S. van Houten: Phys. Rev. Letters **2**, 215 (1962); U. Enz, J. F. Fast, S. van Houten, and J. Smith: Philips Research Repts. **17**, 1451 (1962); R. L. Wild, and R. D. Archer: Bull. Am. Phys. Soc. **7**, 440 (1962); R. Didchenko, and F. P. Gortsema: J. Phys. Chem. Solids **24**, 863 (1963).

[340] T. R. McGuire, and M. W. Shafer: J. Appl. Phys. **35**, 984 (1964).

[341] G. Busch: J. Appl. Phys. **38**, 1386 (1967).

[342] N. G. Nerenson, V. E. Olsen, and G. P. Arnold: Phys. Rev. **127**, 2101 (1962).

[343] V. L. Moruzzi, D. T. Teaney: Solid State Comm. **1**, 127 (1963); G. Busch, P. Junod, R. G. Morris, and J. Muheim: Helv. Phys. Acta **37**, 637 (1964).

[344] S. J. Pickart, and H. A. Alperin: Bull. Am. Phys. Soc. **10**, 32 (1965); R. J. Joenk: Bull. Am. Phys. Soc. **11**, 109 (1966); two papers with more detailed magnetic and neutron diffraction investigations by T. R. McGuire, F. Holtzberg, R. J. Joenk, and S. J. Pickart: H. A. Alperin will be publ. in J. Phys. Chem. Solids **29** (1968).

together with specific heat and optical measurements by Busch and collaborators[345] indicate a Néel temperature of 4.6 °K and a ferromagnetic ordering temperature at 2.8 °K in zero magnetic field. The anisotropy field is about 100 oe only. EuTe is antiferromagnetic and the neutron diffraction[346] shows below 7.8 °K a spin order of the second type as found in MnO also. The Néel temperature of 9.8 °K has been derived from the peak in the temperature variation of specific heat[347].

The values for the magnetic transition temperatures depend, of course, on the experimental procedure they are derived from, but also on sample preparation. Deviations from stoichiometry in favor of higher Eu-concentrations increase the magnetic Curie temperature[348]. In EuO single crystals, grown with excess of RE^{3+}-metal, Curie temperatures up to 135 °K have been found[349]. The increase is probably due to interactions via the conduction electrons to be discussed in Sect. 61. The low saturation moment of 6.8 instead of 7 μ_B may indicate Eu^{2+} deficiency. EuO-crystals with a saturation moment of 7 μ_B have a ferromagnetic Curie temperature of 69.4 °K. The paramagnetic Curie-Weiss temperature of EuS can be found to be as high as 26 °K compared to the standard value of 19 °K given in Table 3.

It is interesting to note, that EuO has a saturation magnetization and magneto-optical Faraday-rotation comparable to values found in metallic iron. Therefore, the material is attractive for use in magneto-optical devices operating at cryogenic temperatures, as has been discussed in part F of this article. The low anisotropies and coercive forces of the order of 1—2 oe make the Eu-chalcogenides useful for ultrahigh frequency transformers at very low temperatures[350], where the permeability of most ferrites is blocked by high anisotropies.

The values for the optical absorption edges in Table 3 have some uncertainty because it is difficult to define them for the exponential absorption edges shown in part F, Fig. 27. The given numbers indicate for which photon energies the material becomes practically transparent. All 1:1 Eu-chalcogenides are, at room temperature, good insulators with resistivities of the order of 10^8 ohm. cm or more.

The ferromagnetism of Eu-chalcogenides has attractive features from, at least two points of view. First, the magnetic properties of nearly "ideal" Heisenberg type ferromagnets with short range interactions can be investigated experimentally and compared with predictions of theory. Second, they satisfy conditions for surprisingly large ferromagnetic exchange interactions between RE-ions even in the absence of any conduction electrons.

The theory of Heisenberg type magnets[351] has been used extensively to analyze the magnetic properties of insulating Eu-chalcogenides. For sake of simplicity only two exchange parameters have been used, and exchange interactions over longer distances are neglected. The parameter J_1 describes inter-

---

[345] P. Schwob, and O. Vogt: Phys. Letters **22**, 374 (1966); G. Busch, P. Junod, R. G. Morris, and J. Muheim: Helv. Phys. Acta **37**, 637 (1964); G. Busch: J. Appl. Phys. **38**, 1386 (1967); H. W. White, D. C. McCollum, and J. Callaway: Phys. Letters **25 A**, 388 (1967).

[346] G. Will, S. J. Pickart, H. A. Alperin, and R. Nathans: J. Phys. Chem. Solids **24**, 1679 (1963).

[347] G. Busch, P. Junod, R. G. Morris, and J. Muheim: Helv. Phys. Acta **37**, 637 (1964).

[348] M. W. Shafer: J. Appl. Phys. **36**, 1145 (1965).

[349] M. W. Shafer, and T. R. McGuire: J. Appl. Phys., **39**, 588 (1968).

[350] S. Tansal, and H. Sobol: Rev. Sci. Instr. **34**, 1075 (1963); S. Methfessel: IEEE—Trans. Magn. **1**, 144 (1965).

[351] J. S. Smart: Effective Field Theories of Magnetism, Philadelphia: W. B. Saunders Co. 1966; Magnetism, Vol. 3 (Ed. G. T. Rado, and H. Suhl), New York: Academic Press 1963.

actions of each Eu-ion with its 12 nearest Eu-neighbors and J_2 connects each Eu-ion with its 6 next nearest neighbors via anions. Fig. 31 shows, as an example, the analysis of the magnetic transition temperatures in the molecular field approach[352]. For antiferromagnetic EuTe, both J_1 and J_2 can be derived from the paramagnetic Curie-Weiss temperature and the Néel temperature. For the

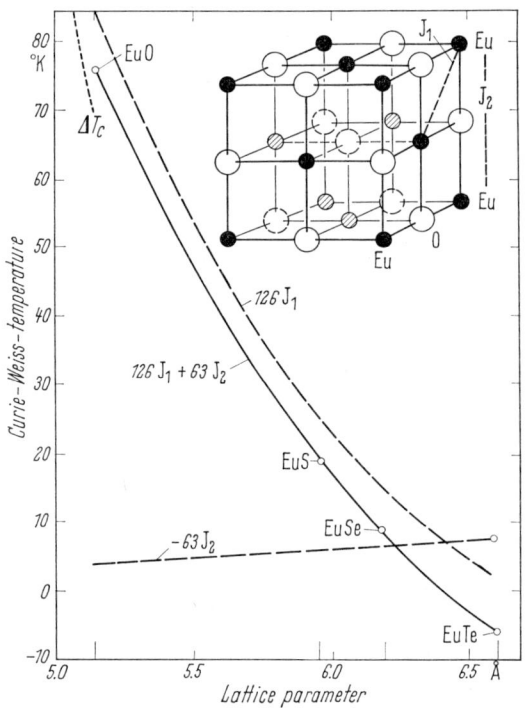

Fig. 31. Comparison of measured paramagnetic Curie-Weiss temperatures with molecular field constants of Table 3. The resulting distance dependence is weaker than expected from high-pressure experiments (------). (From McGuire et al. in Ref. [352].)

second type spin order in EuTe with $z_1 = 12$ nearest and $z_2 = 6$ next nearest Eu-neighbors to each Eu^{2+}-ion in the NaCl-type lattice one obtains for $S = \frac{7}{2}$:

$$k\Theta = \frac{2}{3} S(S+1) \sum z_i J_i = 126 J_1 + 63 J_2, \tag{59.1}$$

$$k T_N = \frac{2}{3} S(S+1)(-z_2 J_2) = -63 J_2. \tag{59.2}$$

For ferromagnetic materials the paramagnetic Curie-Weiss temperature and the ferromagnetic Curie temperature are nearly equal in molecular field theory, and only one parameter can be determined. Since J_2 results probably from a 180°-superexchange interaction via anions, one can assume from comparison with isostructural Mn-chalcogenides, that superexchange may decreases from telluride to oxide by a factor of about two. The large ferromagnetic Curie temperature of EuO is then a consequence of a strong positive nearest neighbor interaction, which exists also in Mn-chalcogenides. There, however, J_1 is negative and not positive as in the Eu-chalcogenides, but it decreases also with increasing cation

---

[352] T. R. McGuire, B. E. Argyle, M. W. Shafer, and J. S. Smart: J. Appl. Phys. 34, 1345 (1963); U. Enz, J. F. Fast, S. van Houten, and J. Smith. J. Appl. Phys. 34, 1257 (1963).

distances. As zeroth approximation one may try the assumption that J_1 depends mainly on interatomic distances which determine the overlap of the coupling wave functions. In this case, one can expect that the pressure dependence of the Curie temperature gives the same distance dependence of the exchange parameters as Fig. 31.

The variation of T_c with pressure[353] as well as the volume variations in neighborhood of the Curie temperature[354] have been measured in EuO and EuS. The dependence of Curie temperature on volume can be derived when the elastic properties are known. One finds $d \ln T_c/d \ln V = -6$ and -5.5 for EuO and EuS, respectively[355]. A power law with the exponent $(-10/3)$ is usually found for the volume dependence of superexchange interactions in insulating materials such as garnets, $3d$-oxides etc.[356]. Since magnetic measurements at high pressure as well as the determination of elastic properties on polycrystalline samples are quite difficult, the reliability of the results should not be overestimated. Fig. 31 includes the distance dependence of the Curie temperature in EuO, measured by McWHAN et al. The exchange parameter J_1 (which makes the largest contribution in EuO) shows a stronger distance dependence than expected from its variation through the compound series. Obviously, the anions influence the cation-cation exchange J_1 not only by their size. This is in qualitative agreement with our discussion of optical properties in Sects. 44, 45. There, the concentration dependence of the optical absorption edge indicated that the overlap of the exited states with neighboring Eu-ions depends on the dielectric constant which determines the effective orbital radii. Since dielectric constant and anion polarizability increase from EuO to EuTe it is reasonable to assume that the increasing cation-cation distances in the NaCl-lattice are partially compensated by increasing effective orbital radii of the overlapping electrons. In any case, the contribution of anion wave functions to J_1 must be small compared to the direct Eu-Eu-interactions. Stronger involvement in the J_1-exchange would result in 90° superexchange, which has, compared to experimental results, the wrong dependence on anion character.

Values for the exchange parameters have been derived also from molecular field analysis of other magnetic properties, such as temperature dependence[357] and field dependence[358] of magnetization, and specific heat[359]. Some typical results have been included into Table 3. A detailed analysis of magnetization data for EuS with application of spin wave theory has been given by CHARAP and BOYD[360].

---

[353] R. STEVENSON, and M. C. ROBINSON: Can. J. Phys. **43**, 1744 (1965); D. B. McWHAN, P. C. SOUERS, and G. JURA: Phys. Rev. **143**, 385 (1966); G. K. SOKOLOVA, K. M. DEMCHUK, K. P. RADIONOV i A. A. SAMOKHALOV: Soviet Phys. JETP **22**, 317 (1966); EuS: P. SCHWOB, and O. VOGT: Phys. Letters **24 A**, 242 (1967); In EuSe the change of J_1, J_2 and anisotropies with pressure changes the spin structure [G. BUSCH, P. SCHWOB, and O. VOGT: Phys. Letters **20**, 602 (1966)]. EuTe (and SmTe) transforms at about 25 kbar in an isostructural high pressure modification containing Eu^{3+} ions [C. J. M. ROOYMANS: Solid State Comm. **3**, 421 (1965); Ber. Bunsenges. phys. Chemie **70**, 1036 (1966)].

[354] B. E. ARGYLE, N. MIYATA, and T. D. SCHULTZ: Phys. Rev. **160**, 413 (1967).

[355] This gives $dJ_1/da_0 = -2.3\ k_B\ °K/\text{Å}$ for EuO.

[356] D. BLOCH: J. Phys. Chem. Solids **27**, 881 (1966).

[357] B. E. ARGYLE: J. Appl. Phys. **36**, 679 (1965); P. JUNOD, and F. LEVY: Phys. Letters **23**, 624 (1966).

[358] G. BUSCH, P. JUNOD, P. SCHWOB, O. VOGT, and F. HULLIGER: Phys. Letters **9**, 7 (1964); I. S. JACOBS, and S. D. SILVERSTEIN: Phys. Rev. Letters **13**, 272 (1964).

[359] V. L. MORUZZI, and D. T. TEANEY: Solid State Comm. **1**, 127 (1963); J. CALLAWAY, and D. C. McCOLLUM: Phys. Rev. Letters **9**, 376 (1962); Phys. Rev. **130**, 1741 (1963); **136**, A 426 (1964); P. J. WOJTOWICZ: J. Appl. Phys. **35**, 991 (1964); B. C. PASSENHEIM, D. C. McCOLLUM, and J. CALLAWAY: Phys. Letters **23**, 634 (1966). H. W. WHITE, D. C. McCOLLUM, and J. CALLAWAY: Phys. Letters **25 A**, 388 (1967).

[360] S. H. CHARAP, and E. L. BOYD: Phys. Rev. **133**, A 811 (1964).

Calhoun and Overmeyer[361] have found in paramagnetic resonance spectra ferromagnetic nearest neighbor exchange interactions between Eu-ions diluted in CaO and SrO. Ferromagnetic and paramagnetic resonance on EuO and EuS single crystals[362] give a spin only state with $g=2$, line width of about 100 to 1,000 oe at 1.5 and 300 °K, respectively. The crystal anisotropy field of the magnetization is $K/M = -190$ oe and smaller than 30 oe for EuO and EuS, respectively. Magnetoelastic and magnetocrystalline properties of EuO single crystals have been extensively investigated by B. E. Argyle et al.[363] and compared with predictions of relevant theories[364]. Irregular variation of crystal anisotropy and crystal field splittings for various anions suggest covalent contributions, for which indications are also found in nuclear resonance[365] and Mössbauer studies[366]. The isomer shift[367] appears to decrease with increasing covalency and indicates that the total s-electron density at the Eu^{151}-nucleus decreases from EuO to EuTe. As a possible explanation, the partial population of $5d$-orbitals in Eu^{2+} by electrons from the ligands has been suggested[368]. The $5d$-electrons screen then the nucleus for the potential acting on the s-electrons.

Molecular field theory gives magnitude and direction of the exchange interactions J_1 and J_2 but leaves open their detailed mechanism. As we have mentioned already, J_2 has probably the character of 180°-superexchange via p-electrons at anion sites, whereas the distance dependence of J_1 supports the assumption of predominantly cation-cation interaction for this exchange parameter. With respect to J_2, the question remains open (as usual in RE-compounds) as to which cation wave functions provide the spin dependent overlap with the anion p-electrons. It can be a direct covalency of the $4f$-electrons, which is usually considered to be negligible. The closed $5s$ and $6p$ shells can be partially polarized[369] and also the covalent mixing of the empty $5d$ orbitals[370] with the p-orbitals at the anions can be spin dependent due to intraatomic $4f$-$5d$-exchange. In general, one has to consider all possible excited states which are sufficiently extended in the lattice to permit partial or complete transfer of magnetic electrons between cations directly or indirectly over interjacent anions. The magnitude of the exchange parameter depends on the value of a transfer integral and the excitation energy (which are derived from the nature of the involved electron states) and, of course, on the magnetic ground state[371].

---

[361] B. A. Calhoun, and J. Overmeyer: J. Appl. Phys. **35**, 989 (1964).
[362] J. F. Dillon, and C. E. Olsen: Phys. Rev. **135**, A 434 (1964); S. von Molnar, and A. W. Lawson: Phys. Rev. **139**, A 1598 (1965).
[363] N. Miyata, and B. E. Argyle: Phys. Rev. **157**, 448 (1967); B. E. Argyle, N. Miyata, and T. D. Schultz: Phys. Rev. **160**, 413 (1967); EuTe: D. S. Rodbell, L. M. Osika, and P. E. Lawrence: J. Appl. Phys. **36**, 666 (1965).
[364] Single-ion molecular field theory: W. P. Wolf: Phys. Rev. **108**, 1152 (1957); Cluster-theory: E. Callen, and H. B. Callen: Phys. Rev. **129**, 578 (1963); **136**, A 1675 (1964); **139**, A 455 (1965).
[365] J. Overmeyer, and R. J. Gambino: Phys. Letters **9**, 108 (1964).
[366] P. Brix, S. Hufner, P. Kienle, and D. Quitman: Phys. Letters **13**, 140 (1964); H. H. Wickman, I. Nowick, J. H. Wernick, D. A. Shirley, and R. B. Frankel: J. Appl. Phys. **37**, 1246 (1966); F. A. Deeney, J. A. Delaney, and V. P. Ruddy: Phys. Letters **25** A, 370 (1967).
[367] A. J. Freeman, and R. E. Watson: Magnetism, Vol. 2A, p. 214 (Ed. G. T. Rado, and H. Suhl). New York: Academic Press 1965.
[368] J. Danon, and A. M. de Graf: J. Phys. Chem. Solids **27**, 1953 (1966).
[369] R. E. Watson, and A. J. Freeman: Phys. Rev. Letters **6**, 277, 388 (1961); J. Phys. Soc. Japan **17**, B 1, 15 (1962); Phys. Rev. **127**, 2058 (1962).
[370] J. B. Goodenough: Magnetism and the Chemical Bond. New York: Whiley-Interscience 1963; S. Methfessel: Z. angew. Phys. **18**, 414 (1965).
[371] For review of simple mechanisms of exchange interactions see J. R. Goodenough: Magnetism and Chemical Bond. New York: J. Wiley 1963.

Of special interest are mechanisms for the ferromagnetic exchange J_1 and several qualitative suggestions have been made (as discussed in section D). A certain preference in the selection of the coupling excited states derives from the nature of the electronic bandstructure one accepts for the Eu-chalcogenides (see discussions in part C and F). When the $4f$-levels are assumed to lie below the valence band (as in Fig. 6) indirect exchange via polarized valence electrons of the Bloembergen-Rowland type (section D) is the preferred mechanism. De Graaf et al.[372] describe the exchange parameters in EuS by interaction via valence electrons of diamagnetic anions which can be excited over a 5.2 eV wide band gap into the conduction band. An exchange interaction of about 0.36 eV is needed between the localized $4f$ electrons and the p-electrons at S^{2-}. This interaction is about three times larger than the $4f$-$5d$ interaction in free Eu-ions[373] and much too large to be acceptable.

When the $4f$-levels lie between the conduction band and the valence band (as in Fig. 24) coupling via d or s electrons is more obvious. Goodenough[374] has suggested for J_1 cation-cation interaction, which is quite abundantly found in insulating compounds where ligand octahedra touch one another along edges. This symmetry favors the overlap of d-wave functions with t_{2g} character at neighboring cation sites. In the NaCl-type lattice of the Eu-chalcogenides the overlap between $5d$-t_{2g} functions from the 12 nearest cation neighbors is quite large; but in purely ionic compounds the $5d$-states are unoccupied. Mixing between $4f$ and $5d$-states are parity forbidden in unperturbed ions. Therefore, Goodenough assumes mixing and overlap between $4f$ and $5d$-orbitals at different cations which leads for the case of overlap between half filled and empty orbitals in third-order perturbation to an effective ferromagnetic exchange parameter of the magnitude[374]:

$$J_1 = +2 b^2 J_{fd}/4 S^2 U^2. \qquad (59.3)$$

Here is b the transfer integral connecting the $4f$ and $5d$ functions at neighboring cations, J_{fd} the intraatomic $4f$-$5d$ exchange, S the total $4f$-spin and U the excitation energy. The formula is in qualitative agreement with the variation of J_1 through the chalcogenide series when U is derived from optical absorption measurements[375]. Smit[376] has considered the possibility of virtual excitation of $4f$-electrons to $5d$-orbitals at the same atom by zero point lattice vibrations as a possible cause of ferromagnetic cation-cation interaction. The result is similar to Goodenough's, when the intraatomic exchange is replaced by the matrix element which provides the $4f$-$5d$ excitation, and gives the right magnitude for J_1 in EuO. However, one immediate consequence of a mechanism involving lattice vibrations is that the Curie temperature should show an isotope effect and that J_1 should increase with temperature. The constant Θ in the Curie-Weiss law of paramagnetic susceptibility should begin to decrease with temperature in the neighborhood of 700 °K. Such effects have *not* been observed. As a third possibility the population of overlapping $5d$-orbitals by p-electrons from anions could be considered for the origin of J_1 as well as of J_2. But then J_1 should increase with increasing anion polarizability from

---

[372] A. M. de Graaf u. S. Strässler: Phys. Kondens. Materie **1**, 13 (1963); A. M. de Graaf, and R. M. Xaver: Phys. Letters **18**, 225 (1965); J. Danon, and A. M. de Graaf: J. Phys. Chem. Solids **27**, 1953 (1966); Indirect exchange via filled bands in insulators has been discussed also by J. Callaway: Nuovo cimento **26**, 625 (1962).

[373] For numbers see discussion of optical properties of chalcogenides in Sect. 48, reference 264.

[374] J. B. Goodenough: Magnetism and Chemical Bond, p. 149, New York: J. Wiley 1963; S. Methfessel: Z. angew. Phys. **18**, 414 (1965).

[375] S. Methfessel, F. Holtzberg, and T. R. McGuire: IEEE Trans. Magn. **2**, 305 (1966).

[376] J. Smit: J. Appl. Phys. **37**, 1455 (1966), and in Atomist. Approach to Nature and Prop. of Materials (J. A. Pask), p. 287, New York: J. Wiley 1967.

EuO to EuTe as J_2 does. However, mixing of magnetic $4f$-electrons into the empty conduction band states can probably give a type of "virtual" Ruderman-Kittel exchange which has the right properties[377].

The magnetic and electrical properties of *actinide chalcogenides and pnictides* have been reviewed recently[378] and discussed in terms of indirect exchange via conduction electrons. As expected from the lesser localization of $5f$ electrons compared to $4f$ electrons the ferromagnetic Curie temperatures are higher (US: $\sim$180 °K, USe: 188 °K, UTe: 104 °K) than in Eu-chalcogenides (EuO $\sim$ 70 °K; EuS: 16 °K).

60. Magnetic interactions in RE-hydrides. In this concluding section on magnetic interactions we mention the possibility of depleting the RE-metals of their conduction electrons by *hydrogenation*. The RE-elements react exothermally and under volume expansion with H_2. The reaction energies are comparable to those found in hydrides of alkali and alkaline earth elements. The phase diagrams are rather complicated with certain similarities to the Pd-H system. Several compounds are formed without well defined compositions. The hydrogen goes as negative ion on lattice interstices and reduces the number of conduction electrons. The face centered cubic dihydrides have nearly temperature independent resistivity, the hexagonal trihydrides are semiconductors[379]. Most dyhydrides are antiferromagnetic with Néel temperatures of the order of 10 to 100 °K. The magnetic properties[380] are rather complicated in addition to the usual crystal field effects on the ground state. The insulating trihydrides give no indication of magnetic order down to 4.2 °K. Since crystal structures are different it cannot be decided clearly if this is connected to the insulating properties or results from the change in crystal symmetry. Nuclear resonance studies on light RE-hydrides have been discussed in terms of Ruderman-Kittel interactions via $5d$-band with varying population[381]. However, the insulating cubic hydride EuH_2 is ferromagnetic below 24 °K without conduction electrons[382] and the coupling mechanism is not understood.

II. Magnetic interaction via semiconducting carriers.

61. Eu-chalcogenides with NaCl-structure. Magnetic interaction via semiconducting carriers has been indicated for several $3d$-compounds but can be identified most clearly in *Eu-chalcogenides*[383]. Their NaCl-type structure has relatively simple ion configurations and superexchange interactions are small. The solid solubility of the insulating Eu-chalcogenides with metallic rocksalt compounds of trivalent RE-elements makes it possible to vary the electrical conductivity by twelve orders of magnitude from insulator to metal in the same crystal structure. In contrast

---

[377] S. METHFESSEL: Z. angew. Phys. **18**, 414 (1965); IEEE Trans. Mag. **1**, 144 (1965).

[378] J. GRUNZWEIG, and M. KUZNIETZ: J. Appl. Phys., **39**, 905 (1968).

[379] G. E. STURDY, and R. N. R. MULFORD: J. Am. Chem. Soc. **78**, 1083 (1956); W. L. KORST, and J. C. WARF: Acta Cryst. **9**, 452 (1956); R. N. R. MULFORD, and C. E. HOLLEY: J. Phys. Chem. **59**, 1222 (1955); A. PEBLER, and W. E. WALLACE: J. Phys. Chem. **66**, 148 (1962); G. G. LIBOWITZ: Solid State Chem. of Binary Metal Hybrides, New York: W. A. Benjamin 1965; W. G. BOS, and K. H. GAYER: J. Nucl. Mater. **18**, 1 (1966). B. STALINSKI: Bull. acad. polon. sci. **5**, 997, 1001 (1957); **7**, 269 (1959).

[380] Y. KUBOTO, and W. E. WALLACE: J. Appl. Phys. **33**, 1348 (1962); **34**, 1348 (1963); J. Chem. Phys. **39**, 1285 (1963); D. E. COX, G. SHIRAN, and W. J. TAKEL: J. Appl. Phys. **3**, 1373 (1967).

[381] J. P. KOPP, and D. S. SCHREIBER: J. Appl. Phys. **38**, 1373 (1967).

[382] R. L. ZANOWICK, and W. E. WALLACE: Phys. Rev. **126**, 537 (1962).

[383] F. HOLTZBERG, T. R. MCGUIRE, S. METHFESSEL, and J. C. SUITS: Phys. Rev. Letters **13**, 18 (1964); Proc. Intern. Conf. Magnetism, Nottingham 1964, p. 470; S. METHFESSEL: Z. angew. Phys. **18**, 414 (1965); F. HOLTZBERG, T. R. MCGUIRE, and S. METHFESSEL: J. Appl. Phys. **37**, 976 (1966).

to many $3d$-compounds the number and ground state of magnetic electrons per Eu-ion does not change during the insulator-metal transition to a first approximation. In the special case of solid solutions between Eu- and Gd-compounds, the $4f^7$ configuration at cation sites even remains unchanged by Gd substitution.

Fig. 32 gives, as an example, the variation of the electrical resistivity at room temperature with Gd-concentration in the solution system $Eu_{1-x}Gd_xSe$. The Gd-concentration x can be taken as a crude measure of the number of electrons induced by excess of charge compensation between cations and anions. However, the actually measured number of conduction electrons has been found to be smaller than the chemical Gd-concentration[384]. For metallic samples with 50% or more GdSe in EuSe the discrepancy amounts to a factor of 2 to 3 only, but for single crystals containing 1% or 5% GdSe the true carrier concentrations (for 1%: 3.1×10^{18} cm^{-3}, for 5%: 3.9×10^{19} cm^{-3}) are smaller by orders of magnitude than the expected values of 1.7×10^{20} cm^{-3} and 8.4×10^{20} cm^{-3}, respectively (see table 6).

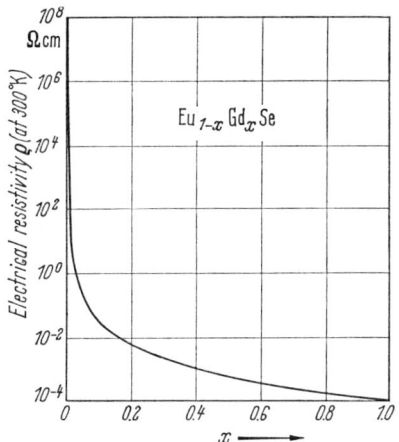

Fig. 32. Variation of electrical resistivity at room temperature (log-scale) in $Eu_{1-x}Gd_xSe$ with GdSe-concentration x. (From Ref. [377].)

The third valence electron of Gd (which remains after the p-shell of the surrounding anions is filled) can occupy, obviously, states of different character. It can be trapped by the higher core charge of the Gd-ion itself, or by lattice defects which have an effective positive charge, such as cation vacancies or anion interstitials. The strong binding to all such local deviations from stoichiometry generally renders the electrons unavailable for electrical transport and magnetic interactions around room temperature and below. The contributions to electrical conductivity must come from states with loosely bound electrons which can be excited into the conduction band at prevailing temperatures or can migrate from Gd^{2+} to Gd^{3+}-impurities.

The numerical ratio of conducting electrons to trapped electrons depends on electron concentration and conditions during crystal preparation and, for high temperatures, on the Boltzmann factor. When the stoichiometry is forced to the metallic side by crystal growth or annealing in closed containers with excess metal vapor, the obtained samples are more conductive than others after annealing in vacuum. The presence of conduction electrons at room temperature can be observed not only electrically but also optically by the related free carrier absorption for red and infrared light.

Fig. 33 shows the variation of the paramagnetic Curie-Weiss temperature of EuSe in solid solutions with GdSe and LaSe as well as accompanying changes in cell dimensions. We remember from Sect. 59 that pure EuSe is an insulator with a positive Curie-Weiss temperature of 9 °K and a metamagnetic spin order caused by nearly equal values of the exchange interactions J_1 and J_2. The metal GdSe is antiferromagnetic, has a Néel temperature of about 50 °K and has electrical properties similar to Gd-metal. LaSe is a diamagnetic metal. The alloying of a few percent of a conducting rocksalt compound into insulating EuSe raises the

---

[384] S. v. MOLNAR, and S. METHFESSEL: J. Appl. Phys. 38, 959 (1967).

paramagnetic Curie temperature steeply. The same effect occurs in EuS and EuTe and when chalcogenides of other trivalent lanthanides such as Tb, Dy, Lu and also Y are used for alloying with Eu-chalcogenides. Traces of lighter elements such as trivalent transition metals, however, might increase the electrical conductivity but do not effect the Curie temperature so drastically. The Curie temperature of EuO can be increased from 69 to 135 °K when the material is made conductive by forcing excess RE^{3+}-ions into it[385]. It is interesting to note

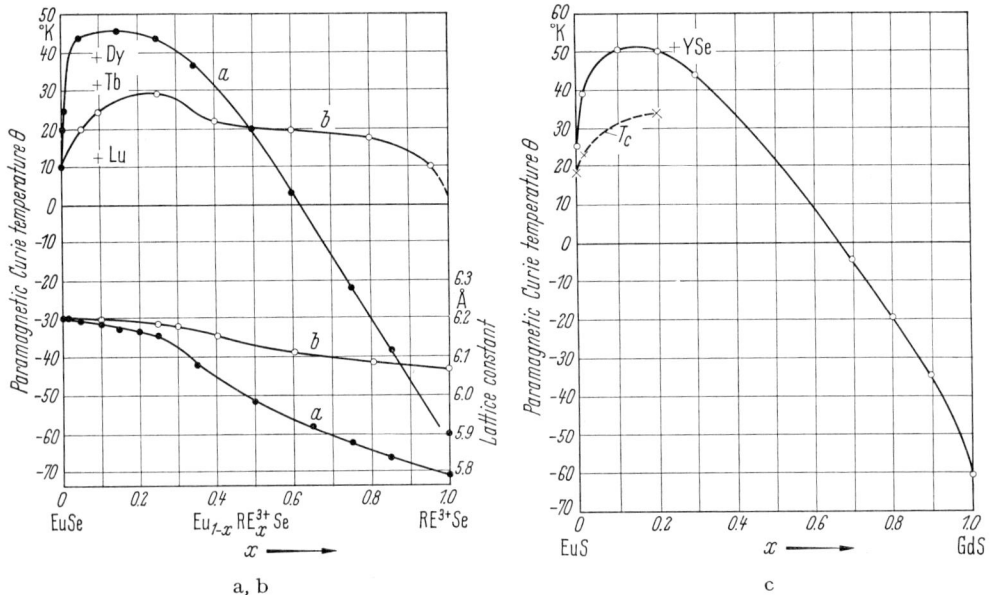

Fig. 33a—c. Variation of the paramagnetic Curie-Weiss temperature Θ, the lattice constant a_0 and the ferromagnetic Curie-temperature T_c with RE^{3+} concentration in Eu-chalcogenides: a) $Eu_{1-x}Gd_xSe$, b) $Eu_{1-x}La_xSe$, c) $Eu_{1-x}Gd_{1-x}S$.
The paramagnetic Curie-Weiss temperatures of some samples containing 10% LuSe, TbSe, DySe and 26% YSe are marked by +. (From Ref. [383].)

that in this case the paramagnetic Curie temperature Θ lies close to the ferromagnetic ordering temperature T_c, whereas in EuS and EuSe T_c is much less effected by conductivity than Θ. The resulting difference between T_c and Θ is unusually large in semiconducting crystals with small RE^{3+} concentrations, and disappears in samples with metallic conductivity[386]. A deviation between T_c and Θ indicates that the ferromagnetic exchange interactions available for ordering below T_c are smaller than those measured as spin interactions in the paramagnetic region at high temperatures. Such a condition can result from chemical inhomogenities in the sample, but may also be a typical feature of the magnetic interaction via conduction electrons in semiconductors. As electrons are trapped at trapping centers, or as the Boltzmann factor changes with temperature, the

---

[385] M. W. SHAFER, and T. R. McGUIRE: J. Appl. Phys. **39**, 588 (1968).
[386] In EuSe is, as an example, for $x=0.01$, $T_c \sim 8$ °K, $\Theta = 24$ °K; for $x=0.15$, $T_c \sim 27$ °K, $\Theta = 46$ °K, for $x = 0.5$, $T_c \sim 20$ °K, $\Theta = 20$ °K. When T_c and Θ are very different from one another the temperature dependence of the magnetization around T_c is flatter than the Brillouinfunction, and T_c is experimentally and in principle difficult to define.

number of conduction electrons participating in magnetic bonds will change. An extreme case of discrepancy between magnetic properties at high and low temperatures is found in certain alloys of EuTe, which is antiferromagnetic in its pure state. Alloying EuTe with metallic antiferromagnetic GdTe increases Θ to a maximum value of $+28\,°K$ at $x=0.2$, although no spontaneous ferromagnetic order occurs at low temperatures over this range of alloying. High fields of the order of 50 koe are required for ferromagnetic alignment[387].

The assumption of *indirect exchange* via conduction electrons as the mechanism for the ferromagnetic increase of Θ is rather plausible. In considering *direct exchange* as an alternative we expect from the distance dependence of J_1 and J_2 in Fig. 31 the effect of alloying on Θ to go qualitatively in the wrong direction and to be quantitatively too small. Ascribing the effect to *additional Eu^{2+}-Gd^{3+} ferromagnetic exchange* also fails, as the diamagnetic RE-elements La, Lu, Y raise the Curie temperature also. Finally, it is possible also to remove conduction electrons from Gd-doped EuS or EuSe by evaporating metallic Eu out of the sample in vacuo while the Gd-concentration remains the same[388]. The insulating, sample has, in spite of its Gd^{3+} contents, the same magnetic properties as found in the pure insulating material. On the other hand, GdSe can be made ferromagnetic when a sufficient number of conduction electrons is removed, as in replacing Gd^{3+} by Na^+ ions[388].

The only important property of impurity atoms as concerns magnetism in these materials is apparently their capability to provide a certain number of electrons to states which are sufficiently delocalized to produce additional exchange between Eu-neighbors. Conduction electrons satisfy, of course, this condition. But electrons in bound impurity states with sufficiently large orbital radii (as found in semiconductors of Ge- or Si-type) must also be considered in this connection.

The similarity of the experimental curve for $Eu_{1-x}Gd_xSe$ in Fig. 33 with the theoretical Curie temperature variation in Fig. 13 suggests coupling via conduction electrons[389]. The first transition between ferromagnetism and antiferromagnetism with increasing electron concentration is theoretically predicted for $\frac{1}{4}$ of an electron per magnetic atom[390]. The corresponding experimental point lies at a Gd-concentration of $x=0.6$. Electrical measurements in this region give 0.43 conduction electron for each Gd ion. Considering this factor, theory and experiment are in fair agreement with one another. However, we must be aware of the possibility that contributions of $5d$-orbitals at the bottom of the conduction band or other irregularities in the band structure invalidate the concept of a parabolic band to a certain degree. The detailed discussion of conduction-particle exchange has been given in section D.

62. Kasuya's impurity model. KASUYA and YANASE[391] have devised an impurity model for the description of magnetic, and electrical properties of doped Eu-chalcogenides. Good agreement with experimental data is obtained especially for semiconducting samples with small impurity concentrations. The localized character of an impurity state elegantly avoids the difficulty confronted by the

---

[387] F. HOLTZBERG, T. R. McGUIRE, and S. METHFESSEL: J. Appl. Phys. **37**, 976 (1966). See also Sect. 82.

[388] F. HOLTZBERG, and S. METHFESSEL: To be published.

[389] The solutions with La-chalcogenides show a more complicated variation of Θ with concentration, since the diamagnetic La-ions influence the magnetic properties not only by increasing conductivity but also by dilution of the spin lattice.

[390] See calculation Appendix of D. MATTIS "The Theory of Magnetism". New York: Harper & Row 1965.

[391] T. KASUYA, and A. YANASE: J. Appl. Phys. **39**, 430 (1968), and papers to be publ. in J. Phys. Soc. Japan; see also the magnetic impurity model of HEIKES and DEGENNES discussed in Sect. 64.

free electron theory in this region where the exchange interactions of the electrons with the localized ion spins can be larger than the Fermi energy. The influence of the magnetic spin order on the conduction electrons becomes then too big to be handled by perturbation theory, as will be seen in chapter H.

In KASUYA'S model the excess electron remains bound to the RE^{3+}-impurity with an energy of the order of 0.5 eV. This impurity state is much deeper than usually found in Ge and Si and extends mostly to the 12 nearest Eu-neighbors only. In addition to the usual properties of such states, we find now an exchange interaction of the impurity electron with the local $4f$-spins in the spin lattice. This i-f-exchange is supposed to be larger than the electron-phonon interaction but weaker than the Coulomb attraction by the impurity atom. The resulting "magnetic impurity state" has similarity with the "magnetic exciton" (Sect. 48) when the Coulomb attraction of the optically excited core is replaced by the core charge of the higher valent impurity ion. The difference in energy of the magnetic impurity state in the disordered and ordered spin lattice (which is analog to the optical red shift of the $4f$—$5d$ transition but has different magnitude) is responsible for the Curie temperature increase within the "molecule" formed by the RE^{3+} ion with its electron and the 12 surrounding Eu^{2+} ions. The Hamiltonian describing the magnetic properties of the "free molecule"

$$H_m = -2s[J_0 S_c + J_1 \Sigma S_m]$$

contains the exchange interaction J_0 of the impurity electron spin s with the spin S_c at the impurity ion, such as Gd^{3+}, in the center of the molecule, and the interaction J_1 with the spins S_m at the surrounding Eu^{2+}-ions, which are included in the molecule. This Hamiltonian can be solved rigorously. For a Gd-molecule including 12 Eu-neighbors KASUYA takes the approximation $J_0/J_1 = 12$. Calculation of the energy levels shows that parallel alignment of the electron spin and the Gd-spin is the more stable configuration. That means, a total spin of $S_0 = \frac{7}{2} + \frac{1}{2} = 4$ has to be assigned to the Gd-ion in the center, which has now a spin configuration similar to the $4f^7 5d^1$ ($^9D_2$) state of Gd^{2+} ions (Table 1). The antiferromagnetic state with $S_0 = 3$ has an energy which is higher by 720 cm^{-1}. For temperatures below $J_1 \sim 25$ °K in EuSe the exchange interaction between the impurity electron and the 12 Eu-neighbors alignes the Eu-spins parallel to one another and the "molecule" obtains a "giant moment". This effect depends only on J_1 and *not* on the character of the central ion. Therefore, it is qualitatively the same for Gd and La doping. The formation of "giant molecules" can be observed experimentally[392] as a quite unusual behavior of the magnetic specific heat in the neighborhood of the Curie temperature. Fig. 34 gives an example.

The magnetic susceptibility measured in applied magnetic fields at a certain temperature derives from several contributions:

a) exchange interactions J between Eu and Gd spins as found in the insulating material (in EuSe approximately 1 cm^{-1} = 1.44 °K).

b) exchange interaction $J_0 \sim 90$ cm^{-1} which couples the impurity electron spin to the Gd-spin in the center of the "molecule".

c) exchange interaction J_1 by which the impurity electron alignes the 12 neighboring Eu-spins to a giant moment (in EuSe about 25 °K).

d) exchange interaction J_2 of the giant moment with other Eu-spins which do not belong to a "molecule" (in EuSe about 18 °K).

---

[392] V. L. MORUZZI, D. T. TEANEY, and B. J. C. VAN DER HOVEN: Phys. Rev. Letters **20**, 719 (1968).

e) exchange interaction J_3 between two impurity states which is analog to the interaction of two hydrogen atoms and determines the relative alignment of giant moments in neighboring "molecules" (in EuSe about 20 °K).

For small impurity concentrations molecular field theory is applicable to the calculation of the susceptibility of the whole system. The observed difference between T_c and Θ is interpreted, according to this theory, as deriving from the assumption that J, J_0, J_1 and J_2 have ferromagnetic nature, whereas J_3 is antiferromagnetic. In the temperature interval between Θ and 300 °K where Θ is determined experimentally, the antiferromagnetic exchange between molecules is not

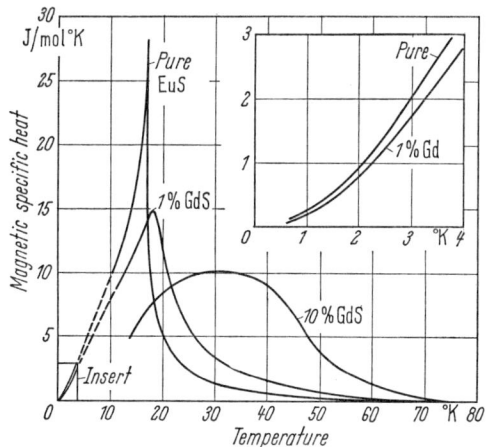

Fig. 34. Temperature dependence of the magnetic part of the specific heat in pure and Gd-doped EuS. In conducting and ferromagnetic samples the specific heat behavior is unusual compared to other magnetic materials. The reason probably lies in the formation of spin clusters with giant magnetic moment or in non-uniform exchange interactions varying with the local density of charge carriers [V. L. MORUZZI, D. T. TEANEY, and B. J. C. VON DER HOEVEN: Phys. Rev. Letters 20, 719 (1968).]

important since molecules do not exist here. In the temperature range around and below the ferromagnetic ordering temperature T_c, the antiferromagnetic interaction J_3 competes effectively with the ferromagnetic exchange and lowers T_c below the value expected from the high temperature susceptibility. Kasuya concludes from the antiferromagnetic nature of J_3 and the numerical agreement of $12\,J_1 \sim 209$ cm^{-1} with the s-f-exchange in the free Eu$^+$-ion (see Sect. 48, ref. 264), that the impurity electrons have s-character[393]. He expects J_3 to be ferromagnetic and $T_c \sim \Theta$ for impurity electrons with d-character. The extension of the theory to large impurity concentrations (where the interaction between overlapping molecules is expected to become more complicated) was not available at the time of writing (1967—1968).

63. RE^{3+}-chalcogenides with Th_3P_4-structure. The magnetic properties of the *2:3 chalcogenides* of RE^{3+} ions with S, Se give another striking example of ferromagnetic interaction via conduction electrons. We noted in Sect. 58 that those compounds crystallize in the b. c. c. Th_3P_4-structure with sufficient defects D in the cation lattice to provide charge compensation in the 2:3 composition: $RE^{3+}_{\frac{8}{3}} D_{\frac{1}{3}} X^{2-}_4$.

---

[393] Note the contrast to the APW-calculations of EuS in Fig. 6. The d-bands X_3 and Γ'_{25} lie below the s-band Γ_1 at the bottom of the conduction band.

The cation defects do not show observable order and can be filled continously by solid solution of the insulating 2:3 compound with the metallic $RE^{3+}X^{2-}$ compound[394]. The electrical conductivity increases by several orders of magnitude with increasing cation excess and changes its character from semiconducting to metallic.

In Gd-chalcogenides[395] the increase in conductivity is accompanied by the occurence of ferromagnetism with Curie temperatures up to 100 °K, as is shown in Table 4 for the solid solution system $Gd_2Se_3 + xGdSe$, as an example[396]. Compared to the Curie temperature of pure Gd-metal (~300 °K) a value of 100 °K is remarkably high for compounds which contain more nonmagnetic anions than magnetic cations and which results from solid solution of two anti-

Table 4. *Properties of solid solutions between Gd_2Se_3 and $GdSe$*[395]
n = number of excess electrons per RE^{3+}; a_0 = lattice constant; ϱ = resistivity at R.T.; Θ = paramagnetic Curie-Weiss-temperature; μ_{eff} = paramagnetic moment per RE-ion.

	Gd_2Se_3	n pro RE^{3+}	a_0 Å	$\varrho_{R.T.}$ (Ω cm)	Θ °K	μ_{eff}
1	+0,0 GdSe	0,0	8,718	~10^6	−10	7,75
2	+0,05 GdSe	0,02	8,718	3,4	−10	7,63
3	+0,275 GdSe	0,12	8,718	0,2	+3	7,96
4	+0,407 GdSe	0,17	8,718	—	+17	7,68
5	+0,475 GdSe	0,19	8,718	$3,1 \cdot 10^{-2}$	+30	7,93
6	+0,55 GdSe	0,22	8,718	$2,0 \cdot 10^{-3}$	+65	8,01
7	+0,625 GdSe	0,24	8,718	$1,4 \cdot 10^{-3}$	+88	7,74
8	+0,625 EuSe	0,0	8,865	~10	0	7,73
9	+0,625 YSe	0,24	8,706	$8 \cdot 10^{-3}$	+47	7,96

ferromagnetic components. ($Gd_2Se_3: \Theta = -10$ °K, $T_N = 6$ °K, $GdSe: \Theta = -60$ °K, $T_N = 50$ °K). The Gd-rich conductive samples show quite normal dependence of magnetization on applied field and temperature. In contrast to the 1:1 chalcogenides the paramagnetic Curie temperature Θ and the ferromagnetic ordering temperature T_c agree within a few °K.

Table 4 includes also two samples where Gd_2Se_3 has been alloyed with EuSe or YSe instead with GdSe in order to demonstrate the significance of the conductivity to the Curie temperature increase. EuSe has the same magnetic moment per cation as GdSe but is an insulator. Its addition to Gd_2Se_3 increases the spin concentration slightly but does not produce conductivity; the Curie temperature remains unchanged[397]. YSe is a diamagnetic metal with electrical properties

---

[394] It is interesting to note that Sc, the first 3d-element, forms also a wide solubility range between a metallic ScS-compound and semiconducting Sc_2S_3. ScS has a NaCl-type structure. Sc_2S_3 has a defect structure which is closely related to the NaCl-type structure, but the unit cell is twelve times as large due to a complex ordering arrangement of cation voids. (J. P. Dismukes, and J. G. White: Inorg. Chemistry 3, 1220 (1964)).
[395] F. Holtzberg, T. R. McGuire, S. Methfessel, and J. C. Suits: J. Appl. Phys. 35, 1033 (1964); S. Methfessel: Z. angew. Phys. 18, 414 (1965).
[396] The system $Gd_2S_3 + GdS$ has similar properties, but Gd_2Te_3 does not have the Th_3P_4 type structure.
[397] The composition $EuGd_2S_4$ is semiconducting and ferromagnetic below 6 °K [F. Hulliger, and O. Vogt: Phys. Letters 17, 238 (1965); Helv. Phys. Acta 34, 199 (1966)]; compounds have been prepared also with alkaline earth ions filling the cation vacancies [S. W. Kurnick, R. L. Fitzpatrick, and M. F. Merriam: R. E. Research (ed. J. F. Nachman, C. E. Lundin). New York: Gordon and Breach 1962, p. 249; J. Flahaut, L. Domange et M. Patrie: Bull. chem. France 2048 (1962); S. M. Golabi, J. Flahaut et J. Domange. Compt. rend. 259, 820 (1964)].

comparable to GdSe. Its admixture lowers the resistivity to about 10^{-2} ohm cm in this specific sample and increases the Curie temperature to 47 °K.

The application of Kasuya's impurity model to the conductivity induced ferromagnetism in sesquichalcogenides is difficult. The excess Gd-atoms, used for doping, disappear unrecognizably in the statistically distributed holes of the defect Gd-lattice as indicated by the constancy of the lattice parameter through the entire concentration range, until the Th_3Ph_4-type structure breaks down at the largest Gd concentration given in Table 4. Therefore, the charge irregularities necessary for binding conduction electrons into impurity states are difficult to assign and the use of a conduction band model appears to be more appropriate.

There is a certain temptation to link phenomenologically, the remarkable increase in Curie temperature to the existence of an unusually large homogeneity range in the RE-chalcogen compounds, and to seek a common cause for both in some peculiarities of crystal symmetry, orbital overlap etc. Experience indicates that the outer d-orbitals may be responsible for both effects. Indeed, comparison of the crystal symmetry in lattices of the NaCl and Th_3P_4-type supports the thesis that the orientation in the crystalline field of the lowest energy $5d$-orbitals and their overlap at neighboring cations may account for the observed phenomena[398]. The Gd—Gd distance in GdSe and Gd_2Se_3 has about the same value of 4.08 Å and the coordination of the nearest three Se-neighbors in Gd_2Se_3 is nearly identical to the Gd-position in GdSe with respect to one of the faces of the Se-octahedra in the NaCl-type structure (Sect. 58). The $5d$-orbitals have enough direct overlap between neighboring cations to form a band in which large numbers of excess electrons can be delocalized, and large deviation from charge compensation becomes tolerable. Similar effects occur in oxides of the first $3d$-elements where large cation size provides sufficient overlap of $3d$-orbitals. With increasing overlap the Gd-ions deviate increasingly from their divalent state with $4f^75d^1$ configuration in the ^9D-groundstate. The contribution of $5d$ electrons to atomic moment decreases while the exchange interaction via overlapping $5d$-states increases. In this picture, the difference in Curie temperature in the insulating and conducting state is ascribed qualitatively to the difference in character of the RE-elements considered to be $4f^7$-ions in the first case and $5d^1$-ions in the latter.

64. Conducting exchange in $3d$-compounds with NaCl-structure. Intensive search for ferromagnetic exchange via conduction carriers in *$3d$-transition element compounds* has been made by HEIKES and his group at Westinghouse, and in the Philips Research Laboratories. Introduction of ions of different valency at equivalent lattice sites can produce metallic-type conductivity in the $3d$ band of originally insulating compounds[399], and may give rise to ferromagnetic double-exchange (Sect. 28).

In MnO, CoO, NiO and CuO partial substitution of the $3d$-cations by monovalent Li up to 0.35, 0.20, 0.15 or 0.02 parts, respectively, gives single phase materials with the NaCl-type crystal structure[400]. Mixed states of divalent and trivalent $3d$-ions and electrical conductivity have been found but the character

---

[398] S. METHFESSEL: Z. angew. Phys. **18**, 414 (1965); F. HOLTZBERG, and S. METHFESSEL: J. Appl. Phys. **37**, 1433 (1966). Sect. 59.

[399] J. H. DE BOER, and E. J. W. VERWEG: Proc. Phys. Soc. (London) **49**, 59 (1937); N. F. MOTT: Proc. Phys. Soc. (London) A **62**, 416 (1949); E. J. W. VERWEY: Semiconducting materials. London: Butterworth 1951; E. J. W. VERWEY, P. W. HAAIJMAN, F. C. ROMEIJN, and G. W. VAN OOSTERHOUT: Phil. Res. Repts. **5**, 173 (1950).

[400] W. D. JOHNSTON, and R. R. HEIKES: J. Am. Chem. Soc. **78**, 3255 (1956); R. R. HEIKES, and W. D. JOHNSTON: J. Chem. Phys. **26**, 582 (1957); W. D. JOHNSTON, R. R. HEIKES, and D. SESTRICH: J. Phys. Chem. Solids **7**, 1 (1958).

of all samples was semiconducting rather than metallic. None of the samples studied became ferromagnetic[401]. It was assumed that the double exchange interaction is pre-empted when the charge carriers are trapped at lattice sites by formation of electrostatic polarons; the electron degeneracy has negligible effect under those conditions. Electron motion has been described in these materials as a classical diffusion process activated by thermal lattice fluctuations[402]. Since the strength of the electron trapping decreases with increasing electronic polarizability of the lattice[403], it was conjectured by HEIKES et al. that Se at the anion sites might produce the desired conditions.

Preliminary investigations of the system $Li_xMn_{1-x}Se$ in the concentration range $0 \leq x \leq 0.1$ give a complex magnetic behavior[404] which has been attributed to the superposition of ferromagnetic double exchange onto antiferromagnetic superexchange[405] of the pure MnSe. For small Li concentrations up to $x=0.05$ the spin order of the pure MnSe is retained but the Néel temperature lowered to 83 °K. For $x=0.07$ the same type of ordering sets in at 73 °K but at 45 °K the spin direction changes abruptly and rotates out of the (111)planes. At 20 °K a small remanent magnetic moment of about 0.5 μ_B per Mn-ion is observable. The composition $Li_{0.1}Mn_{0.9}Se$ becomes ferromagnetic with 0.7 μ_B per Mn-ion below 110 °K. At 70 °K the ferromagnetism disappears again and an antiferromagnetic order of the third kind appears. The transition at 70 °K is accompanied with thermal hysteresis and tetragonal distortion.

HEIKES explains the occurence of spontaneous magnetization with small moments at low temperatures in these compositions by assuming that the conduction carriers freeze out below about 130 °K and concentrate in the neighborhood of the Li$^+$ ions. The 3d-hole which is needed to provide the charge compensation for the more strongly electropositive Li$^+$-ion[406] is shared by the 12 nearest Mn-neighbors located at identical lattice sites. The trapped 3d-hole combines the nearest Mn-neighbors to an entity which is quite analogous to Kasuya's magnetic impurity state for conduction electrons in doped Eu-chalcogenides (Sect. 62).

DE GENNES[407] has shown, that double-exchange in such a localized state will couple the 12 nearest Mn-spins together into a ferromagnetic spin cluster with a net localized moment. For small Li-concentrations these localized moments have no direct interaction with one another and behave paramagnetically above the ordering temperature of the remaining lattice. Below the Néel point the local moments

---

[401] Some spontaneous magnetization has been found in $Li_xN_{1-x}O$ for $0.3 \leq x \leq 0.5$ [J. B. GOODENOUGH, D. G. WICKHAM, and W. J. CROFT: J. Phys. Chem. Solids **5**, 107 (1958); N. PARAKIS, J. WUCHER et G. PARRAVANO: Compt. rend. **248**, 2306 (1959); however, it was due not to double exchange but to ferrimagnetism resulting from preferential ordering of the Li$^+$ ions into alternate(111)planes, which modifies the antiferromagnetic Ni—O—Ni superexchange].

[402] In Part H we refer to more recent investigations on electrical properties of NiO and CoO single crystals which have indicated 3d-conduction in a narrow band instead of hopping. Earlier results obtained on sintered samples appear to be misleading because of grain-boundary effects.

[403] N. F. MOTT, and R. W. GURNEY: Electronic Processes in Ionic Crystals. Oxford: Clarendon Press 1948.

[404] R. R. HEIKES, T. R. McGUIRE, and R. J. HAPPEL: Phys. Rev. **121**, 703 (1961); S. J. PICKART, R. NATHANS, and G. SHIRANE: Phys. Rev. **121**, 707 (1961).

[405] Pure MnSe has below $T_N \sim 173$ °K an antiferromagnetic order of the second type (i. e. ferromagnetic(111)-layers with alternating magnetization) and the atomic moment of 5.0 μ_B of the ^6S state of Mn^{2+}. The value of T_N is uncertain because of thermal hysteresis due to crystallographic phase changes [R. LINDSAY: Phys. Rev. **84**, 569 (1959)]. $\Theta = -250$ °K.

[406] The second ionization potential of Li is 76 eV compared to a third ionization potential of about 35 eV for transition metal ions.

[407] P. G. DE GENNES: Phys. Rev. **118**, 141 (1960).

are exchange-coupled with the host lattice and will be constrained into the overall antiferromagnetic structure of the crystal. With increased Li-concentration the double exchange interactions between spin clusters become more prominent and produce a field-dependent increase of susceptibility. The competition between ferromagnetic double exchange coupling between clusters and antiferromagnetic exchange in the host lattice leads to spin canting at lower temperatures. In the $Li_{0.1}Mn_{0.9}Se$ sample the cluster concentration is large enough for the clusters to lose their localized character, and double exchange becomes dominant throughout the entire lattice; the paramagnetic Curie temperature is now $+55$ °K compared to -250 °K in pure MnSe. The competition with antiferromagnetic superexchange results in spin canting or in ordering of the third type[408], below 71 °K. At 71 °K the sample changes more or less abruptly to a state of ferromagnetic ordering and behaves, with increasing temperature, like a normal ferromagnet with a Curie point of 110 °K.

GOODENOUGH[408] has pointed out, that the occurence of ferromagnetism in these materials is not a sufficient argument for double-exchange. Rules of superexchange predict ferromagnetic interaction between Mn^{2+} and Mn^{3+} ions via Se-anions when the Mn^{3+} ion is in the low spin state $t_{2g}^4 e_g^0$. If one assumes sufficient sensitivity of the crystal field splitting to thermal lattice expansion, the transition at 71 °K can be understood as the result of a transition from the high spin to the low spin state of Mn^{3+}. At low temperatures the high spin state can be stabilized by Jahn-Teller distortion[409]. If single crystals were available magnetic anisotropy measurements could be used to distinguish between the superexchange and double exchange scheme.

KARPENKO and BERDYSHEV[410] describe again the ferromagnetism in $Li_xMn_{1-x}Se$ as an excellent example of indirect exchange via free conduction electron carriers produced by thermal excitation in a semiconductor. Using unpublished information emanating from the Westinghouse Research Laboratories group (from which they learned that below the ferromagnetic-antiferromagnetic phase transition at 71 °K the sample with $x=0.1$ is a semiconductor, with activation energy[411] of 0.01 eV, and a metallic conductor with $n \doteq 10^{17}$—10^{18} cm^{-3} above the temperature at which the phase transition takes place) these authors assumed that the temperature dependence of the (ferromagnetic) indirect exchange between nearest-neighbor Mn ions was responsible both for the magnetic phase transition and the Mott transition. A loose modification of the formula (24.7) adapted for nearest-neighbors, yields the effective ferromagnetic bond as calculated by these authors [with $n(T)$ and $p(T)$ the temperature-dependent electron and hole concentrations and m_e, m_h the respective carrier masses]:

$$J_c = J^2 \frac{aS\hbar^2}{24(kT)^2} \left[\frac{n(T)}{m_e} + \frac{p(T)}{m_h} \right]. \tag{64.1}$$

Here J is the s-d exchange coupling constant, S is the d-shell spin, and a is the separation between nearest neighboring magnetic atoms. To the indirect ex-

---

[408] J. B. GOODENOUGH: Magnetism and the Chemical Bond, p. 238. New York: Interscience-Wiley 1963.

[409] Similar transitions have been found for Co^{3+} in $LaCoO_3$: J. B. GOODENOUGH: J. Phys. Chem. Solids 6, 287 (1958); R. R. HEIKES, R. C. MILLER, and R. MAZELSKY: Physica 30, 1600 (1964).

[410] B. V. KARPENKO, and A. A. BERDYSHEV: Sov. Phys. Solid State 5, 2494 (1964); relevant theories are discussed in chapter D, Sects. 24 and 26.

[411] Compared to 0.4 eV in pure MnSe.

change parameter we must add the antiferromagnetic superexchange J_s to obtain:

$$J_{\text{eff}} \doteq \sqrt{2z}(2S J_s - J_c) \qquad (64.2)$$

in which $z =$ number of n. n. These formulas have some heuristic value, as they indicate that at low temperature (where $n(T)$ and $p(T)$ vanish exponentially) the dominant mechanism is the antiferromagnetic superexchange, whereas the situation may reverse at increasing temperature. The analysis needs further justification, however, before leading to a quantitatively acceptable theory.

The relationship between the transition to ferromagnetism and free carrier concentration shows rough agreement with experimental data[412], although a final conclusion on the validity of the indirect exchange model in $Li_{1-x}Mn_xSe$ needs more experimental studies on single crystals, which are not available at present.

65. Ferromagnetic Cr-chalcogen-spinels. In the earlier Sect. 56 we already discussed the role of double exchange in perovskites. It was difficult to separate in these compounds the conductivity-induced ferromagnetism with convincing clearness from ferromagnetic superexchange, from electron-ordering effects and Jahn-Teller distortions. But there was undoubtedly a certain contribution of double exchange to the stabilization of ferromagnetic order at certain chemical compositions. In 1964, the discovery of ferromagnetic spinels at the Philips Laboratories brought new aspects into the discussion. LOTGERING had found that the normal spinels $CuCr_2S_4$, $CuCr_2Se_4$ and $CuCr_2Te_4$ have metallic p-type conductivity (about 10^{-2} to 10^{-4} ohm cm at R. T.) and show normal ferromagnetic order below Curie temperatures of 420, 460 and 365 °K, respectively[413]. In the sulphide and selenide the paramagnetic Curie-Weiss temperatures lie quite close to the ferromagnetic ordering temperature. This is a characteristic for ferromagnets with negligible negative interactions. The saturation moments per chemical formula unit at 4 °K are 4.58 and 4.94 μ_B. LOTGERING has suggested, that the magnetic properties can be explained by the assumption that the Cu, at the tetrahedral A-sites of the spinel structure, is monovalent in the diamagnetic $3d^{10}$-configuration. Thus, one half of the Cr-ions at the octahedral B-sites must be in the tetra-valent state with $3d^2$ and one half in the trivalent $3d^3$ configuration: $Cu^+|Cr^{3+}Cr^{4+}|X_4$. This concept is consistent with an observed magnetic moment of nearly $3+2=5 \mu_B$ per formula unit. The random distribution of Cr-ions with different valency over identical octahedral sites results in metallic conductivity in the $3d$-band of Cr and ferromagnetic double exchange[414]. It is interesting to note, that the combination of ferromagnetic ordering with an integral number of Bohr magnetons per atomic moment and metallic conduction with nearly one carrier per magnetic ion has been found to occur only in RE-materials, in the

---

[412] For $a \sim 10^{-8}$ cm, $S=2$, $m_e = m_h = m$ and $J = 10^{-12} - 10^{-13}$ erg, one obtains $J_c = An(T)$ with $A = 10^{-29} - 10^{-31}$ at 4 °K; $10^{-30} - 10^{-32}$ at 20 °K and $10^{-31} - 10^{-33}$ at 70 °K. In neighborhood of the transition temperature, $T = 70$ °K, is $n(T) = 10^{17} - 10^{19}$ cm^{-3}. Therefore, the conduction carrier exchange, $J_c = 10^{-14} - 10^{-15}$ erg, may exceed the antiferromagnetic superexchange, $J_s = T_N/k = 10^{-14} - 10^{-15}$ erg.

[413] F. K. LOTGERING: Proc. Intern. Conf. Magnetism, Nottingham 1964, p. 533, Solid State Comm. **2**, 55 (1964); the existence of the compounds has been shown by H. HAHN, C. DE LORENT, and B. HARDER: Z. anorg. Chem. **283**, 138 (1956); H. HAHN, and F. SCHRÖDER: Z. anorg. Chem. **269**, 135 (1952).

[414] Other chalcogen spinels such as $FeCr_2S_4$ and $CoCr_2S_4$ are clearly ferrimagnetic in support of LOTGERING's statement, that Cu is the only transition metal which forms stable monovalent ions with chalcogens, or of the later discussed statement of GOODENOUGH, that there are no Cr^{4+} ions in thio spinels.

perovskite manganites and CuCr-chalcogen spinels and in CrO_2 with rutile structure[415].

GOODENOUGH[416] has objected to the assumption that Cu^+ can be stabilized against Cr^{4+} in chalcogenides[417]. He prefers the formal valency assignment $Cu^{2+}Cr_2^{3+}X_4$ which had been rejected by LOTGERING because it would require a ferrimagnetic ordering of the Cr^{3+} moments of 3 μ_B and of the Cu^{2+} moments of 1 μ_B in order to provide the observed moment of 5 μ_B per formula unit instead of $1 + 2 \times 3 = 7 \mu_B$. GOODENOUGH, however, derives the moment reduction from

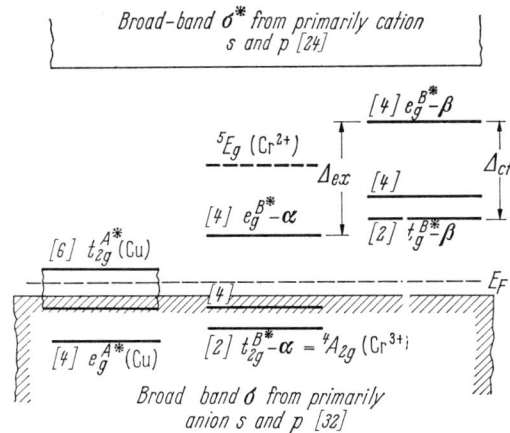

Fig. 35. Schematic electron energy diagram for antibonding d^*-states in normal AB_2X_4-spinels with $A = Cu^{2+}$, $B = Cr^{3+}$. The exchange splitting Δ_{ex} separates one-ion states with spin directions α and β. The crystal field of the octahedral B-site is Δ_{cf} (including some trigonal component). δ_{cf} is the tetrahedral crystal-field splitting of the A-site. Numbers in brackets refer to total degeneracy per molecule. These are obtained by multiplying the number of orbitals per atom contributing to a band or level by the number of atoms per molecule and then doubling to include spin degeneracy. 2 Cr^{3+} provide six d-electrons per molecule which are parallel spin aligned and fill up the t_{2g}^{B*}-α band. The diagram neglects the level splitting by electron-electron repulsion which is probably larger than the crystal field splitting. [J. B. GOODENOUGH: Coll. Intern. C. N. R. S. No. 157, Orsay, Oct. 1965 (Editions C. N. R. S. 1967) and private comm.]

polarized conduction electrons in delocalized t_{2g}-orbitals at the Cu-ions (Fig. 35). Cu^{2+} has in tetrahedral coordination the 3 d^9-electron configuration $eg^4 t_{2g}^5$ with one unoccupied t_{2g}-orbital. Orbitals with t_{2g} symmetry can form in tetrahedral ligand

---

[415] $T_c = 121$ °C, saturation moment about 2 μ_B per Cr^{4+} ion: C. GUILLAUD, A. MICHEL, J. BERNARD et M. FALLOT: Compt. rend. 219, 58 (1944); K. SIRATORI, and S. IIDA: J. Phys. Soc. Japan 15, 210 (1960); D. S. RODBELL, J. M. LOMMEL, and R. C. DE VRIES: J. Phys. Soc. Japan 21, 2430 (1966); J. B. GOODENOUGH: Magnetism and Chemical Bond. New York: Wiley 1963; Bull. Soc. Chim. France 1965, p. 1200. The electrical resistivity in sintered powders is about 10^{-1} ohm cm at RT and shows a pronounced maximum at the Curie temperature due to grain effects [B. KUBOTA, and E. HIROTA: J. Phys. Soc. Japan 16, 345 (1960); D. S. CHAPIN, J. A. KAFALAS, and J. M. HONIG: J. Phys. Chem. 69, 1402 (1965); N. NAKAYAMA, E. HIROTA, and T. NISHIKAWA: J. Am. Ceram. Soc. 49, 52 (1966)]. Single crystals have a RT resistivity of 2.5×10^{-4} cm with the temperature dependence usually found in metallic ferromagnets (D. S. RODBELL, J. M. LOMMEL, and R. C. DE VRIES: J. Phys. Soc. Japan 21, 2430 (1966)].

[416] J. B. GOODENOUGH: Coll. Intern. C. N. R. S., No. 157, Orsay, Oct. 1965 (Editions C. N. R. S., 1967).

[417] The II. ionization energy of Cu is 20.29 eV, whereas Cr has the IV. ionization energy of about 50 eV.

coordination strong σ-bonds with the chalcogen ions. As a consequence, the localization of the t_{2g}-electrons is broken up by cation-anion-cation interactions. The crystal field approach has to be replaced by the band theory description of a narrow band with one hole per molecule and with metallic conductivity. However, the e_g-orbitals of the Cu^{2+} ions at the A-sites and the t_{2g}^3-electrons of Cr^{3+} at the octahedral B-sites remain localized. There is some covalent mixing of anion orbitals into the Cr^{3+}-orbitals which gives the crystal field splitting and the nearest neighbor $Cr^{3+}-Cr^{3+}$ superexchange interactions. The 90° cation-anion-cation superexchange is expected to be ferromagnetic and the nearest neighbor cation-cation superexchange antiferromagnetic[418]. The Cu-conduction electrons are supposed to be polarized antiparallel[419] to the spin of the $^4A_{2g}$-ground state of Cr^{3+}. This polarization contributes by double exchange or Ruderman-Kittel exchange (depending on band width) to the ferromagnetic $Cr^{3+}-Cr^{3+}$ interaction in addition to the ferromagnetic superexchange. The resulting spin density of the conduction electrons is distributed over the lattice. For complete polarization with $-1\ \mu_B$ it can provide the right reduction of the observed magnetic moment.

Fig. 35 shows schematically the energetical position of the ligand field levels with respect to the conduction and valence band and to the Fermi-energy E_F in the Cu-t_{2g}-band in Goodenough's model. One-electron levels are used not only for the description of the broad bonding σ and antibonding σ^* bands but also for the narrow d-bands of the A-site copper atoms. The localized-electron d-states at the B-site chromium ions are shown as multi-electron ligand field levels. Fig. 35. shows in addition to the ground state levels an excited Cr^{2+} state which can be important for the interpretation of the magnetic shift of the optical absorption edge (Sect. 48). The mixture of localized levels (applying to individual molecules) and band states (which belong to the whole crystal) within one energy level diagram is, of course, not very satisfying if one believes that the local state and the band state are different mathematical descriptions of a singular physical situation. Goodenough, however, has considered here both states to be of different thermodynamic character[420]. In this case, energy diagrams of the type shown in Fig. 35 appear to be not only unavoidable but even characteristic for the situation in magnetic materials.

The problem of the copper valency, Cu^+ or Cu^{2+}, and the question of conductivity induced ferromagnetism by $Cr^{3+}-Cr^{4+}$ double exchange or by Cu^{2+} conduction electrons have been attacked by neutron diffraction, by NMR and by preparation of a large number of normal chalcogen spinels with a variety of cation replacements on the A and B sites as well as anion substitution. Neutron diffraction[421] on $CuCr_2Se_4$ and $CuCr_2Te_4$ shows no moment at the Cu-ions. The magnitude of the scattering moments is equal for all Cr-ions and compatible only with the assumption that all chromium atoms are in the trivalent state. From this, one expects a magnetic moment of 6 μ_B per molecule and not 5 μ_B as observed

---

[418] J. B. Goodenough: Magnetism and the Chemical Bond. New York: Interscience-Wiley 1963.

[419] The conduction electron polarization in rare-earth materials is usually parallel to the local 4f-spins. In $CuCr_2S_4$ a net antiparallel-spin polarization is anticipated. The $A-B$ superexchange between localized d-electrons in spinels is always found to be negative. The molecular exchange field acting on the t_{2g}-band electrons at the Cu from the Cr-spins may be assumed to be negative also, as far as the band is narrow and consists of d-functions mainly.

[420] Recent review: J. B. Goodenough: J. Appl. Phys. **39**, 403 (1968). Remarks about transitions between localized and band states are made in Sects. 7, 12, 13, 72 and in other parts of this article.

[421] C. Colominas: Phys. Rev. **153**, 558 (1966); M. Robbins, H. W. Lehmann, and J. G. White: J. Phys. Chem. Solids **28**, 897 (1967).

in magnetic measurements. NMR of the nuclei Cu^{63} and Cu^{65} in the selenide and telluride [422] gives in the paramagnetic temperature region a positive Knight shift proportional to the paramagnetic magnetization. This means that the internal magnetic field at the Cu-nuclei has the same direction as the applied field. The hyperfine field at the Cu-nuclei in the ferromagnetic state of $CuCr_2Se_4$ without applied field varies with temperatures proportional to the ferromagnetic saturation magnetization and extrapolates to $+71$ koe at $0\,°K$. The existence and magnitude of the hyperfine field and its consistency with the paramagnetic Knight shift contradict the assumption of strictly diamagnetic Cu^+ ions. There must be some polarized conduction or valence electrons at the copper sites. The hyperfine field [423] in $CuCr_2Te_4$ has only about half the value of that in $CuCr_2Se_4$, related perhaps to the Curie temperature difference for both compounds. These results are consistent with Goodenough's model assuming that the partially filled t_{2g}-band of Cu^{2+} has an antiferromagnetic polarization of $1\,\mu_B$ with respect to the Cr^{3+}-spins which is not sufficiently localized to be seen by neutron diffraction [424].

66. Mixed Cr-spinels. A large number of normal spinels have been prepared with a variety of atoms at the A, B or X-sites and some of their properties are given in Table 5 and Fig. 36. Much qualitative information about electron levels in the rather complicated spinel lattice [425] has been derived from attempts to relate the electrical and magnetic properties of all these isostructural compounds consistently with the chemical characteristics of the substitutional atoms. Obviously, there is no reliable relationship between ferromagnetism and conductivity. The ferromagnetic Curie-temperature is more sensitive to the replacement of Cu by other nonmagnetic ions than to variations in electrical conductivity. When the copper at the A-sites is replaced by strictly divalent cations with filled d-shell, such as Cd, Zn, Hg, the compounds become nonmetals but remain ferromagnetic with reduced Curie temperature. Since Mn, Co, Fe reduce the free carrier concentration also, the location of the partially filled d-levels of the A-atoms in the energy scheme with respect to the broad valence and conduction bands [426] appears to be an important factor in the electrical properties.

Lotgering [427] has adopted his model $Cu^+Cr_2X_4$ to the newer experimental results by abandoning the concept of $Cr^{3+} - Cr^{4+}$ conductivity and double exchange. Instead, the admixture of narrow t_{2g}-band states of Cr^{3+} into the top of the filled valence band has been suggested. In the new model for $CuCr_2X_4$ the metallic p-type conductivity is due to holes in the broad valence band (Fig. 37a). The holes come from the overlap of the valence band with a narrow (~ 0.1 eV wide) t_{2g}-band of Cr which has 12 levels for two Cr-atoms per molecule and is occupied for a fraction of $5/12$. (1 electron goes to Cu^+). In contrast to Goodenough's model in Fig. 35 the Cr^{3+}-t_{2g}-states are not filled and located below the top of the valence band. Therefore, the Fermi energy moves into the valence band and hole conductivity becomes possible. A number of $\delta \approx 0.1$ holes per molecule would give about 8×10^{20} charge carriers per cm^3 and a calculated moment of $5.0 + \delta = 5.1\,\mu_B$ in good agreement with Hall measurements and ob-

---

[422] P. R. Locher: Solid State Comm. 5, 185 (1967).

[423] H. Yokoyama, R. Watanabe, and S. Chiba: J. Phys. Soc. Japan 22, 659 (1967).

[424] J. B. Goodenough: Solid State Comm. 5, 577 (1967).

[425] The band structure of the nonmagnetic spinel $CdIn_2S_4$ has been calculated approximately by W. Rehwald [Phys. Rev. 155, 861 (1967)].

[426] W. Albers, and C. Haas: Phys. Letters 8, 300 (1964); W. Albers, G. van Aller, and C. Haas: Coll. Intern. C. N. R. S. No. 157, Orsay 1965 (Editions C. N. R. S. 1967) p. 19. More details in chapter H, Sect. 78.

[427] F. K. Lotgering, and R. R. van Stapele: Solid State Comm. 5, 143 (1967); J. Appl. Phys. 39, 417 (1968).

Table 5. *Properties of chalcogen-spinels* (s. c. = semiconductor).

	a_0 (Å)	T_c (°K)	θ (°K)	μ_B/Mol	C_M	ϱ_{300} (ohm·cm)	Literature	
a) ACr_2S_4								
$CuCr_2S_4$	9.822	420—390	+425—390	4.58	2.36—2.40	10^{-3}	5), 15), 17)	p-type metal
$ZnCr_2S_4$	9.988	$T_N=18$	+18	—	3.34	5×10^{10}—3	1), 2), 5), 14), 15), 19),	s.c., weak ferromagnet
$CdCr_2S_4$	10.244	84.5—97	+135—156	5.15—5.55	3.2—3.8	2×10^3	2), 3), 6), 9), 12), 13), 19)	s.c., mob: <0,5 cm²/V·sec absorpt. edge: 1.37eV, magnetic blueshift
$HgCr_2S_4$	10.237	36—60	+137—142	5.35—5.46	3.62		2), 3), 10)	s.c. metamagn. helix below 25°K
$Hg_{1-x}Cd_xCr_2S_4$		T_c increases monoton, goes through max. metamagn. disappears with increasing x					4)	
$MnCr_2S_4$	10.110	~100°K	0		2.50	10^3—10^9	1), 5), 15)	ferrimagnetic s.c.
$CoCr_2S_4$	9.93				3.54	10^4	1), 5), 15)	s.c., ferrimagnetic
$FeCr_2S_4$	9.995	$T_N=195$		1.59		1.8—20	1), 5), 15)	p-type s.c., ferrimagn.
$Fe_{1-x}Cu_xCr_2S_4$							7)	3 band s.c., mobil: 10 cm²/Vsec
b) ACr_2Se_4								
$CuCr_2Se_4$	10.331	434—466	+433—465	4.94		~10^{-4}	15), 19), 21)	p-type metal
$ZnCr_2Se_4$	10.443	$T_N=20$ helical	+115	—		~25	1), 14), 15), 16) 19), 20)	s.c., absorp. edge: 1.28eV redshift in H-field
$CdCr_2Se_4$	10.755	129.5—142	+190—210	5.4—5.98	3.66—4.48	2×10^3	2), 3), 6), 8), 9), 11), 12), 13), 15), 17), 19)	s.c., absorp. edge: 1.29eV— 1.32 ev., magnetic redshift, mobil:7 cm²/Vsec
$HgCr_2Se_4$	10.753	106—120	+192—200	5.4—5.64	3.34—3.79		2), 3), 12), 15), 17), 19), 24	s.c.
$Zn_{1-x}Cd_xCr_2Se_4$		$0 < x \leq 0.4$: $T_N=20°$K; $0.4 < x \leq 1$: T_c, θ vary linearly with x					4)	
c) ACr_2Te_4								
$CuCr_2Te_4$	11.137	365	400	4.93	2.90		15)	metallic
d) CuB_2S_4								
$CuTi_2S_4$	9.994	temp. independ. paramagnetic				4×10^{-4}	5), 18)	metal
CuV_2S_4	9.808	complicated susceptibility				6×10^{-4}	5), 18)	metal
$CuCo_2S_4$	9.461	temp. independ. paramagnetic				4×10^{-4}	5), 18)	p-type metal
$CuRh_2S_4$		temp. independ. paramagnetic				5×10^{-3}	5), 18), 25), 26)	metal, superc. $T<4.8$ °K
$CuCrTiS_4$		<4,4	0—25		1.97		5), 17), 18)	n-type or p-type s.c.

Mixed Cr-spinels.

e) CuB_2Se_4

$CuRh_2Se_4$	10.259		temp. independ. paramagnetic		3.2×10^{-3}	[15], [17], [18], [21], [25], [26]	metal, mobil: 4,5 cm²/V·sec, superc.
$CuCrRhSe_4$		255—270	255	2.1	1.1	10^{-3}	[17], [18]	temp. indep. resistivity

f) $CuCr_2X_4$

$CuCr_2Se_3Cl$	10.377	magnetic at R.T.	4,40			[22]	hygroscopic at R.T.
$CuCr_2Se_3Br$	10.416	274	345	5.25		[22], [23]	s.c.
$CuCr_2Se_{4-x}Br_x$		a_0, T_c and vary linearly with x				[22], [23]	s.c. to metallic
$CuCr_2Te_3I$	11.125	294		4.10		[22]	
$CuCr_2Te_{4-x}I_x$		a_0 has max. at $x=0.4$, T_c varies irregularly with x				[22]	

[1] W. Albers, G. van Aller, and C. Haas: Coll. Intern. C.N.R.S. No. 157 (Editions C.N.R.S. 1967).
[2] P. K. Baltzer, H. W. Lehmann, and M. Robbins: Phys. Rev. Letters **11**, 493 (1965); **15**, 493 (1965).
[3] F. K. Baltzer, P. J. Wojtowicz, M. Robbins, and E. Lopatin: Phys. Rev. **151**, 367 (1966).
[4] P. K. Baltzer, M. Robbins, and P. J. Wojtowicz: J. Appl. Phys. **38**, 953 (1967).
[5] R. J. Bouchard, P. A. Russo, and A. World: Inorg. Chem. **4**, 685 (1965).
[6] G. Busch, B. Magyar, and P. Wachter: Phys. Letters **23**, 438 (1966).
[7] G. Haake, and L. C. Beegle: J. Appl. Phys. **39**, 656 (1968).
[8] C. Haas, A. M. J. G. van Run, P. F. Bongers, and W. Albers: Solid State Comm. **5**, 657 (1967).
[9] G. Harbeke, and H. Pinch: Phys. Rev. Letters **17**, 1090 (1966).
[10] J. M. Hastings, and L. M. Corliss: J. Appl. Phys. **39**, 632 (1968).
[11] R. C. Le Craw, H. von Philipsborn, and M. D. Sturge: J. Appl. Phys. **38**, 965 (1967).
[12] H. W. Lehmann, and M. Robbins: J. Appl. Phys. **37**, 1389 (1966).
[13] H. W. Lehmann, and G. Harbeke: J. Appl. Phys. **38**, 946 (1967).
[14] F. K. Lotgering: Philips Res. Rep. **11**, 190, 337 (1956).
[15] F. K. Lotgering: Intern. Conf. Magnetism, Nottingham 1964, p. 533; Solid State Comm. **2**, 55 (1964).
[16] F. K. Lotgering: Solid State Comm. **3**, 347 (1965).
[17] F. K. Lotgering, and R. P. van Stapele: Solid State Comm. **5**, 143 (1967).
[18] F. K. Lotgering, and R. P. van Stapele: J. Appl. Phys. **39**, 417 (1968).
[19] N. Menyuk, K. Dwight, J. Arnott, and W. Wold: J. Appl. Phys. **37**, 1387 (1966).
[20] R. Plumier: Compt. rend. **260**, 3348 (1965); J. Appl. Phys. **37**, 964 (1966).
[21] M. Robbins, H. W. Lehmann, and J. G. White: J. Phys. Chem. Solids **28**, 897 (1967).
[22] M. Robbins, P. K. Baltzer, and E. Lopatin: J. Appl. Phys. **39**, 662 (1968).
[23] J. G. White, and M. Robbins: J. Appl. Phys. **39**, 664 (1968).
[24] V. B. Berger, J. I. Budnick, and T. J. Burch: J. Appl. Phys. **39**, 658 (1968).
[25] M. H. van Maaren, G. M. Schaeffer, and F. K. Lotgering: Phys. Letters **25 A**, 238 (1967).
[26] M. Robbins, R. H. Willens, and R. C. Miller: Solid State Comm. **5**, 933 (1967).

served magnetic properties. The dominant contribution to the ferromagnetic interaction is attributed in this model to indirect coupling of Cr-spins via the holes in the valence band. (The filled $3d^{10}$ states of monovalent Cu are not shown in the figure 37.)

When Cr is replaced by Co or Rh one obtains p-type metals[428] with temperature independent magnetic susceptibility (Table 5). The Rh or Co t_{2g}-states are sup-

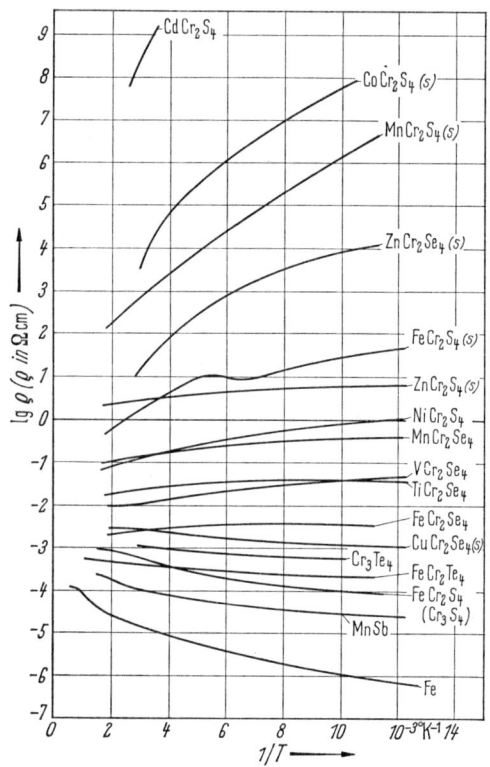

Fig. 36. Resistivity versus reciprocal temperature for ACr_2X_4-chalcogenides. Compounds with normal spinel structure are marked by s. The others have the cubic Cr_3S_4 type structure which can be described as a defect NiAs structure with ordered vacancies. Curves for Fe and MnSb are given for comparison. [From W. ALBERS, G. VAN ALLER, and C. HAAS: Coll. Intern. C. N. R. S. No. 157, Orsay, Oct. 1965 (Editions C. N. R. S. 1967).]

posed to lie deeper, below the Fermi energy. Therefore, they are filled, i. e. Rh and Co occur only in the trivalent state without holes corresponding to Rh^{4+} or Co^{4+} ions. The valence band contains one very mobile hole per molecule (corresponding to a formal valency of 1.75 per S-ion). The magnetic susceptibility is a composite of the PAULI susceptibility of the free holes and VAN VLECK susceptibility of the $3d$-ions, with the largest contribution possibly from the latter. For $CuTi_2S_4$ it is assumed, that Ti is in the tetravalent diamagnetic state. The conduction band contains 1 electron/mol and large electrical conductivity and Pauli

---

[428] $CuRh_2S_4$ and $CuRh_2Se_4$ are superconducting spinels below 4.8 or 3.5 °K, respectiveley (M. H. VAN MAAREN, G. M. SCHAEFFER, and F. K. LOTGERING: Phys. Letters **25 A**, 238 (1967); M. ROBBINS, R. H. WILLENS, and R. C. MILLER: Solid State comm. **5**, 933 (1967).

susceptibility is observable[429] due to the high density of states of the Ti^{3+} band at the Fermi level (Fig. 37c). Partial substitution of Cr by Rh^{3+} or Ti^{4+} lowers or raises the Fermi level according to the concentration. $CuCrRhSe_4$, as an example, is p-type metallic and ferromagnetic because the position of the Fermi level is similar to that in $CuCr_2Se_4$. The compound $CuCrTiS_4$, however, behaves quite differently. The energy scheme is that of a semiconductor with filled valence and

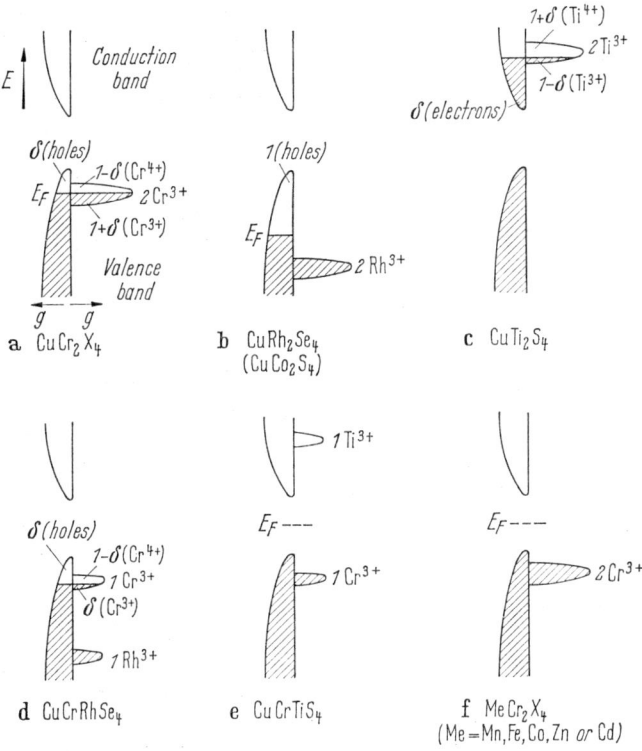

Fig. 37 a—f. The energy level scheme for several chalcogen-spinels in the model of F. K. LOTGERING and R. P. VAN STAPELE [J. Appl. Phys. **39**, 417 (1968); Solid State Comm. **5**, 143 (1967)]. Shown is the density of states as function of energy using the negative abscissa for the band states in the broad valence and conduction band and the positive axis for the narrow $3d$-band states.

empty conduction band. There is no magnetic coupling between Cr-spins. With variation of the Cr:Ti ratio the conductivity decreases by orders of magnitude and changes its character from p- to n-type.

The substitution of Cu in $CuCr_2X_4$ by Zn, Cd or Hg results in semiconducting ferromagnets with rather high ordering temperatures[430]. The filled narrow A^{2+} level lies somewhere below the Fermi level and all Cr^{3+}-states as well as the valence

---

[429] GOODENOUGH, however, assignes conductivity and PAULI-susceptibility to the partially filled t_{2g}-band of Ti^{3+} at B-sites. Ti^{3+} has only one d-electron in contrast to Cr^{3+} in Fig. 35, which has a filled t_{2g}-band.

[430] F. K. LOTGERING: Intern. Conf. on Magnetism, Nottingham 1964, p. 533; Solid State Comm. **2**, 55 (1965); P. K. BALTZER, H. W. LEHMANN, and M. ROBBINS: Phys. Rev. Letters **15**, 493 (1965); N. MENYUK, K. DWIGHT, J. ARNOTT, and W. WOLD: J. Appl. Phys. **37**, 1387 (1966).

band are filled. The low temperature resistivity decreases by four orders of magnitude if a small amount of 0.2% to 5% Cu is substituted for A^{2+}, because monovalent Cu produces holes in the valence band. GOODENOUGH[431] has argued against this interpretation. The A^{2+}-ions which have been selected for the proof of monovalent copper and Cr^{3+}-hole conductivity have in common (in contrast to Cu) that their t_{2g}-antibonding bands are all filled and cannot sustain free carrier motion. The experimental facts can be explained as well by the assumption that the introduction of Cu^{2+} creates shallow impurity levels which, at sufficient concentration, provide conductivity. Then one understands also why substitution of Ag^+, which should make holes in the valence band like Cu^+ in Lotgering's model, actually does influence the conductivity only very little[432]. In order to adapt Goodenough's model to more detailed experimental facts one considers, as in Fig. 35, a certain mixing of the t_{2g}-states of the A-cations into the top of the valence band instead of the mixing of the B-cation states in Lotgering's model.

ROBBINS et al.[433] have eliminated the possibility of ambiguous valence states for Cu and Cr by substituting *halides* for the chalcogen anions. In the compound $CuCr_2Se_3Br$, as an example, the total negative valence of the anions is 7 and the valence assignments $Cu^+[Cr_2^{3+}]Se_3^{2-}Br^-$ (with diamagnetic Cu and a moment of $6\,\mu_B$ at the B-sites) or $Cu^{2+}[Cr^{3+}Cr^{2+}]Se_3Br$ (with $1\,\mu_B$ at the A-sites and $7\,\mu_B$ at the B-sites) are, in principle, conceivable. The compounds which have been investigated are semiconducting and ferromagnetic with surprisingly high Curie temperatures (Table 5). In solid solutions of the type $CuCr_2Se_{4-x}Br_x$ the lattice parameter, the magnetic moment per molecule and the Curie temperature vary linearly with x between the values given in table 5 for the compounds with $x=0$ and $x=1$. Neutron diffraction results[434] support the valence attribution $Cu^+[Cr_2^{3+}]Se_3Br$. Unfortunately, this result is compatible with both models[435] for the metallic $CuCr_2Se_4$. Nevertheless, $CuCr_2Se_3Br$ and $CuCr_2Te_3I$ are, at present, the ferromagnetic semiconductor with the highest Curie temperature above room temperature.

The magnetic order and transition temperature of *insulating compounds* can be analyzed in the Heisenberg model assuming strong ferromagnetic superexchange between nearest neighbor B-cations and weaker longer-ranged interactions, which can play an important role in determining the ground-state spin configurations as greater numbers of atoms are involved. The sign of the exchange interactions can be obtained from the Goodenough-Kanamori-rules[436]. There

---

[431] J. B. GOODENOUGH: Solid State Comm. **5**, 577 (1967).

[432] 5% Ag decrease the p-type resistivity at R. T. by about two orders of magnitude. Differences in conductivity can be caused also by carrier trapping around lattice defects. The concentration of such defects depends on preparation conditions and the physico-chemical properties of the starting materials, and can be different for Cu and Ag compounds.

[433] M. ROBBINS, P. K. BALTZER, and E. LOPATIN: J. Appl. Phys. **39**, 662 (1968).

[434] J. G. WHITE, and M. ROBBINS: J. Appl. Phys. **39**, 664 (1968).

[435] A. W. SLEIGHT [Mat. Res. Bull. **2**, 1107 (1967)] has analyzed neutron and X-ray data of $CuCr_2X_4$ and $CuCr_2Se_3Br$ for interatomic distances. The observed Cu-anion distances in $CuCr_2X_4$ are found to be about 5% smaller than calculated from Pauling's covalent radii for Cu^+ in tetrahedral coordination, but in $CuCr_2Se_3Br$ the values agree within 1—2%. This has been interpreted as a support to the Goodenough model. P. M. RACCAH, R. J. BOUCHARD, and A. WOLD [J. Appl. Phys. **37**, 1436 (1966)] have found, that metallic $CuCr_2S_4$ has an irregularly large U-parameter of 0.384 compared to the semiconducting spinels with Co, Zn, Mn or Fe at the A-sites. The U-parameter is the only free parameter in the spinel structure and measures the anion sublattice distortion from cubic close-packing.

[436] For detailed discussion see P. K. BALTZER, P. J. WOJTOWICZ, M. ROBBINS, and E. LOPATIN: Phys. Rev. **151**, 367 (1966); J. Phys. Chem. Solids **28**, 2423 (1967); F. K. LOTGERING: Proc. Intern. Conf. on Magnetism, Nottingham 1964, p. 533; Solid State Comm. **3**, 347 (1965); G. BLASSE, and J. F. FAST: Philips Res. Rept. **18**, 393 (1963); E. F. BERTAUT: J. Phys. Radium **25**, 516 (1964); K. DWIGHT, and N. MENYUK: J. Appl. Phys. **39**, 660 (1968); Phys. Rev. **163**, 435 (1967). Present article, page 451.

are two principal superexchange interactions between each magnetic B-site cation and its *six nearest* B-site neighbors:

a) direct cation-cation superexchange which is always antiferromagnetic for two identical cations and depends strongly on distance.

b) 90° cation-anion-cation superexchange which is ferromagnetic between two Cr^{3+} with filled t_{2g}-states. Its strength is less sensitive to cell size and increases from oxide to telluride with increasing covalency.

A multiplicity of antiferromagnetic exchange paths connects each magnetic cation with a large number of more distant neighbors, especially its *30 next nearest* B-site neighbors.

1) 180° cation-anion-cation superexchange, decreasing from oxide to telluride with increasing lattice size,

2) cation-anion-anion-cation superexchange which follows the same rules as the cation-anion-cation superexchange, but is weaker,

3) cation B-anion-cation A-anion-cation B superexchange which is antiferromagnetic with a strength depending on the nature of the A-site cation.

BALTZER et al.[437] have given a simplified 2 parameter theory for the magnetic order and critical temperatures which is convenient for the analysis of experimental data. One parameter J represents the ferromagnetic B-anion-B 90°-superexchange, between nearest neighbors. The other parameter K takes into consideration the antiferromagnetic B-anion-A-anion-B superexchange as only interaction to the next-nearest neighbors[438].

The electrical conductivity of $CdCr_2S_4$ or $CdCr_2Se_4$ can be increased by several orders of magnitude by doping[439] with Ag, In or Ga. Doping with Au or Ag produces p-type conductivity with temperature dependent mobility of about 30 cm²/v.sec. Ga and In doping gives n-type conductivity with much lower mobility. There is no change of magnetic properties with conductivity in both cases. The n-type material has strong magneto-resistance effects around the ferromagnetic Curie temperature (which can be described by critical scattering), whereas the p-material shows no irregularities in this temperature range beside a change in slope of the resistivity-temperature curve.

H. Electrical properties of magnetic semiconductors.

67. General aspects. In chapter C we discussed the theory of electrical transport, and here we shall investigate the experimental situation. Charge transport measurements are supposed to give information about the motion of electrons in partly filled bands. Since electron transport results from acceleration of all electrons by applied electrical, magnetic fields or temperature gradients, considerations about the nature of the band structure are a natural starting point for interpretation or classification of electrical properties. Going beyond this, the fundamental difficulty and challenge in the study of magnetic materials lies in the problem of how to introduce into the one-electron picture the correlations which are responsible for the appearance of atomic magnetic moments and their ordering throughout the crystal. Since this problem has not been solved in suffi-

---

[437] P. K. BALTZER, P. J. WOJTOWICZ, M. ROBBINS, and E. LOPATIN: Phys. Rev. **151**, 367 (1966); J. Phys. Chem. Solids **28**, 2433 (1967).

[438] Examples for J/k and K/k in °K, respectively: $CdCr_2S_4$: 11.8; -0.33; $CdCr_2Se_4$: 14.0; -0.10; $HgCr_2S_4$: 13; -0.6; $HgCr_2Se_4$: 15.8; -0.51.

[439] H. W. LEHMANN, and G. HARBEKE: J. Appl. Phys. **38**, 946 (1967); C. HAAS, A. M. J. G. VAN RUN, P. F. BONGERS, and W. ALBERS: Solid State Comm. **5**, 657 (1967); Al, Dy, Ti and Ni in $CdCr_2Se_4$ are electrically inactive. H. W. LEHMANN: Phys. Rev. **163**, 488 (1967); C. HAAS: Phys. Rev. **168**, 531 (1968).

cient generality by a rigorous theoretical treatment, the interpretation of transport properties in magnetic materials requires a great amount of intuition. By weighing a large variety of chemical, crystallographic, magnetic, optical and electrical data against one another one tries to design qualitative energy level schemes (such as shown in Figs. 24, 35, 37, 39) which give optimal agreement with most experimental facts.

In this situation, the unreliability of electrical measurements on sintered ceramics can have disastrous effects. Compounds which have been known to be semiconductors with well understood properties became metals when the investigations were repeated on single crystalline samples. NiO was considered and discussed as the classical example of a hopping-type semiconductor for more than a decade. In single crystals, however, it is a narrow band semiconductor as we discuss in Sect. 80. The majority of older ceramic samples had uncertain stoichiometry, large concentrations of accidental impurities and lattice effects and contained impure, semiconducting grain boundaries. Their properties were very little related to intrinsic material properties. Recent improvements in techniques for growing single crystals from chemically complicated materials give hope that the rigid material standards requisite in semiconductor physics for many years can be obtained for magnetic semiconductors also. Until then, all discussions and surveys of this field have necessarily a rather preliminary character.

I. The phase diagram of correlated electrons.

68. Choice of parameters. GOODENOUGH[440] has developed in several papers a phenomenological electron phase diagram which is quite instructive for the classification of magnetic materials according to the degree of electron correlation required for consistent interpretation of electrical, magnetic and optical properties. Fig. 38 shows this diagram for the case of a single electron per interacting orbital, corresponding to a half-filled band[441]. The magnetic ordering temperature T_N is used as a measure of the strength of the electron-electron interactions. This is, of course, a rather arbitrary selection from the large variety of experimental parameters related to electron correlations but it has the advantage of simplicity in definition and experimental determination. The horizontal axis gives the electron transfer matrix element b which represents the energy gain by electron delocalization due to transfer to neighboring atoms in the lattice. An electron which can extend its orbital by including wave functions from other atoms has, of course, lower kinetic energy. The ligand field model of molecules relates b to the square of the covalent mixing parameter which is proportional to the transferred part of atomic wave functions from adjacent atoms. The meaning of the transfer integral b in the band model is very complex and can include many parameters such as electron-electron repulsions, electron-phonon interactions etc, which are not too well understood. In the following qualitative considerations, b is considered to be proportional to the band width W which one obtains from tight binding calculations of the band structure with omission of all electron-electron and electron-phonon interactions. The transition from the localized description to the band model is supposed to take place at a critical value b_c. Spontaneous magnetic moments appear for $b < b_m$.

---

[440] J. B. GOODENOUGH: Czech. J. Phys. B **17**, 304 (1967); Mater. Research Bull. **2**, 37, 165 (1967); J. Appl. Phys. **39**, 403 (1968).
[441] J. B. GOODENOUGH, J. M. LONGO, and J. A. KAFALAS: To be published in Mater. Research Bull.

69. Conductivity by localized electrons, for $b \ll b_c$. We have discussed in earlier parts of this article the properties of electrons in atoms and crystals which are relevant to magnetic semiconductors; the phase diagram gives a convenient summary. At the left side of b_c, for $b \to 0$, the crystal lattice is described by an array of isolated atoms without any overlap of magnetic electrons. We consider here only one single-electron state, such as $4f$ or $5d$, and neglect more extended orbitals with higher energy, at the same atom, which can have larger overlaps in the same crystal. (The case of crystals with two partially filled electron levels will be discussed where relevant).

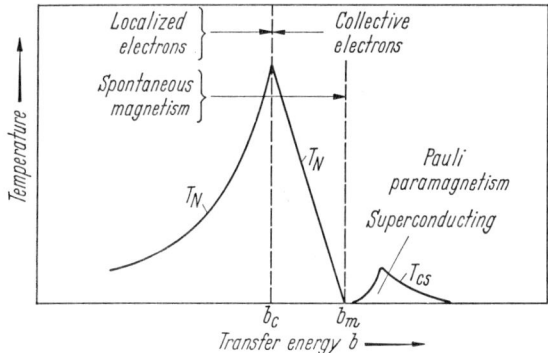

Fig. 38. Schematic phase diagram for electrons relates physical properties of materials to the bandwidth and amount of delocalization of magnetic electrons. (J. GOODENOUGH: To be published).

In this part of the phase diagram, the electron energy scheme is dominated by the ionization energies between different valence states of the magnetic cations as modified by the electrostatic Madelung-potential of the crystal and its dielectric constant[442]. The Hund-rule interactions between electrons at the same atom and the Stark effect of the electrostatic crystal field are much smaller effects. Each electron is restricted to remain with a certain spin direction at a specific atomic site. There is no charge transport through a lattice of identical atoms with an integral number of electrons per atom at temperatures low compared to the separation of the energy levels[443]. When the electron number is nonintegral, i. e. some atoms are for some reasons in a higher valence state, hole conductivity can occur by tunneling of electrons over the short distances between atoms. The hole mobility is expected to vary exponentially with temperature. A thermally activated diffusion process appears to be a suitable model. Since the mean time of an electron residing at an individual atom is long compared to the lattice relaxation time an appreciable activation energy arises from local lattice polarization around the different valent atom (Landau-trapping[444]). The properties of the $4f$-electrons in paramagnetic, insulating salts are believed to approximate the situation at this extreme end of the electron phase diagram. Conduc-

---

[442] An example is the energy level diagram of NiO designed by F. J. MORIN (in N. B. HANNAY: Semiconductors. New York: Reinhold 1959) and S. VAN HOUTEN [J. Phys. Chem. Solids **17**, 7 (1960); Halbleiterprobleme **4**, 118 (1961), Ed. F. SAUTER. Braunschweig: Vieweg & Sohn].

[443] Heat transport can occur at lower temperatures than charge transport by phonon, exciton, and magnon modes. The lack of electrical conductivity is related to the energy gap of the second kind, discussed in the section on the Mott transition.

[444] L. D. LANDAU: Physik. Z. Sowjet **3**, 664 (1933); N. F. MOTT, and P. W. GURNEY: Electronic Process in Ionic Crystals. London: Oxford Univ. Press 1940.

tivity by $4f$-electrons in lattices with different valent atoms has been reported for Pr_2O_{3+x} to have mobilities of the order of 10^{-5} to 10^{-6} cm^2/V · sec.[445].

70. Localized electrons with $b \to b_c$. The d-electrons in most transition element compounds have larger b-values than discussed in Sect. 69. The ground state of the crystal can no longer be derived from free atom states in a purely electrostatic crystal field, but molecular orbital theory has to be applied to the molecul comprising the magnetic cation and the surrounding ligands (Fig. 39). The admixture λ between d-orbitals and ligand orbitals[446] stabilizes the ligand orbitals as bonding states

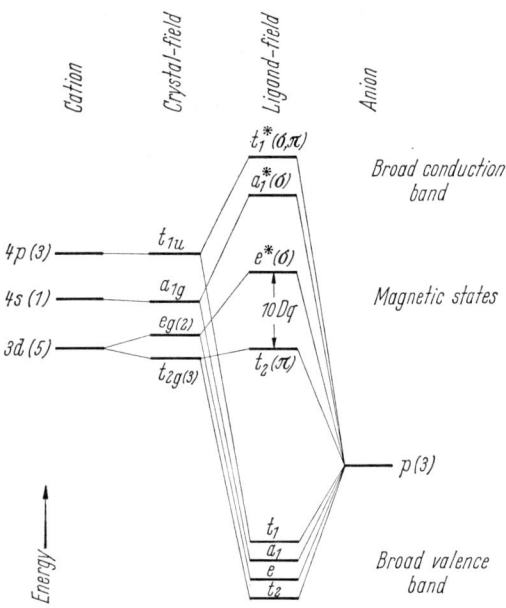

Fig. 39. Schematic diagram of the electron energy levels in a molecule containing a $3d$-ion surrounded by six ligand ions in octahedral coordination. The free ion levels of cation and anion are shifted relative to one another according to the difference of their electronegativities given in Fig. 45.

which are filled by the valence electrons from the anions. The d-electrons from the cations have to go into the antibonding states which have more localized character. Since the overlap depends on orbital and crystalline symmetry the covalency removes the degeneracy of the d-orbitals and produces crystal field splittings[447]. In octahedral ligand configuration, as an example, the d-orbitals with e_g-symmetry form strong σ-bonds with ligand p-orbitals and are more destabilized than the t_{2g}-orbitals which extend mainly between the ligands and form weak π-bonds. This difference in destabilization gives a crystal field splitting 10 Dq of about 1—2 eV in most $3d$-compounds which decreases with the anion-cation distance and difference in electronegativity (Fig. 39). Spin-orbit interactions lift

---

[445] J. M. Honig, A. A. Cella, and J. C. Cornwell: Rare Earth Research 2 (Ed. K. S. Vorres), p. 555. New York: Gordon and Breach 1964.

[446] λ^2 is proportional to the fraction to which p-electrons are transferred and proportional to b.

[447] P. W. Anderson: Solid State Physics **14**, 99 (1963); Magnetism **1**, 26 (1963); Phys. Rev. **115**, 2 (1959); S. Sugano, and R. G. Shulman: Phys. Rev. **129**, 2481 (1963); **130**, 506, 512, 517 (1963).

the threefold degeneracy of partially filled t_{2g} orbitals. Any accidentally remaining degeneracy of partially filled states can be further reduced by Jahn-Teller distortion of the crystalline environment. With more than two outer electrons or holes, intra-atomic exchange separates electron states with different spin alignment, unless the crystal field splittings are larger than this Hund-rule splitting.

The theory of superexchange, described in chapter D, shows that the ground-state energy of a lattice with magnetic molecules can be further reduced by short range spin order. The transfer from d-electrons beyond the ligands to the nearest cations is considered as second order perturbation. This electron transfer from one cation via the intermediate anion to the next cation is subject to Coulomb repulsion U and intra-atomic exchange at the transferred-to cation. The strength of superexchange and the magnetic ordering temperature increase proportional to b^2/U in this part of the diagram. Since superexchange is an exciton excitation between cations of equal valency it does not produce net charge transport.

71. Electron properties in a double-exchange coupled spin lattice. For cations with different valency the interaction between molecules becomes a first order perturbation. Superexchange is replaced by double exchange. Actual charge transport occurs from the lower to the higher valent cation. When the transfer integral b is still small compared to the intra-atomic exchange splitting of the one-electron states, then its value depends critically on the angle ϑ between the spins of the interacting cations $b = b_0 \cos \vartheta/2$. (see Sect. 28) The cation-cation overlap or the width of the "d-band" are large when spins are parallel and narrow for antiparallel spins. Hoping of electrons to neighbors with preferably parallel spin results in ferromagnetic spin alignment at low temperatures. Since double exchange between different valent ions is always accompanied by the super-exchange its actual effect on the spin ordering depends on the latter. In anti-ferromagnetic spin lattices it produces spin cantings. In ferromagnetic materials the Curie temperature increases above the value resulting from superexchange alone. Examples of both cases were given in chapter G.

In his paper on double exchange DE GENNES considers[448] two possibilities for the character of the "ZENER"-electrons responsible for double exchange in anti-ferromagnetic lattices: free carriers which fill the bottom of a band[449] which is derived by tight binding approximation as a linear combination of localized wave functions; bound electrons in an impurity state around the nonmagnetic impurity atom which induces the valence change of the magnetic cation.

The free carriers will restrict their motion through the lattice to ferromagnetic layers or chains if superexchange permits such spin arrangements. This type of conductivity is impossible without spin distortion in materials where superexchange produces two alternating sublattices so that each atom is surrounded by opposite spins of the other sublattice only. Two possibilities have to be considered now for the free carrier behavior. They can keep their freedom of motion and cant all spins uniformly through the whole crystal, or they can form bound or nearly bound states, where a large local spin distortion is built up by the localization of the electron in this area. The carrier becomes self-trapped by the distortions of the spin lattices it produces in its immediate environment. The resulting centers are able to move only slowly and their interaction with phonons and magnons will be so strong that they will fall always into bound states. The total energy of trapped carriers increases proportionally to the carrier concentration while the

---

[448] P. G. DE GENNES: Phys. Rev. **118**, 141 (1960).
[449] The band width must be at least as large as the intra atomic exchange, i.e. 0.1 to 1 eV, in order to avoid incorrect predictions about magnetic properties such as deviations of the paramagnetic susceptibility from the Curie-Weiss law.

energy corresponding to uniform spin canting by free carriers is found to vary with the square of carrier concentration. Therefore, the self-trapped configuration is always more stable for small carrier concentrations. The motion of carriers through an excited spin lattice at finite temperatures is extremely complicated because of the feedback of the carrier properties on the statistical behavior of the spin lattice by double exchange.

Electrons which are bound by Coulomb attraction to impurity atoms with a positive effective charge cant the spins of the neighboring magnetic ions toward a more parallel alignment. The magnetic impurity states have a local magnetic moment which can interact with the moments of other impurity states by direct overlap of distorted regions or by long reaching "wings" in which superexchange interactions smooth out the local perturbation of the spin lattice. The magnetic properties of Li doped MnSe, discussed in Sect. 64, are an example of this case.

The effect of double exchange by free carriers, self-trapped carriers or bound impurity states on magnetic properties is so similar that other criteria such as neutron-diffraction or specific heat measurements are needed for selection of the most appropriate description for specific materials. Effects of magnetic order on conductivity are expected to derive from the variation of the transfer integral or "band width" with spin order in the case of free carriers. Self-trapped carriers and impurity trapped carriers have a "magnetic" contribution to their activation energy for motion as free carriers or for hopping to the next unoccupied impurity state, respectively. This activation energy depends on the variation of the spin order with increasing temperature and can be modified by applied magnetic fields. In practice, however, the resultant effects are always seriously perturbed by the strong interactions of the slow carriers with the lattice phonons which have been neglected in the foregoing treatment.

72. The breakdown of the localized electron-model for $b \sim b_c$. When we go in the electronic phase diagram from the region of superexchange and double exchange toward larger b-values, i.e. to smaller interatomic distances and more covalent anions, the validity of the localized electron model breaks down very fast. The effect of the increasing overlap is augmented by the exponential decrease of the electron-electron repulsion U, with increasing b and electron number, because more mobil electrons can more easily screen local charge variations. For values of b larger than b_c the screening of the core potentials is sufficiently perfect and the electrons can move nearly freely everywhere in the crystal. The band model is now the superior starting approximation compared to the local description. The Mott-problem, that is the question how sharp the value of b_c is defined in real materials and how abruptly physical properties change at b_c, has been discussed in chapter C.

At present, the majority of solid state physicists believe that in transition metal semiconductors the appearance of the Fermi surface should have the character of a thermodynamic transition where at least some observable physical properties must vary their magnitude abruptly[450]. The real lattice, however, has so many free parameters that it may be able to obliterate (or in some cases, magnify) the expected abrupt changes in electronic properties by lattice distortions or changes in crystalline structure.

---

[450] GOODENOUGH gives the following empirical estimate: For carrier mobilities larger than 0.1 cm²/V·sec at R.T. the jump frequency in the thermally activated diffusion model becomes larger than the frequency of the lattice vibrations ($\sim 10^{12}$ sec^{-1}). With mobilities of more than 0.5 cm²/V·sec the band model for free carriers satisfies the Uncertainty Principle. Thus electrons having mobilities of the order of 0.1—0.5 cm²/V·sec at room temperature are characteristic of the problematic transitional region [J. B. GOODENOUGH: J. Appl. Phys. **39**, 403 (1968)].

Experiments to observe the transitions from the localized to band electron states are usually designed to increase the overlap λ by applying hydrostatic pressure to materials where the transfer integral b between local states is expected to be slightly smaller than b_c. In principle, the variation in b with spin order in the case of double exchange could cause also such transitions when ferromagnetic order is induced by sufficiently strong applied magnetic fields. One expects under such circumstances negative magneto-resistance, i. e. resistivity decrease with increasing field strength, in contrast to the positive magnetoresistance effects usually observed in semiconductors (see also Sect. 81).

73. Broad band model for $b \gg b_c$. We begin the discussion of the area on the right side of b_c in the phase diagram with very large b-values compared to b_c and b_m. Here we find the nonmagnetic broad-band metals, such as alkali-metals, and broad-band semiconductors of the germanium and silicon type. Electrons in partially filled bands run freely through such lattices by jumping from orbital to orbital at successive atoms during a time which is negligibly short compared to the lattice relaxation time. Electron mobilities are of the order of 10^2—10^3 cm^2/V·sec. The interaction with thermal lattice vibrations is a small perturbation. The electron-electron repulsion which could occur at a specific lattice site is well screened by easy redistribution of all the other electrons. In electron transport the charge carrier always finds orbitals at the next atom to which it can jump into without increase of its energy. The spin response to applied magnetic fields is described by the very small, temperature independent Pauli-susceptibility. There are no spin interactions beyond those dictated by the Pauli principle and no spontaneous magnetic moments.

The problem is to re-introduce the neglected electron-electron interactions for partially filled bands with finite band width $W \sim b$. One critical parameter is the intra-atomic exchange integral J_a which gives the energy difference of two electrons with parallel or antiparallel spin in two different orbitals at the same atom, discussed in chapter D. This is of the order of 0.1 to 1 eV, and HUND's rules say that the parallel spin state has lower energy. J_a can be thought of as a correction to the electrostatic repulsion U which two electrons experience when they jump into the same atom. For d-orbitals in insulating compounds a value for U of about 10 eV is consistent with their magnetic properties. In broad-band metals with sufficient carrier concentration the repulsion energy U can be reduced practically to zero by screening. The magnitude of U compared to the band width and to the intra-atomic exchange energy in real materials is an important parameter for the description of their physical properties.

74. Electron correlations in magnetic metals. The weakest observable interaction shown in Fig. 38 is superconductivity. With decreasing band width superconductivity is quenched[451]. The magnetic susceptibility increases above the Pauli-susceptibility and becomes temperature dependent. The electronic specific heat is exchange enhanced[452]. Below b_m spontaneous magnetic moments appear which have a nonintegral value per atom. This is the region of the ferromagnetic $3d$-transition metals. The ferromagnetic exchange between conduction electron spins is here comparable in magnitude to the increase in kinetic energy which is required to bring the electron with parallel spin in the next higher orbital in order to satisfy the Pauli-principle. The balance is in favor of ferromagnetism when the

---

[451] N. F. BERK, and J. R. SCHRIEFFER: Phys. Rev. Letters **17**, 433 (1966).
[452] S. DONIACH, and S. ENGELSBERG: Phys. Rev. Letters **17**, 750 (1966); E. BUCHER, W. F. BRINKMAN, J. P. MAITA, and H. J. WILLIAMS: Phys. Rev. Letters **18**, 1125 (1967); Several papers in J. Appl. Phys. **39**, 545—570, 956—965 (1968).

states at the Fermi energy E_F lie close together and the density of states $N(E_F)$ is high. The condition for ferromagnetism of band electrons[453] is

$$2JN(E_F) > 1 \tag{74.1}$$

and for nickel and iron one finds the product to have the value of 1.23 and 1.11, respectively[454]. The ferromagnetic order shifts the spin-down band to higher energies relative to the spin-up band. The larger electron occupation of the spin-up band produces the magnetic moment which has, in most cases, a nonintegral value per atom. The energy separation of the bands, i. e. the energy required to reverse the spin direction of an electron without changing its orbital or wave number, has been estimated to lie between 0.4 eV and 2 eV for nickel[455] with some preference to the low side.

The Coulomb repulsion between the electrons of antiparallel spin in the same state, U, results in an enhanced spin susceptibility and so may reduce the critical magnitude of J required for ferromagnetism. And indeed several authors have pointed to the possibility that not the intra-atomic exchange, but the electron-electron repulsion U is the important parameter for the occurence of ferromagnetism in Nickel[456], although we have seen (chapter C) that is is more related to the Mott transition, and possibly, to antiferromagnetism than to ferromagnetism. The repulsion results in energy separation between spin-up and spin-down bands of about 0.5 eV in ferromagnetic nickel.

The discussion of the relationship of band structure to magnetic, electric or optical properties of transition metal compounds and their alloys is out of the scope of this article and we refer to some recent review articles[457]. The details of the electronic phase diagram in this region can be found in publications by Goodenough and Penn[458].

75. Spin disorder resistivity. The most versatile description of the influence of magnetic ordering on the electrical conductivity of band electrons is given by the spin-wave scattering theory[459], which is discussed in chapter C, sections 16—21,

---

[453] J. C. Slater: Phys. Rev. **49**, 537 (1936); **82**, 538 (1951); E. C. Stoner: Proc. Roy. Soc. (London A **165**, 372 (1938); A **169**, 339 (1939); Rep. Progr. Phys. **11**, 43 (1947); E. P. Wohlfarth: Proc. Roy. Soc. (London) A **195**, 434 (1949); Proc. Intern. Conf. Magnetism, Nottingham 1964, p. 51; J. Friedel, G. Leman, and S. Olszewski: J. Appl. Phys. **32**, 325 S (1961); D. C. Mattis: Phys. Rev. **132**, 2521 (1963) and "Theory of Magnetism". New York: Harper & Row 1965.

[454] M. Shimizu, and A. Katsuki: Phys. Letters **8**, 7 (1964).

[455] J. G. Slater: Phys. Rev. **49**, 537 (1936); J. Appl. Phys. **39**, 761 (1968); J. H. Van Vleck: Rev. Mod. Phys. **25**, 220 (1953); J. Friedel: Advances in Phys. **3**, 446 (1955); J. C. Phillips: Phys. Rev. **133**, A 1020 (1964); J. Appl. Phys. **39**, 755 (1968); H. Ehrenreich, H. R. Phillip, and D. J. Olechna: Phys. Rev. **131**, 2469 (1963).

[456] J. H. van Vleck: Rev. Mod. Phys. **25**, 220 (1953); J. Kanamori: Proc. Theoret. Phys. (Kyota) **30**, 275 (1963); M. C. Gutzwiller: Phys. Rev. Letters **10**, 159 (1963); Phys. Rev. **134**, A 923 (1964); J. Hubbard: Proc. Roy. Soc. (London) A **276**, 238 (1963); A **277**, 237 (1964); A **281**, 401 (1964); Proc. Phys. Soc. (London) **84**, 455 (1964).

[457] For example: C. Herring: Magnetism 4 [Ed. G. T. Rado, and H. Suhl). New York: Academic Press 1966]; N. F. Mott: Advances in Phys. **13**, 325 (1964); D. C. Mattis: The Theory of Magnetism. New York: Harper and Row 1965; S. Doniach: J. Appl. Phys. **39**, 751 (1968).

[458] J. B. Goodenough: J. Appl. Phys. **39**, 403 (1968); D. R. Penn: Phys. Rev. **142**, 350 (1966).

[459] T. Kasuya: Progr. Theoret. Phys. (Kyoto) **16**, 58 (1956); **22**, 227 (1959); J. Friedel, and P. De Gennes: J. Phys. Chem. Solids **4**, 71 (1958); J. Kondo: Progr. Theoret. Phys. (Kyoto) **27**, 772 (1962); T. van Peski-Tinbergen, and A. J. Dekker: Physica **29**, 917 (1963); R. J. Weiss, and A. S. Marotta: J. Phys. Chem. Solids **9**, 302 (1959); I. Mannari: Progr. Theoret. Phys. (Kyoto) **22**, 335 (1959); D. A. Goodings: Phys. Rev. **132**, 542 (1963). Review Articles: S. V. Vonsovski: J. Phys. Soc. Japan **17**, 45 (1961); N. F. Mott: Advances in Phys. **13**, 325 (1964); B. R. Coles: Advances in Phys. **7**, 40 (1963); C. Haas: Phys. Rev. **168**, 531 (1968).

and particularly 19. The electrons are assumed to interact with the magnetic moments by intra-atomic exchange only. Therefore, it is not very critical which model one uses for the justification of magnetism, and the theory applies equally well to 3d-metals and RE-materials. The intra-atomic exchange is considered in the scattering approximation to be small compared to the band width and Fermi energy so that the effect of magnetic order can be treated as a small perturbation on the motion of the electrons. When all spins are ordered ferromagnetically at low temperature with the same periodicity as the ion lattice, the exchange interaction adds a spin dependent term to the lattice potential which polarizes the spins of the conducting electrons by a certain amount but does not change their mobility. With increasing temperature spin waves with decreasing wave length are excited. The spinwaves interact with the electron motion and shorten the electron mean free path. The situation is analogous to the scattering of neutrons in a magnetic lattice.

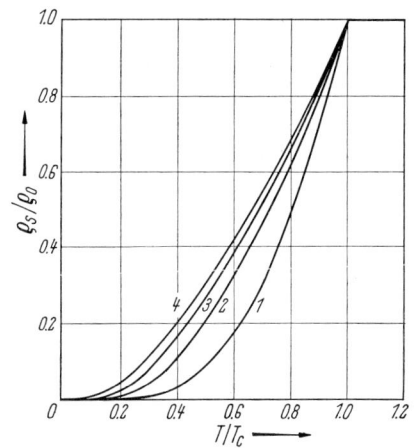

Fig. 40. Relative temperature dependence of the spin disorder resistivity. Curves 1 to 4 are calculated for atomic spins of $S = \frac{1}{2}, \frac{3}{2}, \frac{5}{2}$ and $\frac{7}{2}$, respectively using simple molecular-field theory. [T. VAN PESKI-TINBERGER, and A. J. DEKKER: Physica 29, 917 (1963)].

The spinwave part ϱ_m of the resistivity is additive with the phonon part ϱ_p and the residual resistivity ϱ_0 and can be separated out from dc-measurements by the different temperature dependence. The residual resistivity ϱ_0 is temperature independent[460], ϱ_b follows the Grüneisen-formula[461]. The magnetic part ϱ_m increases as T^2 at low temperatures and the theoretical temperature dependence is shown in Fig. 40 for several values of the scattering spin moments. When molecular-field theory is used for the temperature dependence of the magnetic ordering, the disorder resistivity reaches its largest value at the Curie temperature and remains constant with further increase of temperature. The high temperature limit of the magnetic disorder resistivity has the value[462]:

$$\varrho_m^\infty = 4.3 \times 10^{-4} (10^{23}/N) (m J/E_F)^2 S(S+1) \text{ ohm} \cdot \text{cm} \tag{75.1}$$

(N = number of conduction electrons per cm³; m = relative electron mass; J = effective intra-atomic exchange in eV; E_F = Fermi energy in eV; S = atomic spin).

Fig. 41 gives the experimental curves for the relative temperature dependence of ϱ_m in several materials with quite different magnetic and electrical properties. Gadolinium metal has localized spins and approximates best the assumptions of the theory. The magnetism in iron is due to exchange between itinerant band electrons. MnSb and Cr_3Te_4 are metallic compounds with superexchange ferromagnetism.

---

[460] Y. A. ROCHER [J. Phys. Radium 22, 367 (1961)] has discussed magnetic contributions to the residual resistivity by local irregularities in the spin alignment around nonmagnetic impurity atoms.

[461] E. GRÜNEISEN: Ann. d. Phys. 16, 530 (1933); $\varrho_p \sim T^5$ for temperatures low compared to the Debye-temperature and proportional to T at high temperatures.

[462] T. KASUYA: Progr. Theoret. Phys. (Kyoto) 16, 58 (1956) For Gd-metal, as an example, one obtains with $N = 3.07 \times 10^{22}$ cm^{-3}, $J = 0.157$ eV, $E_F = 4.4/\text{meV}$, $\varrho_m^\infty = 120 \times 10^{-6}$ m³ ohm · cm compared to an experimental value of about 130×10^{-6} ohm · cm.

MnTe is an antiferromagnetic p-type semiconductor with an energy gap of about 1 eV and 10^{18}—10^{19} cm^{-3} carriers in a broad valence band. The temperature dependence of resistivity in some broad band semiconductors containing localized moments appears to be a superposition of the mobility variations due to spin scattering onto the change in carrier concentration by thermal excitations. Fig. 42 gives examples[463].

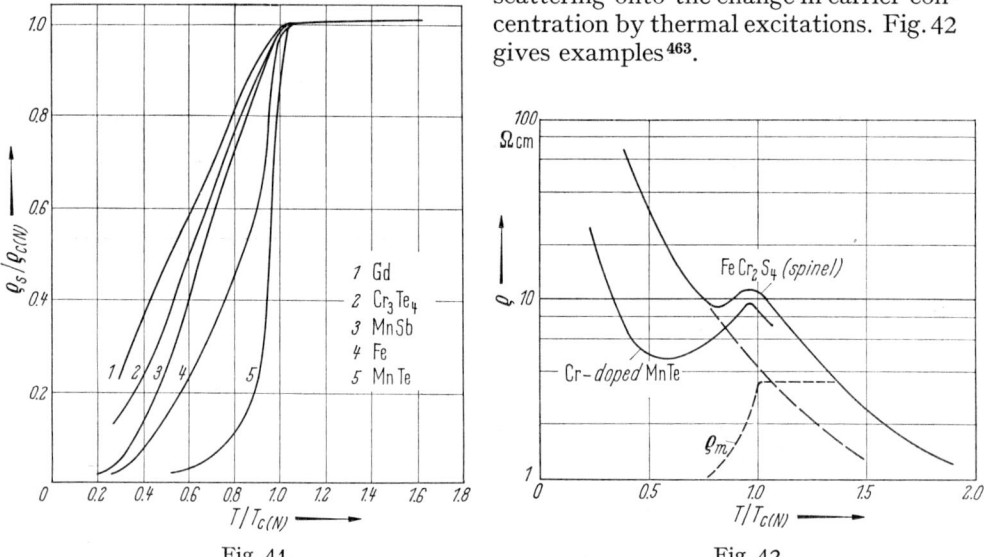

Fig. 41.

Fig. 42.

Fig. 41. Experimental measurements of $\varrho_m/\varrho_m^\infty$ as a function of T/T_c for the ferromagnetic RE-metal Gd ($\varrho_m^\infty = 1.3 \times 10^{-4}$ cm, $T_c = 296$ °K), for ferromagnetic, metallic compounds MnSb ($\varrho_m^\infty = 2 \times 10^{-4}$ ohm cm, $T_c = 587$ °K and Cr$_3$Te$_4$ ($\varrho_m^\infty = 2 \times 10^{-3}$ ohm cm, $T_c = 350$ °K), for the ferromagnetic 3d-metal Fe ($\varrho_m^\infty = 8 \times 10^{-5}$ ohm cm, $T_c = 1030$ °K) and for the antiferromagnetic semiconductor MnTe ($\varrho_m^\infty = 3$ ohm cm, $T_N = 310$ °K). (From W. Albers, G. van Aller, and C. Haas: Coll. Intern. C. N. R. S. No. 157, Editions C. N. R. S. 1967).

Fig. 42. In broad band semiconductors the magnetic disorder resistivity ϱ_m can be superposed on the temperature dependent resistivity $\varrho = \varrho_0 + \varrho_L + \varrho_t \cdot \exp(-E/kT)$. Examples are FeCr$_2S_4$ and Cr-doped MnTe. (From W. Albers, G. van Aller, and C. Haas: Coll. Intern. C. N. R. S. No. 157, Editions C. N. R. S. 1967.)

A more accurate treatment of the temperature dependence of the magnetic disorder resistivity has to include the spin-spin correlations around and above the Curie-temperature[464] which had been omitted so far. The influence of spin correlations on conductivity is negligible when the wave length of the Fermi-electrons $\lambda = 2\pi/k_F$ is smaller than the double distance $2d$ between magnetic ions. For sufficiently long wave length, $k_F d \ll \pi$, a phenomenon analogous to critical opalescence occurs. The resistivity shows a cusp-like peak at the Curie temperature (Fig. 8 gives an example). The peak amplitude decreases sharply with increasing values of $k_F d$ and disappears a little above $k_F d = 2$. The theory of De Gennes and Friedel[464] includes electron interactions with long range spin correlations only. Fisher and Langer[465] have pointed out that dominant contributions to magnetic resistivity can come from short-range spin fluctuations. Then, the slope of the resistivity curve $d\varrho/dT$ is expected to vary like the

[463] Another example may be CrSb: T. Suzuoka: J. Phys. Soc. Japan 12, 1344 (1957).

[464] P. G. De Gennes, and J. Friedel: J. Phys. Chem. Solids 4, 71 (1958); Y. A. Rocher: Advances in Phys. 11, 233 (1962); C. Haas, A. M. J. G. van Run, P. F. Bongers, and W. Albers: Solid State Comm. 5, 651 (1967).

[465] M. E. Fisher, and J. S. Langer: Phys. Rev. Letters 20, 665 (1968).

magnetic specific heat with temperature, i. e. like $(T/T_c - 1)^{-\alpha}$. The magnetocaloric energy and ϱ_m should have the same kind of singularity around the magnetic ordering temperature.

In transition metals and alloys where the narrow d-levels are embedded in the broad conduction band of s-like electrons, the scattering of highly mobile s-electrons into narrow d-orbitals by phonon interactions can contribute to electrical resistivity also. A discussion of such effects in metals can be found in a recent review by Mott[466].

76. Adaptation of the band model to strong electron interactions. Our discussion of the qualitative electron phase diagram in Fig. 38 comes now to the region closer to b_c where the band width b becomes comparable to or smaller than electron-electron, electron-magnon, electron-phonon interactions and where those effects can no longer be treated as small perturbations. In order to prevent complicated many-body problems in the description of the electrical properties one adheres to the band model and tries to experiment with suitable corrections. Our parameter b is then the band width as obtained from a standard band structure calculation by the LCAO or tight binding method. It contains the crystal symmetry, the core potentials, orbital overlaps between adjacent atoms, difference in electronegativity of cations and anions. Every one of these parameters can be responsible for small values of b in partially filled bands of specific materials. Small b-values can produce, at most, small electron mobilities but are not sufficient to explain the variety of electron properties observed in transition metal compounds. The divergence between theoretical prediction and experimental observation is especially marked in semiconducting compounds where the irregular transport properties of narrow band carriers are not masked by the presence of s-electrons as they are in metals.

The following corrections have been suggested in order to adapt the band model to the reality of experimental observations:

a) *Exchange splittings:* Intra-atomic exchange is of the order of 0.1—1 eV and can split a sufficiently narrow band of a ferromagnetic material into two separated bands with spin-up or spin-down states only. In the case of an accurately half-filled band the material could be an insulator with a saturation magnetization equal to the free ion value. With increasing temperature, the energy gap decays as a function of magnetic short range order and metallic conductivity should occur somewhat above the Curie temperature. For antiferromagnetic and ferrimagnetic materials, the existence of magnetic sublattices at low temperatures can effect the bandstructure because of a reduction in crystal symmetry. The bands will split, forming one subband for each magnetic sublattice[467]. This can explain the insulating character of $3d$-compounds with certain electron numbers. MnO and NiO, as examples, are antiferromagnetic with two sublattices. Therefore, the d-band splits into two subbands, each containing 5 states. The

---

[466] N. F. Mott: Advances in Phys. **13**, 325 (1964); see also D. A. Goodings: Phys. Rev. **132**, 542 (1963).

[467] J. C. Slater: Phys. Rev. **82**, 538 (1951); T. Matsubara, and T. Yokota: Proc. Int. Conf. Theor. Phys. Japan 1953 (Sci. Council Japan, Tokyo 1954); J. Des Cloiseaux: J. Phys. Radium **20**, 606, 751 (1959); F. J. Morin: Phys. Rev. Letters **3**, 34 (1959). — The splitting of the conduction band in RE-metals by helical spin order and its consequences on magnetic and electrical properties has been treated in the following papers: A. R. Mackintosh: Phys. Rev. Letters **9**, 90 (1962); Phys. Letters **4**, 140 (1963); P. G. De Gennes, and D. Saint James: Solid State Comm. **1**, 62 (1963); H. Miwa: Progr. Theoret. Phys. (Tokyo) **28**, 208 (1962); R. J. Elliot, and F. A. Wedgwood: Proc. Phys. Soc. (London) **81**, 846 (1963); **84**, 63 (1964); J. Callaway: Proc. Int. Conf. Phys. Semicond. Exeter 1962, p. 582.

five d-electrons in Mn^{2+} occupy accurately one subband. Ni^{2+} has 8 electrons. If the split of the d-orbitals by crystal symmetry in t_{2g} and e_g-orbitals is taken into account, one obtains an empty e_g-band, all other bands filled. However, there are many other insulating transition element compounds where this electron and orbital arithmetic is not successful (e. g. CoO).

The most problematic point in the model is the temperature dependence of resistivity. The materials should have a semiconductor to metal transition at the magnetic ordering temperature which has not been observed. But the underlying assumption that magnetic order disappears completely at the Curie temperature or Néel temperature, is derived from molecular field theory which is very unreliable on this point. It is well known that spin-spin correlations persist up to two to three times the magnetic ordering temperature. This could be connected with a certain amount of short-range spin order which maintains the gap locally for narrow band electrons. The electron-phonon interaction can maintain the energy gap without long range order. The situation is then quite similar to the electron-magnon interactions discussed under d below.

b) *Electron-electron repulsion:* If the Coulomb repulsion between two electrons of opposite spin at the same ion is larger than the band width b, the tenfold d-band, as an example, splits up in ten subbands each containing a single electron state only [468]. Fig. 43 shows the splitting of an s-band, as an example. The consequences of the appearance of an energy gap of the second type has been discussed in chapter C. The one-band model is able to explain the insulating ground state in

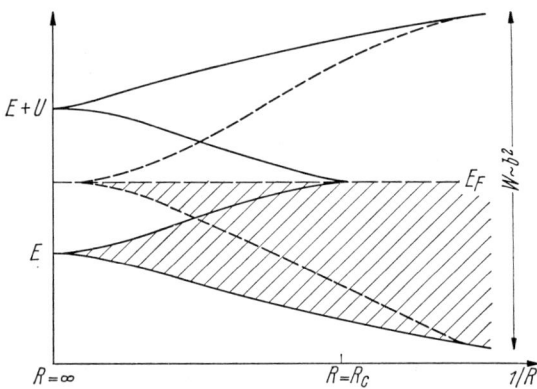

Fig. 43. Schematic illustration how an s-band varies with inverse interatomic distance R, when the electron-electron repulsion U is considered (full lines) or omitted (dotted lines). For $R > R_c$ the energy U splits the band into two subbands each containing one single electron state and separated from one another by a forbidden gap $E_g = (U^2 + \frac{1}{4} W^2)^{\frac{1}{2}} - \frac{1}{2} W$. A material with one s-electron per unit cell would be a metal for $R < R_c$ and an insulator for $R > R_c$. Screening by electrons from other atoms can move the center of the subbands faster together with decreasing R than indicated in the figure. [J. HUBBARD: Proc. Roy. Soc. (London) A **277**, 237 (1964).]

---

[468] N. F. MOTT: Proc. Phys. Soc. (London) A **62**, 416 (1949); Nuovo cimento, Suppl. **7**, 312 (1958); Phil. Mag. **6**, 287 (1961); P. W. ANDERSON: Solid State Phys. **14**, 99 (1963); M. C. GUTZWILLER: Phys. Rev. Letters **10**, 159 (1963); Phys. Rev. **134**, A 923 (1964); J. HUBBARD: Proc. Roy. Soc. (London) A **276**, 238 (1963); A **277**, 237 (1964); A **281**, 401 (1964); Transition Metal Compounds (Ed. E. R. SCHATZ). New York: Gordon and Breach 1964; Proc. Roy. Soc. (London) A **285**, 542 (1965); A **296**, 82 (1966); A **296**, 100 (1966); W. KOHN: Phys. Rev. **133**, A 171 (1964); G. KEMENY: Ann. Phys. (N. Y.) **32**, 69, 404 (1965). — Most recent survey article. D. ADLER: Solid State Phys. **21** (1968) (in press).

most transition metal compounds with partially filled bands. There should be a Mott-transition into a metallic state when inter-atomic distances are decreased and band width enlarged by applying hydrostatic pressure, by crystalline phase transitions etc. The expected transitions have a smooth character but secondary effects such as the electron-phonon interactions (strain) can make them abrupt (see chapter C). The most serious deficiency of the one-band theory, in its present form, is the omission of interactions between electrons at different atoms. The mutual screening of electrons can reduce the effective value of the Coulomb repulsion in actual materials radically below the free ion value, which is of the order of 10 eV. The description of the transport properties is complicated and has not been treated in sufficient detail. In insulators the electron moving to the adjacent atom remains bound to the hole it left behind (Mott-trapping[469]).

c) *Electron-phonon interactions:* In narrow bands the motion of the carriers through the lattice is slow. The mean time of an electron at each ion core is long enough to permit the surrounding lattice ions to adapt to the momentary charge distribution. The resulting lattice distortion forms a positive image charge which binds the electron. The strength of the interaction depends on the polarizibility of the ion lattice and can be characterized by the difference between the static dielectric constant and the square of the refraction index (i. e. the high frequency dielectric constant).

In weakly polarizable crystals, the electron charge is only partly screened by ion dislocations in a large volume. The electron forms, with the polarization cloud, a "large polaron" which interacts weakly with phonons and moves trough the lattice as a pseudo-particle with large effective mass or small band width[470]. In crystals with large dielectric constant the radius becomes small and the "small polaron" has a narrow pseudo-particle band width, and strong interaction with phonons. The pseudo-particle band width of the small polaron decreases with increasing temperature and transport at high temperature becomes very similar to diffusion[471]. When the radius of the polaron becomes comparable to interatomic distances the interaction with the phonons in the crystal lattice is so strong that the trapped electron will not participate in conductivity (Landau-trapping).

---

[469] N. F. Mott: Phil. Mag. **6**, 287 (1961); G. Kemeny: Ann. Phys. (N. Y.) **32**, 404 (1965); D. C. Langreth: Phys. Rev. **143**, 707 (1966); G. W. Pratt, and L. G. Caron: J. Appl. Phys. **39**, 485 (1968).

[470] H. Fröhlich, and N. F. Mott: Proc. Roy. Soc. (London) A **171**, 496 (1939); F. E. Low, and D. Pines: Phys. Rev. **98**, 414 (1955); S. J. Pekar: Fortschr. Physik **1**, 367 (1954).

[471] R. R. Heikes, and W. D. Johnston: J. Chem. Phys. **26**, 582 (1957); Phys. Rev. **99**, 1232 (1955); J. Yamashita, and T. Kurosawa: Phys. Chem. Solids **5**, 34 (1958); J. Phys. Soc. Japan **15**, 802, 1211 (1960); G. H. Jonker: Phys. Chem. Solids **9**, 165 (1959); G. L. Sewell: Phys. Mag. **3**, 1361 (1958); R. R. Heikes, A. A. Marududin et R. C. Miller: Ann. phys. **8**, 733 (1963); T. Holstein: Ann. Phys. (N. Y.) **8**, 325, 343 (1959); L. Friedman, and T. Holstein: Ann. Phys. (N. Y.) **21**, 494 (1963); L. Friedman: Phys. Rev. **131**, 2445 (1963); **135**, A 233 (1964); M. I. Klinger: Soviet Phys. Solid State **4**, 2260 (1963); Phys. Status Solidi **3**, 805 (1963); **11**, 499 (1965); I. G. Lang i Y. A. Firsov: Soviet Phys. JETP **16**, 1301 (1963); **18**, 262 (1964); H. G. Reik: Phys. Letters **5**, 236 (1963); Solid State Comm. **1**, 67 (1963); E. L. Nagaev: Soviet Phys. Solid State **4**, 1611 (1963); Y. A. Firsov: Soviet Phys. Solid State **5**, 1566 (1964); J. Schnakenberg: Z. Physik **185**, 123 (1965); K. Schotte: Z. Physik **196**, 393 (1966); J. Appel: Phys. Rev. **141**, 506 (1966); H. G. Reik, and D. Hesse: J. Phys. Chem. Solids **28**, 581 (1967); V. N. Bogomolov, E. K. Kodinov, D. N. Mirlin i Y. A. Firsov: Soviet Phys. Solid State **9**, 2077 (1967). — Review articles: G. H. Jonker u. S. van Houten: Halbleiterprobleme **6**, 118 (Ed. W. Schottky). Braunschweig: Vieweg 1961; E. R. Schatz: Transition Metal Compounds. New York: Gordon and Breach 1964; J. Appel: Solid State Phys. **21** (1968) (in print); D. Adler: Solid State Phys. **21** (1968) (in print).

HEIKES[472] has considered the influence of spin order on the hopping mobility in formula (21.3). He assumes that antiparallel spin alignment can not change the activation energy but reduces only the number of neighboring sites into which the electron can hop in moments of favorable lattice excitations. The probability of parallel spins is $\frac{1}{2}(M - M/M_0)^2$ when M and M_0 are sublattice magnetizations at T and 0 °K, respectively. With the crude approximation $(M/M_0)^2 = 1 - (T/T_c)^2$ for the temperature dependence of the magnetization, one finds for the hopping mobility a factor of T for $T < T_c$ and of T^{-1} for $T > T_c$, which produces a change in the slope of the resistivity-temperature dependence at T_c.

The materials to which the hopping theory applies belong actually in the phase diagram to the left side of b_c where the description by localized atomic orbitals is an alternative possibility. The difference between both models lies in the concept of the mechanism which reduces the electron mobility through the crystal. It is the Coulomb interaction of the electron with the environment in one case and with the ion core in the other. At present, it is customary to distinguish between thermally activated hopping and polaron band conductivity, although a unified theory is conceivable.

Until a few years ago, the hopping model was considered to give the adequate description for many 3d-semiconductors, especially oxides. However, recent investigations of pure single crystals have clearly indicated that the agreement between theory and experiment was fortuitous. The earlier experimental data were not related to intrinsic material properties but modified by grain boundary effects. Even in materials such as NiO, where a quite small b-value can be expected for the d-electrons, the electrical properties are best described by a narrow band model with exchange and correlation splittings. The electron-phonon interaction appears now to be a much weaker perturbation of electron motion than expected before (Sect. 80).

d) *Electron-magnon-interactions.* The carriers in the earlier mentioned model of magnetic disorder scattering did not contribute actively to exchange forces between atomic magnetic moments. If the magnetic ordering temperature is observed to depend on the presence of conducting electrons the feed-back of the electron behavior on the spin order has to be considered. The situation is qualitatively analogous to the formation of polarons by electrostatic interactions between moving carriers and the surrounding ion lattice. The intra-atomic exchange between conduction electron spin and ion spins polarizes the spin lattice to the extent of producing ferromagnetic order in the ground state at zero temperature. With increasing temperature the spin lattice goes through a spectrum of excited states. The spin excitations are uniform throughout the crystal when the band width of the coupling electrons is large compared to the intra-atomic exchange. Electrons in a narrow band can be sufficiently slow that the spin lattice relaxes around a momentary position in the lattice. The electron becomes trapped in a spin cluster with locally increased magnetic order. The resulting "magnetic polaron" can be large or small depending on the differential magnetic susceptibility of the spin lattice for electrons[473]. In the ferromagnetic lattice at 0 °K the band width of this pseudo-particle is equal to the band width b of the undressed electron. It decreases with increasing temperature as in the case of the electrostatic polaron. Unfortunately, the analogy between magnetic and electrostatic polarons can not be carried up to higher temperatures as magnons have more complicated properties than phonons (e.g. one can not superpose them linearly). Around the

---

[472] R. HEIKES; Ann. Phys. **8**, 733 (1963) and in: Transition Metal Compounds. (Ed. E. SCHATZ), p. 1. New York: Gordon and Breach 1964; J. APPEL: Phys. Rev. **141**, 506 (1966).

[473] Theory of magnetic polaron or spin polaron is given in chapter C., Eqs. (21.4) and (21.7).

Curie temperature, where the magnetic susceptibility is at a maximum, magnetic polarons may be quite small. The conductivity will then occur by thermal diffusion.

The magnetic polarons may be trapped at lattice defects or impurity atoms; such behavior of impurity electrons in a ferromagnetic lattice has been discussed by KASUYA in order to interpret the experimental properties of doped Eu-chalcogenides (see C and G). An occupied impurity state is surrounded by a cluster of local spins with relatively high order. This spin polarization lowers the electron energy and provides an activation energy when the electron moves to an other impurity site which is empty and has, therefore, no surrounding polarization cloud. To this activation energy one adds contributions from chemical fluctuations, electrostatic lattice distortions etc. It disappears in the ferromagnetic lattice at 0 °K and reaches its maximum around the Curie temperature. For impurity electrons or narrow band electrons in antiferromagnetic lattices, the concept of narrow band electrons with correlation and exchange corrections merges with the De Gennes-theory of double-exchange and hopping between localized magnetic atoms (Sect. 71).

e) *Symmetry distortions*: The energy of a crystal may be lowered by symmetry distortions which split narrow bands into occupied and empty states[474]. This is analogue to the Jahn-Teller effect for atomic states. The induced reductions in translational symmetry introduce planes of energy discontinuity at the Fermi surface. ADLER and BROOKS[475] have shown that the resulting energy gap can explain the properties of materials such as V_2O_3, VO, VO_2 which are semiconductors at low temperatures and go by a first order phase transition into the metallic state at high temperatures[476], as we have already discussed in sect. 12 *et seq.*, part C.

II. Electrical properties of 3d-semiconductors.

77. Qualitative approaches to the band structure problem. The 3d-compounds surpass any other group of materials in the perplexing diversity of their electrical properties. At present, band structure calculations cannot derive convincing interpretations from first principles[477]. Therefore, it is challenging to anticipate the further development of band theory by designing qualitative rules which can order the electrical, optical and magnetic properties of all 3d-compounds consistently as functions of known parameters such as atomic number or size of cations and anions, differences in their electronegativity, number of outer electrons available for chemical bonding, lattice symmetry, orbital hybridization etc. Such schemes have been suggested by GOODENOUGH, FRIEDEL, MOSER, SUCHET and others[478].

---

[474] J. B. GOODENOUGH: Mater. Bull. **2**, 37, 165 (1967); Coll. Intern. C. N. R. S. No. 157, Orsay 1965 (Editions C. N. R. S. 1967).
[475] D. ADLER, and H. BROOKS: Phys. Rev. **155**, 826 (1967).
[476] J. FEINLEIB, and W. PAUL: Phys. Rev. **155**, 841 (1967); D. ADLER, J. FEINLEIB, H. BROOKS, and W. PAUL: Phys. Rev. **155**, 851 (1967); D. ADLER: Solid State Phys. **21** (1968) (in print); Proc. Int. Conf. Metal-Nonmetal transitions, San Francisco 1968, will appear in Rev. Mod. Phys., Oct. 1968.
[477] Most recent review on band structure calculations in magnetic semiconductors: J. C. SLATER: J. Appl. Phys. **39**, 761 (1968); Physics Today **21**, 61 (1968).
[478] J. B. GOODENOUGH: Magnetism and the Chemical Bond. New York: Wiley-Interscience 1963; Bull. Soc. Chim. France 1965, p. 1200; Coll. Intern. C. N. R. S. No. 157, Orsay 1965 (Editions C. N. R. S. 1967); J. Appl. Phys. **37**, 1415 (1966); **39**, 403 (1968); Phys Rev., **164**, 785 (1967); Czech. J. Phys. B **17**, 304 (1967); Mater. Research Bull. **2**, 37, 165 (1967); J. FRIEDEL: J. Phys. Radium **23**, 501, 692 (1962); Bull. Soc. Chim. France 1965, p. 1186; E. MOSER, and W. B. PEARSON: Phys. Rev. **101**, 1608 (1956); Progr. Semiconductors

The discussion starts usually with a qualitative design of the molecular-orbital diagram of the cation and its ligands (Fig. 39). The set of one-electron orbitals includes the split into separate d^1, d^2 etc. orbitals by Coulomb repulsion, the separation between different spin states by intra-atomic exchanges and ligand field effects. Then one investigates the possibility that the sharp levels of the isolated atoms are broadened by inter-atomic interactions. The occurrence of conductivity in stoichiometric samples requires sufficiently strong interactions that one-electron states coalesce into a band containing more than one state per transition metal atom. There are interactions between adjacent cations and cation-anion-cation interactions. In the first case the cation-cation distance is a crucial parameter. The second, more indirect interaction is mainly determined by orbital hybridization and covalent overlap with ligand orbitals. GOODENOUGH has suggested for "Class 1"-oxides a semi-empirical formula for the critical distance R_c, below which bands are formed by cation-cation overlap:

$$R_c^{3d} = \{3.20 - 0.05\, m - 0.03\,(Z - Z_{Ti}) - 0.04\, S(S+1)\}\,\text{Å}$$

$$R_c^{4d} = R_c^{3d} + 0.88\,\text{Å}; \qquad R_c^{5d} = R_c^{3d} + 1.36\,\text{Å}.$$

The first term is the effective d-orbital radius in the crystal which depends also on the orbital overlap and covalent mixing with ligand anions. It is assumed to be the same for all oxids independent of crystal symmetry. The second and third term describes the ion contraction in the $3d$-element series with increasing atomic number Z with reference to Ti and with increasing cationic valency m of the M^{m+} ion, respectively. The last term reflects the intra-atomic exchange stabilization for a localized magnetic state versus a metallic band state. Examples of Class 1 oxides are TiO and VO. In the NaCl-type structure the lower lying t_{2g}-orbitals bridge cations which are closer than R_c. The resulting narrow t_{2g}-band is filled with 3 electrons in VO which is a semiconductor at low temperatures. TiO has only two electrons and is metallic at all temperatures. The bands in "Class 2" oxides are formed by strong π or σ bonding of the d-levels with ligand orbitals. Examples can be found among bronzes with rutile, perovskite or ReO_3-type structure.

78. Position of the d-levels relative to E_F. ALBERS et al.[479] have proposed the classification of magnetic materials into eight groups corresponding to the relative position of the d-levels in the band structure of the outer electrons. In Fig. 44 the valence bands contain essentially the p-electrons of the anions, the empty conduction band states correspond approximately to the ionic configuration $(3d^n)4s$ of the transition metal cation. The separation between valence and conduction band is given by the difference in electronegativity of anions and cations. The forbidden gap in isostructural compound series should vary like the distance between the points for the anions and cations in the electronegativity diagram of Fig. 45.

5, 103 (1960); F. HULLIGER, and E. MOSER: J. Phys. Chem. Solids **26**, 429 (1965); Progr. Solid State Chem. **2**, 330 (1965); F. HULLIGER: Helv. Phys. Acta **32**, 615 (1959); J. Phys. Chem. Solids **26**, 639 (1965); J. P. SUCHET: J. Phys. Radium **23**, 487 (1962); Chimie Physique des Semiconducteurs. Paris: Dunod 1962, engl. transl.: New York: van Nostrand 1965; Compt. rend. **255**, 2080 (1962); **256**, 2563 (1963); **258**, 4486 (1964); **260**, 5239 (1965); Phys. Status Solidi **2**, 167 (1962); G. BLASSE: Bull. Chim. Soc. France 1965, p. 1212; W. ALBERS, and C. HAAS: Phys. Letters **8**, 300 (1964); Coll. Intern. C.N.R.S. No. 157, p. 19, Orsay 1965 Orsay 1965 (Editions C.N.R.S. 1967).

[479] W. ALBERS, and C. HAAS: Phys. Letters **8**, 300 (1964); W. ALBERS, G. VAN ALLER, and C. HAAS: Coll. Intern. C.N.R.S. No. 157, Orsay 1965 (Editions C.N.R.S. 1967), p. 19.

Sect. 78. Position of the d-levels relative to E_F.

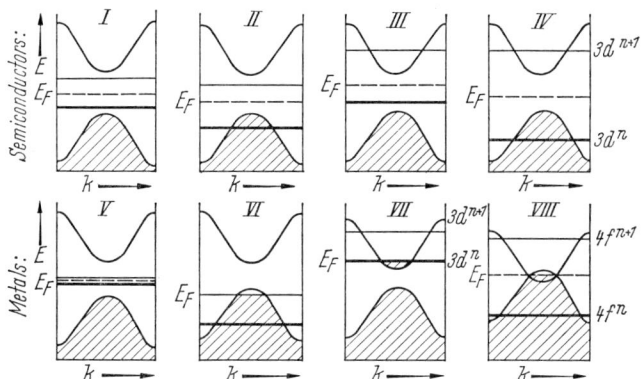

Fig. 44. Electrical and magnetic properties of magnetic materials are defined by the position of magnetic d or f-levels with respect to the filled valence and empty conduction bands k. The Fermi energy E_F separates filled and empty d and f levels. Group I to IV are semiconductors, V—VII metals. Details are given in the text. (From W. ALBERS, G. VAN ALLER, and C. HAAS: Coll. Intern. C. N. R. S. No. 157, Editions C. N. R. S. 1967.)

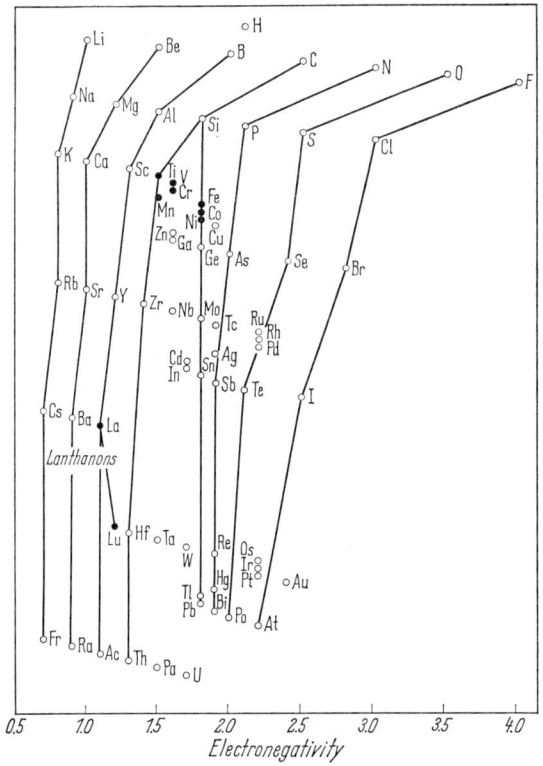

Fig. 45. The difference in electronegativity between magnetic cations (full circle) and anions defines together with stereo-chemistry the amount of covalent orbital mixing and the width of the forbidden gap between valence and conduction band states. (From L. PAULING: The Nature of the Chemical Bond. Ithaca: Cornell 1948.)

The magnetic d-levels lie somewhere below the empty conduction states. Fig. 44 includes the highest occupied $3d^n$ state which defines the magnetic moment and the lowest empty next state $3d^{n+1}$ which may contribute to conductivity. The position of the d-states relative to the states $3d^n 4s$ in the conduction band can be derived approximately by a thermo-chemical Born-Haber cycle[480]. One takes the $4s$ or a $5d$-electron away from the cation paying for the II and III ionization potential, respectively. Transfer into the p-shell of the anion gains the electron-affinity. The lattice energy is taken into account by subtraction of the Madelung-potential[481]. Furthermore, corrections are introduced for the dielectric lattice polarization[482] and the electrostatic crystal field splitting. The influence of covalent orbital overlap and elastic effects due to different size of the divalent and trivalent cations are neglected. The procedure has been carried through for NiO by MORIN and VAN HOUTEN[483]. The position of the obtained energy levels is quantitatively not very reliable. But for compound series with different cations in equal crystal symmetry a relationship can be expected between the position of the $3d$-levels and the variation of the II. and III. ionization potentials within the $3d$-element series (Fig. 46).

The electrical and magnetic properties of transition element compounds are characterized in Fig. 44 by certain combinations of the above mentioned anion and cation parameters. The materials in groups I to IV are semiconductors, i. e. insulators in the pure stoichiometric state at low temperatures and conductors at high temperature or with suitable impurities or deviations from stoichiometry. Materials in group V to VIII are magnetic metals. The following characteristics are typical of each group:

Group I. Magnetic electron levels are separated from the valence and conduction band. Since there is a gap between $3d^n$ and $3d^{n+1}$ levels, electron transport occurs only by hopping or in a narrow polaron band. The atomic magnetic moment has an integral number of spins. Examples: NiO etc. The gap between valence and conduction band is about 5 to 10 eV in $3d$-oxides.

Group II. Transport in the conducting state is carried by a combination of narrow band electrons and broad band holes. The d-levels are markedly widened by hybridization with the p-orbitals. Examples may be the semiconducting Mn-chalcogenides where the stability of the $3d$-electrons in the half-filled shell combines with the relatively low electronegativity of the chalcogens. (The other chalcogenides are metals.) Hole conductivity can produce indirect exchange such as in the Cr-chalcogen-spinels or in the Mn-perovskites which have been discussed in chapter G. It is typical that the electron mobility in n-type $CdCr_2Se_4$ is by two orders of magnitude smaller than the hole mobilities in p-type material (see Sect. 66).

Group III. Conductivity by excitation of electrons into the broad conduction band or by holes in the narrow d-band with mobilities below 1 cm^2/V·sec.

Group IV. Both electrons and holes move in broad bands. This is an alternative model for MnTe etc. to group II.

---

[480] M. BORN: Verh. deut. Phys. Ges. **21**, 13 (1919); F. HABER: ibid. p. 750; M. BORN: Atomtheorie des festen Zustandes. Leipzig: Teubner 1923; L. PAULING: Nature of the Chemical Bond, Ithaka: Cornell Univ. Press 1960; O. K. RICE: Electronic Structure and Chemical Binding, New York: Mc Graw-Hill 1940, or other physico-chemical textbooks.

[481] J. SHERMAN: Chem. Revs. **11**, 93 (1932); T. C. WADDINGTON: Adv. Inorg. Radiochem. **1**, 158 (1959); This handbook, Vol. VII-1 (1955); Landolt-Börnstein, Vol. I-4, 537 (1955).

[482] W. JOST: J. Chem. Phys. **1**, 466 (1933); N. F. MOTT, and N. J. LITTLETON: Trans. Faraday Soc. **34**, 485 (1938); E. S. RITTNER, R. A. HUTNER, and F. K. DU PRÉ: J. Chem. Phys. **17**, 198 (1949); **18**, 379 (1950).

[483] F. J. MORIN: in Semiconductors (Ed. N. B. HANNAY), New York: Reinhold 1959; S. VAN HOUTEN: J. Phys. Chem. Solids **17**, 7 (1960); G. H. JONKER u. S. VAN HOUTEN: Halbleiterprobleme **6**, 118 (1961).

Group V. These materials are metals with temperature independent carrier concentration because the filled and empty $3d$-states coalesce into a single band. Splitting of this narrow band by distortion of crystal symmetry or by magnetic ordering can produce abrupt metal to semiconductor transitions. Examples are TiO, VO etc.[484] and perhaps also[485] MnAs. A similar situation may exist with respect to the $5d$-levels in Eu-chalcogenides producing the giant magneto-resistance effect to be discussed in a later section.

Group VI. The d-electron levels are broadened by hybridization with the broad valence band. Some valence electrons are transferred into the empty $3d^{n+1}$ level. Variation in the magnetic moment and p-type metallic conductivity with large mobility and large carrier concentration result. The spin-polarization of the holes produces ferromagnetic indirect exchange. Examples are the CuCr-spinels corresponding to Lotgering's interpretation in Fig. 37.

Group VII. The filled d-levels lie in the conduction band and are broadened by interaction with broad band states. FRIEDEL[486] has estimated the width of the d-band at the Fermi level to be about $\frac{1}{3}(E_F - E_c)$ with E_c as energy at the bottom of the conduction band. In transition metals with large Fermi energies around 7 eV the d-band is quite wide in contrast to compounds with smaller conduction electron densities. The value of the atomic magnetic moment is non-integral, corresponding to a mixture of $3d^n$ and $3d^{n+1}$ ions. Examples are the magnetic transition metals and their alloys, pnictides such as MnSb, chalcogenides such as Cr_3Te_4, CrTe, $FeCr_2Te_4$ etc. In most transition metals, however, the gap between valence and conduction bands may be closed as in group VIII.

Group VIII. Refers to the RE-metals with widely spaced $4f$-levels which are completely filled or empty. In addition to the $4f$-levels one may consider the $5d$-levels located around the Fermi level as in Group VII. The empty $4f^{n+1}$ levels can widen by interaction with the holes above the Fermi level. When their position is close enough to E_F, as in Ce, Gd, the wider $4f^{n+1}$ level can become partially filled by conduction electrons and contribute a small nonintegral magnetic moment[487].

Admittedly crude, this classification of transition element compounds into eight groups demonstrates qualitatively the important parameters. In application to specific materials one must remember that the covalency effects are omitted and that these can have a dominating influence on the character of d-orbitals. The general experimental experience indicates the following trend.

Compounds containing strongly electronegative anions, such as fluorine, provide for the magnetic d-electrons a situation which is very similar to that described by localized electron models. All anions to the left of a vertical line somewhere between oxygen and nitrogen in Fig. 45 form strongly covalent compounds with $3d$-elements and application of the bandmodel with some of the corrections of sect. 76 is justified. Exceptions are some Mn-compounds, such as MnS, MnSe and MnTe which are semiconductors because of the high stability of the half-filled $3d$-shell in Mn^{2+}. FRIEDEL has argued[488] that the hydrides, borides, carbides and ni-

---

[484] F. J. MORIN: Phys. Rev. Letters **3**, 34 (1959); J. Appl. Phys. **32**, 2195 (1961); J. B. GOODENOUGH: Phys. Rev. **117**, 1442 (1962). More recent measurements indicate that VO does not exhibit a semiconductor-metal transition: J. B. GOODENOUGH: J. Appl. Phys. **39**, 409 (1968).

[485] J. B. GOODENOUGH: Solid State Res. MIT Lincoln Lab. 1963.

[486] J. FRIEDEL: J. Phys. Radium **23**, 501, 692 (1962); Bull. Soc. Chim. France 1965, p. 1186.

[487] Y. A. ROCHER: J. Phys. Chem. Solids **23**, 1621 (1962); Advances in Phys. **11**, 233 (1962). A. BLANDIN: J. Appl. Phys. **39**, 1285 (1968).

[488] J. FRIEDEL: Bull. Soc. Chim. France 1965, p. 1186.

trides behave more like alloys than compounds. The 3 d-atoms take the role of anions and fill holes in their d-band by valence electrons from the elements H, B, C or N.

79. The intermediate character of the 3d-element oxides. The 3 d-oxides are obviously in the intermediate region where the transition between the localized Heitler-London and band descriptions must occur somewhere and somehow. A critical parameter is here the stability of the 3 d-shell which increases with increasing atom number in the series (Fig. 46) and is modified by the covalent overlap with anion wave functions. It should prove interesting to study in detail the influence of ion properties and stereochemistry on the band structure and on the transition to local electron states. Unfortunately, the present experimental data on electrical properties of 3 d-oxides is rather chaotic. The majority of the earlier measurements on sintered ceramics have lost their credibility since the paramount significance of grain boundaries, of uncertain stoichiometries, and of the large concentrations of accidental impurities and lattice defects in the investigated samples has been realized. New data on single crystals or specially prepared ceramics give, at present, a fresh and clean start. Under those circumstances there is no need to add here another summary of material properties to the available review articles [489].

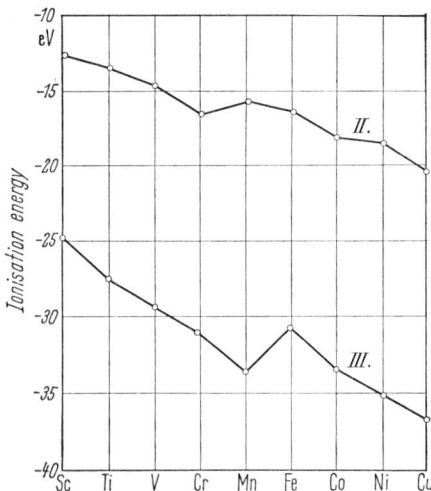

Fig. 46. The variation of the II. and III. ionization energy through the 3d-element series gives a crude measure for the separation of the 3 d^n-states from the conduction states 3 $d^n 4s$ and for the increasing stability of the 3 d^n state with increasing atomic number. (From C. E. MOORE: Atomic energy levels. Circ. Nat. Bur. Standards 467, May 1958.)

The most recent discussion of properties of transition metal oxides is given by ADLER [490]. The summarizing diagram in Fig. 47 is reproduced from one of his recent lectures on this topic [491]. It demonstrates well the wide spectrum of electrical properties found in transition element oxides. The elements at the beginning of the 3 d-series such as Ti, V and Co have, like the 4d and 5d elements Nb, Mo and Re, rather loosely bound d-electrons with relatively large orbital radii. The compounds TiO, ReO$_2$ are good metals with temperature independent Pauli-susceptibility; CrO$_2$, and R$_2$O$_2$ are ferromagnetic and diamagnetic metals, respectively. The compounds NiO, CoO, MnO, MnS have the same NaCl-type structure as the metal TiO, but with 10^{20} times larger room temperature resistivities, they are among the best insulators, owing to the high stability of their d-shell. All those insulating materials show antiferromagnetic ordering at low temperature. The electrical resistivity shows, at best, a small knee at the Néel temperature which is usually interpreted

---

[489] E. J. VERWEY: Semiconducting Materials. London: Butterworth 1951; J. B. GOODENOUGH: Magnetism and the Chemical Bond, New York: Wiley-Interscience 1963; F. J. MORIN, in Semiconductors (Ed. N. B. HANNAY), New York: Reinhold 1959; Bell. System Tech. J. **37**, 1047 (1958); G. H. JONKER u. S. VAN HOUTEN: Halbleiterprobleme **6**, 118 (1960); E. R. SCHATZ: Transition Metal Compounds. New York: Gordon and Breach 1964.

[490] D. ADLER: Solid State Phys. **21** (1968) (in press); Proc. Int. Conf. on Metal-Nonmetal Transitions, San Francisco 1968, to be published in Rev. Mod. Phys., Oct. 1968.

[491] We are indebted to Prof. ADLER for the permission to use this unpublished diagram.

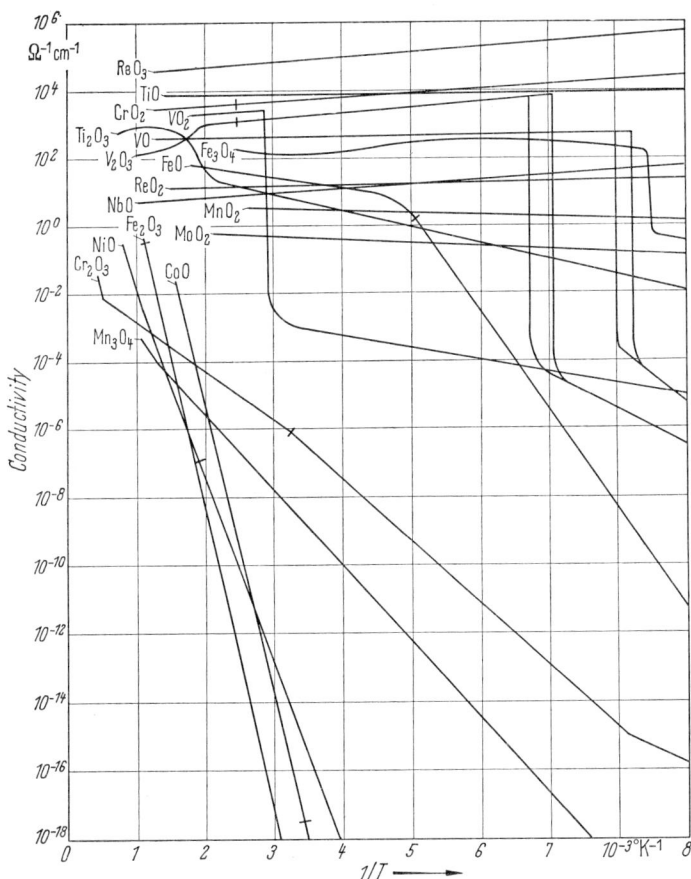

Fig. 47. Electrical conductivity as a function of reciprocal temperature for some examples of transition metal oxides. When magnetic ordering occurs in the plotted temperature range, the transition temperature is marked by a small cross line. (From D. ADLER: Private communication.) VO has been found in more recent measurements to show no semiconductor to metal transition [see J. B. GOODENOUGH: J. Appl. Phys. **39**, 409 (1968).]

as a variation in the activation energy for hopping. Between metals and insulators is a large group of compounds which show a semiconductor to metal transition with variation in temperature: VO, VO_2, V_2O_3, V_6O_{13}, V_3O_5, V_4O_7, Ti_2O_3, Ti_3O_5, CrS, FeS, NiS and probably many more. These transitions are usually not connected to the onset of magnetic ordering but to crystalline distortions. Ti_2O_3 and VO_2, as examples, have a distinct transition in their resistivity curves, but no magnetic ordering can be observed. VO and V_2O_3 have complicated magnetic properties. NiS is the only example where resistivity changes at the magnetic ordering temperature are not accompanied by large crystalline distortions. The nonmetal-metal transitions in magnetic semiconductors have been discussed recently in several papers by a group at Harvard and reviewed by ADLER[492].

[492] See chapter C; D. ADLER, and H. BROOKS: Phys. Rev. **155**, 826 (1967); J. FEINLEIB, and W. PAUL: Phys. Rev. **155**, 841 (1967); D. ADLER, J. FEINLEIB, H. BROOKS, and W. PAUL: Phys. Rev. **155**, 851 (1967); D. ADLER: Solid State Phys. **21** (1968) (in print); D. ADLER: Proc. Inter. Conf. on Metal-Nonmetal Transitions, San Francisco 1968, to be published in Rev. Mod. Phys. Oct. 1968.

80. Recent developments. The electrical properties of semiconducting anti ferromagnets had been considered until some years ago as being well described by the thermally activated hopping of carriers between localized electron states[493]. NiO with Li-doping was the classic example for this model and had been investigated in great details. The resistivity was described with a temperature independent carrier concentration with a temperature dependent mobility of 10^{-2} to 10^{-8} cm^2/V·sec. The absence of a detectable Hall effect gave additional evidence for the validity of the hopping model. The situation has changed entirely, as recently Hall mobilities of about 0.1 to 0.6 cm^2/V·sec have been reported for single crystals and carefully prepared NiO ceramics with Li-doping[494]. The Hall mobility decreases exponentially with temperature proportional to the resistivity as expected for temperature dependent carrier concentration with constant mobility, in contrast to the hopping model. Following the new results, NiO is a narrow band or polaron band semiconductor. The conducting states are formed by indirect overlap of $3d$-orbitals via anions and have an estimated width of about 0.4 eV. The effective carrier mass is of the order of 10. Near the Néel temperature, $T_N \sim 525$ °K, the spin disorder scattering changes carrier mobility and produces the knee in the resistivity-temperature curve. At temperatures below 200 to 300 °K impurity conductivity through partially compensated acceptor levels dominates. It is typical for the sample purity in transition metal compounds that the impurity conductivity is reduced but also present in undoped samples. At higher temperatures the conductivity by thermally excited holes in the $3d$-band is the dominating feature. The activation energy is between 0.1 and 0.3 eV depending on Li concentration.

It has been demonstrated that the deviating results of earlier measurements originated from grain boundary effects[495]. This development has brought quite new aspects into recent discussion of transport properties in magnetic semiconductors. One has to realize that the electron-phonon interactions in narrow bands can produce appreciable dressing of the electrons as polarons, but are probably not strong enough to localize the electrons as visualized in the hopping model. The band model (with some of the corrections we have discussed earlier) appears to be absolutely sufficient, subject to the solution of many problems concerning the Coulomb repulsions, phonon and magnon interactions, and their effects on electrical conductivity. The preparation and investigation of samples with reliable stoichiometry and purity for a larger number of materials is of paramount significance and will be decisive for further progress in this field.

---

[493] F. J. MORIN: Phys. Rev. **83**, 1005 (1951); **93**, 1195 (1954); R. R. HEIKES, and W. D. JOHNSTON: J. Chem. Phys. **26**, 582 (1957); R. R. HEIKES, A. A. MARADUDIN et R. C. MILLER: Ann. phys. **8**, 733 (1963); S. VAN HOUTEN: J. Phys. Chem. Solids **17**, 7 (1960); E. R. SCHATZ: Transition Metal Compounds. New York: Gordon and Breach 1964.

[494] P. V. ZHUZE i I. A. SHELUKH: Soviet Phys. Solid State **5**, 1278 (1963); M. KSDENZOV, N. L. ANSELM, L. L. VASILEA i V. M. LATSHEVA: Soviet Phys. Solid State **5**, 1116 (1963); M. ROILOS, and P. NAGELS: Solid State Comm. **2**, 285 (1964); S. KOIDE: J. Phys. Soc. Japan **20**, 123 (1965); A. J. SPRINGTHORPE, I. G. AUSTIN, and B. A. SMITH: Solid State Comm. **3**, 143 (1965); Phys. Letters **21**, 20 (1966); A. J. BOSMAN, H. J. VAN DAAL, and G. F. KNUVERS: Phys. Letters **19**, 372 (1965); A. J. BOSMAN, and C. CREVECOEUR: Phys. Rev. **144**, 763 (1966); P. NAGELS, and M. DENAYER: Solid State Comm. **5**, 193 (1967); I. A. AUSTIN, A. J. SPRINGTHORPE, B. A. SMITH, and C. E. TURNER: Proc. Phys. Soc. (London) **90**, 157 (1967); H. J. VAN DAAL, and A. J. BOSMAN: Phys. Rev. **158**, 736 (1967). See also review by D. ADLER: Solid State Phys. **21** (1968) (in press).

[495] M. NACHMAN, L. N. COJOCARN, and L. V. RIBCO: Phys. Status Solidi **8**, 773 (1965); A. J. BOSMAN, and C. CREVECOEUR: Phys. Rev. **144**, 763 (1966). A phenomenological analysis of the grain-boundary effects and their dependence on magnetic ordering via magnetostriction has been given for LaMnO$_3$ by J. VOLGER: Physica **20**, 49 (1954).

III. Rare earth chalcogenides.

81. Comparison with $3d$-semiconductors. One novel feature of the RE-semiconductors is that $5d$-orbitals provide an extra set of narrow but empty electron states which lie above the extremely narrow magnetic $4f$-levels. When the $5d$-bands are assumed to be narrow compared to intraatomic exchange (several 0.1 eV) a small concentration of electrons in those bands can be expected to produce relatively large correlation between electrical properties and magnetic spin order, especially at low temperatures where electron-phonon interactions are unimportant. Similar effects can be obtained from electrons trapped in narrow impurity states, as has been discussed in earlier parts of this article. It is not yet decided clearly which effects are dominant in the large magnetoresistance effects observed in semiconducting Eu-chalcogenides.

A qualitative comparison between $3d$- and RE-materials suggests the importance of the $5d$-orbital for the properties of the latter and their similarity to the $3d$-electrons. The band structure of Gd-metal, as an example, was found in Sect. 41 to be very similar to Ni with a large density of d-electrons at the Fermi-energy. The APW-calculations for EuS in Fig. 6 give a conduction band which has a structure quite similar to that calculated for isostructural TiO, but is somewhat narrower[496]. In the preceding section TiO has been shown to be a narrow band metal with two itinerant $3d$-electrons which give Pauli susceptibility but no atomic magnetic moment. The mechanical and physico-chemical properties of the metals with NaCl-type structure at the beginning of the $3d$-series[497] are quite similar to those of the isostructural conducting RE^+-compounds with one itinerant electron per cation. The high melting points, hardness, brittleness and wide range of homogeneity are common characteristics. Consequently, the characteristics of metallic NaCl-type compounds, such as CeS, ThS, US, have been assigned to itinerant $5d$- or $6d$-electrons, respectively[498]. The overlap of $3d$-wave functions and the resulting electrical properties discussed in Sect. 79 did depend on the electronegativity of the anions. Most compounds with anions to the left of O in Fig. 45 had band-like d-states. The situation with RE^{3+} compounds appears to be similar also in this respect. All oxides and fluorides exist only as insulating compounds with balanced anion and cation charges. Anions with smaller electronegativity than oxygen permit formation of metallic compounds with less sharp stoichiometry. This supports the assumption that the conditions for the formation of a narrow d-band capable of containing a non-integral number of delocalized d-electrons are very similar for $3d$-electrons in transition elements and for $5d$-electrons in RE-elements. The symmetry of the d-wave functions in the NaCl-type structure suggests a picture of RE- and transition element compounds in this structure as covalent semiconductors in which d-bands of t_{2g}-symmetry are superposed on a mostly ionic lattice[499]. The Eu-chalcogenides with their large magneto-resistive effects appear then to be a variant of a d-band material, such as TiO, where the d-band is subject to relatively strong intra-atomic exchange forces from $\frac{7}{2}$ spins at each cation site. The

---

[496] TiO: V. ERN, and A. C. SWITENDICK: Phys. Rev. **137**, A 1927 (1965).

[497] For a review see. R. KIEFFER u. F. BENESOVSKY: Hartstoffe. Berlin-Göttingen-Heidelberg: Springer 1963.

[498] For example: E. D. EASTMAN, L. BREWER, L. A. BROMLEY, P. W. GILLES, and N. L. LOFGREN: J. Am. Chem. Soc. **72**, 4019 (1950).

[499] In MnO, however, there is only very little electron-density in the diagonal direction as seen in electron density maps obtained from neutron diffractions: N. M. OLEKHNOVICH, N. N. SIROTA, in: Chemical bonds in semiconductors and solids (Ed. N. N. SIROTA), transl. from Russ. Consultant Bureau, New York 1967.

possibility arises that a magnetically induced metal-semiconductor transition can occur in the neighborhood of the magnetic ordering temperature.

82. Giant magneto-resistance effects. In section 61 we have described general properties of Eu-chalcogenides which are made electrically conductive by replacing a certain part of Eu^{2+}-cations with trivalent RE-ions. Fig. 33 shows the variation of magnetic Curie-Weiss temperature with increasing RE^{3+} concentration when the electrical resistivity was decreased by many orders of magnitude. Indirect exchange via conduction electrons appeared to give a suitable description of the magnetic properties of samples with larger Gd^{3+}-concentrations. Difficulties occurred at small carrier concentrations because the Fermi energy of the conducting electrons became comparable to or smaller than the intra-atomic exchange which could no longer be treated as a small perturbation.

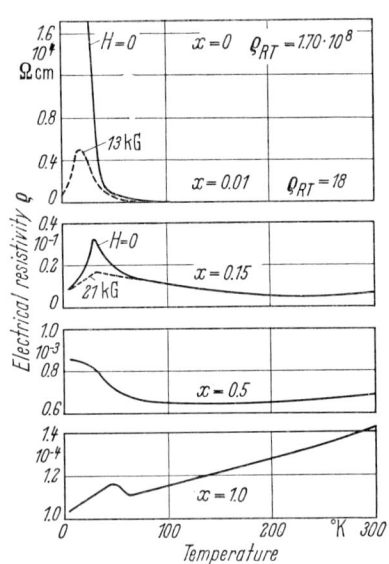

Fig. 48. Temperature dependence of resistivity in polycrystalline $Eu_{1-x}Gd_xSe$ samples for various Gd concentrations x. (From F. HOLTZBERG, T. R. MCGUIRE, S. METHFESSEL, J. C. SUITS: Proc. Int. Conf. on Magnetism, Nottingham 1964, p. 470.)

Fig. 48 surveys the temperature dependence of resistivity in four polycrystalline samples of EuSe diluted with different parts x of GdSe[500]. The pure GdSe is an antiferromagnetic metal with electrical properties very similar to Gd-metal. Its Néel temperature at about 50 °K is marked by a resistivity peak due to spin disorder scattering. The sample of $Eu_{1-x}Gd_xSe$ with $x=0.5$ is still metallic with an electron concentration of about 5×10^{21} cm^{-3}, i.e. 0.3 to 0.4 electrons per molecule. A positive Curie-Weiss temperature and ferromagnetic Curie temperature lie close together at 20 °K. The resistivity increase with decreasing temperature below 100 °K does not vary with applied magnetic field and is, therefore, considered to be related not to magnetic scattering but rather to impure grain boundaries or other unexplained effects. The sample with $x=0.15$ has semiconducting properties with a field-dependent peak in resistivity at its Curie temperature of 27 °K (The paramagnetic Curie-Weiss temperature is remarkably higher, $\Theta = 46$ °K). The resistivity peak at the magnetic Curie temperature of the sample with $x=0.01$ has an amplitude of more than 10^8 ohm·cm which lies outside the limits of Fig. 48. The peak amplitude is unusually sensitive to applied magnetic fields. The peak is remarkably reduced by an applied field of 13×10^3 gauss, as an example, but remains still to more than one order of magnitude higher than the low temperature resistivity.

In Table 6 the carrier concentration and activation energies for selenide samples with various Gd-concentrations are listed. The two samples with smallest Gd-concentrations show a large deviation between the nominal carrier concentration derived from chemical composition and the true concentration calculated

---

[500] F. HOLTZBERG, T. R. MCGUIRE, S. METHFESSEL, and J. C. SUITS: Proc. Int. Conf. Magnetism, Nottingham 1964; S. METHFESSEL: Z. angew. Phys. **18**, 414 (1965).

from resistivity and Hall measurements. This deviation depends strongly on preparation conditions and can be varied to a certain amount by suitable annealing treatment as already discussed in 61. An interesting and not yet clearly answered question concerns if and how the change in electrical properties around $x = 0.3$ is related to the irregular variation of the lattice constant in this concentration range, as shown in Fig. 33. It is conceivable that the relatively enlarged lattice constant for low x-values is a direct consequence of the large concentration of lattice defects which trap most of the excess electrons, and should, therefore, depend on preparation conditions. But it is also possible that for larger concentrations the metallic bonding due to the excess electrons begins to make a markable contribution to the stability of the crystal.

Table 6. *Electrical properties of $Eu_{1-x}Gd_xSe$ samples at room temperature*
[from S. von Molnar and S. Methfessel: J. Appl. Phys. **38**, 959 (1967)]

Sample	$n_{nom.}^{(a)}$ (cm^{-3})	$n_{true}^{(b)}$ (cm^{-3})	$\mu^{(c)}$ (cm^2/vsec)	q (ev)$^{(d)}$	$\Theta^{(e)}$	$T_c^{(f)}$
$Eu_{.99}Gd_{.01}Se$ *	1.7×10^{20}	3.10×10^{18}	15	3.2×10^{-2}	20	8
$Eu_{.95}Gd_{.05}Se$ *	8.4×10^{20}	3.93×10^{19}	12	1.4×10^{-2}	35	—
$Eu_{.65}Gd_{.35}Se$	6.2×10^{21}	1.04×10^{21}	0.4	2.1×10^{-3}	28	27
$Eu_{.50}Gd_{.50}Se$	9.3×10^{21}	4.0×10^{21}	1.6	1.7×10^{-3}	24	24

* single crystal.
(a) Carrier concentration caluclated from the nominal chemical composition and experimental lattice parameter.
(b) Carrier concentration according to the relation $R = 1/ne$. ($R = -$ Hall constant).
(c) Mobility according to the relation $\mu = |R|/\rho$.
(d) Activation energy calculated from the temperature dependence of resistivity.
(e) Paramagnetic Curie temperature.
(f) Ferromagnetic Curie temperature.

In Fig. 8 the temperature dependence of resistivity in a $Eu_{0.95}Gd_{0.05}S$ sample with about 2×10^{20} cm^{-3} carriers was shown to be in good agreement with critical scattering theory. It is interesting to note, that the behavior of the resistivity in the neighborhood of the Curie temperature in this sample is quite similar to that measured in n-conducting ferromagnetic spinels, doped with In or Ga[501]. In the spinel, however, the Curie temperature is found not to depend on the electron concentration.

When the carrier concentration is reduced to about 4×10^{19} cm^{-3} the peak at the Curie temperature rises very strongly, as shown in Fig. 49 for the single crystal $Eu_{0.95}Gd_{0.05}Se$ listed in Table 6. The resistivity above the Curie temperature varies exponentially with the reciprocal temperature as usual in semiconductors. The slope of the curve in Fig. 49 corresponds to an activation energy of about 10^{-2} eV. However, the measurements of Hall effect, Seebeck voltage and optical free carrier absorption for red light indicate that the resistivity maximum at T_c is a *pure mobility effect* at constant carrier concentrations[502]. The anomalous Hall effect related to the spontaneous magnetization is negligible

---

[501] C. Haas, A. M. J. G. van Run, P. F. Bongers, and W. Albers: Solid State Comm. **5**, 657 (1967); H. W. Lehmann: Phys. Rev. **163**, 488 (1967).
[502] It is conceivable, that the lowering of the lowest unoccupied states with increasing magnetic order, which gives the magnetic red shift of the optical absorption edge, decreases the separation between occupied and unoccupied states. Then the activation energy and carrier concentration would become a function of spin order. Contributions of this effect to electrical conductivity have been observed in Cr-spinels [H. W. Lehmann: Phys. Rev. **163**, 488 (1967)] but not in Eu-chalcogenides.

compared to that usually found in ferromagnetic metals[503]. Further remarkable points are: a) the strong resistivity decrease with spin alignment forced by applied magnetic fields which results in a "giant" negative magneto-resistance, b) the persistence of magnetic effects on resistivity up to temperatures of 2 to 3 times T_c and c) the fact that the resistivity in the ordered spin state at low temperatures is actually smaller than at room temperature (a behavior usually found only in metals).

Fig. 50 shows the resistivity curve for EuS containing 5% LaS and about 8×10^{18} cm^{-3} conduction electrons[504]. The peak at the Curie temperature extends

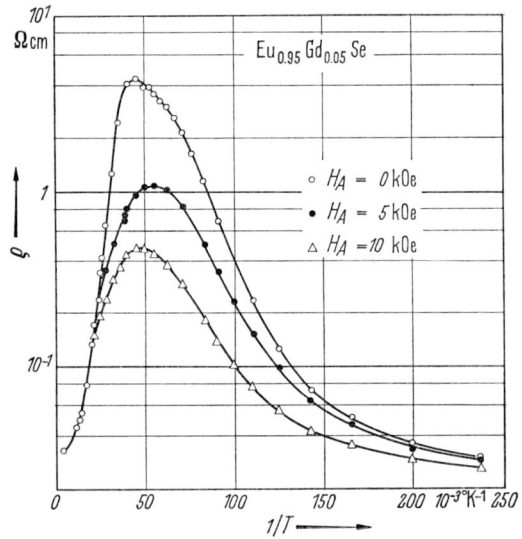

Fig. 49. The dependance of log ϱ on reciprocal temperature in a single crystal of Eu$_{0.95}$Gd$_{0.05}$Se without and with applied magnetic field. [From S. von Molnar, and S. Methfessel: J. Appl. Phys. **38**, 959 (1967).]

now over more than seven orders of magnitude and is clearly beyond the scope of any "weak coupling" approach which treats intraatomic exchange as a small perturbation compared to bandwidth and electron-phonon interactions[505] (e. g. discussion in sect 21 in C). Remarkably the resistivity effect is rather independent of the nature of the impurity atoms, since La^{3+} is diamagnetic and Gd^{2+} has a $\frac{7}{2}$ spin as Eu^{2+}. The Figs. 51 and 52 demonstrate how the character of magnetic spin ordering at low temperatures modifies the resistivity temperature dependence. In Eu$_{0.99}$Gd$_{0.01}$Se a vestige of the complicated spin structure of pure EuSe (which disappears for higher conduction electron concentrations) causes the resistivity to rise again at low temperatures. Applied magnetic fields of sufficient strength overcome this type of spin order and the associated rise in resistivity.

---

[503] There is another possibility that we actually observe a two band conductivity with the Hall effect coming from carriers in a broad band which is very weakly coupled to the spin disorder. The absence of an anomalous Hall effect may also support the Kasuya model's assumption of s-character for the conduction electrons (Sects. 62, 84).

[504] We are indebted to Dr. S. von Molnar for the permission to use unpublished data.

[505] The irregular electrical properties of Eu-chalcogenides have been reported first by J. C. Suits: Bull. Am. Phys. Soc. **8**, 381 (1963) for impure EuSe and by R. R. Heikes, and C. W. Chen [Physics **1**, 159 (1964); C. W. Chen, F. Carter, and R. R. Heikes. J. Appl. Phys. **36**, 1160 (1965)] for La-doped EuS.

$Eu_{0.8}Gd_{0.2}Te$ has a positive paramagnetic Curie-Weiss temperature of about 28 °K and a Curie temperature of about 10 °K. But the magnetization has extremely high saturation fields of about 4.5×10^4 oe below the Curie temperature. This was interpreted by the formation of canted or ferromagnetic spin clusters around the excess electrons in the antiferromagnetic layer structure[506]. The temperature dependence of the resistivity indicates the trapping of the conduction electrons in such spin clusters.

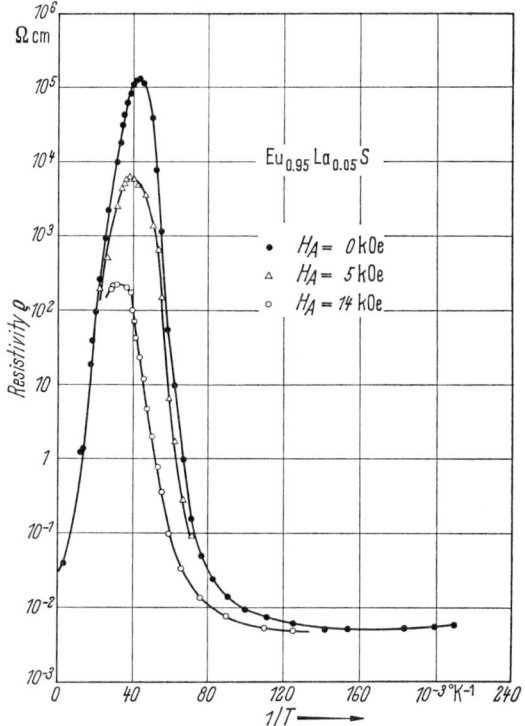

Fig. 50. Magnetic resistivity peak in single crystalline EuS containing 5% diamagnetic LaS (S. von Molnar: Private communication). The applied field H_A is the parameter.

The unusually strong effects of magnetic spin order on the electrical resistivity for small carrier concentrations below about 10^{19} cm^{-3} correspond to reductions in carrier mobility from about 10 cm^2/V·sec at room temperature and in the ferromagnetic state to less than 10^{-8} cm^2/V·sec near the Curie temperature T_c. Mobility variations of this magnitude cast doubt upon the validity of free electron theory in this concentration range. There must be a magnetically caused transition from band states into local or magnetic polaron states in neighborhood of the Curie temperature.

83. Exchange effects on 5d-electrons. We have already mentioned the description of doped Eu-chalcogenides by two alternatives, of having a narrow 5d-conduction band in the forbidden gap below the conduction band, or of introducing impurity electron states within the forbidden energy gap. In the first case the excess electrons are removed from the RE^{3+} donors and reside in bound states of large radius which

---

[506] F. Holtzberg, T. R. McGuire, and S. Methfessel: J. Appl. Phys. **37**, 976 (1966).

condense out of the one-electron t_{2g}-band of the Eu^{2+}-ions. In principle, the magnetic polaron theory applies to the narrowing of a one-electron spin-up band which occurs when some of the Eu-spins become reversed by thermal excitation. The band modification by spin disorder is most drastic when for small conduction electron concentrations the Ruderman-Kittel oscillations in the conduction electron polarization have longer wave length than the thermal fluctuations in the spin

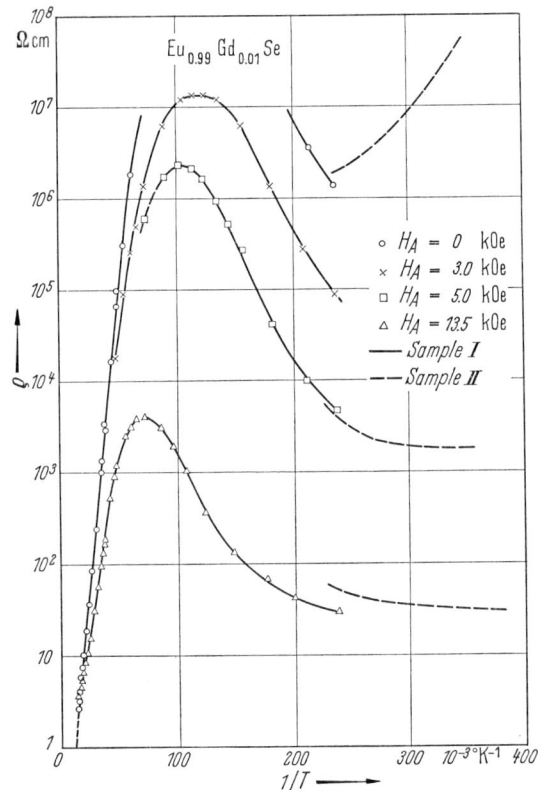

Fig. 51. The electrical resistivity of EuS single crystals containing 1% GdSe increases again when the helical spin order develops at low temperatures. The spin order can be modified by applied magnetic fields. [From S. von Molnar, and S. Methfessel: J. Appl. Phys. **38**, 959 (1967).]

lattice[507]. The mutual electron interaction is then very ineffective and if the Fermi energy is small compared to the Hund's rule exchange the electrons can be trapped by the spin alignment they produce themselves in their environment. The trapping energy would be greatest at the Curie temperature. The resistivity could drop exponentially with temperature above T_c because of thermal excitation of electrons out of the trapped state. Since the trapped electrons are not fixed relative to the lattice but merely have their mobility reduced by the dressing with the

---

[507] The effect of spin order on the energy of the 5d-band has been demonstrated in pure EuS and EuSe not only by the red shift of the optical absorption edge but also by electron tunneling from Al-electrodes into these materials. The barrier voltage decreases by about 25% to 35% when the spin order increases below the Curie-temperature or in applied fields [L. Esaki, P. J. Stiles, and S. von Molnar: Phys. Rev. Letters **19**, 852 (1967)].

surrounding spin cloud, this will not be observable as variation of carrier concentration in Hall measurements etc. At present, the question remains open, whether the spin-polaron can be "small" enough to explain the observed resistivity peaks in the neighborhood of T_c.

84. Magnetic breakdown of impurity bands. The impurity models deny that the excess electrons from RE^{3+} additions can be ionized into the $5d$-band of the Eu-lattice at lower temperatures. The Coulomb attraction from the RE^{3+} core is

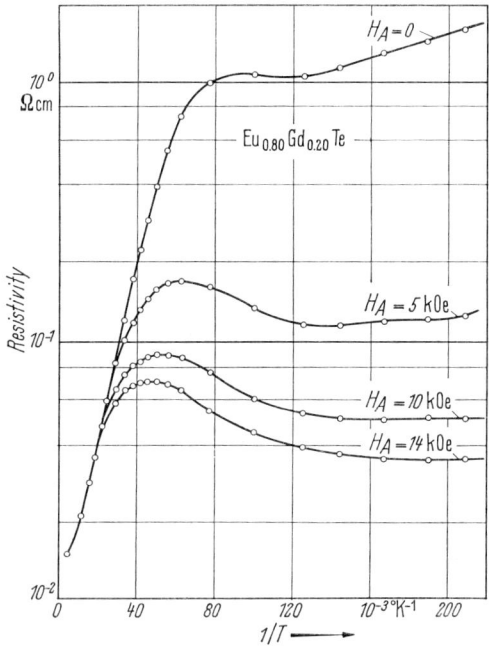

Fig. 52. The temperature dependence of the resistivity of EuTe containing 20% GdTe reflects the antiferromagnetic order and its modification by applied magnetic fields. (From S. von Molnar: Private communication.)

considered to be the longest-range perturbation and forms a bound state which is subject to perturbation effects of the intra-atomic exchange and electron-phonon interactions. It is difficult to estimate the position of these impurity states within the forbidden gap of the Eu-chalcogenide lattice. Since the third ionization energies of La and Eu have about the same value in Table 1, a Born-Haber cycle would bring the $5d$-electron level of La^{2+} at about the same energy as the $4f^7$-level of Eu^{2+} in the lattice, i.e. about 1 to 2 eV below the empty conduction band. However, further corrections such as larger covalent destabilization, dielectric lattice polarization by the larger core charge and the elastic energy due to the larger size of La^{2+} compared to Eu^{2+} can increase the energy of the $5d$-impurity level appreciably and can merge it into the $5d$-band of the Eu^{2+} ions[508].

---

[508] The free Eu-atom has the configuration $4f^7 6s^2$, and the empty $5d$-states about 1 eV above the ground state. But the larger destabilization of the s-electrons by covalent overlap with anion orbitals compared to the weakly covalent t_{2g}-orbitals which are additionally stabilized by cation-cation overlap can result in low lying d-levels as shown in the band structure of EuO in Fig. 6. In La and Gd the $5d$ electrons are more strongly bound than the $6s$-electrons.

Impurity bands in semiconductors are associated with strongly correlated states[509]. The electron-electron repulsion is larger than the band width and the band contains only one or two electron states per impurity atom (depending if the exchange interaction is larger or smaller than the band width). The width of the impurity band depends on the average distance between donors, i. e. on the impurity concentration. For small concentrations the overlap between excess electron wave functions on neighboring donors is negligible. Conductivity can take place by thermal excitation into the conduction band or by hopping from lower to higher valent impurity atoms. The activation energy for impurity hopping includes two parts[510]: the separation energy of the hole from the lattice defect (which produced the hole by binding or compensating one electron) and the electron-phonon interaction.

The separation energy varies with the distance of the donor impurity orbitals from the compensating acceptor lattice defects and fluctuates locally with the concentration of the lattice defects. It disappears when the overlap between impurity wave functions becomes comparable to the amplitude of the local fluctuations in Coulomb attractions from donor states with and without compensating lattice defects in its neighborhood. Kasuya and Yanase[510] have estimated that this occurs when about one percent of cations are replaced by higher valent impurities.

The contribution of the electron-phonon interaction to the hopping activation energy arises from the Landau-trapping of the carriers in the electrostatic polarization cloud surrounding them in the lattice. If the trapping energy is large compared to that of an optical phonon the lattice distortion can not follow the charge transfer and an activation energy for transport is observed which increases exponentially with temperature. For small trapping energy the carrier and its polarization cloud can move through the lattice as a small or large polaron. When the orbital overlap is sufficiently large at higher impurity concentrations both contributions to the activation energy disappear and the impurity band sustains quasimetallic conductivity without activation energy.

The number of carriers in the impurity band depends on the amount of compensation by lattice defects, interstitials with effective positive charge etc. which is measured by the deviation between the nominal and the true carrier concentration in Table 6. Materials with a true carrier concentration of accurately one per impurity atom would be insulators in the ferromagnetic state if the band width is smaller than intra-atomic exchange splitting, otherwise they are metals[511].

Heikes and Chen[512] have interpreted qualitatively the resistivity temperature dependence in La-doped EuS by spin-polaron effects on the impurity band. The band width is supposed to decrease exponentially when the magnetic order decreases with increasing temperature, as has been suggested by Wolfram and Callaway[513]. In the neighborhood of the Curie temperature the band width be-

---

[509] Impurity banding in semiconductors has been discussed in many papers, such as: T. Kasuya: J. Phys. Soc. Japan **13**, 1096 (1958); T. Kasuya, and S. Koide: J. Phys. Soc. Japan **13**, 1287 (1958); N. F. Mott, and W. D. Twose: Advances in Phys. **10**, 107 (1961); H. Fritzsche: J. Phys. Chem. Solids **6**, 69 (1958); T. Matsubara, and Y. Tohozawa: Progr. Theoret. Phys. (Kyoto) **26**, 739 (1961); K. Takeyama: J. Phys. Soc. Japan **23**, 1013 (1967).
[510] T. Kasuya, and A. Yanase: To be published Rev. Mod. Phys., Oct. 1968.
[511] The validity of the critical scattering theory of De Gennes-Friedel for the sample in Fig. 8 indicates that the modification of the impurity band by intra-atomic exchange must be negligibly small in this case.
[512] R. R. Heikes, and C. W. Chen: Phys. **1**, 159 (1964).
[513] T. Wolfram, and J. Callaway: Phys. Rev. **127**, 1605 (1962), but see chapter C.

comes so small that the electrons are trapped by phonons in localized polaron states on lattice sites. At higher temperatures the resistivity decreases as $\varrho = \varrho_0 T e^{q/KT}$, when electrons are thermally excited to hop from lattice site to lattice site. The experimentally determined activation energy increases from 0.016 to 0.028 eV with increasing La-concentration from 0.01 to 0.1 parts. This surprising variation is unexplained and contradicts the results for EuSe in Table 6. In the ferromagnetic state below T_c the conductivity is supposed to occur in a wide impurity band without activation energy.

KASUYA and YANASE[514] discuss the electrical properties of the $Eu_{0.99}Gd_{00.1}Se$ sample of Fig. 51 in more detail. The intraatomic exchange is assumed to be larger than both the band width and the electron-phonon interaction but smaller than the separation of the impurity state from the conduction band. The conduction band is assumed to have mainly 6s-character at its bottom and an electron mobility of about 10^3 cm/V·sec is estimated. In the ferromagnetic state at low temperatures the conductivity occurs in a narrow impurity band without observable activation energy. The carriers are holes, produced by compensation of every tenth impurity electron. The holes are supposed to move in an impurity band on the Gd^{3+} ions which lies about 0.6 eV below the bottom of the broad conduction band. The orbitals of the impurity electrons are rather concentrated around the Gd^{3+} ions and extend only to the nearest Eu^{2+} neighbors[515]. When the spin lattice becomes disordered with increasing temperature the presence of electrons at occupied RE^{3+} impurities polarizes the spins at the neighbor Eu-ions more than around an empty impurity site or elsewhere in the lattice. The neighbors of an occupied impurity state thus form a magnetic "molecule" with a giant moment and enhanced magnetic interaction, as already discussed in chapter G. The gain in s—f exchange resulting from the higher spin order around the occupied state requires an activation for hopping to an empty site in order to align there the surrounding spins. The related activation energy is strongly temperature dependent, disappearing at very low and very high temperatures because the spin order becomes equal around both the occupied and empty sites, (the alignment is ferromagnetic or perfectly random, respectively). The maximum value is expected to occur around the Curie temperature of the sample where the occupied impurity sites are clearly distinguished from the remaining lattice by the spin order of their environment. The temperature dependence of the magnetic activation energy is related to the transition from band conductivity at low temperatures to hopping at high temperatures, in good agreement with the experiment. The large negative magnetoresistance effect is explained as the consequence of improved spin alignment around empty sites by the magnetic field which reduces the activation energy[516]. The transition from band conductivity to hopping conductivity should be observable in the Hall effect as a reduction in the number of mobile carriers. The temperature independence of the Hall effect in other Eu-chalcogenide samples has been interpreted as indication for two band conductivity. With increasing temperature carriers are excited over the 0.6 eV gap into the conduction band where they have a large mobility. Agreement with experimental data at room temperature is obtained for 6×10^{14} cm^{-3} electrons with a mobility of about 10^3 cm^2/V·sec in the s-type conduction band. This electron conductivity and the hole conductivity in the impurity band are additive.

---

[514] T. KASUYA, and A. YANASE: To be published Rev. Mod. Phys., Oct. 1968.

[515] The occupation ratio is roughly 2:6:2 for the central atom, the nearest neighbor Eu-atoms and the rest of the lattice.

[516] It may be viewed also as resulting from a decrease of the differential susceptibility in the presence of finite magnetic fields [see Eqs. (21.6) and (21.7)].

The presently available experimental data are not sufficient for a final decision whether the magnetically induced transition from band to local states is a property of impurity centers or of the intrinsic outer $5d$-orbitals in RE-materials. Independent of such details, the electrical and magnetic properties of Eu-chalcogenides demonstrate important consequences of exchange for the electron band structure in magnetic semiconductors. This is a valuable complement to those semiconducting $3d$ element compounds in which the magnetic properties have been found to be a minor perturbation to the single-electron band structure, compared to electrostatic interactions of electrons with one another and with the ionic crystal lattice.

Acknowledgements. We are indebted to many colleagues for valuable discussions. We thank D. ADLER, E. CALLEN, T. KASUYA and S. VON MOLNAR for permission to use unpublished work. J. D. AXE, G. BURNS, H. CALLEN, A. J. FREEMAN, M. J. FREISER, J. B. GOODENOUGH, J. F. JANAK and T. KASUYA have read large parts of the manuscript and have kindly contributed their valuable comments. One of us (D. C. M) thanks the department of Physics at Yale University, particularly Professors W. LAMB and V. HUGHES, for their hospitality during the Fall of 1967.

Sachverzeichnis.

(Deutsch-Englisch.)

Bei gleicher Schreibweise in den beiden Sprachen sind die Wörter jeweils einfach aufgeführt.

Abfall, exponentieller, *exponential decay* 21, 22.
Absorption 6, 12, 476.
—, Magneto-, *magnetic absorption* 482.
Absorptionslinie, *absorption line* 217.
Absorptionsspektrum von Eu^{2+}-Ionen, *absorption spectrum of Eu^{2+} ions* 474.
adiabatische Entmagnetisierung, *adiabatic demagnetization* 1.
adiabatische Näherung, *adiabatic approximation* 43, 51.
adiabatische Suszeptibilität, *adiabatic susceptibility* 5, 10, 17.
akustische paramagnetische Resonanz, *acoustic paramagnetic resonance* 91.
akustischer Zweig, *acoustic branch* 37.
alkylsubstituiertes Benzol, *alkyl-substituted benzene* 176.
Allylradikal, *allyl radical* 189.
Alterungseffekt, *aging effect* 10.
Anderson-Mechanismus, Kramers-, *Kramers-Anderson mechanism* 379.
angeregte Konfiguration, *excited configuration* 321.
angeregtes Triplettsystem, *excited triplet system* 210.
Anion 185.
—, Anthracen-, *anthracene anion* 156.
—, Benzosemichinon, *benzosemiquinone anion* 180.
—, Butadien-, *butadiene anion* 160.
—, Cyclooctatetraen-, *cyclo-octatetrene anion* 176.
—, Phenyl-, *benzene anion* 186.
Anionradikal, *anion radical* 155.
anisotrope Kopplung zwischen Elektronenspin und Magnetfeld, *anisotropic coupling between electron spin and magnetic field* 133.
Anisotropie in Relaxationszeiten, *anisotropy in relaxation times* 63.
— in Spinbahnkopplung, *anisotropy in spin-orbit coupling* 64.
— in der Verformung, *anisotropy in the strain* 85.
— in Zustandsfunktionen, *anisotropy in state functions* 63.
anomaler Magnetowiderstand, *giant magneto-resistance effect* 554.
Anregung, Triplett-, *triplet excitation* 219.
Anthracene 185.
Anthracenanion, *anthracene anion* 156.

antiferromagnetische Kopplung, *antiferromagnetic coupling* 267.
Antiferromagnetismus, Bandmodell des, *band model of antiferromagnetism* 417.
—, kompensierter, *compensated antiferromagnetism* 267.
—, nichtkompensierter, *not-compensated antiferromagnetism* 267.
Atom, α-Kohlenstoff-, *α-carbon atom* 159.
—, β-Kohlenstoff-, *β-carbon atom* 159.
— in Festkörpern, *atom in solids* 245.
—, Liganden-, *ligand atom* 132.
atomares Energieniveau, *atom energy level* 401.
Aufspaltung, Bahn-, *orbital splitting* 131.
—, Bahnniveau-, *orbital level splitting* 129.
—, elektronische, *electronic splitting* 107, 108, 133, 136, 139.
—, Hyperfein-, *hyperfine splitting* 14, 99, 104, 108, 133, 135, 139, 140, 148.
—, Kristallfeld-, *crystal field splitting* 407, 534.
—, Manganathyperfein-, *manganate hyperfine splitting* 140.
—, Superhyperfein-, *superhyperfine splitting* 132, 184, 255.
—, Winkelverteilung der elektronischen, *angular variation of electronic splitting* 107, 140.
—, zeitabhängige Hyperfein-, *temperature-dependent hyperfine splitting* 198.
Aufspaltungseffekt, Stark-, *Stark splitting effect* 108, 127.
Aufspaltungsfaktor, Landé-, *Landé splitting factor* 126.
Aufspaltungskonstante, ^{13}C-, *^{13}C splitting constant*, 165.
—, ^{19}F-Fluor-, *^{19}F fluorine splitting constant* 167.
—, ^{14}N-, *^{14}N splitting constant* 166.
—, Vorzeichen der Hyperfein-, *sign of the hyperfine splitting constant* 168.
Aufspaltungsparameter, D, *splitting parameter, D* 137.
Aufspaltungstensor, Hyperfein-, *hyperfine splitting tensor*, 147, 148, 187.
Austausch, Deuterium-, *exchange, deuterium* 188.
—, direkter, *direct exchange* 432.
—, Doppel-, *double exchange* 351, 364, 367, 411, 446, 450, 520, 535.
—, indirekter, *indirect exchange* 410, 430, 440, 487, 515, 521.

Austausch, kinetischer, *kinetic exchange* 270, 330, 370, 432.
—, potentieller, *potential exchange* 270, 330, 370.
—, Spin, *spin exchange* 199, 203.
—, Super-, *super exchange* 336.
Austauschintegral, Definition für 2 Elektronen, *exchange integral, definition for 2 electrons* 266, 272, 275.
—, — für 3 Elektronen, *exchange integral, definition for 3 electrons* 314.
—, direktes, Vorzeichen, *exchange integral, direct sign* 306.
—, Einfluß der Elektron-Kern-Wechselwirkung, *exchange integral, influence of electron nuclear interaction* 308, 310.
—, H_2-Problem, *exchange integral, H_2 problem* 276, 334.
—, Heisenbergsches, *Heisenberg exchange integral* 277, 330.
—, indirektes, halbempirische Regeln zur Abschätzung, *exchange integral, indirect, semiempirical rules for the estimation* 379.
—, —, im Kristall, *exchange integral, indirect, in the crystal* 366.
—, —, Vorzeichen, *exchange integral, indirect sign* 373.
—, —, für 3 Zentren und 4 Elektronen, *exchange integral, indirect, for 3 centres and 4 electrons* 343, 347, 349, 359.
—, —, zahlenmäßige Bestimmungen, *exchange integral, indirect, numerical calculations* 381.
— zwischen orthogonalen Funktionen, *exchange integral between orthogonal functions* 276, 350.
Austauschmechanismus, nicht-Kramersscher, *non-Kramers exchange mechanism* 269.
Austausch, Mechanismus des indirekten, *mechanism of indirect exchange* 268, 337.
Austauschwechselwirkung, *exchange interaction* 14, 20, 34, 35, 71, 73, 75, 77.
—, biquadratische, *biquadratic exchange interaction* 312.
—, direkte, *direct exchange interaction* 268, 286.
—, indirekte, *indirect exchange interaction* 268, 335.
—, —, Andersons neuer Lösungsweg, *exchange interaction, indirect, Anderson's new approach* 269, 368.
—, —, im Kristall, *exchange interaction, indirect, in the crystal* 365.
—, paramagnetische Relaxation durch modulierte, *paramagnetic relaxation by modulated exchange interaction* 71.
Auswahlregeln, *selection rules* 459.

Bahn, d-, *d orbital* 127, 128, 129.
—, $3d$-, $3d$ orbital 127.
—, e_g-, e_g *orbital* 128.
—, π-, π *orbital* 127.
—, σ-, σ *orbital* 127.
—, s-, *s orbital* 128.
—, t_{2g}-, t_{2g} *orbital* 128.

Bahn, $4f$-, $4f$ *orbital* 454.
—, Einelektronenmolekül-, *one-electron molecular orbital* 378.
—, nichtlokalisierte, *delocalised orbital* 99.
—, nichtlokalisierte Molekül-, *delocalised molecular orbital* 95, 101, 126.
—, orthogonale Molekül-, *orthogonal molecular orbital* 321.
Bahnaufspaltung, *orbital splitting* 131.
Bahndrehimpuls, Unterdrückung des, *quenching of the orbital angular momentum*, 36.
Bahnen, räumliche Verteilung von $3d$-, *spatial distribution of $3d$ orbitals* 127.
Bahnenergieniveau, *orbital energy level* 129.
Bahn-Gitterpotential, *orbit-lattice potential* 48.
Bahn-Gitterwechselwirkung, *orbit-lattice interaction* 48.
Bahnniveauaufspaltung, *orbital level splitting* 129.
Bahnradius der SE, *orbital radius of RE* 452.
Bahnwechselwirkung, Spin-, *spin-orbit interaction* 35, 462.
Bandbreite, *bandwidth* 109, 113.
Bandmodell des Antiferromagnetismus, *band model of antiferromagnetism* 417.
Bandtheorie, *band theory* 397.
Bandstruktur, *band structure* 392.
— von Eu-Chalkogeniden, *band structure of Eu chalcogenides* 477, 511.
—, EuS-, *band structure for $Eu^{++}S$* 418.
— für SE-Metalle, *band structure for RE metals* 472.
Beweglichkeit von $3d$, $4f$-Elektronen, *mobility of $3d$, $4f$ electrons* 398.
Benzol, *benzene* 219.
—, alkylsubstituiertes, *alkyl-substituted benzene* 176.
Benzosemichinon, *benzo-semiquinone* 104.
—, Dibutyl-, *dibutyl-benzo-semiquinone* 106.
Benzosemichinonanion, *benzosemiquinone anion* 180.
Bestrahlungsschaden, *irradiation damage* 95, 98.
Bethe-Slater-Kurve, *Bethe-Slater curve* 307.
Bindung, kovalente, *covalent bonding* 67, 69.
—, Kristall-, *crystal bonding* 67.
biochemisch, *biochemical* 95.
Biradikal, *biradical* 205, 226.
Bloch-Gleichung, *Bloch equation* 8.
Bloch-Welle, *Bloch wave* 321.
Bloembergen-Rowland-Kopplung, *Bloembergen-Rowland coupling* 411.
Bogolyubov, Störungstheorie nach, *Bogolyubov's perturbation theory* 322, 365.
Bohrsches Magneton, *Bohr magneton* 95, 404.
Bolton-Beziehung, Colpa-, *Colpa- Bolton relation* 158.
Boltzmann-Faktor, *Boltzmann factor* 96.
Boltzmann-Relaxation, Maxwell-, *Maxwell-Boltzmann relaxation*, 30.
Boltzmann-Verteilung, *Boltzmann distribution* 72.
Born-Haber cycle 548.
Brillouin function 405.

Sachverzeichnis.

Brons-van Vleck-Formel, *Brons-van Vleck form* 74, 81.
Brückenschaltung, *bridge circuit* 11.
Butadienanion, *butadiene anion* 160.

Casimir und Du Pré, Theorie von, *theory of Casimir and Du Pré* 4.
—- und Du Pré-Formalismus, *Casimir and Du Pré formalism* 77.
—- und Du Pré-Relaxationskurve, *Casimir and Du Pré relaxation curve* 7.
^{13}C-Aufspaltungskonstante, ^{13}C *splitting constant* 165.
CdS 255.
Chalkogenid, Eu-, *Eu chalcogenide* 476, 481, 506, 512, 554.
—, Gd-, *Gd-chalcogenide* 518.
Chalkogenide, Eu-, magnetische Rotverschiebung, *Eu chalcogenides, magnetic redshift* 483.
Chalkogeniden, Bandstruktur von Eu-, *bandstructure of Eu chalcogenides* 477, 511.
chemischer Übergang, *chemical transfer* 202.
CH-Gruppe, *CH group* 101, 102.
CH$_2$ 101.
CH$_2$-Gruppe, *CH$_2$ group* 102.
CH$_3$-Gruppe, *CH$_3$ group* 101, 102, 103.
Chlorbenzosemichinon, *chlorobenzosemiquinone* 105.
Chlorsuperhyperfeinstruktur, *chlorine superhyperfine structure* 252.
Chrom-Halogen, *chromium halide* 481.
Cl$_2^-$-Spektrum, *Cl$_2^-$ spectrum* 241.
Colpa-Bolton-Beziehung, *Colpa-Bolton relation* 158.
Condon-Effekt, Franck-, *Franck-Condon effect*, 80.
Coulombsche Wechselwirkung, *Coulomb interaction* 35.
Creation operator 84.
CrBr$_3$ 267.
Cr-Chalkogen-Spinell, *Cr-chalcogen spinel* 522.
Cr-Spinelle, magnetische Rotverschiebung, *Cr spinels, magnetic red shift* 487.
Curie-Gesetz, *Curie's law* 2.
Curie-Konstante, *Curie constant* 2, 3, 405.
Curie-Temperatur, *Curie temperature* 265.
Curie-Weiss-Temperatur, *Curie-Weiss temperature* 409.
Cyclooctatetraenanion, *cyclo-octatetrene anion* 176.

Davydov splitting 483.
d-Bahn, d *orbital* 127, 128, 129.
d-Wellenfunktion, d-*wave function* 132.
3d-Bahn, 3d *orbital* 127.
3d-Oxid, 3d-*element oxide* 550.
Debye-Grenzfrequenz, *Debye cut-off frequency* 82.
Debye-Relaxation, *Debye relaxation* 21, 42.
Debye-Temperatur, *Debye temperature* 42.
Debye-Verteilung, *Debye distribution* 38, 49, 80.
Deuteriumatom, *deuterium atom* 246.

Deuteriumaustausch, *deuterium exchange* 188.
Dichtematrix-Operator, *density matrix operator* 2, 15.
Dibutylbenzosemichinon, *dibutylbenzosemiquinone* 106.
Diffusionsvorgang, *diffusion process* 75.
Diphenylpicrylhydrazyl, DPPH 182.
dipolare Wechselwirkung, *dipolar interaction* 73.
Dipol-Dipol-Effekt, *dipole-dipole effect* 122.
Dipol-Dipol-Operator, *dipole-dipole operator* 207.
Dipol-Dipol-Wechselwirkung, *dipole-dipole interaction* 125.
—, magnetische, *magnetic dipole-dipole interaction* 14, 34, 35, 75.
Dipol-Resonanzkurve, Verbreiterung der, *dipole broadening* 122.
Dipolübergang, elektrischer, *electric dipole transition*, 37, 459.
—, magnetischer, *magnetic dipole transition* 37.
Diracscher Spin-Hamilton-Operator, *Dirac spin Hamiltonian* 266.
Dirac-van Vleck-Hamilton-Operator, *Dirac-van Vleck Hamiltonian* 267.
direkter Prozeß, *direct process* 39, 121.
direkter Relaxationsprozeß, *direct process of relaxation* 39.
— — in Seltene-Erde-Salzen, *direct relaxation process in rare-earth salts* 48.
direkter Vorgang in Nicht-Kramers-Salzen, *direct process in non-Kramers salts* 48.
Dispersion 12.
Doppelaustausch, *double exchange* 351, 364, 367, 411, 446, 450, 520, 535.
Doppelresonanz, *double resonance* 142, 144.
Dublett, Kramers-, *Kramers doublet* 131.
Du Pré, Theorie von Casimir und, *theory of Casimir and Du Pré* 4.
— -Formalismus, Casimir- und, *Casimir and Du Pré formalism* 77.
— -Relaxationskurve, Casimir- und, *Casimir and Du Pré relaxation curve* 7.
Donatorzentrum, *electron donor centre* 80.
DPPH, diphenylpicrylhydrazyl 182.

EuI$_2$ 267.
EuO 267.
EuS 267.
EuSe 267.
Eu$_2$SiO$_4$ 267.
Eckhart theorem, Wigner- 48.
e_g-Bahn, e_g *orbital* 128.
Einelektronenmolekülbahn, *one-electron molecular orbit* 378.
Einelektronenübergangsintegral, *one-electron transition integral* 330.
Einelektronhyperfeinkonstante, *one-electron hyperfine constant* 149.
Electron-electron repulsion 537.
elektrische Leitfähigkeit in Thio-Spinellen, *electrical conductivity in thio-spinels* 531.
elektrische Quadrupolwechselwirkung, *electric quadrupole interaction* 135, 136.

elektrischer Dipolübergang, *electric dipole transition* 37, 459.
elektrischer Quadrupolübergang, *electric quadrupole transition* 37.
elektrisches Feld, inneres, *internal electric field* 108.
elektrisches Kernquadrupol, *electric quadrupole, nuclear* 140.
elektrisches Quadrupolmoment, *electric quadrupole moment* 135, 141.
Elektromagnet, *electromagnet* 98, 119.
Elektron, hydratisiertes, *hydrated electron* 238.
—, ungepaartes, *unpaired electron* 94, 98, 100, 109, 137.
Elektron-Elektron-Wechselwirkung, *electron-electron interaction* 289, 341, 306.
Elektronen, Beweglichkeit von 3d, 4f-, *mobility of 3d, 4f electrons* 398.
—, Dichte ungepaarter, *unpaired electron density* 155, 163.
—, kovalente Bindung von 4f-, *covalent bonding of 4f electrons*, 470.
—, Zentrum für ungerade, *odd- electron centre* 259.
Elektronenpolarisation, Leitungs-, *conduction electron polarization* 433, 490.
Elektronenbeweglichkeit, *electron mobility* 397.
Elektronenkorrelation, *electron correlation* 284, 398, 400, 402, 532.
Elektronenmasse, effektive, *effective mass* 398.
Elektronenresonanz, paramagnetische, *electron paramagnetic resonance (E.P.R.)* 36.
Elektronenschalen, Radius äußerer, *radius of electron shells* 389.
Elektronenspin und Magnetfeld, anisotrope Kopplung zwischen, *anisotropic coupling between electron spin and magnetic field*, 133.
Elektronenspinresonanz, *electron spin resonance* 9, 94.
— von Gasen, *electron spin resonance of gases* 256.
Elektronenspinresonanzspektrometer, *electron spin resonance spectrometer* 112.
Elektronenübergang, intramolekularer, *intramolecular electron transfer* 203.
Elektronenwechselwirkung, *electron interaction* 541.
Elektronenzentrum, lokalisiertes, *local electron centre*, 78.
elektronische Aufspaltung, *electronic splitting* 107, 108, 133, 136, 139.
— —, Winkelverteilung der, *angular variation of electronic splitting* 107, 140.
elektronisches Phasendiagramm, *electron phase diagram* 532.
elektronisches Quadrupol, *electronic quadrupole* 136.
Elektron Kern Doppelresonanz, *ENDOR (Electron Nuclear DOuble Resonance)* 142, 143, 144, 229, 236, 241, 247, 252, 253, 262.

Elektron-Magnon-Wechselwirkung, *electron-magnon interaction* 398, 544.
Elektron-Phonon-Wechselwirkung, *electron-phonon interaction* 398, 543.
Emission, induzierte, *stimulated emission* 121.
— eines Phonons, spontane, *spontaneous emission of a phonon* 46.
—, spontane, *spontaneous emission* 46.
Empfänger, phasenempfindlicher, *phase sensitive detector* 113, 114, 117.
Empfängerrauschen, *detector noise* 109, 110.
Energiegap, zweites, *second gap* 413.
Energieniveau, Bahn-, *orbital energy level* 129.
— in SE-Atomen, *energy level for lanthanide atoms* 458.
Energy reservoir, exchange- 22.
entarteter Zustand, zweifach, *doubly degenerate state* 36.
Entartung, Kramers-, *Kramers degeneracy* 36, 40, 69, 72.
—, Zeitumkehr-, *time-reversal degeneracy* 43.
Entmagnetisierung, adiabatische, *adiabatic demagnetization*, 1.
Enzym, *enzyme* 95.
ESR von Gasen, *ESR of gases* 256.
ESR-Spektrometer, *ESR spectrometer* 112.
Eu-Chalkogenid, *Eu chalcogenide* 476, 481, 506, 512, 554.
Eu-Chalkogenide, magnetische Rotverschiebung, *Eu chalcogenides, magnetic red shift* 483.
Eu-Chalkogeniden, Bandstruktur von, *band-structure of Eu chalcogenides*, 477, 511.
Exchange-energy reservoir 22.
Exchange narrowing 123, 125.
Exciton-magnon absorption 483.
exponentieller Abfall, *exponential decay* 21, 22.

Faraday-Rotation in Granaten, *Faraday rotation in garnets* 481.
Farbzentrum, *colour centre* 72, 78.
α-Fe_2O_3 268.
Fermi-Gas, *Fermi gas* 266.
Fermi-Kontaktpotential, *Fermi contact potential* 72.
Fermi-Wechselwirkung, *Fermi interaction* 72.
Ferrimagnetismus, *ferrimagnetism* 265, 267.
Ferrit, *ferrite* 335.
Ferromagnetismus, *Ferromagnetism* 265, 266.
— von Bandelektronen, *Ferromagnetism of band electrons* 538.
Festkörpermaser, *solid-state maser* 2.
Festkörpern, Atom in, *atom in solids* 245.
^{19}F-Fluoraufspaltungskonstante, ^{19}F *fluorine splitting constant* 167.
flacher Akzeptor, *shallow acceptor impurity* 254.
flacher Donator, *shallow donor impurity* 253.
Fluor-Superhyperfeinstruktur, *fluorine super-hyperfine structure* 252.
Fock-Verfahren, beschränktes Hartree-, *restricted Hartree-Fock method* 352.
— —, uneingeschränktes Hartree-, *unrestricted Hartree-Fock method* 352, 369.

Fourier-Spektrum, *Fourier spectrum* 90.
Franck-Condon-Effekt, *Franck-Condon effect* 80.
freien Radikalen, Übergang mit, *free-radical transient* 142.
freien Radikals, g-Wert eines, *free radical g-value* 126.
freies Radikal, *free radical* 77, 94, 95, 98, 116, 119, 126, 151.
— —, stabiles, *stable free radical* 182.
Fremdatome, magnetische, *magnetic impurity state* 516.
Fremdatompaar, *impurity pair* 254.
Fremdion, *impurity ion* 78.
Frequenzkontrolle, automatische, *automatic frequency control* 108, 116, 117.
Frequenzmodulation, Hoch-, *high-frequency modulation*, 113.
Füllfaktor, *filling factor* 109.
Funkelrauschen, *flicker noise* 109.
F-Zentrum, *F-centre* 78, 144, 228.
—, abgewandeltes, *modified F-centre* 233.
F-Zentrum-Wechselwirkung, *F-centre interaction* 231.

Gauss'sche Kurvenform, *Gaussian line shape* 24, 123.
Gauss'sche Linie, *Gaussian line* 124.
Gauss'sche-Lorentz-Kurvenform, *Gaussian-Lorentz line shape* 26.
Gd-Chalkogenid, *Gd-chalcogenide* 518.
g-Faktor, *g factor* 115.
g-Komponente, *g component* 147.
g-Parameter 95.
g tensor 146.
g-Wert, *g value* 99, 115, 120, 126, 131, 132, 133, 136, 139.
— eines freien Radikals, *free radical g value* 126.
Germanium 254.
Giant moment 516.
Giacometti-Nordio-Pavan-Beziehung, *Giacometti-Nordio-Pavan relation* 159.
Gitterfehler in Halbleitern, *defects in semiconductors* 253.
— in Isolatoren, *defects in insulating solids* 228.
Gitter-Hamilton-Operator, *Lattice Hamiltonian* 38.
Gitteroperator, Spin-, *spin-lattice operator* 30.
Gitterpotential, Bahn-, *orbit- lattice potential* 48.
Gitterrelaxation, Spin-, *spin-lattice relaxation* 1, 30, 56, 59, 78, 120.
Gitterrelaxationsoperator, effektiver Spin-, *effective spin-lattice relaxation operator*, 56.
Gitterrelaxationszeit, Spin-, *spin- lattice relaxation time*, 6, 9, 120, 121, 132.
Gitterstörung, Spin-, *spin-lattice perturbation*, 34.
Gittertemperatur, *lattice temperature* 31.
Gitterwechselwirkung, Spin-, *spin- lattice interaction*, 4, 48, 64, 99, 120, 124.
Gitterzustand, *lattice state* 38.

Goodenough-Kanamori-Regeln, *Goodenough-Kanamori rules* 451.
Goodenough-Mechanismus, *Goodenough mechanism* 379.
Granat, *garnet* 497.
Granaten, Faraday-Rotation in, *Faraday rotation in garnets* 481.
Granatstruktur, *garnet structure* 253.
Grundzustand, Triplett, *ground state triplet* 221.

H 241.
Haber cycle, Born- 548.
Halbleiter, *semiconductor* 95.
—, magnetischer, *magnetic semiconductor* 389, 396.
Halbleitern, Gitterfehler in, *defects in semiconductors* 253.
Hamilton-Operator, *Hamiltonian* 3, 14, 20, 235.
—, Diracscher Spin-, *Dirac spin Hamiltonian* 266.
—, Dirac-van Vleck-, *Dirac-van Vleck Hamiltonian* 267.
—, Gitter-, *lattice Hamiltonian* 38.
—, Heisenberg-, *Heisenberg Hamiltonian* 267.
—, Spin-, *spin Hamiltonian* 4, 14, 39, 56, 74, 132, 135, 136, 137, 141, 170, 227, 227, 232, 242.
Hamilton-Operator-Matrix, *Hamiltonian matrix* 208.
Hartree-Fock-Verfahren, beschränktes, *restricted Hartree-Fock method* 352.
—, uneingeschränktes, *unrestricted Hartree-Fock method* 352, 369.
Hartree operator 67.
Heisenberg-Hamilton-Operator, *Heisenberg Hamiltonian* 267.
Heisenberg-Magnet, *Heisenberg magnet* 507.
Heisenbergsches Austauschintegral, *Heisenberg exchange integral* 277, 330.
Heitler-London-Methode, *Heitler-London method* 271.
Heitler-London-Modell, *Heitler-London model* 409.
Heitler-London-Theorie, asymptotisch exakte Lösung, *Heitler-London theory, asymptotically exact solution* 317.
— für den Kristall, *Heitler-London theory for the crystal* 287.
HMO-Methode, *HMO method* 175.
HMO-Näherung, *HMO approach* 147, 151, 159.
Hohlraum, Resonanz-, *resonance cavity* 109.
Hohlraumresonator, *cavity resonator* 97, 111, 112, 113.
Hopping 410, 427.
Hückel-Molecular-Orbital-Näherung, *Hückel Molecular Orbital approximation (HMO)* 147, 151.
Hundsche Regel, *Hund's rule* 377, 404.
Hyperfeinaufspaltung, *hyperfine splitting* 14, 99, 104, 108, 133, 135, 139, 140, 148.
—, Super-, *super-hyperfine splitting* 132, 184, 255.

Hyperfeinaufspaltung, zeitabhängige, *temperature-dependent hyperfine splitting* 198.
Hyperfeinaufspaltungskonstante, Vorzeichen der, *sign of the hyperfine splitting constant* 168.
Hyperfeinaufspaltungstensor, *hyperfine splitting tensor* 147, 148, 187.
Hyperfeineffekt zweiter Ordnung, *hyperfine effect, second-order* 149.
Hyperfeinkonstante, Einelektron-, *one-electron hyperfine constant* 149.
Hyperfeinkopplung, *hyperfine coupling* 149.
Hyperfeinkopplungsparameter, *hyperfine coupling parameter* 149, 160, 169.
Hyperfeinlinie, *hyperfine line* 103, 116.
—, Super-, *super-hyperfine line* 103.
Hyperfeinmuster, Super-, *super-hyperfine pattern* 107.
Hyperfeinstruktur, *hyperfine structure* 101, 132.
—, Chlorsuper-, *chlorine super-hyperfine structure* 252.
—, Fluor-Super-, *fluorine super-hyperfine structure* 252.
—, Mechanismus der, *mechanism of hyperfine structure* 162.
Hyperfeinstrukturaufspaltung, *hyperfine structure splitting* 14.
Hyperfeinwechselwirkung, *hyperfine interaction* 100, 102, 132, 143.
—, anisotrope, *anisotropic hyperfine interaction* 187.
—, Modulation der, *modulation of hyperfine interaction* 194.
— zweiter Ordnung, *second-order of hyperfine interaction* 147/148.
Hyperkonjugation, *hyperconjugation* 172, 176.
H-Zentrum, *H centre* 241, 244.

Impulssättigung, *pulse saturation* 10, 12.
induzierte Emission, *stimulated emission* 46, 96.
inneres elektrisches Feld, *internal electric field* 108.
Integral, Überlappungs-, *overlap integral* 276, 278, 311, 323, 354, 359, 362.
inverser Orbach-Relaxationsprozeß, *inverted Orbach relaxation process* 46, 65.
Ionenanordnung, *arrangement of ions* 267.
Ionenenergieniveau, *ionic energy level* 35.
Ionenwechselwirkung, *interionic interaction* 72.
Ionenzustand, *ionic state* 282, 293.
Ionisierungsarbeit für SE, *ionization energy for RE* 458.
Ionisationsenergie, *ionization energy* 294, 319, 382.
Isolator 265.

Jahn-Teller-Effekt, *Jahn-Teller effect* 69, 70.
Jahn-Teller theorem 36.

Kalorimeterverfahren, *calorimeter method* 12.
Kanamori-Regeln, Goodenough-, *Goodenough-Kanamori rules* 451.
Kationradikal, *cation radical* 155.
KCl 144.
Keim, *nucleus* 99, 101.
—, Kupfer-, *copper nucleus* 100.
—, Mangan, *manganese nucleus* 138.
kernmagnetische Resonanz, *nuclear magnetic resonance* 2, 119, 142.
Kernmagneton, *nuclear magneton* 99.
Kernmoment, magnetisches, *nuclear magnetic moment* 94.
Kernquadrupol, elektrisches, *nuclear electric quadrupole* 140.
Kernquadrupolrelaxation, *nuclear quadrupole relaxation* 82.
Kernspin, *nuclear spin* 99, 100.
Kernspinresonanz, *nuclear spin resonance* 9.
Kernwechselwirkung, Mehr-, *multi-nuclear interaction* 104.
Kittel-Funktion, Ruderman-, *Ruderman-Kittel function*, 434.
Klystron 97, 108, 110, 115, 116.
α-Kohlenstoffatom, *α-carbon atom* 159.
β-Kohlenstoffatom, *β-carbon atom* 159.
Kohlenwasserstoff, ,,alternant", *hydrocarbon, "alternant"* 154.
Konfigurationswechselwirkung, *configuration interaction* 337.
Kontaktpotential, Fermi-, *Fermi contact potential* 72.
Konzentration, *concentration* 70.
Konzentrationseffekt, *concentration effect* 78.
Koordination, achtfache, *eight-fold coordination* 69.
—, neunfache, *nine-fold coordination* 47.
—, sechsfache, *six-fold coordination* 36.
Kopplung, Anisotropie in Spinbahn-, *anisotropy in spin-orbit coupling* 64.
—, antiferromagnetische, *antiferromagnetic coupling* 267.
—, Bloembergen-Rowland-, *Bloembergen-Rowland coupling* 411.
— zwischen Elektronenspin und Magnetfeld, anisotrope, *anisotropic coupling between electron spin and magnetic field* 133.
—, Hyperfein-, *hyperfine coupling* 149.
—, Russel-Saunders-, *Russel-Saunders coupling* 402.
—, Spinbahn-, *spin orbit coupling* 36, 47, 79, 126, 129, 130, 132, 259, 460.
—, Spin-Phonon-, *spin-phonon coupling* 91.
Kopplungsparameter, Hyperfein-, *hyperfine coupling parameter* 149, 160, 168.
Korrelation, Elektronen-, *electron correlation* 284, 398, 400, 402, 532.
—, π-σ-correlation 164.
Korrelationszeit, *correlation time* 90.
kovalente Bindung, *covalent bonding* 67, 69.
— — von 4f-Elektronen, *covalent bonding of 4f electrons* 470.
Kovalenz, *covalency* 65, 82, 383.
—, 4f-, *covalency of the 4f electrons* 510.
Kramers-Anderson-Mechanismus, *Kramers-Anderson mechanism* 379.
Kramers-Dublett, *Kramers doublet* 131.

Kramers-Entartung, *Kramers degeneracy* 36, 40, 69, 72.
Kramers-Kronig-Beziehung, *Kramers-Kronig relation* 19.
Kramers-Salz, *Kramers salt* 39, 40.
—, Nicht, *non-Kramers salt* 39, 40, 41.
—, Raman-Prozeß in Nicht-, *Raman process in non-Kramers salt* 50, 41.
Kramersscher Austauschmechanismus, nicht-, *non-Kramers exchange mechanism* 269.
Kramers theorem 36.
Kristall, konzentriertes, *concentrated crystal* 65.
—, Molekül-, *molecular crystal* 68.
—, Wahl der Wellenfunktion des, *choice of the wave function for the crystal* 290, 300, 304.
Kristallbindung, *crystal bonding* 67.
Kristalldetektor, *crystal detector* 98, 112.
Kristallfeld, *crystal field* 22, 34, 35, 47, 107, 131, 408, 463, 466.
— in Salzen Seltener-Erden, *crystal field in rare-earth salts* 47.
Kristallfeldaufspaltung, *crystal field splitting* 407, 534.
Kristallfeldparameter, *crystal field parameter* 47, 60, 62, 63.
Kristallfeldpotential, *crystal field potential* 48.
Kristallfeldtheorie, schwaches Kristallfeld, *crystal field theory, weak crystal field* 374.
—, starkes Kristallfeld, *crystal field theory strong crystal theory* 376.
Kristallrauschen, Überschuß, *excess crystal noise* 110.
Kristallspektrum, SE-, *crystal spectrum in RE* 463.
kritische Adernabstände, *critical distance* 399.
Kronig-Beziehung, Kramers-, *Kramers-Kronig relation* 19.
Kupferkeim, *copper nucleus* 100.

Lage der d-Niveaus, *position of the d-levels* 546.
Landau trapping 311, 543.
Landé-Aufspaltungsfaktor, *Landé splitting factor* 126, 404.
Laplace-Gleichung, *Laplace's equation* 47.
Larmor-Frequenz, *Larmor frequency* 8, 22, 24.
Lavesphasen, *Laves phase* 492.
LCAO-Methode, *LCAO* 67, 151, 378, 409.
—, MO-, *MO-LCAO* 321.
Lebensdauer des Phonons, *life-time of the phonon* 86, 87.
Leitungselektronenpolarisation, *conduction electron polarization* 433.
Ligandenatom, *ligand atom* 132.
Ligandenfeldtheorie, *ligand field theory* 127, 270, 379, 406.
Ligandenkern, *ligand nucleus* 132.
Li_xMn_1 520.
Linear combination of atomic orbitals (LCAO) 67, 151, 378, 409.
Linienform für nichtausgerichtete Systeme, *line shape for non-oriented systems* 214.
Linienbreite, *line width* 98.

Linienbreitenanomalie, *line width anomaly* 192.
Linienbreiteneffekt, *line width effect* 192.
Liouville operator 16.
lokalisierte Fehlstelle, *localized defect* 82.
lokalisiertes Elektronenmodell, *localized electron model* 400.
Löwdin, Störungstheorie nach, *Löwdin's perturbation theory* 344, 349, 361.
London-Methode, Heitler-, *Heitler-London method* 271.
London-Modell, Heitler-, *Heitler-London model* 409.
longitudinale Relaxationszeit, *longitudinal relaxation time* 8.
longitudinales Phonon, *longitudinal phonon* 82.
Lorentz-Kurvenform, *Lorentzian line shape* 26, 123, 124.
—, Gauss'sche-, *Gaussian-Lorentz line shape* 26.

Madelung potential 548.
Magic-T 119.
Magnet, supraleitender, *superconducting magnet* 120.
Magnetfeldmodulation, *magnetic field modulation* 110, 112, 113, 115.
magnetische Dipol-Dipol-Wechselwirkung, *magnetic dipole-dipole interaction* 14.
magnetische Fremdatome, *magnetic impurity state* 516.
magnetische spezifische Wärme, *magnetic specific heat* 10.
magnetische Wechselwirkung über halbleitende Träger, *magnetic interaction via semiconducting carriers* 512 .
— — in SE-Elementen, *magnetic interaction in RE elements* 491.
magnetischer Dipolübergang, *magnetic dipole transition* 37.
magnetischer Halbleiter, *magnetic semiconductor* 391, 396.
magnetischer Moment von SE, *magnetic moment of RE* 490.
magnetisches Hochfrequenzspektrum, *magnetic high-frequency spectrum* 136.
magnetisches Kernmoment, *nuclear magnetic moment* 94.
magnetisches Moment, *magnetic moment* 94, 99.
magnetisches Polaron, *magnetic polaron* 399, 419, 421.
magnetisches Quasiteilchen, *magnetic quasiparticle* 368.
Magnetoabsorption, *magnetic absorption* 482.
magnetomechanisches Verhältnis, *magnetomechanical ratio* 8.
Magneton, Kern-, *nuclear magneton* 99.
Magnetonnummer, effektive, *effective magneton number* 405.
Magneto-Optik, *magneto-optic* 480.
Magnon absorption, exciton- 483.
Magnonendrag, *magnon drag* 430.

Magnon-Wechselwirkung, Elektron-, *electron-magnon interaction* 398.
Malonsäure, *malonic acid* 259.
Malonsäureradikal, *malonic acid radical* 165, 187.
Manganat, *manganate* 141.
Manganathyperfeinaufspaltung, *manganate hyperfine splitting* 140.
Manganfluorosilikat, *manganese fluosilicate* 138.
Mangankern, *manganese nucleus* 138, 141.
Maser 108.
—, Festkörper-, *solid-state maser* 2.
Mastergleichung, verallgemeinerte, *generalised master equation* 18, 23.
Mattuck's und Strandberg's effektiver Störungsoperator, *Mattuck's and Strandberg's effective perturbation operator* 56.
Maxwell-Boltzmann-Relaxation, *Maxwell-Boltzmann relaxation* 30.
McConnel-Gleichung, *McConnel equation* 162, 165.
Mechanismus des indirekten Austauschs, *mechanism of indirect exchange* 268, 337.
— — —, Doppelübergang, d_1-p, $p'-d_2$, *mechanism of indirect exchange, double transition*, d_1-p, $p'-d_2$ 351, 359, 364.
— — —, Polarisationsmechanismus, *mechanism of indirect exchange, polarization mechanism* 352.
— — —, Übergang in die besetzte d-Bahn, $p-d'$, *mechanism of indirect exchange, transition to the occupied orbital* 348, 359, 369, 373.
— — —, — in die leere d-Bahn, $p-d'$, *transition to the vacant orbital* 337, 371, 373.
— — —, zweifache Anregung, $p-d_1$, $p-d_2$, *mechanism of indirect exchange, twofold excitation*, $p-d_1$, $p-d_2$ 349.
Mehrkernwechselwirkung, *multi-nuclear interaction* 104.
Mehrniveausysteme, Relaxation in, *relaxation in multilevel systems* 65.
Mehrspinprozeß, *multi-spin process* 74, 75.
Methylen, *methylene* 221.
Methylgruppe, *methyl group* 106.
Mikrowellenbrücke, *microwave bridge* 112, 113, 116.
Mn-Perowskit, *Mn perovskite* 498.
Mn^{2+} ion 137, 252.
Mn^{55} nucleus, Mn^{55}-*Kern* 137.
MnO 267, 268, 335, 380, 382.
Modulation, Hochfrequenz-, *high-frequency modulation* 113.
—, Magnetfeld-, *magnetic field modulation* 110, 112, 113, 115.
Modulationsverbreiterung, *modulation broadening* 117.
MO-LCAO-Methode, *MO-LCAO* 321.
Molekülbahn, *molecular orbital* 409, 470.
—, Einelektronen-, *one-electron molecular orbital* 378.
—, nichtlokalisierte, *delocalized molecular orbital* 95, 101, 126.

Molekülbahn, orthogonale, *orthogonal molecular orbital* 321.
Molekülkristall, *molecular crystal* 68.
—, paramagnetische Relaxation in einem, *paramagnetic relaxation in a molecular crystal* 2.
Moment, magnetisches, *magnetic moment* 94, 99.
—, — Kern-, *nuclear magnetic moment* 94.
Motional narrowing 122, 125.
Mott localization 411.
Mott trapping 399, 543.
Mott-Übergang, *Mott problem* 536.
—, *Mott transition* 397, 413, 414, 543.
—, kritischer Abstand, *Mott transition, critical distance* 397.
$4f$-Multiplett, $4f$ *multiplet* 464.
M-Zentrum, M *centre* 235.

NaCl-Verbindung, metallische, *metallic NaCl-type compound* 505.
Näherung, adiabatische, *adiabatic approximation* 43, 51.
—, nichtadiabatische, *non-adiabatic approximation* 43, 52.
Naphthalentriplett, *naphthalen triplet* 207, 224.
Narrowing, exchange 123, 125.
—, motional, *narrowing* 122, 125.
^{14}N-Aufspaltungskonstante, ^{14}N *splitting constant* 166.
nephelauxeter Effekt, *nephelauxetic effect* 471, 504.
nichtadiabatische Näherung, *non-adiabatic approximation* 43, 52.
Nicht-Kramers-Salz, *non-Kramers salt* 39, 40, 41.
—, Raman-Prozeß in, *Raman process in non Kramers salt* 41, 50.
Nicht-Kramers-Salze, direkter Vorgang in, *direct process in non-Kramers salts* 48.
nicht-Kramersscher Austauschmechanismus, *non-Kramers exchange mechanism* 269.
Nichtorthogonalitätsintegral, *non-orthogonality integral* 276, 278, 311, 323, 354, 359, 362.
Nichtorthogonalitätskatastrophe, *non-orthogonality catastrophe* 290.
NiO 268, 552.
Nordio-Pavan-Beziehung, Giacometti-, *Giacometti-Nordio-Pavan relation* 159.

O_2 ion 248.
Oktaederfeld, *octahedral field* 129.
Oktaedersymmetrie, *octahedral symmetry* 129.
Operator, Vernichtungs-, *annihilation operator* 38.
Operatoräquivalent, *operator equivalent* 466.
optischer Zweig, *optical branch* 37.
optisches Phonon, *optical phonon* 80, 81.
optisches Spektrum, *optical spectrum* 404.
Orbach-Relaxationsprozeß, *Orbach process of relaxation* 43, 51.
—, inverser, *inverted Orbach process of relaxation* 46, 65.

Orbach-Relaxationsprozeß in Seltene-Erde-Salzen, *Orbach relaxation process in rare-earth salts* 51.
Orbital, bindendes, *bonding orbital* 132.
—, lockerndes, *antibonding orbital* 132.
—, nichtbindendes, *nonbonding orbital* 177.
Orbital quenching 408.
Orthoferrit, *orthoferrite* 498.
Overhauser-Effekt, *Overhauser effect* 142, 143.
Overhauser-Verfahren, *Overhauser technique* 142.

π-Bahn, π *orbital* 127.
π-Radikal, π-*radical* 151.
—, anorganisches, *inorganic* π-*radical* 190.
— in bestrahlten organischen Kristallen, π-*radical in irradiated organic crystals* 187.
π-σ-Korrelation, π-σ *correlation* 164.
π-σ-Wechselwirkung, π-σ *interaction* 163.
paramagnetische Elektronenresonanz, *electron paramagnetic resonance (E.P.R.)* 36.
paramagnetische Relaxation, *paramagnetic relaxation* 2.
— — in lokalen Elektronenzentren, *paramagnetic relaxation in local electron centres* 78.
— — durch modulierte Austauschwechselwirkung, *paramagnetic relaxation by modulated exchange interaction* 71.
— — in Molekülkristallen, *paramagnetic relaxation in molecular crystals* 2.
— —, Stereochemie und, *stereochemistry and paramagnetic relaxation* 68.
paramagnetische Resonanz, *paramagnetic resonance* 2, 8, 145.
— —, akustische, *acoustic paramagnetic resonance* 91.
paramagnetische Sättigung, *paramagnetic saturation* 47.
paramagnetische Suszeptibilität, *paramagnetic susceptibility* 404.
paramagnetische Relaxation, WALLERs Mechanismus der, *Waller's mechanism of paramagnetic relaxation* 65, 70, 74, 77.
paramagnetisches Ion, *paramagnetic ion* 131.
paramagnetisches Resonanzsättigungsexperiment, *paramagnetic resonance saturation experiment* 28.
paramagnetisches Zentrum in Silizium, *paramagnetic centre in silicon* 253.
Paramagnetismus, *paramagnetism* 2.
Parameter, D, Aufspaltungs-, *splitting parameter*, D 137.
—, bottleneck 86.
—, g-, *g parameter* 95.
—, Phononen-, *phonon parameter* 63.
—, Relaxations-, *relaxation parameter* 21.
—, Sättigungs-, *saturation parameter* 32, 86.
Parität, *parity* 30, 37, 47.
Pavan-Beziehung, Giacometti-Nordio-, *Giacometti-Nordio-Pavan relation* 159.
Perinaphthenylradikal, *perinaphthenyl radical* 162, 171.
Perowskit, *perovskite* 253, 497.
—, Mn-, *Mn perovskite* 498.

Phenylanion, *benzene anion* 186.
Phonon 37.
— bottleneck 7, 46, 74, 86, 90.
—, Debye-Verteilung für ein, *Debye distribution for a phonon* 38.
—, longitudinales, *longitudinal phonon* 82.
—, optisches, *optical phonon* 80, 81.
—, transversales, *transverse phonon* 82.
Phononendrag, *phonon drag* 430.
Phononenparameter, *phonon parameter* 63.
Phononenrelaxation, *phonon relaxation* 88.
Phononenrelaxationsprozeß in Seltene-Erde-Salzen, Zwei-, *two-phonon relaxation process in rare-earth salts* 50.
Phononenschwingungstyp, lokaler, *local phonon mode* 82.
Phononenspektrum, reelles, *phonon spectrum, real* 85.
Phononensystem, *phonon system* 4.
—, Relaxation im, *relaxation in the phonon system* 80.
Phononenzustandsdichte, *phonon density of states* 30.
Phonon-Kopplung, Spin-, *spin-phonon coupling* 91.
Phonon-Phonon-Wechselwirkung, *phonon-phonon interaction* 87.
Phonon, Lebensdauer des, *life-time of the phonon* 86, 87.
—, Matrixelement eines, *matrix element of a phonon* 30, 38.
—, spontane Emission eines, *spontaneous emission of a phonon* 46.
Phonon-System, Spin-, *spin-phonon system* 34.
—, Zustandsfunktionen des Spin-, *state functions of the spin-phonon* 34.
Phonon-Wechselwirkung, Elektron-, *electron-phonon interaction* 398.
—, Spin-, *spin-phonon interaction* 84, 85.
Phthalocyanin, Kupfer-, *copper phthalocyanine* 106.
Polarisation, Leitungselektronen-, *conduction electron polarization* 433, 490.
—, Spin-, *spin polarization* 174, 311, 352, 380.
Polaron 398.
—, großes, *large polaron* 543.
—, kleines, *small polaron* 543.
—, magnetisches, *magnetic* 399, 419, 421.
Polarzustand, *polar state* 282, 293.
potentieller Austausch, *potential exchange* 270, 330, 370.
Prinzip des detaillierten Gleichgewichts, *principle of detailed balance* 31.
Problem der 3 Zentren und 4 Elektronen, nichtorthogonale Bahnen, *problem of 3 centres and 4 electrons, non-orthogonal orbitals* 359.
— — — — —, orthogonale Bahnen, *orthogonal orbitals* 336.
Projektionsoperator, *projection operator* 325.
Proton, α- 187.
—, β- 188.

Protonenübergang, *proton transfer* 202.
Pulssättigungsverfahren, *pulse saturation technique* 10, 12, 32.

Quadrupol, elektrisches Kern-, *nuclear electric quadrupole* 140.
—, elektronisches, *electronic quadrupole* 136.
Quadrupolauswahlregel, *quadrupolar selection rule* 57.
Quadrupolmoment, elektrisches, *electric quadrupole moment* 135, 141.
Quadrupolrelaxation, Kern-, *nuclear quadrupole relaxation* 81, 82.
Quadrupolübergang, elektrischer, *electric quadrupole transition* 37.
Quadrupolwechselwirkung, elektrische, *electric quadrupole interaction* 135, 136.
—, Kern-, *nuclear quadrupole interaction* 34, 35, 72.
Quantelung, zweite, *second quantization* 294, 322, 369.
Quarzkristallstandard, *quartz crystal standard* 115.
Quasi-Raman-Prozeß, *Quasi-Raman process* 1.
Quasiteilchen, magnetisches, *magnetic quasiparticle* 368.

Radialintegral, *radial integral* 460.
Radikal, Allyl-, *allyl radical* 189.
—, Anion-, *anion radical* 155.
— in bestrahlten nichtausgerichteten Festkörpern, *radical in non-oriented irradiated solids* 192.
—, freies, *free radical* 77, 94, 95, 98, 116, 119, 126, 151.
—, Kation-, *cation radical* 155.
—, Malonsäure, *malonic acid radical* 165, 187.
—, monozyklisches, *monocyclic radical* 177.
—, π-, *π-radical* 151.
—, —, anorganisches, *inorganic π-radical* 190.
—, —, in bestrahlten organischen Kristallen, *π-radical in irradiated organic crystals* 187.
—, Perinaphthenyl-, *perinaphthenyl radical* 162, 171.
—, σ-, *σ-radical* 204.
—, Semichinon-, *semiquinone radical* 104.
—, stabiles freies, *stable free radical* 182.
—, Vinyl-, *vinyl radical* 205.
Radikalen, Übergang mit freien, *free-radical transient* 142.
Radikals, g-Wert eines freien, *free-radical g-value* 126.
Radius äußerer Elektronenschalen, *radius of electron shells* 389.
räumliche Verteilung von 3d-Bahnen, *spatial distribution of 3d orbitals* 127.
Raman-Prozeß in Nicht-Kramers-Salzen, *Raman process in non-Kramers salts* 50.
—, Quasi-, *quasi-Raman process* 1.
—, Übergangswahrscheinlichkeit für, *transition probability for Raman process* 41.
Raman-Relaxationsprozeß, *Raman process of relaxation* 41, 53, 121.

Raman-Relaxationsprozeß in Seltene-Erde-Salzen, *Raman relaxation process in rare-earth salts* 50.
Rauschen, *noise* 109, 110.
—, Empfänger-, *detector noise* 109, 110.
—, Funkel-, *flicker noise* 109.
—, Überschußkristall-, *excess crystal noise* 110.
Relaxation, Debye 21, 42.
— in Ionen im S-Zustand, *Relaxation in S-state ions* 64.
—, Kernquadrupol-, *nuclear quadrupole relaxation* 81, 82.
— in lokalen Elektronenzentren, paramagnetische, *paramagnetic relaxation in local electron centres* 78.
— in Mehrniveausystemen, *relaxation in multilevel systems* 65.
—, Maxwell-Boltzmann-, *Maxwell-Boltzmann relaxation* 30.
— durch modulierte Austauschwechselwirkung, paramagnetische, *paramagnetic relaxation by modulated exchange interaction* 71.
— in Molekülkristallen, paramagnetische, *paramagnetic relaxation in molecular crystals* 2.
—, paramagnetische, *paramagnetic relaxation* 2.
—, Phononen-, *phonon relaxation* 88.
— im Phononensystem, *relaxation in the phonon system* 80.
— durch Spinbahnkopplung, *relaxation due to spin-orbit coupling* 47.
—, Spingitter-, *spin-lattice relaxation* 1, 30, 56, 59, 78, 120.
—, Spin-Spin-, *spin-spin relaxation* 1, 9, 13, 72.
—, Stereochemie und paramagnetische, *stereochemistry and paramagnetic relaxation* 68.
—, Waller's Mechanismus der paramagnetischen, *Waller's mechanism of paramagnetic relaxation* 65, 70, 74, 77.
Relaxationserscheinung, *relaxation phenomenon* 120.
Relaxationsfunktion, *relaxation function* 14, 15.
Relaxationskurve, Casimir- und Du Pré-, *Casimir and Du Pré relaxation curve* 7.
Relaxationsoperator, effektiver Spingitter-, *effective spin-lattice relaxation operator* 56.
Relaxationsparameter, *relaxations parameter* 21.
Relaxationsprozeß, direkter, *direct process of relaxation* 39.
—, Orbach-, *Orbach process of relaxation* 43, 51.
—, Raman-, *Raman relaxation process* 41, 53, 121.
— in Seltene-Erde-Salzen, direkter, *direct relaxation process in rare-earth salts* 48.
— —, Orbach-, *Orbach relaxation process in rare-earth-salts* 48.
— —, Raman-, *Raman relaxation process in rare-earth salts* 50.

Sachverzeichnis. 573

Relaxationsprozeß in Seltene-Erde-Salzen, Zweiphononen-, *two-phononon relaxation process in rare-earth salts* 50.
Relaxationszeit, longitudinale, *longitudinal relaxation time* 8.
—, Spingitter-, *spin-lattice relaxation time* 6, 9, 120, 121, 132.
—, transversale, *transversal relaxation time* 8.
Relaxationszeiten, Anisotropie in, *anisotropy in relaxation times* 63.
Resonanz, akustische paramagnetische, *acoustic paramagnetic resonance* 91.
—, Doppel-, *double resonance* 142, 144.
—, Elektronenspin-, *electron spin resonance* 9, 94.
Resonanz, kernmagnetische, *nuclear magnetic resonance* 2, 119, 142.
—, Kernspin-, *nuclear spin resonance* 9.
—, paramagnetische, *paramagnetic resonance* 2, 8, 145.
—, paramagnetische Elektronen-, *electron paramagnetic resonance (E.P.R.)* 36.
Resonanzhohlraum, *resonance cavity* 109.
Resonanz- und Nichtresonanzverfahren, *resonant and non-resonant technique* 79.
Resonanzsättigungseffekt, *resonance saturation effect* 124.
Resonanzsättigungsexperiment, paramagnetisches, *paramagnetic resonance saturation experiment* 28.
Resonanzsättigungsverfahren, *resonance saturation technique* 10.
Rotverschiebung, magnetische, *magnetic red shift* 524.
—, —, Cr-Spinelle, *Cr spinels, magnetic red shift* 487.
—, —, Eu-Chalkogenide, *Eu chalcogenides, magnetic red shift* 483.
Rowland-Kopplung, Bloembergen-, *Bloembergen-Rowland coupling* 411.
Ruderman-Kittel-Funktion, *Ruderman-Kittel function* 434.
Russel-Saunders-Kopplung, *Russel-Saunders coupling* 402.
Rutil, *rutile* 253.
R-Zentrum, *R-centre* 146, 236.

Sättigung, *Saturation* 109, 121, 125, 143.
—, Impuls-, *pulse saturation* 10, 12.
— im stationären Zustand, *steady state saturation* 10.
—, paramagnetische, *paramagnetic saturation* 47.
Sättigungseffekt, *saturation effect* 124.
—, Resonanz-, *resonance saturation* 124.
Sättigungsexperiment, *saturation experiment* 9.
—, paramagnetisches Resonanz-, *paramagnetic resonance saturation* 28.
Sättigungsparameter, *saturation parameter* 32, 86.
Sättigungsverfahren, *saturation technique* 31.
—, Puls-, *pulse saturation technique* 10, 12, 32.
—, Resonanz-, *resonance saturation technique* 10.

Saunders-Kopplung, Russel-, *Russell-Saunders coupling* 402.
s-Bahn, *s orbital* 128.
Schaden, Bestrahlungs-, *irradiation damage* 95, 98.
Schwebungsfrequenzverfahren, *beat-frequency method* 12.
Schwingung, thermische, *thermal vibration* 121.
Schwingungsform, normale, *normal vibrational mode* 48.
Seltene Erden (SE), Bahnradius der, *orbital radius of rare earths (RE)* 452.
—, Grundzustand von 4f-Elektronen in, *ground state of 4f-electrons in RE* 455.
—, Ionisierungsarbeit für, *ionization energy for RE* 458.
—, magnetischer Moment von, *magnetic moment of RE* 490.
—, Schmelzpunkt der, *melting point of RE* 456.
SE-Atomen, Energieniveau in, *energy level for lanthanide atoms* 458.
SE-Chalkogenid, *RE-chalcogenide* 553.
SE-Elementen, magnetische Wechselwirkung in, *magnetic interaction in RE-elements* 491.
SE-Hydrid, *RE-hydride* 512.
SE-Kristallspektrum, *crystal spectrum in RE* 463.
SE-Metalle, Bandstruktur, *RE metals, band structure* 472.
SE-Pniktid: magnetische Eigenschaften, *RE-pnictide, magnetic properties* 495.
SE-Sesquioxid, *RE-sesquioxide* 501.
SE-Spektrum, *RE-spectrum* 459.
SE-III-V-Verbindungen, chemische Eigenschaften, *III-V-compounds of RE^{3+} with pnictides* 493.
SE, Chemie der, *RE chemistry* 393.
SE-Salzen, direkter Relaxationsprozeß in, *direct relaxation process in RE salts* 48.
—, Kristallfeld in, *crystal field in RE salts* 47.
—, Orbach-Relaxationsprozeß in *Orbach relaxation process in RE salts* 51.
—, Raman-Relaxationsprozeß in, *Raman relaxation process in RE salts* 50.
—, Zweiphononenrelaxationsprozeß in, *two-phonon relaxation process in RE salts* 50.
Semichinon, *Semiquinone* 180, 202.
—, Benzo-, *benzo- semiquinone* 104.
Semichinonradikal, *Semiquinone radical* 104.
Silizium, *Silicon* 253.
—, paramagnetisches Zentrum in, *paramagnetic centre in silicon* 253.
Siliziumkristallgleichrichter, *silicon crystal rectifier* 98.
Singulett-Triplett-Aufspaltung, *singlet-triplet splitting* 266, 272.
Slater-Kurve, Bethe-, *Bethe-Slater curve* 307.
Spektrometer, Elektronenspinresonanz-, *electron spin resonance spectrometer* 112.
—, ESR, *ESR spectrometer* 112.
Spektrum, magnetisches Hochfrequenz-, *magnetic high-frequency spectrum* 136.

Spektrum, reelles Phononen-, *real phonon spectrum* 85.
spezifische Wärme bei konstanter Feldstärke, *specific heat at constant field* 5, 17.
— — — Magnetisierung, *specific heat at constant magnetization* 5, 18.
Spin, Kern-, *nuclear spin* 99, 100.
— und Magnetfeld, anisotrope Kopplung zwischen Elektronen-, *anisotropic coupling between electron spin and magnetic field* 133.
Spin operator 266, 279, 299.
Spinaustausch, *spin exchange* 199, 203.
Spinbahnkopplung, *spin orbit coupling* 36, 47, 79, 126, 129, 130, 132, 259, 460.
—, Anisotropie in, *anisotropy in spin orbit coupling* 64.
—, Relaxation durch, *relaxation due to spin orbit coupling* 47.
Spinbahnkopplungskoeffizient, *spin orbit coupling coefficient* 121, 131.
Spinbahnwechselwirkung, *spin-orbit interaction* 35, 462.
Spindichte, *spin density* 161, 162, 164, 173, 180, 225.
—, negative, *negative spin density* 189.
Spindisorderwiderstand, *spin disorder resistivity* 538.
Spinecho, *spin echo* 263.
Spinell, *spinel* 253.
—, Cr-, $CdCr_2SE_4$ *spinel* 481.
—, Cr-Chalkogen-, *Cr-chalcogen spinel* 522.
Spinelle, Cr-, magnetische Rotverschiebung, *Cr spinels, magnetic red shift* 487.
— Thio-, elektrische Leitfähigkeit in, *electrical conductivity in thio-spinels* 531.
—, Superaustauschwechselwirkung in Thio-, *superexchange interaction in thio-spinels* 531.
Spinfluktuationen, Trägerstreuung durch, *scattering by spin fluctuations* 424.
Spingitteroperator, *spin-lattice operator* 30.
Spingitterrelaxation, *spin-lattice relaxation* 1, 30, 56, 59, 78, 120.
Spingitterrelaxationsoperator, effektiver, *effective spin-lattice relaxation operator* 56.
Spingitterrelaxationszeit, *spin-lattice relaxation time* 6, 9, 120, 121, 132.
Spingitterstörung, *spin-lattice perturbation* 34.
Spingitterwechselwirkung, *spin-lattice interaction* 4, 48, 64, 99, 120, 124.
Spin-Hamilton-Operator, *spin Hamiltonian* 4, 14, 39, 56, 74, 132, 133, 135, 136, 137, 141, 170, 227, 232, 242.
—, Diracscher, *Dirac spin Hamiltonian* 266.
Spin-Phonon-Kopplung, *spin-phonon coupling* 91.
Spin-Phonon-System, *spin-phonon system* 34.
—, Zustandsfunktionen des, *state functions of the spin phonon system* 34.
Spin-Phonon-Wechselwirkung, *spin-phonon interaction* 84, 85.
Spinpolarisation, *spin polarization* 174, 311, 352, 380.

Spinprozeß-, Mehr-, *multi-spin process* 74, 75.
Spinresonanz, Elektronen-, *electron spin resonance* 9, 94.
—, Kern-, *nuclear spin resonance* 9.
Spin-Spin-Relaxation, *spin-spin relaxation* 1, 9, 13, 72.
Spin-Spin-Relaxationszeit, *spin-spin relaxation time* 120.
Spin-Spin-Wechselwirkung, *spin-spin interaction* 99, 120, 122.
Spinsystem, *spin system* 4, 5.
Spintemperatur, *spin temperature* 6, 72, 82, 86.
—, negative, *negative spin temperature* 87.
spontane Emission, *spontaneous emission* 46.
stabiles freies Radikal, *stable free radical* 182.
Stark-Aufspaltungseffekt, *Stark effect splitting* 108, 127.
Störungsoperator, effektiver, *effective perturbation operator* 39.
Störungstheorie nach Bogolyubov, *Bogolyubov's perturbation theory* 322, 365.
— erster Ordnung, *first-order perturbation theory* 66, 79.
— höherer Ordnung, *higher order perturbation theory* 74.
— nach Löwdin, *Löwdin's perturbation theory* 344, 349, 361.
— zweiter Ordnung, *second-order perturbation* 50, 66, 68, 80.
strahlungsinduzierte Gitterfehler in Festkörpern, *radiation-induced defects in solids* 145.
strahlungsinduziertes Zentrum, *radiation-induced centre* 254.
Strandbergs effektiver Störungsoperator, Mattucks und, *Mattuck and Strandberg's effective perturbation operator* 56.
Sturm-Liouville theorem 281.
Superaustausch, *superexchange* 336, 410, 448, 535.
Superaustauschwechselwirkung in Thio-Spinellen, *superexchange interaction in thio-spinels-* 531.
Superhyperfeinaufspaltung, *superhyperfine splitting* 132, 184, 255.
Superhyperfeinlinie, *superhyperfine line* 103.
Superhyperfeinmuster, *superhyperfine pattern* 107.
Superhyperfeinstruktur, Chlor-, *chlorine superhyperfine structure* 252.
—, Fluor-, *fluorine superhyperfine structure* 252.
supraleitender Magnet, *superconducting magnet* 120.
Suszeptibilität, adiabatische, *adiabatic susceptibility* 5, 10, 17.
—, komplexe, *complex susceptibility* 6, 19.
—, paramagnetische, *paramagnetic susceptibility* 404.
—, statische, *static susceptibility* 2, 5, 108/109, 125.
Symmetrie, Oktaeder-, *octahedral symmetrie* 129.

Symmetriegruppe, kubische, *cubic symmetry group* 375.
S-Zustand, Relaxation in Ionen im, *relaxation in S-state ions* 64.
S-Zustandsion, *S-state ion* 70, 71.
σ-Bahn, *σ orbital* 127.
σ-Radikal, *σ-radical* 204.

T, magic- 119.
t_{2g}-Bahn, t_{2g} *orbital* 128.
Teller-Effekt, Jahn-, *Jahn- Teller effect* 69, 70.
Teller theorem, Jahn- 36.
Temperatur, Curie-Weiss-, *Curie-Weiss temperature*, 409.
—, Debye-, *Debye temperature* 42.
—, Gitter-, *lattice temperature* 31.
—, negative Spin-, *negative spin temperature* 87.
—, Spin-, *spin temperature* 6, 72, 82, 86.
pole transition 37.
temperaturabhängiger Moment, *temperature-dependent moment* 406.
tetragonale Verzerrung, *tetragonal distortion* 130.
Tetraederfeld, *tetrahedral field* 129, 130.
thermische Schwingung, *thermal vibration* 121.
Thermokraft, *thermoelectric force* 429.
Th_3P_4-Verbindung, Th_3P_4 *compound* 502.
tiefer Donator, *deep donor impurity* 254.
Trägerstreuung durch Spinfluktuationen, *scattering by spin fluctuations* 424.
Transporteigenschaft, *transport property* 412.
Transportintegral, *transport integral* 42, 66.
Transposition 290, 301, 313.
Transversale Relaxationszeit, *transversal relaxation time* 8.
transversales Phonon, *transverse phonon* 82.
Triplett, Naphthalen-, *naphthalene triplet* 207, 224.
Triplettanregung, *triplet excitation* 219.
Triplett-Aufspaltung, Singulett-, *singlet-triplet splitting* 266, 272.
Triplettgrundzustand, *ground state triplet* 221.
Triplettsystem, angeregtes, *excited triplet system* 210.
Triplettzustandssystem, *triplet state system* 205.

Übergang, chemischer, *chemical transfer* 202.
—, elektrischer Dipol-, *electric dipole transition* 37, 459.
—, elektrischer Quadrupol-, *electric quadrupole transition* 37.
—, $4f$-$5d$-, $4f$-$5d$ *transition* 472.
—, intramolekularer Elektronen-, *intramolecular electron transition* 203.
—, magnetischer Dipol-, *magnetic dipole transition* 37.
—, Metall-Isolator-, *metal-insulator transition* 400.
—, Protonen-, *proton transition* 202.
—, verbotener, *forbidden transition* 136.

Übergangsintegral, Einelektronen-, *one electron transition integral* 330.
Übergangswahrscheinlichkeit, *transition probability* 30, 33, 76, 81, 90.
— für den direkten Vorgang, *transition probability for the direct process* 40.
— für Orbach-Prozesse, *transition probability for Orbach processes* 43.
— für Raman-Prozesse, *transition probability for Raman processes* 41.
Überlappung, *overlap* 362.
Überlappungsintegral, *overlap integral* 276, 278, 311, 323, 354, 359, 362.
Überschußkristallrauschen, *excess crystal noise* 110.
Überschußladung, *excess charge* 158.
Unterdrückung des Bahndrehimpulses, *quenching of the orbital angular momentum* 36.
U-Zentrum, *U centre* 247.

V_1-Zentrum, V_1 *centre* 239.
V_F-Zentrum, V_F *centre* 241.
V_K-Zentrum, V_K *centre* 240, 241, 243.
V_{OH}-Zentrum, V_{OH} *centre* 245.
V_t-Zentrum, V_t *centre* 241.
Van Vleck-Aufhebung, *van Vleck-cancellation* 37, 52, 57, 66.
Van Vleck-Formel, Brons-, *Brons- van Vleck form*, 74, 81.
Van Vleck-Hamilton-Operator, Dirac-, *Dirac-van Vleck Hamiltonian* 267.
Van Vleck's Theorie, *van Vleck's theory* 56.
Vektormodell für mehrere Elektronen mit Bahnentartung, *vector model for more than one electron with orbital degeneracy* 304.
— — — ohne Bahnentartung, *vector model for more than one electron without orbital degeneracy* 300.
— bei orthogonalisierten Wellenfunktionen, *vector model with orthogonal atomic wave functions* 321.
— für ein s-Elektron pro Atom, *vector model for one s-electron per atom* 296.
verbotene Zone, *energy gap* 413.
verbotener Übergang, *forbidden transition* 136.
verbreiterte Linie, homogen, *homogeneously broadened line* 125.
— —, inhomogen, *inhomogeneously broadened line* 125.
Verbreiterung der Dipol-Resonanzkurve, *broadening, dipole* 125.
—, Modulations-, *modulation broadening* 117.
verdünntes System, *dilute system* 72.
Verformung, *strain* 48, 83.
Verformungsanisotropie, *strain anisotropy* 85.
Verformungsexperiment, statisches, *static strain experiment* 92.
Verformungsoperator, *strain operator* 30, 49.
Vernichtungsoperator, *annihilation operator* 38.

Verschiebung, chemische, *chemical shift* 169.
Vertauschung, faktorisierbare, *factorable permutation* 291.
—, intraatomare, *intraatomic permutation* 301.
—, nichtfaktorisierbare, *non-factorable permutation* 291.
Vertauschungsgruppe, *permutation group* 295, 303, 305, 313.
Verzerrung, tetragonale, *tetragonal distortion* 130.
Vielniveausystem, *multi-level system* 65.
Vinylradikal, *vinyl radical* 205.
vorübergehendes Zwischenprodukt, *transient intermediate* 98.

Wärme bei konstanter Feldstärke, spezifische, *specific heat at constant field*, 5, 17.
— bei konstanter Magnetisierung, spezifische, *specific heat at constant magnetization* 5, 18.
—, magnetische spezifische, *magnetic specific heat* 10.
—, Peltier, *Peltier heat* 429.
Wärmeübertragungskoeffizient, *heat transfer coefficient* 6, 30.
Waller's Mechanismus der paramagnetischen Relaxation, *Waller's mechanism of paramagnetic relaxation* 65, 70, 74, 77.
Wasserstoffatom, *hydrogen atom* 246, 259.
Wasserstoffmolekül, asymptotisch exakte Lösung, *hydrogen molecule, asymptotically exact solution* 283.
—, Behandlung mit Bogolyubov-Methode, *hydrogen molecule, treatment by Bogolyubov-method* 332.
—, Heitler-London-Theorie, *hydrogen molecule Heitler-London theory* 271.
—, Orthogonalisierung der Einelektronenfunktionen, *hydrogen molecule, orthogonalization of one-electron functions* 280.
Wechselwirkung, anisotrope Hyperfein-, *anisotropic hyperfine interaction* 187.
—, Austausch-, *exchange interaction* 14, 20, 34, 35, 71, 73, 75, 77.
—, Bahn-Gitter-, *orbit-lattice interaction* 48.
—, Coulombsche, *Coulomb interaction* 35.
—, dipolare, *dipolar interaction* 73.
—, Dipol-Dipol-, *dipole-dipole interaction* 125.
—, direkte Austausch-, *direct exchange interaction* 268, 286.
—, elektrische Quadrupol-, *electric quadrupole interaction* 135, 136.
—, Elektron-Elektron, *electron-electron interaction* 289, 306, 341.
—, Elektronen-, *electron interaction* 541.
—, Elektron-Magnon-, *electron-magnon interaction* 398, 544.
—, Elektron-Phonon-, *electron-phonon interaction* 398, 543.
—, Fermi-, *fermi interaction* 72.
—, F-Zentrum, *F-centre interaction* 231.
—, über halbleitende Träger, magnetische, *magnetic interaction via semiconducting carriers* 512.

Wechselwirkung, Hyperfein-, *hyperfine interaction* 100, 102, 132, 143.
—, indirekte Austausch-, *indirect exchange interaction* 268, 335.
—, Ionen-, *interionic interaction* 72.
—, Kernquadrupol-, *nuclear quadrupole interaction* 34, 35, 72.
—, Konfigurations-, *configuration interaction* 337.
—, magnetische Dipol-Dipol-, *magnetic dipole-dipole interaction* 14, 34, 35, 75.
—, Mehrkern-, *multi-nuclear interaction* 104.
—, Modulation der Hyperfein-, *modulation of hyperfine interaction* 194.
—, π-σ-, π-σ *interaction* 163.
—, paramagnetische Relaxation durch modulierte Austausch-, *paramagnetic relaxation by modulated exchange interaction* 71.
—, Phonon-Phonon-, *phonon-phonon interaction* 87.
—, relativistische, *relativistic interaction* 286.
—, in SE-Elementen, magnetische, *magnetic interaction in RE-elements* 491.
—, Spinbahn-, *spin-orbit interaction* 35, 462.
—, Spingitter-, *spin-lattice interaction* 4, 48, 64, 99, 120, 124.
—, Spin-Phonon-, *spin-phonon interaction* 84, 85.
—, Spin-Spin-, *spin-spin interaction* 99, 120, 122.
—, in Thio-Spinellen, Superaustausch-, *superexchange interaction in thio-spinels* 531.
—, zweiter Ordnung, Hyperfein-, *second-order hyperfine interaction* 147/148.
Weiss-Temperatur, Curie-, *Curie-Weiss temperature* 409.
Wellenfunktion, d-, d- *wave function* 132.
— des Kristals, Wahl der, *choice of the wave function for the crystal* 290, 300, 304.
Wertigkeitswechsel, *valency change* 142.
Wigner-Eckhart theorem 48.
Winkelabhängigkeit, *angular variation* 135, 140.

XY_6 complex 56, 68, 69, 81.

Zeeman-Effekt, elektronischer, *electronic Zeeman effec* 35, 79, 126.
Zeeman-Energiereservoir, *Zeeman-energy reservoir* 22.
Zeeman-Kerneffekt, *Zeeman effect, nuclear* 35, 79.
zeitabhängige Hyperfeinaufspaltung, *temperature-dependent hyperfine splitting* 198.
Zeitumkehr, *time-reversal* 66.
Zeitumkehr-Entartung, *time-reversal degeneracy* 43.
Zeitumkehr-Symmetrie, *time-reversal symmetry* 30, 36.
Zentren, Mehrfach-F-, F-*aggregate centres* 235.
Zentrum, abgewandeltes F-; *modified F-centre* 233.
—, Donator-, *electron donor centre* 80.

Zentrum, F-, *F centre* 78, 144, 228.
—, H-, *H centre* 241, 244.
—, lokalisiertes Elektronen-, *local electron centre* 78.
—, M-, *M centre* 235.
— für ungerade Elektronen, *odd-electron centre* 259.
—, V_1-, V_1 *centre* 239.
—, V_K'-, V_K' *centre* 241.
—, V_K-, V_K *centre* 240, 241, 243.
—, V_{OH}-, V_{OH} *centre* 245.
—, V_t-, V_t *centre* 241.
Zentrum-Wechselwirkung, F-, *F- centre interaction* 231.
ZnS 255.
Zone, verbotene, *energy gap* 413.

Zustand, Sättigung im stationären, *steady state saturation* 10.
—, zweifach entarteter, *doubly degenerate state* 36.
Zustandsdichte, Phononen-, *phonon density of states* 30.
Zustandsfunktionen, *state functions* 30.
—, Anisotropie in, *anisotropy in state functions* 63.
— der Elektronen, *state functions of the electrons* 59.
— des Spin-Phonon-Systems, *state functions of the spin-phonon system* 34.
Zweig, akustischer, *acoustic branch* 37.
—, optischer, *optical branch* 37.
Zwischengitterstelle, *interstitial centre* 245.

Subject Index.

(English-German.)

Where English and German spelling of a word is identical the German version is omitted.

Absorption 6, 12, 476.
— line, *Absorptionslinie* 217.
—, magnetic, *Magnetoabsorption* 482.
— spectrum of Eu^{2+} ions, *Absorptionsspektrum von Eu^{2+}-Ionen* 474.
Acoustic branch, *akustischer Zweig* 37.
— paramagnetic resonance, *akustische paramagnetische Resonanz* 91.
Adiabatic approximation, *adiabatische Näherung* 43, 51.
— demagnetization, *adiabatische Entmagnetisierung* 1.
— susceptibility, *adiabatische Suszeptibilität* 5, 10, 17.
Aging effect, *Alterungseffekt* 10.
Alkyl-substituted benzene, *alkylsubstituiertes Benzol* 176.
Allyl radical, *Allylradikal* 189.
Anderson mechanism, Kramers-, *Kramers-Anderson-Mechanismus* 379.
Angular variation, *Winkelabhängigkeit* 135, 140.
Anion 185.
—, anthracene, *Anthracenanion* 156.
—, benzene, *Phenylanion* 186.
—, benzosemiquinone, *Benzosemichinonanion* 180.
—, butadiene, *Butadienanion* 160.
—, cyclo-octatetrene, *Cyclooctatetraenanion* 176.
— radical, *Anionradikal* 155.
Anisotropic coupling between electron spin and magnetic field, *anisotrope Kopplung zwischen Elektronenspin und Magnetfeld* 133.
Anisotropy in relaxation times, *Anisotropie in Relaxationszeiten* 63.
— in spin-orbit coupling, *in Spinbahnkopplung* 64.
— in state functions, *in Zustandsfunktionen* 63.
— in the strain, *in der Verformung* 85.
Annihilation operator, *Vernichtungsoperator* 38.
Anthracene 185.
— anion, *Anthracenanion* 156.
Antiferromagnetic coupling, *antiferromagnetische Kopplung* 267.
Antiferromagnetism, band model of, *Bandmodell des Antiferromagnetismus* 417.
—, compensated, *kompensierter Antiferromagnetismus* 267.

Antiferromagnetism, not-compensated, *nichtkompensierter Antiferromagnetismus* 267.
Approximation, adiabatic, *adiabatische Näherung* 43, 51.
—, non-adiabatic, *nichtadiabatische Näherung* 43, 52.
Atom, α-carbon, α-*Kohlenstoffatom* 159.
—, β-carbon, β-*Kohlenstoffatom* 159.
— energy level, *atomares Energieniveau* 401.
—, ligand, *Ligandenatom* 132.
— in solids, *Atom in Festkörpern* 245.

Band model of antiferromagnetism, *Bandmodell des Antiferromagnetismus* 417.
— structure, *Bandstruktur* 392.
— —, RE metals, *Bandstruktur, SE-Metalle* 472.
— — for Eu^{++}S, *EuS-Bandstruktur* 418.
— — of Eu chalcogenides, *Bandstruktur von Eu-Chalkogeniden* 477, 511.
— theory, *Bandtheorie* 397.
Bandwidth, *Bandbreite* 109, 113.
Beat-frequency method, *Schwebungsfrequenzverfahren* 12.
Benzene, *Benzol* 219.
—, alkyl-substituted, *alkylsubstituiertes Benzol* 176.
— anion, *Phenylanion* 186.
Benzo-semiquinone, *Benzosemichinon* 104.
Benzosemiquinone anion, *Benzosemichinonanion* 180.
—, dibutyl-, *Dibutylbenzosemichinon* 106.
Bethe-Slater curve, *Bethe-Slater-Kurve* 307.
Biochemical, *biochemisch* 95.
Biradical, *Biradikal* 205, 226.
Bloch equation, *Bloch-Gleichung* 8.
— wave, *Bloch-Welle* 321.
Bloembergen-Rowland coupling, *Bloembergen-Rowland-Kopplung* 411.
Bogolyubov's perturbation theory, *Störungstheorie nach Bogolyubov* 322, 365.
Bohr magneton, *Bohrsches Magneton* 95, 404.
Bolton relation, Colpa-, *Colpa-Bolton-Beziehung* 158.
Boltzmann distribution, *Boltzmann-Verteilung* 72.
— factor, *Boltzmann-Faktor* 96.
— relaxation, Maxwell-, *Maxwell-Boltzmann-Relaxation* 30.
Bonding, covalent, *kovalente Bindung* 67, 69.
—, crystal, *Kristallbindung* 67.
Born-Haber cycle 548.

Branch, acoustic, *akustischer Zweig* 37.
—, optical, *optischer Zweig* 37.
Bridge circuit, *Brückenschaltung* 11.
Brillouin function 405.
Broadening, dipole, *Verbreiterung der Dipol-Resonanzkurve* 122.
Broadened line, homogeneously, *homogen verbreiterte Linie* 125.
— —, inhomogeneously, *inhomogen verbreiterte Linie* 125.
Broadening, modulation, *Modulationsverbreiterung* 117.
Brons-van Vleck form, *Brons-van Vleck-Formel* 74, 81.
Butadiene anion, *Butadienanion* 160.

CdS 255.
CH_2 101.
CH group, *CH-Gruppe* 101, 102.
CH_2 group, *CH_2-Gruppe* 102.
CH_3 group, *CH_3-Gruppe* 101, 102, 103.
Cl_2 spectrum, *Cl_2-Spektrum* 241.
^{13}C splitting constant, *^{13}C-Aufspaltungskonstante* 165.
$CrBr_3$ 267.
Cr-chalcogen spinel, *Cr-Chalkogen-Spinell* 522.
Cr-spinels, magnetic red shift, *Cr-Spinelle, magnetische Rotverschiebung* 487.
Calorimeter method, *Kalorimeterverfahren* 12.
Carbon atom, α-, *α-Kohlenstoffatom* 159.
— —, β-, *β-Kohlenstoffatom* 159.
Casimir and Du Pré formalism, *Casimir- und Du Pré-Formalismus* 77.
— and Du Pré relaxation curve, *Casimir und Du Pré-Relaxationskurve* 7.
— and Du Pré, theory of, *Theorie von Casimir und Du Pré* 4.
Cation radical, *Kationradikal* 155.
Cavity, resonance, *Resonanzhohlraum* 109.
— resonator, *Hohlraumresonator* 97, 111, 112, 113.
Centre, electron donor, *Donatorzentrum* 80.
—, F, *F-Zentrum* 78, 144, 228.
—, H, *H-Zentrum* 241, 244.
— interaction, F-, *F-Zentrum-Wechselwirkung* 231.
—, interstitial, *Zwischengitterstelle* 245.
—, local electron, *lokalisiertes Elektronenzentrum* 78.
—, M-, *M-Zentrum* 235.
—, modified, F-, *abgewandeltes F-Zentrum* 233.
—, odd-electron, *Zentrum für ungerade Elektronen* 259.
—, V_1, *V_1-Zentrum* 239.
—, V_F, *V_F-Zentrum* 241.
—, V_K, *V_K-Zentrum* 240, 241, 243.
—, V_{OH}, *V_{OH}-Zentrum* 245.
—, V_t, *V_t-Zentrum* 241.
Centres, F-aggregate, *Mehrfach-F-Zentren* 235.
Chalcogenide, Eu, *Eu-Chalkogenid* 476, 481, 506, 512, 554.
—, Gd-, *Gd-Chalkogenid* 518.

Chalcogenides, band structure of Eu, *Bandstruktur von Eu-Chalkogeniden* 477, 511.
—, Eu, magnetic redshift, *Eu-Chalkogenide, magnetische Rotverschiebung* 483.
Charge, excess, *Überschußladung* 158.
Chemical transfer, *chemischer Übergang* 202.
Chlorine superhyperfine structure, *Chlor-superhyperfeinstruktur* 252.
Chlorobenzosemiquinone, *Chlorbenzosemichinon* 105.
Chromium halide, *Chrom-Halogen* 481.
Colour centre, *Farbzentrum* 72, 78.
Colpa-Bolton relation, *Colpa-Bolton-Beziehung* 158.
Concentration, *Konzentration* 70.
— effect, *Konzentrationseffekt* 78.
Condon effect, Franck-, *Franck-Condon-Effekt* 80.
Conduction electron polarization, *Leitungselektronenpolarisation* 433.
Configuration interaction, *Konfigurationswechselwirkung* 337.
Contact potential, Fermi, *Fermi-Kontaktpotential* 72.
Coordination, eight-fold, *achtfache Koordination* 69.
—, nine-fold, *neunfache Koordination* 47.
—, six-fold, *sechsfache Koordination* 36.
Copper nucleus, *Kupferkeim* 100.
Correlation, π-σ, *π-σ-Korrelation* 164.
—, electron, *Elektronenkorrelation* 284, 398, 400, 402, 532.
— time, *Korrelationszeit* 90.
Coulomb interaction, *Coulombsche Wechselwirkung* 35.
Coupling, anisotropy in spin-orbit, *Anisotropie in Spinbahnkopplung* 64.
—, antiferromagnetic, *antiferromagnetische Kopplung* 267.
— between electron spin and magnetic field, anisotropic, *anisotrope Kopplung zwischen Elektronenspin und Magnetfeld* 133.
—, Bloembergen-Rowland, *Bloembergen-Rowland-Kopplung* 411.
—, hyperfine, *Hyperfeinkopplung* 149.
— parameter, hyperfine, *Hyperfeinkopplungsparameter* 149, 160, 168.
—, Russel-Saunders, *Russel-Saunders-Kopplung* 402.
—, spin orbit, *Spinbahnkopplung* 36, 47, 79, 126, 129, 130, 132, 259, 460.
—, spin-phonon, *Spin-Phonon-Kopplung* 91.
Covalency, *Kovalenz* 65, 82, 383.
— of the 4f electrons, *4f-Kovalenz* 510.
Covalent bonding, *kovalente Bindung* 67, 69.
— — of 4f electrons, *kovalente Bindung von 4f-Elektronen* 470.
Creation operator 84.
Critical distance, *kritische Adernabstände* 399.
Crystal bonding, *Kristallbindung* 67.
—, choice of the wave function for the, *Wahl der Wellenfunktion des Kristalls* 290, 300, 304.
—, concentrated, *konzentriertes Kristall* 65.

Crystal detector, *Kristalldetektor* 98, 112.
— field, *Kristallfeld* 22, 34, 35, 47, 107, 131, 408, 463, 466.
— — parameter, *Kristallfeldparameter* 47, 60, 62, 63.
— — potential, *Kristallfeldpotential* 48.
— — in rare-earth salts, *Kristallfeld in Salzen seltener Erden* 47.
— — splitting, *Kristallfeldaufspaltung* 407, 534.
— — theory, strong crystal field, *Kristallfeldtheorie, starkes Kristallfeld* 376.
— — —, weak crystal field, *Kristallfeldtheorie, schwaches Kristallfeld* 374.
—, molecular, *Molekülkristall* 68.
— noise, excess, *Überschußkristallrauschen* 110.
— spectrum in RE, *SE-Kristallspektrum* 463.
Curie constant, *Curie-Konstante* 2, 3, 405.
Curie's law, *Curie-Gesetz* 2.
Curie temperature, *Curie-Temperatur* 265.
Curie-Weiss temperature, *Curie-Weiss-Temperatur* 409.
Cyclo-octatetrene anion, *Cyclooctatetraeanion* 176.

d orbital, *d-Bahn* 127, 128, 129.
d-wave function, *d-Wellenfunktion* 132.
$3d$-element oxide, $3d$-*Oxid* 550.
$3d$ orbital, $3d$-*Bahn* 127.
DPPH, diphenylpicrylhydrazyl 182.
Damage, irradiation, *Bestrahlungsschaden* 95, 98.
Davydov splitting 483.
Debye cut-off frequency, *Debye-Grenzfrequenz* 82.
— distribution, *Debye-Verteilung* 38, 49, 80.
— relaxation, *Debye-Relaxation* 21, 42.
— temperature, *Debye-Temperatur* 42.
Decay, exponential, *exponentieller Abfall* 21, 22.
Deep donor impurity, *tiefer Donator* 254.
Defect in insulating solids, *Gitterfehler in Isolatoren* 228.
— in semiconductors, *Gitterfehler in Halbleitern* 253.
Degeneracy, Kramers, *Kramers-Entartung* 36, 40, 69, 72.
—, time-reversal, *Zeitumkehr-Entartung* 43.
Degenerate state, doubly, *zweifach entarteter Zustand* 36.
Demagnetization, adiabatic, *adiabatische Entmagnetisierung* 1.
Density matrix operator, *Dichtematrix-Operator* 2, 15.
— of states, phonon, *Phononenzustandsdichte* 30.
Detector noise, *Empfängerrauschen* 109, 110.
—, phase sensitive, *phasenempfindlicher Empfänger* 113, 114, 117.
Deuterium atom, *Deuteriumatom* 246.
— exchange, *Deuteriumaustausch* 188.
Diffusion process, *Diffusionsvorgang* 75.
Dibutyl-benzosemiquinone, *Dibutylbenzosemichinon* 106.

Dilute system, *verdünntes System* 72.
Diphenylpicrylhydrazyl, DPPH 182.
Dipolar interaction, *dipolare Wechselwirkung* 73.
Dipole broadening, *Verbreiterung der Dipol-Resonanzkurve* 122.
Dipole-dipole effect, *Dipol-Dipol-Effekt* 122.
— interaction, *Dipol-Dipol-Wechselwirkung* 125.
— —, magnetic, *magnetische Dipol-Dipol-Wechselwirkung* 14, 34, 35, 75.
— operator, *Dipol-Dipol-Operator* 207.
Dipole transition, electric, *elektrischer Dipolübergang* 37, 459.
— —, magnetic, *magnetischer Dipolübergang* 37.
Dirac spin Hamiltonian, *Diracscher Spin-Hamilton-Operator* 266.
Dirac-van Vleck Hamiltonian, *Dirac-van Vleck-Hamilton-Operator* 267.
Direct process, *direkter Prozeß* 39, 121.
— — in non-Kramers salts, *direkter Vorgang in Nicht-Kramers Salzen* 48.
— — of relaxation, direkter Relaxationsprozeß 39.
— relaxation process in rare-earth salts, *direkter Relaxationsprozeß in Seltene-Erde-Salzen* 48.
Dispersion 12.
Distortion, tetragonal, *tetragonale Verzerrung* 130.
Double exchange, *Doppelaustausch* 351, 364, 367, 441, 446. 450, 520, 535.
— resonance, *Doppelresonanz* 142, 144.
Doublet, Kramers, *Kramers-Dublett* 131.
Du Pré formalism, Casimir and, *Casimir- und Du Pré-Formalismus* 77.
— relaxation curve, Casimir and, *Casimir- und Du Pré-Relaxationskurve* 7.
—, theory of Casimir and, *Theorie von Casimir und Dupré* 4.

EuI_2 267.
EuS 267.
EuO 267.
EuSe 267.
Eu_2SiO_3 267.
e_g orbital, e_g-*Bahn* 128.
ENDOR (Electron Nuclear DOuble Resonance), *Elektron-Kern-Doppelresonanz* 142, 143, 144, 229, 236, 241, 247, 252, 253, 262.
ESR of gases, *ESR von Gasen* 256.
ESR spectrometer, *ESR-Spektrometer* 112.
Eu chalcogenides, magnetic red shift, *Eu-Chalkogenide, magnetische Rotverschiebung* 483.
Eckhart theorem, Wigner-, 48.
Electric dipole transition, *elektrischer Dipolübergang* 37, 459.
Electric field, internal, *inneres elektrisches Feld* 108.
— quadrupole interaction, *elektrische Quadrupolwechselwirkung* 135, 136.

Electric quadrupole moment, *elektrisches Quadrupolmoment* 135, 141.
— quadrupole, nuclear, *elektrisches Kernquadrupol* 140.
— quadrupole transition, *elektrischer Quadrupolübergang* 37.
Electrical conductivity in thio-spinels, *elektrische Leitfähigkeit in Thio-Spinellen* 531.
Electromagnet, *Elektromagnet* 98, 119.
Electron centre, local, *lokalisiertes Elektronenzentrum* 78.
— —, odd-, *Zentrum für ungerade Elektronen* 259.
— correlation, *Elektronenkorrelation* 284, 398, 400, 402, 532.
Electrons, covalent bonding of 4f, *kovalente Bindung von 4f-Elektronen* 470.
Electron density, unpaired, *Dichte ungepaarter Elektronen* 155, 163.
— donor centre, *Donatorzentrum* 80.
Electron-electron interaction, *Elektron-Elektron-Wechselwirkung* 289, 306, 341.
— repulsion 537.
Electron, hydrated, *hydratisiertes Elektron* 238.
— interaction, *Elektronenwechselwirkung* 541.
Electron-magnon interaction, *Elektron-Magnon-Wechselwirkung* 398, 544.
Electrons, mobility of $3d, 4f$, *Beweglichkeit von 3d,4f-Elektronen* 398.
Electron model, localized, *lokalisiertes Elektronenmodell* 400.
— nuclear double resonance (ENDOR), *Elektron-Kern-Doppelresonanz* 142, 143, 144, 229, 236, 241, 247, 252, 253, 262.
— paramagnetic resonance (E.P.R.), *paramagnetische Elektronenresonanz* 36.
— phase diagram, *elektronisches Phasendiagramm* 532.
Electron-phonon interaction, *Elektron-Phonon-Wechselwirkung* 398, 543.
Electron polarization, conduction, *Leitungselektronenpolarisation* 433, 490.
— shells, radius of, *Radius äußerer Elektronenschalen* 389.
— spin and magnetic field, anisotropic coupling between, *anisotrope Kopplung zwischen Elektronenspin und Magnetfeld* 133.
— spin resonance, *Elektronenspinresonanz* 9, 94.
— — — of gases, *von Gasen* 256.
— — — spectrometer, *Elektronenspinresonanzspektrometer* 112.
— transfer, intramolecular, *intramolekularer Elektronenübergang* 203.
—, unpaired, *ungepaartes Elektron* 94, 98, 100, 109, 137.
Electronic quadrupole, *elektronisches Quadrupol* 136.
— splitting, *elektronische Aufspaltung* 107, 108, 133, 136, 139.
— —, angular variation of, *Winkelverteilung der elektronischen Aufspaltung* 107, 140.

Emission of a phonon, spontaneously, *spontane Emission eines Phonons* 46.
—, spontaneous, *spontane Emission* 46.
—, stimulated, *induzierte Emission* 121.
Energy gap, *verbotene Zone* 413.
— level for lanthanide atoms, *Energieniveau in SE-Atomen* 458.
— —, orbital, *Bahnenergieniveau* 129.
— reservoir, exchange- 22.
Enzyme, *Enzym* 95.
Eu chalcogenide, *Eu-Chalkogenid* 476, 481, 506, 512, 554.
Eu chalcogenides, bandstructure of, *Bandstruktur von Eu-Chalkogeniden* 477, 511.
Excess charge, *Überschußladung* 158.
— crystal noise, *Überschußkristallrauschen* 110.
Exchange, deuterium, *Deuteriumsaustausch* 188.
—, direct, *direkter Austausch* 432.
—, double, *Doppelaustausch* 351, 364, 367, 411, 446, 450, 520, 535.
Exchange-energy reservoir 22.
Exchange, indirect, *indirekter Austausch* 410, 430, 440, 487, 515, 521.
— interaction, indirect, Anderson's new approach, *indirekte Austauschwechselwirkung, Andersons neuer Lösungsweg* 269, 368.
— integral between orthogonal functions, *Austauschintegral zwischen orthogonalen Funktionen* 276, 350.
— —, definition for 2 electrons, *Austauschintegral, Definition für 2 Elektronen* 266, 272, 275.
— —, definition for 3 electrons, *Austauschintegral, Definition für 3 Elektronen* 314.
— —, direct, sign, *Austauschintegral, direktes, Vorzeichen* 306.
— —, H_2 problem, *Austauschintegral, H_2-Problem* 276, 334.
— —, Heisenberg, *Heisenbergsches Austauschintegral* 277, 330.
— —, indirect, for 3 centres and 4 electrons, *Austauschintegral, indirektes, für 3 Zentren und 4 Elektronen* 343, 347, 349, 358.
— —, indirect, in the crystal, *Austauschintegral, indirektes, im Kristall* 366.
— —, influence of electron nuclear interaction, *Austauschintegral, Einfluß der Elektron-Kern-Wechselwirkung* 308, 310.
— —, indirect, numerical calculations, *Austauschintegral, indirektes, zahlenmäßige Bestimmungen* 381.
— —, indirect, semiempirical rules for the estimation, *Austauschintegral, indirektes, halbempirische Regeln zur Abschätzung* 379.
— —, indirect, sign, *Austauschintegral, indirektes, Vorzeichen* 373.
Exchange interaction, *Austauschwechselwirkung* 14, 20, 34, 35, 71, 73, 75, 77.
— —, biquadratic, *biquadratische Austauschwechselwirkung* 312.
— —, direct, *direkte Austauschwechselwirkung* 268, 286.

Exchange interaction, indirect, *indirekte Austauschwechselwirkung* 268, 335.
— —, indirect, in the crystal, *indirekte Austauschwechselwirkung, im Kristall* 365.
— —, paramagnetic relaxation by modulated, *paramagnetische Relaxation durch modulierte Austauschwechselwirkung* 71.
—, kinetic, *kinetischer Austausch* 270, 330, 370, 432.
—, mechanism of indirect, *Mechanismus des indirekten Austausches* 268, 337.
— mechanism, non-Kramers, *nicht-Kramersscher Austauschmechanismus* 269.
— narrowing 123, 125.
—, potential, *potentieller Austausch* 270, 330, 370.
—, spin, *Spinaustausch* 199, 203.
—, super, *Superaustausch* 336.
Excitation, triplet, *Triplettanregung* 219.
Excited configuration, *angeregte Konfiguration* 321.
— triplet system, *angeregtes Triplettsystem* 210.
Exciton-magnon absorption 483.
Exponential decay, *exponentieller Abfall* 21, 22.

F-centre, *F-Zentrum* 78, 144, 228.
F-centre interaction, *F-Zentrum-Wechselwirkung* 231.
F-centre, modified, *abgewandeltes F-Zentrum* 233.
^{19}F fluorine splitting constant, ^{19}F-*Fluoraufspaltungskonstante* 167.
Faraday rotation in garnets, *Faraday-Rotation in Granaten* 481.
α-Fe$_2$O$_3$ 268.
Fermi contact potential, *Fermi-Kontaktpotential* 72.
— gas, *Fermi-Gas* 266.
— interaction, *Fermi-Wechselwirkung* 72.
Ferrimagnetism, *Ferrimagnetismus* 265, 267.
Ferrite, *Ferrit* 335.
Ferromagnetism, *Ferromagnetismus* 265, 266.
— of band electrons, *Ferromagnetismus von Bandelektronen* 538.
Filling factor, *Füllfaktor* 109.
Flicker noise, *Funkelrauschen* 109.
Fluorine superhyperfine structure, *Fluor-Superhyperfeinstruktur* 252.
Fock method, restricted Hartree-, *beschränktes Hartree-Fock-Verfahren* 352.
— —, unrestricted Hartree-, *uneingeschränktes Hartree-Fock-Verfahren* 352, 369.
Forbidden transition, *verbotener Übergang* 136.
Fourier spectrum, *Fourier-Spektrum* 90.
Franck-Condon effect, *Franck-Condon-Effekt* 80.
Free radical, *freies Radikal* 77, 94, 95, 98, 116, 119, 126, 151.
— radical g-value, *g-Wert eines freien Radikals* 126.
— radical, stable, *stabiles freies Radikal* 182.

Free-radical transient, *Übergang mit freien Radikalen* 142.
Frequency control, automatic, *automatische Frequenzkontrolle* 108, 116, 117.
— modulation, high-, *Hochfrequenzmodulation* 113.

g component, *g-Komponente* 147.
g factor, *g-Faktor* 115.
g parameter, *g-Parameter* 95.
g tensor 146.
g value, *g-Wert* 99, 115, 120, 126, 131, 132, 133, 136, 139.
— —, free radical, *g-Wert eines freien Radikals* 126.
Gd chalcogenide, *Gd-Chalkogenid* 518.
Gap, energy, *verbotene Zone* 413.
—, second, *zweites Energiegap* 413.
Garnet, *Granat* 497.
Garnets, Faraday rotation in, *Faraday-Rotation in Granaten* 481.
Garnet structure, *Granatstruktur* 253.
Gaussian line, *Gauss'sche Linie* 124.
— — shape, *Gauss'sche Kurvenform* 24, 123.
Gaussian-Lorentz line shape, *Gauss'sche-Lorentz-Kurvenform* 26.
Germanium 254.
Giant magneto-resistance effect, *anomaler Magnetowiderstand* 554.
— moment 516.
Giacometti-Nordio-Pavan relation, *Giacometti-Nordio-Pavan-Beziehung* 159.
Goodenough-Kanamori rules, *Goodenough-Kanamori-Regeln* 451.
Goodenough mechanism, *Goodenough-Mechanismus* 379.
Ground state triplet, *Triplettgrundzustand* 221.

H 241.
H centre, *H-Zentrum* 241, 244.
HMO approach, *HMO-Näherung* 147, 151, 159.
HMO method, *HMO-Methode* 175.
Haber cycle, Born- 548.
Hamiltonian, *Hamilton-Operator* 3, 14, 20, 235.
—, Dirac spin, *Diracscher Spin-Hamilton-Operator* 266.
—, Dirac-van Vleck-, *Dirac-van Vleck-Hamilton-Operator* 267.
—, Heisenberg, *Heisenberg-Hamilton-Operator* 267.
—, lattice, *Gitter-Hamilton-Operator* 38.
— matrix, *Hamilton-Operator-Matrix* 208.
—, spin, *Spin-Hamilton-Operator* 4, 14, 39, 56, 74, 132, 133, 135, 136, 137, 141, 170, 227, 232, 242.
Hartree-Fock method, restricted, *beschränktes Hartree-Fock-Verfahren* 352.
— —, unrestricted, *uneingeschränktes Hartree-Fock-Verfahren* 352, 369.
Hartree operator 67.
Heat at constant field, specific, *spezifische Wärme bei konstanter Feldstärke* 5, 17.

Heat at constant magnetization, specific, *spezifische Wärme bei konstanter Magnetisierung* 5, 18.
—, Peltier *Peltier-Wärme* 429.
—, magnetic specific, *magnetische spezifische Wärme* 10.
— transfer coefficient, *Wärmeübertragungskoeffizient* 6, 30.
Heisenberg exchange integral, *Heisenbergsches Austauschintegral* 277, 330.
— Hamiltonian, *Heisenberg-Hamilton-Operator* 267.
— magnet, *Heisenberg-Magnet* 507.
Heitler-London method, *Heitler-London-Methode* 271.
— model, *Heitler-London-Modell* 409.
— theory, asymptotically exact solution, *Heitler-London-Theorie, asymptotisch exakte Lösung* 317.
— — for the crystal, *für den Kristall* 287.
Hopping 410, 427.
Hückel Molecular Orbital approximation (HMO), *Hückel-Molecular-Orbital-Näherung* 147, 151.
Hund's rule, *Hundsche Regel* 377, 404.
Hydrocarbon, "alternant", *Kohlenwasserstoff, „alternant"* 154.
Hydrogene atom, *Wasserstoffatom* 246, 259.
Hydrogene molecule, asymptotically exact solution, *Wasserstoffmolekül, asymptotisch exakte Lösung* 283.
— —, Heitler-London theory, *Wasserstoffmolekül, Heitler-London-Theorie* 271.
— —, orthogonalization of one-electron functions, *Wasserstoffmolekül, Orthogonalisierung der Einelektronenfunktionen* 280.
— —, treatment by BOGOLYUBOV's method, *Wasserstoffmolekül, Behandlung mit Bogolyubov-Methode* 332.
Hyperconjugation, *Hyperkonjugation* 172, 176.
Hyperfine constant, one-electron, *Einelektronhyperfeinkonstante* 149.
— coupling, *Hyperfeinkopplung* 149.
— coupling parameter, *Hyperfeinkopplungsparameter* 149, 160, 168.
— effect, second-order, *Hyperfeineffekt zweiter Ordnung* 149.
— interaction, *Hyperfeinwechselwirkung* 100, 102, 132, 143.
— —, anisotropic, *anisotrope Hyperfeinwechselwirkung* 187.
— —, modulation of, *Modulation der Hyperfeinwechselwirkung* 194.
— —, second-order, *Hyperfeinwechselwirkung zweiter Ordnung* 147/148.
— line, *Hyperfeinlinie* 103, 116.
— —, super-, *Superhyperfeinlinie* 103.
— pattern, super-, *Superhyperfeinmuster* 107.
— splitting, *Hyperfeinaufspaltung* 14, 99, 104, 108, 133, 135, 139, 140, 148.
— splitting constant, sign of the, *Vorzeichen der Hyperfeinaufspaltungskonstante* 168.

Hyperfine splitting, super-, *Superhyperfeinaufspaltung* 132, 184, 255.
— —, temperature-dependent, *zeitabhängige Hyperfeinaufspaltung* 198.
— splitting tensor, *Hyperfeinaufspaltungstensor* 147, 148, 187.
— structure, *Hyperfeinstruktur* 101 132.
— —, chlorine super-, *Chlorsuperhyperfeinstruktur* 252.
— —, fluorine super-, *Fluor-Superhyperfeinstruktur* 252.
— —, mechanism of, *Mechanismus der Hyperfeinstruktur* 162.
— structure splitting, *Hyperfeinstrukturaufspaltung* 14.

Impurity ion, *Fremdion* 78.
— pair, *Fremdatompaar* 254.
— state, magnetic, *magnetische Fremdatome* 516.
Integral, overlap, *Überlappungsintegral* 276, 278, 311, 323, 354, 359, 362.
Interaction, π-σ, π-σ-*Wechselwirkung* 163.
—, anisotropic hyperfine, *anisotrope Hyperfeinwechselwirkung* 187.
—, configuration, *Konfigurationswechselwirkung* 337.
—, Coulomb, *Coulombsche Wechselwirkung* 35.
—, dipolar, *dipolare Wechselwirkung* 73.
—, dipole-dipole, *Dipol-Dipol-Wechselwirkung* 125.
—, direct exchange, *direkte Austauschwechselwirkung* 268, 286.
—, electric quadrupole, *elektrische Quadrupolwechselwirkung* 135, 136.
—, electron, *Elektronenwechselwirkung* 541.
—, electron-electron, *Elektron-Elektron-Wechselwirkung* 289, 306, 341.
—, electron-magnon, *Elektron-Magnon-Wechselwirkung* 398, 544.
—, electron-phonon, *Elektron-Phonon-Wechselwirkung* 398, 543.
—, exchange, *Austauschwechselwirkung* 14, 20, 34, 35, 71, 73, 75, 77.
—, F-centre, *F-Zentrum-Wechselwirkung* 231.
—, Fermi, *Fermi-Wechselwirkung* 72.
—, hyperfine, *Hyperfeinwechselwirkung* 100, 102, 132, 143.
—, indirect exchange, *indirekte Austauschwechselwirkung* 268, 335.
—, interionic, *Ionenwechselwirkung* 72.
—, magnetic dipole-dipole, *magnetische Dipol-Dipol-Wechselwirkung* 14, 34, 35, 75.
—, modulation of hyperfine, *Modulation der Hyperfeinwechselwirkung* 194.
—, multi-nuclear, *Mehrkernwechselwirkung* 104.
—, nuclear quadrupole, *Kernquadrupolwechselwirkung* 34, 35, 72.
—, orbit-lattice, *Bahn-Gitterwechselwirkung* 48.
—, paramagnetic relaxation by modulated exchange, *paramagnetische Relaxation durch modulierte Austauschwechselwirkung* 71.

Interaction, phonon-phonon, *Phonon-Phonon-Wechselwirkung* 87.
— in RE-elements, magnetic, *magnetische Wechselwirkung in SE-Elementen* 491.
—, relativistic, *relativistische Wechselwirkung* 286.
—, second-order hyperfine, *Hyperfeinwechselwirkung zweiter Ordnung* 147/148.
— via semiconducting carriers, magnetic, *magnetische Wechselwirkung über halbleitende Träger* 512.
—, spin-lattice, *Spingitterwechselwirkung* 4, 48, 64, 99, 120, 124.
—, spin-orbit, *Spinbahnwechselwirkung* 35, 462.
—, spin-phonon, *Spin-Phonon-Wechselwirkung* 84, 85.
—, spin-spin, *Spin-Spin-Wechselwirkung* 99, 120, 122.
— in thio-spinels, superexchange, *Superaustauschwechselwirkung in Thio-Spinellen* 531.
Internal electric field, *inneres elektrisches Feld* 108.
Interstitial centre, *Zwischengitterstelle* 245.
Inverted Orbach relaxation process, *inverser Orbach-Relaxationsprozeß* 46, 65.
Ionic energy level, *Ionenenergieniveau* 35.
— interaction, inter-, *Ionenwechselwirkung* 72.
— state, *Ionenzustand* 282, 293.
Ionization energy, *Ionisationsenergie* 294, 319, 382.
— — for RE, *Ionisierungsarbeit für SE* 458.
Ions, arrangement of, *Ionenanordnung* 267.
Irradiation damage, *Bestrahlungsschaden* 95, 98.
Isolator 265.

Jahn-Teller effect, *Jahn-Teller-Effekt* 69, 70.
— theorem 36.

KCl 144.
Kanamori rules, Goodenough-, *Goodenough-Kanamori-Regeln* 451.
Kittel function, Ruderman-, *Ruderman-Kittel-Funktion* 434.
Klystron 97, 108, 110, 115, 116.
Kramers-Anderson mechanism, *Kramers-Anderson-Mechanismus* 379.
Kramers degeneracy, *Kramers-Entartung* 36, 40, 69, 72.
Kramers doublet, *Kramers-Dublett* 131.
Kramers exchange mechanism, non-, *nicht-Kramersscher Austauschmechanismus* 269.
Kramers-Kronig relation, *Kramers-Kronig-Beziehung* 19.
Kramers salt, *Kramers-Salz* 39, 40.
— —, non-, *Nicht-Kramers-Salz* 39, 40, 41.
— —, Raman process in non-, *Raman-Prozeß in Nicht-Kramers-Salz* 41, 50.
Kramers theorem 36.
Kronig relation, Kramers-, *Kramers-Kronig-Beziehung* 19.

LCAO, *LCAO-Methode* 67, 151, 378, 409.
—, MO-, *MO-LCAO-Methode* 321.
Li_xMn_1 520.
Landau trapping 411, 543.
Landé splitting factor, *Landé-Aufspaltungsfaktor* 126, 404.
Lanthanide atoms, energy level for, *Energieniveau in SE-Atomen* 458.
Laplace's equation, *Laplace-Gleichung* 47.
Larmor frequency, *Larmor-Frequenz* 8, 22, 24.
Lattice Hamiltonian, *Gitter-Hamilton-Operator* 38.
— interaction, spin-, *Spingitterwechselwirkung* 4, 48, 64, 99, 120, 124.
— operator, spin-, *Spingitteroperator* 30.
— perturbation, spin-, *Spingitterstörung* 34.
— potential, orbit-, *Bahn-Gitterpotential* 48.
— relaxation operator, effective spin-, *effektiver Spingitterrelaxationsoperator* 56.
— relaxation, spin-, *Spingitterrelaxation* 1, 30, 56, 59, 78, 120.
— relaxation time, spin-, *Spingitterrelaxationszeit* 6, 9, 120, 121, 132.
— state, *Gitterzustand* 38.
— temperature, *Gittertemperatur* 31.
Laves phase. *Lavesphasen* 492.
Life-time of the phonon, *Lebensdauer des Phonons* 86, 87.
Ligand atom, *Ligandenatom* 132.
— field theory, *Ligandenfeldtheorie* 127, 270, 379, 406.
— nucleus, *Ligandenkern* 132.
Line shape for non-oriented systems, *Linienform für nichtausgerichtete Systeme* 214.
— width, *Linienbreite* 98.
— — anomaly, *Linienbreitenanomalie* 192.
— — effect, *Linienbreiteneffekt* 192.
Linear combination of atomic orbitals LCAO 67, 151, 378, 409.
Liouville operator 16.
Localized defekt, *lokalisierte Fehlstelle* 82.
Localized electron model, *lokalisiertes Elektronenmodell* 400.
Löwdin's perturbation theory, *Störungstheorie nach Löwdin* 344, 349, 361.
London method, Heitler-, *Heitler-London-Methode* 271.
— model, Heitler-, *Heitler-London-Modell* 409.
Longitudinal phonon, *longitudinales Phonon* 82.
— relaxation time, *longitudinale Relaxionszeit* 8.
Lorentz line shape, Gaussian-, *Gaußsche-Lorentz-Kurvenform* 26.
Lorentzian line shape, *Lorentz-Kurvenform* 26, 123, 124.

M centre, *M-Zentrum* 235.
Mn perovskite, *Mn-Perowskit* 498.
Mn^{2+} ion 137, 252.
Mn^{55} nucleus 137.
MnO 267, 268, 335, 380, 382.
MO-LCAO, *MO-LCAO-Methode* 321.

Madelung potential 548.
Magic-T 119.
Magnet, superconducting, *supraleitender Magnet* 120.
Magnetic absorption, *Magnetoabsorption* 482.
— dipole-dipole interaction, *magnetische Dipol-Dipol-Wechselwirkung* 14.
— dipole transition, *magnetischer Dipolübergang* 37.
— field modulation, *Magnetfeldmodulation* 110, 112, 113, 115.
— high-frequency spectrum, *magnetisches Hochfrequenzspektrum* 136.
— impurity state, *magnetische Fremdatome* 516.
— interaction in RE elements, *magnetische Wechselwirkung in SE-Elementen* 491.
— — via semiconducting carriers, *magnetische Wechselwirkung über halbleitende Träger* 512.
— moment, *magnetisches Moment* 94, 99.
— — of RE, *magnetisches Moment von SE* 490.
— —, nuclear, *magnetisches Kernmoment* 94.
— polaron, *magnetisches Polaron* 399, 419, 421.
— quasiparticle, *magnetisches Quasiteilchen* 368.
— semiconductor, *magnetischer Halbleiter* 391, 396.
— specific heat, *magnetische spezifische Wärme* 10.
Magneto-optic, *Magneto-Optik* 480.
Magnetomechanical ratio, *magnetomechanisches Verhältnis* 8.
Magneton, nuclear, *Kernmagneton* 99.
— number, effective, *effektive Magnetonnummer* 405.
Magnon absorption, exciton- 483.
— drag, *Magnonendrag* 430.
— interaction, electron-, *Elektron-Magnon-Wechselwirkung* 398.
Malonic acid, *Malonsäure* 259.
— acid radical, *Malonsäureradikal* 165, 187.
Manganate, *Manganat* 141.
— hyperfine splitting, *Manganathyperfeinaufspaltung* 140.
Manganese fluosilicate, *Manganfluorosilikat* 138.
— nucleus, *Mangankern* 138, 141.
Maser 108.
—, solid-state, *Festkörpermaser* 2.
Mass, effective, *effektive Elektronenmasse* 398.
Master equation, generalised, *verallgemeinerte Mastergleichung* 18, 23.
Mattuck and Strandberg's effective perturbation operator, *Mattucks und Strandbergs effektiver Störungsoperator* 56.
Maxwell-Boltzmann relaxation, *Maxwell-Boltzmann-Relaxation* 30.
McConnel equation, *McConnel-Gleichung* 126, 165.
Mechanism of indirect exchange, *Mechanismus des indirekten Austauschs* 268, 337.

Mechanism of indirect exchange, double transition, d_1-p, $p'-d_2$, *Mechanismus des indirekten Austauschs, Doppelübergang*, d_1-p, $p'-d_2$ 351, 359, 364.
— — —, polarization mechanism, *Polarisationsmechanismus* 352.
— — —, transition to the occupied orbital, *Übergang in die besetzte d-Bahn*, $p-d'$ 348, 359, 369, 373.
— — —, — to the vacant orbital, *in die leere d-Bahn*, $p-d'$ 337, 371, 373.
— — —, twofold excitation, $p-d_1$, $p-d_2$, *zweifache Anregung*, $p-d_1$, $p-d_2$ 349.
Methylene, *Methylen* 221.
Methyl group, *Methylgruppe* 106.
Microwave bridge, *Mikrowellenbrücke* 112, 113, 116.
Mobility, *Elektronenbeweglichkeit* 397.
— of $3d$, $4f$ electrons, *Beweglichkeit von $3d$, $4f$-Elektronen* 398.
Mode, normal vibrational, *normale Schwingungsform* 48.
Modulation broadening, *Modulationsverbreiterung* 117.
—, high-frequency, *Hochfrequenzmodulation* 113.
—, magnetic field, *Magnetfeldmodulation* 110, 112, 113, 115.
Molecular crystal, *Molekülkristall* 68.
— —, paramagnetic relaxation in a, *paramagnetische Relaxation in einem Molekülkristall* 2.
— orbit, one-electron, *Einelektronenmolekülbahn* 378.
— —, orthogonal, *orthogonale Molekülbahn* 321.
— orbital, *Molekülbahn* 409, 470.
— —, delocalised, *nichtlokalisierte Molekülbahn* 95, 101, 126.
Moment, magnetic, *magnetisches Moment* 94, 99.
—, nuclear magnetic, *magnetisches Kernmoment* 94.
Motional narrowing 122, 125.
Mott localization 411.
— Problem, *Mott-Übergang* 536.
— trapping 399, 543.
— transition, *Mott-Übergang* 397, 413, 414, 543.
— —, critical distance, *Mott-Übergang, kritischer Abstand* 397.
Multi-level system, *Vielniveausystem* 65.
— systems, relaxation in, *Relaxation in Mehrniveausystemen* 65.
Multi-nuclear interaction, *Mehrkernwechselwirkung* 104.
Multi-spin process, *Mehrspinprozeß* 74, 75.
Multiplet, $4f$, *$4f$-Multiplett* 464.

NiO 268, 552.
^{14}N splitting constant, *^{14}N-Aufspaltungskonstante* 166.
NaCl-type compound, metallic, *metallische NaCl-Verbindung* 505.

Naphthalen triplet, *Naphthalentriplett* 207, 224.
Narrowing, exchange 123, 125.
—, motional 122, 125.
Nephelauxetic effect, *nephelauxeter Effekt* 471, 504.
Noise, *Rauschen* 109, 110.
—, detector, *Empfängerrauschen* 109, 110.
—, excess crystal, *Überschußkristallrauschen* 110.
—, flicker, *Funkelrauschen* 109.
Non-adiabatic approximation, *nichtadiabatische Näherung* 43, 52.
Non-Kramers exchange mechanism, *nicht-Kramersscher Austauschmechanismus* 269.
Non-Kramers salt, *Nicht-Kramers-Salz* 39, 40, 41.
— —, Raman process in, *Raman-Prozeß in Nicht-Kramers-Salz* 50, 41.
— —, direct process in, *direkter Vorgang in Nicht-Kramers-Salzen* 48.
Non-orthogonality catastrophe, *Nichtorthogonalitätskatastrophe* 290.
— integral, *Nichtorthogonalitätsintegral* 276, 278, 311, 323, 354, 359, 362.
Nordio-Pavan relation, Giacometti-, *Giacometti-Nordio-Pavan-Beziehung* 159.
Nuclear electric quadrupole, *elektrisches Kernquadrupol* 140.
— interaction, multi-, *Mehrkernwechselwirkung* 104.
— magnetic moment, *magnetisches Kernmoment* 94.
— magnetic resonance, *kernmagnetische Resonanz* 2, 119, 142.
— magneton, *Kernmagneton* 99.
— quadrupole relaxation, *Kernquadrupolrelaxation* 82.
— spin, *Kernspin* 99, 100.
— spin resonance, *Kernspinresonanz* 9.
Nucleus, *Keim* 99, 101.
—, copper, *Kupferkeim* 100.
—, manganese, *Mangankern* 138.

O_2 ion 248.
Octahedral field, *Oktaederfeld* 129.
Octahedral symmetry, *Oktaedersymmetrie* 129.
Odd-electron centre, *Zentrum für ungerade Elektronen* 259.
One-electron hyperfine constant, *Einelektronhyperfeinkonstante* 149.
— molecular orbit, *Einelektronenmolekülbahn* 378.
— transition integral, *Einelektronenübergangsintegral* 330.
Operator, annihilation, *Vernichtungsoperator* 38.
— equivalent, *Operatoräquivalent* 466.
Optical branch, *optischer Zweig* 37.
— phonon, *optisches Phonon* 80, 81.
— spectrum, *optisches Spektrum* 404.
Orbach process, inverted, *inverser Orbach-Relaxationsprozeß* 46, 65.

Orbach process of relaxation, *Orbach-Relaxationsprozeß* 43, 51.
— —, transition probability for, *Übergangswahrscheinlichkeit für Orbach-Prozesse* 43.
— relaxation process in rare-earth salts, *Orbach-Relaxationsprozeß in Seltene-Erde-Salzen* 51.
Orbit interaction, spin-, *Spinbahnwechselwirkung* 35, 462.
Orbit-lattice interaction, *Bahn-Gitterwechselwirkung* 48.
— potential, *Bahn-Gitterpotential* 48.
Orbit, one-electron molecular, *Einelektronenmolekülbahn* 378.
—, orthogonal molecular, *orthogonale Molekülbahn* 321.
Orbital angular momentum, quenching of the, *Unterdrückung des Bahndrehimpulses* 36.
—, antibonding, *lockerndes Orbital* 132.
—, bonding, *bindendes Orbital* 132.
—, d, d-*Bahn* 127, 128, 129.
—, $3d$, $3d$-*Bahn* 127.
—, delocalised, *nichtlokalisierte Bahn* 99.
—, delocalised molecular, *nichtlokalisierte Molekülbahn* 95, 101, 126.
—, e_g, e_g-*Bahn* 128.
—, energy level, *Bahnenergieniveau* 129.
—, $4f$, $4f$-*Bahn* 454.
— level splitting, *Bahnniveauaufspaltung* 129.
—, nonbonding, *nichtbindendes Orbital* 177.
—, π, π-*Bahn* 127.
— quenching 408.
— radius of RE, *Bahnradius der SE* 452.
—, σ, σ-*Bahn* 127.
—, s, s-*Bahn* 128.
Orbitals, spatial distribution of $3d$, *räumliche Verteilung von $3d$-Bahnen* 127.
Orbital, t_{2g}, t_{2g}-*Bahn* 128.
— splitting, *Bahnaufspaltung* 131.
Orbitals, spatial distribution of $3d$, *räumliche Verteilung von $3d$-Bahnen* 127.
Orthoferrite, *Orthoferrit* 498.
Overhauser effect, *Overhauser-Effekt* 142, 143.
— technique, *Overhauser-Verfahren* 142.
Overlap, *Überlappung* 362.
— integral, *Überlappungsintegral* 276, 278, 311, 323, 354, 359, 362.

π orbital, π-*Bahn* 127.
π-radical, π-*Radikal* 151.
—, inorganic, *anorganisches π-Radikal* 190.
—, in irradiated organic crystals, *in bestrahlten organischen Kristallen* 187.
π-σ correlation, π-σ-*Korrelation* 164.
π-σ interaction, π-σ-*Wechselwirkung* 163.
Paramagnetic centre in silicon, *paramagnetisches Zentrum in Silizium* 253.
— ion, *paramagnetisches Ion* 131.
— relaxation, *paramagnetische Relaxation* 2.
— — in local electron centres, *paramagnetische Relaxation in lokalen Elektronenzentren* 78.

Paramagnetic relaxation by modulated exchange interaction, *paramagnetische Relaxation durch modulierte Austauschwechselwirkung* 71.
— — in molecular crystals, *paramagnetische Relaxation in Molekülkristallen* 2.
— —, stereochemistry and, *Stereochemie und paramagnetische Relaxation* 68.
— —, Waller's mechanism of, *Waller's Mechanismus der paramagnetischen Relaxation* 65, 70, 74, 77.
Paramagnetic resonance, *paramagnetische Resonanz* 2, 8, 145.
— —, acoustic, *akustische paramagnetische Resonanz* 91.
— —, electron (E.P.R.), *paramagnetische Elektronenresonanz* 36.
— — saturation experiment, *paramagnetisches Resonanzsättigungsexperiment* 28.
— saturation, *paramagnetische Sättigung* 47.
— susceptibility, *paramagnetische Suszeptibilität* 404.
Paramagnetism, *Paramagnetismus* 2.
Parameter, bottleneck 86.
—, D, splitting, *Aufspaltungsparameter*, D 137.
—, g, *g-Parameter* 95.
—, phonon, *Phononenparameter* 63.
—, relaxation, *Relaxationsparameter* 21.
—, saturation, *Sättigungsparameter* 32, 86.
Parity, *Parität* 30, 37, 47.
Pavan relation, Giacometti-Nordio-, *Giacometti-Nordio-Pavan-Beziehung* 159.
Perinaphthenyl radical, *Perinaphthenylradikal* 162, 171.
Permutation group, *Vertauschungsgruppe* 295, 303, 305, 313.
—, factorable, *faktorisierbare Vertauschung* 291.
—, intraatomic, *intraatomare Vertauschung* 301.
—, non-factorable, *nichtfaktorisierbare Vertauschung* 291.
Perovskite, *Perowskit* 253, 497.
—, Mn, *Mn-Perowskit* 498.
Perturbation operator, effective, *effektiver Störungsoperator* 39.
— —, Mattuck and Strandberg's effective, *Mattucks und Strandbergs effektiver Störungsoperator* 56.
— theory, Bogolyubov's, *Störungstheorie nach Bogolyubov* 322, 365.
— —, first-order, *Störungstheorie erster Ordnung* 66, 79.
— —, higher order, *Störungstheorie höherer Ordnung* 74.
— —, Löwdin's, *Störungstheorie nach Löwdin* 344, 349, 361.
— —, second-order, *Störungstheorie zweiter Ordnung* 50, 66, 68, 80.
Phonon 37.
— bottleneck 7, 46, 74, 86, 90.
— coupling, spin-, *Spin-Phonon-Kopplung* 91.

Phonon, Debye distribution, *Debye-Verteilung für ein Phonon* 38.
— density of states, *Phononenzustandsdichte* 30.
— drag, *Phononendrag* 430.
— interaction, electron-, *Elektron-Phonon-Wechselwirkung* 398.
— —, spin-, *Spin-Phonon-Wechselwirkung* 84, 85.
Phonon, life-time of the, *Lebensdauer des Phonons* 86, 87.
—, longitudinal, *longitudinales Phonon* 82.
—, matrix element, *Matrixelement eines Phonons* 30, 38.
— mode, local, *lokaler Phononenschwingungstyp* 82.
—, optical, *optisches Phonon* 80, 81.
— parameter, *Phononenparameter* 63.
Phonon-phonon interaction-, *Phonon-Phonon-Wechselwirkung* 87.
Phonon relaxation, *Phononenrelaxation* 88.
— — process in rare-earth salts, two-, *Zweiphononenrelaxationsprozeß in Seltene-Erde-Salzen* 50.
— spectrum, real, *reelles Phononenspektrum* 85.
—, spontaneous emission of a, *spontane Emission eines Phonons* 46.
— system, *Phononensystem* 4.
— —, relaxation in the, *Relaxation im Phononensystem* 80.
— —, spin-, *Spin-Phonon-System* 34.
— —, state functions of the spin-, *Zustandsfunktionen des Spin-Phonon-Systems* 34.
—, transverse, *transversales Phonon* 82.
Phthalocyanine, copper, *Kupferphthalocyanin* 106.
Polar state, *Polarzustand* 282, 293.
Polarization, conduction, electron, *Leitungselektronenpolarisation* 433, 490.
—, spin, *Spinpolarisation* 174, 311, 352, 380.
Polaron 398.
—, large, *großes Polaron* 543.
—, magnetic, *magnetisches Polaron* 399, 419, 421.
—, small, *kleines Polaron* 543.
Position of the d-levels, *Lage der d-Niveaus* 546.
Potential exchange, *potentieller Austausch* 270, 330, 370.
Principle of detailed balance, *Prinzip des detaillierten Gleichgewichts* 31.
Problem of 3 centres and 4 electrons, non-orthogonal orbitals, *Problem der 3 Zentren und 4 Elektronen, nichtorthogonale Bahnen* 359.
— — — — —, orthogonal orbitals, *Problem der 3 Zentren und 4 Elektronen, orthogonale Bahnen* 336.
Projection operator, *Projektionsoperator* 325.
Proton, α- 187.
—, β- 188.
— transfer, *Protonenübergang* 202.

Pulse saturation, *Impulssättigung* 10, 12.
— — technique, *Impulssättigungsverfahren* 10, 12, 32.

Quadrupolar selection rule, *Quadrupolauswahlregel* 57.
Quadrupole, electronic, *elektronisches Quadrupol* 136.
— interaction, electric, *elektrische Quadrupolwechselwirkung* 135, 136.
— —, nuclear, *Kernquadrupolwechselwirkung* 34, 35, 72.
— moment, electric, *elektrisches Quadrupolmoment* 135, 141.
—, nuclear electric, *elektrisches Kernquadrupol* 140.
— relaxation, nuclear, *Kernquadrupolrelaxation* 81, 82.
— transition, electric, *elektrischer Quadrupolübergang* 37.
Quantization second, *zweite Quantelung* 294, 322, 369.
Quartz crystal standard, *Quarzkristallstandard* 115.
Quasi-Raman process, *Quasi-Raman-Prozeß* 1.
Quasiparticle, magnetic, *magnetisches Quasiteilchen* 368.
Quenching of the orbital angular momentum, *Unterdrückung des Bahndrehimpulses* 36.

R-centre, *R-Zentrum* 146, 236.
Radial integral, *Radialintegral* 460.
Radiation-induced centre, *strahlungsinduziertes Zentrum* 254.
— defects in solids, *strahlungsinduzierte Gitterfehler in Festkörpern* 145.
Radical, allyl, *Allylradikal* 189.
—, anion, *Anionradikal* 155.
—, cation, *Kationradikal* 155.
—, free, *freies Radikal* 77, 94, 95, 98, 116, 119, 126, 151.
— g-value, free, *g-Wert eines freien Radikals* 126.
—, malonic acid, *Malonsäureradikal* 165, 187.
—, monocyclic, *monozyklisches Radikal* 177.
— in non-oriented irradiated solids, *Radikal in bestrahlten nichtausgerichteten Festkörpern* 192.
—, π-, *π-Radikal* 151.
—, —, inorganic, *anorganisches π-Radikal* 190.
—, —, in irradiated organic crystals, *π-Radikal in bestrahlten organischen Kristallen* 187.
—, perinaphthenyl, *Perinaphthenylradikal* 162, 171.
—, σ-, *σ-Radikal* 204.
—, semiquinone, *Semichinonradikal* 104.
—, stable free, *stabiles freies Radikal* 182.
— transient, free, *Übergang mit freien Radikalen* 142.
—, vinyl, *Vinylradikal* 205.

Radius of electron shells, *Radius äußerer Elektronenschalen* 389.
Raman process in non-Kramers salts, *Raman-Prozeß in Nicht-Kramers-Salzen* 50.
— —, quasi-, *Quasi-Raman-Prozeß* 1.
— — of relaxation, *Raman-Relaxationsprozeß* 41, 53, 121.
— processes, transition probability for, *Übergangswahrscheinlichkeit für Raman-Prozesse* 41.
— relaxation process in rare-earth salts, *Raman-Relaxationsprozeß in Seltene-Erde-Salzen* 50.
Rare-earth (RE)-chalcogenide, *Seltene Erde (SE)-Chalkogenid* 553.
RE chemistry, *Chemie der Seltenen Erden* 393.
RE, crystal spectrum in, *SE-Kristallspektrum* 463.
RE-elements, magnetic interaction in, *magnetische Wechselwirkung in SE-Elementen* 491.
RE, ground state of 4f-electrons in, *Grundzustand von 4f-Elektronen in Se* 455.
RE-hydride, *SE-Hydrid* 512.
RE, ionization energy for, *Ionisierungsarbeit für SE* 458.
—, magnetic moment of, *magnetischer Moment von SE* 490.
—, melting point of, *Schmelzpunkt der SE* 456.
RE metals, band structure, *SE-Metalle, Bandstruktur* 472.
RE, orbital radius of, *Bahnradius der SE* 452.
RE^{3+} with pnictides, III-V-compounds of, *SE-III-V-Verbindungen, chemische Eigenschaften* 493.
RE-pnictide, *SE-Pniktid: magnetische Eigenschaften* 495.
RE salts, crystal field in, *Kristallfeld in SE-Salzen* 47.
— —, direct relaxation process in, *direkter Relaxationsprozeß in SE-Salzen* 48.
— —, Orbach relaxation process in, *Orbach-Relaxationsprozeß in SE-Salzen* 51.
— —, Raman relaxation process in, *Raman-Relaxationsprozeß in SE-Salzen* 50.
— —, two-phonon relaxation process in, *Zweiphononenrelaxationsprozeß in SE-Salzen* 50.
RE-sesquioxide, *SE-Sesquioxyd* 501.
RE-spectrum, *SE-Spektrum* 459.
Red shift, magnetic, *magnetische Rotverschiebung* 524.
— —, —, Cr spinels, *magnetische Rotverschiebung, Cr-Spinelle* 487.
— —, —, Eu chalcogenides, *Eu-Chalkogenide* 483.
Relaxation curve, Casimir and DuPré, *Casimir- und DuPré-Relaxationskurve* 7.
—, Debye, *Debye-Relaxation* 21, 42.
—, direct process of, *direkter Relaxationsprozeß* 39.

Relaxation due to spin-orbit coupling, *Relaxation durch Spinbahnkopplung* 47.
— function, *Relaxationsfunktion* 14, 15.
— in local electron centres, paramagnetic, *paramagnetische Relaxation in lokalen Elektronenzentren* 78.
—, Maxwell-Boltzmann, *Maxwell-Boltzmann-Relaxation* 30.
— by modulated exchange interaction, paramagnetic, *paramagnetische Relaxation durch modulierte Austauschwechselwirkung* 71.
— in molecular crystals, paramagnetic, *paramagnetische Relaxation in Molekülkristallen* 2.
— in multilevel systems, *Relaxation in Mehrniveausystemen* 65.
—, nuclear quadrupole, *Kernquadrupolrelaxation* 81, 82.
— operator, effective spin-lattice, *effektiver Spingitterrelaxationsoperator* 56.
—, Orbach process of, *Orbach-Relaxationsprozeß* 43, 51.
—, paramagnetic, *paramagnetische Relaxation* 2.
— parameter, *Relaxationsparameter* 21.
— phenomenon, *Relaxationserscheinung* 120.
—, phonon, *Phononenrelaxation* 88.
— in the phonon system, *im Phononensystem* 80.
— process in rare-earth salts, direct, *direkter Relaxationsprozeß in Seltene-Erde-Salzen* 48.
— — — —, Orbach, *Orbach-Relaxationsprozeß in Seltene-Erde-Salzen* 48.
— — — —, Raman, *Raman-Relaxationsprozeß in Seltene-Erde-Salzen* 50.
— — — —, two-phonon, *Zweiphononenrelaxationsprozeß in Seltene-Erde-Salzen* 50.
—, Raman process of, *Raman-Relaxationsprozeß* 41, 53, 121.
—, spin-lattice, *Spingitterrelaxation* 1, 30, 56, 59, 78, 120.
—, spin-spin, *Spin-Spin-Relaxation* 1, 9, 13, 72.
— in S-state ions, *Relaxation in Ionen im S-Zustand* 64.
—, stereochemistry and paramagnetic, *Stereochemie und paramagnetische Relaxation* 68.
— times, anisotropy in, *Anisotropie in Relaxationszeiten* 63.
— time, longitudinal, *longitudinale Relaxationszeit* 8.
— —, spin-lattice, *Spingitterrelaxationszeit* 6, 9, 120, 121, 132.
— —, transversal, *transversale Relaxationszeit* 6.
—, Waller's mechanism of paramagnetic, *Waller's Mechanismus der paramagnetischen Relaxation* 65, 70, 74, 77.
Resonance, acoustic paramagnetic, *akustische paramagnetische Resonanz* 91.

Resonance cavity, *Resonanzhohlraum* 109.
—, double, *Doppelresonanz* 142, 144.
—, electron paramagnetic (E.P.R.), *paramagnetische Elektronenresonanz* 36.
—, electron spin, *Elektronenspinresonanz* 9, 94.
—, nuclear magnetic, *kernmagnetische Resonanz* 2, 119, 142.
—, nuclear spin, *Kernspinresonanz* 9.
—, paramagnetic, *paramagnetische Resonanz* 2, 8, 145.
— saturation effect, *Resonanzsättigungseffekt* 124.
— saturation experiment, paramagnetic, *paramagnetisches Resonanzsättigungsexperiment* 28.
— saturation technique, *Resonanzsättigungsverfahren* 10.
Resonant and non-resonant technique, *Resonanz- und Nichtresonanzverfahren* 76.
Rowland coupling, Bloembergen-, *Bloembergen-Rowland-Kopplung* 411.
Ruderman-Kittel function, *Ruderman-Kittel-Funktion* 434.
Russel-Saunders coupling, *Russel-Saunders-Kopplung* 402.
Rutile, *Rutil* 253.

s orbital, *s-Bahn* 128.
S-state ion, *S-Zustandsion* 70, 71.
S-State ions, relaxation in, *Relaxation in Ionen im S-Zustand* 64.
σ-radical, *σ-Radikal* 204.
σ orbital, *σ-Bahn* 127.
Saturation, *Sättigung* 109, 121, 125, 143.
— effect, *Sättigungseffekt* 124.
— —, resonnance, *Resonanzsättigungseffekt* 124.
— experiment, *Sättigungsexperiment* 9.
— —, paramagnetic resonance, *paramagnetisches Resonanzsättigungsexperiment* 28.
—, paramagnetic, *paramagnetische Sättigung* 47.
— parameter, *Sättigungsparameter* 32, 86.
—, pulse, *Impulssättigung* 10, 12.
—, steady state, *Sättigung im stationären Zustand* 10.
— technique, *Sättigungsverfahren* 31.
— —, pulse, *Impulssättigungsverfahren* 10, 12, 32.
— —, resonance, *Resonanzsättigungsverfahren* 10.
Saunders coupling, Russel-, *Russel-Saunders-Kopplung* 402
Scattering by spin fluctuations, *Trägerstreuung durch Spinfluktuationen* 424.
Selection rules, *Auswahlregeln* 459.
Semiconductor, *Halbleiter* 95.
—, magnetic, *magnetischer Halbleiter* 389, 396.
Semiconductors, defects in, *Gitterfehler in Halbleitern* 253.
Semiquinone, *Semichinon* 180, 202.
—, benzo-, *Benzosemichinon* 104.
— radical, *Semichinonradikal* 104.

Shallow acceptor impurity, *flacher Akzeptor* 254.
— donor impurity, *flacher Donator* 253.
Shift, chemical, *chemische Verschiebung* 169.
Silicon, *Silizium* 253.
— crystal rectifier, *Siliziumkristallgleichrichter* 98.
—, paramagnetic centre in, *paramagnetisches Zentrum in Silizium* 253.
Singlet-triplet splitting, *Singulett-Triplett-Aufspaltung* 266, 272.
Slater curve, Bethe-, *Bethe-Slater-Kurve* 307.
Solid-state maser, *Festkörpermaser* 2.
Solids, atom in, *Atom in Festkörpern* 245.
Spatial distribution of $3d$ orbitals, *räumliche Verteilung von 3d-Bahnen* 127.
Specific heat at constant field, *spezifische Wärme bei konstanter Feldstärke* 5, 17.
— — at constant magnetization, *bei konstanter Magnetisierung* 5, 18.
Spectrometer, ESR, *ESR-Spektrometer* 112.
—, electron spin resonance, *Elektronenspinresonanzspektrometer* 112.
Spectrum, magnetic high-frequency, *magnetisches Hochfrequenzspektrum* 136.
—, real phonon, *reelles Phononenspektrum* 85.
Spin density, *Spindichte* 161, 162, 164, 173, 180, 225.
— —, negative, *negative Spindichte* 189.
— disorder resistivity, *Spindisorderwiderstand* 538.
— echo, *Spinecho* 263.
— exchange, *Spinaustausch* 199, 203.
— fluctuations, scattering by, *Trägerstreuung durch Spinfluktuationen* 424.
— Hamiltonian, *Spin-Hamilton-Operator* 4, 14, 39, 56, 74, 132, 133, 135, 136, 137, 141, 170, 227, 232, 242.
— —, Dirac, *Diracscher Spin-Hamilton-Operator* 266.
Spin-lattice interaction, *Spingitterwechselwirkung* 4, 48, 64, 99, 120, 124
— operator, *Spingitteroperator* 30.
— perturbation, *Spingitterstörung* 34.
— relaxation, *Spingitterrelaxation* 1, 30, 56, 59, 78, 120.
— relaxation operator, effective, *effektiver Spingitterrelaxationsoperator* 56.
— relaxation time, *Spingitterrelaxationszeit* 6, 9, 120, 121, 132.
Spin and magnetic field, anisotropic coupling between electron, *anisotrope Kopplung zwischen Elektronenspin und Magnetfeld* 133.
—, nuclear, *Kernspin* 99, 100.
— operator 266, 279, 299.
Spin-orbit coupling, *Spinbahnkopplung* 36, 47, 79, 126, 129, 130, 132, 259, 460.
— —, anisotropy in, *Anisotropie in Spinbahnkopplung* 64.
— coupling coefficient, *Spinbahnkopplungskoeffizient* 121, 131.
— coupling, relaxation due to, *Relaxation durch Spinbahnkopplung* 47.

Spin-orbit interaction, *Spinbahnwechselwirkung* 35, 462.
Spin-phonon coupling, *Spin-Phonon-Kopplung* 91.
— interaction, *Spin-Phonon-Wechselwirkung* 84, 85.
— system, *Spin-Phonon-System* 34.
— —, state functions of the, *Zustandsfunktionen des Spin-Phonon-Systems* 34.
Spin polarization, *Spinpolarisation* 174, 311, 352, 380.
— process, multi-, *Mehrspinprozeß* 74, 75.
— resonance, electron, *Elektronenspinresonanz* 9, 94.
— —, nuclear, *Kernspinresonanz* 9.
Spin-spin interaction, *Spin-Spin-Wechselwirkung* 99, 120, 122.
— relaxation, *Spin-Spin-Relaxation* 1, 9, 13, 72.
— relaxation time, *Spin-Spin-Relaxationszeit* 120.
Spin system, *Spinsystem* 4, 5.
— temperature, *Spintemperatur* 6, 72, 82, 86.
— —, negative, *negative Spintemperatur* 87.
Spinel, *Spinell* 253.
— $CdCr_2SE_4$, *Cr-Spinell* 481.
—, Cr-chalcogen, *Cr-Chalkogen-Spinell* 522.
Spinels, Cr, magnetic red shift, *Cr-Spinelle, magnetische Rotverschiebung* 487.
—, electrical conductivity in thio-, *elektrische Leitfähigkeit in Thio-Spinellen* 531.
—, superexchange interaction in thio-, *Superaustauschwechselwirkung in Thio-Spinellen* 531.
Splitting, angular variation of electronic, *Winkelverteilung der elektronischen Aufspaltung* 107, 140.
— constant, ^{13}C, ^{13}C-*Aufspaltungskonstante* 165.
— —, ^{14}N, ^{14}N-*Aufspaltungskonstante* 166.
— —, ^{19}F fluorine, ^{19}F-*Fluoraufspaltungskonstante* 167.
— —, sign of the hyperfine, *Vorzeichen der Hyperfeinaufspaltungskonstante* 168.
—, crystal field, *Kristallfeldaufspaltung* 407, 534.
—, electronic, *elektronische Aufspaltung* 107, 108, 133, 136, 139.
— factor, Landé, *Landé-Aufspaltungsfaktor* 126.
—, hyperfine, *Hyperfeinaufspaltung* 14, 99, 104, 108, 133, 135, 139, 140, 148.
—, manganate hyperfine, *Manganathyperfeinaufspaltung* 140.
—, orbital, *Bahnaufspaltung* 131.
—, orbital level, *Bahnniveauaufspaltung* 129.
— parameter, D, *Aufspaltungsparameter, D* 137.
—, Stark effect, *Stark-Aufspaltungseffekt* 108, 127.
—, superhyperfine, *Superhyperfeinaufspaltung* 132, 184, 255.
—, temperature-dependent hyperfine, *zeitabhängige Hyperfeinaufspaltung* 198.

Subject Index.

Splitting tensor, hyperfine, *Hyperfeinaufspaltungstensor* 147, 148, 187.
Spontaneous emisssion, *spontane Emission* 46.
Stable free radical, *stabiles freies Radikal* 182.
Stark effect splitting, *Stark-Aufspaltungseffekt* 108, 127.
State, doubly degenerate, *zweifach entarteter Zustand* 36.
— functions, *Zustandsfunktionen* 30.
— —, anisotropy in, *Anisotropie in Zustandsfunktionen* 63.
— — of the electrons, *Zustandsfunktionen der Elektronen* 59.
— — of the spin-phonon system, *Zustandsfunktionen des Spin-Phonon-Systems* 34.
Steady state saturation, *Sättigung im stationären Zustand* 10.
Stimulated emission, *induzierte Emission* 46, 96.
Strain, *Verformung* 48, 83.
— anisotropy, *Verformungsanisotropie* 85.
— experiment, static, *statisches Verformungsexperiment* 92.
— operator, *Verformungsoperator* 30, 49.
Strandberg's effective perturbation operator, Mattuck and, *Mattucks und Strandbergs effektiver Störungsoperator* 56.
Sturm-Liouville theorem 281.
Superexchange, *Superaustausch* 336, 410, 448, 535.
— interaction in thio-spinels-, *Superaustauschwechselwirkung in Thio-Spinellen* 531.
Superconducting magnet, *supraleitender Magnet* 120.
Superhyperfine line, *Superhyperfeinlinie* 103.
— pattern, *Superhyperfeinmuster* 107.
— splitting, *Superhyperfeinaufspaltung* 132, 184, 255.
— structure, chlorine, *Chlorsuperhyperfeinstruktur* 252.
— —, fluorine, *Fluor-Superhyperfeinstruktur* 252.
Susceptibility, adiabatic, *adiabatische Suszeptibilität* 5, 10, 17.
—, complex, *komplexe Suszeptibilität* 6, 19.
—, paramagnetic, *paramagnetische Suszeptibilität* 404.
—, static, *statische Suszeptibilität* 2, 5, 108, 109, 125.
Symmetry group, cubic, *kubische Symmetriegruppe* 375.
—, octahedral, *Oktaedersymmetrie* 129.

T, magic-, 119.
t_{2g} orbital, t_{2g}-*Bahn* 128.
Th_3P_4 compound, Th_3P_4-*Verbindung* 502.
Teller effect, Jahn-, *Jahn-Teller-Effekt* 69, 70.
— theorem, Jahn- 36.
Temperature, Curie-Weiss, *Curie-Weiss-Temperatur* 409.
—, Debye, *Debye-Temperatur* 42.

Temperature-dependent hyperfine splitting, *zeitabhängige Hyperfeinaufspaltung* 198.
— moment, *temperaturabhängiger Moment* 406.
Temperature, lattice, *Gittertemperatur* 31.
—, spin, *Spintemperatur* 6, 72, 82, 86.
—, negative spin, *negative Spintemperatur* 87.
Tetragonal distortion, *tetragonale Verzerrung* 130.
Tetrahedral field, *Tetraederfeld* 129, 130.
Thermal vibration, *thermische Schwingung* 121.
Thermoelectric force, *Thermokraft* 429.
Time-reversal, *Zeitumkehr* 66.
— degeneracy, *Zeitumkehr-Entartung* 43.
— symmetry, *Zeitumkehr-Symmetrie* 30, 36.
Transfer, chemical, *chemischer Übergang* 202.
—, intramolecular electron, *intramolekularer Elektronenübergang* 203.
—, proton, *Protonenübergang* 202.
Transient intermediate, *vorübergehendes Zwischenprodukt* 98.
Transition, electric dipole, *elektrischer Dipolübergang* 37, 459.
—, electric quadrupole, *elektrischer Quadrupolübergang* 37.
—, 4f-5d, *4f-5d-Übergang* 472.
—, forbidden, *verbotener Übergang* 136.
— integral, one-electron, *Einelektronenübergangsintegral* 330.
—, magnetic dipole, *magnetischer Dipolübergang* 37.
—, metal-insulator, *Metall-Isolatorübergang* 400.
Transition probability, *Übergangswahrscheinlichkeit* 30, 33, 76, 81, 90.
— — for the direct process, *Übergangswahrscheinlichkeit für den direkten Vorgang* 40.
— — for Orbach processes, *Übergangswahrscheinlichkeit für Orbach-Prozesse* 43.
— — for Raman processes, *Übergangswahrscheinlichkeit für Raman-Prozesse* 41.
Transport integral, *Transportintegral* 42, 66.
— property, *Transporteigenschaft* 412.
Transposition 290, 301, 313.
Transversal relaxation time, *transversale Relaxationszeit* 8.
Transverse phonon, *transversales Phonon* 82.
Triplet excitation, *Triplettanregung* 219.
Triplet, ground state, *Triplettgrundzustand* 221.
—, naphthalene, *Naphthalentriplett* 207, 224.
— splitting, singlet-, *Singulett-Triplett-Aufspaltung* 266, 272.
— state system, *Triplettzustandssystem* 205.
— system, excited, *angeregtes Triplettsystem* 210.

U centre, U-*Zentrum* 247.

V_1 centre, V_1-*Zentrum* 239.
V_F centre, V_F-*Zentrum* 241.
V_K centre, V_K-*Zentrum* 240, 241, 243.

V_{OH} centre, V_{OH}-*Zentrum* 245.
V_t centre, V_t-*Zentrum* 241.
Valency change, *Wertigkeitswechsel* 142.
Van Vleck cancellation, *Van Vleck-Aufhebung* 37, 52, 57, 66.
Van Vleck form, Brons-, *Brons-van-Vleck-Formel* 74, 81.
Van Vleck Hamiltonian, Dirac-, *Dirac-van-Vleck-Hamilton-Operator* 267.
Van Vleck's theory, *Van Vlecks Theorie* 56.
Vector model for more than one electron with orbital degeneracy, *Vektormodell für mehrere Elektronen mit Bahnentartung* 304.
— — — — — without orbital degeneracy, *Vektormodell für mehrere Elektronen ohne Bahnentartung* 300.
— — for one s-electron per atom, *Vektormodell für ein s-Elektron pro Atom* 296.
— — with orthogonalized atomic wave functions, *Vektormodell bei orthogonalisierten Wellenfunktionen* 321.
Vibrational mode, normal, *normale Schwingungsform* 48.

Vibration, thermal, *thermische Schwingung* 121.
Vinyl radical, *Vinylradikal* 205.

Waller's mechanism of paramagnetic relaxation, *Waller's Mechanismus der paramagnetischen Relaxation* 65, 70, 74, 77.
Wave function, d-, *d-Wellenfunktion* 132.
— — for the crystal, choice of the, *Wahl der Wellenfunktion des Kristalls* 290, 300, 304.
Weiss temperature, Curie-, *Curie-Weiss-Temperatur* 409.
Wigner-Eckhart theorem 48.

XY_6 complex 56, 68, 69, 81.

ZnS 255.
Zeeman effect, electronic, *elektronischer Zeeman-Effekt* 35, 79, 126.
— —, nuclear, *Zeeman-Kerneffekt* 35, 79.
Zeeman-energy reservoir, *Zeeman-Energiereservoir* 22.

QC
21
H327
v.18,pt1

MAY 18 1969